T0136780

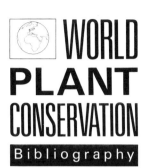

WORLD
PLANT
CONSERVATION
Bibliography

WORLD PLANT CONSERVATION

Bibliography

compiled by

Royal Botanic | Threatened Plants Unit
Gardens Kew | World Conservation
          | Monitoring Centre

1990

THE ROYAL BOTANIC GARDENS, KEW

The mission of the Royal Botanic Gardens, Kew, is to ensure better management of the Earth's environment by increasing our knowledge and understanding of the plant kingdom: the basis of life on Earth.

Whenever possible, the Royal Botanic Gardens, Kew, will endeavour to reduce and reverse the rate of destruction of the world's plant species and their habitats.

WORLD CONSERVATION MONITORING CENTRE

The World Conservation Monitoring Centre is a joint venture between the three partners in the World Conservation Strategy: IUCN – The World Conservation Union, UNEP – United Nations Environment Programme, and WWF – World Wide Fund for Nature (Formerly World Wildlife Fund). Its mission is to support conservation and sustainable development through the provision of information on the world's biological diversity.

© :   World Conservation Monitoring Centre
and
Royal Botanic Gardens, Kew
1990

ISBN :   0 947643 24 9
Cover design :   Stevens Richardson Design Associates for Media Resources
Typesetting :   Royal Botanic Gardens, Kew
Printed by :   Whitstable Litho Printers Ltd.

CONTENTS

PREFACE

There is an ever increasing pressure on the natural resources of planet Earth by mankind – as the human population continues to increase, so does the need for more food, fuelwood, fodder, fibres, timber and other plant products – either grown in cultivation or taken directly from nature. Mankind relies on plants to survive, and it is ironic, therefore, that his actions are threatening the very future of these life sustaining plant resources. The daily threats to plant diversity include worldwide destruction and degradation of habitats, uncontrolled harvesting from "the wild", the spread of monocultures and the insidious effects of pollutants, such as acid rain. Additionally, it now seems likely that man's activities are contributing to major climatic changes, with uncalculated but probably locally calamitous future consequences for various plant communities.

Many people can join in the battle to rescue plants or manage the world's vegetation in better ways. Whoever they are and wherever they work, they all require information to be effective. Institutions such as the Royal Botanic Gardens, Kew, and the World Conservation Monitoring Centre have been at the fore in providing the basic tools of appropriate botanical information to aid decision makers on a global scale. The present bibliography provides an excellent and expanded compilation of the references to be found in WCMC's pioneering volume *Plants in Danger; what do we know?**, which provides much useful information on relevant subjects in the plant conservation field.

The present collaborative bibliography provides points of reference for those interested in conservation action or theory, in both richer, more studied countries with extensive plant conservation literatures, and poorer countries where little work has been done. In the former, the conservationist can feel lost in a sea of information: the bibliography will serve to pinpoint major published works to examine. In poorer countries, there is often a shortage of knowledge on what has previously been achieved. The bibliography will alert activists to existing references and a concerned effort can then be made to obtain the most relevant literature.

Hopefully, this volume will be the first of regular productions by the teams at the Royal Botanic Gardens, Kew, and the Threatened Plants Unit at WCMC who have produced a most valuable and worthwhile publication as a result of their collaboration.

Alan Hamilton
International Plants Conservation Officer
World Wide Fund for Nature

* *Davis, S.D. et al. (1986). Plants in danger; what do we know? Cambridge. IUCN Conservation Monitoring Centre. xlv, 461p.*

ACKNOWLEDGEMENTS

A book of this type, covering such an extensive literature, and generated from a computer held database that has been developed over a number of years, could not have been compiled without the help of a large number of individuals, both within the Royal Botanic Gardens, Kew and the World Conservation Monitoring Centre.

We must single out Stephen Davis to whom we give special thanks for his meticulous care in the day-to-day supervision and overall management of the project within WCMC. Suzy Dickerson has also played a key role in having senior editorial responsibility for the project within Kew. In addition, the bibliography has benefited greatly from Stephen Droop's expertise and patience in proof-reading and database correction.

In preparing this bibliography for publication, we would also especially like to thank the following: Christine Leon (editorial control, data-input); Noel McGough (editorial control, liaison); Susan Wickens (library checking), without whom the work would never have seen the light of day.

We thank the many staff of the Royal Botanic Gardens, Kew, who gave of their time to help proof-read the text and translate many citation titles, specifically: Sally Dawson, Diana Polhill, Emma Powell, Kathleen Rattue, Brian Stannard, Maria Wallace. Thanks also go to Milan Svanderlik of Media Resources and the staff of the Typing Pool. We also thank Bill Loder for handling the type-setting of the final text, moving the text through various computer systems so successfully.

Within the staff of the World Conservation Monitoring Centre we thank Richard Fiennes, Robert Madams, Judy Sheppard, Saskia Hartwig and Jana Zantovska. Special thanks also go to Duncan Mackinder, Al Blake and other members of the Computer Services Unit of the WCMC who helped develop and maintain the software for the database who provided technical help in formatting the final text.

We also thank Hugh Synge for writing the original software and managing the database in its early days. The bibliography has also benefited from the help of numerous individuals who have submitted reference lists to us over the years. Particular thanks go to Bob DeFillips at the IUCN office in the Smithsonian Institute, Washington, D.C., for the addition of many references for the Pacific; to Janie Villa-Lobos, at the same institution, whose monthly *Biological Conservation Newsletter* is a valuable source of information on recently published material; and to Norman Myers and David Given for the inclusion of many relevant papers.

Dr R. Pellew
Director
World Conservation Monitoring Centre
Cambridge

Professor G.Ll. Lucas
Deputy Director
Royal Botanic Gardens
Kew

INTRODUCTION

AIM OF THIS BIBLIOGRAPHY

The relatively recent explosion in 'green' thinking and policy making worldwide, together with the concern at the rate of loss of both species and habitats, has led to a growing demand for up-to-date facts and figures about plant species and vegetation types and how they should be conserved.

Fortunately, much has been published on this subject in recent years, yet there has been no obvious route for the conservation and development community to gain access to this information quickly and easily. The relevant literature tends to be markedly multi-disciplinary, and is thinly scattered across such varied fields as botany, forestry, ecology, horticulture, planning, economics, politics and environmental law. Moreover, at the same time, a sizeable volume of the literature is of an ephemeral nature and much is repetitive. In this respect, the plant conservation literature is quite unlike the plant taxonomic literature, to which plant taxonomists gain easy access through the *Kew Record*, published regularly by the Royal Botanic Gardens, Kew. By comparison, the selection of references for the *World Plant Conservation Bibliography* has been by no means straightforward.

Furthermore, although a number of new journals have appeared in recent years devoted almost exclusively to conservation – for example, *Ambio, Biological Conservation, Environmental Conservation* and *Conservation Biology* – much essential plant conservation information continues to be found only in rather obscure or little known journals.

Accessing this information is therefore not only time-consuming but can often be very difficult, especially for individuals or small organizations with no direct links to the larger scientific libraries.

Aware, therefore, of an increasing demand for quick and easy access to this vast pool of information, the World Conservation Monitoring Centre and the Royal Botanic Gardens, Kew, have together produced this somewhat ambitious publication entitled *World Plant Conservation Bibliography*. Its primary aim is to help fulfil a very obvious need of conservationists and policy makers for a catalogue of 'significant' references on the conservation of plant diversity. Its precise coverage is described in more detail later. Clearly, the broad scope of the subject dictates that a bibliographic compilation such as this is a rather open-ended task and one that will never be complete. We make no claims, therefore, about the comprehensiveness of this volume, but we hope it will prove useful to those who need to gain access to the literature and in this way indirectly promote conservation at local, national and international levels.

The present bibliography is the result of a number of years of collaborative effort by WCMC and the Royal Botanic Gardens, Kew. Staff at both institutions have been engaged in screening the literature for relevant material, and the resultant database maintained and developed by WCMC as part of its data collecting activities. The role of Kew in the development of this bibliography illustrates how the resources of a modern botanic garden can be harnessed to provide working documents for plant conservation.

This volume is designed to complement WCMC's earlier publication, prepared by its Threatened Plants Unit entitled *Plants in Danger; What do we know? (1986)*, which is a more structured work containing selected references to the plant conservation literature. Users may find it helpful, therefore, to refer to these two publications together.

Every effort has been made to ensure that the citations are sufficiently clear and complete for users to identify them quickly. However, in order to make the World Plant Conservation Bibliography available as soon as possible, some minor inconsistencies, such as journal abbreviations, may still be present. Special care, however, has been taken to ensure that subject cross-references are as complete as possible in order to indicate the coverage of any one reference within the very multi-disciplinary field of conservation. By choosing to arrange the references by geography, in addition to including indices on plant families and individual plant taxa, we hope that we have presented the information in a way that will be most easily accessible.

WHAT DOES THE BIBLIOGRAPHY CONTAIN?

This volume contains over 10,000 reference citations to published literature considered relevant to international, national or local plant conservation. As a result, the user will find a mixture of sometimes highly specific papers, for example on the conservation and population status of an individual threatened plant species, to more general papers tackling conservation strategies and policies at a national, regional or international level.

Most citations date from the late 1970s onwards. Earlier publications may also be cited where they describe vegetation cover and rates of habitat loss, since these may provide valuable perspectives for conservationists and land managers on which to gauge present-day changes.

To assist in our selection of references from the potentially vast field of relevant literature, we chose to adopt the following set of criteria:

References on the following are *included*:

- Rare and threatened plants and their conservation, including papers outlining the status of a species that is threatened nationally or locally, but which may be common on a world scale. (Includes a few papers on lower plants and fungi, but coverage for these groups is very sparse);

- Erosion and conservation of plant genetic resources, *in situ* and/or *ex situ*, especially accounts relating to individual crops and their wild relatives, as well as medicinal and other economic plants;

- Threats to plant life, including invasive and introduced species where they pose a threat, acid rain, etc.;

- Extent and loss of habitats and vegetation;

- Floristic and vegetation accounts of centres of plant diversity and/or endemism, including papers using satellite imagery;

- A selection of forestry papers, especially those covering forest policy and management for conservation;

- Protected areas, especially those for which plant species inventories and accounts of their vegetation have been compiled;

- Role and work of botanical institutions, especially botanic gardens, in plant conservation;

- *Ex situ* plant collections where held for conservation purposes (e.g. gene banks);

- Conservation thinking and policy, especially background papers on conservation in a country or region;

- Reviews of the legal basis for conservation relevant to plants;

- Bibliographies on the above topics.

HISTORICAL BACKGROUND

The origin of this bibliography dates back to the mid-1970s and the work of Gren Lucas and Hugh Synge at the Royal Botanic Gardens, Kew. At that time, plant conservation references were recorded in a card index which was later to be maintained and developed by IUCN's Threatened Plants Committee Secretariat, now the Threatened Plants Unit (TPU) of the World Conservation Monitoring Centre.

By 1979, Dawn Scott of the Royal Botanic Gardens' library staff began a more systematic screening of the literature for plant conservation material. References found during indexing of new accessions and in preparing the *Kew Record*, an annual taxonomic bibliography, were collated in the monthly *Conservation Current Awareness List (CCAL)*, which was produced by the Kew Library from January 1979 to January 1981. Coverage was worldwide, with the exception of the United Kingdom. However, *CCAL* was never widely available outside of Kew; its main use was by Kew's own staff and visiting scientists.

With the purchase of a computer system in 1981, TPU were able to begin managing citations to plant conservation references in a computer datafile, using software written by Hugh Synge, later to be up-dated by computer staff in the WCMC. Initially, back issues of *CCAL* were incorporated together with additional references as they were screened by the Kew Library staff. Additional sources of information included commercial abstracting databases (such as Agricola and CABS) together with recently published regional bibliographies on aspects of plant conservation. Here, we acknowledge the valuable bibliography published by the New York Botanical Garden Library entitled *Endangered Plant Species of the World and their Endangered Habitats: A Compilation of the Literature* (*1978, 1985*) and *A Bibliography of Plant Conservation in the Pacific Islands: Endangered Species, Habitat Conversion, Introduced Biota* (*1987*), by R.A. DeFilipps.

The database has now become an important research tool. Its contents have greatly expanded beyond threatened plants to cover topics such as protected areas, the role of botanic gardens in conservation, the loss of habitats, other types of threats to plant life, and more general works on conservation.

Aware of the increasing demand for easier access to this information, the IUCN/WWF Joint Plant Advisory Group, during its 1988 meeting at Kew, recommended that the bibliography should be made more widely available. An opportunity to implement this recommendation came in the summer of 1989 when the Royal Botanic Gardens, Kew, found the funds for its publication. Since then, RBG, Kew, and WCMC, have worked in close collaboration to prepare it for publication. The present volume is the result.

THE FUTURE

As it is intended to produce supplements and revisions, as time and finance permit, we warmly invite users to send corrections and additional references. Such collaboration will be greatly appreciated as this will help to improve the coverage and quality of future editions. Please forward any communication to either of the addresses given below:

Threatened Plants Unit
World Conservation Monitoring Centre
219c Huntingdon Road
Cambridge, CB3 0DL

Conservation Unit
Economic and Conservation
  Section (ECOS)
Royal Botanic Gardens
Kew, Richmond
Surrey, TW9 3AB

HOW TO USE THIS BIBLIOGRAPHY

Reference citations are grouped into three sections: *General, Regional* and *Country.*

References included under the *General* section are those which tend to be geographically non-specific. The *Regional* section includes those references covering more than one country but confined to a particular geographical region of the world. For example, a paper dealing with ecological guidelines for the Middle East would be cited under the Regional heading *North Africa and the Middle East* rather than be cited under each country to which it was relevant. 18 regions are recognised in this volume, as listed in the Contents page.

The *Country* section contains references specific to individual countries. Where a publication is relevant to a small number of countries, then the reference is repeated under the appropriate country headings. Country names used are those adopted by the International Standards Organization (ISO).

Large countries are subdivided into their respective states or provinces. Thus, a paper dealing with threatened plants of western China would appear under the country heading *China.* However, a paper on threatened plants of Yunnan and Sichuan would be found under the appropriate province headings.

Occasionally, a paper may be listed under both *Regional* and *Country* or state headings, where the subject matter seems relevant to more than one geographical entity. A case in point would be a general paper on the conservation of Pacific Island floras which included detailed case studies from Hawaii. Such a reference would be cited under the Regional heading *Pacific Islands* as well as under *U.S. – Hawaii.*

Within each section, references are arranged in alphabetical order of author. For a full list of countries and states, and commonly used synonyms, see the *Geographical Index.*

Format of Reference Citations

References are cited as follows:

> **Citation number**   **Author (Date). Title. Title translation (if original not in English). Publication details. Notes, e.g. language key; Cross-references.**

Some explanation is necessary for some of these items:

Citation Number

All references are numbered in the left hand column according to their numerical order of appearance in the bibliography. This number is designed to help users find specific references quickly when using the subject indices given at the end of the book.

Title Translation

We have provided English translations, or in some cases, transliterations, for Polish and Russian titles. Translations are given in round brackets after the foreign title. Some English translations from German and Spanish have also been made in cases where titles are long and complex.

Where references were screened from secondary sources and abstracts, it has often only been possible to include a translated title. In these cases, the English translation is given in square brackets.

Because of the limitations of the computer system, accents have sadly had to be omitted.

Language key

Following each citation, language abbreviations are given for all papers not written in English. The language code of the main text is given first, followed in brackets by the codes for other langauges, if used. The latter usually refer to abstracts or summaries. Abbreviations are as follows:

Afrikaans	Af	Italian	It
Albanian	Al	Japanese	Ja
Arabic	Ara	Kazakh	Ka
Armenian	Arm	Korean	Ko
Azerbaijani	Az	Latvian	Latv
Belorussian	Be	Lithuanian	Li
Bulgarian	Bu	Macedonian	Mac
Catalan	Ca	Mongolian	Mong
Chinese	Ch	Norwegian	No
Czech	Cz	Polish	Pol
Danish	Da	Portuguese	Por
Dutch	Du	Rumanian	Rum
English	En	Russian	Rus
Estonian	Es	Serbo-Croatian	Se
Finnish	Fi	Slovakian	Slk
Flemish	Fl	Slovenian	Sln
French	Fr	Spanish	Sp
German	Ge	Swedish	Sw
Georgian	Geo	Thai	Th
Greek	Gre	Turkish	Tu
Hebrew	He	Turkmenian	Turkm
Hindi	Hi	Ukrainian	Uk
Hungarian	Hu	Vietnamese	We
Indonesian	In		

Notes

Where the subject coverage of a reference is unclear from its title, sometimes we have added a phrase or short sentence as a guide to users. References containing illustrations and/or maps are indicated by the use of *'Illus.'* and/or *'Maps'*. Illustrations are assumed to be black and white unless *'Col. illus.'* is used.

Cross-References

The following cross-reference codes to broad *subject areas* covered by the reference, are added in brackets after each citation, as appropriate:

B – Bibliography (i.e. whether the reference itself is a bibliography)
E – Economic Plants and/or Genetic Resources
G – Botanic Gardens and Cultivation (includes seed banks and all aspects of *ex situ* plant conservation)
L – Legislation (includes relevance to existing and proposed laws and conventions)
P – Protected Areas (includes nature reserves and national parks, as well as proposals for their establishment)
T – Threatened Plants

Sample citation 1:

Trautmann, W., Korneck, D. (1978). Zum Gefahrdungsgrad der Pflanzenformationen in der Bundesrepublik Deutschland. (The threat to vegetation formations in the German Federal Republic.) Veroff. Natursch. Landschaftspfl. Baden-Wurttemberg, 11: 35-40. Ge. Illus. (T)

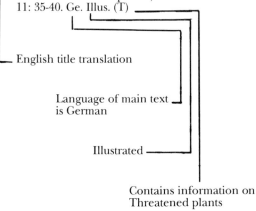

English title translation

Language of main text is German

Illustrated

Contains information on Threatened plants

Sample citation 2:

Kondratiuk, I., Ivashyn, D.S. Burda, R.I. (1978) [Study of the flora and vegetation of the Stanitsa-Lugansk branch of the Lugansk State Reservation]. *Introd. Aklim. Rosl. Ukr. Akad. Nauk. URSS*, 13: 18-22. Uk (Rus). (P)

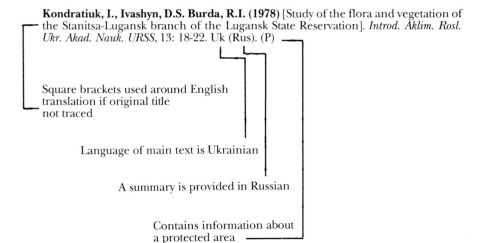

Square brackets used around English translation if original title not traced

Language of main text is Ukrainian

A summary is provided in Russian

Contains information about a protected area

Index

In addition to the cross-reference codes appearing after each reference citation, the user may also access the bibliography by the three indices for *Countries, Families* or *Species* given at the back of the book.

An explanatory note on the use of the indices is given at the beginning of each.

IUCN RED DATA BOOK CATEGORIES

The following IUCN Red Data Book categories and their codes are occasionally referred to in the notes following reference citations:

Extinct (Ex)

Taxa which are no longer known to exist in the wild after repeated searches of their type localities and other known or likely places.

Endangered (E)

Taxa in danger of extinction and whose survival is unlikely if the causal factors continue operating.

Included are taxa whose numbers have been reduced to a critical level or whose habitats have been so drastically reduced that they are deemed to be in immediate danger of extinction.

Vulnerable (V)

Taxa believed likely to move into the Endangered category in the near future if the causal factors continue operating.

Included are taxa of which most or all the populations are *decreasing* because of over-exploitation, extensive destruction of habitat or other environmental disturbance; taxa with populations that have been seriously *depleted* and whose ultimate security is not yet assured; and taxa with populations that are still abundant but are *under threat* from serious adverse factors throughout their range.

Rare (R)

Taxa with small world populations that are not at present Endangered or Vulnerable, but are at risk.

These taxa are usually localized within restricted geographical areas or habitats or are thinly scattered over a more extensive range.

Indeterminate (I)

Taxa *known* to be Extinct, Endangered, Vulnerable or Rare but where there is not enough information to say which of the four categories is appropriate.

Candidate (C)

Taxa whose status is being assessed and which are suspected but not definitely known to belong to any of the above categories.

Not Threatened (nt)

Neither rare nor threatened.

Insufficiently Known (K)

Taxa that are suspected but not definitely known to belong to any of the above categories, following assessment, because of the lack of information.

LEGAL CONVENTIONS

WHC: *Convention concerning the Protection of the World Cultural and Natural Heritage* (Unesco, Paris, 1972). This convention provides for the designation of areas of 'outstanding universal value' as world heritage sites, with the principle aim of fostering international cooperation in safeguarding these important areas. Sites, which must be nominated by the signatory nation responsible, are evaluated for their world heritage quality before being declared by the international World Heritage Committee. The convention entered into force 17 December 1975. For each party adopting the convention since August 1975 the convention enters into force four months after the date of adoption.

CITES: *Convention on International Trade in Endangered Species of Wild Fauna and Flora.* CITES is an international agreement designed to prohibit the international trade in an agreed list of currently endangered species and to control and monitor the international trade in additional species that might otherwise become endangered. The Convention works by issuance of import and export licences by designated government Management Authorities, who are advised by designated Scientific Authorities.

RAMSAR: *Convention on Wetlands of International Importance especially as Waterfowl Habitat* (Ramsar, Iran, 1971). An international treaty providing the framework for international cooperation for the conservation of wetland habitats. The Convention places general obligations on contracting party states relating to the conservation of wetlands throughout their territory, with special obligations pertaining to those wetlands which have been designated in a 'List of Wetlands of International Importance'. The Convention entered into force 21 December 1975. For each party adopting the Convention since August 1975 the Convention enters into force four months after the date of adoption.

BERNE: *Convention on the Conservation of European Wildlife and Natural Habitats* (Berne, Switzerland, 1979). An international treaty whose aim is to conserve wild flora and fauna and their natural habitats, especially, those species and habitats whose conservation requires the co-operation of several Council of Europe Member States. The Convention places general obligations on contracting parties in respect of wild flora, fauna and their habitats and more specific obligations pertaining to named species and various methods of their exploitation according to four Appendices. The Convention entered into force 19 September 1979.

ACRONYMS

AABGA	American Association of Botanic Gardens and Arboreta
AFA	American Forest Association
AIDEC	Association for European Industrial Development and Economic Cooperation (Netherlands)
ANPWS	Australian National Parks and Wildlife Service
AOS	American Orchid Society
ASB	Association of Southeastern Biologists (USA)
ASEAN	Association of South East Asian Nations
ASPO	American Society of Planning Officials
AVIS	Arid Vegetation Information System
BES	British Ecological Society
BIOTROP	Regional Centre for Tropical Biology (Indonesia)
BLM	Bureau of Land Management
BRC	Biological Records Centre (UK)
BSBI	Botanical Society of the British Isles
BSE	Botanical Society of Edinburgh (UK)
BSI	Botanical Survey of India
CNPPA	IUCN Commission on National Parks and Protected Areas
CNPS	California Native Plant Society
CNRS	Centre National de la Recherche Scientifique
CSIR	Council for Scientific and Industrial Research
CSIRO	Commonwealth Scientific and Industrial Research Organization
CSSA	Cactus and Succulent Society of America
DSIR	Department of Science and Industrial Research (New Zealand)
ECE	Economic Commission for Europe
EMBRAPA	Empresa Brasileira de Pesquisa Agropecuària
ESCAP	Economic and Social Commission for Asia and the Pacific
EUCARPIA	European Association for Research on Plant Breeding
FAO	Food and Agriculture Organization of the United Nations
FEEMA	Fundacao Estadual de Engenhacia do Meio Ambiente
GEMS	Global Environmental Monitoring System (UNEP)
GENIS	Genetic Resources Information System
IBP	International Biological Programme
IBP/CT	International Biological Programme Conservation Section
IBRD	International Bank for Reconstruction and Development
ICARDA	International Centre for Agricultural Research in the Dry Areas
ICASALS	International Centre for Arid and Semi-Arid Land Studies
ICOM	International Council of Museums
ICONA	Instituto Nacional para la Conservacion de la Naturaleza
ICPB	International Council for Bird Preservation
ICRISAT	International Crop Research Institute for the Semi-Arid Tropics (India)
IIED	International Institute for Environment and Development
INEAC	Institut National pour l'Etude Agronomique du Congo
INIA	Instituto Nacional de Investigaciones Agrarias
IOS	International Organization for Succulent Plant Study
IRRI	International Rice Research Institute
IUBS	International Union for Biological Sciences
IUCN	International Union for Conservation of Nature and Natural Resources
IUFRO	International Union of Forest Research Organizations
LIPI	Indonesian Institute of Sciences
MAB	Man and the Biosphere Programme (Unesco)
MIT	Massachussetts Institute of Technology
MNS	Malayan Nature Society
NAPRALERT	Natural Products Alert (USA)
NCC	Nature Conservancy Council (UK)
NOAA	National Oceanic and Atmospheric Administration (USA)
NPCA	National Parks and Conservation Association
NRDC	Natural Resources Defense Council
OCS	The Oceanic Society (USA)

ODA	Overseas Development Administration (UK)
OECD	Organization for Economic Cooperation and Development
OPTIMA	Organization for the Phyto-taxonomic Investigation of the Mediterranean Area
ORSTOM	Office de la Recherche Scientifique et Technique Outre-Mer (Office for Scientific and Technical Research Overseas)
RGS	Royal Geographical Society
RSNC	Royal Society for Nature Conservation
SCAR	Scientific Committee on Antarctic Research
SEAMEO	Southeast Asian Regional Center for Tropical Biology
SPREP	South Pacific Regional Environment Programme
TNC	The Nature Conservancy (U.S.A.)
TPC	IUCN Threatened Plants Committee Secretariat
TPU	WCMC Threatened Plants Unit (formerly TPC, and the IUCN Threatened Plants Unit)
UCLA	University of California at Los Angeles (USA)
UNCTAD	United Nations Conference on Trade and Development
UNDP	United Nations Development Programme
UNEP	United Nations Environment Programme
Unesco	United Nations Educational, Scientific and Cultural Organization
UNFPA	United Nations Fund for Population Activities
USAID	US Agency for International Development
USDA	United States Department of Agriculture
USFS	US Department of the Interior, Fish and Wildlife Service
WCI	World Conservation International (USA)
WCMC	World Conservation Monitoring Centre
WRI	World Resources Institute
WWF	World Wide Fund for Nature (formerly World Wildlife Fund)

GENERAL REFERENCES

GENERAL

1 **Abramovitz, J.N. (1989).** A survey of U.S.-based efforts to research and conserve biological diversity in developing countries. Washington, D.C., World Resources Institute. 71p. Includes lists of projects and organizations.

2 **Adams, B. (1982).** Third World EIA: a whole new ball game. *Ecos*, 3(3): 30-35. Environmental impact assessment.

3 **Adams, W.M., Rose, C.I. (1978).** The selection of reserves for nature conservation. Discussion Paper in Conservation No. 20. London, University College. 34p. (P)

4 **Adamus, P.R., Garrett, C.C. (1978).** Evaluating species for protection in natural areas. *Biol. Conserv.*, 13: 165-178. (P)

5 **Akeroyd, J.R. (1988).** *Biarum davisii* and its conservation cultivation. *Bot. Gard. Conserv. News*, (2): 18-19. Illus., map. (G/T)

6 **Aldrich, J.W. (1975).** Our wildflowers and a program to protect them. *The Conservationist*, 29(5): 23-29. (T)

7 **Aldrich, R.C. (1979).** Remote sensing of wildlife resources: a state-of-the-art review. *In* U.S. Department of Agriculture Forest Service. General Technical Report. Fort Collins, CO, Rocky Mountain Forest and Range Experiment Station. 56p. (P)

8 **Allan, D. (1985).** Threatened plants and animals of the world. *Envir. Conserv.* 12(1): 79-81. (T)

9 **Allen, D.E. (1987).** Changing attitudes to nature conservation: the botanical perspective. *Biol. J. Linn. Soc.*, 32(2): 203-212.

10 **Allen, E.B. (1988).** The reconstruction of disturbed arid lands: an ecological approach. Boulder, CO, Westview Press. 267p.

11 **Allen, R. (1975).** Conservation and the new economic order. *IUCN Bull.*, n.s., 6(7): 25-26.

12 **Allen, R. (1980).** How to save the world. *The Ecologist*, 10(6/7): 190-194. Illus.

13 **Allen, R. (1980).** How to save the world: strategy for world conservation. London, Kogan Page. 150p.

14 **Allen, R.P. & P. (1982).** The case for *in situ* conservation of crop genetic resources. *Nature and Resources*, 18(1): 15-20. Crop. (E)

15 **Alleweldt, G. (1983).** Collection, conservation and evaluation of genetic resources in grapevine. *Bull. Oiv*, 56(624): 91-103. Fr. Gene banks. (E)

16 **Almaca, C.I. (1978).** Ecologie humaine et conservation de la nature. *Bol. Soc. Port. Cienc. Nat.*, 18: 91-92. Fr.

17 **Altieri, M., Merrick, L. (1987).** *In situ* conservation of crop genetic resources through maintenance of traditional farming systems. *Econ. Bot.*, 41(1): 86-96. (E)

18 **Altieri, M.A., Merrick, L.C. (1988).** Agroecology and *in situ* conservation of native crop diversity in the Third World. *In* Wilson, E.O., ed. Biodiversity. Washington, D.C., National Academy Press. 361-369. Crop genetic resources. (E/G)

19 **Altieri, M.A., Merrick, L.C., Anderson, M.K. (1987).** Peasant agriculture and the conservation of crops and wild plant resources. *Conserv. Biol.*, 1(1): 49-58. (Sp). Illus. (E/G)

1

GENERAL

20 **Anca, A. (1979).** Cresterea aportului padurilor la protectia mediului inconjurator, sarcina actuala si de perspectiva. (Contribution of forests to present and future environmental protection.) *Rev. Padur. Ind. Lemn. Celul Hirtie Silvic. Expl. Padur.*, 94(4): 191.

21 **Anderson, M.P. (1909).** The passing of the wildflowers. *J. New York Bot. Gard.*, 10: 147-163. (T)

22 **Andrews, S., Natkin, M., comps (1983).** The world environment handbook: a directory of government natural resource management agencies in 144 countries. New York, World Environment Center. xiv, 130p.

23 **Angiboust, A. (1987).** U.I.P.P.: 5e Congres national de la protection des plantes. La protection de l'environnement recherche du zero de pollution. *Rev. Hort.*, 278: 4-5. Fr.

24 **Anishetty, N.M., Ayad, W.G., Toll, J. (1981).** Directory of germplasm collections: 3. Cereals. IV *Sorghum* and millets. AGB: IBPGR/81/55. Rome, IBPGR. vi, 37p. (E/G)

25 **Anon. (1972).** Vanishing plantlife. *Ecol. Today*, 2(1): 8. (T)

26 **Anon. (1973).** Ecologists plan 22 studies of endangered species. *Sci. News*, (1973): 87. (T)

27 **Anon. (1975).** Convention on International Trade in Endangered Species of Wild Fauna and Flora. International trade in endangered species of wild fauna and flora: Convention done in Washington March 3, 1973 ... entered into force July 1, 1975. *In* United States treaties and other international agreements. 27(2): 1087-1361. (L)

28 **Anon. (1975).** Wild Life Protection Bill. *BSE News*, 17: 2. (L)

29 **Anon. (1975).** Wildflowers in distress. *Sci. News*, 107: 338.

30 **Anon. (1976).** Endangered plants: plant-lovers needed. *Nat. Parks Conserv. Mag.*, 50(8): 27. (T)

31 **Anon. (1978).** Conference on the practical role of botanic gardens in the conservation of rare and threatened plants. (Konferenz uber die Rolle botanischer Garten bei der Erhaltung seltener und bedrohter Pflanzen.) *Gartn.-Bot. Brief*, 58: 5-8. En, Ge. (G/T)

32 **Anon. (1978).** Desert vegetation has potential uses, but ecological issues need to be considered. *ICASALS Newsletter*, 11(4): 1, 2. Arid Lands Conference on Plant Resources, Texas Tech. Univ., 8-15 October 1978. (E)

33 **Anon. (1978).** Genetische variatie via de (genen)bank. *Tuinderij*, 18(16): 16-19. Du. Gene bank. (G)

34 **Anon. (1978).** Unesco's programmes on natural resources and environment for 1979-80. *Nat. Resources*, 14(4): 13-28.

35 **Anon. (1979).** A plea for plants. *IUCN Bull.*, 10(2): 9, 16. Illus. (T)

36 **Anon. (1979).** Extinct species rise from the grave. *New Sci.*, 84(1176): 94. Illus. (G/T)

37 **Anon. (1979).** La conservazione delle foreste nel mondo. *Nat. Montagna*, 26(2): 10-14. VIII Congresso Forestale Mondiale, Jakarta, 16-28 October 1978. It.

38 **Anon. (1979).** New sites on the World Heritage List. *Nat. Resources*, 15(4): 21-23. (P)

39 **Anon. (1979).** Red for danger. *Conserv. Rev.*, 18: 4-5. Col. illus. *The IUCN Plant Red Data Book*, 1978. (T)

40 **Anon. (1979).** The attack on flower power. *IUCN Bull.*, 10(2): 12-13. Illus. (T)

GENERAL

41 **Anon. (1979).** What is CITES and why is it important for plants? *Threatened Pl. Commit. Newsl.*, 4: 6-12. (L/T)

42 **Anon. (1979).** Why national parks are important to you! *Victoria's Resources*, 21(1): 39-41. Illus. (P)

43 **Anon. (1980).** "Extinct" species found in botanic gardens. *Threatened Pl. Committ. Newsl.*, 5: 1-2. (G/T)

44 **Anon. (1980).** Endangered species information request. *AABGA Newsl.*, 70: 12. (T)

45 **Anon. (1980).** International consultation on wildlife resources for rural development 7-11 July 1980, Hyderabad, India. *Tigerpaper*, 7(3): 1-27. Illus.

46 **Anon. (1980).** Plant material: endangered species information request. *AABGA Newsl.*, 72: 3. (T)

47 **Anon. (1980).** Survival or extinction. *Mon. Bull. Alp. Gard. Club B.C.*, 23(1): 10-11. (T)

48 **Anon. (1980).** The strategy at a glance: executive summary. *IUCN Bull.*, 11(4): 38-39. Illus. World Conservation Strategy.

49 **Anon. (1981).** Appendix 3. The IUCN Red Data Book categories. *In* Synge, H., *ed.* The biological aspects of rare plant conservation. Chichester, Wiley. 531-540. Proceedings of International Conference, King's College, Cambridge, 14-19 July 1980. (T)

50 **Anon. (1981).** Convention on International Trade in Endangered Species of Wild Fauna and Flora (CITES) comes of age. *Tigerpaper*, 8(3): 23-25. Illus. From *IUCN Bull.*, Jan/Feb 1981. (L/T)

51 **Anon. (1981).** Endangered species. *Canad. Bot. Ass. Bull.*, 14(4): 42. (T)

52 **Anon. (1981).** International Carnivorous Plant Society Seed Bank July 27, 1981. *Carniv. Pl. Newsl.*, 10(3): 59. (G)

53 **Anon. (1981).** Ist diese Welt noch zu retten? *Gartn.-Bot. Brief*, 68: 7-9. Ge.

54 **Anon. (1981).** Termine des Ausverkaufs. *Gartn.-Bot. Brief*, 68: 15. Ge, En.

55 **Anon. (1981).** The conservation of lower plants: report from a panel discussion. *In* Synge, H., *ed.* The biological aspects of rare plant conservation. Chichester, Wiley. 125-137. Proceedings of International Conference, King's College, Cambridge, 14-19 July 1980. (T)

56 **Anon. (1981).** World Conservation Strategy: a year on. *IUCN Bull.*, 12(3-4): 24.

57 **Anon. (1981).** XIII International Botanical Congress, Sydney, Australia. 21-28 August 1981. *Threatened Pl. Commit. Newsl.*, 8: 1-3. Resolutions on plant conservation. (P)

58 **Anon. (1982).** A proposal for an I.O.S. resolution on the international trade convention. *IOS Bull.*, 4(1): 11. (L)

59 **Anon. (1982).** Endangered species. *AABGA Newsl.*, 96: 3. (T)

60 **Anon. (1982).** Endangered species. *BioScience*, 32(8): 652. (L/T)

61 **Anon. (1982).** Red Data species. *Threatened Pl. Commit. Newsl.*, 10: 11-12. (T)

62 **Anon. (1982).** Trade in endangered species. *Cact. Succ. J.* (U.K.), 37(1): 20. (L/T)

63 **Anon. (1982).** Trade in rare plants. To what extent are wild plant species threatened by trade? *Threatened Pl. Commit. Newsl.*, 10: 7-11. (T)

64 **Anon. (1982).** UN acts to save the world's seeds. *New Sci.*, 92(1284): 786. Illus. (E/G)

GENERAL

65 **Anon. (1983).** House leader calls for special hearing on germplasm system. *Diversity*, 4: 3.

66 **Anon. (1983).** How the World Heritage Convention works. *Ambio*, 12(3-4): 140-144. (L/P)

67 **Anon. (1983).** Kew's out to save wild orchids. *J. Wellington Orchid Soc.*, 7(7): 147-148. (G)

68 **Anon. (1983).** NPGS network news. First National Germplasm Conference hailed a success. *Diversity*, 4: 6. National Plant Germplasm System. (E)

69 **Anon. (1983).** Newly gathered orchids to be banned in LBOS shows. *Asian Orch. Ornamentals*, 3(12): 15. Los Banos Orchid Society. (T)

70 **Anon. (1983).** Plant breeding research forum warns government inaction may cause future food shortages. *Diversity*, 4: 4-5. (E)

71 **Anon. (1983).** The Bali declaration. *Envir. Conserv.*, 10(1): 73. World National Parks Congress, Bali, 11-22 Oct., 1982. (P)

72 **Anon. (1983).** WWF formulating plans for the Year of the Plant. *GC & HTJ*, 193(24): 5. (T)

73 **Anon. (1984).** Action plan for biosphere reserves. *Nature & Resources*, 20(4): 11-22. (P)

74 **Anon. (1984).** Conserving biological diversity. *Pl. Sci. Bull.*, 30(2): 9-10.

75 **Anon. (1984).** IUCN/WWF Plants Programme. *Threatened Pl. Newsl.*, 13: 4. (E/G/L/P/T)

76 **Anon. (1984).** Orchids surviving extinction threat. *World Wildl. News*, no. 28: 5. Illus. (T)

77 **Anon. (1984).** WWF launches Plants Campaign. *Threatened Pl. Newsl.*, 13: 2-3. (E/G/L/P/T)

78 **Anon. (1985).** Areas of critical environmental concern. Part II. 1951-1970. *Wilderness*, 48(168): 31-39.

79 **Anon. (1985).** Code of conduct for growers and collectors of succulent plants. *Cact. Succ. J.* (U.K.), 3(1): 3-6. (G)

80 **Anon. (1985).** The top twelve plants. *Spec. Surv. Commis., Newsl.*, 5: 18-19. Illus. (T)

81 **Anon. (1987).** Threatened protected areas of the world, additions for 1986. *Parks*, 12(1): 2. Lists 23 additions to the register updated annually by IUCN's Protected Areas Data Unit. Current register totals 74 sites worldwide. (P)

82 **Anon. (1988).** A role for seed-bank surveys in conservation? *Oryx*, 22(3): 139. Soil seed banks may contain rare species. (T)

83 **Anon. (1988).** Medicinal plants. *Threatened Pl. Newsl.*, 20: 2-3. Includes the Chiang Mai Declaration. (E/T)

84 **Anon. (1988).** Twelve critically endangered species. *Species*, 10: 22-24. Illus. Includes 3 plant taxa. (T)

85 **Anon. (1989).** Reports from the Reunion congress. Botanic gardens meet in La Reunion. *Bot. Gard. Conserv. News*, 1(4): 11-16. Illus. Reports on the Second International Botanic Gardens Conservation Congress, Reunion, 24-28 April 1989. (G)

86 **Apel, J. (1981).** Exactly how modern is the international seed exchange ? *Threatened Pl. Commit. Newsl.*, 8: 7-9. (G)

GENERAL

87 **Appleman, F.J. (1955).** Natuurebescherming. *Techtona*, 43: 219-235.

88 **Arbuckle, J.G., Brown, M.A., Bryson, N.S., eds (1985).** Environmental Law Handbook. 8th ed. Rockville, Maryland, Government Institutes, Inc. 586p. (L)

89 **Arnklit, F. (1978).** Setting up a practical small seed bank. *Gartn.-Bot. Brief*, 58: 46-47. (Ge). Botanical Garden of Copenhagen. (G)

90 **Ashton, P.S. (1987).** Biological considerations in *in situ* vs *ex situ* plant conservation. *In* Bramwell, D., Hamann, O., Heywood, V., Synge, H., *eds.* Botanic gardens and the World Conservation Strategy. London, Academic Press. 117-130. (Sp). (G/P)

91 **Ashton, P.S. (1988).** Conservation of biological diversity in botanical gardens. *In* Wilson, E.O., *ed.* Biodiversity. Washington, D.C., National Academy Press. 269-278. (E/G)

92 **Aslin, R.G., Gallaher, H.G. (1978).** Planting trees for conservation: a step-by-step approach. *C. Kansas State University*, 596: 6. Illus.

93 **Astley, D., Innes, N.L. (1982).** Genetic resources of *Allium* species: a global report. AGPG/IBPGR/81/77. Rome, IBPGR. iii, 38p. Illus. (E/T)

94 **Attonn, H. (1988).** Foundation group's efforts geared to some rare isle species, last hope for many plants. *Notes Waimea Arbor.*, 15(1): 10-11. Illus. (T)

95 **Ave, J., King, V., de Wit, J. (1983).** West Kalimanton, a bibliography. Leiden, Koninklijk Instituut voor Taal-Land-en Volkenkunde (Biographical Series 13.) 1855 refs.

96 **Ayad, G., Anishetty, N.M. (1980).** Directory of germplasm collections: 1. Food legumes. Rome, IBPGR. 22p. (E/G)

97 **Ayensu E.S. (?)** The plight of the world's vanishing plants. Washington, D.C., Smithsonian Institution. 12p. (T)

98 **Ayensu, E.S. (1976).** International cooperation among conservation-orientated botanical gardens and institutions. *In* Simmons, J.B., *et al.*, *eds.* Conservation of threatened plants. NATO Conference Series 1: Ecology, vol. 1. New York and London, Plenum Press. 259-269. Proc. Conf. on the Functions of Living Plant Collections in Conservation and Conservation-Orientated Research and Public Education, Kew, 2-6 Sept. 197 (G)

99 **Ayensu, E.S. (1978).** Calling the roll of the world's vanishing plants. *Smithsonian*, 9(8): 122-129. (T)

100 **Ayensu, E.S. (1979).** International convention on endangered species. *IUBS Newsl.*, 15: 3-4. CITES, San Jose, Costa Rica, 19-30 March 1979. (L/T)

101 **Ayensu, E.S. (1981).** International approaches to the conservation of wild orchids. *In* Proceedings of the orchid symposium, 13th International Botanical Congress, Sydney, 1981. Orchid Society of New South Wales. 27-34. Col. illus. (L/P/T)

102 **Ayensu, E.S. (1983).** The world's diminishing plant resources. *In* Jain, S.K., Mehra, K.L., *eds.* Conservation of tropical plant resources. Howrah, Botanical Survey of India. 19-28. (T)

103 **Ayensu, E.S., DeFilipps, R.A. (1981).** Smithsonian Institution endangered flora computerized information. *In* Morse, L.E., Henifin, M.S., *eds.* Rare plant conservation: geographical data organization. New York, The New York Botanical Garden. 111-122. (T)

104 **Ayensu, E.S., DeFilipps, R.A. (1981).** The international regulation of trade in endangered species of wild orchids. *Amer. Orch. Soc. Bull.*, 50(8): 959-967. (L/T)

GENERAL

105 **Ayensu, E.S., Heywood, V.H., Lucas, G.L., DeFilipps, R.A. (1984).** Our green and living world: the wisdom to save it. Cambridge, Cambridge University Press. 255p. Col. illus. Account of plant diversity, emphasizing economic values of plants; the need for and goals of conservation. (E/T)

106 **Aymonin, G.G. (1979).** Strategies pour la sauvegarde des especes vegetales: quelques aspects recents. *Bull. Soc. Étude Sci. Nat. Beziers, n.s.,* 7(48): 24-37. Fr. (T)

107 **Aymonin, G.G. (1980).** Quelques considerations sur la notion de "plantes en danger". *Bull. Soc. Bot. France, Lett. Bot.,* 127(1): 105-110. Fr. (T)

108 **Aymonin, G.G. (1986).** Connaitre les plantes protegees. Alpes-massifs centrales. Paris, France, La Federation Francaise des Societes de protection de la nature avec le concours du Ministre de l'Environnement. 44p. Fr. (T)

109 **Babareka, Ia. Z., Shapira, D.K., Naryzhnaia, T.I., Starkova, N.Iu. (1984).** [Rare form of *Berberis vulgaris* L.]. Vestsi Akademii navuk BSSR. Seryia biialagichnykh navuk = Isvestiia Akademii nauk BSSR. Seriia biologicheskikh nauk. Minsk "Navuka i tekhinika", (5): 7-9. Be (En, Rus). (T)

110 **Bachmura, F.T. (1971).** Economics of vanishing species. *Nat. Resources J.,* 11(4): 674-692. (E/T)

111 **Baerends, G.P. (1979).** Natuurbescherming emotioneel ot rationeel? *Natura,* 76(1): 1-6. Illus.

112 **Balick, M., ed. (1988).** The palm - tree of life: biology, utilization and conservation. *Advances in Economic Botany,* 6: 1-282. (E)

113 **Balzer, J.L. (1984).** Development and conservation: co-existence through rational planning. *In* McNeely, J.A., Miller, K.R., *eds.* National parks, conservation and development. The role of protected areas in sustaining society. Washington, D.C., Smithsonian Institution Press. 719-721.

114 **Banks, L. (1980).** Saving gardens and garden plants. *Garden* (London), 105(3): 89-93. Col. illus. (G/T)

115 **Barabe, D. (1979).** Role et organisation des collections de plantes vivantes. *Gartn.-Bot. Brief,* 61: 5-17. Fr (Ge). Illus. (G)

116 **Barber, J.C., Krugman, S.L. (1974).** Preserving forest-tree germplasm. *Amer. For.,* 80(10): 8-11, 42-43. (E/G)

117 **Barbey, P. (1981).** Des refuges pour plantes rares. *Schweizer Naturschutz,* 47(3): 30. Fr. (G/T)

118 **Barbier, R. (1977).** A propos des refuges a orchidees. *Orchidophile,* 30: 868-871. Fr. (P/T)

119 **Bark-Signon, I., Patzke, E. (1986).** Schutzenswerte Gebiete im Raum Duren: 2. Die Drover Heide. (TK 5205/3). *Gottinger Flor. Rundbr.,* 19(2): 110-116. Ge.

120 **Baskin, J.M. & C.C. (1978).** The seed bank in a population of an endemic plant species and its ecological significance. *Biol. Conserv.,* 14(2): 125-130. (G)

121 **Batisse, M. (1982).** The biosphere reserve: a tool for environmental conservation and management. *Envir. Conserv.,* 9(2): 101-111. Illus. (P)

122 **Batisse, M. (1984).** Biosphere reserves throughout the world: current situation and perspectives. *In* Unesco-UNEP. Conservation, science and society. Paris, Unesco. v-xii. Contributions to the First International Biosphere Reserve Congress, Minsk, Byelorussia/USSR, 26 September - 2 October 1983. (P)

123 **Bayer, M.B. (1983).** Comments on the IOS registration scheme and conservation. *IOS Bull.,* 4(2): 65. (G/T)

GENERAL

124 **Beaumont, P. (1989).** Environmental management and development of drylands. London, Routledge. 505p.

125 **Beckner, J. (1979).** Are orchids endangered? *Amer. Orchid Soc. Bull.*, 48(10): 1010-1017. (T)

126 **Beebe, S. (1984).** A model for conservation. *Nat. Conserv. News*, 34(1): 4-7.

127 **Bekele, E. (1986).** The interfaces of ecology, genetics, and biosystematics in conservation. *Symb. Bot. Upsal.*, 26(2): 101-113.

128 **Bell, C.R. (1984).** The role of plant propagation in taxonomy and conservation. *Combined Proceedings - International Plant Propagators' Society*, (1983 publ. 1984) 33: 170-175. (G)

129 **Bellamy, D. (1979).** The role of the media in conservation. *In* Synge, H., Townsend, H., *eds.* Survival or extinction. Proceedings of a conference held at the Royal Botanic Gardens, Kew, 11-17 September 1978. Kew, Bentham-Moxon Trust. 175-177.

130 **Beloussova, L.S., Denisova, L.V. (1983).** Redkie rasteniya mira. (Rare plants of the world.) Moscow, Lesnaya Prom-st'. 340p. Rus. Col. illus. (T)

131 **Bender, M. (1984).** Evergreen tree listed as endangered. *Endangered Spec. Tech. Bull.*, 9(2): 4-5. (T)

132 **Bennett, D. (1988).** Against a triage strategy for threatened species. *Buzzworm*, 1(1): 30-31. (T)

133 **Benseler, R.W. (1987).** Conservation ethics, animals, and rare plant protection. *In* Elias, T.S., *ed.* Conservation and management of rare and endangered plants. Proceedings from a conference, Sacramento, California, 5-8 November 1986. Sacramento, California Native Plant Society. 623-626. Rare plant protection requires a commitment to ecosystem conservation. (T)

134 **Benson, L. (1977).** How do you preserve an ecosystem? *Fremontia*, 5(2): 3-7. (P)

135 **Berg, K., Bittman, R. (1988).** Rediscovery of the Humboldt milk-vetch. *Fremontia*, 16(1): 13-14. (T)

136 **Bernays, E.M. (1892).** Proposition dans le but de preserver les especes en voie de disparition. *Bull. Soc. Roy. Bot. Belgique*, 31: 158-162. Fr.

137 **Bettencourt, E., Konopka, J., Damania, A.B. (1989).** Directory of germplasm collections: 1. Food legumes. Rome, IBPGR. 190p. (E/G)

138 **Beyer, I. (1979).** The Botanic Gardens Conservation Co-ordinating Body. *Threatened Pl. Commit. Newsl.*, 4: 1-5. (G/T)

139 **Beyer, I., Synge, H., Worth, W. (1983).** Botanic garden news. *Threatened Pl. Newsl.*, 12: 5-7. Botanic Gardens Conservation Co-ordinating Body: use of computerised botanic garden records. (G)

140 **Beyer, R.I. (1981).** TPC Botanic Gardens Conservation Co-ordinating Body. *Threatened Pl. Commit. Newsl.*, 8: 5-7. (G)

141 **Bishop, R.C. (1977).** The economics of endangered species: alternative approaches to public decisions. Madison, Wisconsin, College of Agricultural and Life Sciences. University Dept of Agricultural Economics staff papers series. (T)

142 **Bishop, R.C. (1978).** Endangered species and uncertainty: the economics of a safe minimum standard. *Amer. J. Agric. Econ.*, 60(1): 10-18. (T)

143 **Biswas, M.R. (1978).** U.N. Conference on desertification in retrospect. *Envir. Conserv.*, 5(4): 247-262.

GENERAL

144 **Blackie, J.R., Ford, E.D., Horne, J.E.M., Kinsman, D.J.J., Last, F.T., Moorhouse, P. (1980).** Environmental effects of deforestation: an annotated bibliography. Occasional Paper No. 10. Ambleside, U.K., Freshwater Biological Association. 173p. (B)

145 **Blower, J. (1984).** National parks for developing countries. *In* McNeely, J.A., Miller, K.R., *eds.* National parks, conservation and development. The role of protected areas in sustaining society. Washington, D.C., Smithsonian Institution Press. 722-727. (P)

146 **Boardman, R. (1981).** International organization and the conservation of nature. London, Macmillan. 215p. (L)

147 **Bobisud, L.B., Neuhaus, R.J. (1975).** Pollinator constancy and survival of rare species. *Oecologia*, 21(3): 263-272. (G/T)

148 **Boecklen, W.J. (1986).** Optimal design of nature reserves: consequences of genetic drift. *Biol. Conserv.*, 38(4): 323-338. (P)

149 **Bolgiano, C. (1985).** How to map the future. *Amer. Forests*, 91(10): 22-24. Illus., maps.

150 **Bond, C. (1978).** Wild plant racketeers. *African Wildlife*, 32: 29-32.

151 **Bongers, W. (1981).** Nature conservation today. *Tigerpaper*, 8(2): 30-31. Illus.

152 **Bonnefous, E. (1970).** L'homme ou la nature? Hachette. xiii, 462p. Fr.

153 **Borisov, V.A., Beloussova, L.S., Vinokirov, A.A. (1985).** [Protected natural territories of the world.] Moscow, Agropromizdat. Rus. (P)

154 **Bormann, F.H. (1985).** Air pollution and forests: an ecosystem perspective. *BioScience*, 35(7): 434-441.

155 **Bourdoux, P. (1982).** Protection et conservation de la nature, collections de reference, quelques suggestions susceptibles d'en ameliorer l'efficacite! *Cactus* (Brussels), 6(5): 98-104, 113-115. Fr. (G/T)

156 **Bouvarel, P. (1978).** Contribution of forestry research to the conservation of the environment. *In* Proceedings of the 7th World Forestry Congress, Buenos Aires, 4-18 October 1972, vols 4 and 5. Buenos Aires, Instituto Forestal Nacional. 4689-4712. En, Sp, Fr.

157 **Bouvarel, P. (1985).** Un equilibre fragile et toujours menace. (A balance that is fragile and always threatened.) *Rev. Forest. Francaise*, 37 (Special number): 141-144. Fr.

158 **Bradshaw, A.D. (1976).** Disturbed lands: their reclamation and development as nature reserves. *In* Canadian Botanical Association. Natural areas: Proceedings of a Symposium at the Thirteenth Annual Meeting of the Canadian Botanical Association, Bishop's University, Lennoxville, Quebec, 6-10 June. 30-38. Quebec, CBA. (Fr). Illus. (P)

159 **Bradshaw, M.E., Doody, J.P. (1978).** Plant population studies and their relevance to nature conservation. *Biol. Conserv.*, 14(3): 223-242. (T)

160 **Brady, N.C. (1988).** International development and the protection of biological diversity. *In* Wilson, E.O., *ed.* Biodiversity. Washington, D.C., National Academy Press. 409-418.

161 **Bramwell, D. (1975).** El jardin botanico y la conservacion de la naturaleza. *Aguauro*, 48: 4-6. Sp. Illus. (G)

162 **Bramwell, D. (1984).** Biosystematics and conservation. *In* Grant, W.F., *ed.* Plant systematics. Toronto, Academic Press. 633-641.

GENERAL

163 **Bramwell, D. (1988).** Are botanic gardens doing the right conservation research? *Species*, 11: 30-31. (G/T)

164 **Bramwell, D., Hamann, O., Heywood, V., Synge, H., eds (1987).** Botanic gardens and the World Conservation Strategy. Proceedings of an international conference, 26-30 November 1985, Las Palmas de Gran Canaria. London, Academic Press. 367p. (G/T)

165 **Bramwell, D., ed. (1979).** Plants and islands. London, Academic Press. 459p. Illus., maps. (G/P/T)

166 **Brandenburg, W.A., Oost, E.H., Vooren, J.G.V.D. (1982).** Taxonomic aspects of the germplasm conservation of cross-pollinated cultivated plants. *In* Porceddu, E., Jenkins, G., *eds.* Seed regeneration in cross-pollination species: proceedings of the C.E.C./EUCARPIA Seminar, Nyborg, Denmark, 15-17 July 1981. Rotterdam, A.A. Balkema. 33-41. (E)

167 **Brenan, J.P.M. (1981).** Plants for man: their diversity, codification and exploitation. *International Relations*, 7(1): 1005-1020. (E)

168 **Brickell, C. (1977).** Conserving cultivated plants. *The Garden* (U.K.), 102(5): 197-201. (G)

169 **Brickell, C. (1982).** Conservation of plants in gardens. *In* Evans, A., *ed.* Alpines '81, Report of the Fifth International Rock Garden Plant Conference and Show, 13-16 April 1981, Nottingham. 59-75. (G/T)

170 **Brickell, C., Sharman, F. (1986).** The vanishing garden: a conservation guide to garden plants. London, John Murray and the Royal Horticultural Society. 261p. Illus., col. illus. (G)

171 **Brightman, F., ed. (1982).** Natural history book reviews: an international bibliography. Berkhamsted, AB Academic Publishers. 90p. (B)

172 **Broggi, M.F. (1981).** Establishment of protected areas. *Naturopa*, 38: 26-28. Illus. (P)

173 **Brondegaard, V.J. (1984).** Zur Gefahrdung der kulturgeschichtlichen Wildpflanzen. *Palmengarten*, 48(3): 154-162. Ge. Illus. (G)

174 **Brown, A.G., Palmberg, C.M., eds (1978).** Third World consultation on forest tree breeding. Session 1. Exploration, utilization and conservation of gene resources. Canberra, CSIRO. 112p. (Fr, Sp). Part of a collective document. (E)

175 **Brown, L.R., Wolf, E.C. (1984).** Soil erosion: quiet crisis in the world economy. Worldwatch Paper No. 60. Washington, D.C., Worldwatch Institute. 49p.

176 **Brown, L.R., et al. (1987).** State of the world 1987. New York and London, Norton.

177 **Brown, R.A. (1981).** Plants in the computer. *Amer. Hort.*, 52(4): 36-42. (T)

178 **Brown, S., Larmer, P., Thomas, A., Wall, S. (1985).** Why save endangered species: an ethical perspective. *Endangered Species Tech. Bull. Reprint*, 2(6): 3-4. (T)

179 **Brown, W.L. (1983).** Genetic diversity and genetic vulnerability - an appraisal. *Economic Botany*, 37(1): 4-12. Crop. (E)

180 **Brumback, W. (1983).** Propagating endangered plants: theory and practice. *Wild Flower News and Notes*, 1: 4-5. (G/T)

181 **Brumback, W.E. (1980).** Endangered plant species and botanic gardens. *In* Longwood Program Seminars, vol.12. Newark, University of Delaware. 65-71. (G/T)

182 **Brush, S.B. (1989).** Rethinking crop genetic resource conservation. *Conservation Biology*, 3(1): 19-29. (E)

GENERAL

183 **Bruton, D.F. (1973).** Endangered species - so what. *Trail and Landscape*, 7(3): 72-73. (T)

184 **Bruton, M.N. (1980).** The conservation of diversity. *East. Cape Nat.*, 69: 29-30.

185 **Buchanan, R.A. (1979).** Edge disturbance in natural areas. *Austral. Parks*, August: 39-43.

186 **Buchinger, M. (1969).** Saving natural areas. Carson City, NV, Foresta Institute for Ocean and Mountain Studies. 41p. (P)

187 **Buchinger, M. (1978).** Regional problems in the development of national parks. *In* Proceedings of the 7th World Forestry Congress, Buenos Aires, 4-18 October 1972, vol. 3: Conservation and recreation. Buenos Aires, Instituto Forestal Nacional. 4039-4055. (Sp, Fr). (P)

188 **Budowski, G. (1974).** Forest plantations and nature conservation. *IUCN Bull., n.s.,* 5(7): 25-26. (P)

189 **Budowski, G. (1974).** Priorities in conservation action. *IUCN Bull., n.s.,* 5(3): 9.

190 **Budowski, G. (1975).** Conservation problems in the Third World. *IUCN Bull., n.s.,* 6(10): 37.

191 **Budowski, G. (1976).** The global problems of conservation and the potential role of living collections. *In* Simmons, J.B., *et al.*, *eds.* Conservation of threatened plants. NATO Conference Series 1: Ecology, vol. 1. New York and London, Plenum Press. 9-13. Proc. Conference on the Functions of Living Plant Collections in Conservation and Conservation-Orientated Research and Public Education, Kew, 2-6 September 1975. (G)

192 **Bugnot, P. (1978).** Trends in legislation. *Naturopa*, 31: 33-35. (L)

193 **Bunyard, P. (1985).** World climate and tropical forest destruction. *The Ecologist*, 15(3): 125-136.

194 **Burdick, M. (1978).** Creating natural communities: how public gardens can help. *Bull. Amer. Assoc. Bot. Gard. Arbor.*, 12(1): 29-32. (G)

195 **Burhenne, W.E. (1984).** The permanence of conservation institutions. *In* McNeely, J.A., Miller, K.R., *eds.* National parks, conservation and development. The role of protected areas in sustaining society. Washington, D.C., Smithsonian Institution Press. 58-59. (L)

196 **Burley, F.W. (1988).** Monitoring biological diversity for setting priorities in conservation. *In* Wilson, E.O., *ed.* Biodiversity. Washington, D.C., National Academy Press. 227-230.

197 **Burns, D. (1987).** Runway and treadmill deforestation. Gland, IUCN. 20p. Effects of logging practices.

198 **Burton, R. (1989).** Sending plants home. *The Garden* (U.K.), August: 369-372. Illus. Re-introductions. (G/T)

199 **Buttrick, S.C. (1979).** Ecology Forum No. 31 - Lower but not lowly. *Nat. Conserv. News*, 29(2): 25-26, 30.

200 **Butts, G.F. (1978).** Preservation of tree land corners. *Proc. Amer. Congr. Surv. Mapp.*, 1978: 54-56.

201 **Cahn, R. & P. (1985).** Saved but threatened. *Audubon*, 87(3): 48-51. World threatened protected areas. (P)

202 **Cain, S.A., Sokolov, V., Smith, F., Curry-Lindahl, K., Carvalho, J.C., Frankel, O., Scott, P.M. (1970).** Preservation of natural areas and ecosystems and protection of rare and endangered species. *In* UNESCO, Proc. Symp. Use & Conserv. Biosphere, Nat. Resour. Res. vol. X. (P/T)

GENERAL

203 **Cairns, J. (1988).** Increasing diversity by restoring damaged ecosystems. *In* Wilson, E.O., *ed.* Biodiversity. Washington, D.C., National Academy Press. 333-343. Illus.

204 **Cairns, J., Dickson, K.L., Herricks, E.E., eds (1977).** Recovery and restoration of damaged ecosystems. Proceedings of the International Symposium on the Recovery of Damaged Ecosystems, Virginia Polytechnic Institute and State University, Blacksburg, Virginia, 23-25 March 1975. Charlottesville, University Press of Virginia. x, 531p. Conservation and management.

205 **Cairns, J., jr (1980).** The recovery process in damaged ecosystems. Ann Arbor, MI, Ann Arbor Science. 167p. (P)

206 **California Native Plant Society (1982).** Species diversity of those under threat. *Calif. Native Pl. Soc. Bull.*, 12(1): 7. (T)

207 **Campbell, F. (1986).** The nuts and bolts of wild orchid protection. *Garden* (U.S.), 10(2): 2-4. (T)

208 **Campbell, F.T., Tarr, J. (1980).** The international trade in plants is still unregulated. *Nat. Parks Conserv. Mag.*, 54(4): 11-15. (L/T)

209 **Canadian Wildlife Service. (1973).** Convention on International Trade in Endangered Species (CITES) of Wild Fauna and Flora. CITES Report No. 1. Ontario, Environment Ottawa. (L/T)

210 **Candolle, R. de (1981).** A plea for the world's forests. *Envir. Conserv.*, 8(1): 2-3.

211 **Candolle, R. de (1983).** World charter for nature. *Envir. Conserv.*, 10(1): 67-68.

212 **Carp, E., ed. (1972).** International conference on the conservation of wetlands and waterfowl, Ramsar, Iran, 1971. Slimbridge, U.K., Intl. Wildfowl Res. Bur. 303p.

213 **Carpenter, R.A., ed. (1983).** Natural systems for development: what planners need to know. New York, Macmillan. Illus.

214 **Carpenter, S.L., Kennedy, W.J.D. (1981).** Environmental conflict management: new ways to solve problems. *Mountain Res. Devel.*, 1(1): 65-70. (Fr, Ge).

215 **Carter, C., comp. (1976).** Endangered species (plants). Tracer Bulletin TB 76-6. Washington, D.C., Library of Congress. 12p. (B/T)

216 **Cartman, J. (1979).** Propagation and conservation. *Canterbury Bot. Soc. J.*, 13: 33-36. Illus. (G)

217 **Castri, F. di, Hadley, M. (1984).** Comparative ecological research and representative natural areas. *In* Unesco-UNEP. Conservation, science and society. Paris, Unesco. 301-311. Contributions to the First International Biosphere Reserve Congress, Minsk, Byelorussia/USSR, 26 September - 2 October 1983. Illus. (P)

218 **Castri, F. di, Robertson, J. (1982).** The biosphere reserve concept: 10 years after. *Parks*, 6(4): 1-6. Illus. 10th anniversary of Unesco MAB Programme. (P)

219 **Cathey, H.M. (1976).** Endangered plants; an editorial. *Amer. Hort.*, 55(6): 2-3. (T)

220 **Catling, P.M. (1979).** Some thoughts on endangered and threatened plants. *Canad. Bot. Ass. Bull.*, 12(3): 49-51. (T)

221 **Catling, P.M. (1980).** Endangered orchids? - another viewpoint. *Amer. Orchid Soc. Bull.*, 49(3): 220-222. (T)

222 **Catling, P.M., Whiting, R.E. (1976).** Orchids, fragile and endangered. *Ontario Naturalist*, 16(3): 20-30. (T)

223 **Chang, T.T. (1985).** Preservation of crop germplasm. *Iowa State J. Res.*, 59(4): 365-378, 457-496. (E/G)

GENERAL

224 **Channon, G. (1974).** Save the species! *Amer. Orch. Soc. Bull.*, 43: 123-124. (T)

225 **Chapman, G.P. (1988).** An approach to desert containment and retrieval. *Biologist*, 35(4): 217-220.

226 **Chapman, V.J. (1967).** Conservation of maritime vegetation and the introduction of submerged freshwater aquatics. *Micronesica*, 3: 31-35.

227 **Chavez, M.E. (1988).** Utria: the bay of beauty becomes a national park. *Nature Conservancy News*, Spring: 3, 10. (P)

228 **Chazdon, R.L. (1988).** Conservation-conscious collecting: concerns and guidelines. *Principes*, 32: 13-17.

229 **Cherfas, J. (1979).** Planting the seeds for survival. *New Sci.*, 81(1145): 751. (T)

230 **Cherfas, J. (1987).** Ecology invades a new environment. *New Sci.*, 116(1583): 42-46. Illus. Deals with acid rain, tropical forests, diversity and stability debate.

231 **Chiariello, N. (1976).** Plant endangerment and ecological stability. *Conserv. News*, 41(13): 10-13. (T)

232 **Child, G.S. (1984).** FAO and protected area management: where do we go from here? *In* McNeely, J.A., Miller, K.R., *eds.* National parks, conservation and development. The role of protected areas in sustaining society. Washington, D.C., Smithsonian Institution Press. 685-688. Tables. (P)

233 **Chomchalow, N. (1983).** Methods of conservation of medicinal and aromatic plants. *IBPGR Region. Committ. Southeast Asia Newsl.*, 7(4): 6. (E)

234 **Chudovska, O. (1979).** Die Konservierung von tropischen Orchideen fur Herbare. *Vierteljahrsschr. Naturforsch. Ges. Zurich*, 124(3): 199-203. Ge (En). Illus. (T)

235 **Chudovska, O. (1979).** Ein neues Konservierungsverfahren fur Sukkulenten. *Kakt. And. Sukk.*, 30(5): 106-107. Ge. Illus. (T)

236 **Clapham, A.R., ed. (1980).** The IBP survey of conservation sites: an experimental study. International Biological Programme 24. Cambridge, Cambridge University Press. 344p. (P)

237 **Clark, T., Crete, R., Cada, J. (1989).** Designing and managing successful endangered species recovery programs. *Environ. Manage.*, 13(2): 159-170. (T)

238 **Claver, I. (1978).** The role of forestry schools in the face of the new demands for the conservation of the environment. Human activities of nature. *In* Proceedings of the 7th World Forestry Congress, Buenos Aires, 4-18 October 1972, vol. 3: Education. Buenos Aires, Instituto Forestal Nacional. 3333-3363. En, Sp, Fr.

239 **Clements, M.A. (1984).** The conservation of orchids using symbiotic methods at the Royal Botanic Gardens, Kew. *Orchid Res. Newsl.*, 3: 5. (G)

240 **Coblentz, B.E. (1978).** The effects of feral goats (*Capra hircus*) on island ecosystems. *Biol. Conserv.*, 13(4): 279-286.

241 **Commonwealth Science Council (1986).** Biological diversity and genetic resources of underexploited plants. Summary report. International workshop held at Royal Botanic Gardens, Kew, Surrey, U.K. 20-24 October 1986. CSC Technical Bulletin Publication Series No. 206. London, Commonwealth Science Council. 42p. (E)

242 **Commonwealth Science Council (1986).** Biological diversity and genetic resources techniques and methods: mass propagation using tissue culture and vegetative methods. International workshop held at Kuala Lumpur, Malaysia, 14-25 July 1986. London, Commonwealth Science Council. 139p. (E/G)

GENERAL

243 **Conacher, A.J. (1978).** Resources and environmental management: some fundamental concepts and definitions. *Search*, 9(12): 437-441. Illus.

244 **Connally, E.H. (1981).** Saving our last frontier. *Nation. Parks*, 55(3): 5-9. Illus., map.

245 **Conover, A. (1987).** Expedition to a "Lost World". *Int. Wildlife*, 17(3): 4-11.

246 **Consiglio Nazionale delle Richerche (1981).** Problemi scientifici e tecnici della conservazione del patrimonio vegetale. *Optima Leafl.*, 114: 146p. It. Illus., col. illus., maps.

247 **Constantz, G. (1977).** Should we save all endangered species? *Frontiers*, 41(4): 39. (T)

248 **Corley, M.F.V. (1979).** Notes on some rare Dicranaceae (*Oncophorus riparius, Oncophorus sardous, Arctoa anderssonii*). *J. Bryol.*, 10(4): 383-386. (T)

249 **Corner, E.J.H. (1979).** Conservation of the plant world - purpose, action, education, maintenance. *In* Lebedev, D.V., *ed.* Proceedings of the International Botanical Congress, Leningrad, 3-10 July 1975. Leningrad, Komarov Botanical Institute. 65-72. (T)

250 **Cosijn, R. (1981).** Forest reserves and nature conservation. *Nederl. Bosbouw Tijdschr.*, 53(11/12): 333-342. Du. (P)

251 **Costantino, I.N. (1978).** Desarrollo internacional de los parques nacionales. (International development of national parks.) *In* Proceedings of the 7th World Forestry Congress, Buenos Aires, 4-18 October 1972, vol. 3: Conservation and recreation. Buenos Aires, Instituto Forestal Nacional. 4112-4117. Sp. (P)

252 **Council on Environmental Quality (1978).** Environmental quality. 9th Annual report of the Council on Environmental Quality. Washington, D.C., U.S. Govt Print. Off. n.p.

253 **Coupland, R.T., ed. (1979).** Grassland ecosystems of the world: analysis of grasslands and their uses. Cambridge, Cambridge University Press. 401p. Illus., map. (E)

254 **Cowan, B. (1978).** Landscaping amid rare plants. *Fremontia*, 5(4): 32-33. (G)

255 **Cowan, McT. (1973).** Vanishing species: habitat change and reconciling conflict. *In* IUCN. Twelfth technical meeting, Banff, Alberta, Canada, 12-15 September 1972. 321-333. (T)

256 **Cracknell, M.P. (1979).** Protecting genetic resources. *IFAP News*, 28(10): 1, 2. (E/T)

257 **Cragg, J.B. (1968).** The theory and practice of conservation. *In* IUCN Commission on Ecology. Proceedings of technical meeting, Ankara-Bursa-Istanbul, 9-16 October 1967. 25-35.

258 **Creech, J.L. (1978).** Institutional plant collections - long-range concepts. *Bull. Amer. Assoc. Bot. Gard. Arbor.*, 12(4): 105-109. Illus. (G)

259 **Cribb, P.J. (1985).** Crisis for the world's wild orchids. *Country Life*, 177(4568): 566-568. Illus., col. illus. (T)

260 **Cribb, P.J. (1986).** The world wide orchid crisis. *Kew Mag.*, 3(1): 4-8. Illus. (T)

261 **Crisp, P., Astley, D. (1985).** Genetic resources in vegetables. *Progress in Plant Breeding*, 1: 281-310.

262 **Cromarty, A.S., Ellis, R.H., Roberts, E.H. (1983).** The design of seed storage facilities for genetic conservation. AGPG:IBPGR/82/23. Rome, IBPGR Secretariat. v, 96p. Illus. Replaces AGPE:IGPGR/76/25. Seed bank. (E/G)

GENERAL

263 **Crompton, B. (1976).** Rare grasses in a vanishing habitat. *Fremontia*, 4(3): 22-23. (T)

264 **Cronk, Q., et al. (1988).** Conservation and biodiversity. *Threatened Pl. Newsl.*, no. 19: 2-3. (E)

265 **Cronquist, A. (1971).** Adapt or die. *Bull. Jard. Bot. Nat. Belg.*, 41(1): 135-144.

266 **Crovello, T.J. (1976).** Use of computers to determine the endangered status of plant species. *Proc. Indiana Acad. Sci.*, 85: 352-353. (T)

267 **Crovello, T.J. (1977).** Computers as an aid to solving endangered species problems. *In* Prance, G.T., Elias, T.S., *eds.* Extinction is forever. Proceedings of a symposium at the New York Botanical Garden, 11-13 May 1976. New York, New York Botanical Garden. 337-346. (T)

268 **Crovello, T.J. (1981).** The literature as a rare plant information resource. *In* Morse, L.E., Henifin, M.S., *eds.* Rare plant conservation: geographical data organization. New York, New York Botanical Garden. 83-93. (T)

269 **Croze, H. (1984).** Global monitoring and biosphere reserves. *In* Unesco/UNEP. Conservation, science and society. Paris, Unesco. 361-370. Contributions to the First International Biosphere Reserve Congress, Minsk, Byelorussia/USSR, 26 September - 2 October 1983. (P)

270 **Croze, H. (1984).** Monitoring within and outside protected areas. *In* McNeely, J.A., Miller, K.R., *eds.* National parks, conservation and development. The role of protected areas in sustaining society. Washington, D.C., Smithsonian Institution Press. 628-633. (P)

271 **Cullen, J. (1976).** The use of records systems in the planning of botanic gardens collections. *In* Simmons, J.B., *et al., eds.* Conservation of threatened plants. NATO Conference Series 1: Ecology, vol. 1. New York and London, Plenum Press. 95-103. Proc. Conference on the Functions of Living Plant Collections in Conservation and Conservation-Orientated Research and Public Education, Kew, 2-6 September 1975. (G)

272 **Cullen, J. (1978).** Programs and policies with respect to maintenance of record system. *Bull. Amer. Assoc. Bot. Gard. Arbor.*, 12(4): 123-129. Illus. (G)

273 **Cullen, J., Donald, D., Shaw, R., Wymer. E., Wyse Jackson, P. (1988).** The cultivation and propagation of threatened plants: a proposal for the documentation of botanic garden methods. Kew, IUCN Botanic Gardens Conservation Secretariat. 27p. (G)

274 **Cullen, J., Lear, M., Mackinder, D.C., Synge, H. (1987).** Principles and standards for the computerisation of garden record schemes, as applied to conservation, with proposals for an International Transfer Format. *In* Bramwell, D., Hamann, O., Heywood, V., Synge, H., *eds.* Botanic gardens and the World Conservation Strategy. London, Academic Press. 301-337. (Sp). (G)

275 **Curry-Lindahl, K. (1972).** Conservation for survival. New York, William Morrow. n.p.

276 **Curry-Lindahl, K. (1978).** WWF och IUCN - stotte-pelare for internationell naturvard. (WWF and IUCN - pillars for international nature conservation.) *Sveriges Nat.*, 69(4): 305-310. Sw (En). Illus.

277 **Cutright, N.J. (1978).** A partial bibliography on rare and endangered species. *Vermont Inst. Nat. Sci. Publ.*, 5: 1-16. (B/T)

278 **Dagenbach, H. (1978).** Uber die Nachzucht des Speierlings (*Sorbus domestica* L.) ein Beitrag zur Erhaltung einer vom Aussterben bedrohten Baumart. *Veroff. Natursch. Landschaftspfl. Baden-Wurttemberg*, 47/48: 191-203. Illus. (T)

279 **Dahlberg, K.A. (1983).** Crop germplasm conservation. *Science*, 221(4609): 414. (E/G)

GENERAL

280 **Dalfelt, A. (1984).** The role and constraints of international development agencies in promoting effective management of protected areas. *In* McNeely, J.A., Miller, K.R., *eds.* National parks, conservation and development. The role of protected areas in sustaining society. Washington, D.C., Smithsonian Institution Press. 692-697. (P)

281 **Dalmas, J.-P. (1989).** Les actions de protection de la flore dans les Parcs Nationaux. *In* Chauvet, M., *ed.* Plantes sauvages menacees de France. Bilan et protection. Actes du colloque de Brest, 8-10 octobre 1987. Paris, Bureau des ressources genetiques. 277-286. Fr. (P)

282 **Damania, A.B., Williams, J.T. (1980).** Directory of germplasm collections: 2. Root crops. Rome, International Board for Plant Genetic Resources. 54p. Illus. (E/G)

283 **Dasmann, R. (1979).** Biosphere exploitation and conservation: a world picture. *In* Hedberg, I., *ed.* Systematic botany, plant utilization and biosphere conservation. Stockholm, Almqvist & Wiksell International. 155-157.

284 **Dasmann, R. (1984).** Biosphere reserves and human needs. *In* Unesco-UNEP. Conservation, science and society. Paris, Unesco. 509-513. Contributions to the First International Biosphere Reserve Congress, Minsk, Byelorussia/USSR, 26 September - 2 October, 1983. (P)

285 **Dasmann, R.F. (1973).** A system for defining and classifying natural regions for purposes of conservation: a progress report. IUCN Occasional Paper No. 7. Morges, IUCN. 47p. Maps.

286 **Dasmann, R.F. (1973).** Classification and use of protected natural and cultural areas. IUCN Occasional Paper No. 4. Morges, IUCN. 24p. (P)

287 **Dasmann, R.F. (1976).** The threatened world of nature. Horace M. Albright Conservation Lectureship No. 16. University of California, Department of Forestry and Conservation. 14p.

288 **Dasmann, R.F. (1984).** Environmental conservation. New York, U.S.A. 486p. Illus.

289 **Dasmann, R.F. (1984).** The relationship between protected areas and indigenous peoples. *In* McNeely, J.A., Miller, K.R., *eds.* National parks, conservation and development. The role of protected areas in sustaining society. Washington, D.C., Smithsonian Institution Press. 667-671. (P)

290 **Dassler, H.-G., Borititz, S. (1976).** Einfluss von Luftverunreinigungen auf die Vegetation: Ursachen, Wirkungen, Gegenmassnahmen. (Influence of air pollution on vegetation: causes, effects, counter-measures.) Jena Fischer. 189p. Ge. Illus., col. illus.

291 **Datta, S.C. (1983).** Conservation of genetic resources and endangered plants. *Sci. Cult.*, 49(6): 153-156. (E/G/L/T)

292 **Davey, A.J., Jefferies, R.L. (1981).** Approaches to the monitoring of rare plant populations. *In* Synge, H., *ed.* The biological aspects of rare plant conservation. Chichester, Wiley. 219-232. Proceedings of International Conference, King's College, Cambridge, 14-19 July 1980. (T)

293 **Davidson, D.A., ed. (1986).** Soil erosion and conservation. Longman. Includes information on vegetation management.

294 **Davis, A.M., Glick, T.F. (1978).** Urban ecosystems and island biogeography. *Envir. Conserv.*, 5(4): 299-305. Illus.

295 **Davis, G.M. (1977).** Rare and endangered species: a dilemma. *Frontiers*, 41(4): 12-14. (T)

296 **Davis, S.D., Droop, S.J.M., Gregerson, P., Henson, L., Leon, C.J., Lamlein Villa-Lobos, J., Synge, H., Zantovska, J. (1986).** Plants in danger: what do we know? Cambridge, IUCN Conservation Monitoring Centre. xlv, 461p. (B/G/L/T)

GENERAL

297 **Davy, A.J., Jefferies, R.L. (1981).** Approaches to the monitoring of rare plant populations. *In* Synge, H., *ed.* The biological aspects of rare plant conservation. Chichester, Wiley. 219-232. Proceedings of International Conference, King's College, Cambridge, 14-19 July 1980. (T)

298 **Deschatres, R. (1982).** Plantes rares, plantes menacees, plantes protegees. *Rev. Scient. Bourb.*, 3-24. Fr. (T)

299 **Diamond, J.M. (1975).** The island dilemma: lessons of modern biogeographic studies for the design of natural reserves. *Biol. Conserv.*, 7: 129-146. (P)

300 **Diamond, J.M. (1987).** Extant unless proven extinct? Or, extinct unless proven extant? *Conserv. Biol.*, 1(1): 77-79. Discussion on application of Red Data Book categories to poorly known species; deals with birds but relevant to other groups.

301 **Dickinson, S. (1968).** Further thoughts on orchid conservation. *Orchidata*, 8: 88-89. (T)

302 **Dittberner, P.L., et al. (1981).** The use of the Plant Information Network (PIN) in rare plant conservation. *In* Morse, L.E., Henifin, M.S., *eds.* Rare plant conservation: geographical data organization. New York, New York Botanical Garden. 149-165. (T)

303 **Dixit, R.D. (1986).** Tree ferns: an urgent need of conservation. *Indian Fern J.*, 3(1-2): 42-45. Illus. (T)

304 **Dodson, C.H. (1979).** Orchids, a threatened family? *Bull. Marie Selby Bot. Gard.*, 6(5): 38-40. Illus. (T)

305 **Donald, D. (1982).** Plant introductions in conservation. *In* Pinder, N., *ed.* Conservation and introduced species: a discussion meeting held at University College London on 18 April 1980. Discussion Paper in Conservation No. 30. London, UCL. 50-59. British Association of Nature Conservationists in conjunction with Ecology and Conservation Unit of UCL.

306 **Dotto, L. (1986).** Planet in jeopardy: environmental consequences of nuclear war. Chichester, Wiley. 134p.

307 **Douglas, J.S., Hart, R. A. de J. (1984).** Forest farming: towards a solution to problems of world hunger and conservation. London, Intermediate Technology Publications. 2nd ed. xxi, 207p. Illus. Includes bibliography.

308 **Douglas, J.S., Hart, R.A. de J. (1976).** Forest farming: towards a solution to problems of world hunger and conservation. London, Robinson and Watkins. 197p.

309 **Droste zu Huelshoff, B. von (1984).** How Unesco's Man and the Biosphere programme is contributing to human welfare. *In* McNeely, J.A., Miller, K.R., *eds.* National parks, conservation and development. The role of protected areas in sustaining society. Washington, D.C., Smithsonian Institution Press. 689-691. (P)

310 **Droste, B. von, Gregg, W.P. jr. (1985).** Biosphere reserves: demonstrating the value of conservation in sustaining society. *Parks*, 10(3): 2-5. Illus. (P)

311 **Droste, B. von, Vernhes, J.R. (1984).** Biosphere reserves and world heritage sites: relationships and perspectives. *In* Unesco-UNEP. Conservation, science and society. Paris, Unesco. 242-245. Contributions to the First International Biosphere Reserve Congress, Minsk, Byelorussia/USSR. 26 September - 2 October, 1983. Table. (P)

312 **Drury, W.H. (1974).** Rare species. *Biol. Conserv.*, 6: 162-169. (T)

313 **Drury, W.H. (1980).** Rare species of plants. *Rhodora*, 82(829): 3-48. Illus., map. (T)

314 **DuMond, D.M. (1973).** A guide for the selection of rare, unique and endangered plants. *Castanea*, 38: 387-395. (T)

16

GENERAL

315 **Dubos, R. (1972).** Conservation, stewardship, and the human heart. *Audubon*, 74(5): 20-28.

316 **Duffey, E. (1980).** The forest world: the ecology of the temperate woodlands. London, Orbis. 120p. Col. illus. National parks as sanctuaries on p. 112-117. (P)

317 **Dugan, P. (1986).** New moves in wetland campaign. *WWF News*, 39: 3. (P)

318 **Duke, J.A. (1976).** Economic appraisal of endangered plant species. *Phytologia*, 34(1): 21-27. (E/L/T)

319 **Dunbar, R. (1984).** Scapegoat for a thousand deserts. *New Sci.* 104(1430):30-33. Illus. Goats.

320 **Dunford, C. (1979).** IUCN's new programme for the world's drylands. *IUCN Bull.*, 10(9): 86-87. Illus. Extract from paper *IUCN Programme for Drylands*.

321 **Dunsterville, G.C.K. (1975).** A letter to orchid conservationists. *Amer. Orch. Soc. Bull.*, 44: 883-885.

322 **Dunsterville, G.C.K. (1975).** Love us, but please do not raze us! *Orquidea*, 5: 55-58.

323 **Dunsterville, G.C.K. (1985).** Conservation in an overpopulated world. *Amer. Orch. Soc. Bull.*, 54(10): 1189-1193. Illus. (T)

324 **Durrell, L. (1986).** State of the ark: an atlas of conservation action. London, Bodley Head. 224p. Illus., maps. Biological resources, species extinctions, deforestation. (E/P/T)

325 **Duvigneaud, P. (1979).** Noosphere et avenir de la vegetation du globe. (Noosphere and the future of the vegetation of the globe.) *In* Lebedev, D.V., *ed.* Proceedings of the International Botanical Congress, Leningrad, 3-10 July 1975. Leningrad, Komarov Botanical Institute. 72-92. Fr.

326 **Duyvendak, R., Luesink, B. (1979).** Preservation of genetic resources in grasses. *In* Harten, A.M. van, Zeven, A., *eds.* Broadening the genetic base of crops. Wageningen, Netherlands. 67-73. (E)

327 **Eckholm, E. (1978).** Disappearing species: the social challenge. Worldwatch Paper No. 22. Washington, D.C., Worldwatch Institute. 38p. Social costs of species extinctions. (T)

328 **Eckholm, E. (1978).** The age of extinction. *The Futurist*, 12(5): 289-300.

329 **Eckholm, E. (1978).** Wild species vs. man: the losing struggle for survival. *The Living Wilderness*, 42(142): 10-22.

330 **Eckholm, E. (1979).** The dispossessed of the Earth: land reform and sustainable development. Worldwatch Paper No. 30. Washington, D.C., Worldwatch Institute. 48p.

331 **Eckholm, E.P. (1976).** Losing ground: environmental stress and world food prospects. Oxford, Pergamon Press. 223p. Soil erosion.

332 **Ehrenberg, C. (1977).** Bevarande, skotsel och fornyande av naturliga bestand som genresurser. (Conservation, management and regeneration of natural stands as gene resources.) *Rapp. Uppsatser, Inst. Skogsgenet.*, 24: 93-98. (E/G)

333 **Ehrenfeld, D.W. (1976).** The conservation of non-resources. *Amer. Sci.*, 64(6): 648-656.

334 **Ehrlich, P. (1981).** Who cares about the death of species? *New Sci.*, 91(1267): 480-481. (T)

335 **Ehrlich, P. (1988).** The loss of diversity: causes and consequences. *In* Wilson, E.O., *ed.* Biodiversity. Washington, D.C., National Academy Press. 21-27. Includes estimates of extinction rates.

GENERAL

336 **Ehrlich, P.R. & A.H. (1981).** Extinction: the causes and consequences of the disappearance of species. New York, Random House. 305p.

337 **Ehrlich, P.R. (1980).** The strategy of conservation, 1980-2000. *In* Soule, M.E., Wilcox, B.A., *eds.* Conservation biology: an evolutionary-ecological perspective. Massachusetts, Sinauer Associates. 329-344.

338 **Ehrlich, P.R. (1982).** Human carrying capacity, extinctions, and nature reserves. *BioScience*, 32(5): 331-333. (P)

339 **Ehrlich, P.R., Mooney, H.A. (1983).** Extinction, substitution, and ecosystem services. *BioScience*, 33(4): 248-254.

340 **Eiberle, K. (1979).** Beziehungen waldbewohnender Tierarten zur Vegetationsstruktur. (Relations between forest fauna and vegetation structure.) *Schweiz. Zeitschr. Fortwesen*, 130(3): 201-224. Ge (Fr).

341 **Eidsvik, H.K. (1978).** The park planning process. *Tigerpaper*, 5(4): 3-8. (P)

342 **Eidsvik, H.K. (1980).** National parks and other protected areas: some reflections on the past and prescriptions for the future. *Environ. Conserv.*, 7(3): 185-190. (P)

343 **Eidsvik, H.K. (1984).** Evolving a new approach to biosphere reserves. *In* Unesco-UNEP. Conservation, science and society. Paris, Unesco. 73-80. Contributions to the First International Biosphere Reserve Congress, Minsk, Byelorussia/USSR, 26 September - 2 October 1983. (P)

344 **Einarsson, E. (1978).** Where there's a will there's a way. *Naturopa*, 31: 19. Illus.

345 **Ekeland, K. (1988).** [The preservation of ancient monuments and the conservation of biological species]. *Sven. Bot. Tidskr.*, 82(6): 490-498. Sw (En). (T)

346 **Elias, P. (1978).** Zachrana ohrozenych taxonov v botanickych zahradach. (Rescue of endangered taxa in botanical gardens.) *Pamatky Prir.*, 4: 236-239. Cz. (G/T)

347 **Elias, T.S. (1976).** Extinction is forever. *Gard. Jour.*, 26: 52-55.

348 **Elliott, D.K., ed. (1986).** The dynamics of extinction. Chichester, Wiley. x, 294p. 12 papers presented at conference on mass extinctions, Northern Arizona Univ., 1983.

349 **Elliott, R. (1979).** Should we preserve species? *Habitat*, 7(5): 9-10. Illus. (T)

350 **Elliott, Sir H., ed. (1974).** Second World Conference on National Parks. Yellowstone and Grand Teton National Parks, U.S.A., September 18-27, 1972. Morges, IUCN. 504p. Illus. (P)

351 **Ellis, R.H., Roberts, E.H., Whitehead, J. (1980).** A new, more economic and accurate approach to monitoring the viability of accessions during storage in seed banks. *Pl. Genet. Resourc. Newsl.*, 41: 3-18. (Fr, Sp). (G)

352 **Emanuelsson, U. (1988).** Tva allvarliga naturvardsproblem ar 2010. (Two serious nature conservancy problems in 2010.) *Sven. Bot. Tidskr.*, 82(6): 411-416. Sw (En).

353 **Emonds, G. (1981).** Guidelines for national implementation of the Convention on International Trade in Endangered Species of Wild Fauna and Flora. Gland, IUCN. xi, 148p. IUCN Environment Policy and Law Paper, no. 17. (L/T)

354 **Encendencia, M.E.M. (1977).** Forest ecosystem and its conservation. *Visca Review*, 2: 10-11.

355 **Ern, H. (1980).** Biological aspects of rare plant conservation. *Gartn. Bot. Brief*, 65: 59-60. Ge. (T)

356 **Esquinas-Alcazar, J.T., Gulick, P.J. (1983).** Genetic resources of *Cucurbitaceae*. AGPG: IBPGR/82/48. Rome, IBPGR. v, 101p. Illus. maps. (E)

GENERAL

357 **Esser, K. (1976).** Genetic factors to be considered in maintaining living plant collections. *In* Simmons, J.B., *et al., eds.* Conservation of threatened plants. NATO Conference Series 1: Ecology, vol. 1. New York and London, Plenum Press. 185-198. Illus. Proc. Conference on the Functions of Living Plant Collections in Conservation and Conservation-Orientated Research and Public Education, Kew, 2-6 September 1975. (G)

358 **Evans, A. (1982).** Conservation of plants in the wild. *In* Evans, A., *ed.* Alpines '81, Report of the Fifth International Rock Garden Plant Conference and Show, 13-16 April 1981, Nottingham. Alpine Garden Society. 53-58. (T)

359 **Everard, B. (1975).** Vanishing flowers of the world. World pictorial map series. London, WWF and Bartholomew. (T)

360 **FAO (1975).** The methodology of conservation of forest genetic resources, report on a pilot study. Misc. 75/8. Rome, FAO/UNEP. 117p.

361 **FAO (1989).** Plant genetic resources: their conservation *in situ* for human use. Rome, FAO. 38p. Illus. Crop genetic resources. (E)

362 **FAO Forestry Department (1986).** Databook on endangered tree and shrub species and their provenances. Forestry Paper No. 77. Rome, FAO. 524p. Datasheets on 81 species. (E/T)

363 **Faber, D. (1962).** 20 'lost' plants. *Bartonia*, 31: 7-10. (T)

364 **Fairbrothers, D.E. (1975).** Rare and endangered plants. *BioScience*, 25(9): 590. (T)

365 **Falk, D. (1987).** Exploring seed storage of endangered plants. *Center Plant Conserv. Newsl.*, 2(3): 7. Seed bank. (G/T)

366 **Falk, D., McMahan, L. (1988).** Endangered plant conservation: managing for diversity. *Natural Areas J.*, 8: 91-99. (T)

367 **Farmer, R.E. (1974).** Supplement to activity analysis: propagation of rare and endangered plants. Norris, TN, Tennessee Valley Authority. n.p. (G/T)

368 **Farnsworth, N.R. (1982).** The potential consequences of plant extinction in the United States on the current and future availability of prescription drugs. *In* Prescott-Allen, C. & R., *eds.* The first resource; wild species in the American economy. New Haven, Yale University Press. (E)

369 **Farnsworth, N.R. (1988).** Screening plants for new medicines. *In* Wilson, E.O., *ed.* Biodiversity. Washington, D.C., National Academy Press. 83-97. Medicinal plants; includes discussion on NAPRALERT database. (E)

370 **Favre, D.S. (1989).** International trade in endangered species: a guide to CITES. Dordrecht, Martinus Nijhoft. 415p. (E/L/T)

371 **Fay, M.F. (1988).** Micropropagation at the Royal Botanic Gardens Kew. *Bot. Gard. Conserv. News*, 1(3): 42-45. Illus. (G/T)

372 **Ferreira, J., Hillyard, D. (1987).** Genetic conservation issues in land restoration: open forum discussion. *In* Elias, T.S., *ed.* Conservation and management of rare and endangered plants. Proceedings from a conference, Sacramento, California, 5-8 November 1986. Sacramento, California Native Plant Society. 523-524.

373 **Filion, L., Villeneuve, P., eds (1977).** Ecological reserves and nature conservation. Papers presented at a conference at Iles-de-la-Madeleine, Quebec, 17-18 June 1976. Rapport, Conseil Consultatif des Reserves Ecologiques, Quebec, No. 1. vii, 45p. Fr, En. (P)

374 **Fisher, A.C. (1982).** Economic analysis and the extinction of species. Berkeley, Dept of Energy and Resources, University of California.

375 **Fisher, J. (1969).** Wildlife in danger. New York, Viking. 368p. (T)

GENERAL

376 **Fitter, R. (1986).** Wildlife for man: how and why we should conserve our species. London, Collins. 223p. (E/T)

377 **Florescu, I.I. (1979).** Padurea gradinarita si protectia mediului. (Selective forests and environmental protection.) *Rev. Padur. Ind. Lemn. Celul. Hirtie Silvic. Expl. Padur.*, 94(4): 227-228.

378 **Folch i Guillen, R. (1979).** Conservation de la nature et amenagement forestier: possibilites de cooperation et danger de conflits. *Webbia*, 34(1): 73-85. Fr (It).

379 **Force, J.E. (1984).** A research planning strategy to assess the impacts of air pollutants on forest resources. *Bull. G.B. Forest. Comm.*, (61): 74-80. Illus.

380 **Ford-Lloyd, B., Jackson, M. (1986).** Plant genetic resources: an introduction to their conservation and use. Victoria, Edward Arnold. v, 152p. Illus., maps. (E)

381 **Forman, L., Bridson, D. (1989).** Conservation and the herbarium. *In* Forman, L., Bridson, D., *eds.* The herbarium handbook. Kew, Royal Botanic Gardens. 203-204. Herbarium specimens can be an important data source on threatened plants. (T)

382 **Forster, R.R. (1973).** Planning for man and nature in national parks: reconciling perpetuation and use. IUCN Publications New Series No. 26. Morges, IUCN. 84p. Illus. (P)

383 **Fosberg, F.R. (1973).** Temperate zone influence on tropical forest land use: a plea for sanity. *In* Meggers, B.J., Ayensu, E.S., Duckworth, W.D., *eds.* Tropical forest ecosystems in Africa and South America: a comparative review. Washington, D.C., Smithsonian Institution Press. 345-350.

384 **Fosberg, F.R. (1979).** Whither terrestrial ecosystems? Preservation of the habitat of man. *In* Larsen, K., Holm-Nielsen, L.B. *eds.* Tropical botany. London, Academic Press. 63-93.

385 **Fosberg, F.R. (1981).** An overview of natural area preservation in the world. *Heritage Record*, 12: 6p. (P)

386 **Fosberg, F.R. (1983).** The human factor in the biogeography of oceanic islands. *Compt. Rend. Soc. Biogeogr.*, 59(2): 147-190. Mentions introduced species; deforestation.

387 **Foster, J. (1988).** Protected landscapes. Summary Proceedings of an International Symposium, Lake District, United Kingdom, 5-10 October 1987. Gland, IUCN, Countryside Commission and Countryside Commission for Scotland. 80p. Illus. (P)

388 **Fox, A.M. (1979).** Of nature and man and the nature of man. *Victoria's Resources*, 21(1): 3-9. Illus.

389 **Fox, D.B. (1983).** The conservation of lilies and Liliaceae. *Roy. Hort. Soc. Lily Group Bull.*, 50: 49. (G/T)

390 **Frankel, O.H. (1970).** Genetic conservation in perspective. *In* Frankel, O.H., Bennett, E., *eds.* Genetic resources in plants - their exploration and conservation. Philadelphia, PA, F.A.Davis. 469-489. (E/G)

391 **Frankel, O.H. (1970).** Genetic conservation of plants useful to man. *Biol. Conserv.*, 2: 162-169. (E/G)

392 **Frankel, O.H. (1974).** Genetic conservation: our evolutionary responsibility. *Genetics*, 78: 53-65. (G)

393 **Frankel, O.H. (1976).** The time scale of concern. *In* Simmons, J.B., *et al.*, *eds.* Conservation of threatened plants. NATO Conference Series 1: Ecology, vol. 1. New York and London, Plenum Press. 245-248. Proc. Conference on the Functions of Living Plant Collections in Conservation and Conservation-Orientated Research and Public Education, Kew, 2-6 September 1975. (T)

GENERAL

394 **Frankel, O.H. (1981).** Genetic erosion and genetic conservation. *In* XIII International Botanical Congress, Sydney, Australia, 21-28 August 1981: abstracts. Sydney, Australian Academy of Science. 216.

395 **Frankel, O.H. (1983).** Genetic principles of *in situ* preservation of plant resources. *In* Jain, S.K., Mehra, K.L., *eds.* Conservation of tropical plant resources. Howrah, Botanical Survey of India. 55-65. (T)

396 **Frankel, O.H., Bennett, E. (1970).** Genetic resources in plants - their exploration and conservation. Philadelphia, PA, F.A. Davis. 554p. (E/G)

397 **Frankel, O.H., Bennett, E., eds (1970).** Genetic resources. Introduction. *In* Frankel, O.H., Bennett, E., *eds.* Genetic resources in plants - their exploration and conservation. Philadelphia, PA, F.A. Davis. 7-17. (G)

398 **Frankel, O.H., Soule, M.E. (1981).** Conservation and evolution. Cambridge, Cambridge University Press. viii, 327p. Illus.

399 **Frankel, O.H., Soule, M.E. (1981).** The role of botanical gardens in conservation. *In* Frankel, O.H., Soule, M.E., *eds.* Conservation and evolution. Cambridge, Cambridge University Press. 163-174. (G)

400 **Frankel, O.H., ed. (1973).** Survey of crop genetic resources in their centres of diversity. Rome, FAO. 164p. (E)

401 **Franklin, J.F., Sokolov, V.E., Gunin, P.D., Herrmann, R., Puzachenko, Y.V., Wiersma, G.B. (1984).** Similar biosphere reserves and principles for their selection. *In* Unesco-UNEP. Conservation, science and society. Paris, Unesco. 377-383. Contributions to the First International Biosphere Reserve Congress, Minsk, Byelorussia/USSR, 26 September - 2 October 1983. (P)

402 **Freedman, B. (1989).** Environmental ecology. The impacts of pollution and other stresses on ecosystem structure and function. London, Academic Press. Includes effects of pollution and acidification on forests.

403 **Fuente, G. de la (1978).** Proteger la naturaleza. (Protection of nature.) *In* Proceedings of the 7th World Forestry Congress, Buenos Aires, 4-18 October 1972, vol. 3: Conservation and recreation. Buenos Aires, Instituto Forestal Nacional. 4122-4125. Sp (En, Fr).

404 **Fuller, D., Fitzgerald, S., eds (1987).** Conservation and commerce of cacti and other succulents. Washington, D.C., TRAFFIC (U.S.A.), World Wildlife Fund. 264p. (L/T)

405 **Game, M. (1980).** Best shape for nature reserves. *Nature*, 287(5783): 630-632. (P)

406 **Garratt, K. (1984).** The relationship between adjacent lands and protected areas: issues of concern for the protected area manager. *In* McNeely, J.A., Miller, K.R., *eds.* National parks, conservation and development. The role of protected areas in sustaining society. Washington, D.C., Smithsonian Institution Press. 65-71. (P)

407 **Garrett-Jones, C. (1981).** World Conservation Strategy: a vital issue evaded? *Conserv. News*, 81: 7-8.

408 **George, T.E. (1969).** Conservation by cultivation. *Victorian Nat.*, 86: 67-70. (G)

409 **Gerasimov, I.P., Preobrazhenskii, V.S. (1979).** [National parks as a form of territory utilization and management for recreation and tourist activities]. *Izv. Akad. Nauk SSSR*, 5: 19-24. Rus. (P)

410 **Ghiselin, J. (1973).** Wilderness and the survival of species. Declining populations lose options in recessive genes. *Living Wilderness*, Winter Issue: 22-27. 1973-1974. (T)

411 **Ghosh, R.C. (1983).** Forest management and its role in conservation. Howrah, Botanical Survey of India. v, 73-81. (E/P/T)

GENERAL

412 **Giacomini, V. (1978).** La protezione della diversita nella natura. *Nat. Montagna*, 25(4): 5-11.

413 **Gibson, T., et al. (1981).** International trade in plants. Washington, TRAFFIC (USA). xxv, 99p. (L)

414 **Gillette, R. (1973).** Endangered species moving toward a ceasefire. *Science*, 179: 1107-1109. (L/T)

415 **Girardon, J. (1988).** Semences: la loi du plus fort. *L'Expresse Aujourd'hui*, 10 June: 22-27. Col. illus., map. (E/G)

416 **Given, D.R. (1977).** Threatened plants and horticulture. *Roy. New Zealand Inst. Hort. Ann. J.*, 5: 5-11. (G/T)

417 **Given, D.R. (1983).** Conservation of Pteridophyta: development of the International Association of Pteridology - Threatened Plants Committee Project. *Pac. Sci. Congr. Proc.*, 15(1-2): 83. (T)

418 **Given, D.R. (1983).** Conservation of island floras. *In* Jain, S.K., Mehra, K.L., *eds.* Conservation of tropical plant resources. Howrah, Botanical Survey of India. 82-100. (T)

419 **Given, D.R. (1985).** Botanic gardens and the conservation of threatened plants. *Roy. New Zealand Inst. Hort. Ann. J.*, 12: 47-51. (G/T)

420 **Given, D.R. (1986).** Botanic gardens and conservation: the road from Las Palmas. *Ann. J. Roy. N.Z. Instit. Hort.*, 14: 3-9. (G)

421 **Given, D.R. (1987).** Plant variety rights: some implications for conservation. *Tuatara*, 29(1-2): 13-18. (E/G)

422 **Given, D.R. (1987).** What a conservationist requires of *ex situ* collections. *In* Bramwell, D., Hamann, O., Heywood, V., Synge, H., *eds.* Botanic gardens and the World Conservation Strategy. London, Academic Press. 105-118. (Sp). (G)

423 **Given, D.R. (1988).** Greening the botanic gardens. Botanic gardens and conservation. *For. Bird*, 19(3): 28-30. (G/T)

424 **Given, D.R. (1989).** The Botanic Gardens Conservation Secretariat and its role in plant conservation. *Ann. J. Roy. N.Z. Instit. Hort.* (G)

425 **Given, D.R., Jermy, A.C. (1985).** Conservation of pteridophytes: a postcript. *Proc. Roy. Soc. Edinburgh*, Sect. B, 86: 435. (T)

426 **Given, D.R., ed. (1983).** Conservation of plant species and habitats. Wellington, Nature Conservation Council. 128p. Illus. A symposium held at 15th Pacific Science Congress, Dunedin, New Zealand, February 1983. (P/T)

427 **Given, D.R., ed. (1984).** Background to the symposium. *In* Conservation of plant species and habitats. Proceedings of a symposium held at 15th Pacific Science Congress, Dunedin, New Zealand, February 1983. Wellington, Nature Conservation Council. 1-4.

428 **Gladstones, J.S., Francis, C.M., Collins, W.J. (1981).** Genetic resources and plant breeding. *J. Agric. West. Austral.*, 22(4): 126-130. Concerned with the collection of subtropical forage and pasture plants and temperate annual pasture legumes, including species of *Trifolium, Medicago, Ornithopus* and *Lupinus*. (E)

429 **Glass, C., Foster, R. (1978).** Propagation of succulents and the new conservation laws. *Pac. Hort.*, 39(4): 3-6. Illus. (G/L)

430 **Goldmith, E. (1982).** The retreat from Stockholm. *Ecologist*, 12(3): 98-100. Illus. Refers to UN Conference on the Human Environment, Stockholm, 1972.

431 **Goldsmith, E. (1980).** World Ecological Areas Programme: a proposal. *Envir. Conserv.*, 7(1): 27-29. (P)

GENERAL

432 **Good, R.B., Leigh, J.H. (1983).** The criteria for the assessment of the conservation status of rare and threatened plants. *Pac. Sci. Congr. Proc.*, 15(1-2): 86. (T)

433 **Goodall, D.W., Perry, R.A., eds (1979).** Arid-land ecosystems: structure, functioning and management, 1. Cambridge, Cambridge University Press. 881p. Illus., maps.

434 **Goodland, R. (1984).** The World Bank, environment, and protected areas. *In* McNeely, J.A., Miller, K.R., *eds*. National parks, conservation and development. The role of protected areas in sustaining society. Washington, D.C., Smithsonian Institution Press. 698-705. (P)

435 **Goodland, R.J.A. (1987).** The World Bank's wildlands policy: a major new means of financing conservation. *Conservation Biology*, 1(3): 210-213.

436 **Grabherr, G., Kusstatscher, K., Mair, A. (1985).** Zur vegetationsokologischen Aufbereitung aktueller Naturschutzprobleme im Hochgebirge. (On the vegetational-ecological processing of current conservation problems in high mountains.) *Verh. Zool.-Bot. Ges. Osterr.*, 123: 269-291. Ge. Map.

437 **Grainger, A. (1982).** Desertification. London, Earthscan. 98p.

438 **Grainger, A. (1989).** The threatening desert: controlling desertification. London, Earthscan. 288p.

439 **Grant, W.F., ed. (1984).** Plant biosystematics. Toronto, Academic Press. Illus., maps. Tropical forests. Includes contributions on endangered species. (T)

440 **Grant, W.S., Robinson, E.R., Ferrar, A.A. (1988).** Population genetics and the management of small populations. *S. Afr. J. Sci.*, 84(11): 868-870.

441 **Green, B. (1981).** What's wrong with conservation? *Ecos*, 2(4): 26-29.

442 **Green, P.S. (1973).** Plants; wild and cultivated. Conference on horticulture and field botany held at the Royal Horticultural New Hall, Vincent Square, London, Sept. 2-3, 1974. BSBI Conference Report No. 13. Hampton, U.K., Classey. 231p. (G)

443 **Green, T. (1988).** Endangered species: another opinion. *Amer. Orchid Soc. Bull.*, 57(1): 29-30. (T)

444 **Gregg, W.P. (1984).** Building science programmes to support the multiple roles of biosphere reserves. *In* Unesco-UNEP. Conservation, science and society. Paris, Unesco. 312-319. Contributions to the First International Biosphere Reserve Congress, Minsk, Byelorussia/USSR, 26 September - 2 October 1983. (P)

445 **Grenell, P. (1987).** Innovative programmatic approaches to resource conservation. *In* Elias, T.S., *ed*. Conservation and management of rare and endangered plants. Proceedings from a conference, Sacramento, California, 5-8 November 1986. Sacramento, California Native Plant Society. 401-403. Discusses wetland and watershed programmes and land management based on California State Coastal Conservancy programme.

446 **Grey-Wilson, C., Knees, S., Mathew, B. (1987).** Conservation and the plant hunter. *Garden* (U.K.), 112(1): 30-35. Col. illus. Covers present international legislation on plant collecting. (L/T)

447 **Griggs, R.F. (1940).** The ecology of rare plants. *Bull. Torrey Bot. Club*, 67: 575-594. (T)

448 **Grodzinsky, A.M. (1988).** Conservation of endangered species *ex situ* in botanic gardens: problems and goals. *Newsl. Int. Assoc. Bot. Gard. Europ. Medit. Div.*, 10: 6-10. (G/T)

449 **Gunn, A.S. (1980).** Why should we care about rare species? *Envir. Ethics*, 2(1): 17-37. (T)

GENERAL

450 **Gyorgy, L. (1978).** Ashabad 1978: Hogyan tovabb a nemzetkozi kermesvetvedelemben. *Buvar*, 33(11): 482-483. Illus.

451 **Haas, H. (1978).** Pilzverbreitung und Pilzschutz. (Distribution and conservation of fungi.) *Veroff. Landesst. Natursch. Landschaftspfl. Baden-Wurttemberg*, 11: 155-160. (T)

452 **Hagsater, E. (1966).** Can there be a different view on orchids and conservation? *Amer. Orch. Soc. Bull.*, 45: 18-21. (T)

453 **Hagsater, E. (1986).** Can there be a different view on orchids and conservation? *Amer. Orchid Soc. Bull.*, 55(3): 268-271. (T)

454 **Hagsater, E., Stewart, J. (1986).** Estrategias para la conservacion de orquideas. (Strategies for orchid conservation). *Orquidea* (Mexico), 10(1): 213-227. En, Sp. (T)

455 **Hagsater, E., Stewart, J. (1986).** Orchid conservation at the international level. *Orchid Rev.*, 94(1108): 56-60. Col. illus., illus. (T)

456 **Hagsater, E., Stewart, J. (1986).** Orchid conservation at the international level. *Orchid Rev.*, 94(1110): 123-125. (T)

457 **Hales, D.F. (1984).** The World Heritage Convention: status and directions. *In* McNeely, J.A., Miller, K.R., *eds.* National parks, conservation and development. The role of protected areas in sustaining society. Washington, D.C., Smithsonian Institution Press. 744-750. (L/P)

458 **Halffter, G. (1985).** Biosphere reserves: conservation of nature for man. *Parks*, 10(3): 15-18. Illus. (P)

459 **Hall, A.V. (1988).** Can you rescue threatened species? *Veld & Flora* (Kirstenbosch), 14(3): 100-101. (T)

460 **Hall, J.B. (1983).** Positive management for strict natural reserves: reviewing effectiveness. *Forest Ecol. Manage.*, 7(1): 57-66. Forestry conservation. (P)

461 **Hamann, O. (1985).** The IUCN/WWF Plants Conservation Programme 1984-1985. *Vegetatio*, 60(3): 147-149. (T)

462 **Hamann, O. (1986).** The plants programme: what has it done. *Species*, 7: 14-16. IUCN Threatened Plants Programme. (T)

463 **Hamann, O. (1987).** The IUCN/WWF Plants Conservation Programme in action. *In* Bramwell, D., Hamann, O., Heywood, V., Synge, H., *eds.* Botanic gardens and the World Conservation Strategy. London, Academic Press. 31-43. (Sp). (E/G/T)

464 **Hamann, O. (1988).** IUCN-WWF's internationale program for plantebevarelse. (The IUCN-WWF Plants Conservation Programme.) *Svensk Bot. Tidskr.*, 82(6): 466-470. Sw (En). (T)

465 **Hamann, O., Leon, C., Synge, H. (1984).** International Union for Conservation of Nature and Natural Resources. Botanic Gardens and IUCN: a vital link for plant conservation. *Rep. Bot. Inst. Univ. Aarhus*, 10: 25-32. (G/T)

466 **Hanemann, W.M. (1988).** Economics and the preservation of biodiversity. *In* Wilson, E.O., *ed.* Biodiversity. Washington, D.C., National Academy Press. 193-199.

467 **Hanson, J., Williams, J.T., Freund, R. (1985).** Institutes conserving crop germplasm: the IBPGR global network of genebanks. AGPG: IBPGR/84/18. Rome, IBPGR. iii, 25p. (E/G)

468 **Harcourt, A.H., Pennington, H., Weber, A.W. (1986).** Public attitudes to wildlife and conservation in the Third World. *Oryx*, 20(3): 152-154. Paper presented at a symposium 'Current Issues in Primate Conservation' London, Dec. 1985 organised by FFPS and the Primate Society of G.B.

GENERAL

469 **Hardjosentono, P., Hehuwat, F., Soemarmo, B. (1984).** Involvement of politicians in the development of parks and protected areas. *In* McNeely, J.A., Miller, K.R., *eds.* National parks, conservation and development. The role of protected areas in sustaining society. Washington, D.C., Smithsonian Institution Press. 245-248. (L/P)

470 **Harlan, J.R. (1972).** Genetic conservation of plants that feed the world. *Nat. Parks Conserv. Mag.,* 46(10): 15-17. (E)

471 **Harlan, J.R. (1983).** The scope for collection and improvement of forage plants. *In* McIvor, J.G., Bray, R.A., *eds.* Genetic resources of forage plants. East Melbourne, CSIRO. 3-14. (E)

472 **Harley, J.L. (1978).** The objectives of conservation. *Unasylva,* 30(121): 25-28. Illus.

473 **Harper, J.L. (1981).** The meanings of rarity. *In* Synge, H., *ed.* The biological aspects of rare plant conservation. Chichester, Wiley. 189-203. Proceedings of International Conference, King's College, Cambridge, 14-19 July 1980. (T)

474 **Harrington, J.F. (1970).** Sampling gene pools. *In* Frankel, O.H., Bennett, E., *eds.* Genetic resources in plants - their exploration and conservation. Philadelphia, PA, F.A. Davis. 501-521. (E/G)

475 **Harris, L.D. (1986).** The fragment forest: island biogeography, theory and the preservation of biotic diversity. Chicago, University of Chicago Press. 211p.

476 **Harrison, E.A. (1975).** Endangered species: a bibliography with abstracts. USDC/NTIS Microf. PS-75/881. Springfield, VA. 44p. (B)

477 **Harrison, J. (1983).** Maintaining a database on the world's protected areas. *Parks,* 7(4): 3-5. Illus. (P)

478 **Harrison, J. (1984).** An international data bank on biosphere reserves and the need for standardization. *In* Unesco-UNEP. Conservation, science and society. Paris, Unesco. 371-376. Contributions to the First International Biosphere Reserve Congress, Minsk, Byelorussia/USSR, 26 September - 2 October 1983. (P)

479 **Harrison, J., Karpowicz, Z., Leon, C. (1986).** Monitoring environmental conservation: towards an integral global overview. *In* World Meteorological Organization. Environmental pollution monitoring and research programme. No. 45. Vol. 3. 95-105.

480 **Harrison, J., Miller, K., McNeely, J. (1982).** The world coverage of protected areas: development goals and environmental needs. *Ambio,* 11(5): 238-245. Maps. (P)

481 **Harrison, J., Miller, K., McNeely, J. (1984).** The world coverage of protected areas: development goals and environmental needs. *In* McNeely, J.A., Miller, K.R., *eds.* National parks, conservation and development. The role of protected areas in sustaining society. Washington, D.C., Smithsonian Institution Press. 24-33. Tables. (P)

482 **Harrison, S. (1988).** Local extinction, metapopulations and endangered species. *End. Species Update,* 5(5): 10. (T)

483 **Harvey, B., Knamiller, G.W. (1981).** Development and conservation: a global dilemma. *Envir. Conserv.,* 8(3): 199-205.

484 **Hastey, E.L. (1978).** BLM's plant preservation policies. *Fremontia,* 5(4): 25-27. (L/T)

485 **Hauri, R., et al. (1978).** Neue Naturschutzgebiete. *In* Naturschutzinspektorat des Kantons Bern Bericht 1977. *Mitt. Naturforsch. Ges. Bern,* 35: 33-173. Ge. Illus.

GENERAL

486 **Hawkes, J.G. (1976).** Sampling gene pools. *In* Simmons, J.B., *et al., eds.* Conservation of threatened plants. NATO Conference Series 1: Ecology, vol. 1. New York and London, Plenum Press. 145-154. Proc. Conference on the Functions of Living Plant Collections in Conservation and Conservation-Orientated Research and Public Education, Kew, 2-6 September 1975. (G)

487 **Hawkes, J.G. (1978).** The taxonomist's role in the conservation of genetic diversity. *In* Street, H.E., *ed.* Essays in plant taxonomy. New York, Academic Press. 125-142. (G)

488 **Hawkes, J.G. (1978).** Zachowanie i wykorzystanie roslinnych zasobow genowych. (The conservation and utilisation of plant genetic resources.) *Biul. Inst. Hodowli Aklim. Roslin,* 133: 139-145. Pol (Rus, En). (E/G)

489 **Hawkes, J.G. (1987).** A strategy for seed banking in botanic gardens. *In* Bramwell, D., Hamann, O., Heywood, V., Synge, H., *eds.* Botanic gardens and the World Conservation Strategy. London, Academic Press. 131-149. (Sp). (G)

490 **Hawkes, J.G., Williams, J.T., Croston, R.P. (1983).** A bibliography of crop genetic resources. Rome, IBPGR. 442p. Comprehensive bibliography. (B/E/G/P/T)

491 **Hayes, S. (1977).** Plants are wildlife too! *WWF Newsl. (Canada),* Christmas: 1-2. (T)

492 **Heard, B. (1983).** Conservation: a matter of conscience. *Amer. Nurseryman,* 158(10): 27-29. Illus.

493 **Hedberg, I., ed. (1979).** Systematic botany, plant utilization and biosphere conservation. Stockholm, Almqvist & Wiksell International. 157p. Proceedings of a symposium held in Uppsala in commemoration of the 500th anniversary of the University of Stockholm. (T)

494 **Hedberg, O. (1977).** Conservation of the plant world - a global scheme. *In* Bicentenary celebration of C.P. Thunberg's visit to Japan: Tokyo, Kyoto and Nagasaki, 17-25 May 1976. Tokyo, Royal Swedish Embassy & Botanical Society of Japan. 17-22. Illus. (T)

495 **Helliwell, D.R. (1976).** The extent and location of nature conservation areas. *Envir. Conserv.,* 3(4): 255-258. (P)

496 **Hendee, J.C., Stankey, G.H., Lucas, R.C. (1978).** Wilderness management. Miscellaneous Publications No. 1365. U.S. Department of Agriculture, Forest Service. 381p. Illus. (P)

497 **Henifin, M.S. (1981).** Planning field work on rare or endangered plant populations. *In* Morse, L.E., Henifin, M.S., *eds.* Rare plant conservation: geographical data organization. New York, The New York Botanical Garden. 309-312. (T)

498 **Henifin, M.S., et al. (1981).** Guidelines for the preparation of status reports on rare and endangered species survey. *In* Morse, L.E., Henifin, M.S., *eds.* Rare plant conservation: geographical data organization. New York, The New York Botanical Garden. 261-282. (T)

499 **Henke, H. (1981).** Cornerstones for survival. *Naturopa,* 38: 7-9. Illus. (P)

500 **Hepper, F.N. (1969).** The conservation of rare and vanishing species of plants. *In* Fisher, J., *et al., eds.* Wildlife in danger. New York, Viking Press. 352-360. (T)

501 **Herbst, D. (1977).** Vanishing plants. *Water Spectrum,* 9(4): 20-26. (T)

502 **Heslop-Harrison, J. (1973).** The plant kingdom: an exhaustible resource? *Trans. Bot. Soc. Edinb.,* 42: 1-15. (T)

503 **Heslop-Harrison, J. (1974).** Genetic resource conservation: the end and the means. *J. Roy. Soc. Arts,* 123: 157-169. (G)

GENERAL

504 **Heslop-Harrison, J. (1974).** Postscript: the Threatened Plants Committee. *In* Hunt, D.R., *ed.* Succulents in peril. *Int. Org. Succ. Plant Study Bull. Suppl.*, 3(3): 30-32. (T)

505 **Heslop-Harrison, J. (1975).** Man and the endangered plant. *IUCN Intl. Yearb.*, 1975: 103-106. (T)

506 **Heslop-Harrison, J. (1976).** Reproductive physiology (transcript). *In* Simmons, J.B., *et al.*, *eds.* Conservation of threatened plants. NATO Conference Series 1: Ecology, vol. 1. New York and London, Plenum Press. 199-205. Proc. Conference on the Functions of Living Plant Collections in Conservation and Conservation-Orientated Research and Public Education, Kew, 2-6 September 1975. Includes conservation problems. (G)

507 **Heslop-Harrison, J., Thimann, K.V., Kolesnikov, B.P. (1979).** Presentation of the resolutions. *In* Lebedev, D.V., *ed.* Proceedings of the International Botanical Congress, Leningrad, 3-10 July 1975. Leningrad, Komarov Botanical Institute. 109-113. All-Congress Symposium "Conservation of the Plant World", 7 July 1975. (T)

508 **Heywood, V.H. (1987).** The changing role of the botanic garden. *In* Bramwell, D., Hamann, O., Heywood, V., Synge, H., *eds.* Botanic gardens and the World Conservation Strategy. London, Academic Press. 3-18. A review of the history of the role of botanic gardens and their emerging role in conservation. (Sp). (G)

509 **Heywood, V.H. (1988).** Botanic gardens and germplasm conservation. *Bot. Gard. Conserv. News*, 1(3): 15-16. Illus. (E/G)

510 **Heywood, V.H. (1988).** Rarity: a privilege and a threat. *In* Greuter, W., Zimmer, B., *eds.* Proceedings of the XIV International Botanical Congress. Konigstein, Koeltz. 277-290. (G/T)

511 **Heywood, V.H. (1989).** Patterns, extents and modes of invasions by terrestrial plants. *In* Drake, J.A., *et al.*, *eds.* Biological invasions. Wiley. 31-52. Analysis of the threat to natural communities from introduced plants, and the part played by botanic gardens. (G)

512 **Higgs, A.J. (1981).** Island biogeography theory and nature reserve design. *J. Biogeogr.*, 8(2): 117-124. (P)

513 **Higgs, A.J., Usher, M.B. (1980).** Should nature reserves be large or small? *Nature*, 285(5766): 568-569. (P)

514 **Higler, L.W.G. (1971).** Water management in nature reserves. *In* Duffey, E., Watt, A.S., *eds.* The scientific management of animal and plant communities for conservation. The 11th Symposium of the BES, University of East Anglia, Norwich, 7-9 July 1970. Oxford, Blackwell. 311-315. (P)

515 **Hilgert, H.J. (1981).** Das "Washingtoner-Artenschutzubereinkommen" und die Kakteen-Gesellschaften. *Kakt. And. Sukk.*, 32(7): 148-149. Ge. The Washington Species Protection Treaty and the importation of protected plants. (L/T)

516 **Hillgarter, F.W. (1978).** Die Walderneuerung ist ein Bestandteil der Waldpflege. (Forest renovation is a part of forest tending.) *Mitt. Forst. Bundes-Versuchsanstalt Wien*, 124: 229-237. Ge. Includes plant protection.

517 **Hills, L.D. (1983).** Seeds of hope. *Ecologist*, 13(5): 175-178. Illus., maps. Advocates vegetable sanctuaries. (E/P/T)

518 **Hills, L.D. (1983).** The conservation of vegetables. *Threatened Pl. Newsl.*, 11: 10-12. (E/G)

519 **Hinrichsen, D. (1986).** Multiple pollutants and forest decline. *Ambio*, 15(5): 258-265. Col.illus.

GENERAL

520 **Hintum, T.J.L. van (1988).** GENIS: a fourth generation information system for the database management of genebanks. *Pl. Genet. Resourc. Newsl.*, 75/76: 13-15. (E)

521 **Hofmann, M. (1976).** Index systems for special collections as illustrated by the Orchidaceae. *In* Simmons, J.B., *et al.*, *eds.* Conservation of threatened plants. NATO Conference Series 1: Ecology, vol. 1. New York and London, Plenum Press. 119-122. Proc. Conference on the Functions of Living Plant Collections in Conservation and Conservation-Orientated Research and Public Education. (G)

522 **Hogner, Dorothy Childs (1977).** Endangered plants. New York, NY, T.Y. Crowell. 83p. Juv. lit. (T)

523 **Holden, C. (1987).** World Bank launches new environment policy. *Science*, 236: 769.

524 **Holdgate, M.W. (1987).** Changing habitats of the world. *Oryx*, 21: 149-159.

525 **Holdgate, M.W., Woodman, M.J., eds (1978).** The breakdown and restoration of ecosystems. Nato Conference Series: 1. Ecology, vol. 3. New York, London, Plenum Press. xi, 496p.

526 **Holland, M. (1979).** AVIS: a prototype arid vegetation information system. *Desert Plants*, 1(2): 71-76.

527 **Holmgren, A.H. (1979).** Strategies for preservation of rare plants. *In* Wood, S.L., *ed.* The endangered species: a symposium. (T)

528 **Holsinger, K., Gottlieb, L.D. (1989).** The conservation of rare and endangered plants. *Trends Ecol. Evol.*, 4(7): 193-194. (T)

529 **Holt, R.A. (1983).** Exotic species control: an island perspective. *Nature Conserv. News*, 33(4): 23-24. Introduced species.

530 **Holzner, W., Werger, M.J.A., Ikusima, I., eds (1983).** Man's impact on vegetation. Geobotany 5. The Hague, Junk. xxi, 370p. Illus., maps.

531 **Hondelmann, W. (1976).** Seed banks. *In* Simmons, J.B., *et al.*, *eds.* Conservation of threatened plants. NATO Conference Series 1: Ecology, vol. 1. New York and London, Plenum Press. 212-224. Proc. Conference on the Functions of Living Plant Collections in Conservation and Conservation-Orientated Research and Public Education, Kew, 2-6 September 1975. (G)

532 **Hood, L. (1976).** What's next for endangered plants? *Fremontia*, 4(1): 14-16. (T)

533 **Hoose, P.M. (1981).** Building an ark: tools for the preservation of natural diversity through land protection. Covelo, CA, Island Press. 300p. (P)

534 **Hoover, W.S. (1981).** Endangered *Begonia* habitats: can we help? *Begonian*, 49: 132-134, 143. Illus.

535 **Hoover, W.S. (1982).** An update on endangered *Begonia* habitats. *Begonian*, 49: 33, 42. (T)

536 **Horich, C.K. (1979).** Al alcance nuestro: la perpetuacion de orquideas escasas; un topico actual de gran transcendencia. (Within our reach: the preservation of rare orchids; a present day topic of great importance.) *Orquideologia*, 13(3): 248-260. Sp, En. Illus. Also published in French in *Orchidophile*, 40: 1486-1489, 1980 entitled: "Le but a atteindre: la perpetuation des orchidees rares". (T)

537 **Housley, R.M. (1985).** Managing wilderness today for the future. Ogden, Utah, U.S. Department of Agriculture, Forest Service. General Tech. Report INT 182: 9-12. Forest fires, wildlife conservation and use of wilderness. (P)

538 **Howard, A. (1978).** The legal position of rare plants today. *Fremontia*, 5(4): 17-21. (L/T)

GENERAL

539 **Howden, C. (1978).** A gene bank for plants. *Hort. New Zealand*, 6: 12. Wakehurst Place gene bank. (G)

540 **Hoyt, E. (1988).** Conserving the wild relatives of crops. Rome, IBPGR and Gland, WWF and IUCN. 45p. Col. illus., maps. (E/G/P)

541 **Hrapko, J.O. (1981).** Botany reference list: rare and endangered flowering plants. Edmonton, Alberta, Provincial Museum of Alberta. 10p. (B/T)

542 **Huettl, R.F. (1988).** "New type" forest declines and restabilization/revitalization strategies. A programmatic focus. *Water Air Soil Pollut.*, 41(1-4): 95-111.

543 **Humphrey, S. (1988).** The habitat conservation plan. *Conserv. Biol.*, 2(3): 240-244.

544 **Hunt, D.R. (1982).** Whatever happened to conservation? *Cact. Succ. J.* (U.K.), 44(2): 31-35. (L)

545 **Hunt, D.R., ed. (1974).** Succulents in peril. *Suppl. Intl. Org. Succ. Plant Study Bull.*, 3(3): 32p. (T)

546 **Hunt, G., Fifield, R. (1987).** Remote sensing and the whole world picture show. *New Sci.*, 115(1574): 46-51. Illus., col. illus., maps.

547 **Hunt, P.F. (1969).** Conservation of orchids. *IUCN Bull.*, n.s., 2(10): 76. (T)

548 **Hutcherson, K. (1976).** Endangered species: the law and the land. *J. Forest.*, 74(1): 31-34. (L/T)

549 **Huxley, A. (1974).** The ethics of plant collecting. *J. Roy. Hort. Soc.*, 99: 242-249. (T)

550 **Huxley, A. (1985).** Green inheritance: the World Wildlife Fund book of plants. London, Collins-Harvill and New York, Anchor Press and Doubleday. 193p. Illus. Col. illus. Maps. (E/P/T)

551 **Huxley, A. (1985).** The ethics of plant collecting. *Quart. Bull. Alp. Gard. Soc.*, 53(2): 114-123. Col. illus. (T)

552 **Huxley, A. (1986).** The world in tatters: crisis-time for plant life. *Country Life*, 180(4657): 1608-1611. Col. illus. (T)

553 **Huxley, A. (1987).** Botanists strike back: the danger of plant smuggling. *Country Life*, 181(19): 96-97. Col. illus.

554 **IBPGR (1981).** Revised priorities among crops and regions. AGP:IBPGR/81/34. Rome, IBPGR. 18p. (E)

555 **IBPGR (1982).** Working Group on the Genetic Resources of Citrus, Tsukuba, Japan, 4-6 November 1981: report. AGPG/IBPGR/82/7. Rome, IBPGR. iii, 13p. (E/G)

556 **IBPGR (1983).** Conservation. *Ann. Rep. Int. Board Pl. Genet. Resourc. 1982*: 64-72 (1982). Col. illus. (E/G)

557 **IBPGR (1983).** Genetic resources of *Vitis* species. AGPG: IBPGR/83/52. Rome, IBPGR, Working Group on *Vitis* Genetic Resources. iii, 11p. (E)

558 **IBPGR (1983).** Practical constraints affecting the collection and exchange of wild species and primitive cultivars. AGPG/IBPGR/83/49. Rome, IBPGR. 11p. (E/G)

559 **IBPGR (1985).** Conservation. *In* International Board for Plant Genetic Resources. Annual report 1983. IPBGR/84/61. Rome, IBPGR. 73-83. Col. illus. (E)

560 **IBPGR (1985).** Ecogeographical surveying and *in situ* conservation of crop relatives. Rome, IBPGR. 27p. Report of an IBPGR Task Force, 30 July-1 August, 1984, Washington, D.C. (E/G/T)

561 **IBPGR (1987).** Seed conservation. *Ann. Rep. Int. Board Pl. Genet. Resourc. 1987*: 27-41. Illus., maps. (E/G)

GENERAL

562 **IUCN (1971).** Convention on export, import and transit of certain species of wild animals and plants. *IUCN Bull., n.s.,* 2(19): 162, 166. (L)

563 **IUCN (1973).** A working system for classification of world vegetation. IUCN Occasional Paper No. 5. Morges, IUCN. 21p.

564 **IUCN (1973).** Convention on International Trade in Endangered Species of Wild Fauna and Flora. *IUCN Bull., n.s.,* 4(3): 12. Spec. Suppl. (L/T)

565 **IUCN (1973).** Convention to control trade in threatened wildlife. *IUCN Bull., n.s.,* 4(3): 9. (L/T)

566 **IUCN (1973).** Twelfth technical meeting, Banff, Alberta, Canada, 12-15 September 1972. Conservation for development. IUCN Publications New Series No. 28. Morges, IUCN. 383p.

567 **IUCN (1974).** Biotic provinces of the world: further development of a system for defining and classifying natural regions for purposes of conservation. IUCN Occas. Paper No. 9). Morges, IUCN. (P)

568 **IUCN (1975).** Conservation of endangered species. *IUCN Yearbook,* 1974: 23-35. (T)

569 **IUCN (1976).** Plant conservation. *IUCN Bull., n.s.,* 7(1): 3.

570 **IUCN (1976).** Twelfth general assembly, Kinshasa, Zaire, 8-18 September, 1975. Proceedings. IUCN Publications New Series, Supplementary Paper No. 44-E. Morges, IUCN. 297p. (IUCN Publications new series, Suppl. paper no.44-E).

571 **IUCN (1977).** Thirteenth (extraordinary) General Assembly. Geneva, Switzerland 19-21 April, 1977. Proceedings. Morges, IUCN. 203p.

572 **IUCN (1977).** Threatened plants. *IUCN Bull., n.s.,* 8(10): 61. (T)

573 **IUCN (1979).** Proceedings of the 14th Session of the General Assembly of IUCN and 14th IUCN Technical Meeting. Ashkhabad, 26 September-5 October 1978. Morges, IUCN. 155p.

574 **IUCN (1985).** Threatened natural areas, plants and animals of the world. *Parks,* 10(1): 15-17. Illus., map. (P/T)

575 **IUCN (1986).** Botanic gardens and the World Conservation Strategy. Kew, U.K., IUCN Conservation Monitoring Centre. 13p. Lists the recommendations made at an international conference, Las Palmas, 26-30 November 1985. (G)

576 **IUCN (1987).** The IUCN Botanic Gardens Conservation Strategy: a summary. *In* Bramwell, D., Hamann, O., Heywood, V., Synge, H., *eds.* Botanic gardens and the World Conservation Strategy. London, Academic Press. xxxvii-xxxix. (G)

577 **IUCN Botanic Gardens Conservation Secretariat (1987).** The International Transfer Format (ITF) for botanic garden plant records. Pittsburgh, Hunt Institute for Botanical Documentation. (G)

578 **IUCN Botanic Gardens Conservation Secretariat (1989).** The Botanic Gardens Conservation Strategy. Gland, WWF and IUCN-BGCS. 60p. Map. (G)

579 **IUCN Botanic Gardens Conservation Secretariat (in co-operation with the Threatened Plants Unit of WCMC) (1988).** Carnivorous plants of the world. A preliminary list of the conservation species; questionnaire and preliminary report on their occurrence in botanic gardens. Kew, IUCN Botanic Gardens Conservation Secretariat. 9p. (G/T)

580 **IUCN Botanic Gardens Conservation Secretariat (in co-operation with the Threatened Plant Unit of WCMC) (1989).** Conifers of conservation importance: a preliminary world list. Questionnaire and preliminary report on their occurrence in botanic gardens. Kew, IUCN Botanic Gardens Conservation Secretariat. 21p. (G/T)

GENERAL

581 **IUCN Botanic Gardens Conservation Secretariat (in co-operation with the Threatened Plants Unit of WCMC) (1989).** Extinct plant species of the world. Survey, questionnaire and preliminary report on their occurrence in botanic gardens. Kew, IUCN Botanic Gardens Conservation Secretariat. Unpaginated. (G/T)

582 **IUCN Commission on National Parks and Protected Areas (1984).** Categories, objectives and criteria for protected areas. *In* McNeely, J.A., Miller, K.R., *eds.* National parks, conservation and development. The role of protected areas in sustaining society. Washington, D.C., Smithsonian Institution Press. 47-53. (P)

583 **IUCN Conservation Monitoring Centre (1985).** The 1985 United Nations list of national parks and protected areas. Cambridge, IUCN. 174p. (P)

584 **IUCN Conservation Monitoring Centre (1987).** Directory of wetlands of international importance: sites designated under the Convention on Wetlands of International Importance especially as Waterfowl Habitat. Gland, and Cambridge, IUCN, 445p. Prepared for the Third Conference of Contracting Parties to the Convention, RAMSAR, Canada conference, 1987. Includes details of vegetation and noteworthy plants for many sites. (P)

585 **IUCN Conservation Monitoring Centre (1987).** Protected landscapes: experience around the world. Gland, IUCN. 404p. Prepared for the International Symposium on Protected Landscapes, Grange-over-Sands, England, 5-10 October 1987. (P)

586 **IUCN Threatened Plants Unit (1983).** The botanic gardens list of threatened tree ferns. Kew, IUCN Botanic Gardens Conservation Co-ordination Body Report No. 6. Lists plants in cultivation in botanic gardens. (G/T)

587 **IUCN Threatened Plants Unit (1984).** The IUCN/WWF Plants Conservation Programme 1984-85. Kew, Threatened Plants Unit. 29p.

588 **IUCN Threatened Plants Unit (1987).** Centres of plant diversity. A guide and strategy for their conservation. An outline for a book being prepared by the Joint IUCN-WWF Plants Conservation Programme and IUCN Threatened Plants Unit. Kew, IUCN. 40p. Includes candidate list of centres of plant diversity and endemism; draft regional accounts on floras and sample data sheets on a number of sites. (P/T)

589 **IUCN, UNEP, WWF (1980).** World Conservation Strategy: living resource conservation for sustainable development. Morges, IUCN, WWF, UNEP. Illus., maps.

590 **Iltis, H.H. (1969).** The conservation of nature: whose problem? *BioScience*, 23: 248-253.

591 **Iltis, H.H. (1972).** Shepherds leading sheep to slaughter: the extinction of species and the destruction of ecosystems. *Amer. Biol. Teach.*, 34(4): 201-205.

592 **Iltis, H.H. (1974).** Flowers and human ecology. *In* Selmes, C., *ed.* New movements in the study and technology of biology. London, Maurice Temple Smith. n.p.

593 **Iltis, H.H. (1988).** Serendipity in the exploration of biodiversity: what good are weedy tomatoes? *In* Wilson, E.O., *ed.* Biodiversity. Washington, D.C., National Academy Press. 98-105. Illus. Value of wild genetic resources. (E/T)

594 **Ingelog, T. (1988).** Praktisk floravard idag och imorgon. (Practical conservation of the flora, today and tomorrow.) *Svensk Bot. Tidskr.*, 82(6): 503-506. Sw (En). Illus. (T)

595 **Inskipp, T. (1984).** Plants working group meeting on CITES implementation. *TRAFFIC Bull.*, 6(1): 11-12. (L)

596 **International Orchid Commission (1984).** Conservation: state of the art. *Austral. Orchid Rev.*, 49(3): 183-184. (T)

31

GENERAL

597 **Ives, J., Pitt, D.C., eds (?)** Deforestation. Social dynamics in watersheds and mountain systems. London, Routledge. 224p.

598 **Jackson, M., Ford-Lloyd, B.V., Parry, M.J., eds (1990).** Climatic change and plant genetic resources. Belhaven Press. 242p. Covers effects of climate change on vegetation, agriculture and implications for genetic resource conservation. (E)

599 **Jackson, P. (1978).** WWF and national parks. *Parks*, 3(3): 9-11. Illus. (P)

600 **Jackson, P. (1983).** World charter for nature. *Nat. Malaysiana*, 8(2): 8-9. Col. illus.

601 **Jacobs, M. (1977).** Gardens, species, universities. *In* Stone, B.C., *ed.* The role and goals of tropical botanic gardens. Proceedings, symposium held at Rimba Ilmu Botanic Garden, Univ. of Malaya, Kuala Lumpur, August 1974. Kuala Lumpur, Rimba Ilmu Univ. Malaya. (G)

602 **Jacobs, M. (1977).** It is the genera of threatened plants that need attention. *Flora Males. Bull.*, 30: 2828-2830. (T)

603 **Jacobs, M. (1980).** Vanishing species: things wrong between man and nature. *Flora Males. Bull.*, 33: 3427-3433. (T)

604 **Jacobs, P., Munro, A., eds (1987).** Conservation with equity: strategies for sustainable development. Proceedings of the Conference on Conservation and Development: Implementing the World Conservation Strategy, Ottawa, Canada, 31 May - 5 June 1986. Cambridge, IUCN. xvii, 466p. (B/L)

605 **Jacquiot, C. (1981).** Les reserves biologiques forestieres, element important de notre patrimoine scientifique. *Compt. Rend. Seanc. Acad. Agric. France*, 67(2): 208-212. Fr (En). (P)

606 **Jaeger, R.G. (1974).** Competitive exclusion of comments on survival and extinction of species. *BioScience* 24(1): 33-38.

607 **Jain, H.K. (1982).** Plant breeders' rights and genetic resources. *Indian J. Genet. Pl. Breed.*, 42(2): 121-128. Crop. (E/L)

608 **Jain, J.K., ed. (1986).** Combating desertification in developing countries. Country reports prepared for the United Nations Conference on Desertification. Scientific Reviews on Arid Zone Research No. 4. Jodhpur, Scientific Publishers. xiv, 315p. Maps.

609 **Janes, F. (1980).** Conservation of orchid species not endemic to G.B. *Orchid Rev.*, 88(1042): 127. (T)

610 **Jarvis, C. (1981).** Protected areas: plant lists. *Threatened Pl. Commit. Newsl.*, 8: 14-15. News of IUCN activities to promote and collate plant lists for protected areas. (P)

611 **Jarvis, C., Leon, C., Oldfield, S. (1981).** Appendix 2. Bibliography of Red Data Books and threatened plants lists. *In* Synge, H., *ed.* The biological aspects of rare plant conservation. Chichester, Wiley. 513-529. Proceedings of International Conference, King's College, Cambridge, 14-19 July 1980. (B/T)

612 **Jeffrey, C. (1970).** Conservation - the fourfold way. Potters Bar, The Conservation Society. 48p.

613 **Jenkins, D.W. (1975).** At last - a brighter outlook for endangered plants. *Nat. Parks Conserv. Mag.*, 49(1): 13-17. (T)

614 **Jenkins, R.E. (1973).** The preservation of ecosystems. *Atlant. Nat.*, 28(2): 44, 47. (P)

615 **Jenkins, R.E. (1975).** Endangered plant species: a soluble ecological problem. *Nat. Conserv. News*, 25(4): 20-21. (T)

GENERAL

616 **Jenkins, R.E. (1981).** Rare plant conservation through elements-of-diversiy information. *In* Morse, L.E., Henifin, M.S., *eds.* Rare plant conservation: geographical data organization. New York, The New York Botanical Garden. 33-40. (T)

617 **Jenkins, R.E. (1988).** Information management for the conservation of biodiversity. *In* Wilson, E.O., *ed.* Biodiversity. Washington, D.C., National Academy Press. 231-239. Illus., map. Includes discussion on TNC Natural Heritage Data Centers.

618 **Jenkins, R.E., Bedford, W.B. (1973).** The use of natural areas to establish environmental baselines. *Biol. Conserv.*, 5(3): 168-174. (P)

619 **Jenny, J. (1981).** Acquisition bill for: Endangered Wildlife and Unique Ecosystems. *Calif. Native Pl. Soc. Bull.*, 11(1): 8. (L/T)

620 **Jensen, D.B. (1987).** Concepts of preserve design: what we have learned. *In* Elias, T.S., *ed.* Conservation and management of rare and endangered plants. Proceedings from a conference, Sacramento, California, 5-8 November 1986. Sacramento, California Native Plant Society. 595-603. Reserve size, fragmentation, isolation and edge effects influence population extinctions. (P)

621 **Jianming, J. (1987).** Protecting biological resources to sustain human progress. *Ambio*, 16(5): 262-266. Col. illus. (E)

622 **Johnsen, I. (1988).** [Surveillance of natural habitats]. *Sven. Bot. Tidskr.*, 82(6): 417-418. Sw (En).

623 **Johnson, B. (1983).** Forestry crisis: what else is new? *IUCN Bull.*, 14(10-12): 122-123. Illus.

624 **Johnson, D.V. (1988).** Worldwide endangerment of useful palms. *Adv. Econ. Bot.*, 6: 268-273. (E/T)

625 **Johnson, M.K. (1979).** Review of endangered species policies and legislation. *Wildl. Soc. Bull.*, 7(2): 79-93. (L/T)

626 **Joint IUCN-WWF Plants Conservation Programme (1988).** Biodiversity: the key role of plants. Kew, IUCN and WWF. 13p. Includes section on saving biodiversity. (E/L/T)

627 **Jones, G.E. (1987).** The conservation of ecosystems and species. Natural Environment: Problems and Management Series. London, Routledge. 288p.

628 **Jones, J.G., Beardsley, W.G., Countryman, D.W., Schweitzer, D.L. (1978).** Estimating economic costs of allocating land to wilderness. *Forest Science*, 24(3): 410-422. (P)

629 **Jong, P.C. de (1979).** The role of collections of wild species as seed orchards for the cultivation of unusual plants. *In* Synge, H., Townsend, H., *eds.* Survival or extinction. Proceedings of a conference held at the Royal Botanic Gardens, Kew, 11-17 September 1978. Kew, Bentham-Moxon Trust. 215-217. (G)

630 **Jonsell, B. (1988).** Taxonomi och floravard - vad behover vi veta om arters variation relationer? (Taxonomy and conservation of the flora ?-what do we need to know about variation and relations among species?) *Svensk Bot. Tidskr.*, 82(6): 403-409. Sw (En). Illus., maps. (T)

631 **Joyce, C. (1989).** Dying to get on the list. *New Sci.*, 123(1684): 42-47. Col. illus. CITES appendices. (E/L/T)

632 **Joye, C. (1978).** Problemes d'amenagement du territoire et de conservation de la nature, les methodes d'evaluation des sites et des paysages. *Nat. Mosana*, 31(1): 1-17.

GENERAL

633 **Julien, M.H., ed. (1987).** Biological control of weeds: a world catalogue of agents and their target weeds. 2nd ed. Farnham Royal, Slough, CAB International. Includes list of invasive species, their status and control.

634 **Kafton, D.L. (1977).** Gene conservation of commercial forest trees. *California Agriculture*, 31(9): 30-31. (E/G)

635 **Kakarya, P. (1979).** Park, reserves and traditional communities. *In* Second South Pacific Conference on National Parks and Reserves, Proceedings, vol. 1. 123-129. (P)

636 **Kassas, M. (1983).** The global biosphere: conservation for survival. *Mazingira*, 7(2): 2-13. Illus.

637 **Kasteele, F.S.C. Stoop van de (1974).** Conservation of wild *Lilium* species. *Biol. Conserv.*, 6(1): 26-31. (T)

638 **Kaufman, P.B., Lacroix, J.D., eds (1979).** Plants, people and environment. New York, Macmillan. xii, 542p. Illus.

639 **Kaule, G. (1978).** Upravlenie torfyanymi bolotami v tselyakh okhrany prirody. (Management von Mooren fur den Naturschutz?) (The management of peat wetlands as a whole with the protection of nature.) *Telma*, 8: 197-200. Rus (Ge, En).

640 **Keddy, P.A. (1983).** Transplanting rare plants to protect them: a plant ecologist's perspective. *Canad. Bot. Ass. Bull.*, 16(2): 13-15. (T)

641 **Kemp, R.H. (1978).** Exploration, utilization and conservation of genetic resources. *Unasylva*, 30(119-120): 10-16, 47-41. (E/G)

642 **Kempf, C., Piantanida, T. (1987).** Les forets meurent aussi. Strasbourg, Editions Bueb and Reumaux. 171p. Fr.

643 **Kennedy, G. (1975).** Orchids and conservation - a different view. *Amer. Orch. Soc. Bull.*, 44: 401-405. (T)

644 **Keystone Center (1988).** Final report of the Keystone International Dialogue on Plant Genetic Resources. Session I: *Ex situ* conservation of plant genetic resources. 15-18 August 1988. Keystone, Colorado, The Keystone Center. 46p. (E)

645 **Kilborn, J. (1977).** Endangered house plants. *Horticulture*, 55(1): 36-40. (G/T)

646 **Kilburn, P.D. (1961).** Endangered relict trees. *Nat. Hist.*, 70(10): 56-63. (T)

647 **Killian, R. (1981).** Selected natural diversity bibliography. Arlington, VA, The Nature Conservancy. 8p. (B)

648 **King, F.W. (1974).** International trade and endangered species. London, Zoological Society of London. 456p. (L)

649 **Kira, T. (1978).** Conservation of plant world as part of the human environment. *In* Bicentenary Celebration of C.P. Thunberg's visit to Japan. Tokyo, Kyoto and Nagasaki, 17-25 May 1976. Tokyo, Royal Swedish Embassy & Botanical Society of Japan. 23-27. Illus. (T)

650 **Kirkpatrick, C.M., ed. (1978).** Wildlife and people. West Lafayette, IN, Purdue Research Foundation. vii, 191p. Illus., maps. Proc. 1978 John S. Wright Forestry Conference, Purdue University, 23-24 February 1978.

651 **Kjellqvist, E. (1988).** Genebankens roll: bevarandearbetet. (The role of the gene bank in the conservation work.) *Svensk Bot. Tidskr.*, 82(6): 473-476. Sw (En). Col. illus. (G/T)

652 **Kleinschmit, J. (1978).** The criteria for the conservation of genetic resources of trees. *In* Proceedings of the 8th World Forestry Congress, Jakarta, 16-28 October 1978: Forestry for quality of life. II, 6. (E/G)

GENERAL

653 **Klemm, C. de (1983).** Les forets et la strategie mondiale de la conservation. (Forests and the World Conservation Strategy.) *In* Prieur, M., *ed.* Forets et environnement: en droit compare et international. Paris, Presses Universitaires de France. 261-279. Fr. (L)

654 **Klemm, C. de (1984).** Protecting wild genetic resources for the future: the need for a world treaty. *In* McNeely, J.A., Miller, K.R., *eds.* National parks, conservation and development. The role of protected areas in sustaining society. Washington, D.C., Smithsonian Institution Press. 661-666. (E/L)

655 **Klemm, C. de (1985).** Preserving genetic diversity: a legal review. *Landscape Planning*, 12: 221-238. Reviews legal and planning tools to preserve genetic diversity outside protected areas. (L/T)

656 **Klemm, C. de (1989).** Les instruments juridiques de protection de la flore sauvage: droit international et exemples etrangers. *In* Chauvet, M., *ed.* Plantes sauvages menacees de France. Bilan et protection. Actes du colloque de Brest, 8-10 octobre 1987. Paris, Bureau des ressources genetiques. 259-268. Fr. (L)

657 **Klemm, C. de (?)** Wild plant conservation and the law. IUCN. A detailed study of national and international legislation relating to plants. In prep. (B/L/T)

658 **Knees, S., Cheek, M. (1988).** Changes in regulations affecting international trade in carnivorous plants. *Carniv. Pl. Newsl.*, 17(2): 45-46. (T)

659 **Koeppel, H.-W. von, Arnold, F. (1981).** Schriftenreihe fur Landschaftsplege und Naturschutz. Bonn-Bad Godesburg. 186p. Ge. Maps; illus. Extensive bibliography. (B/P)

660 **Kolesnikov, B.P., Semenova-Tyan-Shanskaya, A.M., Dyrenkov, S.A. (1977).** [Protection of the world of plants at the 12th International Botanical Conference, Leningrad, 3-10 July 1975]. *Bot. Zhurn.*, 62(12): 1792-1807. Rus. (T)

661 **Koller, G.L., Brown, R.A. (1978).** Collecting plants - why bother? *Bull. Amer. Assoc. Bot. Gard. Arbor.*, 12(2): 52-58. (G)

662 **Konzak, C.F., Dietz, S.M. (1969).** Documentation for the conservation, management and use of plant genetic resources. *Econ. Bot.*, 23: 299-308. (E/G)

663 **Koopowitz, H. (1986).** A gene bank to conserve orchids. *Amer. Orchid Soc. Bull.*, 55(3): 247-250. Col. illus. Gene bank. (G/T)

664 **Koopowitz, H. (1986).** Conservation problems in the *Amaryllidaceae. Herbertia*, 42: 21-25. (T)

665 **Koopowitz, H. (1987).** Prospects for effective long-term conservation of the Orchidaceae. Part 1 from a paper delivered at the Tokyo World Orchid Conference in March, 1987. *Orchid Advocate*, 8(6): 212-214. (T)

666 **Koopowitz, H., Kaye, H. (1983).** Plant extinction: a global crisis. Washington, D.C., Stone Wall Press. 239p. (T)

667 **Koopowitz, H., Kaye, H. (1986).** Orchids: myths and realities: part 2. The ethics of collection. *Orchid Advocate*, 12(1): 9-12. (T)

668 **Koopowitz, H., Kaye, H. (1986).** Orchids: myths and realities: part 3. The politics of orchid collecting. *Orchid Advocate*, 12(2): 61-66. (T)

669 **Koopowitz, H., Ward, R. (1984).** A technological solution for the practical conservation of orchid species. *Orchid Advocate*, 10(2): 43-45. Illus. (T)

670 **Kornas, J. (1982).** Man's impact upon the flora: processes and effects. *Mem. Zool*, 37: 11-30. (Pol, Rus). (T)

671 **Kovacs, M., ed. (1985).** Pollution control and conservation. Chichester, Ellis Harwood. 398p.

GENERAL

672 **Kowarik, I., Sukopp, H. (1984).** Auswirkungen von Luftverunreinigungen auf die spontane Vegetation (Farn- und Blutenpflanzen). (Effects of air pollution on the spontaneous vegetation ferns and flowering plants.) *Angew. Bot.*, 58(2): 157-170. Ge (En). Map. (T)

673 **Krugman, S.L., Phares, R.E. (1978).** Use and management of biosphere reserves of the Man and Biosphere Program for environmental monitoring and conservation. *In* Proceedings of the 8th World Forestry Congress, Jakarta, 16-28 October 1978: Forestry for quality of life. II, 16. (P)

674 **Kubitzki, K. (1977).** The problem of rare and of frequent species: The monographer's view. *In* Prance, G.T., Elias, T.S., *eds.* Extinction is forever. Proceedings of a symposium at the New York Botanical Garden, 11-13 May 1976. New York, New York Botanical Garden. 331-336. Map, table. (T)

675 **Kummel, F. (1980).** Schutz und Erhaltung der Sukkulenten: eine Aufgabe die jeden angeht! *Kakt. And. Sukk.*, 31(1): 2-7. Illus. (T)

676 **Kummel, F. (1981).** Schutz and Erhaltung der Sukkulenten. *Gartn.-Bot. Brief.*, 68: 44-52. Ge. Illus. (T)

677 **Kunkele, S. (1979).** Internationaler Artenschutz: Referat vor der Gesellschaft fur Okologie anlasslich der 9. Jahreshauptversammlung am 19.09.1979. *Mitteilungsbl. Arbeitskr. Heim. Orch. Baden-Wurttemberg*, 11(3): 223-234. Ge.

678 **La Verne Smith, L. (1980).** Laws and information needs for listing plants. *Rhodora*, 82(829): 193-199. (L/T)

679 **Labous, P. (1983).** *Crataegus* preserved. *GC & HTJ*, 192(28): 17. Illus. Project to save rare species in Oxford University parks. (G/T)

680 **Lacate, D.S. (1973).** Collecting, storing and evaluating data for nature conservation purposes. *In* Costin, A.B., Groves, R.H., *eds.* Nature conservation in the Pacific. Proc. Symposium A-10, XII Pacific Science Congress, Canberra, August-September 1971. IUCN and Australian Nat. Univ. Press. 13-25.

681 **Lachaux, C. (1980).** Les parcs nationaux. Paris, Puf. 128p. National parks of the world. (P)

682 **Lamb, R. (1979).** World without trees. London, Wildwood House. 221p.

683 **Lamlein, J. (1984).** Botanic Gardens Conservation Co-ordinating Body. *Endangered Species Tech. Bull. Reprint*, 1(6): 3-4. (G)

684 **Lamlein, J. (1984).** WWF-US Plants Programme. *Threatened Pl. Newsl.*, 13: 7. (E/T)

685 **Lamotte, M., Hadley, M. (1984).** Biosphere reserves in savannah regions: building the research-conservation connection. *In* Unesco-UNEP. Conservation, science and society. Paris, Unesco. 44-58. Contributions to the First International Biosphere Reserve Congress, Minsk, Byelorussia/USSR, 26 September - 2 October 1983. Tables, map. (P)

686 **Lancaster, R. (1980).** Gardeners to the rescue. *Country Life*, 168(4330): 550-551. Illus. (G)

687 **Landry, J., et al. (1979).** A rating system for threatened and endangered species of wildlife. *New York Fish and Game J.*, 26(1): 11-21. (T)

688 **Lapin, P. (1984).** The role of botanic gardens in the conservation of rare and endangered plant species. *Rep. Bot. Inst. Univ. Aarhus*, 10: 16-24. (G/P/T)

689 **Lapin, P.I. (1978).** Puti okhrany i obogashcheni i a rastitel'nosti. (Methods for the protection and enrichment of vegetation.) Moscow, Znanie. 62p.

GENERAL

690 **Larsen, K., Morley, B., Ern, H. (1987).** The role of the International Association of Botanic Gardens (IABG) in conservation world-wide. *In* Bramwell, D., Hamann, O., Heywood, V., Synge, H., *eds.* Botanic gardens and the World Conservation Strategy. London, Academic Press. 277-284. (Sp). (G)

691 **Larson, J.S. (?)** Managing woodland and wildlife habitat in and near cities. *In* Larson, J.S., *ed.* Trees and forests in an urbanizing environment. Publication No. 17. Amherst, University of Massachusetts. 125-127.

692 **Lasserre, P. (1979).** Coastal lagoons: sanctuary ecosystems, cradles of culture, targets for economic growth. *Nat. Resources*, 15(4): 2-21. Wetlands.

693 **Laumonier, Y., Purnajaya (1987).** The use of ecological vegetation maps in conservation and management of endangered plants and animals. *In* Santiapillai, C., Ashby, K.R., *eds.* Proceedings of the Symposium on the Conservation and Management of Endangered Plants and Animals. Bogor, Indonesia, 18-20 June 1986. Bogor, SEAMEO-BIOTROP. 217-224. (T)

694 **Laux, D.W., Sicker, W., Blumenbach, D. (1979).** Bibliographie der Pflanzenschutz - Literatur. *Biol. Bund. Land- Forstwirtsch. Berlin-Dahlem*, 15(1): xxxiii, 209p. (B)

695 **Lavarack, P.S. (1979).** Orchid conservation - a matter of conscience. *Orchadian*, 6(3): 69-70. (T)

696 **Leeper, E.M. (1978).** Local botanists collect data to save endangered plants. *BioScience*, 28: 229. (T)

697 **Lehr, J.H. (1974).** The Torrey symposium on endangered plant species. *Bull. Torrey Bot. Club*, 101(4): 213. (T)

698 **Leich, H.H. (1972).** The environment conference in Stockholm (June 5-16, 1972). *Appalachia*, 38(13): 118-124.

699 **Leitzell, T.L. (1983).** Extinction, evolution, and environment management. Center for Philosophy and Public Policy, Univ. Maryland Working Paper PS-2. 15p. (P)

700 **Lek, C.W. (1968).** Conservation of habitats. *IUCN Publ., n.s.*, 10: 337-339.

701 **Lenco, M. (1982).** Remote sensing and natural resources. *Nat. Resources*, 18(2): 2-9. Illus., maps.

702 **Leon, C. (1989).** L'UICN et la conservation des plantes au niveau international. *In* Chauvet, M., *ed.* Plantes sauvages menacees de France. Bilan et protection. Actes du colloque de Brest, 8-10 octobre 1987. Paris, Bureau des Ressources Genetiques. 3-9. (G/T)

703 **Leroy, J.F. (1967).** Un chapitre d'ethnobotanique: la conservation des especes vegetales. *J. Agric. Trop. Bot. Appl.*, 14(12): 511-525. (T)

704 **Lesouef, J.-Y. (1983).** Conserver aujourd'hui pour utiliser demain. *Penn Bed*, 14(112): 69-78. Fr. Illus., maps. (T)

705 **Lesouef, Y. (1968).** The protection of rare plants by cultivation. *J. Agric. Trop. Bot. Appl.*, 15(7/8): 328-329. (G/T)

706 **Lewin, R. (1984).** Parks: how big is big enough? *Science*, 225: 611-612. (P)

707 **Lewis, R.R., III, ed. (1982).** Creation and restoration of coastal plant communities. Boca Raton, FL, CRC Press. 232p.

708 **Lindley, M. (1980).** World Conservation Strategy sets priorities for global resources. *Nature*, 284(5752): 113-114. Illus.

GENERAL

709 **Lindqvist, O.V. (1984).** Bringing biosphere reserves into the economy: what is needed? *In* Unesco-UNEP. Conservation, science and society. Paris, Unesco. 486-491. Contributions to the First International Biosphere Reserve Congress, Minsk, Byelorussia/USSR, 26 September - 2 October 1983. (P)

710 **Linnington, S., Smith, R.D. (1987).** Deferred regeneration. A manpower-efficient technique for germplasm conservation. *Pl. Genet. Resource. Newsl.*, 70: 2-12. Gene bank. (G)

711 **Linton, R.M. (1970).** Terracide. Boston, MA, Little Brown. 376p.

712 **Lipske, M. (1979).** The withering wreath. *Defenders*, 53(6): 298-305. (T)

713 **Logofet, D.O., Svirezhev, Y.M. (1984).** Modelling of population and ecosystem dynamics under reserve conditions. *In* Unesco/UNEP. Conservation, science and society. Paris, Unesco. 331-339. Contributions to the First International Biosphere Reserve Congress, Minsk, Byelorussia/USSR, 26 September - 2 October 1983. Illus. (P)

714 **Lopez Serrano, F. (1975).** Proteccion a la naturaleza: arboles para la vida. *Vida Silvestre*, 13: 60-64. Col. illus.

715 **Love, A. (1977).** Genetical background of biological conservation. *In* Studia Phytologica in Honorem Jubilantis. A.O. Harvat. Pecs, MTA Pecsi Bizottsaga. 85-88.

716 **Lovejoy, T.E. (1976).** We must decide which species will go forever. *Smithsonian*, 7(4): 52-59. (T)

717 **Lovejoy, T.E. (1977).** Conservation: can systematics provide the answer? *Assoc. Syst. Coll. Newsl.*, 5(1): 4-5.

718 **Lovejoy, T.E. (1981).** Discontinuous wilderness: minimum areas for conservation. *Tigerpaper*, 8(2): 13-16. Illus. (P)

719 **Lovejoy, T.E. (1984).** Biosphere reserves: the size question. *In* Unesco-UNEP. Conservation, science and society. Paris, Unesco. 146-151. Contributions to the First International Biosphere Reserve Congress, Minsk, Byelorussia/USSR, 26 September - 2 October 1983. (P)

720 **Lucas, G. (1979).** On the Convention on International Trade in Endangered Species of Wild Fauna and Flora. *In* Lebedev, D.V., *ed.* Proceedings of the International Botanical Congress, Leningrad, 3-10 July 1975. Leningrad, Komarov Botanical Institute. 107-108. (L/T)

721 **Lucas, G. (1979).** Organizations and contacts for conservation throughout the world. *In* Synge, H., Townsend, H., *eds.* Survival or extinction. Proceedings of a conference held at the Royal Botanic Gardens, Kew, 11-17 September 1978. Kew, Bentham-Moxon Trust. 15-23.

722 **Lucas, G. (1984).** Plants: a kingdom at risk. *IUCN Bull.*, 15(1-3): 1, 10-11. (T)

723 **Lucas, G., Synge, H. (1978).** The IUCN plant Red Data Book. Switzerland, IUCN. 540 p.

724 **Lucas, G.Ll. (1974).** The Convention on Trade in Endangered Species. *In* Hunt, D.R., *ed.* Succulents in peril. Suppl. *Intl. Org. Succ. Plant Study Bull.*, 3(3): 13-15. (L/T)

725 **Lucas, G.Ll. (1976).** Conservation: recent developments in international co-operation and legislation. *In* Simmons, J.B., *et al.*, *eds.* Conservation of threatened plants. NATO Conference Series 1: Ecology, vol. 1. New York and London, Plenum Press. 271-277. Proc. Conference on the Functions of Living Plant Collections in Conservation and Conservation-Orientated Research and Public Education, Kew, 2-6 September 1975. (L)

GENERAL

726 **Lucas, G.Ll. (1977).** Conservation: recent developments in international cooperation. *In* Prance, G.T., Elias, T.S., *eds.* Extinction is forever. Proceedings of a symposium at the New York Botanical Garden, 11-13 May 1976. New York, New York Botanical Garden. 356-359. (L)

727 **Lucas, G.Ll. (1979).** Extinction and conservation: a world picture. *In* Hedberg I., *ed.* Systematic botany, plant utilization and biosphere conservation. Stockholm, Almqvist & Wiksell International. 127-130. (T)

728 **Lucas, G.Ll. (1979).** The Threatened Plants Committee of IUCN and island floras. *In* Bramwell, D., *ed.* Plants and islands. London, Academic Press. 423-430. (Sp). (T)

729 **Lucas, G.Ll. (1984).** The survival of species genetic diversity. *In* McNeely, J.A., Miller, K.R., *eds.* National parks, conservation and development. The role of protected areas in sustaining society. Washington, D.C., Smithsonian Institution Press. 56-57.

730 **Lucas, G.Ll., Synge, H. (1977).** The IUCN Threatened Plants Committee and its work throughout the world. *Envir. Conserv.*, 4(3): 179-187. (T)

731 **Lucas, G.Ll., Wickens, G.E. (1985).** Arid lands plants: the data crisis. *Arid Lands Newsl.*, 23: 10. (T)

732 **Lucas, P.H.C. (1984).** How protected areas can help meet society's evolving needs. *In* McNeely, J.A., Miller, K.R., *eds.* National parks, conservation and development. The role of protected areas in sustaining society. Washington, D.C., Smithsonian Institution Press, 72-77. (P)

733 **Ludwig-Maximilians-Universitat Munchen (1978).** Symposium 100 Jahre Forstwissenschaft in Munchen vom 25-27 Oktober 1978. (Symposium on 100 years of Forestry Science in Munich, 25-27 October 1978.) *Forsch.-berichte Forst. Forsch. Munch.*, 42: 352p. Includes forestry protection.

734 **Lukashev, V.K. & K.I. (1979).** Nauchnye osnovy okhrany okruzhayushchei sredy. BSSR, Vyssheishaya Shkola, 1980. (Principles of protecting the environment, scientific basis.) *Novye Knigi SSSR*, 9: 55. Rus.

735 **Luks, Yu.A., Kryukova, I.V. (1980).** O sozdanii edinoi deistvennoi sistemy zapovednosti redkikh i ischezayushchikh rastenii. (Contribution to the constructing of a united effective reservation system of rare and vanishing plants.) *Bot. Zhurn.*, 65(5): 737-747. Rus. (P/T)

736 **Luks, Yu.A., Kryukova, I.V. (1980).** [Creation of single effective reservation system of rare and vanishing plants.] *Bot. Zhurn.*, 65(5): 737-747. Rus. (P/T)

737 **Luoma, J.R. (1987).** Forests are dying but is acid rain really to blame? *Audobon*, 89(2): 36-51. Acid rain.

738 **Lyons, G. (1972).** Conservation: a waste of time? *Cact. Succ. J.* (U.S.), 44: 173-177.

739 **Lyster, S. (1985).** International wildlife law. England, Grotius. xxiii, 470p. Analysis of the international treaties and conventions protecting wildlife, including texts of 12 of the most important ones. (L)

740 **MacBryde, B. (1980).** Why are so few endangered plants protected? *Amer. Hort.*, 59(5): 29-33. Col. illus. (L/T)

741 **MacBryde, B. (1982).** Approaches and priorities for conserving endangered plants. *Amer. Ass. Adv. Sci. Abstr. Pap. Nation. Meeting*, 148:64. (T)

742 **MacBryde, B. (1984).** Assessment of vulnerable native plants updated. *Endangered Species Tech. Bull.*, 8(12): 1, 6-8. (T)

743 **MacBryde, B., Altevogt, R. (1977).** Endangered plant species. *In* McGraw Hill Yearbook of Science and Technology. New York, NY, McGraw-Hill. (T)

GENERAL

744 **MacBryde, B., McMahan, L. (1981).** Plants protected by the Convention on International Trade in Endangered Species of Wild Fauna and Flora: a list of plants reported in trade, including common names and synonyms. Washington, D.C., International Convention Advisory Commission. 207p. (E/L/T)

745 **MacFarland, C. (1984).** Relating the biosphere reserve to other protected area categories. *In* Unesco-UNEP. Conservation, science and society. Paris, Unesco. 196-203. Contributions to the First International Biosphere Reserve Congress, Minsk, Byelorussia/USSR, 26 September - 2 October 1983. Tables. (P)

746 **MacKenzie, D. (1983).** World control of food genes wrecked. *New Sci.*, 100(1386): 645. (E)

747 **MacKenzie, J., El-Ashry, M. (1988).** Ill winds: air pollution's toll on trees and crops. Washington, D.C., World Resources Institute. 71p. Acid rain. (E)

748 **MacLean, D.C. (1982).** Vegetation surveys to evaluate fluoride effects on plants. *In* Murray, F., *ed.* Fluoride emmissions: their monitoring and effects on vegetation and ecosystems. Sydney, Academic Press. 153-156. Pollution.

749 **Machlis, G.E., Tichnell, M.S. (1987).** Economic development and threats to national parks: a preliminary analysis. *Envir. Conserv.*, 14(2): 151-156. (P)

750 **Madison Audubon Society (1982).** The threat of global extinctions: a symposium. 30 January 1982. Wisconsin-Madison, Herbarium and Department of Botany, University of Wisconsin-Madison. 40p. (T)

751 **Maesen, L.J.G. van der, Pundir, R.P.S. (1984).** Availability and use of wild *Cicer* germplasm. *Pl. Genet. Resource. Newsl.*, 57: 19-24. (Fr, Sp). (E/G)

752 **Magnanini, A. (1972).** Nature conservation through national parks development and endangered species protection. Morges, IUCN. 14p. (P/T)

753 **Maheshwari, J.K. (1980).** Taxonomic aspect of plant conservation. *In* Nair, P.K.K., *ed.* Modern trends in plant taxonomy. New Delhi, Vikas. 173-181. (T)

754 **Mahler, W.F. (1983).** The role of plant succession in the extinction of plant species. *Sida Contrib. Bot.*, 10(2): 191. (T)

755 **Makhdoum, M.F., Khorasani, N. (1988).** Differences between environmental impacts of logging and recreation in mature forest ecosystems. *Envir. Conserv.*, 15(2): 137-142, 166.

756 **Malik, A. (1984).** Protected areas and political reality. *In* McNeely, J.A., Miller, K.R., *eds.* National parks, conservation and development. The role of protected areas in sustaining society. Washington, D.C., Smithsonian Institution Press. 10-11. (P)

757 **Manion, P.D. (1988).** Pollution and forest ecosystems. *In* Greuter, W., Zimmer, B., *eds.* Proceedings of the XIVth International Botanical Congress. Konigstein, West Germany, Koeltz. 405-421.

758 **Margules, C., Higgs, A.J., Rafe, R.W. (1982).** Modern biogeographic theory: are there any lessons for nature reserve design? *Biol. Conserv.*, 24(2): 115-128. (P)

759 **Margules, C., Usher, M.B. (1981).** Criteria used in assessing wildlife conservation potential: a review. *Biol. Conserv.*, 21(2): 79-109.

760 **Marsh, H. (1987).** Orchid conservation project. *Newsl. Scott. Orchid Soc.*, 22: 20. (T)

761 **Martin, M. (1965).** Rare succulents. *Nat. Cact. Succ. J.* (U.K.), 20(2): 23-24. (T)

762 **Massey, J.R., Whitson, P.D. (1977).** Species biology: definition, direction, data and decisions. *In* Conference on endangered plants in the southeast. Proceedings May 11-13, 1976, Asheville, North Carolina. Asheville, NC, Southeast Forest Exp. Station. 88-94.

GENERAL

763 **Massey, J.R., Whitson, P.D. (1980).** Species biology, the key to plant preservation. *Rhodora*, 82(829): 97-103. (T)

764 **Mather, A.S. (1990).** Global forest resources. Belhaven Press. 224p. Illus. Extent of forests, current utilisation and future trends. (E)

765 **Mathews, W.H. (1971).** Man's impact on terrestrial and oceanic ecosystems. Cambridge, MA, MIT Press. 540p.

766 **Mathisen, J.E. (1974).** Dilemma of the orchids - orchid conservation. *Amer. Orch. Soc. Bull.*, 43: 1043-1048. (T)

767 **Matiuk, I.S. (1979).** [Overall utilization of reed and nature conservation]. *Lesnoe Khoz.*, 7: 53-54. Rus. *Phragmites communis.* (T)

768 **Matsuo, T., ed. (1975).** Gene conservation: exploration, collection, preservation and utilization of genetic resources. JIBP Synthesis, 5. Tokyo, Univ. of Tokyo Press. 229p. (E/G)

769 **Maudsley, P. (1981).** A threatened flora: the dangers of plant extinction discussed at an international meeting in Canberra. *GC & HTJ*, 190(20): 21-22. Illus. (T)

770 **May, R.M. (1975).** Island biogeography and the design of wildlife preserves. *Nature*, 254: 177-178. (P)

771 **Mayo, A. (1986).** 60,000 plants may become endangered in the next 40 years. *Species*, 6: 4-5. (T)

772 **McCormick, J. (1983).** National parks come of age. *Ecos*, 4(1): 40-41. (P)

773 **McCormick, J. (1985).** Acid Earth: the global threat of acid pollution. London, Earthscan. 199p. Acid rain.

774 **McCormick, J.F. (1977).** Endangered species - questions of science, ethics and law. *In* Conference on endangered plants in the southeast. Proceedings 11-13 May, 1976, Asheville, North Carolina. Asheville, NC, Southeast Forest Exp. Station. 81-97. (L/T)

775 **McCoy, E.D. (1983).** The application of island-biogeographic theory to patches of habitat: how much land is enough? *Biol. Conserv.*, 25(1): 53-61. (P)

776 **McCrone, J.D. (1984).** Cluster biosphere reserves. *In* Unesco-UNEP. Conservation, science and society. Paris, Unesco. 208-213. Contributions to the First International Biosphere Reserve Congress, Minsk, Byelorussia/USSR, 26 September - 2 October, 1983. (P)

777 **McDavid, S.J. (1977).** Endangered species. *Fernback Quarterly*, 3(2): 1-3. (T)

778 **McEachern, J., Towle, E.L. (1974).** Ecological guidelines for island development. IUCN Publications new series No. 30. Morges, IUCN. 65p.

779 **McFarlane, B.S., Louguet, P. (1980).** Planning for threatened and endangered plant species. *J. Soil Water Conserv.*, 35(5): 221-223. (T)

780 **McIvor, J.G., Bray, R.A., eds (1983).** Genetic resources of crop forage plants. CSIRO, East Melbourne. (E)

781 **McMahan, L. (1982).** Terrestrial orchids in trade. *TRAFFIC (U.S.A.) Newsl.*, 4(1): 6. (L/T)

782 **McMahan, L., Walter, K. (1987).** Where the rare plants are. *Center Plant Conserv. Newsl.*, 2(3): 6. (T)

783 **McNaught, K. (1981).** Town margin nature reserves. *Nat. Cambridge*, 24: 20-22. Illus. (P)

GENERAL

784 **McNeely, J. (1987).** Our common future: IUCN and the Brundtland Commission. *IUCN Bull.*, 18(10-12): 5-6. Illus. Reviews IUCN's efforts to implement the findings of the Brundtland Commission.

785 **McNeely, J., Scriabine, R. (1983).** Parks for people. *Ambio*, 12(1): 51-53. Illus. World Parks Congress, Bali, Oct. 1982. (P)

786 **McNeely, J.A. (1982).** Protected areas have come of age. *Ambio*, 11(5): 236-237. Col. illus. (P)

787 **McNeely, J.A. (1983).** IUCN National Parks and protected areas: priorities for action. *Envir. Conserv.*, 10(1): 13-21. (P)

788 **McNeely, J.A. (1984).** Biosphere reserves and human ecosystems. *In* Unesco-UNEP. Conservation, science and society. Paris, Unesco. 492-498. Contributions to the First International Biosphere Reserve Congress, Minsk, Byelorussia/USSR, 26 September - 2 October 1983. (P)

789 **McNeely, J.A. (1984).** Introduction: protected areas are adapting to new realities. *In* McNeely, J.A., Miller, K.R., *eds*. National parks, conservation and development. The role of protected areas in sustaining society. Washington, D.C., Smithsonian Institution Press. 1-7. (P)

790 **McNeely, J.A. (1984).** The World Heritage Convention: protecting natural and cultural wonders of global importance - a slide presentation. *In* McNeely, J.A., Miller, K.R., *eds*. National parks, conservation and development. The role of protected areas in sustaining society. Washington, D.C., Smithsonian Institution Press. 735-736. (E/L/P)

791 **McNeely, J.A. (1988).** Economics and biological diversity: developing and using economic incentives to conserve biological resources. Gland, IUCN. 256p. (E/P)

792 **McNeely, J.A. (1989).** Conserving biological diversity. A decision-maker's guide. *IUCN Bull.*, 20(4-6): 6-7. Illus. (E)

793 **McNeely, J.A., Miller, K.R. (1984).** Recommendations of the World National Parks Congress. *In* McNeely, J.A., Miller, K.R., *eds*. National Parks, conservation and development. The role of protected areas in sustaining society. Washington, D.C., Smithsonian Institution Press. 765-776. (P)

794 **McNeely, J.A., Miller, K.R., Reid, W.V.C., Mittermeier, R.A., Werner, T.B. (1989).** Conserving the world's biological diversity. IUCN, World Resources Institute, WWF-US and The World Bank. 250p. Illus. (B/L)

795 **McNeely, J.A., Miller, K.R., eds (1984).** National parks, conservation and development: the role of protected areas in sustaining society. Washington, D.C., Smithsonian Institution Press. 838p. (P)

796 **McNeely, J.A., Navid, D., eds (1984).** Conservation, science and society: the proceedings of the First International Congress on Biosphere Reserves. Minsk, Byelorussia. 600p. (P)

797 **McNeely, J.A., Thorsell, J.W., eds (1987).** Guidelines for development of terrestrial and marine national parks for tourism and travel. Madrid, World Tourism Organization. 29p. (P)

798 **McNeill, J. (1981).** CITES: Convention on International Trade in Endangered Species. *Canad. Bot. Ass. Bull.*, 14(4): 38-39. (L/T)

799 **Medwecka-Kornas, A. (1981).** The floristic and phytosociological definition and description of conservation sites. *In* Synge, H., *ed*. The biological aspects of rare plant conservation. Chichester, Wiley. 431-445. Proceedings of International Conference, King's College, Cambridge, 14-19 July 1980. (P)

GENERAL

800 **Mehra, K.L., Sastrapradja, S. (1985).** Proceedings of the International Symposium on South East Asian Plant Genetic Resources. Jakarta, Indonesia, 20-24 August 1985. Bogor, Lembaga Biologi Nasional - LIPI. 211p. Graphs, tables. (E/G)

801 **Meijer, W. (1973).** Endangered plant life. *Biol. Conserv.*, 5(3): 163-167. Illus. (T)

802 **Melville, R. (1970).** Plant conservation and the Red Book. *Biol. Conserv.*, 2: 185-188. (T)

803 **Melville, R. (1970).** Plant conservation in relation to horticulture. *J. Roy. Hort. Soc.*, 95(11): 473-480. (G)

804 **Melville, R. (1970).** Red Data Book: volume 5. Angiospermae. Morges, IUCN Survival Service Commission. 12p, 68 leaves. (T)

805 **Melville, R. (1971).** Conservation of orchids. *Orch. Rev.*, 79: 21-22. (T)

806 **Mengesha, M.H., Rao, K.E.P. & S.A. (1982).** Sorghum and millets genetic resources at ICRISAT. *Pl. Genet. Resource. Newsl.*, 51: 21-26. (Fr, Sp). (E/G)

807 **Miasek, M.A., Long, C.R. (1978).** Endangered plant species of the world and their endangered habitats: a compilation of the literature. New York, Library of the New York Botanical Garden. 46p. (B/T)

808 **Miasek, M.A., Long, C.R., comps (1985).** Endangered plant species of the world and their endangered habitats: a compilation of the literature. Plant bibliography no. 6. New York, Library of the New York Botanical Garden. 154p. 1647 refs. (B/T)

809 **Miegroet, M. van (1978).** The forester and environmental problems. *In* Proceedings of the 7th World Forestry Congress, 4-18 October 1972, vol. 3: Education. Buenos Aires, Instituto Forestal Nacional. 3398-3425. En, Sp, Fr.

810 **Millar, C.I. (1987).** The California forest germplasm conservation project: a case for genetic conservation of temperate tree species. *Conservation Biology*, 1(3): 191-193. (E/G)

811 **Miller, K.R. (1982).** Parks and protected areas: considerations for the future. *Ambio*, 11(5): 315-317. (P)

812 **Miller, K.R. (1984).** Biosphere reserves and the global network of protected areas. *In* Unesco-UNEP. Conservation, science and society. Paris, Unesco. 3-13. Contributions to the First International Biosphere Reserve Congress, Minsk, Byelorussia/USSR, 26 September - 2 October 1983. Tables. (P)

813 **Miller, K.R. (1984).** The Bali Action Plan: a framework for the future of protected areas. *In* McNeely, J.A., Miller, K.R., *eds.* National parks, conservation and development. The role of protected areas in sustaining society. Washington, D.C., Smithsonian Institution Press. 756-764. (P)

814 **Miller, K.R. (1984).** The natural protected areas of the world. *In* McNeely, J.A., Miller, K.R., *eds.* National parks, conservation and development. The role of protected areas in sustaining society. Washington, D.C., Smithsonian Institution Press. 20-23. (P)

815 **Miller, P. (1985).** The impacts of air pollution on forest resources. Forestry research west - United States Department of Agriculture, Forest Service. Fort Collins, Colo. The Service: 1-5. Illus.

816 **Miller, R., Bratton, S., White, P. (1987).** A regional strategy for reserve design and placement based on an analysis of rare and endangered species distribution patterns. *Biol. Conserv.*, 39(4): 255-268. (P)

817 **Miller, R.S., Botkin, D.B. (1974).** Endangered species models and predictions. *Amer. Sci.*, 62(2): 172-180. (T)

GENERAL

818 **Mills, S. (1983).** Shades of reasons for protecting wildlife. *New Sci.*, 98(1361): 685-687. Illus.

819 **Mittermeier, R. (1986).** Species and genetic resources. *IUCN Bull.*, 17(1-3): 24-25. (E)

820 **Mlinsek, D. (1978).** Research aspects in the evaluation and analysis of environmental impacts of forestry. *In* Proceedings of the 8th World Forestry Congress, Jakarta, 16-28 October 1978. Forestry for quality of life. II, 14.

821 **Mlot, C. (1989).** Blueprint for conserving plant diversity. *BioScience*, 39(6): 364-368. (T)

822 **Moir, W.W.G. (1967).** Final report of the conservation committee to the American Orchid Society. *Orchidata*, 7: 106-110. (T)

823 **Moir, W.W.G. (1970).** Conservation of native orchids. *Amer. Orch. Soc. Bull.*, 39: 425-426. (T)

824 **Moir, W.W.G. (1983).** The particular variants. *Orchid Rev.*, 91(1078): 244-245. About the over-collection of exceptional variants. (E/T)

825 **Monod, Th. (1968).** La conservation des habitats: problemes de definitions et de choix. *In* Hedberg, I. & O., *eds.* Conservation of vegetation in Africa south of the Sahara. Symposium Proceedings, at 6th plenary meeting of AETFAT, Uppsala. *Acta Phytogeogr. Suec.*, 54: 32-35. Fr. Habitats and biotopes. (P)

826 **Mooney, P.R. (1983).** The law of the seed: another development and plant genetic resources. *Devel. Dialogue*, 1-2: 172p. Illus., maps. (E/G/L)

827 **Moore, D.M. (1983).** Human impact on island vegetation. *Geobotany*, 1983(5): 237-246. Illus.

828 **Moore, H.E., jr (1979).** Endangerment at the specific and generic levels in palms. *Principes*, 23(2): 47-64. (T)

829 **Moore, P. (1987).** A thousand years of death. *New Sci.*, 113(1542): 46-48. Peatlands.

830 **Moritz, H. (1978).** Nationalparkplanung. (National park planning.) *Natur und Land*, 64(6): 200-201. Ge. (P)

831 **Morse, L.E., Henifin, M.S., Ballman, J.C., Lawyer, J.I. (1981).** Geographical data organization in botany and plant conservation: a survey of alternative strategies. *In* Morse, L.E., Henifin, M.S., *eds.* Rare plant conservation: geographical data organization. New York, New York Botanical Garden. 9-29. (T)

832 **Morse, L.E., Henifin, M.S., eds (1981).** Rare plant conservation: geographical data organization. New York, New York Botanical Garden. v, 377p. Illus., maps. (T)

833 **Moseley, J., Thelen, K., Miller, K. (1976).** National parks planning. A manual with annotated examples. FAO Forestry Paper No. 6. Rome, FAO Forest Resources Division. vii, 42p. (P)

834 **Mueller-Dombois, D. (1988).** Forest decline and dieback - a global ecological problem. *Trends Ecol. Evol.*, 3(11): 310-312.

835 **Muenscher, W.C. (1937).** Why our native wild flowers need protection. *Wild Flower*, 14: 72-75. (L/T)

836 **Mugnozza, G.T.S. (1981).** [Genetic erosion: magnitude of the phenomenon and ways of intervening.] *Ann. Accad. Agric. Torino*, 124: 1-15. It. Maps. Crop. Gene banks.

837 **Mukherjee, S.K. (1983).** Conservation of germplasm resources of horticultural crops: fruits. *In* Jain, S.K., Mehra, K.L., *eds.* Conservation of tropical plant resources. Howrah, Botanical Survey of India. 159-163. (E/G)

44

GENERAL

838 **Mull, W.P. (1975).** Magnificent minutiae. *Defenders*, 50(6): 487-490. Effect of introduced species.

839 **Muller, C. (1980).** Conservation a long terme des graines de sapins et son incidence sur le comportement des semis en pepinieries. *Seed Sci. Techn.*, 8(1): 103-118. Fr (En, Ge). Illus. (G)

840 **Munro, D. (1984).** Global sharing and self-interest in protected areas conservation. *In* McNeely, J.A., Miller, K.R., *eds.* National parks, conservation and development. The role of protected areas in sustaining society. Washington, D.C., Smithsonian Institution Press. 672-676. (P)

841 **Munro, D.A. (1979).** A strategy for the conservation of wild living resources. *ICOM*, 2(4): 6-15.

842 **Munro, D.A. (1979).** Towards a world strategy of conservation. *Envir. Conserv.*, 6(3): 169-170.

843 **Muresan, G., Furnica, H., Copaceanu, D. (1985).** Nouvelles technologies d'istexploitation des forets et leur correlation avec les exigences de la loi concernant la conservation du patrimoine forestier. *In* Proceedings of the 8th World Forestry Congress, Jakarta, 16-28 October 1978: Forestry for industrial development. Fr. (L)

844 **Murphy, D.D. (1989).** Conservation and confusion: wrong species, wrong scale, wrong conclusions. *Conservation Biology*, 3(1): 82-84. Discusses habitat fragmentation, size of reserves and extinction rates. (P)

845 **Murray, M., Oldfield, S. (1984).** Choosing a top ten. *IUCN Bull.*, 15(7/9): 79-82. Illus. (T)

846 **Murtopo, A. (1984).** Keynote address: the World Conservation Strategy and the developing world. *In* McNeely, J.A., Miller, K.R., *eds.* National parks, conservation and development. The role of protected areas in sustaining society. Washington, D.C., Smithsonian Institution Press. 678-680.

847 **Myers, N. (1976).** An expanded approach to the problem of diappearing species. *Science*, 193(4249): 198-202.

848 **Myers, N. (1977).** Garden of Eden to weed patch: the Earth's vanishing genetic heritage. *NRDC Newsl.*, 6(1): 1-15. (T)

849 **Myers, N. (1978).** Disappearing legacy. *Nature Canada*, 7(4): 41-54.

850 **Myers, N. (1979).** The sinking ark. A new look at the problem of disappearing species. Oxford, Pergamon Press. xiii, 307p. Includes deforestation and tropical forests. (E/T)

851 **Myers, N. (1980).** The problem of disappearing species: what can be done? *Ambio*, 9(5): 229-235. Illus. (T)

852 **Myers, N. (1981).** The Earth's vanishing genetic heritage. *Calif. Native Pl. Soc. Bull.*, 11(6): 6.

853 **Myers, N. (1983).** A priority-ranking strategy for threatened species? *The Environmentalist*, 3: 97-120. (T)

854 **Myers, N. (1983).** A wealth of wild species: storehouse for human welfare. Boulder, Colorado, Westview Press. xiii, 272p. Illus. (E)

855 **Myers, N. (1983).** By saving wild species, we may be saving ourselves. *Nat. Conserv. News*, 33(6): 6-13.

856 **Myers, N. (1984).** Eternal values of the parks movement and the Monday morning world. *In* McNeely, J.A., Miller, K.R., *eds.* National parks, conservation and development. The role of protected areas in sustaining society. Washington, D.C., Smithsonian Institution Press. 656-660. (P)

GENERAL

857 **Myers, N. (1984).** Genetic resources in jeopardy. *Ambio*, 13(3): 171-174.

858 **Myers, N. (1984).** Plants - an embarrassment of choices. *IUCN Bull.*, 15(1-3): 16-18.

859 **Myers, N. (1985).** The Gaia atlas of planet management. London, Pan Books and New York, Doubleday. 272p. Illus. Global overview of natural resources and environmental problems, including deforestation, species extinctions.

860 **Myers, N. (1985).** The plant kingdom - a material underpinning of our daily lives is grossly threatened. *J. Roy. Soc. Arts*, 134: 38-44. (E/T)

861 **Myers, N. (1986).** A treaty on endangered species habitat protection: a proposal. *In* Kennedy, M., Burton, R., eds. "A threatened species conservation strategy for Australia". Policies for the future. Manly, N.S.W., Ecofund. 49-51. (L)

862 **Myers, N. (1986).** Economics and ecology in the international arena: the phenomenon of "linked linkages". *Ambio*, 15(5): 296-300. Illus.

863 **Myers, N. (1987).** The extinction spasm impending: synergisms at work. *J. Conserv. Biol.*, 1(1): 14-21.

864 **Myers, N. (1988).** Draining the gene pool: the causes and consequences of genetic erosion. *In* Kloppenberg, J., ed. Seeds and sovereignty: debate over the use and control of plant genetic resources. Durham, NC, Duke Univ. Press. 90-113. (E)

865 **Myers, N. (1988).** Favoring a triage strategy for threatened species. *Buzzworm*, 1(1): 28-29. (T)

866 **Myers, N. (1988).** Mass extinction - profound problem, splendid opportunity. *Oryx*, 22(40): 205-210. (T)

867 **Myers, N. (1989).** A mass extinction episode ahead of us: predictable or preventable? *In* Western, D., Pearl, M., eds. Conservation for the twenty-first century. Oxford, Oxford Univ. Press. 42-49. (T)

868 **Myers, N. (1989).** Extinction rates past and present. *BioScience*, 39: 39-41. (T)

869 **Myers, N. (1989).** Loss of biological diversity and its potential impact on agriculture and food production. *In* Pimentel, D., Hall, C., eds. Food and natural resources. Academic Press. 49-67. (E/T)

870 **Myers, N. (1989).** The environmental basis of sustainable development. *In* Schramm, G., Warford, J.J., eds. Environmental management and economic development. Baltimore, John Hopkins Univ. Press. 57-68.

871 **Myers, N. (1989).** The future of forests. *In* Friday, L., Laskey, R.A., eds. The fragile environment. Cambridge, Cambridge Univ. Press. 22-40. Deforestation.

872 **Myers, N., Ayensu, E.S. (1983).** Reduction of biological diversity and species loss. *Ambio*, 12(2): 72-74. Col. illus. (P/T)

873 **Nabhan, G.P. (1979).** Who is saving the seeds to save us? *Mazingara*, 9: 54-59. (E)

874 **Nageswara Rao, P.P. (1980).** Importance of habitat in the conservation of wildlife. *Sci. Cult.*, 46(8): 292-293.

875 **Nagy, I.V. (1975).** Kornyezetvedelmi Vilagnap. (World day of environmental protection.) *Buvar*, 30(7): 290. Hu.

876 **National Parks and Conservation Association (1975).** How to save a wildflower: NPCA progress report on endangered plants. *Nat. Parks Conserv. Mag.*, 49(4): 10-14. (T)

877 **National Research Council Committee on Germplasm Resources (1978).** Conservation of germplasm resources: an imperative. Washington, D.C., National Academy of Sciences. 118p. (E/G)

GENERAL

878 **Nature Conservancy Council (1980).** Nature conservation and agriculture. Bibliography Series No. 1. Banbury, NCC. 1979 addendum 1980. 12p. (B)

879 **Navarro Valdivielso, B. (1987).** The botanic gardens as a vehicle for environmental education. *In* Bramwell, D., Hamann, O., Heywood, V., Synge, H., *eds.* Botanic gardens and the World Conservation Strategy. London, Academic Press. 59-65. (Sp). (G)

880 **Nayar, M.P. (1984).** Extinction of species and concept of rarity in plants. *J. Econ. Taxon. Bot.*, 5(1): 1-6. (T)

881 **Nayar, N.M. (1983).** Conservation of plantation crops genetic resources. *In* Jain, S.K., Mehra, K.L., *eds.* Conservation of tropical plant resources. Howrah, Botanical Survey of India. 246-253. (E/G/T)

882 **Negrutiu, F. (1979).** Padurile de agrement in contextul actiunilor privind protectia mediului inconjurator. (Recreational forests within the plans for environmental protection.) *Rev. Padur. Ind. Lemn. Celul. Hirtie Silvic. Expl. Padur.*, 94(4): 229-230. Rum.

883 **Nelson, J.R. (1987).** Rare plant surveys: techniques for impact assessment. *In* Elias, T.S., *ed.* Conservation and management of rare and endangered plants. Proceedings from a conference, Sacramento, California, 5-8 November 1986. Sacramento, California Native Plant Society. 159-166. (T)

884 **Nerem, R.S., Holz, R.K., Helfert, M.R., Tapley, B.D. (1985).** Vegetation change detection from NOAA polar orbiting satellites. *Geol. J.*, 11(4): 313-320. Illus.

885 **New York State Agric. Exper. Station (1979).** New methods of germplasm preservation by quick-freezing plant cell cultures. New York State Agric. Exper. Station Annual Rept. No. 98. 1-11. (G)

886 **Ng, N.Q. (1982).** Genetic resources programme of IITA. *Pl. Genet. Resource. Newsl.*, 49: 26-31. (Fr, Sp). Map. International Institute of Tropical Agriculture. (E/G)

887 **Nicholson, R. (1988).** Arnold Arboretum focuses on growing rare woody plants. *Center for Plant Conservation*, 3(1): 3. (G)

888 **Nickerson, N.H. (1980).** Habitat protection. *Rhodora*, 82(829): 201-206.

889 **Nicot, J. (1973).** Les champignons dans la destruction de la nature. Les mycologues et la protection de l'environnement. *Rev. Mycol.*, 37(1-2): 96-99. Fr. (T)

890 **Nikles, D.G. (1980).** Realized and potential gains from using and conserving genetic resources of *Araucaria. In* Int. Union of Forestry Research Organizations. Forestry problems of the genus *Araucaria*. IUFRO meeting, Curitiba, Parana (Brazil), 21-28 October 1979. Parana, Fundacao de Pesquisas Florestais do Parana. 87-95. (E/T)

891 **Nikolaev, G.V. (1978).** [Conservation of drug plants]. *Lesnoe Khoz.*, 9: 66-70. Rus. (E)

892 **Nilsson, O. (1988).** Da botaniska tradgardarnas roll: bevarandearbetet. (The role of botanic gardens in the conservation work.) *Sven. Bot. Tidskr.*, 82(6): 471-472. Sw (En). (G/T)

893 **Noirfalise, A., Segers, M. (1979).** Chasse et conservation de la nature. *Ann. Gembloux*, 85: 179-185. Fr.

894 **Nommsalu, F. (1979).** Uueneb looduskaitse - strateegia. (Nature conservation strategy is renewed.) *Eesti Loodus*, 22(4): 212-216. Es (Rus, En). Illus. 14th Assembly of IUCN, Ashkhabad, 1978.

895 **Norse, E.A. (1987).** International lending and the loss of biological diversity. *J. Conservation Biology*, 1(3): 259-260.

GENERAL

896 **Norton, B.G. (1983).** On the inherent danger of undervaluing species. Working Paper PS-3. College Park, MD, Center for Philosophy and Public Policy, University of Maryland. 28p.

897 **Noss, R.F. (1988).** The longleaf pine landscape of the southeast: almost gone and almost forgotten. *End. Species Update*, 5(5): 1-8.

898 **Obata, J. (1976).** Cultivating an "extinct" species. *Newsl. Hawaii. Bot. Soc.*, 15(2): 35-37. (G/T)

899 **Oedekoven, K. (1980).** The vanishing forest. *Envir. Law Pol.*, 6: 184-185. (L/T)

900 **Ohgane, E., Maeda, M. (1978).** Consideration of forest protection from a technical theoretical view point. *Nippon Ringaku Kaishi Nippon Ringakkai*, 60(9): 323-326. (Ja).

901 **Ojeda, E.M., Rojas, C.S., Llanes, L.F.P., Banos, M.G. (1978).** Indice de proyectos en desarrollo en ecologia de zonas aridas, vol.1. (Index of current research in arid zones ecology.) Xalapa, Instituto de Investigaciones sobre Recursos Bioticos, A.C. vii, 97p. Sp (En, Fr).

902 **Olaczek, R. (1985).** [IUCN categories of threat to plant and animal species.] *Chronmy Przyr. Ojczysta*, 41(6): 5-21. Pol (En). Illus. (T)

903 **Oldfield, S. (1980).** Conservation of carnivorous plants. *Carniv. Pl. Soc. J.*, 4: 15-16. (T)

904 **Oldfield, S. (1981).** Convention on Trade in Endangered Species. *Threatened Pl. Committ. Newsl.*, 8: 3-5. Several species of insectivorous plants added to the Appendices of CITES at New Delhi, 1981. (L/T)

905 **Oldfield, S. (1981).** The trade in tree ferns. *TRAFFIC Bull.*, 3(3/4): 30. (L)

906 **Oldfield, S. (1983).** Orchids and conservation legislation. *Orchid Rev.*, 91(1077): 220-221. (L/T)

907 **Oldfield, S. (1983).** Threat from wild bulb trade. *GC & HTJ*, 194(17): 15. Illus. (T)

908 **Oldfield, S. (1984).** Conservation of rare and endangered bulbs. *Kew Mag.*, 1(1): 23-29. Illus. (T)

909 **Oldfield, S. (1984).** Improving CITES for plants. *Threatened Pl. Newsl.*, 13: 16-17. Illus. Trade. (L/T)

910 **Oldfield, S. (1984).** Protected plants. *Cact. Succ. J.* (U.K.), 2(3): 83. (E/L/T)

911 **Oldfield, S. (1985).** The laundered-bulb trade is a dirty business. *Garden* (New York), May/June: 2-4. (L/T)

912 **Oldfield, S. (1985).** Trade in endangered species. *Aroideana*, 6(3): 83-84. (T)

913 **Oldfield, S. (1985).** Whither international trade in plants? *New Sci.*, 106(1454): 10-11. Illus. (L/T)

914 **Oldfield, S. (1986).** Botanical trafficking. *Nation. Council Conserv. Pl. Gard. Newsl.*, 8: 27-28. (T)

915 **Olembo, R. (1984).** UNEP and protected areas. *In* McNeely, J.A., Miller, K.R., *eds.* National parks, conservation and development. The role of protected areas in sustaining society. Washington, D.C., Smithsonian Institution Press. 681-684. (P)

916 **Olembo, R.J. (1978).** Environmental issues in forest and wildland management. *In* Proceedings of the 8th World Forestry Congress, Jakarta, 16-28 October 1978: Forestry for quality of life. I, 13.

GENERAL

917 **Olsen, O., Arnklit, F. (1979).** Setting up a practical small seed bank. *In* Synge, H., Townsend, H., *eds.* Survival or extinction. Proceedings of a conference held at the Royal Botanic Gardens, Kew, 11-17 September 1978. Kew, Bentham-Moxon Trust. 185-188. Seed bank. (G)

918 **Ormazabal, C. (1986).** Preservacion de recursos fitogeneticos in situ a traves de parques nacionales y otras areas protegidas. Importancia, avances, limitaciones y proyeccion futura. *Boletin Tecnico,* 16: 29p. Sp.

919 **Orpet, E.O. (1981).** Recoleccion de orquideas. (Collecting orchids.) *Orquideologia,* 15(1): 88-91, 100-103. Sp (En). Col. illus.

920 **Ostapko, V.M. (1982).** [Polymorphism and plant conservation.] Introduktsiia ta aklimatizatsiia roslyn na Ukraini. Kyiv, Naukova dumka. 21. 8-15. Uk (Rus). Illus.

921 **Otto, V. (1982).** The brink of extinction. *Fiddlehead Forum,* 9(3): 17. *Diplazium laffanianum.* (T)

922 **Ovington, J.D. (1984).** Ecological processes and national park management. *In* McNeely, J.A., Miller, K.R., *eds.* National parks, conservation and development. The role of protected areas in sustaining society. Washington, D.C., Smithsonian Institution Press. 60-64. Tables. (P)

923 **Owen, J.S. (1972).** Some thoughts on management of national parks. *Biol. Conserv.,* 4(4): 241-246. (P)

924 **Palmberg, C. (1981).** A vital fuelwood gene pool is in danger. *Unasylva,* 33(133): 22-30. Illus. FAO project. (E)

925 **Palmberg, C. (1987).** Conservation of genetic resources of woody species. Rome, FAO. 16p., annexes. Mimeo. Paper prepared for Simposio Sobre Silvicultura y Mejoramiento Geneti co. Cief. Buenos Aires, 6-10 April 1987. Includes lists of commercially exploited timber trees. (E/T)

926 **Palmer, A.R. (1982).** A computer-based natural resource mapping system for use on small nature reserves. *In* AETFAT Synopses 10th Congress: the origin, evolution and migration of African floras, Pretoria, 18-23 January 1982. Pretoria. 82p. Abstract. (P)

927 **Park, C. (1982).** EIA: scope and significance. *Ecos,* 3(3): 2-5. Environmental impact assessment.

928 **Park, C.C. (1987).** Acid rain: rhetoric and reality. London, Routledge. 284p. Illus.

929 **Parker, D. (1986).** Could the amateur collector do more for conservation? *Cact. Succ. J.* (U.K.), 4(3): 76-78.

930 **Patrascoiu, N. (1979).** Consideratii privind amenajarea padurilor cu functii speciale de protectie a mediului inconjurator. (Thoughts on forest management with the special role of the environment in protection.) *Rev. Padur. Ind. Lemn. Celul. Hirtie Silvic. Expl. Padur.,* 94(4): 235-238.

931 **Pattison, G.A. (1988).** The National Council for the Conservation of Plants and Gardens (NCCPG) in the U.K. *Bot. Gard. Conserv. News,* 1(3): 38-39. (G/T)

932 **Pavlik, B.M. (1987).** Autecological monitoring of endangered plants. *In* Elias, T.S., *ed.* Conservation and management of rare and endangered plants. Proceedings from a conference, Sacramento, California, 5-8 November 1986. Sacramento, California Native Plant Society. 385-390. (T)

933 **Pawlick, T. (1985).** A killing rain: the global threat of acid precipitation. San Francisco, CA, Sierra Club Books. 206p. (T)

934 **Paylore, P., Mabbutt, J.A. (1980).** Desertification: a world bibliography update, 1976-80. Tucson, University of Arizona. 196p. (B)

GENERAL

935 **Pearce, F. (1987).** Acid rain. *New Sci.*, 116(1585): 4p. suppl. Illus., map. Includes effects on trees; mainly Europe.

936 **Pearsall, S.H. (1984).** *In absentia* benefits of nature preserves: a review. *Envir. Conserv.*, 11(1): 3-10. (P)

937 **Pellet, H., et al. (1978).** Endangered species project. *Landscape Hort. Res. Bull,* 29. (T)

938 **Pellew, R. (1988).** WCMC - the new look for CMC. *Species,* 11: 5-6. Describes the work and databases of the World Conservation Monitoring Centre.

939 **Pena M, Tellez V. (1988).** Orquidea..una voz en extincion? Consideraciones sobre aspectos de conservacion. *In* Simposio sobre diversidad biologica de Mexico. 3-7 Octubre 1988. Oaxtepac, Morelos, Mexico. Mexico, Instituto de Biologia, Universidad Nacional de Mexico. 43-44. Sp. (T)

940 **Penman, S., Rankin, E., Moye, J. (1987).** Rare and endangered native orchid species survey. *Orchadian,* 9(1): 5-9. Map. (T)

941 **Penny, S. (1979).** Nature conservation. London, National Book League. 33p. (B)

942 **Perina, V., Tesar, V. (1983).** K planovani a realizaci obnovnich cilu v imisnich oblastech. (Planning and implementation of forest regeneration aims in the air polluted areas.) *Lesnicka Prace,* 62(9): 389-395. Cz (En, Fr, Ge, Rus). Pollution, forest.

943 **Perrard, O. (1988).** Nature: l'arche de noe se vide. *L'Expresse Aujourd'hui,* 10 June: 28-34. Col. illus. Overview on loss of species and methods of *in situ* conservation. (G/T)

944 **Perring, F.H. (1975).** Plant conservation without botanists. *New Sci.,* 67(959): 194-195.

945 **Perring, F.H. (1976).** The future and electronic data processing. *In* Simmons, J.B., *et al., eds.* Conservation of threatened plants. NATO Conference Series 1: Ecology, vol. 1. New York and London, Plenum Press. 113-117. Proc. Conference on the Functions of Living Plant Collections in Conservation and Conservation-Orientated Research and Public Education, Kew, 2-6 September 1975.

946 **Perring, F.H. (1977).** New records for endangered plants. *BSE News,* 21: 8-9. Procedure for reporting finds. (T)

947 **Perring, F.H. (1983).** Collecting and collating Red Data Book data. *In* Jain, S.K., Mehra, K.L., *eds.* Conservation of tropical plant resources. Howrah, Botanical Survey of India. 164-171. (T)

948 **Peterken, G.F. (1968).** International selection of areas for reserves. *Biol. Conserv.,* 1(1): 55-61. (P)

949 **Peterken, G.F. (1968).** The IBP/CT survey of areas of significance to conservation. *In* Hedberg, I. & O., *eds.* Conservation of vegetation in Africa south of the Sahara. Symposium Proceedings, at 6th plenary meeting of AETFAT, Uppsala. *Acta Phytogeogr. Suec.,* 54: 44-47.

950 **Peters, R.L., Darling, J.D.S. (1985).** The greenhouse effect and nature reserves. *BioScience,* 35(11): 707-717. Illus. (P)

951 **Phillips, D.H. (1981).** Legislating for introduced species: the lessons from plant health. *In* Pinder, N., *ed.* Conservation and introduced species: a discussion meeting held at University College London on 18 April 1980. Discussion Paper in Conservation No. 30. London, UCL. 50-59. British Association of Nature Conservationists in conjunction with Ecology and Conservation Unit of UCL. (L)

952 **Pickett, S.T.A., White, P.S., eds (1985).** The ecology of natural disturbance and patch dynamics. London, Academic Press. 472p.

GENERAL

953　**Pickoff, L.J. (1975).** Our role in conservation. *Cact. Succ. J.* (U.S.), 47(1): 20-22.

954　**Pigram, J. (1983).** Outdoor recreation and resource management. London, Croom Helm; New York, St. Martin's Press. 262p. Illus., maps. See Chapter 7: National Parks and Wilderness. (P)

955　**Pitt, D., ed. (?)** The future of the environment. London, Routledge. 192p.

956　**Plant Conservation Roundtable (1986).** Conservation guidelines. *Natural Areas J.*, 6(3): 31-32. Guidelines for plant collecting. (T)

957　**Plenipotentiary Conference to Conclude an International Convention on Trade in Certain Species of Wildlife, Washington, D.C., 1973. (1973).** World wildlife conference: efforts to save endangered species. Washington, D.C., U.S. Dept. of State. 30p. (L/T)

958　**Plucknett, D.L., Smith, N.J., Williams, N., Anishetty, N.H. (1987).** Gene banks and the world's food. Princeton, New Jersey, Princeton University Press. 247p. Illus. (E/G)

959　**Poel, A.J. van der (1978).** Die Anforderungen der Umweltgestalter, der Landschaftspfleger und der Landschaftsplaner an die Waldbewirtschaftung. (The requirements of the environmentalists, the conservationists and landscape planners with regard to forest management.) *Mitt. Forst. Bund. Versuchs. Wien*, 124: 41-44.

960　**Polunin, N. (1968).** Conservational significance of botanical gardens. *Biol. Conserv.*, 1: 104-105. (G)

961　**Polunin, N. (1983).** Wilderness and the biosphere. *Envir. Conserv.*, 10(4): 281-282.

962　**Polunin, N. (1984).** The Foundation for Environmental Conservation: auspices and objectives. *Envir. Conserv.*, 11(1): 77-78.

963　**Polunin, N., Eidsvik, H.K. (1979).** Ecological principles for the establishment and management of national parks and equivalent reserves. *Envir. Conserv.*, 6(1): 21-26. Illus. (P)

964　**Poore, D. (1984).** Introduction: Biosphere Reserves for science and monitoring. *In* Unesco/UNEP. Conservation, science and society. Paris, Unesco. 247-250. Contributions to the First International Biosphere Reserve Congress, Minsk, Byelorussia/USSR, 26 September - 2 October 1983. (P)

965　**Poore, M.E.D. (1982).** Ecology and conservation: a practitioner's view. *New Phytol.*, 90(3): 405-417.

966　**Poore, M.E.D. (1983).** Replenishing the world's forests: why replenish? World forests, past, present and future. *Commonw. Forest. Rev.*, 62(3): 163-168. (Fr, Sp).

967　**Postel, S. (1984).** Air pollution, acid rain, and the future of forests. Worldwatch Paper 58. Washington, D.C., Worldwatch Institute. 54p. Acid rain.

968　**Pough, R.H. (1981).** The need for species diversity. *Defenders*, 56(6): 37.

969　**Prance, G.T. (1982).** Extinction: the size of the problem. *Amer. Assoc. Adv. Sci. Abstr. Pap. Nation. Meeting*, 148: 63. (T)

970　**Prescott-Allen, C. & R. (1986).** The forest resources. New Haven and London, Yale University Press. 529p. (E)

971　**Prescott-Allen, R. & C. (1982).** The case for *in situ* conservation of crop genetic resources. *Nat. Resources*, 18(1): 15-20. Illus. (E/P)

972　**Prescott-Allen, R. & C. (1982).** What's wildlife worth? Economic contributions of wild plants and animals to developing countries. London, International Institute for Environment and Development/Earthscan Paperbacks. 90p. Illus. Gene. (E)

GENERAL

973 **Prescott-Allen, R. & C. (1983).** Park your genes. *Ambio*, 12(1): 37-39. Illus. (E/P/T)

974 **Prescott-Allen, R. & C. (1984).** Park your genes: protected areas as *in situ* genebanks for the maintenance of wild genetic resources. *In* McNeely, J.A., Miller, K.R., *eds.* National parks, conservation and development. The role of protected areas in sustaining society. Washington, D.C., Smithsonian Institution Press. 634-638. (E/P)

975 **Prescott-Allen, R. & C. (1988).** Genes from the wild. Using wild genetic resources for food and raw materials. 2nd ed. London, Earthscan. 111p. Illus. (E)

976 **Prescott-Allen, R. (1982).** Conservation and development: the need for a world conservation strategy. *Conserv. News*, 83: 16-17.

977 **Preston, D.J. (1975).** Endangered plants. *Amer. For.*, 81(4): 8-11. (T)

978 **Preston, F.W. (1948).** The commonness and rarity of species. *Ecology*, 29: 254-283. (T)

979 **Prieur, M. (1984).** Forets et environnement: en droit et international. Limoges, Paris Presses. 289p. Fr (Sp). (E/L/P/T)

980 **Principe, P.P. (1985).** The value of biological diversity among medicinal plants. Paris, Environment Directorate, OECD. (E)

981 **Principe, P.P. (?)** Valuing diversity of medicinal plants. *In* Proceedings of the International Consultation on the Conservation of Medicinal Plants, Chiang Mai, Thailand, March 1988. Academic Press. In press. Medicinal plants. (E)

982 **Protopopova, V.V. (1980).** [The first all-union symposium: protection and cultivation of orchids.] *Ukr. Bot. Zh.*, 37(6): 102-103. Rus. (G)

983 **Puri, G.S., Meher-Homji, V.M., Gupta, R.K., Puri, S. (1983).** Forest ecology. 2nd edition. Volume 1: phytogeography and forest conservation. New Delhi, Oxford & IBH Publishing Co. xxii, 549p. Illus., maps, key.

984 **Quinn, J.F., Hastings, A. (1987).** Extinction in subdivided habitats. *Conservation Biology*, 1(3): 198-208. Population biology applied to the problems of extinction rates and reserve design. (P/T)

985 **Rabinowitz, D. (1981).** Seven forms of rarity. *In* Synge, H., *ed.* The biological aspects of rare plant conservation. Chichester, Wiley. 205-217. Proceedings of International Conference, King's College, Cambridge, 14-19 July 1980. (T)

986 **Rabinowitz, D. (1985).** Biologists attitudes towards rare species. *Plant Sci. Bull.*, 31(6): 41-42. (T)

987 **Rabotnov, T.A. (1981).** The coenopopulation as an approach to the conservation of rare plant species. *In* Synge, H., *ed.* The biological aspects of rare plant conservation. Chichester, Wiley. 505-506. Abstract of paper. Proceedings of International Conference, King's College, Cambridge, 14-19 July 1980. (T)

988 **Rackham, O., et al. (1983).** Trees in the 21st. century. Berkhamsted, A B Academic Publishers. 133p. Illus. Maps. (E)

989 **Radford, A.E., et al. (1981).** Natural heritage - classification, inventory and information. Chapel Hill, NC, Univ. of North Carolina. 485p.

990 **Rahmani, A.R. (1980).** Conservation of plant species. *Tigerpaper*, 7(1): 24-26. Illus. (T)

991 **Raiford, W.N. (1988).** Social forestry: an answer to deforestation. *World Dev.*, 1(5): 12-15. (E)

992 **Ralls, K., Brownwell, R., jr (1989).** Protected species - research permits and the value of basic research. *BioScience*, 39(6): 394-396. (T)

GENERAL

993 **Ramsay, W. (1976).** Priorities in species preservation. *Environ. Affairs*, 5(4): 595-616. (T)

994 **Ratcliffe, D.A. (1981).** Introduced species in forestry and their relevance to nature conservation. *In* Pinder, N., *ed.* Conservation and introduced species: a discussion meeting held at University College, London on 18 April 1980. Discussion Paper in Conservation No. 30. London, UCL. 32-37. British Association of Nature Conservationists in conjunction with Ecology and Conservation Unit of UCL. (E)

995 **Ratcliffe, D.A. (1981).** Why protection? *Naturopa*, 38: 4-6. Illus.

996 **Raup, D.M. (1988).** Diversity crises in the geological past. *In* Wilson, E.O., *ed.* Biodiversity. Washington, D.C., National Academy Press. 51-57. Relevance to predictions of current and future extinction rates. Illus.

997 **Rauschert, S. (1981).** Nomenklaturregeln und Naturschutz. (Regulations of nomenclature and nature preservation.) *Arch. Naturschutz Landschaftsforsch.*, 21(3): 175-177. Ge. (L)

998 **Raven, P.H. (1976).** Ethics and attitudes. *In* Simmons, J.B., *et al.*, *eds.* Conservation of threatened plants. NATO Conference Series 1: Ecology, vol. 1. New York and London, Plenum Press. 155-179. Proc. Conference on the Functions of Living Plant Collections in Conservation and Conservation-Orientated Research and Public Education, Kew, 2-6 September 1975. (T)

999 **Raven, P.H. (1981).** Research in botanical gardens. *Bot. Jahrb. Syst. Pflanzengesch. Pflanzengeogr.*, 102(1-4): 53-72. Incl. function as plant repositories for threatened species. Pres. at the Tercentenary Symp. of the Berlin Botanical Garden -research progress in bot. gdns, on Balkan flora & genus *Bromus*. 10-13 Sept. 1979. (G)

1000 **Raven, P.H. (1986).** 60,000 plants under threat. *Threatened Pl. Newsl.*, 16: 2-3. (T)

1001 **Raven, P.H. (1987).** The scope of the plant conservation problem world-wide. *In* Bramwell, D., Hamann, O., Heywood, V., Synge, H., *eds.* Botanic gardens and the World Conservation Strategy. London, Academic Press. 19-29. Describes current rates of loss of plant-rich habitats and estimates that 60,000 vascular plant species are at risk in the tropics and sub-tropics. (Sp). (T)

1002 **Raven, P.H., Dahlgren, R. (1981).** Vart ansvar infor en utdoende natur. (Our responsibilities towards a vanishing nature.) *Svensk. Bot. Tidskr.*, 75(1): 7-18. Sw (En). Illus.

1003 **Ray, G.C., Hayden, B.P., Dolan, R. (1984).** Development of a biophysical coastal and marine classification system. *In* McNeely, J.A., Miller, K.R., *eds.* National parks, conservation and development. The role of protected areas in sustaining society. Washington, D.C., Smithsonian Institution Press. 39-46. Maps.

1004 **Read, M. (1990).** Progress for plant conservation in Lausanne. *Oryx*, 24: 3-4. Outcome of plant proposals considered at 1989 CITES meeting. (L/T)

1005 **Read, R.W., Desautels, P.E. (1983).** CITES: orchids as a family endangered by legislation? *Amer. Orchid Soc. Bull.*, 52(1): 15-21. Col. illus. Convention on International Trade in Endangered Species of Wild Fauna and Flora. (L/T)

1006 **Reckin, J. (1978).** Vorstellungen uber die Nutzung des Botanischen Gartens aus der Sicht des Biologielehrers. *Gartn.-Bot. Brief*, 58: 44-45. (G)

1007 **Reid, W.V., Miller, K.R. (1989).** Keeping options alive: the scientific basis for conserving biodiversity. World Resources Institute. 128p. (B/E/G/L/P/T)

1008 **Reisner, M. (1978).** What are species good for? *Frontiers*, 42(4): 24-27.

1009 **Rekas, A.M.B. (1976).** Computerized information systems for threatened and endangered species. *ASB Bull.*, 23(3): 144-149. (T)

GENERAL

1010 **Repetto, R. (1988).** The forest for the trees? Government policies and the misuse of forest resources. Washington, D.C., World Resources Institute. 105p.

1011 **Reppenhagen, W. (1980).** Conservation by propagation. *Cact. Succ. J.* (U.K.), 42(3): 71-74. Illus., col. illus. (G)

1012 **Reveal, J.L. (1981).** The concept of rarity and population threats in plant communities. *In* Morse, L.E., Henifin, M.S., *eds.* Rare plant conservation: geographical data organization. New York, New York Botanical Garden. 41-47. (T)

1013 **Reveal, J.L., Broome, C.R. (1979).** Plant rarity real and imagined. *Nat. Conserv. News*, 29(2): 4-8. Col. illus. (T)

1014 **Reyes, M.R. (1981).** Natural forests need attention. *Canopy*, 7(4): 1, 5-9. Illus.

1015 **Reynolds, E.R.C., Wood, P.J. (1977).** Natural versus man-made forests as buffers against environmental deterioration. *Forest Ecol. Manage.*, 1(2): 83-96.

1016 **Rhoads, W.A. (1978).** Principles, problems and practicalities of reporting on endangered plants. *In* National Environmental Research Park Symposium: Natural Resource Inventory Charact. and Analysis 1, 1977: 133-147. (T)

1017 **Ricciuti, E.R. (1979).** Plants in danger. New York, Harper & Row. 86p. Juv. lit. (T)

1018 **Rich, B. (1985).** Multilateral development banks: their role in destroying the global environment. *Ecologist*, 15: 56-68. Deforestation.

1019 **Richards, J., Tucker, E., eds (1988).** Deforestation in the twentieth century. Durham, NC, Duke University Press. 321p.

1020 **Richardson, R.C. (1984).** To be, or not to be: endangered? *J. Orchid Soc. Gr. Brit.*, 33(4): 93-94. (T)

1021 **Richter, J. (1981).** Washington commentary: the world's disappearing forests. *Span*, 24(2): 75. Deforestation.

1022 **Rickett, H.W. (1977).** Plants, too, are endangered. *Nation. Wildl.*, 12(3): 28. (T)

1023 **Riedl, O., Zachar, D., eds (1984).** Forest amelioration. Amsterdam, Elsevier. 623p. Illus.

1024 **Riether, W. (1980).** Moglichkeiten des aktiven Schutzes heimischer Orchideenvorkommen. *Naturschutzarb. Naturk. Heimatforsch. Sachsen*, 22: 10-18. Ge.

1025 **Riney, T. (1964).** A rapid field technique and its application for describing conservation status and trends in semi-arid pastoral lands. *Afr. Soils*, 8(2): 159-258.

1026 **Ripley, S.D., Lovejoy, T.E. (1977).** Threatened and endangered species. *Dodo*, 14: 9-22. (T)

1027 **Ripley, S.D., Lovejoy, T.E. (1979).** Threatened and endangered species. *Habitat*, 7(4): 21-28. Illus., col. illus. (T)

1028 **Roberts, E.H., Ellis, R.H. (1977).** Prediction of seed longevity at sub-zero temperatures and genetic resources conservation. *Nature*, 268: 431-433. (G)

1029 **Robinette, G.O. (1972).** Plants, people and environmental quality. Washington, D.C., U.S. Dept of the Interior, National Park Service. xi, 136p. Illus.

1030 **Rock, B.N., Vogelman, J.E., et al. (1986).** Remote detection of forest damage. *Bioscience*, 36(7): 439-445.

1031 **Rodas, M.A.F. (1981).** In order for the forest to survive its inhabitants must survive first. *Ceres*, 14(4): 20-23.

GENERAL

1032 **Rogers, G. (1988).** A species that nearly disappeared. *Missouri Bot. Gard. Bull.*, 76(5): 7. (T)

1033 **Rolston III, H. (1985).** Duties to endangered species. *BioScience*, 35(11): 718-726. Illus. Conservation ethics. (T)

1034 **Romagnesi, H. (1979).** Quelques especes rares ou nouvelles de Macromycetes: 3. *Inocybe.* (Some rare or new species of Macromycetes: 3. *Inocybe.*) *Sydowia*, 8: 349-365. Fr. Illus. (T)

1035 **Roome, N. (1981).** The evaluation of nature conservation benefits. Gloucs. Papers in Local & Rural Planning No. 10. Gloucester, Dept of Town & Country Planning, Gloucs. College of Arts & Techn. 36p.

1036 **Roush, G.J. (1983).** On saving diversity. *Bull. Pac. Trop. Bot. Gard.*, 13(1): 8-16. Illus. (E/T)

1037 **Rowley, G.D. (1979).** Conservation in the glasshouse - status symbols for succulents. *Cact. Succ. J.* (U.K.), 34(3): 63-64. Illus. (G)

1038 **Royal Botanic Gardens, Kew (1984).** Plants for cultivation in arid lands. Bibliography No. 8. Kew, Library and Archives Division, Royal Botanic Gardens. 4p. 1982, 2nd revision 1984. (B/E/G)

1039 **Ruchti, S. (1985).** Naturschutz und Forstwirtschaft. (Nature protection and forestry.) *Schweiz. Zeitsch. fur Forstwesen*, 136(1): 55-70. Ge.

1040 **Rude, K. (1987).** Saving wetlands worldwide. *End. Species Update*, 4(9-10): 1-3. (P)

1041 **Russow, L.-M. (1981).** Why do species matter? *Envir. Ethics*, 3(2): 101-112.

1042 **Ryder, O.A. (1980).** Monitoring genetic variation in endangered species. *In* University of British Columbia. Second international congress of systematic and evolutionary biology, Vancouver, B.C. Canada, July 17-24, 1980. Vancouver, Univ. of British Columbia. 48. (T)

1043 **Rysin, L.P. (1979).** Lesnye rezervaty. (Forest reserves.) *Lesovedenie*, 1: 3-10. (En). (P)

1044 **Sagoff, M. (1980).** On the preservation of species. *Columbia J. Envir. Law*, 7(1): 33-67. (L)

1045 **Sagoff, M. (1982).** We have met the enemy and he is us or conflict and contradiction in environmental law. *Environmental Law*, 12: 283-315. (L)

1046 **Sahavacharin, O. (1980).** Tissue culture for conservation of perennial crops. *Kasetsart J.*, 14(1): 1-7. (E/G)

1047 **Sajeva, M. (1986).** Contributo alla protezione delle specie di piante succulente minacciate da estinzione. (Contribution to the protection of succulent plants threatened with extinction). *Piante Grasse*, 6(4): 124-125. It. (T)

1048 **Sajeva, M. (1987).** Protection of endangered succulent plants at the botanic garden of Palermo. *In* Bramwell, D., Hamann, O., Heywood, V., Synge, H., *eds.* Botanic gardens and the World Conservation Strategy. London, Academic Press. 351-352. (G/T)

1049 **Salm, R.V., Clark, J.R. (1984).** Marine and coastal protected areas: a guide for planners and managers. Gland, IUCN. 302p. Illus., maps. Based on the workshops on managing coastal and marine protected areas, World Congress on National Parks, Bali, Indonesia, October 1982. (L/P)

1050 **Samstag, T. (1984).** Obstacles to saving plants in peril. *Times*, 31 Mar.: [1p.]. (E/T)

1051 **Sand, P. (1978).** Trends in international environment law. *Unasylva*, 29(116): 26-28. Illus. (L)

GENERAL

1052 **Sand, P.H. (1980).** Combatting the trade in endangered species through CITES. *Unasylva*, 31(125): 32-35 (1979 publ. 1980). (L/T)

1053 **Sanwai, M. (1988).** Community forestry: policy issues, institutional arrangements, and bureaucratic reorientation. *Ambio*, 17(5): 342-346. Col. illus.

1054 **Sayer, J., Wachtel, P.S. (1984).** Saving the plants that save us - the programme: the campaign. *IUCN Bull.*, 15(1-3): 12. (E/G/T)

1055 **Schemel, H.-J. (1984).** Landschaftserhaltung durch Tourismus? (Landscape conservation through tourism?). *Gart. Landsch.*, 94(11): 24-27. En, Ge. Illus.

1056 **Schlechter, R. (1985).** Die Orchideen: ihre Beschreibung, Kultur und Zuchtung. 3, vollig neubearteitete Auflage. Band 2, 11-u-12. Lieferung, Bogen 41-46. Berlin, Hamburg, Verlag Paul Parey. 641-727. Ge. Illus.

1057 **Schofield, E.K. (1981).** Endangered plants. *Current Energy & Ecology: the Continuing Guide to Environmental Education*, 3(7): 17-19. (T)

1058 **Schonewald-Cox, C.M., Chambers, S.M., MacBryde, B., Thomas, W.L., eds. (1983).** Genetics and conservation: a reference for managing wild animal and plant populations. London, Benjamin/Cummings Publishing Co. xxii, 722p. Maps. Chrom. nos. (T)

1059 **Schoolcraft, G.D. (1988).** Ash Valley: a newly designated research natural area. *Fremontia*, 16(1): 15-18.

1060 **Schoser, G. (1982).** Zukunftsziel: Schutzsammlung fur sukkulente Pflanzen im Palmengarten Frankfurt? *Palmengarten*, 46(2): 53-54. Ge. (G)

1061 **Schott, L. (1987).** The destruction of wild orchids. *Bol. Soc. Dominicana Orquideologia*, 3(3): 60-71. En (Sp). Illus. (T)

1062 **Schreiber, R., Newman, J. (1988).** Acid precipitation effects on forest habitats: implications for wildlife. *Conserv. Biol.*, 2(3): 249-259. Acid rain.

1063 **Schreiner, K.M. (1976).** Critical habitat: what it is - and is not. *Endangered Species Tech. Bull.*, 1(2): 1, 4.

1064 **Schreiner, K.M., Ruhr, C.E. (1974).** Progress in saving endangered species. *Trans. North Amer. Wildl. Nat. Res. Conf.*, 39: 127-135. (T)

1065 **Schulze, E.D. (1989).** Air pollution and forest decline in a spruce (*Picea abies*) forest. *Science*, 244(4906): 776-782.

1066 **Schwabe-Braun, A., International Society for Vegetation Science (1981).** Vegetation as anthropo-ecological subject. Endangered vegetation and its conservation. Vaduz, Cramer. xvi, 662p. Ge, En, Fr, (En). Illus., maps.

1067 **Scoby, D.R. (1971).** Environmental ethics. Minneapolis, MN, Burgess. 239p.

1068 **Scott, J.M., Csuti, B., Estes, J.E., Anderson, H. (1989).** Status assessment of biodiversity protection. *Conservation Biology*, 3(1): 85-87.

1069 **Sebek, V. (1983).** Bridging the gap between environmental science and policy-making: why public policy often fails to reflect current scientific knowledge. *Ambio*, 12(2): 118-120. (L)

1070 **Segnestam, M. (1982).** Varldsstrtegin for naturvard - 2 ar senare. (World Conservation Strategy: 2 years on.) *Sveriges Nat.*, 3: 24. Sw (En).

1071 **Seigler, D.S., ed. (1977).** Crop resources. New York, Academic Press. xi, 233p. (E)

1072 **Semenova-Tyan-Shanskaya, A.M., Boch, M.S. (1978).** [On the establishment of the section "Protection of the world of plants" in the Scientific Council for the problem "Biological bases of rational usage, transformation and protection of the plant world"]. *Bot. Zhurn.*, 63(6): 917-919. Rus. (T)

GENERAL

1073 **Seyberth, M. (1978).** Vereinbarkeit oder Konflikt zwischen Land-, Forst- und Wasserwirtschaft mit den Grundsatzen des Schutzes der Natur und der Erhaltung der naturlichen Lebensgrundlagen. *Allg. Forstzeitung,* 89(8): 250-252. Ge. Illus.

1074 **Shabecoff, P. (1978).** New battles over endangered species. *New York Times,* (Sec. 6), 4 June: 38-42, 44. (L)

1075 **Shaw, R.L. (1976).** Future: integrated international policies. *In* Simmons, J.B., *et al., eds.* Conservation of threatened plants. NATO Conference Series 1: Ecology, vol. 1. New York and London, Plenum Press. 39-47. Proc. Conference on the Functions of Living Plant Collections in Conservation and Conservation-Orientated Research and Public Education, Kew, 2-6 September 1975. (G/T)

1076 **Sheehan, T. (1981).** Orchidophilos. *Florida Orchidist,* 24(4): 160-162. Illus.

1077 **Sherbrooke, W.C., Paylore, P. (1973).** World desertification: cause and effect. A literature review and annotated bibliography. Arid Lands Resource Information Paper No. 3. Tucson, University of Arizona, Office of Arid Lands Studies. iv, 168p. (B)

1078 **Silva, M.E., et al. (1986).** A bibliographical list of coastal and marine protected areas: a global survey. Woods Hole Oceanographic Institution Technical report WHOI-86-11. 156p. (B/P)

1079 **Simberloff, D. (1978).** Our fragile evolutionary heritage: islands and their species. *Nat. Conserv. News,* 28(4): 4-10. Illus., col. illus.

1080 **Simberloff, D. (1978).** Sizes and shapes of wildlife refuges. *Frontiers,* 42(4): 28-32. Illus. (P)

1081 **Simberloff, D. (1983).** What a species needs to survive. *Nat. Conserv. News,* 33(6): 18-22.

1082 **Simberloff, D. (1984).** The next mass extinction? *Garden* (New York), 8(2): 2-8, 32. Col. illus., maps. (T)

1083 **Simberloff, D., Cox, J. (1987).** Consequences and costs of conservation corridors. *Conserv. Biol.,* 1(1): 63-71. (Sp). (P)

1084 **Simberloff, D.S., Abele, L.G. (1976).** Island biogeography theory and conservation practice. *Science,* 191: 285-286. (P)

1085 **Simmons, I.G. (1979).** Biogeography: natural and cultural. London, Edward Arnold. 400p. Illus., maps. Includes introductions and extinctions. (T)

1086 **Simmons, J.B. (1976).** Preface. *In* Simmons, J.B., *et al., eds.* Conservation of threatened plants. NATO Conference Series 1: Ecology, vol. 1. New York and London, Plenum Press. ix-xii. Proc. Conference on the Functions of Living Plant Collections in Conservation and Conservation-Orientated Research and Public Education, Kew, 2-6 September 1975. (G)

1087 **Simmons, J.B. (1976).** Present: the resource potential of existing living plant collections. *In* Simmons, J.B., *et al., eds.* Conservation of threatened plants. NATO Conference Series 1: Ecology, vol. 1. New York and London, Plenum Press. 27-38. Proc. Conference on the Functions of Living Plant Collections in Conservation and Conservation-Orientated Research and Public Education, Kew, 2-6 September 1975. Map. (G)

1088 **Simmons, J.B. (1979).** The collection, establishment and distribution of natural source plant material. *In* Synge, H., Townsend, H., *eds.* Survival or extinction. Proceedings of a conference held at the Royal Botanic Gardens, Kew, 11-17 September 1978. Kew, Bentham-Moxon Trust. 75-84. (G)

GENERAL

1089 **Simmons, J.B., Beyer, R.I., Brandham, P.E., Lucas, G.Ll., Parry, V.T.H., eds. (1976).** Conservation of threatened plants. NATO Conference Series 1: Ecology, vol. 1. New York and London, Plenum Press. xvi, 336p. Proceedings of the Conference on the Functions of Living Plant Collections in Conservation and Conservation-Orientated Research and Public Education, Kew, 2-6 September 1975. (G/T)

1090 **Simon, J.L. (1986).** Disappearing species, deforestation and data. *New Sci.*, 110(1508): 60-63. Illus., col. illus. (T)

1091 **Simon, J.L., Wildavsky, A. (?)** On species loss, the absence of data, and risks to humanity. *In* Simon, J.L., Kahn, H., eds. The resourceful Earth: a response to Global 2000. Oxford, Basil Blackwell. 171-183. Questions estimates of extinction rates.

1092 **Singh, R.B. (1982).** Diversity conserved through unity. *IBPGR Region. Committ. Southeast Asia Newsl.*, 6(2): 3-5. (E)

1093 **Singh, R.B. (1982).** Genebanks and plant introductions. *IBPGR Region. Committ. Southeast Asia Newsl.*, 6(2): 3. (G)

1094 **Skarapanau, S.H. (1981).** [Intensification of crop farming and problems of environment and conservation]. *Vestsi Akademii Navuk BSSR*, 1981(1): 11-15. Be (En, Rus).

1095 **Slatyer, R. (1983).** The origin and evolution of the World Heritage Convention. *Ambio*, 12(3-4): 138-140. (L/P)

1096 **Slatyer, R. (1984).** The World Heritage Convention: introductory comments. *In* McNeely, J.A., Miller, K.R., eds. National parks, conservation and development. The role of protected areas in sustaining society. Washington, D.C., Smithsonian Institution Press. 734. (L/P)

1097 **Sloet van Oldruitenborgh, C.J.M. (1981).** [Diversity in landscape, vegetation and flora as a basis for a nature protection model in an old culture landscape]. *In* Schwabe-Braun, A., *ed.* Vegetation als anthropo-okologischer Gegenstand (Rinteln, 5-8 April 1971). Gefahrdete Vegetation und ihre Erhaltung (Rinteln, 27-30 March 1972). Vaduz, J. Cramer. 495-501. Ge (Fr, En).

1098 **Smale, T. (1984).** A sideways look at conservation. *Cact. Succ. J.* (U.K.), 2(2): 49-51. Illus. A criticism of CITES. (L/T)

1099 **Smith, E.L. (1980).** Laws and information needs for listing plants. *Rhodora*, 82: 193-199. (L/T)

1100 **Smith, N. (1981).** Wood: an ancient fuel with a new future. Worldwatch Paper No. 30. Washington, D.C., Worldwatch Institute. 48p. Includes forest cover statistics; fuelwood crisis. (E)

1101 **Smith, R.L. (1976).** Ecological genesis of endangered species: the philosophy of preservation. *Ann. Rev. Ecol. Syst.*, 7: 33-35. (T)

1102 **Soerianegara, I. (1978).** Forest management and conservation of forest resources management. *In* Proceedings of the 8th World Forestry Congress, Jakarta, 16-28 October 1978: Forestry for quality of life. II, 9.

1103 **Soil Conservation Society of America (1982).** Resource conservation glossary. 3rd Ed. Iowa, Soil Conservation Society of America. 193p.

1104 **Sokolov, V.E., Gunin, P.D., Drozdov, A.V., Puzachenko, Y.G. (1984).** Siting criteria for biosphere reserves. *In* Unesco-UNEP. Conservation, science and society. Paris, Unesco. 133-138. Contributions to the First International Biosphere Reserve Congress, Minsk, Byelorussia/USSR, 26 September - 2 October 1983. Tables. (P)

GENERAL

1105 **Sokolov, V.S., Puzachenko, Y.G., Skulkin, V.S. (1984).** The concept of analogous biosphere reserves. *In* Unesco-UNEP. Conservation, science and society. Paris, Unesco. 204-207. Contributions to the First International Biosphere Reserve Congress, Minsk, Byelorussia/USSR, 26 September - 2 October 1983. (P)

1106 **Soule, M.E. (1984).** Applications of genetics and population biology, the what, where and how of nature reserves. *In* Unesco-UNEP. Conservation, Science and Society. Paris, Unesco. 252-264. Contributions to the First International Biosphere Reserve Congress, Minsk, Byelorussia/USSR, 26 September - 2 October 1983. Illus. Tables. (G/P)

1107 **Soule, M.E. (1985).** What is conservation biology? *BioScience*, 35(11): 727-734. Illus.

1108 **Soule, M.E. ed. (1986).** Conservation biology: the science of scarcity and diversity. Sunderland, MA, Sinauer Assoc. 584p.

1109 **Soule, M.E., Kohm, K.A., eds (1989).** Research priorities for conservation biology. Washington, D.C., Island Press. 97p.

1110 **Soule, M.E., Wilcox, B.A., eds (1980).** Conservation biology; an evolutionary-ecological perspective. Massachussetts, Sinauer Associates. xv, 395p.

1111 **Soule, M.E., ed. (1987).** Viable populations for conservation. Cambridge, Cambridge Univ. Press. 189p.

1112 **Specht, R.L., ed. (1981).** Heathlands and related shrublands: analytical studies. Ecosystems of the World 9B. Amsterdam, Elsevier. 385p. Includes section on conservation, 231-259.

1113 **Spellerberg, I.F. (1981).** Ecological evaluation for conservation. Studies in Biology No. 133. London, Edward Arnold. 59p. Illus., map.

1114 **Squire, S. (1984).** Frozen assets. *Nation. Wildl.*, 22(5): 6-13. (G)

1115 **Srun, S.M. (1983).** An overview of the world's deforestation. *Canopy*, 9(8): 3-5. Illus. (E/L)

1116 **St. John, T.V. (1987).** Mineral acquisition in native plants. *In* Elias, T.S., *ed.* Conservation and management of rare and endangered plants. Proceedings from a conference, Sacramento, California, 5-8 November 1986. Sacramento, California Native Plant Society. 529-535. Includes symbionts and mineral nutrition of rare and endangered plants which may have a bearing on cultivation and transplanting success. (G/T)

1117 **Stacey, S. (1988).** Inventory time at God's own drug store. *BBC Wildlife*, 6(7): 378-381. Col. illus. Conservation of medicinal plants. (E)

1118 **Stankey, G.H. (1982).** The role of management in wilderness and natural-area preservation. *Envir. Conserv.*, 9(2): 149-155. (P)

1119 **Staritsky, G. (1980).** *In vitro* storage of aroid germplasm. *Pl. Genet. Resource. Newsl.*, 42: 25-27. (Fr, Sp). Illus. (G)

1120 **Stebbins, G.L. (1942).** The genetic approach to problems of rare and endemic species. *Madrono*, 6(6): 241-258. (T)

1121 **Stebbins, G.L. (1972).** The scientific and aesthetic value of plants and animals in unusual places. *Calif. Nat. Pl. Soc. Newsl.*, 7(4): 3. (G)

1122 **Stebbins, G.L. (1979).** Rare species as examples of plant evolution. *In* Wood, S.L., *ed.* The endangered species: a symposium. Great Basin Naturalist Memoir No. 3. Provo, UT, Brigham Young Univ. (T)

1123 **Stebbins, G.L. (1980).** Rarity of plant species: a synthetic viewpoint. *Rhodora*, 82(829): 77-86. (T)

59

GENERAL

1124 **Steele, A. (1975).** Species orchid seed - conservation and distribution. *Amer. Orch. Soc. Bull.*, 44: 514-515. (G/T)

1125 **Steubing, L. (1984).** Monitoring methodology of bioindicators of immission load. *In* Unesco-UNEP. Conservation, science and society. Paris, Unesco. 411-426. Contributions to the First International Biosphere Reserve Congress, Minsk, Byelorussia/USSR, 26 September - 2 October 1983. Tables, illus.

1126 **Stewart, J. (1986).** Orchid conservation at the international level. *Amer. Orchid Soc. Bull.*, 55(3): 242-246. (T)

1127 **Stewart, J. (1987).** Conserving orchids. *Species*, 9: 28-30. (T)

1128 **Stopp, G.H., jr (1985).** Acid rain: a bibliography of research annotated for easy access. Metuchen, NJ, Scarecrow Press. 174p. Acid rain. (B)

1129 **Stowe, J.S. (1976).** The significance of natural areas. *In* Canadian Botanical Association. Natural areas: Proceedings of a Symposium at the Thirteenth Annual Meeting of the Canadian Botanical Asociation, Bishop's University, Lennoxville, Quebec, 6-10 June. CBA. 3-7. (Fr). (P)

1130 **Strom, A.A. (1979).** Impressions of a developing conservation ethic, 1870-1930. *Parks and Wildlife*, 2(3-4): 45-53. Illus.

1131 **Stuckey, R.L. (1975).** Selected list of references on the preservation of natural areas and rare organisms. Columbus, Ohio, College of Biological Sciences, Ohio State University. 6p. (B/P/T)

1132 **Stuckey, R.L. (1976).** Additions to selected list of references on the preservation of natural areas and rare organisms. Columbus, Ohio, College of Biological Sciences, Ohio State University. 3p. (B/P/T)

1133 **Stumm, W. (1986).** Water, an endangered ecosystem. *Ambio*, 15(4): 210-217. (T)

1134 **Sukendar (1983).** [Notes on orchid germplasm.] *Bul. Kebun Raya*, 6(1): 15-17. In.

1135 **Sun, M. (1985).** Possible acid rain woes in the west. *Science*, 228(4695): 34-35. Illus.

1136 **Swanson, G.A., ed. (1979).** The mitigation symposium. Gen. Tech. Rep. RM-65. U.S. Department of Agriculture Forest Service. 669p. This grandiose project contains 109 contributed papers. Titles and summaries are to be found in *Ecological Abstracts* 1981/1, 118-127. (P)

1137 **Swindells, P. (1981).** Who will conserve? Philip Swindells looks at the changing face of the botanic garden and the likely effect this may have in the future. *GC & HTJ*, 190(10): 34. (G)

1138 **Symoens, J.J. (1979).** Reserves naturelles, parcs nationaux, parcs naturels: essai de mise au point. *Nat. Belg.*, 60(1): 2-43. Fr. Illus. (P)

1139 **Synge, A.H.M. (1977).** Conservation: endangered plant life. *In* Huxley, A., *ed.* The encyclopedia of the plant kingdom. London, Salamander. 88-97. Illus. (T)

1140 **Synge, A.H.M. (1979).** Botanic gardens and island plant conservation. *In* Bramwell, D., *ed.* Plants and islands. London, Academic Press. 379-390. (Sp). (G/T)

1141 **Synge, H. (1979).** Threatened trees and the "IUCN Plant Red Data Book". *Int. Dendrol. Soc. Yearbook*, (1978): 22-25. (T)

1142 **Synge, H. (1983).** The conservation of vegetables. *Threatened Pl. Newsl.*, 11: 10-12. (E/G)

1143 **Synge, H. (1983).** The identification of threatened floras around the world: the role of IUCN. *In* Jain, S.K., Mehra, K.L., *eds.* Conservation of tropical plant resources. Howrah, Botanical Survey of India. 225-229. (T)

GENERAL

1144 **Synge, H. (1984).** Gardening for conservation. *IUCN Bull.*, 15(1-3): 15-16. (G)

1145 **Synge, H. (1986).** Plants in danger: what do we know? *IUCN Bull.*, 17(1-3): 46. (T)

1146 **Synge, H. (1987).** Introduction. *In* Bramwell, D., Hamann, O., Heywood, V., Synge, H., *eds.* Botanic gardens and the World Conservation Strategy. London, Academic Press. xxi-xxxv. Describes the work of the IUCN Botanic Gardens Co-ordinating Body and the role of botanic gardens in conservation. (G)

1147 **Synge, H. (1988).** The Joint IUCN-WWF Plants Conservation Programme: achievements 1984-1987 and activities planned 1988-1990. Kew, IUCN and WWF. 27p. (E/G/P/T)

1148 **Synge, H., Townsend, H., eds (1979).** Survival or extinction. Kew, Bentham-Moxon Trust. 250p. Illus., maps. Proceedings: The Practical Role of Botanic Gardens in the Conservation of Rare and Threatened Plants, Kew, 11-17 September 1978. (G/T)

1149 **Synge, H., ed. (1981).** The biological aspects of rare plant conservation. Chichester, Wiley. 558p. Proceedings of International Conference, King's College, Cambridge, 14-19 July 1980. Maps. (P/T)

1150 **Synge, H., ed. (1983).** Plants on CITES: problems and opportunities. *Threatened Pl. Newsl.*, (11): 2-7. (L)

1151 **Synge, P.M. (1979).** Plant conservation in 1978: a brief report. *Int. Dendrol. Soc. Yearbook*, (1978): 17-21. (T)

1152 **Sytnik, K. (1978).** [A praising of the ancient science of botany]. *Nauka Pol.*, 7: 73-77. Rus. Problems of flora conservation. (T)

1153 **Szijj, J. (1972).** Some suggested criteria for determining the international importance of wetlands in the western palearctic. *In* Carp, E., *ed.* International conference on the conservation of wetlands and waterfowl, Ramsar, Iran, 1971. Slimbridge, U.K., International Wildfowl Res. Bur.

1154 **Takayanagi, K. (1980).** Seed storage and viability tests. *In* Tsunoda, S., Hinata, K., Gomez-Campo, C., *eds. Brassica* crops and wild allies: biology and breeding. Tokyo, Scientific Society Press. 303-321. Includes preservation of germplasm seeds. (E/G)

1155 **Talbot, L.M. (1978).** Endangered species: international dimensions of the problem. *Frontiers*, 42(4): 16-19. Illus. (T)

1156 **Talbot, L.M. (1980).** A world conservation strategy. *J. Roy. Soc. Arts*, 128(5288): 493-510.

1157 **Talbot, L.M. (1983).** IUCN in retrospect and prospect. *Envir. Conserv.*, 10(1): 5-11.

1158 **Talbot, L.M. (1984).** The role of protected areas in the implementation of the World Conservation Strategy. *In* McNeely, J.A., Miller, K.R., *eds.* National parks, conservation and development. The role of protected areas in sustaining society. Washington, D.C., Smithsonian Institution Press. 15-16. (P)

1159 **Tendron, G., Pruvost, P., Roumeguere, J.Y., Yon, D. (1984).** Principes de gestion des espaces naturels. Dijon, Cah. AIDEC. 17p. Fr. (T)

1160 **Terborgh, J. (1974).** Preservation of natural diversity: the problem of extinction prone species. *BioScience*, 24(12): 715-722. (T)

1161 **Tesliuk, N.K. (1984).** [Determination of the nature conservation age structure of forests.] *Lesnoe Khoz.*, 12: 43-45. Rus.

1162 **Thacher, P.S. (1978).** The United Nations Environment Programme and the ecological world. *Envir. Conserv.*, 5(4): 241-245.

GENERAL

1163 **Thacher, P.S. (1984).** Peril and opportunity: what it takes to make our choice. *In* McNeely, J.A., Miller, K.R., *eds*. National parks, conservation and development. The role of protected areas in sustaining society. Washington, D.C., Smithsonian Institution Press. 12-14.

1164 **Thamen, R.R. (1986).** Trees, conflict resolution, and peace: the preservation of trees as a precondition for environmental and social stability. *In* Maas, J.P., Stewart, R.A.C., *eds*. Toward a world of peace: people create alternatives. Suva, University of the South Pacific. xiii, 601p.

1165 **The Nature Conservancy (1974).** The Nature Conservancy Preserve directory. Arlington, VA, The Nature Conservancy. 154p. (P)

1166 **Thelen, K.D., Child, G.S. (1984).** Biosphere reserves and rural development. *In* Unesco-UNEP. Conservation, science and society. Paris, Unesco. 470-477. Contributions to the First International Biosphere Reserve Congress, Minsk, Byelorussia/USSR, 26 September - 2 October 1983. (P)

1167 **Thibodeau, F. (1987).** Germplasm collecting for public gardens. *The Public Garden*, 2(1): 10-13. Illus. (G)

1168 **Thibodeau, F.R., Falk, D.A. (1985).** Saving the pieces. *Restor. Manage. Notes*, 2(2): 71-72. Illus.

1169 **Thibodeau, F.R., Falk, D.A. (1987).** Building a national *ex situ* conservation network - the U.S. Center for Plant Conservation. *In* Bramwell, D., Hamann, O., Heywood, V., Synge, H., *eds*. Botanic gardens and the World Conservation Strategy. London, Academic Press. 285-294. (Sp). Map. (G)

1170 **Thomas, G. (1973).** The role of the private garden. *In* Green, P.S., *ed*. Plants; wild and cultivated. Conference on horticulture and field botany held at the Royal Hort. New Hall, Vincent Square, London, Sept. 2-3, 1974. Hampton, U.K., Classey. 59-65. (G)

1171 **Thomas, J.W., Salwasser, H. (1989).** Bringing conservation biology into a position of influence in natural resource management. *Conservation Biology*, 3(2): 123-127. (P)

1172 **Thomas, K. (1983).** Man and the natural world. New York, Pantheon. 426p.

1173 **Thomas, W.L. (1965).** Man's role in changing the face of the earth. Chicago, IL, University of Chicago Press. 1193p.

1174 **Thompson, A.R. (1970).** A conservation regime for the north - what have the lawyers to offer? *In* Fuller, W.A., Kevan, P.G., *eds*. Productivity and conservation in northern circumpolar lands. Proceedings of a conference, Edmonton, Alberta, 15-17 October 1969. Morges, IUCN. 301-320. (L)

1175 **Thompson, P. (1975).** Should botanic gardens save rare plants? *New Sci.*, 68(979): 636-638. (G/T)

1176 **Thompson, P.A. (1970).** Seed banks as a means of improving the quality of seed lists. *Taxon*, 19(1): 59-62. (G)

1177 **Thompson, P.A. (1972).** The role of the botanic garden. *Taxon*, 21(1): 115-119. (G)

1178 **Thompson, P.A. (1973).** Seeds in the bank. *Bull. Min. Agric. Fish. Food*, 17: 71-73. (G)

1179 **Thompson, P.A. (1973).** Techniques for the storage of species and ecotypes. *In* Hawkes, J.G., Lange, W., *eds*. European and regional gene banks. Proceedings of a conference on European and regional gene banks, Izmir, Turkey, April 10-15, 1972. Wageningen, Eucarpia. 65-66. (G)

1180 **Thompson, P.A. (1974).** The use of seed banks for conservation of populations of species and ecotypes. *Biol. Conserv.*, 6(1): 15-19. (G)

GENERAL

1181 **Thompson, P.A. (1975).** The collection, maintenance, and environmental importance of the genetic resources of wild plants. *Envir. Conserv.*, 2(3): 223-228. (G)

1182 **Thompson, P.A. (1976).** Factors involved in the selection of plant resources for conservation as seed in gene banks. *Biol. Conserv.*, 10: 159-167. (G)

1183 **Thompson, P.A. (1976).** Seeds everlasting. *Garden* (U.K.), 101: 105-109. (G)

1184 **Thompson, P.A. (1979).** Factors involved in the selection of plant resources for conservation as seed in gene banks. *Gartn.-Bot. Brief*, 62: 5-13. (E/G)

1185 **Thompson, P.A. (1979).** Preservation of plant resources in gene banks within botanic gardens. *In* Synge, H., Townsend, H., *eds.* Survival or extinction. Proceedings of a conference held at the Royal Botanic Gardens, Kew, 11-17 September 1978. Kew, Bentham-Moxon Trust. 179-184. Gene bank. (G)

1186 **Thompson, P.A., Brown, G.E. (1972).** The seed unit at the Royal Botanic Gardens, Kew. *Kew Bull.*, 26(3): 445-446. (G)

1187 **Thompson, P.A., Smith, R.D., Dickie, J.B., Sanderson, R.H., Probert, R.J. (1981).** Collection and regeneration of populations of wild plants from seed. *Biol. Conserv.*, 20: 229-245. (G)

1188 **Thorhaug, A. (1986).** Review of seagrass restoration efforts. *Ambio*, 15(2): 110-117.

1189 **Thorsell, J. (1985).** Parks on the borderline. *IUCN Bull.*, 16(110-12): 128-130. (P)

1190 **Thorsell, J. (1985).** Parks that promote peace. *WWF News*, 38: 2. International border parks. (P)

1191 **Thorsell, J. (1989).** Why do we need national parks? *TODY*, 1(2): 6-7. (P)

1192 **Thrupp, A. (1981).** The peasant view of conservation. *Ceres*, 14(4): 31-34. Illus.

1193 **Tijmens, W.J. (1979).** Plants in the landscape. *Newsl. Soc. Protect. Envir.*, 9(2): 3-6.

1194 **Tinker, J. (1979).** Wildlife in the middle. *New Sci.*, 82(1152): 263-265. Illus.

1195 **Toll, J., Sloten, D.H. van (1982).** Directory of germplasm collections: 4. Vegetables. Rome, IBPGR. v, 187p. Illus. (E/G)

1196 **Tomaselli, R. (1977).** Why and wherefore of botanical gardens. *Naturopa*, 31: 37-38. (G)

1197 **Tomlinson, D. (1980).** Time for unity: international wildlife conservation. *Country Life*, 168(4344): 2045-2046. Illus.

1198 **Townsend, H. (1979).** The potential and progress of the Technical Propagation Unit at the Royal Botanic Gardens, Kew. *In* Synge, H., Townsend, H., *eds.* Survival or extinction. Proceedings of a conference held at the Royal Botanic Gardens, Kew, 11-17 September 1978. Kew, Bentham-Moxon Trust. 189-193. Micropropagation. (G)

1199 **Trzyna, T.C., ed. (1989).** World directory of environmental organizations. A handbook of national and international organizations and programs - governmental and non-governmental - concerned with protecting the Earth's resources. California, Sacramento, California Institute of Public Affairs; IUCN. 175p.

1200 **Tsitsin, N.V. (1976).** [The role of botanical gardens in preservation of the vegetation.] *Bull. Main Bot. Gard. Soviet Acad. Sci.*, 100: 6-13. Rus. (G)

1201 **Tsunoda, S., Hinata, K., Gomez-Campo, C. (1980).** Preservation of genetic resources. *In* Tsunoda, S., Hinata, K., Gomez-Campo, C., *eds. Brassica* crops and wild allies: biology and breeding. Tokyo, Scientific Society Press. 339-354. (E/G)

GENERAL

1202 **Turner, J.S. (1966).** The decline of the plants. *In* Marshal, A.J., *ed.* The great extermination. London, Heinemann. 221p. (T)

1203 **Tuxen, R. (1978).** L'importance de la phytosociologie. *In* Filion, L., Villeneuve, P., *eds.* Les reserves ecologiques et la protection de la nature. Les realisations nord-americanes, europeenes et japonaises. Quebec, Conseil Consultatif des Reserves Ecologiques. Rapport No. 1. 23-24. Textes presentes au colloque des Iles-de-la-Madeleine les 17 et 18 Juin 1976. Fr. (P)

1204 **U.K. Secretary of State for Foreign & Commonwealth Affairs (1980).** Revised appendices I, II and III to the Convention on International Trade in Endangered Species of Wild Fauna and Flora. Done at Washington on 3 March 1973. Presented to Parliament. April 1980. London, HMSO. 31p. (L/T)

1205 **U.S. Congress Office of Technology Assessment (1987).** Technologies to maintain biological diversity. Washington, D.C., U.S. Govt Printing Office. 334p. Illus., tables, graphs. Status and trends of natural resources and environmental problems. (T)

1206 **U.S. Department of State (1973).** World wildlife conference, efforts to save endangered species. Washington, D.C., U.S. Govt Print. Office. 30p. (T)

1207 **U.S. Department of State. Council of Environmental Quality (1980).** The global 2000 report to the President. Entering the twenty-first century, 3 vols. Washington, D.C., U.S. Govt Print. Office. Analysis of global trends in population, resources and the environment, including food and agriculture, forestry and loss of genetic resources. (E/G)

1208 **U.S. Department of the Interior. Fish and Wildlife Service (1985).** Endangered and threatened wildlife and plants: public hearings and reopening of comment period on proposed Endangered status for *Abutilon menziesii*, *Hibiscadelphus distans*, and *Scaevola coriacea*. *Fed. Register*, 50(202): 42196-42197. (L/T)

1209 **Udvardy, M.D.F. (1975).** A classification of the biogeographical provinces of the world. Prepared as a contribution to Unesco's Man and the Biosphere Programme Project No. 8. IUCN Occasional Paper No. 18. Morges, IUCN. 48p.

1210 **Udvardy, M.D.F. (1984).** A biogeographical classification system for terrestrial environments. *In* McNeely, J.A., Miller, K.R., *eds.* National parks, conservation and development. The role of protected areas in sustaining society. Washington, D.C., Smithsonian Institution Press. 34-38. Tables.

1211 **Udvardy, M.D.F. (1984).** The IUCN/Unesco system of biogeographic provinces in relation to the biosphere reserves. *In* Unesco-UNEP. Conservation, science and society. Paris, Unesco. 16-19. Contributions to the First International Biosphere Reserve Congress, Minsk, Byelorussia/USSR, 26 September - 2 October 1983. Tables. (P)

1212 **Unesco (1970).** Use and conservation of the biosphere. Proceedings of the Intergovernmental Conference of Experts on the Scientific Basis for Rational Use and Conservation of the Resource of the Biosphere, 1968. Paris, Unesco. 272p.

1213 **Unesco (1974).** International Working Group on Project 3. Impact of human activities and land use practices on grazing lands: savanna, grassland (from temperate to arid areas). Final report. Man and the Biosphere [MAB] Report Series No. 25. Hurley, Unesco. 97p.

1214 **Unesco (1976).** Expert panel on Project 4. Impact of human activities on the dynamics of arid and semi-arid zone ecosystems, with particular attention to the effects of irrigation. Programme on Man and the Biosphere [MAB] Report Series No. 29. Paris, Unesco. 44p.

GENERAL

1215 **Unesco-UNEP. (1984).** Conservation, science and society. Paris, Unesco. Contributions to the First International Biosphere Congress, Minsk, Byelorussia/USSR, 26 September - 2 October 1983. (P/T)

1216 **Unesco/UNEP (1974).** Task Force on criteria and guidelines for the choice and establishment of biosphere reserves. Final report. Programme on Man and the Biosphere [MAB] Report Series No. 19. Paris, Unesco. 61p. Includes diagrams for zoning systems. (P)

1217 **Unkel, W. (1986).** Conservation of plant diversity in forest service planning. *Fremontia*, 13(4): 16-17. (P/T)

1218 **Unninanayar, S. (1984).** The world climate programme and biosphere reserves. *In* Unesco-UNEP. Conservation, science and society. Paris, Unesco. 427-434. Contributions to the First International Biosphere Reserve Congress, Minsk, Byelorussia/USSR, 26 September - 2 October 1983. Illus., maps. (P)

1219 **Uolters, S.M. (1976).** [The role of botanical gardens in preservation of rare and vanishing plant species.] *Byull Glavn Bot. Sada*, 100: 24-26. Rus. (G/T)

1220 **Usher, M.B. (1973).** Biological management and conservation; ecological theory, application and planning. London, Chapman and Hall. 394p.

1221 **Usher, M.B. (1979).** Changes in the species-area relations of higher plants on nature reserves. *J. Appl. Ecol.*, 16(1): 213-215. Illus. (P)

1222 **Usher, M.B. (1988).** Biological invasions of nature reserves: a search for generalisations. *Biol. Conserv.*, 44(1-2): 119-135. Invasive species. (P)

1223 **Usher, M.B., Kruger, F.J., Macdonald, I.A.W., Loope, L.L., Brockie, R.E. (1988).** The ecology of biological invasions into nature reserves: an introduction. *Biol. Conserv.*, 44(1-2): 1-8. Map. Invasive species. (P)

1224 **Vakhrameeva, M.G., Denisova, L.V. (1980).** Dynamics of cenopopulations of 3 species of the family Orchidaceae. *Moscow Univ. Biol. Sci. Bull.*, 35(1): 51-55. Includes conservation.

1225 **Van Helden, P.D. (1981).** The contribution of small areas to conservation. *Veld & Flora*, 67(1): 25-27. (P)

1226 **Varley, J.A. (1979).** Physical and chemical soil factors affecting the growth and cultivation of endemic plants. *In* Synge, H., Townsend, H., *eds.* Survival or extinction. Proceedings of a conference held at the Royal Botanic Gardens, Kew, 11-17 September 1978. Kew, Bentham-Moxon Trust. 199-205. (G/T)

1227 **Vasquez de la Parra, R. (1978).** A tree. *In* Proceedings of the 7th World Forestry Congress, Buenos Aires, 4-18 October 1972. vol. 3: Conservation and Recreation. Buenos Aires, Instituto Forestal Nacional. 3697-3701. On the need for conservation.

1228 **Vaughan, D.A., Hymowitz, T. (1983).** Progress in wild perennial soyabean characterization. *Pl. Genet. Resource. Newsl.*, no. 56: 7-12. (Fr,Sp). (E/G)

1229 **Vent, W. (1978).** Zur Bedeutung und Entwicklung Botanischer Garten. *Gleditschia*, 6: 9-18. Ge. (G)

1230 **Verney, R. (1984).** Towards a conservation ethic. *J. Roy. Soc. Arts*, 132(5336): 501-512.

1231 **Vernhes, J.R. (1983).** US-MAB Symposium: learning to conserve genetic resources. *Ambio*, 12(1): 34-37. Illus. (E/G/T)

1232 **Vernon, R.W. (1979).** The Wheeler orchid collection and species bank. *Amer. Orchid Soc. Bull.*, 48(7): 689-690. Col. illus. (G)

1233 **Vick, R. (1989).** Conservation: some possible directions. *Can. Pl. Conserv. Prog. Newsl.*, 4(1): 11-14.

GENERAL

1234 **Vitousek, P.M. (1988).** Diversity and biological invasions of oceanic islands. *In* Wilson, E.O., *ed.* Biodiversity. Washington, D.C., National Academy Press. 181-189. Introduced and invasive species.

1235 **WWF (1972).** 20,000 plants in danger. *Biol. Conserv.* 4: 255. (T)

1236 **WWF-U.S. (?)** Future in the wild. A conservation handbook. Washington, D.C., WWF-U.S. 24p. Col. illus.

1237 **Wachacki, B. (1974).** Technology to the rescue of endangered trees. *Rocz. Sekc. Dendrol. Pol. Bot.*, 28: 147-151. (T)

1238 **Walker, J. (1982).** Botanic defence. *GC & HTJ*, 191(1): 26. Illus. (G/T)

1239 **Wallace, G.D. (1975).** Suggestions for preventing extinction of rare and endangered plants. *AABGA Bull.*, 9(3): 67-68. (G/T)

1240 **Walsh, J. (1981).** Germplasm resources are losing ground. *Science*, 214: 421-423. (E)

1241 **Walters, S.M. (1973).** The role of botanic gardens in conservation. *J. Roy. Hort. Soc.*, 98(7): 311-315. (G)

1242 **Walters, S.M. (1981).** Priorities in rare species conservation for the 1980s. *In* Synge, H., *ed.* The biological aspects of rare plant conservation. Chichester, Wiley. xxv-xxviii. Proceedings of International Conference, King's College, Cambridge, 14-19 July 1980. (T)

1243 **Wang, X. (1985).** [The 12 most endangered plant species of the world.] *Guihaia*, 5(4): 413-414. Ch. (T)

1244 **Wang, X.P. (1988).** [The achievements and future tasks of the Joint IUCN-WWF Plants Conservation Programme]. *Guihaia*, 8(4): 365-370.

1245 **Ward, B. (1988).** Progress for a small planet. Rev. ed. London, Earthscan. 320p.

1246 **Warren, A., Goldsmith, F.B., eds. (1983).** Conservation in perspective. Chichester, Wiley. xiv, 474p. Maps.

1247 **Webb, L. (1966).** The rape of the forests. *In* Marshal, A.J., *ed.* The great extermination. London, Heinemann. 156-205.

1248 **Webb, W.M. (1925).** Plant protection. *J. Bot.*, 63: 273-275. (P)

1249 **Weber, F. (1976).** Des montagnes a soulever. Lausanne, Zurich, Ex-Libris. 288p. Fr. Illus.

1250 **Webster, B. (1982).** Genetic researchers anxious lest rare plant types vanish. *New York Times*, 25 January. (T)

1251 **Weinberg, J.H. (1975).** Botanocrats and the fading flora. *Sci. News*, 108(6): 92-95. (T)

1252 **Went, F.W., Babu, V.R. (1978).** Plant life and desertification. *Envir. Conserv.*, 5(4): 263-272. Illus.

1253 **Wentzel, K.F. (1983).** Maximale SO2-Konzentrations - Werte zum Schutze der Walder. (Maximally allowable SO2 concentrations for protection of forests.) *Aquilo. Ser. Botanica*, 19(5): 167-176. Ge (En). Illus. Literature review. Air quality standards, pollution.

1254 **Western, D., Pearl, M., eds (1989).** Conservation for the twenty-first century. Oxford University. Discusses future trends and techniques for conservation.

GENERAL

1255 **Westhoff, V. (1971).** The dynamic structure of plant communities in relation to the objectives of conservation. *In* Duffey, E., Watt, A.S., eds. The scientific management of animal and plant communities for conservation. The 11th Symposium of the BES, University of East Anglia, Norwich, 7-9 July 1970. Oxford, Blackwell. 3-14.

1256 **Westoby, J. (1983).** Who's deforesting whom? *IUCN Bull.*, 14(10-12): 124-125.

1257 **Westoby, J. (1987).** The purpose of forests. Oxford, Basil Blackwell. 343p. (E)

1258 **Westoby, J., Blackwell, B. (1989).** Introduction to world forestry. Oxford, New York. 228p. (E)

1259 **Wetterberg, G.B. (1984).** The exchange of wildlands technology: a management agency perspective. *In* McNeely, J.A., Miller, K.R., eds. National parks, conservation nd development. The role of protected areas in sustaining society. Washington, D.C., Smithsonian Institution Press. 728-732. (P)

1260 **Whelan, T. (1988).** Debt for conservation agreement. *IUCN Bull.*, 19(1-3): 8. (P)

1261 **White, P.S., Bratton, S.P. (1980).** After preservation: philosophical and practical problems of change. *Biol. Conserv.*, 18(4): 241-255. (P)

1262 **Whitehead, E.E., et al., eds (1988).** Arid lands: today and tomorrow. Proceedings of an international research and development conference, Tucson, Arizona, USA. 20-25 October, 1988. Boulder, CO, Westview Press and London, Belhaven Press. xlv, 1435p. Illus., maps.

1263 **Whitson, P.D., Massey, J.R. (1981).** Information systems for studying the population status of threatened and endangered plants. *In* Morse, L.E., Henifin, M.S., eds. Rare plant conservation: geographical data organization. New York, The New York Botanical Garden. 217-236. (T)

1264 **Widen, B. (1988).** [Population studies and conservation management of plants]. *Sven. Bot. Tidskr.*, 82(6): 449-457. Sw (En). (T)

1265 **Wilcox, B., Murphy, D. (1985).** Conservation strategy: the effects of fragmentation on extinction. *Amer. Naturalist*, 125(6): 879-887.

1266 **Wilcox, B.A. (1980).** Insular ecology and conservation. *In* Soule, M.E., Wilcox, B.A., eds. Conservation biology: an evolutionary and ecological perspective. Massachusetts, Sinauer Associates. 95-117.

1267 **Wilcox, B.A. (1984).** *In situ* conservation of genetic resources: determinants of minimum area requirements. *In* McNeely, J.A., Miller, K.R., eds. National parks, conservation and development. The role of protected areas in sustaining society. Washington, D.C., Smithsonian Institution Press. 639-647. (E/P)

1268 **Wilkes, G. (1985).** Germplasm conservation toward the year 2000: potential for new crops and enhancement of present crops. *In* Yeatman, C.W., Kafton, D., Wilkes, G., eds. Plant genetic resources. A conservation imperative. Boulder, CO, Westview Press. 131-161. (E/G/P)

1269 **Wilkins, C.P., Bengochea, T., Dodds, J.H. (1982).** The use of *in vitro* methods for plant genetic conservation. *Outlook Agric.*, 11(2): 67-72. Illus. (G)

1270 **Wilkins, C.P., Dodds, J.H. (1983).** The application of tissue culture techniques to plant genetic conservation. *Science Progress*, 68(270): 259-284. (G)

1271 **Williams, J.T. (1982).** Genetic conservation of wild plants. *Nat. Resources*, 18(1): 14-15. (G)

1272 **Williams, J.T. (1983).** International plant genetic resources programmes. *In* Jain, S.K. Mehra, K.L., eds. Conservation of tropical plant resources. Howrah, Botanical Survey of India. 236-239. (G)

GENERAL

1273 **Williams, J.T. (1987).** Vavilov's centres of diversity and the conservation of genetic resources. *Pl. Genet. Resourc. Newsl.*, 72: 6-8. (E)

1274 **Williams, J.T., Creech, J.L. (1987).** Genetic conservation and the role of botanic gardens. *In* Bramwell, D., Hamann, O., Heywood, V., Synge, H., *eds.* Botanic gardens and the World Conservation Strategy. London, Academic Press. 161-173. (Sp). (E/G)

1275 **Wilson, E.O. (1980).** Species extinction: no. 1. Problem of the eighties. *Calif. Native Pl. Soc. Bull.*, 10(3): 8. Excerpt from *Harvard Magazine Inc.*, Jan-Feb. 1980. (T)

1276 **Wilson, E.O. (1985).** The biological diversity crisis. *BioScience*, 35(11): 700-706. Illus. (E/T)

1277 **Wilson, E.O. (1985).** The biological diversity crisis: a challenge to science. *Issues Sci. Tech.*, Autumn 1985: 20-29.

1278 **Wilson, E.O. (1988).** Biodiversity. National Academy of Sciences, Washington, D.C. 521p. Chapters covering rates of forest destruction, species diversity, useful species. (E/T)

1279 **Wilson, E.O. (1988).** The current state of biological diversity. *In* Wilson, E.O., *ed.* Biodiversity. Washington, D.C., National Academy of Science Press. 3-18. Numbers of described species; extinction rates; tropical rain forests as centres of biodiversity. (T)

1280 **Wilson, E.O., ed. (1988).** Biodiversity. Washington, D.C., National Academy of Science Press. 521p. Many chapters; covers species numbers and rates of extinction; deforestation rates.

1281 **Winter, J. (1981).** Conservation of indigenous flora. *Park Admin.*, 34(1): 65-71. Illus.

1282 **Winterhoff, W. (1978).** Gefahrdung und Schutz von Pilzen. (Endangerment and conservation of fungi.) *Veroff. Landesst. Natursch. Landschaftspfl. Baden-Wurttemberg*, 11: 161-167. Ge. (T)

1283 **Wirth, W. (1986).** Uber die internationale und nationale Gesetzgebung zum Arten-und Naturschutz und den Besitz von Orchideen. (About the international and national legislation on species and natural protection and possession of orchids.) *Ber. Arbeitskr. Heim. Orch.*, 3(1): 145-155. Ge. (L/T)

1284 **Withers, L.A. (1980).** The cryo preservation of higher plant tissue and cell cultures, an overview with some current observations and future thoughts. *Cryoletters*, 1(8): 239-250. (G)

1285 **Withers, L.A. (1982).** Plant genetic conservation: recalcitrant seed and tissue culture. *Biol. Internat.*, 5: 2-9. (E/G)

1286 **Withers, L.A., Williams, J.T., eds (?)** Crop genetic resources; the conservation of difficult material. Proceedings of an International Workshop held at the University of Reading, UK, 8-11 September 1980. International Union of Biological Sciences, Series B42. Illus. (E/G)

1287 **Wochok, Z.S. (1981).** The role of tissue culture in preserving threatened and endangered plant species. *Biol. Conserv.*, 20(2): 83-89. (G/T)

1288 **Wolf, E.C. (1985).** Challenges and priorities in conserving biological diversity. *Envir. Awareness*, 8(3): 67-79. (E/G)

1289 **Wolf, E.C. (1986).** Beyond the Green Revolution: new approaches for Third World agriculture. Worldwatch Paper No. 73. Washington, D.C., Worldwatch Institute. 46p. (E)

GENERAL

1290 **Wolf, E.C. (1987).** On the brink of extinction: conserving the diversity of life. Worldwatch Paper No. 78. Washington, D.C., Worldwatch Institute. 54p. Tables on biodiversity; extinction rates. (T)

1291 **Wood, D.A. (1977).** Endangered species: a bibliography on the world's rare, endangered and recently extinct wildlife and plants. Stillwater, Oklahoma State University. 85p. Environmental Series No. 3. 1149 entries. (B/T)

1292 **Woods, A. (1982).** The propagation and distribution of rare and endangered plants. *Threatened Pl. Commit. Newsl.*, 9: 14-15. (G/T)

1293 **Woodwell, G.M. (1977).** The challenge of endangered species. *In* Prance, G.T., Elias, T.S., *eds.* Extinction is forever. Proceedings of a symposium at the New York Botanical Garden, 11-13 May 1976. New York, New York Botanical Garden. 5-10. Table of net primary productivity of the world's major ecosystems. Discusses stabilizing the relationship between man and resources. (T)

1294 **Woodwell, G.M. et al. (1983).** Global deforestation: contribution to atmospheric carbon dioxide. *Science*, 222(4628): 1081-1086.

1295 **World Commission on Environment and Development (1987).** Our common future. Oxford, Oxford Univ. Press. 400p. Covers environmental problems; role of international economy; institutional and legal changes for a sustainable future. (L)

1296 **World Forestry Congress (1978).** Third Technical Commission: Conservation and recreation. *In* Proceedings of the 7th World Forestry Congress, Buenos Aires, 4-18 October 1972, vol. 3. Buenos Aires, Instituto Forestal Nacional. 3433-4313. En, Sp, Fr.

1297 **World Resources Institute, IIED (1986).** World resources 1986. An assessment of the resource base that supports the global economy. New York, Basic Books. 353p. Illus., maps. Chapters on population, agriculture, forests, wildlife, pollution. (L/P)

1298 **World Resources Institute, IIED, UNEP (1988).** World resources 1988-89. An assessment of the resource base that supports the global economy. New York, Basic Books. 372p. Illus., maps. Chapters on population, agriculture, forests, wildlife, pollution, data tables for 146 countries. (L/T)

1299 **Worthington, E.B. (1982).** World campaign for the biosphere. *Envir. Conserv.*, 9(2): 93-100.

1300 **Wright, D.F. (1977).** A site evaluation for use in the assessment of potential nature reserves. *Biol. Conserv.*, 11: 293-305. (P)

1301 **Wyse Jackson, P. (1988).** Botanic gardens unite for action. *Species*, 11: 10. Describes the work of the IUCN Botanic Gardens Conservation Secretariat. (G)

1302 **Wyse Jackson, P. (1989).** What role can Field Clubs play in conservation? *In* Wyse Jackson, P., Akeroyd, J.R., Moriarty, C., *eds.* In the field of the naturalists. Dublin, Dublin Naturalists' Field Club. 22-30.

1303 **Yeatman, C.W., Kafton, D., Wilkes, G., eds (1985).** Plant genetic resources. A conservation imperative. Boulder, CO, Westview Press. 164p. (E/G/P/T)

1304 **Young, E. (1981).** Nature conservation in the urban environment. *Park Admin.*, 34(1): 56-65.

1305 **Young, J.L. (1974).** Seed banks. *Amer. Orch. Soc. Bull.*, 43(2): 124-125. (G)

1306 **Young, V. (1979).** Importance of islands as reserves. *In* Second South Pacific Conference on National Parks and Reserves, Proceedings, vol. 1. 130-139. (P)

GENERAL

1307 **Zakaria, M.B. (1987).** Conservation of medicinal and economic plants in retrospective of their chemical constituents. *In* Santiapillai, C., Ashby, K.R., *eds.* Proceedings of the Symposium on the Conservation and Management of Endangered Plants and Animals. Bogor, Indonesia, 18-20 June 1986. Bogor, SEAMEO-BIOTROP. 87-91. Illus. Medicinal. (E)

1308 **Zamula, N.T., Kharkevich, S.S., Egorova, L.N. (1978).** [Fourth session of Far Eastern Regional Scientific Council for the problem "Biological bases for rational usage, transformation and protection of the plant world", May 30-June 3 1977, Blagoveshchensk]. *Bot. Zhurn.*, 63(6): 912-917. Rus. (T)

1309 **Zimmerman, J.H. (1961).** Conservation of rare plants and animals. *WI Acad. Rev.*, (1961): 6-11. (T)

REGIONAL REFERENCES

ANTARCTICA

1310 **Anon. (1983).** Plan to flatten islands for airstrip. *Oryx*, 17: 144. Adelia Land, Antarctica. Includes five islands: Cuvier, Lion, Pollux, Zeus & Buffon.

1311 **Broady, P., Given, D.R., Greenfield, L.G., Thompson, K. (1987).** The biota and environment of a rare Antarctic habitat: steam warmed ground on volcanic Mount Melbourne, Northern Victoria Land. *Polar Biology*, 7: 97-113. Site is now an SSSI; includes rare moss flora. (P)

1312 **Elworthy, J. (1984).** Keynote address: the Antarctic Realm. *In* McNeely, J.A., Miller, K.R., *eds.* National parks, conservation and development. The role of protected areas in sustaining society. Washington, D.C., Smithsonian Institution Press. 364-368. (P)

1313 **Given, D. (1987).** Plants in the Antarctic. *IUCN Bull.*, 18(10-12): 11. Illus. Report on field visit in 1984.

1314 **Given, D.R. (1981).** Flora of offshore and onlying islands. *In* XIII International Botanical Congress, Sydney, Australia, 21-28 August 1981: abstracts. Sydney, Australian Academy of Science. 215.

1315 **Given, D.R. (1987).** Plants in the Antarctic. *Species*, 8: 25. Important bryophyte community on Mount Melbourne.

1316 **IUCN (1984).** Conservation and development of Antarctic ecosystems. Gland, IUCN. 36p.

1317 **IUCN/SCAR Working Group (1986).** Conservation in the Antarctic. Report of the Joint IUCN/SCAR Working Group on long-term conservation in the Antarctic. Gland, Switzerland, IUCN and International Council of Scientific Unions Scientific Committee on Antarctic Research. 50p. Maps, graphs. (L)

1318 **Lucas, P.H.C. (1982).** International agreement on conserving the Antarctic environment. *Ambio*, 11(5): 292-295. Illus., col. illus., maps. (L/P)

1319 **Lucas, P.H.C. (1984).** Finding ways and means to conserve Antarctica. *In* McNeely, J.A., Miller, K.R., *eds.* National parks, conservation and development. The role of protected areas in sustaining society. Washington, D.C., Smithsonian Institution Press. 369-375. Map.

1320 **Oldfield, S. (1987).** Fragments of paradise. A guide for conservation action in the U.K. Dependent Territories. Oxford, Pisces Publications; for the British Association for Nature Conservationists. 192p.

ANTARCTICA

1321 **Pingitore, E.J. (1982).** Especies interesantes de La Tierra del Fuego e Islas del Antartico Sur. *Bol. Soc. Hort. Argentina*, 38: 10-12. Sp. Tentative list of 38 threatened species. (T)

1322 **Smith, R.I.L. (1988).** Destruction of Antarctic terrestrial ecosystems by a rapidly increasing fur seal population. *Biol. Conserv.*, 45: 55-72.

1323 **Thom, D.A. (1984).** Future directions for the Antarctic Realm. *In* McNeely, J.A., Miller, K.R., *eds*. National parks, conservation and development. The role of protected areas in sustaining society. Washington, D.C., Smithsonian Institution Press. 412-416. (P)

ARCTIC

1324 **Polunin, N. (1970).** Botanical conservation in the Arctic. *Biol. Conserv.*, 2(3): 197-205. Map. (T)

1325 **Roots, E.F. (1985).** The Northern Science Network: regional co-operation for research and conservation. *Nature and Resources*, 21(2): 2-10. Illus. Map. Circumpolar. (P/T)

CARIBBEAN ISLANDS

1326 **Anon. (1979).** Activities of international organisations. *Caribbean Conserv. News*, 1(19): 19-20.

1327 **Anon. (1981).** Conservation on the islands. *IUCN Bull.*, 12(5-6): 29.

1328 **Anon. (1981).** IUCN's Caribbean conservation initiative. *IUCN Bull.*, 12(5-6): 26.

1329 **Anon. (1981).** The CCA: a "mini" IUCN in the Caribbean. *IUCN Bull.*, 12(5-6): 28.

1330 **Anon. (1981).** The IUCN Caribbean Strategy. *IUCN Bull.*, 12(5-6): 28. Illus.

1331 **Chinnery, L. (1979).** Biological records centres for the Caribbean. *Caribbean Conserv. News*, 1(17): 17-18.

1332 **Densmore, D. (1986).** Endangered plants of our Caribbean Islands: a unique flora faces unique problems. *Endangered Species Tech. Bull.*, 11(3): 3-4. (T)

1333 **FAO (1985).** Intensive multiple-use forest management in the tropics. Analysis of case studies from India, Africa, Latin America and the Caribbean. Rome, FAO Forestry Paper 55. 180p. (E)

1334 **FAO (1986).** Report on natural resources for food and agriculture in Latin America and the Caribbean. FAO Environment and Energy Paper 8. Rome, FAO. 102p. (E)

1335 **Geoghegan, T. (1985).** Public participation and managed areas in the Caribbean. *Parks*, 10(1): 12-14. Illus. (P)

1336 **Heywood, V.H. (1979).** The future of island floras. In Bramwell, D., *ed*. Plants and islands. London, Academic Press. 431-441. (Sp). (P/T)

1337 **Howard, R.A. (1977).** Conservation and the endangered species of plants in the Caribbean Islands. *In* Prance, G.T., Elias, T.S., *eds*. Extinction is forever. Proceedings of a symposium at the New York Botanical Garden, 11-13 May 1976. New York, New York Botanical Garden. 105-114. The nature of the vegetation, location of endangered species, the nature of the present threats and attempts at conservation. (L/P/T)

1338 **IUCN (1981).** Conserving the natural heritage of Latin America and the Caribbean. The planning and management of protected areas in the Neotropical Realm. Proceedings of the 18th working session of IUCN's Commission on National Parks and Protected Areas, Lima, Peru, 21-28 June 1981. Gland, IUCN. 329p. (P)

CARIBBEAN ISLANDS

1339 **IUCN (1981).** IUCN/WWF projects in the Caribbean. *IUCN Bull.*, 12(5-6): 36. Illus.

1340 **IUCN Botanic Gardens Conservation Secretariat (1988).** Rare and threatened palms of the New World; questionnaire and preliminary report on their occurrence in botanic gardens. Part 2 - The Caribbean. Kew, IUCN Botanic Gardens Conservation Secretariat. 5p. (G/T)

1341 **IUCN Commission on National Parks and Protected Areas (1982).** IUCN Directory of Neotropical protected areas. Dublin, Tycooly International. 436p. (Sp). Maps. (P)

1342 **Johnson, D. (1987).** Conservation status of wild palms in Latin America and the Caribbean. *Principes*, 31(2): 96-97. (T)

1343 **Johnson, T.H. (1988).** Biodiversity and conservation in the Caribbean: profiles of selected islands. ICBP Monograph No. 1. Cambridge, U.K., ICBP. 144p. Data on flora, fauna, vegetation, geography, projects and recommendations for selected islands. (T)

1344 **Lawyer, J.I., et al. (1979).** A guide to selected current literature on vascular plant floristics for the contiguous U.S., Alaska, Canada, Greenland and the U.S. Caribbean and Pacific Islands. New York, New York Botanical Garden. 138p. Suppl. 1: Research on floristic information synthesis: a report to the Division of Natural History, Nat. Park Service, U.S. Dept of Interior. Washington, D.C. (B)

1345 **Lugo, A.E., Schmidt, R., Brown, S. (1981).** Tropical forests in the Caribbean. *Ambio*, 10(6): 318-324. Col. illus. (T)

1346 **Matos, C. (1979).** Where to begin. *Caribb. Conserv. News*, 1(16): 12-16.

1347 **Munson, V. (1979).** Conference on environmental management and economic growth in the smaller Caribbean Islands (17-21 September 1979). *Caribb. Conserv. News*, 1(19): 7-9.

1348 **Nietschmann, B. (1984).** Biosphere reserves and traditional societies. *In* Unesco-UNEP. Conservation, science and society. Paris, Unesco. 499-508. Contributions to the First International Biosphere Reserve Congress, Minsk, Byelorussia/USSR, 26 September - 2 October 1983. Map. (P)

1349 **Oldfield, S. (1987).** Fragments of paradise. A guide for conservation action in the U.K. Dependent Territories. Oxford, Pisces Publications; for the British Association for Nature Conservationists. 192p.

1350 **Putney, A.D., Jackson, I., Renard, Y. (1984).** The Eastern Caribbean natural area management programme: a regional approach to research and development for conservation action. *In* McNeely, J.A., Miller, K.R., *eds.* National parks, conservation and development. The role of protected areas in sustaining society. Washington, D.C., Smithsonian Institution Press. 608-615. Map. (P)

CENTRAL AMERICA

1351 **Anon. (1979).** First Regional Conference of Central American Nongovernmental Conservation Associations, 4-7 December 1978, Guatemala City. *Caribbean Conserv. News*, 1(16): 10.

1352 **Boom, B. (1987).** Libertad for Central America's forests. *Garden* (U.S.), 11(1): 3-5. Tropical forest.

1353 **Budowski, G. (1965).** The choice and classificaiton of natural habitats in need of preservation in Central America. *Turrialba*, 15(3): 238-246. (P)

1354 **Budowski, G. (1977).** A strategy for saving wild plants: experience from Central America. *In* Prance, G.T., Elias, T.S., *eds.* Extinction is forever. Proceedings of a symposium at the New York Botanical Garden, 11-13 May 1976. New York, New York Botanical Garden. 368-373. (T)

CENTRAL AMERICA

1355 **Budowski, G., MacFarland, C. (1984).** Keynote address: the Neotropical Realm. *In* McNeely, J.A., Miller, K.R., *eds*. National parks, conservation and development. The role of protected areas in sustaining society. Washington, D.C., Smithsonian Institution Press. 552-560. Table. (P)

1356 **Denevan, W. (1982).** Causes of deforestation and forest woodland degradation in tropical Latin America. Washington, D.C., Office of Technology Assessment.

1357 **Dod, D. (1977).** Orquideas terrestres: primitivas pero fascinantes. *Bol. Soc. Dominican Orquideologia*, 1(4): 41-46. Sp.

1358 **Dodson, C.H. (1968).** Conservation of orchids. *In* IUCN. Conservation in Latin America. Proceedings San Carlos de Bariloche, Argentina, 27 March-2 April 1968. Morges, IUCN. 170. (T)

1359 **Dourojeanni, M.J. (1984).** Future directions for the Neotropical Realm. *In* McNeely, J.A., Miller, K.R., *eds*. National parks, conservation and development. The role of protected areas in sustaining society. Washington, D.C., Smithsonian Institution Press. 621-625. (P)

1360 **FAO (1985).** Intensive multiple-use forest management in the tropics. Analysis of case studies from India, Africa, Latin America and the Caribbean. Rome, FAO Forestry Paper 55. 180p. (E)

1361 **FAO (1986).** Report on natural resources for food and agriculture in Latin America and the Caribbean. FAO Environment and Energy Paper 8. Rome, FAO. 102p. (E)

1362 **FAO (1986).** Some medicinal forest plants of Africa and Latin America. FAO Forestry Paper No. 67. Rome, FAO. 252p. (E)

1363 **Farnworth, E.G., Golley, F.B., eds. (1974).** Fragile ecosystems: evaluation of research and application in the neotropics: a report of the Institute of Ecology, June 1973. New York, Springer-Verlag. 258p.

1364 **Ffolliott, P.F., Halffter, G. (1980).** Social and environmental consequences of natural resources policies, with special emphasis on biosphere reserves. *In* USDA Forest Service. Proceedings of the International Seminar, 8-13 April, 1980, Durango, Mexico. General technical report. Fort Collins, Rocky Mountain Forest and Range Experiment Station. 56p. (P)

1365 **Fuller, K.S., Swift, B. (1984).** Latin American wildlife trade laws (Leyes del Comercio de Vida Sylvestre en America Latina). Washington, D.C., WWF-US. En (Sp). Mimeograph. Accounts for each country in South and Middle America on wildlife laws, with species lists; mostly faunal but flora to be covered in more detail in revised version in preparation. (L/T)

1366 **Gentry, A. (1979).** Extinction and conservation of plant species in tropical America: a phytogeographical perspective. *In* Hedberg, I., *ed*. Systematic botany, plant utilization and biosphere conservation. Stockholm, Almqvist & Wiksell International. 110-126. Illus., maps. (T)

1367 **Gilbert, L.E. (1981).** Food web organization and the conservation of neotropical diversity. *In* Jordan, C.F., *ed*. Tropical ecology. Stroudsberg, Pennsylvania, Hutchinson Ross Publishing Company. 113-135. (Benchmark Papers in Ecology, vol. 10.) Reprinted from Soule, M.E., Wilcox, B.A., *eds*. Conservation biology: an evolutionary-ecology perspective. Sunderland, Massachussetts, Sinauer Associates. 11-33 (1980).

1368 **Goodland, R., (1977).** Panel discussion. *In* Prance, G.T., Elias, T.S., *eds*. Extinction is forever. Proceedings of a symposium at the New York Botanical Garden, 11-13 May 1976. New York, New York Botanical Garden. 360-367. Panel discussion on important differences between conservation in Southeast Asia and Latin America. (E/L/P/T)

1369 **Hartshorn, G.S. (1980).** Neotropical forest dynamics. *Biotropica*, 12(2) suppl.: 23-30. (Sp).

CENTRAL AMERICA

1370 **Hartshorn, G.S. (1983).** Wildlands conservation in Central America. *In* Sutton, S.L., Whitmore, T.C., Chadwick, A.C., *eds.* Tropical rain forest: ecology and management. Oxford, Blackwell Scientific. 423-444. Special Publications Series of the British Ecological Society No. 2. Maps.

1371 **Hernandez, M.O. (1968).** The disappearance of valuable native orchids in Latin America. IUCN Publications, New Series No. 13. Morges, IUCN. 168-169. (T)

1372 **IUCN (1981).** Conserving the natural heritage of Latin America and the Caribbean. The planning and management of protected areas in the Neotropical Realm. Proceedings of the 18th working session of IUCN's Commission on National Parks and Protected Areas, Lima, Peru, 21-28 June 1981. Gland, IUCN. 329p. (P)

1373 **IUCN Commission on National Parks and Protected Areas (1982).** IUCN Directory of Neotropical protected areas. Dublin, Tycooly International. 436p. (Sp). Maps. (P)

1374 **IUCN Threatened Plants Unit (1986).** The botanic gardens list of rare and threatened species of Central America. Kew, IUCN Botanic Gardens Co-ordinating Body Report No. 18. 36p. Lists plants in cultivation in botanic gardens. (G/T)

1375 **Janzen, D.H. (1974).** The deflowering of Central America. *Nat. Hist.,* (1974): 47-51. (T)

1376 **Janzen, D.H. (1988).** Tropical dry forests: the most endangered major tropical ecosystem. *In* Wilson, E.O., *ed.* Biodiversity. Washington, D.C., National Academy Press. 130-137.

1377 **Jimenez, J.A. (1984).** A hypothesis to explain the reduced distribution of the mangrove *Pelliciera rhizophorae* Tr. and Pl. *Biotropica*, 16(4): 304-308. Maps. (T)

1378 **Johnson, D. (1987).** Conservation status of wild palms in Latin America and the Caribbean. *Principes,* 31(2): 96-97. (T)

1379 **La Bastille, A. (1979).** Facets of wildland conservation in Central America. *Parks,* 4(3): 1-5. Illus. (P)

1380 **Linares, O. (1976).** Garden hunting in the American tropics. *Human Ecology,* 4(4): 331-349. (G)

1381 **MacFarland, C., Barborak, J.R., Morales, R. (1984).** Training personnel for biosphere reserves and other managed wildlands and watersheds: Catie's experience in Central America. *In* Unesco-UNEP. Conservation, science and society. Paris, Unesco. 605-612. Contributions to the First International Biosphere Reserve Congress, Minsk, Byelorussia/USSR, 26 September - 2 October 1983. (P)

1382 **Marshall, N.T. (1989).** Parlor palms. Increasing popularity threatens Central American species. *TRAFFIC* (U.S.A.), 9(3): 1-3. Illus. *Chamaedorea* spp. (E/L/T)

1383 **Miller, K.R. (1975).** Ecological guidelines for the management and development of national parks and reserves in the American humid tropics. *In* IUCN. The use of ecological guidelines for development in the American humid tropics. Proceedings, Caracas, Venezuela, 20-22 February 1974. IUCN Publications New Series No. 31. Morges, IUCN. 91-105. (P)

1384 **Milton, J.P. (1975).** The ecological effects of major engineering projects. *In* IUCN. The use of ecological guidelines for development in the American humid tropics. Proceedings, Caracas, Venezuela, 20-22 February 1974. Switzerland, IUCN New Series No. 31. 207-221.

1385 **Munoz Pizarro, C. (1975).** Especies vegetales que se extinguen en nuestro pais. *In* Capurro, L., Vergara, R., *eds.* Present y futuro del medio humano. Mexico, Edit. Cont. CECSA. Capitulo XI. 161-179. Sp. (T)

CENTRAL AMERICA

1386 **Myers, N. (1981).** The hamburger connection: how Central America's forests become North America's hamburgers. *Ambio*, 10(1): 3-8. Illus., map. Deforestation.

1387 **Nations, J., Komer, D. (1982).** Rainforest, cattle and the hamburger society. Austin, Center for Human Ecology. Deforestation.

1388 **Nations, J.D., Komer, D.I. (1983).** Central America's tropical rainforests: positive steps for survival. *Ambio*, 12(5): 232-238. Illus., col. illus.

1389 **New York Botanical Garden (1976).** Symposium on threatened and endangered species of plants in the Americas and their significance in ecosystems today and in the future. Program and Abstracts. New York, New York Botanical Garden. 55p. (T)

1390 **Pan American Union (1967).** La Convencion para la Proteccion de la Flora, de la Fauna y de las Bellezas Escenicas Naturales de los Estados Americanos: listas de especies de fauna y flora en vias de Extincion en los estados miembros. Washington, D.C., Organization of American States. 48p. Sp. (T)

1391 **Prance, G.T. (1976).** Threatened and endangered plants in the Americas. *BioScience*, 26(10): 633-634. (T)

1392 **Rajanaidu, N. (1983).** *Elaeis oleifera* collection in South and Central America. *Pl. Genet. Resource. Newsl.*, 56: 42-51. En (Fr, Sp). Maps. (E/T)

1393 **Ravenna, P. (1977).** Neotropical species threatened and endangered by human activity in the Iridaceae, Amaryllidaceae and allied bulbous families. *In* Prance, G.T., Elias, T.S., *eds*. Extinction is forever. Proceedings of a symposium at the New York Botanical Garden, 11-13 May 1976. New York, New York Botanical Garden. 257-266. Illus. (T)

1394 **Read, M. (1989).** Bromeliads threatened by trade. *Kew Mag.*, 6(1): 22-29. Illus. (T)

1395 **Vale, T.R. (1985).** What kind of conservationist? *J. Geogr.*, 84(6): 239-241. Central America, attitudes to deforestation. Tropical forest. (T)

1396 **Villa-Lobos, J., comp. (?)** Threatened plants of Central America. Washington D.C., Smithsonian Institution. c.400p. A collaborative project between IUCN, the Smithsonian Institution and the World Conservation Monitoring Centre. In prep. (B/T)

1397 **Wadsworth, F.H. (1975).** Natural forests in the development of the American humid tropics. *In* IUCN. The use of ecological guidelines for development in the American humid tropics. Proceedings, Caracas, Venezuela, 20-22 February 1974. Morges, IUCN. 129-138.

1398 **Watters, R.F. (1975).** Shifting agriculture - its past, present and future. *In* IUCN. The use of ecological guidelines for development in the American humid tropics. Proceedings, Caracas, Venezuela, 20-22 February 1974. 77-86.

1399 **Wille, C. (1985).** The hamburger connection. *Audubon Action*, 3(3): 2. Destruction of Central American forests for raising cattle. Deforestation.

EUROPE

1400 **Aird, I.A. (1980).** Conservation. *Cyclamen Soc. J.*, 4(1): 7. Conservation of *Cyclamen*. (T)

1401 **Alcamo, J., Amann, M., Hettelingh, J.-P., Holmberg, M., Hordijk, L., Kamari, J., Kauppi, L., Kornal, G., Makela, A. (1987).** Acidification in Europe: a simulation model for evaluating control strategies. *Ambio*, 16(5): 232-245. Illus., col. illus., maps. Acid rain; forest soil acidification; lake acidification; modelling.

EUROPE

1402 **Allium Working Group (1984).** European cooperative programme for the conservation and exchange of crop genetic resources. Report of a working group held at the Research Centre for Agrobotany, Tapioszele, Hungary, 29-31 May 1984. Rome, International Board for Plant Genetic Resources. iii, 11p. (AGPG: IBPGR/84/89). (E/G)

1403 **Andersson, F. (1983).** Acid rain. *Naturopa*, 43: 14-18. Illus., col. illus.

1404 **Anon. (1978).** Resolution (77)6 on the conservation of rare and threatened plants in Europe. *Naturopa*, no. 31: 22. European Ministerial Conferences on the Environment (Vienna 1973, Brussels 1976). (L/T)

1405 **Anon. (1980).** EEC environmental policies. *Conserv. News*, 77: 3-5. (L)

1406 **Anon. (1980).** Protect the Mediterranean. *IUCN Bull.*, 11(9/10): 81-83, 95. Illus.

1407 **Anon. (1981).** Siebenstern (*Trientalis*): es war einmal. *Gartn.-Bot. Brief*, 69: 48-49. Ge. Illus. (G)

1408 **Anon. (1983).** Kuidas kaitsta haruldasi taimi? 1. (How should rare plants be protected? - part 1.) *Eesti Loodus*, 26(3): 159-164. Es. Illus., col. illus. (T)

1409 **Anon. (1983).** Kuidas kaitsta haruldasi taimi? 2. (How should rare plants be protected? - part 2.) *Eesti Loodus*, 26(4): 210-218. Es (Rus, En). Illus., col. illus. (T)

1410 **Anon. (1983).** Pluies acides: maladie de la foret. *Ardenne et Gaume*, 128: 2-3. Fr. Acid rain.

1411 **Anon. (1987).** European forest damage. *Acid News*, 3: 6-7. Illus., map. Acid rain.

1412 **Anon. (1988).** National strategies for protection of flora, fauna and their habitats. New York, United Nations. (L/P/T)

1413 **Ashmore, M., Tickle, A. (1987).** Acid rain: never pure and rarely simple. *Trends Ecol. Evol.*, 2(3): 58-59. Illus. Pollution.

1414 **Aulitzky, H. (1974).** Endangered alpine regions and disaster prevention measures. Nature and Environment Series No. 6. Strasbourg, Council of Europe. vi, 115p. (T)

1415 **Aymonin, G.G. (1973).** Especes menacees. *Bull. Soc. Etude Sci. Nat. Beziers, n.s., t.1*, 42: 15-20. Fr. (T)

1416 **Aymonin, G.G. (1973).** La disparition des populations vegetales. *Le Courrier de la Nature*, 25: 1-7. (T)

1417 **Aymonin, G.G. (1975).** La regression des vegetaux hygrophiles en Europe: anipleur et signification. *In* Comptes Rendus du 100e Congres National des Societes Savantes. Fascicule II. Paris, Bibliotheque Nationale. 271-282. (T)

1418 **Bailey, R.H., Dunning, R.J. (1984).** The Society and the conservation of *Cyclamen*. *Cyclamen Soc. J.*, 8(1): 9-10. (T)

1419 **Baum, P. (1981).** An ecological action plan for nature conservation in western Europe. *Threatened Pl. Commit. Newsl.*, 8: 16-18.

1420 **Baumann, H., Dafni, A., Golz, P., Greuter, W., Reinhard, H.R., Tigges, M. (1981).** OPTIMA-Project "Kartierung des Mediterranen Orchideen" 2. Orchideenforschung und Naturschutz im Mittelmeergebiet Internationales Artenschutzprogramm. *Beih. Veroff. Natursch. Landschaftspfl. Baden-Wurttemberg*, 19: 1-190. Ge (En). Col. illus., maps.

1421 **Bonneau, M., Landamann, G. (1988).** Europe's deteriorating forests. What ails the forest? *Recherche* (Paris), 19(205): 1542-1553.

1422 **Borhidi, A. (1984).** Role of mapping the flora of Europe in nature conservation. *Norrlinia*, 2: 87-98. Maps.

EUROPE

1423 **Broussalis, P. (1978).** Where have all the flowers gone? *Naturopa*, 31: 4-6. Illus. (T)

1424 **Caldecott, J. (1988).** Climbing towards extinction. *New Sci.*, 118 (1616): 62-66. Col. illus. (T)

1425 **Campbell, F. (1987).** Alert to gardeners. *TRAFFIC* (U.S.A.), 7(4): 11. Over-collecting of bulbs. (G/T)

1426 **Carbiener, R. (1978).** Forests or just trees? *Naturopa*, 31: 10-13. Illus. (T)

1427 **Cerovsky, J., comp. (1986).** Nature conservation in the socialist countries of East Europe. Vrchlabi, Administration of the Krkonose (Giant Mountains) National Park. 116p. Illus., maps. Covers Bulgaria, Czechoslovakia, Hungary, German Democratic Republic, Poland, Romania & U.S.S.R.

1428 **Chester, P.F. (1983).** Perspectives on acid rain. *J. Roy. Soc. Arts*, 131(5326): 587-603. Maps.

1429 **Cizek, K., Sofron, J., Vondracek, M. (1985).** Statni prirodni rezervace Jaleri vrch v planickem breberi. (Naturschutzgebiet Jeberi vrch im Planitzer Kamm.) *Zprary Muz. Zapad. Kraje, Prir.*, 30-31: 15-17. Cz (Ge).

1430 **Clauser, F. (1983).** Exploitation versus conservation. *Naturopa*, 43: 26-28. Illus.

1431 **Clement, J.L. (1977).** Aspect juridique de la protection des orchidees en Europe. *Orchidophile*, 29: 823-828. Fr. Illus. (L/T)

1432 **Commission of the European Communities (1977).** State of the environment: first report. Brussels, EEC. 261p. Illus., maps.

1433 **Cook, C.D.K. (1976).** Autecology. *In* Simmons, J.B., *et al.*, eds. Conservation of threatened plants. NATO Conference Series 1: Ecology, vol. 1. New York and London, Plenum Press. 207-210. Proc. Conference on the Functions of Living Plant Collections in Conservation and Conservation-Orientated Research and Public Education, Kew, 2-6 September 1975. Incl. ecology of rare wetland plants. (T)

1434 **Council of Europe (1977).** Selective bibliography on wetlands. Strasbourg, Council of Europe. 37p. (B/P/T)

1435 **Council of Europe (1977).** The heathlands of western Europe. Nature and Environment Series No. 12. Strasbourg, Council of Europe. 11p. (B/P/T)

1436 **Council of Europe (1981).** Alluvial forests in Europe. Nature and Environment Series No. 22. Strasbourg, Council of Europe. 65p. Illus. Maps. (P/T)

1437 **Council of Europe (1981).** Convention on the Conservation of European Wildlife and Natural Habitats. *Council Europe Newsl.*, 81(4/5): 1-2. (L)

1438 **Council of Europe (1981).** Dry grasslands of Europe. Nature and Environment Series No. 21. Strasbourg, Council of Europe. 56p. Maps. (P/T)

1439 **Council of Europe (1981).** National activities: protecting the Mediterranean forest. *Council of Europe Newsl.*, 81(8/9): 2.

1440 **Council of Europe (1981).** Symposium: tourism, environmental conservation and environmental education. Report of European Committee for the Conservation of Nature and Natural Resources, Malta, 3-10 October 1981: Report. Strasbourg, Council of Europe. 53p. Illus. (P/T)

1441 **Council of Europe (1982).** Selective bibliography on legislations on nature conservation and the environment. Strasbourg, Council of Europe. 65p. (B/L/P/T)

EUROPE

1442 **Council of Europe (1982).** Study protection and management of natural genetic resources: the experience of the Council of Europe. Strasbourg, Council of Europe, European Committee for the Conservation of Nature and Natural Resources: Steering Committee. 65p. (G/T)

1443 **Council of Europe (1983).** Council of Europe. Nature parks: a European challenge. *Council of Europe Newsl.*, 83(11): 1. (P)

1444 **Council of Europe (1984).** Campaign on the Water's Edge. *Council of Europe Newsl.*, 84(4): 4p.

1445 **Council of Europe (1984).** Coastal areas, river banks and lake shores: their planning and management in compatibility with the ecological balance. 4th European Ministerial Conference on the Environment: Report presented by the Delegation of Greece. Athens, 25-27 April 1984. Council of Europe. 104p. (P/T)

1446 **Council of Europe (1984).** Conservation de la nature en Europe; vingt annees d'activites. Strasbourg, Council of Europe, Comite Europeen pour la Conservation de la Nature et des Ressources Naturelles. 111p. Fr. Maps. Brief notes on the situation in Council of Europe countries. (P/T)

1447 **Council of Europe (1984).** Council of Europe documents on nature conservation and the natural environment. Strasbourg, Council of Europe. 39p. (B/G/L/P/T)

1448 **Council of Europe (1984).** Salt marshes in Europe. Nature and Environment Series No. 30. Strasbourg, Council of Europe. 178p. Illus. (P/T)

1449 **Council of Europe (1984).** Selective bibliography on National Parks, Regional Nature Parks, Nature Reserves and other protected reserves, landscapes and sites. Documentation Series No. 15. Strasbourg, Council of Europe. 64p. (B/P/T)

1450 **Council of Europe (1984).** Selective bibliography on the water's edge. Documentation Series No. 12. Strasbourg, Council of Europe, European Information Centre for Nature Conservation. 119p. (B/L/P/T)

1451 **Council of Europe (1984).** Survey of the activities undertaken in the field of nature conservation by the Council of Europe since the third ministerial conference. 4th Ministerial Conference on the Environment, Athens 25-27 April. Strasbourg, Council of Europe. 27p. (L/P/T)

1452 **Council of Europe (1985).** Catalogue of data banks in the field of nature conservation. 2nd Colloquy on Computer Applications in the Field of Nature Conservation, Strasbourg, 26-27 February 1985. Strasbourg, Council of Europe. 109p. (B/P/T)

1453 **Council of Europe (1985).** Selective bibliography on acid rain and its effects. Strasbourg, Council of Europe, European Information Centre for Nature Conservation. 30p. En, Fr, Ge. (B/P/T)

1454 **Council of Europe. Division of Environment and Natural Resources (1984).** Council of Europe work on protected areas. *In* McNeely, J.A., Miller, K.R., *eds*. National parks, conservation and development. The role of protected areas in sustaining society. Washington, D.C., Smithsonian Institution Press. 706-711. (P)

1455 **Cutrera, A., et al. (1987).** European environmental yearbook 1987. London, DocTer International U.K. 815p. Maps. Nature conservation, protection of the environment, town and country planning in Belgium, Denmark, Federal Republic of Germany, France, Greece, Ireland, Italy, Luxembourg, Netherlands, Portugal, Spain, UK (L)

1456 **Delforge, P., Tyteca, D. (1982).** Quelques orchidees rares ou critiques d'Europe Occidentale. *Bull. Soc. Roy. Bot. Belg.*, 115(2): 271-288. Fr (En). Illus., maps. (T)

1457 **Der Deutsche Naturschutzring (1980).** Moore: Bedeutung, Schutz, Regeneration. (Wetlands: significance, protection, regeneration.) Bonn, Deutscher Naturschutzring. 21p. Ge. Col. illus. (P)

EUROPE

1458 **Dierschke, H. (1978).** Monotony! *Naturopa*, no. 31: 29-32. Illus.

1459 **Dijkema, K.S. (1984).** La vegetation halophile. *In* Council of Europe. Conservation de la nature en Europe. Strasbourg, Council of Europe. 31-32. Fr. Saltmarshes. (P/T)

1460 **Doody, J.-P. (1984).** Les dunes. *In* Council of Europe. Conservation de la Nature en Europe. Strasbourg, Council of Europe. 33. Fr. (P/T)

1461 **Duffey, E. (1982).** National parks and reserves of western

Europe. London, Macdonald. 288p. Illus., col. illus., maps. (P)

1462 **Eiberle, K. (1978).** Wald und zoologischer Naturschutz. (Forest and wildlife conservation.) *Mitt. Eidgen. Anst. Forstl. Versuchsw.*, 54(4): 418-426. Ge (En, Fr, It).

1463 **Einarsson, E., et al. (1974).** Listes des especies vegetales menacees en Europe. Doc. CE/Nat./74 (101), Conseil de l'Europe. 40p. (T)

1464 **Ellenberg, H. (1985).** Veranderungen der Flora Mitteleuropas unter dem Einfluss von Dungung and Immissionen. (Changes in the flora of Central Europe under the influence of fertilisers and air pollution.) *Schweiz. Zeitschr. Forstw.*, 136(1): 19-39. Ge (Fr). Forest, pollution.

1465 **European Information Centre for Nature Conservation (1981).** Threatened European alluvial forests. *Envir. Conserv.*, 8(4): 306.

1466 **FAO, UNDP (1981).** Directory of crop genetic resources institutions in European countries. RER/80/005/ECP/GR/Directory 1981. Rome, FAO. 212p. (E)

1467 **Fessler, A. (1981).** Erfahrungen mit Vermehrungskulturen in Botanischen Garten (alphabetische). Hinwiese zur Anzucht einiger gefahrdeter Wildpflanzenarten. (Experience with propagation techniques in botanic gardens.) *Gartn. Bot. Brief*, 69: 19-28. Ge. (G)

1468 **Fischer, F. (1978).** Forstwirtschaft und Landschaftsschutz. (Forestry and environmental conservation.) *Mitt. Eidgenoss. Anst. Forstl. Versuchsw.*, 54(4): 519-530. Ge (En, Fr, It).

1469 **Fogle, H.W., Winters, H.F. (1981).** North American and European fruit and tree nut germplasm resources inventory. *U.S.D.A. Misc. Publ.*, 1406: 732p. (E)

1470 **Frey, W., Kurschner, H. (1985).** Final report of the section of botany, supporting period 1974-1985. Tubingen atlas of the Near and Middle East. Tubingen University, Special Research Division 19. Berlin. 28p. Maps.

1471 **Gammell, A. (1982).** The Berne Convention. *Ecos*, 3(4): 15-16. (L)

1472 **Garve, E. (1988).** Stand des Niedersachsischen Pflanzenarten-Erfassungsprogramms und Bericht von den Gelandetreffen. *Flor. Rundbr.*, 21(2): 134-146. Ge (En). Maps.

1473 **Gay, P. (1984).** Proposals for improved mechanisms to develop the biosphere reserve system as an integrated network. *In* Unesco-UNEP. Conservation, science and society. Paris, Unesco. 81-84. Contributions to the First International Biosphere Reserve Congress, Minsk, Byelorussia/USSR, 26 September - 2 October 1983. (P)

1474 **Gehu, J.-M. (1977).** La protection de la nature et le concept de reserve ecologique en Europe. (Nature conservation and the concept of ecological reserves in Europe.) *In* Filion, L., Villeneuve, P., *eds*. Ecological reserves and nature conservation. Rapport, Conseil Consultatif des Reserves Ecologiques, Quebec, No. 1: 14-19. Papers presented at a conference at Iles-de-la-Madeleine, Quebec, 17-18 June 1976. Fr. (P)

1475 **Gehu, J.-M. (1978).** 80,000 kms of coastline. *Naturopa*, 31: 14-16. Illus.

EUROPE

1476 **Gehu, J.-M. (1984).** Les pelouses sur calcaire. *In* Council of Europe. Conservation de la Nature en Europe. Strasbourg, Council of Europe. 27-28. Fr. Map. (P/T)

1477 **Gimingham, C.H. (1981).** Conservation: European heathlands. *In* Specht, R.L. *et al., eds.* Heathlands and related shrublands. Analytical studies. Ecosystems of the World 9B. Oxford, Elsevier. 249-259.

1478 **Gissy, M. (1985).** Europe's salt marshes: urgent need for protective network of reserves. *Envir. Conserv.*, 12(4): 371-372. (P)

1479 **Glenny, M. (1987).** Living in a socialist smog. *New Sci.*, 115(1579): 41-44. Illus. Overview of acid rain problem; mentions effects on forests.

1480 **Gomez-Campo, C. (1972).** Preservation of west Mediterranean members of the cruciferous tribe Brassiceae. *Biol. Conserv.*, 4(5): 355-360.

1481 **Gomez-Campo, C. (1978).** The OPTIMA Seeds and Living Material Commission. *Gartn. Bot. Brief*, 58: 29. (Ge). Abstract. (G)

1482 **Gomez-Campo, C. (1979).** The role of seed banks in the conservation of Mediterranean flora. *Webbia*, 34(1): 101-107. (It). (G)

1483 **Gomez-Campo, C. (1981).** The 'Artemis' project for seed collection of Mediterranean endemics. *Threatened Pl. Commit. Newsl.*, 8: 9-11. (G)

1484 **Goodwillie, R. (1980).** European peatlands. Nature and Environment Series No. 19. Strasbourg, Council of Europe. 75p. Maps.

1485 **Goodwillie, R. (1982).** Peatlands: living relics. *Naturopa*, 42: 18.

1486 **Goodwillie, R. (1984).** Les tourbieres. *In* Council of Europe. Conservation de la Nature en Europe. Strasbourg, Council of Europe. 25. Fr. Peatlands. (P/T)

1487 **Greuter, W. (1979).** Mediterranean conservation as viewed by a plant taxonomist. *Webbia*, 34(1): 87-99. (It). (T)

1488 **Greuter, W. (1981).** Probleme des Naturschutzes im Mittelmeergebiet. *Beih. Veroff. Naturschutz Landschaftspflege Baden-Wurttemberg*, 19: 155-158. Ge.

1489 **Grube, A. (1986).** Erganzungen zu "Verbreitung und Gefahrdung der Orchideen in Hessen". (Further material on the "distribution and threat to orchids in Hessen".) *Ber. Arbeitskr. Heim. Orch.*, 3(1): 58-63. Ge. (T)

1490 **Guillard, J. (1983).** The green mantle of Europe: forestry policies. *Naturopa*, 43: 20-23. Illus.

1491 **Halliday, G., Elkington, T. (1981).** Threats and conservation. *In* Halliday, G., Malloch, A., *eds.* Wild flowers; their habitats in Britain and Northen Europe. s.l., Peter Lowe. 165-173. Col. illus. (T)

1492 **Hauser, A., et al. (1978).** Forest and European countryside. 5th European Seminar on Applied Ecology - Forest and Landscape Conservation, 3-7 October 1977. *Mitteil. Eidgenoss. Anst. Forstl. Versuchsw.*, 54(4): 365-564. En, Ge, Fr. Illus.

1493 **Hawkes, J.G., Lange, W., eds. (1973).** European and regional gene banks. Wageningen, Netherlands, Eucarpia. 109p. Proceedings of a Conference of Eucarpia, Izmir, Turkey 1972. (G)

1494 **Hawkes, J.G., ed. (1978).** Conservation and agriculture. London, Duckworth. xvi, 284p.

1495 **Heywood, V.H. (1971).** Preservation of the European flora. The taxonomist's role. *Bull. Jard. Bot. Nat. Belg.*, 41(1): 153-166. (T)

1496 **Heywood, V.H. (1984).** Co-operation between the IUCN Species Survival Commission and European-Mediterranean Botanic Gardens in conservation strategies. *Rep. Bot. Inst. Univ. Aarhus*, 10: 33-39. (G/T)

EUROPE

1497 **Heywood, V.H. (1987).** The conservation of medicinal and aromatic plants - the need for a cooperative approach. *In* Mota, M., Bafta, J., *eds.* EUCARPIA, International Symposium on Conservation of Genetic Resources of Aromatic and Medicinal Plants, Oeiras, Potugal, 1984. 25-37. (E/T)

1498 **Hummel, F.C. (1982).** In the forests of the E.E.C. *Unasylva*, 34(138): 2-16. Illus. (E/T)

1499 **Hummel, F.C. (1983).** Trees in the evolution of the European landscape. *In* Arboricultural Association & Int. Assoc. Arboriculture. Trees in the 21st century. Berkhamstead, AB Academic Publishers. 23-33. (T)

1500 **IUCN (1976).** A new phase for Europe's plants. *IUCN Bull., n.s.*, 7(10): 55. (T)

1501 **IUCN Conservation Monitoring Centre (1982).** Threatened plants, amphibians and reptiles, and mammals (excluding marine species and bats) of the European Economic Community. IUCN Conservation Monitoring Centre. c. 400p. Species conservation data-sheets. (L/P/T)

1502 **IUCN Threatened Plants Committee Secretariat (1977).** List of rare, threatened and endemic plants in Europe. Nature and Environment Series No. 14. Strasbourg, Council of Europe. iv, 286p. (T)

1503 **IUCN Threatened Plants Committee Secretariat (1979).** The botanic gardens list of rare and threatened species of Europe recorded in cultivation. Kew, IUCN Botanic Gardens Co-ordinating Body Report No. 1. (G/T)

1504 **IUCN Threatened Plants Unit (1983).** List of rare, threatened and endemic plants in Europe, 2nd Ed. Nature and Environment Series No. 27. Strasbourg, Council of Europe. 357p. Lists over 2000 threatened vascular taxa, with IUCN conservation categories. (L/T)

1505 **IUCN Threatened Plants Unit (1984).** The botanic gardens list of rare and threatened species of Europe recorded in cultivation (update). Kew, IUCN Botanic Gardens Co-ordinating Body Report No. 12. Updated version of IUCN Botanic Gardens Co-ordinating Body Report No. 1, 1979. (G/T)

1506 **Jenik, J. (1984).** Man-made biosphere reserves? Re-examination of criteria. *In* Unesco-UNEP. Conservation, science and society. Paris, Unesco. 214-220. Contributions to the First International Biosphere Reserve Congress, Minsk, Byelorussia/USSR, 26 September - 2 October 1983. Maps. (P)

1507 **Jordanov, D., et al., eds (1975).** Problems of the Balkan flora and vegetation. Sofia, Bulgarian Academy of Sciences. Proceedings of the First International Symposium on Balkan Flora and Vegetation, Vorna, 7-14 June 1973. 441p. (T)

1508 **Klemm, C. de (1975).** The integrated management of the European wildlife heritage. Nature and Environment Series No. 9. Strasbourg, Council of Europe. 40p.

1509 **Knapp, R. (1971).** Influence of indigenous animals on the dynamics of vegetation in conservation areas. *In* Duffey, E., Watt, A.S., *eds.* The scientific management of animal and plant communities for conservation. The 11th Symposium of the BES, University of East Anglia, Norwich, 7-9 July 1970. Oxford, Blackwell. 387-390. (P)

1510 **Koester, V. (1980).** Nordic countries' legislation on the environment with special emphasis on conservation: a survey. Gland, IUCN. 44p. (L)

1511 **Kornas, J. (1976).** Decline of the European flora - facts, comments and forecasts. *Phytocoenosis* (Warsaw), 5: 173-185. (T)

1512 **Kornas, J. (1983).** Man's impact upon the flora and vegetation in Central Europe. *Geobotany*, 1983(5): 277-286. (T)

1513 **Kovanda, M. (1982).** *Dianthus gratianopolitanus*: variability, differentiation, relationships. *Preslia*, 54: 223-242. (T)

EUROPE

1514 **Labatut, P. (1977).** Protection des orchidees d'Europe? connais pas! *Orchidophile*, 28: 792. Fr. (T)

1515 **Lauriola, P.R. (1979).** Mediterranean landscapes. *Naturopa*, 33: 21-23. Illus. Protection through planning.

1516 **Lazare, J.-J., Miradalles, J., Villar, L. (1987).** *Cypripedium calceolus* L. (Orchidaceae) en el Pirineo. *An. Jardin Bot. Madrid*, 43(2): 375-382. Sp (En). Illus., map. Rediscovery in Pyrenees; distribution data. (T)

1517 **Leon, C. (1981).** Endangered species within the European Economic Community. *Threatened Pl. Commit. Newsl.*, 8: 16. (T)

1518 **Leon, C. (1982).** Council of Europe's Berne Convention comes into force. *Threatened Pl. Commit. Newsl.*, 10: 6-7. (L/T)

1519 **Leon, C.J., Lucas, G.L. (1982).** Can we preserve our flora? *Naturopa*, no. 42: 21. Illus. (L/T)

1520 **Lichtenthaler, H.K., Buschmann, C. (1984).** Das Waldsterben aus botanischer Sicht: Verlauf, Ursachen und Massnahmen. (Forest dying from botanical point of view.) Karlsruhe Braun, 80p. Ge. Illus., col. illus. Effects of air pollution on forests of Central Europe.

1521 **Loening, U.E. (1982).** Forests in developing countries: role of the European Community. *Conserv. News*, 83: 16.

1522 **Lucas, G.Ll. (1980).** We can do it. *Naturopa*, 34/35: 21. Col. illus. Actions needed to save threatened plants and habitats. (T)

1523 **Lucas, G.Ll. (1982).** Extension of the work of the Threatened Plants Committee (TPC). *In* Heywood, V.H., Clark, R.B., eds. Taxonomy in Europe. Amsterdam, North Holland. 17-18. (T)

1524 **Materna, J. (1983).** Bewirtschaftung des immissionsgeschadigten und - gefahrdeten Fichtenwaldes. (Management of spruce forests harmed and endangered by emissions [*Picea*].) *Forstwiss. Forsch. Beih. Fortswiss. Zentralbl.*, 1983(38): 44-46. Ge.

1525 **Matthews, J.D. (1983).** Replenishing the world's forests: the British Isles and Europe. *Commonw. For. Rev.*, 62(3): 179-180 (Fr, Sp). (P/T)

1526 **Moberg, R. (1988).** [Aspects of protecting cryptogams]. *Sven. Bot. Tidskr.*, 82(6): 401-402. Sw (En). (T)

1527 **Morandini, R. (1977).** Problems of conservation, management and regeneration of Mediterranean forests: research priorities. *MAB Technical Notes*, 2: 73-79.

1528 **Neuhausl, R. (1984).** Umweltgemasse naturliche Vegetation, ihre Kartierung und Nutzung fur den Umweltschutz. (Natural vegetation consistent with the environmental conditions, its mapping and application for environmental protection.) *Preslia*, 56(3): 205-212. Ge (Cz, En). Illus.

1529 **Newbold, C. (1984).** Les zones humides. *In* Council of Europe. Conservation de la Nature en Europe. Strasbourg. 34-35. Fr. (P/T)

1530 **Nie, H.W. de (1987).** The decrease in aquatic vegetation in Europe and its consequences for fish populations. EIFAC Occas. Paper No. 19. Rome, European Inland Fisheries Advisory Commission, FAO. 52p. Illus.

1531 **Nilsson, J. (1986).** So much will nature stand. *Acid News*, 3-4: 9-11. Illus. Tolerance levels of forests to pollution levels.

1532 **Noirfalise, A. (1978).** Protection et restoration des forets en Europe. *Bull. Soc. Roy. Forest. Belgique, n.s.*, 85(3): 123-133. Fr.

EUROPE

1533 **Noirfalise, A. (1979).** Man's use of the forest. *Naturopa*, 33: 26-28. Illus.

1534 **Nusslein, F. (1978).** The economic value of wildlife production in forests and woodlands of Europe. *In* Proceedings of the 7th World Forestry Congress, Buenos Aires, 4-18 October 1972, vol. 3: Conservation and recreation. Buenos Aires, Instituto Forestal Nacional. 3967-3987. En, Sp, Fr. (E)

1535 **Offner, H. (1976).** Am Rande vermerkt: Eindrucke von der Vorbereitung und Durchfuhrung der dritten Generalversammlung der Foderation der Natur- und Nationalparke Europas vom 23. bis 27. September 1976 in Plitvice. *Naturschutz- und Naturparke*, 83: 15-20. Ge. Illus. (P)

1536 **Offner, H. (1976).** Naturparke in Europa. *Naturschutz- und Naturparke*, 80: 5-11. Illus. (P)

1537 **Offner, H. (1977).** Europaische Zusammenarbeit in Natur- und Nationalparken. *Naturschutz- und Naturparke*, 86: 37-39. (P)

1538 **Oldfield, S. (1984).** EEC ruling on CITES. *Threatened Pl. Newsl.*, 13: 17. Trade. (L/T)

1539 **Oldfield, S. (1984).** The conservation of European orchids. *Orchid Rev.*, 92(1090): 255. (T)

1540 **Oldfield, S. (1987).** Mires in danger in Western Europe. *Oryx*, 21(4): 229-232. Illus. (T)

1541 **Olivier, L. (1984).** La reintroduction de plantes indigenes disparues ou menacees de disparition en Region Mediterraneenne un des vecteurs de la strategie conservatoire. *Compt. Rend. Seanc. Soc. Biogeogr.*, 60(1): 51-66. Fr (En). (T)

1542 **Ozenda, P. (1984).** La carte de la vegetation. *In* Council of Europe. Conservation de la Nature en Europe. Strasbourg, Council of Europe. 40-41. Fr. (P/T)

1543 **Ozenda, P.G. (1982).** The vegetation map. *Naturopa*, 42: 22-23. Illus.

1544 **Pahlsson, L. (1979).** Basic classification for nature conservation in the Nordic countries. *In* Hedberg, I., ed. Systematic botany, plant utilization and biosphere conservation. Stockholm, Almqvist & Wiksell International. 132-136. Map.

1545 **Pahlsson, L. (1984).** Reference areas with representative types of nature in the nordic countries, and the proposed ECE system of representative ecological areas. *In* Unesco-UNEP. Conservation, science and society. Paris, Unesco. 233-241. Contributions to the First International Biosphere Reserve Congress, Minsk, Byelorussia/USSR. 26 September - 2 October 1983. Maps, diagrams. (P)

1546 **Pearce, F. (1986).** The strange death of Europe's trees. *New Sci.*, 112(1537): 41-45. Illus., map.

1547 **Pearce, F. (1986).** Unravelling a century of acid pollution. *New Sci.*, 111(1527): 23-24. Illus. Effects of acid rain on forests and lakes.

1548 **Pignatti, S. (1983).** Human impact in the vegetation of the Mediterranean basin. *Geobotany*, 1983(5): 151-161. Illus.

1549 **Pigott, C.D. (1981).** The status, ecology and conservation of *Tilia platyphyllos* in Britain. *In* Synge, H., ed. The biological aspects of rare plant conservation. Chichester, Wiley. 305-317. Illus., maps. Proceedings of International Conference, King's College, Cambridge, 14-19 July 1980. (T)

1550 **Pitschmann, H. (1978).** Mountain flora: a rupture of nature. *Naturopa*, 31: 23-25. Illus.

1551 **Plas-Haarsma, M. van der (1987).** Cyclamen in trade. *TRAFFIC Bull.*, 9(1): 11. Includes trade statistics for 12 species. (T)

EUROPE

1552 **Poore, D., Gryn-Ambroes, P. (1980).** Nature conservation in Northern and Western Europe. Gland, Switzerland, UNEP, IUCN, WWF. 408p.

1553 **Postel, S. (1984).** Air pollution, acid rain and the future of forests. 5. *Amer. Forests*, 90(11): 46-47. Pollution.

1554 **Postel, S. (1987).** New summary: ten million hectares of forest damaged. *Acid News*, 3-4: 16-18. Extent of forest damage caused by acid rain.

1555 **Pysek, A. (1987).** Prvni zkusenosti s pestovanim ohrozeneho drhu lebedy ruzoue (*Atriplex rosea* L.). (Erste Erfahrungen mit dem Anbau der bedrohten Arb Rosea Melde (*Atriplex rosea* L.) *Zpravy Muz. Zapad. Kraje, Prir.*, 34-35: 19-22. Cz (Ge).

1556 **Quezel, P. (1977).** Forests of the Mediterranean basin. *MAB Technical Notes*, 2: 9-29. Includes protection. (P)

1557 **Quezel, P. (1982).** The Mediterranean maquis. *Naturopa*, 42: 15. Illus. (T)

1558 **Quezel, P. (1984).** A system of terrestrial biosphere reserves for the Mediterranean. *In* Unesco-UNEP. Conservation, science and society. Paris, Unesco. 23-32. Contributions to the First International Biosphere Reserve Congress, Minsk, Byelorussia/USSR, 26 September - 2 October 1983. Tables. (P)

1559 **Quezel, P. (1984).** Le maquis mediterraneen. *In* Council of Europe. Conservation de la Nature en Europe. Strasbourg, Council of Europe. 24. Fr. (P/T)

1560 **Rabbinge, R., Loomis, R.S. (1979).** There is another way. *Naturopa*, 33: 32-34. Illus.

1561 **Ramin, I. von (1979).** Botanic gardens and the conservation of European orchids. *In* Synge, H., Townsend, H., eds. Survival or extinction. Proceedings of a conference held at the Royal Botanic Gardens, Kew, 11-17 September 1978. Kew, Bentham-Moxon Trust. 207-210. (G)

1562 **Read, M. (1989).** Grown in Holland? Brighton, Fauna and Flora Preservation Society. 12p. Illus., maps. European bulb trade. (E/G/L/T)

1563 **Richard, J.L. (1978).** Le role de la foret et du forestier dans la protection de la flore. *Mitt. Eidgenoss. Anstalt Forstl. Versuchsw.*, 54(4): 411-417. Fr (En, Ge, It). (T)

1564 **Riedel, W., Heydemann, B. (1980).** Schutz von Flora und Fauna und ihrer naturlichen Lebensraume. Sankelmark, Akademic Sankelmark. 136p. Ge.

1565 **Romagnesi, H. (1980).** Quelques especes rares ou nouvelles de Macromycetes. VIII. Russulacees. *Bull. Trimest. Soc. Mycol. Fr. Paris, La Societe.*, 96(3): 297-314. (T)

1566 **Rose, C., Neville, M. (1985).** Tree dieback survey. London, Friends of the Earth. 28p. Col. illus. Maps. Pub. undated, but presumed 1985. (T)

1567 **Rosencranz, A. (1986).** The acid rain controversy in Europe and North America: a political analysis. *Ambio*, 15(1): 19-21.

1568 **Runge, F. (1988).** Anderungen der Flora des Naturachutzgebietes 'Huronensee' bei Munster wahrend der letzen 62 Jahre. *Flor. Rundbr.*, 21(2): 95-97. Ge (En).

1569 **Scheifele, M. (1979).** Europe's forests. *Naturopa*, 33: 24-25. Illus.

1570 **Schutt, P. (1978).** Die gegenwartige Epidemie des Tannensterbens: ihre geographische Verbreitung im nordlichen Teil des naturlichen Areals von *Abies alba*. (The present epidemic of fir decline: its geographical distribution in the northern part of the natural range of *Abies alba*.) *Europ. J. Forest Path.*, 8(3): 187-190. (T)

1571 **Schutt, P. (1983).** So stirbt der Wald: Schadbilder und Krankheitsverlauf. (Thus dies the forest.) Munchen, BLV Verlagsgesellschaft. 95p. Ge (En). Col. illus., maps. Air pollution.

EUROPE

1572 **Sealey, J.L. (1981).** Cyclamen for western gardens. *Pac. Hort.*, 42(4): 21-26. Illus., col. illus. (G/T)

1573 **Seidenfaden, G. (1984).** La convention relative a la conservation de la vie sauvage et du milieu naturel de l'Europe. *In* Council of Europe. Conservation de la Nature en Europe. Strasbourg, Council of Europe. 50-53. Fr. Berne Convention. (L/P/T)

1574 **Seybold, S. (1974).** Das Red Data Book und die Roten Listen. *Mitteilungsbl. Arbeitskr. Heim. Orch. Baden-Wurttemberg*, 6(2): 10-13. Ge. (T)

1575 **Shrenk, W.J. (1981).** Rare orchids in western Europe: 1. Vanishing beauties and inconspicuous ducklings. *Amer. Orchid Soc. Bull.*, 50(10): 1185-1194. Illus., col. illus. (T)

1576 **Stangl, J., Veselsky, J. (1978).** *Inocybe appendiculata* Kuhner (Beitrage zur Kenntnis seltenerer Inocyben: 13.) *Ceska Mykol.*, 32(3): 161-166. Ge (Cz). Illus.

1577 **Stewart, J. (1989).** The conservation of European orchids. Strasbourg, Council of Europe. 63p. (T)

1578 **Synge, H. (1980).** Endangered monocotyledons in Europe and South West Asia. *In* Brickell, C.D., Cutler, D.F., Gregory, M., *eds.* Petaloid monocotyledons: horticultural and botanical research. Linnean Society Symposium Series No. 8. London, Academic Press. 199-206. Illus. (T)

1579 **Tendron, G., et al. (1984).** L'Environnement et la foret. Droit et economie de l'environnement. *Publications specialisees.* Trevoux: 9-10. Fr. (T)

1580 **Tendron, G., et al. (1984).** Les recommendations et les chartes. *In* Council of Europe. Conservation de la nature en Europe. Vingt annees d'activites. Strasbourg, Council of Europe. 48-49. Fr. (P/T)

1581 **Terrasson, F. (1984).** The image of nature, a determinant for strategy and action. Gland, IUCN Commission on Environmental Planning. 14p. (T)

1582 **Terrasson, F., Tendron, G. (1975).** Evolution and conservation of hedgerow landscapes in Europe. Nature and Environment Series No. 8. Strasbourg, Council of Europe. ii, 43p. Illus., maps.

1583 **Terrasson, M.F. (1984).** La bocage europeen. *In* Council of Europe. Conservation de la nature en Europe. Vingt annees d'activites. Strasbourg, Council of Europe. 23. Fr. (P/T)

1584 **Theunert, R. (1988).** Bemerkugen Zuden Gefasspflanzen der Vohrumer Torgrube. *Flor. Rundbr.*, 21(2): 107-109. Ge (En).

1585 **Thirgood, J.V. (1981).** Man and the Mediterranean forest. A history of resource depletion. London, Academic Press. 194p.

1586 **Tigges, M. (1981).** Grundzuge eines internationalen Artenschutzprogrammes. *Beih. Veroff. Natursch. Landschaftspfl. Baden-Wurttemberg*, 19: 158-189. Ge.

1587 **Tomaselli, R. (1977).** Degradation of the Mediterranean maquis. *MAB Technical Notes*, 2: 33-72. Maps.

1588 **Tooze, S. (1984).** European forests. Pollution, pathogens and pests. *Nature* 307, 12 Jan.: 97. (T)

1589 **Townsend, P. (1979).** Study tour of European 'natur' parks. Castleton, Peak National Park Study Centre. 106p. Illus., maps. (P)

1590 **Trautmann, W. (1984).** Les plantes vasculaires menacees. *In* Council of Europe. Conservation de la nature en Europe. Strasbourg, Council of Europe. 20-22. Fr. Table. (T)

EUROPE

1591 **Trautmann, W. (1984).** Prerequisites for a representative network of biosphere reserves for Europe. *In* Unesco-UNEP. Conservation, science and society. Paris, Unesco. 20-22. Contributions to the First International Biosphere Reserve Congress, Minsk, Byelorussia/USSR, 26 September - 2 October 1983. (P)

1592 **Unesco (1974).** Expert panel on Project 2. Ecological effects of different land uses and management practices on temperate and mediterranean forest landscapes. Final report. Programme on Man and the Biosphere [MAB] Report Series No. 19. Paris, Unesco. 80p.

1593 **Unesco (1977).** Mediterranean forests and maquis: ecology, conservation and management. *MAB Technical Notes*, 2: 79p.

1594 **Unesco (1979).** Programme on man and the biosphere. Workshop on biosphere reserves in the Mediterranean region: development of a conceptual basis and a plan for the establishment of a regional network. Final report. MAB Report No. 45. Paris, Unesco. 62p. (P)

1595 **Vanesse, R. (1984).** Les landes a bruyere. *In* Council of Europe. Conservation de la Nature en Europe: vingt annees d'activites. Strasbourg, Council of Europe. 26. Fr. Heathlands. (P/T)

1596 **Vent, W. (1982).** Zur Dendroflora der Nordhemisphare. *Gleditschia*, 9: 57-76 (1981 publ. 1982). Ge (En). (T)

1597 **Veronesi, P. (1980).** Ora di conservazione della Natura se ne parla anche al Parlamento Europeo. *Nat. Montagna*, 27(3): 243-245. It.

1598 **Walters, S.M. (1971).** Index to the rare, endemic vascular plants of Europe. *Boissiera*, 19: 87-89. (T)

1599 **Walters, S.M. (1976).** The conservation of threatened vascular plants in Europe. *Biol. Conserv.*, 10: 31-41. (T)

1600 **Walters, S.M. (1977).** The role of European botanic gardens in the conservation of rare and threatened plant species. *Gartn.-Bot. Brief*, 51: 2-20. (G/T)

1601 **Walters, S.M. (1979).** Conservation of the European flora: *Aldrovanda vesiculosa* L., a documented case-history of a threatened species. *In* Hedberg, I., *ed.* Systematic botany, plant utilization and biosphere conservation. Stockholm, Almqvist & Wiksell International. 76-82. Maps. (T)

1602 **Walters, S.M. (1979).** The role of Mediterranean botanic gardens in plant conservation. *Webbia*, 34(1): 109-116. (It). (G/T)

1603 **Walters, S.M. (1981).** Aims and methods in mapping the endemic vascular plants of the Balkan peninsula. *In* Mapping the Flora of the Balkan Peninsula. Sofia. (T)

1604 **Waycott, A. (1983).** National parks of Western Europe. Southampton, Inklon Publications. 159p. Illus., maps. (P)

1605 **Westhoff, V. (1978).** No biotype no protection. *Naturopa*, no. 31: 7-9. Illus.

1606 **Wirth, H., ed. (1981).** Nature reserves in Europe. London, Jupiter Books. 330p. Illus., col. illus. (P)

1607 **Woike, S. (1969).** Beitrag zum Vorkommen von *Coleanthus subtilis* (Tratt.) Seidl (Feines Scheidenblutgras) in Europa. *Folia Geobot.Phytotax.*, 402-413. Ge. Map (T)

1608 **Wolkinger, F. (1982).** Dry grasslands. *Naturopa*, 42: 18-19. Illus. (T)

1609 **Wood, C. (1982).** Implications of the European EIA initiative. *Ecos*, 3(3): 25-29. Environmental impact assessment. (L)

1610 **Wood, J., Clements, M., Muir, H. (1984).** Plants in peril. 2. *Kew Mag.*, 1(3): 139-142. Illus. *Cypripedium calceolus*. (T)

EUROPE

1611 **Wood, J.J. (1985).** Orchids of Europe: *Orchis italica* Poir. & *Orchis spitzellii* Sauter ex W. Koch. *Orchid Digest*, 49(3): 113-116. Illus. (T)

1612 **World Conservation Monitoring Centre (1989).** European plants in peril. A review of threatened plants in the European Community. Cambridge, WCMC. 53p. Illus., maps. Data sheets and distribution maps for selected species, including lower plants. (T)

1613 **Yon, D. (1980).** Evolution des forets alluviales en Europe. Facteurs de destruction et elements strategiques de conservation. *In* La vegetation des forets alluviales. IX Colloques Phytosociologiques. Strasbourg, Vaduz. Fr. Illus. (T)

1614 **Yon, D. (1982).** A natural asset by the waterside: alluvial forests. *Naturopa*, 42: 20. Col. illus. (T)

1615 **Yon, M.D. (1984).** Les forets alluviales. *In* Council of Europe. Conservation de la Nature en Europe. Strasbourg, Council of Europe. 29-30. Fr. Map. (P/T)

1616 **Ziegler, H. (1988).** Deterioration of forests in Central Europe. *In* Greuter, W., Zimmer, B., *eds.* Proceedings of the XIV International Botanical Congress, Berlin, 24 July - 1 August 1987. 423-444. Illus., maps. (T)

INDIAN OCEAN ISLANDS

1617 **Bliss-Guest, P. (1983).** Environmental stress in the East African Region. *Ambio*, 12(6): 290-295. Col. illus, maps.

1618 **Carlquist, S. (1965).** Island life: a natural history of the islands of the world. New York, Natural History Press. 451p. Origin, evolution and adaptations of island flora and fauna; Galapagos and Hawaiian Islands well covered.

1619 **Commonwealth Science Council (1988).** Biological diversity and genetic resources techniques and methods: conservation biology. Executive summary. Regional training workshop held at Royal Botanic Gardens, Peradeniya, Sri Lanka, 16-21 May 1988. London, Commonwealth Science Council. 22p. (E)

1620 **Finn, D. (1983).** Land use and abuse in the East African region. *Ambio*, 12(6): 296-301. Col. illus., maps.

1621 **Fosberg, F.R., Sachet, M.-H. (1972).** Status of floras of western Indian Ocean Islands. *In* IUCN. Comptes Rendus de la Conference Internationale sur la Conservation de la Nature et de ses Ressources a Madagascar, 1970. Publications IUCN Nouvelle Serie 36. Morges, IUCN. 152-155. (T)

1622 **Hedberg, I. (1979).** Possibilities and needs for conservation of plant species and vegetation in Africa. *In* Hedberg, I., *ed.* Systematic botany, plant utilization and biosphere conservation. Stockholm, Almqvist & Wiksell International. 83-104. (T)

1623 **Heywood, V.H. (1979).** The future of island floras. In Bramwell, D., *ed.* Plants and islands. London, Academic Press. 431-441. (Sp). (P/T)

1624 **IUCN (1970).** Eleventh Technical Meeting Papers and Proceedings, New Delhi, India, Vol. II. Problems of threatened species. IUCN Publications New Series No. 18. Morges, IUCN. 132p. (T)

1625 **IUCN Botanic Gardens Conservation Secretariat (1989).** Rare and threatened plants of the western Indian Ocean Islands; *ex situ* conservation in botanic gardens. Kew, IUCN Botanic Gardens Conservation Secretariat. 25p. (G/T)

1626 **IUCN Threatened Plants Unit (1984).** The botanic gardens list of endemic species of the western Indian Ocean. Kew, IUCN Botanic Gardens Conservation Co-ordination Body Report No. 10. 13p. Lists plants in cultivation in botanic gardens. (G/T)

INDIAN OCEAN ISLANDS

1627 **Keraudren, M. (1968).** Synthese regionale. Discussion. *In* Hedberg, I. & O., *eds.* Conservation of vegetation in Africa south of the Sahara. Symposium Proceedings, at 6th plenary meeting of AETFAT, Uppsala. *Acta Phytogeogr. Suec.*, 54: 279-282. Fr. Overview of Indian Ocean islands. (P/T)

1628 **Kundaeli, J.N. (1983).** Making conservation and development compatible. *Ambio*, 12(6): 326-331. Col. illus.

1629 **Lesouef, J.-Y. (1979).** The endangered plants of the oceanic islands: their cultivation in the Stangalarc'h Conservatory (Brest). *OPTIMA Newsl.*, 819: 16. (G/T)

1630 **MacKinnon, J. & K. (1986).** Review of the protected areas system in the Indo-Malayan realm. Gland, Switzerland and Cambridge, U.K., IUCN and UNEP. 284p. Illus., maps. (P)

1631 **Melville, R. (1970).** Endangered plants and conservation in the islands of the Indian Ocean. *In* Papers and proceedings of the IUCN 11th technical meeting, New Delhi, India, 25-28 November 1969. Switzerland, IUCN. 103-107. (T)

1632 **Sachet, M.-H., Fosberg, F.R. (1971).** Island bibliographies supplement. Washington, D.C., Pacific Science Board, Nat. Academy of Sciences. 427p. Islands worldwide. (B/P/T)

MEDITERRANEAN

1633 **Anon. (1980).** Protect the Mediterranean. *IUCN Bull.*, 11(9/10): 81-83, 95. Illus.

1634 **Avishai, M. (1985).** The role of Mediterrannean botanic gardens in the maintenance of living conservation-oriented collections. *In* Gomez-Campo, C. *ed.* Plant conservation in the Mediterranean Area. Dordrecht, Dr W. Junk. 221-236. Diagram. (G/T)

1635 **Bacar, H. (1984).** A regional approach to marine and coastal protected areas: the Mediterranean Sea. *In* McNeely, J.A., Miller, K.R., *eds.* National parks, conservation and development. The role of protected areas in sustaining society. Washington, D.C., Smithsonian Institution Press. 438-441. (P)

1636 **Barbero, M., et al. (1989).** Menaces pesant sur la flore mediterraneenne francaise. *In* Chauvet, M., *ed.* Plantes sauvages menacees de France. Bilan et protection. Actes du colloque de Brest, 8-10 octobre 1987. Paris, Bureau des Ressources Genetiques. 11-21.

1637 **Baumann, H., Dafni, A., Golz, P., Greuter, W., Reinhard, H.R., Tigges, M. (1981).** OPTIMA-Project "Kartierung des Mediterranen Orchideen" 2. Orchideenforschung und Naturschutz im Mittelmeergebiet Internationales Artenschutzprogramm. *Beih. Veroff. Natursch. Landschaftspfl. Baden-Wurttemberg*, 19: 1-190. Ge (En). Col. illus., maps.

1638 **Boulos, L. (1985).** The arid eastern and south-eastern Mediterranean region. *In* Gomez-Campo, C., *ed.* Plant conservation in the Mediterranean area. Dordrecht, Dr W. Junk. 124-140. Illus. (T)

1639 **Gomez-Campo, C. (1972).** Preservation of west Mediterranean members of the cruciferous tribe Brassiceae. *Biol. Conserv.*, 4(5): 355-360.

1640 **Gomez-Campo, C. (1978).** The OPTIMA Seeds and Living Material Commission. *Gartn. Bot. Brief*, 58: 29. (Ge). Abstract. (G)

1641 **Gomez-Campo, C. (1978).** The Organization for the Phyto-Taxonomic Investigation of the Mediterranean Area (OPTIMA). *In* Synge, H., Townsend, H., *eds.* Survival or extinction. Proceedings of a conference held at the Royal Botanic Gardens, Kew, 11-17 September 1978. Kew, Bentham-Moxon Trust. 195-197.

1642 **Gomez-Campo, C. (1979).** The role of seed banks in the conservation of Mediterranean flora. *Webbia*, 34(1): 101-107. (It). (G)

MEDITERRANEAN

1643 **Gomez-Campo, C. (1981).** The 'Artemis' project for seed collection of Mediterranean endemics. *Threatened Pl. Commit. Newsl.*, 8: 9-11. (G)

1644 **Gomez-Campo, C. (1985).** Seed banks as an emergency conservation strategy. *In* Gomez-Campo, C., *ed.* Plant conservation in the Mediterranean area. Dordrecht, Netherlands, Dr W. Junk. 237-247. Illus. Diagrams. (G/T)

1645 **Gomez-Campo, C. (1985).** The conservation of Mediterranean plants: principles and problems. *In* Gomez-Campo, C., *ed.* Plant conservation in the Mediterranean area. Dordrecht, Dr W. Junk. 3-8. (T)

1646 **Gomez-Campo, C., ed. (1985).** Plant conservation in the Mediterranean area. Geobotany Series 7. Dordrecht, Dr. W. Junk Publishers. 269p. Illus., maps. (T)

1647 **Greuter, W. (1979).** Mediterranean conservation as viewed by a plant taxonomist. *Webbia*, 34(1): 87-99. (It). (T)

1648 **Heywood, V.H. (1984).** Co-operation between the IUCN Species Survival Commission and European-Mediterranean Botanic Gardens in conservation strategies. *Rep. Bot. Inst. Univ. Aarhus*, 10: 33-39. (G/T)

1649 **Heywood, V.H. (1987).** The conservation of medicinal and aromatic plants - the need for a cooperative approach. *In* Mota, M., Bafta, J., *eds.* EUCARPIA, International Symposium on Conservation of Genetic Resources of Aromatic and Medicinal Plants, Oeiras, Potugal, 1984. 25-37. (E/T)

1650 **IUCN (1959).** Animaux et vegetaux rares de la Region Mediterraneenne. (Rare animals and plants of the Mediterranean Region.) Proceedings of the IUCN 7th Technical Meeting, 11-19 September 1958, Athens, vol. 5. Brussels, IUCN. 209p. Fr (En, Sp). (T)

1651 **IUCN Conservation Monitoring Centre (1982).** Threatened plants, amphibians and reptiles, and mammals (excluding marine species and bats) of the European Economic Community. IUCN Conservation Monitoring Centre. c. 400p. Species conservation data-sheets. (L/P/T)

1652 **IUCN Threatened Plants Committee Secretariat (1977).** List of rare, threatened and endemic plants in Europe. Nature and Environment Series No. 14. Strasbourg, Council of Europe. iv, 286p. (T)

1653 **Lauriola, P.R. (1979).** Mediterranean landscapes. *Naturopa*, 33: 21-23. Illus. Protection through planning.

1654 **Le Houerou, H.N. (1981).** Impact of man and his animals on mediterranean vegetation. *In* Castri, F. di, Goodall, D.W., Specht, R.L., *eds.* Ecosystems of the world. 11. Mediterranean-type shrublands. Amsterdam, Elsevier. 479-517.

1655 **Leon, C., Lucas, G., Synge, H. (1985).** The value of information in saving threatened Mediterranean plants. *In* Gomez-Campo, C., *ed.* Plant conservation in the Mediterranean area. Dordrecht, Dr W. Junk. 177-196. (T)

1656 **Malato-Beliz, J. (1976).** Relations entre agriculture et conservation de la vegetation naturelle dans la region Mediterraneene. *Collana Verde*, 39: 269-290. Fr.

1657 **Mooney, H.A. (1988).** Lessons from Mediterranean-climate regions. *In* Wilson, E.O., *ed.* Biodiversity. Washington, D.C., National Academy of Science Press. 157-165. Discussion of biodiversity of tropical and temperate zones.

1658 **Morandini, R. (1977).** Problems of conservation, management and regeneration of Mediterranean forests: research priorities. *MAB Technical Notes*, 2: 73-79.

1659 **Noirfalise, A. (1979).** Man's use of the forest. *Naturopa*, 33: 26-28. Illus.

1660 **Noy-Meir, Anikster, Y., Waldman, M., Ashri, A. (1988).** Population dynamics research for *in situ* conservation: wild wheat in Israel. *Pl. Genet. Resource. Newsl.*, 75/76: 9-ll. Illus. *Triticum dicoccoides*. (E)

MEDITERRANEAN

1661 **Olivier, L. (1979).** Multiplication and re-introduction of threatened species of the littoral dunes in Mediterranean France. *In* Synge, H., Townsend, H., *eds*. Survival or extinction. Proceedings of a conference held at the Royal Botanic Gardens, Kew, 11-17 September 1978. Kew, Bentham-Moxon Trust. 91-93. (G/T)

1662 **Olivier, L. (1984).** La reintroduction de plantes indigenes disparues ou menacees de disparition en Region Mediterraneenne un des vecteurs de la strategie conservatoire. *Compt. Rend. Seanc. Soc. Biogeogr.*, 60(1): 51-66. Fr (En). (T)

1663 **Ozenda, P. (1984).** La carte de la vegetation. *In* Council of Europe. Conservation de la Nature en Europe. Strasbourg, Council of Europe. 40-41. Fr. (P/T)

1664 **Pignatti, S. (1983).** Human impact in the vegetation of the Mediterranean basin. *Geobotany*, 1983(5): 151-161. Illus.

1665 **Pons, A., Quezel, P. (1985).** The history of the flora and vegetation and past and present human disturbance in the Mediterranean. *In* Gomez-Campo, C., *ed*. Plant conservation in the Mediterranean Area. Dordrecht, Dr W. Junk. 25-43. Diagrams. (T)

1666 **Quezel, P. (1977).** Forests of the Mediterranean basin. *MAB Technical Notes*, 2: 9-29. Includes protection. (P)

1667 **Quezel, P. (1982).** The Mediterranean maquis. *Naturopa*, 42: 15. Illus. (T)

1668 **Quezel, P. (1984).** A system of terrestrial biosphere reserves for the Mediterranean. *In* Unesco-UNEP. Conservation, science and society. Paris, Unesco. 23-32. Contributions to the First International Biosphere Reserve Congress, Minsk, Byelorussia/USSR, 26 September - 2 October 1983. Tables. (P)

1669 **Quezel, P. (1984).** Le maquis mediterraneen. *In* Council of Europe. Conservation de la Nature en Europe. Strasbourg, Council of Europe. 24. Fr. (P/T)

1670 **Quezel, P. (1985).** Definition of the Mediterranean area and the origin of its flora. *In* Gomez-Campo, C., *ed*. Plant conservation in the Mediterranean Area. Dordrecht, Dr W. Junk. 9-24. Maps. Tables. (T)

1671 **Quezel, P., Barbero, M. (1985).** Carte de la vegetation potentielle de la Region Mediterraneenne. Feuille no. 1: Mediterrannee Orientale. Paris, Editions du CNRS. Fr. Includes notes on vegetation zones. (P)

1672 **Ruiz de la Torre, J. (1985).** Conservation of plant species within their native ecosystems. *In* Gomez-Campo, C. *ed*. Plant conservation in the Mediterranean area. Dordrecht, Dr W. Junk. 197-219.

1673 **Snogerup, S. (1985).** The Mediterranean islands. *In* Gomez-Campo, C., *ed*. Plant conservation in the Mediterranean Area. Dordrecht, Dr W. Junk. 160-173. Illus. Map. (T)

1674 **Synge, H. (1980).** Endangered monocotyledons in Europe and South West Asia. *In* Brickell, C.D., Cutler, D.F., Gregory, M., *eds*. Petaloid monocotyledons: horticultural and botanical research. Linnean Society Symposium Series No. 8. London, Academic Press. 199-206. Illus. (T)

1675 **Thirgood, J.V. (1981).** Man and the Mediterranean forest. A history of resource depletion. London, Academic Press. 194p.

1676 **Tomaselli, R. (1977).** Degradation of the Mediterranean maquis. *MAB Technical Notes*, 2: 33-72. Maps.

1677 **Torre, J. Ruiz de la (1985).** Conservation of plant species in their native ecosystems. *In* Gomez-Campo, C., *ed*. Plant conservation in the Mediterranean area. Dordrecht, Dr W. Junk. 197-219. Diagrams. (T)

MEDITERRANEAN

1678 **Unesco (1974).** Expert panel on Project 2. Ecological effects of different land uses and management practices on temperate and mediterranean forest landscapes. Final report. Programme on Man and the Biosphere [MAB] Report Series No. 19. Paris, Unesco. 80p.

1679 **Unesco (1977).** Mediterranean forests and maquis: ecology, conservation and management. *MAB Technical Notes*, 2: 79p.

1680 **Unesco (1979).** Programme on man and the biosphere. Workshop on biosphere reserves in the Mediterrranean region: development of a conceptual basis and a plan for the establishment of a regional network. Final report. MAB Report No. 45. Paris, Unesco. 62p. (P)

1681 **Walters, S.M. (1979).** The role of Mediterranean botanic gardens in plant conservation. *Webbia*, 34(1): 109-116. (It). (G/T)

1682 **Walters, S.M. (1981).** Aims and methods in mapping the endemic vascular plants of the Balkan peninsula. *In* Mapping the Flora of the Balkan Peninsula. Sofia. (T)

MIDDLE ASIA, INDOCHINA AND JAPAN

1683 **Anon. (1980).** Dr. Ashton returns from the endangered tropical forests of the Far East. *Plant Sciences, Newsl. Arnold Arbor.*, Fall/Winter 1980: 2-3. Illus. (T)

1684 **Anon. (1987).** Blue Vanda. Conservation of a rare orchid through micropropagation. *Bot. Gard. Conserv. News*, 1(1): 47-49. Illus. (G/T)

1685 **Awasthi, D.D. (1982).** Conservation and economic importance of Himalayan flora. *In* Paliwal, G.S., *ed.* The vegetational wealth of the Himalayas. Papers of the symposium at the Dept of Botany, Univ. of Garhwal, 1-6 October 1979. Delhi, Puja Publishers. 450-453. Illus. (E)

1686 **Bahaguna, S. (1989).** Chipko movement. The role of the local communities in upland conservation. *Forest News*, 2(4): 5-7. Supplement to *Tigerpaper*, 15(4).

1687 **Bahuguna, S.L. (1982).** Chipko Andolan: a novel movement. *In* Paliwal, G.S., *ed.* The vegetational wealth of the Himalayas. Papers of the symposium at the Dept of Botany, Univ. of Garhwal, 1-6 October 1979. Delhi, Puja Publishers. 3-11. Illus. Himalayas.

1688 **Baitulin, I.O., Bykov, B.A., Roldugin, I.I. (1985).** K organizatsii Zailiskogo prirodnogo natsional'nogo parka v severnom Tyan'-Shane. (On the organization of the Zailisk National Nature Park in North Tyan' Shan'.) *Izv. Akad. Nauk Kaz. SSR, Biol.*, 6: 10-13. Rus (Ka). (P)

1689 **Bhattacharyya, U.C., Malhotra, C.L. (1984).** The danger of vanishing orchid species in north west Himalaya. *In* Paliwal, G.S., *ed.* The vegetational wealth of the Himalayas. Delhi. 249-255. (T)

1690 **Bisht, N.S. (1982).** Problems of forest conservation in Himalayas. *In* Paliwal, G.S., *ed.* The vegetational wealth of the Himalayas. Papers of the symposium at the Dept of Botany, Univ. of Garhwal, 1-6 October 1979. Delhi, Puja Publishers. 417-421. Illus.

1691 **Chowdhery, H.J., Wadhwa, B.M. (1983).** Is *Microgynoceium tibeticum* Hook.f. (Chenopodiaceae) a rare species? *Indian J. Forestry*, 6(3): 248-249. Illus. Discussion on distribution and status of *Microgynoecium tibeticum*. (T)

1692 **Commonwealth Science Council (1988).** Biological diversity and genetic resources techniques and methods: conservation biology. Executive summary. Regional training workshop held at Royal Botanic Gardens, Peradeniya, Sri Lanka, 16-21 May 1988. London, Commonwealth Science Council. 22p. (E)

1693 **Davis, P.H., Harper, P.C., Hedge, I.C., eds. (1971).** Plant life of South-West Asia. Edinburgh, Botanical Society of Edinburgh. x, 335p.

MIDDLE ASIA, INDOCHINA AND JAPAN

1694 **Davis, S. (1985).** Conserving S.E. Asian plant genetic resources. *Threatened Pl. Newsl.*, 15: 8-9. Review of an International Symposium held in Jakarta on 20-24 August 1985. Use of wild relatives in breeding of *Lycopersicon, Citrus* and *Mangifera.* (E/P)

1695 **Dzharmaganbetov, T.Zh. (1984).** K okhrane redkikh i nakhodyashchikhcya pod ugrozoi ischeznoveniya vidov rastenii v Gur'evskoi oblasti. *In* Izuch. i okhrana zapoved. ob'ektov. Alma-Ata. 84-85. Rus. Note in *Ref. Zhurn., Biol.*, 4(2): V736 (1985).

1696 **ESCAP (1985).** State of the environment in Asia and the Pacific. Vol. 1: Summary. Bangkok, ESCAP.

1697 **FAO Regional Office for Asia and the Pacific (1985).** Dipterocarps of South Asia. Bangkok, FAO. 321p. (Rapa Monograph 1985/4). (E/T)

1698 **FAO/UNEP (1981).** Tropical Forest Resources Assessment Project (in the framework of GEMS). Forest resources of tropical Asia. Part 1: Regional synthesis. Rome, FAO. 475p. En, Fr. Tropical forests; deforestation.

1699 **FAO/UNEP (1981).** Tropical Forest Resources Assessment Project (in the framework of GEMS). Forest resources of tropical Asia. Part II: Country briefs. Rome, FAO. 586p. En, Fr. Tropical forests; deforestation.

1700 **Frey, W., Kurschner, H. (1985).** Final report of the section of botany, supporting period 1974-1985. Tubingen atlas of the Near and Middle East. Tubingen University, Special Research Division 19. Berlin. 28p. Maps.

1701 **Griggs, T. (1981).** The green gene banks of Asia. *Asia 2000* (Hong Kong), 1(3): 24-27. (G)

1702 **Gupta, K.M., Bandhu, D., eds (1979).** Man and forests: a new dimension in the Himalaya. New Delhi, Today and Tomorrow's Publishers.

1703 **Gupta, R., Sethi, K.L. (1983).** Conservation of medicinal plants resources in the Himalayan region. *In* Jain, S.K., Mehra, K.L., *eds.* Conservation of tropical plant resources. Howrah, Botanical Survey of India. 101-109. (E)

1704 **Hamilton, L.S., ed. (1983).** Forest and watershed development in Asia and the Pacific. London, Westview. 560p.

1705 **Hepper, F.N. (1977).** The practical importance of plant ecology in arid zones. *In* Dalby, D., Harrison Church, R.J., Bezzaz, F., *eds.* Drought in Africa 2. London, International African Institute. 105-106.

1706 **Holloway, C.W. (1976).** Conservation of threatened vertebrates and plant communities in the Middle East and South West Asia. *In* Proceedings of an International Meeting on Ecological Guidelines for the use of Natural Resources in the Middle East and South West Asia, Persepolis, Iran, 24-30 May 1975. IUCN Publ. New Series No. 34. Morges, IUCN. 179-188. (T)

1707 **Holloway, R. (1989).** Doing development: governments, NGOs and the rural poor in Asia. London, Earthscan. 192p.

1708 **IUCN (1976).** Proceedings of an International Meeting on Ecological Guidelines for the use of Natural Resources in the Middle East and South West Asia, held at Persepolis, Iran, 24-30 May 1975. IUCN Publications New Series No. 34. Morges, IUCN. 231p. Maps.

1709 **IUCN Commission on National Parks and Protected Areas (1985).** The Corbett Action Plan for protected areas of the Indomalayan Realm. *In* Thorsell, J.W., *ed.* Conserving Asia's natural heritage. The planning and management of Protected Areas in the Indomalayan Realm. Proc. of the 25th Working Session of IUCN CNPPA. Gland, IUCN. 219-242. (P)

MIDDLE ASIA, INDOCHINA AND JAPAN

1710 **Ives, J.D. (1981).** Applied mountain geoecology. *In* Lall, J.S., *ed.* The Himalaya: aspects of change. New Delhi, India International Centre, Oxford University Press. 377-402. A general article discussing conditions in the Andes and Rockies as well as the Himalayas. (P/T)

1711 **Jain, S.K., Mehra, K.L. (1983).** Conservation of tropical plant resources. Proceedings of the Regional Workshop on Conservation of Tropical Plant Resources in South East Asia, New Delhi, March 8-12, 1982. Howrah, Botanical Survey of India. v, 253p. Illus. Contains papers on some non-tropical areas, e.g. Himalayas, Kashmir. South-east Asia is also loosely applied. (T)

1712 **Jungius, H. (1983).** The role of indigenous flora and fauna in rangeland management systems of the arid zones in western Asia. *J. Arid. Envir.*, 6(2): 75-85. (E)

1713 **Kamakhina, G.L. (1986).** Endemichnye i redkie vidy rastenii uchastka Babazo Kopetdagskogo Zapovednika. (Endemic and rare species of plants in the area of the Babazo Kopetdag Nature Reserve). *In* Priroda Tsentr. Kopetdaga. Ashkhabad. 24-39. Rus. (P/T)

1714 **Kiew, R. (1988).** Portraits of threatened plants 16. *Phoenix paludosa* Roxb. *Malay. Naturalist*, 42(1): 16. Illus. (T)

1715 **Kostermans, A.J.G.H., ed. (1987).** Proceedings of the Third Round Table Conference on Dipterocarps. Jakarta, Unesco. 657p. Papers presented at an international conference held at Mulawarman Univ., Samarinda, E. Kalimantan, 16-20 April 1985. Includes conservation. (E/T)

1716 **MacKinnon, J. & K. (1986).** Review of the protected areas system in the Indo-Malayan realm. Gland, Switzerland and Cambridge, U.K., IUCN and UNEP. 284p. Illus., maps. (P)

1717 **MacKinnon, J. (1985).** Outline of methodolgy for preparation of a review of the protected areas system of the "Indomalayan Realm". *In* Thorsell, J.W., *ed.* Conserving Asia's natural heritage. The planning and management of protected areas in the Indomalayan Realm. Proc. of the 25th working session of IUCN's CNPPA. Gland, IUCN. 53-59. (P)

1718 **Maheshwari, J.K. (1971).** The baobab tree: disjunctive distribution and conservation. *Biol. Conserv.*, 4(1): 57-60. (T)

1719 **Maheshwari, J.K. (1982).** Biosphere reserves in the Himalayas. *In* Paliwal, G.S., *ed.* The vegetational wealth of the Himalayas. Papers of the symposium at the Dept of Botany, Univ. of Garhwal, 1-6 October 1979. Delhi, Puja. 439-449. Illus. (P)

1720 **Malhotra, C.L. (1982).** Economic exploitation of rare north-western Himalayan plants. *In* Paliwal, G.S., *ed.* The vegetational wealth of the Himalayas. Papers of the symposium at the Dept of Botany, Univ. of Garhwal, 1-6 October 1979. Delhi, Puja. 221-225. Illus. Himalayas. (E/T)

1721 **Mascarenhas, A.F., comp. (?)** Biological diversity and genetic resources techniques and methods: tissue culture directory for Asia-Pacific. CSC Technical Publication Series No. 249. London, Commonwealth Science Council. 12p. (E/G)

1722 **Maunder, M. (1988).** Plants in peril, 3. *Ulmus wallichiana* Planchon (Ulmaceae). *Kew Mag.*, 5(3): 137-140. Illus. (T)

1723 **McNeely, J.A. (1987).** How dams and wildlife can co-exist: natural habitats, agriculture, and major water resource development projects in tropical Asia. *J. Conservation Biology*, 1(3): 228-238. Maps. (P)

1724 **Miri, P.K. (1980).** Genetic resources of wheat. *Indian J. Gen. Pl. Breed.*, 40(1): 26-34. (E)

MIDDLE ASIA, INDOCHINA AND JAPAN

1725 **Muradov, K., Govorukhina, V. (1988).** Nauchnye rekomendatsii akademika N.I. Vavilova o ratsional'nom ispol'zovanii i okhrane rastitel'nogo mira Kopetdaga. (Scientific recommendations of Academician N.I. Varilova on the efficient use of and protection of the plant world of the Kopet Dag.) *Izv. Akad. Nauk Turkm. SSR, Biol. Nauk*, 2: 23-27. Rus. (T)

1726 **Numata, M. (1973).** Further case studies in selecting and allocating land for nature conservation: South-East Asia. *In* Costin, A.B., Groves, R.H., *eds*. Nature conservation in the Pacific. Proc. Symposium A-10, XII Pacific Science Congress, Canberra, August-September 1971. IUCN and Australian Nat. Univ. Press. 66-71. (P)

1727 **Paliwal, G.S., ed. (1982).** The vegetational wealth of the Himalayas. Papers contributed at the symposium held between 1-6 October 1979, at the Department of Botany, University of Garhwal. Delhi, Puja Publishers. xv, 566p. Illus. Himalayas.

1728 **Plas-Haarsma, M. van der (1987).** Cyclamen in trade. *TRAFFIC Bull.*, 9(1): 11. Includes trade statistics for 12 species. (T)

1729 **Poore, D. (1976).** Conservation of vegetation, flora and fauna as a part of land use policy. *In* Proceedings of an International Meeting on Ecological Guidelines for the use of Natural Resources in the Middle East and South West Asia, Persepolis, Iran, 24-30 May 1975. IUCN Publ. New Series No. 34. Morges, IUCN. 215-223.

1730 **Pradhan, U.C. (1975).** Conservation of eastern Himalayan orchids. Problems and prospects. Part I. *Orch. Rev.*, 83: 314-317. (T)

1731 **Pradhan, U.C. (1975).** Conservation of eastern Himalayan orchids. Problems and prospects. Part II. *Orch. Rev.*, 83: 345-347. (T)

1732 **Pradhan, U.C. (1975).** Conservation of eastern Himalayan orchids. Problems and prospects. Part III. *Orch. Rev.*, 83: 374. (T)

1733 **Raizada, M.B. (1982).** Plant resources of the Indian subcontinent. *In* Paliwal, G.S., *ed.* The vegetational wealth of the Himalayas. Papers of the symposium at the Dept of Botany, Univ. of Garhwal, 1-6 October 1979. Delhi, Puja Publishers. 233-240. Illus.

1734 **Remanandan, P. (1983).** The wild gene pool of *Cajanus* at ICRISAT, present and future. Proceedings of the International Workshop on Pigeonpeas, Patancheru, 15-19 December 1980, Vol 2. Patancheru, ICRISAT. 29-38. (E)

1735 **Sale, J.B. (1985).** Regional and international cooperation in protected area management in Indomalaya. *In* Thorsell, J.W., *ed.* Conserving Asia's natural heritage. The planning and management of protected areas in the Indomalayan Realm. Proc. of the 25th working session of IUCN's CNPPA. Gland, IUCN. 203-207. (P)

1736 **Sale, J.B. (1985).** Wildlife research in the Indomalayan Realm. *In* Thorsell, J.W., *ed.* Conserving Asia's natural heritage. The planning and management of protected areas in the Indomalayan Realm. Proc. of the 25th working session of IUCN's CNPPA. Gland, IUCN. 137-149.

1737 **Sealey, J.L. (1981).** Cyclamen for western gardens. *Pac. Hort.*, 42(4): 21-26. Illus., col. illus. (G/T)

1738 **Singh, R.B. (1980).** Genetic resources of cotton in Asia. *IBPGR Region. Committ. Southeast Asia Newsl.*, 5(1): 9-11. (E/G)

1739 **Singh, S. (1985).** An overview of the conservation status of the national parks and protected areas of the Indomalayan Realm. *In* Thorsell, J.W., *ed.* Conserving Asia's natural heritage. The planning and management of protected areas in the Indomalayan Realm. Proc. of the 25th working session of IUCN's CNPPA. Gland, IUCN. 1-5. (P)

MIDDLE ASIA, INDOCHINA AND JAPAN

1740 **Snidvongs, K. (1984).** Future directions for the Indomalayan Realm. *In* McNeely, J.A., Miller, K.R., *eds.* National parks, conservation and development. The role of protected areas in sustaining society. Washington, D.C., Smithsonian Institution Press. 206-210. (P)

1741 **Spears, J. (1988).** Preserving biological diversity in the tropical forests of the Asian region. *In* Wilson, E.O., *ed.* Biodiversity. Washington, D.C., National Academy Press. 393-402. Deforestation rates; solutions to deforestation; role of development agencies.

1742 **Taber, R.D. (1976).** New developments in wildlife conservation for South West Asia and the Middle East. *In* Proceedings of an International Meeting on Ecological Guidelines for the use of Natural Resources in the Middle East and South West Asia, Persepolis, Iran, 24-30 May 1975. IUCN Publ. New Series No. 34. Morges, IUCN. 196-204.

1743 **Thorsell, J.W. (1985).** Threatened protected areas of the Indomalayan Realm. *In* Thorsell, J.W. *ed.* Conserving Asia's natural heritage. The planning and management of protected areas in the Indomalayan Realm. Proc. of the 25th working session of IUCN's CNPPA. Gland, IUCN. 43-49. (P)

1744 **Thorsell, J.W., ed. (1985).** Conserving Asia's natural heritage. The planning and management of protected areas in the Indomalayan Realm. IUCN. Proceedings of the 25th Working Session of IUCN's Commission on National Parks and Protected Areas. Corbett National Park India, 4-8 Feb. 1985. Gland, IUCN. 242p. (P)

1745 **Uniyal, B.P., et al. (1982).** Economic exploitation of rare Himalayan plants. *In* Paliwal, G.S., *ed.* The vegetational wealth of the Himalayas. Papers of the symposium at the Dept of Botany, Univ. of Garhwal, 1-6 October 1979. Delhi, Puja Publishers. 219-245. Illus. Himalayas. Medicinal. (E/T)

1746 **Whitmore, T.C. (1975).** Tropical rain forests of the Far East. Oxford, Clarendon Press. 282p. Ecology, classification and distribution of forest types, impact of man. Revised edition 1984. (E/P)

NEW WORLD

1747 **American Horticultural Society (1982).** North American horticulture: a reference guide. New York, C. Scribner's Sons. xvi, 367p. Includes conservation organizations.

1748 **Anon. (1982).** Prickly but imperilled species: the hardy cactus may be doomed by rustlers and smugglers. *Aloe, Cact. Succ. Soc. Zimbabwe Quart. Newsl.,* 50: 27-29. (T)

1749 **Anon. (1982).** The end of a locality or to hell with cacti: let's play ball. *Aloe, Cact. Succ. Soc. Zimbabwe Quart. Newsl.,* 50: 31-32.

1750 **Anon. (1984).** Arnold Arboretum joins forces with other gardens to save the rarest plants. *Plant Sciences,* 4(3): 3. (G)

1751 **Anon. (1986).** Scientists clone endangered cacti. *Amer. Horticulturist,* 65(11): 1-2. Illus. (G/T)

1752 **Anon. (1987).** Scientists clone endangered cacti. *Wildflower,* 3(2): 43-44. (T)

1753 **Anon. (1988).** Cactus theft. *Oryx,* 22(3): 149. Plant collecting threat. (T)

1754 **Barthlott, W. (1980).** Kakteen von Ausrottung bedroht. *Kakt. And. Sukk.,* 31(7): 214. Ge. (T)

1755 **Benson, L. (1975).** Cacti - bizarre, beautiful, but in danger. *Nat. Parks Conserv. Mag.,* 49(7): 17-21. (T)

NEW WORLD

1756 **Benson, L. (1977).** Preservation of cacti and management of the ecosystem. *In* Prance, G.T., Elias, T.S., *eds.* Extinction is forever. Proceedings of a symposium at the New York Botanical Garden, 11-13 May 1976. New York, New York Botanical Garden. 283-300. Illus. (P)

1757 **Bourdoux, P. (1981).** Protection de la nature. Conservation des especes sauvages ou inconscience Americaine (suite). *Cactus* (Brussels), 5(1): 20-22. Fr. (T)

1758 **Budowski, G., Miller, K.R. (1973).** North and Latin America: comparisons and contrasts in evaluating land for nature conservation. *In* Costin, A.B., Groves, R.H., *eds.* Nature conservation in the Pacific. Proc. Symposium A-10, XII Pacific Science Congress, Canberra, August-September 1971. Morges, IUCN. 27-38. (P)

1759 **Campbell, F.T. (1980).** Cactus rustlers cash in as rare cacti succumb. *Not Man Apart*, May: 10-11. (T)

1760 **Canfield, J. (1976).** Endangered species of the Americas: report on a symposium at the New York Botanical Garden. *Univ. Wash. Arbor. Bull.*, 39(4): 4-8. Illus. (T)

1761 **Carey, R.H. (1980).** Safeguarding the cacti. *Garden* (New York), 4(5): 4-7, 31. (T)

1762 **Connor, E.F. (1986).** The role of Pleistocene forest refugia in the evolution and biogeography of tropical biotas. *Trends Ecol. Evol.*, 1(6): 165-168. Maps. Rain forest.

1763 **Dickenson, R.E. (1984).** Keynote address: the Nearctic Realm. *In* McNeely, J.A., Miller, K.R., *eds.* National parks, conservation and development. The role of protected areas in sustaining society. Washington, D.C., Smithsonian Institution Press. 492-495. (P)

1764 **Dransfield, J., Johnson, D., Synge, H., comps (1988).** The palms of the New World: a conservation census. Gland and Cambridge, IUCN. xv, 30p. IUCN-WWF Plants Conservation Programme, Publication No. 2. (E/T)

1765 **Eidsvik, H.K. (1984).** Future directions for the Nearctic Realm. *In* McNeely, J.A., Miller, K.R., *eds.* National parks, conservation and development. The role of protected areas in sustaining society. Washington, D.C., Smithsonian Institution Press. 546-549. (P)

1766 **FAO/UNEP (1981).** Tropical Forest Resources Assessment Project (in the framework of GEMS). Los recursos forestales de la America tropical. Part I: Regional synthesis; Part II: Country briefs. Rome, FAO. 86, 343p. Sp. Tropical forests; deforestation. Both parts bound in one volume.

1767 **Fadiman, A. (1987).** Dr. Plotkin's jungle pharmacy. *Life*, 10(6): 15-19. Medicinal plants, ethnobotany. (E)

1768 **Fogle, H.W., Winters, H.F. (1981).** North American and European fruit and tree nut germplasm resources inventory. *U.S.D.A. Misc. Publ.*, 1406: 732p. (E)

1769 **Folkerts, G.W. (1977).** Endangered and threatened carnivorous plants of North America. *In* Prance, G.T., Elias, T.S., *eds.* Extinction is forever. Proceedings of a symposium at the New York Botanical Garden, 11-13 May 1976. New York, New York Botanical Garden. 301-313. Illus. (T)

1770 **Gentry, A.H. (1982).** Patterns of neotropical plant species diversity. *Evolutionary Biology*, 15: 1-84.

1771 **Hoffmann, W. (1982).** Brauchen wir Kakteen - Wildpflanzen? *Kakt. And. Sukk.*, 33(2): 41. Ge. (T)

1772 **Hubbell, S.P., Foster, R.B. (1983).** Diversity of canopy trees in a neotropical forest and implications for conservation. *In* Sutton, S.L., Whitmore, T.C., Chadwick, A.C., *eds.* Tropical rain forest: ecology and management. Oxford, Blackwell.

NEW WORLD

1773 **IRRI (1978).** Proceedings of the Workshop on the Genetic Conservation of Rice. A survey of rice genetic resources and conservation in Africa and the Americas. Abstracts of reports. Los Banos, IRRI. 15-21. (E/G)

1774 **IUCN (1981).** Conserving the natural heritage of Latin America and the Caribbean. The planning and management of protected areas in the Neotropical Realm. Proceedings of the 18th working session of IUCN's Commission on National Parks and Protected Areas, Lima, Peru, 21-28 June 1981. Gland, IUCN. 329p. (P)

1775 **IUCN Botanic Gardens Conservation Secretariat (1988).** Rare and threatened palms of the New World; questionnaire and preliminary report on their occurrence in botanic gardens. Kew, IUCN Botanic Gardens Conservation Secretariat. 16p. (G/T)

1776 **Ives, J.D. (1981).** Applied mountain geoecology. *In* Lall, J.S., *ed.* The Himalaya: aspects of change. New Delhi, India International Centre, Oxford University Press. 377-402. A general article discussing conditions in the Andes and Rockies as well as the Himalayas. (P/T)

1777 **Machlis, G.E., Neumann, R.P. (1987).** The state of national parks in the Neotropical Realm. *Parks*, 12(2): 3-8. Illus. (P)

1778 **Mickel, J.T. (1977).** Rare and endangered pteridophytes in the New World and their prospects for the future. *In* Prance, G.T., Elias, T.S., *eds.* Extinction is forever. Proceedings of a symposium at the New York Botanical Garden, 11-13 May 1976. New York, New York Botanical Garden. 323-328. (T)

1779 **Nabhan, G. (1985).** Native crop diversity in Aridoamerica: conservation of regional gene pools. *Econ. Bot.*, 39(4); 387-399. (E)

1780 **Nabhan, G.P. (1989).** Enduring seeds: native American agriculture and wild plant conservation. San Francisco, North Point Press. 225p. Includes chapters on loss of genetic diversity; techniques for plant conservation. (E/G)

1781 **New York Botanical Garden (1976).** Symposium on threatened and endangered species of plants in the Americas and their significance in ecosystems today and in the future. Program and Abstracts. New York, New York Botanical Garden. 55p. (T)

1782 **Oldfield, S. (1984).** Les Cactacees: une famille menacee par le commerce. *Cactus* (Brussels), 8(5): 112-115. Fr. Trade. (T)

1783 **Oldfield, S. (1984).** The Cactaceae, a family threatened by trade. *Oryx*, 18: 148-151. Illus. (T)

1784 **Oldfield, S. (1984).** The commercial collection of cacti, a conservation problem. *Kew Mag.*, 1(4): 181-187. Illus. Trade. (T)

1785 **Oldfield, S. (1985).** Conservation and the trade. *Cact. Succ. J.* (U.K.), 3(4): 89-90. (T)

1786 **Oldfield, S. (1985).** The western European trade in cacti and other succulents. *TRAFFIC Bull.*, 7(3/4): 44-57. Illus. Numerous tables, and specific and generic names of threatened succulents. (T)

1787 **Parsons, J.J. (1975).** The changing nature of New World tropical forests since European colonization. *In* IUCN. The use of ecological guidelines for development in the American humid tropics. Proceedings, Caracas, Venezuela, 20-22 February 1974. Morges, IUCN. 28-38.

1788 **Prance, G.T. (1976).** Threatened and endangered plants in the Americas. *BioScience*, 26(10): 633-634. (T)

NEW WORLD

1789 **Prance, G.T., Elias, T.S., eds (1977).** Extinction is forever. Proceedings of a symposium entitled Threatened and Endangered Species of Plants in the Americas and their Significance in Ecosystems Today and in the Future. New York, New York Botanical Garden. 437p. Illus., maps. Essays on state of knowledge on this topic for most parts of the Americas. Extensive appendices. (E/L/P/T)

1790 **Simerda, I.B. (1989).** Effective propagation methods for endangered cacti. *Bot. Gard. Conserv. News*, 1(4): 41-46. Illus. (G/T)

1791 **Uhl, C., Parker, G. (1986).** Is a one-quarter pound hamburger worth a half-ton rain forest? *Interciencia*, 11(5): 213. (T)

1792 **Weber, W. (1982).** Vanished bromeliads. *J. Bromeliad Soc.*, 32(5): 215-219. Illus. *Vriesea lancifolia, V. recurvata.* (T)

1793 **Wells, S.M., Pyle, R.M., Collins, N.M., comps (1983).** Threatened communities. *In* IUCN. The IUCN invertebrate Red Data Book. Gland and Cambridge, IUCN. 559-615. Data sheets on important sites and communities, many of which are centres of plant diversity and endemism. (P)

NORTH AFRICA AND MIDDLE EAST

1794 **Anon. (1980).** Protect the Mediterranean. *IUCN Bull.*, 11(9/10): 81-83, 95. Illus.

1795 **Anon. (1981).** Conservation rediscovered: learning from the Bedouin. *IUCN Bull.*, 12(3-4): 14. Illus.

1796 **Boughey, A.S. (1960).** Man and the African environment. *Proc. Trans. Rhod. Sci. Assoc.*, 48: 8-18.

1797 **Cloudsley-Thompson, J.L. (1984).** Key environments: Sahara Desert. Oxford, Pergamon Press. x, 348p. Illus., maps.

1798 **Croft, J. (1966).** Vanishing Africa. *Defenders Wildl. News*, 41(2): 117-120. (T)

1799 **Frey, W., Kurschner, H. (1985).** Final report of the section of botany, supporting period 1974-1985. Tubingen atlas of the Near and Middle East. Tubingen University, Special Research Division 19. Berlin. 28p. Maps.

1800 **Gomez-Campo, C. (1978).** The OPTIMA Seeds and Living Material Commission. *Gartn. Bot. Brief*, 58: 29. (Ge). Abstract. (G)

1801 **Gomez-Campo, C. (1979).** The role of seed banks in the conservation of Mediterranean flora. *Webbia*, 34(1): 101-107. (It). (G)

1802 **Gomez-Campo, C. (1981).** The 'Artemis' project for seed collection of Mediterranean endemics. *Threatened Pl. Commit. Newsl.*, 8: 9-11. (G)

1803 **Greuter, W. (1979).** Mediterranean conservation as viewed by a plant taxonomist. *Webbia*, 34(1): 87-99. (It). (T)

1804 **Hepper, F.N. (1977).** The practical importance of plant ecology in arid zones. *In* Dalby, D., Harrison Church, R.J., Bezzaz, F., eds. Drought in Africa 2. London, International African Institute. 105-106.

1805 **Hoffmann, L. (1968).** Project MAR: its principles and objectives; its special significance for the Near and Middle East Region. *In* IUCN Comission on Ecology. Proceedings of a Technical Meeting on Wetland Conservation, Ankara-Bursa-Istanbul, 9-16 October 1967. Morges, IUCN.

1806 **Holloway, C.W. (1976).** Conservation of threatened vertebrates and plant communities in the Middle East and South West Asia. *In* Proceedings of an International Meeting on Ecological Guidelines for the use of Natural Resources in the Middle East and South West Asia, Persepolis, Iran, 24-30 May 1975. IUCN Publ. New Series No. 34. Morges, IUCN. 179-188. (T)

NORTH AFRICA AND MIDDLE EAST

1807 **IUCN (1976).** Proceedings of an International Meeting on Ecological Guidelines for the use of Natural Resources in the Middle East and South West Asia, held at Persepolis, Iran, 24-30 May 1975. IUCN Publications New Series No. 34. Morges, IUCN. 231p. Maps.

1808 **IUCN Threatened Plants Unit (1986).** The botanic gardens list of rare and threatened species of North Africa. Kew, IUCN Botanic Gardens Co-ordinating Body Report No. 17. 9p. Lists plants in cultivation in botanic gardens. (G/T)

1809 **Jungius, H. (1983).** The role of indigenous flora and fauna in rangeland management systems of the arid zones in western Asia. *J. Arid. Envir.*, 6(2): 75-85. (E)

1810 **Kernick, M. (1978).** Forage genetic resources in N. Africa, Near and Middle East -a special area of concern. *Pl. Genet. Resource. Newsl.*, 33: 9-14. (Fr, Sp). Erosion caused by overgrazing; priority lists of indigenous grasses, legumes, browse shrubs and trees. (E)

1811 **Leach, G., Mearns, R. (1989).** Beyond the woodfuel crisis people, land and trees in Africa. London, Earthscan. 320p. (E)

1812 **Lock, J.M. (1989).** Legumes of Africa; a checklist. Kew, Royal Botanic Gardens. 619p. Provides IUCN conservation status for all taxa. (E/T)

1813 **Lucas, G. (1980).** Deux cas remarquables de taxa menaces: Cyprinodontidae nord-africains, *Araucaria* neo-caledoniens. *Compte R. Seances Soc. Biogeogr.*, 56(489): 51-52. Fr. (T)

1814 **Maunder, M. (1986).** Plants in peril, 8. *Kew Mag.*, 3(2): 88-90. Illus. *Cupressus dupreziana.* (T)

1815 **Miri, P.K. (1980).** Genetic resources of wheat. *Indian J. Gen. Pl. Breed.*, 40(1): 26-34. (E)

1816 **Mirimanian, K.P. (1976).** Measures for protection and rational land utilization in arid mountain regions. *In* Proceedings of an International Meeting on Ecological Guidelines for the use of Natural Resources in the Middle East and South West Asia, Persepolis, Iran, 24-30 May 1975. IUCN Publ. New Series No. 34. Morges, IUCN. 224-226. (P)

1817 **Morandini, R. (1977).** Problems of conservation, management and regeneration of Mediterranean forests: research priorities. *MAB Technical Notes*, 2: 73-79.

1818 **Nicholson, R.G. (1986).** Collecting rare conifers in North Africa. *Arnoldia*, 46(1): 20-29. Illus.

1819 **Noy-Meir, Anikster, Y., Waldman, M., Ashri, A. (1988).** Population dynamics research for *in situ* conservation: wild wheat in Israel. *Pl. Genet. Resource. Newsl.*, 75/76: 9-11. Illus. *Triticum dicoccoides.* (E)

1820 **Olivier, L. (1984).** La reintroduction de plantes indigenes disparues ou menacees de disparition en Region Mediterraneenne un des vecteurs de la strategie conservatoire. *Compt. Rend. Seanc. Soc. Biogeogr.*, 60(1): 51-66. Fr (En). (T)

1821 **Pignatti, S. (1983).** Human impact in the vegetation of the Mediterranean basin. *Geobotany*, 1983(5): 151-161. Illus.

1822 **Poore, D. (1976).** Conservation of vegetation, flora and fauna as a part of land use policy. *In* Proceedings of an International Meeting on Ecological Guidelines for the use of Natural Resources in the Middle East and South West Asia, Persepolis, Iran, 24-30 May 1975. IUCN Publ. New Series No. 34. Morges, IUCN. 215-223.

1823 **Prior, J., Tuohy, J. (1987).** Fuel for Africa's fires. *New Sci.*, 115(1571): 48-51. Illus., col. illus. Trees for fuelwood. (E)

1824 **Quezel, P. (1977).** Forests of the Mediterranean basin. *MAB Technical Notes*, 2: 9-29. Includes protection. (P)

NORTH AFRICA AND MIDDLE EAST

1825 **Quezel, P. (1984).** A system of terrestrial biosphere reserves for the Mediterranean. *In* Unesco-UNEP. Conservation, science and society. Paris, Unesco. 23-32. Contributions to the First International Biosphere Reserve Congress, Minsk, Byelorussia/USSR, 26 September - 2 October 1983. Tables. (P)

1826 **Saussay, C. du (1984).** La protection des forets en droit Africain. (Protection of forests in African law.) *In* Prieur, M., *ed.* Forets et environnement: en droit compare et international. Paris, Presses Universitaires de France. 147-164. Fr. (L)

1827 **Seidenfaden, G. (1984).** La convention relative a la conservation de la vie sauvage et du milieu naturel de l'Europe. *In* Council of Europe. Conservation de la Nature en Europe. Strasbourg, Council of Europe. 50-53. Fr. Berne Convention. (L/P/T)

1828 **Synge, H. (1980).** Endangered monocotyledons in Europe and South West Asia. *In* Brickell, C.D., Cutler, D.F., Gregory, M., *eds.* Petaloid monocotyledons: horticultural and botanical research. Linnean Society Symposium Series No. 8. London, Academic Press. 199-206. Illus. (T)

1829 **Taber, R.D. (1976).** New developments in wildlife conservation for South West Asia and the Middle East. *In* Proceedings of an International Meeting on Ecological Guidelines for the use of Natural Resources in the Middle East and South West Asia, Persepolis, Iran, 24-30 May 1975. IUCN Publ. New Series No. 34. Morges, IUCN. 196-204.

1830 **Thirgood, J.V. (1981).** Man and the Mediterranean forest. A history of resource depletion. London, Academic Press. 194p.

1831 **Tomaselli, R. (1977).** Degradation of the Mediterranean maquis. *MAB Technical Notes*, 2: 33-72. Maps.

1832 **Unesco (1977).** Mediterranean forests and maquis: ecology, conservation and management. *MAB Technical Notes*, 2: 79p.

1833 **Unesco (1979).** Programme on man and the biosphere. Workshop on biosphere reserves in the Mediterrranean region: development of a conceptual basis and a plan for the establishment of a regional network. Final report. MAB Report No. 45. Paris, Unesco. 62p. (P)

1834 **Walters, S.M. (1979).** The role of Mediterranean botanic gardens in plant conservation. *Webbia*, 34(1): 109-116. (It). (G/T)

OLD WORLD

1835 **Ahmad, A.M. (1979).** Genetic resources of tropical forest trees - conservation options. *Malay. Nature J.*, 33(1): 25-38. (E)

1836 **Anon. (1974).** Conference on plant protection in tropical and sub-tropical areas, November 4-15, 1974, Manila, Philippines. Eschborn, Germany, Federal Agency for Economic Cooperation. 342p. (T)

1837 **Carp, E., comp. (1980).** Directory of wetlands of international importance in the western Palearctic. Nairobi, UNEP and Gland, WWF. 506p. Maps.

1838 **Chang, T.T. (1982).** The IRRI germplasm bank. *IBPGR Region. Committ. Southeast Asia Newsl.*, 6(2): 5. *Oryza* gene bank. (G)

1839 **Chang, T.T. (1983).** Exploration and collection of the threatened genetic resources and the system of documentation of rice genetic resources at IRRI. Howrah, Botanical Survey of India. v, 48-54. (E/G/T)

1840 **Chang, T.T. (1984).** Conservation of rice genetic resources: luxury or necessity? *Science*, 224(4646): 251-256. (E/G)

1841 **Chang, T.T. (1985).** Crop history and genetic conservation: rice - a case study. *Iowa State J. Res.*, 59(4): 425-455, 457-496. Illus., maps. *Oryza sativa, Oryza glaberrima.* (E)

OLD WORLD

1842 **Chang, T.T., Adair, C.R., Johnston, T.H. (1982).** The conservation and use of rice genetic resources. *Adv. Agron.*, 35: 37-91. (E)

1843 **Denton, R., Chang, T.T., eds (1983).** Rice germplasm conservation workshop held at the International Rice Research Institute, Philippines, 25-26 April 1983. Laguna, Philippines, IRRI. v, 109p. (E)

1844 **IRRI (1978).** Proceedings of the Workshop on the Genetic Conservation of Rice. Recommendations of the workshop on rice genetic conservation. Los Banos, IRRI. 26-35. (E)

1845 **Ramade, F. (1984).** Keynote address: the Palaearctic Realm. *In* McNeely, J.A., Miller, K.R., *ed.* National parks, conservation and development. The role of protected areas in sustaining society. Washington, D.C., Smithsonian Institution Press. 418-425. Tables. (P)

1846 **Ranjitsinh, M.K. (1984).** Keynote address: the Indomalayan Realm. *In* McNeely, J.A., Miller, K.R., *eds.* National parks, conservation and development. The role of protected areas in sustaining society. Washington, D.C., Smithsonian Institution Press. 148-153. (P)

1847 **Robertson, J. (1985).** The World Heritage Convention and the international biosphere reserve network of The Man and the Biosphere (MAB) programme: their status in the Indomalayan Realm. *In* Thorsell, J.W., *ed.* Conserving Asia's natural heritage. The planning and management of protected areas in the Indomalayan Realm. Proc. of the 25th working session of IUCN's CNPPA. Gland, IUCN. 208-216. (P)

1848 **Saharia, V.B. (1985).** Manpower planning and training needs for protected area management in the Indomalayan Realm. *In* Thorsell, J.W., *ed.* Conserving Asia's natural heritage. The planning and management of protected areas in the Indomalayan Realm. Proc. of the 25th working session of IUCN's CNPPA. Gland, IUCN. 125-129. (P)

1849 **Segnestam, M. (1984).** Future directions for the Western Palaearctic Realm. *In* McNeely, J.A., Miller, K.R., *eds.* National parks, conservation and development. The role of protected areas in sustaining society. Washington, D.C., Smithsonian Institution Press. 486-490. (P)

1850 **Singh, R.B. (1983).** Collection and conservation of wild species of rice. *IBPGR Region. Committ. Southeast Asia Newsl.*, 7(2-3): 21. (E/G)

1851 **Spooner, B., Mann, H.S., eds (1982).** Desertification and development: dryland ecology in social perspective. London, Academic Press. xx, 407p. Arid. (E)

1852 **Ward, S., Harrison, J. (1985).** Monitoring protected areas in the Indomalayan Realm. *In* Thorsell, J.W., *ed.* Conserving Asia's natural heritage. The planning and management of protected areas in the Indomalayan Realm. Proc. of the 25th working session of IUCN's CNPPA. Gland, IUCN. 5-8. (P)

1853 **Wells, S.M., Pyle, R.M., Collins, N.M., comps (1983).** Threatened communities. *In* IUCN. The IUCN invertebrate Red Data Book. Gland and Cambridge, IUCN. 559-615. Data sheets on important sites and communities, many of which are centres of plant diversity and endemism. (P)

PACIFIC ISLANDS

1854 **Anon. (1953).** Our heritage in the Pacific. Washington, D.C., Pacific Science Board of the National Academy of Sciences - National Research Council. 13p.

1855 **Anon. (1958).** The vegetation of Micronesia. Engineer Intelligence Study No. 257. Washington, D.C., U.S. Geological Survey, Military Geology Branch. 160p. Includes causes of denudation of vegetation in all island groups.

PACIFIC ISLANDS

1856 **Anon. (1979).** Second South Pacific Conference on National Parks and Reserves. Sydney. 2 vols, 128, 176p. Vol. 1: Verbatim transcript; Vol. 2: Formal papers presented. (P)

1857 **Bowman, R.I. (1963).** The scientific need for island reserve areas. *In* Scientific use of natural areas symposium. XVI International Congress of Zoology. Miami, Coconut Grove. 60-76. Field Research Projects, Natural Areas Studies No. 2. (P)

1858 **Bryant, J.J. (1989).** Environmental education in the South Pacific: towards sustainable development. *The Environmentalist*, 9(1): 45-54. Illus.

1859 **Byrne, J.E., ed. (1979).** Literature review and synthesis of information on Pacific island ecosystems. Washington, D.C., U.S. Fish and Wildlife Service, Office of Biological Services.

1860 **Carew-Reid, J. (1984).** The South Pacific Regional Environment Program. *Ambio*, 13(5-6): 377.

1861 **Carlquist, S. (1965).** Island life: a natural history of the islands of the world. New York, Natural History Press. 451p. Origin, evolution and adaptations of island flora and fauna; Galapagos and Hawaiian Islands well covered.

1862 **Chapman, V.J. (1969).** Conservation of island ecosystems in the south-west Pacific. *Biol. Conserv.*, 1: 159-165. Descriptions of various islands with emphasis on species and vegetation types needing protection. (P/T)

1863 **Cheatham, N.H. (1968).** Forestry and conservation in the Trust Territory of the Pacific Islands. *S. Pacif. Bull.*, 18(4): 38-41, 47.

1864 **Clapp, R.B., Sibley, F.C. (1971).** Notes on the vascular flora and terrestrial vertebrates of Caroline Atoll, Southern Line Islands. *Atoll Res. Bull.*, 145: 1-18.

1865 **Commonwealth Science Council (1987).** Biological diversity and genetic resources: life support species. Summary report. International workshop on maintenance and evaluation of life support species in Asia and the Pacific region, New Delhi, 4-7 April 1987. London, Commonwealth Science Council. 25p. (E)

1866 **Connell, J. (1984).** Islands under pressure - population growth and urbanization in the South Pacific. *Ambio*, 13(5-6): 306-308, 310-312.

1867 **Connell, J. (1986).** Population, migration, and problems of atoll development in the South Pacific. *Pacific Studies*, 9(2): 41-58.

1868 **Costin, A.B., Groves, R.H., eds (1973).** Nature conservation in the Pacific. Proceedings of Symposium A-10, XII Pacific Science Congress, August-September 1971, Canberra, Australia. IUCN New Series No. 25. Morges, IUCN and Canberra, Australian National University Press. 337p. Illus., maps.

1869 **Cumberland, K.B. (1963).** Man's role in modifying island environments in the southwest Pacific, with special reference to New Zealand. *In* Fosberg, F.R., ed. Man's place in the island ecosystem: a symposium. Honolulu, Bishop Museum Press. 186-206.

1870 **Dahl, A.L. (1980).** Regional ecosystems survey of the South Pacific area. South Pacific Commission Tech. Paper No. 179. Noumea, South Pacific Commission and IUCN. 99p. Maps.

1871 **Dahl, A.L. (1984).** Future directions for the Oceanian Realm. *In* McNeely, J.A., Miller, K.R., eds. National parks, conservation and development. The role of protected areas in sustaining society. Washington, D.C., Smithsonian Institution Press. 359-362. (P)

1872 **Dahl, A.L. (1984).** Oceania's most pressing environmental concerns. *Ambio*, 13(5-6): 296-299. Col. illus., map.

PACIFIC ISLANDS

1873 **Dahl, A.L. (1986).** Review of the protected areas system in Oceania. Gland, Switzerland and Cambridge, U.K., IUCN and UNEP. 239p. Maps. Data sheets on each island, including land area, population, species of conservation interest, ratings on conservation status, species richness, pressures, vulnerability, human impact. (P/T)

1874 **Dahl, A.L., Baumgart, I.L. (1982).** The state of the environment in the South Pacific. *In* SPREP Conference Human Environment, Report. Noumea, New Caledonia, South Pacific Commission. 47-71.

1875 **Daly, K. (1989).** Eradication of feral goats from small islands. *Oryx*, 23(2): 71-75. Eradication of introduced species.

1876 **DeFilipps, R.A. (1987).** A bibliogaphy of plant conservation in the Pacific islands: endangered species, habitat conversion, introduced biota. *Atoll Res. Bull.*, 311: 195p. Comprehensive coverage; includes references on invasive species. (B/E/G/P/T)

1877 **Dodge, E.S. (1976).** Islands and empires: Western impact on the Pacific and East Asia. Minneapolis, University of Minnesota Press. 350p.

1878 **Doran, E. (1959).** Handbook of selected Pacific islands. California, Point Mugu, Pacific Missile Range. 223p. Includes discussion of deforestation.

1879 **Douglas, G. (1970).** Draft check list of Pacific oceanic islands. *Micronesica*, 5(2): 327-463. Remarks on land use; conservation status. (T)

1880 **Douglas, G. (1973).** Review of IBP/CT survey of oceanic islands. *In* Costin, A.B., Groves, K.H., *eds.* Nature conservation in the Pacific. Proc. Symposium A-10, XII Pacific Science Congress, Canberra, August-September 1971. IUCN and Australian Nat. Univ. Press. 203-207.

1881 **ESCAP (1985).** State of the environment in Asia and the Pacific. Vol. 1: Summary. Bangkok, ESCAP.

1882 **Eaton, P. (1985).** Land tenure and conservation: protected areas in the South Pacific. SPREP Topic Review No. 17. Noumea, New Caledonia, South Pacific Commission. (P)

1883 **Eaton, P. (1986).** Tenure and taboo: customary rights and conservation in the South Pacific. *In* South Pacific Commission. Report of the Third South Pacific National Parks and Reserves Conference, Apia, 1985. Vol. II: Collected key issue and case study papers. Noumea, SPC. 114-134.

1884 **Elliott, H.F.I. (1973).** Past, present and future conservation status of Pacific islands. *In* Costin, A.B., Groves, K.H., *eds.* Nature conservation in the Pacific. Proc. Symposium A-10, XII Pacific Science Congress, Canberra, August-September 1971. IUCN and Australian Nat. Univ. Press. 217-227.

1885 **Ely, C.A., Clapp, R.B. (1973).** The natural history of Laysan Island, northwestern Hawaiian Islands. *Atoll Res. Bull.*, 171: 1-361.

1886 **Falanruw, M.V.C. (1984).** People pressure and management of limited resources on Yap. *In* McNeely, J.A., Miller, K.R., *eds.* National parks, conservation and development. The role of protected areas in sustaining society. Washington, D.C., Smithsonian Institution Press. 348-354.

1887 **Fischer, J.L. & A.M. (1957).** The eastern Carolines. Washington, D.C., Pacific Science Board. 274p. (T)

1888 **Fosberg, F.R. (1950).** The problem of rare and vanishing plant species. *In* Proc. Papers International Technical Conference, Protection of Nature, Lake Success, NY, 1949. 502-504. (T)

1889 **Fosberg, F.R. (1953).** A conservation program for Micronesia. *In* Proc. Seventh Pacific Science Congress, 1949. Wellington. 4. 670-673.

PACIFIC ISLANDS

1890 **Fosberg, F.R. (1953).** Vegetation of central Pacific atolls: a brief summary. *Atoll Res. Bull.*, 23: 1-26.

1891 **Fosberg, F.R. (1954).** The protection of nature in the islands of the Pacific. VIII Congres International de Botanique, 1954. *Compt. Rend. Seances*, 21-27: 104-117. Paris, 1954-1957. (P)

1892 **Fosberg, F.R. (1959).** Conservation situation in Oceania. *In* Proc. Ninth Pacific Science Congress, 7: 30-31.

1893 **Fosberg, F.R. (1959).** Long-term effects of radioactive fallout on plants? *Atoll Res. Bull.*, 61: 1-11.

1894 **Fosberg, F.R. (1963).** Grazing animals and the vegetation of oceanic islands. *In* Unesco. Symposium on the Impact of Man on Humid Tropics Vegetation (Goroka, Papua New Guinea). Djakarta. 168-169.

1895 **Fosberg, F.R. (1972).** Man's effects on island ecosystems. *In* Farvar, M.T., Milton, J.P., *eds.* The careless technology: ecology and international development. New York, Natural History Press. 869-880.

1896 **Fosberg, F.R. (1973).** On present condition and conservation of forests in Micronesia. *In* Pac. Sci. Assoc. Standing Comm. on Pac. Bot. Symposium: Planned utilization of the lowland tropical forests, August 1971. Bogor, Indonesia.

1897 **Fosberg, F.R. (1973).** Past, present and future conservation problems of oceanic islands. *In* Costin, A.B., Groves, K.H., *eds.* Nature conservation in the Pacific. Proc. Symposium A-10, XII Pacific Science Congress, Canberra, August-September 1971. IUCN and Australian Nat. Univ. Press. 209-215. Includes effects of introduced and invasive species.

1898 **Fosberg, F.R. (1973).** Vascular plants - widespread island species. *In* Costin, A.B., Groves, K.H., *eds.* Nature conservation in the Pacific. Proc. Symposium A-10, XII Pacific Science Congress, Canberra, August-September 1971. IUCN and Australian Nat. Univ. Press. 167-169.

1899 **Fosberg, F.R. (1984).** Phytogeographic comparison of Polynesia and Micronesia. *In* Radovsky, F.J., Raven, P.H., Sohmer, S.H., *eds.* Biogeography of the tropical Pacific. Bishop Museum Special Publication No. 72. Assoc. Systematics Collectors & Bernice P Bishop Museum. 33-44. Includes man's impact on vegetation.

1900 **Hamilton, L.S., ed. (1983).** Forest and watershed development in Asia and the Pacific. London, Westview. 560p.

1901 **Harris, D.R. (1962).** Invasions of oceanic islands by alien plants. *Trans. Inst. Brit. Geogr.*, 31: 67-82. Introduced species.

1902 **Heyerdahl, T. (1963).** Prehistoric voyages as agencies for Melanesian and South American plant and animal dispersal to Polynesia. *In* Barrau, J., *ed.* Plants and the migrations of Pacific peoples. Honolulu, Hawaii, Bishop Museum Press. 23-35. Introduced species. (T)

1903 **Heywood, V.H. (1979).** The future of island floras. In Bramwell, D., *ed.* Plants and islands. London, Academic Press. 431-441. (Sp). (P/T)

1904 **Holdgate, M.W., Nicholson, E.M. (1967).** An international conservation programme for the Pacific Islands. *Micronesica*, 3(1): 51-54.

1905 **Holdgate, M.W., Wace, N.M. (1961).** The influence of man on the floras and faunas of southern islands. *Polar Record*, 10(68): 475-493.

1906 **IBPGR (1980).** Genetic resources of the Far East and the Pacific Islands. *IBPGR Region. Committ. Southeast Asia Newsl.*, 5(1): 12-13. (E)

PACIFIC ISLANDS

1907 **Johannes, R.E. (1984).** Traditional conservation methods and protected marine areas in Oceania. *In* McNeely, J.A., Miller, K.R., *eds.* National parks, conservation and development. The role of protected areas in sustaining society. Washington, D.C., Smithsonian Institution Press. 344-347. (P)

1908 **King, W.B. (1973).** Conservation status of Central Pacific islands. *Wilson Bull.*, 85: 89-103.

1909 **Knott, N.P. (1973).** Further case studies in selecting and allocating land for nature conservation: Micronesia, a multiple land-capability inventory method. *In* Costin, A.B., Groves, K.H., *eds.* Nature conservation in the Pacific. Proc. Symposium A-10, XII Pacific Science Congress, Canberra, August-September 1971. IUCN and Australian Nat. Univ. Press. 61-66.

1910 **Lamoureux, C.H. (1961).** Botanical observations on Leeward Hawaiian atolls. *Atoll Res. Bull.*, 79: 1-10.

1911 **Lamoureux, C.H. (1963).** The flora and vegetation of Laysan Island. *Atoll Res. Bull.*, 97: 1-12.

1912 **Lamoureux, C.H. (1963).** Vegetation of Laysan. *Proc. Hawaii. Acad. Sci.*, 37: 22.

1913 **Lamoureux, C.H. (1964).** The Leeward Hawaiian Island. *Newsl. Hawaii. Bot. Soc.*, 3(2): 7-11.

1914 **Lawyer, J.I., et al. (1979).** A guide to selected current literature on vascular plant floristics for the contiguous U.S., Alaska, Canada, Greenland and the U.S. Caribbean and Pacific Islands. New York, New York Botanical Garden. 138p. Suppl. 1: Research on floristic information synthesis: a report to the Division of Natural History, Nat. Park Service, U.S. Dept of Interior. Washington, D.C. (B)

1915 **Laycock, G. (1970).** Haunted sands of Laysan. *Audubon*, 72(2): 42-49. Vegetation has been destroyed by rabbits. (T)

1916 **Lesouef, J.-Y. (1979).** The endangered plants of the oceanic islands: their cultivation in the Stangalarc'h Conservatory (Brest). *OPTIMA Newsl.*, 819: 16. (G/T)

1917 **Marten, K.D. (1985).** Tropical forestry in Melanesia and some Pacific islands. *In* Dahl, A.L., Carew-Reid, J., *eds.* Environment and resources in the Pacific. Regional Seas Reports and Studies No. 69. Geneva, UNEP. 115-128.

1918 **Mascarenhas, A.F., comp.** (?) Biological diversity and genetic resources techniques and methods: tissue culture directory for Asia-Pacific. CSC Technical Publication Series No. 249. London, Commonwealth Science Council. 12p. (E/G)

1919 **Melville, R. (1979).** Endangered island floras. *In* Bramwell, D., *ed.* Plants and islands. London, Academic Press. 361-378. (Sp). (P/T)

1920 **Mueller-Dombois, D. (1973).** Natural area system development for the Pacific region, a concept and symposium. Island Ecosystems IRP/IBP Hawaii, Technical Report No. 26. Honolulu, University of Hawaii. 55p. (P)

1921 **Mueller-Dombois, D. (1983).** Canopy dieback and successional processes in Pacific forests. *Pacific Science*, 37(4): 317-325.

1922 **Mueller-Dombois, D. (1984).** Zum Baumgruppensterben in pazifischen Inselwaldern. *Phytocoenologia*, 12(1): 1-8. Forest dieback.

1923 **Mull, M.E. (1975).** Comments on natural resources management plan. *Elepaio*, 35(11): 127-131. Discusses re-establishment of endemic species.

1924 **Murdock, G.P. (1963).** Human influences on the ecosystems of high islands of the tropical Pacific. *In* Fosberg, F.R., *ed.* Man's place in the island ecosystem: a symposium. Honolulu, Bishop Museum Press. 145-154.

PACIFIC ISLANDS

1925 **Newell, L.A. (1986).** Demographics and mangrove resources in the eastern Carolines. *Amer. Pacif. Forest. News*, July: 5-7.

1926 **Nicholson, E.M. (1969).** Draft check list of Pacific oceanic islands. *Micronesica*, 5(2): 327-463. Provides details of vegetation cover, threats. (T)

1927 **Nicholson, E.M., Eldredge, L.C., eds. (1970).** International Biological Programme Technical Meeting on Conservation of Pacific Islands held at Koror, Palau and Guam in November, 1968. Proceedings. *Micronesica*, 5(2): 1-496.

1928 **Nietschmann, B. (1984).** Biosphere reserves and traditional societies. *In* Unesco-UNEP. Conservation, science and society. Paris, Unesco. 499-508. Contributions to the First International Biosphere Reserve Congress, Minsk, Byelorussia/USSR, 26 September - 2 October 1983. Map. (P)

1929 **Owen, R.P. (1979).** A conservation program for the Trust Territory. *Micronesian Reporter*, 27(1): 22-28.

1930 **Pacific Information Centre (1983).** Environmental issues in the South Pacific: a preliminary bibliography. Suva, Fiji, University of the South Pacific Library. ix, 64p. (B)

1931 **Pulea, M. (1984).** Environmental legislation in the Pacific region. *Ambio*, 13(5-6): 369-371. (L)

1932 **Pulea, M., Va'ai, A.V.S. (1983).** Review of international and regional conventions relevant to the environmental management of the South Pacific region. Noumea, New Caledonia, South Pacific Commission. 26p. (L)

1933 **Radovsky, F.J., Raven, P.H., Sohmer, S.H., eds (1984).** Biogeography of the tropical Pacific. B.P. Bishop Museum Special Publication No. 72. 221p.

1934 **Ranjitsinh, M.K. (1979).** Forest destruction in Asia and the South Pacific. *Ambio*, 8(5): 192-201. Illus., col. illus.

1935 **Rappaport, R.A. (1963).** Aspects of man's influence upon island ecosystems, alteration and control. *In* Gressitt, J.L., *ed.* Pacific basin biogeography, Honolulu, Bishop Museum Press. 155-174. Includes discussion on introduced species. (P/T)

1936 **Raynal, J. (1979).** Three examples of endangered nature in the Pacific Ocean. *In* Hedberg, I., *ed.* Systematic botany, plant utilization and biosphere conservation. Stockholm, Almqvist & Wiksell International. 145-150.

1937 **Reboul, J.L. (1975).** Deux exemples d'introductions malheureuses pour la nature polynesienne. *Te Natura o Polynesia*, 2: 14-20. Fr. (T)

1938 **Robbins, R.G. (1972).** Vegetation and man in the south-west Pacific and Papua New Guinea. *In* Ward, R.G., *ed.* Man in the Pacific islands: essays on geographical change in the Pacific islands. London, Clarendon Press. x, 339p.

1939 **Routley, R. & V. (1980).** Destructive forestry in Melanesia and Australia. *Ecologist*, 10(1-2): 56-67. Illus.

1940 **SPREP (1985).** Action strategy for protected areas in the South Pacific region. Noumea, New Caledonia, South Pacific Commission. 24p.

1941 **Sachet, M.-H. (1957).** The vegetation of Melanesia: a summary of the literature. *Proc. Eighth Pacific Science Congress*, 4: 35-47. (B)

1942 **Sachet, M.-H., Fosberg, F.R. (1955).** Island bibliographies: Micronesian botany, land development and ecology of coral atolls, vegetation of tropical Pacific islands. Washington, D.C., Nat. Academy of Sciences, Nat. Research Council. 577p. (B/G)

1943 **Sachet, M.-H., Fosberg, F.R. (1971).** Island bibliographies supplement. Washington, D.C., Pacific Science Board, Nat. Academy of Sciences. 427p. Islands worldwide. (B/P/T)

PACIFIC ISLANDS

1944 **Singh, B. (1984).** Keynote address: the Oceanian Realm. *In* McNeely, J.A., Miller, K.R., *eds.* National parks, conservation and development. The role of protected areas in sustaining society. Washington, D.C., Smithsonian Institution Press. 310-314. (P)

1945 **Skottsberg, C. (1953).** Report of the standing committee for the protection of nature in and around the Pacific for the years 1939-1948. Proc, 7th. Pac. Sci. Conf. 4: 586-612. Notes on vegetation and nature protection on various island groups. (P/T)

1946 **Sloth, B. (1988).** Nature legislation and nature conservation as part of tourism development in the island Pacific. Suva, Fiji, Tourism Council of the South Pacific. 82p. Pacific Regional Tourism Development Programme. (L)

1947 **Sorensen, J. (1974).** Remote oceanic islands: approaches to conservation of an international resource. Berkeley, University of California. 26p.

1948 **South Pacific Commission (1986).** Report of the Third South Pacific National Parks and Reserves Conference, Apia, 1985. Vol. II. Noumea, SPC. Collected key issue and case study papers. (P)

1949 **Steenis, C.G.G.J. van (1965).** Man and plants in the tropics: an appeal to Micronesians for the preservation of nature. *Micronesica*, 2: 61-65.

1950 **Stoddart, D.R. (1968).** Catastrophic human intervention with coral atoll ecosystems. *Geography*, 53(1): 25-40.

1951 **Stoddart, D.R. (1968).** Isolated island communities. *Science J.*, 4(4): 32-38. Includes map indicating threats.

1952 **Stoddart, D.R. (1975).** Scientific importance and conservation of central Pacific islands. London, Dept of Education and Science, and Foreign and Commonwealth Office. 28p. Report to the Southern Zone Research Committee, Royal Society.

1953 **Stoddart, D.R. (1976).** Conservation and scientific importance of central Pacific Islands. Report to the Southern Zone Research Committee, Royal Society, Dept of Education and Science, and Foreign and Commonwealth Office. 28p.

1954 **Sykes, W.R. (1983).** Conservation on South Pacific Islands. *In* Given, D.R., *ed.* Conservation of plant species and habitats. Wellington, New Zealand, Nature Conservation Council. 37-42. Illus. (P/T)

1955 **Talbot, L.M. (1969).** Highlights of conservation in the International Biological Programme in the Asia-Pacific region. *Malaysian Forest.*, 32: 391-394.

1956 **Taylor, R.H. (1968).** Introduced mammals and islands: priorities for conservation and research. *Proc. New Zealand Ecol. Soc.*, 15: 61-67.

1957 **Thaman, R.R. (1974).** *Lantana camara*: its introduction, dispersal and impact on islands of the tropical Pacific Ocean. *Micronesica*, 10(1): 17-39. Invasive species.

1958 **Thamen, R.R. (1986).** Microparks in the Pacific islands. The relevance of traditional and modern small scale conservation areas in the Pacific islands. *In* South Pacific Commission. Report of the Third South Pacific National Parks and Reserves Conference, Apia, 1985. Vol. II. Collected key issue and case study papers. Noumea, SPC. 215-236. (P)

1959 **Trust Territory of the Pacific Islands (1976).** Adopted regulations, Title 45: Fish, shellfish and game, chapter 5: Endangered species. *Territorial Register*, 2(1): 4 December. (T)

1960 **Vasiliev, N.G., Manko, Y.Y. (1983).** Protection of forests in the northwestern Pacific region. *Pac. Sci. Congr. Proc.*, 15(1-2): 245. (P)

PACIFIC ISLANDS

1961 **Venkatesh, S., Va'ai, S., Pulea, M. (1983).** An overview of environmental protection legislation in the South Pacific countries. Noumea, New Caledonia, South Pacific Commission. 63p. (L)

1962 **Wace, N.M. (1960).** The botany of the southern oceanic islands. *Proc. Roy. Soc.*, B152: 475-490. Discusses the vulnerability of oceanic island floras to competition from introduced species.

1963 **Wace, N.M. (1978).** The character of oceanic islands and the problem of their rational use and conservation. Morges, IUCN.

1964 **Watson, J.S. (1961).** Feral rabbit populations on Pacific islands. *Pacific Sci*, 15(4): 591-593. Discusses rabbit damage to vegetation.

1965 **Wenkam, R. (1971).** Micronesian parks: a proposal. *Micronesian Reporter*, 19(3): 9-22. Illus. (P)

1966 **Whistler, W.A. (1981).** A naturalist in the south Pacific: north to Tokelau. *Bull. Pacific Trop. Bot. Gard.*, 11(2): 29-37. Includes observations on rare species and introduced plants. (T)

1967 **Wiens, H.J. (1962).** Atoll environment and ecology. New Haven, Yale Univ. Press. 532p. Includes discussion on introduced plants; pressures on resources.

1968 **Williams, J.T. (1988).** Identifying and protecting the origins of our food plants. *In* Wilson, E.O., ed. Biodiversity. Washington, D.C., National Academy Press. 240-247. Discusses crop origins and germplasm use; IBPGR program; *in situ* and *ex situ* conservation. (E/G)

1969 **Wodzicki, K. (1981).** Some nature conservation problems in the South Pacific. *Biol. Conserv.*, 21(1): 5-18. Map.

PANTROPICAL

1970 **Ahmad, A.M. (1979).** Genetic resources of tropical forest trees - conservation options. *Malay. Nature J.*, 33(1): 25-38. (E)

1971 **Almeda, F., Pringle, C.M., ed. (1988).** Tropical rainforests: diversity and conservation. San Francisco, California Academy of Sciences and Pacific Division, American Association for the Advancement of Science. xiii, 306p. Illus., col. illus., maps.

1972 **Alvim, P. de T. (1977).** The balance between conservation and utilization in the humid tropics with special reference to Amazonian Brazil. *In* Prance, G.T., Elias, T.S., eds. Extinction is forever. Proceedings of a symposium at the New York Botanical Garden, 11-13 May 1976. New York, New York Botanical Garden. 347-352. (E)

1973 **Anon. (1974).** Conference on plant protection in tropical and sub-tropical areas, November 4-15, 1974, Manila, Philippines. Eschborn, Germany, Federal Agency for Economic Cooperation. 342p. (T)

1974 **Anon. (1980).** Save the rainforests. *IUCN Bull.*, 11(5): 17-18. Illus. Tropical forest.

1975 **Anon. (1981).** Cycad Society seed bank. *Cycad Newsl.*, 4(3): 16. (G)

1976 **Anon. (1981).** Mangroves: conservation on the waterfront. *IUCN Bull.*, 12(3-4): 18.

1977 **Anon. (1981).** Working through the eleventh hour. Every week the remaining tropical lowland forest diminishes by an area about the size of Delaware. *Missouri Bot. Gard. Bull.*, 69(5): 1, 3-6.

1978 **Anon. (1988).** Tropical rain forests. A plan for action. *For. Bird*, 19(3): 28-30.

1979 **Arthur, A.J. (1962).** Symposium on the Impact of Man on Humid Tropics Vegetation, Goroka, Papua-New Guinea, September, 1960. Canberra, A.J. Arthur, Commonwealth Govt. Printer. 401p.

PANTROPICAL

1980 **Ashton, P.S. (1981).** Techniques for the identification and conservation of threatened species in tropical forests. *In* Synge, H., *ed.* The biological aspects of rare plant conservation. Chichester, Wiley. 155-164. Proceedings of International Conference, King's College, Cambridge, 14-19 July 1980. (T)

1981 **Ashton, P.S. (1981).** Tropical botanical gardens: meeting the challenge of declining resources. *Longwood Programme Seminars*, 13: 55-57. (G/P/T)

1982 **Baker, D. (1987).** Remote future for Third World satellite data. *New Sci.*, 116(1583): 48-51. Illus. Remote-sensing, deforestation.

1983 **Balick, M.J. (1987).** The economic utilization of the babassu palm: a conservation strategy for sustaining tropical forest resources. *J. Wash. Acad. Sci.*, 77(4): 215-223. Illus., map. (E)

1984 **Barkhuisen, B.P. (1975).** The cycad garden of Unisa/Die broodboomtuin van Unisa. University of South Africa, Pretoria. 80pp. En, Af. (G)

1985 **Batten, M. (1983).** The rush is on to study jungles. *Internat. Wildl.*, 13(3): 16-19.

1986 **Bax, J., Hooghiemstra, R., Witte, H. de (1986).** Tropenbos: initiating a global research programme to conserve and develop tropical rain forests. Utrecht, Tropenbos. Illus. Proposal to set up global research programme in rain forests.

1987 **Beardsley, T. (1986).** Tropical rain forests: ecologists unite for diversity. *Nature*, 323(6085): 193. Covers parks. (P)

1988 **Beckett, G. (1984).** WWF/IUCN tropical forest campaign 1982-1984. *Int. Dendrol. Soc. Year Book 1983*: 97-106. Illus.

1989 **Beusekom, C.F. van, et al., eds (1987).** Tropenbos, wise utilization of tropical rain forest lands. Tropenbos Scientific Series 1. The Hague, Min. Ed. & Sci.

1990 **Bruijnzeel, P.S. (1986).** Environmental impacts of (de)forestation in the humid tropics. A watershed perspective. *Wallaceana*, 46: 3-13. Illus. (T)

1991 **Brune, A. (1988).** Conservation of tropical orchids in their natural habitat. *Amer. Orchid Soc. Bull.*, 57(10): 1127-1131. (T)

1992 **Budowski, G. (1981).** Why save tropical rain forests? Some arguments for campaigning conservationists. *In* Jordan, C.F., *ed.* Tropical ecology. Benchmark Papers in Ecology, vol. 10. Stroudsberg, PA, Hutchinson Ross. 324-333. Reprinted from *Amazoniana*, 4: 529-538 (1976). (T)

1993 **Burger, W. (1977).** Cloud forests. *Field Mus. Nat. Hist. Bull.* (Chicago), 48(10): 11-16. Col. illus.

1994 **Burger, W. (1989).** Tropical forests and the number of species on planet Earth. *Field Mus. Nat. Hist. Bull.*, 60(5): 9-14.

1995 **Burley, F.W. (1988).** The Tropical Forestry Action Plan: recent progress and new initiatives. *In* Wilson, E.O., *ed.* Biodiversity. Washington, D.C., National Academy Press. 403-408.

1996 **Burley, J., Styles, B.T., eds (1975).** Tropical trees: variation, breeding and conservation. Symposium on variation, breeding and conservation of tropical forest trees, Oxford University, 1975. Linnean Society Symposium Series No. 2. New York, Academic Press. 243p.

1997 **Campbell, D.G., Hammond, H.D., eds (1989).** Floristic inventory of tropical countries. The status of plant systematics, collections, and vegetation, plus recommendations for the future. New York, New York Botanical Garden. 545p. Maps. Floristics, vegetation, centres of endemism, threatened areas, extinction rates.

1998 **Caufield, C. (1982).** Tropical moist forests: the resource, the people, the threat. London, IIED/Earthscan Paperbacks. 67p. Illus., maps.

PANTROPICAL

1999 **Caufield, C. (1985).** In the rainforest. New York, Alfred A. Knopf. 304p. Tropical forest; deforestation.

2000 **Chadwick, A.C., Sutton, S.L. eds (1984).** Tropical rain-forest: the Leeds symposium. Leeds, Leeds Philosophical and Literary Society. Illus.

2001 **Christensen, B. (1983).** Mangroves - what are they worth? *Unasylva*, 35(139): 2-15. Illus. Includes report on Malaysia's mangrove management.

2002 **Colchester, M. (1988).** The vanishing jungle. Ecologists make friends with economists. *The Economist*, 15 October: 25-28. Illus., map. Deforestation.

2003 **Collins, M. (1984).** Removing the nail from the tropical forest's coffin. *Oryx*, 18: 1-2.

2004 **Collins, M. (1987).** International protection of cycads. *Bull. Fairchild Trop. Gard.*, 42(3): 28-29. Illus. (T)

2005 **Collins, N.M. (1988).** The conservation and management of genetic resources. *In* McDermott, M.J., *ed.* The future of the tropical rain forest. Proceedings of an international conference at St. Catherine's College, Oxford, 27-28 June 1988. Oxford, Oxford Forestry Institute. 9-12. (E)

2006 **Committee on Selected Biological Problems in the Humid Tropics (1982).** Ecological aspects of development in the humid tropics. Washington, D.C., National Academy Press. ix, 297p. Illus. Includes chapter on germplasm and conservation of genetic resources. (E)

2007 **Cromie, W. (1980).** World's rain forests vanishing: 'progress' threatens to make millions of animal, plant species extinct. *Pittsburgh Press*, 13 June: 1B. (T)

2008 **Davidson, J. (1985).** Economic use of tropical moist forests. Commission on Ecology Paper No. 9. Gland, IUCN. 28p. Illus., map. (E)

2009 **Davidson, J. (1987).** Bioenergy tree plantations in the tropics. Ecological implications and impacts. Commission on Ecology Paper No. 12. Gland, IUCN. 47p. Illus., maps. (E)

2010 **Davis, S. (1983).** World action against jungle destruction. *Threatened Pl. Newsl.*, 11: 16-17.

2011 **Denslow, J., Padoch, C., eds (1988).** People of the tropical rain forest. Berkeley, CA California Press. 232p.

2012 **Eckholm, E. (1985).** U.N. and aid groups seek to save dwindling Third World forests. *New York Times*, July 29: A-11.

2013 **Elfring, C. (1984).** Can technology save tropical forests? *BioScience*, 34(6): 350-352. Illus. (E/T)

2014 **Elliott, G.K. (1989).** The future of the tropical forest: the timber trade views. *In* McDermott, M.J., *ed.* The future of the tropical rain forest. Proceedings of an international conference at St. Catherine's College, Oxford, 27-28 June 1988. Oxford, Oxford Forestry Institute. 62-63. (E)

2015 **Emmons, L. (1989).** Tropical rain forests: why they have so many species and how we may lose this biodiversity without cutting a single tree. *Orion Nat. Quart.*, 8(3): 8-14.

2016 **Enabor, E.E. (1982).** Economics of tropical forest resources conservation. *In* Srivastava, P.B.L. *et al.*, *eds.* Tropical forests: source of energy through optimisation and diversification. Proceedings International Forestry Seminar, 11-15 Nov. 1980, Serdang, Selangor, Malaysia. Kuala Lumpur, Forestry Dept HQ. (E)

PANTROPICAL

2017 **Erwin, T.L. (1988).** The tropical forest canopy: the heart of biotic diversity. *In* Wilson, E.O., *ed.* Biodiversity. Washington, D.C., National Academy Press. 123-129. Map, illus. Discussion mainly about invertebrates, but figures have relevance to current debate on biodiversity and extinction rates.

2018 **Esteban, I.D. (1980).** Tropical rain forest management: ecological and environmental aspects. *Canopy*, 6(12): 3-6. Illus.

2019 **Esteban, I.D. (1981).** Tropical rain forest management: ecological and environmental aspects. *Canopy*, 7(1): 3-5; 7(2): 3-6. Illus.

2020 **Ewel, J. (1980).** Tropical succession: manifold routes to maturity. *Biotropica*, 12(2) suppl.: 2-7.

2021 **FAO (1977).** Conservation of tropical rain forests I. *Tigerpaper*, IV(1): 20-23. Illus., map. Based on the work of T.C. Whitmore.

2022 **FAO (1977).** Conservation of tropical rain forests II. *Tigerpaper*, 4(2): 6-9. Based on the work of T.C. Whitmore.

2023 **FAO (1982).** Conservation and development of tropical forest resources. FAO Forestry Paper No. 37. Proceedings and recommendations of the 2nd meeting of experts on tropical forests sponsored by FAO/UNEP/Unesco, Rome, 12-15 Jan. 1982. Rome, FAO. ix, 122p. (E)

2024 **FAO (?)** Forest genetic resources information - no. 14. Rome, FAO. 48p. Maps. Includes data sheets on timber trees. (E)

2025 **FAO Forest Resources Division (1984).** A guide to *in situ* conservation of genetic resources of tropical woody species. Rome, FAO. 196p. Maps. (E/P/T)

2026 **FAO Forestry Department (1986).** FAO's Tropical Forestry Plan. *Unasylva*, 38(2): 37-64. Illus.

2027 **FAO, World Resources Institute, The World Bank, UNDP (?)** The Tropical Forestry Action Plan. FAO, IBRD, WRI, UNDP. 32p. Illus., maps. (E)

2028 **Fisher, J. (1980).** Our diminishing rain forest. *Bull. Fairchild Trop. Gard.*, 35(3): 10-15. Illus.

2029 **Fontaine, R. (1981).** What is really happening to tropical forests? *Ceres*, 14(4): 15-19. Illus.

2030 **Forsyth, A., Miyata, K. (1984).** Tropical nature. New York, Charles Scribner's Sons. 248p.

2031 **Fosberg, F.R. (1977).** Tropical floristic botany - concepts and status - with special attention to tropical islands. *In* Larsen, K., Holm-Nielsen, L.B., *eds.* Tropical botany. London, Academic Press. 89-105. Mentions introduced and invasive species.

2032 **Garwood, N.C., Janos, D.P., Brokaw, N. (1979).** Earthquake-caused landslides: a major disturbance to tropical forests. *Science*, 205(4410): 997-999. Illus.

2033 **Gentry, A.H. (1986).** Endemism in tropical versus temperate plant communities. *In* Soule, M.E., *ed.* Conservation biology: the science of scarcity and diversity. Sunderland, Massachussetts, Sinauer Assoc. 153-181.

2034 **Giliomee, J.H. (1982).** The disappearance of the tropical rain forest. *J. Dendrol.*, 2(1-2): 50-54. Map. (E/T)

2035 **Gleissman, S.R., Garcia, R., Amador, M. (1981).** The ecological basis for the application of traditional technology in the management of tropical ecosystems. *Agro-ecosystems*, 7: 173-185.

2036 **Goldsmith, E. (1980).** World Ecological Areas Programme: a proposal to save the world's tropical rain forests. *Ecologist*, 10(1/2): 2-4. Tropical forest.

PANTROPICAL

2037 **Gomez-Pompa, A., Butanda C., A. (1977).** Indice de proyectos en desarrollo en ecologia tropical, Vol. 2. (Index of current tropical ecology research.) Xalapa, Instituto de Investigaciones sobre Recursos Bioticos, A.C. iii, 279p. (En).

2038 **Gomez-Pompa, A., Vasquez-Yanes, C., Guevara, S. (1972).** The tropical rainforest: a non-renewable resource. *Science*, 177: 762-765.

2039 **Goreau, T.J., De Mello, W.Z. (1988).** Tropical deforestation: some effects on atmospheric chemistry. *Ambio*, 17(4): 275-281. Col. illus.

2040 **Gradwohl, J., Greenberg, R. (1988).** Saving the tropical forests. London, Earthscan. 207p. Illus., maps. (P)

2041 **Grainger, A. (1980).** The state of the world's tropical forests. *Ecologist*, 10(1/2): 6-54. Illus., maps.

2042 **Grainger, A. (1983).** Improving the monitoring of deforestation in the humid tropics. *In* Sutton, S.L., *et al*. Tropical rain forest: ecology and management. Oxford, Blackwell. 387-395.

2043 **Grainger, A. (1984).** Quantifying changes in forest cover in the humid tropics: overcoming current limitations. *J. World Forest. Res. Management*, 1: 3-63. Tropical forest; deforestation.

2044 **Green, K.M. (1983).** Using Landsat to monitor tropical forest ecosystems: realistic expectations of digital processing technology. *In* Sutton, S.L., *et al.*, eds. Tropical rain forest: ecology and management. Oxford, Blackwell. 397-409.

2045 **Guldager, P. (1975).** *Ex situ* conservation stands in the tropics. *In* The methodology of conservation of forest genetic resources: report on a pilot study. Rome, FAO. 85-92. (E/G)

2046 **Guppy, N. (1983).** The case for an Organization of Timber Exporting Countries (OTEC). *In* Sutton, S.L., *et al.*, eds. Tropical rain forest: ecology and management. Oxford, Blackwell. 459-463. (E)

2047 **Guppy, N. (1984).** Tropical deforestation: a global view. *Foreign Affairs*, 1984: 928-965. A long and important review of problems that arise or may arise from deforestation of the tropics. (T)

2048 **Hadley, M., Lanly, J.-P. (1983).** Tropical forest ecosystems: identifying differences, seeking similarities. *Nat. Resources*, 19(1): 2-19. Illus. Includes assessment of status. (E)

2049 **Haffer, J. (1982).** General aspects of refuge theory. *In* Prance, G.T., *ed*. Biological diversification in the tropics. New York, Columbia Univ. Press. 6-25.

2050 **Hall, J.B. (1989).** Priorities and trends in tropical rain forest research. *In* McDermott, M.J., *ed*. The future of the tropical rain forest. Proceedings of an international conference at St. Catherine's College, Oxford, 27-28 June 1988. Oxford, Oxford Forestry Institute. 80-81.

2051 **Harrison, J. (1986).** How much is protected? *IUCN Bull.*, 17(1-3): 22-23. Tropical forest. (T)

2052 **Henderson-Sellers, A. (1988).** Tropical deforestation and global climatic change. *In* McDermott, M.J., *ed*. The future of the tropical rain forest. Proceedings of an international conference at St. Catherine's College, Oxford, 27-28 June 1988. Oxford, Oxford Forestry Institute. 13-15.

2053 **Hervey, R.J. (1983).** Simultaneous exploitation and preservation of tropical forests. *Interciencia*, 7(6): 361. (T)

2054 **Heywood, V.H. (1981).** Forest destruction - can we stop it in time? *Oryx*, 16(1): 27-30. (T)

2055 **Holden, C. (1980).** Rain forests vanishing. *Science*, 208(4442): 378.

PANTROPICAL

2056 **Holm-Nielson, L.B., Balslev, H., Nelson, I., eds (1989).** Tropical forests: botanical dynamics, speciation and diversity. London, Academic Press. 400p. Proceedings of a conference held at the University of Aarhus, Denmark, 8-10 August 1988.

2057 **Hoover, W.S. (1983).** As rain forests come tumbling down, will species begonias come through? *Begonian*, 50: 40-42. Illus. (T)

2058 **Horich, C.K. (1980).** La destruccion de los bosques tropicales: causa grave de la desaparicion de su flora orquideofila. (The destruction of the tropical forest: a major cause of the disappearance of orchid floras.) *Orquidea*, 7(4): 265-276. Sp (En). Illus. (T)

2059 **Howard, B. (1989).** The role of the Overseas Development Administration in the future of the tropical rain forest. *In* McDermott, M.J., *ed*. The future of the tropical rain forest. Proceedings of an international conference at St. Catherine's College, Oxford, 27-28 June 1988. Oxford, Oxford Forestry Institute. 70-76.

2060 **Hpay, T. (1986).** The International Tropical Timber Agreement: its prospects for tropical timber trade, development and forest management. IUCN/IIED Tropical Forest Policy Paper No. 3. Cambridge, IUCN, London, IIED-Earthscan and WWF. 20p. (E)

2061 **Huguet, L. (1983).** Que penser de la "disparition" des forets tropicales. *Bois Forets Trop.*, 1(195): 7-22. Fr.

2062 **Huguet, L. (1983).** Replenishing the world's forests: the future of the world's tropical forests. *Commonw. Forest. Rev.*, 62(3): 195-200. (Fr, Sp).

2063 **IUCN (1964).** The ecology of man in the tropical environment. Ninth technical meeting, Nairobi, 17-20 September 1963. (L'ecologie de l'homme dans le milieu tropical.) IUCN Publications New Series No. 4. Morges, IUCN. 355p. Maps.

2064 **IUCN Threatened Plants Committee Secretariat (1980).** The botanic gardens list of cycads. Kew, IUCN Botanic Gardens Conservation Co-ordination Body Report No. 3. 14p. Interim report of plants in cultivation in botanic gardens. (G/T)

2065 **Iltis, H.H. (1981).** The tropical forests - so rich, so fragile, so irreplaceable - is extermination or preservation to be their fate? *Envir. Educ. Rep.*, 9: 2-3.

2066 **Iltis, H.H. (1983).** Tropical forests: what will be their fate? *Environment*, 25(10): 55-60. (T)

2067 **International Earthcare Center (1983).** Fact-sheet on tropical rain-forests. *Envir. Conserv.*, 10(1): 71-73.

2068 **International Tropical Timber Organization (1989).** The role of the International Tropical Timber Organization. *In* McDermott, M.J., *ed*. The future of the tropical rain forest. Proceedings of an international conference at St. Catherine's College, Oxford, 27-28 June 1988. Oxford, Oxford Forestry Institute. 52-57. (E)

2069 **Jackson, P. (1983).** The tragedy of our tropical rain forests. *Ambio*, 12(5): 252-254. Illus., col. illus.

2070 **Jacobs, M. (1980).** Significance of the tropical rain forests on 12 points. *BioIndonesia*, 7: 75-94. Map.

2071 **Jain, S.K., Mehra, K.L. (1983).** Conservation of tropical plant resources. Proceedings of the Regional Workshop on Conservation of Tropical Plant Resources in South East Asia, New Delhi, March 8-12, 1982. Howrah, Botanical Survey of India, v, 253p. Illus. Contains papers on some non-tropical areas, e.g. Himalayas, Kashmir. South-east Asia is also loosely applied. (T)

2072 **Janzen, D.H. (1972).** The uncertain future of the tropics. *Nat. Hist.*, (1972) (1): 80-94.

PANTROPICAL

2073 **Johns, A.D. (1983).** Tropical forest primates and logging: can they co-exist? *Oryx*, 17: 114-118. Illus. Map. (T)

2074 **Johnson, B. (1983).** Rain forests and foreign policies: a look at Britain's impact. *In* Sutton, S.L., Whitmore, T.C., Chadwick, A.C., *eds*. Tropical rain forest: ecology and management. Oxford, Blackwell Scientific. 477-485. Special Publications Series of the British Ecological Society No. 2.

2075 **Johnson, B. (1984).** The forestry crisis: what must be done. *Ambio*, 13(1): 48-49. Tropical forests.

2076 **Johnson, B. (1985).** Chimera or opportunity? An environmental appraisal of the recently concluded International Tropical Timber Agreement. *Ambio*, 14(1): 42-44. Illus. Deforestation, tropical forests. (L)

2077 **Johnstone, B. (1987).** Japan saps the world's rain forests. *New Sci.*, 114(1554): 18.

2078 **Joyce, C. (1986).** Species are the spice of life. *New Sci.*, 112(1529): 20-21. Illus. Biodiversity; tropical forests.

2079 **Kemp, R.H., Burley, J., Keiding, H., Nickles, D.G. (1978).** International cooperation in the exploration, conservation and development of tropical and sub-tropical forest gene resources. *In* Proceedings of the 7th World Forestry Congress, Buenos Aires, 4-18 October 1972, vol. 4 and 5. Buenos Aires, Instituto Forestal Nacional. 5083-5105. En, Sp, Fr. (E)

2080 **Kennedy, M.G. (1980).** The struggle for the world's rainforest. *Habitat* (Australia), 8(2): 8-9. Illus.

2081 **Kitching, R., Schofield, C. (1986).** Every pitcher tells a story. *New Sci.*, 109(1492): 48-50. Col. illus. Complex communities of invertebrates within traps of tropical pitcher plants can reveal health of the environment. (P)

2082 **Knees, S., Gardner, M. (1983).** Mahogany: an endangered resource? *Threatened Pl. Newsl.*, 12: 12-13. (P/T)

2083 **Knees, S.G., Gardner, M.F. (1983).** Mahoganies: candidates for the Red Data Book. *Oryx*, 17: 88-92. Illus. (E/T)

2084 **Lanly, J.P. (1983).** Assessment of the forest resources of the tropics. *Forestry Abstr.*, 44(6): 287-318. Deforestation. Tropical forests. Review article, describing the work of FAO/UNEP Tropical Forest Resources Assessment Project.

2085 **Lanly, J.P., Clement, J. (1979).** Present and future natural forest and plantation areas in the tropics. *Unasylva*, 31(123): 12-20. (E)

2086 **Lewin, R. (1986).** Damage to tropical forests, or why were there so many kinds of animals? *Science*, 234: 149-150.

2087 **Ling, C.Y. (1989).** The Tropical Forestry Action Plan: people or profits? A response from a non-governmental perspective. *In* McDermott, M.J., *ed*. The future of the tropical rain forest. Proceedings of an international conference at St. Catherine's College, Oxford, 27-28 June 1988. Oxford, Oxford Forestry Institute. 33-36. (E)

2088 **Ljungman, L.L. (1988).** The Tropical Forestry Action Plan. *In* McDermott, M.J., *ed*. The future of the tropical rain forest. Proceedings of an international conference at St. Catherine's College, Oxford, 27-28 June 1988. Oxford, Oxford Forestry Institute. 20-22. (E)

2089 **Longman, K.A., Jenik, J. (1987).** Tropical forest and its environment. 2nd ed. Harlow, Longman. 347p. Illus., maps. Rain forest ecology; deforestation; management.

PANTROPICAL

2090 **Lovejoy, T.E. (1982).** Designing refugia for tomorrow. *In* Prance, G.T., *ed.* Biological diversification in the tropics: Proc. 5th. Int. Symp. of the Assoc. for Trop. Biology, Macuto Beach, Caracas, Venezuela, 8-13 Feb., 1979. NY, Columbia Univ. Press. 673-680. (P)

2091 **Lovejoy, T.E., Bierregaard, R.O., Rankin, J.M., Schubart, H.O.R. (1983).** Ecological dynamics of tropical forest fragments. *In* Sutton, S.L., *et al.*, *eds.* Tropical rain forest: ecology and management. BES Special Publ. 2. Oxford, Blackwell. 377-384.

2092 **Lucas, S.A. (1985).** Cycads: the living fossils. *Bull. Pac. Trop. Bot. Gard.*, 15(1): 1-10. *Cycas*, endangered species (T)

2093 **Lugo, A., Brown, S. (1984).** Conserving tropical rainforest ecosystems and assigning priorities. *In* Unesco-UNEP. Conservation, science and society. Paris, Unesco. 37-43. Contributions to the First International Biosphere Reserve Congress, Minsk, Byelorussia/USSR, 26 September - 2 October 1983. Table. (P)

2094 **Lugo, A.E. (1981).** Research on the global role of tropical forests relative to the carbon balance. U.S. Department of Agriculture, Forest Service. Southern Forest Experimental Station. 1980 Annual Letter, Inst. Tropical Forestry, Rio Pedras, Puerto Rico (June 1981). 31-32. (T)

2095 **Lugo, A.E. (1988).** Estimating reductions in the diversity of tropical forest species. *In* Wilson, E.O., *ed.* Biodiversity. Washington, D.C., National Academy Press. 58-70. Extinction rates; deforestation; tropical forests.

2096 **Lugo, A.E. (1988).** The future of the forest. Ecosystem rehabilitation in the tropics. *Environment*, Washington, D.C., 30(7): 16-20, 41-45. Tropical forest.

2097 **Lugo, A.E., Brown, S. (1982).** Conversion of tropical moist forests: a critique. *Interciencia*, 7(2): 89-93.

2098 **Lugo, A.E., Brown, S. (1982).** Rebuttal to the "Response to the Lugo-Brown critique by Myers". *Interciencia*, 7(6): 360. Deforestation. (T)

2099 **Mabberley, D.J. (1983).** Tropical rain forest ecology. Glasgow, Blackie. 156p. Illus., maps. Includes deforestation.

2100 **MacKinnon, J. & K., Child, G., Thorsell, J. (1986).** Managing protected areas in the tropics. Gland, IUCN. 295p. Illus., maps. Rain forest; deforestation; includes genetic resource conservation. (L/P)

2101 **Macdonald, I.A.W., Frame, G.W. (1988).** The invasion of introduced species into nature reserves in tropical savannas and dry woodlands. *Biol. Conserv.*, 44(1-2): 67-93. Invasive species. Includes 5 case studies. (P)

2102 **Maclean, J.T. (1985).** Ecology of tropical rainforests. Quick Bibliography Series -National Agricultural Library (U.S.). Beltsville, Maryland. 16p. Extensive bibliography. (B)

2103 **McDermott, M.J., ed. (1988).** The future of the tropical rain forest. Proceedings of an international conference held in St. Catherine's College, Oxford, 27-28 June 1988. Oxford, Oxford Forestry Institute. 110p. (E)

2104 **Medina, E. (1982).** Deforestation of tropical forests. *Interciencia*, 7(6): 357. (T)

2105 **Meijer, W. (1980).** A new look at the plight of tropical rain-forests. *Envir. Conserv.*, 7(3): 203-206. Map.

2106 **Mergen, F., ed. (1981).** Tropical forests: utilization and conservation. New Haven, CT, Yale School of Forestry and Environmental Studies. 199p.

2107 **Miller, K.R. (1973).** Conservation and development of tropical rain forest areas. *In* Elliott, H.F.I., *ed.* Twelfth technical meeting, Banff, Alberta, Canada, 12-15 September 1972. Morges, IUCN. 259-270. (T)

PANTROPICAL

2108 **Misra, R. (1970).** Save the tropical ecosystems. *Intecol Bull.*, 2: 29.

2109 **Mohan Ram, H.Y. (1983).** Role of botanic gardens in the conservation of tropical plants. *In* Jain, S.K., Mehra, K.L., *eds.* Conservation of tropical plant resources. Howrah, Botanical Survey of India. 154-158. (G)

2110 **Moore, H.E., jr (1977).** Endangerment at the specific and generic levels in palms. *In* Prance, G.T., Elias, T.S., *eds.* Extinction is forever. Proceedings of a symposium at the New York Botanical Garden, 11-13 May 1976. New York, New York Botanical Garden. 267-282. Illus. (T)

2111 **Moore, P. (1986).** What makes rainforests so special? *New Sci.*, 111(1522): 38-40. Illus., col. illus. Biodiversity and rain forests.

2112 **Morgan, F., Vincent, J.R. (1987).** Natural management of tropical moist forests: silvicultural and management prospects of sustained utilization. New Haven, Yale School of Forestry.

2113 **Myers, N. (1978).** Conservation of forest animal and plant genetic resources in tropical rainforests. *In* Proceedings of the 8th World Forestry Congress, Jakarta, 16-28 October 1978: Forestry for quality of life. II, 13. (E/G)

2114 **Myers, N. (1978).** Forests for people. *New Sci.*, 80(1134): 951-953. Illus. Exploitation of world's tropical forests.

2115 **Myers, N. (1979).** Tropical rain forests: whose hand is on the axe? *Nat. Parks Conserv. Mag.*, 53(11): 9-13. Illus., map.

2116 **Myers, N. (1980).** Conversion of tropical moist forests. Washington, D.C., National Academy of Sciences. ix, 205p.

2117 **Myers, N. (1981).** Conservation needs and opportunities in tropical moist forests. *In* Synge, H., *ed.* The biological aspects of rare plant conservation. Chichester, Wiley. 141-154. Proceedings of International Conference, King's College, Cambridge, 14-19 July 1980. (P)

2118 **Myers, N. (1982).** Depletion of tropical moist forests: a comparative review of rates and causes in the three main regions. *Acta Amazonica*, 12(4): 745-758. (Por).

2119 **Myers, N. (1982).** Response to the Lugo-Brown critique of "Conservation of tropical moist forests". *Interciencia*, 7(6): 358-360. Deforestation.

2120 **Myers, N. (1984).** The primary source: tropical forests and our future. New York, London, W.W. Norton. xiv, 399p. Illus., maps. (T)

2121 **Myers, N. (1985).** Tropical deforestation and species extinctions: the latest news. *Futures*, 17: 451-463. (T)

2122 **Myers, N. (1988).** Threatened biotas: "hot spots" in tropical forests. *Environmentalist*, 8(3): 187-208.

2123 **Myers, N. (1988).** Tropical deforestation and remote sensing. *Forest Ecology and Management*, 23: 215-225.

2124 **Myers, N. (1988).** Tropical forest species: going, going, going.... *Scientific American*, 259: 132.

2125 **Myers, N. (1988).** Tropical forests and their species: going, going ...? *In* Wilson, E.O., *ed.* Biodiversity. Washington, D.C., National Academy Press. 28-35.

2126 **Myers, N. (1988).** Tropical forests: a storehouse for human welfare. *In* Almeda, F., Pringle, C.M., *eds.* Tropical rainforests: diversity and conservation. San Francisco, California Academy Press. 13-27.

2127 **Myers, N. (1988).** Tropical forests: much more than stocks of wood. *J. Trop. Ecol.*, 4(2): 208-220. (E)

2128 **Myers, N. (1988).** Tropical forests: why they matter to us. *Geog. Rev.*, 1: 16-19.

PANTROPICAL

2129 **Myers, N. (1989).** The greenhouse effect: a tropical forestry response. *Biomass*, 18: 73-78.

2130 **Myers, N. (1989).** Tropical deforestation and climatic change. *Envir. Conserv.*, 15: 293-298.

2131 **Myers, N. (1989).** Tropical forests and climate. *In* Berger, A., *ed.* Climate and geo-sciences. Dordrecht, Netherlands, D. Reidel.

2132 **Myers, N. (1989).** Tropical forests and life on Earth. *In* Head, S., Heinzman, R., *eds.* Lessons of the rainforest. San Francisco, Sierra Club Books.

2133 **Nations, J., Komer, D. (1982).** Rainforest, cattle and the hamburger society. Austin, Center for Human Ecology. Deforestation.

2134 **Nectoux, F. (1895).** Timber! An investigation of the UK tropical timber industry. London, Friends of the Earth. (E)

2135 **Nectoux, F., Dudley, N. (1987).** A hardwood story: an investigation into European influence in tropical forest loss. London, Friends of the Earth. (E)

2136 **Ng, F.S.P. (1983).** Ecological principles of tropical lowland rain forest conservation. *In* Sutton, S.L., *et al.*, *eds.* Tropical rain forest: ecology and management. Oxford, Blackwell. 359-375.

2137 **Norman, C. (1985).** Virgin rain forest reprieved. *Science*, 227(4684): 273.

2138 **Odum, W.E. (1976).** Ecological guidelines for tropical coastal development. IUCN New Series No. 42. Morges, IUCN. 60p.

2139 **Office of Technology Assessment (1984).** Technologies to sustain tropical forest resources. Washington, D.C., Office of Technology Assessment.

2140 **Oldfield, M.L. (1981).** Tropical deforestation and genetic resources conservation. *In* Sutlive, V.H., Altshuler, N., Zamora, M.D., *eds.* Blowing in the wind: deforestation and long-range implications. Williamsburg, Virginia, Dept of Anthropology, College of William & Mary. 277-345. Studies in Third World Societies No. 14. (E/G/T)

2141 **Oldfield, S. (1981).** The endangered cycads. *Oryx*, 16: 45. (T)

2142 **Oldfield, S. (1985).** More protection for cycads? *Oryx*, 19(3): 132. (T)

2143 **Oldfield, S. (1985).** The western European trade in cacti and other succulents. *TRAFFIC Bull.*, 7(3/4): 44-57. Illus. Numerous tables, and specific and generic names of threatened succulents. (T)

2144 **Oldfield, S. (1988).** Buffer zone management in tropical moist forests. Case studies and guidelines. Gland, IUCN. 49p. Illus., maps. (P)

2145 **Oldfield, S. (1988).** Rare tropical timbers. Gland, Switzerland and Cambridge, U.K., IUCN. 37p. Illus. (E/T)

2146 **Oldfield, S. (1989).** The tropical chainsaw massacre. *New Sci.*, 123(1683): 54-57. Col. illus. (E)

2147 **Ooi, S.C., Rajanaidu, N. (1979).** Establishment of oil palm genetic resources -theoretical and practical considerations. *Malay. Appl. Biol.*, 8(1): 15-28. (Mal). (E)

2148 **Osborne, R. (1985).** Cycads and the law. *Encephalartos*, 4: 18. (L/T)

2149 **Palmberg, C. (1983).** Conservation of variation in tropical tree species. *Pl. Genet. Resource. Newsl.*, 55: 28-31. (Fr, Sp). Illus. (E/G)

2150 **Perry, R.T. (1982).** The moist tropical forest: its conversion and protection. *The Environmentalist*, 2(2): 117-132. Illus., maps. Detrimental effects of slash and burn agriculture.

PANTROPICAL

2151 **Plotkin, M.J. (1988).** New agricultural and industrial products from the tropics. *WWF Reports,* June/July 1988: 7-10. Illus. (E)

2152 **Plotkin, M.J. (1988).** The outlook for new agricultural and industrial products from the tropics. *In* Wilson, E.O., *ed.* Biodiversity. Washington, D.C., National Academy Press. 106-116. Value of wild genetic resources; lists potential crop plants. (E)

2153 **Plumwood, V., Routley, R. (1982).** World rainforest destruction: the social factors. *Ecologist,* 12(1): 4-7, 9-22. Illus., map.

2154 **Poore, D. (1974).** Saving tropical rain forests. *IUCN Bull. New Ser.,* 5(8): 29-30.

2155 **Poore, D. (1976).** Ecological guidelines for development in tropical rain forests. Morges, IUCN. viii, 39p. Illus.

2156 **Poore, D. (1978).** Ecological guidelines for development in tropical rain forests - part I. *Tigerpaper,* 5(2): 11-15. Illus. (T)

2157 **Poore, D. (1978).** Ecological guidelines for development in tropical rain forests - part II. *Tigerpaper,* 5(3): 18-22. Illus. (T)

2158 **Poore, D., Sayer, J. (1987).** The management of tropical moist forest lands: ecological guidelines. Gland, Switzerland and Cambridge, U.K., IUCN. v, 63p. (P)

2159 **Poore, M.E.D. (1983).** Driving forces for destruction. *People,* 10(1): 14-17. Illus. Tropical forests. A discussion of the UNEP Gems Report.

2160 **Prance, G.T. (1982).** Forest refuges: evidence from woody angiosperms. *In* Prance, G.T., *ed.* Biological diversification in the tropics. New York, Columbia University Press. 137-158. Centres of plant endemism.

2161 **Prance, G.T. (1984).** Ethnobotany aids in managing tropical rainforests. *Nat. Res. Techn. Bull.,* 6-3, 8. (E)

2162 **Prance, G.T., ed. (1982).** Biological diversification in the tropics. Proceedings of the Fifth International Symposium of the Association for Tropical Biology, held at Macuto Beach, Caracas, Venezuela, February 8-13, 1979. New York, Columbia Univ. Press. 714p. Illus., maps. (P)

2163 **Putz, F.E. (1988).** Blueprint for saving tropical forest. *Garden,* 12(2): 2-5, 32. Col. illus. (L)

2164 **Rao, Y.S., Chandrasekharan, C. (1983).** The state of forestry in Asia and the Pacific. *Unasylva,* 35(140): 11-21. Deforestation.

2165 **Raven, P.H. (1976).** The destruction of the tropics. *Missouri Bot. Gard. Bull.,* 64(9): 5-6. Illus. (T)

2166 **Raven, P.H. (1978).** The destruction of the tropics. *Bull. Amer. Assoc. Bot. Gard. Arbor.,* 12(1): 2-3. Deforestation.

2167 **Raven, P.H. (1988).** Our diminishing tropical forests. *In* Wilson, E.O., *ed.* Biodiversity. Washington, D.C., National Academy Press. 119-122. Deforestation.

2168 **Raven, P.H. (1988).** Tropical floristics tomorrow. *Taxon,* 37(3): 549-560. Includes discussion of species diversity and endemism.

2169 **Real, H.G. (1982).** A tool to save tropical ecosystems. *Garden* (New York), 6(6): 2-4, 32. Illus.

2170 **Richards, P.W. (1975).** Doomsday for the world's tropical rain forests? *Unesco Courier,* 32: 16-24.

2171 **Robinson, M. (1985).** Alternatives to destruction: investigations into the use of tropical forest resources with comments on repairing the effects of destruction. *Envir. Profess.,* 7: 232-239. (T)

PANTROPICAL

2172 **Roche, L. (1978).** Community forestry and the conservation of plants and animals. *In* Proceedings of the 8th World Forestry Congress, Jakarta, 16-28 October 1978: Forestry for quality of life. II, 19. (E)

2173 **Roche, L. (1979).** Forestry and the conservation of plants and animals in the tropics. *Forest Ecol. Management*, 2(2): 103-122. Illus. (T)

2174 **Roche, L., Dourojeanni, M.J. (1984).** A guide to *in situ* conservation of genetic resources of tropical woody species. Rome, FAO. (E)

2175 **Ross, M.S., Donovan, D.G. (1986).** Land clearing in the humid tropics, based on experience in the conversion of tropical moist forests in South East Asia. IUCN/IIED Tropical Forest Policy Paper No. 1. Cambridge, IUCN, London, IIED-Earthscan and WWF. 19p.

2176 **Rubinoff, I. (1982).** Tropical forests: can we afford not to give them a future? *Ecologist*, 12(6): 253-258. (T)

2177 **Rubinoff, I. (1983).** A strategy for preserving tropical forests. *In* Sutton, S.L., Whitmore, T.C., Chadwick, A.C., *eds.* Tropical rain forest: ecology and management. Special Publ. Series of the British Ecological Society No. 2. Oxford, Blackwell. 465-476.

2178 **Saenger, P., Hegerl, E.J., Davis, J.D.S., eds (1981).** First report on the global status of mangrove ecosystems. Gland, IUCN, Commission on Ecology. v, 132p. Maps. Contains lists of threatened plants. (T)

2179 **Sahabat Alam Malaysia (?)** Proceedings of the Conference on 'Forest Resources Crisis in the Third World', 6-8 September 1986. Penang, Sahabat Alam Malaysia. Deforestation. (E)

2180 **Salati, E., Vose, P.B. (1983).** Depletion of tropical rain forests. *Ambio*, 12(2): 67-71. Illus. (T)

2181 **Salati, E., Vose, P.B. (1984).** Depletion of tropical rain forests. *Trees S. Afr.*, 35(3-4): 50-57. Illus.

2182 **Schmidt, R.C. (?)** Current programmes of tropical rain forest management. Proceedings of an International Workshop on Rain Forest Regeneration and Management, 24-28 November 1989. Guri, Venezuela. In press.

2183 **Schoser, G. (1977).** The conservation of tropical orchids. *In* Stone, B.C., *ed.* The role and goals of tropical botanic gardens. Proceedings, symposium held at Rimba Ilmu Botanic Garden, Univ. of Malaya, Kuala Lumpur, August 1974. Kuala Lumpur, Rimba Ilmu Univ. 175-179. (T)

2184 **Secrett, C. (1986).** Rain forest: protecting the planet's richest resource London, Friends of the Earth. 90p. Illus., maps. (1986?).

2185 **Shiva, V. (1987).** Forestry crisis and forestry myths. A critical review of tropical forests. A call for action. Malaysia, World Rainforest Movement.

2186 **Simberloff, D. (1986).** Are we on the verge of mass extinction in tropical rain forests? *In* Elliott, D.K., *ed.* Dynamics of extinction. New York, Wiley. 165-180. (T)

2187 **Skorupa, J.P., Kasenene, J.M. (1984).** Tropical forest management: can rates of natural treefalls help guide us? *Oryx*, 18(2): 26-101. Illus., map.

2188 **Soepadmo, E. (1977).** Conservation of wild fruit tree species. *In* Stone, B.C., *ed.* The role and goals of tropical botanic gardens. Proceedings, symposium held at Rimba Ilmu Botanic Garden, Univ. of Malaya, Kuala Lumpur, August 1974. Kuala Lumpur, Rimba Ilmu Univ. 207-210. (E/G)

2189 **Sommer, A. (1976).** Attempt at an assessment of the world's tropical forests. *Unasylva*, 28(112-113): 5-24.

PANTROPICAL

2190 **Spears, J. (1988).** The Tropical Forestry Action Plan: environmental concerns and donor response. *In* McDermott, M.J., *ed.* The future of the tropical rain forest. Proceedings of an international conference at St. Catherine's College, Oxford, 27-28 June 1988. Oxford, Oxford Forestry Institute. 23-27. (E)

2191 **Spears, J.S. (1980).** Can farming and forestry coexist in the tropics? *Unasylva*, 32(128): 2-12. Illus. (T)

2192 **Spears, J.S. (1980).** Can the wet tropical forest survive? *Commonw. Forest. Rev.*, 58: 165-180.

2193 **Spears, J.S. (1983).** Replenishing the world's forests. Tropical reforestation: an achievable goal? *Commonw. Forest. Rev.*, 62(3): 201-217.

2194 **Stadtmuller, T. (1989).** Cloud forests in the humid tropics: a bibliographic review. Tokyo, United Nations University. 81p. (B)

2195 **Steinhart, P. (1983).** Trouble in the tropics. *Nat. Wildl.*, 22(1): 16-20.

2196 **Steinlin, H.J. (1982).** Monitoring the world's tropical forests. *Unasylva*, 34(137): 2-8. Illus. (T)

2197 **Stoel, T. (1980).** IUCN rainforest statement: averting 'a major setback to development'. *IUCN Bull.*, 11(5): 22-26. (T)

2198 **Sutlive, V.H., Altshuler, N., Zamora, M.D., eds (1981).** Where have all the flowers gone? Deforestation in the Third World. Studies in Third World Societies No. 13. Williamsburg, Virginia, Dept of Anthropology, College of William and Mary. xi, 277p. En, Fr. Maps. (L/P)

2199 **Sutlive, V.H., Altshuler, N., Zamora, M.D., eds (1986).** Blowing in the wind: deforestation and long-range implications. Studies in Third World Societies No. 14. Williamsburg, Virginia, Dept of Anthropology, College of William & Mary. 514p.

2200 **Sutton, S.L., Whitmore, T.C., Chadwick, A.C. (1983).** Tropical rain forest: ecology and management. Oxford, Blackwell. 498p.

2201 **Talbot, J.J., Pettinger, L.R. (1980).** Use of remote sensing for monitoring deforestation in tropical and sub-tropical latitudes. *Ciencia Interamericana*, 21: 63-71.

2202 **Tangley, L. (1986).** Saving tropical forests. *BioScience*, 36(1): 4-8. Tropical forests: a call for action.

2203 **Tangley, L. (1988).** Studying (and saving) the tropics. *BioScience*, 38(6): 375-385. Organization for Tropical Studies.

2204 **Terborgh, J. (1986).** Keystone plant resources in the tropical forest. *In* Soule, M.E., *ed.* Conservation biology: the science of scarcity and diversity. Sunderland, MA, Sinauer. 330-344.

2205 **Thomson, K. (1989).** The Friends of the Earth Tropical Rain Forest Campaign. *In* McDermott, M.J., *ed.* The future of the tropical rain forest. Proceedings of an international conference at St. Catherine's College, Oxford, 27-28 June 1988. Oxford, Oxford Forestry Institute. 58-56. (E)

2206 **Tinker, J. (1979).** In place of forests. *New Sci.*, 82(1151): 170. Tropical forests.

2207 **Train, R. (1986).** The threat to tropical forests. *Atlantic Naturalist*, 36: 3-4.

2208 **U.S. Department of State (1980).** The world's tropical forests: a policy, strategy, and program for the United States. Washington, D.C., U.S. Govt Print. Office. 53p.

2209 **U.S. Department of State. Agency for International Development (1978).** Proceedings of the U.S. Strategy Conference on Tropical Deforestation, June 12-14, 1978, Washington, D.C. Washington, D.C., U.S. Department of State. 78p.

PANTROPICAL

2210 **U.S. General Accounting Office (1982).** Changes needed in U.S. assistance to deter deforestation in developing countries: report to the Congress. Washington, D.C., U.S. General Accounting Office. 56p. Illus.

2211 **UNCTAD. Information Unit (1984).** International agreement on tropical timber adopted. *Envir. Conserv.*, 11(2): 182. (E/L)

2212 **UNEP/FAO (1982).** The Global Environment Monitoring System. GEMS PAC Info Series 3: The global assessment of tropical forest resources. Nairobi, UNEP/FAO. 14p. Extent of tropical forest and deforestation rates.

2213 **Unesco (1974).** International working group on Project 1. Ecological effects of increasing human activities on tropical and sub-tropical forest ecosystems. Final report. Programme on Man and the Biosphere [MAB] Report Series No. 16. Rio de Janeiro, Unesco. 96p.

2214 **Unesco (1976).** Scientific problems of the humid tropical zone deltas and their implications. Proceedings of the Dacca Symposium, 1964. Dacca? 422p.

2215 **Unesco/UNEP/FAO (1978).** Tropical forest ecosystems: a state-of-knowledge report. Natural Resources Research XIV. Paris, Unesco. 3 vols. 683p. Maps.

2216 **Urquhart, T. (1987).** Save the birds - why bother? *New Sci.*, 115(1567): 55-58. Illus., col. illus., map. Birds at risk could help pinpoint threatened forests and centres of endemism.

2217 **Villa Lobos, J. (1987).** SI/MAB biological diversity progam. *Biol. Conserv. Newsl.*, 49: 1-2.

2218 **Vogel, E.F. de (1976).** Tropical orchids as an endangered plant group. *Flora Males. Bull.*, 29: 2602-2604. (T)

2219 **WWF (1983).** WWF launch Tropical Forest Campaign. *Threatened Pl. Commit. Newsl.*, 10: 1-4.

2220 **WWF (1988).** ITTO Tropical forest conservation and the International Tropical Timber Organization. Position Paper No. 1. Gland, WWF. 24p. Maps. (E)

2221 **Wells, S.M., Pyle, R.M., Collins, N.M., comps (1983).** Threatened communities. *In* IUCN. The IUCN invertebrate Red Data Book. Gland and Cambridge, IUCN. 559-615. Data sheets on important sites and communities, many of which are centres of plant diversity and endemism. (P)

2222 **White, P.T. (1983).** Nature's dwindling treasures - rain forests. *Natl. Geog.*, 163(1): 2-47.

2223 **Whitelock, L.M. (1978).** The twilight of the cycads (endangered species). *Garden* (New York), 2(5): 6-10. (T)

2224 **Whitmore, T.C. (1975).** Conservation review of tropical rain forests, general considerations and Asia. Switzerland, IUCN. 116p. Introduction to forest types and their distribution, protected area coverage and conservation priorities for rain forest countries from India east to Polynesia. (E/P)

2225 **Whitmore, T.C. (1979).** Tree conservation in the tropical rain forest. *Int. Dendrol. Soc. Yearbook*, (1979): 16-19. Illus.

2226 **Whitmore, T.C. (1980).** 17. The conservation of tropical rain forest. *In* Soule, M.E., Wilcox, B.A., eds. Conservation biology: an evolutionary and ecological perspective. Massachusetts, Sinauer Associates. 303-318.

2227 **Wickens, G., Goodin, J.R., Field, D.V., eds (1985).** Plants for arid lands: proceedings of the Kew International Conference on Economic PLants for Arid Lands, held in the Jodrell Laboratory, Royal Botanic Gardens, Kew, England, 23-27 July 1984. London, Allen & Unwin. xiv, 452p. Illus., map. (E/T)

PANTROPICAL

2228 **Wilcox, B.A. (1988).** Tropical deforestation and extinction. *In* IUCN Conservation Monitoring Centre. 1988 Red List of threatened animals. Gland, Switzerland and Cambridge, U.K., IUCN. v-x.

2229 **Williams, J.T. (1977).** Conservation of genetic resources in the tropics. *In* Stone, B.C., *ed.* The role and goals of tropical botanic gardens. Proceedings, symposium held at Rimba Ilmu Botanic Garden, Univ. of Malaya, Kuala Lumpur, August 1974. Kuala Lumpar, Rimba Ilmu Universiti. 187-200. Map. (E/G)

2230 **Winterbottom, B. (1989).** Environmental implications of the Tropical Forestry Action Plan. *In* McDermott, M.J., *ed.* The future of the tropical rain forest. Proceedings of an international conference at St. Catherine's College, Oxford, 27-28 June 1988. Oxford, Oxford Forestry Institute. 28-32.

2231 **Wong, K.M. (1984).** Tropical forests. Can we cope with the dwindling resource? *Wallaceana*, W35 March: 3-7. Illus.

2232 **World Resources Institute (1985).** Tropical forests: a call for action. Report of an International Task Force convened by the World Resources Institute, the World Bank and the United Nations Development Programme. Washington, D.C., World Resources Institute. 3 vols; 47, 55, 22p. Illus., maps.

2233 **World Resources Institute, World Bank (1985).** Accelerated Action Plan for Tropical Forests. Washington, D.C., World Resources Institute and World Bank.

2234 **Worrall, J. (1979).** Trees: a Third World crisis. *Newsl. Soc. Protect. Environ.*, 9(2): 7-8. (E)

2235 **Wyatt-Smith, J. (1987).** The management of tropical moist forest for the sustained production of timber: some issues. IUCN/IIED Tropical Forest Policy Paper No. 4. Cambridge, IUCN, London, IIED-Earthscan and WWF. 20p. Map. (E)

2236 **Yeom, F. (1984).** Lesser known tropical wood species: how bright is their future? *Unasylva*, 36(146): 2-16. (E)

SOUTH AMERICA

2237 **Allegretti, M.H. (?)** Extractive reserves: an alternative for reconciling development and environmental conservation in Amazonia. *In* Anderson, A.B. *ed.* Alternatives to deforestation: steps toward sustainable use of the Amazon rainforest. New York, Columbia University Press. In press.

2238 **Alvim, P. de T. (1978).** Floresta amazonica: equilibrio entre utilizacao e conservacao. (Amazon forest: balance between utilization and conservation.) *Cienc. Cult.*, 30(1): 9-16. Por (En). (E)

2239 **Anderson, A.B., Prance, G.T., Albuquerque, de B.M. (1975).** A vegetacao lenhosa da Campina da Reserva Biologica INPA-SUFRAMA (Manaus-Caracarai, km 62). *Acta Amazonica*, 5(3): 225-246. (P)

2240 **Anderson, A.B., ed. (?)** Alternatives to deforestation: steps toward the sustainable use of the Amazon rainforest. New York, Columbia University Press. In press. (E)

2241 **Armitage, F.B., Joustra, P.A., Ben Salem, B. (1980).** Genetic resources of tree species in arid and semi-arid areas: a survey for the improvement of rural living in Latin America, Africa, India and Southwest Asia. Rome, FAO. vi, 118p. (E)

2242 **Balick, M.J. (1985).** Useful plants of Amazonia: a resource of global importance. *In* Prance, G.T., Lovejoy, T.E., *eds.* Key environments: Amazonia. Oxford, Pergamon. 339-368. (E)

2243 **Barrett, S.W. (1980).** Conservation in Amazonia. *Biol. Conserv.*, 18(3): 209-235. Maps.

SOUTH AMERICA

2244 **Bazan, F. (1968).** International coordination of the conservationist laws. *In* IUCN. Conservation in Latin America. Proceedings, San Carlos de Bariloche, Argentina, 27 March-2 April 1968. Morges, IUCN. 253-256. (L)

2245 **Budowski, G., MacFarland, C. (1984).** Keynote address: the Neotropical Realm. *In* McNeely, J.A., Miller, K.R., *eds.* National parks, conservation and development. The role of protected areas in sustaining society. Washington, D.C., Smithsonian Institution Press. 552-560. Table. (P)

2246 **Buschbacher, R.J. (1986).** Tropical deforestation and pasture development. *BioScience*, 36(1): 22-28. Illus. Amazon river basin.

2247 **Caldevilla, G.M. (1978).** Other forest products: a source of foreign exchange. *In* Proceedings of the 7th World Forestry Congress, Buenos Aires, 4-18 October 1972, vol. 3: Conservation and recreation. Buenos Aires, Instituto Forestal Nacional. 3879-3896. En, Sp, Fr. Includes conservation. (E)

2248 **Castro, J.P.C. de (1981).** 'Vereda' vegetation and its legal protection. *Brasil Florestal*, 11(46): 39-54. Por. (L)

2249 **Clement, C.R., Chavez Flores, W.B. (1983).** Review of genetic erosion of Amazon perennial crops. *Pl. Genet. Resourc. Newsl.*, 55: 21-23. (E)

2250 **Denevan, W. (1982).** Causes of deforestation and forest woodland degradation in tropical Latin America. Washington, D.C., Office of Technology Assessment.

2251 **Dodson, C.H. (1968).** Conservation of orchids. *In* IUCN. Conservation in Latin America. Proceedings San Carlos de Bariloche, Argentina, 27 March-2 April 1968. Morges, IUCN. 170. (T)

2252 **Dourojeanni, M.J. (1978).** Economic values of wild production in the forest regions of Latin America. *In* Proceedings of the 7th World Forestry Congress, Buenos Aires, 4-18 October 1972, vol. 3: Conservation and recreation. Buenos Aires, Instituto Forestal Nacional. 3897-3920. (E)

2253 **Dourojeanni, M.J. (1984).** Future directions for the Neotropical Realm. *In* McNeely, J.A., Miller, K.R., *eds.* National parks, conservation and development. The role of protected areas in sustaining society. Washington, D.C., Smithsonian Institution Press. 621-625. (P)

2254 **Eden, M.J. (1990).** Ecology and land management in Amazonia. Belhaven Press. 256p. Illus. (E)

2255 **FAO (1985).** Intensive multiple-use forest management in the tropics. Analysis of case studies from India, Africa, Latin America and the Caribbean. Rome, FAO Forestry Paper 55. 180p. (E)

2256 **FAO (1986).** Report on natural resources for food and agriculture in Latin America and the Caribbean. FAO Environment and Energy Paper 8. Rome, FAO. 102p. (E)

2257 **FAO (1986).** Some medicinal forest plants of Africa and Latin America. FAO Forestry Paper No. 67. Rome, FAO. 252p. (E)

2258 **Farnworth, E.G., Golley, F.B., eds. (1974).** Fragile ecosystems: evaluation of research and application in the neotropics: a report of the Institute of Ecology, June 1973. New York, Springer-Verlag. 258p.

2259 **Fearnside, P.M. (1985).** Agriculture in Amazonia. *In* Prance, G.T., Lovejoy, T.E., *eds.* Key environments: Amazonia. Oxford, Pergamon. 393-418.

2260 **Fearnside, P.M. (1989).** Forestry management in Amazonia: the need for new criteria in evaluating development options. *Forsest Ecology and Management*, in press.

SOUTH AMERICA

2261 **Forero, E. (1987).** 80,000 plants in South America: the case for creating more botanic gardens. *In* Bramwell, D., Hamann, O., Heywood, V., Synge, H., *eds.* Botanic gardens and the World Conservation Strategy. London, Academic Press. 227-237. (Sp). (G)

2262 **Freese, C.J.S.C. (1986).** Prioridades biologicas de conservacion en los Andes tropicales. *Parks*, 11(2-3): 8-11. Sp (En). Maps.

2263 **Fuller, K.S., Swift, B. (1984).** Latin American wildlife trade laws (Leyes del Comercio de Vida Sylvestre en America Latina). Washington, D.C., WWF-US. En (Sp). Mimeograph. Accounts for each country in South and Middle America on wildlife laws, with species lists; mostly faunal but flora to be covered in more detail in revised version in preparation. (L/T)

2264 **Gentry, A. (1979).** Extinction and conservation of plant species in tropical America: a phytogeographical perspective. *In* Hedberg, I., *ed.* Systematic botany, plant utilization and biosphere conservation. Stockholm, Almqvist & Wiksell International. 110-126. Illus., maps. (T)

2265 **Gilbert, L.E. (1981).** Food web organization and the conservation of neotropical diversity. *In* Jordan, C.F., *ed.* Tropical ecology. Stroudsberg, Pennsylvania, Hutchinson Ross Publishing Company. 113-135. (Benchmark Papers in Ecology, vol. 10.) Reprinted from Soule, M.E., Wilcox, B.A., *eds.* Conservation biology: an evolutionary-ecology perspective. Sunderland, Massachussetts, Sinauer Associates. 11-33 (1980).

2266 **Goodland, R., (1977).** Panel discussion. *In* Prance, G.T., Elias, T.S., *eds.* Extinction is forever. Proceedings of a symposium at the New York Botanical Garden, 11-13 May 1976. New York, New York Botanical Garden. 360-367. Panel discussion on important differences between conservation in Southeast Asia and Latin America. (E/L/P/T)

2267 **Goodland, R., Irwin, H.S., Tillman, G. (1978).** Ecological development for Amazonia. *Cienc. Cult.*, 30(3): 275-289.

2268 **Goodland, R.J., Irwin, H.S. (1975).** Amazon jungle: green hell to red desert? New York, Elsevier. 155p.

2269 **Hartshorn, G.S. (1980).** Neotropical forest dynamics. *Biotropica*, 12(2) suppl.: 23-30. (Sp).

2270 **Herrera, R., Jordan, C.F., Medina, E., Klinge, H. (1981).** How human activities disturb the nutrient cycles of a tropical rainforest in Amazonia. *Ambio*, 10(2-3): 109-114. Illus., map.

2271 **Hueck, K. (1978).** Los bosques de Sudamerica. Ecologia, composicion e importancia economica. Eschborn, F.R.G., G.T.Z. 476p. Sp

2272 **IUCN (1968).** Conservation in Latin America. Proceedings of the Latin American Conference on Renewable Natural Resources, San Carlos de Bariloche, Argentina, 27 March-2 April 1968. IUCN Publications New Series No. 13. Morges, IUCN. 517p.

2273 **IUCN (1975).** The use of ecological guidelines for development in the American humid tropics. Proceedings of International Meeting held at Caracas, Venezuela, 20-22 February 1974. IUCN Publications New Series No. 31. Morges, IUCN. 249p. Maps.

2274 **IUCN (1981).** Conserving the natural heritage of Latin America and the Caribbean. The planning and management of protected areas in the Neotropical Realm. Proceedings of the 18th working session of IUCN's Commission on National Parks and Protected Areas, Lima, Peru, 21-28 June 1981. Gland, IUCN. 329p. (P)

SOUTH AMERICA

2275 **IUCN Commission on Ecology (1983).** Ecological structures and problems of Amazonia. Proceedings of a symposium organised by the Dept of Biological Sciences of Federal Univ. of Sao Carlos and IUCN, Sao Carlos, Brazil, 18 March 1982. Gland, Switzerland, IUCN. 79p. Illus., maps. (E/P)

2276 **IUCN Commission on National Parks and Protected Areas (1982).** IUCN Directory of Neotropical protected areas. Dublin, Tycooly International. 436p. (Sp). Maps. (P)

2277 **Iltis, H.H., comp. (1978).** Extinction or preservation: what biological future for the South American tropics? Madison, University of Wisconson, Dept of Botany. 208p.

2278 **Ives, J.D. (1981).** Applied mountain geoecology. *In* Lall, J.S., *ed.* The Himalaya: aspects of change. New Delhi, India International Centre, Oxford University Press. 377-402. A general article discussing conditions in the Andes and Rockies as well as the Himalayas. (P/T)

2279 **Janzen, D.H. (1988).** Tropical dry forests: the most endangered major tropical ecosystem. *In* Wilson, E.O., *ed.* Biodiversity. Washington, D.C., National Academy Press. 130-137.

2280 **Jimenez, J.A. (1984).** A hypothesis to explain the reduced distribution of the mangrove *Pelliciera rhizophorae* Tr. and Pl. *Biotropica*, 16(4): 304-308. Maps. (T)

2281 **Johnson, D. (1987).** Conservation status of wild palms in Latin America and the Caribbean. *Principes*, 31(2): 96-97. (T)

2282 **Jordan, C.F. (1982).** Amazon rain forests. *Amer. Sci.*, 70(4): 394-401. Illus. (T)

2283 **Kozarik, J.M. (1978).** Watershed and forest management. Spheres of action and interrelationships. *In* Proceedings of the 7th World Forestry Congress, Buenos Aires, 4-18 October 1972, vol. 3: Conservation and recreation. Buenos Aires, Instituto Forestal Nacional. 3583-3603. En, Sp, Fr.

2284 **Lamlein, J. (1982).** Plant conservation in tropical America: the TPC-SI programme. *Threatened Pl. Commit. Newsl.*, 9: 6. *Puya raimondii.* (T)

2285 **Lamlein, J. (1984).** Latin American conservation network. *Threatened Pl. Newsl.*, 13: 6-7.

2286 **Laroche, R.C.M. (1978).** Contribuicao ao conhecimento da ecologia de floresta pluvial tropical e sua conservacao. *Rodriguesia*, 30(4): 105-107. Por.

2287 **Linares, O. (1976).** Garden hunting in the American tropics. *Human Ecology*, 4(4): 331-349. (G)

2288 **Lovejoy, T.E. (1979).** Conservation beyond our borders. *Nat. Conserv. News*, 29(4): 4-7.

2289 **Mares, M.A. (1986).** Conservation in South America: problems, consequences and solutions *Science*, 233: 734-739.

2290 **Meganck, R.A., Geobel, J.M. (1979).** Les parcs d'Amerique Latine et leurs problemes. *Parks* (USA), 4(2): 4-8. Fr. (P)

2291 **Miller, K.R. (1975).** Ecological guidelines for the management and development of national parks and reserves in the American humid tropics. *In* IUCN. The use of ecological guidelines for development in the American humid tropics. Proceedings, Caracas, Venezuela, 20-22 February 1974. IUCN Publications New Series No. 31. Morges, IUCN. 91-105. (P)

2292 **Miller, K.R. (1978).** Planning national parks for ecodevelopment: methods and cases from Latin America. Ann Arbor, University of Michigan, Center for Strategic Wildland Management Studies, School of Natural Resources. ix, 624p. 2 vols. Illus. Maps. Spanish language version also available. (P)

SOUTH AMERICA

2293 **Milton, J.P. (1975).** The ecological effects of major engineering projects. *In* IUCN. The use of ecological guidelines for development in the American humid tropics. Proceedings, Caracas, Venezuela, 20-22 February 1974. 207-221. Switzerland, IUCN New Series No. 31.

2294 **Morales, H.L. (1984).** For a self-sustained development. *In* Unesco-UNEP. Conservation, science and society. Paris, Unesco. 478-485. Contributions to the First International Biosphere Reserve Congress, Minsk, Byelorussia/USSR, 26 September - 2 October 1983. Tables. (P)

2295 **Munoz Pizarro, C. (1975).** Especies vegetales que se extinguen en nuestro pais. *In* Capurro, L., Vergara, R., *eds.* Present y futuro del medio humano. Mexico, Edit. Cont. CECSA. Capitulo XI. 161-179. Sp. (T)

2296 **Myers, N. (1979).** Islands of conservation. *New Sci.*, 83(1169): 600-602. Illus., map. Amazonian forest.

2297 **Nations, J., Komer, D. (1982).** Rainforest, cattle and the hamburger society. Austin, Center for Human Ecology. Deforestation.

2298 **New York Botanical Garden (1976).** Symposium on threatened and endangered species of plants in the Americas and their significance in ecosystems today and in the future. Program and Abstracts. New York, New York Botanical Garden. 55p. (T)

2299 **Ospina Hernandez, M. (1968).** The disappearance of valuable native orchids in Latin-America. *In* IUCN. Conservation in Latin America. Proceedings, San Carlos de Bariloche, Argentina, 27 March-2 April 1968. Morges, IUCN. 168-169. (T)

2300 **Pires, J.M., Prance, G.T. (1985).** The vegetation types of the Brazilian Amazon. *In* Prance, G.T., Lovejoy, T.E., *eds.* Key environments: Amazonia. Oxford, Pergamon. 109-145.

2301 **Plotkin, M.J. (1985).** Standardized format for conservation and ethnobotanical data in Tropical South America Project. *Taxon*, 34(1): 120-121. (E)

2302 **Prance, G., Lovejoy, T., eds (1985).** Key environments: Amazonia. Oxford, Pergamon. 442p. Illus., maps. (T)

2303 **Prance, G.T. (1976).** Threatened and endangered plants in the Americas. *BioScience*, 26(10): 633-634. (T)

2304 **Prance, G.T. (1978).** Conservation problems in the Amazon Basin. *In* Schofield, E.A., *ed.* Earthcare: global protection of natural areas. Proceedings of the XIV Biennial Wilderness Conference. 191-207.

2305 **Prance, G.T. (1985).** The increased importance of ethnobotany and under-exploited plants in a changing Amazon. *In* Hemming, J., *ed.* Change in the Amazon basin. Vol.1: Man's impact on forests and rivers. Cambridge, Cambridge University Press. 129-136. (E)

2306 **Prance, G.T. (1986).** The Amazon: paradise lost? *In* Kaufman, L., Mallort, K., *eds.* The last extinction. Cambridge, Mass., MIT Press. 62-106.

2307 **Prance, G.T. (1986).** The conservation and utilization of the Amazon rainforest. *Revta. Acad. Colomb. Cienc. Exact. Fisic. Nat.*, 16(60): 117-127. (E)

2308 **Prance, G.T. (1989).** Economic prospects for tropical rainforest ethnobotany. *In* Browder, J.O., *ed.* Fragile hands on Latin America. Boulder, Westview Press. 61-74. (E)

2309 **Prance, G.T., Balee, W., Boom, B.M., Carneiro, R.L. (1987).** Quantitative ethnobotany and the case for conservation in Amazonia. *Conservation Biology*, 1: 296-310.

2310 **Rajanaidu, N. (1983).** *Elaeis oleifera* collection in South and Central America. *Pl. Genet. Resource. Newsl.*, 56: 42-51. En (Fr, Sp). Maps. (E/T)

SOUTH AMERICA

2311 **Rands, R.J. (1975).** Phragmipediums - and their future. *Amer. Orch. Soc. Bull.*, 44: 235-238.

2312 **Ravenna, P. (1977).** Neotropical species threatened and endangered by human activity in the Iridaceae, Amaryllidaceae and allied bulbous families. *In* Prance, G.T., Elias, T.S., *eds.* Extinction is forever. Proceedings of a symposium at the New York Botanical Garden, 11-13 May 1976. New York, New York Botanical Garden. 257-266. Illus. (T)

2313 **Read, M. (1989).** Bromeliads threatened by trade. *Kew Mag.*, 6(1): 22-29. Illus. (T)

2314 **Schultes, R.E. (1979).** The Amazonia as a source of new economic plants. *Econ. Bot.*, 33(3): 259-266. (E)

2315 **Smith, N. (1983).** New genes from wild potatoes. *New Sci.*, 98(1359): 558-560, 564-565. Illus., col. illus. International Potato Centre, Lima. (E)

2316 **Timoni, J.L., et al. (1980).** Conservacao genetica da *Araucaria angustifolia* (Bert.) O. Ktze. *In* Int. Union of Forestry Research Organizations. Forestry problems of the genus *Araucaria*. IUFRO meeting, Curitiba, Parana (Brazil), 21-28 October 1979. Fundacao de Pesquisas Florestais do Parana. 115-118. Por (En). (T)

2317 **Uhl, C. (1984).** You can keep a good forest down. *Nat. Hist.*, 92(4): 71-79. Tropical rain forest.

2318 **Wadsworth, F.H. (1975).** Natural forests in the development of the American humid tropics. *In* IUCN. The use of ecological guidelines for development in the American humid tropics. Proceedings, Caracas, Venezuela, 20-22 February 1974. Morges, IUCN. 129-138.

2319 **Walker, C.C. (1981).** Asclepiads under threat. *Asklepias*, no. 24: 41. (T)

2320 **Watters, R.F. (1975).** Shifting agriculture - its past, present and future. *In* IUCN. The use of ecological guidelines for development in the American humid tropics. Proceedings, Caracas, Venezuela, 20-22 February 1974. 77-86.

2321 **Wetterberg, G.B., Castro, C.S.D., Quintao, A.T.B., Porto, E.R. (1978).** Estado atual dos parques nacionais et reservas equivalentes na America do Sul, 1978. (Present state of national parks and equivalent reserves in South America, 1978.) *Brasil Florest.*, 9(36): 11-36. Por. Illus., maps. (P)

2322 **Wetterberg, G.B., Prance, G.T., Lovejoy, T.E. (1981).** Conservation progress in Amazonia: a structural review. *Parks*, 6(2): 5-10. (P)

2323 **Wetterberg, G.B., et al. (1976).** Uma Analise de Prioridades em Conservacao da Natureza na Amazonia. Serie Tecnica No. 8. Brasilia, Instituto Brasileiro de Desenvolvimento Florestal. 62p. Por.

2324 **Williams, J.T. (1981).** Cacao genetic resources in Latin America. *Pl. Genet. Resourc. Newsl.*, 45: 20-22. (Fr, Sp). (E)

2325 **Woodwell, G.M., Houghton, R.A., Stone, T.A., Nelson, R.F., Kavalick, W. (1987).** Deforestation in the tropics: new measurements in the Amazon Basin, using Landsat and NOAA advanced very high resolution radiometer imagery. *J. Geophysical Research*, 92: 2157-2163.

2326 **Wright, R.M. (1979).** Small steps, new paths; a review of conservation efforts south of the border. *Nat. Conserv. News*, 29(4): 14-19.

SOUTH ATLANTIC AND SOUTHERN OCEAN IS.

2327 **Carlquist, S. (1965).** Island life: a natural history of the islands of the world. New York, Natural History Press. 451p. Origin, evolution and adaptations of island flora and fauna; Galapagos and Hawaiian Islands well covered.

SOUTH ATLANTIC AND SOUTHERN OCEAN IS.

2328 **Clark, M.R., Dingwall, P.R. (1985).** Conservation of islands in the Southern Ocean: a review of the protected areas of Insulantarctica. Gland and Cambridge, IUCN. 188p. Illus., maps. Directory of information on physical and biological features of the southern islands; conservation status, problems and priorities; details of administration and management of protected areas. (P/T)

2329 **Heywood, V.H. (1979).** The future of island floras. In Bramwell, D., *ed*. Plants and islands. London, Academic Press. 431-441. (Sp). (P/T)

2330 **IUCN Threatened Plants Unit (1983).** The botanic gardens list of threatened plants of the South Atlantic and Southern Ocean Islands. Kew, IUCN Botanic Gardens Conservation Co-ordination Body Report No. 7. Lists plants in cultivation in botanic gardens. (G/T)

2331 **Meurk, C.D., Lee, W.G., Foggo, M.N. (1983).** Conservation on subantarctic islands: a botanical perspective. *Pac. Sci. Congr. Proc.*, 15(1-2): 164. (T)

2332 **Oldfield, S. (1987).** Fragments of paradise. A guide for conservation action in the U.K. Dependent Territories. Oxford, Pisces Publications; for the British Association for Nature Conservationists. 192p.

2333 **SCAR/IUCN (1986).** Conservation of subantarctic islands. The biological basis for conservation of subantarctic islands. Report of the Joint SCAR/IUCN Workshop at Paimpont, France, 12-14 September 1986. SCAR/IUCN. 32p. Illus., maps. (L/P)

2334 **Sachet, M.-H., Fosberg, F.R. (1971).** Island bibliographies supplement. Washington, D.C., Pacific Science Board, Nat. Academy of Sciences. 427p. Islands worldwide. (B/P/T)

SOUTH EAST ASIA

2335 **Alphonso, A.G. (1978).** Orchid species of South East Asia and the need for their conservation. *In* Soon, T.E., *ed*. Orchids. Singapore, Times Periodicals. 62-65. Illus., col. illus. (T)

2336 **Anon. (1980).** Dr. Ashton returns from the endangered tropical forests of the Far East. *Plant Sciences, Newsl. Arnold Arbor.*, Fall/Winter 1980: 2-3. Illus. (T)

2337 **Anon. (1981).** Genetic resources of east Asia and the Pacific. *Pl. Genet. Resource. Newsl.*, 45: 48-49. (Fr, Sp). (E)

2338 **Anon. (1981).** Keeping the stoves of the Third World burning. *IUCN Bull.*, 12(3-4): 15-16. Illus. (E)

2339 **Anon. (1982).** Miscellaneous information: conservation. *Flora Males. Bull.*, 35: 3757-3767. Including some national parks. (P)

2340 **Anon. (1988).** Rare species of orchid found. *Bull. Orchid Soc. S.E. Asia*, 6(6): 5. *Vanda hookeriana*. News clipping from *New Straits Times* June 7, 1988. (T)

2341 **Bompard, J.M., Kostermans, A.J.G.H. (1985).** Preliminary results of an IUCN/WWF sponsored project for conservation of wild *Mangifera* species in situ in Kalimantan (Indonesia). Montpellier, Laboratoire de Botanique Tropicale and Bogor, BIOTROP. Col. illus., maps. Mimeo. (E/P/T)

2342 **Borja, E.B. (1981).** Saving Asia's forests. *Canopy*, 7(12): 6. Illus.

2343 **Chomchalow, N. (1982).** Medicinal and aromatic plants germplasm conservation in ESCAP region. *IBPGR Region. Committ. Southeast Asia Newsl.*, 7(4): 5. (E/G/T)

2344 **Chomchalow, N. (1983).** Genetic diversity and genetic erosion in ESCAP region. *IBPGR Region. Committ. Southeast Asia Newsl.*, 7(4): 5-6. (E/G)

2345 **Comber, J.B. (1981).** What can we do about conservation ? *In* Comber, J., *ed*. Wayside orchids of Southeast Asia. Kuala Lumpur, Heinemann Asia. vii, 28p. Col. illus. (T)

SOUTH EAST ASIA

2346 **Commonwealth Science Council (1987).** Biological diversity and genetic resources: life support species. Summary report. International workshop on maintenance and evaluation of life support species in Asia and the Pacific region, New Delhi, 4-7 April 1987. London, Commonwealth Science Council. 25p. (E)

2347 **Commonwealth Science Council (1988).** Biological diversity and genetic resources techniques and methods: conservation biology. Executive summary. Regional training workshop held at Royal Botanic Gardens, Peradeniya, Sri Lanka, 16-21 May 1988. London, Commonwealth Science Council. 22p. (E)

2348 **Dahl, A.L. (1984).** Oceania's most pressing environmental concerns. *Ambio*, 13(5-6): 296-299. Col. illus., map.

2349 **Davidson, J., Tho Y.P., Bijleveld, M. (1983).** The future of tropical rain forests in South East Asia. Commission on Ecology Papers No. 10. Gland, IUCN. 127p. Proceedings of a symposium organised by the Forest Research Institute, Kepong, Malaysia and the IUCN Commission on Ecology held in Kepong, Malaysia, 1-2 September 1983. Illus., maps. Rain forest, deforestation.

2350 **Davis, S. (1983).** The work of I.U.C.N. in promoting plant conservation in South East Asia and the Pacific region. *In* Given, D.R., *ed.* Conservation of plant species and habitats. Wellington, Nature Conservation Council. 117-120. (E/P/T)

2351 **Dransfield, J. (1981).** The biology of Asiatic rattans in relation to the rattan trade and conservation. *In* Synge, H., *ed.* The biological aspects of rare plant conservation. Chichester, Wiley. 179-186. Proceedings of International Conference, King's College, Cambridge, 14-19 July 1980. (E/T)

2352 **ESCAP (1985).** State of the environment in Asia and the Pacific. Vol. 1: Summary. Bangkok, ESCAP.

2353 **FAO/UNEP (1981).** Tropical Forest Resources Assessment Project (in the framework of GEMS). Forest resources of tropical Asia. Part 1: Regional synthesis. Rome, FAO. 475p. En, Fr. Tropical forests; deforestation.

2354 **FAO/UNEP (1981).** Tropical Forest Resources Assessment Project (in the framework of GEMS). Forest resources of tropical Asia. Part II: Country briefs. Rome, FAO. 586p. En, Fr. Tropical forests; deforestation.

2355 **Fortes, M.D. (1988).** Mangrove and seagrass beds of East Asia: habitats under stress. *Ambio*, 17(3): 207-213. Col. illus., map.

2356 **Goodland, R., (1977).** Panel discussion. *In* Prance, G.T., Elias, T.S., *eds.* Extinction is forever. Proceedings of a symposium at the New York Botanical Garden, 11-13 May 1976. New York, New York Botanical Garden. 360-367. Panel discussion on important differences between conservation in Southeast Asia and Latin America. (E/L/P/T)

2357 **Griggs, T. (1981).** The green gene banks of Asia. *Asia 2000* (Hong Kong), 1(3): 24-27. (G)

2358 **Hamilton, L.S., ed. (1983).** Forest and watershed development in Asia and the Pacific. London, Westview. 560p.

2359 **IBPGR (1980).** Genetic resources of the Far East and the Pacific Islands. *IBPGR Region. Committ. Southeast Asia Newsl.*, 5(1): 12-13. (E)

2360 **IBPGR Regional Committee for South East Asia (1985).** Botanic gardens and plant genetic resources conservation. *IBPGR South East Asia Region. Committ. Newsl.*, 9(4): 3-4. (E/G/T)

2361 **IBPGR/IRRI (1978).** Proceedings of the Workshop on the Genetic Conservation of Rice. Laguna, Philippines, IRRI. 54p. Illus. (E)

2362 **IBPGR/IRRI Rice Advisory Committee (1982).** Conservation of the wild rices of tropical Asia. *Pl. Genet. Resource. Newsl.*, 49: 13-18. (Fr, Sp). Illus. (E)

SOUTH EAST ASIA

2363 **IUCN (1970).** Eleventh Technical Meeting Papers and Proceedings, New Delhi, India, Vol. II. Problems of threatened species. IUCN Publications New Series No. 18. Morges, IUCN. 132p. (T)

2364 **IUCN (1975).** The use of ecological guidelines for development in tropical forest areas of South East Asia. Papers and proceedings of the regional meeting at Bandung, Indonesia, 28 May-1 June 1974. IUCN Publications New Series No. 32. Morges, IUCN. 185p. Illus.

2365 **IUCN Commission on National Parks and Protected Areas (1985).** The Corbett Action Plan for protected areas of the Indomalayan Realm. *In* Thorsell, J.W., *ed.* Conserving Asia's natural heritage. The planning and management of Protected Areas in the Indomalayan Realm. Proc. of the 25th Working Session of IUCN CNPPA. Gland, IUCN. 219-242. (P)

2366 **Ingram, G.B. (1987).** Conservation of wild plants in crop genepools and their intraspecific variation: current needs and opportunities in the moist forest of SE Asia. *In* Santiapillai, C., Ashby, K.R., *eds.* Proceedings of the Symposium on the Conservation and Management of Endangered Plants and Animals. Bogor, Indonesia, 18-20 June 1986. Bogor, SEAMEO-BIOTROP. 63-85. Crop relatives, gene bank. (E/G/P)

2367 **Jacobs, M. (1977).** What a botanist can contribute to conservation in Malesia. *Flora Males. Bull.*, 30: 2826-2827.

2368 **Jacobs, M. (1979).** A plea for S.E. Asia's forests. *Habitat*, 7(4): 8-13, 31. Col. illus.

2369 **Jacobs, M. (1979).** Botanical philosophy on the selection of rain forest reserves in Malesia. *Flora Males. Bull.*, 32: 3247-3250. Tropical forest. (P)

2370 **Jain, S.K., Mehra, K.L. (1983).** Conservation of tropical plant resources. Proceedings of the Regional Workshop on Conservation of Tropical Plant Resources in South East Asia, New Delhi, March 8-12, 1982. Howrah, Botanical Survey of India, v, 253p. Illus. Contains papers on some non-tropical areas, e.g. Himalayas, Kashmir. South-east Asia is also loosely applied. (T)

2371 **Johnson, A. (1968).** Rare plants and the community in South East Asia. IUCN Publications New Series No. 10. Morges, IUCN. 340-343. (T)

2372 **Jong, K., ed. (1977).** Transactions of the Fifth Aberdeen-Hull Symposium on Malesian Ecology: biological aspects of plant genetic resource conservation in South-East Asia. Hull, University of Hull, Department of Geography. xiv, 95p. (Fr, Ge, In). (E)

2373 **Kemp, R.H., Whitmore, T.C. (1978).** International co-operation for the conservation of tropical and sub-tropical forest genetic resources exemplified by South East Asia. *In* Proceedings of the 8th World Forestry Congress, Jakarta, 16-28 October 1978: Forestry for quality of life. II, 8. (E)

2374 **Kiew, R. (1988).** Portraits of threatened plants 16. *Phoenix paludosa* Roxb. *Malay. Naturalist*, 42(1): 16. Illus. (T)

2375 **Kostermans, A.J.G.H. (1987).** Rare trees of Malesia. *In* Santiapillai, C., Ashby, K.R., *eds.* Proceedings of the Symposium on the Conservation and Management of Endangered Plants and Animals. Bogor, Indonesia, 18-20 June 1986. Bogor, SEAMEO-BIOTROP. 147-150. (T)

2376 **Kostermans, A.J.G.H., ed. (1987).** Proceedings of the Third Round Table Conference on Dipterocarps. Jakarta, Unesco. 657p. Papers presented at an international conference held at Mulawarman Univ., Samarinda, E. Kalimantan, 16-20 April 1985. Includes conservation. (E/T)

2377 **Lee, D. (1980).** The sinking ark: environmental problems in Malaysia and Southeast Asia. Kuala Lumpur, Heinemann. 85p.

SOUTH EAST ASIA

2378 **MacKinnon, J. & K. (1986).** Review of the protected areas system in the Indo-Malayan realm. Gland, Switzerland and Cambridge, U.K., IUCN and UNEP. 284p. Illus., maps. (P)

2379 **MacKinnon, J. (1979).** Conservation of wildlife and forests in Southeast Asia. *Tigerpaper*, 6(4): 2-5. Illus.

2380 **MacKinnon, J. (1985).** Outline of methodolgy for preparation of a review of the protected areas system of the "Indomalayan Realm". *In* Thorsell, J.W., *ed.* Conserving Asia's natural heritage. The planning and management of protected areas in the Indomalayan Realm. Proc. of the 25th working session of IUCN's CNPPA. Gland, IUCN. 53-59. (P)

2381 **Maheshwari, J.K. (1970).** The need for conservation of flora and floral provinces in southeast Asia. IUCN New Series No. 18. Switzerland, IUCN. 89-94.

2382 **Mascarenhas, A.F., comp.** (?) Biological diversity and genetic resources techniques and methods: tissue culture directory for Asia-Pacific. CSC Technical Publication Series No. 249. London, Commonwealth Science Council. 12p. (E/G)

2383 **McNeely, J.A. (1987).** How dams and wildlife can co-exist: natural habitats, agriculture, and major water resource development projects in tropical Asia. *J. Conservation Biology*, 1(3): 228-238. Maps. (P)

2384 **Meijer, W. (1973).** Devastation and regeneration of lowland dipterocarp forests in Southeast Asia. *BioScience*, 23(9): 528-533. Illus., map. Deforestation. (E/T)

2385 **Meijer, W. (1982).** *Rafflesia.* La plus grande fleur du monde, menacee d'extinction? *Terre Vie*, 36(2): 297-304. Fr (En). Illus., map. (P/T)

2386 **Meijer, W. (1982).** Plant refuges in the Indo-Malesian region. *In* Prance, G.T., *ed.* Biological diversification in the tropics. Proc. 5th Int. Symp. of the Assoc. for Tropical Biology, Macuto Beach, Caracas, Venezuela, Feb. 8-13, 1979. NY, Columbia Univ. Press. 576-586. Centres of species diversity and endemism.

2387 **Mirimanian, K.P. (1976).** Measures for protection and rational land utilization in arid mountain regions. *In* Proceedings of an International Meeting on Ecological Guidelines for the use of Natural Resources in the Middle East and South West Asia, Persepolis, Iran, 24-30 May 1975. IUCN Publ. New Series No. 34. Morges, IUCN. 224-226. (P)

2388 **Nectoux, F., Kuroda, Y. (1989).** Timber from the South Seas: an analysis of Japan's tropical timber trade and its environmental impact. Gland, WWF International. (E)

2389 **Nicholls, F.G. (1968).** Regulation and co-ordination of collections of flora and fauna. *In* Talbot, L.M., *ed.* Conservation in tropical southeast Asia. Morges, Switzerland, IUCN. 347-348. (L)

2390 **Numata, M. (1973).** Further case studies in selecting and allocating land for nature conservation: South-East Asia. *In* Costin, A.B., Groves, R.H., *eds.* Nature conservation in the Pacific. Proc. Symposium A-10, XII Pacific Science Congress, Canberra, August-September 1971. IUCN and Australian Nat. Univ. Press. 66-71. (P)

2391 **Oldfield, S. (1985).** Disappearing pitcher plants. *Oryx*, 19: 71-72. (T)

2392 **Pardo, R. (1979).** Southeast Asia: the disappearing forests. *Amer. Forests*, 85(4): 46-49, 54-56. Illus. Deforestation; tropical forest.

2393 **Poore, D. (1974).** Ecological guidelines for development in tropical forest areas of South East Asia. Switzerland, IUCN. 33p.

2394 **Qureshi, I.M., Kaul, O.N. (1970).** Some endangered plants and threatened habitats in southeast Asia. Switzerland, IUCN. 115-126. (IUCN New Series no. 18.) (T)

SOUTH EAST ASIA

2395 **Ranjitsinh, M.K. (1979).** Forest destruction in Asia and the South Pacific. *Ambio*, 8(5): 192-201. Illus., col. illus.

2396 **Rao, A.N. (1983).** Conservation of plant wealth in Southeast Asia. *Pac. Sci. Congr. Proc.*, 15(1-2): 193-194. (T)

2397 **Rao, A.N., Keng, H., Wee, Y.C. (1983).** Problems in conservation of plant resources in South East Asia. *In* Jain, S.K., Mehra, K.L., *eds*. Conservation of tropical plant resources. Howrah, Botanical Survey of India. 181-204. Illus. (E/T)

2398 **Rao, S. (1983).** Conservation of plant wealth in south east Asia. *In* Given, D.R., *ed*. Conservation of plant species and habitats. Wellington, Nature Conservation Council. 43-57. (P/T)

2399 **Ross, M.S., Donovan, D.G. (1986).** Land clearing in the humid tropics, based on experience in the conversion of tropical moist forests in South East Asia. IUCN/IIED Tropical Forest Policy Paper No. 1. Cambridge, IUCN, London, IIED-Earthscan and WWF. 19p.

2400 **Rubeli, K., et al. (1978).** Special symposium issue on role and management of national parks. *Malayan Nat. J.*, 29(4): 213-343 (1976 publ. 1978). (P)

2401 **Sagarik, R. (1988).** Local customs and the conservation of wild orchids. *Malay Orchid Rev.*, 22: 62-64. (T)

2402 **Sale, J.B. (1985).** Regional and international cooperation in protected area management in Indomalaya. *In* Thorsell, J.W., *ed*. Conserving Asia's natural heritage. The planning and management of protected areas in the Indomalayan Realm. Proc. of the 25th working session of IUCN's CNPPA. Gland, IUCN. 203-207. (P)

2403 **Sale, J.B. (1985).** Wildlife research in the Indomalayan Realm. *In* Thorsell, J.W., *ed*. Conserving Asia's natural heritage. The planning and management of protected areas in the Indomalayan Realm. Proc. of the 25th working session of IUCN's CNPPA. Gland, IUCN. 137-149.

2404 **Santiapillai, C., Ashby K.R., eds (1987).** Proceedings of the Symposium on the Conservation and Management of Endangered Plants and Animals. Bogor, Indonesia, 18-20 June 1986. Bogor, SEAMEO-BIOTROP. 246p. Maps. (E/G/L/T)

2405 **Sastrapradja, S., Kartawinata, K., Adisoemarto, S., Tarumingkeng, R.C. (1978).** The conservation of forest animal and plant genetic resources. *In* Proceedings of the 8th World Forestry Congress, Jakarta, 16-28 October 1978: Forestry for quality of life. III, 47. (E)

2406 **Singh, R.B. (1980).** Genetic resources of cotton in Asia. *IBPGR Region. Committ. Southeast Asia Newsl.*, 5(1): 9-11. (E/G)

2407 **Singh, R.B. (1981).** Conservation of plant genetic resources in Southeast Asia. *In* Fourth International Sabrao Congress, 4-8 May 1981, at Universiti Kebangsaan Malaysia and Federal Hotel, Kuala Lumpur. Bangkok, IBPGR/FAO. (E)

2408 **Singh, R.B. (1983).** Conservation of plant genetic resources: a cooperative effort in South East Asia. *In* Jain, S.K., Mehra, K.L., *eds*. Conservation of tropical plant resources. Howrah, Botanical Survey of India. 211-224. (E)

2409 **Singh, S. (1985).** An overview of the conservation status of the national parks and protected areas of the Indomalayan Realm. *In* Thorsell, J.W., *ed*. Conserving Asia's natural heritage. The planning and management of protected areas in the Indomalayan Realm. Proc. of the 25th working session of IUCN's CNPPA. Gland, IUCN. 1-5. (P)

2410 **Snidvongs, K. (1984).** Future directions for the Indomalayan Realm. *In* McNeely, J.A., Miller, K.R., *eds*. National parks, conservation and development. The role of protected areas in sustaining society. Washington, D.C., Smithsonian Institution Press. 206-210. (P)

SOUTH EAST ASIA

2411 **Soepadmo, E. (1978).** The role of tropical botanic gardens and arboreta in the conservation of rare and threatened plant genetic resources in S.E. Asia. *Gartn.-Bot. Brief*, 58: 30. (Ge). Abstract. (G)

2412 **Soepadmo, E. (1979).** The role of tropical botanic gardens in the conservation of threatened valuable plant genetic resources in South East Asia. *In* Synge, H., Townsend. H., *eds.* Survival or extinction. Proceedings of a conference held at the Royal Botanic Gardens, Kew, 11-17 September 1978. Kew, Bentham-Moxon Trust. 63-74. Gene. (E/G/T)

2413 **Soepadmo, E. (1983).** Forest and man: an ecological appraisal. Kuala Lumpur, University of Malaya. 43p. Maps. Tropical forest. Deforestation. Includes map of existing forest areas. (E)

2414 **Spears, J. (1988).** Preserving biological diversity in the tropical forests of the Asian region. *In* Wilson, E.O., *ed.* Biodiversity. Washington, D.C., National Academy Press. 393-402. Deforestation rates; solutions to deforestation; role of development agencies.

2415 **Spencer, J.E. (1977).** Shifting cultivation in Southeast Asia. Berkeley, Univ. of California Press.

2416 **Steenis, C.G.G.J. van (1971).** Plant conservation in Malesia. *Bull. Jard. Bot. Nat. Belg.*, 41: 189-202. (T)

2417 **Talbot, L.M. (1969).** Highlights of conservation in the International Biological Programme in the Asia-Pacific region. *Malaysian Forest.*, 32: 391-394.

2418 **Talbot, L.M. (1973).** Conservation problems in South-east Asia. *In* Costin, A.B., Groves, K.H., *eds.* Nature conservation in the Pacific. Proc. Symposium A-10, XII Pacific Science Congress, Canberra, August-September 1971. IUCN and Australian Nat. Univ. Press. 255-262.

2419 **Talbot, L.M., ed. (1968).** Conservation in tropical southeast Asia. Morges, IUCN. (L)

2420 **Thorsell, J.W. (1985).** Threatened protected areas of the Indomalayan Realm. *In* Thorsell, J.W. *ed.* Conserving Asia's natural heritage. The planning and management of protected areas in the Indomalayan Realm. Proc. of the 25th working session of IUCN's CNPPA. Gland, IUCN. 43-49. (P)

2421 **Thorsell, J.W., ed. (1985).** Conserving Asia's natural heritage. The planning and management of protected areas in the Indomalayan Realm. IUCN. Proceedings of the 25th Working Session of IUCN's Commission on National Parks and Protected Areas. Corbett National Park India, 4-8 Feb. 1985. Gland, IUCN. 242p. (P)

2422 **Wagner, J.P. (1985).** The "scandalwood". *Hawaii*, 2(2) (Issue No. 4): 51-52. Exploitation of sandalwood has had devastating effects. (T)

2423 **Walters, S.M. (1983).** The role of botanic gardens in plant conservation in South-East Asia. *In* Jain, S.K. Mehra, K.L., *eds.* Conservation of tropical plant resources. Howrah, Botanical Survey of India. 230-235. (G/T)

2424 **White, A. (1984).** Vulnerable marine resources, coastal reserves, and pollution: a Southeast Asian perspective. *In* McNeely, J.A., Miller, K.R., *eds.* National parks, conservation and development. The role of protected areas in sustaining society. Washington, D.C., Smithsonian Institution Press. 170-174. (P)

2425 **Whitmore, T.C. (1975).** Conservation review of tropical rain forests, general considerations and Asia. Switzerland, IUCN. 116p. Introduction to forest types and their distribution, protected area coverage and conservation priorities for rain forest countries from India east to Polynesia. (E/P)

2426 **Whitmore, T.C. (1975).** Tropical rain forests of the Far East. Oxford, Clarendon Press. 282p. Ecology, classification and distribution of forest types, impact of man. Revised edition 1984. (E/P)

SOUTH EAST ASIA

2427 **Whitmore, T.C. (1980).** Utilization, potential and conservation of *Agathis*, a genus of tropical Asian conifers. *Econ. Bot.*, 34(1): 1-12. Illus. (E)

2428 **Whitmore, T.C. (1984).** A vegetation map of Malesia at scale 1:5 million. *J. Biogeography*, 11: 461-471. Maps. 10 forest types depicted; deforested areas and conservation areas over 200 sq. km shown; explanatory text. (P)

2429 **Williams, J.T., et al., eds (1975).** South East Asian plant genetic resources. Proceedings of a symposium on South East Asian genetic resources held at Kopo, Cisarua, Bogor, Indonesia, March 20-22, 1975. Bogor, IBPGR. 272p. (E)

SOUTHERN AFRICA

2430 **Acocks, J.P.H. (1975).** Veld types of South Africa, 2nd. ed. *Mem. Bot. Surv. S. Afr.*, 40: 128p. Includes vegetation map 1:1,500,000.

2431 **Anon. (1979).** Conservation through cultivation. *Veld & Flora* (Kirstenbosch), 65(3): 65. (G)

2432 **Armitage, F.B., Joustra, P.A., Ben Salem, B. (1980).** Genetic resources of tree species in arid and semi-arid areas: a survey for the improvement of rural living in Latin America, Africa, India and Southwest Asia. Rome, FAO. vi, 118p. (E)

2433 **Boddington, G. (1987).** Urbanisation or conservation? *Veld & Flora* (Kirstenbosch), 73(2): 71-73. Col. illus. (T)

2434 **Boughey, A.S. (1960).** Man and the African environment. *Proc. Trans. Rhod. Sci. Assoc.*, 48: 8-18.

2435 **Edwards, D. (1974).** Survey to determine the adequacy of existing conserved areas in relation to vegetation types. A preliminary report. *Koedoe*, 17: 2-37. (P)

2436 **Hall, A.V., Winter, B. de, Arnold, T.H. (1981).** Threatened plants in Southern Africa. *In* XIII International Botanical Congress, Sydney, Australia, 21-28 August 1981: abstracts. Sydney, Australian Academy of Science. 214. (T)

2437 **Hall, A.V., Winter, B. de, Fourie, S.P., Arnold, T.H. (1984).** Threatened plants in southern Africa. *Biol. Conserv.*, 28(1): 5-20. Map. (T)

2438 **Hall, A.V., Winter, M. de & B. de, Oosterhout, S.A.M. van (1980).** Threatened plants of Southern Africa: a report of the Committee for Terrestrial Ecosystems National Programme for Environmental Sciences. South African National Scientific Programmes Report No. 45. Pretoria, CSIR. 244p. (T)

2439 **Hedberg, I. & O., eds. (1968).** Conservation of vegetation in Africa south of the Sahara. *Acta Phytogeogr. Suec.* (Uppsala), 54: 320p. Col. illus., maps. Proceedings of a symposium held at the 6th Plenary meeting of the "Association pour l'Etude Taxonomique de la Flore d'Afrique Tropicale" (AETFAT) in Uppsala, Sept. 12th-16th, 1966. (P/T)

2440 **Hedberg, I. (1979).** Possibilities and needs for conservation of plant species and vegetation in Africa. *In* Hedberg, I., ed. Systematic botany, plant utilization and biosphere conservation. Stockholm, Almqvist & Wiksell International. 83-104. (T)

2441 **Hedberg, O. (1978).** Nature in utilization and conservation of high mountains in eastern Africa (Ethiopia to Lesotho). *In* The use of high mountains of the world. Wellington, N.Z., Department of Lands and Survey and IUCN. 42-56. Illus., map.

2442 **Hoffman, M.T., Everard, D.A. (1987).** Neglected and abused: the eastern Cape subtropical thickets. *Veld & Flora* (Kirstenbosch), 73(2): 43-45. Map. (T)

2443 **Huntley, B.J. (1978).** Ecosystem conservation in southern Africa. *In* Werger, M.J.A., ed. Biogeography and ecology of Southern Africa. The Hague, W. Junk. 1333-1384. (P)

SOUTHERN AFRICA

2444 **Huntley, B.J., Ellis, S. (1982).** The conservation status of vegetation types of Southern Africa. *In* AETFAT Synopses 10th Congress: the origin, evolution and migration of African floras, Pretoria, 18-23 January 1982. Abstract. Pretoria. 82.

2445 **IUCN (1981).** Conserving Africa's natural heritage: the planning and management of protected areas in the Afrotropical Realm. Gland, IUCN. 271p. Proceedings on the 17th meeting of IUCN's Commission on National Parks and Protected Areas, Garoua, Cameroon, 17-23 November 1980. (P)

2446 **IUCN Threatened Plants Committee Secretariat (1982).** The botanic gardens list of rare and threatened plants for Southern Africa. Kew, IUCN Botanic Gardens Conservation Co-ordination Body Report No. 5. Lists plants in cultivation in botanic gardens. (G/T)

2447 **IUCN Threatened Plants Unit (1985).** The botanic gardens list of succulent euphorbias and aloes of Africa (including Madagascan succulents update). Kew, IUCN Botanic Gardens Co-ordinating Body Report No. 15. Lists plants in cultivation in botanic gardens. (G/T)

2448 **Leach, G., Mearns, R. (1989).** Beyond the woodfuel crisis: people, land and trees in Africa. London, Earthscan. 320p. (E)

2449 **Lock, J.M. (1989).** Legumes of Africa; a checklist. Kew, Royal Botanic Gardens. 619p. Provides IUCN conservation status for all taxa. (E/T)

2450 **Low, B. (1987).** Non-government conservation: clarifying future roles in urbanizing Greater Cape Town. *Veld & Flora* (Kirstenbosch), 73(2): 68-71. (L)

2451 **Lucas, G.Ll. (1979).** The problems of species conservation in Africa - an appeal. *In* Kunkel, G., *ed.* Taxonomic aspects of African economic botany; proceedings of the IX Plenary meeting of AETFAT, Las Palmas de Gran Canaria, 18-23 March 1978. Las Palmas, Excmo Ayuntamiento. 245-247.

2452 **McMahan, L. (1984).** Protecting Africa's 'Halfmens'. *Endangered Species Tech. Bull. Reprint*, 1(5): 1. Arborescent aloes. (T)

2453 **Nicholson, H.B. (1977).** The cedars of south and central Africa. A conservation tragedy. *Trees S. Afr.*, 29(2/4): 61-63. (T)

2454 **Piearce, G.D. (1986).** How to save the Zambezi teak forests. *Unasylva*, 38(2): 29-36. (T)

2455 **Poppendieck, H.-H. (1976).** Mesembryanthemums and the problems of their cultivation. *In* Simmons, J.B., *et al.*, *eds.* Conservation of threatened plants. NATO Conference Series 1: Ecology, vol. 1. New York and London, Plenum Press. 55-60. Proc. Conference on the Functions of Living Plant Collections in Conservation and Conservation-Orientated Research and Public Education, Kew, 2-6 September 1975. (G)

2456 **Richards, D.J. (1980).** Cycads plundered. *Aloe Cact. Succ. Rhodesia Quart. Newsl.*, 46: 7. (T)

2457 **Rowley, G. (1978).** Ecologie, conservation ou sauvetage de la nature. *Cactus* (Brussels), 2(6): 120-123. Fr. *Haworthia graminifolia.* (T)

2458 **Saussay, C. du (1984).** La protection des forets en droit Africain. (Protection of forests in African law.) *In* Prieur, M., *ed.* Forets et environnement: en droit compare et international. Paris, Presses Universitaires de France. 147-164. Fr. (L)

2459 **Thresher, P. (1972).** African national parks and tourism - an interlinked future. *Biol. Conserv.*, 4(4): 279-284. (P)

2460 **Turner, I. (1987).** In search of *Zamia wallisi. Encephalartos*, 11: 20-21. (T)

2461 **White, F. (1981).** The history of the Afromontane archipelago and the scientific need for its conservation. *Afr. J. Ecol.*, 19: 33-54. En (Fr). Maps.

SOUTHERN AFRICA

2462 **Witkowski, E.T.F., Walker, B.H. (1981).** Towards an ecological basis for the management of Klaserie Nature Reserve. *Forum Bot.*, 19(2): 37. Abstract. Paper presented at SAAB Annual Congress, January 1981. South African Association of Botanists. (P)

TROPICAL AFRICA

2463 **Allan, W. (1968).** Soil resources and land use in Tropical Africa. *In* Hedberg, I. & O., *eds.* Conservation of vegetation in Africa south of the Sahara. Symposium Proceedings, at 6th plenary meeting of AETFAT, Uppsala. *Acta Phytogeogr. Suec.*, 54: 9-12.

2464 **Anderson, D., Grove, R. (1988).** Conservation in Africa - peoples, policies and practice. Cambridge, Cambridge University Press. 365p.

2465 **Anon. (1979).** Getting better - or getting worse? *African Wildlife*, 33(1): 8-11. Illus. Extracts from IUCN Bulletin. 14th General Assembly of IUCN, Ashkhabad. (T)

2466 **Anon. (1983).** Coffins and forests. *Oryx*, 17: 141 Swiss coffin manufacturer has switched from a rare tropical hardwood, abachi, from West Africa to local poplar wood. (E/T)

2467 **Armitage, F.B., Joustra, P.A., Ben Salem, B. (1980).** Genetic resources of tree species in arid and semi-arid areas: a survey for the improvement of rural living in Latin America, Africa, India and Southwest Asia. Rome, FAO. vi, 118p. (E)

2468 **Aubreville, A. (1968).** Synthese regionale. *In* Hedberg, I. & O., *eds.* Conservation of vegetation in Africa south of the Sahara. Symposium Proceedings, at 6th plenary meeting of AETFAT, Uppsala. *Acta Phytogeogr. Suec.*, 54: 134-136. Fr. A general overview of conditions in tropical African territories. (P/T)

2469 **Aubreville, A.M. (1985).** The disappearance of the tropical forests of Africa. *Unasylva*, 37(146): 18-28.

2470 **Ayensu, E.S. (1984).** Keynote address: the Afrotropical Realm. *In* McNeely, J.A., Miller, K.R., *eds.* National parks, conservation and development. The role of protected areas in sustaining society. Washington, D.C., Smithsonian Institution Press. 80-86. (P)

2471 **Bechtel, H. (1979).** Ostafrika in Farbe. Ein Reisfuhrer fur Naturfreunde. Stuttgart, Kosmos. 71p. Ge. Col. illus., map. National parks and reserves in tropical East Africa. (P)

2472 **Bliss-Guest, P. (1983).** Environmental stress in the East African Region. *Ambio*, 12(6): 290-295. Col. illus, maps.

2473 **Boudet, G. (1972).** Desertification de l'Afrique tropicale seche. *Adansonia, ser. 2,* 12(4): 505-524. Fr.

2474 **Boughey, A.S. (1960).** Man and the African environment. *Proc. Trans. Rhod. Sci. Assoc.*, 48: 8-18.

2475 **Brown, L.R., Wolf, E.C. (1985).** Reversing Africa's decline. Worldwatch Paper No. 65. Washington, D.C., Worldwatch Institute. 81p. Includes deforestation rates; reforestation programmes.

2476 **Cloudsley-Thompson, J.L. (1984).** Key environments: Sahara Desert. Oxford, Pergamon Press. x, 348p. Illus., maps.

2477 **Commonwealth Science Council (1987).** Biological diversity and genetic resources techniques and methods: conservation biology. Summary report. Regional training workshop held at Lusaka, Zambia, 16-22 May 1987. London, Commonwealth Science Council. 35p. (E)

2478 **Cribb, P.J. (1983).** Orchids and conservation in tropical Africa and Madagascar. *Bothalia*, 14(3 & 4): 1013-1014. (Fr.) (T)

TROPICAL AFRICA

2479 **Croft, J. (1966).** Vanishing Africa. *Defenders Wildl. News*, 41(2): 117-120. (T)

2480 **Curry-Lindahl, K. (1968).** Les parcs nationaux commes archives de recherches et documentation de l'evolution des especes. *In* Hedberg, I. & O., *eds.* Conservation of vegetation in Africa south of the Sahara. Symposium Proceedings, at 6th plenary meeting of AETFAT, Uppsala. *Acta Phytogeogr. Suec.*, 54: 35-38. Fr. On national parks and biotopes. (P)

2481 **Curry-Lindahl, K. (1968).** Zoological aspects on the conservation of vegetation in Tropical Africa. *In* Hedberg, I. & O., *eds.* Conservation of vegetation in Africa south of the Sahara. Symposium Proceedings, at 6th plenary meeting of AETFAT, Uppsala. *Acta Phytogeogr. Suec.*, 54: 25-32.

2482 **Darling, P. (1989).** The changing Sahara. London, Earthscan. 352p.

2483 **Davies, B.R. (1983).** SCOPE project on African wetlands. *S. Afr. J. Sci.*, 79(7): 257-258. Map. Scientific Committee on Problems of the Environment.

2484 **Davis, R.K. (1978).** National parks as the basis for generating rural economic activity. *In* Proceedings of the 7th World Forestry Congress, Buenos Aires, 4-18 October 1972, vol. 3: Conservation and recreation. Buenos Aires, Instituto Forestal Nacional. 4056-4080. En (Sp, Fr). (P)

2485 **Devoe, N.N. (1982).** Forestry issues in West Africa. *In* Stock, M., Force, J.E., Ehrenreich, D., *eds.* Women in natural resources. Proceedings of a conference, University of Idaho, Moscow, Idaho, 8-9 March, 1982. 49-63. Maps. Silvicultural practices on the conservation of tropical moist forest, personnel issues. (E)

2486 **FAO (1985).** Intensive multiple-use forest management in the tropics. Analysis of case studies from India, Africa, Latin America and the Caribbean. Rome, FAO Forestry Paper 55. 180p. (E)

2487 **FAO (1986).** Some medicinal forest plants of Africa and Latin America. FAO Forestry Paper No. 67. Rome, FAO. 252p. (E)

2488 **FAO/UNEP (1981).** Tropical Forest Resources Assessment Project (in the framework of GEMS). Forest resources of tropical Africa. Part 1: Regional synthesis. Rome, FAO. 108p. En, Fr.

2489 **FAO/UNEP (1981).** Tropical Forest Resources Assessment Project (in the framework of GEMS). Forest resources of tropical Africa. Part II: Country briefs. Rome, FAO. 586p. En, Fr. Tropical forests; deforestation.

2490 **Finn, D. (1983).** Land use and abuse in the East African region. *Ambio*, 12(6): 296-301. Col. illus., maps.

2491 **Foley, G., Buren, A. van (1980).** Substitutes for wood: turning to coal and other approaches to easing the pressure on fuelwood resources. *Unasylva*, 32(130): 11-24. Illus. (E)

2492 **Fryer, G. (1972).** Conservation of the Great Lakes of east Africa: a lesson and a warning. *Biol. Conserv.*, 4(4): 256-262. (P)

2493 **Gaisler, J. (1983).** Baborsky narodni park a jeho endemicky brhlik Sitta ledanti. *Ziva*, 31(3): 110-112. Cz. Illus. (P)

2494 **Gbile, Z.O., Soladoye, M.O., Adesina, S.K. (1988).** Plants in traditional medicine in West Africa. *In* Goldblatt, P., Lowry, P.P., *eds.* Modern systematic studies in African botany. Proc. Eleventh Plenary Meeting, AETFAT, Missouri Bot. Gard., 10-14 June 1985. Missouri, Missouri Botanic Garden. 343-349. Medicinal plants. (E)

2495 **Geerling, C. (1985).** The status of the woody species of the Sudan and Sahel zones of West Africa. *Forest Ecol. Manage.*, 13: 247-255. (T)

137

TROPICAL AFRICA

2496 **Gilbert, V.C. (1984).** Cooperative regional demonstration projects: environmental education in practice. *In* Unesco-UNEP. Conservation, science and society. Paris, Unesco. 566-572. Contributions to the First International Biosphere Reserve Congress, Minsk, Byelorussia/USSR, 26 September - 2 October 1983. (P)

2497 **Goodier, R. (1968).** Nature conservation and forest clearance in Africa with special reference to some ecological considerations of tsetse control. *In* Hedberg, I. & O., *eds*. Conservation of vegetation in Africa south of the Sahara. Symposium Proceedings, at 6th plenary meeting of AETFAT, Uppsala. *Acta Phytogeogr. Suec.*, 54: 20-25.

2498 **Gorse, J. (1987).** Desertification in the Sahelian and Sudanian zones of West Africa. *Unasylva*, 37: 2-18.

2499 **Greathead, D.J. (1968).** Biological control of *Lantana* - a review and discussion of recent developments in East Africa. *Pest Articl. News Summ.*, Sect. C, 14: 167-175. Illus. Invasive species.

2500 **Happold, D.C.D., ed. (1971).** Wildlife conservation in West Africa. Proceedings of a Symposium held at the University of Ibadan, Nigeria, 2 April 1970. Morges, IUCN. 60p.

2501 **Hedberg, I. & O., eds. (1968).** Conservation of vegetation in Africa south of the Sahara. *Acta Phytogeogr. Suec.* (Uppsala), 54: 320p. Col. illus., maps. Proceedings of a symposium held at the 6th Plenary meeting of the "Association pour l'Etude Taxonomique de la Flore d'Afrique Tropicale" (AETFAT) in Uppsala, Sept. 12th-16th, 1966. (P/T)

2502 **Hedberg, I. (1979).** Possibilities and needs for conservation of plant species and vegetation in Africa. *In* Hedberg, I., *ed*. Systematic botany, plant utilization and biosphere conservation. Stockholm, Almqvist & Wiksell International. 83-104. (T)

2503 **Hedberg, O. (1978).** Nature in utilization and conservation of high mountains in eastern Africa (Ethiopia to Lesotho). *In* The use of high mountains of the world. Wellington, N.Z., Department of Lands and Survey and IUCN. 42-56. Illus., map.

2504 **Hepper, F.N. (1968).** Regional synthesis. *In* Hedberg, I. & O., *eds*. Conservation of vegetation in Africa south of the Sahara. Symposium Proceedings, at 6th plenary meeting of AETFAT, Uppsala. *Acta Phytogeogr. Suec.*, 54: 98-105. Discussion of vegetation in West Africa. Answers 3 questions: what requires to be conserved in West Africa, what has been achieved and what should be done now? (P/T)

2505 **Hepper, F.N. (1977).** The practical importance of plant ecology in arid zones. *In* Dalby, D., Harrison Church, R.J., Bezzaz, F., *eds*. Drought in Africa 2. London, International African Institute. 105-106.

2506 **Huntley, B.J. (1988).** Conserving and monitoring biotic diversity: some African examples. *In* Wilson, E.O., *ed*. Biodiversity. Washington, D.C., National Academy Press. 248-260. Includes section on centres of endemism and species richness. (P/T)

2507 **IRRI (1978).** Proceedings of the Workshop on the Genetic Conservation of Rice. A survey of rice genetic resources and conservation in Africa and the Americas. Abstracts of reports. Los Banos, IRRI. 15-21. (E/G)

2508 **IUCN (1963).** Conservation of nature and natural resources in modern African states. Report of a Symposium organized by CCTA and IUCN and held under the auspices of FAO and Unesco at Arusha, Tanganyika, September 1961. IUCN Publications New Series No. 1. Morges, IUCN. 367p. Illus.

2509 **IUCN (1976).** Proceedings of a meeting on the creation of a coordinated system of national parks and reserves in eastern Africa. Seronera Lodge, Serengeti National Park, Tanzania, 14-19 October 1974. IUCN Publications New Series, Supplementary Paper No. 45. Morges, IUCN. 205p. Maps. (P)

TROPICAL AFRICA

2510 **IUCN (1981).** Conserving Africa's natural heritage: the planning and management of protected areas in the Afrotropical Realm. Gland, IUCN. 271p. Proceedings on the 17th meeting of IUCN's Commission on National Parks and Protected Areas, Garoua, Cameroon, 17-23 November 1980. (P)

2511 **IUCN (?)** La conservation des ecosystemes forestiers d'Afrique centrale. IUCN, Tropical Forest Programme Series. 124p. In prep.

2512 **IUCN Commission on National Parks and Protected Areas (1986).** Review of the protected areas system in the Afrotropical realm. Gland, Switzerland and Cambridge, IUCN and UNEP. 259p. Illus., 5 maps. (P)

2513 **IUCN Threatened Plants Unit (1985).** The botanic gardens list of succulent euphorbias and aloes of Africa (including Madagascan succulents update). Kew, IUCN Botanic Gardens Co-ordinating Body Report No. 15. Lists plants in cultivation in botanic gardens. (G/T)

2514 **IUCN/UNEP (1987).** The IUCN Directory of Afrotropical protected areas. Gland, Switzerland and Cambridge, U.K., IUCN. xix, 1034p. Illus., maps. (P)

2515 **Kio, P.R.O. (1982).** Factors and policies affecting forest resources use and conservation in Africa. *In* Vogt, F., *ed.* Energy conservation and use of renewable energies in the bio-industries. 2: Proc., 2nd int. seminar, Trinity College, Oxford, 6-10 September 1982. Oxford, Pergamon Press. 425-432. (E)

2516 **Kundaeli, J.N. (1983).** Making conservation and development compatible. *Ambio*, 12(6): 326-331. Col. illus.

2517 **Lamprey, H.F. (1975).** The distribution of protected areas in relation to the needs of biotic community conservation in eastern Africa. IUCN Occas. Paper No. 16. Morges, IUCN. 85p. Maps. Presented at Regional Meeting on Coordinated System of National Parks and Reserves in Eastern Africa, Seronera Lodge, Serengeti National Park, Tanzania, 14-19 October 1974. (P)

2518 **Leach, G., Mearns, R. (1989).** Beyond the woodfuel crisis: people, land and trees in Africa. London, Earthscan. 320p. (E)

2519 **Ledant, J.P. (1984).** La reduction de biomasse vegetale en Afrique de l'Ouest. 1. Apercu general. (Reduction of plant biomass in West Africa. 1. General survey.) *Ann. Gembl.*, 90(4): 195-216. Fr.

2520 **Lock, J.M. (1989).** Legumes of Africa; a checklist. Kew, Royal Botanic Gardens. 619p. Provides IUCN conservation status for all taxa. (E/T)

2521 **Lucas, G.Ll. (1979).** The problems of species conservation in Africa - an appeal. *In* Kunkel, G., *ed.* Taxonomic aspects of African economic botany; proceedings of the IX Plenary meeting of AETFAT, Las Palmas de Gran Canaria, 18-23 March 1978. Las Palmas, Excmo Ayuntamiento. 245-247.

2522 **Lusigi, W.J. (1984).** Future directions for the Afrotropical Realm. *In* McNeely, J.A., Miller, K.R., *eds.* National parks, conservation and development. The role of protected areas in sustaining society. Washington, D.C., Smithsonian Institution Press. 137-146. Map. (P)

2523 **Madams, R. (1989).** The Threatened Plants Unit of the World Conservation Monitoring Centre (WCMC): its research and activities in tropical Africa. *Trop. Afr. Bot. Gard. Bull.*, 1: 19-20. (T)

2524 **Maheshwari, J.K. (1971).** The baobab tree: disjunctive distribution and conservation. *Biol. Conserv.*, 4(1): 57-60. (T)

2525 **McMahan, L. (1984).** Protecting Africa's 'Halfmens'. *Endangered Species Tech. Bull. Reprint*, 1(5): 1. Arborescent aloes. (T)

2526 **Milburn, M. (1984).** Dragon's blood in east and west Africa and the Canary Islands. *Africa* (Rome, Instituto Italo-Africano), 39(3): 486-493. (T)

TROPICAL AFRICA

2527 **Myers, N. (1982).** Forest refuges and conservation in Africa with some appraisal of survival prospects for tropical moist forests throughout the biome. *In* Prance, G.T., *ed.* Biological diversification in the tropics. Proc. 5th Int. Symp. of the Assoc. for Trop. Biology, Macuto Beach, Caracas, Venezuela, 8-13 Feb. 1979. NY, Columbia Univ. Press. 658-672. (E/P/T)

2528 **Norton Griffiths, M., Ryden, P., eds (1989).** The IUCN Sahel studies, 1989. Gland, IUCN. 178p. Illus., map. Covers population, natural resource management, pastoral conservation, fuelwood and conservation areas. Also published in French. (E/P)

2529 **Phillips, J.F.V. (1968).** The influence of fire in Trans-Saharan Africa. *In* Hedberg, I. & O., *eds.* Conservation of vegetation in Africa south of the Sahara. Symposium Proceedings, at 6th plenary meeting of AETFAT, Uppsala. *Acta Phytogeogr. Suec.,* 54: 13-20.

2530 **Poppendieck, H.-H. (1976).** Mesembryanthemums and the problems of their cultivation. *In* Simmons, J.B., *et al., eds.* Conservation of threatened plants. NATO Conference Series 1: Ecology, vol. 1. New York and London, Plenum Press. 55-60. Proc. Conference on the Functions of Living Plant Collections in Conservation and Conservation-Orientated Research and Public Education, Kew, 2-6 September 1975. (G)

2531 **Prior, J., Tuohy, J. (1987).** Fuel for Africa's fires. *New Sci.,* 115(1571): 48-51. Illus., col. illus. Trees for fuelwood. (E)

2532 **Remanandan, P. (1983).** The wild gene pool of *Cajanus* at ICRISAT, present and future. Proceedings of the International Workshop on Pigeonpeas, Patancheru, 15-19 December 1980, Vol 2. Patancheru, ICRISAT. 29-38. (E)

2533 **Ruedin, Y.-M. (1987).** Africa..."Great things are done when men and mountains meet...". *Development Forum,* 15(3): 8-9. Illus., map. Report on the United Nations University and Unesco programme to conserve Africa's mountain ecosystems.

2534 **Saussay, C. du (1984).** La protection des forets en droit Africain. (Protection of forests in African law.) *In* Prieur, M., *ed.* Forets et environnement: en droit compare et international. Paris, Presses Universitaires de France. 147-164. Fr. (L)

2535 **Smithsonian Peace Corps Environmental Program (1977).** Peace Corps volunteer conference on West African Parks and Wildlife, Niamey, Niger, June 20-24, 1977 - Proceedings. Washington, D.C., Smithsonian Institution. 152p. Illus., maps. (P)

2536 **Steencroft, M. (1968).** Education towards the conservation of nature in tropical Africa. *In* Hedberg, I. & O., *eds.* Conservation of vegetation in Africa south of the Sahara. Symposium Proceedings, at 6th plenary meeting of AETFAT, Uppsala. *Acta Phytogeogr. Suec.,* 54: 38-44.

2537 **Thresher, P. (1972).** African national parks and tourism - an interlinked future. *Biol. Conserv.,* 4(4): 279-284. (P)

2538 **Tybirk, K., Lawesson, J.E., Nielsen, I., eds (1989).** Sahel Workshop 1989 University of Aarhus. Risskov, Denmark, University of Denmark. 75p. Illus., maps.

2539 **Verdcourt, B. (1968).** Regional synthesis. *In* Hedberg, I. & O., *eds.* Conservation of vegetation in Africa south of the Sahara. Symposium Proceedings, at 6th plenary meeting of AETFAT, Uppsala. *Acta Phytogeogr. Suec.,* 54: 186-192. Overview of problems associated with tropical East Africa. (P/T)

2540 **Verdcourt, B. (1968).** Why conserve natural vegetation? *In* Hedberg, I. & O., *eds.* Conservation of vegetation in Africa south of the Sahara. Symposium Proceedings, at 6th plenary meeting of AETFAT, Uppsala. *Acta Phytogeogr. Suec.,* 54: 1-9.

2541 **Western, D., Henry, W. (1979).** Economics and conservation in Third World (East African) national parks. *BioScience,* 29(7): 414-418. Illus. (P)

TROPICAL AFRICA

2542 **White, F. (1981).** The history of the Afromontane archipelago and the scientific need for its conservation. *Afr. J. Ecol.*, 19: 33-54. En (Fr). Maps.

2543 **Wickens, G.E. (1989).** The Sahel - a double disaster. *Disaster Management*, 2(2): 100-101. Illus. Desertification.

2544 **Wild, H. (1968).** Regional synthesis - discussion. *In* Hedberg, I. & O., eds. Conservation of vegetation in Africa south of the Sahara. Symposium Proceedings, at 6th plenary meeting of AETFAT, Uppsala. *Acta Phytogeogr. Suec.*, 54: 232-233. Overviews the protection of vegetation in southern tropical Africa. (P/T)

COUNTRY REFERENCES

AFGHANISTAN

2545 **McNeely, J.A., Thorsell, J.W., eds (1985).** People and protected areas in the Hindukush-Himalaya. Kathmandu, ICIMOD. 250p. (P)

2546 **Sayer, J.A. (1979).** Afghanistan's efforts to manage wildlife and environment. *Unasylva*, 31(123): 4-42.

2547 **Sayer, J.A. (1979).** Conservation in Afghanistan. *Tigerpaper*, 6(2-3): 41-42. Illus.

ALBANIA

2548 **Jordanov, D., et al., eds (1975).** Problems of the Balkan flora and vegetation. Sofia, Bulgarian Academy of Sciences. Proceedings of the First International Symposium on Balkan Flora and Vegetation, Vorna, 7-14 June 1973. 441p. (T)

2549 **Kuzmanov, B. (1981).** Balkan endemism and the problem of species conservation, with particular reference to the Bulgarian flora. *Bot. Jahrb. Syst.*, 102(1-4): 255-270. Illus., maps. (T)

ALGERIA

2550 **Dobr, J. (1988).** *Cupressus dupreziana. Threatened Pl. Newsl.*, 20: 8. Expedition Tarout '81 reintroduction project. (T)

2551 **Faurel, L. (1959).** Plantes rares et menacees d'Algerie. *In* Animaux et vegetaux rares de la region Mediterraneenne. Proceedings of the IUCN 7th Technical Meeting 11-19 September 1958, Athens, vol. 5. Brussels, IUCN. 140-155. Fr. Includes lists of rare or threatened plants in different parts of Algeria. (T)

2552 **IUCN Threatened Plants Committee Secretariat (1980).** First preliminary draft of the list of rare, threatened and endemic plants for the countries of North Africa and the Middle East. Kew, IUCN Threatened Plants Committee Secretariat. 170p. (T)

2553 **Mathez, J., Quezel, P., Raynaud, C. (1985).** The Maghreb countries. *In* Gomez-Campo, C., ed. Plant conservation in the Mediterranean Area. Dordrecht, Dr W. Junk. 141-157. Illus., maps. (T)

2554 **Quezel, P. (1964).** L'endemisme dans la flore d'Algerie. *Compt. Rend. Somme. Seanc. Soc. Biogeogr.*, 361: 137-149. Fr.

2555 **Stevenson, A.C., Skinner, J., Hollis, G.E., Smart, M. (1988).** The El Kala National Park and environs, Algeria: an ecological evaluation. *Envir. Conserv.*, 15(4): 335-348. Includes conservation evaluation of plant communities. (P)

ALGERIA

2556 **Tatole, V. (1985).** Rezervatii naturale si parcuri nationale in Algeria. (Reserves naturelles et parcs nationaux en Algerie.) *Ocrot. Nat.*, 29(2): 152. Rum (Fr). (P)

AMERICAN SAMOA

2557 **Graf, D.F. (1972).** American Samoa - annual environmental report. Pago Pago, American Samoa. Office of the Governor. 17p.

2558 **Nelson, R.E. (1964).** A look at the forests of American Samoa. U.S. Forest Service, Research Note PSW-53. 14p.

2559 **Sachet, M.-H. (1954).** A summary of information on Rose Atoll. *Atoll Res. Bull.*, 29: 1-25. Includes conservation.

2560 **Trotman, I.G. (1979).** Un premier parc national dans les Samoa occidentales. *Parcs* (USA), 3(4): 5-8. Fr. (P)

2561 **U.S. Congress, Office of Technology Assessment (1987).** Integrated renewable resource management for U.S. insular areas. Washington, D.C., U.S. Govt Printing Office. 443p. Illus., maps. OTA-F-325. Includes description of vegetation, remaining coverage, threatened species. (P/T)

2562 **Watters, R.F. (1960).** The nature of shifting cultivation: a review of recent research. *Pacific Viewpoint*, 1: 59-99. Impact of shifting cultivation.

2563 **Whistler, W.A. (1980).** The vegetation of Eastern Samoa. *Allertonia*, 2(2): 45-190. Extent and condition of vegetation.

ANGOLA

2564 **Huntley, B.J. (1979).** Angola. *In* Hedberg, I., *ed.* Systematic botany, plant utilization and biosphere conservation: Proc. of a sym. held in Uppsala in commemoration of the 500th ann. of the Univ. Stockholm, Almqvist & Wiksell Int'l. 99.

2565 **Smirnova, E.S., Dementeva, V.S. (1980).** *Welwitschia mirabilis.* Moskva, Nauka. 163-168. Rus. Illus. (G/T)

2566 **Teixeira, J.B. (1966).** A conservacao da vegetacao das suas especies en Angola. *Proteccao da Natureza Nova Ser.*, 8: 6-15. (T)

2567 **Teixeira, J.B. (1968).** Angola. *In* Hedberg, I. & O., *eds.* Conservation of vegetation in Africa south of the Sahara. Symposium Proceedings, at 6th plenary meeting of AETFAT, Uppsala. *Acta Phytogeogr. Suec.*, 54: 193-197. Map. 3 National parks, 2 nature reserves, 41 forest reserves and a number of game reserves, created by law some years ago, described. (L/P)

ARGENTINA

2568 **Barroso, J. (1978).** Linea de accion para la planificacion de los parques y reservas nacionales y monumentos naturales. *In* Proceedings of the 7th World Forestry Congress, Buenos Aires, 4-18 October 1972, vol. 3: Conservation and recreation. Buenos Aires, Instituto Forestal Nacional. 4089-4094. Sp (En, Fr). Ref. in *Forest. Abs.*, 41: 4974 (1980). (P)

2569 **Barroso, J. (1978).** Proyecto de la Republica Argentina para un centro de capacitacion en parques nacionales. (An Argentinian project for a training centre on national parks.) *In* Proceedings of the 7th World Forestry Congress, Buenos Aires, 4-18 October 1972, vol. 3: Conservation and recreation. Buenos Aires, Instituto Forestal Nacional. 4081-4085. Sp (En, Fr). (P)

2570 **Brucher, H. (1979).** Uber seltene und wenig bekannte Wildkartoffeln aus dem ariden Westen Argentiniens. (Rare and insufficiently known wild potatoes [*Solanum* sect. Tuberarium] from arid west Argentina.) *Angew. Bot.*, 53(1/2): 1-14. Ge. Illus. (T)

ARGENTINA

2571 **Cabrera, A.L. (1977).** Threatened and endangered species in Argentina. *In* Prance, G.T., Elias, T.S., *eds.* Extinction is forever. Proceedings of a symposium at the New York Botanical Garden, 11-13 May 1976. New York, New York Botanical Garden. 245-247. Examples of problems faced by Argentina's endangered species and plant communities, and protection of the natural vegetation. (P/T)

2572 **Correa Luna, H. (1978).** Problemas que se presentaron durante el desarrollo de la politica parquistica desde 1966 (VI Congresso Forestal Mundial). (Problems arising in the development of national park policies from 1966 to 1972.) *In* Proceedings of the 7th World Forestry Congress, Buenos Aires, 4-18 October 1972, vol. 3: Conservation and recreation. Buenos Aires, Instituto Forestal Nacional. 4100-4104. Sp (En, Fr). (P)

2573 **Correa Luna, H., Cafferata, R. (1978).** Proyecto de creacion del Parque Nacional Talampaya, La Rioja. (The creation of Talampaya National Park, La Rioja, Argentina.) *In* Proceedings of the 7th World Forestry Congress, Buenos Aires, 4-18 October 1972, vol. 3: Conservation and recreation. Buenos Aires, Instituto Forestal Nacional. 4105-4111. Sp (En, Fr). (P)

2574 **Debenedetti, L (1978).** Analisis del desarrollo de la politica del servicio nacional de parques nacionales. (Review of the development of the policies of the national parks service.) *In* Proceedings of the 7th World Forestry Congress, Buenos Aires, 4-18 October 1972, vol. 3: Conservation and recreation. Buenos Aires, Instituto Forestal Nacional. 4118-4121. Sp (En, Fr). (P)

2575 **Dimitri, M.J. (1977).** Pequena flora ilustrada de los parques nacionales andino-patagonicos. 2nd ed. Buenos Aires, Servicio Nacional de Parques Nacionales. xiii, 122p. (P)

2576 **Gardner, M.F., Knees, S.G. (1987).** *Fitzroya cupressoides*: the tree and its future. *Int. Dendrol. Soc. Year Book*, 1987: 78-82 (1987 publ. 1988). (T)

2577 **Halloy, S. (1987).** Anconquija - Argentina. *Biol. Conserv. Newsl.*, 48: 1-2. Mentions medicinal plants and timber trees. (E)

2578 **Holdgate, M.W. (1968).** Conservation in the *Nothofagus* forest zone. *In* IUCN. Conservation in Latin America. Proceedings, San Carlos de Bariloche, Argentina, 27 March-2 April 1968. 122-132.

2579 **Luti, R. (1984).** They survive under the southern winds: wildlife protection in northern coastal Patagonia. *In* McNeely, J.A., Miller, K.R., *eds.* National parks, conservation and development. The role of protected areas in sustaining society. Washington, D.C., Smithsonian Institution Press. 561-564. (P)

2580 **Perada, J. (1978).** Nuevo equilibrio biologico en el sud de Neuquen. (New biological equilibrium in the south of Neuquen, Argentina.) *In* Proceedings of the 7th World Forestry Congress, Buenos Aires, 4-18 October 1972, vol. 3: Conservation and recreation. Buenos Aires, Instituto Forestal Nacional. 4143-4145. Por.

2581 **Pigretti, E.A. (1984).** Legislacion forestal y medio ambiente en la Argentina. (Forest and environmental legislation in Argentina.) *In* Prieur, M., *ed.* Forets et environnement: en droit compare et international. Paris, Presses Universitaires de France: 199-204. Sp. (L)

2582 **Pingitore, E.J. (1976).** The Republic of Argentina tree ferns. *Los Angeles Int. Fern Soc.*, 3(10): 198-203; 3(11): 222-225; 3(12): 246-249. Includes list of 8 endangered and 2 rare species. (T)

2583 **Pingitore, E.J. (1981).** Especies vegetales en vias de extincion de la Republica Argentina. *Bol. Soc. Hort. Argentina*, 37: 10-13. Sp. Tentative list of 69 threatened species. (T)

2584 **Pingitore, E.J. (1982).** Especies interesantes de La Tierra del Fuego e Islas del Antartico Sur. *Bol. Soc. Hort. Argentina*, 38: 10-12. Sp. Tentative list of 38 threatened species. (T)

ARGENTINA

2585 **Pingitore, E.J. (1983).** Rare palms in Argentina. *Principes*, 26(1): 9-18. Illus., maps. (T)

2586 **Reid, R. (1988).** "Extinct" cereal rediscovered. *Threatened Pl. Newsl.*, 19: 6. Illus. (E/T)

2587 **Sanchis Munoz, J.R. (1968).** Integration and co-ordination of legislation and national and province or state conservationist action. *In* IUCN. Conservation in Latin America. Proceedings, San Carlos de Bariloche, Argentina, 27 March-2 April 1968. Morges, IUCN. 256-261. (L)

2588 **Sota, E.R. de la (1977).** The problems of threatened and endangered plant species and plant communities in Argentina. *In* Prance, G.T., Elias, T.S., *eds.* Extinction is forever. Proceedings of a symposium at the New York Botanical Garden, 11-13 May 1976. New York, New York Botanical Garden. 240-244. (T)

2589 **Sota, E.R. de la (1979).** Argentina: the conservation of endemic and threatened plant species within botanic gardens. *In* Synge, H., Townsend, H., *eds.* Survival or extinction. Proceedings of a conference held at the Royal Botanic Gardens, Kew, 11-17 September 1978. Kew, Bentham-Moxon Trust. 95-99. (G/T)

2590 **Valentini, J.A. (1978).** Relacion bosque - agricultura - ganaderia en el Manejo de los suelos del Parque Chaqueno Humedo. *In* Proceedings of the 7th World Forestry Congress, Buenos Aires, 4-18 October 1972, vol. 3: Conservation and recreation. Buenos Aires, Instituto Forestal Nacional. 3693-3696. Sp. (P)

AUSTRALIA

2591 **Althofer, G.W. (1981).** Rare or endangered species of plants. *Burrendong Arbor. Brigge*, 55: 9-11. *Eucalyptus, Grevillea.* (T)

2592 **Althofer, G.W. (1984).** Rare or endangered species of native plants. *Burrendong Arbor. Brigge*, 66: 9-12. (T)

2593 **Anon. (1978).** Eighteenth annual meeting of Australian Orchid Council: Gold Coast Conservation Project. *Austral. Orchid. Rev.*, 43(4): 192. (T)

2594 **Anon. (1979).** How to go bush in style (Stony Range Flora Reserve). *Good Gardening*, 37: 45-47. Illus. (P)

2595 **Anon. (1979).** National parks - Australia. *Parks and Wildlife*, 2(3-4): 92-160. Illus., col. illus., maps. (P)

2596 **Anon. (1982).** A national conservation strategy. *Austral. Parks Recr.*, May: 52. From a seminar held in Canberra, 30 November-3 December 1981.

2597 **Anon. (1982).** The Australian Orchid Foundation orchid seed bank. *Amer. Orchid Soc. Bull.*, 51(6): 622. (G)

2598 **Anon. (1983).** New Australian reserves. *Oryx*, 17: 144. (P)

2599 **Anon. (1984).** Conservation of orchids in Kiutpo Forest. *Native Orchid Soc. S. Austral. J.*, 8(3): 30. (T)

2600 **Australia: Department of Home Affairs and Environment (1982).** Towards a national conservation strategy: a discussion paper: national conservation strategy for Australia, living resource conservation for sustainable development. Canberra, Department of Home Affairs and Environment. 71p.

2601 **Australia: Department of Home Affairs and Environment (1983).** A national conservation strategy for Australia: summary record of a conference held in Canberra, 10-13 June 1983. Canberra, Department of Home Affairs and Environment. 66p.

AUSTRALIA

2602 **Australian and New Zealand Association for the Advancement of Science (1979).** A vanishing heritage: the problem of endangered species and their habitats. Wellington, Nature Conservation Council. 273p. (T)

2603 **Barnaby, W. (1983).** Australia: environment down under. *Ambio*, 12(1): 27-33. Illus. Deforestation, water pollution, erosion. (L)

2604 **Bates, B. (1988).** Australian notes: conservation of orchids, real versus perceived. *Newsl. New Zealand Native Orchid Group*, 26: 7-8. (T)

2605 **Bates, R. (1983).** Destruction of native vegetation in Spring Gully Conservation Park by sheep. *S. Austral. Nat.*, 58(1): 12-14. Map. (P/T)

2606 **Bates, R. (1983).** Our disappearing orchids: *Caladenia latifolia*. *Native Orchid Soc. S. Austral. J.*, 7(10): 93. Illus. (T)

2607 **Beardsell, D. & C. (1982).** Australia's earth orchids: aground. *Garden* (New York), 6(5): 10-14. Col. illus. (T)

2608 **Bennett, J.W. (1982).** Valuing the existence of a natural ecosystem. *Search*, 13(9-10): 232-235. Illus. (P)

2609 **Brandenburg, J.P. (1981).** Bush in the urban environment. *Austral. Parks Recr.*, Nov.: 7-10. Illus. (P)

2610 **Buckley, R. (1981).** Endangered plants in central Australia. *In* Synge, H., *ed.* The biological aspects of rare plant conservation. Chichester, Wiley. 504-505. Proceedings of International Conference, King's College, Cambridge, 14-19 July 1980. Paper reviews recent assessments of Australian rare plants. (T)

2611 **Buckley, R. (1982).** Use and conservation of Central Australian dunefields. *Biol. Conserv.*, 22(3): 197-205. Map.

2612 **Carr, G. (1982).** Some conservation problems in the Australian flora. Report No. 2, Australian Flora Foundation. 29-30. (T)

2613 **Caufield, C. (1983).** Rainforests can cope with careful logging. *New Sci.*, 99(1373): 631. Illus. (E/T)

2614 **Chippendale, G.M. (1963).** The relic nature of some Central Australian plants. *Trans. Roy. Soc. Aust.*, 86: 31-34. (T)

2615 **Christensen, P.E.S. (1981).** Clearfelling and native fauna in south-west forests. *Forest Focus*, 24: 10-23.

2616 **Clements, M.A. (1985).** The conservation of Australian orchids. *In* Tan, Kiat, *ed.* Proceedings of the Eleventh World Orchid Conference, Miami, Fl. 138-141. (T)

2617 **Clements, M.A., Boden, R.W. (1981).** The propagation from seed of the endangered orchid *Diuris punctata* J.E.Sm. var. *albo-violacea* Rupp ex Dockrill. *In* XIII International Botanical Congress, Sydney, Australia, 21-28 August 1981: abstract. Sydney, Australian Academy of Science. 340. (G/T)

2618 **Coaldrake, J.E. (1973).** Conservation problems of coastal sand and open-cast mining. *In* Costin, A.B., Groves, A.K., *eds.* Nature conservation in the Pacific. Proc. Symposium A-10, XII Pacific Science Congress, Canberra, August-September 1971. IUCN and Australian Nat. Univ. Press. 299-314. Map.

2619 **Connell, S., Lamont, B. (1988).** Rare and endangered Matchstian *Banksia*. *Austral. Nat. Hist.*, 22(8): 354-355. Col. illus., maps. (T)

2620 **Costin, A.B., Groves, R.H., eds (1973).** Nature conservation in the Pacific. Proceedings of Symposium A-10, XII Pacific Science Congress, August-September 1971, Canberra, Australia. IUCN New Series No. 25. Morges, IUCN and Canberra, Australian National University Press. 337p. Illus., maps.

AUSTRALIA

2621 **Crisp, M.D. (1985).** Conservation of the genus *Daviesia*. Australian National Botanic Gardens Occasional Publication No. 6. Canberra, Australian Government Publishing Service. v, 33p. Illus., map. (T)

2622 **Douglas, I. (1975).** Pressures on Australian rain-forests. *Envir. Conserv.*, 2(2): 109-119. Illus, maps. (P)

2623 **Dowe, J. (1989).** The unexpected rediscovery of *Carpoxylon macrospermum*. *Principes*, 33(2): 63-67. (T)

2624 **Dunphy, M. (1979).** The deforestation of Australia. *Habitat*, 7(4): 14-18.

2625 **Dunphy, M.J. (1979).** The bushwalking conservation movement, 1914-1965. *Parks and Wildlife*, 2(3-4): 54-64. Illus., map.

2626 **Fairley, A. (1981).** The observer's book of national parks of Australia. Sydney, Methuen and London, Warne. xii, 168p. Col. illus., maps. (P)

2627 **Fox, A.M. (1984).** People and their park: an example of free running socio-ecological succession. *In* McNeely, J.A., Miller, K.R., *eds*. National parks, conservation and development. The role of protected areas in sustaining society. Washington, D.C., Smithsonian Institution Press. 290-295. (P)

2628 **Fry, I., Benson, J. (1986).** Australia's threatened plants. *In* Kennedy, M., Burton, R., *eds*. "A threatened species conservation strategy for Australia". Policies for the future. Manly, N.S.W., Ecofund. 19-31. (T)

2629 **Fry, I., Kennedy, M. (1986).** Correlating habitats with high priority threatened species. *In* Kennedy, M., Burton, R., *eds*. "A threatened species conservation strategy for Australia". Policies for the future. Manly, N.S.W., Ecofund. 39-42. Covers threatened plants and animals. (T)

2630 **Gardner, J.E., Nelson, J.G. (1981).** National parks and native peoples in northern Canada, Alaska and northern Australia. *Envir. Conserv.*, 8(3): 207-215. Maps. (P)

2631 **Giles, J. (1979).** Research in national parks. *Parks and Wildlife*, 2(3-4): 86-88. Illus. (P)

2632 **Gill, A.M. (1977).** Management of fire-prone vegetation for plant species conservation in Australia. *Search*, 8(1-2): 20-26. Illus. (T)

2633 **Gill, A.M. (1979).** Fire in the Australian landscape. *Landscape Planning*, 6(3/4): 343-357. Illus. Includes national parks. (P)

2634 **Given, D.R. (1981).** Flora of offshore and onlying islands. *In* XIII International Botanical Congress, Sydney, Australia, 21-28 August 1981: abstracts. Sydney, Australian Academy of Science. 215.

2635 **Good, R.B., Lavarack, P.S. (1981).** The status of Australian plants at risk. *In* Synge, H., *ed*. The biological aspects of rare plant conservation. Chichester, Wiley. 81-91. Proceedings of International Conference, King's College, Cambridge, 14-19 July 1980. Maps. (T)

2636 **Groves, R.H. (1979).** The status and future of Australian grasslands. *New Zealand J. Ecol.*, 2: 76-81.

2637 **Groves, R.H. (1982).** Changing directions in research on Australian rare plants. *In* Groves, R.H., Ride, W.D.L., *eds*. Species at risk: research in Australia. Berlin and New York, Springer-Verlag. 175-179. (T)

2638 **Groves, R.H., Ride, W.D.L., eds (1982).** Species at risk: research in Australia. Symposium on the Biology of Rare and Endangered Species in Australia, Canberra, Australia, 25-26 November 1981. Berlin and New York, Springer-Verlag. 216p. Illus., map. (T)

AUSTRALIA

2639 **Harris, J.A. (1982).** Rare or endangered plants. *Burrendong Arbot. Brigge*, 58: 1-5. Editorial. *Eucalyptus bakeri, E. behriana, E. watsoniana, Grevillea leiophylla.* (T)

2640 **Hartley, W., Leigh, J. (1979).** Plants at risk in Australia. Occasional Paper No. 3. Canberra, Australian National Parks and Wildlife Service. 80p. (T)

2641 **Heberle, R.L. (1983).** West Australian terrestrial orchid seeds. *Austral. Orchid Rev.*, 48(1): 35-36. (T)

2642 **Hollowoway, C. (1979).** IUCN. The Red Data Book and some issues of concern to the identification and conservation of threatened species. *In* Tyler, M.J., *ed.* The status of Australasian wildlife. Canberra. 1-12. (T)

2643 **Hopper, S.D., Campbell, N.A., Moran, G.R. (1981).** Morphometric and allozyme variation in the rare south-west Australian species *Eucalyptus caesia*. *In* XIII International Botanical Congress, Sydney, Australia, 21-28 August, 1981, Abstracts. Sydney, Australian Academy of Science. 124. (T)

2644 **Horning, D.S., jr, Mitchell, A.D. (1982).** Relative resistance of Australian native plants to fluoride. *In* Murray, F., *ed.* Fluoride emissions: their monitoring and effects on vegetation and ecosystems. New York, Academic Press. 157-176.

2645 **Hornsby, P. (1980).** Conserving orchids at Yundi. *Native Orchid Soc. S. Austral. J.*, 4(3): 3. (T)

2646 **IUCN (1986).** IUCN's list of top 24 threatened animals and plants. *Biol. Conserv. Newsl.*, 44: 1-6. (T)

2647 **IUCN Threatened Plants Unit (1983).** The botanic gardens list of rare and threatened plants of Australia. Kew, IUCN Botanic Gardens Conservation Co-ordination Body Report No. 9. Lists plants in cultivation in botanic gardens. (G/T)

2648 **Johnstone, D.A. (1984).** Future directions for the Australian Realm. *In* McNeely, J.A., Miller, K.R., *eds.* National parks, conservation and development. The role of protected areas in sustaining society. Washington, D.C., Smithsonian Institution Press. 301-308. Table. (P)

2649 **Kelleher, G. (1984).** Conserving the Great Barrier Reef through zoning. *In* Unesco-UNEP. Conservation, science and society. Paris, Unesco. 152-158. Contributions to the First International Biosphere Reserve Congress, Minsk, Byelorussia/USSR, 26 September - 2 October 1983. Map. (P)

2650 **Kelleher, G., Kenchington, R. (1984).** Australia's Great Barrier Reef Marine Park: making development compatible with conservation. *In* McNeely, J.A., Miller, K.R., *eds.* National parks, conservation and development. The role of protected areas in sustaining society. Washington, D.C., Smithsonian Institution Press. 267-273. Map. (P)

2651 **Kennedy, M. (1988).** Conservation advances in Australia. *Species*, no. 10: 14. Announcement of National Endangered Species Program. (L/T)

2652 **Kennedy, M., Burton, R., eds (1986).** "A threatened species conservation strategy for Australia". Policies for the future. Manly, N.S.W., Ecofund. Illus., maps. (L/T)

2653 **Kennedy, M., Fry, I. (1986).** Legislative proposals. *In* Kennedy, M., Burton, R., *eds.* "A threatened species conservation strategy for Australia". Policies for the future. Manly, N.S.W., Ecofund. 52-57. Endangered species habitat protection legislation; includes model of the Endangered Species Habitat Act for New South Wales. (L)

2654 **Kerr, R. (1981).** Have you seen *Cryptanthemis slateri* lately? *Austral. Orchid Rev.*, 46(2): 88-91. Illus. (T)

2655 **Lawn, R.J., Cottrell, A. (1988).** Wild mungbean and its relatives in Australia. *Biologist*, 35(6): 267-273. Illus., map. (E)

AUSTRALIA

2656 **Leigh, J., Boden, R. (1979).** Australian flora in the Endangered Species Convention - CITES. Spec. Publ. No. 3. Canberra, Australian National Parks and Wildlife Service. v, 93p. Col. illus. (T)

2657 **Leigh, J., Boden, R. (1980).** Australian endangered plant species. *Austral. Pl.*, 10(84): 357-358. (T)

2658 **Leigh, J., Boden, R., Briggs, J. (1984).** Extinct and endangered plants of Australia. Melbourne, Macmillan. 369p. Includes detailed case studies of 76 species presumed extinct and 203 which are endangered. (T)

2659 **Leigh, J., Briggs, J., Hartley, W. (1981).** Rare or threatened Australian plants. Spec. Publ. No. 7. Canberra, Australian National Parks and Wildlife Service. 178p. Illus., maps. (T)

2660 **Leigh, J.H., Briggs, J.D., Hartley, W. (1972).** The conservation status of Australian plants. *In* Groves, R.H., Ride, W.D.L., eds. Species at risk: research in Australia. New York, Springer-Verlag. 13-25. (T)

2661 **Ling, J.K. (1984).** Kaiserstuhl Conservation Park dedicated. *S. Austral. Nat.*, 58(3): 49-50. Illus. (P)

2662 **Maconochie, J.R. (1981).** General plant ecology and biology of the Australian arid zone. *North. Terr. Bot. Bull.*, 4: 1-19. Includes list of rare and/or endemic plant species in Central Australia. (T)

2663 **Mascarenhas, A.F., comp. (?)** Biological diversity and genetic resources techniques and methods: tissue culture directory for Asia-Pacific. CSC Technical Publication Series No. 249. London, Commonwealth Science Council. 12p. (E/G)

2664 **McComb, A.J., Lake, P.S. (1988).** The conservation of Australian wetlands. Chipping Norton, Surrey Beatty and World Wildlife Fund Australia. viii, 196p. Col. illus., maps.

2665 **McCraith, G. (1981).** Pollinate to conserve. *Austral. Orchid Rev.*, 46(2): 101. Seed bank at Australian Orchid Foundation, Sydney. (G)

2666 **McLaren, R. (1981).** Preserve our flora: selection of best forms. *Austral. Pl.*, 11(87): 185. Col. illus. *Anthocercis albicans.* (T)

2667 **McMichael, D.F. (1982).** What species, what risk? *In* Groves, R.H., Ride, W.D.L., eds. Species at risk: research in Australia. Berlin and New York, Springer-Verlag. 163-274. (T)

2668 **McMichael, D.F., Gare, N.C. (1984).** Keynote address: the Australian Realm. *In* McNeely, J.A., Miller, K.R., eds. National parks, conservation and development. The role of protected areas in sustaining society. Washington, D.C., Smithsonian Institution Press. 258-266. (P)

2669 **Melville, R. (1971).** Endangered angiosperms and conservation in Australia. *Bull. Jard. Bot. Belg.*, 41: 145-152. (T)

2670 **Melville, R. (1973).** Relict plants in the Australian flora and their conservation. *In* Costin, A.B., Groves, K.H., eds. Nature conservation in the Pacific. Proc. Symposium A-10, XII Pacific Science Congress, Canberra, August-September 1971. IUCN and Australian Nat. Univ. Press. 83-90. Illus. (T)

2671 **Moriarty, O. (1979).** A challenge to the system: different, better and more parks. *Austral. Parks & Recreation*, Nov.: 47-49. Illus. (P)

2672 **Morrison, D., Auld, T., Gallagher, K. (1983).** The distribution, ecology and conservation status of *Acacia suaveolens*. *Victorian Nat.*, 100(4): 140-145. Illus., map. (T)

2673 **Mosley, J.G. (1979).** People and parks. *Habitat*, 7(4): 3-5. Col. illus. (P)

AUSTRALIA

2674 **Mosley, J.G. (1984).** Protected areas and environmental planning in Australia: the continuing evolution of a diverse range of protected areas. *In* McNeely, J.A., Miller, K.R., *eds.* National parks, conservation and development. The role of protected areas in sustaining society. Washington, D.C., Smithsonian Institution Press. 274-282. Tables. (P)

2675 **Murdoch, B. (1979).** Letters to the Editor: conservation of native orchids. *Orchadian*, 6(6): 129. (T)

2676 **Nevill, J. (1979).** Preserving native wildlife in the rural landscape. *Habitat*, 7(1): 22-26. Col. illus.

2677 **New South Wales. Forestry Commission (1984).** Logging and conservation: the private collector and the commercial picker. *In* Proceedings of Australasian Native Orchid Society anniversary workshop, Naamaroo, Lane Cove, 7-8 July 1984. 4p.

2678 **New South Wales. National Parks and Wildlife Service (1984).** Conservation of native orchids and licensing of private growers and commercial pickers. *In* Proceedings of Australasian Native Orchid Society anniversary workshop, Naamaroo, Lane Cove, 7-8 July 1984. 5p. Trade. (G/T)

2679 **Newsome, A.E. (1973).** The adequacy and limitations of flora conservation for fauna conservation in Australia and New Zealand. *In* Costin, A.B., Groves, R.H., *eds.* Nature conservation in the Pacific. Proc. Symposium A-10, XII Pacific Science Congress, Canberra, August-September 1971. IUCN and Australian Nat. Univ. Press. 93-110. Illus., maps. (P)

2680 **Palmer, A. (1982).** Rare Australian orchid saved from extinction. *J. Wellington Orchid Soc.*, 6(1): 9. Seed propagated at National Botanic Gardens. *Diuris punctata* var. *albo-violacea.* (T)

2681 **Parsons, R.F., Browne, J.H. (1982).** Causes of plant species rarity in semi-arid southern Australia. *Biol. Conserv.*, 24(3): 183-192. Map. (T)

2682 **Pederick, L.A. (1976).** Conservation of gene resources for the improvement of native species in Australia. *Austral. Forest.*, 39(2): 113-120. (E)

2683 **Phillips, G.P. (1979).** Mining and national parks. *Austral. Parks & Recreation*, (Nov.): 9-12. (P)

2684 **Pryor, L.D. (1973).** Nature conservation in relation to modern trends in Australian forestry. *In* Costin, A.B., Groves, A.K., *eds.* Nature conservation in the Pacific. Proc. Symposium A-10, XII Pacific Science Congress, Canberra, August-September 1971. IUCN and Australian Nat. Univ. Press. 321-326.

2685 **Pryor, L.D. (1981).** Australian endangered species: eucalypts. Special Publication No. 5. Canberra, Australian National Parks and Wildlife. 139p. Illus., map. (T)

2686 **Recher, H.F. (1978).** Wildlife conservation: a case for managing forests as ecosystems. *In* Proceedings of the 8th World Forestry Congress, Jakarta, 16-28 October 1978: Forestry for quality of life. II, 9.

2687 **Routley, R. & V. (1980).** Destructive forestry in Melanesia and Australia. *Ecologist*, 10(1-2): 56-67. Illus.

2688 **Saunders, D., Hobbs, R. (1989).** Corridors for conservation. *New Sci.*, 121(1649): 63-68. Illus. Discusses the importance of "wildlife corridors" for conservation.

2689 **Slatyer, R. (1983).** Conservation of Australia's biological resources. *Austral. J. Ecol.*, 8(1): 1-2. (L/T)

2690 **Smithers, C. (1978).** A preliminary list of the ferns of Tuglo Wildlife Refuge. *Hunter Nat. Hist.*, 10(4): 189-191. Illus. (P)

AUSTRALIA

2691 **Specht, R.L. (1981).** Conservation of vegetation types. *In* Groves, R.H., *ed.* Australian vegetation. Cambridge, Cambridge Univ. Press. 393-410. Maps. (P)

2692 **Specht, R.L. (1981).** Conservation: Australian heathlands. *In* Specht, R.L., *ed.* Heathlands and related shrublands: analytical studies. Ecosystems of the World 9B. Amsterdam, Elsevier. 235-240.

2693 **Specht, R.L., Roe, E.M., Boughton, V.H., eds. (1974).** Conservation of major plant communities in Australia and Papua New Guinea. Australian J. Bot. Supp. Series 7. 667p. Detailed assessment of conservation status of major plant communities. (T)

2694 **Stocker, G.C. (1981).** The conservation of plants in northern Australia. *In* XIII International Botanical Congress, Sydney, Australia, 21-28 August 1981: abstracts. Sydney, Australian Academy of Science. 215. (T)

2695 **Strom, A.A. (1979).** Some events in nature conservation over the last forty years. *Parks and Wildlife,* 2(3-4): 65-73. Illus.

2696 **Teare, L., Davis, N.P. (1982).** A survey of endangered succulents in Australia. *Calindrinia,* 2: 72-77. (T)

2697 **Terry, P. (1982).** Satellite spots elusive orchid. *Nature Orchid Soc. S. Austral. J.,* 6(7): 63-64. (T)

2698 **Tisdell, C.A. (1983).** An economists's critique of the World Conservation Strategy, with examples from the Australian experience. *Envir. Conserv.,* 10(1): 43-52.

2699 **Townsend, D. (1979).** Management of parks - a planning perspective. *Parks and Wildlife,* 2(3-4): 89-91. (P)

2700 **Tracey, J.G. (1981).** Australia's rain forest: where are the rare plants and how do we keep them? *In* Synge, H., *ed.* The biological aspects of rare plant conservation. Wiley. 165-178. Proceedings of International Conference, King's College, Cambridge, 14-19 July 1980. Map. (T)

2701 **WWF (1984).** World Wildlife Fund Australia Conservation Programme 1984. Sydney, WWF-Australia. 27p. Illus. Lists WWF-Australia projects. (T)

2702 **WWF-Australia (1986).** Conservation Programme 1983. Project 45: propagation of threatened Australian plants in Victorian reserves. Sydney, WWF-Australia. p.17. Project aimed at re-establishing endangered or vulnerable plant species in suitable habitats. (P/T)

2703 **Walker, K. (1983).** Death of the Murray. *Austral. Nat. Hist.,* 21(1): 23-28. Illus., col. illus. (P)

2704 **Walsh, G. (1983).** Some logic at last. *Orchadian,* 7(9): 219-220. Rainforest logging. Originally published in *The Orchidophile,* Sydney ANOS Group Bulletin, Jan. 1983.

2705 **Walsh, G. (1983).** The toothpick: tool of conservation, or of confusion? *Orchadian,* 7(9): 207-209. Author assists orchid pollination and introduces species beyond their natural range. (T)

2706 **Walsh, G.F. (1982).** Is this conservation? *Orchadian,* 7(6): 141. *Dendrobium x gracillimum.* (L/T)

2707 **Webb, L.J. (1981).** Current problems in the preservation of the remaining tropical and subtropical rainforests of Australia. *In* XIII International Botanical Congress, Sydney, Australia, 21-28 August 1981: abstracts. Sydney, Australian Academy of Science. 105. (T)

AUSTRALIA

2708 **Webb, L.J., Tracey, J.G., Kikkawa, J., Williams, W.T. (1973).** Techniques for selecting and allocating land for nature conservation in Australia. *In* Costin, A.B., Groves, R.H., *eds.* Nature conservation in the Pacific. Proc. Symposium A-10, XII Pacific Science Congress, Canberra, August-September 1971. IUCN and Australian Nat. Univ. Press. 39-52. (P)

2709 **Williams, O.B. (1981).** Monitoring changes in populations of desert plants. *In* Synge, H., *ed.* The biological aspects of rare plant conservation. Chichester, Wiley. 233-240. Proceedings of International Conference, King's College, Cambridge, 14-19 July 1980. (T)

2710 **Womersley, J.S., comp. (1974).** Conservation of primitive, rare, and endangered species. *In* Specht, R.L., *et al., eds.* Conservation of major plant communities in Australia and Papua New Guinea. Australian J. Bot. Suppl. No. 7: 594. (T)

2711 **Wrigley, J.W. (1978).** Success with growing native Australian endangered plants. *Gartn.-Bot. Brief*, 58: 32. Abstract. (G/T)

2712 **Wrigley, J.W. (1979).** Australia: the cultivation of native endangered plants in Canberra Botanic Gardens. *In* Synge, H., Townsend, H., *eds.* Survival or extinction. Proceedings of a conference held at the Royal Botanic Gardens, Kew, 11-17 September 1978. Kew, Bentham-Moxon Trust. 101-105. (G/T)

AUSTRALIA - Coral Sea Islands Territory

2713 **McMichael, D.F., Talbot, F.H. (1970).** Conservation of islands and coral reefs of the Great Barrier Reef system, the islands of the Coral Sea, and Norfolk and Lord Howe Islands. *Micronesia*, 5(2): 493-496.

AUSTRALIA - NSW - Lord Howe Island

2714 **Anon. (1979).** Conservation of Lord Howe Island's terrestrial and marine environments. *The Lord Howe Island Signal*, 2(26): 1-2. (P)

2715 **Dodson, J.R. (1982).** Modern pollen rain and recent vegetation history on Lord Howe Island: evidence of human impact. *Rev. Palaeobot. Palynol.*, 38(1/2): 1-21. Illus., maps.

2716 **Green, P.S. (1979).** Observations on the phytogeography of the New Hebrides, Lord Howe Island and Norfolk Island. *In* Bramwell, D., *ed.* Plants and islands. London, Academic Press. 41-53. (Sp). Map.

2717 **McMichael, D.F., Talbot, F.H. (1970).** Conservation of islands and coral reefs of the Great Barrier Reef system, the islands of the Coral Sea, and Norfolk and Lord Howe Islands. *Micronesia*, 5(2): 493-496.

2718 **Melville, P.S. (1970).** Notes relating to the floras of Norfolk and Lord Howe islands, I. *J. Arnold Arb.*, 51(2): 204-220. Illus. (P/T)

2719 **Moore, H.E. (1966).** Palm hunting around the world. IV. Lord Howe Island. *Principes*, 10: 13-21. (T)

2720 **Pickard, J. (1973).** An annotated botanical bibliography of Lord Howe Island. *Contrib. N.S.W. Nat. Herb.*, 4: 470-491. (B)

2721 **Pickard, J. (1980).** The palm seed industry on Lord Howe Island. *Principes*, 24(1): 3-13. (E/T)

2722 **Pickard, J. (1982).** The effect of feral goats (*Capra hircus* L.) on the vegetation of Lord Howe Island. *Austral. J. Ecol.*, 1(2): 103-113. Illus., map. (P/T)

2723 **Pickard, J. (1983).** Rare or threatened vascular plants of Lord Howe Island. *Biol. Conserv.*, 27(2): 125-139. Maps. (T)

AUSTRALIA - NSW - Lord Howe Island

2724 **Pickard, J. (1984).** Exotic plants on Lord Howe Island: distribution in space and time, 1853-1981. *J. Biogeography,* 11: 28p. Maps. Effects of introduced species. (P/T)

2725 **Recher, H.F., Clark, S.S. (1983).** A biological survey of Lord Howe Island with recommendations for the conservation of the island's wildlife. *Biol. Conserv.,* 6: 263-273.

2726 **Smith, N., et al. (1977).** Lord Howe Island. Sydney, The Australian Museum. 42p. Includes discussion on conservation and exploitation.

AUSTRALIA - New South Wales

2727 **Althofer, G.W. (1982).** Rare or endangered species of native plants. *Burrendong Arbor. Brigge,* 60: 11. *Halgania preissiana, Hakea bakerana, Hemigenia cuneifolia.* (T)

2728 **Althofer, G.W. (1983).** Rare or endangered species of native plants. *Burrendong Arbor. Brigge,* 63: 11. *Boronia* sp. (undescribed, close to *B. ruppii*). (T)

2729 **Anderson, R.H. (1941).** The effect of settlement upon the New South Wales flora. *Proc. Linn. Soc. New South Wales,* 66(1-2): v-xxiii. (T)

2730 **Anon. (1978).** Orchids at Jervis Bay Annex National Botanic Gardens. *Austral. Orchid Rev.,* 44(3): 151. (G)

2731 **Anon. (1979).** Royal National Park. *Habitat,* 7(2): 3-6. Illus., col. illus. (P)

2732 **Anon. (1979).** The Kosciusko environment. *Habitat,* 7(3): 12-19. Illus., col. illus. Edited abstract from Costin, A., Wimbush, D., Gray, M. and Totterdell, C., *Kosciusko Alpine Flora,* Melbourne, CSIRO Division of Plant Industry.

2733 **Anon. (1983).** Arboretum wins award. *Oryx,* 17: 144. Burrendong Arboretum, New South Wales, Australia. (G/T)

2734 **Barker, W.R., Morrison, S.P. (1989).** *Hakea pulvinifera* L. Johnson (Proteaceae): a rediscovered species under threat. *J. Adelaide Bot. Gard.,* 11(2): 175-178. (T)

2735 **Barton, L. (1982).** *Dendrobium falcorostrum* and logging. *Orchadian,* 7(5): 106. (T)

2736 **Baur, G.N., comp. (1981).** Forest preservation in State Forests of New South Wales. Sydney, Forestry Commission of New South Wales. 78p. Map. (P)

2737 **Benson, J. (1988).** Conservation of flora in western New South Wales. *Nation. Parks J.,* 32(3): 16-22. (T)

2738 **Benson, J.S. (1987).** The effect of 200 years of European settlement on the vegetation of New South Wales, Australia: an overview. Paper presented at XIVth International Botanical Congress, West Berlin, July 1987. New South Wales, National Parks and Wildlife Service. 47p. Maps. Extent of original forest and impact of settlement; threatened communities. (P/T)

2739 **Brickhill, J. (1988).** Conservation of mallee in western New South Wales. *Nation. Parks J.,* 32(3): 34-37.

2740 **Brooker, M.I.H. (1979).** Notes on red gums, grey gums and *Eucalyptus pumila* Cambage - a rare species with obscure affinities. *Austral. Forest Res.,* 9(4): 265-276. Illus. (T)

2741 **Butt, L. (1987).** Cycads of Australia: *Macrozamia miquelii. Encephalartos,* 10: 30. Illus. (T)

2742 **Butt, L. (1988).** Cycads of Australia. Small cycads of N.S.W. continued. *Encephalartos,* 14: 17. Illus. (T)

2743 **Clemens, J., Franklin, M.H. (1980).** A description of coastal heath at North Head, Sydney Harbour National Park: impact of recreation and other disturbance since 1951. *Austral. J. Bot.,* 28(4): 463-478. Maps. (P)

AUSTRALIA - New South Wales

2744 **Goldstein, W. (1978).** Watch out for those protected plants. *Forest Timber*, 14(3): 17-21. Illus. (L/T)

2745 **Gray, M. (1979).** A review of the rare and endangered species, *Rutidosis leptorrhynchoides* F. Muell. *Austral. Syst. Bot. Soc. Newsl.*, 20: 2-5. (T)

2746 **Harrison, M. (1983).** Cultivation based on habitat observation: 8. *Dendrobium falcorostrum* Fitz G. *Orchadian*, 7(10): 227-229. (G/T)

2747 **Johnstone, D.A. (1984).** Public participation in reserve management in New South Wales. *In* Unesco-UNEP. Conservation, science and society. Paris, Unesco. 540-547. Contributions to the First International Biosphere Reserve Congress, Minsk, Byelorussia/USSR, 26 September - 2 October 1983. (L/P)

2748 **Keane, P.A., Wild, A.E.R., Rogers, J.H. (1979).** Trampling and erosion in alpine country. *J. Soil Conserv. Serv. N.S.W.*, 35(1): 7-12.

2749 **Macdonald, I.A.W., Graber, D.M., DeBenedetti, S., Groves, R.H., Fuentes, E.R. (1988).** Introduced species in nature reserves in Mediterranean-type climatic regions of the world. *Biol. Conserv.*, 44(1-2): 37-66. Invasive species. Includes 7 case studies. (P)

2750 **Martin, R. (1980).** The Warrumbungle National Park. *Austral. Pl.*, 10(84): 376-382. Illus. (P)

2751 **McAlpine, D.K. (1982).** Saving the Illawarra greenhood. *Orchadian*, 7(4): 75. Illus. (T)

2752 **McMichael, D.F. (1973).** Further case studies in selecting and allocating land for nature conservation: New South Wales. *In* Costin, A.B., Groves, R.H., *eds.* Nature conservation in the Pacific. Proc. Symposium A-10, XII Pacific Science Congress, Canberra, August-September 1971. IUCN and Australian Nat. Univ. Press. 53-56. Map. (P)

2753 **Morgan, G., Terry, J. (1988).** Nature conservation in western New South Wales. *Nation. Parks J.*, 32(3): 6-14.

2754 **Pratt, N. (1979).** Kosciusko National Park - conserved for what? *Austral. Parks & Recreation*, Nov.: 27-31. Illus. (P)

2755 **Pressey, R.L. (1988).** Wetlands of western New South Wales: characteristics, conservation and management. *Natl Parks J.*, 32(3): 23-30.

2756 **Smith, J. (1981).** The distribution and conservation status of a rare conifer, *Microstrobos fitzgeraldii* (Podocarpaceae). *Cunninghamia*, 1(1): 125-128. Map. (T)

2757 **Somerville, J. (1988).** Conservation of river red gums in New South Wales. *Nation. Parks J.*, 32(3): 31-33. (T)

2758 **Washington, H. (1979).** Wollemi - a new park for all seasons. *Habitat*, 7(4): 6. Illus., col. illus. (P)

2759 **Wrigley, J. (1987).** Sunraysia Oasis Botanical Gardens, Australia. *Bot. Gard. Conserv. News*, 1(1): 9. (G)

AUSTRALIA - Northern Territory

2760 **Braithwaite, R.W., Dudzinski, M.L., Ridpath, M.G., Parker, B.S. (1984).** The impact of water buffalo on the monsoon forest ecosystem in Kakadu National Park. *Austral. J. Ecol.*, 9(4): 309-322. Illus. (P)

2761 **Brown, G. (1988).** Conserving a rare palm, *Ptychosperma bleeseri* Burret, at Darwin Botanic Gardens, Northern Territories, Australia. *Bot. Gard. Conserv. News*, 1(3): 14. Illus. Sometimes treated as a synonym of *Carpentaria acuminata*. (G/T)

2762 **Chinner, D.W. (1979).** N.T. arid zone parks. *Austral. Parks & Recreation*, (Nov.): 15-19. Illus. (P)

AUSTRALIA - Northern Territory

2763 **Chuk, M. (1982).** The status and ecology of *Acacia peuce* in the northern territory. Conservation Commission of the Northern Territory. Technical report 2. 30p. Illus. (T)

2764 **Fox, A. (1983).** Kakadu is aboriginal land. *Ambio*, 12(3-4): 161-166. Illus., col. illus. (P)

2765 **Hill, M. (1983).** Kakadu National Park and the aboriginals: partners in protection. *Ambio*, 12(3-4): 158-160, 162-167. Col. illus., maps. (P)

2766 **James, S.H. (1982).** The relevance of genetic systems in *Isotoma petraea* to conservation practice. *In* Groves, R.H., Ride, W.D.L., *eds.* Species at risk: research in Australia. New York, Springer-Verlag. 63-71. (T)

2767 **Lonsdale, M., Braithwaite, R. (1988).** The shrub that conquered the bush. *New Sci.*, 120(1643): 52-55. Col. illus., map. *Mimosa pigra* from Central America threatens native vegetation; invasive species. (P)

2768 **Macdonald, I.A.W., Frame, G.W. (1988).** The invasion of introduced species into nature reserves in tropical savannas and dry woodlands. *Biol. Conserv.*, 44(1-2): 67-93. Invasive species. Includes 5 case studies. (P)

2769 **McBride, J.R., Lewis, H.T. (1984).** Occurrence of fire scars in relation to the season and frequency of surface fires in eucalyptus forests of the Northern Territory, Australia (Kakadu National Park). *Forest Sci.*, 30(4): 970-976. (P)

2770 **McPhee, H. (1985).** Conservation at Ayers Rock. *Austral. Pl.*, 13(103): 115. (P)

2771 **Walker, J. (1985).** Towards an expert system for fire management at Kakadu National Park. Canberra, CSIRO, Institute of Biological Resources, Division of Water and Land Resources. 169p. Illus., maps. (P)

2772 **Williams, A.R. (1984).** Changes in *Melaleuca* forest density on the Magela floodplain, Northern Territory, between 1950 and 1975. *Austral. J. Ecol.*, 9(3): 199-202. Maps.

AUSTRALIA - Queensland

2773 **Anon. (1979).** Corio Bay: a place of beauty and natural abundance. *Habitat*, 7(1): 12-15. Col. illus.

2774 **Bailes, J.W. (1983).** Native orchid conservation at Currumbin Sanctuary. *Austral. Orchid Rev.*, 48(4): 248-249. Illus. (P)

2775 **Batianoff, E.N. (1987).** Plants of the Sunshine Coast, Queensland (Noosa National Park to MC/Codum). Society for Growing Australian Plants, Queensland Region. 61p. Maps. (P)

2776 **Bird, C. & A. (1983).** On course to Cooktown. *Launceston Nat.*, 16(5-6): 6-7. Visit to Carnarvon National Park. (P)

2777 **Boyland, D.E. (1970).** Ecological and floristic studies in the Simpson Desert National Park, South Western Queensland. *Proc. Roy. Soc. Queensland*, 82(1): 1-16. Illus. (P)

2778 **Butt, L. (1987).** Cycads of Australia: *Macrozamia miquelii*. *Encephalartos*, 10: 30. Illus. (T)

2779 **Churchett, G. (1982).** All in a day's walk: the flora and fauna of the Lamington National Park and other parks in South East Queensland. Athelstone, S. Australia, G. Churchett. 147p. Illus.; maps. Popular guide to plants and animals. (P)

2780 **Edgecombe, J. (1980).** Tangeera-Green Island National Park: profile of a Barrier Reef tourist island. *Habitat* (Australia), 8(2): 20-23. Illus., col. illus. (P)

2781 **Gasteen, J. (1979).** The discovery that caused a land rush. *Habitat*, 7(5): 6-8. Illus. Dawson Valley.

AUSTRALIA - Queensland

2782 **Greenwood, R.H. (1968).** Pressure on land and resources in Queensland. *Proc. Roy. Soc. Queensland,* 80: v-xiv. (P)

2783 **Harrison, M. (1983).** Cultivation based on habitat observation: 8. *Dendrobium falcorostrum* Fitz G. *Orchadian,* 7(10): 227-229. (G/T)

2784 **James, S.H. (1982).** The relevance of genetic systems in *Isotoma petraea* to conservation practice. *In* Groves, R.H., Ride, W.D.L., eds. Species at risk: research in Australia. New York, Springer-Verlag. 63-71. (T)

2785 **Keto, A.I. (1984).** The conservation status of the rainforests of the wet tropics of north-east Queensland. Brisbane, Rainforest Conservation Society of Queensland.

2786 **Lavarack, P.S., Stanton, J.P. (1977).** Vegetation of the Jardine River catchment and adjacent coastal areas. *Proc. Roy. Soc. Queensl.,* 88: 39-48. Illus., map. (T)

2787 **McCabe, J. (1979).** The Iwasaki Resort proposal - some background on a resource conflict. *Habitat,* 7(1): 16-17.

2788 **Nietschmann, B. (1984).** Indigenous island peoples, living resources and protected areas. *In* McNeely, J.A., Miller, K.R., eds. National parks, conservation and development. The role of protected areas in sustaining society. Washington, D.C., Smithsonian Institution Press. 333-343. Map, table. (P)

2789 **Perner, J. (1988).** Visiting *Macrozamia platyrachis. Encephalartos,* 14: 26-27. Illus. Notes on distribution, population size and threats. (G/T)

2790 **Queensland Department of Forestry (1983).** Conservation status of plant communities and areas of biological significance. *In* Rainforest research in north Queensland. Queensland, Department of Forestry. 36. Map. (E)

2791 **Queensland Department of Forestry (1983).** Minor forms of flora. *In* Rainforest research in north Queensland. Queensland, Department of Forestry. 34-36. Includes conservation status of plant communities.

2792 **Queensland Department of Forestry (1983).** Rainforest research in north Queensland. Queensland, Department of Forestry. vii, 52p. Illus., col. illus., maps. (E)

2793 **Rainforest Conservation Society of Queensland (1986).** Tropical rainforests of north Queensland: their conservation significance. A report to the Australian Heritage Commission by the Rainforest Conservation Society of Queensland. Australian Heritage Commission Special Publication Series No. 3. Canberra, Australian Govt Publishing Service. x, 195p. Illus., col. illus., maps.

2794 **Stocker, G.C. (1981).** Regeneration of a north Queensland rain forest following felling and burning. *Biotropica,* 13(2): 86-92.

2795 **Thomas, M.B., McDonald, W.J.F. (1987).** Rare and threatened plants of Queensland. Brisbane, Dept of Primary Industries. 68p. Maps. (P/T)

2796 **Winter, J.W., Bell, F.C., Pahl, L.I., Atherton, R.G. (1987).** Rainforest clearfelling in north-eastern Australia. *Proc. Roy. Soc. Queensland,* 98: 41-57. Deforestation.

AUSTRALIA - South Australia

2797 **Anon. (1977).** Australia (South Australia). Wildlife Regulations, 1975, made under the National Parks and Wildlife Act, 1972-1974. *Food Agric. Legis.,* 26(1): 117-119. (L/P)

2798 **Anon. (1987).** Rediscovery of a *Veronica. Oryx,* 21(1): 58. (P/T)

2799 **Bates, R. (1977).** No. 2 in a series on rare or endemic South Australian orchids. *S. Austral. Nat.,* 51(4): 66. Illus. (T)

AUSTRALIA - South Australia

2800 **Bates, R. (1977).** No. 3 in a series on rare or endemic South Australian orchids. *S. Austral. Nat.*, 51(4): 67. Illus. (T)

2801 **Bates, R. (1981).** Endangered orchid species of South Australia. *Native Orchid Soc. S. Austral. J.*, 5(3): 26-27. (T)

2802 **Bates, R. (1984).** Effects of "Ash Wednesday" bushfires on orchids in the Adelaide hills. *Orchadian*, 7(12): 286. (T)

2803 **Beesley, P.L., Boden, R.W., Crisp, M.D., Rymer, J.L., Sessions, S.G. (1981).** An endangered species of Compositae in south eastern Australia. *In* XIII International Botanical Congress, Sydney, Australia, 21-28 August 1981: abstracts. Sydney, Australian Academy of Science. 309. (T)

2804 **Conacher, A.J. (1977).** Conservation and geography: the case of the Manjimup woodchip industry, southwestern Australia. *Austral. Geog. Stud.*, 15(2): 104-122.

2805 **Crossen, T.I. (1978).** A new concept in park design and management. *Biol. Conserv.*, 15(2): 105-125. Maps. Black Hill Native Flora Park, Adelaide. (P)

2806 **Davies, R.J-P. (1982).** The conservation of major plant associations in South Australia. Adelaide, Conservation Council of South Australia Inc. 367p. Illus., maps. Detailed analysis of plant associations. Index of conservation priorities.

2807 **Davies, R.J-P. (1983).** Surviving examples of South Australia's most threatened plant associations. Adelaide, Conservation Council of South Australia Inc. 43p. Analysis of plant communities including occurrence of threatened species. (T)

2808 **Davies, R.J.-P. (1986).** Threatened plant species of the Mount Lofty Ranges and Kangaroo Island Regions of South Australia. Adelaide, Conservation Council of South Australia. 174p. Illus., maps. (G/L/P/T)

2809 **Hopper, S.D. (1981).** Recent developments in flora conservation in south Western Australia. *In* XIII International Botanical Congress, Sydney, Australia, 21-28 August 1981: abstracts. Sydney, Australian Academy of Science. 216.

2810 **James, S.H. (1982).** The relevance of genetic systems in *Isotoma petraea* to conservation practice. *In* Groves, R.H., Ride, W.D.L., eds. Species at risk: research in Australia. New York, Springer-Verlag. 63-71. (T)

2811 **Jessop, J.P. (1977).** Endangered species in the South Australian native vascular flora. *J. Adelaide Bot. Gard.*, 1: 135-140. (T)

2812 **Margules, C. (1978).** The status of rare plant alliances, birds and mammals in South Australia. Technical Memorandum 78/23. Canberra, CSIRO, Division of Land Use Research. 33p. Maps. (T)

2813 **Preiss, K. (1980).** Ferguson Conservation Park. *S. Austral. Nat.*, 54(4): 52-57. (P)

2814 **Specht, R.L. (1961).** Flora conservation in South Australia: 1. The preservation of plant formations and associations recorded in South Australia. *Trans. Roy. Soc. S. Australia*, 85: 177-196. Map. Lists major national parks and reserves in South Australia. (P/T)

2815 **Wotton, D. (1980).** Nixon-Skinner Conservation Park. *S. Austral. Nat.*, 54(3): 40-42. Illus. (P)

AUSTRALIA - Tasmania

2816 **Anon. (1978).** Rare and endangered plant species. *Launceston Nat.*, 12(1-2): 7. (T)

2817 **Baidya, K.N. (1983).** Tasmanian forest regeneration burning: consideration for ecology or economy? *Tigerpaper*, 10(3): 21-30. Illus., maps.

2818 **Baidya, K.N. (1983).** The South-West Tasmania wilderness crisis. *Envir. Conserv.*, 10(1): 59-61. Illus. Map. (T)

AUSTRALIA - Tasmania

2819 **Bosworth, P. (1982).** Pressure for exploitation in South West Tasmania's wilderness. *Ambio*, 11(5): 268-273. Illus., col. illus., maps. (P/T)

2820 **Bosworth, P.K. (1984).** Increasing pressures for resources exploitation in an area of high nature conservation value, Southwest Tasmania. *In* McNeely, J.A., Miller, K.R., *eds*. National parks, conservation and development. The role of protected areas in sustaining society. Washington, D.C., Smithsonian Institution Press. 283-289. Maps, tables. (P)

2821 **Brown, J.J., et al. (1977).** Conservation of flora in Tasmania. The status of plant species which are primitive, endemic or of geographic significance. Wildlife Technical Report No. 77/4. Hobart, National Parks and Wildlife Service. (T)

2822 **Candolle, R. de (1983).** Tasmania's proposed dams in its South-West Wilderness. *Envir. Conserv.*, 10(1): 68. (P)

2823 **Dragun, A.K. (1983).** Hydroelectric development and wilderness conflict in South-West Tasmania. *Envir. Conserv.*, 10(3): 197-204. Illus., map.

2824 **Felton, K. (1978).** Land reserved in Tasmania for nature conservation. *Austral. Forest.*, 41(3): 146-152. (P)

2825 **Halliwell, B. (1979).** A rare dwarf conifer. *Mon. Bull. Alp. Gard. Club B.C.*, 22(1): 11. (T)

2826 **Halliwell, B. (1981).** *Pratia irrigua*: a gardens success story. *Threatened Pl. Commit. Newsl.*, 8: 12. (G/T)

2827 **Innes, C. (1987).** *Isophysis*. *The Garden* (U.K.), 112(1): 14-15. Illus. (T)

2828 **Kirkpatrick, J. (1980).** Development and wilderness in Tasmania. *Habitat* (Australia), 8(2): 12-14. Maps.

2829 **Kirkpatrick, J.B. (1983).** An iterative method for establishing priorities for the selection of nature reserves: an example from Tasmania. *Biol. Conserv.*, 25(2): 127-134. Maps. (P)

2830 **Kirkpatrick, J.B. (1986).** Conservation of plant species, alliances and associations of the Treeless High Country of Tasmania, Australia. *Biol. Conserv.*, 37(1): 43-57. Map. (T)

2831 **Kirkpatrick, J.B., Dickinson, K.J.M. (1982).** Recent destruction of natural vegetation in Tasmania. *Search*, 13(7-8): 186-187. Map.

2832 **Kirkpatrick, J.B., Harwood, C.E. (1983).** Conservation of Tasmanian macrophytic wetland vegetation. *Pap. Proc. Roy. Soc. Tasmania*, 117: 5-20. Map. Species list. (T)

2833 **Law, G. (1984).** The Tasmanian wilderness campaign. *Ecos*, 5(3): 15-19. Illus., map.

2834 **Lee, N. (1982).** Atmospheric emissions, legislation and monitoring at the Comalco Bay Bell Smelter. *In* Murray, F., *ed*. Fluoride emissions: their monitoring and effects on vegetation and ecosystems. Sydney, Academic Press. 31-44. Illus., vegetation survey.

2835 **Ogden, J., Powell, J.A. (1979).** A quantitative description of the forest vegetation on an altitudinal gradient in the Mount Field National Park, Tasmania and a discussion of its history and dynamics. *Austral. J. Ecol.*, 4(3): 293-325. Illus. (P)

2836 **Read, J., Hill, R.S. (1983).** Rainforest invasion onto Tasmanian old-fields. *Austral. J. Ecol.*, 8(2): 149-161. Illus, maps.

2837 **Shepherd, R.R., Winkler, C.B., Jones, R. (1975).** The conservation area in land management - physical and administrative aspects of the management of the central plateau of Tasmania. *Proc. Ecol. Soc. Austral.*, 9: 267-284. (P)

AUSTRALIA - Tasmania

2838 **Thompson, P. (1980).** Tasmania's last wild rivers: choosing power or glory. *Habitat* (Australia), 8(2): 15-19. Illus., col. illus. (L/P)

2839 **WWF-Australia (1983).** Project 14: the conservation of the Huon pine. *WWF-Australia Newsl.*, 14: 1. (E/T)

AUSTRALIA - Tasmania - Macquarie Island

2840 **Brothers, N.P., Copson, G.R. (1988).** Macquarie Island flora and fauna - interpreting progress and predictions for the future. *In* Banks, M.R., Smith, S.J., eds. Macquarie Island. *Pap. Proc. Roy. Soc. Tasmania*, 122(1). 318p. Papers presented at the Macquarie Island symposium, Hobart, 11-15 May 1987.

2841 **Costin, A.B., Moore, D.M. (1960).** The effects of rabbit grazing on the grasslands of Macquarie Island. *J. Ecol.*, 48: 729-732.

2842 **Knox, G.A. (1973).** Conservation and research on the offshore islands of New Zealand. *In* Costin, A.B., Groves, K.H., eds. Nature conservation in the Pacific. Proc. Symposium A-10, XII Pacific Science Congress, Canberra, August-September 1971. IUCN and Australian Nat. Univ. Press. 229-239. Maps.

AUSTRALIA - Victoria

2843 **Bird, E.C.F. (1979).** Sites of special scientific interest. *Victorian Nat.*, 96(1): 21-28. Illus. Includes botanical sites.

2844 **Campbell, A.G. (1952).** The dolorous story of Wilson's Promontory National Park. *J. Wild Life Preserv. Soc. Austral.*, 2(3): 32-34. (P)

2845 **Cropper, S. (1987).** Ecological notes and suggestions for conservation of a recently discovered site of *Lepidium hyssopifolium* Desv. (Brassicaceae) at Bolwarrah, Victoria, Australia. *Biol. Conserv.*, 41(4): 269-278. Illus., map. (T)

2846 **FitzSimons, P.F. (1979).** The case of community involvement: conservation of coastal dunes. *Victoria's Resources*, 21(3): 5-7. Illus.

2847 **Garnet, J.R. (1956).** A bill for Victorian national parks. *Victorian Nat.*, 72(12): 187. (L/P)

2848 **James, S.H. (1982).** The relevance of genetic systems in *Isotoma petraea* to conservation practice. *In* Groves, R.H., Ride, W.D.L., eds. Species at risk: research in Australia. New York, Springer-Verlag. 63-71. (T)

2849 **May, R. (1979).** Park management in Victoria - the National Parks Service answers the challenge. *Victoria's Resources*, 21(1): 21-27, 38. Illus., col. illus. (P)

2850 **Muell, F. (1979).** A review of the rare and endangered species *Rutidosis leptorrhynchoides*. *Austral. Syst. Bot. Soc. Newsl.*, 20: 2-5. (T)

2851 **Parsons, R.F., Scarlett, N.H., Stuwe, J. (1981).** A register of rare and endangerd native plants in Victoria. *Threatened Pl. Committ. Newsl.*, 7: 22-23. Outline of a project to survey and document rare and threatened plants. (T)

2852 **Saunders, D.S. (1979).** National parks - the Victorian scene. *Victoria's Resources*, 21(1): 10-14. Illus., col. illus. (P)

2853 **Scarlett, N.H., Parsons, R.F. (1982).** Rare plants of the Victorian plains. *In* Groves, R.H., Ride, W.D.L., eds. Species at risk: research in Australia. New York, Springer-Verlag. 89-105. (T)

2854 **Smith, I., et al. (1979).** Spaciousness and variety: Victoria's new parks. *Victoria's Resources*, 21(1): 15-20. Illus., col. illus. (P)

2855 **Stuwe, J. (1980).** Rare and endangered Victorian plants: 1. *Acacia enterocarpa*. *Victorian Nat.*, 97(4): 157-158. (T)

AUSTRALIA - Victoria

2856 **Stuwe, J. (1981).** Rare and endangered Victorian plants: 2. *Brachycome muelleroides.* *Victorian Nat.*, 98(5): 216-218. Illus., map. (T)

2857 **Stuwe, J. (1982).** Rare and endangered Victorian plants: 3. *Acacia glandulicarpa.* *Victorian Nat.*, 99(2): 62-65. Illus. (T)

2858 **Wrigley, J. (1987).** Sunraysia Oasis Botanical Gardens, Australia. *Bot. Gard Conserv. News*, 1(1): 9. (G)

AUSTRALIA - Western Australia

2859 **Anon. (1979).** Flora conservation. *S. W.A.N.S.*, 9(1): 15. Brief notes. (T)

2860 **Anon. (1979).** Nature reserves. *S. W.A.N.S.*, 9(1): 31-32. (P)

2861 **Anon. (1982).** Satellite finds *Rhizanthella gardneri*. *Austral. Orchid Rev.*, 47(3): 208. (T)

2862 **Anon. (1988).** New park protects karri forest. *Oryx*, 22(3): 152. Shannon Karri National Park. (P)

2863 **Brown, J.M., Hopkins, A.J.M. (1983).** The Kwongan (sclerophyllous shrublands) of Tutanning Nature Reserve, Western Australia. *Austral. J. Ecol.*, 8(1): 63-73. Maps. (P)

2864 **Burbidge, A.A. (1978).** The wildlife of the proposed Wadana Nature Reserve, near Yuna, Western Australia. Report No. 32. Western Australia, Dept of Fisheries & Wildlife. 55p. (P)

2865 **Burbidge, A.A., Fuller, P.J., Cashin, K. (1980).** Wildlife of the proposed Toolonga Nature Reserve. Report No. 39. Western Australia, Dept of Fisheries & Wildlife. 63p. (P)

2866 **Fox, J.E.D., Downes, S., Maslin, B.R. (1980).** The vascular plants of Yalgorup National Park. *West Austral. Herb. Res. Notes*, 31: 1-18. Maps. (P)

2867 **George, A.S. (1979).** *Hakea aculeata* (Proteaceae), a rare and endangered new species from Western Australia. *Nuytsia*, 2(6): 375-377. Illus. (T)

2868 **Griffin, E.A., Hnatiuk, R.J., Hooper, S.D. (1982).** Flora conservation values of vacant Crown land south of Mount Adams, Western Australia. *West Austral. Herb. Res. Notes*, 7: 31-47. Illus., maps. Recommendation that land should become a nature reserve. (P/T)

2869 **Harris, J.A. (1981).** Western Australia: release of Crown Land for agricultural use. *Burrendong Arbor. Brigge*, 56: 1-4.

2870 **Hawkeswood, T.J. (1984).** *Calothamnus accedens* T.J. Hawkeswood (Myrtaceae), a rare and endangered new species from Western Australia. *Nuytsia*, 5(2): 305-310. (T)

2871 **Heddle, E.M., Havel, J.J., Loneragan, O.W. (1980).** Focus on northern Jarrah forest conservation and recreation areas. *Forest Focus*, 22: 3-19. Col. illus., map. (P)

2872 **Hnatiuk, R.J., Hopkins, A.J.M. (1980).** Western Australian species-rich kwongan (sclerophyllous shrubland) affected by drought. *Austral. J. Bot.*, 28(5-6): 573-585.

2873 **Hopper, S. (1983).** Orchid conservation in Western Australia. *In* Robinson, R., ed. Proceedings of the 8th Australian Orchid Conference, Townsville, 27 August-4 September 1983. Townsville Orchid Society. 123-126. Illus. (T)

2874 **Hopper, S.D. (1981).** Recent developments in flora conservation in south Western Australia. *In* XIII International Botanical Congress, Sydney, Australia, 21-28 August 1981: abstracts. Sydney, Australian Academy of Science. 216.

2875 **Hopper, S.D. (1983).** Applied plant systematics: case studies in the conservation of rare Western Australia flora. *Austral. Syst. Bot. Soc. Newsl.*, 35: 1-6. (T)

AUSTRALIA - Western Australia

2876 **Hopper, S.D., Campbell, N.A., Moran, G.R. (1981).** *Eucalyptus caesia*, a rare mallee of granite rocks from south-western Australia. *In* Groves, R.H., *ed.* Species at risk research in Australia. Proc., Symp. on biology of rare and endangered species in Australia, Canberra, 25-26 Nov. 1981. Berlin, Springer Verlag. 46-61. Illus., maps. Conservation, geographic distribution, taxonomy. (T)

2877 **James, S.H. (1982).** The relevance of genetic systems in *Isotoma petraea* to conservation practice. *In* Groves, R.H., Ride, W.D.L., *eds.* Species at risk: research in Australia. New York, Springer-Verlag. 63-71. (T)

2878 **Macdonald, I.A.W., Graber, D.M., DeBenedetti, S., Groves, R.H., Fuentes, E.R. (1988).** Introduced species in nature reserves in Mediterranean-type climatic regions of the world. *Biol. Conserv.*, 44(1-2): 37-66. Invasive species. Includes 7 case studies. (P)

2879 **Marchant, N.G., Keighery, G.J. (1979).** Poorly collected and presumable rare vascular plants in Western Australia. *Kings Park Research Notes*, 5: 1-103. (T)

2880 **McComb, A.J., Loneragan, W.A. (1980).** Cannington Swamp. *Carniv. Pl. Newsl.*, 9(3): 63-64. Important site for carnivorous plants.

2881 **McDermott, P. (1982).** *Kennedia beckxiana*: an endangered Australian species. *Vitis* (Glasnevin), 1(2): 15. (T)

2882 **McKenzie, N.L., Burbidge, A.A., eds (1979).** The wildlife of some existing and proposed nature reserves in the Gibson, Little Sandy and Great Victoria Deserts in Western Australia. *Wildlife Res. Bull., West Austral.*, 8: 36p. Illus. (P)

2883 **Monk, D., Hnatiuk, R.J., George, A.S. (1979).** Vegetation survey of Frank Hann National Park. *West. Austral. Herb. Res. Notes*, 2: 23-49. Maps. (P)

2884 **Patrick, S.J., Hopper, S.D. (1982).** A guide to the gazetted rare flora of Western Australia: supplement I. Report No. 54. Perth, Department of Fisheries and Wildlife. 94p. Illus.; maps. (L/T)

2885 **Rye, B.L. (1982).** Geographically restricted plants of southern Western Australia. Rep. Dept Fish. Wildl. West. Austral. No. 49. Perth. 63p. Map. (T)

2886 **Rye, B.L. (1982).** Rare Western Australian plants 1: Caesia. Leaflet issued by the Department of Fisheries and Wildlife, Perth, South Australia. 2p. Col. illus., map. *Eucalyptus caesia*. (T)

2887 **Rye, B.L. (1982).** Rare Western Australian plants 2: Green honeysuckle. Leaflet issued by the Department of Fisheries and Wildlife, Perth, South Australia. 2p. Col. illus., map. *Lambertia rariflora*. (T)

2888 **Rye, B.L. (1982).** Rare Western Australian plants 3: Fitzgerald eremophila. Leaflet issued by the Department of Fisheries and Wildlife, Perth, South Australia. 2p. Col. illus., map. *Eremophila denticulata*. (T)

2889 **Rye, B.L. (1982).** Rare Western Australian plants 4: Good's banksia. Leaflet issued by the Department of Fisheries and Wildlife, Perth, South Australia. 2p. Col. illus., map. *Banksia goodii*. (T)

2890 **Rye, B.L. (1982).** Rare Western Australian plants 5: Lesueur hakea. Leaflet issued by the Department of Fisheries and Wildlife, Perth, South Australia. 2p. Col. illus., map. *Hakea megalosperma*. (T)

2891 **Rye, B.L. (1982).** Rare Western Australian plants 6: Mogumber bell. Leaflet issued by the Department of Fisheries and Wildlife, Perth, South Australia. 2p. Col. illus., map. *Darwinia carnea*. (T).

2892 **Rye, B.L. (1982).** Rare Western Australian plants 7: Augusta kennedia. Leaflet issued by the Department of Fisheries and Wildlife, Perth, South Australia. 2p. Col. illus., map. *Kennedia macrophylla*. (T)

AUSTRALIA - Western Australia

2893 **Rye, B.L. (1982).** Rare Western Australian plants 8: Underground orchid. Leaflet issued by the Department of Fisheries and Wildlife, Perth, South Australia. 2p. Col. illus., map. *Rhizanthella gardneri.* (T)

2894 **Rye, B.L., Hopper, S.D. (1981).** A guide to the gazetted rare flora of Western Australia. *Rep. Dept Fish. Wildl. West. Austral.,* 42: 1-211. (T)

2895 **Rye, B.L., Hopper, S.D. (1982).** Misapplication of the aboriginal name "Gungurru" to *Eucalyptus caesia* Benth. and notes on the species' distribution. *J. Roy. Soc. West. Austral.,* 65: 93-95. Illus., maps. (T)

2896 **Rye, B.L., Hopper, S.D., Watson, L.E. (1980).** Commercially exploited vascular plants native in Western Australia: census, atlas and preliminary assessment of conservation status. *Rep. Dept Fish. Wildl. West. Austral.,* 40: 5-367. Maps. (E)

2897 **Shugg, H.B. (1979).** Fire control and nature reserves. *S.W.A.N.S.,* 9(1): 9-11. (P)

2898 **Speck, N.H., Baird, A.M. (1984).** Vegetation of Yule Brook Reserve near Perth, Western Australia. *J. Roy. Soc. West. Australia,* 66(4): 147-162. Illus., maps. (P)

2899 **Williams, O.B. (1978).** Desertification in the pastoral rangelands of the Gascoyne Basin, Western Australia. *Search,* 9(7): 257-261. Illus., map.

AUSTRIA

2900 **Anon. (1981).** The last location of *Artemisia laciniata. Naturopa,* 38: 9. Illus. (T)

2901 **Bach, H. (1978).** Karntner Naturschutzhandbuch, Vol. 1. Karntner, Klagenfurt. 779p. Ge. Illus. Includes threatened and protected plants, and threatened habitats in the Province of Karnten. (P/T)

2902 **Barker, M.L. (1982).** Comparison of parks, reserves and landscape protection in three countries of the eastern Alps. *Envir. Conserv.,* 9(4): 275-285. Maps. (P)

2903 **Breiteneder, K., ed. (1978).** Wald- Forstwirtschaft?-Naturschutz. (Forests, forestry and nature conservation.) *Allgem. Forstzeit.,* 89(8): 247-268. Ge. A series of articles on forests in Austria.

2904 **Buchwald, K. (1978).** [Research on the safeguarding of rural cultural landscapes in the highlands of southern Tyrol: 1. Economic and scenic problems in the areas of highland farmers and methods for their safeguarding]. *Landschaft. Stadt.,* 10(4): 153-164. Ge. Illus.

2905 **Conrad, K. (1979).** Zur Geschichte des Alpenparkes in den Hohen Tauern. *Naturschutz- und Naturparke,* 93: 41-44. Illus. (P)

2906 **Council of Europe (1981).** National activities: Austria. *Council of Europe Newsl.,* 81(6/7): 2. (P)

2907 **Council of Europe (1984).** Danube forests under sentence of death. *Council of Europe Newsl.,* 84(3): 1.

2908 **Gluck, P. (1984).** [State action in the ecology crisis.] *Allgem. Forstzeit.,* 95(5): 123, 125-128. Ge. Pollution damage to forests. (L)

2909 **Grims, F. (1979).** Ein Fundort von *Diphasium issleri* (Rouy) Holub im Kobernausser-Wald, Oberosterreich. *Linzer Bio. Beitr.,* 11(2): 279-285. Ge. Illus. (T)

2910 **Harti, H. (1979).** Die Natur- und Landschaftschutzgebiete Karnten aus der sicht des Biologen. (Prirodnye i landshaftnye rezervaty v karintii s tochki zreniya biologa). (Nature and landscape reserves in Carinthia from the viewpoint of a biologist.) *Poroc Vzhodnoalp. Dinar. Dr. Preuc. Veget.,* 14: 173-177 (1978 publ. 1979). Ge (Sln, It). (P)

AUSTRIA

2911 **Kasy, F. (1976).** Naturschutzgebiete im ostlichen Osterreich als Refugien bermerkenswerter thermophiler Pflanzen- und Schmetterlingsarten. (Sanctuaries in eastern Austria as refugial-areas of remarkable thermophile species of plants and Lepidoptera.) Mitteleurop. Trocken-Standorte Pflanz.-Tierokol. Sicht. 63-72. Ge (En). Paper at annual symposium, Graz. (P)

2912 **Kux, S., Kasperowski-Schmid, E., Katzmann, W. (1981).** Naturschutz - Empfehlungen zur Umweltgestaltung und Umweltpflege II. Wien, Osterreichisches Bundesinstitut fur Gesundheitswesen. 125p. Ge. Illus. Includes principles and problems of nature conservation and countryside management; species protection; habitat protection; lists threatened animals, plants and protected areas. (P/T)

2913 **Mayer, H., Zukrigl, K. (1980).** Naturwaldreservate in Osterreich. (Natural forest reserves in Austria.) *Allgem. Forstzeit.*, 91(8): 215-216. Ge. Present and planned reserves of virgin or semi-natural forests are listed, with details of area, altitude and forest types. (P)

2914 **Meurer, M. (1979).** Vegetationskundliche Sukzessions - untersuchungen einer Brandstelle im Sudtiroler Langental. *Jahrb. Ver. Schutz Bergwelt*, 44: 155-170. Ge. Illus.

2915 **Niesslein, E. (1978).** Coordination of environment conservation with forest economic objectives. *In* Proceedings of the 7th World Forestry Congress, Buenos Aires, 4-18 October 1972, vols 4-5. Buenos Aires, Instituto Forestal Nacional. 5658-5670.

2916 **Niklfeld, H. (1986).** Rote Listen Gefahrdeter Pflanzen Osterreichs. Wien, Bundesministerium fur Gesundheit und Umweltschutz. 202p. Ge. Col. illus., maps. (T)

2917 **Olga, B. (1984).** Vadon termo orchideainkrol: 2. *Cypripedium calceolus* L. *Orchidean*, 7: 4-7. Hu. Illus. (T)

2918 **Plank, S. (1975).** Gesetzlich Gesschutzte Pflanzen in Osterreich. Graz, Ludwig-Boltzmann-Institut fur Umweltwissenschaften und Natur schutz. 50p. (L)

2919 **Plank, S. (1979).** Seltene oder bemerkenswerte Porlinge aus der Steiermark: 1. (Rare or noteworthy Polyporaceae from Styria.) *Mitt. Naturwiss. Ver. Steiermark*, 109: 163-173. Ge. Illus., maps. (T)

2920 **Scheiring, H. (1978).** Forstwirtschaft und Naturschutz - Gegensatz oder Gemeinsamkeit. (Forest management and nature protection - opposition or unity.) *Allgem. Forstzeit.*, 89(8): 254-255. Ge. Illus. (P)

2921 **Steinbach, J. (1978).** Naturschutz: Land-und Forstwirtschaft. (Nature protection: agriculture and forestry.) *Allgem. Forstzeit.*, 89(8): 263-265. Illus. (P)

2922 **Stemberger, T. (1978).** Land und Forstwirtschaft - Naturschutz. (Agriculture and forestry - nature protection.) *Allgem. Forstzeit.*, 89(8): 258-260. Illus. (P)

2923 **Toepfer, A. (1981).** A priceless heritage. *Naturopa*, 38: 21-22. Illus. (P)

2924 **Traxler, G. (1978).** Verschollene und gefahrdete Gefasspflanzen im Burgenland: Rote Liste bedrohter Gefasspflanzen. (Extinct and endangered vascular plants in Burgenland: Red list of threatened vascular plants.) *Natur und Umwelt im Burgenland*, 1: 1-24. Ge. Lists 619 regionally threatened flowering plants in Burgenland. (T)

2925 **Traxler, G. (1980).** Zur Roten Liste der Gefasspflanzen des Burgenlandes. Beitrage, Erganzungen und Berichtigungen (I)-(IV). (About the Red List of vascular plants in Burgenland. Additions, completions and corrections (I)-(IV).) *Natur und Umwelt im Burgenland*, 3(1): 9-14; 4(1): 22-25; 5(112): 3,4 (1980-1982) and *Volk und Heimat*, 3: 42-43 (1984). Ge. (T)

AUSTRIA

2926 **Traxler, G. (1982).** Liste der Gefasspflanzen des Burgenlandes. (List of vascular plants in the Burgenland.) *Veroffent. Internat. Clusius-Forschungsges. Gussing*, 6: 1-32. Ge. Checklist; includes conservation categories. (T)

2927 **WWF (1985).** Dam construction threatens Danube. *WWF News*, 46: 1-7.

2928 **Weilharter, R. (1984).** Allgemeine Forstschutzprobleme in Oberosterreich. (General forest protection problems in Upper Austria.) *Allgem. Forstzeit.*, 95(4): 96-98. Ge (En). Illus., maps.

2929 **Weiskirchner, O. (1979).** Rote Liste bedrohter Farn- und Blutenpflanzen in Salzburg. (Red List of threatened ferns and flowering plants in Salzburg.) Salzburg, Amt der Salzburger Landesregierung, Naturschutzreferat. 41p. Ge. Lists c. 720 taxa. (T)

2930 **Wolkinger, F. (1978).** Vorschlage fur einen Steppen-National-park Neusiedlersee. *Natur und Land*, 64(6): 203-211. Ge. Illus., map. (P)

2931 **Zierl, H. (1978).** Naturschutz und Waldwirtschaft aus der Sicht eines mitteleuropaischen Nationalparks. (Nature protection and forest management from the viewpoint of a central European national park.) *Allgem. Forstzeit.*, 89(8): 262-263. Illus. (P)

2932 **Zimmerman, A., Kniely, G., Maurer, W., Melzer, H. (?)** Atlas zur liste verschollener und gefahrdeter farn- blutenpflanzen fur die Steiermark. Graz. Ge. In prep. Distribution maps of species treated in Zimmermann and Kniely, 1980. (T)

2933 **Zimmermann, A., Kniely, G. (1980).** Liste verschollener und gefahrdeter farn- und blutenpflanzen fur die Steiermark. (List of missing and endangered ferns and flowering plants for Steiermark.) *Mitt. Inst. Umweltwiss. Naturschutz*, 3: 3-29. Ge. Lists over 540 taxa including not threatened endemics. (T)

2934 **Zukrigl, K. (1979).** Probleme des Vegetationsschutzes in Waldern dargestellt an Beispielen aus Osterreich. (Problems of conserving vegetation in forests, illustrated by examples from Austria.) *Phytocoenologia*, 6: 532-543. (En).

2935 **Zukrigl, K. (1983).** Forest reserves in Austria. *Zeitschrift fur okologie, natur und umweltschutz*, 5(2): 20-27. Ge. (P)

BAHAMAS

2936 **Attrill, R. (1979).** Bahamian wildlife - a historic perspective. *Caribbean Conserv. News*, 1(16): 16-20.

2937 **Byrne, R. (1980).** Man and the variable vulnerability of island life: a study of recent vegetation change in the Bahamas. *Atoll Res. Bull.* 240. 200p. Illus., maps.

2938 **Campbell, D.G. (1978).** The ephemeral islands: a natural history of the Bahamas. London, Macmillan. viii, 151p. Illus., col. illus., map.

2939 **Popenoe, J. (1984).** Threatened plants in the Bahamas. *Threatened Pl. Newsl.*, 13: 11. (T)

2940 **Sauleda, R.P., Adams, R.M. (1979).** *Encyclia inaguensis* Nash ex Britton and Millspaugh - a rare orchid from the Bahama Islands and the Caicos group. *Bull. Amer. Orchid Soc.*, 48(3): 257-260. Illus. (T)

BANGLADESH

2941 **FAO (1977).** Sundarban forest - Bangladesh. *Tigerpaper*, 4(2): 13-15. Illus. Tropical rain forest.

2942 **Gittins, S.P., Akonda, A.W. (1982).** What survives in Bangladesh? *Tigerpaper*, 9(4): 5-11. Illus., map. (P)

BANGLADESH

2943 **Khan, M.A.R. (1985).** Future conservation directions for Bangladesh. *In* Thorsell, J.W., *ed.* Conserving Asia's natural heritage. The planning and management of protected areas in the Indomalayan Realm. Proc. of the 25th working session of IUCN's CNPPA. Gland, IUCN. 114-122.

2944 **Sarker, N.M., Fazlul Huq, A.K.M. (1985).** Protected areas of Bangladesh. *In* Thorsell, J.W., *ed.* Conserving Asia's natural heritage. The planning and management of protected areas in the Indomalayan Realm. Proc. of the 25th working session of IUCN's CNPPA. Gland, IUCN. 36-38. (P)

BELGIUM

2945 **Anon. (1979).** Conservation de la nature. *Nat. Mosana*, 32(3): 159-160.

2946 **Anon. (1979).** Conservation de la nature. *Rapp. Act. Jard. Bot. Nat. Belg.* (Meise), 1979: 95. Fr.

2947 **Anon. (1979).** Conservation de la nature: Arboretum Kalmthout. *Zoo Anvers*, 3: 112-114. Illus. (G)

2948 **Baere, D. de, Mahieu, R. (1984).** Het Goorken en de Lokkerse Dammen (Arendonk, Belgie). 1. (Introduction and phytosociological survey.) *Bull. Soc. Roy. Bot. Belg.*, 117(2): 328-340. Du (En). Maps. (P)

2949 **Bary-Lenger, A., Evrard, R., Tathy, P. (1979).** La foret: ecologie - gestion - economie - conservation. Liege, Vaillant-Carmanne S.A. 619p. Fr.

2950 **Cassimans, C., Wavrin, H. de, Herman, R., Stenuit, J., Woue, L. (1978).** La foret - le parc naturel. Chapelle-lez-Herlaimont, Centre d'Education pour la Protection de la Nature. 32p. Fr. (P)

2951 **Council of Europe (1981).** National activities: Belgium. *Council of Europe Newsl.*, 81(6/7): 2. Plans for Viroin-Hermeton Nature Park and parks in Walloon region (Rurbusch, Attert, Ave-et-Auffe). (P)

2952 **Council of Europe (1984).** Benelux: nature conservation and protection. *Council of Europe Newsl.*, 84(2): 2.

2953 **D'Hose, R., Langhe, J.E. de (1974).** Nieuwe groeiplaatsen van zeldzame planten in Belgie. II. (New locations of rare plants in Belgium. II.) *Bull. Soc. Roy. Bot. Belg.*, 107(1): 107-114. Fl. First of numerous papers. (T)

2954 **D'Hose, R., Langhe, J.E. de (1975).** Nieuwe groeiplaatsen van zeldzame planten in Belgie. III. (New locations of rare plants in Belgium. III.) *Bull. Soc. Roy. Bot. Belg.*, 108: 35-45. Fl (Fr, Ge). (T)

2955 **D'Hose, R., Langhe, J.E. de (1976).** Nieuwe groeiplaatsen van zeldzame planten in Belgie. IV. (New locations of rare plants in Belgium. IV.) *Bull. Soc. Roy. Bot. Belg.*, 109: 29-41. Fl (Fr, Ge). (T)

2956 **D'Hose, R., Langhe, J.E. de (1977).** Nieuwe groeiplaatsen van zeldzame planten in Belgie. V. (New locations of rare plants in Belgium. V.) *Bull. Soc. Roy. Bot. Belg.*, 110: 20-28. Fl (Fr, Ge). (T)

2957 **D'Hose, R., Langhe, J.E. de (1978).** Nieuwe groeiplaatsen van zeldzame planten in Belgie. VI. (New locations of rare plants in Belgium. VI.) *Bull. Soc. Roy. Bot. Belg.*, 111(1): 19-26. Fl (Fr, Ge). (T)

2958 **D'Hose, R., Langhe, J.E. de (1979).** Nieuwe groeiplaatsen van zeldzame Planten in Belgie. VII. (New locations of rare plants in Belgium. VII.) *Bull. Soc. Roy. Bot. Belg.*, 112(1): 21-34. Fl (Fr, Ge). Maps. (T)

2959 **D'Hose, R., Langhe, J.E. de (1980).** Nieuwe groeiplaatsen van zeldzame planten in Belgie. VIII. (New locations of rare plants in Belgium. VIII.) *Bull. Soc. Roy. Bot. Belg.*, 113: 119-125 Fl (Fr, Ge). (T)

BELGIUM

2960 **D'Hose, R., Langhe, J.E. de (1981).** Nieuwe groeiplaatsen van zeldzame planten in Belgie. IX. (New locations of rare plants in Belgium. IX.) *Bull. Soc. Roy. Bot. Belg.*, 114(1): 41-48. Fl (Fr, Ge). (T)

2961 **D'Hose, R., Langhe, J.E. de (1982).** Nieuwe groeiplaatsen van zeldzame planten in Belgie. X. (New locations of rare plants in Belgium. X.) *Bull. Soc. Roy. Bot. Belg.*, 115(2): 289-296. Fl (Fr, Ge). (T)

2962 **D'Hose, R., Langhe, J.E. de (1983).** Nieuwe groeiplaatsen van zeldzame planten in Belgie. XI. (New locations of rare plants in Belgium. XI.) *Bull. Soc. Roy. Bot. Belg.*, 116(2): 195-200. Fl (Fr, Ge). (T)

2963 **D'Hose, R., Langhe, J.E. de (1984).** Nieuwe groeiplaatsen van zeldzame planten in Belgie. XII. (New locations of rare plants in Belgium. XII.) *Bull. Soc. Roy. Bot. Belg.*, 117(2): 351-358. Fl (Fr, Ge). (T)

2964 **D'Hose, R., Langhe, J.E. de (1986).** Nieuwe groeiplaatsen van zeldzame planten in Belgie. XIII. (New locations of rare plants in Belgium. XIII.) *Bull. Soc. Roy. Bot. Belg.*, 118: 165-171 Fl (Fr, Ge). (T)

2965 **D'Hose, R., Langhe, J.E. de (1987).** Nieuwe groeiplaatsen van zeldzame planten in Belgie. XIV. (New locations of rare plants in Belgium. XIV.) *Bull. Soc. Roy. Bot. Belg.*, 119: 153-160. Fl (Fr, Ge). (T)

2966 **D'Hose, R., Langhe, J.E. de (1987).** Nieuwe groeiplaatsen van zeldzame planten in Belgie. XV. (New locations of rare plants in Belgium. XV.) *Bull. Soc. Roy. Bot. Belg.*, 120: 106-110. Fl (Fr, Ge). (T)

2967 **Delvosalle, L., Demaret, F., Lambinon, J., Lawalree, A. (1969).** Plantes rares, disparues ou menacees de disparition en Belgique: l'appauvrissement de la flore indigene. Ministere de l'Agriculture, Service des Reserves Naturelles domaniales et de la Conservation de la Nature, No. 4. 129p. Fr. Maps. Lists over 300 extinct and threatened vascular plants, and 148 threatened bryophytes; describes threats to the flora. (T)

2968 **Delvosalle, L., Vanhecke, L. (1982).** Essai de notation quantitative de la rarefaction d'especes aquatiques et palustres en Belgique entre 1960 et 1980. *In* Symoens, J.J., Hooper, S.S., Compere, P., eds. Studies on aquatic vascular plants. Proc. Int. Colloqu. on Aquatic Vascular Plants, Brussels, 23-25 January 1981. Roy. Bot. Soc. Belg. 403-409. Fr (En). Maps. Wetlands. (T)

2969 **Duvigneaud, J. (1983).** Reserves et parcs naturels de Wallonie. *Nat. Mosana*, 36(1): 1-3. Fr. (P/T)

2970 **Duvigneaud, J., Coulon, F. (1980).** Les sites Dolomitiques de Belgique, hier et aujourd'hui. Problemes de la preservation de leur flore et de leur vegetation. *Nat. Mosana*, 33(1): 10-25. Fr. (P/T)

2971 **Duvigneaud, J., Meriaux, J.L., Speybroeck, D. van (1982).** La conservation des pelouses calcaires de Belgique et du nord de la France. Metz, Institut Europeen d'Ecologie de Metz et Entente nationale pour la protection de la nature. 42p. Fr. Heathland.

2972 **Duvigneaud, J., et al. (1983).** Conservation de la nature: la fange de l'Abime, a Willerzie, est devenue reserve naturelle: la protection de la tourbiere des Hauts Buttes, dans le departement des Ardennes: creation d'une nouvelle reserve d'Ardenne et Gaume a Theux. *Nat. Mosana*, 36(1): 24-25. Fr. Peatlands. (P/T)

2973 **Fabri, R. (1983).** *Bromus grossus* s.l. et *B. secalinus* en Belgique et au Grand-Duche de Luxembourg. *Bull. Soc. Roy. Bot. Belg.*, 116(2): 207-223. Fr (En). Illus., maps. (T)

2974 **Gryseels, M., Hermy, M. (1981).** Derelict marsh and meadow vegetation of the Leiemeersen at Oostkamp (Prov. west-Flanders, Belgium). *Bull. Soc. Roy. Bot. Belg.*, 114(1): 125-139. Map. Includes "Evaluation & management for nature conservation purposes".

BELGIUM

2975 **Hermy, M., Viane, R., Vanhercke, L. (1981).** *Thelypteris limbosperma* (All.) H.P.Fuchs en *Pilularia globulifera* L. in Het Bos van Houthulst (Staden, Houthulst; W.-VL). *Dumortiera*, 21: 25-?. Fl (En). (T)

2976 **Lawalree, A. (1971).** L'appauvrissement de la flore belge. *Bull. Jard. Bot. Nat. Belg.*, 41: 167-171. Fr.

2977 **Lawalree, A. (1981).** Plantes sauvages protegees en Belgique. Meise, Jardin Botanique National de Belgique. 32p. Col. illus. Describes habitats and threats of 64 protected species.

2978 **Nesterov, Ya S., Krasheninnik, N.V. (1978).** Rastitel'nye resursy Belgii. (The plant resources of Belgium.) *Trudy Prikl. Bot. Genet. Selek.*, 61(2): 71-78. Rus. (T)

2979 **Noirfalise, A. (1971).** La conservation des biocoenoses en Belgique. *Bull. Jard. Bot. Nat. Belg.*, 41: 219-230.

2980 **Noirfalise, A. (1979).** Belgium's nature reserves. *Nature and Nationalpark*, 17(64): 11-13. En, Ge, Fr. (P)

2981 **Petit, J. (1979).** Chromique de la Montagne Saint-Pierre: 2. Une Liste Rouge de plantes menacees. *Rev. Vervietoise Hist. Nat.*, 36(7-9): 54-57. Fr. (T)

2982 **Stieperaere, H. (1985).** *Viola lactea* SM and *V. persicifolia* Schreiber, two neglected violets of the Belgian flora. *Bull. Soc. Roy. Bot. Belg.*, 118: 157-164. Illus. (T)

2983 **Tips, W. (1977).** The Walenboscomplex; a conservation site of national importance in Brabant, Belgium. *Biol. Conserv.*, 11(4): 243-250. Illus., maps. (P)

2984 **Vanhecke, L., Charlier, G. (1982).** The regression of aquatic and marsh vegetation and habitats in the north of Belgium between 1904 and 1980: some photographic evidence. *In* Symoens, J.J., Hooper, S.S., Compere, P., eds. Studies on aquatic vascular plants. Brussels, Societe Royale de Botanique de Belgique. 410-411. Proceedings of the international colloquium on aquatic vascular plants, 23-25 January 1981, Brussels.

BELIZE

2985 **Anon. (1979).** The Belize Audubon Society. *Caribbean Conserv. News*, 1(16): 10-12.

2986 **Anon. (1984).** Last mahogany forest in Belize may fuel a power station. *Oryx*, 18(2): 67-68. (E/T)

2987 **Caufield, C. (1983).** British aid may fuel Belize forest scheme. *New Sci.*, 100(1385): 560. Illus., map.

2988 **D'Arcy, W.G. (1977).** Endangered landscapes in Panama and Central America: the threat to plant species. *In* Prance, G.T., Elias, T.S., eds. Extinction is forever. Proceedings of a symposium at the New York Botanical Garden, 11-13 May 1976. New York, New York Botanical Garden. 89-104. Maps. The flora and deforestation today. (P/T)

BENIN

2989 **Adjanohoun, E. (1968).** Le Dahomey. *In* Hedberg, I. & O., eds. Conservation of vegetation in Africa south of the Sahara. Symposium Proceedings, at 6th plenary meeting of AETFAT, Uppsala. *Acta Phytogeogr. Suec.*, 54: 86-91. Fr. Map, illus. Describes vegetation, associations already protected and recommends additional areas for protection. (P)

2990 **Adjanohoun, E.J. (1979).** Possibilities and needs for conservation of plant species and vegetation in Africa. Appendix: preliminary lists of rare and threatened species in African countries. The Republic of Benin. *In* Hedberg, I., ed. Systematic botany, plant utilization and biosphere conservation. Stockholm, Almqvist & Wiksell International. 91-92. (T)

BENIN

2991 **Huonto-Hotegbe, T. (1971).** The national parks and game reserves of Dahomey. *In* Happold, D.C.D., *ed.* Wildife conservation in West Africa. Proceedings of a Symposium held at the University of Ibadan, Nigeria, 2 April 1970. Morges, IUCN. 25-27. (P)

2992 **Osborne, R. (1988).** *Encephalartos barteri. Encephalartos*, 14: 8-16. Illus., map. Data sheet; includes conservation. (G/T)

2993 **Paradis, G., Houngnon, P. (1977).** La vegetation de l'istaire classee de la lama dans la mosaique foret-savane du Sud-Benin (ex Sud-Dahomey). *Bull. Mus. Nat. Hist. Nat., Bot.*, 34: 169-167. Fr. (P)

BERMUDA

2994 **Hayward, S.J., Gomez, V.H., Sterrer, W., eds. (1981).** Bermuda's delicate balance: people and environment. Bermuda, The Bermuda National Trust. 402p.

2995 **Oldfield, S. (1987).** Fragments of paradise. A guide for conservation action in the U.K. Dependent Territories. Oxford, Pisces Publications; for the British Association for Nature Conservationists. 192p.

2996 **Phillips, B.R. (1980).** Saving an endangered plant in Bermuda. *Threatened Pl. Committ. Newsl.*, 5: 6-7. (T)

2997 **Tiberi, J.D. (1982).** Nature study and conservation: invasion force continues. *Mon. Bull. Dept Agric. Fish. Bermuda*, 53(6): 48-50.

BHUTAN

2998 **Deb, D.B., Raghavan, R.S. (1983).** A rare species of *Agapetes* D. Don ex G. Don (Ericaceae). *Bull. Bot. Surv. India*, 24(1-4): 171-173 (1982 publ. 1983). Illus. (T)

2999 **Mahat, G. (1985).** Protected areas of Bhutan. *In* Thorsell, J.W., *ed.* Conserving Asia's natural heritage. The planning and management of protected areas in the Indomalayan Realm. Proc. of the 25th working session of IUCN's CNPPA. Gland, IUCN. 26-29. Map. (P)

3000 **Numata, M. (1987).** Vegetation, plant industry and nature conservation in Bhutan. *In* Ohsawa, M., *ed.* Life zone ecology of the Bhutan Himalaya. Japan, Chiba Univ. Illus., map. (L/P)

3001 **Pradhan, U.C. (1975).** Conservation of eastern Himalayan orchids. Problems and prospects. Part I. *Orch. Rev.*, 83: 314-317. (T)

3002 **Pradhan, U.C. (1975).** Conservation of eastern Himalayan orchids. Problems and prospects. Part II. *Orch. Rev.*, 83: 345-347. (T)

3003 **Pradhan, U.C. (1975).** Conservation of eastern Himalayan orchids. Problems and prospects. Part III. *Orch. Rev.*, 83: 374. (T)

3004 **Rustomji, N.K. (1986).** Sikkim, Bhutan and India's north-eastern borderlands. *In* Lall, J.S., *ed.* The Himalayas: aspects of change. New Delhi, India International Centre; Oxford, Oxford University Press. 236-252. (P/T)

3005 **Sargent, C. (1985).** The forests of Bhutan. *Ambio*, 14(2): 75-80. Illus., maps. Himalaya, deforestation.

BOLIVIA

3006 **Arce, S.J.P., Estenssoro, C.S., Ergueta, S.P. (1987).** Diagnostico del estado de la flora, fauna y communidades importantes para la conservacion. Bolivia, La Paz, Centro de Datos para la Conservacion. 98p. (T)

BOLIVIA

3007 **Centro de datos para la conservacion (1989).** Sintesis de la problematica de conservacion ambiental en Bolivia, en base a las subregiones naturales. La Paz, CDC. 45p.

3008 **Estenssoro, S. (1987).** Lista preliminar de plantas especiales CDC-Bolivia. La Paz, CDC. 45p. (T)

3009 **Marconi, M., Estenssoro, S., Ergueta, P., Arce, P. (1989).** Principales ecoregiones de Bolivia y prioridades. La Paz, CDC. 34p.

3010 **Margolis, K. (1989).** Ecosystem conservation in the Beni. *Orion Nat. Quart.*, 8(3): 15-16.

3011 **Redwood, S. (1987).** Going for gold in Bolivia. *New Sci.*, 115(1574): 41-43. Col. illus. Deforestation, soil erosion in Amazon due to mining.

3012 **Riesgo Reguera, A. (1977).** La reserve nacional de Saja. *Vida Silvestre*, 22: 101-115. Col. illus. (P)

3013 **Villa-Lobos, J. (1988).** Debt swaps. *Threatened Pl. Newsl.*, 19: 4-5. Conservation organizations assume part of nation's debt in return for land protection. (P)

BOTSWANA

3014 **Anon. (1987).** Palm decline in Botswana. *Oryx*, 21(1): 52. Extract from *Kalahari Conservation Soc. Newsl.*, 13. (T)

3015 **Anon. (1987).** Plant threatens Okavango. *Oryx*, 21(1): 52. *Salvia molesta* threatens Okavango Delta.

3016 **Carter, J.M. (1983).** The development of wildlife management areas. *In* Which way Botswana's wildlife? Proceedings of the Symposium of the Kalahari Conservation Society, Gaborone, 15-16 August 1983. 21-25. (P)

3017 **Cook, H.J. (1983).** The struggle against environmental degradation: Botswana's experience. *Desertification Control*, 8: 9-15.

3018 **Field, D.I. (1979).** Botswana. *In* Hedberg, I., ed. Systematic botany, plant utilization and biosphere conservation: Proc. of a sym. held in Uppsala in commemoration of the 500th ann. of the Univ. Stockholm, Almqvist & Wiksell Int'l. 99.

3019 **Hannah, L., Wetterberg, G., Duvall, L. (1988).** Botswana biological diversity assessment. Washington, D.C., Agency for International Development. 72p. (T)

3020 **Parris, R., Bothama, J. du P., Waanders, E., Boshoff, A.F. (1977).** Preliminary map of the south-western Kalahari Desert. *Koedoe*, 20: 163-165.

3021 **Skarpe, C. (1983).** Cattle grazing and the ecology of the western Kalahari. *In* Which way Botswana's wildlife? Proceedings of the Symposium of the Kalahari Conservation Society, Gaborone, 15-16 August 1983. 21-25.

3022 **White, R. (1978).** A working plan for the Gaborone Wildlife Reserve for the years 1977/78 to 1981/82. Gaborone, Department of Wildlife, National Parks and Tourism. 27p. Maps. (P)

3023 **Wild, H. (1968).** Bechuanaland Protectorate. *In* Hedberg, I. & O., eds. Conservation of vegetation in Africa south of the Sahara. Symposium Proceedings, at 6th plenary meeting of AETFAT, Uppsala. *Acta Phytogeogr. Suec.*, 54: 198-202. Map. Associations already protected and those proposed. Lists plant species in various areas of the country. (P/T)

3024 **Williamson, D., Williamson, J. (1985).** Botswana's fences and the depletion of Kalahari wildlife. *Parks*, 10(2): 5-7. Map, illus. Fences interrupt migration routes of animals, which are threatened as is the grass cover of the soil. (T)

BRAZIL

3025 **Almeida, R.F. de (1979).** Subsidos para una nova legislacao florestal. *In* Sociedade Botanica do Brasil. Resumo dos Trabalhos de XXX Congresso Nacional de Botanica 21 a 27 de Janeiro de 1979. Campo Grande-MS. 149-150. Abstract. (L)

3026 **Alves, L. da C. (1982).** *Scaevola plumieri* Vahl: uma especie em extincao. Flora, especies raras ou ameacadas de extincao: 1. *Cadernos FEEMA Ser. Trab. Techn.*, 18: 7-11. Por. Illus. (T)

3027 **Alvim, P. de T. (1977).** The balance between conservation and utilization in the humid tropics with special reference to Amazonian Brazil. *In* Prance, G.T., Elias, T.S., *eds.* Extinction is forever. Proceedings of a symposium at the New York Botanical Garden, 11-13 May 1976. New York, New York Botanical Garden. 347-352. (E)

3028 **Anderson, A.B., Gely, A. (?)** Extractivism and forest management by rural inhabitants in the Amazon estuary. *In* Posey, D.A., Balee, W., *eds.* Natural resource management by indigenous and old societies in Amazonia. New York, New York Botanical Society. In press. Harvesting the floodplain forests.

3029 **Andrade, A.G. de & J.C. de, Carauta, J.P.P. (1982).** *Bumelia obtusifolia* Roem. & Schult. var. *excelsa* (DC) Miq. (Sapotaceae), quixabeira ameacada de extincao. Flora, especies raras ou ameacadas de extincao: 2. *Cadernos FEEMA Ser. Trab. Techn.*, 18: 1-9. Por. Illus. (T)

3030 **Andrade, A.G. de, Carauta, J.P.P., Andrade, J.C. de (1981).** *Bumelia obtusifolia* Roem. & Schult. var *excelsa* (DC) Miq. (Sapotaceae) - ameacada de extincao. *Bradea*, 3(28): 221-228. Por. Illus. (T)

3031 **Anon. (1968).** List of Brazilian species of animals and plants in danger of extinction. *Braz. Found. Conserv. Nat. Boletin.* Info. 3. (T)

3032 **Anon. (1978).** Conservacao de natureza: IBDF submete ao M.A. a regulamentacao do codigo florestal. (Conservation of natural resources: the Brasilian Forest Department Institute submits the forestry code regulation to the Ministry of Agriculture.) *Brasil Florestal* (Brasilia), 9(33): 6-16. Por. (L)

3033 **Anon. (1979).** Brasil tem regulamento para os parques e mais tres reservas. (Brazil has regulation for parks and three more reserves.) *Brasil Florestal*, 9(39): 11-28. (En). Illus., map. (P)

3034 **Anon. (1979).** Criadas tres novas unidades de conservacao. *Brasil Florestal*, 9(38): 6-17. Por (En). Illus.

3035 **Anon. (1980).** Amazonia - what price development? *IUCN Bull.*, 11(5): 27. Illus.

3036 **Anon. (1980).** Botanico ingles alerta para devastacao da flora brasileira. *O Globo*, 1p. (T)

3037 **Anon. (1980).** Botanicos denunciam desmatamento no Sul. *Correio da Bahia*, 2p.

3038 **Anon. (1980).** Brazil rethinks its entire Amazon strategy. *IUCN Bull.*, 11(5): 28-31. Illus.

3039 **Anon. (1980).** Ingles estuda flora do Brasil. *Correio da Bahia*, 1p.

3040 **Anon. (1980).** The Jari project - profit and loss. *IUCN Bull.*, 11(5): 29.

3041 **Anon. (1983).** Aspects of forestry in Brazil. *Allgem. Forstzeit.*, 6/7: 133-181. Ge. Illus., col. illus. (E)

3042 **Anon. (1983).** Brazilian disaster. *Oryx*, 17: 143. Road put through Araguaia National Park, and other N.P.s threatened, Pakas Novas N.P., Rio Trombetas Reserve, Aparados da Sierra N.P. (P)

3043 **Anon. (1984).** Dam will destroy virgin forest in Brazil. *Oryx*, 18(2): 68.

BRAZIL

3044 **Aoki, H. (1982).** Consideracoes sobre a preservacao dos Cerrados. (Reflections on conservation of the Cerrados.) *Silvicult. Sao Paulo,* 16A(1): 372-384. Por. Literature review. Forests, natural resources, phytosociology, vegetation. (P)

3045 **Barham, J., Caufield, C. (1984).** The problems that plague a Brazilian dam. *New Sci.,* 104(1425): 10. Illus.

3046 **Benton, P. (1972).** Fight for the drylands: struggle and achievement in Brazil. London, Collins. 188p.

3047 **Bevaart, C. (1981).** Ecology and development in the rain forests of the Amazon Basin. Leiden, State University. 187p. Ge. Illus.

3048 **Binswanger, H. (1989).** Brazilian policies that encourage deforestation in the Amazon. Washington, D.C., World Bank. 24p.

3049 **Brandao, L.G., et al. (1982).** Brasil Florestal, ano 2.000: diretrizes estrategicas para o Setor Florestal Brasileiro. *Brasil Florest.,* 12(50): 7-33. Por (En). Illus.

3050 **Buenos, C.N. de (1982).** *Phillostyllon brasiliense* Benth. & J.D. Hooker (Ulmaceae): especie ameacada de extincao. Flora, especies raras ou ameacadas de extincao: 2. *Cadernos FEEMA Ser. Trab. Techn.,* 18: 11-15. Por. Illus. (T)

3051 **Burman, A. (1989).** A note on the threatened woody bamboo taxa in Brazil. *Bot. Gard. Conserv. News,* 1(5): 23-24. Illus. (E/T)

3052 **Cabral, A.H.D. (1983).** O peculiar interesse municipal e a tutela juridica de areas verdes. (Local municipal interest and the judicial protection of "green areas".) *Brasil Florest.,* 13(53): 5-23. Por. (En). Illus. (L/P)

3053 **Camara, I.de G. (1983).** Tropical moist forest conservation in Brazil. *In* Sutton, S.L., Whitmore, T.C., Chadwick, A.C., *eds.* Tropical rain forest: ecology and management. Oxford, Blackwell Scientific. 413-421. Special Publications Series of the British Ecological Society No. 2. Maps.

3054 **Carauta, J.P.P., Castro, M.W. de (1982).** Plantas em perigo de extincao: *Dorstenia.* Flora, alguns estudos: 1. *Cadernos FEEMA Ser. Trab. Tecn.,* 1: 29-65. Por (En). Illus. 16 spp. of *Dorstenia.* (T)

3055 **Carauta, J.P.P., Coimbra Filho, A.F. (1982).** *Pavonia alnifolia* Saint-Hilaire, quetea ameacada de extincao. Flora, especies raras ou ameacadas de extincao: 2. *Cadernos FEEMA Ser. Trab. Techn.,* 18: 27-34. Por. Illus. (T)

3056 **Carauta, J.P.P., Lins, E.A.M. (1982).** *Ficus lanuginosa* Cas., molemba: uma especie ameacada de extincao. Flora, especies raras ou ameacadas de extincao: 1. *Cadernos FEEMA Ser. Trab. Techn.,* 18: 31-35. Por. Illus. (T)

3057 **Carauta, J.P.P., et al. (1981).** Conservacao da flora: salvemos o que resta! *Atas Soc. Bot. Brasil Rio de Janeiro,* 1(1): 3-7. Por (En). (T)

3058 **Carvalho, J.C.M. (1968).** Lista das especies de animais e plantas ameacadas de extincao no Brasil. (List of plant and animal species threatened with extinction in Brazil). *Fund. Brasil. Conserv. Natureza, Bol. Inform.,* 3: 11-16. Por. 13 species listed. (T)

3059 **Casari, M.B. (1982).** *Halophila decipiens* Ostenfeld var. *pubescens* Hartog (Hydrocharitaceae): ameacada de extincao. Flora, especies raras ou ameacadas de extincao: 1. *Cadernos FEEMA Ser. Trab. Techn.,* 18: 1-6. Por. Illus. (T)

3060 **Casari, M.B., et al. (1980).** Nove especies ameacadas ou em perigo de desaparecimento no Brasil. Resumos do 31 Congresso Nacional de Botanica. Ilheus, Sociedade Bota nica do Brasil. 123. Por.

3061 **Caufield, C. (1982).** Brazil, energy and the Amazon. *New Sci.,* 96(1329): 240-243. Illus., map. Hydroelectric power schemes threaten vast areas of tropical forest.

BRAZIL

3062 **Caufield, C. (1982).** Defoliant clouds a rainforest's future. *New Sci.*, 95(1320): 539. Illus. Threat to Amazonian rainforests.

3063 **Cavalcanti, D.F. (1981).** Plantas em extincao no Brasil. *Fund. Brasil. Conserv. Natureza, Bol. Inform.*, 16: 115-119. Por. (T)

3064 **Clement, C.R., Muiler, C.H., Flores, W.B.C. (1982).** Recursos geneticos de especies frutiferas nativas da Amazonia Brasileira. *Acta Amazonica*, 12(4): 677-695. Por (En). Illus. (E/G)

3065 **Colchester, M. (1989).** Indian development in Amazonia: risks and strategies. *The Ecologist*, 19(6): 249-254. Illus.

3066 **Coutinho, L.M. (1980).** As queimadas e seu papel ecologico. *Brasil Florest.*, 10(44): 7-23. Por. Illus.

3067 **Coutinho, S. da C. (1982).** A conservacao de germoplasma de especies florestais. *In* Resumos do XXXIII Congresso Nacional de Botanica, 24 a 30 de janeiro de 1982. Maceio, Sociedade Botanicas do Brasil. 27. Por. (E)

3068 **Coutinho, S.C., ed. (1990).** Vegetacao do Jari; conservacao *in situ* de recursos geneticos florestais. Rio de Janeiro, Jari, Companhia Florestal Monte Dourado. c.150p. In prep.

3069 **Croat, T. (1985).** A new collection of the rare *Alloschemone occidentalis* (Poepp.) Engl. & Krause. *Aroideana*, 8(3): 80-82. (T)

3070 **Denevan, W.M. (1973).** Development and the imminent demise of the Amazon rain forest. *Profess. Geogr.*, 25: 130-135.

3071 **Duveen, D.I. (1983).** *Laelia sincorana* Schltr.: the appearance, disappearance and reappearance of a showy Brazilian orchid. *Orchid Digest*, 47(4): 135-137. Illus. (T)

3072 **Eden, M.J. (1982).** Silvicultural and agroforestry developments in the Amazon Basin of Brazil. *Commonw. Forest. Rev.* 61(3): 195-202. (Fr, Sp). Illus., map. (E)

3073 **Eyde, R.H., Olson, S.L. (1983).** The dead trees of Ilha da Trindade. *Bartonia*, 49: 32-51. (T)

3074 **Fearnside, P. (1986).** Spatial concentration of deforestation in the Brazilian Amazon. *Ambio*, 15(2): 72-79. Tropical forest.

3075 **Fearnside, P., Ferreira, G. de Lima (1985).** Amazonian forest reserves, fact or fiction? *The Ecologist*, 15(5/6): 297-299. (P)

3076 **Fearnside, P.M. (1982).** Deforestation in the Brazilian Amazon: how fast is it occurring? *Interciencia*, 7(2): 82-88. Map.

3077 **Fearnside, P.M. (1982).** Rebuttal to the Lugo-Brown critique of "Deforestation of the Amazon". *Interciencia*, 7(6): 362. Deforestation.

3078 **Fearnside, P.M. (1987).** Deforestation and international economic development projects in Brazilian Amazonia. *Conservation Biology*, 1(3): 214-221. Map.

3079 **Fearnside, P.M. (1989).** A prescription for slowing deforestation in Amazonia. *Environment*, 31(4): 17-20, 39-40.

3080 **Fearnside, P.M. (1989).** Deforestation in Brazilian Amazonia: the rates and causes of forest destruction. *The Ecologist*, 19(6): 214-218. Illus., map.

3081 **Fearnside, P.M. (1989).** The charcoal of Carajas: a threat to the forests of Brazil's eastern Amazon region. *Ambio*, 18(2): 141-143. Col. illus. (T)

3082 **Filho, A.C., et al. (1983).** Flora fanerogamica da Reserva do Parque Estadual das Fontes do Ipiranga (Sao Paulo, Brasil). (Phanerogam flora of the Fontes do Ipiranga State Park Reserve (Sao Paulo, Brazil).) *Hoehnea*, 10: 24-124. Por. Illus. Various papers each covering an individual plant family. (P)

BRAZIL

3083 **Filho, L.E. de M. (1982).** Projeto Parque das Dunas. Por. XXXII Congresso Nacional de Botanica, 25 a 31 de Janeiro de 1981, Teresina, Piaui. *An. Soc. Bot. Brasil*: 331 (1981 publ. 1982). Por. (P)

3084 **Fonece, G.A.B. da (1985).** The vanishing Brazilian Atlantic forest. *Biol. Conserv.*, 34(1): 17-34. Maps. Tropical forest; deforestation. (T)

3085 **Freire, G.V. (1982).** *Talisia esculenta* Radlk., pitomba: especie vulneravel. Flora, especies raras ou ameacadas de extincao: 2. *Cadernos FEEMA Ser. Trab. Techn.*, 18: 41-45. Por. Illus. (T)

3086 **Gentchujnicov, I.D. (1980).** 1. Contribuicao para o estudo ecologico de campos naturais: Humaita, AM. *In* Sociedade Botanica do Brasil. XXXI Congresso Nacional de Botanica, 20 a 27 de janeiro de 1980. Itabuna, Ilheus. Bahia, Soc. Bot. Brasil. 130. Por. Abstract.

3087 **Giacometti, D.C. (1982).** Conservacao e manejo de recursos geneticos vegetais. *In* Resumos do XXXIII Congresso Nacional de Botanica, 24 a 30 de janeiro de 1982. Maceio, Sociedade Botanica do Brasil. 26. Por. (E)

3088 **Giacometti, D.C., Coradin, L. (1985).** Conservacao e manejo de recursos geneticos vegetais no Brasil. (Management and conservation of plant genetic resources in Brazil). *In* Anais do XXXIII Congresso Nacional de Botanica, Maceio, 24-30 Janeiro de 1982. Brasilia, EMBRAPA. 201-218. Por (En). Maps. (E)

3089 **Giacometti, D.C., Pinta da Cunha, M.A., Coredin, L. (1980).** A preservacao dos recursos geneticos vegetais no Brasil. *Bol. Bot. Latinoam.*, 6: 10-13. (E/T)

3090 **Goodland, R.J., Irwin, H.S. (1977).** Amazonian forest and cerrado: development and environmental conservation. *In* Prance, G.T., Elias, T.S., *eds.* Extinction is forever. Proceedings of a symposium at the New York Botanical Garden, 11-13 May 1976. New York, New York Botanical Garden. 214-233. Maps, graphs. (P/T)

3091 **Goodland, R.J.A. (1980).** Environmental ranking of Amazonian development projects in Brazil. *Envir. Conserv.*, 7(1): 9-26. Illus. (P)

3092 **Gusmao, I. de C., Mittermeier, R.A. (1984).** Genetic diversity, endemism and protected areas: a case study of the endangered primates of Brazil's atlantic forest region. *In* McNeely, J.A., Miller, K.R., *eds.* National parks, conservation and development. The role of protected areas in sustaining society. Washington, D.C., Smithsonian Institution Press. 572-574. Map. (P)

3093 **Hartshorn, G.S. (1982).** Take the profits and run. *Garden* (New York), 6(1): 25-31. Illus. Amazon forests. (E)

3094 **Hecht, S. (1985).** Cattle ranching in Amazonia: political and ecological considerations. *In* Schmink, M, Wood, C., *ed.* Frontier expansion in Amazonia. Gainesville, Univ. of Florida Press.

3095 **Hecht, S.B. (1989).** The sacred cow in the green hell: livestock and forest conversion in the Brazilian Amazon. *The Ecologist*, 19(6): 229-234. Illus.

3096 **Hildyard, N. (1989).** Amazonia: the future in the balance. *The Ecologist*, 19(6): 207-210.

3097 **IUCN (1988).** Amazonian forest destruction. *Species*, 11: 14. Brazilian Institute for Space Research detects burning of 204,608 sq. km of Amazon forest in 1987.

3098 **IUCN Commission on Ecology (1983).** Ecological structures and problems of Amazonia. Proceedings of a symposium organised by the Dept of Biological Sciences of Federal Univ. of Sao Carlos and IUCN, Sao Carlos, Brazil, 18 March 1982. Gland, Switzerland, IUCN. 79p. Illus., maps. (E/P)

BRAZIL

3099 **Instituto Brasileiro de Desenvolvimento Florestal (1978).** Fotointerpretacion y mapeo de las reservas de *Araucaria angustifolia* (Bert.) O.Ktze. del sur del Brazil. *In* Proceedings of the 7th World Forestry Congress, Buenos Aires, 4-18 October 1972, vol. 2: Silviculture. Buenos Aires, Instituto Forestal Nacional. 2362-2366. (En, Fr). (P)

3100 **Jannuzzi, C.M.L., Almeida, H.A. de (1982).** *Dorstenia gracilis* Car. & alii, um caiapia raro. Flora, especies raras ou ameacadas de extincao: 1. *Cadernos FEEMA Ser. Trab. Techn.*, 18: 43-47. Por. Illus. (T)

3101 **Janzen, D.H. (1972).** The uncertain future of the tropics. *Nat. Hist.*, (1972) (1): 80-94.

3102 **Johns, A.D. (1988).** Economic development and wildlife conservation in Brazilian Amazonia. *Ambio*, 17(5): 302-306. Col. illus., maps.

3103 **Jordan, C.F., ed. (1987).** Amazonian rain forests. Ecosystem disturbance and recovery. Ecological Studies No. 60. New York, Springer-Verlag. 133p.

3104 **Kenneally, K.F. (1988).** The Amazon rainforest - another threat to its survival. *Austral. Syst. Bot. Soc. Newsl.*, 57: 3-5. Map; tropical rain forest.

3105 **Kirkbridge, J.H., jr (1982).** Reconhecimento das especies em perigo de extincao. *In* Resumos do XXXIII Congresso Nacional de Botanica, 24 a 30 de janeiro de 1982. Maceio, Sociedade Botanica do Brasil. 28. Por. (T)

3106 **Klein, R.M. (1980).** Aspectos ecologicos do pinheiro brasileiro. *In* Sociedade Botanica do Brasil. XXXI Congresso Nacional de Botanica, 20 a 27 de janeiro de 1980. Itabuna, Ilheus. p.114. Abstract.

3107 **Klein, R.M. (1981).** Fitofisionomania, importancia e recursos da vegetacao do Parque Estadual da Serra do Tabuleiro. (Phytophysionomy, importance and vegetation resources of the Serra do Tabuleiro State Park.) *Sellowia*, 33(33): 5-54. Por (Ge). Maps. (P)

3108 **Lamlein, J. (1982).** Plant conservation in tropical America: a remarkable achievement for Brazil. *Threatened Pl. Commit. Newsl.*, 9: 6-7. (P)

3109 **Laroche, R.C. (1977).** Uso racionalizado dos recursos florestais e sua productividade. *Rodriguesia*, 29(43): 143-145. Por. Includes nature conservation and reserves. (P)

3110 **Laroche, R.C.M. (1980).** Contribuicao ao conhecimento da ecologia da floresta pluvial tropical e sua conservacao: 2. *Rodriguesia*, 32(53): 117-120. Por (Fr). Illus.

311! **Liddell, R. (1980).** Collections and conservation of Brazilian orchids. *In* Sukshom, M.R., ed. Proceedings of the 9th World Orchid Conference. Thailand, Amarin Press. 283-285. (T)

3112 **Lins, E.A.M., Hoffmann, C. (1983).** *Brosimum glazioui* Taubert, marmelinho: uma especie rara. *Cadernos FEEMA Ser. Trab. Techn.*, no. 18: 35-39. Por. Illus. Flora, especies raras ou ameacadas de extincao: 2. (T)

3113 **Lobo, M. da G.A. (1982).** *Sipolisia languginosa* Glaziou: uma especie ameacada de extincao. Flora, especies raras ou ameacadas de extincao: 1. *Cadernos FEEMA Ser. Trab. Techn.*, 18: 21-24. Por. (T)

3114 **Longo, A.N. (1979).** Papel da iniciativa privada na preservacao da flora em Blumenau. *In* Sociedade Botanica do Brasil. Resumo dos trabalhos do XXX Congresso Nacional de Botanica 21-27 de Janeiro de 1979. Campo Grande - MS. 151. Abstract. (T)

3115 **Louro, R.P., Santiago, L.J. (1984).** A regiao de Barra de Marica, RJ, e a importancia de sua preservacao. *Atas Soc. Bot. Brasil Rio de Janeiro*, 2(15):109-120. Por (En).

BRAZIL

3116 **Lovejoy, T.E. (1982).** Hope for a beleaguered paradise. *Garden* (New York), 6(1): 32-36. Illus., col. illus., map. Amazon forests.

3117 **Lugo, A.E., Brown, S. (1983).** Deforestation in the Brazilian Amazon. *Interciencia*, 7(6): 361-362. Deforestation. (T)

3118 **Luna Lugo, A. (1981).** [Some aspects of management of productive tropical forests in relation to protection of the environment.] *In* Chavarria, M., *ed.* Simposio Internacional Sobre las Ciencias Forestales y su Contribucion al Desarrollo de la America Tropical. San Jose, Costa Rica. 35-39. Sp (En).

3119 **Lutzenberger, J.A. (1982).** The systematic demolition of the tropical rain forest in the Amazon. *Ecologist*, 12(6): 248-252. (E/T)

3120 **Machado, A.D. (1977).** The Brazilian humid tropics program. *In* Prance, G.T., Elias, T.S., *eds.* Extinction is forever. Proceedings of a symposium at the New York Botanical Garden, 11-13 May 1976. New York, New York Botanical Garden. 353-355. Objectives of the programme.

3121 **Machado, E. de F. (1978).** Diagnostico do subsistema de conservacao e preservacao de recursos naturais renovaveis. *In* Instituto Brasileiro de Desenvolvimento Florestal. Forestry development and planning: projections for the period 1979-1985. Por. Appraisal of forest sector of Brazil: Colecao. Brasil, Secretario. (P)

3122 **Machado, P.A.L. (1984).** Les forets et l'environnement au Bresil. (Forests and environment in Brazil.) *In* Prieur, M., *ed.* Forets et environnement: en droit compare et international. Paris, Presses Universitaires de France. 205-217. Fr. (L)

3123 **Mahar, D. (1989).** Government policies and deforestation in Brazil's Amazon region. Washington, D.C., World Bank. 56p.

3124 **Malingreau, J.-P., Tucker, C.J. (1988).** Large-scale deforestation in the southeastern Amazon Basin of Brazil. *Ambio*, 17(1): 49-55. Illus., col. illus., maps. Deforestation.

3125 **Martinelli, G. (1980).** *Worsleya rayneri* (J.D. Hooker) Traub et Moldenke (Amaryllidaceae) - uma especie ameacada de extincao. *In* Sociedade Botanica do Brasil. XXXI Congresso Nacional de Botanica, 20 a 27 de janeiro de 1980. Itabuna, Ilheus. 125. Abstract. (T)

3126 **Martinelli, G., Leme, E. (1986).** Rare bromeliads from Brazil, No. 2 *Vriesea triligulata. J. Bromeliad Soc.*, 36(1): 18-20. (T)

3127 **Mayo, S.J., Fevereiro, V.P.B. (1982).** Mata de Pau Ferro: a pilot study of the Brejo forest of Paraiba, Brazil. Kew, RBG/Bentham-Moxon Trust. 29p. (Por). Illus., maps.

3128 **Meijer, W. (1973).** Endangered plant life in Brazil. *Biol. Conserv.*, 5: 147. (T)

3129 **Melo Carvalho, J.C. de (1968).** A conservacao da natureza e recursos naturais no mundo e no Brasil. *An. Soc. Bot. Brasil*, 187-205. Por. XIX Congresso Nacional de Botanica, Fortaleza, 21-29 January, 1968.

3130 **Mittermeier, R. (1986).** Atlantic forest: now for the good news. *IUCN Bull.*, 17(1-3): 30. Tropical forest.

3131 **Molion, L.C.B. (1989).** The Amazonian forests and climatic stability. *The Ecologist*, 19(6): 211-213. Illus.

3132 **Mori, S.A., Boom, B.M., Prance, G.T. (1981).** Distribution patterns and conservation of eastern Brazilian coastal forest tree species. *Brittonia*, 33(2): 233-245. Maps.

3133 **Myers, N. (1989).** Ecology of the Amazonian rainforest. *In* Johnson, C., Knowles, R., Colchester, M., *eds.* Rainforests: land-use options for Amazonia. Oxford, Oxford Univ. Press. 9-12.

BRAZIL

3134 **Nassar, N.M.A. (1978).** Conservation of the genetic resources of Cassava (*Manihot esculenta*). Determination of wild species localities with emphasis on probable origin. *Econ. Bot.*, 32(3): 311-320. (E)

3135 **Neto, H.M. (1980).** *Eugenia copacabanensis* Kiaerskon, uma especie ameacada de extincao. *In* Sociedade Botanica do Brasil. XXXI Congresso Nacional de Botanica, 20 a 27 de janeiro de 1980. Itabuna, Ilheus. 121-122. Abstract. (T)

3136 **Ogawa, H.Y., Mota, I.S.D., Favrin, L.J.B., Valentino, R.A.L., Andrade, W.J.D. (1982).** [Forest inventory of Sao Paulo State; natural vegetation cover of two subregions of Paraiba Valley.] *Silvicultura em Sao Paulo*, 16A(1): 441-446. Por (En).

3137 **Padua, M.T.J. (1978).** Conservacao da natureza: o patrimonio natural e o mais nobre legado que podemos deixar para nossos filhos. (Conservation of natural resources: a natural heritage is the noblest legacy that we can leave to our children.) *Brasil Florestal* (Brasilia), 9(34):6-11. Por. Illus., map. National forest parks. (P)

3138 **Padua, M.T.J. (1986).** Pantanal Matogrossense. *Parks*, 11(2-3): 15-19. Illus., maps.

3139 **Padua, M.T.J., Melo Carvalho, J.C. de (1979).** New action in the field of conservation of nature in Brazil. *Envir. Conserv.*, 6(3): 224.

3140 **Padua, M.T.J., Quintao, A.T.B. (1982).** Parks and biological reserves in the Brazilian Amazon. *Ambio*, 11(5): 309-314. Maps. (P)

3141 **Padua, M.T.J., Quintao, A.T.B. (1984).** A system of national parks and biological reserves in the Brazilian Amazon. *In* McNeely, J.A., Miller, K.R., eds. National parks, conservation and development. The role of protected areas in sustaining society. Washington, D.C., Smithsonian Institution Press. 565-571. Maps. (P)

3142 **Pascoli, A.H.G. (1982).** *Helicostylis tomentosa* (Poep. & Endl.) Rusby (Moraceae): especie ameacada de extincao. Flora, especies raras ou ameacadas de extincao: 1. *Cadernos FEEMA Ser. Trab. Techn.*, 18: 25-29. Por. Illus. (T)

3143 **Pereira, B.A.S. (1982).** Especies ornamentais nativas da Bacia do Rio Sao Bartolomeu, Distrito Federal. (Native ornamental species of the Sao Bartolomeu river basin, Federal District Brazil, conservation, flora, forested areas.) *Brasil Florest.*, 12(51): 19-28. Por (En). Illus.

3144 **Peters, C.M., Balick, M.J., Kahn, F., Anderson, A.B. (1989).** Oligarchic forests of economic plants in Amazonia: utilization and conservation of an important tropical resource. *Conservation Biology*, 3(4): 341-349. Illus. (E)

3145 **Pires, J.M. (1977).** The Amazon forest: a natural heritage to be preserved. *In* Prance, G.T., Elias, T.S., eds. Extinction is forever. Proceedings of a symposium at the New York Botanical Garden, 11-13 May 1976. New York, New York Botanical Garden. 158-194. Maps. Basic information and observations of the complex Brazilian Amazonian ecosystems. (P/T)

3146 **Posey, D.A. (1985).** Indigenous management of tropical forest ecosystems: the case of the Kayapo Indians of the Brazilian Amazon. *Agroforestry Systems*, 3(2): 139-158. Illus. (E)

3147 **Posey, D.A. (1989).** Alternatives to forest destruction: lessons from the Mebengokre Indians. *The Ecologist*, 19(6): 241-244. Illus. Sustainable use of forest resources.

3148 **Prance, G.T. (1977).** The phytogeographic subdivisions of Amazonia and their influence on the selection of biological reserves. *In* Prance, G.T., Elias, T.S., eds. Extinction is forever. Proceedings of a symposium at the New York Botanical Garden, 11-13 May 1976. New York, New York Botanical Garden. 195-213. Maps. (P)

BRAZIL

3149 **Prance, G.T. (1979).** Exploitation and conservation in Brazil. *In* Hedberg, I., *ed.* Systematic botany, plant utilization and biosphere conservation. Stockholm, Almqvist & Wiksell International. 146-149.

3150 **Prance, G.T. (1982).** The Amazon: Earth's most dazzling forest. *Garden* (New York), 6(1): 2-10. Col. illus., map.

3151 **Prance, G.T. (1985).** The changing forests. *In* Prance, G.T., Lovejoy, T.E., *eds.* Key environments: Amazonia. Oxford, Pergamon. 146-165. Maps.

3152 **Quintao, A.T.B. (1977).** Planejamento local e areas de desenvolvimento. (Local planning and areas of development in national parks.) *Brasil Florest.*, 8(32): 6-13. Por (En). Illus. (P)

3153 **Quintao, A.T.B. (1983).** Evolucao do conceito de Parques Nacionais e sua relacao com o processo de desenvolvimento. (The evolution of the national parks concept in relation with the development process.) *Brasil Florest.*, 13(54): 13-28. Por (En). Illus., map. (P)

3154 **Rankin, J.M. (1985).** Forestry in the Brazilian Amazon. *In* Prance, G.T., Lovejoy, T.E., *eds.* Key environments: Amazonia. Oxford, Pergamon. 369-392. (E)

3155 **Rodrigues, V. (1980).** Nucleo de desertificacao de Gilbues, Pi. *In* Sociedade Botanica do Brasil. XXXI Congresso Nacional de Botanica, 20 a 27 de janeiro de 1980. Itabuna, Ilheus. 115. Por. Abstract.

3156 **Rodrigues, W.A. (1979).** Especies amazonicas preservadas no Tropical Hotel de Manaus. *In* Sociedade Botanica do Brasil. Resumo dos trabalhos de XXX Congresso Nacional de Botanica 21 a 27 Janeiro de 1979. Campo Grande - MS. 153-154. Por. Abstract. (T)

3157 **Salati, E., Vose, P.B.F. (1984).** Amazon basin: a system in equilibrium. *Science*, 225(4658): 129-138. Amazonia.

3158 **Santos, A.A. dos (1982).** *Dicksonia sellowiana* (Presl.) Hooker; samambaiacu: ameacado de extincao. Flora, especies raras ou ameacadas de extincao: 2. *Cadernos FEEMA Ser. Trab. Techn.*, 18: 21-26. Por. Illus. (T)

3159 **Santos, M. dos (1982).** *Dorstenia fischeri* Bureau: caiapia ameacado de extincao. Flora, especies raras ou ameacadas de extincao: 2. *Cadernos FEEMA Ser. Trab. Techn.*, no. 18: 17-20. Por. (T)

3160 **Schubart, H.O.R. (1982).** Fundamentos ecologicos para o manejo florestal na Amazonia. (Ecological bases for forest management in the Amazon region.) *Silvicult. Sao Paulo*, 16A(2): 713-731. Por. Illus. Conservation of forest resources.

3161 **Schwartz, T. (1989).** The Brazilian forest peoples' movement. *The Ecologist*, 19(6): 245-247. Illus.

3162 **Silva, S.O., Alves, E.J., Cunha, G.A.P. da, Sampaio, J.M.M., Sobrinho, A.P.C., Luna, J.V.U. (1982).** Preservacao e utilizacao de recursos geneticos vegetais de Batata-Doce, Citrus, frutas tropicais e mandioca. *In* Resumos do XXXIII Congresso Nacional de Botanica, 24 a 30 de janeiro de 1982. Maceio, Sociedade Botanica do Brasil. 87. Por. (E)

3163 **Sioli, H. (1985).** The effects of deforestation in Amazonia. *Geog. J.*, 15(2): 197-203.

3164 **Smith, N.J.H. (1982).** Rainforest corridors: the Transamazon colonization scheme. Berkeley, Univ. of California Press. Deforestation.

3165 **Sokol, S., Szczepka, M.Z. (1983).** Parki narodowe w Brazylii. (Brazilian national parks international nature conservation.) *Chronmy Przyr. Ojczysta*, 39(6): 100-109. Pol. Maps. (P)

BRAZIL

3166 **Taylor, K.I. (1988).** Deforestation and Indians in Brazilian Amazonia. *In* Wilson, E.O., *ed.* Biodiversity. Washington, D.C., National Academy Press. 138-144. Tropical forests; deforestation. Includes discussion on resource management techniques used by the Yanomami and Kayapo tribes and effects of shifting cultivation. (E)

3167 **Teixeira, A.R. (1982).** Proposta a criacao de um Instituto Nacional de Ecologia e Conservacao da Natureza, a ser sediado em Brasilia. *An. Soc. Bot. Brasil*: 325-330 (1981 publ. 1982). Por. XXXII Congresso Nacional de Botanico, 25 a 31 de Janeiro de 1981, Teresina, Piaui.

3168 **Thome, J.W. (1977).** A politica de conservacao dos recursos naturais: 2. Exame do sistema educacional relativo a conservacao da natureza e da rede institucional educativa. *Natureza Revista*, 3: 52-59. Por.

3169 **Thome, J.W. (1977).** A politica de conservacao dos recursos naturais: formulacao das necessidades de pesquisa e preparo de pessoal para o inventario e avaliacao dos recursos naturais. *Natureza Revista*, 2: 54-57. Por.

3170 **Thome, J.W. (1978).** A politica de conservacao dos recursos naturais: 3. O conservacionismo, uma filosofia vivencial? *Natureza Revista*, 5: 42-45. Por.

3171 **Thomson, K., Dudley, N. (1989).** Transnationals and oil in Amazonia. *The Ecologist*, 19(6): 219-224. Illus.

3172 **Toledo Filho, L. de, Silva, J. de A., Ferreira, J.M. (1983).** Estrategia para utilizacao das Florestas Nacionais das Regioes Sul e Sudeste. (The national forests of the southern and southeastern regions of Brazil.) *Brasil Florest.*, 13(54): 5-12. Por (En). Illus., map.

3173 **Treece, D. (1989).** The militarization and industrialization of Amazonia: the Calha Norte and Grande Carajas programmes. *The Ecologist*, 19(6): 225-228. Map.

3174 **Uhl, C. (1988).** Restoration of degraded lands in the Amazon basin. *In* Wilson, E.O., *ed.* Biodiversity. Washington, D.C., National Academy Press. 326-332.

3175 **Uhl, C., et al. (1989).** Disturbance and regeneration in Amazonia: lessons for sustainable land-use. *The Ecologist*, 19(6): 235-240. Illus.

3176 **Veening, W. (1989).** Is er hoop voor Amazonie? *Panda*, 11: 8-10. Du. Col. illus., maps. Deforestation.

3177 **Vianna, M.C. (1982).** *Vochysia oppugnata* (Vell.) Warm., canela-santa: uma Vochysiaceae ameacada. Flora, especies raras ou ameacadas de extincao: 1. *Cadernos FEEMA Ser. Trab. Techn.*, 18: 13-20. Por. Illus. (T)

3178 **Villa-Lobos, J.L. (1985).** Atlantic forest protected. *Threatened Pl. Newsl.*, 15: 16-17. (P)

3179 **Woodwell, G.M., Houghton, R.A., Stone, T.A., Nelson, R.F., Kavalick, W. (1987).** Deforestation in the tropics: new measurements in the Amazon Basin, using Landsat and NOAA advanced very high resolution radiometer imagery. *J. Geophysical Research*, 92: 2157-2163.

BRAZIL - Alagoas State

3180 **Andrade-Lima, D. de (1977).** Preservation of the flora of northeastern Brazil. *In* Prance, G.T., Elias, T.S., *eds.* Extinction is forever. Proceedings of a symposium at the New York Botanical Garden, 11-13 May 1976. New York, New York Botanical Garden. 234-239. Map. Vegetation types, their problems, and preservation. (P/T)

BRAZIL - Maranhao State

3181 **Oren, D.C. (1987).** Grande Carajas, international financing agencies, and biological diversity in southeastern Brazilian Amazonia. *Conservation Biology*, 1(3): 222-227. Map.

BRAZIL - Mato Grosso State

3182 **Alho, C.J.R., Lacher, jr, T.E., Goncalves, H.C. (1988).** Environmental degradation in the Pantanal ecosystem. *BioScience*, 38(3): 164-171. (P)

BRAZIL - Minas Gerais State

3183 **Saturnino, H.M. & M.A.C., Ferreira, M.B. (1978).** Algumas consideracoes sobre exportacao e importacao de plantas ornamentais em Minas Gerais. *An. Soc. Bot. Brasil*, 213-217 (1977 publ. 1978). Por (En). 28 Congresso Nacional de Botanica, 23-30 Janeiro 1977. (L)

BRAZIL - Para State

3184 **Anderson, A., Gely, A., Strudwick, J., Sobel, G.L., Pinto, M.G.C. (1985).** Um sistema agroforestal na varzea do estuario Amazonica (Ilha das Oncas, Municipio de Barcarena, Estado de Para.). *Acta Amazonica*, 15: 195-224. Harvesting the floodplain forests.

3185 **Oren, D.C. (1987).** Grande Carajas, international financing agencies, and biological diversity in southeastern Brazilian Amazonia. *Conservation Biology*, 1(3): 222-227. Map.

BRAZIL - Paraiba State

3186 **Andrade-Lima, D. de (1977).** Preservation of the flora of northeastern Brazil. *In* Prance, G.T., Elias, T.S., *eds.* Extinction is forever. Proceedings of a symposium at the New York Botanical Garden, 11-13 May 1976. New York, New York Botanical Garden. 234-239. Map. Vegetation types, their problems, and preservation. (P/T)

BRAZIL - Rio de Janeiro State

3187 **Carauta, J.P.P., Valente, M. da C. (1976).** A vegetacao da Pedra da Gavea, Parque Nacional da Tijuca, Rio de Janeiro. *An. Soc. Bot. Brazil*, 231-239. Por (En). Illus. XXV Congresso Nac. de Bot., Mossoro, Rio de Janeiro, 20-26 January 1974. (P)

3188 **Carauta, J.P.P., Vianna, M.C., Araujo, D.S.D.D., Oliveira, R.F.D. (1978).** The vegetation of "Poco das Antas" Biological Reserve. *Bradea*, 2(46): 299-305. (En). Map. (P)

3189 **Casari, M.B. (1982).** Especies de *Anthurium* (Araceae) raras ou ameacadas no Estado do Rio de Janeiro. (Rare or endangered *Anthurium* (Araceae) species in the State of Rio de Janeiro.) Flora, alguns estudos: 1. *Cadernos FEEMA Ser. Trab. Techn.*, 1: 17-27. Por (En). Illus. (T)

3190 **Pereira Carauta, J., Souza Ferreira de Rocha, E. de (1988).** Conservacao da flora no trecho fluminense da bacia hidrografica do Rio Paraiba do Sul. *Albertoa*, 1(11): 85-86. Por (En). (T)

BRAZIL - Rondonia Federal Territory

3191 **Dourojeanni, M.J. (1986).** Rondonia: disaster or tough lesson? *IUCN Bull.*, 17(1-3): 31-32.

3192 **Fearnside, P.M., Salati, E., (1985).** Explosive deforestation in Rondonia, Brazil. *Envir. Conserv.*, 12(4): 355-356. (T)

BRAZIL - Sao Paulo State

3193 **Bertoni, J.E.D.A., Stubblebine, W.H., Martins, F.R., Leitao Filho, H.D.F. (1982).** Nota previa: comparacao fitossociologica das principais especies de florestas de terra firme e ciliar na reserva estadual de Porto Ferreira. (Phytosociological comparison of the main forest species in terra firme and gallery forests of the state reserve of Porto Ferreira.) *Silvicult. Sao Paulo*, 16A(1): 563-571. Por (En). (P)

BRAZIL - Sao Paulo State

3194 **Marino, M.C. (1989).** The Sao Paulo Institute of Botany: research and conservation. *Bot. Gard. Conserv. News,* 1(4): 39-40. (G)

3195 **Mattos, J.R. & N.F. (1982).** Contribuicao ao conhecimento da flora do Parque Estadual de Campos do Jordao, SP. (Study of the flora of the Campos do Jordao State Park, Sao Paulo State.) *Silvicult. Sao Paulo,* 16A(1): 647-662. Por (En). (P)

3196 **Nello, H. do A., Lima, W. de P. (1978).** Urban pressure on the forest: the example of Sao Paulo. *In* Proceedings of the 7th World Forestry Congress, Buenos Aires, 4-18 October 1972, vol. 3: Conservation and recreation. Buenos Aires, Instituto Forestal Nacional. 3702-3712. (En, Sp, Fr).

3197 **Noffs, M.D.S., Baptista-Noffs, L.J. (1982).** Mapa da vegetacao do Parque Estadual da Ilha do Cardoso - as principais formacoes. (Vegetation map of the Ilha do Cardoso State Park - the main formations.) *Silvicult. Sao Paulo,* 16A(1): 620-628. Por (En). Maps. (P)

3198 **Por, F.D., Imperatriz-Fonseca, V.L. (1984).** The Jureia Ecological Reserve, Sao Paulo, Brazil: facts and plans. *Envir. Conserv.,* 11(1): 67-70. Illus., maps. (P)

BRAZIL - Sergipe State

3199 **Andrade-Lima, D. de (1977).** Preservation of the flora of northeastern Brazil. *In* Prance, G.T., Elias, T.S., *eds.* Extinction is forever. Proceedings of a symposium at the New York Botanical Garden, 11-13 May 1976. New York, New York Botanical Garden. 234-239. Map. Vegetation types, their problems, and preservation. (P/T)

BRITISH INDIAN OCEAN TERRITORY

CHAGOS ARCHIPELAGO (Brit Indian Oc Terr)

3200 **Oldfield, S. (1987).** Fragments of paradise. A guide for conservation action in the U.K. Dependent Territories. Oxford, Pisces Publications; for the British Association for Nature Conservationists. 192p.

BRITISH VIRGIN ISLANDS

3201 **Anon. (1979).** The Virgin Islands Conservation Society. *Caribbean Conserv. News,* 1(18): 14-15.

3202 **Ayensu, E.S., DeFilipps, R.A. (1978).** Endangered and threatened plants of the United States. Washington, D.C., Smithsonian Institution and World Wildlife Fund Inc. 403p. Lists 90 'Extinct', 839 'Endangered' and 1211 'Threatened' taxa for the continental U.S. Also covers Hawaii, Puerto Rico and Virgin Is. (T)

3203 **Little, E.L., jr, Woodbury, R.O. (1980).** Rare and endemic trees of Puerto Rico and the Virgin Islands. U.S. Department of Agriculture, Forest Service Conserv. Res. Rep., 27. Washington, D.C., U.S. Govt Print. Off. 26p. (T)

3204 **Little, E.L., jr, Woodbury, R.O., Wadsworth, F.H. (1976).** Flora of Virgin Gorda (British Virgin Islands). U.S. Department of Agriculture, Forest Service Res. Pap., ITF-21. Washington, D.C., U.S. Govt Print. Off. 36p. (Sp). Illus. Remnants of tropical deciduous forest in Gorda Peak National Park. (P)

3205 **Rogers, C. (1985).** Toward a Lesser Antillean biosphere reserve. *Parks,* 10(3): 22-24. Illus. (P)

3206 **Willan, R.L. (1958).** Forestry development in the British Virgin Islands. Rome, FAO. 26p.

BRUNEI DARUSSALAM

3207 **Bailes, C.P. (1984).** Orchids of Borneo and their conservation. *Malay. Orchid Rev.,* 18: 20-27. Illus., col. illus. (T)

BULGARIA

3208 **Aleksandrov, A.H. (1978).** [Study and conservation of the genetic resources of forest trees in Bulgaria]. *Gorskostopanska Nauka*, 15(4): 34-44. Bul (Rus, En). (E/T)

3209 **Alexiev, A., Simenova, A., Shtilianova, E. (1985).** [Threatened and rare decorative plants and problems related to their cultivation]. *In* International Symposium "Protection of natural areas and the genetic material they contain". 23-28 Sept. 1985. Sofia, Unesco. Rus (En). (G/T)

3210 **Andreev, N., Nikolov, V. (1985).** [Endemic and relict taxa and vegetational communities in the biosphere reservation "Cervenata Stena"]. *In* International Symposium "Conservation of natural areas and of the genetic material they contain". 23-28 Sept. 1985. Sofia, Unesco. Rus (En). Map. (G/P)

3211 **Bondev, I.A., Ganchev, S.P., Slavova, L.M., Boyadjiiski. (1979).** New and rare species of higher plants in the Osogov Mountains; southwestern Bulgaria. *Fitologiya*, 11: 71-73. (T)

3212 **Boscaiu, N. (1985).** Cartea Rosie a Republicii Populare Bulgare: vol. 1 Plante. (Le Livre Rouge de la Republique Populaire Bulgare: 1. Plantes). *Ocrot. Nat. Mediul. Inconjurat.*, 29(1): 59-62. Rum. (T)

3213 **Bulgaria. Bulgarska Akademiya na Naukite (1984).** Chervena kniga na NR Bulgariya: izcheznali, zastrasheni ot izchezvane i redki rasteniya i zhivotni: tom 1. Rasteniya. (Red Data Book of the Peoples's Republic of Bulgaria: v.1. Plants.) Sofiya, Izdatelstvo na Bulgarskata Akademiya na Naukite. 447p. Bul (En, Rus). Illus., col. illus., maps. (T)

3214 **Carter, F.W. (1978).** Nature reserves and national parks in Bulgaria. *Espace Geographique*, 7(1): 69-72. (Fr). (P)

3215 **Dimitrov, D. (1977).** Rare plant species of the Bulgarian Black Sea coast. *Priroda*, 26(3): 95-96. (T)

3216 **Dimitrov, D. (1978).** The oldest forest reservation in Bulgaria. *Priroda*, 27(40): 80-82. Illus. (P)

3217 **Filizov, D. (1978).** [Bistritsa Forest Reservation]. *Gorsko Stopanstvo*, 34(12): 42-44. Bu. Illus. (P)

3218 **Horeanu, C. (1979).** Ocrotirea naturii in R.P. Bulgaria. (La protection de la nature en Bulgarie.) *Ocrot. Nat.*, 23(2): 179. Rum (Fr).

3219 **Ianovici, V. (1979).** Protectia mediului inconjurator si prezervarea cadrului biologic in Tarile. Balcanice. (La protection de l'environnement et la conservation du cadre biologique dans les pays Balkaniques.) *Ocrot. Nat.*, 23(1): 5-7. Rum (Fr). Conference at Bucharest on 18-22 September 1978.

3220 **Jordanov, D., et al., eds (1975).** Problems of the Balkan flora and vegetation. Sofia, Bulgarian Academy of Sciences. Proceedings of the First International Symposium on Balkan Flora and Vegetation, Vorna, 7-14 June 1973. 441p. (T)

3221 **Kasov, D. (1978).** Projektovanie lesnych parkov. (Protecting forest parks.) *Les* (Bratislava), 34(10): 399-402. Illus. (P)

3222 **Kostov, K., Lozanov, I. (1985).** National genetic resources programme in Bulgaria. *Pl. Genet. Resourc. Newsl.*, 62: 28-31. List of collected plant resources in Bulgaria from 1956-1983. (E/G)

3223 **Kozuharov, S. (1975).** On the endemism in the Bulgarian flora. *In* Jordanov, D., *et al., eds.* Problems of the Balkan flora and vegetation. Proc. of the First International Symposium on Balkan Flora and Vegetation. Sofia, Bulgarian Academy of Sciences. 162-168. (T)

3224 **Kruscheva, R., Pirbanov, R. (1978).** Album of protected and rare plants. Sofia, Zemizdat. 117p. Bu. (T)

BULGARIA

3225 **Kuzmanov, B. (1978).** [About the "Red book of rare Bulgarian plants".] *Phytology* (Bulgarian Academy of Sciences), 9: 17-32. Bu (En). Lists 150 rare Bulgarian plants. (T)

3226 **Kuzmanov, B. (1981).** Balkan endemism and the problem of species conservation, with particular reference to the Bulgarian flora. *Bot. Jahrb. Syst.*, 102(1-4): 255-270. Illus., maps. (T)

3227 **Kuzmanov, B. (1981).** Mapping and protection of the threatened plants in the Bulgarian flora. *In* Velcev, V.I., Kozuharov, S.I., *eds*. Mapping the flora of the Balkan Peninsula. 247p. (T)

3228 **Kuzmanov, B., Ivanoeva, S., Georgieva, S. (1985).** [*Achillea urumoffii*, Hall - Bulgarian endemic for the Red Data Book of the People's Republic of Bulgaria]. *In* International Symposium "Conservation of natural areas and of the genetic material they contain". 23-28 Sept. 1985. Sofia, Unesco. Bu (En). (T)

3229 **Kuznetsov, V.S. (1979).** Problemy sokhraneniya i razvitiya parkov, lesoparkov i rekreatsionnykh zon na Chernomorskom poberezh'e Bolgarii. (Problems of conservation and development of parks, forest-parks and recreation zones at the Black Sea coast of People's Republic of Bulgaria.) *Byull. Gos. Nikit. Bot. Sada*, 3(40): 17-20. Rus (En). (P)

3230 **Martinek, V. (1979).** Stav a zamereni ochrany lesu v Bulharsku. (Status and trends of forest protection in Bulgaria.) *Lesn. Cesk. Akad. Zemed.*, 25(4): 375-380. Cz.

3231 **Mel'nik, V.I. (1987).** Krasnaya kniga Narodnoi Respubliki Bolgarii. (The Red Book of the People's Republik of Bulgaria.) *Byull. Glavn. Bot. Sada* (Moscow), 146: 87-88. Rus. (T)

3232 **Meshinev, T., Todorova, O. (1985).** [Flora of the National Park of "Etara"]. *In* International Symposium "Conservation of natural areas and the genetic material they contain". 25-28 Sept. 1985. Sofia, Unesco. Rus (En). (G/P)

3233 **Nedialkov, M. (1984).** Biosphere reserves in Bulgaria and their relation to other protected zones. *In* Unesco-UNEP. Conservation, science and society. Paris, Unesco. 229-232. Contributions to the First International Biosphere Reserve Congress, Minsk, Byelorussia/USSR, 26 September - 2 October 1983. (P)

3234 **Stanev, S. (1975).** Zvezdi gasnat v planinata. Razkazi za nashite redki rasteniya. (The stars are becoming extinct in the mountains. Stories about our rare plants.) Zemizdat. 132p. Illus. (T)

3235 **Stefanov, B. (1936).** Remarks upon the causes determining the relict distribution of plants. *Spis. Bulg. Acad. Sci.*, 53: 133-179. (T)

3236 **Stefanov, B. (1978).** [Plants that are very rare in Bulgaria or that have recently disappeared and the cause of their decline]. *Gorskost. Nauka*, 15(6): 3-10. Bu (Rus, Ge). (T)

3237 **Stefanov, B., Bankov, M. (1978).** [Floristic communication about certain plants with rare habitats in Bulgaria]. *Gorskost. Nauka*, 15(5): 96-97. Bu (En, Rus). (T)

3238 **Stoilov, D., Nostev, V., Rangelova, M., Minkovska, R., Pavlova, E. (1979).** Bulgaria. Protected Natural Sites. (Naturschutz Objekte Bulgarien.) Committee on Environmental Protection at the Council of Ministers of the People's Republic of Bulgaria, Scientific Research Centre for the protection of Natural Environment and Water Resources. (P)

3239 **Stoilov, D., et al. (1981).** Protected natural sites in the People's Republic of Bulgaria. Sofia, Committee on Environmental Protection, Council of Ministers of the People's Republic of Bulgaria. 31p. Translated from Bulgarian by I. Saraouleva. (P)

BULGARIA

3240 **Velchev, V., Bondev, I., Kozuharov, S. (1975).** The problem of protection of the natural flora and vegetation in Bulgaria. *In* Jordanov, D., *et al., eds.* Problems of Balkan flora and vegetation. Sofia, Bulgarian Academy of Sciences. 431-435. Proceedings of the 1st International Symposium on Balkan Flora and Vegetation, Varna, June 7-14 1973.

3241 **Velchev, V., Kozuharov, S., Bondev, I., Kuzmanov, B., Markova, M. (1984).** Red Data Book of the People's Republic of Bulgaria. Volume 1. Plants. Sofia, Bulgarian Academy of Sciences. 447p. Bul (En). Illus., maps. Describes 763 threatened species; includes data on distribution, habitats and ecology. (T)

3242 **Velchev, V., Stoeva, M. (1985).** Population approach to the investigation of the threatened and rare species in Bulgarian flora in connection of their conservation. *In* International Symposium "Conservation of natural areas and the genetic material they contain". Sofia, Unesco. 182. (G/T)

3243 **Vykhodtsevsky, N.M., Simeonovsky, M.Iv. (1978).** Kratkii istoricheskii ocherk okhrany prirody i Bolgarii. (A brief historical essay on nature conservation in Bulgaria.) *Byull. Mosk. Obshch. Ispyt. Prir., Biol.*, 83(4): 161-165. Rus.

BURKINA FASO

3244 **Panzer, K.F., Rhody, B. (1981).** Applicability of large-scale aerial photography to the inventory of natural resources in the Sahel of Upper Volta. *USDA Forest Service General Technical Report WO.* 1981(28): 287-299. En (Sp). Illus. Presented at the Arid Lands Resources Inventories workshop, La Paz, Mexico, 1980.

BURMA

3245 **Han, S. (1984).** Better than machines: elephants in Burma. *In* McNeely, J.A., Miller, K.R., *eds.* National parks, conservation and development. The role of protected areas in sustaining society. Washington, D.C., Smithsonian Institution Press. 175-177. Table. Use of elephants for timber extraction causes less damage to forests.

3246 **Richards, S.V. (1981).** Conservation and utilization of pasture legumes in Burma. *IBPGR Region. Committ. Southeast Asia Newsl.*, 5(3): 5. (E)

BURUNDI

3247 **INEAC (1954).** Carte des Sols et de la Vegetation du Congo, du Rwanda et du Burundi. Brussels, INEAC. Fr. A series of vegetation and soil maps covering Zaire Rwanda and Burundi in c. 25 parts, published between 1954 and c. 1970. Each map is accompanied by a descriptive memoir, & several maps are to dif. scales.

3248 **Lewalle, J. (1968).** Burundi. *In* Hedberg, I. & O., *eds.* Conservation of vegetation in Africa south of the Sahara. Symposium Proceedings, at 6th plenary meeting of AETFAT, Uppsala. *Acta Phytogeogr. Suec.*, 54: 127-130. Fr. Illus. Lists of plants needing protection. (P/T)

CAMEROON

3249 **Burnett, G.W. (1984).** Cameroon: national parks and nation-building. *Envir. Conserv.*, 11(1): 70-72. Map. (P)

3250 **Depierre, D. (1978).** Les biocenoses au Cameroun: richesse et fragilite. (The biocenoses of Cameroun: their richness and fragility.) *Bois Forets Trop.*, 182: 72-76. Illus.

3251 **Gartlan, J.S. (1985).** The Korup Management Plan: conservation and development in the Ndian Division of Cameroon. Gland, IUCN/WWF. 179p. (P)

CAMEROON

3252 **Gartlan, J.S. (1986).** The biological and historical importance of the Korup forest. *In* Gartlan, S., Macleod, H., *eds.* Proceedings of the Workshop on Korup National Park, Mundemba, Ndian Division, South-West Province, Cameroon, August 1986. 28-35. (P)

3253 **Gartlan, S. (1988).** Conservation et utilisation rationnelle des ecosystemes forestiers en Afrique centrale; rapport national Cameroun. Gland, Switzerland, IUCN. 178p. Fr. Maps.

3254 **Gartlan, S. (1989).** The Korup National Park, Cameroon. *Trop. Afr. Bot. Gard. Bull.*, 1: 14-15. (P)

3255 **Gartlan, S., Macleod, H., eds (1986).** Proceedings of the workshop on Korup National Park, Mundemba, Ndian Division, South-West Province, Cameroon. Gland, WWF/IUCN. (P)

3256 **Gartlan, S., Momo, D. (1986).** Korup: new approach to conservation. *IUCN Bull.*, 17: 1-3. (P)

3257 **Hepper, F.N. (1988).** A botanical garden for conserving the forest in Cameroon. *Bot. Gard. Conserv. News*, 1(2): 16-17. Illus. (G/L/T)

3258 **IUCN (1989).** La conservation des ecosystemes forestiers du Cameroun. IUCN, Tropical Forest Programme Series No.10. 196p. Illus.

3259 **Letouzey, R. (1968).** Cameroun. *In* Hedberg, I. & O., *eds.* Conservation of vegetation in Africa South of the Sahara. Symposium Proceedings, at 6th plenary meeting of AETFAT, Uppsala. *Acta Phytogeogr. Suec.*, 54: 115-121. Fr. Illus., map. Lists of species for protection. Forest preserves. (P/T)

3260 **Macleod, H.L., Numbem, S.T., Thomas, D.W. (1986).** The conservation of Oku Mountain forest, Cameroon. ICBP Study Report No. 15. Cambridge, International Council for Bird Preservation. 90p. Maps. Includes vegetation, deforestation, provisional species list.

3261 **McKee, D. (1988).** *Cecropia peltata*, an introduced neotropical pioneer tree, replacing *Musanga cecropioides* in southwestern Cameroon. *Biotropica*, 20: 62-264. Map. Invasive species poses a threat.

3262 **Sanford, W.W. (1970).** Conservation of west African orchids II: The Republic of Cameroon. *Biol. Conserv.*, 3: 47-50. (T)

3263 **Stark, M.A., Hudson, R.J. (1985).** Plant communities' structure in Benoue National Park, Cameroon: a cluster association analysis. *Afr. J. Ecol.*, 23(1): 21-27. (Fr). Forest. (P)

3264 **Thomas, D.W. (1986).** The botanical uniqueness of Korup and its implications for ecological research. *In* Gartlan, S., Macleod, H., *eds.* Proceedings of the Workshop on Korup National Park, Mundemba, Ndian Division, South-West Province, Cameroon, August 1986. Gland, WWF. 36-40. (P)

CANADA

3265 **Achuff, P.L. (1989).** Plant conservation in the prairie provinces. *Can. Pl. Conserv. Prog. Newsl.*, 4(1): 2-3. (T)

3266 **Adams, W.T. (1981).** Population genetics and gene conservation in Pacific Northwest conifers. *In* Scudder, G.G.E., Reveal, J.L., *eds.* Evolution today: proceedings of the second International Congress of Systematic and Evolutionary Biology. Pittsburgh. Hunt Inst. Bot. Documentation. 401-415.

3267 **Allen, G.M., Oldham, M.J. (1985).** *Plantago cordata* Lam. (Heart-leaved plantain) still survives in Canada. *The Plant Press* (Field Botany in Ontario), 3(3): 94-96. (T)

3268 **Ambrose, J. (1987).** Conservation of Canada's rare plant resources. *Wildflower*, 3(3): 40-41. Illus. Includes role of botanic gardens. (G/P/T)

CANADA

3269 **Ambrose, J.P., Kevan, P.G., Gadawski, R.M. (1985).** Hop tree (*Ptelea trifoliata*) in Canada: population and reproduction biology of a rare species. *Canad. J. Bot.*, 63(1): 1928-1935. (T)

3270 **Anon. (1981).** Canada's forest: prime source of national wealth: can be saved. *Trees S. Afr.*, 32(4): 103-104.

3271 **Argus, G.W. (1975).** The conservation of Canadian rare and endangered plants. Pap. Midwest Wildl. Conf., 7p. (T)

3272 **Argus, G.W. (1976).** The conservation of Canadian rare and endangered plants. *In* Mosquin, T., Suchal, C., *eds* Canada's threatened species and habitats. Proc. of the symposium on Canada's threatened species and habitats, held in Ottawa, May 20-24, 1976. 139-143. (T)

3273 **Argus, G.W. (1977).** Canada. *In* Prance, G.T., Elias, T.S., *eds.* Extinction is forever. Proceedings of a symposium at the New York Botanical Garden, 11-13 May 1976. New York, New York Botanical Garden. 17-29. Present knowledge of rare and endangered plants in Canada including legislative action within each province and other action by various organizations. (L/P/T)

3274 **Argus, G.W. (1978).** List of Canadian flora affected by CITES (Convention on International Trade in Endangered Species of Wild Fauna and Flora). CITES Rep. No. 4. Ottawa, Canadian Wildlife Service, Environment Canada. 14p. (L/T)

3275 **Argus, G.W. (1981).** Status reports on endangered, threatened and rare Canadian plants. *Canad. Bot. Ass. Bull.*, 14(2): 15. (T)

3276 **Argus, G.W. (1983).** Endangered species. *Amer. Assoc. Bot. Gard. Arbor. Inc. Newsl.*, 101: 3. (T)

3277 **Argus, G.W., McNeill, J. (1974).** Conservation of evolutionary centres in Canada. *In* Maini, J.S., Carlisle, A., *eds.* Conservation in Canada: a conspectus. Canadian Forest Service Publication 1340. Dept of Environment. 131-141.

3278 **Barabe, D., Cornellier, N., Laliberte, S. (1985).** La conservation artificielle des plantes rares. *Nat. Canad.*, 112(2): 275-281. Fr (En). (G/T)

3279 **Baskerville, G.L. (1988).** Redevelopment of a degrading forest system. *Ambio*, 17(5): 314-322. Col. illus.

3280 **Benson, L. (1982).** The cacti of the United States and Canada. Stanford, CA, Stanford University Press. ix, 1044p. Illus., col. illus., maps, keys. Includes conservation, 242-247. (T)

3281 **Bossert, E. (1982).** Les plantes en danger. *Foret Conservation*, 49(2): 22-28. (T)

3282 **Brumback, W.E. (1981).** Endangered plant species programs for botanic gardens with examples from North American institutions. Newark, Univ. of Delaware. 232p. M.Sc. thesis. (G/T)

3283 **Carruthers, J.A. (1970).** Present national parks and future needs in Canada. *In* Fuller, W.A., Kevan, P.G., *eds.* Productivity and conservation in northern circumpolar lands. Proceedings of a conference, Edmonton, Alberta, 15-17 October 1969. Morges, IUCN. 297-299. (P)

3284 **Craig, I.L., Fedak, G. (1985).** *Hordeum arizonicum* threatened with extinction. *Cereal Res. Comm.*, 13(2/3): 269-271. (E/T)

3285 **Currah, R.S., Smreciu, E.A., Seymour, P.N.D. (1986).** A national plant conservation programme for Canadian botanic gardens. *Canad. Bot. Ass. Bull.*, 19(3): 29-30. (G)

3286 **Currah, R.S., Smreciu, E.A., Seymour, P.N.D. (1987).** A national plant conservation programme for Canadian botanic gardens. *In* Bramwell, D., Hamann, O., Heywood, V., Synge, H., *eds.* Botanic gardens and the World Conservation Strategy. London, Academic Press. 295-299. (Sp). (G)

CANADA

3287 **Dojcsak, G. (1976).** Kanada nemzeti parkjai. (The national parks of Canada.) *Buvar*, 31(5): 211-214. Hu. Illus., map. (P)

3288 **Dube, D.E. (1979).** Fire management in national parks. *Inform Rep. NOR-X North. Forest Res. Centre*, 215: 78-79. (P)

3289 **Dummler, H. (1977).** Ein Reisebericht: vom 12. Internationalen Seminar uber Nationalparke und gleichwertige Schutzgebiete com 2. August bis 1. September 1977 in Kanada, USA und Mexico. *Naturschutz- und Naturparke*, 87: 1-8. Ge. Illus., map. (P)

3290 **Eidsvik, H. (1983).** Under joint responsibility: the Kluane-Wrangell-St. Elias World Heritage Site. *Ambio*, 12(3-4): 191-196. Col. illus., map. (P)

3291 **Fenge, T. (1982).** Towards comprehensive conservation of environmentally significant areas in the northwest territories of Canada. *Envir. Conserv.*, 9(4): 305-313. Map. (L/P)

3292 **Forster, R. (1989).** Conserving national horticultural resources. *Can. Pl. Conserv. Prog. Newsl.*, 4(1): 11-14. (G)

3293 **Francis, G. (1984).** Environmental education in biosphere reserves: some Canadian experience. *In* Unesco-UNEP. Conservation, science and society. Paris, Unesco. 601-604. Contributions to the First International Biosphere Reserve Congress, Minsk, Byelorussia/USSR, 26 September - 2 October 1983. (P)

3294 **Gardner, J.E., Nelson, J.G. (1981).** National parks and native peoples in northern Canada, Alaska and northern Australia. *Envir. Conserv.*, 8(3): 207-215. Maps. (P)

3295 **Gimbarzevsky, P. (1977).** L'Anse aux Meadows National Historic Park. Integrated survey of biophysical resources. Information Report No. FMR-X-99. Ottawa, Forest Management Research Institute. viii, 113p. (P)

3296 **Gimbarzevsky, P. (1978).** Integrated survey of biophysical resources in national parks. Ottawa, Forest Management Institute. 21p. (P)

3297 **Haber, E. (1986).** Rare and endangered plants of Canada: report of the plants subcommittee, the Committee on the Status of Endangered Wildlife in Canada (COSEWIC). *Canad. Field-Nat.*, 100(3): 400-403. (T)

3298 **Inglis, J. (1984).** Developing biosphere reserves in northern Canada. *In* Unesco-UNEP. Conservation, science and society. Paris, Unesco. 514-519. Contributions to the First International Biosphere Reserve Congress, Minsk, Byelorussia/USSR, 26 September - 2 October 1983. (P)

3299 **Jana, S. (1982).** Canada: ICARDA collaboration for cereal germplasm conservation. *Pl. Genet. Resource. Newsl.*, 49: 5-10. (Fr, Sp). Map. Includes collecting. *Hordeum spontaneum.* (E)

3300 **Joyce, C. (1987).** Trees and lakes 'need fear no acid'. *New Sci.*, 115(1579): 21. Acid rain.

3301 **Kartesz, J.T. & R. (1977).** The biota of North America: part 1. Vascular plants. Volume 1: rare plants. Pittsburgh, Bonac. iii, 361p. Includes Endangered Species Act, 1973. (L/T)

3302 **Kershaw, L.J., Morton, J.K. (1976).** Rare and potentially endangered species in the Canadian flora. A preliminary list of vascular plants. *Canad. Bot. Assoc. Bull.*, 9(2): 26-30. An abbreviated version of a computerised list of the rare and endangered vascular plants in Canada compiled at the University of Waterloo and last updated in 1978. (T)

3303 **Lamont, S.M. (1978).** Notes on the plants of the Douglas Provincial Park seepage areas. *Blue Jay*, 36(4): 190-192. Illus. (P)

CANADA

3304 **Lapalme, L.A. (1980).** Les parcs nationaux d'un reseau national. (National Parks: the planning of a national resource.) *Geoscope,* 11(1): 1-27. Fr. Conservation of Canadian forest and mineral resources. (P)

3305 **Lawyer, J.I., et al. (1979).** A guide to selected current literature on vascular plant floristics for the contiguous U.S., Alaska, Canada, Greenland and the U.S. Caribbean and Pacific Islands. New York, New York Botanical Garden. 138p. Suppl. 1: Research on floristic information synthesis: a report to the Division of Natural History, Nat. Park Service, U.S. Dept of Interior. Washington, D.C. (B)

3306 **Looman, J., Best, K.F. (1979).** Budd's flora of the Canadian Prairie Provinces. Hull, Quebec, Canadian Government Publishing Centre. 863p. Illus., map, key. Distinguishes some of the rare plants. (T)

3307 **Lopoukhine, N. (1985).** Resource management in national parks. *Forest. Chron.,* 61(5): 374-377. Maps. (P)

3308 **MacKenzie, D. (1984).** Canada counts its lost trees. *New Sci.,* 101(1395): 8. Illus.

3309 **Maini, J.S., Carlisle, A. (1974).** Conservation in Canada: a conspectus. Dept of Environment, Canadian Forest Service Publication 1340.

3310 **McKeating, G.B., Bowman, I. (1977).** Endangered species. *Ontario Fish and Wildlife Rev.,* 16(4): 1-24. (T)

3311 **McLean, A. (1976).** Protection of vegetation in ecological reserves in Canada. *Canad. Field-Nat.,* 90: 144-148. (P)

3312 **Middleton, J., Merriam, G. (1985).** The rationale for conservation: problems from a virgin forest. *Biol. Conserv.,* 33(2): 133-146.

3313 **Mondor, C., Kun, S. (1982).** The long struggle to protect Canada's vanishing prairie. *Ambio,* 11(5): 286-291. Col. illus., maps. (P)

3314 **Mondor, C., Kun, S. (1984).** The Lone Prairie: protecting natural grasslands in Canada. *In* McNeely, J.A., Miller, K.R., *eds.* National parks, conservation and development. The role of protected areas in sustaining society. Washington, D.C., Smithsonian Institution Press. 508-517. Illus.; maps. (P)

3315 **Morton, J.K. (1972).** The role of botanic gardens in conservation of species and genetic material. *In* Rice, P.F., *ed.* Proceedings of Symposium on A National Botanical Garden System for Canada. Tech. Bull. No.6. Hamilton, Ontario, Royal Botanic Gardens. 46-54. (G)

3316 **Morton, J.K., ed. (1976).** Proceedings of the Symposium: Man's impact on the Canadian flora. *Canad. Bot. Assoc. Bull.,* 9(Suppl.)(1): 30p.

3317 **Mosquin, T., Suchal, C., eds (1977).** Canada's threatened species and habitats. Proceedings of the symposium on Canada's threatened species and habitats, held in Ottawa, May 20-24, 1976. Canadian Nature Federation Publication No. 6. Ottawa, Canadian Nature Foundation. 185p. (T)

3318 **Nelson, J.G. (1976).** Man's impact on the western Canadian landscape. Toronto, McLelland & Stewart. 205p.

3319 **Nelson, J.G. (1984).** Living with exploitation in the subarctic and arctic of Canada. *In* McNeely, J.A., Miller, K.R., *eds.* National parks, conservation and development. The role of protected areas in sustaining society. Washington, D.C., Smithsonian Institution Press. 527-533.

3320 **Nelson, J.G., Needham, R.D., Nelson, S.H., Scace, R.C., eds (1979).** The Canadian national parks: today and tomorrow. Conference II: Ten years later. Waterloo, Ontario, Studies in land use history and landscape change No. 7. xix, 1-525; xix, 526-838. Proceedings of a conference at Banff, Alberta on 8-13 October, 1978. (P)

CANADA

3321 **Packer, J.G., Vitt, D.H. (1974).** Mountain Park: a plant refugium in the Canadian Rocky Mountains. *Can. J. Bot.*, 52: 1393-1409. (P)

3322 **Peterson, E.B. (1976).** Regulatory aspects of ecological reserves in Canada. *In* Canadian Botanical Association. Natural areas: Proceedings of a Symposium at the 13th Annual Meeting of the Canadian Botanical Association, Bishop's University, Lennoxville, Quebec, 6-10 June. CBA. 8-14. (Fr). (P)

3323 **Peterson, J.S. (1982).** 1981 Rocky Mountain Regional Rare Plant Conference. *Green Thumb*, 39(1): 24-26. Illus. (T)

3324 **Postel, S. (1984).** Air pollution, acid rain and the future of forests. 5. *Amer. Forests*, 90(11): 46-47. Pollution.

3325 **Reed, F.L.C. (1983).** Replenishing the world's forests: forest renewal in Canada. *Commonw. Forest. Rev.*, 62(3): 169-177. En (Fr, Sp).

3326 **Revel, R.D. (1981).** Conservation in northern Canada: International Biological Programme conservation sites revisited. *Biol. Conserv.*, 21(4): 263-287. Maps; proposed establishment of new national parks. (L/P)

3327 **Rice, P.F., ed. (1972).** Proceedings of Symposium on A National Botanical Garden System for Canada. Tech. Bull. No.6. Hamilton, Ontario, Royal Botanic Gardens. (G)

3328 **Rosencranz, A. (1986).** The acid rain controversy in Europe and North America: a political analysis. *Ambio*, 15(1): 19-21.

3329 **Saltykovskaya, L.V. (1980).** Conservation of natural resources and the environment in East Canada. *Izv. Akad. Nauk SSSR, Ser. Geogr.*, 2: 85-94. Rus. (P)

3330 **Scudder, G.C.E. (1979).** Present patterns in the fauna and flora of Canada. *In* Danks, H.V., ed. Canada and its insect fauna. *Mem. Entomol. Soc. Can.*, 108: 87-179.

3331 **Shay, J.M. (1974).** A survey of the rare and endangered plants of Canada. Ottawa, Ont., Ottawa Mus. Nat. Sci. n.p. (T)

3332 **Singleton, N. (1977).** Endangered species legislation in Canada. *In* Mosquin, T., Suchal, C., eds. Canada's threatened species and habitats. Proceedings of symposium. Publication No. 6. Ottawa, Ontario, Canadian Nature Federation. 10-21. (L)

3333 **Soper, J.H. (1979).** Nature conservation in Canada. *In* Hedberg, I., ed. Systematic botany, plant utilization and biosphere conservation. Stockholm, Almqvist & Wiksell International. 143-146.

3334 **Taschereau, P.M. (1985).** The status of ecological reserves in Canada. Nova Scotia, Dalhousie Univ. 120p. (P)

3335 **Taylor, R.L. (1983).** The activities of COSEWIC: Committee on the Status of Endangered Wildlife in Canada. *In* Given, D.R., ed. Conservation of plant species and habitats. Welington, Nature Conservation Council. 29-36. (T)

3336 **U.S. Department of the Interior. Fish and Wildlife Service (1988).** Endangered and threatened wildlife and plants; determination of Threatened status for *Cirsium pitcheri*. *Fed. Register*, 53(137): 27137-27141. (L/T)

3337 **U.S. Department of the Interior. Fish and Wildlife Service (1989).** Endangered and threatened wildlife and plants; determination of Threatened status for eastern and western prairie fringed orchids. *Fed. Register*, 54(187): 39857-39862. *Platanthera leucophaea, P. praeclara*. (L/T)

3338 **WWF (1974).** Symposium on Endangered and Threatened Species of North America, Washington, D.C., 1974. Proceedings of the Symposium on Endangered and Threatened Species of North America. Washington, D.C., World Wildlife Fund. 339p. (T)

CANADA

3339 **Walker, J. (1970).** The influence of man on vegetation at Churchill. *In* Fuller, W.A., Kevan, P.G., eds. Productivity and conservation in northern circumpolar lands. Proceedings of a conference, Edmonton, Alberta, 15-17 October 1969. Morges, IUCN. 266-269.

3340 **Waller, D.M., O'Malley, D.M., Gawler, S.C. (1987).** Genetic variation in the extreme endemic *Pedicularis furbishiae* (Scorphulariaceae). *Conserv. Biol.*, 1(4): 335-340. Illus. (T)

3341 **Wiley, L. (1969).** Rare wild flowers of North America. 2nd ed. rev. Portland, OR, n.p. 501p. (T)

CANADA - Alberta

3342 **Achuff, P.L., Corns, I.G.W. (1985).** Plants new to Alberta from Banff and Jasper National Parks. *Canad. Field-Nat.*, 99(1): 94-98. Maps. (P)

3343 **Argus, G.W., White, D. (1976).** A preliminary list of the rare plants of Alberta. Ottawa Museum of Natural Science. 15p. (T)

3344 **Argus, G.W., White, D.J. (1978).** The rare vascular plants of Alberta. *Syllogeus*, 17: 47p. En, Fr. (T)

3345 **Cline, D., Erdman, K., Pearce, W. (1984).** The Dinosaur World Heritage Site: responsible management in Canada. *In* McNeely, J.A., Miller, K.R., eds. National parks, conservation and development. The role of protected areas in sustaining society. Washington, D.C., Smithsonian Institution Press. 741-743. (P)

3346 **Cowley, M., Lieff, B.C. (1984).** Extending the biosphere reserve by involving local people in western Canada. *In* Unesco-UNEP. Conservation, science and society. Paris, Unesco. 548-552. Contributions to the First International Biosphere Reserve Congress, Minsk, Byelorussia/USSR, 26 September - 2 October 1983. (P)

3347 **Fairbarns, M., Loewen, V., Bradley, C. (1987).** The rare vascular flora of Alberta. Volume 1: A summary of the taxa occurring in the Rocky Mountain natural region. Edmonton, Alberta Forestry, Lands and Wildlife. 27p. (T)

3348 **Holland, W.D., Coen, G.M. (1983).** Ecological (biophysical) land classification of Banff and Jasper National Parks. Edmonton, Alberta Institute of Pedology, 2 vols. Includes bibliographies. Vol. 1 - summary, Vol. 2 - soil and vegetation resources. (P)

3349 **Kuijt, J. (1982).** A flora of Waterton Lakes National Park. Edmonton, University of Alberta Press. xxiv, 684p. Illus., col. illus. (P)

3350 **Miller, M. (1978).** Perspectives for fire management in Alberta provincial parks and wilderness areas. *Inform. Rep. NOR-X North. Forest Res. Centre*, 210: 34-36. (P)

3351 **Packer, J.G., Bradley, C.E. (1978).** A checklist of the rare vascular plants of Alberta with maps. Edmonton, Alberta, Recreation, Parks and Wildlife. Maps. (T)

3352 **Packer, J.G., Bradley, C.E. (1984).** A checklist of the rare vascular plants in Alberta. Prov. Mus. Alta. Nat. Hist. Occ. Paper No. 5. (T)

3353 **Price, M.F. (1983).** Management planning in the Sunshine area of Canada's Banff National Park. *Parks*, 7(4): 6-10. Illus., map. (P)

3354 **Ryhanen, H.M. (1978).** Forest protection in Alberta, 1977. *Inform. Rep. NOR-X North. Forest Res. Centre*, 210: 69-72. Illus.

3355 **Tande, G.F. (1979).** Fire history and vegetation pattern of coniferous forests in Jasper National Park, Alberta. *Canad. J. Bot.*, 57(18): 1912-1931. (Fr). Illus., maps. (P)

CANADA - Alberta

3356 **Walker, B.D., Kojima, S., Holland, W.D., Coen, G.M. (1978).** Land classification of the Lake Louise study area, Banff National Park. *Inf. Rep., North. Forest Res. Centre NOR-X*, 160. 121p. Illus., map. (P)

3357 **Wallis, C. (1977).** Preliminary lists of the rare flora and fauna of Alberta. Edmonton, Alberta, Alberta Provincial Parks. 22p. (T)

3358 **Wallis, C. (1987).** The rare vascular flora of Alberta. Volume 2: A summary of the taxa occurring in the grasslands, parkland and boreal forest. Edmonton, Alberta Forestry, Lands and Wildlife. 10p. Map. (T)

3359 **Wallis, C., Allen, L. (1987).** Assessment and monitoring of rare plants in Alberta, Canada. *In* Elias, T.S., *ed.* Conservation and management of rare and endangered plants. Proceedings from a conference, Sacramento, California, 5-8 November 1986. Sacramento, California Native Plant Society. 579-585. (T)

3360 **Wallis, C., Bradley, C., Fairbarns, M., Loewen, V. (1987).** The rare vascular flora of Alberta. Volume 3: Species summary sheets. Edmonton, Alberta Forestry, Lands and Wildlife. Unpaginated. Map. (T)

3361 **Wallis, C., Fairbarns, M., Loewen, V., Bradley, C. (1987).** The rare vascular flora of Alberta. Volume 4: Bibliography. Edmonton, Alberta Forestry, Lands and Wildlife. Unpaginated. (B/T)

CANADA - British Columbia

3362 **Anon. (1983).** New park for Canada. *Oryx*, 17: 142. Valhalla Mountain Range, on western edge of the Kootenays in British Columbia to be a 60,000 ha shoreline to Mountain Peak Provincial Park. (P)

3363 **Anon. (1987).** Reserve for phantom orchid. *Oryx*, 21(2): 120. Extract from *Nature Canada*, 15(4). Reserve established in British Columbia for *Cephalanthera austinae*. (T)

3364 **Guppy, G.A. (1977).** Endangered plants in British Columbia *Davidsonia*, 8: 24-30. (T)

3365 **Haber, E., Soper, J.H. (1980).** Vascular plants of Glacier National Park, British Columbia, Canada. *Syllogeus*, 24: 34p. Illus. Checklist of plants. (P)

3366 **Heriot, J.E. (1983).** Endangered wild flowers of B.C. *B.C. Naturalist*, 21(3): 13-14. (T)

3367 **Inselberg, A.E., Klinka, K., Ray, C. (1982).** Ecosystems of MacMillan Park on Vancouver Island. Victoria, B.C., Ministry of Forests, Land Management Report. 113p. Illus. (P)

3368 **Krajina, V.J. (1973).** Further case studies in selecting and allocating land for nature conservation: British Columbia. *In* Costin, A.B., Groves, R.H., *eds.* Nature conservation in the Pacific. Proc. Symposium A-10, XII Pacific Science Congress, Canberra, August-September 1971. IUCN and Australian Nat. Univ. Press. 57-61. Map. (P)

3369 **Krajina, V.J. (1976).** Progress of ecological reserves in British Columbia. *In* Canadian Botanical Association. Natural areas: Proceedings of a Symposium at the 13th Annual Meeting of the Canadian Botanical Association, Bishop's University, Lennoxville, Quebec, 6-10 June. 15-21. (Fr). Illus. (P)

3370 **Roemer, H.L., Pojar, J., Joy, K.R. (1988).** Protected old-growth forests in coastal British Columbia. *Nat. Areas J.*, 8(3): 146-159. (P)

3371 **Soper, J.H., Szczawinski, A.F. (1976).** Mount Revelstoke National Park: wildflowers. Natural History Series No. 3. Victoria, British Columbia Provincial Museum. xv, 96p. Illus., col. illus., map. (P)

CANADA - British Columbia

3372 **Stace-Smith, R., et al. (1980).** Threatened and endangered species and habitats in British Columbia and the Yukon. Victoria, B.C., B.C. Ministry of Environment, Fish and Wildlife Branch. 302p. Proc. of the Symposium co-sponsored by Fed. of B.C. Naturalists Inst. of Environmental Studies, Douglas College, B.C. Min. of Environ., Fish and Wildlife Branch held in Richmond, B.C. March 8 and 9, 1980. (T)

3373 **Straley, G., Taylor, R., Douglas, G. (1985).** The rare vascular plants of British Columbia. *Syllogeus* (Ottawa, National Museums of Canada), 59: 165p. (T)

3374 **Sullivan, T.P. (1979).** Virgin Douglas fir forest on Saturna Island, British Columbia. *Canad. Field-Nat.*, 93(2): 126-131. Illus. Saturna Island Ecological Reserve. (P)

3375 **Taylor, R.L. (1982).** Plant introduction scheme from the University of British Columbia Botanical Garden. *Bull. Amer. Assoc. Bot. Gard. Arbot.*, 16(3): 115-118. (G)

3376 **Taylor, R.L., Straley, G.B. (1985).** The rare vascular plants of British Columbia. *Syllogeus*, 59. (T)

3377 **Taylor, T. (1973).** Endangered species. *Bull. British Columbia Bot. Club*, 47-48. (T)

CANADA - Manitoba

3378 **Argus, G.W., White, D. (1975).** A preliminary list of the rare plants of Manitoba. Ottawa Mus. Nat. Sci., 15p. (T)

3379 **Haddon, B.D. (1984).** Reporting and summarizing forestry change data Manitoba pilot study. Information report PI-X 36. National Forestry Institute. Chalk River, Ontario, Canadian Forestry Service. 26p. Illus., maps.

3380 **Hall, R.J. (1984).** Use of large-scale aerial photographs in regeneration assessments. Information report NOR-X.264. Edmonton, Northern Forest Research Centre. 31p. (Fr). Illus., maps.

3381 **Johnson, K. (1975).** Preservation of native plants. *Manitoba Nature*, Winter: 20-23.

3382 **Krivda, W. (1982).** Cottonwood: a rare tree near the Pas, Manitoba. *Blue Jay*, 40(2): 78-79. Illus. *Populus deltoides*. (T)

3383 **Shay, J.M. (1974).** Preliminary list of endangered and rare species in Manitoba. Winnipeg, Man., University of Manitoba, Botany Dept. n.p. (T)

3384 **White, D.J., Johnson, K.L. (1980).** The rare vascular plants of Manitoba. *Syllogeus*, 27: 1-52. (Fr). Maps. (T)

CANADA - New Brunswick

3385 **Day, R.T. (1983).** A survey and census of the endangered furbish lousewort, *Pedicularis furbishiae*, in New Brunswick. *Canad. Field-Nat.*, 97(3): 325-327. Maps. (T)

3386 **Hinds, H.R. (1983).** The rare vascular plants of New Brunswick. *Syllogeus*, 50: 38p. En, Fr. Maps. (T)

3387 **Hirvonen, R., Madill, R.J. (1978).** Fundy National Park, NB (New Brunswick) and the proposed western extension. Integrated resource survey. Information Report No. FMR-X-105. Ottawa, Ontario, Forest Management Institute. xi, 225p. (Fr). Illus. (P)

CANADA - Newfoundland

3388 **Robertson, A., Roberts, B.A. (1982).** Checklist of the alpine flora of Western Brook Pond and Deer Pond areas, Gros Morne National Park. *Rhodora*, 84(837): 101-115. (Fr). Illus. (P)

CANADA - Northwest Territories

3389 **Cody, W.J., Scotter, G.W., Talbot, S.S. (1979).** Additions to the vascular plant flora of Nahanni National Park, Northwest Territories. *Nat. Canad.*, 106(4): 439-450. (Fr). (P)

3390 **Jones, G., Bright, P. (1977).** An annotated list of higher plant species observed or collected in the general area of the Auyuittuq National Park. *In* Clifton College, Westminster School. Baffin 1977: the Joint Clifton and Westminster expedition to Baffin Island. 97-107. (P)

CANADA - Nova Scotia

3391 **Argus, G.W., White, D. (1975).** A preliminary list of the rare plants of Nova Scotia. Ottawa Museum of Natural Science. 9p. (T)

3392 **Hinds, H.R. (1984).** Additions to the flora of Cape Breton Highlands National Park, Nova Scotia. *Rhodora*, 86(845): 67-71. (P)

3393 **Isnor, W. (1981).** Provisional notes on the rare and endangered plants and animals of Nova Scotia. Curatorial Report No. 46. Halifax, Nova Scotia, Nova Scotia Museum. Maps. Notes on identification, distribution, habitat and vulnerability for 82 vascular plants. (T)

3394 **Keddy, P. (1985).** Lakeshores in the Tusket River Valley, Nova Scotia: distribution and status of some rare species, including *Coreopsis rosea* Nutt. and *Sabatia kennedyana* Fern. *Rhodora*, 87(851): 309-320. (T)

3395 **Keddy, P. (1987).** Two nationally significant botanical sites in Nova Scotia now protected. *Canad. Bot. Ass. Bull.*, 20(1): 9-10. (P)

3396 **Maher, R.V., White, D.J., Argus, G.W., Keddy, P.A. (1978).** The rare vascular plants of Nova Scotia. *Syllogeus*, 18: 1-38. (Fr). (T)

3397 **Roland, A.E. (1975).** Checklist of the vascular plants of Kejimkujik National Park. Unpublished. 81p. (P)

3398 **Wisheu, I.C., Keddy, P.A. (1989).** The conservation and management of a threatened coastal plain plant community in eastern North America (Nova Scotia, Canada). *Biol. Conserv.*, 48(3): 229-238.

CANADA - Ontario

3399 **Alexander, M.E. (1978).** Reconstructing the fire history of Pukaskwa National Park. *Inform. Rep. NOR-X North. Forest Res. Centre*, 210: 4-11. Map. (P)

3400 **Ambrose, J.D. (1980).** A regional rare woody plant program. *In* Second International Congress of Systematic and Evolutionary Biology, the University of British Columbia, Vancouver, Canada, 17-24 July 1980. p.409. Abstract. (T)

3401 **Ambrose, J.D. (1987).** A rare plant programme for southern Ontario. *In* Bramwell, D., Hamann, O., Heywood, V., Synge, H., *eds.* Botanic gardens and the World Conservation Strategy. London, Academic Press. 341-343. (T)

3402 **Argus, G.W., White, D.J. (1977).** The rare vascular plants of Ontario. *Syllogeus*, 14. (T)

3403 **Argus, G.W., White, D.J. (1983).** Atlas of the rare vascular plants of Ontario. Ottawa, Botany Div., National Museum of Natural Sciences. 74p. (T)

3404 **Argus, G.W., et al., eds. (1982).** Atlas of the rare vascular plants of Ontario. Ottawa, Ontario, National Museum of Natural Sciences. First two parts edited by G.W. Argus and D.J. White (1982,1983), 3rd part by G.W. Argus and C.J. Keddy (1984), 4th and final part by G.W. Argus and K. Pryer (in prep.). Supersedes Argus and White, 1977. (T)

CANADA - Ontario

3405 **Fernald, M.L. (1935).** Critical plants of the upper Great Lakes region of Ontario and Michigan. *Rhodora*, 37: 197-222, 238-262, 272-301, 324-341. (T)

3406 **Hirvonen, R. (1978).** Georgian Bay Islands National Park. Information Report No. FMR-X-117. Ottawa, Ontario, Forest Management Institute. viii, 115p. (Fr). Illus. (P)

3407 **U.S. Department of the Interior. Fish and Wildlife Service (1978).** Endangered species: Great Lakes Region. Twin Cities, MN, Fish and Wildlife Service. (L/T)

3408 **U.S. Department of the Interior. Fish and Wildlife Service (1988).** Endangered and threatened wildlife and plants; determination of Threatened status for *Hymenoxys acaulis* var. *glabra* (Lakeside daisy). *Fed. Register*, 53(121): 23742-23745. (L/T)

3409 **U.S. Department of the Interior. Fish and Wildlife Service (1988).** Endangered and threatened wildlife and plants; determination of Threatened status for *Solidago houghtonii* (Houghton's goldenrod). *Fed. Register*, 53(137): 27134-27137. (L/T)

3410 **U.S. Department of the Interior. Fish and Wildlife Service (1988).** Endangered and threatened wildlife and plants; proposal to determine *Iris lacustris* (Dwarf lake iris) to be a Threatened species. *Fed. Register*, 52(233): 46334-46336. (L/T)

3411 **U.S. Department of the Interior. Fish and Wildlife Service (1989).** Endangered and threatened wildlife and plants; Threatened status for *Phyllitis scolopendrium* var. *americana* (American hart's-tongue). *Fed. Register*, 54(134): 29726-29730. (L/T)

3412 **Zander, R.H. (1976).** Floristics and environmental planning in western New York and adjacent Ontario: distribution of legally protected plants and plant sanctuaries. Buffalo, NY, Buffalo Society of Natural Sciences. 47p. (L/P/T)

CANADA - Quebec

3413 **Anon. (1982).** Pourquoi les parcs nationaux? *Milieu* (Quebec), 25: 6-8. Fr. Col. illus. (P)

3414 **Bouchard, A., Barabe, D., Dumais, M., Hay, S. (1981).** La preservation des plantes rares au Quebec. *Gartn. Bot. Brief*, 68: 53-54. Fr. (T)

3415 **Bouchard, A.D., Barabe, D., Dumais, M., Hay, S. (1983).** The rare vascular plants of Quebec. *Syllogeus*, 48. 79p. En, Fr. (T)

3416 **Brien, Y., Kiendl, E., Le Demezet, M. (1978).** Parcs, reserves et protection de la nature au Quebec. *Penn Bed*, 91: 211-221. Illus. (P)

3417 **Couillard, L., Grondin, P. (1986).** La vegetation des milieux humides du Quebec. (The vegetation of the wetlands of Quebec). Les Publications du Quebec, Gouvernement du Quebec. 377p. Fr. Illus., maps.

3418 **Gauvin, G., Jolicoeur, H. (1979).** Pourquoi constituer un parc de conservation avec les Grands Jardins. Quebec, Direction de la Recherche Faunique. v, 9p. Fr. Illus. map. (P)

3419 **Guillebon, E. de (1978).** L'istaccueil du public en foret: les centres d'interpretation de la nature au Quebec. *Rev. Forest. Franc.*, 30(6): 479-482. Fr. (P)

3420 **Lamoureux, G. (1972).** Inventaire floristique preliminaire du Parc National de la Maurice. Unpubl., 34p. (P)

3421 **Lemieux, G. (1976).** Some problems in the establishment of ecological reserves in Quebec. *In* Canadian Botanical Association. Natural areas: Proceedings of a Symposium at the 13th Annual Meeting of the Canadian Botanical Association, Bishop's University, Lennoxville, Quebec, 6-10 June, 1976. Quebec, Canadian Botanical Association. 22-29. (Fr). Illus. (P)

CANADA - Quebec

3422 **Lemieux, G. (1977).** Les reserves ecologiques dans les contextes canadiens et quebecois. *In* Filion, L., Villeneuve, P., *eds.* Ecological reserves and nature conservation. Rapport, Conseil Consultatif des Reserves Ecologiques, Quebec. 1. 4-8. Papers presented at a conference at Iles-de-la-Madeleine, Quebec, 17-18 June 1976. Fr. (P)

3423 **Mercier, J.C. (1985).** Protection of the forest: a social and economic imperative. *Forest. Chron.*, 61(5): 378-381.

CANADA - Saskatchewan

3424 **Argus, G.W., White, D. (1975).** A preliminary list of the rare plants of Saskatchewan. Ottawa Museum of Natural Science. 14p. (T)

3425 **Harms, V.L. (1977).** Further records of the rare dragon's mouth (swamp-pink) orchid, in Saskatchewan. *Blue Jay*, 35(3): 138-141. Illus. (T)

3426 **Harms, V.L. (1978).** Additional records for some rare or uncommon native orchids in Saskatchewan. *Blue Jay*, 36(3): 130-136. Illus., maps. (T)

3427 **Harms, V.L. (1978).** The white floating marsh marigold, *Caltha natans*, in Saskatchewan. *Blue Jay*, 36(4): 186-188. Illus. Rare plant. (T)

3428 **Hooper, D.F. (1982).** Rare plants found near Hudson Bay, Saskatchewan. *Blue Jay*, 40(2): 69-72. Illus. (T)

3429 **Hudson, J.H. (1977).** Rare and endangered native plant species in Saskatchewan. south of lat. 55. *Blue Jay*, 35(3): 126-137. Illus. (T)

3430 **Lamont, S.M. (1983).** Big Gully Creek Sanctuary - a memorial to those who worked to protect it. *Blue Jay*, 41(2): 67-68. (P)

3431 **Maher, R.V., Argus, G.W., Harms, V.L., Hudson, J.H. (1979).** The rare vascular plants of Saskatchewan. (Les plantes vasculaires rares de la Saskatchewan.) *Syllogeus*, 20: 55-25. (Fr). Maps. (T)

CANADA - Yukon Territory

3432 **Douglas, G.W., Argus, G.W., Dickson, H.L., Brunton, D.F. (1981).** The rare vascular plants of the Yukon. Les plantes vasculaires rares du Yukon. Ottawa, National Museum of Natural Sciences. 61, [35], 64p. *Syllogeus*, 28. (Fr). Maps. Contribution to Unesco Program on Man and the Biosphere. (T)

3433 **Stace-Smith, R., et al. (1980).** Threatened and endangered species and habitats in British Columbia and the Yukon. Victoria, B.C., B.C. Ministry of Environment, Fish and Wildlife Branch. 302p. Proc. of the Symposium co-sponsored by Fed. of B.C. Naturalists Inst. of Environmental Studies, Douglas College, B.C. Min. of Environ., Fish and Wildlife Branch held in Richmond, B.C. March 8 and 9, 1980. (T)

CANADIAN ARCTIC

3434 **Cody, W.J. (1963).** Some rare plants from the Mackenzie District, N.W.T. *Canad. Field-Nat.*, 77: 226-228. (T)

3435 **Cody, W.J. (1979).** Vascular plants of restricted range in the continental Northwest Territories, Canada. *Syllogeus*, 23: 1-57.

3436 **Fuller, W.A., Kevan, P.G., eds (1970).** Productivity and conservation in northern circumpolar lands. Proceedings of a conference, Edmonton, Alberta 15-17 October 1969. IUCN New Series No. 16. Morges, IUCN. 344p. Illus., map.

CAPE VERDE

3437 **Grandvaux Barbosa, L.A. (1968).** L'Archipel du Cap-Vert. *In* Hedberg, I. & O., *eds.* Conservation of vegetation in Africa south of the Sahara. Symposium Proceedings, at 6th plenary meeting of AETFAT, Uppsala. *Acta Phytogeogr. Suec.*, 54: 94-97. Fr. Diagram. Discusses vegetation and protection of plant species on each island. (T)

3438 **Kammer, F. (1979).** The influence of man on the vegetation of Macaronesia. *In* Berichte der Internationalen Symposien der Internationalen Vereinigung fur Vegetatsionskunde Herausgegeben von Reinhold Tuxen, 20-23 March 1978. 601-616. Illus. (T)

3439 **Kammer, F. (1982).** Beitrage zu einer kritischen Interpretation der rezenten und fossilen Gefasspflanzenflora und Wirbeltierfauna der Azoren, des Madeira-Archipels, der Ilhas Selvagens, der Kanarischen Inseln und der Kapverdischen Inseln, mit einem Ausblick auf Artenschwund in Makaronesien. Freiburg im Breisgau. 179p. Ge.

3440 **Malato-Beliz, J. (1983).** Proposition pour l'elaboration d'un catalogue des plantes endemiques, menacees et disparues en Macaronesie. *In* Comunicacoes apresentadas ao 2 Congresso Internacional pro flora Macaronesica, Funchal, 19-25 de Junho de 1977. Funchal. 437-440. Fr. (T)

CENTRAL AFRICAN REPUBLIC

3441 **Barber, K.B., Buchanan, S.A., Galbreath, P.F. (1980).** An ecological survey of the St Floris National Park, Central African Republic. Washington, D.C., U.S. Dept of the Interior. xi, 161p. Maps. (P)

3442 **Buchanan, S.A., Schacht, W.H. (1979).** Ecological investigations in the Manovo-Gounda-St Floris National Park, submitted to the Ministre des Eaux, Forets, Chasses et Peches. Bangui. 38p. Map. (P)

3443 **Fabregues, B.P. de (1981).** The national wildlife reserve of Manovo-Gounda-Saint Floris (C.A.R.). Vegetation and wildlife. *Rev. Elev. Med. Vet. Pays Trop.*, 34(2): 221-?. (P)

3444 **Guigonis, G. (1968).** Republique Centrafricaine. *In* Hedberg, I. & O., *eds.* Conservation of vegetation in Africa south of the Sahara. Symposium Proceedings, at 6th plenary meeting of AETFAT, Uppsala. *Acta Phytogeogr. Suec.*, 54: 107-111. Fr. Map. Describes climate, vegetation, protection of nature, and associations requiring protection. (P/T)

3445 **IUCN (?)** La conservation des ecosystemes forestiers de la Republique centrafricaine. IUCN, Tropical Forest Programme Series. c.200p. In prep.

CHAD

3446 **Gaston, A. (1980).** La vegetation du Tchad (nord-est et sud-est du Lac Tchad): evolutions recentes sous des influences climatiques et humaines. Maisons-Alfort, Institut d'Elevage et de Medecine Veterinaire des Pays Tropicaux. Fr. Colour map 1:1,000,000 covering about a quarter of Chad with unpublished descriptive thesis of 333p. (T)

3447 **Gillet, H. (1968).** Tchad et Sahel Tchadien. *In* Hedberg, I. & O., *eds.* Conservation of vegetation in Africa south of the Sahara. Symposium Proceedings, at 6th plenary meeting of AETFAT, Uppsala. *Acta Phytogeogr. Suec.*, 54: 54-58. Fr. Describes vegetation types and aspects of flora protection, including threatened species. (T)

CHILE

3448 **Anon. (1984).** A tree loses some protection. *Oryx*, 18(2): 113. (T)

CHILE

3449 **Bahre, C.J. (1979).** Destruction of the natural vegetation of north-central Chile. University of California Publications in Geography No. 23. Berkeley, University of California Press. x, 117p. Illus., maps. (T)

3450 **Bruhin, D. (1985).** The two endemic palms of Chile. *Int. Dendrol. Soc. Yearbook,* 1984: 119-122. (T)

3451 **Chile Corporacion Nacional Forestal (1983).** Chile: sus parques nacionales y otras areas naturales. Madrid, Instituto de Cooperacion Iberoamericana and Instituto de la Caza Fotografica y Ciencias de la Naturaleza. 224p. Sp. Col. illus., maps. (P)

3452 **Corporacion Nacional Forestal (1974).** Plan de manejo Parque Nacional Fray Jorge. Santiago, Org. Naciones Unidas para la Agric. y la Aliment. Officina Regional para America Latina. Vegetacion: 5-8. Sp. (P)

3453 **Corporacion Nacional Forestal (1975).** Plan de manejo Parque Nacional Torres del Paine. Santiago, Org. Naciones Unidas para la Agric. y la Aliment. Officina Regional para America Latina. Plants: 9-10. Sp. (P)

3454 **Corporacion Nacional Forestal (1985).** Simposio flora nativa arborea y arbustiva de Chile amenazadas de extincion. Ministerio de Agricultura. Santiago, Chile. 80p. (T)

3455 **Cowley, J. (1985).** Plants in peril, 5. *Kew Mag.,* 2(2): 285-288. *Tecophilaea cyanocrocus.* (T)

3456 **Endlicher, W., Mackel, R. (1985).** Natural resources, land use and degradation in the coastal zone of Arauco and the Nahuelbuta Range, central Chile. *Geojournal,* 11(1): 43-60. Illus., maps.

3457 **Gajardo, M.R., Serra, V. (1979).** [Floristic composition of *Nothofagus* forests in relation to altitude (at Niblinto, Malleco Forest Reserve, Chile).] *Ciencias Forestales,* 1(3): 29-38. Sp. (P)

3458 **Gardner, M.F., Knees, S.G. (1987).** *Fitzroya cupressoides:* the tree and its future. *Int. Dendrol. Soc. Year Book,* 1987: 78-82 (1987 publ. 1988). (T)

3459 **Holdgate, M.W. (1968).** Conservation in the *Nothofagus* forest zone. *In* IUCN. Conservation in Latin America. Proceedings, San Carlos de Bariloche, Argentina, 27 March-2 April 1968. 122-132.

3460 **Macdonald, I.A.W., Graber, D.M., DeBenedetti, S., Groves, R.H., Fuentes, E.R. (1988).** Introduced species in nature reserves in Mediterranean-type climatic regions of the world. *Biol. Conserv.,* 44(1-2): 37-66. Invasive species. Includes 7 case studies. (P)

3461 **Marticorena, C. (1980).** Threatened plants and areas of Chile. Universidad de Concepcion. List of threatened plants of the continent and Islas of Mas a Tierra, Mas Afuera, Santa Clara, San Felix and San Ambrosio. (T)

3462 **Merino Cuevas, R. (1978).** Los parques nacionales y su contribucion al desarrollo economico nacional. *Cienc. Forest.,* 1(1): 13-17. Illus. (P)

3463 **Munoz Pizarro, C. (1967).** La extincion de especies vegetales en Chile. *In* La conservacion de la naturaleza y la prensa en la America Lalina. Mexico, Instituto Mexicano de Recursos Naturales Renovables. 75-82. Sp. (T)

3464 **Munoz Pizarro, C. (1971).** Chile: plantas en extincion. Santiago, Editorial Universitaria. 248p. Sp. Illus., col. illus. (T)

3465 **Munoz Pizarro, C. (1977).** Threatened and endangered species of plants in Chile. *In* Prance, G.T., Elias, T.S., eds. Extinction is forever. Proceedings of a symposium at the New York Botanical Garden, 11-13 May 1976. New York, New York Botanical Garden. 251-253. Plant list. (T)

CHILE

3466 **Munoz Pizarro, C. (1980).** I. Areas naturales: localidas y regiones de Chile dignas de proteccion. II. La extincion de especies vegetales. *In* 2a Jorn. Latinoam. de Parques Nacionales. SAG, Minist. Agric., Via del Mar. 23p. Sp. (P/T)

3467 **Oldfield, S. (1983).** Plants and CITES: *Fitz-Roya* back in trade! *Threatened Pl. Newsl.,* 12: 13. (T)

3468 **Oltremari, J.V., Jackson, R.G. (1985).** Chile's National Parks: present and future. *Parks* 10(2): 1-4. Maps, illus. Table of 34 National Parks. (P)

3469 **Osorio, O.M., Cerda, M.L., Donoso, M.M., Peredo, L.H., Gara, B.R. (1978).** Programa para la proteccion de los bosques nacionales. (Programme for the protection of national forests.) *In* International Union of Forest Research Organizations. Meeting of IUFRO working groups 2.06.12 & 2.07.07. Pests and diseases of tropical pines. Medellin, 3-14 September 1978. Bogota, IUFRO. 13p. Sp. (P)

3470 **Poblete, I.C., ed. (1985).** Flora nativa arborea y arbustiva de Chile amenazada de extincion. Santiago, Ministerio de Agricultura. Siumposio. 80p. Sp. Illus., col. illus. (T)

3471 **Putney, A.D. (1971).** Interpretive master plan: Puyehue National Park, Chile. Univ. of Washington. Checklist of plant species, 130-140. (P)

3472 **Ramirez, C. (1980).** Conservacion de la vegetacion nativa en tierras bajas valdivianas. *Medio Ambiente* (Valdivia), 4(2): 82-89. Sp.

3473 **Ramirez, G.C. (1978).** Estudio floristico y vegetacional del Parque Nacional Tolhuaca. (Floristic and vegetational study of the Tolhuaca National Park.) *Publ. Ocas. Mus. Nac. Hist. Nat. Santiago,* 23. Sp. (P)

3474 **Reid, R. (1988).** "Extinct" cereal rediscovered. *Threatened Pl. Newsl.,* 19: 6. Illus. (E/T)

3475 **Republica de Chile, Corporacion Nacional Forestal (1985).** Actao: simposio flora arborea y arbustiva de Chile amenazada de extincion, 27 al 30 de Agosto de 1985. Santiago, Chile, Corporacion Nacional Forestal. 80p. Sp. Illus., col. illus.

3476 **Schlegel Sachs, F.M. (1982).** Especies Chilenas Amenazadas. Univ. Austral de Chile. List of threatened plants including Ex:9, E:53, V:15, R:42. (T)

3477 **Skottsberg, C. (1961).** The status of conservation in Chile, Juan Fernandez, and Easter Island. *Proc. Eighth Pacific Science Congr.,* 6: 128-131. (P/T)

3478 **Veblen, T.T., Delmastro, R.J., Schlatter, J.E. (1976).** The conservation of *Fitzroya cupressoides* and its environment in southern Chile. *Envir. Conserv.,* 3: 291-301. (T)

CHILE - Easter Island

3479 **Anon. (1984).** Easter enigma solved. *IUCN Bull.,* 15(1-3): 6. Deforestation led to soil erosion and collapse of civilisation. (P/T)

3480 **Browne, M.W. (1987).** New findings reveal ancient abuse of lands. *New York Times, Science Times,* C1-C3. (P)

3481 **Corporacion Nacional Forestal (1976).** Plan de manejo Parque Nacional Rapa Nui. Santiago, Org. Nacional Unidas para la Agric. y la Aliment. Officina Regional para America Latina. Plants: 9-10. Sp. (P)

3482 **Elton, C.S. (1958).** The ecology of invasions by animals and plants. London, Methuen. 181p. Introduced species. (P)

3483 **Godley, E. (1989).** The supposed Easter Island *Sophora* in Cristchurch [*sic*], New Zealand. *Bot. Gard. Conserv. News,* 1(4): 37-38. (G/T)

3484 **Havas, V. (1985).** Next stop, Easter Island. *Islands,* 5(6): 12. Discusses threat of airstrip construction. (P)

CHILE - Easter Island

3485 **Lobin, W., Barthlott, W. (1988).** *Sophora toromiro* (Leguminosae); the lost tree of Easter Island. *Bot. Gard. Conserv. News,* 1(3): 32-34. Illus. (G/P/T)

3486 **Skottsberg, C. (1961).** The status of conservation in Chile, Juan Fernandez, and Easter Island. *Proc. Eighth Pacific Science Congr.,* 6: 128-131. (P/T)

3487 **Skottsberg, C.J.F. (1920).** The natural history of the Juan Fernandez and Easter Island. Uppsala, Almqvist and Wiksell. 3 vols, 1920-1956. (P)

3488 **Weimarck, G. (1984).** Conservation work with *Sophora toromiro,* the tree of Easter Island. *Rep. Bot. Instit. Univ. Aarhus,* 10: 40-42. (G/P/T)

CHILE - Juan Fernandez

3489 **Anon. (1968).** Chile: the Juan Fernandez Islands. *IUCN Bull.,* 2(8): 61. Introduced species threaten native flora. (P)

3490 **Anon. (1987).** *Thyrsopteris elegans. IUCN Bull.,* 18(1-3): 12-13. Discusses threat posed to this and other species by erosion. (P/T)

3491 **Barrau, J. (1960).** The sandalwood tree. *South Pacific Bull.,* 10(4): 39, 63. Recounts the history of decimation; lists species of *Santalum* in Oceania.

3492 **Bruhin, D. (1985).** The two endemic palms of Chile. *Int. Dendrol. Soc. Yearbook,* 1984: 119-122. (T)

3493 **Corporacion Nacional Forestal (1976).** Plan de manejo Parque Nacional Juan Fernandez. Santiago, Org. Nacional Unidas para la Agric. y la Aliment. Officina Regional para America Latina. Plants: 6-10. Sp. (P)

3494 **IUCN (1986).** IUCN's list of top 24 threatened animals and plants. *Biol. Conserv. Newsl.,* 44: 1-6. (T)

3495 **IUCN Threatened Plants Unit (1984).** The botanic gardens list of rare and threatened species of the Galapagos and Juan Fernandez Islands. Kew, IUCN Botanic Gardens Conservation Co-ordinating Body Report No. 11. 25p. Lists plants in cultivation in botanic gardens. (G/P/T)

3496 **Kunkel, G. (1968).** Robinson Crusoe's Island. *Pacific Discovery,* 21(1): 1-8. Illus., map. (T)

3497 **Marticorena, C. (1980).** Threatened plants and areas of Chile. Universidad de Concepcion. List of threatened plants of the continent and Islas of Mas a Tierra, Mas Afuera, Santa Clara, San Felix and San Ambrosio. (T)

3498 **Munoz Pizarro, C. (1969).** El Archipielago de Juan Fernandez y la conservacion de sus recursos naturales renovables. *Bol. Acad. Ciencias Instituto Chile,* 1(2): 83-103. Sp. Illus. (T)

3499 **Nishida, H. & M. (1979).** The vegetation of the Mas a Tierra (Robinson Crusoe) Island, Juan Fernandez. *In* Nishida, M. *ed.* A report of the palaeobotanical survey to southern Chile by a grant-in-aid for overseas scientific survey, 1979. Japan, Chiba University. 41-48. Map. (P/T)

3500 **Perry, R. (1984).** Juan Fernandez Islands: a unique botanical heritage. *Envir. Conserv.,* 11(1): 72-76. Illus. (P/T)

3501 **Sanders, R.W., Stuessy, T.F., Marticorena, C. (1982).** Recent changes in the flora of the Juan Fernandez Islands, Chile. *Taxon,* 31(2): 284-289. Map. (T)

3502 **Skottsberg, C. (1918).** The islands of Juan Fernandez. *Geogr. Rev.,* 5(5): 362-383. Illus., maps. Includes description of vegetation.

3503 **Skottsberg, C. (1961).** The status of conservation in Chile, Juan Fernandez, and Easter Island. *Proc. Eighth Pacific Science Congr.,* 6: 128-131. (P/T)

CHILE - Juan Fernandez

3504 **Skottsberg, C. (1962).** Report of the subcommittee on nature protection. *Proc. Ninth Pacific Science Congr.*, 4: 29-38. Discusses destruction of vegetation on Masatierra, Santa Clara and Masafuera. (P)

3505 **Skottsberg, C.J.F. (1920).** The natural history of the Juan Fernandez and Easter Island. Uppsala, Almqvist and Wiksell. 3 vols, 1920-1956. (P)

3506 **Sparre, B. (1968).** Plants from Robinson Crusoe's island. *Taxon*, 22(1): 171. Over-collecting. (P/T)

3507 **Stuessy, T.F., Sanders, R.W., Matthei, O.R. (1983).** *Juania australis* revisited in the Juan Fernandez Islands, Chile. *Principes*, 27(2): 71-74. Illus., map. (P/T)

3508 **Stuessy, T.F., Sanders, R.W., Silva, M. (1984).** Phytogeography and evolution of the flora of the Juan Fernandez Islands: a progress report. *In* Radovsky, F.J., Raven, P.H., Somer, S.H., *eds.* Biogeography of the tropical Pacific. Honolulu, Bishop Museum Special Publ. No. 72: 55-69. Introduced animals have caused decline in native flora. (P/T)

3509 **Stuessy, T.F., Silva O. & M. (1983).** The evolution of the flora of the Juan Fernandez Islands. *Rep. Chilean Univ. Life*, 15: 3-6. Illus. (P)

CHINA

3510 **Anon. (1980).** China's environmental consciousness. *Unasylva*, 32(129): 35.

3511 **Anon. (1982).** Plant conservation in China. *Threatened Pl. Commit. Newsl.*, 10: 15. (L/T)

3512 **Anon. (1982).** The Ding Hu Shan Forest Ecosystem Reserve. Guangzhou, Popular Science Press. 236p. Ch. Illus., maps. Subtropical forest; tropical forest. (E/P)

3513 **Anon. (1983).** Plant conservation in China. *Oryx*, 17: 142. (L/T)

3514 **Anon. (1987).** Magnolia saved. *Oryx*, 21: 187. Report from *China Daily*, 1 September 1986. Baohua magnolia protected. (T)

3515 **Campbell, J.N.J. (1984).** Giant panda conservation and bamboo forest destruction. *In* Veziroglu, T.N., *ed.* The biosphere: problems and solutions. Amsterdam, Elsevier Science Publisher. 599. Illus.

3516 **Caufield, C. (1984).** A window on China's environment. *New Sci.*, 101(1391): 28-29. Illus.

3517 **Chang, P. (1981).** [Protecting forests and developing forestry]. Pei-Ching, Chung-Kuo Lin Yeh Ch'u Pan She. 46p. Ch. Afforestation.

3518 **Chaoyu, J. (1982).** Crop germplasm resources conservation system in China. *Pl. Genet. Resource. Newsl.*, 51: 2-5. (Fr, Sp). (E)

3519 **Chen, J.Y., Huang, Z.F. (1988).** [A preliminary study of the introduction and preserving of the rare and endangered plants]. *Guihaia*, 8(2): 179-189. Ch (En). (T)

3520 **Chen, S.-P., et al. (1988).** [The mangrove of South China Sea]. *Guihaia*, 8: 215-224. Ch (En). Illus. Distribution, conservation and management.

3521 **Fu, L.G., Chen, S.Z. (1981).** Discovery and designation of *Cathaya argyrophylla*. *Plant Journal (Zhiwu Zazhi)*, 4: 42p. Ch. (T)

3522 **Glaeser, B. (1982).** Okologie und Umweltschutz in der VR China: Eindrucke und Erfahrungsberichte einer Umweltdelegation. (Ecology and environmental protection in the People's Republic of China.) Bochum, West Germany, Studien Verlag Brockmeyer. 341p. Ge. Illus., maps.

CHINA

3523 **Hanxi, Y. (1984).** Research programmes in nature reserves of China. *In* Unesco-UNEP. Conservation, science and society. Paris, Unesco. 351-353. Contributions to the First International Biosphere Reserve Congress, Minsk, Byelorussia/USSR, 26 September - 2 October 1983. (P)

3524 **Hoey-Smith, J.R.P. van (1973).** Two very rare conifers *Microbiota decussata* and *Cathaya argyrophylla*. *IDS Year Book 1972*: 50-54. Illus. (T)

3525 **Hsiung, W. (1983).** Replenishing the world's forests: forestry progress in China. *Commonw. Forest. Rev.*, 62(3): 191-193. (Fr, Sp).

3526 **IUCN (1986).** IUCN's list of top 24 threatened animals and plants. *Biol. Conserv. Newsl.*, 44: 1-6. (T)

3527 **Jiang, C.Y. (1982).** Crop germplasm resources conservation system in China. *Pl. Genet. Resourc. Newsl.*, 51: 2-5. Crop. (E)

3528 **Lan, H. (1980).** A new great wall for China. *Canopy*, 6(9): 12-13. Illus.

3529 **Lewis, W. (1979).** Ginseng of northeastern China. *In* Hensley, D.L., Alexander, S., Roberts, C.R., *eds.* Proceedings of the First National Ginseng Conference. Kentucky, Governor's Council on Agriculture. 69-77. Illus., maps. *Panax ginseng, Acanthopanax santicosus*. (E/T)

3530 **Lianhe, D., Xincheng, Z. (1982).** [Suggestion to protect the community composed of rare tree species of Houhe Forest Area in Wufeng County, Hubei Province]. *Acta Phytoecol. Geobot. Sin.*, 6(1): 84-85. Ch. Illus. 15 species mentioned. (T)

3531 **Liu, J.-Y. (1980).** [Investigation of the vegetation and landscape of Chushan bamboo forest recreation area]. *Quart. J. Chinese Forest.*, 13(4): 129-154. Ch (En). Illus., maps. (P)

3532 **Luk, S.-H. (1983).** Recent trends of desertification in the Maowusu Desert, China. *Envir. Conserv.*, 10(3): 213-224. Illus., maps.

3533 **Ma, X., Yang, Z., Xu, Z., Tao, G. (1988).** [A study of the causes threatening *Pellacalyx yunnanensis*. A species receiving priority protection in China]. *Acta Bot. Yunnanica*, 10(3): 311-316. Ch (En). Illus. (T)

3534 **MacKinnon, J. & K. (1986).** Review of the protected areas system in the Indo-Malayan realm. Gland, Switzerland and Cambridge, U.K., IUCN and UNEP. 284p. Illus., maps. (P)

3535 **National Environment Protection Bureau, Botanical Institute of Chinese Academy of Sciences (1987).** [The list of rare and endangered plants protected in China]. Beijing, Academy Press. 96p. Ch. (T)

3536 **Seidensticker, J., Eisenberg, J.F., Simons, R. (1984).** The Tangjiake, Wanghang and Fengtongzhai giant panda reserves and biological conservation in the People's Republic of China. *Biol. Conserv.*, 28(3): 217-251. Illus., maps. (P)

3537 **Smil, V. (1983).** Deforestation in China. *Ambio*, 12(5): 226-231. Illus., col. illus., map.

3538 **Smil, V. (1984).** Eco-troubles in the People's Republic. *Garden* (New York), 8(1): 12-23. Illus., col. illus., maps.

3539 **Song, K.S. (1985).** Sishionbana nature conservation area: the Indomalayan area of China. *In* Thorsell, J.W., *ed.* Conserving Asia's natural heritage. The planning and management of protected areas in the Indomalayan Realm. Proc. of the 25th working session of IUCN's CNPPA. Gland, IUCN. 32-35. Map. (P)

3540 **Thorhaug, A. (1978).** Conservation concern in the People's Republic of China. *Envir. Conserv.*, 5(4): 245-246.

3541 **UNEP (1977).** Conservation measures in China. UNEP booklet series No. 1. Nairobi, UNEP. 38p.

199

CHINA

3542 **WWF (1980).** Major conservation research centre for China. *Envir. Conserv.*, 7(3): 177-178.

3543 **Wang Xianpu (1986).** Protected natural areas in China. *Arnoldia*, 46(4): 38-45. (P)

3544 **Wang, C.-H. (1980).** Recent researches on relationship between air pollution and plants in China. Nanjing, Jiangsu, Institute of Botany. 14p.

3545 **Wang, H.-P. (1980).** Nature conservation in China: the present situation. *Parks*, 5(1): 1-10. Illus.

3546 **Wang, H.-P., Jin, X. (1986).** Plant conservation in China. *Species, Newsl. SSC*, 7: 5-6. (T)

3547 **Wang, H.-P., Xiaobai, J. (1986).** Beijing Botanic Garden - its contribution to conservation. *Species*, 7: 29-30. (G)

3548 **Wang, M.-J., Wu, S.-P., Chen, Y.-H., Shen, J.-Y. (1986).** Introduction experiments on the rare and endangered tree species from central and north subtropical area of eastern China. *Bull. Nanjing Bot. Gard.*, 1986: 39-48. Ch (En). (G/T)

3549 **Wang, X. (1986).** Protected natural areas in China. *Arnoldia*, 46(4): 38-45. (P)

3550 **Wang, X., Jin, X., Sun, C. (1986).** *Burretiodendron hsienmu* Chun & How: its ecology and its protection. *Arnoldia*, 45(4): 46-51. Illus., map. (T)

3551 **Wang, X.-P. (1986).** (On the threatened protected areas and their relieving countermeasure.) *Guihaia*, 6(1-2): 141-145. Ch (En). (P)

3552 **Wang, Y. (1987).** Nature conservation regions in China. *Ambio*, 16(6): 326-331. Col. illus., maps. (P/T)

3553 **Wang, Z.H., He, D.Q., Song, S.D., Chen, S.P. & D.R., Tu, M.Z. (1982).** The vegetation of Ding Hu Shan Biosphere Reserve. *In* Anon. The Ding Hu Shan Forest Ecosystem Reserve. Guangzhou, Popular Science Press. 77-141. Ch. Illus., map. Tropical; subtropical forest; coniferous forest; grasslands. (P)

3554 **Xiaogui, S. (1982).** [Suggestion on the establishment of natural reserves in Shandong Province]. *Acta Phytoecol. Geobot. Sin.*, 6(1): 85-86. Ch. (P)

3555 **Xiyang, T. (1988).** Living treasures: an odyssey through China's extraordinary nature reserves. Bantam Books. 208p. (P)

3556 **Xu, Z., Tao, G. (1987).** [Discussion on the method of systematic assessment to regional threatened plants and their prior conservation.] *Acta Bot. Yunnanica*, 9(2): 193-202. Ch (En). (T)

3557 **Yu, M. (1986).** [Some aspects of botanical studies on environment protection in China]. *Acta Bot. Austro Sinica*, 2: 193-198. Ch (En).

3558 **Zhao, D., Sun, B. (1986).** Air pollution and acid rain in China. *Ambio*, 15(1): 2-5.

3559 **Zhu, J., Reng, R., Jiang, Z. (1983).** (Base our forestry on ecological-economic principles and foster strategical thinking of simultaneous protection of environment and exploitation of wood: more about the new strategy of forestry development in our country). *J. Nanjing Techn. Coll. Forest Prod.*, no. 2: 24-31. Ch (En).

CHINA - Anhui Province

3560 **He Shan-an, Yang Z.-B., Wang, M.-J., Zhong, S.-X., Shen, J.-Y., Tao, J.-C. (1987).** Investigation and introduction of some rare and endangered species in Nanjing Botanical Garden, Mem. Sun Yat-Sen. *In* Bramwell, D., Hamann, O., Heywood, V., Synge, H., *eds*. Botanic gardens and the World Conservation Strategy. London, Academic Press. 255-260. (Sp). Includes list of rare and endangered taxa with distributions. (G/T)

CHINA - Fujian Province

3561 **Wang, X., Cao, K. (1987).** Endangered camellia in China. *Species*, 9: 24. (P/T)

CHINA - Guangdong - Hainan Island

3562 **Chen, S., et al. (1986).** [Mangrove and its reservation in Hainan Island]. *Ecol. Sci.*, 1986/1: 12-19. Ch (En). Illus.

CHINA - Guangxi Autonomous Region

3563 **Chen, Y. (1985).** [An endemic and endangered genus *Heteroplexis* Chang from Guangxi]. *Guihaia*, 5(4): 337-343. Ch (En). Illus., map. Key. (T)

3564 **Wang, X.P. (1983).** [Discussion on the management and administration of the protective area, Da Ming Shan, Wu-Ming County, Guangxi in the sense and role of national park.] *Guihaia*, 3(2): 145-151. Ch. Map. (P)

CHINA - Hubei Province

3565 **Bartholemew, B., Boufford,, D.E., Spongberg, S.A. (1983).** *Metasequoia glyptostroboides* - its present status in central China. *J. Arnold Arbor.*, 64(1): 105-128. Illus. (T)

3566 **Bartholemew, B., et al. (1983).** The 1980 Sino-American botanical expedition to western Hubei Province, People's Republic of China. *J. Arnold Arbor.*, 64(1): 1-103. Illus. Deforestation.

3567 **Wang, S., et al. (1988).** [The present state of preservation of the precious, rare and threatened plants in Hubei, as well as the proposition for studying further these plants]. *J. Wuhan Bot. Res.*, 6(3): 285-298. Ch (En). (T)

CHINA - Hunan Province

3568 **Guo, C., Ma, P. (1989).** Nanyue Arboretum, Hunan Province, China. *Bot. Gard. Conserv. News*, 1(4): 17-19. Includes list of rare species in cultivation. (G/T)

CHINA - Jiangsu Province

3569 **He Shan-an, Yang Z.-B., Wang, M.-J., Zhong, S.-X., Shen, J.-Y., Tao, J.-C. (1987).** Investigation and introduction of some rare and endangered species in Nanjing Botanical Garden, Mem. Sun Yat-Sen. *In* Bramwell, D., Hamann, O., Heywood, V., Synge, H., *eds.* Botanic gardens and the World Conservation Strategy. London, Academic Press. 255-260. (Sp). Includes list of rare and endangered taxa with distributions. (G/T)

3570 **Liu, F.-X., Huang, Z.-Y. (1981).** [The threatened precious vegetation resources in Jiangsu province: the Nanmus and mixed Nanmus forests]. *J. Nanjing Techn. Coll. Forest Prod.*, 4: 90-96. Ch (En). (T)

3571 **Yang, Z.-B., Huang, Z.-Y., Zong, S.-X. (1985).** [Synopsis on the vegetation of Yuntaishan Natural Protected Areas, Jiangsu Province.] *Bull. Nanjing Bot. Gard. Mem. Sun Yat Sen*, 1983: 130-132. (1983 publ. 1985). Ch. (P)

CHINA - Jiangxi Province

3572 **Wang, X., Cao, K. (1987).** Endangered camellia in China. *Species*, 9: 24. (P/T)

CHINA - Jilin Province

3573 **Harrison, M. (1984).** The Changbai Shan Nature Reserve. *Herbal Rev.*, 9(4): 19-21. Illus. (P)

CHINA - Shaanxi Province

3574 **Wu Zhenhgi, Fu Qing (1988).** Second batch precious rare and endangered plants of China from Shaanxi. *Shaanxi Forest Sci. Tech.*, 2: 19-22. Ch. (T)

CHINA - Shandong Province

3575 **Wang, X., Cao, K. (1987).** Endangered camellia in China. *Species*, 9: 24. (P/T)

CHINA - Sichuan Province

3576 **Anon. (1982).** Recent news from China. Amazing primeval forest is a "living laboratory". *Mountain Res. Devel.*, 2(2): 228-229. (P)

3577 **Tangjun, T., ed. (1984).** The rare plants and flowers of western Sichuan. Chengdu, Sichuan, Academia Sinica. 108p. Col. illus. (T)

3578 **Wang Xianpu, Fu Yenfeng (1987).** Establishment of rare plant centre for Sichuan, China. *Bot. Gard. Conserv. News*, 1(1): 14. (G)

CHINA - Yunnan Province

3579 **Cribb, P.J. (1984).** The golden slipper orchid of Yunnan. *Garden* (London), 109(9): 352-353. Col. illus. (T)

3580 **WWF (1986).** Protecting the last tropical forests in China. *WWF Monthly Report*, March 1986: 65-67. Illus., map. Covers Xishuangbanna protected areas. (P)

3581 **Xu, Z. (1987).** The work of Xishuangbanna Tropical Tropical Botanical Garden in conserving plants of the Yunnan tropics. *In* Bramwell, D., Hamann, O., Heywood, V.H., Synge, H., *eds*. Botanic gardens and the World Conservation Strategy. London, Academic Press. 239-253. (Sp). Map. (G/P/T)

3582 **Zou, S.Q. (1988).** The vulnerable and endangered plants of Xishuangbanna Prefecture, Yunnan Province, China. *Arnoldia*, 48(2): 2-7. Illus. (T)

CHINA - Zhejiang Province

3583 **He Shan-an, Yang Z.-B., Wang, M.-J., Zhong, S.-X., Shen, J.-Y., Tao, J.-C. (1987).** Investigation and introduction of some rare and endangered species in Nanjing Botanical Garden, Mem. Sun Yat-Sen. *In* Bramwell, D., Hamann, O., Heywood, V., Synge, H., *eds*. Botanic gardens and the World Conservation Strategy. London, Academic Press. 255-260. (Sp). Includes list of rare and endangered taxa with distributions. (G/T)

3584 **Wang, X., Cao, K. (1987).** Endangered camellia in China. *Species*, 9: 24. (P/T)

CHRISTMAS ISLAND

3585 **Gare, N.C. (1979).** Conservation on Christmas Island, Indian Ocean. *Tigerpaper*, 6(2-3): 38-40. Illus. (P)

3586 **Hicks, J. (1982).** Christmas Island: a new Australian national park. *Parks*, 7(1): 1-4. Illus., map. (P)

3587 **Mitchell, B.A. (1974).** The forest flora of Christmas Island. *Commonw. Forest. Rev.*, 53(1): 19-29. Illus. (P/T)

3588 **Powell, D. & Covacevich, J. (1983).** Lister's Palm, *Arenga listeri*, on Christmas Island: a rare or vulnerable species? *Principes*, 27(2): 89-93. Illus. (T)

COLOMBIA

3589 **Anon. (1984).** Colombian marsh polluted. *Oryx*, 18(2): 112. San Silvestre marsh. World Environmental Report, 30 October 1983. (P)

COLOMBIA

3590 **Blackstone, J. (1979).** Vanishing forest. *Vole*, 2(8): 7-8.

3591 **Boh, L. (1984).** Conservation and ethnobotany in Colombia. *Threatened Pl. Newsl.*, 13: 5. (E/G)

3592 **Bunyard, P. (1989).** Guardians of the forest: indigenist policies in the Colombian Amazon. *The Ecologist*, 19(6): 255-258. Illus., map.

3593 **Denevan, W.M. (1973).** Development and the imminent demise of the Amazon rain forest. *Profess. Geogr.*, 25: 130-135.

3594 **Fernandez-Perez, A. (1977).** The preparation of the endangered species list of Colombia. *In* Prance, G.T., Elias, T.S., *eds.* Extinction is forever. Proceedings of a symposium at the New York Botanical Garden, 11-13 May 1976. New York, New York Botanical Garden. 117-127. (T)

3595 **Gentry, A.H. (1982).** Phytogeographic patterns as evidence for a Choco refuge. *In* Prance, G.T., *ed.* Biological diversification in the tropics. New York, Columbia Univ. Press. 112-136. Centre of plant diversity and endemism.

3596 **Gentry, A.H. (1986).** Species richness and floristic composition of Choco region plant communities. *Caldesia*, 15: 71-91. Centre of plant diversity and endemism.

3597 **Goodland, R.J., Irwin, H.S. (1977).** Amazonian forest and cerrado: development and environmental conservation. *In* Prance, G.T., Elias, T.S., *eds.* Extinction is forever. Proceedings of a symposium at the New York Botanical Garden, 11-13 May 1976. New York, New York Botanical Garden. 214-233. Maps, graphs. (P/T)

3598 **Herrera-MacBryde, O. (1988).** New Colombian parks. *Threatened Pl. Newsl.*, 20: 5-6. Map. (P)

3599 **McMahan, L. (1982).** Colombian plant import stats staggering. *TRAFFIC (U.S.A.) Newsl.*, 4(3 & 4): 9. Trade. (L)

3600 **Meganck, R.A. (1978).** Battle for a Colombian park. *Oryx*, 14(4): 352-358. Illus., maps. (P)

3601 **Norstog, K.J. (1985).** Exploring for *Zamia wallisii*. *Bull. Fairchild Trop. Gard.*, 40(1): 6-17. Illus. (T)

3602 **Ospina, M. (1969).** Colombian orchids and their conservation. *In* Corrigan, M.J., *ed.* Proceedings of the 6th World Orchid Conference, Sydney, 1971. 95-98. (T)

3603 **Prance, G.T. (1977).** The phytogeographic subdivisions of Amazonia and their influence on the selection of biological reserves. *In* Prance, G.T., Elias, T.S., *eds.* Extinction is forever. Proceedings of a symposium at the New York Botanical Garden, 11-13 May 1976. New York, New York Botanical Garden. 195-213. Maps. (P)

3604 **Prance, G.T. (1979).** La Amazonia Colombiana: la oportunidad para un programa racional de investigacion, conservacion y aprovechamento. *In* Seminario sobre los recursos naturales renovables y el desarollo regional Amazonico. Bogota, Instituto Geografico Agustin Codazzi. 153-169.

3605 **Quaintance, C.W. (1971).** The vanishing flora of Colombia. *Biol. Conserv.*, 3: 145-147. (T)

3606 **Salati, E., Vose, P.B.F. (1984).** Amazon basin: a system in equilibrium. *Science*, 225(4658): 129-138. Amazonia.

3607 **Terborgh, J., Winter, B. (1983).** A method for siting parks and reserves with special reference to Colombia and Ecuador. *Biol. Conserv.*, 27(1): 45-58. Maps. (P)

COMOROS

3608 **Baumer, M. (1978).** La conservation et la valorisation des resources ecologiques dans les iles des Comores, de Maurice, de la Reunion, des Seychelles. Paris, Agence de Cooperation Culturelle et Technique. 92p. Fr.

CONGO

3609 **Farron, C. (1968).** Congo-Brazzaville. *In* Hedberg, I. & O., *eds.* Conservation of vegetation in Africa south of the Sahara. Symposium Proceedings, at 6th plenary meeting of AETFAT, Uppsala. *Acta Phytogeogr. Suec.*, 54: 112-115. Fr. Map. (P/T)

3610 **Germain, R. (1968).** Congo-Kinshasa. *In* Hedberg, I. & O., *eds.* Conservation of vegetation in Africa south of the Sahara. Symposium Proceedings, at 6th plenary meeting of AETFAT, Uppsala. *Acta Phytogeogr. Suec.*, 54: 121-125. Fr. Discusses the different biotopes within each province. (P/T)

3611 **IUCN (?)** La conservation des ecosystemes forestiers du Congo. IUCN, Tropical Forest Programme Series. c.200p. In prep.

COOK ISLANDS

3612 **Merlin, M.M. (1985).** Woody vegetation in the upland region of Rarotonga, Cook Islands. *Pac. Sci.*, 39(1): 81-99.

COSTA RICA

3613 **Allen, W.H. (1988).** Biocultural restoration of a tropical forest. *BioScience*, 38(3): 156-161. (P)

3614 **Anon. (1981).** Costa Rica takes the lead in national parks. *IUCN Bull.*, 12(5-6): 32-33. Illus. (P)

3615 **Anon. (1988).** Costa Rica swaps debt for national park. *Oryx*, 22(1): 58. (P)

3616 **Atwood, J.T. (1987).** The vascular flora of La Selva Biological Station, Costa Rica -Orchidaceae. *Selbyana*, 10(1): 76-145. (P)

3617 **Bates, D.M. (1987).** The Robert and Catherine Wilson Botanical Garden at Las Cruces, Costa Rica. *Bot. Gard. Conserv. News*, 1(1): 22-28. Illus. (G)

3618 **Beebe, S. (1984).** A model for conservation. *TNC News*, 34: 4-7.

3619 **Boaz, M.A., Mendoza, R. (1981).** The national parks of Costa Rica. Madrid, Incafo SA. 310p. Col. illus., maps. (P)

3620 **Boza, M.A. (1974).** Costa Rica: a case study of strategy in the setting up of national parks in a developing country. *In* IUCN. Second World Conference on National Parks, Yellowstone and Grand Teton National Parks, U.S.A., 18-27 September 1972. 183-192. Map. (P)

3621 **Boza, M.A., Bonilla, A. (1978).** Los parques nacionales de Costa Rica. Madrid, INCAFO & CIC. 224p. Sp. Coleccion "La naturaleza en Iberoamerica". Col. illus., map. (P)

3622 **Boza, M.A., Lemieux, G.H. (1978).** Les parcs nationaux du Costa Rica; bilan strategique actuel et amenagement futur. *Rev. Geogr. Inst. Pan-Amer. Geogr. Hist.* (Mexico), 86/87: 431-357. Sp. (P)

3623 **Carr, A. (1982).** Tropical forest conservation and estuarine ecology. *Biol. Conserv.*, 23(4): 247-259. Maps. Tortuguero, Costa Rica.

3624 **Caufield, C. (1982).** Costa Rica's national parks survive threats. *New Sci.*, 96(1328): 144. Maps. (L/P)

3625 **Cherfas, J. (1986).** How to grow a tropical forest. *New Sci.*, 112(153): 26-27. (P)

COSTA RICA

3626 **Cherfas, J. (1987).** Costa Rica swaps its debt for forests in Guanacaste. *New Sci.*, 115(1572): 22. (P)

3627 **Cifuentes, M., Craig, M., Morales, R. (1984).** Strategic planning of national or regional systems of biosphere reserves: a methodology and case study from Costa Rica. *In* Unesco-UNEP. Conservation, science and society. Paris, Unesco. 93-120. Contributions to the First International Biosphere Reserve Congress, Minsk, Byelorussia/USSR, 26 September - 2 October 1983. Tables, maps. (P)

3628 **D'Arcy, W.G. (1977).** Endangered landscapes in Panama and Central America: the threat to plant species. *In* Prance, G.T., Elias, T.S., *eds.* Extinction is forever. Proceedings of a symposium at the New York Botanical Garden, 11-13 May 1976. New York, New York Botanical Garden. 89-104. Maps. The flora and deforestation today. (P/T)

3629 **Gomez, L.D., ed. (1986).** Vegetacion de Costa Rica. Apuntes para una biogeografia Costarricense. San Jose, Editorial Universidad Estatal a Distancia. 327p., 118p. 2 vols.

3630 **Gonzales, R. (1979).** Establecimiento y desarrollo de reservas forestales en Costa Rica. (Designation and development of forest reserves in Costa Rica.) *Agron. Costarricense*, 3(2): 161-166. Sp (En). (P)

3631 **Grayum, M., Nevers, G. de (1988).** New and rare understory palms from the Peninsula de Osa, Costa Rica, and adjacent regions. *Principes*, 32(3): 101-114. (T)

3632 **Hammel, B.E. (1986).** The vascular flora of La Selva Biological Station, Costa Rica - Cecropiaceae. *Selbyana*, 9(1): 192-195. (P)

3633 **Hammel, B.E. (1986).** The vascular flora of La Selva Biological Station, Costa Rica - Cyclanthaceae. *Selbyana*, 9(1): 196-202. (P)

3634 **Hammel, B.E. (1986).** The vascular flora of La Selva Biological Station, Costa Rica - Lauraceae. *Selbyana*, 9(1): 218-233. (P)

3635 **Hammel, B.E. (1986).** The vascular flora of La Selva Biological Station, Costa Rica - Marantaceae. *Selbyana*, 9(1): 234-242. (P)

3636 **Hammel, B.E. (1986).** The vascular flora of La Selva Biological Station, Costa Rica - Moraceae. *Selbyana*, 9(1): 243-259. (P)

3637 **IUCN (1986).** IUCN's list of top 24 threatened animals and plants. *Biol. Conserv. Newsl.*, 44: 1-6. (T)

3638 **Janzen, D.H. (1983).** No park is an island: increase in interference from outside as park size decreases. *Oikos*, 41(3): 402-410. (Rus). (P)

3639 **Janzen, D.H. (1986).** Guanacaste National Park. *Parks*, 11(4): 3-12. Illus., maps. (P)

3640 **Janzen, D.H. (1986).** Guanacaste National Park: tropical, ecological and cultural restoration. San Jose, Costa Rica, Editorial Universidad Estatal a Distancia. 103p. Illus., maps. (P)

3641 **Lamlein, J. (1982).** Plant conservation in tropical America: reserve in Costa Rica expanded. *Threatened Pl. Commit. Newsl.*, 9: 7. (P)

3642 **Lawton, R., Dryer, V. (1980).** The vegetation of the Monteverde Cloud Forest Reserve. *Brenesia*, 18: 101-116. (Sp). (P)

3643 **MacFarland, C., Morales, R., Barborak, J.R. (1984).** Establishment, planning and implementation of a national wildlands system in Costa Rica. *In* McNeely, J.A., Miller, K.R., *eds.* National parks, conservation and development. The role of protected areas in sustaining society. Washington, D.C., Smithsonian Institution Press. 592-600. Map. (P)

COSTA RICA

3644 **Mathieu, P. (1978).** Ecosistema, vegetacion y problematica humana en el medio tropical; e caso de Costa Rica. (Ecosystem, vegetation and the human problem in the tropics; the case of Costa Rica.) *Rev. Geogr. Inst. Pan-Amer. Geogr. Hist.* (Mexico), 86/87: 275-305. Sp. Illus., map. Destruction & conservation of forests.

3645 **Morales, R., Barborak, J.R., MacFarland, C. (1984).** Planning and managing a multi-component biosphere reserve, the case of the La Amistad/Talamanca Range/Bocas de Toro Wildlands Complex of Cosa Rica and Panama. *In* Unesco-UNEP. Conservation, science and society. Paris, Unesco. 168-177. Contributions to the First International Biosphere Reserve Congress, Minsk, Byelorussia/USSR, 26 September - 2 October, 1983. Map. (P)

3646 **Murphy, J. (1986).** Growing a forest from scratch. *Time,* 128(26): 65. (P)

3647 **Omang, J. (1987).** In the tropics, still rolling back the rain forest primeval. *Smithsonian,* 17(12): 56-67. Deforestation.

3648 **Putney, A.D. (1974).** Interpretative plan Poas Volcano National Park (Costa Rica). San Jose, National Parks Foundation and FAO. 57-68. Tech. Working Doc. no. 12, FAO Project RLAT/TF 199. (P)

3649 **Sader, S.A., Joyce, A.T. (1988).** Deforestation rates and trends in Costa Rica, 1940 to 1983. *Biotropica,* 20(1): 11-19. Maps. Deforestation.

3650 **Silberman, M. (1981).** Costa Rica parks planning: chaos gives way to order. *IUCN Bull.,* 12(7-9): 50-51. Illus. (P)

3651 **Simons, P. (1988).** Costa Rica's forests are reborn. *New Sci.,* 120(1635): 43-47. Col. illus., maps. Tropical forest. (P)

3652 **Torres, H. (1987).** La Amistad Biosphere Reserve. Conservation in the name of friendship. IUCN Special Report: Central America. *IUCN Bull.,* 18(10-12): 6. Illus. (P)

3653 **Torres, H., Hurtado, L., eds (1987).** Parque international de La Amistad - plan general de manejo y desarrollo. Madrid, Instituto de Cooperacion Iberoamericano. (P)

3654 **Torres, H., Hurtado, L., eds (1987).** Reserva la Biosfera de La Amistad: una estrategia para su conservacion y desarrollo. Turrialba, Costa Rica, Tropical Agricultural Research and Training Center. (P)

3655 **U.S. Department of the Interior. Fish and Wildlife Service (1983).** Endangered and threatened wildlife and plants: proposed Endangered status for *Jatropha costaricensis* (Quemador del Pacifico). *Fed. Register,* 48(137): 32525-32526. (L/T)

3656 **U.S. Department of the Interior. Fish and Wildlife Service (1984).** Endangered and threatened wildlife and plants: determination of Endangered status for *Jatropha costaricensis* (Costa Rican jatropha). *Fed. Register,* 49(146): 30199-30201. (L/T)

3657 **U.S. Department of the Interior. Fish and Wildlife Service (1984).** Protection becomes final for four plants. *Endangered Species Tech. Bull.,* 9(8): 3-5. Illus. (L/T)

3658 **Villa-Lobos, J. (1988).** Debt swaps. *Threatened Pl. Newsl.,* 19: 4-5. Conservation organizations assume part of nation's debt in return for land protection. (P)

3659 **Wilbur, R.L. (1986).** The vascular flora of La Selva Biological Station, Costa Rica - introduction. *Selbyana,* 9(1): 191. (P)

3660 **Woolliams, K.R. (1985).** Endangered *Heliconia:* how serious a problem? *Notes Waimea Arbor.,* 12(1): 5-8. (T)

COTE D'IVOIRE

3661 **Adjanohoun, E., Ake Assi, L., Guillaumet, J.L. (1968).** La Cote d'Ivoire. *In* Hedberg, I. & O., *eds.* Conservation of vegetation in Africa south of the Sahara. Symposium Proceedings, at 6th plenary meeting of AETFAT, Uppsala. *Acta Phytogeogr. Suec.*, 54: 76-81. Fr. Map, illus. Describes vegetation, outlines protected plant sites, recommends additional sites for protection. (P)

3662 **Ake Assi, L. (1988).** Especes rares et en voie d'extinction de la flore de la Cote d'Ivoire. *In* Goldblatt, P., Lowry, P.P., *eds.* Modern systematic studies in African botany. Proc. Eleventh Plenary Meeting, AETFAT, Missouri Bot. Gard., 10-14 June 1985. Missouri, Missouri Bot. Garden. 461-463. Map. (T)

3663 **Anon. (1978).** Ecologie. *In* Janvier, G., Person, G., *eds.* Bibliographie de la Cote d'Ivoire: Sciences de la Terre, Sciences de la Vie (1970-1976). 114-121. Annales de L'Universite d'Abidjan, vol. hors. ser. (B)

3664 **Anon. (1981).** SOS: save the Tai. *IUCN Bull.*, 12(3-4): 10-11. Illus. (P)

3665 **Bousquet, B. (1978).** Un parc de foret dense en Afrique: le Parc National de Tai (Cote-d'Ivoire). (A park in the African high forest: the Tai National Park [Ivory Coast].) *Bois Forets Trop.*, 179: 27-46. Fr (En, Sp). Illus. (P)

3666 **Bousquet, B. (1978).** Un parc de foret dense en Afrique; le Park National de Tai: 2. (A park in the African high forest: the Tai National Park [Ivory Coast]: 2.) *Bois Forets Trop.*, 180: 23-37. Fr (En, Sp). Illus. (P)

3667 **Dosso, H., Guillaumet, J.L., Hadley, M. (1981).** The Tai project: land use problems in a tropical rain forest. *Ambio*, 10(2-3): 120-125. Illus., maps. (P)

3668 **Lamotte, M. (1983).** The undermining of Mount Nimba. *Ambio*, 12(3-4): 174-179. Illus., col.illus., maps. (P)

3669 **Lanly, J.P. (1969).** Regression de la foret dense en Cote d'Ivoire. *Bois Forets Trop.*, 127: 45-59. Fr. Tropical forest.

3670 **Rahm, U. (1973).** Propositions pour la creation du Parc National Ivoirien de Thai (TAI). IUCN Occasional Papers No. 3. Switzerland, IUCN. 47p. Fr. Illus., map. (P)

3671 **Reynolds, D.W. (1981).** Comoe National Park: an investigation of the ecology of the northeastern region and recommendations for its future development. Ivory Coast, Minister of Water & Forests. 55p. Maps. (P)

3672 **Roth, H.H. (1984).** We all want the trees: resource conflict in the Tai National Park, Ivory Coast. *In* McNeely, J.A., Miller, K.R., *eds.* National parks, conservation and development. The role of protected areas in sustaining society. Washington, D.C., Smithsonian Institution Press. 127-129. Table. (P)

CUBA

3673 **Anon. (1984).** Castro's *Erythrina. Threatened Pl. Newsl.*, 13: 12-13. (G/T)

3674 **Borhidi, A., Muniz, O. (1983).** Catalogo de plantas Cubanas amenazadas o extinguidas. La Habana, Acad. Ciencias de Cuba. 85p. Sp. Lists 959 species of gymnosperms and flowering plants threatened or extinct, including 832 endemics, with their distribution by provinces and assignment into categories -noncompatible with IUCN categories. (T)

3675 **Eremeev, A.G. (1978).** Klassifikatsiya lesov Respubliki Kuba po narodnokhozyaistvennomu znacheniyu. (Classification of the forests of Cuba according to economic importance.) *Lesnoe Khoz.*, 2: 93-95. Rus. Includes protection, national parks, nature reserves. (P)

3676 **IUCN Botanic Gardens Conservation Secretariat (in co-operation with the Threatened Plants Unit of WCMC) (1989).** Rare and threatened plants of Cuba: *ex situ* conservation in botanic gardens. Kew, IUCN Botanic Gardens Conservation Secretariat. 37p. (G/T)

CUBA

3677 **Jhones, C.C., Oliver, P.H., Iniguez, L. (1988).** Aspectos botanicos y valores para la conservacion de la naturaleza de la llanura costera del norte de la Provincia de Las tunas, Cuba. *Acta Botanica Cubana*, 61: 26p. Sp. Includes species checklist. (P)

3678 **Leiva, A. (1988).** The National Botanic Garden of Cuba. *Bot. Gard. Conserv. News*, 1(3): 20-24. Describes some of the rare plant collections. (G/T)

3679 **Risco Rodriguez, E. del (1982).** La conservacion de la naturaleza y los jardines botanicos. *Rev. Jard. Bot. Nacion. Univ. Habana*, 3(1): 167-195. Sp (En). (G)

3680 **Samek, V. (1968).** La proteccion de la naturaleza en Cuba. *Ser. Transform. Natur.*, Acad. Cienc. Cuba, 7: 1-23.

3681 **Samek, V., Risco, E. del (1983).** Ochrana prirody na Kube. *Ziva*, 31(6): 205-206. Cz. Illus.

3682 **U.S. Department of the Interior. Fish and Wildlife Service (1983).** Endangered and threatened wildlife and plants: proposed Endangered status for *Cereus robinii* (Key tree cactus). *Fed. Register*, 48(147): 34483-34486. (L/T)

3683 **U.S. Department of the Interior. Fish and Wildlife Service (1984).** Endangered and threatened wildlife and plants: final rule to determine *Cereus robinii* (Key tree-cactus) to be an Endangered species. *Fed. Register*, 49(140): 29234-29237. (L/T)

CYPRUS

3684 **Anon. (1984).** Cyprus moves to protect wetland. *Oryx*, 18(2): 107. (P)

3685 **IUCN Threatened Plants Committee Secretariat (1980).** First preliminary draft of the list of rare, threatened and endemic plants for the countries of North Africa and the Middle East. Kew, IUCN Threatened Plants Committee Secretariat. 170p. (T)

3686 **Ionnides, O. (1973).** Nature conservation in Cyprus. *Nature in Focus*, 14: 16-17.

3687 **Zahariadi, C.A. (1980).** Deux taxons nouveaux ou rares d'*Ornithogalum* de la mediterranee orientale et quelques considerations sur la notion de l'endemisme. *Ann. Mus. Goulandris*, 4: 249-261. Fr. Illus. (T)

CZECHOSLOVAKIA

3688 **Anon. (1984).** Barevne plany Pruhonickeho Parku (K obrazku na zadni strane obalky). *Ziva*, 32(1): 11. Cz. Map.

3689 **Augustin, P., Bratislava, Z. (1979).** K problematike vyrubi stromov a krikov pri narusovani ekologickej rovnovahy. (Problem of felling trees and bushes and disturbance to the ecological equilibrium.) *Zahradnictvo*, 4(2): 83-84. Rus. Forest conservation.

3690 **Balatova-Tulackova, E. (1980).** Ubersichte der Vegetationseinheiten der Wiesen im Naturschutzgebiet Zd'arske vrchy: 1. (Survey of the meadow plant communities in the protected area of the Zd'arske vrchy hills: 1.) *Preslia*, 52(4): 311-331. Ge (En, Cz). Grassland. (P)

3691 **Balatova-Tulackova, E. (1983).** Feuchtwiesen des Landschaftsschutzgebietes Jiserske hory. I. (Wet meadow communities of the protected landscape region Jizerske hory Mountains. I.) *Folia Geobot. Phytotax.*, 18(2): 113-136. Ge (En). Illus. Montane, grassland. (P)

3692 **Balhar, R. (1978).** Podil statni ochrany prirody na reseni rekreace v Severomoravskem kraji. (State nature conservation and solution of recreation in the northern Moravia region.) *Pamatky a Priroda*, 4: 227-229. Illus.

3693 **Bares, I., Sehnalova, J. (1981).** Preservation of land-races of cultivated plants in Czechoslovakia. *Kulturpflanze*, 29: 67-77. (E)

CZECHOSLOVAKIA

3694 **Bosackova, E. (1972).** [Actual state and protection of the low moor vegetation in Zitnij Ostrov.] Bratislava, Slovensky ustav pamiatkovej starostlivosi ochrany prirody. 82p. Rus.

3695 **Bosackova, E. (?)** Chranene rastliny na Slovensku a podmienky ich ochrany. (Protected plants in Slovakia and protective measures.) Vydalo VPL pre Poverenictvo SNR pre kulturu a informacie. 6p. (T)

3696 **Bucek, A., Vicek, V. (1984).** Promena krajiny jinomoravskych udolnich niv. (Changes to the valley meadows of the South Moravian region.) *Ziva*, 32(4):122-124. Cz. Illus. Map.

3697 **Burda, J. (1978).** Z chranenych uzemi na Jicinsku. (From protected regions in the Jicin area.) *Pamatky a Priroda*, 6: 367-368. Cz. Illus. (P)

3698 **Capek, M., Jancarik, V. (1979).** [Protection of forests in the Czechoslovak Socialist Republic]. *Gorsko Stopanstvo*, 35(2): 38-43. Bul. Illus.

3699 **Cerovsky, J. (1979).** Czechoslovakia - promotion of protected plants. *Threatened Pl. Commit. Newsl.*, 4: 18. (T)

3700 **Cerovsky, J. (1981).** Zasady viberudruhu pro ochranu. (Principles for selection of protected species.) *In* Holub, J., ed. Mizejici flora a ochrana fytogenofondu v CSSR. (The vanishing flora and protection of the gene pool in Czechoslovakia.) Studie CSAV, 20. Prague, Academia. 17-22. Cz (En). (T)

3701 **Cerovsky, J. (1982).** Botanisch-okologische Probleme des Arten-Schutzes in der CSSR unter Berucksichtigung der praktischen Naturschutzarbeit. (Botanical and ecological problems of species preservation in the CSSR with regard to practical conservation work.) *Berichte der ANL*, 6: 90-92. (T)

3702 **Cerovsky, J. (1984).** Education for the protection of nature: a national plan for Czechoslovakia. *In* Unesco-UNEP. Conservation, science and society. Paris, Unesco. 577-584. Contributions to the First International Biosphere Reserve Congress, Minsk, Byelorussia/USSR, 26 September - 2 October 1983. (P)

3703 **Cerovsky, J., Holub, J., Prochazka, F. (1979).** [Red list of the flora in the Czech Socialist Republic]. *Pamatky Prir.*, 4(6): 361-378. Cz. Illus. (T)

3704 **Cerovsky, J., Podhazska, Z. (1981).** Registrace kriticky ohrozenych drulu vyssich rostlin v CSR. (Registration of critically endangered plant species in the Czech Socialist Republic.) *Pamatky Prir.*, 6: 577-583. Cz. (T)

3705 **Cherven, Yu.L. (1980).** Zatsita na gorite v CSSR. (Die Schutz des Waldes. in CSSR.) *Nauc. Trud. Lesoteh. Inst. Sofia*, 21: 113-127 (1979). Cz (Ge).

3706 **Chmelar, J. (1981).** Priciny ustupu nekterych druhu vrb v kulturnich oblastech. (The agents responsible for retreat of some willow species in culture areas.) *In* Holub, J., ed. Mizejici flora a ochrana fytogenofondu v CSSR. (The vanishing flora and protection of the gene pool in Czechoslovakia.) Studie CSAV, 20. Prague, Academia. 75-76. Cz (En). (T)

3707 **Chytil, J. (1983).** Inventarizacni pruzkum CHPV "Udoli oslavy a Chvojnice". *Ziva*, 31(3): 117-118. Cz. Illus., map.

3708 **Council of Europe (1980).** National activities: Tribon Basin. *Council of Europe, Newsl. Nat.*, 7/12: 4. Southern Bohemia near Austrian frontier.

3709 **Deyl, C. (1982).** *Nasturtium microphyllum* (Boenn.) Reichenb. a jine vzacne rostliny na Hane u Hrdiboric. (*Nasturtium microphyllum* (Boenn.) Reichenb. and other rare plants from lowland Hana [Central Moravia]). *Zpravy Cesk. Bot. Spol. CSAV*, 17(1): 53-56. Cz (En). (T)

3710 **Dobrkovsky, F. (1978).** Cholticky Zamecky Park. (Caste Park in Chiltice.) *Zahradnictvo*, 3(10): 476-478. Cz. Illus. (P)

CZECHOSLOVAKIA

3711 **Dostal, L. (1978).** Najblizsie ulohy ochrany prirody v ochrane rastlinstva na vychodnom Slovensku. (Forthcoming tasks on plant protection in eastern Slovakia.) *Acta Bot. Slov. Acad. Sci. Slov., A. Taxon. Geobot.,* 4: 51-61. Rus (En). (T)

3712 **Elias, P. (1979).** Zriedkavejsie rastliny zeleznicnych komunikacii na zapadnom Slovensku: 1. (Rarer plant species in railways of West Slovakia: 1.) *Biol. Bratisl.,* 34(1): 67-70. Cz (En). (T)

3713 **Elias, P. (1981).** O ochrana genofondu ovocnych drevin. (On the protection of fruit tree genetic diversity). *In* Holub, J., *ed.* Mizejici flora a ochrana fytogenofondu v CSSR. (The vanishing flora and protection of the gene pool in Czechoslovakia.) Studie CSAV, 20. Prague, Academia. 125-127. Cz (En). (E/G/T)

3714 **Elias, P. (1981).** [Rarer plant species on railways of West Slovakia, Czechoslovakia: 2.] *Biol. Bratisl.,* 36(1): 73-78. Slk (En, Rus). (T)

3715 **Fajmonova, E. (1979).** Waldgesellschaften des Verbandes Vaccinio-Piceion Br.-Bl. 1939 im Naturschutzgebiet und in der Schutzzone Slovensky Raj. (Forest associations of Vaccinio-Piceion Br.-Bl. 1939 in conservation and protected zone of the Slovensky Raj.) *Acta Facult. Rer. Nat. Univ. Comen. Bot.,* 27: 53-75. Ge (Rus, Slk). (P)

3716 **Fials, K., Kvet, J. (1978).** Dynamic balance between plant species in South Moravian reedswamps. *In* Duffey, E., Watt, A.S., *eds.* The scientific management of animal and plant communities for conservation. The 11th Symposium of the BES, University of East Anglia, Norwich, 7-9 July, 1970. 241-269. Illus., map.

3717 **Futak, J. (1981).** Endemicke rastliny Slovenska. (Endemic plants of Slovakia.) *In* Holub, J., *ed.* Mizejici flora a ochrana fytogenofondu v CSSR. (The vanishing flora and protection of the gene pool in Czechoslovakia.) Studie CSAV, 20. Prague, Academia. 45-48. Cz (En). (T)

3718 **Hadac, E. (1981).** Endemicke rostliny CSR. (Endemic plants of the Czech Socialist Republic.) *In* Holub, J., *ed.* Mizejici flora a ochrana fytogenofondu v CSSR. (The vanishing flora and protection of the gene pool in Czechoslovakia.) Studie CSAV, 20. Prague, Academia. 41-44. Cz (En). (T)

3719 **Hajkova, A. (1987).** Poznamky k vyskytu nekterych ohrozenych druhii v Beskydech a Podbeskydse pahorkatine. (Bemerkungen zum Auftreten einiger gefahrdeten Arten in den Beskiden und im Hugelland unterhalb der Beskiden). *Cas. Slezsk. Muz., Vedy Prir., A,* 36(1): 89-90. Cz (Ge, Rus). Illus. (T)

3720 **Havlickova, J., Rydlo, J. (1982).** Ohrozena lokalita *Dactylorhiza incarnata* ve Strednim Polabi. (An endangered locality of *Dactylorhiza incarnata* in Central Elbe Lowland (Central Bohemia.) *Zprav. Cesk. Bot. Spol.,* 17(2): 142-144. Cz (En). (T)

3721 **Hendrych, R. (1977).** Zanikle nebo nezvestne rostliny nasi kveteny. (Extinct or missing plants of our flora.) *Ziva,* 25(3): 84-85. (T)

3722 **Hendrych, R. (1981).** Bemerkungen zum Endemismus in der Flora der Tschechoslowakei. (Observations on endemism in the flora of Czechoslovakia). *Preslia,* 53: 97-120. Ge (En). (T)

3723 **Holub, J. (1981).** Ochrana fytogenofondu z hlediska taxonomickeho a fytogeografickeho. (Protection of the floristic diversity from the aspect of taxonomy and phytogeography). *In* Holub, J., *ed.* Mizejici flora a ochrana fytogenofondu v CSSR. (The vanishing flora and protection of the gene pool in Czechoslovakia.) Studie CSAV, 20. Prague, Academia. 27-39. Cz (En). (T)

3724 **Holub, J., Prochazka, F., Cerovsky, J. (1979).** Seznam vykynulych, endemickych a ohrozenych taxonu vyssich rostlin kveteny CSR (1. verze). (List of extinct, endemic and threatened taxa of vascular plants of the flora of the Czech Socialist Republic - first draft.) *Preslia,* 51(3): 213-237. Cz (En). (T)

CZECHOSLOVAKIA

3725 **Holub, J., ed. (1981).** Mizejici flora a ochrana fytogenofondu v CSSR. (The vanishing flora and protection of the gene pool in Czechoslovakia.) Studie CSAV, 20. Prague, Academia. 174p. Cz. Proceedings from a conference. (T)

3726 **Holubickova, B. (1981).** *Pinus uncinata,* priklad ruznych typu ohrozeni druhu. (*Pinus uncinata,* ein Beispiel von verschiedenen.) *In* Holub, J., *ed.* Mizejici flora a ochrana fytogenofondu v CSSR. (The vanishing flora and protection of the gene pool in Czechoslovakia.) Studie CSAV, 20. Prague, Academia. 71-73. Cz (Ge). (T)

3727 **Horakova, J. (1978).** [Co-operation of museums with the assets of state nature protection and state conservation care]. *Pamatky a Priroda,* 1: 27-31. Cz.

3728 **Hron, F. (1981).** Cesky svaz ochrancu prirody a dobrovolna ochrana prirody v CSR. *Ziva,* 27(2): 42-43. Cz.

3729 **Husak, S. (1981).** Mizejici okrasne typy nasi kveteny vlivem hortikultury. (Decorative native plants of the Czechoslovak flora threatened by the influence of horticulture.) *In* Holub, J., *ed.* Mizejici flora a ochrana fytogenofondu v CSSR. (The vanishing flora and protection of the gene pool in Czechoslovakia.) Studie CSAV, 20. Prague, Academia. 105-109. Cz (En). Collectors. (E/G/T)

3730 **Jatiova, M., Lapacek, V. (1978).** Chranena krajinna oblast Palava. (The protected Palava country area.) *Ziva,* 26(1): 15-16. Cz. (P)

3731 **Jirgle, J. Kucera, J., Tichy, J., Materna, J. (1983).** Problematika Krusnych hor. (Problems of the Ore Mountains.) *Zpravy Lesnickeho Vyzkumu,* 28(1): 6-16. Cz.

3732 **Kantor, J (1983).** Postaveni ceskoslovenskeho lesniho semenarstvi a slechteni lesnich drevin v obdobi ohrozeni lesu imisemi. (Czechoslovak forest tree breeding in relation to threats from air pollution.) *Lesnicka Prace,* 62(10): 458-463. Cz (En, Rus). Illus.

3733 **Kociova, M. (1981).** Ustup jedle v slovenskych Karpatok a opatrenie na jej zachranu. (Retreat of the fir in Slovak Carpathians and provisions for its rescue.) *In* Holub, J., *ed.* Mizejici flora a ochrana fytogenofondu v CSSR. (The vanishing flora and protection of the gene pool in Czechoslovakia.) Studie CSAV, 20. Prague, Academia. 67-69. Cz (En). (T)

3734 **Kolbek, J. (1985).** Malo znama rostlinna spolecenstva Chranene krajinne oblasti Krivoklatsko. (Less known natural plant communities of nature reserve Krivoklatsko.) *Preslia,* 57(2): 151-169. Cz (En, Ge). (P)

3735 **Kopriva, F. (1978).** Navrh na rekonstrukci cenne lesostepni lokality ve Statni Priorodni Rezervaci Karlstejn. (Proposal for reconstruction of valuable forest-steppe land in the Karlstejn State National Park.) *Pamatky a Priroda,* 6: 372-374. Illus. (P)

3736 **Korpel, S. (1975).** Rezervacia Kasivarovska Dubina. (The Kasivarova Oakwood Reserve.) *Zborn. Vedeck. Prac Lesn. Fak. Vsld Vo Zvolene,* 17(11): 19-52. Illus. (P)

3737 **Krippel, E. (1978).** [State National Park Abrod calls for help]. *Pamatky a Priroda,* 6: 369-371. Sl. Illus. (P)

3738 **Krix, K. (1979).** Sit statnich prirodnich rezervaci a typologie lesu. (System of state natural parks and forest typology.) *Pamatky a Priroda,* 4(4): 225-233. Cz (En, Ge, Rus). Illus., map. (P)

3739 **Kriz, Z. (1979).** Dreviny moravskych parku a moznosti zlepseni produkce sadovnickeno skolkarstvi. *Ziva,* 27(1): 11-15. Cz. Illus.

3740 **Kriz, Z. (1981).** Uloha botanickych zahrad a arboret v zachrane genofondu domacich a cizich drevin. (Aufgabe der botanischen Garten und Arboreten bei der Retung des Genofondes der heimischen und fremden Geholze.) *In* Holub, J., *ed.* Mizejici flora a ochrana fytogenofondu v CSSR. (The vanishing flora and protection of the gene pool in Czechoslovakia.) Studie CSAV, 20. Prague, Academia. 61-66. Cz (Ge). (G/T)

CZECHOSLOVAKIA

3741 **Kubat, K. (1981).** Ohrozene druhy severozapadnich Cech. (Threatened species in north-west Bohemia.) *In* Holub, J., *ed.* Mizejici flora a ochrana fytogenofondu v CSSR. (The vanishing flora and protection of the gene pool in Czechoslovakia.) Studie CSAV, 20. Prague, Academia. 133-137. Cz. (T)

3742 **Kubikova, J. (1977).** The vegetation of Prokop Valley Nature Reserve in Prague. *Folia Geobot. Phytotax.*, 12(2): 167-199. (P)

3743 **Kubikova, J., Manych, J. (1979).** Soucasna kvetena statni prirodni rezervace Prokopske udoli v Praze. (Present flora of the Prokop Valley Nature Reserve in Prague.) *Zpravy Cesk. Bot. Spol. CSAV*, 14(1): 37-58. Cz (En). Maps. (P)

3744 **Kucera, M. (1979).** Jifovce v Pruhonickem parku. *Ziva*, 27(2): 49-51. Illus., col. illus.

3745 **Kuhn, F. (1981).** Ochrana genofondu mistnich plodin v CSSR. (Der schutz des genofondes der lokalfruchte in der CSSR.) *In* Holub, J., *ed.* Mizejici flora a ochrana fytogenofondu v CSSR. (The vanishing flora and protection of the gene pool in Czechoslovakia.) Studie CSAV, 20. Prague, Academia. 117-120. Cz (Ge). (G/T)

3746 **Lednicky, V. (1978).** Znecisteni ovzdusi v Chranene krajinne oblasti Beskydy. *Ziva*, 26(5): 164-165. Cz.

3747 **Lochman, J. (1977).** Biological forest protection against the damage done by red deer. *Commun. Inst. Forest. Cechosloveniae*, 10: 57-69. (Cs, Rus).

3748 **Madar, Z. (1984).** Regles juridiques relatives a la protection de la foret en Tchecoslovaquie. (Judicial regulations relative to forest protection in Czechoslovakia.) *In* Prieur, M., *ed.* Forets et environnement: en droit compare et international. Paris, Presses Universitaires de France. 139-145. Fr. (L)

3749 **Magic, D. (1980).** (K probleme zashchity trebuyushchikh okhrany territorii i organizatsiya sistemy zashchity.) K otazke starostlivosti o chranene uzemie a tvorby ich systemu. *In* Zb. ref. z 3-go Zjazdu Slov. bot. spolecn pri SAV, Zvolen. 123-128. Slk (En).

3750 **Magic, D., Bosackova, E., Krejca, J., Usak, O. (1979).** Atlas chranenych rastlin: obrazkova kvetena Slovenska. (Field-guide of protected plants.) Obzor, Bratislava. 260p. Slk. (T)

3751 **Maglocky, S. (1980).** Cerveny zoznam ohrozenych druhov rastlin SSR. (The Red List of the threatened plants of the Slovak Socialist Republic.) *In* Zbornik referatov 3 zjasdu Slovenskej Botanickej Spolvcnosti, Zvolen 30 June - 5 July 1980. 149-150. Slk. (T)

3752 **Maglocky, S. (1983).** Zoznam vyhynutych endemickych a ohrozenych taxonov vyssich rastlin flory Slovenska. (List of extinct, endemic and threatened taxa of vascular plants of the flora of Slovakia.) *Biologia, Ser. A: Bot.*, 38(3): 825-852. Slk (En, Rus). (T)

3753 **Malik, J. (1983).** K otazce vlivu znecisteni ovzdusi na plodivost smrku v horach. (Influence of air pollution on the fructification of spruce in the mountains.) *Lesnicka Prace*, 62(9): 400-404. Cz (En, Fr, Ge, Rus). Pollution, forest.

3754 **Mann, J. (1978).** Nalezy vzachejsich druhu hub v ololi Zamberka v. roce 1977. (Collections of very rare species of mushrooms near Zamberk in 1977.) *Mykol. Sborn.*, 55(3): 69-71. Illus. (T)

3755 **Marsakova, M., Kuznikova, H. (1978).** Chranene krajinne oblasti ve statistice. (Protected landscape areas in statistics.) *Pamatky a Priroda*, 1: 33-36. Cz. Illus. (P)

CZECHOSLOVAKIA

3756 **Moldan, B., Stepanek, P. (1984).** Biogeochemical balance as a prerequisite to ecological stability: a model for the Krivoklatsko Biosphere Reserve, Czechoslovakia. *In* Unesco-UNEP. Conservation, Science and Society. Paris, Unesco. 340-346. Contributions to the First International Biosphere Reserve Congress, Minsk, Byelorussia/USSR, 26 September - 2 October 1983. (P)

3757 **Molikova, M. (1978).** (Beitrag zur botanischen Charakeristik der Naturschutzgebiete im Bezirk Benesov [Mittelbohmen].) *Bohemia Centralis*, 7: 71-82. Cz (Ge).

3758 **Moravec, J., et al. (1983).** Rostlinna spolecenstva ceske socialisticke republiky a jejich ohrozeni. (Red list of plant communities of the Czech Socialist Republic and their endangerment.) Litomerice, Severoceskon Prirodou, Priloha 1983/1. 110p. Cz. (T)

3759 **Mucina, L. (1978).** [The Tematin Hills National State Reservation]. *Pamatky a Priroda*, 1: 55-56. Sl. Illus. (P)

3760 **Nevrly, M. (1978).** Zmizi housenkovec roketovy z nasi kveteny. (Will *Beckmannia cruciformis* disappear from our flora?) *Pamatky a Priroda*, 1: 49-50. Cz. Illus. (T)

3761 **Novy, P. (1978).** Pripravovana chranena uzemi prirody okrescu Tachov. (Prepared protected national parks in the Tachov area.) *Pamatky a Priroda*, 4: 230-235. Cz. Illus., map. (P)

3762 **Petricek, V. (1978).** Statni Prirodni Rezervace Provodinske Kameny. (The Provodinske Kameny State National Reservation.) *Pamatky a Priroda*, 1: 59-60. Cz. Illus. (P)

3763 **Petricek, V., Kolbek, J. (1979).** [Vegetation of Natural Reservation Belysov]. *Zprav. Cesk. Bot. Spol.*, 14(1): 61-68. Cz (Ge). (P)

3764 **Petricek, V., Kolbek, J. (1985).** Vapnomilne bory na Ustecku utociste vzacnych druhu rostlin. (Lime-loving woods at Usteck: a refuge for rare species of plants.) *Ziva*, 34(1): 5-7. Cz. Illus., maps. (T)

3765 **Piekos-Mirkowa, H. (1980).** [Distribution of rare taxa of vascular plants in the area of the Tatra Mts.] *Chronmy Przyr. Ojczysta*, 36(3): 34-44. Pol (En). (T)

3766 **Pitoniak, P., Petrik, A., Dzubinova, L'., Uhlirova-Simekova, J., Fajmonova, E. (1978).** Flora a vegetacia chranenej krajinnej oblasti Slovensky Raj. (The flora and vegetation of the Slovensky Raj protected landscape region.) *Biologicke Prace*, 24(6): 136p. Slk (Rus, En). (P)

3767 **Plansky, B. (1978).** Hledame dalsi naleziste vzacnych hub. (Seeking new localities of rare mushrooms.) *Mykol. Sborn.*, 55(1/2): 19-20. Cz. Illus. *Floccularia rickenii*, *Boletus separans*. (T)

3768 **Plansky, B. (1978).** Hledame dalsi vzacne houby v Ceskoslovansku. (Seeking new localities of rare mushrooms in Czechoslovakia.) *Mykol. Sborn.*, 55(3): 75-76. Cz. Illus. (T)

3769 **Plesnik, P. (1976).** Die Vegetationsstufen in der Slowakei. *PM*, 17(4): 257-276. Ge. Illus. (T)

3770 **Potucek, O. (1981).** Problemy ochrany genofondu kulturnich odrud na prikladu obilovin. (Probleme des genofonds kultursorten am beispiel von getreide.) *In* Holub, J., ed. Mizejici flora a ochrana fytogenofondu v CSSR. Praha, Academia. 121-123. Cz (Ge). Gene. (G/T)

3771 **Prochazka, F. (1978).** Nove pozoruhodne naleziste chranenych a vzacnych rostlin na kralovehradecku. (New remarkable sites of protected and rare plants in the region of Hradec Kralove.) *Pamatky a Priroda*, 1: 46-48. Cz. Illus. (L/T)

3772 **Prochazka, F. (1978).** Vstavac zvrhly - vzacny zastupce Ceske kveteny. (*Orchis hybrida*, a rare representative of Czech flora.) *Ziva*, 26(1): 16. Cz. (T)

CZECHOSLOVAKIA

3773 **Prochazka, F. (1979).** Prstnatec russowuv (*Dactylorhiza russowii*) - nove rozeznany a vyhynuly druh Ceskoslovenske kveteny. (The Russian orchid (*Dactylorhiza russowii*) - newly recognised and extinct species of the Czechoslovak flora (reclassified from *Orchis traunsteineri*).) *Preslia*, 51(3): 245-254. Cz (En, Ge). Illus. (T)

3774 **Prochazka, F. (1980).** Soucasne zmeny vychodoceske flory a poznamky k rozsireni chranenych druhu rostlin. Hradec Kralove, Krajske Muzeum Vychodnich Cech. 134p. Cz. Maps.

3775 **Prochazka, F. (1981).** Ohrozeni a ochrana ceskoslovenskych orchideji. (Gefahrdung und Schutz der tschechoslowakischen Orchideen.) *In* Holub, J., *ed.* Mizejici flora a ochrana fytogenofondu v CSSR. Praha, Academia. 99-103. Cz (Ge). (T)

3776 **Prochazka, F. (1981).** Vyhynule druhy ceskoslovenske flory. (Extinct species in the Czechoslovak flora.) *In* Holub, J., *ed.* Mizejici flora a ochrana fytogenofondu v CSSR. Studies CSAV, 20. Prague, Academia. 13-15. Cz (En). (T)

3777 **Prochazka, F., Cerovsky, J., Holub, J. (1983).** Chranene a ohrozene druhy kveteny CSR. (Protected and endangered species of the flora of the CSR (Czech Socialist Republic). Prague, Central House of Young Pioneers and Youth. 103p. Cz. A study and method handbook for leaders of young pioneer naturalist groups. (P/T)

3778 **Randuska, D., Krizo, M. (1983).** Chranene rastliny. (Protected plants.) Bratislava, Priroda. 430p. Cz. (T)

3779 **Resovska, Z. (1978).** O planovane rekonstrukcii Horskeho Parku v Bratislave. (Planned reconstruction of Mountain Park in Bratislava.) *Zahradnictvo*, 3(8): 378-379. Cz. Illus. Forest conservation. (P)

3780 **Rogalewicz, V., Bares, I. (1987).** Czechoslovak information system for genetic resources. *Pl. Genet. Res. Newsl.*, 72: 14-16. (E)

3781 **Rudiger, A. (1978).** Die Mala Fatra: ein Landschaftsschutzgebiet in der Slowakei. *Naturschutz- und Naturparke*, 89: 33-38. Ge. Illus.

3782 **Rudiger, A. (1979).** Der moorreiche Bohmerwald: ein Landschaftsschutzgebiet im Herzen Europas. *Naturschutz- und Naturparke*, 94: 37-44. Ge. Illus.

3783 **Ruzicka, I. (1978).** Ochrana zbytku priorozenych bucin v Jihlavskych Vrsich. (Conservation of remains of natural beech forests in the Jihlava Mountains.) *Pamatky a Priroda*, 1: 51-54. Illus.

3784 **Rybar, P. (1979).** Protected plant and animal species of East Bohemia. Pardubice, Kraj. stiredisko st. pambatkovbe pbeice a ochrany pirbirody Vbychodoiceskbeho kraje. 174p. (T)

3785 **Samek, V. (1981).** Problematika ochrany genofondu lesnich drevin. (Preservation of gene pool in forest trees). *In* Holub, J., *ed.* Mizejici flora a ochrana fytogenofondu v CSSR. Praha, Academia. 51-55. Cz (En). (G/T)

3786 **Samoril, K. (1980).** [Preservation of the living environment. Vol. 1.] Prague, Manual Publications, ROH Reference Library. 76p. Cz.

3787 **Saul, J. (1985).** Kvetena chranene prirodni pamatky Oburky-Trestenec na Brnensku. (Flora des geschuzen Naturdenkmales Oburky-Trestenec in der Umgebung von Brno (Mittelmahren).) *Zpravy Cesk. Bot. Spol. CSAV*, 20(2): 145-149. Cz. (T)

3788 **Sediva-Novakova, J., Dvorak, F. (1983).** Rogsireni vzacnych, ohrozenych a jinak fytogeograficky pogoruhodnych druhu rostlin na Jihlavsku. (Distribution of rare, threatened and phytogeographically remarkable plants in the Jihlava district, West Moravia.) *Zpravy Cesk. Bot. Spol. CSAV*, 18(3): 214-218. Cz (En). (T)

CZECHOSLOVAKIA

3789 **Sedlacek, J., Dvorak, F. (1983).** Vzacne a ohrozene druhy jihovychodni casti Hustopecske Pahorkatiny. (Rare and endangered species in the southeast part of Hustopecska Pahorkatina.) *Zpravy Cesk. Bot. Spol. CSAV*, 18(1): 61-66. Cz (Ge). Southern Moravia. (T)

3790 **Sindelar, J. (1983).** Uchovani genofondu lesnich drevin z oblasti ohrozenych prumyslovymi imisemi. (Preservation of the genetic resources of forest tree species as threatened by industrial pollution.) *Lesnicka Prace*, 62(10): 443-452. Cz (En, Fr, Ge, Rus). (E)

3791 **Sindelar, J. (1984).** [Measures for conservation, reproduction and better use of the genefund of European larch in CSR with respect to the Jeseniky Larch Lesnictvi.] *Lesn. Ceskosl. Akad. Zemed.* 30(7): 569-587. Cz (En, Rus). Gene, mountain forest. *Larix decidua.*

3792 **Sindelar, J. (1984).** [Measures for rescue and reproduction of the genetic resources of forest from the polluted regions.] *Lesnicka Prace*, 63(9): 392-401. Cz (En, Fr, Ge, Rus). Illus.

3793 **Sindelarova, J. (1978).** Lesy v tvorbe a ochrane zivortniho prostredi. (Studijni Zprava). (The significance of forests in environment formation and protection.) *Studijni Informace*, 4: 84. Cz. Illus.

3794 **Skalicky, V. (1981).** Otazky ustupu a vymirani plevelu. (Die Fragen des Ruckganges und des Aussterbens von Unkrautern.) *In* Holub, J., ed. Mizejici flora a ochrana fytogenofondu v CSSR. Praha, Academia. 83-88. Cz (Ge). (T)

3795 **Skalicky, V., Jenik, J. (1974).** [Conservation of the flora and vegetation of the Bohemian Karst]. *Bohem. Centr.*, 3: 101-140. Cz (En). (T)

3796 **Smejkal, M. (1981).** Antropogenni zmeny flory na prikladu zdarskych vrchu. (Anthropogene Veranderungen der Flora am Beispiel von Zdarske Vrchy-Saarer Bergland.). *In* Holub, J., ed. Mizejici flora a ochrana fytogenofondu v CSSR. Praha, Academia. 129-132. Cz (En). (T)

3797 **Somsak, L. (1977).** Ohrozene a zriedkave taxony horskych a vysokohorskych poloh Slovenska. (The threatened and rare taxa of the mountain range of Slovakia.) Bratislava. Sl. (T)

3798 **Somsak, L. (1979).** Czechoslovakia: the role of botanic gardens in the conservation of rare and threatened plants in Slovakia. *In* Synge, H., Townsend, H., eds. Survival or extinction. Proceedings of a conference held at the Royal Botanic Gardens, Kew, 11-17 September 1978. Kew, Bentham-Moxon Trust. 107-112. (G/T)

3799 **Somsak, L., Slivka, D. (1981).** Chranene rastliny Slovenska. (Protected plants in Slovakia.) 2nd Ed. Bratislava. Illus. (T)

3800 **Sourkova, M. (1981).** *Bupleurum rotundifolium* - jeho drivejsi a soucasne rozsireni v Ceskoslovensku. (*Bupleurum rotundifolium* - seine ehamlige und rezente Verbreitung in der Tschechoslowakei.) *In* Holub, J., ed. Mizejici flora a ochrana fytogenofondu v CSSR. (The vanishing flora and protection of the gene pool in Czechoslovakia.) Studie CSAV, 20. Prague, Academia. 89-94. Cz (Ge). (T)

3801 **State Institute for the Care of Historical Monuments and Nature Conservation (1987).** Conservation of nature and natural environment in the Czechoslovak Socialist Republic. Prague, State Institute for the Care of Historical Monuments and Nature Conservation. 13p. Illus. (P/T)

3802 **Stodola, J. (1983).** Kotvice Plovouci - *Trapa natans* L. (Water Chestnut - *Trapa natans* L.) *Skalnicky*, 7: 29-30. Cz. Illus. (T)

3803 **Stolina, M. (1978).** Ochrana lesa a ochrana zivotneho prostredia cloveka. (Protection of the forest and environment.) *Lesn. Cesk. Akad. Zemed.*, 24(10): 829-842. Cz (En, Ge, Rus). Illus.

215

CZECHOSLOVAKIA

3804 **Stolina, M. (1982).** Vyznamna uloha - ochrana horskych lesov. (Important task -protection of mountain forests.) *Les*, 38(8): 365-370. Sl.

3805 **Supuka, J. (1978).** K problematike dendrologickej struktury parkovych Lesov. (Problems of dendrologic structure of park forests.) *Zahradnictvo*, 3(1): 42-44. Cz. Illus. (P)

3806 **Supuka, J., Vrestiak, P. (1977).** [Role of the Mlynany Arboretum in solving problems relating to the development and conservation of the environment]. *Zahradnictvo*, 2(8): 374-375. Cz. (G)

3807 **Svoboda, A.M. (1981).** Ochrana a vyuzivani genofondu drevin. (Protection and exploitation of gene pool in woody plants.) *In* Holub, J., *ed.* Mizejici flora a ochrana fytogenofondu v CSSR. Praha, Academia. 57-59. Cz (En).

3808 **Tesar, V. (1983).** Znecisteni ovzdusi a lesni hospodarstvi na Trutnovsku. (Air pollution and forest management in the Trutnov area.) *Zpravy Lesnickeho Vyzkumu*, 28(1): 24-26. Cz. Pollution.

3809 **Tilschova, T. (1978).** Vyskyt vzacnejsi ciruvky. (Occurrence of the rather rare *Tricholoma constrictum.*) *Mykol. Sborn.*, 55(1/2): 8-9. Cz. Illus. (T)

3810 **Vacek, S. (1983).** Ekologie poskozeni horskych ochrannych lesu. (Ecology of damage to mountain protection forests.) *Zpravy Lesn. Vyzk.*, 28(1): 29-33. Cz. Illus., maps.

3811 **Vagenknecht, V. (1981).** Ohrozenost rostlin lidskou cinnosti na prikladu chranenych druhu rostlin na Slovensku. (Bedrohung der Pflanzen durch Menschentatigkeit auf dem Beispiel von geschutzten Arten in der Slowakie.) *In* Holub, J., *ed.* Mizejici flora a ochrana fytogenofondu v CSSR. Praha, Academia. 23-26. Cz (Ge). (T)

3812 **Vaneckova, L. (1981).** Vymizele a mizejici druhy Moravskeho krasu. (Extinct and endangered species in the Moravian Karst.) *In* Holub, J., *ed.* Mizejici flora a ochrana fytogenofondu v CSSR. Studie CSAV, 20. Praha, Academia. 139-141. Cz (En). (T)

3813 **Vanek, J. (1980).** Dreviny ve statni prirodni reservaci Babiccino udoli a v jejim ochrannem pasmu. (Die Holzarten im Naturschutzgebiet "Babiccino udoli" und in ihre nachste Ungebung [sic].) *Prace Studie*, 12: 73-81. Cz (Ge, Rus). (P)

3814 **Vesely, J. (1961).** Chranene rostliny. (Protected plants.) 2nd Ed. Praha, Orbis. 85p. Cz. Col. illus. Lists protected species and their conservation status. (T)

3815 **Vesely, J. (1961).** Priroda Ceskoslovenska, juji vyvoj a ochrana. (Nature in Czechoslovakia, its development and conservation.) Bratislava, Osveta. 146p. Cz.

3816 **Voloscuk, I. (1980).** (Okhrana lesnykh ekosistem) Ochrana lesnych ekosystemov. (Protection of forest ecosystems.) *In* Zb. ref. 2 3-go Zjazdu Slov. bot. spolecn pm SAV Zvolen. 117-121. Slk (Rus, En).

3817 **Voloscuk, I. (1983).** Starostlivost o chranene uzemia na lesnom podnom fonde. (Care of protected territories on forest land.) *Lesnicky Casopis*, 29(6): 525-532. Slk (En). Illus. (P)

3818 **Vrestiak, P. (1978).** Parkove lesy v sustave rekreacnej zelene miest. (Park forests in the system of recreational green areas.) *Zahradnictvo*, 3(6): 284-285. Cz. Illus. (P)

3819 **Wiltowski, J. (1979).** Ochrona przyrody w Slowacji. (Nature conservation in Slovakia.) *Chronmy Przyr. Ojczysta*, 35(1): 46-52. Cz (En). Illus.

3820 **Zahar, D., Mracek, Z. (1979).** [Landscape creation and environmental conservation in the Czechoslovak Socialist Republic]. *Gorsko Stopanstvo*, 35(2): 33-37. Bul. Illus.

CZECHOSLOVAKIA

3821 **Zahradnik, J., Mrkacek, Z. (1984).** Chranene druhy rostlin a zivocichu na uzemi pramenu Cidliny a nejblizsiho okoli. (Protected plants and animals in the Cidlina river spring area and its surroundings.) *Cas. Slov. Nar. Mur. Prir. Vedy*, 153(3): 165-171. Cz. Illus., map. (T)

3822 **Ziabek, I. (1984).** Hospodareni v lesich postihovanych imisemi. (Forest management in forests damaged by pollution.) *Lesnicka Prace*, 63(2): 51-55. Cz (En, Fr, Ge, Rus).

DENMARK

3823 **Arnklit, F. (1982).** Erfahrungen mit einer Samenbank im Botanischen Garten Kopenhagen. *Gartn.-Bot. Brief*, 72: 15-22. Ge. Illus. (G)

3824 **Council of Europe (1983).** Native protection laws strengthened. *Council of Europe Newsl.*, 83(8-9): 4. (L/T)

3825 **Faurholdt, N. (1986).** Knaelaebe (*Epipogium aphyllum*) genfundet i Danmark. *Urt*, 10: 3-5. Da (En). Illus. (T)

3826 **Fisenne, O. von (1979).** Mons Klint: das danische Naturschutzgebiet - ein Paradies fur Naturfreunde und Naturforscher. *Naturschutz- und Naturparke*, 92: 37-41. Illus.

3827 **Fredningsstyrelsen (1980).** Status over den Danske plante - og dyreverden. (Status of the Danish plant and animal world.) Copenhagen, Denmark, Fredningsstyrelsen. 456p. Da (En). Illus., maps. Papers presented at a symposium in Copenhagen, 18-20 April 1980. (T)

3828 **Frisen, R. (1988).** [Cooperation for protection of species and biotopes in the Nordic countries]. *Sven. Bot. Tidskr.*, 82(6): 499-501. Sw (En). (P)

3829 **Gravesen, P. (1976).** Forelobig oversigt over botaniske lokaliteter, 4 vols. 1976-1983. Kobenhavn, Miljoministeriets Fredningsstyrelse i Samarbejde med Dansk Botanisk Forening. Da. Col. illus., maps. Describes hundreds of botanical localities and assesses their conservation value as part of a long-term monitoring programme; covers flowering plants, mosses, fungi, lichens and algae. (T)

3830 **Holme, A. (1977).** En tabt botanisk lokalitet - eller en ny? *URT*, 3: 78-82. Illus.

3831 **Hylgaard, T. (1980).** Recovery of plant communities on coastal sand-dunes disturbed by human trampling. *Biol. Conserv.*, 19(1): 15-25.

3832 **Jensen, H.E. (1981).** Naturpark pa Avedore Holme? *URT*, 3: 67-71. Da. Illus. (P)

3833 **Lojtnant, B. (1977).** Truede danske planter. *URT*, 3: 72-74. Da. Illus., maps. *Orchis ustulata*.

3834 **Lojtnant, B. (1979).** Danmarks orchideer. Aarhus, Naturhistorisk Museum, Arhus. 30. Da. Illus. Gives lists and status of Danish orchids. (T)

3835 **Lojtnant, B. (1980).** Status over den danske flora. *In* Moller, H.S., *et al.*, eds. Status over den Danske Plante - og Dyreverden. Proceedings of a Symposium 18-20 April 1980. Miljoministeriet, Fredningsstyrelsen. 327-340. Da (En). Describes conservation status and threats to the flora. (T)

3836 **Lojtnant, B. (1984).** Otte artseksempler. (Eight species examples.) *In* Lojtnant, B., *ed.* Spredningsokologi. (Dispersal ecology.) Denmark, Naturfredningsradet og Fredningsstyrelsen, 84: 39-42. Col. illus. (T)

3837 **Lojtnant, B. (1985).** Rodliste over Danmarks karplanter. Dansk Botanisk Forening. 24p. Da. Illus. A revised threatened plant list of Danish higher plants. (T)

DENMARK

3838 **Lojtnant, B. (1986).** Truede planter og dyr i Danmark - ensamling rodlister. (Threatened plants and animals of Denmark - a collection of Red Lists.) Kobenhavn, Fredningsstyrelsen & Landbrigsministeriets Vildtforvaltning. 55p. Da. Col. illus. (T)

3839 **Lojtnant, B., Worsoe, E. (1977).** Forelobig status over den Danske flora. Reports from the Botanical Institute University of Aarhus, No. 2. 341p. Da (En). Illus. Detailed survey of status of over 200 native vascular plants in Denmark. (T)

3840 **Lojtnant, B., Worsoe, E. (1977).** Threatened and vulnerable Danish vascular plants 5, *Najas marina* and 6, *Herminium monorchis. Flora Fauna*, 83(3-4): 51-56. (T)

3841 **Lojtnant, B., Worsoe, E. (1977).** Truede dansk planter. *URT*, 4: 105-107. Da. Illus., maps. *Schoenus ferrugineus.* (T)

3842 **Lojtnant, B., Worsoe, E. (1977).** Truede og sarbare danske karplanter: 1. *Kaskelot.*, 32: 27-30. Da. Illus., col. illus., maps. *Gymnadenia albida* - diminishing population. (T)

3843 **Lojtnant, B., Worsoe, E. (1978).** Truede danske planter: 7. Gulstenbraek - *Saxifraga hirculus* L. *URT*, 3: 85-88. Da. Illus., maps. (T)

3844 **Lojtnant, B., Worsoe, E. (1980).** Slaegten *Thesium* (Nalebaeger)- uddod i Danmark? *Flora Fauna*, 86 (3/4): 65-71. Da (En). Illus. (T)

3845 **Lojtnant, B., Worsoe, E. (1981).** *Pulmonaria angustifolia* (Sky-Blue Lungwort) threatened with extinction in Denmark. Endangered and vulnerable Danish vascular plants II. *Flora Fauna*, 87(1): 7-12. (T)

3846 **Lojtnant, B., ed. (1984).** Spredningsokogi. (Dispersal ecology.) Denmark, Naturfredningsradet og Fredningsstyrelsen, 84: 111p. Da, En. Illus., maps. Covers island biogeography and design of nature reserves for threatened species, using examples from Denmark and Sweden. (P/T)

3847 **Mehl, J. (1985).** *Dactylorhiza sambucina* L. auf der Ostsee-Insel Bornholm (Danemark): Verbreitung, Okologie, Gefahrdung. *Ber. Arbeitskr. Heim. Orch.*, 2(2): 265-288. Ge. Illus. (T)

3848 **Nielsen, J., Jensen, S.L. (1979).** Reservatet vorso. Urorte skove og opgivne landbrugsarealer. (The Vorso Reserve. Untouched forests and abandoned farmland.) *Tidsskr.-Dansk Skov.*, 64(4): 237-259. Da (En). Illus. (P)

3849 **Nilsson, O. (1979).** Threatened plants in the Nordic countries. *In* Hedberg, I., *ed.* Systematic botany, plant utilization and biosphere conservation: Proceedings of a symposium held in Uppsala in commemoration of the 500th ann. of the Univ. Stockholm, Almqvist & Wiksell International. 72-75. (T)

3850 **Pawley, F. (1983).** Denmark: a bad case of vanishing environment. *Ambio*, 12(5): 277. Illus.

3851 **Petersen, P.M. (1980).** Changes of the vascular plant flora and vegetation in a protected Danish mire, Maglemose, 1913-1979. *Bot. Tidsskr.*, 75(1): 77-88. Illus., map. (P)

3852 **Vuokko, S. (1983).** Uhatut Kasvimme. (Our threatened plants.) Suomen Luonnonsuojelum Tuki Oy, Helsinki. 96p. Fi. Col. illus. Covers Nordic region. (T)

DOMINICA

3853 **Anon. (1970).** Dominica: a chance for a choice. Washington, D.C., The Conservation Foundation. 48p. Some considerations and recommendations on conservation of the island's natural resources.

3854 **Honychurch, P.N. (1978).** Vegetation [of Dominica's national park]. Roseau, Dominica National Park Service. Illus., maps. (P)

DOMINICA

3855 **Thorsell, J.W. (1979).** National parks in developing countries - the Dominican experience. *Bull. Agric. Forest. Alberta*, 2(1): 17-19. Illus., map. (P)

3856 **Thorsell, J.W. (1984).** National parks from the ground up: experience from Dominica, West Indies. *In* McNeely, J.A., Miller, K.R., *eds.* National parks, conservation and development. The role of protected areas in sustaining society. Washington, D.C., Smithsonian Institution Press. 616-620. (P)

3857 **Thorsell, J.W., Wood, G. (1976).** Dominica's Moren Trois Pitons National Park. *Nature Canada*, 5(4): 14-16, 33-34. (P)

DOMINICAN REPUBLIC

3858 **Dod, D. (1981).** El jardin botanico sirve como foco de actividades de conservacionistas. *Bol. Jard. Bot. Nacion. Dr. Rafael M. Moscoso*, 5(2): 6. Sp. (G)

3859 **Dod, D.D. (1980).** Threatened and endangered species information: instructions for collecting orchids from the Dominican Republic. *Amer. Orchid Soc. Bull.*, 49(7): 763. (T)

3860 **Garcia, R.G., Pimentel, B.J. (1986).** Florula de la Reserva Cientifica "Dr Orlando Cruz Franco", Provincia Monte Cristi/Republica Dominicana. (A florula of the scientific Reserve "Dr Orlando Cruz Franco" Prov. Monti Cristi Dominican.) *Moscosoa*, 4: 206-214. Sp (En). Map. (P)

3861 **Gollschalk, M. (1981).** Programa de hibridacion y reproduction por semilla de orquideas nativas. *Bol. Jard. Bot. Nacion. Dr. Rafael M. Moscoso*, 5(2): 10-11. Sp. Illus.

3862 **Jimenez, J. de J. (1978).** Lista tentativa de plantas de la Republica Dominicana que deben protegerse para evitar su extincion. Coloquio Internacional sobre la practica de la conservacion. Santo Domingo. CIBIMA/UASD. Sp. Lists 133 species of threatened flowering plants, of which 49 are endemic. (T)

3863 **Read, R.W., Zanoni, T.A., Mejia, M. (1987).** *Reinhardtia paiewonskiana* (Palmae), a new species for the West Indies. *Brittonia*, 39(1): 20-25. (T)

3864 **U.S. Department of the Interior. Fish and Wildlife Service (1989).** Proposed listings - July 1989. Palo de Rosa (*Ottoschulzia rhodoxylon*). *Endangered Species Tech. Bull.*, 14(8): 5. (L/T)

ECUADOR

3865 **Armstrong, G.D., Macey, A. (1979).** Proposals for a Sangay National Park in Ecuador. *Biol. Conserv.*, 16(1): 43-61. Illus., maps. (P)

3866 **Balslev, H. (1986).** A listing of Ecuadorian palms. *In* Johnson, D.V., *ed.* Final Report WWF 3322. Economic botany and threatened species of the palm family in Latin America and the Caribbean. Part 2. 48-51. (T)

3867 **Brandbyge, J., Azanza, E. (1982).** Conservation in the Ecuadorean Amazon. *Rep. Bot. Inst. Univ. Aarhus*, 5: 14-15. From report on the 5th and 7th Danish-Ecuadorean expedition. (E/P/T)

3868 **Dodson, C.H. (1981).** *Epidendrum ilense*: the saving of a truly endangered species. *Amer. Orchid Soc. Bull.*, 50(9): 1083-1086. Illus. (T)

3869 **Gentry, A.H. (1977).** Endangered plant species and habitats of Ecuador and Amazonian Peru. *In* Prance, G.T., Elias, T.S., *eds.* Extinction is forever. Proceedings of a symposium at the New York Botanical Garden, 11-13 May 1976. New York, New York Botanical Garden. 136-149. Maps. An assessment of threatened and endangered habitats. (P/T)

3870 **Herrera-MacBryde, O. (1987).** Machalilla National Park, Ecuador. *Biol. Conserv. Newsl.*, 51: 1-2. Brief overview; lists 6 species. (P)

ECUADOR

3871 **Herrera-MacBryde, O. (1988).** Machalilla. *Threatened Pl. Newsl.*, 19: 9. Report on master plan; names dominant species of dry secondary forest. (P)

3872 **IUCN (1986).** IUCN's list of top 24 threatened animals and plants. *Biol. Conserv. Newsl.*, 44: 1-6. (T)

3873 **Irawati (1981).** Langkah baru pelestarian anggrek. (A new step in orchid conservation.) *Bul. Kebun Raya*, 5(2): 43-45. In (En). (T)

3874 **MacBryde, B. (1972).** Set-backs to conservation in Ecuador. UNDP/FAO-ECU/71/527. 61p.

3875 **Macey, A., Armstrong, G., Gallo, N., Hall, M.L. (1976).** Sangay, un estudio de las alternativas de Manejo, appendix no. 4 especies de plantas segun las zonas de vida. Quito, World Wildlife Fund. WWF Proyecto PNUD/FAO, ECU/71/527. (P)

3876 **Mantilla Mata, J. (1968).** A conservation programme for Ecuador. *In* IUCN. Conservation in Latin America. Proceedings, San Carlos de Bariloche, Argentina, 27 March-2 April 1968. Morges, IUCN. 392-399.

3877 **Mills, K. (1975).** Flora de la Sierra: un estudio en el Parque Nacional de Cotopaxi 1974/5. (Flora of the Sierra (mountains of Ecuador): a study in Cotopaxi National Park 1974/5.) *Cienc. Nat.*, 16(1): 25-44. Sp. (P)

3878 **Putney, A.D. (1976).** Estrategia prelimina para la conservacion de areas silvestres sobresalientes del Ecuador. UNDP/FAO-ECU/71/527. 61p. Sp.

3879 **Salazar, A.P. (1984).** Ecuadorian strategy for the conservation of wildlands and wildlife. *In* McNeely, J.A., Miller, K.R., *eds.* National parks, conservation and development. The role of protected areas in sustaining society. Washington, D.C., Smithsonian Institution Press. 581-583. (P)

3880 **Salazar, A.P., Huber, R.M., jr (1982).** Ecuador's active conservation program. *Parks*, 6(4): 7-10. Illus.

3881 **Terborgh, J., Winter, B. (1983).** A method for siting parks and reserves with special reference to Colombia and Ecuador. *Biol. Conserv.*, 27(1): 45-58. Maps. (P)

3882 **Veening, W. (1989).** Is er hoop voor Amazonie? *Panda*, 11: 8-10. Du. Col. illus., maps. Deforestation.

3883 **Villa-Lobos, J. (1988).** Debt swaps. *Threatened Pl. Newsl.*, 19: 4-5. Conservation organizations assume part of nation's debt in return for land protection. (P)

ECUADOR - Galapagos

3884 **Acosta-Solis, M. (1963).** Protection and conservation problems on the Galapagos Islands. *Occ. Papers Calif. Acad. Sci.*, 44: 141-146. (P)

3885 **Acosta-Solis, M. (1966).** Problems of conservation and economic development of the Galapagos Islands. *In* Bowman, R.I., *ed.* The Galapagos. Berkeley and Los Angeles, Univ. of California Press. 282-285. (P)

3886 **Adsersen, H. (1976).** A botanist's notes on Pinta. *Not. Galap.*, 24: 26-28. Effects of goats on vegetation. (P)

3887 **Adsersen, H. (1989).** The rare plants of the Galapagos Islands and their conservation. *Biol. Conserv.*, 47(1): 49-77. Map. (P/T)

3888 **Anon. (1978).** Gaining ground in Galapagos. *IUCN Bull.*, 9(5): 27. Recovery of vegetation following elimination of feral goats in parts of the archipelago. (P)

3889 **Anon. (1985).** Wildlife of Galapagos still in danger. *New Sci.*, 106(1459): 7. Illus. Outbreak of fire on Isabela. (P)

3890 **Anon. (1986).** Recovering the Galapagos. *IUCN Bull.*, 17(4-6): 77. Attempts to reduce the threat from introduced species. (P)

ECUADOR - Galapagos

3891 **Black, J.M. (1976).** Galapagos National Park, problems and solutions. *Parks*, 1(1): 2-4. Invasive species. (P)

3892 **Brockie, R.E., Loope, L.L., Usher, M.B., Hamann, O. (1988).** Biological invasions of island nature reserves. *Biol. Conserv.*, 44(1-2): 9-36. Includes 6 case studies. (P)

3893 **Calvopina, L.H. & F. (1980).** Reproductive biology of wild goats and growth and development of vegetation in permanent goat exclosures on Isla San Salvador (Santiago). *In* Charles Darwin Research Station. Annual Report 1980. Santa Cruz Island, Galapogos, Ecuador. 87-97. (P)

3894 **Calvopina, L.H., Vries, T. de (1975).** Estructura de la poblacion de cabras salvajes (*Capra hircus* L.) y los danos causados en la vegetacion de la Isla San Salvador, Galapagos. (Structure of the population of wild goats and damage caused to the vegetation on the island of San Salvador, Galapagos.) *Rev. Universidad Catolica*, 3(8): 219-241. (En).

3895 **Carlquist, S. (1965).** Island life: a natural history of the islands of the world. New York, Natural History Press. 451p. Origin, evolution and adaptations of island flora and fauna; Galapagos and Hawaiian Islands well covered.

3896 **Christian, K.A., Tracy, C.R. (1980).** An update on the status of Isla Santa Fe since the eradication of the feral goats. *Not. Galap.*, 31: 16-17. (P)

3897 **Colinvaux, P.A., Schofield, E.K., Wiggins, I.L. (1968).** Galapagos flora: Fernandina (Narborough) caldera before recent volcanic event. *Science*, 162: 1144-1145. (P)

3898 **Cruz, F. & J., Lawesson, J.E. (1986).** *Lantana camara* L., a threat to native plants and animals. *Not. Galap.*, 43: 10-11. Illus. An aggressive introduced weed. (P/T)

3899 **Curry-Lindahl, K. (1981).** Twenty years of conservation in the Galapagos. Assessment, lessons and future priorities. *Not. Galap.*, 34: 8-9. (P)

3900 **Dawson, E.Y. (1962).** Cacti of the Galapagos Islands and of coastal Ecuador. *Cact. Succ. J.*, (U.S.), 34(4): 99-105. (P/T)

3901 **DeGroot, R.S. (1983).** Tourism and conservation in the Galapagos Islands. *Biol. Conserv.*, 26(4): 291-300. (P)

3902 **DeRoy, T. (1987).** When aliens take over. *International Wildlife*, 17(1): 34-37. Effect of introduced species. (P)

3903 **DeVries, T. (1977).** Como la caza de chivos afecta la vegetacion en las Islas Santa Fe y Pinta, Galapagos. *Rev. Universidad Catolica*, 5(16): 171-181. (P)

3904 **DeVries, T., Black, J. (1983).** Of men, goats and guava: problems caused by introduced species in the Galapagos. *Not. Galap.*, 38: 18-21. (P/T)

3905 **Delderfield, R. (1980).** Cruising the Galapagos. Darwin's wildlife revisited. *Traveller*, 11(4): 16-20. Illus., col. illus., map.

3906 **Dorst, J., et al. (1972).** Conservation. *In* Simkin, T., *et al.*, eds. Galapagos science: 1972 status and needs. Washington, D.C., Smithsonian Institution. 69-74. (P)

3907 **Duffy, D.C. (1981).** Ferals that failed. *Not. Galap.*, 33: 21-22. Introduced species. (P)

3908 **Eckhardt, R.C. (1972).** Introduced plants and animals in the Galapagos Islands. *BioScience*, 22(10): 585-590. (P)

3909 **Eliasson, U. (1968).** On the influence of introduced animals on the natural vegetation of the Galapagos Islands. *Not. Galap.*, 11: 19-21. (P/T)

3910 **Eliasson, U. (1982).** Changes and constancy in the vegetation of the Galapagos Islands. *Not. Galap.*, 36: 7-12. Illus. Includes effects of introduced and invasive species. (P)

ECUADOR - Galapagos

3911 **Fosberg, F.R. (1966).** The volcanic island ecosystem. *In* Bowman, R.I., *ed.* The Galapagos. Berkeley and Los Angeles, University of California Press. 55-61. (P)

3912 **Gilbert, C. (1974).** The Galapagos and man. *Oceans*, 7(2): 40-47. (P)

3913 **Gold, H. (1984).** The Galapagos: seeing what Charles Darwin saw. *Islands*, 4(5): 40-59. (P)

3914 **Grant, P.R. (1981).** Population fluctuations, tree rings and climate. *Not. Galap.*, 33: 12-16. (P/T)

3915 **Groot, R.S. de (1983).** Tourism and conservation in the Galapagos Islands. *Biol. Conserv.*, 26(4): 291-300. Map. (P)

3916 **Hamann, O. (1975).** Vegetational changes in the Galapagos. *Biol. Conserv.*, 7: 37-59. (P)

3917 **Hamann, O. (1979).** Regeneration of vegetation on Santa Fe and Pinta Islands, Galapagos, after the eradication of goats. *Biol. Conserv.*, 15(3): 215-236. (P)

3918 **Hamann, O. (1979).** The survival strategies of some threatened Galapagos plants. *Not. Galap.*, 30: 22-27. Illus., map. (T)

3919 **Hamann, O. (1981).** Plant communities of the Galapagos Islands. *Dansk Botanisk Arkiv*, 34(2): 1-163. Illus., maps. Description of vegetation types with notes on distribution patterns and species endemism. (T)

3920 **Hamann, O. (1984).** Changes and threats to the vegetation. *In* Perry, R., *ed.* Key environments: Galapagos. New York, Pergamon. 115-131. (T)

3921 **Hamann, O. (1984).** Plants introduced into Galapagos - not by man, but El Nino? *Not. Galap.*, 39: 15-19. Introduced species following El Nino event of 1982-83. (P)

3922 **Hamann, O. (1985).** The El Nino influence on the Galapagos vegetation. *In* Fundacion Charles Darwin para las Islas Galapagos. El Nino en las Islas Galapagos: el evento de 1982-1983. Quito, Charles Darwin Foundation. 299-330. Illus. (P)

3923 **Heacox, K. (1984).** El cuidado de las Islas Encantadas. *Americas*, 36(6): 2-5, 46-49. (P)

3924 **Hickman, J. (1985).** The enchanted islands: the Galapagos discovered. Dover, New Hampshire, Tanager Books. 169p. Includes impact of introduced species. (P)

3925 **Hunter, M. (1984).** Galapagos Island opuntias. *Cact. Succ. J.* (U.S.), 56(6): 258-261. Illus. (T)

3926 **IUCN Threatened Plants Unit (1984).** The botanic gardens list of rare and threatened species of the Galapagos and Juan Fernandez Islands. Kew, IUCN Botanic Gardens Conservation Co-ordinating Body Report No. 11. 25p. Lists plants in cultivation in botanic gardens. (G/P/T)

3927 **Jackson, M.H. (1985).** Galapagos: a natural history guide. Calgary, University of Calgary Press. 284p. Illus., col. illus., maps. (P)

3928 **Johnson, M.P., Raven, P.H. (1973).** Species number and endemism: the Galapagos Archipelago revisited. *Science*, 179: 893-895. (P)

3929 **Kastdalen, A. (1982).** Changes in the biology of Santa Cruz Island between 1935 and 1965. *Not. Galap.*, 35: 7-12. (P)

3930 **Kempf, E. (1985).** The Galapagos fire: 50 years' damage. *WWF News*, 37: 8. (P/T)

3931 **Koster, F., Villa, J. (1984).** Science for conservation in the Galapagos. *In* Unesco-UNEP. Conservation, science and society. Paris, Unesco. 289-296. Contributions to the First International Biosphere Reserve Congress, Minsk, Byelorussia/USSR, 26 September - 2 October 1983. (P)

ECUADOR - Galapagos

3932 **Kramer, P. (1973).** Wildlife conservation in the Galapagos Islands (Ecuador). *Nature and Resources*, 9(4): 3-10. Discussion of agricultural practices. Introduced plants. (P)

3933 **Kramer, P. (1974).** Galapagos conservation: present position and future outlook. *Not. Galap.*, 22: 3-5. (P)

3934 **Kramer, P. (1983).** The Galapagos: islands under siege. *Ambio*, 12(3-4): 186-190. Illus., col. illus., map. (P)

3935 **Lawesson, J.E. (1986).** Report on the most threatened endemic plants in Galapagos. Santa Cruz, Charles Darwin Research Station. 10p. Annotated list; maps. (P/T)

3936 **Lawesson, J.E. (1986).** The problems of plant protection in the Galapagos. *Not. Galap.*, 44: 12. (P)

3937 **Lawesson, J.E., Adsersen, H., Bentley, P. (1987).** An updated and annotated check list of the vascular plants of the Galapagos Islands. Reports from the Botanical Institute No. 16. Denmark, University of Aarhus. 74p. (Sp). (P/T)

3938 **Lewin, R. (1978).** Galapagos: gentle giants of the Galapagos. Part 3. *New Sci.*, 79(1114): 334-336. (P)

3939 **Lewin, R. (1978).** Galapagos: the endangered islands. *New Sci.*, 79(1112): 168-172. (P)

3940 **Lewin, R. (1978).** Galapagos: the rise of optimism. Part 2. *New Sci.*, 79(1113): 261-263. (P)

3941 **Loope, L.L., Hamann, O., Stone, C.P. (1988).** Comparative conservation biology of oceanic archipelagoes: Hawaii and Galapagos. *BioScience*, 38(4): 272-282. Illus., maps. Includes discussion on introduced and invasive species. (P)

3942 **McConnell, S. (1979).** The Galapagos National Park. *Town and Country Planning*, 48(9): 305-307. Illus. (P)

3943 **McHugh, D. (1986).** Recovering the Galapagos. Gland, Switzerland, IUCN Press Service. 3p. Programme to eradicate pigs and goats. (P)

3944 **McKie, R. (1987).** The end of the Galapagos we know? *New Sci.*, 114(1565): 70. Illus. (P)

3945 **Perry, R. (1969).** Conservation problems in the Galapagos Islands. *Micronesica*, 5(2): 275-281. (P)

3946 **Perry, R. (1974).** Sunflower trees of the Galapagos. *Not. Galap.*, 22: 11-13. (P/T)

3947 **Perry, R., ed. (1984).** Key environments: Galapagos. New York, Pergamon. 321p. (P/T)

3948 **Porter, D.M. (1979).** Endemism and evolution in Galapagos Islands vascular plants. *In* Bramwell, D., *ed.* Plants and islands. London, New York, Academic Press. 225-256. (P/T)

3949 **Schofield, E.K. (1973).** A unique and threatened flora. *Garden* (U.S.), 23: 68-73. Illus. (P/T)

3950 **Schofield, E.K. (1973).** Annotated bibliography of Galapagos botany, 1836-1971. *Ann. Missouri Bot. Gard.*, 60(2): 461-477. Includes references to articles covering impact of introduced species. (B/P/T)

3951 **Schofield, E.K. (1973).** Galapagos flora: the threat of introduced plants. *Biol. Conserv.*, 5: 48-51. (P)

3952 **Schofield, E.K. (1981).** Hope for the Galapagos. *Garden* (New York), 5(1): 16-21. Illus., col. illus., map. (P)

ECUADOR - Galapagos

3953 **Schofield, E.K. (1989).** Effects of introduced plants and animals on island vegetation: examples from the Galapagos archipelago. *Conservation Biology*, 3(3): 227-238. Illus., map. (P/T)

3954 **Stone, C.P., Loope, L.L., Smith, C.W. (1988).** Conservation biology in the Galapagos archipelago: perspectives from Hawaii. *Elepaio*, 48(1): 1-8. (P)

3955 **Summerhays, S. (1984).** The endangered species of Darwin's Islands. *Envir. Southwest*, 504: 3-6. Illus., col. illus. (P/T)

3956 **Tindle, R.W. (1983).** Galapagos conservation and tourism: eleven years on. *Oryx*, 16: 126-129. Illus. (P)

3957 **Tuoc, L.T. (1983).** Some thoughts on the control of introduced plants. *Not. Galap.*, 37: 25-26. (P)

3958 **Unsworth, W. (1987).** A special place: problems in paradise. *Environment Now*, 1: 45-47. Col. illus. (P)

3959 **Villa, J.L., Ponce, A. (1984).** Islands for people and evolution: the Galapagos. *In* McNeely, J.A., Miller, K.R., *eds.* National parks, conservation and development. The role of protected areas in sustaining society. Washington, D.C., Smithsonian Institution Press. 584-587. (P)

3960 **Vries, T. de (1977).** Como la caza de chivos afecta la vegetacion en las Islas Santa Fe y Pinta, Galapagos. (How the hunting of goats affected the vegetation on the islands of Santa Fe and Pinta, Galapagos.) *Rev. Universidad Catolica*, 5(16): 171-181. Sp (En). Illus. (P)

3961 **Vries, T. de, Black, J. (1983).** On men, goats and guava - problems caused by introduced species in the Galapagos. *Not. Galap.*, (38): 18-21. Illus., maps. (P)

3962 **WWF (1981).** The WWF in the Galapagos. *Not. Galap.*, 34: 4-5. Illus. (P)

3963 **WWF (1986).** A living laboratory for conservation. *WWF News*, 40: 4-5. (P)

3964 **Weber, D. (1971).** Pinta, Galapagos: une ile a sauver. *Biol. Conserv.*, 4(1): 8-12. Efforts to salvage Pinta. (P)

3965 **Werff, H. van der (1979).** Conservation and vegetation of the Galapagos islands. *In* Bramwell, D., *ed.* Plants and islands. London, Academic Press. 391-404. (Sp). Map. (P/T)

EGYPT

3966 **Abdallah, M.S., Sa'ad, F.M. (1980).** Proposals for conservation of endangered species of the flora of Egypt. *Not. Agric. Res. Centre Herb.* (Egypt), 5: 1-12. (T)

3967 **Ayyad, M.A. (1978).** Vivi Tackholm biological symposium: synopses of lectures prepared for a symposium held in Lund, Sweden, in August 1978. A preliminary assessment of the effect of protection on the vegetation of Mediterranean desert ecosystems. *Bot. Notiser*, 131(4): 444-445.

3968 **Ayyad, M.A., El-Kadi, H.F. (1982).** Effect of protection and controlled grazing on the vegetation of a Mediterranean desert ecosystem in northern Egypt. *Vegetatio*, 49(3): 129-139. (E/T)

3969 **El Hadidi, M.N., El-Ghani, M.A., Springuel, I., Hoffman, M.A. (1986).** Wild barley *Hordeum spontaneum* L. in Egypt. *Biol. Conserv.*, 37(4): 291-300. Maps. Wadi Habis populations under threat. (T)

3970 **IUCN (1986).** IUCN's list of top 24 threatened animals and plants. *Biol. Conserv. Newsl.*, 44: 1-6. (T)

EGYPT

3971　**IUCN Threatened Plants Committee Secretariat (1980).** First preliminary draft of the list of rare, threatened and endemic plants for the countries of North Africa and the Middle East. Kew, IUCN Threatened Plants Committee Secretariat. 170p. (T)

3972　**WWF (1986).** A desert heritage in danger. *WWF News*, 39: 3. Gebel Elba Mountain region near the Red Sea coast.

EL SALVADOR

3973　**Anon. (1987).** El Jocotal. Saving the lagoon. IUCN Special Report: Central America. *IUCN Bull.*, 18(10-12): 15. Map. (P)

3974　**Anon. (1987).** Elaboracion del plan y estrategia del sistema nacional de areas silvestres protegidas de El Salvador. Ministerio de Agricultura y Ganaderia. 74p. Sp. (P)

3975　**Berendsohn, W.G. (1988).** The project "Laderas de la Laguna" an approach to conservation at the botanical garden in El Salvador. *Bot. Gard. Conserv. News*, 1(2): 23-24. Illus., map. (G/T)

3976　**D'Arcy, W.G. (1977).** Endangered landscapes in Panama and Central America: the threat to plant species. *In* Prance, G.T., Elias, T.S., eds. Extinction is forever. Proceedings of a symposium at the New York Botanical Garden, 11-13 May 1976. New York, New York Botanical Garden. 89-104. Maps. The flora and deforestation today. (P/T)

3977　**Daugherty, H.E. (1973).** Conservacion ambiental ecologica de El Salvador con recomendaciones para un programa de accion nacional. San Salvador, Artes Grafica Publicitarias. 56p. Sp.

3978　**Daugherty, H.E. (1973).** The Montecristo cloud-forest of El Salvador - a chance for protection. *Biol. Conserv.*, 5(3): 227-230. (P)

3979　**Reyna de Aguilar, M.L. (1981).** Flora en vias de extincion. Servicio de Parques Nacional y Vida Silvestre. Sp. List of threatened trees, bromeliads, orchids and of endemic trees in protected areas. (T)

3980　**Witsberger, D. (1980).** Tree species of El Salvador and their conservation status. Unknown. List of trees of El Salvador with annotations for endemics, species of low population considered rare, and those in Montecristo National Park. (P/T)

EQUATORIAL GUINEA

3981　**Guinea, E. (1968).** Fernando Po. *In* Hedberg, I. & O., eds. Conservation of vegetation in Africa south of the Sahara. Symposium Proceedings, at 6th plenary meeting of AETFAT, Uppsala. *Acta Phytogeogr. Suec.*, 54: 130-132. Fr. Map. List of indigenous plants in various ecospheres. (P/T)

ETHIOPIA

3982　**Ashine, T. (1984).** What the World Heritage Convention has meant to Ethiopia. *In* McNeely, J.A., Miller, K.R., eds. National parks, conservation and development. The role of protected areas in sustaining society. Washington, D.C., Smithsonian Institution Press. 737-740. (P)

3983　**Beals, E.W. (1968).** Ethiopia. *In* Hedberg, I. & O., eds. Conservation of vegetation in Africa south of the Sahara. Symposium Proceedings, at 6th plenary meeting of AETFAT, Uppsala. *Acta Phytogeogr. Suec.*, 54: 137-140. Map. A valuable paper discussing the diverse vegetational zones of Ethiopia together with plant species typical of each, some worth conserving. (P/T)

ETHIOPIA

3984 **Gilbert, M. (1979).** Possibilities and needs for conservation of plant species and vegetation in Africa. Appendix: Preliminary lists of rare and threatened species in African countries. Ethiopia. *In* Hedberg, I., *ed.* Systematic botany, plant utilization and biosphere conservation. Stockholm, Almqvist & Wiksell International. 92-93. Contains 29 endemic succulent taxa - E:1, V:4, R:12, I:12. (T)

3985 **Hillman, J. & S. (1985).** Bale Mountains National Park (Ethiopia). *Swara*, 8(5): 25-27. (P)

3986 **Hillman, J.C. (1986).** Bale Mountains National Park, management plan. Addis Ababa, Ethiopia, Ethiopian Wildlife Conservation Organisation. 2 vols, 250p. (P)

3987 **Hurni, H. (1984).** Environmental problems in Ethiopia. *Anthos*, 23(3): 3-10. (Ge, Fr). Illus., map. Land degradation.

3988 **Hurni, H., Messerli, B. (1981).** Mountain research for conservation and development in Simen, Ethiopia (with map, scale 1:100,000). *Mountain Res. Devel.*, 1(1): 49-54. (Fr, Ge). Illus., map. (P)

3989 **MacKenzie, D. (1987).** Can Ethiopia be saved? *New Sci.*, 115(1579): 54-58. Col. illus. Desertification, deforestation, soil erosion. Deals with the need for sustainable agricultural systems.

3990 **Micael, T. (1983).** Germplasm conservation in Eritrea. *Germplasm Newsl.*, 4: 7-9. (E)

3991 **Verfaille, M. (1978).** The ericaceaous belt of the Semien Mountains National Park, Ethiopia. *Biol. Jaarb. Dodonaea*, 46: 210-223. Illus. (P)

3992 **Viswanathan, T.V. (1986).** Endangered tree species in northern Ethiopia. *Envir. Conserv.*, 13(1): 71-72. Lists 6 tree species. (T)

3993 **Weinert, E. (1984).** Schutz der Natur in Athiopien. (Nature conservation in Ethiopia.) *Arch. Natursch. Landschaftsforsch.*, 24(3): 221-233. Ge (En, Rus). Maps. (P)

FALKLAND ISLANDS (MALVINAS)

3994 **Anon. (1983).** No conservation money for Falklands. *Oryx*, 17: 144.

3995 **Bertrand, K. (1981).** Conservation on Carcass Island. *Warrah*, 1: 5-6.

3996 **Correa Luna, H., et al. (1975).** Campana cientifica en las Islas Malvinas, 1974 (Noviembre 17 a Diciembre 2). *Anal. Soc. Cientif. Argentina*, 199: 51-180. Sp. Articles on conservation, agronomy, physiognomy and fauna by visiting Argentine scientists.

3997 **Davidson, D. & M.C. (1981).** Plant conservation in the Falklands. *Warrah*, 1: 20-21. (T)

3998 **Oldfield, S. (1987).** Fragments of paradise. A guide for conservation action in the U.K. Dependent Territories. Oxford, Pisces Publications; for the British Association for Nature Conservationists. 192p.

3999 **Strange, I. (1984).** Fortress Falklands: but is its wildlife secure? *Oryx*, 18: 14-21. Illus., map.

FALKLAND ISLANDS - South Georgia

4000 **Vogel, M., Remmert, H., Smith, R.I.L. (1984).** Introduced reindeer and their effects on the vegetation and the epigeic invertebrate fauna of South Georgia (Subantarctic). *Oecologia*, 62(1): 102-109. Maps.

FIJI

4001 **Barrau, J. (1960).** The sandalwood tree. *South Pacific Bull.*, 10(4): 39, 63. Recounts the history of decimation; lists species of *Santalum* in Oceania.

FIJI

4002 **Bayliss-Smith, T.P. (1978).** Batiki in the 1970's: satellite of Suva. In Unesco/ UNFPA. Fiji Island reports No. 4. Canberra, Australian National Univ. 67-128. Batiki once tree covered, now a fire-climax vegetation.

4003 **Berry, M.B., Howard, M. (1973).** Fiji forest inventory 1. The environment and forest types. Surbiton, British Overseas Development Administration.

4004 **Brookfield, H.C. (1981).** Man, environment and development in the outer islands of Fiji. *Ambio*, 10(2-3): 59-67.

4005 **Dunlap, R.C., Singh, B.B. (1980).** A national parks and reserves system for Fiji: a plan. Suva, Fiji, National Trust for Fiji, WWF and UNEP. 117p., annex, appendices. Illus. (P)

4006 **Garnock-Jones, P.J. (1978).** Plant communities on Lakeba and southern Vanua Balavu, Lau Group, Fiji. *Bull. Roy. Soc. New Zealand*, 17: 95-117.

4007 **Gorman, M.L., Siwatibau, S. (1975).** The status of *Neoveitchia storckii* (Wendl): a species of palm tree endemic to the Fijian island of Viti Levu. *Biol. Conserv.*, 8(1): 73-76. (T)

4008 **Harley, R.L. (1963).** Agriculture on Rotuma Island. *S. Pacif. Bull.*, 13(2): 57-61, 63. (T)

4009 **Kirkpatrick, J.B., Hassall, D.C. (1981).** Vegetation of the Sigatoka sand dunes, Fiji. *New Zealand J. Bot.*, 19(3): 285-297. (P)

4010 **Lal, P.N. (1984).** Environmental implications of coastal development in Fiji. *Ambio*, 13(5-6): 316-321. Col. illus., maps. Mangrove.

4011 **Mune, T.L., Parham, J.W. (1956).** The declared noxious weeds of Fiji and their control. Bulletin No. 31. Fiji, Department of Agriculture. 73p.

4012 **Paine, R.W. (1934).** The control of Koster's curse (*Clidemia hirta*) on Tavenui. *Agriculture Journal* (Fiji), 7(1): 10-21.

4013 **Parham, B.E.V. (1953).** Notes on the alien flora of Fiji, or the effect of settlement upon the vegetation of Fiji. *Fiji Soc. Sci. Industry*, 2(2): 76-88.

4014 **Siwatibau, S. (1984).** Traditional environmental practices in the South Pacific - a case study of Fiji. *Ambio*, 13(5-6): 365-368. Col. illus., charts.

4015 **Watters, R.F. (1960).** The nature of shifting cultivation: a review of recent research. *Pacific Viewpoint*, 1: 59-99. Impact of shifting cultivation.

FINLAND

4016 **Anon. (1980).** [Summaries of the presentations at the Symposium "Science and Nature Conservation" arranged on 14-15 March 1980 in Helsinki by the Finnish Biological Society "Vanamo" and Societas pro Fauna et Flora Fennica]. *Luonnen Tutkija*, 84(3): 159-161.

4017 **Anon. (1982).** Pirkanmaan uhanalaiset kasvit ja niiden esiintymisalueet. (Threatened plants and their localities in the Province of Tampere.) *Tamper. Seutukaav. Julk.*, Ser. B, 116: 1-22. Fi. (T)

4018 **Anon. (1983).** Kuidas kaitsta haruldasi taimi? 1. (How should rare plants be protected? - part 1.) *Eesti Loodus*, 26(3): 159-164. Es. Illus., col. illus. (T)

4019 **Anon. (1983).** Kuidas kaitsta haruldasi taimi? 2. (How should rare plants be protected? - part 2.) *Eesti Loodus*, 26(4): 210-218. Es (Rus, En). Illus., col. illus. (T)

4020 **Borg, P. (1976).** [On nature conservation in Finland]. *Lounais-Hameen Luonto*, 57/1976: 12-26. Fi (En). (P/T)

4021 **Borg, P. (1979).** IUCN/WWF: luonnonsuojelun kansainvalista yhteistyota. *Suomen Luonto*, 3: 100-101. Fi. Illus. (T)

FINLAND

4022 **Borg, P., Malmstrom, K. (1975).** Suomen uhanalaiset elain-ja kasvilajit. (Threatened animals and plants in Finland.) *Luonnon Tutkija*, 79: 33-43. Lists 62 vascular plant species threatened throughout the country. (L/T)

4023 **Erkamo, V. (1938).** *Alisma gramineum* Gmelin fur Finnland neu. *Mem. Soc. Fauna Flora Fennica*, 13: 4-7. Ge. Illus. (T)

4024 **Frisen, R. (1988).** [Cooperation for protection of species and biotopes in the Nordic countries]. *Sven. Bot. Tidskr.*, 82(6): 499-501. Sw (En). (P)

4025 **Haapanen, A. (1978).** Kulojen esiintyminen Ulvinsalon Luonnonpuistossa. (Forest fires in Ulvinsalo Nature Reserve.) *Silva Fennica*, 12(3): 187-200. (En). (P)

4026 **Haeggstrom, C.-A. & E., Lindgren, L. (1982).** Rapport om fridlysta och sallsynta vaxter pa Aland. (Report on the protected and rare plants on the Aland Islands.) Nato biologiska station. 137p. (T)

4027 **Hakila, R. (1985).** Kokemuksia Satakunnan uhanalaisen kasviston selvittelysta. (A survey of threatened plants in Satakunta, W. Finland.) *Lutukka*, 1(1): 6-8. Fi. (T)

4028 **Hamet-Ahti, L. (1979).** *Cladrastis kentuckea*, kommeakukkainen hernikasvipuu. *Dendrol. Sallskap. Not.*, 1(10): 27-29. Fi (En). The tree thrives in the Botanic Garden of Helsinki. (G/T)

4029 **Hamet-Ahti, L. (1984).** Changes of the northern boreal vegetation and flora in Finland after the Second World War. *Phytocoenologia*, 12(2/3): 359-361. (Sp). Forestry practices.

4030 **Haukioja, M. (1976).** Metsaluonnon hoito kansallis-puistoissa. (Management of forest ecosystems in natural parks.) *Terra*, 38(1): 21-23. Fi (En). (P)

4031 **Jaakkola, E. (1980).** Finland's rich forests. *Naturopa*, no. 34/35: 55.

4032 **Jarvinen, O. (1982).** Conservation of endangered plant populations: single large or several small reserves? *Oikos*, 38(3): 301-307. (Rus). (P/T)

4033 **Jokinen, P.W. (1983).** Virgin forests. *Naturopa*, 43: 12-13. Illus., col. illus.

4034 **Kaakinen, E., Salminen, P., Ulvinen, T. (1979).** Lapin kilmion lettojen tuho. (Fenland loss in the Lapland Triangle.) *Suomen Luonto*, 38: 130-131. Describes plant species on the decline. (T)

4035 **Kaakinen, E., Salminen, P., Ulvinen, T. (1979).** Rikkaat lettomme. (Eutrophic peatlands of Finland). *Suomen Luonto*, 38: 127-129. Fi (En). Col. illus. (T)

4036 **Kaantonen, M. (1982).** Melon voimalan tuhoamaa Nokianvirran varren kasvistoa. (Floristic loss caused by Melo Water Power Station in Nokia, SW Finland.) *Mem. Soc. Fauna Flora Fennica*, 58(3): 91-96. Fi (En). Illus., map. (T)

4037 **Kalliola, R. (1970).** Some features of nature and conservation in Finland. *Biol. Conserv.*, 2(2): 120-124.

4038 **Kauppi, M. (1979).** The exploitation of *Cladonia stellaris* in Finland. *Lichenologist*, 11(1): 85-89. Illus. Map. (T)

4039 **Kemppainen, E. (1985).** Uhanalaisista itiokasveista. (Threatened cryptogams in Finland). *Lutukka*, 2: 31-35. Fi (En). (T)

4040 **Kuitunen, T. & M. (1985).** [Endangered mire vascular plant species in the vegetation of the Kirkisuo mire in southern Finland.] *Suo. Helsinki Suoseura*, 36(1): 21-30. Fi (En). Illus. (T)

4041 **Leikola, M. (1980).** Metsiemme puuvarojen riittavyyden turvaaminen. (Securing the forest resources of Finland for the future.) *Luonnon Tutkija*, 84(4): 171-175. Fi (En). Map.

FINLAND

4042 **Lindgren, I. (1988).** [Management problems in meadow land and wooded pasture - threatened vascular plants in Finland]. *Sven. Bot. Tidskr.*, 82(6): 419-422. Sw (En). (T)

4043 **Lindholm, T. (1987).** Luonnonsuojelualueittemme metsaluonto: mita se on ja millaiseksi se kehittyy. (How natural are the forests in Finnish nature reserves and what is their future?) *Lusnnon Tutkija*, 91(1): 13-19. Fi (En). Illus., maps. (P)

4044 **Makirinta, U. (1964).** Uber das Vorkommen von *Pilularia globulifera* L. in Finland. *Arc. Soc, Vanamo*, 18(3): 149-159. Ge. Illus. Maps. (T)

4045 **Murto, R. (1982).** Tutkimuksia Uudenmaan Laanin Uhanalaisista Kasveista. 1. Tammisaaren ja Inkoon saaristo. (Studies on the threatened plants in the Province of Uusimaa. 1. Archipelago of Tammisaari and Inkoo.) Helsingin yliopiston kasvimuseo. 62p. Fi. (T)

4046 **Nilsson, O. (1979).** Threatened plants in the Nordic countries. *In* Hedberg, I., *ed.* Systematic botany, plant utilization and biosphere conservation: Proceedings of a symposium held in Uppsala in commemoration of the 500th ann. of the Univ. Stockholm, Almqvist & Wiksell International. 72-75. (T)

4047 **Raatikainen, M., Saari, V. (1981).** Luonnonsuojelualueittemne rajauksesta. (Principles for the delimitation of Finnish nature reserves.) *Luonnon Tutkija*, 85(4): 150-154. Fi (En). Illus., map. (P)

4048 **Raatikainen, M., Saari, V. (1982).** Miten luonnonsuojelualueiden rajaukset ovat onnistuneet Vaarunjyrkan lyonnonsuojelualueella ja oulangan kansallispuistossa? (On what basis have the boundaries of Finnish Nature Reserves been decided?) *Luonnon Tutkija*, 86(4): 138-145. Fi (En). Maps. (P)

4049 **Rancken, T. (1973).** Ekenas stads parkskogar. (The woodland parks of Ekenas, Finland.) *Lustgarden*, 54: 34-36. Illus. (P)

4050 **Rassi, P., Vaisanen, R., eds (1987).** Threatened animals and plants in Finland. Helsinki, Govt Printing Centre. 82p. Illus. Summary of the report of the Committee for the Conservation of Threatened Animals and Plants in Finland. (L/T)

4051 **Reunala, A. (1984).** L'ecologie de la foret boreale et les regimes juridiques de la protection des forets en Finlande. (Ecology of the boreal forest and judicial regimes of forest protection in Finland.) *In* Prieur, M., *ed.* Forets et environnement: en droit compare et international. Paris, Presses Universitaires de France: 31-38. Fr. (L)

4052 **Salminen, P. (1979).** Lettojen etsintakuulutus. (Wanted! Quagmires). *Suomen Luonto*, 3/79: 132-133. Fi. Illus. Marshland preservation programme. (T)

4053 **Soyrinki, N., Salmela, R., Suvanto, J. (1977).** Oulangan kansallispuiston metsa - ja suokasvillisuus. (The forest and mire vegetation of the Oulanka National Park, northern Finland.) *Acta Forest. Fennica*, 154: 150p. Fi (En). Map. (P)

4054 **Suominen, J. (1973).** [Search for endangered plants]. *Suomen Luonto*, 32(3): 128-129. Fi. (T)

4055 **Suominen, J. (1974).** Tuloksia uhanalaisten kasvien tiedustelusta. (Results from an enquiry about endangered plants in Finland.) *Suomen Luonto*, 33: 24, 29. Fi. (T)

4056 **Toivonen, H., Vuokko, S. (1972).** Suomen luonnon- ja kansallispuistojen kasvillisuudesta ja kasvistosta. (On the vegetation and flora of the national parks and nature reserves in Finland.) *Eripainos*, 76(4): 95-114. Fi (En). (P)

4057 **Toth, K. (1978).** Nemzeti parkok Finnorszagban. *Buvar*, 33(1): 31-35. Hu. Illus., map.

4058 **Udvardy, M.D.F. (1983).** Finland expands its protected areas. *Parks*, 7(4): 11. Map. (P)

FINLAND

4059 **Uhanalaisten elainten ja kasvien suojelutoimikunta (Commission for the Conservation of Threatened Plants and Animals in Finland). (1986).** Uhanalaisten elainten ja kasvien suojelutoimikunnan mietinto. III Hotade vaxter i Finland (Suomen uhanalaiset kasvit). (Threatened animals and plants in Finland. Vol.3 Threatened plants in Finland.) Helsinki, Ministry of the Environment. 431p. Fi. (T)

4060 **Uotila, P. (1969).** Sormilehtinen kylmankukka -Etela-hameen erikoisuus. *Kaikuja Hameesta*, 12: 291-307. Fi. Illus. Maps. (T)

4061 **Uotila, P. (1981).** Uudenmaan uhanalaiset putkilokasvit. (Threatened vascular plants of Usimaa Province of Keltiniu.) *Pohjoismaiden Uhanalaiset Elaimet ja Kasvit.* Helsinki. Fi. (T)

4062 **Uotila, P. (1983).** Projektet hotade vaxter i Nylands lan. (The threatened plants project in the Province of Uusimaa, S. Finland.) *Mem. Soc. Fauna Flora Fennica*, 59(3): 106-112. Sw (En). Maps. (T)

4063 **Uotila, P. (1988).** [Losses and additions to the vascular flora of Finland]. *Sven. Bot. Tidskr.*, 82(6): 379-384. Sw (En). (T)

4064 **Vuokko, S. (1980).** Uhanalaiset kasvimme. (Conservation of threatened plants.) *Luonnen Tutkija*, 84(3): 106-107. Fi (En). (T)

4065 **Vuokko, S. (1983).** Uhatut Kasvimme. (Our threatened plants.) Suomen Luonnonsuojelum Tuki Oy, Helsinki. 96p. Fi. Col. illus. Covers Nordic region. (T)

FRANCE

4066 **Albarre, G., Wilde, J. de (1982).** L'espace rural wallon: conservation amenagement et developpement regional. *Parcs Nation. Ardenne Gaume*, 37(1): 5-12. Fr. Illus., map.

4067 **Annezo, N. (1989).** Inventaire de la flore menacee du Massif Armoricain. Vers une strategie pour la conservation du patrimoine floristique regional. *In* Chauvet, M., *ed.* Plantes sauvages menacees de France. Bilan et protection. Actes du colloque de Brest, 8-10 octobre 1987. Paris, Bureau des Ressources Genetiques. 47-59. (T)

4068 **Anon. (1981).** A recently established conservation botanic garden in France. *Threatened Pl. Commit. Newsl.*, 8: 11. Conservatoire Botanique de Porquerolles, concentrating on conservation of Mediterranean plant species. (G)

4069 **Anon. (1983).** Camargue is protected as Ramsar text approved in French, German, Russian. *Parks*, 7(4): 22. Previously published in *CNPPA Newsl.*, 21 (1982-83). (L/P)

4070 **Anon. (1984).** Study and management in Camargue Nature Reserve. *Parks*, 9(3/4): 4-7. Reprinted from *IUCN Bulletin*, 15(4-6), 1984. (P)

4071 **Arbez, M. (1987).** Les ressources genetiques forestieres en France. Tome 1: Les coniferes. Paris, Institut National de la Recherche Agronomique and Bureau des Ressources Genetiques. 236p. Fr. Maps. (E)

4072 **Auffret-David, Mme (1979).** The nature and regional park of La Briere. *Nature and National Park*, 17(64): 14, 19. En, Ge, Fr. Illus. (P)

4073 **Auger, R., Laporte-Cru, J. (1982).** A propos de quelques plantes rare en voie d'extension ou de disparition de la flore Girondine. *Bull. Soc. Linn. Bordeaux*, 10(1): 15-21. Fr. (T)

4074 **Aymonin, G.G. (1972).** Quelques rarefactions ou disparitions d'especes vegetales en France. Causes possibles et sonsequences chorologiques. *Compt. Rend. Somm. Soc. Biogeogr. Paris*, 428/430: 49-64. Fr. (T)

FRANCE

4075 **Aymonin, G.G. (1973).** Dynamisme regressif d'especes vegetales en France; Symposium sur les Especes Vegetales Menacees ou en Voie de Disparition, Arc-et-Senans, 13-15 Novembre 1973. Strasbourg, Minist. de l'Environnement, France et Conseil de l'Europe. 16p. Fr. (T)

4076 **Aymonin, G.G. (1973).** Quelques rarefactions et disparitions d'especes vegetales in France. *Compt. Rend. Soc. Biogeogr. Paris*, 430: 49-64. Fr. Causes possibles et consequences chronologiques. (T)

4077 **Aymonin, G.G. (1974).** Etudes sur les regressions d'especes vegetales en France. Rapport No. 1. Especes vegetales considerees comme actuellement disparues du territoire (dans les conditions naturelles). Paris, Societe Botanique de France, Direction de la Protection de la Nature, Ministere de la Qualite de la Vie. Fr. (T)

4078 **Aymonin, G.G. (1974).** Etudes sur les regressions d'especes vegetales en France. Rapport No. 2. Listes preliminaires des especes endemiques et des especes menacees en France (dans les conditions naturelles). Paris, Societe Botanique de France, Direction de la Protection de la Nature, Ministere de la Qualite de la Vie. 58p. Fr. (T)

4079 **Aymonin, G.G. (1975).** Perimetres proteges et regression d'especes vegetales. *In* Colloque National sur les Parcs Naturels Regionaux et les Parcs Nationaux Francais 5-8 Juin 1975. Marseille, Societe d'Ecologie. 245-251. (T)

4080 **Aymonin, G.G. (1977).** Especes vegetales en regression: essai d'analyse du phenomene en France. *In* Protection des especes vegetales et des milieux naturels de nos regions. Bruxelles, Cercles des naturalistes de Belgique: p. A1-A11. Actes du colloque international, organize a l'Universite de l'Etat de Mons, samedi 23 avril 1977. (T)

4081 **Aymonin, G.G. (1977).** Etudes sur les regressions d'especes vegetales en France. Rapport No. 3. Listes generales des especes justifiant des mesures de protection. Paris, Societe Botanique de France, Direction de la Protection de la Nature, Ministere de la Culture et de l'Environnement. 58p. Fr. (T)

4082 **Aymonin, G.G. (1980).** Une estimation du degre de modification des milieux naturels: l'analyse des regressions dans la flore. *Bull. Soc. Bot. France*, 127(2): 187-195. Fr. (T)

4083 **Aymonin, G.G. (1981).** La couverture vegetale en Rouergue et dans les Causses Majeurs: diversite, degradation et sauvegarde. *Bull. Soc. Bot. France, Lett. Bot.*, 128(1-2): 93-102. Fr (En). Illus.

4084 **Aymonin, G.G. (1982).** Sur quelques especes remarquables des complexes boises de Bourgogne et leur situation de regression en Europe. *Bull. Soc. Bot. France Act. Bot.*, 128(3-4): 95-100 (1981 publ. 1982). Fr (En). (T)

4085 **Aymonin, G.G. (1986).** Pyrenees Aquitaine. Connaitre les plantes protegees. Paris, La Federation Francaise des Societes de protection de la nature le concours du Ministere de l'Environnement. 44p. Fr. (T)

4086 **Aymonin, G.G. (1988).** Les plantes protegees. Region Mediterraneenne. Neuchatel, Switzerland, Delachaux & Niestle, Paris, la Federation Francaise de Societes de protection de la nature. 48p. Fr. (T)

4087 **Aymonin, G.G. (1988).** Les plantes protegees. Regions ouest-nord-est. Neuchatel, Switzerland, Delachaux & Niestle, Paris, La Federation Franaise des Societes de Protection de la Nature. 48p. Fr. (T)

4088 **Aymonin, G.G., Keraudren-Aymonin, M. (1971).** L'effritement des richesses vegetales malgaches. *Bull. Soc. Bot. France*, 118(3/4): 275-279. Fr. (T)

4089 **Barbero, M., et al. (1989).** Menaces pesant sur la flore mediterraneenne francaise. *In* Chauvet, M., *ed.* Plantes sauvages menacees de France. Bilan et protection. Actes du colloque de Brest, 8-10 octobre 1987. Paris, Bureau des Ressources Genetiques. 11-21.

FRANCE

4090 **Baron, Y. (1975).** Le "Pinail" premiere reserve biologique du departement de la Vienne? *Bull. Soc. Bot. Centre - Ouest, n.s.,* 6: 47-48. (P)

4091 **Baron, Y. (1989).** Elements pour un bilan de la flore messicole en Poitou-Charentes. *In* Chauvet, M., *ed.* Plantes sauvages menacees de France. Bilan et protection. Actes du colloque de Brest, 8-10 octobre 1987. Paris, Bureau des Ressources Genetiques. 79-86.

4092 **Beguin, D. (1989).** Initiatives conservatoires en France pour la protection du milieu naturel. *In* Chauvet, M., *ed.* Plantes sauvages menacees de France. Bilan et protection. Actes du colloque de Brest, 8-10 octobre 1987. Paris, Bureau des ressources genetiques. 319-324. Fr.

4093 **Binet, P., Provost, M. (1971).** Les plantes rares en Normandie. *Sci. Nat.,* 103: 2-6. Fr. (T)

4094 **Blondeau, P. (1980).** Une nouvelle Reserve Ardenne et Gaume: le "Tienne Breumont" a Nismes. *Parcs Nation. Ardenne Gaume,* 35(1): 4-12. Illus. (P)

4095 **Bock, C., Prelli, R. (1975).** Notice explicative de la carte des groupements vegetaux du Cirque de Chaudefour (Monts Dore). *Arvernia Biol. Bot., n.s.,* 16: 26p. Fr. Illus., map. (T)

4096 **Bon, M. (1979).** Rare, endangered or new *Inocybe* species in northern France. *Beith. Sydowia Ann. Mycol. Ser. II,* 8: 76-97. (T)

4097 **Bonnet, D. (1984).** La reconstitution d'une foret de protection: vers une nouvelle image des forets des alpes seches. (Reconstitution of a protection forest: towards a new image of forests in the dry Alps. *Rev. Forest. Francaise,* 36(6): 459-467. Fr. Illus., maps.

4098 **Boullard, B. (1980).** Parcs botaniques et ecologie. *Bull. Ass. Parcs. Bot. France,* 3: 10-11. Fr. (G)

4099 **Bournerias, J. (1989).** Problemes relatifs de la conservation des Orchidees de la flore francaise. *In* Chauvet, M., *ed.* Plantes sauvages menacees de France. Bilan et protection. Actes du colloque de Brest, 8-10 octobre 1987. Paris, Bureau des ressources genetiques. 151-162. Fr. (T)

4100 **Bournerias, M. (1983).** Especes vegetales protegees, especes et biotopes a proteger dans le Bassin de la Seine et le nord de la France. *Cah. Nat.,* 39(1): 19-36. Fr. (L/T)

4101 **Brasseur, F., et al. (1977).** La vegetation de la Reserve naturelle domaniale des etangs de Luchy. Gembloux, Ministere de l'Agriculture, Administration des Eaux et Forets, Service de la Conservation de la Nature. (Travaux No. 8.) Illus. (P)

4102 **Bussy, J.-C. (1980).** La foret de l'an demain. Paris, Maison Rustique. 127p. Fr. Illus., maps.

4103 **Carie, P. (1981).** Une plante rare du Vaucluse, *Legouzia castellana. Bull. Mens. Soc. Linn. Lyon.,* 50(4): 122. Fr. (T)

4104 **Chasseraud, J. (1978).** Les Bois de Boulogne et de Vincennes, reserves naturelles de Paris. (The Bois de Boulogne and the Bois de Vincennes, the nature reserves of Paris.) *Rev. Forest. Franc.,* 3: 165-180. Fr (En, Ge, Sp). Illus. (P)

4105 **Chauvet, M., ed. (1989).** Plantes sauvages menacees de France. Bilan et protection. Paris, Bureau des Ressources Genetiques. 494p. Proceedings: Actes du colloque de Brest, 8-10 octobre 1987. (L/P/T)

4106 **Clement, B., Touffet, J. (1987).** Les especes vegetales menacees ou protegees des zones humides de Bretagne. *In* Chauvet, M., *ed.* Plantes Sauvages Menacees de France. Actes du colloque de Brest, 8-10 Octobre 1987. Paris, Bureau des Ressources Genetiques. 109-118. Fr. (T)

FRANCE

4107 **Collin, G. (1985).** The Cevennes Biosphere Reserve: integrating traditional uses and ecosystem conservation. *Parks*, 10(3): 12-14. Illus. (P)

4108 **Comite scientifique du parc national de Port-Cros (1978).** Travaux scientifiques du Parc National de Port-Cros. *In* Review in comptes rendus des seances de list academie dist. agriculture de France 65. 4 vols. 866-869. Fr. (P)

4109 **Conservatoire Botanique de Brest (1985).** Plantes menacees du Massif Armoricain. Brest, Conservatoire Botanique de Brest. 60p. Fr. (L/T)

4110 **Council of Europe (1982).** France: protected flora. *Council of Europe Newsl.*, 82(6): 2-3. (L/T)

4111 **Council of Europe (1983).** France: new nature reserve in the Bruges Marshes. *Council of Europe Newsl.*, 83(5): 3. (P/T)

4112 **Council of Europe (1983).** France: new nature reserve on the Morbihan Atlantic coast. *Council of Europe Newsl.*, 83(2): 2. (P/T)

4113 **Council of Europe (1983).** Protection of wetlands: more progress. *Council of Europe Newsl.*, 83(1): 2. (P)

4114 **Council of Europe (1984).** France: peatland submerged. *Council of Europe Newsl.*, 84(2): 3. Faigne de la Lande submerged to provide cheap electricity.

4115 **Council of Europe (1985).** Protected marine areas. The example of France: appraisal and prospects. Nature and Environment Series No. 31. Strasbourg, Council of Europe. 133p. Illus. (P/T)

4116 **Coutrot, M. (1985).** Comment la foret est-elle surveillee et protegee? (How is the forest monitered and protected?) *Rev. Forest. Franc.*, 37 (Special number): 127-132. Fr.

4117 **Daunas, R. (1977).** La protection des especes vegetales en France: plantes rares ou en voie de disparition en Poitou-Charentes et regions limitrophes. *Bull. Soc. Bot. Centre Ouest, n.s.*, 8: 133-138. Fr. (T)

4118 **De Beaufort, F. (1989).** L'etude et la conservation du patrimoine floristique de la France. *In* Chauvet, M., *ed.* Plantes sauvages menacees de France. Bilan et protection. Actes du colloque de Brest, 8-10 octobre 1987. Paris, Bureau des Ressources Genetiques. 1-2.

4119 **De Beaufort, F., Maurin, H. (1989).** Les especes menacees de la flore de France: historique des travaux, valorisation des connaissances et Livre Rouge. *In* Chauvet, M., *ed.* Plantes sauvages menacees de France. Bilan et protection. Actes du colloque de Brest, 8-10 octobre 1987. Paris, Bureau des Ressources Genetiques. 213-232. (T)

4120 **Debussche, M. (1978).** Etudes sur la vegetation du Massif Central: 4. La vegetation du Parc National des Cevennes cartographie a moyenne echelle. *Bull. d'Ecol.*, 9(1): 7-18. Fr. Maps. (P)

4121 **Dejean, R., Godron, M., Romane, F. (1981).** Amorces d'une etude ecologique dans le Parc National des Cevennes. *In* Schwabe-Braun, A., *ed.* Vegetation als anthropo-okologischer gegenstand. Vaduz, Liechtenstein, J. Cramer. 83-96. Fr (En, Ge). Maps. (P)

4122 **Demoly, J.-P. (1983).** Les especes taxonomiquement isolees parmi les phanerophytes exotiques recenses en France. *Bull. Ass. Parcs Bot. France*, 6: 17-20. Fr. (T)

4123 **Descoings, B.H. (1988).** Le statut des especes vegetales protegees dans le department de l'Ardeche. *Bull. Mens. Soc. Linn. Lyon.*, 57(6): 177-200. Fr. Map. (T)

FRANCE

4124 **Dobrowolski, J. (1978).** O regionalnym parku Camargue oraz parku narodowym w pirenejach Wschodnich we Francji. (The regional park of Camargue and the national park in the east Pyrenees in France.) *Chronmy Przyr. Ojczysta*, 34(3): 97-100. Illus. (P)

4125 **Doyen, A. (1977).** Le parc naturel regional des Landes de Gascogne. Les loisirs en foret landaise. (The regional natural park of the Landes of Gascony: recreation in the Landes Forest.) *Bull. Soc. Roy. Forest. Belg.*, 84(5): 270-288. Fr (Du). (P)

4126 **Dumont, S. (1978).** Pourquoi et comment organiser l'accueil du public en foret. *Rev. Forest. Franc.*, 30(1): 59-69. Fr. Illus.

4127 **Dupont, P. (1983).** Remarques sur les especes vegetales protegees ou meritant de l'etre en Loire-Atlantique et en Vendee. *Bull. Soc. Sci. Nat. Ouest France*, 5(2): 94-105. Fr. (L/T)

4128 **Dupont, P. (1989).** Quelques problemes de protection des especes vegetales. Exemples en Loire-Atlantique et en Vendee. Remarques sur les responsabilites individuelles et collectives. *In* Chauvet, M., *ed.* Plantes sauvages menacees de France. Bilan et protection. Actes du colloque de Brest, 8-10 octobre 1987. Paris, Bureau des ressources genetiques. 99-108. Fr. (T)

4129 **Duvigneaud, J. (1978).** Une vallee mosane a proteger: la basse vallee du Bocq (novelle commune du Grand Yvoir, Province de Namur). *Nat. Mosana*, 31(2): 57-83. Fr. Map. Includes plant list. (P)

4130 **Duvigneaud, J. (1981).** L'herborisation generale de la Societe Royale de Botanique de Belgique dans la partie septentrionale de la Lorraine francaise les 6-7 septembre 1980. *Bull. Soc. Roy. Bot. Belg.*, 114(1): 140-154. Fr (En). Includes nature conservation in Lorraine.

4131 **Duvigneaud, J. (1982).** La halde calaminaire du Rocheux a Theux: une nouvelle reserve d'Ardenne et Gaume. *Parcs Nat. Ardenne Gaume*, 37(3): 119-136. Fr. Illus., map. (P)

4132 **Duvigneaud, J. (1983).** La protection de la flore et des milieux de la vie sauvage en France. *Nat. Mosana*, 36(1): 4-7. Fr. (L/T)

4133 **Duvigneaud, J. (1983).** Reserves et parcs naturels de Wallonie. *Nat. Mosana*, 36(1): 1-3. Fr. (P/T)

4134 **Duvigneaud, J., Meriaux, J.L., Speybroeck, D. van (1982).** La conservation des pelouses calcaires de Belgique et du nord de la France. Metz, Institut Europeen d'Ecologie de Metz et Entente nationale pour la protection de la nature. 42p. Fr. Heathland.

4135 **Duvigneaud, J., et al. (1983).** Conservation de la nature: la fange de l'Abime, a Willerzie, est devenue reserve naturelle: la protection de la tourbiere des Hauts Buttes, dans le departement des Ardennes: creation d'une nouvelle reserve d'Ardenne et Gaume a Theux. *Nat. Mosana*, 36(1): 24-25. Fr. Peatlands. (P/T)

4136 **Encke, B.G. (1983).** Taubergiessen - an exemplary Franco-German nature conservation project. *Allgem. Forstzeit.*, 20: 511-515. Ge. Illus., maps. (P)

4137 **Ermen, R. van (1982).** La protection des milieux sensibles dans les zones peri-urbaines. *Parcs Nation. Ardenne Gaume*, 37(1): 26-30. Fr. Illus.

4138 **European Information Centre for Nature Conservation (1982).** Let the forests live. *Envir. Conserv.*, 9(2): 164. (T)

4139 **Favarger, C. (1986).** Endemisme, biosystematique et conservation du patrimonie genetique. *Atti. Ist. Bot. Univ. Pavia*, 5: 5-14. Fr (En). (T)

4140 **Feil, W. (1979).** La fleche de Goulven (Finistere): formation et propositions pour sa protection. *Penn Bed*, 96: 5-10. Illus.

FRANCE

4141 **Galland, J.-P. (1989).** Les instruments juridiques de protection de la flore sauvage en France. *In* Chauvet, M., *ed.* Plantes sauvages menacees de France. Bilan et protection. Actes du colloque de Brest, 8-10 octobre 1987. Paris, Bureau des ressources genetiques. 233-242. Fr. (L)

4142 **Garraud, L. (1981).** Contribution a l'etude de la flore du Parc National de la Vanoise, Vallee de Peisey-Nancroix (Savoie). *Bull. Mens. Soc. Linn. Lyon*, 50(7): 239-240. Fr. Includes plant list. (P)

4143 **Gehu, J.-M. (1989).** Les plantes en rarefaction et en danger sur les quatre littoraux francais: biotopes et chorologie. *In* Chauvet, M., *ed.* Plantes sauvages menacees de France. Bilan et protection. Actes du colloque de Brest, 8-10 octobre 1987. Paris, Bureau des Ressources Genetiques. (T)

4144 **Gehu, J.-M., Wattez, J.-R. (1989).** Les plantes rares et menacees du Nord de la France (Nord, Pas-de-Calais, Somme) dans leurs habitats. *In* Chauvet, M., *ed.* Plantes sauvages menacees de France. Bilan et protection. Actes du colloque de Brest, 8-10 octobre 1987. Paris, Bureau des ressources genetiques. 87-98. (T)

4145 **Gensac, P. (1978).** Observations thermometriques de 1973 a 1976 dans le Parc National de la Vanoise. Consequences biologiques. *Trav. Sci. Parc Nation. Vanoise*, 9: 9-24. Fr (En). (P)

4146 **Gensac, P. (1982).** Methodological studies for a map of the environment of the region Rhone-Alps, France, for purposes of protecting nature. *Doc. Cartogr. Ecol.*, 24: 99-102.

4147 **Giacobbi, F. (1979).** France's nature and regional parks - laboratories for the countryside tomorrow. *Nat. Nation. Park*, 17(64): 13-14. (Ge, Fr). (P)

4148 **Grey-Wilson, C. (1985).** Plants in peril. *Kew Mag.*, 2(1): 232-234. *Saxifraga florulenta*. (T)

4149 **Grison, P. (1985).** Un danger qui menace la foret francaise: les desequilibres biologiques. (A danger threatening French forests: biological unbalances.) *Rev. Forest. Franc.*, 37 (Special number): 29-44. Fr. Illus.

4150 **Gruber, M. (1983).** Contribution a la flore des vallees de Louron et d'Aure (Hautes Pyrenees). *Bull. Soc. Linn. Provence*, 34: 115-120 (1982). Fr (Sp). (T)

4151 **Hubert, C. (1983).** Apercu de la vegetation de la Reserve de Tintigny. *Parcs Nation. Ardenne Gaume*, 38(2): 70-76. Fr. Illus., map. (P)

4152 **Huglin, P., Pouget, R., Truel, P. (1983).** Work on the conservation of genetic resources of grapevine in France. *Bull. Oiv*, 56(625): 163-171. Fr. Gene banks. (E)

4153 **Jovet, P., Aymonin, G.G. (1980).** Phenomenes d'appauvrissement dans une flore locale et leur signification generale: l'exemple du Pays Basque occidental francais. *Compt. Rend. Soc. Biogeogr.*, 489: 31-40. Fr.

4154 **Kardell, L. (1973).** Varldens vackraste ekskogar? Nagra reflektioner efter ett besok i vastra Frankrikes ekskogsdistrikt. (The famous oak forests of western France.) *Lustgarden*, 54: 42-52. No (En). Illus., map.

4155 **Koebele, F. (1980).** Le plan de protection de la foret rhenane. *Rev. Forest. Franc.*, 32(1): 91-98. Illus. (P)

4156 **Lahondere, C. (1979).** L'action de la S.B.C.O. en faveur de la protection des vegetaux et du milieu naturel en Charente-Maritime et en Vendee. *Bull. Soc. Bot. Centre - Ouest, n.s.*, 10: 6-8. Fr.

4157 **Lavagne, A. (1977).** Note sur les especes vegetales endemiques ou rares presentes sur les iles d'Hyeres. *Trav. Sci. Parc Nation. Port-Cros*, 3: 181-189. Fr (En). (T)

FRANCE

4158 **Lavagne, A., Archiloque, A., Borel, L., Devaux, J.P., Moutte, P. (1983).** La vegetation du parc naturel regional du Queyras. Commentaires de la carte phyto-ecologique au 50,000 degrees. *Biol. Ecol. Medit.* (Aix-en-Provence Editions de l'Universite de Provence), 10(3): 175-248. Fr (En). Illus., maps. (P)

4159 **Le Brun, P. (1959).** Plantes rares et menacees de la France Mediterraneene. *In* IUCN. Animaux et vegetaux rares de la Region Mediterraneene. Proceedings of the IUCN 7th Technical Meeting, vol. 5. Brussels, IUCN. 103-111. Fr. (T)

4160 **Lecointe, A. (1978).** Bryophytes rare, meconnues ou nouvelles pour la Normandie. *Bull. Soc. Linn. Normandie*, 106: 85-112. Fr. Illus., maps. (T)

4161 **Leon, C. (1988).** Action in France. *Threatened Pl. Newsl.*, 19: 19. Recommendations of the colloquy "Plantes Sauvage Menacees de France: bilan et protection", 8-10 October 1987.

4162 **Lesouef, J., Olivier, L. (1989).** Bilan de la flore endemique et sub-endemique de France. *In* Chauvet, M., *ed*. Plantes sauvages menacees de France. Bilan et protection. Actes du colloque de Brest, 8-10 octobre 1987. Paris, Bureau des ressources genetiques. 119-128. Fr. (T)

4163 **Marchenay, P. (1981).** Ethnobotanique et conservation genetique: l'exemple des arbres fruitiers. *J. Agric. Trad. Bot. Appl.* (Jaiba), 28(2): 85-158. Fr (En). Illus., map. Genetic resources; conservation plants in France. (E)

4164 **Meriaux, J.-L. (1976).** Vegetation de la Mare a Goniaux (Parc Nature Regional de St. Amand-Raismes). *Bull. Soc. Bot. Nord France*, 28-29(1-2): 15-18. Fr (En). (P)

4165 **Meriaux, J.-L. (1982).** Especes rares ou menacees des biotopes lacustres et fluviatiles du nord de la France. *In* Symoens, J.J., Hooper, S.S., and Compere, P., *eds*. Studies on aquatic vascular plants. Brussels, Societe Royale de Botanique de Belgique. 398-402. Fr. Proceedings of the International Colloquium on Aquatic Vascular Plants, 23-25 January 1981, Brussels. (T)

4166 **Molinier, R. (1976).** Especes indigenes disparues ou menacees de disparition prochaine en Basse-Provence. *Bull. Soc. Linn. Provence*, 29: 21-29. Fr. Includes list of endangered species. (T)

4167 **Moreau, B. (1978).** Le parc naturel regional du Morvan. Th. vet./Lyon. Lyon. 90p. Fr. Illus. (P)

4168 **Moutte, P. (1979).** Une zone naturelle a proteger: les etangs de Villepay a Frejus (Var.) *Ann. Soc. Sci. Nat. Archeol. Toulon Var*, 31: 119-127. Fr. Map.

4169 **Muller, S., Leroux, P. (1989).** Apercu de la flore menacee de Lorraine. Les actions de protection engagees. *In* Chauvet, M., *ed*. Plantes sauvages menacees de France. Bilan et protection. Actes du colloque de Brest, 8-10 octobre 1987. Paris, Bureau des Ressources Genetiques. 31-45. (T)

4170 **Nel, J. (1982).** Sur la destruction de la faune et de la flore du Massif de la Canaille (La Ciotat), Bouches du Rhone. *Alexanor*, 12(8): 373-379. Fr. Illus., maps. Fire damage; includes lists of vegetation and Lepidoptera.

4171 **Nicole, C-H. (1974).** Cinq ans de lutte pour la sauvegarde des plantes menacees a la Vallee de Joux. *La Nature Vaudoise*, 3: 4-5. Fr. (T)

4172 **Olivier, L. (1978).** Reproduction and reintroduction in a natural environment of endangered species pertaining to shore dunes in the Mediterranean south of France. *Gartn.-Bot. Brief*, 58: 28. (Ge). Abstract. (T)

4173 **Olivier, L. (1979).** Multiplication and re-introduction of threatened species of the littoral dunes in Mediterranean France. *In* Synge, H., Townsend, H., *eds*. Survival or extinction. Proceedings of a conference held at the Royal Botanic Gardens, Kew, 11-17 September 1978. Kew, Bentham-Moxon Trust. 91-93. (G/T)

FRANCE

4174 **Parent, G.H. (1980).** Quelques observations floristiques recentes dans le reserves naturelles d'Ardenne et Gaume. *Parcs Nation. Ardenne Gaume*, 35(1): 13-25. Illus. (P)

4175 **Peterson, C., Pochon, F., Barazer, N. le (1979).** Guide des parcs et jardins de France. Paris, Association des parcs botaniques de France, Association des vieilles maisons francaises. 170p. Fr. Includes national parks. (G/P)

4176 **Plaisance, G. (1975).** La gestion de la foret dans les parcs naturels regionaux. Marseille, Faculte des Sciences et Techniques de Saint-Jerome. 451-459. Colloque national sur les parcs naturels regionaux et les parcs nationaux francais, Marseille, 5-8 June 1975. Fr. (P)

4177 **Plaisance, G. (1979).** La gestion de la foret dans les parcs naturels regionaux. (Forest management in regional parks.) *Espaces*, 36: 18-22. Fr. (P)

4178 **Plank, S.M. (1975).** Nationalparke in Frankreich und die Problematik ihrer Erhaltung: 1. La Vanoise, Les Ecrins, Port Cros. *Mitt. Ludwig Boltzmann-Inst. Umweltwiss. Natursch.*, 1: 41-57. Ge. Illus., map. (P)

4179 **Prin, R. (1978).** La foret domaniale de la Perthe (Aube) et sa reserve Botanique. *Cah. Nat.*, 33(3): 57-65 (1977 publ. 1978). (P)

4180 **Provost, M. (1981).** Quelques donnees recentes sur la repartition de certaines plantes vasculaires rares, meconnues ou nouvelles en Basse-Normandie (2eme partie). *Bull. Soc. Linn. Normandie*, 108: 71-84. Fr (En, Ge). (T)

4181 **Provost, M. (1981).** Quelques donnees recentes sur la repartition de certaines plantes vasculaires rares, meconnues ou nouvelles en Basse-Normandie (3eme partie). *Bull. Soc. Linn. Normandie*, 109: 67-83. Fr (En, Ge). Maps. (T)

4182 **Rickard, M. (1980).** In search of the world's rarest fern? *Brit. Pteridol. Soc. Bull.*, 2(2): 87-88. (T)

4183 **Rokita, Z. (1978).** Parki narodowe we Francji. (The national parks in France.) *Chronmy Przyr. Ojczysta*, 34(6): 71-79. Pol. Maps. (P)

4184 **Royer, J.-M. (1971).** Repartition et ecologie de quelques plantes rares de la cote calcaire de Saone-et-Loire. *Bull. Mens. Soc. Linn. Lyon*, 40(8): 243-249. Maps. (T)

4185 **Saintenoy-Simon, J., Bruynseels, G. (1983).** Une oasis dans le desert hesbignon: la reserve de Cortil-Wodon (Fernelmont) reserve naturelle. *Parcs Nation. Ardenne Gaume*, 38(2): 94-107. Fr. Illus., maps. (P)

4186 **Secretariat de la Faune et de la Flore, Paris (1988).** Atlas preliminaire des especes vegetales protegees du Dauphine. *Inventaires de faune et de flore*, 51: 163p. Paris, Museum National d'Histoire Naturelle, Secretariat de la Faune et de la Flore. Illus., maps. (L/P/T)

4187 **Serusiaux, E. (1982).** Reflexions sur la protection du patrimoine naturel wallon. *Parcs Nation. Ardenne Gaume*, 37(1): 13-25. Fr. Illus., map.

4188 **Trotereau, A. (1978).** *Gentiana schroeteri* Wettst. dans le Parc National de la Vanoise, hybride nouveau pour la flore francaise. *Trav. Sci. Parc Nation. Vanoise*, 9: 53-57. Fr (Ge). (P/T)

4189 **Turlot, J.-P. (1982).** La protection juridique des orchidees en France. *Orchidophile*, 13(53): 120-122. Fr. (L/T)

4190 **Wattez, J.R., Gehu, J.M., Bonnot, E. (1979).** Actions engagees par la Societe de Botanique du nord de la France pour la protection de la Nature dans be nord de la France. Role des Societes de Sci. Nat. dans la protection de la nature. Paris, Fed. Francaise Soc. Sci. Nat. 10p. Fr.

FRANCE - Corsica

4191 **Aymonin, G.G. (1975).** La nature Corse: menaces et espoirs (propos preliminaire). *Bull. Soc. Bot. France*, 121: 5-8. Fr.

4192 **Chauvet, M., ed. (1989).** Plantes sauvages menacees de France. Bilan et protection. Paris, Bureau des Ressources Genetiques. 494p. Proceedings: Actes du colloque de Brest, 8-10 octobre 1987. (L/P/T)

4193 **Dierschke, H. (1981).** Vorkommen, Gefahrdung und Erhaltungsmoglichkeiten naturnaher Vegetation auf Korsika. (Occurrence, threat and conservation possibilities of natural vegetation in Corsica.) *In* Schwabe-Braun, A., *ed.* Vegetation als anthropo-okologischer Gegenstand. Vaduz, Liechtenstein, J. Cramer. 521-532. Ge (Fr). Illus.

4194 **Gamisans, J., Conrad, M., Olivier, L., eds (1981).** Inventaire des especes rares ou menacees de la Corse. Hyeres, Conservatoire Botanique de Porquerolles. Fr. Two unpublished reports describe conservation status and habitats of over 300 rare or threatened taxa. (T)

4195 **Muracciole, M., Gamisans, J. (1989).** L'inventaire et la protection des plantes rares en Corse. *In* Chauvet, M., *ed.* Plantes sauvages menacees de France. Bilan et protection. Actes du colloque de Brest, 8-10 octobre 1987. Paris, Bureau des Ressources Genetiques. 23-29. (P/T)

FRENCH GUIANA

4196 **Granville, J.-J. de (1975).** Projets de reserves botaniques et forestieres en Guyane. Cayenne, ORSTOM. 29p. 16 maps. (P)

4197 **Granville, J.-J. de (1982).** Rain forest and xeric flora refuges in French Guiana. *In* Prance, G., *ed.* Biological diversification in the tropics. New York, Columbia Univ. Press. 159-181. Vegetation map.

FRENCH POLYNESIA

4198 **Anon. (1983).** Les chevres et vaches de Rapa favorisent l'erosion de l'ile. *Les Nouvelles*, 21 January: 3p. Research on Rapa on destructive effect of sheep and goats on vegetation.

4199 **Decker, B.G. (1971).** Plants, man and landscape in Marquesan valleys, French Polynesia. *Dissert. Abstr. Int.*, 31(10): 1p.

4200 **Merrill, E.D. (1940).** Man's influence on the vegetation of Polynesia, with special reference to introduced species. Proc. Sixth Pacific Science Congress. 4. Los Angeles, California University Press. 629-639.

4201 **Reboul, J.L. (1975).** Deux exemples d'introductions malheureuses pour la nature polynesienne. *Te Natura o Polynesia*, 2: 14-20. Fr. (T)

4202 **Salvat, B. (1976).** Une programme interdisciplinaire sur les ecosystemes insulaires en Polynesie francaise. *Cah. Pac.*, 19: 397-405. Fr.

FRANCE - Clipperton Island

4203 **Sachet, M.-H. (1963).** History of change in the biota of Clipperton Island. *In* Gressitt, J.L., *ed.* Pacific basin biogeography: a symposium. Honolulu, Bishop Mus. Press. 525-534. Includes discussion on effects of introduced species.

FRANCE - French Polynesia - Gambier Is.

4204 **Huguenin, B. (1974).** La vegetation des Iles Gambier, releve botanique des especes introduites. *Cah. Pac.*, 18(2): 459-471. Fr.

FRENCH POLYNESIA

FRANCE - French Polynesia - Marquesas Is

4205 **D'Arcy, W.G. (1976).** Near extinct plant in Climatron. *Missouri Bot. Gard. Bull.*, 64(3): 5. (T)

4206 **Decker, B.G. (1975).** Unique dry-island biota under official protection in northwestern Marquesas Islands (Isles Marquises). *Biol. Conserv.*, 5(1): 66-67.

4207 **Dening, G. (1982).** The Marquesas. Papeete, Tahiti, Les Editions du Pacifique. 111p.

4208 **Gillett, G.W. (1972).** The critical need for conservation in the Marquesas Islands. *Newsl. Hawaii. Bot. Soc.*, 11(4): 33-36.

4209 **Halle, F. (1978).** Arbres et forets des Iles Marquises. *Cah. Pac.*, 21: 315-357. Fr.

4210 **Heyerdahl, T. (1940).** Marquesas Islands. *Proc. Sixth Pacific Science Congress*, 4: 543-546. Discusses impact of introduced species.

4211 **Sachet, M.-H. (1973).** The discovery of *Lebronnecia kokioides*. *Bull. Pacific Trop. Bot. Gard.*, 3(3): 41-43. (T)

4212 **Sachet, M.-H., Schafer, P.A., Thibault, J.C. (1975).** Mohotani: une ile protegee aux Marquises. *Bull. Soc. Etudes Oceaniennes* 16(6): 557-568. Includes effects of introduced animals.

4213 **Scahafer, P.A. (1977).** La vegetation et l'influence humaine aux Iles Marquises. Montpellier, Academie de Montpellier, Univ. des Sciences et Techniques du Languedoc. 31p.

FRANCE - French Polynesia - Society Is.

4214 **Creutz, E. (1966).** The tiare apetahi of Raiatea. *Garden* (U.S.), 16(4): 142-144. *Apetahi raiateensis.* (T)

4215 **Dodd, E. (1976).** Polynesia's sacred isle. New York, Dodd, Mead & Company. 224p. Data on *Apetahia raiateensis*, endemic to Mt Temehani. (T)

4216 **Fosberg, F.R., Sachet, M.-H. (1985).** Rare, endangered, and extinct Society Island plants. *Nat. Geogr. Soc. Res. Rep.*, 21: 161-165. Approximately 200 out of the total 850 indigenous plant species are rare or extinct. (T)

4217 **Whistler, W.A. (1982).** A naturalist in the south Pacific: in search of the apetahi. *Bull. Pacific Trop. Bot. Gard.*, 12(1): 1-4. *Apetahia raiateensis.* (T)

FRANCE - French Polynesia - Tuamotu Is.

4218 **Danielsson, B. (1984).** Under a cloud of secrecy: the French nuclear tests in the southeastern Pacific. *Ambio*, 13(5-6): 336-341. Threat to environment by nuclear testing.

4219 **Doumenge, F. (1963).** L'ile de Makatea et ses problemes. *Cah. Pac.*, 5: 41-68. Discusses phosphate mining impact.

FRANCE - French Polynesia - Tubuai Is.

4220 **Halle, N. (1980).** Les orchidees de Tubai (Archipel des Australes, Sud Polynesie). *Cahiers Indo-Pacif.*, 11(3): 69-130. Fr. (T)

FRENCH SOUTHERN TERRITORIES

FRENCH SOUTHERN TERRS. - Iles Kerguelen

4221 **Pascal, M. (1980).** L'archipel des Kerguelen. *Courr. Nat.* (France), 69: 26-33. Fr.

GABON

4222 **Cabelle, S. (1979).** Gabon. *In* Hedberg, I., *ed.* Systematic botany, plant utilization and biosphere conservation: Proc. of a sym. held in Uppsala in commemoration of the 500th ann. of the Univ. Stockholm, Almqvist & Wiksell Int'l. 92.

4223 **Halle, N, Le Thomas, A. (1968).** Gabon. *In* Hedberg, I. & O., *eds.* Conservation of vegetation in Africa south of the Sahara. Symposium Proceedings, at 6th plenary meeting of AETFAT, Uppsala. *Acta Phytogeogr. Suec.*, 54: 111-112. Fr. Tables of threatened species. (T)

4224 **IUCN (?)** La conservation des ecosystems forestiers du Gabon. IUCN, Tropical Forest Programne Series. c.200p. In prep.

4225 **McShane, T. (1990).** Conservation before the crisis - an opportunity in Gabon. *Oryx*, 24(1): 9-14. (E/T)

4226 **Nicoll, M., Langrand, O. (1986).** Conservation et utilisation rationelle des ecosystemes forestiers du Gabon. Gland, WWF/IUCN. Fr.

GERMAN DEMOCRATIC REPUBLIC

4227 **Bark-Signon, I., Patzke, E. (1986).** Schutzenswerte Gebiete im Raum Duren: 1. Der Durener Vorbahnhof. (TK 5105/3). *Gottinger Flor. Rundbr.*, 19(2): 104-109. Ge.

4228 **Benkert, D. (1978).** Die verschollenen und vom Aussterben bedrohten Blutenpflanzen und Farne der Bezirke Potsdam, Frankfurt, Cottbus und Berlin. *Gleditschia*, 6: 20-59. Ge. (T)

4229 **Benkert, D. (1978).** Liste der in den brandenburgischen Bezirken erloschenen und gefahrdeten Moose, Farn- und Blutenpflanzen. (List of extinct and endangered mosses, ferns and flowering plants in the Brandenburg District.) *Naturshutz. Berlin Brandenburg*, 14(2/3): 34-80. Ge. Illus. (T)

4230 **Benkert, D. (1980).** Zum Problem des floristischen Artenschutzes in der DDR unter besonderer Berucksichtigung der Geholze. *Naturwiss*, 29(3): 277-284. Ge.

4231 **Benkert, D. (1980).** Zum Problem des floristischen Artenschutzes in der DDR unter besonderer Beruecksichtigung der Gehoelze. (The problem of protecting plant species in the GDR with special reference to woody plants.) *Wiss. Zeitschr. Humboldt Univ. Berlin, Mat.-Naturwiss.*, 29(3): 255-259. Ge (En, Fr, Rus.). (T)

4232 **Benkert, D. (1982).** Vorlaufige Liste der verschollenen und gefahrdeten Grosspilzarten der D.D.R. *Boletus*, 6(2): 21-32. Ge. A preliminary list of missing and endangered fungi. (T)

4233 **Benkert, D. (1984).** Die verschollenen und vom Aussterben bedrohten Blutenpflanzen und Farne der Bezirke Potsdam, Frankfurt, Cottbus and Berlin. (Extinct and threatened vascular plants and ferns in the Districts of Potsdam, Frankfurt, Cottbus and Berlin.) *Gleditschia*, 11: 251-259. Ge. (T)

4234 **Benkert, D., Succow, M., Wisniewski, N. (1981).** Zum Wandel der floristischen Artenmannigfaltigkeit in der DDR. *Gleditshia*, 8: 11-30 (1980 publ. 1981). Ge (En). (T)

4235 **Buschendorf, J. (1980).** "Rote Listen" der in der DDR gefahrdeten Pflanzen- und Tierarten. Eine Anregung zur Mitarbeit. *Biol. Schule*, 29: 490-492. Ge. Red Data List. (T)

4236 **Doll, R., Hernke, E. (1980).** Das Naturschutzgebiet "Degensmoor" bei Wesenberg (Kreis Neustrelitz). *Nat. Natursch. Mecklenburg*, 15: 63-72. Ge.

4237 **Ebel, F. (1979).** Die Bedeutung der Botanischen Garten fur Landeskultur und Naturschutz: Mitteilungen aus dem Botanischen Garten der Sektion Biowissenschaften, Halle, Nr. 40. *Wiss. Zeitschr. Martin-Luther Univ. Halle-Wittenberg, Math. Nat.*, 28(1): 95-105. Illus. (G)

GERMAN DEMOCRATIC REPUBLIC

4238 **Ebel, F., Rauschert, S. (1982).** Die Bedeutung der Botanischen Garten fur die Erhaltung gefahrdeter und vom Aussterben bedrohter heimischer Pflanzenarten. (The importance of botanic gardens for the preservation of native plants which are endangered and threatened by extinction.) *Arch. Naturschutz und Landschaftsforsch.*, 22(3): 187-199. Ge (En). (G/T)

4239 **Endtmann, K.J., Endtmann, M. (1979).** Flora und Geschichte des Pimpinellenberges bei Oderberg/Krs. Eberswalde. *Gleditschia*, 7: 201-223. Ge.

4240 **Fukarek, F. (1980).** Uber die Gefahrdung der Flora der Nordbezirke der DDR. Ge. *Phytocoenologia*, 7: 174-182. English abstract.

4241 **Gontscharow, N.R. (1985).** Okonomische Bewertung komplexer zonaler Pflanzenschutzsysteme. (Economic evaluation of complex zonal plant protection systems.) *Landwirtschaftswiss. Deutsch. Demokr. Rep.*, 39(6): 117-120. Ge (En, Rus). Maps.

4242 **Heinrich, W. (1984).** Bemuhungen um Orchideenkartierung und Orchideenschutz in der Umgebung von Jena (Thuringen). (Efforts to map and protect orchids in the surrounding of Jena (Thuringia.) *Wiss. Zeitschr. Friedrich-Schiller Univ. Jena, Mat. Naturwiss. Beitr. Okol. Land.*, 33(3): 371-397. Ge (En, Rus). Maps. (T)

4243 **Hellberg, F. (1987).** Uber Vorkommen und Verbreitung bemerkenswerter Gefasspflanzen in den Truper Blanken bei Lilienthal. (About vascular plants to be found in Truper Blanken and their propagation). *Abh. Naturwiss. Ver. Bremen*, 40(4): 323-330. Ge (En). Map.

4244 **Hempel, W. (1976).** Rote Liste der ausgestorbenen und gefahrdeten Pflanzenarten der drei sachsischen Bezirke, Teil I. *Naturschutzarb. Naturk. Heimatforsch. Sachsen*, 18(2): 73-83. Ge. Red List. (T)

4245 **Hempel, W. (1977).** Rote Liste der ausgestorbenen und gefahrdeten Pflanzenarten der drei sachsischen Bezirke, Teil II. *Naturschutzarb. Naturk. Heimatforsch. Sachsen*, 19(1): 28-40. Ge. Red List. (T)

4246 **Hempel, W. (1977).** Rote Liste der ausgestorbenen und gefahrdeten Pflanzenarten der drei sachsischen Bezirke, Teil III. *Naturschutzarb. Naturk. Heimatforsch. Sachsen*, 19(2): 76-86. Ge. Red List. (T)

4247 **Hempel, W. (1978).** Verzeichnis der in den drei sachsischen Bezirken (Dresden, Leipzig, Karl-Marx-Stadt) vorkommenden wildwachsenden Farn?-und Blutenpflanzen mit Angabe ihrer Gefahrdungsgrade. (Index of native ferns and flowering plants in 3 districts and their conservation status.) Dresden, Bezirksnaturshutzorganen. 65p. Ge. (T)

4248 **Hildebrand, K. (1978).** Aufgaben der Forstwirtschaft und des Naturschutzes im Berliner Erholungs-wald. (Tasks for forestry and nature protection in the recreation forest of Berlin.) *Sozialist. Forstwirtsch.*, 28(5): 140-141. Ge. (P)

4249 **Huppe, J. (1981).** Entwicklung der Flora im NSG "Kipshagener Teiche" in den letzten 50 Jahren. Eine quantitative Bestandsaufnahme. *Nat. Heimat.*, 41: 67-69. Ge.

4250 **Illig, H. (1981).** Die geschutzten Pflanzenarten der Luckauer Flora: 9. Pflanzen der Moore. *Biol. Stud. Kreis Luckau*, 10: 23-29. Ge.

4251 **Jaeger, K. (1979).** [Forest conservation]. *Naturschutz - und Naturparke*, 92: 33-36. Ge. Illus.

4252 **Jeschke, L. (1977).** Zum Umpflanzen gefahrdeter Pflanzenarten. *Naturschutzarbeit in Mecklenburg*, 20: 18-20. Ge. (T)

4253 **Jeschke, L. (1979).** Zum umpflanzen gefahrdeter Pflanzenarten. *Naturschutzarb. Berlin Brandenburg*, 15: 66-68. Ge. (T)

GERMAN DEMOCRATIC REPUBLIC

4254 **Jeschke, L., et al. (1978).** Liste der in Mecklenburg (Bezirke Rostock, Schwerin und Neubrandenburg) erloschenen und gefahrdeten Farn- und Blutenpflanzen. *Bot. Rundbr. Bez. Neubrandenburg*, 8: 1-29. Ge. Lists over 600 extinct and endangered plant taxa. (T)

4255 **Kubitzki, K. (1983).** Erhaltung der Naturressourcen und Schaffung und Erhaltung produktiver Walder - Aufgabe von hohem politischem Gewicht. (Conservation of natural resources and producing and preserving efficient forests - problem of great political importance.) *Sozial. Forstwirtsch.*, 33(12): 355-356. Ge. Illus.

4256 **Pietsch, W. (1983).** Vegetationsverhaeltnisse und oekologische Situation im NSG "Fenn im Wittenmoor". (Vegetation relationships and ecological situation in the nature reserve "Fenn im Wittenmoor".) *Arch. Naturschutz Landschaftsforsch.*, 23(1): 1-23. Ge (En, Gre). Illus., map. Includes species list. (P/T)

4257 **Pietsch, W. (1984).** Die Vegetationsstruktur im NSG "Kalktuff-Niedermoor" (Vorderrhon). (Vegetation relationships in the nature reserve "Kalktuff-Niedermoor" (Vorderrhon.) *Arch. Naturschutz Landschaftsforsch.*, 24(3): 189-220. Ge (En, Rus).

4258 **Prinke, E. (1982).** Floristische Neufunde aus dem Flaming und dem Baruther Urstromtal als Vorarbeit zu einer Flora des Flamings. *Gleditschia*, 9: 173-193 (1981 publ. 1982). Ge (En).

4259 **Rattey, F. (1984).** Zum Auftreten von einigen atlantischen Florenelementen in der nordwestlichen Altmark. *Gleditschia*, 11: 125-130. Ge (En).

4260 **Rauschert, S. (1978).** Liste der in den Bezirken Halle und Magdeburg erloschenen und gefahrdeten Farn- und Blutenpflanzen (Rote Liste Gefasspflanzen Halle-Magdeburg). *Natursch. Naturkdl. Heimat. Bez. Halle Magdeburg*, 15(1): 1-31. Ge. Red List. (T)

4261 **Rauschert, S. (1979).** Liste der in der Deutschen Demokratischen Republik erloschenen und gefahrdeten Farn-und Blutenpflanzen. Berlin, Kulturbund der DDR. 56p. (T)

4262 **Rauschert, S. (1980).** Liste der in den thuringischen Bezirken Erfurt, Gera und Suhl erloschenen und gefahrdeten Farn- und Blutenpflanzen. *Landschaftspfl. Natursch. Thuringen*, 17(1): 1-32. Ge. (T)

4263 **Rauschert, S., Benkert, D., Hempel, W., Jeschke, L. (1978).** Liste der in der Deutschen Demokratischen Republik erloschenen und gefahrdeten Farn- und Blutenpflanzen. Berlin, Kulturbund der DDR. 56p. Ge. Col. illus. Lists over 500 threatened taxa and their status in individual districts. (T)

4264 **Reichhoff, L. (1977).** Umsetzen von Populationen geschutzter oder seltener Pflanzenarten bei Gefahrdung ihrer Standorte. *Natursch. Naturkdl. Heimat. Bez. Halle Magdeburg*, 14: 33-36. Ge. (T)

4265 **Reichhoff, L., Bohnert, W. (1978).** Zur Pflegeproblematik von Festuco-Brometea-, Sedo-Schleranthetea- und Corynephoreta-Gesellschaften in Naturschutzgebieten im Suden der DDR. (Problems of management of Festuco-Brometea-, Sedo-Schleranthetea- and Corynephoreta-association in nature reserves in south GDR.) *Arch. Naturschutz Landschaftsforsch.*, 18(2): 81-102. Ge (En, Rus). Illus., map. (P)

4266 **Reichhoff, L., Bohnert, W. (1983).** Die Vegetation des Naturschutzgebietes "Vogtei" in der Fuhneaue bei Zorbig. (Vegetation of the nature reserve "Vogtei" in the Fuhne flood plain near Zorbig.) *Arch. Naturschutz Landschaftsforsch.*, 23(3): 181-192. Ge (En, Rus). Illus. (P)

4267 **Schlosser, S. (1982).** Genressourcen fur Forschung und Nutzung. *Natursch. Naturkdl. Heimat. Bez. Halle und Magdeburg*, 19: 1-96. Ger. Col. illus, illus., maps. Contains a series of papers about the potential use of plant genetic resources in the G.D.R. (G)

GERMAN DEMOCRATIC REPUBLIC

4268 **Schmidt, P. (1978).** Der Forstbotanische Garten Tharandt und Probleme der Landeskultur und des Umweltschutzes. (The Tharandt Forest Botanical Garden and problems of cultivation and environmental protection.) *Wiss. Zeitschr.* (Dresden), 27(6): 1349-1351. (G)

4269 **Schmidt, P. (1983).** Moglichkeiten und Grenzen des Artenschutzes fur Geholze aus Sicht botanischer Garten und Arboreten. (Possibilities and limits of conservation of woody plant species from the standpoint of botanic gardens and arboreta.) *Folia Dendrologica*, 10(83): 221-243. Ge (Cz, En). (G/T)

4270 **Tauber, F. (1984).** Ocrotirea naturii in alte tari. Ocrotirea naturii in R.D. Germana: 1. Mod de organizare si de functionare a serviciilor. (La protection de la nature dans la Republique Democrate Allemande). *Ocrot. Nat. Mediul. Inconjurat.*, 28(1): 48-50. Rum (Fr).

4271 **Thomasius, H., ed. (1978).** Wald, Landeskultur und Gesellschaft. (Forest, environmental conservation and society.) 2nd ed. Jena, Fischer Verlag. 466p. Ge. Illus.

4272 **Trauboth, V. (1981).** Pflege von Waldschutzgebieten am Beispiel des NSG "Ibengarten". (Preservation and care of protective forest areas on the example of the nature conservation reserve "Ibengarten".) *Arch. Naturschutz Landschaftsforsch.*, 21(3): 149-158. Ge (En, Rus). (P)

4273 **Wentzel, K.F., Zundel, R. (1984).** Hilfe fur den Wald: Ursachen, Schadbilder, Hilfsprogramme. (Help for forest.) Niedernhausen/Ts, Falken-Verlag. 128p. Ge. Col. illus., col. map.

GERMANY, FEDERAL REPUBLIC OF

4274 **Ammer, U. (1978).** Naturschutz und Erholung in Naturparken. *Naturschutz- und Naturparke*, 91: 21-28. Ge. (P)

4275 **Ammer, U. Mossmer, E.M., Schirmer, R. (1985).** Vitalitat und Schutzbefahigung von Bergwaldbestanden im Hinblick auf das Waldsterben. (Vitality and protective capability of mountainous forest stands in regard to the forest decline.) *Forstwiss. Centralbl.*, 104(2): 122-137. Ge (En). Maps.

4276 **Anon. (1974).** Rote Liste bedrohter Farn- und Blutenpflanzen in Bayern. Schriftenreihe Naturschutz- und Landschaftsplege. Bayer, Landesamt fur Umweltschutz. p.8. Ge. Lists 566 spp. of vascular plants threatened in Bavaria, 27.9[of total. (T)

4277 **Anon. (1976).** Bedrohte Pflanzen und Tiere, vorgestellt vom Bayerischen Staatministerium fur Landesentwicklung und Umweltfragen. *Naturschutz- und Naturparke*, 83: 37-39. Ge. Illus.

4278 **Anon. (1977).** Bedrohte Tiere und Pflanzen. *Naturschutz- und Naturparke*, 87: 51-54. Illus.

4279 **Anon. (1977).** Germany (Federal Republic). Act on nature protection and countryside preservation (Federal Nature Protection Act), 20 December 1976. *Food Agric. Legis.*, 26(2): 114-116. (L)

4280 **Anon. (1978).** Botanischer Garten Freiburg: Informationen Nr. 10. Das System der Blutenpflanzen, ein Stammbaummodell. *Gartn.-Bot. Brief*, 58: 52-55. (G)

4281 **Anon. (1978).** Fragen des Artenschutzes in Baden-Wurttemberg: Referate des gleichnamigen Symposiums uber Veranderungen von Flora und Fauna, Arten- und Biotopschutz, Rasterund Biotopkartierung vom 25-27 October 1977 in Bad Boll. Karlsruhe, Landesantstalt fur Unweltschutz Baden Wurttemberg, Institut fur Okologie und Naturschutz. 503p. Ge. Illus., col. illus. (T)

4282 **Anon. (1979).** Naturparke in der Bundesrepublic Deutschland. *Naturschutz- und Naturparke*, 92: 58-60. (P)

GERMANY, FEDERAL REPUBLIC OF

4283 **Anon. (1982).** "Rote Listen" gefahrdeter Pflanzen- und Tierarten von Berlin (West). *Inform. Berlin Landschaft*, 3(11): 4p. Ge. Red List. (T)

4284 **Anon. (?)** Naturschutz in der Grosstadt. Berlin. Naturschutz und Landschaftsplege in Berlin (West). Heft 2. 24p. Ge. Col. illus. Map. (T)

4285 **Apel, J., Nothdurft, H. (1981).** Erhaltungskulturen in Hamburg. *Gartn.-Bot. Brief*, 69: 29-32. Ge. (G)

4286 **Arnold, W., Bosch, B., Schmid, H. (1982).** Present condition and objectives for the conservation of the juniper heathlands of the Swabian mountains. *Forstwiss. Centralbl.*, 101(5): 311-346. Ge. Col. Illus.

4287 **Ashmore, M., Bell, N., Rutler, J. (1985).** The role of the ozone in forest damage in West Germany. *Ambio*, 14(2): 81-87. Pollution

4288 **Barker, M.L. (1982).** Comparison of parks, reserves and landscape protection in three countries of the eastern Alps. *Envir. Conserv.*, 9(4): 275-285. Maps. (P)

4289 **Barney, G.O., ed. (1977).** The Global 2000 Report to the President: entering the twenty-first century. Washington, D.C., U.S. Govt Printing Office. ix, 47p. (Ge). Maps. Trends in population, resources and environment.

4290 **Bayrisches Landesamt fur Umweltschutz (1987).** Berichte aus dem Bayerischen Landesamt fur Umweltschutz (1). Schriftenreihe. Heft 78. 101p. Ge. (T)

4291 **Belke, E. (1979).** Naturpark Homert. *Naturschutz- und Naturparke*, 94: 1-8. Illus. (P)

4292 **Benkert, D. (1978).** Die verschollenen und vom Aussterben bedrohten Blutenpflanzen und Farne der Bezirke Potsdam, Frankfurt, Cottbus und Berlin. *Gleditschia*, 6: 20-59. Ge. (T)

4293 **Benkert, D. (1984).** Die verschollenen und vom Aussterben bedrohten Blutenpflanzen und Farne der Bezirke Potsdam, Frankfurt, Cottbus and Berlin. (Extinct and threatened vascular plants and ferns in the Districts of Potsdam, Frankfurt, Cottbus and Berlin.) *Gleditschia*, 11: 251-259. Ge. (T)

4294 **Bennert, H.W., Kaplan, K. (1983).** Besonderheiten und Schutzwuerdigkeit der Vegetation und Flora des Landschaftsschutzgebietes Tippelsberg/Berger Muehle in Bochum. (Specialities and protection-merit of vegetation and flora of nature protection region Tippelsberg/Berger Muhle in Bochum.) *Dechenia*, 135(1): 5-14. Ge (En). Illus. (P/T)

4295 **Berger, M. (1985).** Naturschutz im Munsterland auf neuen Wegen. *Natur Heimat*, 45(1): 3-8 Ge.

4296 **Bibelriether, H. (1984).** Nationalpark Bayerischer Wald (D). (Bavarian Forest National Park (D).) *Gart. Landsch.*, 94(11): 35-38. En, Ge. Illus. (P)

4297 **Blab, J., Nowak, E., Sukopp, H., Trautmann, W. (1977).** Rote Liste der gefahrdeten Tiere und Pflanzen in der Bundesrepublik Deutschland. (Red List of threatened plants and animals in the Federal Republic of Germany.) Greven, Kilda. 67p. Ge. Lists 822 threatened vascular plant species in W. Germany, almost 31% of total. (T)

4298 **Blab, J., Nowak, E., Trautmann, W., Sukopp, H. (1984).** Rote Liste der gefahrdeten Tiere und Pflanzen in der Bundesrepublik Deutschland, 4th Ed. Greven, Kilda-Verlag. 270p. Ge. Lists threatened flowering plants, mosses, lichens, fungi, and algae. (T)

4299 **Boberfeld, W.O. von (1989).** [Principles to the protection of natural grassland with regard to a botanical aspect]. *Z. Kulturtech. Landentwicklung*, 30(2): 92-104. Ge (En).

4300 **Bocker, R. (1978).** Vegetations- und Grundwasser-Verhaltnisse im Landschafts -Schutzgebiet Tegeler Fliesstal (Berlin - West). *Verh. Bot. Ver. Prov. Brandenburg*, 114: 164p. Ge. Illus., map.

GERMANY, FEDERAL REPUBLIC OF

4301 **Bottcher, H., Gross, G. (1979).** Der Glindbusch bei Rotenburg (Han.) als Wuchsort gefahrdeter Pflanzenarten. (Glindbusch near Rotenburg (Hanover), as a place for growing endangered plant species.) *Gottinger Flor. Rundbr.*, 13(1): 24-29. Ge. (G/T)

4302 **Braun, G. (1982).** Die Immissionsgefahrdung Bayerischer Walder. (Endangering of Bavarian forests by industrial emmissions.) *Allgem. Forstzeit.*, 37(15): 446-447. Ge. Air pollution.

4303 **Brehm, K. (1979).** Factors affecting the vegetation of fresh water reservoirs on the German coast. *In* Jefferies, R.L., Davy, A.J., *eds.* Ecological processes in coastal environments. British Ecological Society Symposia 19. Oxford, Blackwell Scientific Publications. 603-615. (Fr). Maps.

4304 **Brinkman, H. (1978).** Schutzenwerte Pflanzen und Pflanzengesellschaften der Senne. *Sonderh. Ber. Naturwiss. Ver. Bielefeld*, 33-68. Ge.

4305 **Burrichter, E., Pott, R., Raus, T., Wittig, R. (1980).** Die Hudelandschaft *Borkener Paradies* im Emstal bei Meppen. *Abh. Landesmus. Naturk. Munster Westfalen*, 42(4): 3-69. Ge (En). Illus., map. (P/T)

4306 **Carlsen, C. (1977).** Zum Bundesnaturschutzgesetz. *Naturschutz- und Naturparke*, 87: 28-34. Ge. (P)

4307 **Clauss, V. (1980).** Abbau von oberflachennahen mineralischen Rohstoffen in Schleswig-Holstein - ein Problem auch fur Naturschutz und Landschaftspflege. (Surface mineral workings in Schleswig-Holstein: a problem for both nature conservation and landscape management.) *Forstarchiv*, 51(5): 81-84. Ge. Illus.

4308 **Cordes, H. (1979).** Gefahrdete Pflanzenarten aus der "Roten Liste der Farn- und Blutenpflanzen" - ihre Verbreitung im Bereich der Regionalstelle Bremen Teil 1. *Abh. Naturwiss. Ver. Bremen*, 39: 7-40. Ge (En). Maps. (T)

4309 **Council of Europe (1981).** National activities: Federal Republic of Germany. *Council of Europe Newsl.*, 81(6/7): 3. Additions & losses in protected areas. (P)

4310 **Daiss, H., Hennecke, M., Schneider, P. (1988).** Pflegemassnahmen zur Erhaltung orchideenreicher Trockenstandorte im Schwabischen Wald. (The conservation of orchid-rich, dry habitats in the Rems-Murr-district.) *Mitteilungsbl. Arbeitskr. Heim. Orch. Baden-Wurttemberg*, 20(1): 75-101. Ge (En).

4311 **Deppner, F. (1979).** Waldschutzprobleme bei der Walderneuerung nach den Orkanschaden vom 13. November 1972. (Forest conservation problems in afforestation after hurricane injuries inflicted November 13, 1972.) *Forst und Holzwirt*, 34(1): 15-18. Ge. Illus.

4312 **Der Deutsche Naturschutzring (1980).** Moore: Bedeutung, Schutz, Regeneration. (Wetlands: significance, protection, regeneration.) Bonn, Deutscher Naturschutzring. 21p. Ge. Col. illus. (P)

4313 **Dieterich, H. (1978).** Vegetationsveranderungen in Waldschutzgebieten. (Vegetation changes in forest reserves.) *Veroff. Natursch. Landschaftspfl. Baden-Wurttemberg*, 11: 231-235. Ge. (P)

4314 **Doll, R. (1980).** Der GrosseGollin-See im Kreis Templin. *Feddes Rep.*, 91(1-2): 127-140. Ge (En).

4315 **Duell, R. (1980).** [Grid maps of the flora of Duisburg and environs; with data on the habitat, origin and degree of threat to vascular plants in the Greater Duisberg region since 1800]. Opladen, West Germany, Westdeutscher Verlag. Forschungsberichte des Landes Nordrhein-Westfalen, 2910. Fachgruppe Physik-Chemie-Biologie. Ge. Illus. Maps. (T)

4316 **Dummler, H. (1979).** Gedanken zur Fortentwicklung der deutschen Naturparke. *Naturschutz- und Naturparke*, 94: 33-36. Ge. Illus. (P)

GERMANY, FEDERAL REPUBLIC OF

4317 **Eberle, G. (1979).** Pflanzen unserer Feuchtigkeitsgebiete und ihre Gefahrdung. (Plants of our wetlands and the threat to them.) Frankfurt am Main, Kramer. 236p. Ge. Illus.

4318 **Egli, H. (1976).** Die offentliche Kundegebung 20 Jahre deutscher Naturparke. *Naturschutz- und Naturparke*, 82: 13-26. Ge. (P)

4319 **Encke, B.G. (1983).** Taubergiessen - an exemplary Franco-German nature conservation project. *Allgem. Forstzeit.*, 20: 511-515. Ge. Illus., maps. (P)

4320 **Ertl, J. (1978).** Die Naturschutzpolitik des Bundes. (Nature conservation policy of the Federation.) *Forst und Holzwirt*, 33(13): 277.

4321 **Erz, W. (1970).** Nature conservation and landscape management in the Federal Republic of Germany - dates, facts and figures. Cambridge and Bonn-Bad Godesberg, IUCN Wildlife Trade Monitoring Unit. 24p. Map. (L/T)

4322 **Erz, W., Henke, H. (1977).** Zur Moglichkeit von Nationalparken in der Bundesrepublik Deutschland. (Potential sites for national parks in the German Federal Republic.) *Natur und Landschaft*, 52(1): 3-9. Ge. Ref. in *Forest. Abstr.*, 41: 3836 (1980). (P)

4323 **Evers, F.H. (1983).** Pflege und Behandlung immissionsgefahrdeter Waldbestande. (Care and treatment of emmission-endangered forest stands.) *Forstwiss. Forsch. Beih. Forstwiss. Zentralbl.*, 38: 58-59. Ge.

4324 **Fessler, A. (1978).** Pflanzenreservate oder Neuansiedlung. (Plant reserves or replanting.) *Gartenpraxis*, 10: 478-479. Ge. (P)

4325 **Fingerle, K. (1985).** Landlicher Wegebau und Forderungen des Naturschutzes und der Landschaftspflege. *Natursch. Nordh.*, 8: 29-38. Ge. (T)

4326 **Fink, H.G. (1978).** Vorschlage zur Erhebung fur den Artenschutz erforderlicher zusatzlicher Gelandedaten im Rahmen der Kartierung der Flora der Bundesrepublik Deutschland. *Gottinger Flor. Rundbr.*, 12(4): 128-136. Ge. Map.

4327 **Fink, H.G. (1983).** Literaturdokumentation fur Arten- und Biotopschutz: Pflanzen und Pflanzengesellschaften (Gefasspflanzen/Pteridophyta, Spermatophyta). *Natur und Landschaft*, 58: 242. Ge. Illus. (T)

4328 **Finke, L. (1980).** Okologie und Umweltprobleme. (Ecology and environmental problems.) *Geogr. Rundsch.*, 32 (4): 188-193. Ge (En). Illus. (L/P)

4329 **Fritsche, U., Schebek, L., Schubert, C. (1984).** Rettung fur den Wald: Strategien und Aktionen. (Saving the forests: strategies and campaigns.) Frankfurt am Main, Fischer Taschenbuch Verlag. 191p. Ge. Illus. Air pollution.

4330 **Froment, A. (1978).** Birkengutachten: Birken-Dynamik und Erhaltung der Heidegesellschaften im Naturschutzgebiet Luneburger Heide. *Naturschutz- und Naturparks*, 88: 7-10. Ge. (P)

4331 **Garue, E. (1987).** Stand des niedersachsischen Pflanzenarten-Erfassungsprogramms und Bericht von dem Gelandetreffen 1986. (State of the Lower Saxonian plant species recording programme and report on the area meeting 1986.) *Flor. Rundbr.*, 21(1): 56-68. Ge (En). (T)

4332 **Gayler, W. (1978).** Naturgemasse Waldwirtschaft im Naturschutz. (Natural methods of forest management in nature conservation.) *Forst und Holzwirt*, 33(15): 327-329. Ge.

4333 **Gerstberger, P. (1984).** Seltene und bemerkenswerte Blutenpflanzen aus der Umgebung von Bonn sowie aus der Nordeifel. (Rare and noteworthy flowering plants from the neighbourhood of Bonn and North-Eifel.) *Decheniana*, 137: 62-65. Ge (En). (T)

GERMANY, FEDERAL REPUBLIC OF

4334 **Greenberg, D. (1985).** Fast cars and sick trees. *Int. Wildlife*, 15(4): 22-24. Black Forest is dying. Pollution.

4335 **Gross, J. (1967).** Bedrohte Schonheit. Geschutzte Pflanzen. Hannover, Landbuch-Verlag. 133p. Ge.

4336 **Haeupler, H. (1976).** Die verschollenen und gefahrdeten Gefasspflanzen Niedersachsens, Ursachen ihres Ruckgangs und zeitliche Fluktuation der Flora. (The extinct and endangered species of vascular plants in Lower Saxony). *In* Schrift, V., *ed.* Die verschollenen und gefahrdeten Gefasspflanzen Niedersachsens: Ursachen ihres Ruchgangs". 125-131. Ge (En). Lists 687 threatened plants, 37.5[of total. (T)

4337 **Haeupler, H., Montag, A., Woldecke, K. (1976).** Verschollene und gefahrdete Gefasspflanzen in Niedersachsen. Sonderdruck aus "30 Jahre Naturschutz und Landscaftspflege in Niedersachsen", herausgegeben vom Niedersachsischen Ministerium fur Ehnahrung, Landwirtschaft und Forsten. Ge. Col. illus. (T)

4338 **Haeupler, H., Montag, A., Woldecke, K., Garve, E. (1983).** Rote Liste Gefasspflanzen Niedersachsen und Bremen. *Fachbehorde fur Naturschutz, Merkblatt*, Nr. 18. 34p. Ge. (T)

4339 **Haffner, P., Sauer, E., Wolf, P. (1979).** Rote Liste der im Saarland ausgestorbenen und gefahrdeten hoheren Pflanzen. Wiss. Schr.-R. der Obersten Naturschutzbehorde. Ge. Vol. 1 with appendix: Rote Liste der im Saarland ausgestorbenen und gefahrdeten hoheren Pflanzen. 12p. Red List. (T)

4340 **Harms, K.H., Philippi, G., Seybold, S. (1983).** Verschollene und gefahrdete Pflanzen in Baden-Wurttemberg. Rote Liste der Farne und Blutenpflanzen (Pteridophyta, Spermatophyta), 2nd revision. *Veroff. Natursch. Landschaftspfl. Baden-Wurttemberg*, 32: 160p. Ge. Red List. (T)

4341 **Hasel, K. (1978).** Der forstliche Beitrag fur Naturschutz und Raumordnung auf Grund der neuesten Gesetzgebung in Saarland. (Contribution of forestry to nature protection and land utilization planning and recent legislation in Saarland.) *Forst und Holzwirt*, 33(17): 376-384. Ge. (L)

4342 **Hatzfeldt, H. (1984).** Der Wald stirbt: Forstliche Konsequenzen. (The forest is dying: consequences for forestry.) Karlsruhe Muller. 199p. Ge. Effect of air pollution on forests.

4343 **Hausbeck, P., Schuster, R. (1980).** [Endangered nature, rare plants and animals in northern Oberpfalz.] Regensburg, West Germany, Pustet. 196p. Ge. (T)

4344 **Hecker, U. (1982).** Erfahrungen mit Vermehrungskulturen im Botanischen Garten der Johannes-Gutenberg-Universitat Mainz. *Gartn.-Bot. Brief*, 72: 7-14. Ge. (G)

4345 **Hennig, R. (1982).** Der freiwillige Forstschutz im Sachsenwald. (Voluntary forest protection in the Saxony Forest.) *Allgem. Forstzeit.*, 37(46): 1432. Ge. Illus.

4346 **Heydemann, B. (1980).** Bedeutung der Arten fur Okosysteme als Grundlage des Okosystemschutzes. (Importance of species for ecosystems as basis for ecosystem protection.) *In* Riedel, W., *ed.* Schutz von Flora und Fauna und ihrer naturlichen Lebensraume. Sankelmark, Akademie Sankelmark. 9-48. Ge. Illus. National parks. (P)

4347 **Hiemeyer, F. (1980).** Das Naturschutzgebiet "Stadwald Augsburg". *Ber. Naturwiss. Ver. Schwaben*, 84(1-2): 4-13. Ge. Illus., map.

4348 **Hilbig, W., Illig, H., Large, E. (1985).** Schutz und Erhalturg gefahrdeter Ackerwildpflanzen. *Arch. Naturschutz. Landschaftsforsch.*, 25(2): 73. Ge. (T)

4349 **Hillesheim-Kimmel, U., Karafitat, H. (1978).** Botanische Gutachten uber neue Naturschutzgebiete und schutzwurdige Gebiete im Regierungsbezirk Darmstadt. *Beih. Schriftenr. Inst. Natursch. Darmstadt.*, 27: 27-131. Ge. (P)

GERMANY, FEDERAL REPUBLIC OF

4350 **Hofmann, F. (1978).** Wildbach- und Lawinenverbauung und Naturschutz. (Wild brook and avalanche obstruction and nature protection.) *Allgem. Forstzeit.*, 89(8): 252-253. Ge. Illus. Ref. in *Agricola*, 79 suppl.

4351 **Jaeger, K. (1978).** Sozialbrache und Naturschutz. *Naturschutz- und Naturparke*, 89: 8-10. Ge. Illus. (P)

4352 **Jaeger, K. (1979).** Walderhaltung: zur Forderung von Naturschutz und Landschaftspflege. *Naturschutz- und Naturparke*, 92: 33-36. Ge. (P)

4353 **Jeckel, G. (1981).** Die Vegetation des Naturschutzgebietes "Breites Moor" (Kreis Celle, Nordwestdeutschland). *Tuexenia*, 1: 185-209. Ge.

4354 **Jobst, E. (1979).** Was wird aus unseren Almen? *Jahrb. Ver. Schutz Bergwelt*, 44: 41-59. Ge. Illus.

4355 **Kalbeber, H., Korneck, D., Muller, R., Nieschalk, A. & C., Sauer, H., Seibig, A., eds (1980).** Rote Liste der in Hessen ausgestorbenen, verschollenen und gefahrdeten Farn- und Blutenpflanzen: 2. Wiesbaden, Hessische Landesanstalt fur Umwelt. 46p. Ge. Red List. (T)

4356 **Karl, H. (1978).** Welche Ziele verfolgt der Naturschutz in der bayerischen Rhon? (What are the aims of nature protection in the Bavarian Rhon Mtns.?) *Allgem. Forstzeit.*, 33(51/52): 1527. Ge. Illus.

4357 **Kaule, B. (1979).** Die Trockenrasen des bayerischen voralpinen Hugel- und Moorlandes. *Jahrb. Ver. Schutz Bergwelt*, 44: 223-264. Illus.

4358 **Keister, E. (1985).** Death in the Black Forest. *Smithsonian*, 16(8): 211-230.

4359 **Klein, M. (1983).** Schutzwalder im Saarland. (Reserved forests in Saarland.) *Allgem. Forstzeit.*, 38(3): 56-57. Ge. Illus., maps. (P)

4360 **Knapp, R. (1977).** Die Pflanzenwelt der Rhon: unter besonderer Berucksichtigung der Naturpark-Gebiete. (Plant world of Rhon, with special emphasis on nature park areas.) Fulda, Parseller. 136p. Ge. Illus., maps. (P)

4361 **Knauer, N. (1979).** Konfliktpotential einiger Landschaften aus der Sicht des Naturschutzes. *Kieler Not. Pflanzenk. Schleswig-Holstein*, 11(4): 81-88.

4362 **Koebele, F. (1984).** Der Schutzplan der Rheinwalder. (Protection plan for the Rhine forests). *Allgem. Forstzeit.*, 39(17/18): 426-427. Ge. Illus. (P)

4363 **Kohler, A. (1978).** Gefahrdung und Schutz von Susswasserpflanzen. (Conservation of freshwater vegetation.) *Veroff. Natursch. Landschaftspfl. Baden-Wurttemberg*, 11: 251-257. Ge.

4364 **Konig, W. (1976).** Naturpark oberer Bayerischer Wald. *Naturschutz- und Naturparke*, 82: 1-4. Ge. Illus. (P)

4365 **Koop, H. (1980).** Bosreservaten en de oostfriese hudewalder. [Forest reserves in the East-Friesian pastured forests (Lower Saxony).] *Nederl. Bosbouw Tijdschr.*, 52(4): 113-118. Du. (P)

4366 **Korneck, D., Fink, H.G. (1983).** Erhebung und Zustandserfassung der Populationen seltener und gefahrdeter Gefasspflanzen in der Bundesrepublik Deutschland (Pteridophyta, Spermatophyta). *Natur und Landschaft*, 58: 236-237. Ge. Illus. (T)

4367 **Korneck, D., Lang, W. Reichert, H. (1984).** Rote Liste der Rheinland-Pfalz ausgestorbenen, verschollenen und gefahrdeten Farn- und Blutenpflanzen, 2nd Ed. Ministerium fur soziale Gesundheit und Umwelt. Ge. Red List. (T)

GERMANY, FEDERAL REPUBLIC OF

4368 **Korneck, D., Sukopp, H. (1988).** Rote Liste der in der Bundesrepublik Deutschland ausgestorbenen, verschollenen und gefahrdeten Farn -und Blutenpflanzen und ihre Auswertung fur den Arten- und Biotopschutz. Bonn-Bad Godesberg, Bundesforschungsanstalt fur Naturschutz und Landschaftsokologie. 210p. Ge. (T)

4369 **Korneck, D., Sukopp, H., Trautmann, W. (1978).** Rote Liste der Gefasspflanzen in der Bundesrepublik Deutschland. *In* Olschowy, G., *ed.* Natur- und Umweltschutz in der Bundesrepublik Deutschland. 293-302. Ge. Red List. (T)

4370 **Kowarik, I., Sukopp, H. (1984).** Auswirkungen von Luftverunreinigungen auf die Bodenvegetation von Waldern, Heiden und Mooren. (Effects of air pollution on the ground vegetation of forests, heaths and moors.) *Allgem. Forstzeitschr.*, 39(12): 292-293. Ge.

4371 **Krause, W. (1978).** Gezielte Bodenentblossung und Anlage frischer Wasserflachen als Mittel der Bestandserneuerung in Naturschutzgebieten. (Clearing of soil surface and the establishment of new stretches of water to renew vegetation stands in nature conservation areas.) *Veroff. Natursch. Landschaftspfl. Baden-Wurttemberg*, 11: 245-250. Ge.

4372 **Krems, B. (1983).** Rechtsschutzprobleme bei Waldschaden durch Immissionen. (Legal protection problems in respect of damage to forests from pollution.) *Allgem. Forstzeitschr.*, 38(51/52): 1409-1412. Ge.

4373 **Kremser, W. (1978).** Zielvorstellungen zur Entwicklung und zum Schutz der Walder Niedersachsens. (Objectives in the development and protection of the forests of Lower Saxony.) *Allgem. Forst. Jagd-Zeitung*, 149(8): 150-155. Ge.

4374 **Kroth, W., Gleissner, G. (1978).** Leistungen der bayerischen Forstbetriebe zur Verbesserung der Erholungs- und Schutzfunktionen des Waldes. (Results of Bavarian forestry in improving recreational use and protection of the woods.) Munich, Institut fur Forstpolitik und forstliche Betriebswirtschaftslehre der forstlichen Forschungsanstalt. 165p. Ge. Illus., maps.

4375 **Kuck, J.-J. (1978).** Naturpark Nordschwarzwald. *Naturschutz- und Naturparke*, 88: 17-23. Ge. Illus., map. (P)

4376 **Kuhn, K. (1979).** Landwirtschaftliche Nutzung in Naturschutzgebieten an zwei Beispielen in Hessen. *Nat. Mus.* (Frankfurt), 109(3): 80-87. Ge.

4377 **Kulp, H.-G. (1988).** Verbreitung, Gefahrdung und Schutz seltener Ackerwild Krauter auf Sandboden der Stader Geest und der Nordlichen Weser-Aller-Flachlandes. (Distribution, threat and protection of rare agricultural weeds on sandy soils of the Stader Geest and the north-eastern Weser-Aller-lowland.) *Abh. Naturwiss. Ver. Bremen*, 41(1): 127-136. Ge (En). Illus., map. (T)

4378 **Kunkele, S. (1977).** Zur Verbreitung und Gefahrdung der Orchideen im Raum Albstadt (Schwab. Alb). (Distribution and threats to orchids in the Albstadt Region [Swabian Hills].) *Veroff. Natursch. Landschaftspfl. Baden-Wurttemberg*, 46: 19-48. Ge. Illus., maps. (T)

4379 **Kunkele, S. von, Vogt, A. (1973).** Zur Verbreitung und Gefahrdung der Orchideen in Baden-Wurttemburg. *In* Beiheft zu den veroffentlichungen der Landstelle fur Naturschutz und Landschaftspflege. Baden-Wurttemburg. 8-72. Ge. Illus. Col. illus. (T)

4380 **Kunne, H. (1974).** Rote Liste bedrohter Farn- und Blutenpflanzen in Bayern. (Red List of endangered ferns and flowering plants in Bavaria.) *Schr.-R. Natursch. Landschaftspfl.*, 4: 1-44. Lists 566 species. (T)

4381 **Landesamt fur Naturschutz und Landschaftspflege Schleswig-Holstein (1982).** Rote liste der gefahrdeten Pflanzen und Tiere Schleswig-Holsteins. *Schr.-R. Natursch. Landschaftspfl.*, 5: 149p. Ge. (T)

GERMANY, FEDERAL REPUBLIC OF

4382 **Landesanstalt fur Okologie, Landschaftsentwicklung und Forstplanung NRW (1979).** Rote liste der in Nordrhein-Westfalen gefahrdeten Pflanzen und Tiere. *Schr.-R. der LOLF Nordrhein-Westfalen,* 4: 106p. Ge. (T)

4383 **Landesanstalt fur Umweltschutz Baden-Wurttemberg (1978).** Fragen des Artenschutzes in Baden-Wurttemberg. *Veroff. Naturschutz Landschaftspfl. Baden-Wurttemberg,* 11: 1-502. Ge.

4384 **Lienenbecker, H. (1980).** Die Vegetation des Naturschutzprojektes, Schluchten und Moore am oberen Furlbach. *Ber. Naturwiss. Ver. Bielefeld,* suppl. part 2: 53-74. Ge. Illus.

4385 **Litzelmann, M., Hofmann, K. (1979).** Naturschutz - Gebiet Utzenfluh mit den Teilen grosse und kleine Fluh und Falkenwald. *Mitt. Bad. Landesver. Naturk. Naturschutz,* 12(1-2): 121-130. Ge.

4386 **Lohmeyer, W., Muller, T., Pitzer, E., Sukopp, H. (1972).** Die in der Bundesrepublik Deutschland gefahrdeten Arten von Farn- und Blutenpflanzen. *Gott. Flor. Rundbr.,* 6(4): 91-96. Ge. (T)

4387 **Macher, M., Steubing, L. (1984).** Flechten und Waldschaden im Nationalpark Bayerischer Wald. (Lichens and forest damage in the Bavarian Forest National Park.) *Beitr. Biol. Pflanz.,* 59(2): 191-204. Ge (En). Illus. Maps. Pollution. (P)

4388 **Manegold, F.J. (1978).** Pflanzengesellschaften des Naturschutzgebietes "Apels Teich" Kreis Paderborn. (Plant communities of the "Apels Teich" nature protection area in the Paderborn District.) *Nat. Heimat.,* 38(4): 113-118. Ge. (P)

4389 **Mang, F.W.C. (1981).** Vorarbeiten zu einer Roten Liste fur Hamburg. *Kieler Not. Pflanzenk. Schleswig-Holstein,* 13(1/2): 2-30. Ge. Maps. (T)

4390 **Mecklenburg, G. (1977).** Frankenwald: Naturparkplanung im Zonenrandgebiet. *Naturschutz- und Naturparke,* 85: 23-30. Ge. Illus. (P)

4391 **Merkel, J. (1982).** *Luronium natans* (L.) RAF, ein Neufund in Bayern. *Gottl. Flor. Rundbr.,* 16: 43-48. Ge. (T)

4392 **Meyer, O. (1977).** Naturparkarbeit aus der Sicht der obersten Naturschutzbehorde: Vortrag des rheinland-pfalzischen Staatsministers fur Landwirtschaft, Weinbau und Umweltschutz anlasslich der Jahresversammlung des Verbandes Deutscher Naturparke e. V. am 30. September 1977 in Bad Ems. *Naturschutz- und Naturparke,* 87: 9-13. Ge. (P)

4393 **Minder, R. (1979).** Uber die Luneburger Heide. *Naturschutz- und Naturparke,* 93: 33-39. Ge. Illus., col. illus. (P)

4394 **Montag, A. (1976).** Erfassung schutzwurdiger Gebiete in Niedersachsen. *In* 30 Jahre Naturschutz und Landschaftspflege in Niedersachen. Nieders. Min f. Ernahrung, Landwirtschaft und Forsten. 42-47. Ge. (P)

4395 **Mucke, B. (1983).** Damit der Wald nicht stirbt: Ursachen und Folgen der Waldkatastrophe. (So that the forest does not die.) Munchen, Heyne. 237p. Ge. Col. illus., maps.

4396 **Muller, H. (1979).** Seltene Geholze in Raum Frankfurt am Main. (Rare woody plants in the Frankfurt am Main area.) *Mitt. Deutsch. Dendrol. Ges.,* 70: 195-196 (1978 publ. 1979). Ge. (T)

4397 **Muller, L. (1983).** Stand der immissionsschutzrechtlichen Bestimmungen im Hinblick auf die Bekampfung des Waldsterbens. (Legal definitions of pollution with regard to controlling forest death.) *Allgem. Forstzeitschr.,* 38(51/52): 1407-1408. Ge. Acid rain damage to forests. (L)

4398 **Muller, T. (1972).** Vorschlage zu einer Neufassung der Liste besonders zu schutzender Pflanzenarten in der BRD, *Schr.-R. f. Landschaftspfl. u. Naturschutz,* 7: 19-26. Ge. (T)

GERMANY, FEDERAL REPUBLIC OF

4399 **Muller, T. von, Philippi, G., Seybold, S. (1973).** Vorlaufige "Rote Liste" bedrohter Pflanzenarten in Baden-Wurttemburg. *In* Beiheft zu den Veroffentlichungen der Landestelle fur Naturschutz und Landschaftsplege Baden-Wurttemburg. Baden-Wurttemburg. 74-96. Ge. Red List. (T)

4400 **Muller, T., Kast, D. (1969).** Die geschutzten Pflanzen Deutschlands. Stuttgart, Verlag des Schwabischen Albvereins. Ge. (T)

4401 **Nieschalk, A. & C. (1983).** Hochheiden im Waldecker Upland und angrenzenden westfalischen Sauerland. *Philippia*, 5(2): 127-150. Ge (En). Illus. (T)

4402 **Olschowy, G. (1977).** Zur Entwicklung des Naturschutzes und der Landschaftspflege in Deutschland: Vortrag im Rahmen des Symposium uber Landschaftsplanung am 26. August 1976 in Wageningen/Niederlande aus Anlass der Verabschiedung bon R.J. Benthem aus dem aktiven Dienst. *Naturschutz- und Naturparke*, 86: 25-30. Ge. (P)

4403 **Olschowy, G. (1978).** Potentieller Nationalpark Lange Rhon. *Naturschutz- und Naturparke*, 89: 11-16. Illus. (P)

4404 **Olschowy, G. (1979).** Aus der Arbeit des deutschen Rates fur Landespflege: Naturschutzgebiet Luneburger Heide. *Naturschutz- und Naturparke*, 92: 3-12. Ge. Illus. (P)

4405 **Olschowy, G., Henke, H. (1976).** Potentieller Nationalpark Luneburger Heide. *Naturschutz- und Naturparke*, 82: 31-44. Illus., maps. (P)

4406 **Olschowy, G., ed. (1978).** Natur- und Umweltschutz in der Bundesrepublik Deutschland. (Protection of nature and the environment in the Federal Republic of Germany.) Hamburg, Paul Parey. xvi, 926p. Ge. Illus.

4407 **Ott, W. (1977).** Ein Naturparkkonzept fur Baden-Wurttemberg. *Naturschutz- und Naturparke*, 84: 23-30. Ge. Illus., map. (P)

4408 **Petsch, G. (1982).** Walderhaltung und Waldpflege in Verdichtungsgebieten. (Forest management for conservation.) *Allgem. Forstz.*, 37(41): 1247-1249. Ge. Illus. (P)

4409 **Plochmann, R. (1978).** Bemerkungen zum Waldprogramm des Bundes Naturschutz in Bayern. (Notes on the forest programme of the Nature Conservation Society in Bavaria.) *Forst und Holzwirt*, 33(15): 322-325. Ge.

4410 **Raabe, E.-W. (1975).** Rote Liste der in Schleswig-Holstein und Hamburg vom Aussterben bedrohten hoheren Pflanzen. *Heimat*, 82(7/8): 191-200. Ge. Maps. (T)

4411 **Raabe, E.-W. (1978).** Uber den Wandel unserer Pflanzenwelt in neuerer Zeit. *Kieler Not. Pflanzenk. Schleswig-Holstein*, 10(1-2): 24p. Ge. Illus. (T)

4412 **Raabe, U. (1987).** Die Sumpf-Gansedistel, *Sonchus palustris* L., bei Lemforde, Kreis Diepholz. (The marsh-goose thistle, *Sonchus palustris* L., at Lemforde, district of Diepholz.) *Flor. Rundbr.*, 21(1): 48. Ge (En). (T)

4413 **Raehse, S. (1986).** Zur Flora und Vegetation des Landschaftsschutzgebietes "Kalkberge und Diebachsaue" bei Heiligenrode, Landkreis Kassel. (On the flora and vegetation of the protected area 'Kalkberge and Diebachsaue' near Heiligenrode, Kassel.) *Naturschutz Nordhessen*, 9: 45-66. Ge. Maps. (P)

4414 **Reichhoff, L., Bohnert, W., Knapp, H.D. (1978).** Die Vegetation des Naturschutzgebietes "Tote Taler" - Vegetations-differenzierung im Ubergangsbereich zwischen Wald und Rasen. (The vegetation of the "Tote Taler" Nature Reserve: vegetation differences in the transition zone between forest and grassland.) *Arch. Naturschutz Landschaftsforsch.*, 18(3): 141-150. Ge. (P)

4415 **Reineke, D. (1983).** Der Nutzen von Punktrasterkarten fur den Naturschutz. (The value of distribution dot maps for nature conservation.) *Mitteilungsbl. Arbeitskr. Heim. Orch. Baden-Wurttemberg*, 15(1): 1-10. Ge (En). (P/T)

GERMANY, FEDERAL REPUBLIC OF

4416 **Ringler, A. (1980).** Arten- und Biotopschutz im Alpenvorland. *Jahrb. Ver. Schutz. Bergwelt*, 45: 77-123. Ge.

4417 **Rohr, H. (1977).** Naturpark Elm-Lappwald-Dorm. *Naturschutz- und Naturparke*, 84: 33-38. Ge. Illus. (P)

4418 **Sander, H.J. (1985).** Kaum bekannt, schon gefahrdet. *Ber. Arbeitskr. Heim. Orch.*, 2(1): 142. Ge. (T)

4419 **Savelsbergh, E., Geerlings, J. (1988).** Der ehemalige Noltke-Bahnof, eie schutzenserte Teillandschaft im Sudesflichen stadgebiet von Aachen (Tk 5202/231/232). *Flor. Rundbr.*, 21(2): 110-115. Ge (En).

4420 **Schauer, T. (1977).** Veranderte Waldvegetation in den Waldern des Nationalparks Berchtesgaden. *Jahrb. Ver. Schutze Alpenpfl. Tiere*, 42: 31-52. Ge. Illus. (P)

4421 **Scheller, H. (1978).** Der Speierling, ein seltener Baum unserer Walder und Garten. (*Sorbus domestica*, a rare tree of forests and gardens [in West Germany].) *Gartenpraxis*, 12: 598-601. Ge. (T)

4422 **Schiechtl, H.M. (1983).** Bestanderhaltendes Bauen im Naturschutzgebiet. (Conservation-based construction in a nature reserve.) *Gart. Landsch.*, 11: 857-861. Ge (En). Illus. Construction of Munich long distance water pipeline through Pupplinger Au. and protection of the vegetation. (P)

4423 **Schlenker, G., Schill, G. (1979).** Das Feldflora-Reservat auf dem Beutenlay bei Munsingen. (Field-flora reservation in the region of Munsingen.) *Mitt. Ver Forstliche Standortskunde Forstpflanzenzuechtung*, 27: 55-59. Ge. Illus. (P)

4424 **Schlund, G. (1978).** Naturpark Rhon - seine Entwicklung und Aufgaben. (National Park Rhon - its development and tasks.) *Allg. Forstzeitschr.*, 33(51/52): 1521-1524. Ge. Illus. (P)

4425 **Schmidt-Vogt, H. (1983).** Nadelwald-Okosysteme: Bedeutung, Struktur, Dynamik, Gefahrdung. (Significance, structure, dynamics and endangerment of conifer ecosystems.) *Allgem. Forstzeitschr.*, 38(50): 1355-1359. Ge. Forests. Air pollution damage.

4426 **Schonfelder, P., et al. (1986).** Rote Liste gefahrdeter Farn - und Blutenpflanzen Bayerns. Schriftenreihe. Heft 72. Munchen, Bayerisches Landesamt fur Umweltschutz. 77p. Ge. Maps. (T)

4427 **Schreyer, G., Rausch, V. (1978).** Der Schutzwald in der Alpenregion des Landkreises Miesbach. (Conservation forest in the Alp region of Miesbach District.) Munchen, Bayerisches Statsministerium fur Ernahrung, Landwirtschaft und Forsten. 116p. Ge. Illus., maps.

4428 **Schua, L. (1976).** Das Naturdenkmal Langer See: Nachruf auf ein untergegangenes Naturdenkmal am Rande des Naturparks Spessart. *Naturschutz- und Naturparke*, 83: 21-26. Ge. Illus. (P)

4429 **Schwabe-Braun, A. (1979).** Sigma - Soziologie von Weidfeldern im Schwarzwald: Methodik, Interpretation und Bedeutung fur den Naturschutz. (Synsociological relevees of grazing land complexes in the Black Forest: methods, interpretation and importance for nature protection.) *Phytocoenologia*, 6: 22-31. Ge. Illus.

4430 **Schwabe-Braun, A. (1983).** Die Heustadel-wiesen im Nordbadischen Murgtal. Geschichte - Vegetation - Naturschutz. (Hay meadows in the North Baden valley of the R. Murg. History - vegetation - nature protection.) *Veroff. Natursch. Landschaftspfl. Baden-Wurttemberg*, 55/56: 167-237 (1982 publ. 1983).

4431 **Selchow, E. (1979).** Von der Entwicklung der Naturparke. *Naturschutz- und Naturparke*, 93: 45-46. Ge. Map. (P)

GERMANY, FEDERAL REPUBLIC OF

4432 **Seybold, S. (1978).** Auswertung von Arealkarten fur den Artenschutz am Beispiel der Flora von Wurttemberg. (Evaluation of area maps for plant conservation as shown by the Flora of Wurttemberg.) *Veroff. Naturschutz Landschaftspfl. Baden-Wurttemberg*, 11: 41-53. Ge. Maps. (T)

4433 **Shafer, E. (1980).** The forstmeister: resource manager extraordinary. *J. Soil Water Conserv.*, 35(2): 72-75. Illus. Forest conservation and management.

4434 **Simon, S.J. (1977).** Using the living collection as an educational resource - a German view. *Bull. Amer. Assoc. Bot. Gard. Arbor.*, 11(2): 42-45. (G)

4435 **Springer, S. (1984).** Einige bemerkenswerte Arten im Gebiet des Nationalparks Berchtesgaden. (Some noteworthy species in the Berchtesgaden National Park.) Bericht der Bayerische Botanische Gesellschaft zur Erforschung der heimischen Flora. Munchen, W. Ger. Selbstverlag des Gesellschaft (55): 73-74. Ge. (P)

4436 **Stadler (1979).** Aus deutschen Naturparken. *Naturschutz und Naturparke*, 92(1): 53-57. Ge. (P)

4437 **Stangl, J. (1978).** Zwei seltene Agaricales-Arten in der BR Deutschland. (Two rare agaricales-species in the Federal Republic of Germany.) *Zeitschr. Mykol.*, 44: 271-276. Ge (En). Illus., map. (T)

4438 **Steffen, G. (1983).** Luftverschmutzung: wie gefahrdet sind unsere Walder? (Air pollution: how endangered are our forests?) Dortmund, Rhein-Ruhr Druck Sander. 105p. Ge. Col. illus., maps.

4439 **Steiner, F. (1981).** An authentic site: the last alluvial forests of the Rhine. *Naturopa*, 38: 10-12.

4440 **Stern, H. (1979).** Rettet den Wald. (Save the forest.) Munich, Kindler. 393p. Ge. Illus., col. illus.

4441 **Storm, P.C. (1984).** La protection de la foret en Republique Federale d'Allemagne. (Protection of forests in the German Federal Republic.) *In* Prieur, M., *ed.* Forets et environnement: en droit compare et international. Paris, Presses Universitaires de France. 105-123. Fr. (L)

4442 **Sukkop, H., Auhagen, A., Bennert, W., Bocker, R., Hennig, U., Kunick, W., Kutschkau, H., Schneider, C., Hildemar, S., Zimmerman, F. (1981).** Liste der wildwachsenden Farn- und Blutenpflanzen von Berlin (West). Herausgeber: Landesbeauftragter fur Naturschutz und Landschaftspflege Berlin. 31. Ge. Lists 1396 plant species of W. Berlin, with conservation categories. 68p. (T)

4443 **Sukopp, H. (1966).** Verluste der Berliner Flora wahrend der letzten hundert Jahre. *In* Sonderdruck aus Sitzungberichte der Gesellschaft Naturforschender Freunde zu Berlin N.F. Band VI. 126-136. Ge. (T)

4444 **Sukopp, H. (1972).** Grundzuge eines Programmes fur den Schutz von Pflanzenarten in der Bundesrepublik Deutschland. *In* Sonderdruck aus der Schriftenreihe fur Landschaftspflege und Naturschutz Heft 7, S. Bonn-Bad Godesberg. 67-79. Ge. (T)

4445 **Sukopp, H. (1974).** 'Rote Liste' der in der Bundesrepublik Deutschland gefahrdeten Arten von Farn- und Blutenpflanzen (1. Fassung). *Natur und Landschaft*, 49(12): 315-322. Ge. (T)

4446 **Sukopp, H. (1978).** Schutz fur vom Aussterben bedrohte Pflanzenarten. (The protection of endangered plant species.) *Veroff. Naturschutz Landschaftspfl. Baden-Wurttemberg*, 11: 19-33. Ge. Illus. (T)

4447 **Sukopp, H., Bocker, R., Koster, G.-H. (1974).** Nutzung von Schutzgebieten durch Forschung und Lehre. *Natur und Landschaft*, 49(5): 123-129. Ge. Map. (T)

4448 **Sukopp, H., Elvers, H., eds. (1982).** Rote Liste der gefahrdeten Pflanzen und Tiere in Berlin (West). *Landesentwicklung u. Umweltforschung*, 11: 374p. Ge. (T)

GERMANY, FEDERAL REPUBLIC OF

4449 **Sukopp, H., Trautmann, W. (1979).** Endangerment and conservation of flora and vegetation in the Federal Republic of Germany. *In* Hedberg, I., *ed.* Systematic botany, plant utilization and biosphere conservation. Stockholm, Almqvist & Wiksell International. 136-139. (T)

4450 **Sukopp, H., Trautmann, W. (1981).** Causes in the decline of threatened plants in the Federal Republic of Germany. *In* Synge, H., *ed.* The biological aspects of rare plant conservation. Chichester, Wiley. 113-116. Proceedings of International Conference, King's College, Cambridge, 14-19 July 1980. (T)

4451 **Sukopp, H., Trautmann, W., Korneck, D. (1978).** Auswertung de Roten Liste gefahrdeter Farn- und Blutenpflanzen in der Bundesrepublik Deutschland fur den Arten- und Biotopschutz. *Schriftenr. Vegetationsk.*, 12: 142p. Ge (En). Detailed analysis of 2667 threatened plant taxa, threats, habitats, recommendations. (T)

4452 **Sukopp, H., Trautmann, W., eds (1976).** Veranderungen der Flora und Fauna in der Bundesrepublik Deutschland. Proceedings of a symposium 7-9 October 1975. *Schriftenr. Vegetationsk.*, 10. 409p. Ge (En).

4453 **Sukopp, H., et al. (1981).** Liste der wildwachsenden Farn- und Blutenpflanzen von Berlin (West): mit Angaben zur Gefahrdung der Sippen und Angaben uber Zeitpunkt der Einwanderung in das Gebiet von Berlin (West). Berlin, Landesbeauftragter fur Naturschutz und Landschaftspflege. 68p. Ge. (T)

4454 **Tauber, F. (1981).** Valorificarea listei rosii a pterido-si antofitelor amenitate in R.F. Germania in vederea ocrotirii speciilor si biotopurilor. *Ocrot. Nat.*, 25(1): 114-116. Rum. Red Data List. (T)

4455 **Thomas, S. (1979).** Pflanzensoziologisch-okologische Analyse der Vegetation beweideter Strandwalle im Naturschutzgebiet "Ostufer der Muritz". *Arch. Naturschutz Landschaftsforsch.*, 19(3): 217-229. Ge. Illus. (P)

4456 **Thompson, J.A. (1979).** Impressions of the Luneburg Heath Nature Reserve. *Nature and National Park*, 17(64): 20-21. (Ge, Fr). Illus. (P)

4457 **Toepfer, A. (1976).** Das deutsche Naturpark-programm von 1956 und seine bisherige Erfullung. *Naturschutz- und Naturparke*, 80: 1-22. Ge. Illus. (P)

4458 **Toepfer, A. (1977).** Kunftiger deutscher Nationalpark Luneburger Heide? *Naturschutz- und Naturparke*, 84: 11-14. Ge. (P)

4459 **Toepfer, A. (1979).** 70 Jahre Verein Naturschutzpark. *Naturschutz- und Naturparke*, 93: 9-16. Ge. Illus. (P)

4460 **Toepfer, A. (1979).** Aus dem Naturschutzgebiet Luneburger Heide. *Naturschutz- und Naturparke*, 94: 19-23. Ge. Illus. Regular feature. (P)

4461 **Toepfer, A. (1981).** A priceless heritage. *Naturopa*, 38: 21-22. Illus. (P)

4462 **Trauboth, V. (1981).** Tending of forest protection areas, with the example of the Ibengarten nature conservation area. *Arch. Natursch. Landschaftsforsch.*, 21(3): 149-158. Ge. (P)

4463 **Trautmann, W., Korneck, D. (1978).** Zum Gefahrdungsgrad der Pflanzenformationen in der Bundesrepublik Deutschland. (The threat to plant formations in the German Federal Republic.) *Veroff. Natursch. Landschaftspfl. Baden-Wurttemberg*, 11: 35-40. Ge. Illus. (T)

4464 **Trautmann, W., Wolf, G. (1983).** The importance of forest nature reserves (in West Germany) for conservation area systems. *Schrift. Deutsch. Rat. Landespfl.*, 41: 92-94. Ge. (P)

4465 **Vahrenholt, F., ed. (1984).** Tempo 100: Sofort Hilfe fur den Wald? Reinbeck West Germany Rowohlt. Ge. Illus. Pollution.

GERMANY, FEDERAL REPUBLIC OF

4466 **Vangerow, H.-H. (1978).** Nationalpark Bayerischer Wald Idyll an der Grenze. *Naturschutz- und Naturparke*, 89: 1-7. Ge. (P)

4467 **Volquardts, G. (1982).** Waldbildung und Walderhaltung - Moglichkeiten und Grenzen. *Forstarchiv*, 53(2): 43-46. Ge (En). Illus. Forest conservation.

4468 **Weber, H. (1977).** [Inventory of available and planned nature protection areas in the Diepholz District]. *Inf. Naturschutz. Landschaftspfl. West-Niedersachsen*, 1: 91-93. Ge. (P)

4469 **Weinitschke, H., Thron, R. (1979).** Die Erfassung der Naturausstatt von Naturschutzgebieten und ihre Gruppierung mittels mathematischer Vergleiche. (Recording the natural characteristics of nature reserves and the grouping of the reserves by mathematical comparisons.) *Arch. Natursch. Landschaftsforsch.*, 19(4): 241-245. Ge (En, Rus). (P)

4470 **Wentzel, F. (1978).** Amtlicher und ehrenamtlicher Naturschutz. (Official and honorary nature protection.) *Forst und Holzwirt*, 33(17): 371-376. Ge. Illus.

4471 **Wentzel, K.F. (1978).** Am Naturschutz gescheitert: Chronik des Motodrom-Projektes im Naturpark Hoher Vogelsberg. *Naturschutz- und Naturparke*, 91: 29-32. Ge. Illus. (P)

4472 **Wilmanns, O., Kratochwil, A. (1983).** Naturschutz-bezogene Grundlagen-Untersuchungen im Kaiserstuhl. (Nature conservation studies in Kaiserstuhl.) *Veroff. Natursch. Landschaftspfl. Baden-Wurttemberg*, 34: 39-56. Ge. Illus.

4473 **Winterhoff, W., Haas, H., Knoch, D., Krieglsteiner, G.J., Schwobel, H. (1978).** Vorlaufige Rote Liste der gefahrdeten Grosspilze in Baden-Wurttemberg. (Preliminary Red List of endangered large fungi in Baden-Wurttemberg.) *Veroff. Landesst. Natursch. Landschaftspfl. Baden-Wurttemberg*, 11: 169-178. Ge. (T)

4474 **Wittig, R. (1980).** Vegetation, Flora, Entwicklung, Schutzwurdigkeit und Probleme der Erhaltung des NSG "Westruper Heide" in Westfalen. *Abh. Landesmus. Naturk. Munster Westfalen*, 42(1): 3-30. Ge. Illus., map.

4475 **Wittig, R. (1982).** The effectiveness of the protection of endangered oligotrophic-water vascular plants in nature conservation areas of North Rhine-Westphalia (Fed. Rep. of Germany). *In* Symoens, J.J., Hooper, S.S., Compere, P., eds. Studies on aquatic vascular plants. Proc. Int. Colloqu. on Aquatic Vascular Plants, Brussels, 23-25 January 1981. Roy. Bot. Soc. Belgium. 418-424. Map. (P/T)

4476 **Wittig, R. (1982).** Vegetation, Flora und botanische Bedeutung der Naturschutzgebiete "Wildpferdebahn im Merfelder Bruch", "Schwarzes Venn" und "Sinninger Veen". *Abh. Landesmus. Naturk. Munster Westfalen*, 44(2): 3-34. Ge (En, Fr). Maps. (P)

4477 **Wittig, R. (1983).** Investigation and assessment of the botanical efficiency of conservation in selected nature reserves of Westphalia (Federal Republic of Germany). *Biol. Conserv.*, 25(4): 307-314. (P)

4478 **Wolff-Straub, et al. (1982).** Vom Aussterben bedroht! Farn - und Blutenpflanzen. Der Minister fur Ernahrung, Landwirtschaft und Forsten des Landes Nordrhein-Westfalen. p.87. Ge. Col. illus. (T)

4479 **Wornle, P., Hohmer, H. (1979).** Naturschutz im Landkreis Berchtesgadener Land. *Jahrb. Ver. Schutz Bergwelt*, 44: 9-40. Ge. Illus.

4480 **Wulf, M., Cordes, H. (1988).** Uber die Verbreitung und Gefahrdung ausgewahlter Waldkrauter der Beverstedter Geest. (Threats to some selected forest plants of the Beverstedter Geest.) *Abh. Naturwiss. Ver. Bremen.*, 41(1): 67-82. Ge (En). Map. (T)

4481 **Zang, H. (1978).** Natur- und Landschaftsschutz im Landkreis Goslar mit einer knappen Charakterisierung des Naturparks Herz. *Jahrb. Ver. Schutze Alpenpfl. Tiere*, 43: 181-187. Ge. Illus. (P)

GERMANY, FEDERAL REPUBLIC OF

4482 **Zierl, H. (1979).** Nationalpark Berchtesgaden. (The Berchtesgaden National Park.) *Forstwissenschaftliches Centralblatt*, 98(1): 9-18. Ge. Illus. (P)

GHANA

4483 **Ahn, P.M. (1959).** The principal areas of remaining original forest in western Ghana, and their potential value for agricultural purposes. *J. West Afr. Sci. Assoc.*, 5(2): 91-100. Small-scale vegetation map.

4484 **Asibey, E.O.A., Owusu, J.G.K. (1982).** The case for high-forest national parks in Ghana. *Envir. Conserv.*, 9(4): 293-304. Illus. map. (E/P/T)

4485 **Dorm-Adzobu, C. (1982).** Impact of utilization of natural resources of forest and wooded savanna ecosystems in rural Ghana. *Envir. Conserv.*, 9(2): 157-162. (T)

4486 **Hall, J.B. (1979).** Possibilities and needs for conservation of plant species and vegetation in Africa. Appendix: Preliminary lists of rare and threatened species in African countries. Ghana. *In* Hedberg, I., *ed.* Systematic botany, plant utilization and biosphere conservation. Stockholm, Almqvist & Wiksell International. 88-91. Contains 210 species and infraspecific taxa. (T)

4487 **Lawson, G.W. (1968).** Ghana. *In* Hedberg, I. & O., *eds.* Conservation of vegetation in Africa south of the Sahara. Symposium Proceedings, at 6th plenary meeting of AETFAT, Uppsala. *Acta Phytogeogr. Suec.*, 54: 81-86. Map, illus. Describes vegetation and its conservation, makes proposals for protected areas. (P/T)

4488 **Osborne, R. (1987).** *Encephalartos ghellinckii. Encephalartos*, 12: 16-23. Illus., map. Data sheet; includes conservation. (G/T)

4489 **Osborne, R. (1988).** *Encephalartos barteri. Encephalartos*, 14: 8-16. Illus., map. Data sheet; includes conservation. (G/T)

4490 **Swaine, M.D., Hall, J.B. (1981).** The monospecific tropical forest of the Ghanaian endemic tree *Talbotiella gentii. In* Synge, H., *ed.* The biological aspects of rare plant conservation. Chichester, Wiley. 355-363. Illus., map. Proceedings of International Conference, King's College, Cambridge, 14-19 July 1980. (T)

GREECE

4491 **Alibertis, C. & A. (1987).** Rayon d'espoir en ce qui concerne le Cephalanthere de Crete (*Cephalanthera cucullata* Boiss. et Heldr.). *L'Orchidophile*, 81: 68-70. Fr. Col. illus., map. (T)

4492 **Anon. (1980).** New nature reserves in Greece. *Unasylva*, 32(129): 34. (P)

4493 **Anon. (1988).** Greek law to destroy forests. *Oryx*, 22(3): 141. (L)

4494 **Antipas, B., Muller, G. (1974).** Conservation in Greece. Problems and achievements. *Nature in Focus*, 19: 15, 18-21.

4495 **Arianoutsou-Faraggitaki, M. (1985).** Desertification by overgrazing in Greece; the case of Lesvos Island. *J. Arid Envir.*, 9(3): 236-242. Maps. (T)

4496 **Broussalis, P. (1977).** The protection of the flora in Greece and its problems. *Ann. Mus. Goulandris*, 3: 23-30.

4497 **Broussalis, P. (1987).** [The nature around us.] Athens, The Hellenic Society for the Protection of Nature. 60p. Gre. Col. illus. (P)

4498 **Council of Europe (1981).** National activities: Greece. *Council of Europe Newsl.*, 81(6/7): 3. (L)

4499 **Council of Europe (1983).** Greece: Vai palm-grove protected. *Council of Europe Newsl.*, 83(6): 3. (L/P/T)

GREECE

4500 **Diapoulis, C. (1959).** Conservation measures for the plants of the Greek flora. *In* IUCN. Animaux et Vegetaux Rares de la Region Mediterraneenne. Proceedings of the IUCN 7th Technical Meeting, 11-19 September 1958, Athens, vol. 5. Brussels, IUCN. 189-191. Lists 72 threatened plants. (T)

4501 **Efe, A. (1988).** *Liquidambar orientalis. Threatened Pl. Newsl.*, 20: 9-10. Illus. Exploitation and conservation efforts. (E/T)

4502 **Goulimis, C. (1959).** Report on species of plants requiring protection in Greece and measures for securing their protection. *In* IUCN. Animaux et vegetaux rares de la Region Mediterraneenne. Proceedings of the IUCN 7th Technical Meeting, 11-19 Septermber 1958, Athens, vol. 5. Brussels, IUCN. 168-188. Includes list of threatened plants. (P/T)

4503 **Greuter, W. (1967).** *Phoenix theophrastii* in Crete. *Bauhinia*, 3: 243-250. Ge. (T)

4504 **Greuter, W. (1971).** L'apport de l'homme a la flore spontanee de la Crete. *Boissiera*, 19: 329-337. Fr.

4505 **Haritonidou, P. (1983).** Environmental protection for the Great Marsh at Skinias. Can its vanishing flora and fauna be saved? *Ann. Mus. Goulandris*, 6: 45-52. (Gre). Map. (T)

4506 **Hellenic Society for the Protection of Nature (1979).** Proceedings of a Conference on the Protection of the Flora, Fauna and Biotopes in Greece, 11-13 October 1979. Athens, Hellenic Society for the Protection of Nature. 262p.

4507 **Hoffmann, L., Bauer, W., Muller, G. (1971).** Proposals for nature conservation in Northern Greece. A report submitted in May 1970 to the Ministry of Northern Greece. Morges, IUCN. 40p. IUCN Occasional Paper No. 1. Maps.

4508 **Hoiland, K. (1984).** Diktamnos, *Origanum dictamnus*, en truet medisinplante pa Kreta. (Diktamnos, *Origanum dictamnus*, a vulnerable medicinal herb on Crete.) *Blyttia*, 42: 177-180. No. (E/T)

4509 **Hughes, J.D., Thirgood, J.V. (1982).** Deforestation in ancient Greece and Rome: a cause of collapse. *The Ecologist*, 12(5): 196-207. Illus., maps. (T)

4510 **IUCN (1986).** IUCN's list of top 24 threatened animals and plants. *Biol. Conserv. Newsl.*, 44: 1-6. (T)

4511 **IUCN Threatened Plants Committee Secretariat (1982).** The rare, threatened and endemic plants of Greece. *Ann. Mus. Goulandris*, 5: 69-105. (Gr). Lists over 900 threatened taxa. (T)

4512 **Ianovici, V. (1979).** Protectia mediului inconjurator si prezervarea cadrului biologic in Tarile. Balcanice. (La protection de l'environnement et la conservation du cadre biologique dans les pays Balkaniques.) *Ocrot. Nat.*, 23(1): 5-7. Rum (Fr). Conference at Bucharest on 18-22 September 1978.

4513 **Iatrou, G., Georgiadis, T. (1985).** *Minuartia favargeri*: a new species from Peloponnesus (Greece). *Candollea*, 40(1): 129-138. (Fr). Illus., maps. (T)

4514 **Jordanov, D., et al., eds (1975).** Problems of the Balkan flora and vegetation. Sofia, Bulgarian Academy of Sciences. Proceedings of the First International Symposium on Balkan Flora and Vegetation, Vorna, 7-14 June 1973. 441p. (T)

4515 **Kalopissis, J. (1979).** *Cephalanthera cucullata* Boiss et Heldr., the endangered Cretan *Cephalanthera. Newsl. Hell. Soc. Protect. Nat.*, 18: 26-30. (T)

4516 **Kassioumis, C. (1987).** Nature conservation in Greece (legislation and administration of parks and reserves). *Parks*, 12(2): 14-15. (L/P)

4517 **Kuzmanov, B. (1981).** Balkan endemism and the problem of species conservation, with particular reference to the Bulgarian flora. *Bot. Jahrb. Syst.*, 102(1-4): 255-270. Illus., maps. (T)

GREECE

4518 **Leon, C. (1983).** A plant conservationist in Crete. *Threatened Pl. Newsl.*, 11: 12-14.

4519 **Mavrommatis, G. (1976).** [Ecological conditions of the national park of Samaria, Crete.] *Anakoinoskis I.D.E.*, 4(1): 77-106. Gr (En). Map. (P)

4520 **Micevski, K., Matevski, V. (1983).** [Rare and little-known plants in the flora of Macedonia. I.] *God. Zborn. Ann. Skop. Univ. Biol. Fak.*, 36: 149-153. Mac (Ge). Maps. (T)

4521 **Pavlides, G. (1985).** Geobotanical study of the National Park of Lakes Prespa (NW Greece). Part A: Ecology, flora, phytogeography, vegetation. Thessalonika. 13p. (T)

4522 **Sfikas, G. (1979).** Threatened plants of our mountains. *Nature Bull. Hellen. Soc. Protect.*, 18: 11-14, 38, 42-44. (T)

4523 **Sfikas, G. (1986).** About the flora of Oitis. *Nature Bull. Hellen. Soc. Protect.*, 33: 9-19, 47. Gre (En). An inventory of plants in the national park of Oitis. Col. illus., map. (P/T)

4524 **Sfikas, G. (1988).** Vulnerable plants of Greece. *Nature Bull. Hellen. Soc. Protect.*, 41: 35-37. Gre (En). (T)

4525 **Simmons, J. (1979).** The Cretan *Zelkova. Garden* (London), 104(6): 249-251. Illus. *Zelkova abelicea*, endangered in its native habitat. (T)

4526 **Snogerup, S. (1979).** Cultivation and continued holding of Aegean endemics in an artificial environment. *In* Synge, H., Townsend, H., *eds.* Survival or extinction. Proceedings of a conference held at the Royal Botanic Gardens, Kew, 11-17 September 1978. Kew, Bentham-Moxon Trust. 85-90. (G/T)

4527 **Strid, A. (1986).** *Adonis cyllenea* (Ranunculaceae) and *Helichrysum taenari* (Asteraceae) rediscovered in Peloponnisos. *Ann. Musei Goulandris*, 7: 221-231. (T)

4528 **Strid, A., Papanicolau, K. (1985).** The Greek mountains. *In* Gomez-Campo, C., *ed.* Plant conservation in the Mediterranean Area. Dordrecht, Dr W. Junk. 89-111. (T)

4529 **Tsipouridis, C., Hatziharisis, I.A., Galanopoulou, S., Syrgiannidis, G.D. (1987).** Temperate fruit crop germplasm in Greece (with particular reference to *Prunus*). *Pl. Genet. Resource. Newsl.*, 70: 24-25. (E)

4530 **Tsunis, G.L. (1988).** The Valia-Kalda National Park, Greece. *Oryx*, 22(1): 25-29. Illus., maps. (P/T)

4531 **Voliotis, D.T. (1980).** Neue und seltene Arten (bzw. Unterarten) fur die Griechische Flora aus dem Voras-Gebirge. 2. (New and rare species (or subspecies) for the Greek flora from the Voras Mountains. 2.) *Biol. Gallo-Hellenica.* Ge (En). Illus., maps. Unpaginated. (T)

4532 **Voliotis, D.T. (1981).** Neue und seltene Arten (bzw. Unterarten) fur die Griechische Flora aus dem Voras-Gebirge. 3. (New and rare species (or subspecies) for the Greek flora from the Voras Mountains. 3.) *Acta Bot. Croat.*, 40: 251-256. Ge (En). Illus., maps. (T)

4533 **Voliotis, D.T. (1981).** Neue und seltene Arten (bzw. Unterarten) fur die Griechische Flora aus dem Voras-Gebirge. 4. (New and rare species (or subspecies) for the Greek flora from the Voras Mountains. 4.) *Webbia*, 35(2): 311-321. Ge (En). Illus., maps. (T)

4534 **Voliotis, D.T. (1981).** Neue und seltene Arten (bzw. Unterarten) fur die Griechische Flora aus dem Voras-Gebirge. 6. (New and rare species (or subspecies) for the Greek flora from the Voras Mountains. 6.) *Bot. Chron.*, 1(2): 115-123. Ge (En). Illus., maps. (T)

GREECE

4535 **Voliotis, D.T. (1982).** Neue und seltene Arten (bzw. Unterarten) fur die Griechische Flora aus dem Voras-Gebirge. 5. (New and rare species (or subspecies) for the Greek flora from the Voras Mountains. 5.) *Bauhinia,* 7(3): 155-166. Ge (En, Fr). Illus., maps.

4536 **Voliotis, D.T. (1983).** Neue und seltene Arten (bzw. Unterarten) fur die Griechische Flora aus dem Voras-Gebirge. 1. (New and rare species (or subspecies) for the Greek flora from the Voras Mountains. 1.) *Feddes Repert.,* 94(7/8): 575-589. Ge (En). Illus., maps. (T)

4537 **Yannitsaros, A. (1977).** *Linaria hellenica* Turrill, an endemic plant of Laconia (Greece) that requires protection. *Nature Bull. Hellen. Soc. Protect.,* 12: 34-35. (T)

4538 **Yannitsaros, A. (1978).** *Tulipa goulimyi* Sealy et Turrill and the dangers which threaten it. *Nature Bull. Hellen. Soc. Protect.,* 14: 36-37. (T)

4539 **Yannitsaros, A. (1979).** A preliminary study for the protection of the flora and vegetation of the island of Lesbos. *Nature Bull. Hellen. Soc. Protect.,* 18: 15-25, 44-46. (P)

4540 **Zahariadi, C.A. (1980).** Deux taxons nouveaux ou rares d'*Ornithogalum* de la mediterranee orientale et quelques considerations sur la notion de l'endemisme. *Ann. Mus. Goulandris,* 4: 249-261. Fr. Illus. (T)

GREENLAND

4541 **Anon. (1981).** The national park in north east Greenland. *Newsl. Commis. Sci. Res. Greenland,* 4: 15. (P)

4542 **Bocher, T. (1976).** The arctic glasshouse. *In* Simmons, J.B., *et al., eds.* Conservation of threatened plants. NATO Conference Series 1: Ecology, vol. 1. New York and London, Plenum Press. 61-66. Proc. Conference on the Functions of Living Plant Collections in Conservation and Conservation-Orientated Research and Public Education, Kew, 2-6 September 1975. (G)

4543 **Fredskild, B. (1983).** Greenland botanical survey 1983. *Newsl. Commis. Sci. Res. Greenland,* 8: 20-21. (T)

4544 **Lawyer, J.I., et al. (1979).** A guide to selected current literature on vascular plant floristics for the contiguous U.S., Alaska, Canada, Greenland and the U.S. Caribbean and Pacific Islands. New York, New York Botanical Garden. 138p. Suppl. 1: Research on floristic information synthesis: a report to the Division of Natural History, Nat. Park Service, U.S. Dept of Interior. Washington, D.C. (B)

GUADELOUPE

4545 **Rousseau, L., ed. (1977).** Parc Naturel de Guadeloupe. Basse-Terre. Fr. Includes plant list. (P)

4546 **Sastre, C. (1978).** Plantes menacees de Guadeloupe et de Martinique. 1. Especes altitudinales. *Bull. Mus. Natn. Hist. Nat.,* Paris, 3e ser., 519: 65-93. Fr (En). Description of vegetation, sheets on 13 rare and threatened species with illustrations and habitat photographs. (T)

4547 **Sastre, C. (1982).** S.O.S. orchidees de Guadeloupe. *Orchidophile,* 13(52): 83-90. Fr. Col. illus. (T)

4548 **Stehle, H. (1979).** Premieres observations sur la reconstitution du tapis vegetal sur le volcan de la Soufriere de Guadeloupe apres les eruptions de juillet-aout 1976 et mars 1977 (41 contribution). *Bull. Soc. Bot. France,* 126(3): 349-359. Fr (En).

GUAM

4549 **Anon. (1981).** *Serianthes nelsonii:* an up-date. *Not. Waimea Arbor.,* 8(1): 8-9. (T)

4550 **Drahos, N. (1974).** New specimen of Guam's rarest tree found. *Guam Rail,* 8(9): 5. (T)

GUAM

4551 **Fosberg, F.R. (1960).** The vegetation of Micronesia: 1. General descriptions, the vegetation of the Marianas Islands, and a detailed consideration of the vegetation of Guam. *Bull. Amer. Mus. Nat. Hist.*, 119: 75p. Illus.

4552 **Josiah, S.J. (1983).** Guam's badlands. *Glimpses of Micronesia*, 23(2): 32-35. Includes threats to vegetation.

4553 **Merrill, E.D. (1940).** Man's influence on the vegetation of Polynesia, with special reference to introduced species. Proc. Sixth Pacific Science Congress. 4. Los Angeles, California University Press. 629-639.

4554 **Moore, P.H. (1974).** Guam's flora: rare ferns of Guam. *Guam Rail*, 8(8): 5. (T)

4555 **Moore, P.H. (1980).** Notes on the endangered species of Guam. *Not. Waimea Arbor.*, 7(1): 14-16. Checklist of plants growing at Waimea Arboretum. (G/T)

4556 **Perez, G.S.A. (1975).** Guam conservation priorities. *In* Force, R.W., Bishop, B., eds. The impact of urban centers in the Pacific. Honolulu, Hawaii, Pacific Science Association. 89-96.

4557 **Trust Territory of the Pacific Islands (1976).** Adopted regulations, Title 45: Fish, shellfish and game, chapter 5: Endangered species. *Territorial Register*, 2(1): 4 December. (T)

4558 **U.S. Congress, Office of Technology Assessment (1987).** Integrated renewable resource management for U.S. insular areas. Washington, D.C., U.S. Govt Printing Office. 443p. Illus., maps. OTA-F-325. Includes description of vegetation, remaining coverage, threatened species. (P/T)

GUATEMALA

4559 **Anon. (1988).** Ley de proteccion y mejoramiento del ambiente. Guatemala, Fundacion Defensores de la Naturaleza. 19p.

4560 **D'Arcy, W.G. (1977).** Endangered landscapes in Panama and Central America: the threat to plant species. *In* Prance, G.T., Elias, T.S., eds. Extinction is forever. Proceedings of a symposium at the New York Botanical Garden, 11-13 May 1976. New York, New York Botanical Garden. 89-104. Maps. The flora and deforestation today. (P/T)

4561 **Godoy, J.C. (1987).** Monterrico Nature Reserve: a model of multiple use. IUCN Special Report: Central America. *IUCN Bull.*, 18(10-12): 17. Illus. (P)

4562 **Iltis, H., Kolterman, D., Benz, B. (1986).** Accurate documentation of germplasm: the last Guatemalan teosintes (*Zea*, Gramineae). *Econ. Bot.*, 40(1): 69-77. (E/G/T)

4563 **Nations, J.D., Komer, D.I. (1984).** Conservation in Guatemala: final report, presented to WWF-US. Box 5210, Austin, Texas, Center for Human Ecology. 170p. Mimeo. From WWF Project US-269, Development of a conservation program for Guatemala; extensive report listing conservation organizations, individuals and other useful contacts in Guatemala.

4564 **Rodas Zamora, J., Aguilar Cumes, J. (1980).** Lista de algunas especies vegetales en via en extincion. Guatemala City, INAFOR. Sp. (T)

4565 **U.S. Department of the Interior. Fish and Wildlife Service (1985).** CITES news: Service announces proposals for changes in CITES Appendices. *Endangered Species Tech. Bull.*, 10(1): 11. Proposals by U.S., prepared by TRAFFIC (U.S.), to move *Ceratozamia* from Appendix II to Appendix I. (L/T)

4566 **Veblen, T.T. (1976).** The urgent need for forest conservation in highland Guatemala. *Biol. Conserv.*, 9: 141-154.

4567 **Veblen, T.T. (1978).** Forest preservation in the Western Highlands of Guatemala. *Geogr. Rev.*, 68(4): 417-434. Maps.

GUATEMALA

4568 **Veblen, T.T. (1978).** Guatemalan conifers. *Unasylva*, 29(118): 25-30. Includes conservation. (T)

4569 **Wilkerson, S.J. (1985).** The Usumacinta river: troubles on a wild frontier. *Nat. Geogr.*, 168(4): 514-543. Proposed hydroelectric dams may destroy rain forest between Mexico & Guatemala.

GUINEA

4570 **IUCN (?)** La conservation des ecosystemes forestiers de la Guinee. IUCN, Tropical Forest Programme Series. c.200p.

4571 **Lamotte, M. (1983).** The undermining of Mount Nimba. *Ambio*, 12(3-4): 174-179. Illus., col.illus., maps. (P)

4572 **Schnell, R. (1968).** Guinee. *In* Hedberg, I. & O., *eds.* Conservation of vegetation in Africa south of the Sahara. Symposium Proceedings, at 6th plenary meeting of AETFAT, Uppsala. *Acta Phytogeogr. Suec.*, 54: 69-72. Fr. Illus. Describes plant sites of importance for conservation, some already protected. (P)

GUYANA

4573 **Dalfelt, A. (1978).** Nature conservation survey of the Republic of Guyana. Switzerland, IUCN. 55p.

4574 **Stettler, P.H. (1988).** Artenschutz: Aufgabe botanischer Garten? Bemerkungen am Beispiel des Niederungs-Urwaldes von Guyana. (Species protection - task of botanic gardens? Using the example of the lowland forest in Guyana.) *Wiss. Zeitschr. Friedrich-Schiller-Univ. Jena, Mat. Naturwiss. Brei tr. Okol. Land.*, 37(1): 91-92. Ge. (G/T)

HAITI

4575 **Anon. (1984).** National park for Haiti. *Oryx*, 18(2): 112. Morne la Visite. (P)

4576 **Henderson, A., Aubrey, M. (1989).** *Attalea crassispatha*, an endemic and endangered Haitian palm. *Principes*, 33(2): 88-90. (T)

4577 **Hubbuch, C.E. (1989).** *Attalea crassispatha* a rare palm from Haiti. *Bot. Gard. Conserv. News*, 1(5): 55-57. Illus., map. (G/T)

HONDURAS

4578 **Anon. (1983).** Military road threatens Honduran virgin forest. *Oryx*, 17: 110. (P/T)

4579 **Betancourt, J.A., Hanson, D., Wild, K. (1979).** Honduras' La Tigra: public benefits through conservation. *Parks*, 4(2): 9-11. Illus. (P)

4580 **D'Arcy, W.G. (1977).** Endangered landscapes in Panama and Central America: the threat to plant species. *In* Prance, G.T., Elias, T.S., *eds.* Extinction is forever. Proceedings of a symposium at the New York Botanical Garden, 11-13 May 1976. New York, New York Botanical Garden. 89-104. Maps. The flora and deforestation today. (P/T)

4581 **Direccion General de Recursos Naturales Renovables and Programa Recursos Naturales Renovables (1978).** Estudio preliminar de los recursos naturales y culturales de la cuenca y un plan para el desarrollo de una Reserva de la Biosfera en la region del Rio Platano. *Turrialba*. Appendix 4: Lista de especies observadas de la flora. 102-107. Sp. (P/T)

4582 **Froelich, J.W., Schwerin, K.H. (1983).** Conservation and indigenous human land use in the Rio Platano watershed, northeast Honduras. Research Paper Series, Latin American Institute, University of New Mexico, No. 12: 95p. A survey of the area's cultural characteristics, of its mammals and flora and a summary and recommendations on the future of the reserve. (P)

HONDURAS

4583 **Glick, D., Betancourt, J. (1983).** The Rio Platano Biosphere Reserve: unique resource, unique alternative. *Ambio*, 12(3-4): 168-173. Col. illus., maps. Flora p.169. (P)

4584 **Glick, D.A. (1984).** Management planning in the Platano River Biosphere Reserve Honduras. *In* Unesco-UNEP. Conservation, science and society. Paris, Unesco. 159-167. Contributions to the First International Biosphere Reserve Congress, Minsk, Byelorussia/USSR, 26 September - 2 October 1983. (P)

4585 **Mateo, C.Q. (1987).** La Tigra National Park. A source of concern. IUCN Special Report: Central America. *IUCN Bull.*, 18(10-12): 14. Illus. Forests important for watershed protection. (P)

4586 **Stolze, R.G. (1979).** Ferns new or rare in Honduras. *Brenesia*, 16: 139-141. (Sp). (T)

HONG KONG

4587 **Anon. (1987).** Pledge to save Hong Kong's threatened wetlands. *New Sci.*, 115(1572): 22. (P)

4588 **Chung, H., et al. (1985).** Mangroves of Hong Kong. *Ecol. Sci.*, 1985/2: 1-6.

4589 **Eddie, H.H., comp. (1986).** A bibliography of Hong Kong vascular plants. Hong Kong, WWF-Hong Kong. 29p. (B)

4590 **Oldfield, S. (1987).** Fragments of paradise. A guide for conservation action in the U.K. Dependent Territories. Oxford, Pisces Publications; for the British Association for Nature Conservationists. 192p.

4591 **Talbot, L.M. & M.H. (1965).** Conservation of the Hong Kong countryside. Summary report and recommendation. Hong Kong, Govt Printer. vi, 34p. Includes short sections on forestry, water catchments etc., and recommendations on a system of parks, reserves and recreation areas. (P)

HUNGARY

4592 **Anon. (1975).** Bemutatjukhazank nehany novenyritkasagat Vajda Laszlo felvetelein. (Introducing some Hungarian plant rarities.) *Buvar*, 30(6): 260-261. Hu. Illus. (T)

4593 **Balint, K.E., Terpo, A. (1984).** Threatened local species of restricted area in Hungary. *Folia Dendrologica*, 11: 433-443. (Cz, Rus, Ge). Illus. (T)

4594 **Balogh, J. (1974).** Okologiai szabalyozo rendszerek es a kornyezetvedelem. (The ecological systems of regulation and the protection of our environment.) *Buvar*, 29(1): 3-9. Hu. Illus.

4595 **Berczik, A. (1984).** Adaptability of monitoring systems in the management of biosphere reserves - experiences in Hungary. *In* Unesco-UNEP. Conservation, science and society. Paris, Unesco. 384-388. Contributions to the First International Biosphere Reserve Congress, Minsk, Byelorussia/USSR, 26 September - 2 October 1983. (P)

4596 **Bierbauer, J. (1974).** A gyulaji vadrezervatum. (The wild-reserve of Gyulaj.) *Buvar*, 29(3): 146-152. Hu. Illus., map. (P)

4597 **Borhidi, A., Janossy, D. (1984).** Protected plants and animals in Hungary. *Ambio*, 13(2): 106-108. Illus. (T)

4598 **Buschendorf, J. (1979).** Naturschutz in der Ungarischen Volksrepublik. *Arch. Naturschutz Landschaftsforsch.*, 19(4): 281-283. Ge.

4599 **Buzetzky, G. (1975).** Termeszetvedelmi teruletek vizrendezese. (Water regulation in protected areas.) *Buvar*, 30(9): 406-408. Hu. Maps. (P)

HUNGARY

4600 **Csapody, I. (1982).** Vedett novenyeink. (Our protected plants.) Budapest, Gondolat. 346p. Hu. Col. illus. (T)

4601 **Csima, P., Gerzanics, A. (1984).** Natur- und Umweltschutz in Ungarn. Beitrage der Landschaftsplanung; Forschung und Lehre; internationale Zusammenarbeit. (Nature and environmental conservation in Hungary. The contributions of landscape planning; teaching and research, international cooperation.) *Gart. Landsch.*, 7: 26-29. Ge (En). Illus., maps.

4602 **Hably, L., Nemeth, F., Szerdahelyi, T. (1980).** Floristical data on the nature preservation area of Barcs. *Stud. Bot. Hung.*, 14: 79-81. Plant list. (P)

4603 **Holly, L., Unk, J. (1981).** Preservation of Hungarian land-races as genetic resources. *Kulturpflanze*, 29: 63-65. (E)

4604 **Horvath, I., et al. (1979).** Okologiai kutatasok nemzeti parkjainkban. (Ecological studies in the national parks of Hungary.) *Buvar*, 34(10): 436-441. Hu. Illus. (P)

4605 **Imre, H., Sandor, M., Tibor, S. (1979).** Okologiai kutatasok nemzeti parkjainkban. *Buvar*, 34(10): 436-441. Hu. Illus. (P)

4606 **Istvan, S. (1979).** A Keszthelyi-hegyseg erdekes novenyei. *Buvar*, 39(11): 506-509. Hu. Illus. (P)

4607 **Kereszty, Z. (1985).** Die Kartierung der geschutzten und gefahrdeten Pflanzenarten in Ungarn. (The categorisation of protected and endangered plant species in Hungary.) *Stapfia*, 14: 71-76. Ge. (T)

4608 **Kevey, B. (1984).** Deg parkerdejenek tolgy-koris-szil ligetei. (The oak-ash-elm lowland forest in the Deg park forest.) *Bot. Kozlem.*, 71(1/2): 51-61. Hu (Ge). Illus.

4609 **Kmosko, A., Istvan, M. (1978).** Uj tajvedelmi korzet: Feher-to-Pusztaszer. (A new protected region: Feher-to-Pusztaszer.) *Buvar*, 31(5): 199-203. Hu. Illus., map. (P)

4610 **Kopasz, M. (1975).** Uj tajvedelmi korzetunk: az ocsai osturjanos. (A new Hungarian protected region: the Ocsa virgin marshes.) *Buvar*, 30(11): 492-495. Hu. Illus., map. (P)

4611 **Kopasz, M. (1979).** Bermutatjuk a Vertesi Tajvedelmi Korzetet. *Buvar*, 34(4): 147-151. Hu. Illus., map.

4612 **Kovacs, M. (1975).** Oj tajvedelmi korzet: a Keleti Matra. (A new protected region: eastern Matra.) *Buvar*, 30(2): 51-54. Hu. Illus., map. (P)

4613 **Kovacs, M., Priszter, S. (1974).** Die Anderungen der Flora und der Vegetation in Ungarn im Laufe der letzten Hundert Jahre. *Bot. Kozlemenyek*, 61(3): 185-197. Ge.

4614 **Kovacs, M., Priszter, S. (1974).** Pusztulo novenyvilagunk. (Our perishing flora.) *Buvar*, 29(6): 329-332. Hu. Illus., maps. (T)

4615 **Magda, M.N. (1979).** Pakerdo Zankan. *Buvar*, 39(11): 520-521. Hu. Illus.

4616 **Major, I. (1974).** Termeszeti ertekek es termeszetvedelmi tervek Eszak-Dunantulon. (Treasures of nature and plans for the protection of nature in North-Transdanubia.) *Buvar*, 29(6): 323-328. Hu. Illus.

4617 **Manoliu, A. (1979).** Ocrotirea naturii in R.P. Ungara. (La protection de la nature en Hongrie.) *Ocrot. Nat*, 23(2): 179-181. Rum (Fr).

4618 **Meszoly, C. (1976).** Az erdoallomany fejlesztese es a termeszetvedelem. (Reafforestation and nature protection.) *Buvar*, 31(2): 51-54. Hu. Illus.

4619 **Nemeth, A. (1978).** Valtozo erdovedelem: 1. (Changing trends in forest protection: 1.) *Erfga, Erdogazdasag es Faipar*, 2: 19. Hu.

4620 **Nemeth, F. (1979).** Novenyritkasagok a fovaros Kornyeken. (Rare plants in the vicinity of Budapest.) *Buvar*, 34(9): 394-397. Hu. Illus. (T)

HUNGARY

4621 **Nemeth, F. (1979).** The vascular flora and vegetation on the Szabadszallas-Fulopszallas territory of the Kiskunsag National Park (KNP), I. *Stud. Bot. Hungarica*, 13: 79-105. Map. (P)

4622 **Nemeth, F. (1982).** Termeszet-vedelmi novenyterkepezes a Bukk-fensikon. *Buvar*, 37(4): 147-149. Hu. Col. illus. (T)

4623 **Nemeth, F. (1982).** Vadontermo ritka novenyeink. *Buvar*, 36(7): 291-293. Hu. Col. illus. (T)

4624 **Nemeth, F., Seregelyes, T. (1981).** Hute die Blumen. (Save the wild flowers.) Budapest, National Environment and Nature Conservancy Office. 127p. Col. illus., maps. (T)

4625 **Radetzky, J. (1976).** Vedette valt a Dunantul legnagyobb szikes-tavas videke. (The biggest region of natural lakes in Transdanubia was protected.) *Buvar*, 31(4): 156-159. Hu. Illus., map. (P)

4626 **Rakonczay, Z. (1973).** A Hortobagy, elso nemzeti parkunk. (Hortobagy - our first national park.) *Buvar*, 28(2): 67-74. Hu. Illus. (P)

4627 **Rakonczay, Z. (1973).** Termeszetvedelmunk helyzete es tovabbi terveink. (Nature conservation now and in the future.) *Buvar*, 28(5): 259-264. Hu. Illus.

4628 **Rakonczay, Z. (1976).** Termeszetvedelem es idegenforgalom. (Native protection and tourism.) *Buvar*, 31(3): 98. Hu. Illus.

4629 **Research Centre for Agrobotany of the National Institute for Agricultural Variety Testing (?)** [Establishment and improvement of a gene bank for crop plants, collecton and exchange of Hungarian and foreign plant material, their conservation, evaluation and distribution]. Tapioszele, Research Centre for Agrobotany of the National Institute for Agricultural Variety Testing. 12p. Col. illus. (E/G)

4630 **Reuter, C. (1974).** A Del-Dunantul termeszetvedelmenek eredmenyei es tervei. (Successes and plans for the protection of nature of Del-Dunantul [South Transdanubia].) *Buvar*, 29(4): 195-201. Hu. Illus.

4631 **Sag, G., Gotthard, D. (1985).** Plantes de Hongrie etrangeres a la flore francaise et leur repartition. *Bull. Soc. Bot. Fr.*, 132(4/5): 349-358. Fr (En). (T)

4632 **Simon, T. (1984).** A Bugaci Bioszfera Rezervatum edenyes florajanak termeszet-vedelmi ertekelese. *Abstr. Bot.* (Budapest), 8: 95-100. Hu (En). (P)

4633 **Soo, R. (1976).** Negyed evszazade termeszetvedelmi teruletunk - Batorliget. (Batorliget - a protected region for 25 years.) *Buvar*, 31(5): 195-198. Hu. Illus. (P)

4634 **Szabo, B.H. (1974).** Termeszetvedelmi tervek Eszak - Magyarorszagon. (Plans for the protection of nature in north Hungary.) *Buvar*, 29(1): 10-16. Hu. Illus., map.

4635 **Szakatsits. G. (1975).** A lazberci viztarozo szerepe a kornyezetvedelemben. (The role of the reservoir of Lazberc in the protection of the environment.) *Buvar*, 30(8): 344-347. Hu. Illus., map.

4636 **Szenthe, A. (1979).** Tajvedelmi korzet lett a barcsi osborokas. (Junipers in Barcs [south-west of Somogy] are now protected.) *Buvar*, 29(5): 266-270. Hu. Illus., map. (P)

4637 **Szodfridt, I. (1975).** A visszaszorulo fehernyar vedelmeben. (Defending the declining white poplar.) *Buvar*, 30(2): 74-76. Hu. Illus. (T)

4638 **Szodfridt, I. (1978).** Erdei genvagyonunk megorzese: termeszetes osszetetelu fapopulaciokert. *Buvar*, 33(11): 499-501. Hu. Illus.

HUNGARY

4639 **Szujko-Lacza, J. (1981).** A magyarorszagi nemzeti parkok botanikai kutatasa: 1.A Hortobagy Nemzeti Park florakutatasi eredmenyeirol. (Flora studies in the Hortobagy National Park, HNP, in Hungary.) *Bot. Kozlem.*, 68(1-2): 59-71. Hu (En). Illus. (P)

4640 **Szujko-Lacza, J., ed. (1982).** [The flora of the Hortobagy National Park.] [Natural history of the national parks of Hungary No. 3.] Budapest, Akademiai Kiado. 172p. Hu. Illus., map. (P)

4641 **Terpo, A. (1984).** Nature conservation areas and protected woody plants in Hungary. *Folia Dendrologica*, 11: 419-431. (Cz, Rus, Ge). Map. (P/T)

4642 **Tibor, S. (1988).** A hazai edenyes flora terrieszetvedelri-ertek besorolasa. (Conservation status of the Hungarian vascular flora.) *Abstr. Bot.*, (Budapest), 12: 1-23. Hu (En). (T)

4643 **Tolgyesi, I. (1976).** A Kiskunsagi Nemzeti Park szep nyari viragai. (The beautiful summer flowers in the Kiskunsag National Park.) *Buvar*, 31(4): 171-172. Hu. Illus. (P)

4644 **Tolgyesi, I. (1978).** Gazdalkodas a Kiskungsagi Nemzeti Parkban. *Buvar*, 33(6): 259-262. Hu. Illus. (P)

4645 **Toth, I. (1975).** Gemenc: a Gemenci Tajvedelmi Korzet termeszeti kincsei. (The natural riches of Gemeno protected region.) *Buvar*, 30(10): 435-439. Hu. Illus., map. (P)

4646 **Toth, K. (1975).** Masodik nemzeti parkunk: a Kiskunsagi NP. (The second Hungarian national park: the Kiskunsag National Park.) *Buvar*, 30(1): 8-14. Hu. Illus., map. (P)

4647 **Vajda, L. (1974).** Hazank ritka novenyei kepekben. (Pictures of rare plants of our native country.) *Buvar*, 29(4): 224-225. Hu. Illus. (T)

4648 **Varga, Z. (1980).** Okologiai kutatasok termeszetvedelmi teruleteken. (Ecological studies on areas for nature conservation.) *Buvar*, 35(2): 75. Hu. Illus. (P)

4649 **Vas, M. (1983).** Termeszetvedelmi intezkedesek Hatasai a Kallosemjeni Nagymohoson. (Effects of conservation measures on the Nagymohos swamp near Kallosemjen in northeastern Hungary.) *Bot. Kozlem.*, 70(1/2): 25-35. Hu (En). Maps. (P)

4650 **Voross Laszlo, Z. (1975).** A csodabogyo Baranyaban. (A new plant protected in Baranya: the wonderberry.) *Buvar*, 30(3): 124-125. Hu. Illus. (T)

4651 **Voross Laszlo, Z. (1976).** Szarsomlyo ritka novenyei. (Rare plants in Szarsomlyo.) *Buvar*, 31(5): 207-210. Hu. Illus. (T)

4652 **Waldstein, F. (?)** Descriptiones et icones plantarum rariorum hungariae. (Descriptions and illustrations of rare plants of Hungary.) Vienna, Typis Matthiae Andreae Schmidt. 1802-1812. La. 3 vols. Microreproduction of original. (T)

4653 **Zoltan, R. (?)** A Magyar termeszetvedelem helyzete es tavlati programja. (The present situation and future programme of Hungarian nature conservation.) *Rakonczay, Z*, 62(1-4): 91-95. Hu. (En). (P/T)

ICELAND

4654 **Einarsson, E. (1988).** Trude, sarbare og sjae ldne karplantearter pa Island. (Endangered, vulnerable and rare species of vascular plants in Iceland.) *Svensk Bot. Tidskr.*, 82(6): 389-391. Sw (En). (T)

4655 **Nilsson, O. (1979).** Threatened plants in the Nordic countries. *In* Hedberg, I., *ed.* Systematic botany, plant utilization and biosphere conservation: Proceedings of a symposium held in Uppsala in commemoration of the 500th ann. of the Univ. Stockholm, Almqvist & Wiksell International. 72-75. (T)

ICELAND

4656 **Vuokko, S. (1983).** Uhatut Kasvimme. (Our threatened plants.) Suomen Luonnonsuojelum Tuki Oy, Helsinki. 96p. Fi. Col. illus. Covers Nordic region. (T)

INDIA

4657 **Abraham, Z., Mehrotra, B.N. (1981).** Some observations on endemic species and rare plants of the montane flora of the Nilgiris, south India. *In* Botanical Survey of India. Abstracts of papers for the Seminar on Threatened Plants of India, 14-17 Sept. 1981. Howrah, BSI. 21-22. (T)

4658 **Abraham, Z., Mehrotra, B.N. (1983).** Some observations on endemic species and rare plants of the montane flora of the Nilgiris, South India. *J. Econ. Taxon. Bot.*, 3(3): 863-867 (1982 publ. 1983). (T)

4659 **Agarwal, A., Narain, S., eds (1985).** The state of India's environment 1984-1985: the second Citizen's Report. Center for Science and Environment, New Delhi. 398p.

4660 **Agrawal, S. (1981).** Some rare gentians. *In* Botanical Survey of India. Abstracts of papers for the Seminar on Threatened Plants of India, 14-17 Sept. 1981. Howrah, BSI. 30. (T)

4661 **Agrawal, S., Bhattacharyya, U.C. (1979).** *Sebaea khasiana* C.B. Clarke (Gentianaceae): a rare find from N.W. Himalaya. *Bull. Bot. Surv. India*, 21(1-4): 234-235. Illus. (T)

4662 **Almeida, S.M. (1981).** Field observations and identities of some rare taxa fron western India. *In* Botanical Survey of India. Abstracts of papers for the Seminar on Threatened Plants of India, 14-17 Sept. 1981. Howrah, BSI. 4. (T)

4663 **Anon. (1978).** Identification, survey and conservation of India's endangered plants. *Ann. Rep. Nat. Bot. Res. Inst., Lucknow*, 6-7. (T)

4664 **Anon. (1981).** Keeping the stoves of the Third World burning. *IUCN Bull.*, 12(3-4): 15-16. Illus. (E)

4665 **Anon. (1981).** The moving forests. *Hornbill*, 2: 2. (E)

4666 **Anon. (1982).** Forests mismanaged. *New Sci.*, 96(1333): 490-492. Illus. Extracts from Centre for Science and Environment (Delhi) report. (E/T)

4667 **Anon. (1982).** Plants and the Indian Wildlife (Protection) Act. *Bull. Bot. Surv. India*, 22(1-4): 231-232 (1980 publ. 1982). (L/T)

4668 **Anon. (1983).** Woodman spare thy axe. *Trees S. Afr.*, 33(4): 102-103. (T)

4669 **Ansari, M.Y. (1981).** Some threatened economically important species of Liliaceae in India. *In* Botanical Survey of India. Abstracts of papers for the Seminar on Threatened Plants of India, 14-17 Sept. 1981. Howrah, BSI. 32. (E/T)

4670 **Ansari, M.Y. (1983).** The fragrant *Ceropegia* of nineteenth century. *Bull. Bot. Surv. India*, 24(1-4): 190-192 (1982 publ. 1983). Illus. (T)

4671 **Armitage, F.B., Joustra, P.A., Ben Salem, B. (1980).** Genetic resources of tree species in arid and semi-arid areas: a survey for the improvement of rural living in Latin America, Africa, India and Southwest Asia. Rome, FAO. vi, 118p. (E)

4672 **Arora, R.K., Mehra, K.L., Nayar, E.R. (1983).** Conservation of wild relatives of crop plants in India. NBPGR Sci. Monogr. No. 6. New Delhi, National Bureau of Plant Genetic Resources. 14p. Map. (E)

4673 **Arora, R.K., Nayar, E.R. (1981).** Distribution of wild relatives and related rare species of economic plants in India. *In* Botanical Survey of India. Abstracts of papers for the Seminar on Threatened Plants in India, 14-17 Sept. 1981. Howrah, BSI. 32. (E/T)

INDIA

4674 **Arora, Y.K., Gupta, R.K. (1983).** Native ornamental orchids: conservation and cultivation of endangered and extinct species. *J. Econ. Taxon. Bot.*, 4(2): 393-411. Illus. (G/T)

4675 **Aswal, B.S., Mehrotra, B.N. (1983).** *Delphinium uncinatum* Hook.f. et Thoms. (Ranunculaceae) and *Lilium wallichianum* Schultes f. (Liliaceae): two rare finds from north-west Himalaya. *J. Econ. Taxon. Bot.*, 3(3): 773-775 (1982 publ. 1983). (T)

4676 **Bahadur, K.N., Jain, S.S. (1982).** Rare bamboos of India. *Indian J. Forestry*, 4(4): 280-286. (E/T)

4677 **Bahadur, K.N., Naithani, H.B. (1978).** On a rare Himalayan bamboo. *Indian J. Forestry*, 1(1): 39-43. (E/T)

4678 **Bandyopadhyay, J., Jayal, N.D., Schoettli, U., Singh, C., eds (1985).** India's environment: crises and responses. Dehra Dun, Natraj. 309p. (E)

4679 **Basu, S.K. (1985).** The present status of rattan palms in India - an overview. *In* Wong, K.M., Manokaran, N., *eds.* Proceedings of the Rattan Seminar, 2nd - 4th October 1984, Kuala Lumpur, Malaysia. Kepong, Rattan Information Centre. 77-94. Includes list of species and their distribution. (E/T)

4680 **Basu, S.K. (1986).** Threatened palms of India: some case studies. *J. Econ. Taxon. Bot.*, 7(2): 493-497 (1985 publ. 1986) Illus. (T)

4681 **Bennet, S.S.R., Gaur, R.C. (1981).** A few highly exploited species needing conservation. *In* Botanical Survey of India. Abstracts of papers for the seminar on Threatened Plants of India, 14-17 Sept. 1981. Howrah, BSI. 33-34. (E/T)

4682 **Bharadwaj, D.C. (1984).** Biological and abiological land conservation for India and developing countries. *Biol. Mem.*, 9(1): 1-25.

4683 **Bharadwaj, K., Chandra, V. (1981).** Land conservation: selected flora for afforestation with a new approach. *Biol. Mem.*, 5(2): 150-162 (1980). (E)

4684 **Botanical Survey of India (1983).** Materials for a green book of botanic gardens in India. A preliminary list of rare and threatened plants in cultivation in some botanic gardens in India. Howrah, India, Botanical Survey of India. 6p. (G/T)

4685 **Buchholz, H., Gmelin, W., eds (1979).** Medicinal plants and pharmaceuticals. *In* Science and technology and the future. Proceedings and Joint Report of World Future Studies Conference and DSE-Preconference, Berlin (West), 4-10 May 1979. 1164-1167p. Munchen, Saur. (T)

4686 **Chadha, K.L. (1981).** India: scientists met on orchids. *Chron. Hort.*, 21(1): 13-14. National Symposium on Orchids, India, 24-25 October 1980.

4687 **Chandra, P. (1983).** Observations on the rare and endangered ferns of India. *New Bot.*, 10: 41-47. Lists 49 taxa; notes on distribution and conservation status. (T)

4688 **Chandra, V. (1981).** Notes on endangered *Scleria* species in India. *In* Botanical Survey of India. Abstracts of papers for the Seminar on Threatened Plants of India, 14-17 Sept. 1981. Howrah, BSI. 30-31. (T)

4689 **Chandrabose, M., Nair, N.C., Chandrasekaran, V. (1979).** Rediscovery of two rare and threatened flowering plants of south India. *Bull. Bot. Surv. India*, 21(1-4): 235-237. *Helichrysum perlanigerum, Psychotria barberi*. (T)

4690 **Chandrabose, M., Nair, N.C., Chandrasekaran, V. (1980).** Some rare and interesting plants from south India. *Indian J. Bot.*, 3(2): 176-177. (T)

4691 **Chandrabose, M., Srinivasan, S.R. (1981).** Notes on two rare and interesting plants from S. India. *J. Bombay Nat. Hist. Soc.*, 78(3): 630. (T)

4692 **Cook, C.D.K. (1980).** The status of some Indian endemic plants. *Threatened Pl. Committ. Newsl.*, 6: 17-18. Mentions 5 threatened wetland species. (T)

INDIA

4693 **Daniel, J.C. (1985).** India's wetland resources. *In* Thorsell, J.W. *ed.*, Conserving Asia's natural heritage. The planning and management of protected areas in the Indomalayan Realm. Proc. of the 25th working session of IUCN's CNPPA. Gland, IUCN. 39-42.

4694 **Daniel, J.C. (1985).** Wildlife research in India. *In* Thorsell, J.W. *ed.*, Conserving Asia's natural heritage. The planning and management of protected areas in the Indomalayan Realm. Proc. of the 25th working session of IUCN's CNPPA. Gland, IUCN. 150-153.

4695 **Das Gupta, S.P., ed. (1976).** Atlas of forest resources of India. Calcutta National Atlas Organisation, Department of Science and Technology. Maps.

4696 **Das, C.R. (1983).** *Heterosmilax polyandra* (Liliaceae): a rare endemic to India. *J. Bombay Nat. Hist. Soc.*, 79(3): 715. (T)

4697 **Das, S., Deari, N.C. (1981).** A census of endemic orchids in north-eastern India. *In* Botanical Survey of India. Abstracts of papers for the Seminar on Threatened Plants of India, 14-17 Sept. 1981. Howrah, BSI. 13. (T)

4698 **Datta, A. (1981).** Distribution of some rare ferns in India. *In* Botanical Survey of India. Abstracts of papers for the Seminar on Threatened Plants of India, 14-17 Sept. 1981. Howrah, BSI. 38. (T)

4699 **David, J.E., Gaur, S. (1985).** Data centre for natural resources. *In* Thorsell, J.W. *ed.*, Conserving Asia's natural heritage. The planning and management of protected areas in the Indomalayan Realm. Proc. of the 25th working session of IUCN's CNPPA. Gland, IUCN. 9-10.

4700 **Dayanandan, P. (1978).** To conserve the unique but endangered flora of the Peninsular Hills. *In* The Madras Herbarium (MH) 1853-1978, 125th Anniversary Souvenir, part 2: Symposium on Floristic Studies in Pensinsular India. 47. Abstract. (T)

4701 **Deb, D.B. (1986).** Conservation of a threatened species. *Bull. Bot. Surv. India*, 26(3-4): 225-226 (1984 publ. 1986). Illus. (T)

4702 **Department of Environment (India) (?)** List of national parks, sanctuaries, botanic gardens and zoological parks in India. New Delhi, Dept of Environment. 38p. (G/P)

4703 **Department of the Environment (India). (1983).** A national wildlife action plan. Department of the Environment, Government of India. 19p. (L)

4704 **Deshpande, U.R., Singh, N.P., Raghavan, R.S. (1979).** Rediscovery of the rare grass *Parahyparrhenia bellariensis* (Hack.) Clayton. *Bull. Bot. Surv. India*, 20(1-4): 149-150 (1978 publ. 1979). (T)

4705 **Dun, D. (1981).** Threatened plants of India. Howrah, Botanical Survey of India. 40p. Seminar on Threatened Plants of India, 14-17 Sept. 1981. (E/T)

4706 **FAO (1985).** Intensive multiple-use forest management in the tropics. Analysis of case studies from India, Africa, Latin America and the Caribbean. Rome, FAO Forestry Paper 55. 180p. (E)

4707 **Fernandez, R.R. (1981).** *Hyphaene indica* Becc. - a correction to the caption of the front cover photograph of issue no. 8 of the *Indian Forester*, vol. 106, 1980, with an appeal for protection of the endangered species. *Indian Forester*, 107(7): 461-462. Illus. (T)

4708 **Forestry Division (India). (1983).** Guidelines for preparation of working plans and felling in forests. India, Department of Agriculture and Cooperation (Forestry Division). 10p. Deforestation.

4709 **Gadgil, M. (1982).** Conservation of India's living resources through biosphere reserves. *Current Science*, 51(11): 547-550. (P)

INDIA

4710 **Gadgil, M. (1983).** Conservation of plant resources through biosphere reserves. *In* Jain, S.K., Mehra, K.L., *eds.* Conservation of tropical plant resources. Howrah, Botanical Survey of India. 66-71. (P)

4711 **Gadgil, M. (1984).** An approach to ecodevelopment of Western Ghats. Bangalore, Centre for Ecological Sciences, Indian Institute of Science. 43p.

4712 **Gadgil, M., Meher-Homji, V.M. (1986).** Localities of great significance to conservation of India's biological diversity. *Proc. Indian Acad. Sci. (Anim. Sci./Plant Sci.) Suppl.,* November: 165-180. (P)

4713 **Gadgil, M., Meher-Homji, V.M. (1986).** Role of protected areas in conservation. *In* Chopra, V.L., Khoshoo, T.N., *eds.* Conservation for productive agriculture. New Delhi, Indian Council for Agricultural Research. 143-159. (P)

4714 **Ghosh, R.B. & B., Datta, S. (1977).** Two little known or rare plants from eastern India. *J. Bombay Nat. Hist. Soc.,* 74(3): 564.

4715 **Ghosh, R.C. (1978).** Evaluating and analysing environmental impacts of forests in India. *In* Proceedings of the 8th World Forestry Congress, Jakarta, 16-28 October 1978: Forestry for quality of life. II, 8.

4716 **Goel, A.K., Bhattacharyya, U.C. (1982).** Rediscovery of three rarely collected plants from north-western Himalaya. *Indian J. Forest.,* 5(1): 17-20. (T)

4717 **Gopal, B., Sharma, K.P. (1982).** Studies of wetlands in India with emphasis on structure, primary production and management. *Aquatic Bot.,* 12(1): 81-91.

4718 **Gosh, R.B., Ghosh, B., Datta, S. (1977).** Two little known or rare plants from eastern India (*Dittoceras andersonii, Tofieldia yunnanensis*). *J. Bombay Nat. Hist. Soc.,* 74(3): 564. (T)

4719 **Gosh, R.C. (1983).** Forest management and its role in conservation. *In* Jain, S.K., Mehra, K.L., *eds.* Conservation of tropical plant resources. Howrah, Botanical Survey of India. 72-81.

4720 **Gribbin, J. (1982).** The other face of development. *New Sci.,* 96(1333): 489-495. Illus., maps. Himalaya, deforestation.

4721 **Groombridge, B. (1984).** Sandalwood smuggling in India. *TRAFFIC Bull.,* 5/6: 63. Trade. (L)

4722 **Gupta, B.K. (1981).** Indian cymbopogons: their existence and distribution. *In* Botanical Survey of India. Abstracts of papers for the Seminar on Threatened Plants of India, 14-17 Sept. 1981. Howrah, BSI. 29-30. (T)

4723 **Gupta, B.K., Jain, R.K. (1979).** A plea for the conservation of genetic diversity of Himalayan plants. *Indian J. Forestry,* 2(4): 316-317. (T)

4724 **Gupta, P.C., Dakshini, M.N. (1981).** Rare/endemic taxa in the Botanical Garden and Arboreta of the Forest Research Institute, Dehra Dun. *In* Botanical Survey of India. Abstracts of papers for the Seminar on Threatened Plants of India, 14-17 Sept. 1981. Howrah, BSI. 34. (G/T)

4725 **Gurumurti, K., Purohit, C.K., Rawat, J.S. (1984).** Mangroves - a vanishing ecosystem. *Van Vigyan* (Dehra Dun), 22(1/2): 12-19. Illus. (T)

4726 **Haigh, M. (1984).** Deforestation and disaster in northern India. *Land Use Policy,* July 1984: 187-198.

4727 **Hajra, P.K. (1981).** Is Sahastradhara a threatened type locality? *In* Botanical Survey of India. Abstracts of papers for the Seminar on Threatened Plants of India, 14-17 Sept. 1981. Howrah, BSI. 6. (T)

4728 **Hajra, P.K. (1981).** Plants of north west Himalaya with restricted distribution - a census. *In* Botanical Survey of India. Abstracts of papers for the Seminar on Threatened Plants of India, 14-17 Sept. 1981. Howrah, BSI. 1. (T)

INDIA

4729 **Hajra, P.K. (1983).** Rare, threatened and endemic plants of the western Himalayas: monocotyledons. *Pl. Conserv. Bull.*, 4: 1-13. Annotated list of threatened species, numbers, esp. Orchidaceae. (T)

4730 **Henry, A.N., Swaminathan, M.S. (1979).** Rare or little known plants from south India. *J. Bombay Nat. Hist. Soc.*, 76(2): 373-376. (T)

4731 **Henry, A.N., Swaminathan, M.S. (1982).** Five rare orchids from southern India. *Indian J. Forest.*, 5(1): 78-80. (T)

4732 **Henry, A.N., Swaminathan, M.S. (1983).** On the rediscovery of two rare endemic plants of India. *Bull. Bot. Surv. India*, 24(1-4): 234-235 (1982 publ. 1983). Illus. (T)

4733 **Henry, A.N., Vivekananthan, K., Nair, N.C. (1978).** Rare and threatened flowering plants of south India. *J. Bombay Nat. Hist. Soc.*, 75(3): 684-697. Lists 224 angiosperms. (T)

4734 **Husain, A. (1983).** Conservation of genetic resources of medicinal plants in India. *In* Jain, S.K., Mehra, K.L., *eds*. Conservation of tropical plant resources. Howrah, Botanical Survey of India. 110-117. (E/T)

4735 **IUCN (1970).** Eleventh Technical Meeting Papers and Proceedings, New Delhi, India, Vol. II. Problems of threatened species. IUCN Publications New Series No. 18. Morges, IUCN. 132p. (T)

4736 **India. Ministry of Agriculture. Forestry Division (1983).** Guidelines for preparation of working plans and felling in forests. India, Ministry of Agriculture, Forestry Division. 10p. (E)

4737 **Israel, S., Sinclair, T., eds (1987).** Indian wildlife: Sri Lanka, Nepal. Hong Kong, APA Productions. 362p. Col. illus. Provides overview of vegetation and conservation problems.

4738 **Jain, S.K. (1979).** India: botanic gardens and threatened plants - a report. *In* Synge, H., Townsend, H., *eds*. Survival or extinction. Proceedings of a conference held at the Royal Botanic Gardens, Kew, 11-17 Septeber 1978. Kew, Bentham-Moxon Trust. 113-116. (G/T)

4739 **Jain, S.K. (1980).** Orchids of north-eastern India and their conservation. *Indian Sci. Congr. Assoc. Proc.*, 67(4): 36. Abstract.

4740 **Jain, S.K. (1981).** Conservation of threatened plants in India. *Pl. Conserv. Bull.*, 1: 1-8. (T)

4741 **Jain, S.K. (1981).** Threatened plants in India. *In* XIII International Botanical Congress, Sydney, Australia, 21-28 August 1981: abstracts. Sydney, Australian Academy of Science. 214. (T)

4742 **Jain, S.K. (1983).** Documentation of endangered flora of India. *In* Jain, S.K., Mehra, K.L., *eds*. Conservation of tropical plant resources. Howrah, Botanical Survey of India. 240-245. (T)

4743 **Jain, S.K. (1983).** Materials for a Green Book of botanic gardens in India: a preliminary list of rare and threatened plants in cultivation in some botanic gardens in India. Botanical Survey of India. 6p. Lists 86 threatened species in cultivation in botanic gardens. (G/T)

4744 **Jain, S.K. (1985).** Botanical Survey of India: 1958-1983. *Bull. Bot. Surv. India*, 25(1-4): 252-267 (1983 publ. 1985). Includes review of plant conservation activities.

4745 **Jain, S.K., Rao, R.R. (1983).** An assessment of threatened plants of India. Proceedings of the seminar held at Dehra Dun, 14-17 Sept., 1981. Howrah, Botanical Survey of India. 334p. (T)

4746 **Jain, S.K., Rao, R.R. (1983).** Ethnobotany in India - an overview. Howrah, Botanical Survey of India. 22p. Medicinal. (E)

INDIA

4747 **Jain, S.K., Sastry, A.R.K. (1980).** Threatened plants of India: a state-of-the-art report. New Delhi, Botanical Survey of India and Man and Biosphere Committee. 48p. Col. illus. Short accounts of 134 species. (T)

4748 **Jain, S.K., Sastry, A.R.K. (1981).** Techniques and constraints in survey and conservation of threatened plants and habitats in India. *In* Synge, H., *ed.* The biological aspects of rare plant conservation. Chichester, Wiley. 59-66. Proceedings of International Conference, King's College, Cambridge, 14-19 July 1980. (T)

4749 **Jain, S.K., Sastry, A.R.K. (1982).** National parks and biosphere reserves in India. *In* Souvenir, Silver Jubilee Symposium of the International Society for Tropical Ecology. 50-56. (P)

4750 **Jain, S.K., Sastry, A.R.K. (1982).** Threatened plants and habitats - a review of work in India. *Pl. Conserv. Bull.*, 2: 1-9. (T)

4751 **Jain, S.K., Sastry, A.R.K., comps (1983).** Materials for a catalogue of threatened plants of India. Howrah, Botanical Survey of India. 69p. (T)

4752 **Jain, S.K., Sastry, A.R.K., eds (1984).** Indian Plant Red Data Book, 1. Calcutta, Botanical Survey of India. 367p. Illus. Data sheets on 125 species. (T)

4753 **Jayal, N.D., Lausche, B.J. (1985).** Legislation for biosphere reserves: the Indian experience. *In* Unesco-UNEP. Conservation, science and society. Paris, Unesco. 139-145. Contributions to the First International Biosphere Reserve Congress, Minsk, Byelorussia/USSR, 26 September - 2 October 1983. (P)

4754 **Juyal, N., Bhattacharyya, U.C. (1983).** On a recollection of seven rare Carices from north-western Himalaya. *J. Econ. Taxon. Bot.*, 4(1): 298-302. *Carex fusiformis, C. lehmannii, C. supina, C. tristis, C. microglochin, C. myosurus* vars. *praestans* and *eminens*. (T)

4755 **Kammathy, R.V. (1981).** Rare and endemic taxa of Indian Commelinaceae. *In* Botanical Survey of India. Abstracts of papers for the Seminar on Threatened Plants of India, 14-17 Sept. 1981. Howrah, BSI. 26. (T)

4756 **Kapoor, S.L. (1981).** Observations on some rare Indian *Clematis*. *In* Botanical Survey of India. Abstracts of papers for the Seminar on Threatened Plants of India, 14-17 Sept. 1981. Howrah, BSI. 23. (T)

4757 **Kapoor, S.L., Mitra, R. (1979).** Herbal drugs in Indian pharmaceutical industry. Lucknow, National Botanical Research Institute. 85p. Medicinal. (E)

4758 **Karthikeyan, S. (1981).** Endemic grasses of India with an emphasis on endangered, threatened and rare species. *In* Botanical Survey of India. Abstracts of papers for the Seminar on Threatened Plants of India, 14-17 Sept. 1981. Howrah, BSI. 28. (T)

4759 **Kataki, S.K. (1976).** Indian orchids - a note on conservation. *Amer. Orchid Soc. Bull.*, 46(2): 117-121. Lists threatened orchids. (T)

4760 **Kataki, S.K., Jain, S.K., Sastry, A.R.K. (1984).** Threatened and endemic orchids of Sikkim and north-eastern India. Howrah, Botanical Survey of India. 95p. Descriptions, distributions, illus. of over 100 species. (T)

4761 **Khanna, L.S. (1982).** Forest protection. Dehra Dun, Khanna Bandhu. 206p. Illus.

4762 **Khoshoo, T.N. (1984).** Biosphere reserves: an Indian approach. *In* Unesco-UNEP. Conservation, science and society. Paris, Unesco. 185-189. Contributions to the First International Biosphere Reserve Congress, Minsk, Byelorussia/USSR, 26 September - 2 October, 1983. Table. (P)

4763 **Kiew, R. (1988).** Portraits of threatened plants 16. *Phoenix paludosa* Roxb. *Malay. Naturalist*, 42(1): 16. Illus. (T)

INDIA

4764 **Lajos, S. (1979).** Del-India nemzeti parkjaiban. *Buvar*, 34(9): 406-409. Hu. Illus., map. (P)

4765 **Lall, J.D.S., ed. (1981).** The Himalaya: aspects of change. New Delhi, India International Centre, Oxford Univ. Press. 253p. Col. illus. Maps. (P/T)

4766 **Maheshwari, J.K. (1979).** Conservation of endangered flora. *In* Plant taxonomy in India: a state-of-the-art report. Lucknow, National Botanical Research Institute. 71-76. (T)

4767 **Maheshwari, J.K. (1979).** Conservation of rare plants - Indian scene vis-a-vis world scene. *Bull. Bot. Surv. India*, 19(1-4): 167-173 (1977 publ. 1979). (T)

4768 **Mamgain, S.K., Rao, R.R. (1987).** Notes on distribution of some rare and endangered species of Lactuceae (Asteraceae) in India. *Indian Sci. Congr. Assoc. Proc.*, 3: 124. (T)

4769 **Mehra, K.L, Arora, R.K. (1983).** Collection and conservation of genetic resources of field crops in India. *In* Jain, S.K., Mehra, K.L., *eds.* Conservation of tropical plant resources. Howrah, Botanical Survey of India. 142-153. (E)

4770 **Mehra, K.L., Arora, R.K. (1982).** Plant genetic resources of India: their diversity and conservation. NBPGR Sci. Monogr. No.4. New Delhi, National Bureau of Plant Genetic Resources, xiii, 60p. (E/T)

4771 **Mehrotra, A. (1981).** A few rare Indian grasses. *Bull. Bot. Surv. India*, 21(1-4): 237-238. *Arthraxon lanceolatus, Dichanthium woodrowii, Lophopogon duthiei, L. kingii.* (T)

4772 **Mehrotra, A. (1981).** Studies on endemism and rarity in the family Eriocaulaceae in India. *In* Botanical Survey of India. Abstracts of papers for the Seminar on Threatened Plants of India, 14-17 Sept. 1981. Howrah, BSI. 39. (T)

4773 **Mehrotra, A. (1982).** On the status of some endemics in India: 1. *Bull. Bot. Surv. India*, 22(1-4): 233-234 (1980 publ. 1982).

4774 **Mehrotra, A. (1983).** On the status of some endemics (?) in India: 2. *Bull. Bot. Surv. India*, 24(1-4): 233-234 (1982 publ. 1983). (T)

4775 **Mondal, N.R. (1981).** Endemic taxa in Urticaceae, Moraceae, Ulmaceae & Cannabinaceae. *In* Botanical Survey of India. Abstracts of papers for the Seminar on Threatened Plants of India, 14-17 Sept. 1981. Howrah, BSI. 25-26. (T)

4776 **Mukherjee, P.K. (1981).** Distribution and collection of rare umbellifers in India. *In* Botanical Survey of India. Abstracts of papers for the Seminar on Threatened Plants of India, 14-17 Sept. 1981. Howrah, BSI. 23-24. (T)

4777 **Mukherjee, S.K. (1985).** Systematic and ecogeographic studies of genepools: 1. *Mangifera* L. Rome, IBPGR Secretariat. 86p. Tables. Maps. Extensive bibliography. (B/E/G/T)

4778 **Murthy, S.G. (1985).** Sandalwood: case study of a resource decline. *Garden* (New York), 9(1): 16-19. Illus. Deforestation. (E/L/T)

4779 **Myers, N. (1986).** Environmental repercussions of deforestation in the Himalayas. *J. World Forest Res. Manage.*, 2: 63-72.

4780 **Myrthong, S., Rao, R.R. (1981).** A contribution to the knowledge of threatened plants in India - 3. Some noteworthy cotyledonous species which are rare, endangered or endemic. *In* Botanical Survey of India. Abstracts of papers for the Seminar on Threatened Plants of India, 14-17 Sept. 1981. Howrah, BSI. 15-16. (T)

4781 **Naik, V.N. (1981).** Extinction, evolution and immigration of certain plant species in the flora of Marathwada. *In* Botanical Survey of India. Abstracts of papers for the Seminar on Threatened Plants of India, 14-17 Sept. 1981. Howrah, BSI. 18-19. (T)

INDIA

4782 **Nair, N.C. & V.J., Ansari, R. (1980).** Notes on some rare plants from south India. *Bull. Bot. Surv. India*, 22(1-4): 205-207 (1980 publ. 1982). Illus. (T)

4783 **Nair, N.C. (1978).** Plant conservation: the Indian scene. *J. Kerala Nat. Hist. Soc.*, 2: 17-21.

4784 **Nair, N.C., Ghosh, S.R. (1978).** The rare fern *Asplenium grevillei* Wall. ex Hook. et Grev. (Aspleniaceae). A new find for Peninsular India. *Indian Forester*, 104(12): 819-822. (T)

4785 **Nair, N.C., Ramachandran, V.S. (1982).** A note on the rare plant *Oianthus disciflorus* Hook.f. *Bull. Bot. Surv. India*, 22(1-4): 234-235 (1980 publ. 1982). (T)

4786 **Nair, N.C., Srinivasan, S.R. (1982).** Rediscovery of *Eugenia discifera* Gamble (Myrtaceae) and its lectotypification. *Bull. Bot. Surv. India*, 22(1-4): 232-233 (1980 publ. 1982). (T)

4787 **Nair, N.C., Srinivasan, S.R. (1983).** On the rediscovery of *Koilodepas calycinum* Bedd. (Euphorbiaceae) and *Holcolemma canaliculatum* (Nees ex Steud.) Stapf. ex Hubbard (Poaceae) from south India. *Bull. Bot. Surv. India*, 24(1-4): 241-242 (1982 publ. 1983). (T)

4788 **Naithani, B.D. (1981).** Result of over-exploitation of plant species. *In* Botanical Survey of India. Abstracts of papers for the Seminar on Threatened Plants of India, 14-17 Sept. 1981. Howrah, BSI. 7. (E/T)

4789 **Nayar, M.P. (1979).** *Entada pusaetha* DC. on the verge of depletion. *Bull. Bot. Surv. India*, 21(1-4): 229. Illus. (T)

4790 **Nayar, M.P., Sastry, A.R.K., eds (1987).** Red Data Book of Indian plants. Volume 1. Calcutta, Botanical Survey of India. 367p. Illus., col. illus. (G/T)

4791 **Oza, G.M. (1974).** Indian Doum palm faces extinction. *Biol. Conserv.*, 6(1): 65-67. (T)

4792 **Oza, G.M. (1981).** Save trees, save India. *Envir. Conserv.*, 8(3): 248.

4793 **Oza, G.M., Singh, G. (1981).** A census of some plants for the Red Data Book of India. *In* Botanical Survey of India. Abstracts of papers for the seminar on Threatened Plants of India, 14-17 Sept. 1981. Howrah, BSI. 39-40. (T)

4794 **Padmanaban, P. (1983).** Preservation of rare plants. *Bull. Bot. Surv. India*, 23(3&4): 265-266 (1981). (T)

4795 **Pant, P.C. (1981).** Observations on distribution of some rare plants endemic to India. *In* Botanical Survey of India. Abstracts of papers for the Seminar on Threatened Plants of India, 14-17 Sept. 1981. Howrah, BSI. 38-39. (T)

4796 **Panwar, H.S. (1985).** A study of management requirements in Corbett National Park. *In* Thorsell, J.W., *ed.* Conserving Asia's natural heritage. The planning and management of protected areas in the Indomalayan Realm. Proc. of the 25th working session of IUCN's CNPPA. Gland, IUCN. 169-176. (P)

4797 **Panwar, H.S. (1985).** Protected areas and people - the Indian approach. *In* Thorsell, J.W., *ed.* Conserving Asia's natural heritage. The planning and management of protected areas in the Indomalayan Realm. Proc. of the 25th working session of IUCN's CNPPA. Gland, IUCN. 192-200. (P)

4798 **Pradhan, G.M. (1974).** Orchid conservation in India. *Amer. Orch. Soc. Bull.*, 43: 135-139. (T)

4799 **Pradhan, U.C. (1975).** Conservation of eastern Himalayan orchids. Problems and prospects. Part I. *Orch. Rev.*, 83: 314-317. (T)

4800 **Pradhan, U.C. (1975).** Conservation of eastern Himalayan orchids. Problems and prospects. Part II. *Orch. Rev.*, 83: 345-347. (T)

INDIA

4801 **Pradhan, U.C. (1975).** Conservation of eastern Himalayan orchids. Problems and prospects. Part III. *Orch. Rev.*, 83: 374. (T)

4802 **Pradhan, U.C. (1982).** Himalayan plant Red Data Sheets: 2. *Cymbidium sikkimense* Hook.f. Orchidaceae. *Himalayan Pl. J.*, 1(2): 14, 17-18. Illus., map. (T)

4803 **Pradhan, U.C. (1982).** Himalayan plant Red Data Sheets: 3. *Calanthe alpina* Hook.f. Orchidaceae. *Himalayan Pl. J.*, 1(2): 49-51. Illus. (T)

4804 **Pradhan, U.C. (1983).** Himalayan plant Red Data Sheets. 5: *Nepenthes khasiana* Hook.f. *Himalayan Pl. J.*, 2(4): 50-53. Illus. (T)

4805 **Pradhan, U.C. (1985).** Himalayan plant Red Data Sheets: 7. *Coelogyne treutleri* Hook. f. Treutler's Coelogyne. *Himalayan Pl. J.*, 3(6): 48-49. Illus. (T)

4806 **Pradhan, U.C. (1985).** Himalayan plant Red Data Sheets: 8. *Calanthe whiteana* King & Pantling Orchidaceae. *Himalayan Pl. J.*, 3(7): 80-81. Illus. (T)

4807 **Pradhan, U.C. (1985).** Red Data Sheet on Indian Orchidaceae: 1. *Vanda coerulea. Indian Orchid J.*, 1(2): 54-56. Col. illus. (T)

4808 **Prakash, V., Jain, S.K. (1981).** Tribe Isachneae (Poaceae) - its endemism and rarity in India. *In* Botanical Survey of India. Abstracts of papers for the Seminar on Threatened Plants of India, 14-17 Sept. 1981. Howrah, BSI. 24. (T)

4809 **Pramanik, A., (1981).** Observations on some rare selected legumes. *In* Botanical Survey of India. Abstracts of papers for the Seminar on Threatened Plants of India, 14-17 Sept. 1981. Howrah, BSI. 24. (T)

4810 **Raghavan, R.S. (1981).** Conservation forestry and biosphere reserves along Western Ghats. *Myforest*, 17(4): 71-79. (P)

4811 **Raghavan, R.S., Singh, N.P. (1983).** Endemic and threatened plants of western India. *Plant Conserv. Bull.*, 3: 16p. (T)

4812 **Raghavan, R.S., Singh, N.P. (1984).** An inventory of endemic and vulnerable species of western India deserving conservation. *J. Econ. Taxon. Bot.*, 5(1): 153-164. (T)

4813 **Raizada, M.B. (1979).** Future of floristic studies and preservation and conservation needs of rare, endangered and vanishing plants of India. *Bull. Bot. Surv. India*, 19(1-4): 174-175 (1977 publ. 1979). (T)

4814 **Raizada, M.B. (1980).** Endangered Indian plants. *Indian J. Forest.*, 3(4): 370-371. (T)

4815 **Ramakrishnan, P.S. (1983).** Problems and prospects of conservation of plant resources in the north-eastern hill region of India. *In* Jain, S.K., Mehra, K.L., eds. Conservation of tropical plant resources. Howrah, Botanical Survey of India. 172-180.

4816 **Rao, A.S. (1980).** Endangered species of northeastern India. *Indian Sci. Congr. Assoc. Proc.*, 67(4): 35-36. Abstract. (T)

4817 **Rao, A.S., Srivastava, S.C. (1983).** *Hedychium venustum* Wt.: its type locality and endemism. *Bull. Bot. Surv. India*, 24(1-4): 231-232 (1982 publ. 1983). (T)

4818 **Rao, A.V.N. (1978).** *Acanthephippium bicolor* Lindl. A rare and endangered orchid in south India. *Orchid Rev.*, 86: 275-276. Illus. (T)

4819 **Rao, A.V.N., Banerjee, A.K. (1982).** Miscellaneous notes: 37 Cultivation of endangered plants in south India: 2. *Bentinckia condapanna* Berry ex Roxb. *J. Bombay Nat. Hist. Soc.*, 79(1): 237-239. Illus. (G/T)

4820 **Rao, A.V.N., Banerjee, A.K., Subramaniam, A. (1981).** Cultivation of endangered plants in south India. *J. Bombay Nat. Hist. Soc.*, 78(2): 421-423. Illus. BSI Experimental Garden, Yercaud, Tamil Nadu. (G/T)

INDIA

4821 **Rao, K. (1985).** Legislative and organizational support for protected areas in India. *In* Thorsell, J.W., *ed.* Conserving Asia's natural heritage. The planning and management of protected areas in the Indomalayan Realm. Proc. of the 25th working session of IUCN's CNPPA. Gland, IUCN. 157-161. (L/P)

4822 **Rao, K.S.S., Shah, R.R., Kalyansundaram, N.K., Patel, D.H., Dalal, K.C. (1981).** Urgent need for germplasm collection and conservation of rare, endangered and threatened wild plants of medicinal importance. *In* Botanical Survey of India. Abstracts of papers for the Seminar on Threatened Plants of India, 14-17 Sept. 1981. Howrah, BSI. 33. (E/T)

4823 **Rao, T.K., Satyanarayana Rao, S.V.V., Murthy, V.K. (1985).** Indian forest - an overview. *Indian Forest.*, 111(8): 571-578. Maps.

4824 **Rathore, S.R. (1981).** Endemic & rare species of *Calanthe* R.Br. (Orchidaceae) in India. *In* Botanical Survey of India. Abstracts of papers for the Seminar on Threatened Plants of India, 14-17 Sept. 1981. Howrah, BSI. 27-28. (T)

4825 **Rodgers, W.A. (1985).** Biogeography and protected area planning in India. *In* Thorsell, J.W., *ed.* Conserving Asia's natural heritage. The planning and management of protected areas in the Indomalayan Realm. Proc. of the 25th working session of IUCN's CNPPA. Gland, IUCN. 103-113. Tables; map. (P)

4826 **Rodgers, W.A., Panwar, H.S. (1988).** Planning a wildlife protected area network in India. Vol. 1: The report. Dehra Dun, Establishment of the Wildlife Institute of India, FAO. 341p. (P)

4827 **Rodgers, W.A., Panwar, H.S. (1988).** Planning a wildlife protected area network in India. Vol. 2: State summaries. Dehra Dun, Establishment of the Wildlife Institute of India, FAO. (P)

4828 **Saharia, V.B. (1984).** Human dimensions in wildlife management: the Indian experience. *In* McNeely, J.A., Miller, K.R., *eds.* National parks, conservation and development. The role of protected areas in sustaining society. Washington, D.C., Smithsonian Institution Press. 190-196. (P)

4829 **Saharia, V.B. (1985).** The role of the Wildlife Institute of India. *In* Thorsell, J.W., *ed.* Conserving Asia's natural heritage. The planning and management of protected areas in the Indomalayan Realm. Proc. of the 25th working session of IUCN's CNPPA. Gland, IUCN. 132-136.

4830 **Sahni, K.C. (1969).** Protection of rare and endangered plants in the Indian flora. *Forest Res. Inst.*, 2: 95-102. (T)

4831 **Sahni, K.C. (1970).** Protection of rare and endangered plants in the Indian flora. *IUCN Publ. New Ser.*, 18: 95-102. (T)

4832 **Sahni, K.C. (1979).** Endemic, relict, primitive and spectacular taxa in eastern Himalayan flora and strategies for their conservation. *Indian J. Forest.*, 2(2): 181-190. Illus. (T)

4833 **Sahni, K.C. (1980).** Vanishing Indian taxa and their conservation. *In* Nair, P.K.K., *ed.* Modern trends in plant taxonomy. New Delhi, Vikas. 161-172. (T)

4834 **Santapau, H. (1970).** Endangered plant species and their habitats. *In* IUCN. 11th Technical Meeting Papers and Proceedings, 2. Problems of Threatened Species. IUCN New Series 18. Switzerland, IUCN. 83-88. Includes list of threatened medicinal plants and orchids in need of protection. (T)

4835 **Sastry, A.R.K., Hajra, P.K. (1981).** Some rare & endemic rhododendrons in India - a preliminary study. *In* Botanical Survey of India. Abstracts of papers for the Seminar on Threatened Plants of India, 14-17 Sept. 1981. Howrah, BSI. 24-25. (T)

4836 **Sharma, H.P. (1983).** Status report on indigenous herbal drugs and need for protection measures. New Delhi, Dept of Environment. 8p. Medicinal. (E)

INDIA

4837 **Shiva, V., Bandyopadhyay, J. (1983).** Eucalyptus: a disastrous tree for India. *Ecologist*, 13(5): 184-187. Discusses impact on ecology. (E)

4838 **Shiva, V., Bandyopadhyay, J., Jayal, N.D. (1985).** Afforestation in India: problems and strategies. *Ambio*, 14(6): 329-333. Illus.

4839 **Singh, F. (1982).** Conservation action: conservation of orchids in India. *Hornbill*, (1): 15-17. Illus. (P/T)

4840 **Singh, F. (1985).** Conservation of orchids in India. *Indian Forester*, Special Issue, 3(11): 1010-1022. Illus. (T)

4841 **Singh, J.S., Pandey, U., Tiwari, A.K. (1984).** Man and forests: a central Himalayan case study. *Ambio*, 13(2): 80-86. Illus., maps. Deforestation. (T)

4842 **Singh, P. (1983).** Extended distribution of an endemic species: *Tribulus rajasthanensis* Bandari & Sharma. *Bull. Bot. Surv. India*, 24(1-4): 237-238 (1982 publ. 1983). (T)

4843 **Singh, S. (1985).** Protected areas in India. *In* Thorsell, J.W., *ed.* Conserving Asia's natural heritage. The planning and management of protected areas in the Indomalayan Realm. Proc. of the 25th working session of IUCN's CNPPA. Gland, IUCN. 11-18. (P)

4844 **Sivadasan, M. (1981).** Threatened species of Indian Araceae. *In* Botanical Survey of India. Abstracts of papers for the Seminar on Threatened Plants of India, 14-17 Sept. 1981. Howrah, BSI. 26. (T)

4845 **Srinivasan, K. (1959).** Protection of wild plant life. *Bull. Bot. Surv. India*, 1: 85-89.

4846 **Srivastava, R.C. (1985).** Notes on threatened taxa of Malpighiaceae of India. *J. Econ. Taxon. Bot.*, 6(1): 61-72. Illus. (T)

4847 **Subramanian, K.N. (1982).** Need for the development of seed orchards and germplasm banks of medicinal plants. *Ancient Sci. Life*, 2(2): 98-99. (E/G/T)

4848 **Subramanyam, K. (1969).** Nature conservation - a losing battle. *Indian Forester*, 95: 719-723.

4849 **Subramanyam, K., Henry, A.N. (1970).** Rare or little known plants from south India. *Bull. Bot. Surv. India*, 12(1-4): 1-5. (T)

4850 **Subramanyam, K., Streemadhaven, C.P. (1970).** Endangered plant species and their habitats - a review of the Indian situation. IUCN Publications New Series No. 18. Morges, IUCN. 108-114. (T)

4851 **Swaminathan, M.S. (1981).** Environmental protection in India: problems and prospects. *J. Bombay Nat. Hist. Soc.*, 78(3): 429-435.

4852 **Synge, H. (1982).** Plant conservation and development in India. *Threatened Pl. Commit. Newsl.*, 9: 1-3. (T)

4853 **Synge, H. (1983).** Plant conservation and development in India. *Tigerpaper*, 10(2): 9-10 (T)

4854 **Theuerkauf, W.D. (1989).** The golden tongue: quest for a rare Indian orchid. *Bot. Gard. Conserv. News*, 1(5): 37-39. Illus. *Chrysoglossum maculatum*. (G/T)

4855 **Toky, O.P., Ramakrishnan, P.S. (1982).** Forest wealth of the north-eastern India and its conservation. *In* Paliwal, G.S., *ed.* The vegetational wealth of the Himalayas. Papers of the symposium at the Dept of Botany, Univ. of Garhwal, 1-6 October 1979. Delhi, Puja Publishers. 433-438. Illus. Includes shifting cultivation and vegetation degradation.

4856 **Vajravelu, E. (1983).** Rare, threatened and endemic flowering plants of south India: I. *Pl. Conserv. Bull.*, 4: 14-30. (T)

INDIA

4857 **Vajravelu, E., Bhargavan, P. (1981).** Rare and endemic species re-collected after 50 years or more from south India. *In* Botanical Survey of India. Abstracts of papers for the Seminar on Threatened Plants of India, 14-17 Sept. 1981. Howrah, BSI. 19. (T)

4858 **Vajravelu, E., Bhargavan, P. (1982).** Notes on some rare plants from south India. *J. Econ. Tax. Bot.*, 3(3): 969-973. Illus. (T)

4859 **Vajravelu, E., Gopalan, R. (1982).** Rare and little known plants from south India. *J. Econ. Tax. Bot.*, 3(3): 978-980. Illus. (T)

4860 **Varmah, J.C, Sahni, K.C. (1976).** Rare orchids of the North Eastern Region and their conservation. *Indian Forester*, July 1976: 424-431. Illus. (T)

4861 **Vij, S.P., ed. (1986).** Biology, conservation and culture of orchids: papers presented at a national seminar organized by The Orchid Society of India, held at Punjab University, 3-4 April 1985. New Delhi, Affiliated East-West Press. xvi, 492p. Illus., maps. (G)

4862 **Wadhwa, B.M., Chowdhery, H.J. (1981).** The distribution and phytogeography of some rare western Himalayan plants. *In* Botanical Survey of India. Abstracts of papers for the Seminar on Threatened Plants of India, 14-17 Sept. 1981. Howrah, BSI. 5-6. (T)

4863 **Wagner, J.P. (1985).** The "scandalwood". *Hawaii*, 2(2) (Issue No. 4): 51-52. Exploitation of sandalwood has had devastating effects. (T)

4864 **Wallach, B. (1985).** Deforestation: the view from south India. *Garden* (U.S.), 9(1): 8-15. Col. illus., map.

4865 **Wilson, A. (1983).** Who is destroying India's forests. *Ecologist*, 13(2-3): 100-101. (L)

INDIA - Andaman Islands

4866 **Balakrishnan, N.P. (1977).** Recent botanical studies in Andaman and Nicobar Islands. *Bull. Bot. Surv. India*, 19: 132-138. Lists 136 'rare' and 'endangered' endemic species. (T)

4867 **Balakrishnan, N.P. (1982).** Notes on some little known ferns of Andaman and Nicobar islands. *Bull. Bot. Surv. India*, 22(1-4): 136-140 (1980 publ. 1982). (T)

4868 **Balakrishnan, N.P., Rao, M.K.V. (1983).** The dwindling plant species of Andaman and Nicobar Islands. *In* Jain, S.K., Rao, R.R., eds. An assessment of threatened plants of India. Howrah, Botanical survey of India. 186-210. Lists 110 threatened endemic taxa and 136 threatened non-endemics; notes on distribution. (T)

4869 **Bhargava, N., Premenath, R.K. (1987).** Notes on some rare plants from Andaman Islands. *J. Econ. Tax. Bot.*, 10: 317-320. Illus. (T)

4870 **Botanical Survey of India (?)** Endangered flora of Andaman and Nicobar Islands. Botanical Survey of India, Andaman and Nicobar Circle, Port Blair. 5p. Mimeo. Overview of vegetation and threats to species. (T)

4871 **Singh, V.P., Garge, A., Pathak, S.M., Mall, L.P. (1986).** Mangrove forests of Andaman Islands in relation to human interference. *Envir. Conserv.*, 13(2): 169-172, 160. Illus., map. (T)

4872 **Thothathri, K. (1960).** Studies on the flora of the Andaman Islands. *Bull. Bot. Surv. India*, 2: 357-373. 281 species listed, with notes on distribution and abundance on the islands. (T)

INDIA - Andhra Pradesh State

4873 **Nayar, M.P., Ahmed, M., Raju, D.C.S. (1984).** Endemic and rare plants of Eastern Ghats. *Indian J. Forest.*, 7(1): 35-42. Map. Lists 76 species. (T)

INDIA - Andhra Pradesh State

4874 **Raju, D.C.S., Nayar, M.P., Ahmed, M. (1983).** Endemic and endangered flora of the Eastern Ghats. *Indian Sci. Congr. Assoc. Proc.*, 70(3): 53-54. (T)

INDIA - Arunachal Pradesh Union Terr.

4875 **Chatterjee, A.K., Chandiramani, S.S. (1986).** An introduction to Namdapha Tiger Reserve, Arunachal Pradesh, India. *Tigerpaper*, 13(3): 22-27. Illus., maps. (P)

4876 **Dam, D.P. & S.N. (1982).** *Chrysoglossum erraticum* Hook.f. - a rare orchid from the Kameng District, Arunachal Pradesh. *Bull. Bot. Surv. India*, 22(1-4): 185-187 (1980 publ. 1982). Illus. (T)

4877 **Deb, D.B., Raghavan, R.S. (1983).** A rare species of *Agapetes* D. Don ex G. Don (Ericaceae). *Bull. Bot. Surv. India*, 24(1-4): 171-173 (1982 publ. 1983). Illus. (T)

4878 **Johnsingh, A.J.T. (1985).** Understand, assist, protect and conserve. Flora and fauna of Arunachal Pradesh. *The India Magazine*: 64-71. Illus.

4879 **Lal, J., Pal, G.D. (1987).** Depletion of rare and endangered vascular plants from the flora of Itanagar problem and its solution. *Indian Sci. Congr. Assoc. Proc.*, 3: 124 (T)

4880 **Mudgal, V., Jain, S.K. (1982).** *Coptis teeta* Wall. - local uses, distribution and cultivation. *Bull. Bot. Surv. India*, 22: 179-180. Illus. Medicinal. (E/T)

4881 **Naithani, H.B., Bahadur, K.N. (1981).** Observations on extended distribution of new and rare taxa of north-eastern India with special reference to Arunachal Pradesh. *Indian Forester*, 107(11): 712-724. Illus. (T)

4882 **Pal, G.D. (1983).** *Christensenia aesculifolia* (Blume) Maxon: first report of a poorly known fern from Subansiri District, Arunachal Pradesh, India. *Bull. Bot. Surv. India*, 24(1-4): 180-182 (1982 publ. 1983). Illus. (T)

4883 **Pal, G.D., Abbareddy, N.R (1983).** *Galeola nudifolia* Lour., a rare orchid from Subansiri District, Arunachal Pradesh, India. *Bull. Bot. Surv. India*, 24(1-4): 203-205 (1982 publ. 1983). Illus. (T)

INDIA - Assam State

4884 **Ghosh, S.R. (1982).** The rare and threatened fern *Adiantum soboliferum* Wall. ex Hook: a new find for eastern India. *J. Bombay Nat. Hist. Soc.*, 79(3): 716-717. Key. (T)

4885 **Jain, S.K., Hajra, P.K. (1975).** On the botany of Manas Wild Life Sanctuary in Assam. *Bull. Bot. Surv. India*, 17(1-4): 75-86. Illus. (P)

4886 **Lyngdoh, U.C. (1982).** Forest conservation in Khasi and Jaintia Hills. *Indian Forest.*, 108(5): 380. Letter to editor. (T)

4887 **Pradhan, U.C. (1975).** Conservation of eastern Himalayan orchids. Problems and prospects. Part I. *Orch. Rev.*, 83: 314-317. (T)

4888 **Pradhan, U.C. (1975).** Conservation of eastern Himalayan orchids. Problems and prospects. Part II. *Orch. Rev.*, 83: 345-347. (T)

4889 **Pradhan, U.C. (1975).** Conservation of eastern Himalayan orchids. Problems and prospects. Part III. *Orch. Rev.*, 83: 374. (T)

4890 **Sahni, K.C., Naithani, H.B. (1979).** A rare and spectacular rhododendron from Kameng District. *Indian Forester*, 105(1): 77. (T)

4891 **Songh, N.B., Bhattacharrya, U.C. (1981).** Notes on little known taxa of *Sedum* from Sikkim and Assam Himalaya. *In* Botanical Survey of India. Abstracts of papers for the Seminar on Threatened Plants of India, 14-17 Sept. 1981. Howrah, BSI. 15. (T)

INDIA - Bihar State

4892 **Aitken, J.E., Ganguli, M. (1981).** Another forest disappears in India. *Envir. Conserv.*, 8(3): 228. Destruction of Santhal Parganas forest.

INDIA - Gujarat State

4893 **Gujarat Forest Department (1988).** Wildlife sanctuaries and national parks of Gujarat. *Tigerpaper*, 15(4): 3-5. (P)

4894 **Sabnis, S.D., Rao, K.S.S. (1981).** Observations on some rare or endangered endemics of S.E. Kutch flora. *In* Botanical Survey of India. Abstracts of papers for the Seminar on Threatened Plants of India, 14-17 Sept. 1981. Howrah, BSI. 9. (T)

4895 **Shah, G.L. (1981).** Rare and endemic flora of south-east Gujarat. *In* Botanical Survey of India. Abstracts of papers for the Seminar on Threatened Plants of India, 14-17 Sept. 1981. Howrah, BSI. 7-8. (T)

INDIA - Himachal Pradesh State

4896 **Aswal, B.S., Mehrotra, B.N. (1983).** *Delphinium uncinatum* Hook.f. et Thoms. (Ranunculaceae) and *Lilium wallichianum* Schultes f. (Liliaceae): two rare finds from north-west Himalaya. *J. Econ. Taxon. Bot.*, 3(3): 773-775 (1982 publ. 1983). (T)

4897 **Garson, P.J. (1982).** Conservation of wildlife in Himachal's forests. *Tigerpaper*, 9(4): 27-31. Illus., map. (P)

INDIA - Jammu and Kashmir State

4898 **Aswal, B.S., Mehrotra, B.N. (1983).** *Delphinium uncinatum* Hook.f. et Thoms. (Ranunculaceae) and *Lilium wallichianum* Schultes f. (Liliaceae): two rare finds from north-west Himalaya. *J. Econ. Taxon. Bot.*, 3(3): 773-775 (1982 publ. 1983). (T)

4899 **Dhar, U., Kachroo, P. (1981).** Some remarkable features of endemism in Kashmir Himalayas. *In* Botanical Survey of India. Abstracts of papers for the Seminar on Threatened Plants of India, 14-17 Sept. 1981. Howrah, BSI. 2. (T)

4900 **Gaston, A.J. (1982).** A national park for Kishtwar. *Hornbill*, 4: 10-14. Illus. (P)

4901 **Kapur, S.K. (1983).** Threatened medicinal plants of Jammu & Kashmir. *J. Sci. Res. Pl. Med.*, 4(3): 40-46. (E/T)

4902 **Kapur, S.K., Sarin, Y.K. (1984).** Plant resources exploitation and their utilisation in Trikuta hills of Jammu Province (J. and K. State). *J. Econ. Taxon. Bot.*, 5(5): 1143-1158. Trade. (E)

4903 **Kaul, V. (1983).** Conservation of plant resources in aquatic ecosystems with reference to some aquatic habitats of Kashmir. *In* Jain, S.K., Mehra, K.L., *eds.* Conservation of tropical plant resources. Howrah, Botanical Survey of India. 118-131. Wetlands. (T)

4904 **Kehimkar, I.D. (1982).** Conservation action: fur trade in Kashmir. *Hornbill*, 4:23-24. Illus. Includes plant exports. (E)

4905 **Norberg-Hodge, H. (1981).** Ladakh: development without destruction. *In* Lall, J.S., *ed.* The Himalaya: aspects of change. New Delhi, India International Centre, Oxford University Press. 278-284. (P/T)

4906 **Siddique, M.A.A. (1990).** Endangered medicinal plants of Kashmir Himalaya. *Biol. Conserv. Newsl.*, 81: 1. (E/T)

4907 **Singh, N.B., Bhattacharyya, U.C. (1983).** Relocation of *Pseudosedum lievenii* (Crassulaceae) in Ladakh (J. & K.). *J. Econ. Tax. Bot.*, 4(3): 889-890. Illus. (T)

INDIA - Karnataka State

4908 **Dayanandan, P. (1983).** Conserving the flora of the peninsular hills. *Bull. Bot. Surv. India*, 23(3-4): 250-253 (1981 publ. 1983). Illus. (T)

4909 **Gadgil, M., Sukumar, R. (1986).** Scientific programme for the Nilgiri Biosphere Reserve: report of a workshop, Bangalore, March 7-8, 1986. Bangalore, Centre for Ecological Sciences. 48p. Map. (P)

4910 **Jenik, J., Chandran, M.D. Subash (1988).** Plan for talipot palm. *Threatened Pl. Newsl.*, 19: 7-8. Illus. (E/T)

4911 **Karanth, K.U. (1982).** Bhadra Wildlife Sanctuary and its endangered ecosystem. *J. Bombay Nat. Hist. Soc.*, 79(1): 79-86. Map. (P/T)

4912 **Nair, S.S.C. & P.V., Sharatchandra, H.C., Gadgil, M. (1977).** An ecological reconnaissance of the proposed Jawahar National Park. *J. Bombay Nat. Hist. Soc.*, 74(3): 401-435. (P)

4913 **Naithani, B.D. (1967).** Studies on the flora of Bandipur Reserve Forest, Mysore State. *Bull. Bot. Surv. India*, 8(3-4): 252-263. Annotated list. (P)

4914 **Naithani, B.D. (1978).** Botanising the game sanctuary of Mysore District of Karnataka State. *In* The Madras Herbarium (MH) 1853-1978, 125th Anniversary Souvenir, part 2: Symposium on Floristic Studies in Peninsular India. 23. Abstract. (P)

4915 **Nayar, M.P., Ahmed, M., Raju, D.C.S. (1984).** Endemic and rare plants of Eastern Ghats. *Indian J. Forest.*, 7(1): 35-42. Map. Lists 76 species. (T)

4916 **Raju, D.C.S., Nayar, M.P., Ahmed, M. (1983).** Endemic and endangered flora of the Eastern Ghats. *Indian Sci. Congr. Assoc. Proc.*, 70(3): 53-54. (T)

4917 **Saldanha, C.J. (1984).** Karnataka - state of environment report 1983-1984. Bangalore, Centre for Taxonomic Studies. 191p. Maps, illus.

4918 **Singh, N.P. (1979).** On a recollection of a few rare plants from Karnataka. *Bull. Bot. Surv. India*, 21(1-4): 231-234. Illus. *Crotalaria sandoorensis, Eleiotis trifoliolata, Juncus maritimus.* (T)

4919 **Sunder, S.S., Reddy, A.N.Y., Chalwadi, S.M. (1986).** Western Ghats in Karnataka: its ecological decline and steps for revival. *In* Nair, K.S.S., Gnanaharan, R., Kedharnath, S., *eds.* Ecodevelopment of Western Ghats. Proceedings of a seminar, Peechi, Kerala, 17-18 October 1984. Peechi, Kerala Forest Research Institute. 201-203.

4920 **Tewari, P.K., Janardhanan, K.P. (1983).** *Hopea jacobi* C.E.C. Fischer - a rare dipterocarp in Indian flora. *Indian J. Forestry*, 6(1): 80-81. Illus. (T)

INDIA - Kerala State

4921 **Abraham, A. (1986).** Plant wealth of the Western Ghats and the need for its conservation. *In* Nair, K.S.S., Gnanaharan, R., Kedharnath, S., *eds.* Ecodevelopment of Western Ghats. Proceedings of a seminar, Peechi, Kerala, 17-18 October 1984. Peechi, Kerala Forest Research Institute. 32-35.

4922 **Anon. (1987).** The arboretum of the Tropical Botanic Garden, Trivandrum, India. *Bot. Gard. Conserv. News*, 1(1): 16-19. Illus., map. (G)

4923 **Dayanandan, P. (1983).** Conserving the flora of the peninsular hills. *Bull. Bot. Surv. India*, 23(3-4): 250-253 (1981 publ. 1983). Illus. (T)

4924 **Gadgil, M., Sukumar, R. (1986).** Scientific programme for the Nilgiri Biosphere Reserve: report of a workshop, Bangalore, March 7-8, 1986. Bangalore, Centre for Ecological Sciences. 48p. Map. (P)

INDIA - Kerala State

4925 **Grainger, A. (1982).** Will the death knell sound in Silent Valley? *Ecologist*, 12(4): 185-188. Map. (P/T)

4926 **Henry, A.N., Chandrabose, M.S., Swaminathan, M.S., Nair, N.C. (1984).** Agastyamalai and its environs: a potential area for a biosphere reserve. *J. Bombay Nat. Hist. Soc.*, 81(2): 282-290. (P/T)

4927 **Henry, A.N., Swaminathan, M.S. (1983).** Rare or new *Exacum* L. (Gentianaceae) from southern India. *J. Bombay Nat. Hist. Soc.*, 80(2): 456-459. Illus. (T)

4928 **Hussain, S.A. (1980).** A look at the Kerala forests. *Hornbill*, 1: 25-28.

4929 **Jayson, E.A. (1986).** Ecodevelopment of wildlife sanctuaries in the Western Ghats of Kerala. *In* Nair, K.S.S., Gnanaharan, R., Kedharnath, S., eds. Ecodevelopment of Western Ghats. Proceedings of a seminar, Peechi, Kerala, 17-18 October 1984. Peechi, Kerala Forest Research Institute. 60-66. Map. (P)

4930 **Joseph, J., Chandrasekaran, V. (1982).** An account on the flora and vegetation of Neyyar Wildlife Sanctuary and its vicinity, Trivandrum District, Kerala. *Indian J. Bot.*, 5(2): 143-150. List and description of species. (P)

4931 **Karunakaran, C.K. (1986).** Ecodegradation of Kerala forest: historical facts. *In* Nair, K.S.S., Gnanaharan, R., Kedharnath, S., eds. Ecodevelopment of Western Ghats. Proceedings of a seminar, Peechi, Kerala, 17-18 October 1984. Peechi, Kerala Forest Research Institute. 104-109.

4932 **Kumar, C.S. (1986).** Endemic orchids of Western Ghats. *In* Nair, K.S.S., Gnanaharan, R., Kedharnath, S., *eds.* Ecodevelopment of Western Ghats. Proceedings of a seminar, Peechi, Kerala, 17-18 October 1984. Peechi, Kerala Forest Research Institute. 51-54. (T)

4933 **Menon, A.R.R. (1986).** Forest denudation in Kerala: a case study of the Trichur Forest Division. *In* Nair, K.S.S., Gnanaharan, R., Kedharnath, S., *eds.* Ecodevelopment of Western Ghats. Proceedings of a seminar, Peechi, Kerala, 17-18 October 1984. Peechi, Kerala Forest Research Institute. 94-98. Maps. Deforestation.

4934 **Mohanan, C.N. (1983).** A contribution to the botany of Quilon District, Kerala. *Bull. Bot. Surv. India*, 23(1-2): 60-64 (1981 publ. 1983). Lists endemic and threatened species. (T)

4935 **Mohanan, M., Henry, A.N. (1982).** Rediscovery of three rare and endemic plants of India. *Bull. Bot. Surv. India*, 22: 236-237 (1980 publ. 1982). (T)

4936 **Mohanan, M., Henry, A.N., Nair, N.C. (1982).** Notes on three rare and interesting orchids collected from Trivandrum District, Kerala. *J. Bombay Nat. Hist. Soc.*, 79(1): 234-236. (T)

4937 **Mohanan, M., Henry, A.N., Nair, N.C. (1982).** Some rare and fast disappearing plants discovered in Trivandrum District, Kerala. *Bull. Bot. Surv. India*, 22(1-4): 105-108 (1980 publ. 1982). (T)

4938 **Nair, K.K.N. (1986).** Conservation of the genetic diversity of *Dalbergia* species in the Western Ghats with special reference to Kerala. *In* Nair, K.S.S., Gnanaharan, R., Kedharnath, S., *eds.* Ecodevelopment of Western Ghats. Proceedings of a seminar, Peechi, Kerala, 17-18 October 1984. Peechi, Kerala Forest Research Institute. 47-50. Includes checklist of species and their distribution.

4939 **Nair, K.K.N. (1986).** Tropical wet evergreen forests: a valuable asset of Western Ghats to be preserved. *In* Nair, K.S.S., Gnanaharan, R., Kedharnath, S., *eds.* Ecodevelopment of Western Ghats. Proceedings of a seminar, Peechi, Kerala, 17-18 October 1984. Peechi, Kerala Forest Research Institute. 91-93.

4940 **Nair, K.S.S., Gnanaharan, R., Kedharnath, S., eds (1986).** Ecodevelopment of Western Ghats. Peechi, Kerala Research Institute. 315p. Proceedings of a seminar, Peechi, Kerala, 17-18 October 1984. (P/T)

INDIA - Kerala State

4941 **Nair, N.C., Mohanan, C.N. (1981).** On the rediscovery of four threatened species from the sacred groves of Kerala. *J. Econ. Taxon. Bot.*, 2: 233-235. (P/T)

4942 **Nair, N.C., Sreekumar, P.V., Nair, V.J. (1981).** Some rare and interesting plants from Kerala State. *J. Econ. Taxon. Bot.*, (2): 223-225. (T)

4943 **Nair, N.C., Vajravelu, E., Bhargavan, P. (1980).** The flora of Silent Valley and its conservation. *Indian Sci. Congr. Assoc. Proc.*, 67(4): 38-39. Abstract. (P)

4944 **Nair, N.G. (1986).** Endemic trees of Western Ghats: their importance and conservation. *In* Nair, K.S.S., Gnanaharan, R., Kedharnath, S., *eds.* Ecodevelopment of Western Ghats. Proceedings of a seminar, Peechi, Kerala, 17-18 October 1984. Peechi, Kerala Forest Research Institute. 41-44. (T)

4945 **Nair, P.N. (1986).** Forests of Western Ghats, Kerala. *In* Nair, K.S.S., Gnanaharan, R., Kedharnath, S., *eds.* Ecodevelopment of Western Ghats. Proceedings of a seminar, Peechi, Kerala, 17-18 October 1984. Peechi, Kerala Forest Research Institute. 36-40. Tropical forest.

4946 **Nambiar, V.P.K., Sasidharan, N., Renuka, C., Balagopalan, M. (1985).** Studies on the medicinal plants of Kerala forests. KFRI Research Report No. 42. Peechi, Kerala Forest Research Institute. Destruction of natural habitats threatens supply of drugs. (E)

4947 **Nayar, M.P. (1984).** Silent Valley remains silent. *Threatened Pl. Newsl.*, 13: 7-8. (P)

4948 **Rahmani, A.R. (1980).** Silent Valley: India's last tropical rain forest. *Tigerpaper*, 7(1): 17-19. Illus. (P)

4949 **Ramachandran, K.K., Easa, P.S., Vijayakumaran Nair, P. (1987).** Management of Periyar Tiger Reserve - problems and perspectives. *Tigerpaper*, 14(1): 25-32. Illus., maps. (P)

4950 **Ramachandran, V.S., Nair, V.J., Nair, N.C. (1980).** On some very rare or noteworthy plants from Kerala State. *J. Econ. Taxon. Bot.*, 1(1-2): 93-97. (T)

4951 **Ramakrishnan, P.S. (1984).** The need to conserve Silent Valley and tropical rain-forest ecosystems in India. *Envir. Conserv.*, 11(2): 170-171. Illus. (P/T)

4952 **Renuka, C. (1987).** Rattan resources of Kerala and their conservation. *Rattan Inform. Centre Bull.*, 6(1): 1-3. Map. (E)

4953 **Singh, J.S. & S.P., Saxena, A.K., Rawak, Y.S. (1984).** India's Silent Valley and its threatened rain-forest ecosystems. *Envir. Conserv.*, 11(3): 223-233. Illus., maps. (P/T)

4954 **Subramanian, K.N. (1986).** Conservation and protection of fast-vanishing endemic and rare gene pool of Western Ghats. *In* Nair, K.S.S., Gnanaharan, R., Kedharnath, S., *eds.* Ecodevelopment of Western Ghats. Proceedings of a seminar, Peechi, Kerala, 17-18 October 1984. Peechi, Kerala Forest Research Institute. 45-56.

4955 **Theuerkauf, W.D. (1989).** A refuge for south Indian orchids. *Bot. Gard. Conserv. News*, 1(4): 20-22. Western Ghats. (G)

4956 **Vartak, V.D., Kumbhojkar, M.S., Dabadghao, V. (1986).** Sacred groves - a sanctuary for lofty trees and lianas. *In* Nair, K.S.S., Gnanaharan, R., Kedharnath, S., *eds.* Ecodevelopment of Western Ghats. Proceedings of a seminar, Peechi, Kerala, 17-18 October 1984. Peechi, Kerala Forest Research Institute. 55-59. Describes 25 trees found in sacred groves. (P/T)

INDIA - Madhaya Pradesh State

4957 **Buch, M.N. (1985).** When forests are destroyed can man hope to survive? The flora of Madhya Pradesh. *India Mag.*, (1985): 52-61. Illus.

INDIA - Madhaya Pradesh State

4958 **Kaushik, J.P. (1981).** Some interesting plants of Shivpuri District (M.P.). *In* Botanical Survey of India. Abstracts of papers for the Seminar on Threatened Plants of India, 14-17 Sept. 1981. Howrah, BSI. 11. (T)

4959 **Khare, P.K. (1984).** Stand structure and species composition of a dry tropical forest reserve in central India. *Bull. Bot. Soc.*, 30-31: 8-15. Includes list of species, Gopalpura Forest Reserve. (P)

4960 **Murti, S. (1983).** *Cymbidium macrorhizon* Lindl. - an endangered orchid from Madhya Pradesh. *Bull. Bot. Surv. India*, 24(1-4): 236-237 (1982 publ. 1983). (T)

4961 **Oommachan, M. (1981).** Rare and threatened plants of Pachmarhi Hills (Madhya Pradesh). *In* Botanical Survey of India. Abstracts of papers for the Seminar on Threatened Plants of India, 14-17 Sept. 1981. Howrah, BSI. 9-10. (T)

4962 **Rathakrishnan, N.C., Saran, R. (1983).** Notes on rare plants from Madhya Pradesh. *J. Bombay Nat. Hist. Soc.*, 80(3): 665-667. Nine taxa of flowering plants. (T)

4963 **Shukla, B.K. (1981).** Some threatened and endangered plants of Madhya Pradesh. *In* Botanical Survey of India. Abstracts of papers for the Seminar on Threatened Plants of India, 14-17 Sept. 1981. Howrah, BSI. 10. (T)

INDIA - Maharashtra State

4964 **Ansari, M.Y. (1982).** *Ceropegia panchangiensis* Blatt. et McCann (Asclepiadaceae) -a little known species, rediscovered. *Bull. Bot. Surv. India*, 22(1-4): 199-201 (1980 publ. 1982). Illus. (T)

4965 **Dayanandan, P. (1983).** Conserving the flora of the peninsular hills. *Bull. Bot. Surv. India*, 23(3-4): 250-253 (1981 publ. 1983). Illus. (T)

4966 **Gadgil, M., Vartak, V.D. (1978).** Sacred groves of Maharashtra: an inventory. *J. Indian Bot. Soc.*, 57 suppl.: 65. Abstract. (P)

4967 **Malhotra, S.K., Rao, K.M. (1981).** The vegetation of Nagzira Wildlife Sanctuary and its environs (Maharashtra State). *J. Bombay Nat. Hist. Soc.*, 78(3): 475-486. (P)

4968 **Malhotra, S.K., Rao, K.M. (1982).** The vegetation of Nawegaon National Park and its environs (Maharashtra). *Bull. Bot. Surv. India*, 22(1-4): 1-11 (1980 publ. 1982). (P)

4969 **Sharma, B.D., Kulkarni, B.G. (1982).** Some rare and noteworthy plants from Maharashtra. *Bull. Bot. Surv. India*, 22(1-4): 189-191 (1980 publ. 1982). Illus. (T)

4970 **Vartak, V.D. (1981).** Observations on rare, imperfectly known and endemic plants in the sacred groves of western Maharashtra: I. *In* Botanical Survey of India. Abstracts of papers for the Seminar on Threatened Plants of India, 14-17 Sept. 1981. Howrah, BSI. 20. (P/T)

INDIA - Manipur State

4971 **Pradhan, U.C. (1985).** Himalayan plant Red Data Sheets: 6. *Anoectochilus tetrapterus. Himalayan Pl. J.*, 3(5): 19-21. Illus. (T)

INDIA - Meghalaya State

4972 **Chauhan, A.S. (1981).** Observations on some rare and endangered taxa of Meghalaya. *In* Botanical Survey of India. Abstracts of papers for the Seminar on Threatened Plants of India, 14-17 Sept. 1981. Howrah, BSI. 17. (T)

4973 **Joseph, J., Abbareddy, N.R., Haridasan, K. (1982).** *Gastrodia exilis* Hook., a rare and interesting orchid from Khasi and Jaintia Hills, Meghalaya. *Bull. Bot. Surv. India*, 22: 203-205 (1980 publ. 1982). Illus., map. (T)

INDIA - Meghalaya State

4974 **Kataki, S.K. (1981).** Some rare plants in Khasi and Jaintia hills, Meghalaya. *In* Botanical Survey of India. Abstracts of papers for the Seminar on Threatened Plants of India, 14-17 Sept. 1981. Howrah, BSI. 18. (T)

4975 **Kumar, Y., Rao, R.R. (1981).** Studies on the flora of Balphakram Wildlife Sanctuary in Meghalaya: 2. distributional remarks on certain rare & interesting plant species. *In* Botanical Survey of India. Abstracts of papers for the Seminar on Threatened Plants of India, 14-17 Sept. 1981. Howrah, BSI. 13-14. (P/T)

4976 **Lyngdoh, U.C. (1982).** Forest conservation in Khasi and Jaintia Hills. *Indian Forest.*, 108(5): 380. Letter to editor. (T)

4977 **Rao, R.R., Haridasan, K. (1981).** Threatened plants of Meghalaya - a plea for conservation. *In* Botanical Survey of India. Abstracts of papers for the Seminar on Threatened Plants of India, 14-17 Sept. 1981. Howrah, BSI. 12-13. (T)

4978 **Rao, R.R., Haridasan, K. (1982).** Notes on the distribution of certain rare, endangered or endemic plants of Meghalaya with a brief remark on the flora. *J. Bombay Nat. Hist. Soc.*, 79(1): 93-99. (T)

INDIA - Nagaland State

4979 **Ghosh, S.R. (1982).** The rare and threatened fern *Adiantum soboliferum* Wall. ex Hook: a new find for eastern India. *J. Bombay Nat. Hist. Soc.*, 79(3): 716-717. Key. (T)

INDIA - Nicobar Islands

4980 **Balakrishnan, N.P. (1977).** Recent botanical studies in Andaman and Nicobar Islands. *Bull. Bot. Surv. India*, 19: 132-138. Lists 136 'rare' and 'endangered' endemic species. (T)

4981 **Balakrishnan, N.P. (1982).** Notes on some little known ferns of Andaman and Nicobar islands. *Bull. Bot. Surv. India*, 22(1-4): 136-140 (1980 publ. 1982). (T)

4982 **Balakrishnan, N.P., Rao, M.K.V. (1983).** The dwindling plant species of Andaman and Nicobar Islands. *In* Jain, S.K., Rao, R.R., eds. An assessment of threatened plants of India. Howrah, Botanical survey of India. 186-210. Lists 110 threatened endemic taxa and 136 threatened non-endemics; notes on distribution. (T)

4983 **Basu, S.K. (1986).** Observations on two threatened arecoid palms of Nicobar Islands cultivated at the Indian Botanic Gardens, Howrah. *Bull. Bot. Surv. India*, 26(3-4): 207-210 (1984 publ. 1986). Illus. (T)

4984 **Botanical Survey of India (?)** Endangered flora of Andaman and Nicobar Islands. Botanical Survey of India, Andaman and Nicobar Circle, Port Blair. 5p. Mimeo. Overview of vegetation and threats to species. (T)

4985 **Hore, D.K., Balakrishnan, N.P. (1984).** Orchids of Great Nicobar Island and their conservation. *J. Bombay Nat. Hist. Soc.*, 81(3): 626-635. (T)

INDIA - Orissa State

4986 **Aitken, J.E., Ganguli, M. (1981).** Another forest disappears in India. *Envir. Conserv.*, 8(3): 228. Destruction of Santhal Parganas forest.

4987 **Choudhury, B.P. (?)** Assessment and conservation of medicinal plants of Bhubaneswar and its neighbourhood. *In* Indigenous Medicinal Plants New Delhi, Today & Tomorrow's Printers. 211-219. (E/T)

4988 **Nayar, M.P., Ahmed, M., Raju, D.C.S. (1984).** Endemic and rare plants of Eastern Ghats. *Indian J. Forest.*, 7(1): 35-42. Map. Lists 76 species. (T)

4989 **Raju, D.C.S., Nayar, M.P., Ahmed, M. (1983).** Endemic and endangered flora of the Eastern Ghats. *Indian Sci. Congr. Assoc. Proc.*, 70(3): 53-54. (T)

INDIA - Orissa State

4990 **Saxena, H.O., Brahmam, M. (1981).** Rare and endemic flowering plants of Orissa. *In* Botanical Survey of India. Abstracts of papers for the Seminar on Threatened Plants of India, 14-17 Sept. 1981. Howrah, BSI. 11-12. (T)

INDIA - Rajasthan State

4991 **Ali, S., Vijayan, V.S. (1986).** Keoladeo National Park ecology study. Summary report 1980-1985. Bombay Natural History Society. 186p. Illus., maps. (P)

4992 **Jackson, P. (1987).** World heritage site threatened. *WWF News*, 45: 8. (P)

4993 **Panday, R.P., Shetty, B.V., Malhotra, S.K. (1981).** A preliminary census of rare and threatened plants of Rajasthan. *In* Botanical Survey of India. Abstracts of papers for the Seminar on Threatened Plants of India, 14-17 Sept. 1981. Howrah, BSI. 8-9. (T)

4994 **Sharma, I.K. (1978).** Role of local men in extermination and conservation of local flora and fauna. *In* Proceedings of the 8th World Forestry Congress, Jakarta, 16-28 October 1978: Forestry for quality of life. II, 7.

4995 **Singh, V. (1986).** Threatened taxa and scope for conservation in Rajasthan. *J. Econ. Taxon. Bot.*, 7(3): 573-577 (1985 publ. 1986). (T)

4996 **Sinha, N.K. (1981).** List of national parks and wildlife sanctuaries in Rajasthan. *Geobios* (Jodhpur), 8(2): 78-79. Map. (P)

INDIA - Sikkim State

4997 **Ali, S.M. (1981).** Ecological reconnaissance in eastern Himalaya. *Tigerpaper*, 8(2): 1-3. Illus., map.

4998 **Kataki, S.K., Jain, S.K., Sastry, A.R.K. (1984).** Threatened and endemic orchids of Sikkim and north-eastern India. Howrah, Botanical Survey of India. 95p. Descriptions, distributions, illus. of over 100 species. (T)

4999 **Krishna, B., Chakraborty, P. (1981).** Some threatened plants of Sikkim. *In* Botanical Survey of India. Abstracts of papers for the Seminar on Threatened Plants of India, 14-17 Sept. 1981. Howrah, BSI. 17-18. (T)

5000 **Pradhan, U.C. (1971).** Orchid conservation attempts in Sikkim and east Asia. *Amer. Orch. Soc. Bull.*, 40: 307-308. (T)

5001 **Pradhan, U.C. (1975).** Conservation of eastern Himalayan orchids. Problems and prospects. Part I. *Orch. Rev.*, 83: 314-317. (T)

5002 **Pradhan, U.C. (1975).** Conservation of eastern Himalayan orchids. Problems and prospects. Part II. *Orch. Rev.*, 83: 345-347. (T)

5003 **Pradhan, U.C. (1975).** Conservation of eastern Himalayan orchids. Problems and prospects. Part III. *Orch. Rev.*, 83: 374. (T)

5004 **Pradhan, U.C. (1983).** Himalayan plant Red Data Sheets: 4. *Zeuxine pulchra* King & Pantling. *Himalayan Pl. J.*, 2(3): 17-19. Illus. (T)

5005 **Rustomji, N.K. (1986).** Sikkim, Bhutan and India's north-eastern borderlands. *In* Lall, J.S., *ed*. The Himalayas: aspects of change. New Delhi, India International Centre; Oxford, Oxford University Press. 236-252. (P/T)

5006 **Songh, N.B., Bhattacharrya, U.C. (1981).** Notes on little known taxa of *Sedum* from Sikkim and Assam Himalaya. *In* Botanical Survey of India. Abstracts of papers for the Seminar on Threatened Plants of India, 14-17 Sept. 1981. Howrah, BSI. 15. (T)

INDIA - Tamil Nadu State

5007 **Banerjee, A.K., Rao, A.V.N. (1984).** Miscellaneous notes: 22. Cultivation of *Vernonia shevaroyensis* Gamble (Asteraceae): an endemic and endangered plant in the southern Experimental Garden, Botanical Survey of India. *J. Bombay Nat. Hist. Soc.*, 80(3): 663-664 (1983). (G/T)

5008 **Chandrabose, M., Nair, N.C., Chandrasekaran, V. (1982).** Two rare and threatened flowering plants of South India: rediscovered. *Indian J. Forest.*, 5(2): 159-160. (T)

5009 **Gadgil, M., Sukumar, R. (1986).** Scientific programme for the Nilgiri Biosphere Reserve: report of a workshop, Bangalore, March 7-8, 1986. Bangalore, Centre for Ecological Sciences. 48p. Map. (P)

5010 **Henry, A.N., Chandrabose, M.S., Swaminathan, M.S., Nair, N.C. (1984).** Agastyamalai and its environs: a potential area for a biosphere reserve. *J. Bombay Nat. Hist. Soc.*, 81(2): 282-290. (P/T)

5011 **Henry, A.N., Swaminathan, M.S. (1980).** Rare or little known plants from Kanyakumari District Tamil-Nadu India. *Indian J. Forest.*, 3(2): 140-142. (T)

5012 **Henry, A.N., Swaminathan, M.S. (1981).** On the rediscovery of four rare species of *Symplocos* Jacq. (Symplocaceae) in the Muthukuzhivayal region of Kanyakumari District, Tamil Nadu. *Indian Forester*, 107(11): 700-703. (T)

5013 **Henry, A.N., Swaminathan, M.S. (1983).** Rare or new *Exacum* L. (Gentianaceae) from southern India. *J. Bombay Nat. Hist. Soc.*, 80(2): 456-459. Illus. (T)

5014 **Johnsingh, A.J.T., Joshua, J. (1989).** The threatened gallery forest of the River Tambiraparani, Mundanthurai Wildlife Sanctuary, south India. *Biol. Conserv.*, 47(4): 273-280. Map. (P)

5015 **Lakshmanan, K.K., Rajeswari, M., Jayalakshmi, R., Diwakar, K.M. (1984).** Mangrove forest of Krusadai Island, SE India, and its management. *Envir. Conserv.*, 11(2): 174-176. Illus., maps. (P)

5016 **Matthew, K.M., Britto, J.B., Rani, N. (1981).** Some rare plants of Carnatic area, Tamilnadu. *In* Botanical Survey of India. Abstracts of papers for the Seminar on Threatened Plants of India, 14-17 Sept. 1981. Howrah, BSI. 20-21. (T)

5017 **Nair, K.K.N. (1986).** Extinct and endangered endemic angiosperms of Courtallum (Kuttalam), Tamilnadu State. *J. Econ. Taxon. Bot.*, 7(2): 351-358 (1985 publ. 1986). (T)

5018 **Nayar, M.P., Ahmed, M., Raju, D.C.S. (1984).** Endemic and rare plants of Eastern Ghats. *Indian J. Forest.*, 7(1): 35-42. Map. Lists 76 species. (T)

5019 **Raju, D.C.S., Nayar, M.P., Ahmed, M. (1983).** Endemic and endangered flora of the Eastern Ghats. *Indian Sci. Congr. Assoc. Proc.*, 70(3): 53-54. (T)

5020 **Rathakrishnan, N.C. (1978).** Rare and little known orchids from the erstwhile Presidency of Madras. *In* The Madras Herbarium (MH) 1853-1978, 125th Anniversary Souvenir, part 2: Symposium on Floristic Studies in Peninsular India. 34. Abstract only. (T)

5021 **Rathakrishnan, N.C. (1983).** Rare and little-known orchids from the erstwhile Presidency of Madras. *Bull. Bot. Surv. India*, 23(3-4): 237-239 (1981 publ. 1983). (T)

5022 **Sharma, B.D., Shetty, B.V., Vivekananthan, K., Rathakrishnan, N.C. (1978).** Flora of Mudumalai Wildlife Sanctuary, Tamil Nadu. *J. Bombay Nat. Hist. Soc.*, 75(1): 13-42. Map. (P)

5023 **Shetty, B.V., Vivekananthan, K. (1983).** Endemic primitive temperate elements and the relict vegetation of Kundah Range, Nilgiris, Tamil Nadu. *Bull. Bot. Surv. India*, 23(3-4): 254-264 (1981 publ. 1983).

INDIA - Uttar Pradesh State

5024 **Aswal, B.S., Mehrotra, B.N. (1983).** *Delphinium uncinatum* Hook.f. et Thoms. (Ranunculaceae) and *Lilium wallichianum* Schultes f. (Liliaceae): two rare finds from north-west Himalaya. *J. Econ. Taxon. Bot.*, 3(3): 773-775 (1982 publ. 1983). (T)

5025 **Bhatt, A.B. (1981).** Endangered flora of Garwhal Himalayas and their conservation. *In* Botanical Survey of India. Abstracts of papers for the Seminar on Threatened Plants of India, 14-17 Sept. 1981. Howrah, BSI. 3. Includes medicinal plants. (E/T)

5026 **Biswas, S. (1988).** Rare and threatened taxa in the forest flora of Tehri Garhwal Himalaya and the strategy for their conservation. *Indian J. Forest.*, 11(3): 233-237. (T)

5027 **Gaur, R.D., Semwal, J.K. (1981).** Exploitation and threat to survival of some high altitude plants in Garhwal Himalaya. *In* Botanical Survey of India. Abstracts of papers for the Seminar on Threatened Plants of India, 14-17 Sept. 1981. Howrah, BSI. 6-7. (T)

5028 **Goel, A.K., Bhattacharrya, U.C. (1981).** Rare flowering plants of Garwhal Himalayas and their conservation. *In* Botanical Survey of India. Abstracts of papers for the Seminar on Threatened Plants of India, 14-17 Sept. 1981. Howrah, BSI. 2-3. (T)

5029 **Hegde, S.N., Rao, A.N. (1983).** Further contributions to the orchid flora of Arunachal Pradesh - 1. *J. Econ. Tax. Bot.*, 4(2): 383-392. Annotated list of 30 species, of which 3 are new species, 3 new records for India and 3 new records for N.E. Himalayas. 18 species are noted as rare and/or endangered. (T)

5030 **Joshi, G.C., Pande, N.K., Tiwary, D.N. (1988).** Depleting plant resources: a case study from Kumaun Himalaya. *J. Econ. Taxon. Bot.*, 11(1): 13-16 (1987 publ. 1988).

5031 **Khullar, S.P., Sharma, S.S., Singh, P. (1983).** A little known fern from the Himalaya - *Asplenium nesii* Christ, and the nomenclature of *A. exiguum* Bedd. (Aspleniaceae). *J. Bombay Nat. Hist. Soc.*, 80(1): 262-265. Illus. (T)

5032 **Misra, A. (1978).** Cipako Andolana. Delhi, Gandhi Santi Pratish Nayi. Hindi.

5033 **Pant, P.C. (1986).** Flora of Corbett National Park. Flora of India Series No. 4. Howrah, Botanical Survey of India. 224p. Annotated checklist; introductory chapters on geography, vegetation. (P)

5034 **Pant, P.C., Uniyal, B.P., Prasad, R. (1981).** Additions to the plants of Corbett National Park, U.P. *J. Bombay Nat. Hist. Soc.*, 78(1): 50-53. (P)

5035 **Shah, N.C. (1981).** Endangered medicinal and aromatic taxa of U.P. Himalayas. *In* Botanical Survey of India. Abstracts of papers for the Seminar on Threatened Plants of India, 14-17 Sept. 1981. Howrah, BSI. 1. (E/T)

5036 **Sinha, R.L. (1982).** Industrial potential and planned exploitation of Indian medicinal plants in relation to conservation of hill eco-types in U.P. *In* Paliwal, G.S., *ed.* The vegetational wealth of the Himalayas. Papers of the symposium at the Dept. of Botany, Univ. of Garhwal, 1-6 October 1979. Delhi, Puja Publishers. 241-245. Illus. (E)

INDIA - West Bengal State

5037 **Das, S.N., Roy, S.C. (1984).** A note on the occurrence of a few uncommon plants in W. Bengal. *J. Bombay Nat. Hist. Soc.*, 81(2): 518-520. (T)

5038 **Giri, G.S., Nayar, M.P. (1981).** Some threatened plants of Bengal. *In* Botanical Survey of India. Abstracts of papers for the Seminar on Threatened Plants of India, 14-17 Sept. 1981. Howrah, BSI. 12. (T)

INDIA - West Bengal State

5039 **Kar, R.K., Mandal, J.P. (1982).** Danger of extinction of *Lycopodium clavatum* L. from Darjeeling. *Indian Forester*, 108(6): 460. (E/T)

5040 **Mahapata, A.K. (1978).** A brief survey of some unrecorded less known and threatened plant species of Sundarban of west Bengal India. *Bull. Bot. Soc. Bengal*, 32(1-2): 54-58. (T)

5041 **Mukherjee, A.K. (1984).** Mangrove wealth of Indian Sunderbans: utilisation and conservation. *J. Econ. Taxon. Bot.*, 5(1): 227-230. (E)

5042 **Pradhan, U.C. (1975).** Conservation of eastern Himalayan orchids. Problems and prospects. Part I. *Orch. Rev.*, 83: 314-317. (T)

5043 **Pradhan, U.C. (1975).** Conservation of eastern Himalayan orchids. Problems and prospects. Part II. *Orch. Rev.*, 83: 345-347. (T)

5044 **Pradhan, U.C. (1975).** Conservation of eastern Himalayan orchids. Problems and prospects. Part III. *Orch. Rev.*, 83: 374. (T)

INDONESIA

5045 **Adisoemarto, S., Manan, S., Sutisna, U. (1987).** Activities concerning the efforts in the conservation of endangered wild species of flora and fauna of Indonesia: country report. *In* Santiapillai, C., Ashby, K.R., *eds*. Proceedings of the Symposium on the Conservation and Management of Endangered Plants and Animals, Bogor, Indonesia, 18-20 June 1986. Bogor, SEAMEO-BIOTROP. 225-230. (T)

5046 **Anon. (1980).** Laporan Lokakarya Konservasi Jesnis Flora/Fauna Langka Indonesia, Jakarta, 26-28 November 1979. Bogor, Direktorat Jenderal Kehutanan, Direktorat Perlindungan dan Pengawetan Alam. 468p. In.

5047 **Anon. (1980).** List of endangered wild flora in Indonesia. *In* Laporan Lokakarya Konservasi Jenis Flora/Fauna Langka Indonesia, Jakarta, 26-28 November 1979. Bogor, Direktorat Jenderal Kehutanan, Cirektorat Perlindungan dan Pengawetan Alam. 407. (T)

5048 **Anon. (1983).** Eleven new parks for Indonesia. *Malay. Naturalist*, 37(1): 10. (P)

5049 **Atmosoedarjo, S., Daryadi, L., MacKinnon, J., Hillegers, P. (1984).** National parks and rural communities. *In* McNeely, J.A., Miller, K.R., *eds*. National parks, conservation and development. The role of protected areas in sustaining society. Washington, D.C., Smithsonian Institution Press. 237-244. (P)

5050 **Berding, N., Koike, H. (1980).** Germplasm conservation of the *Saccharum* complex: a collection from the Indonesian archipelago. *Hawaii. Pl. Rec.*, 59(7): 87-176. Maps. (E)

5051 **Blower, J. (1978).** Nature conservation in Indonesia. *Tigerpaper*, V(2): 1-4. Illus., map. (T)

5052 **Blower, J. (1979).** Nature conservation in Indonesia. *Parks*, 3(4): 8-10. Map.

5053 **Bompard, J.M., Kostermans, A. (1986).** Local knowledge too late to save wild mangoes? *IUCN Bull.*, 17(1-3): 26. (E/T)

5054 **Cribb, R. (1988).** The politics of environmental protection in Indonesia. Clayton, Victoria, Centre for Southeast Asian Studies Working Paper No. 48: 35p. Maps. (L)

5055 **Daryadi, L. (1981).** The management and conservation of dipterocarps in Indonesia. *Malay. Forest.*, 44(2-3): 190-192. (E)

5056 **Dransfield, J. (1977).** The Kebun Raya, Bogor, and the conservation of Indonesian palms. *In* Stone, B.C., *ed*. The role and goals of tropical botanic gardens. Proceedings, symposium held at Rimba Ilmu Botanic Garden, Univ. of Malaya, Kuala Lumpur, August 1974. 181-185. (G/T)

INDONESIA

5057 **Duryat, H.M., Lavieren, L.P. van (1984).** Indonesia's experience in training protected areas personnel. *In* McNeely, J.A., Miller, K.R., *eds.* National parks, conservation and development. The role of protected areas in sustaining society. Washington, D.C., Smithsonian Institution Press. 228-232. (P)

5058 **FAO (1980).** National conservation plan for Indonesia. Bogor, UNDP/FAO National Parks Development Project Ins/78/061. 8 vols. Field report. 1-Introduction; 2-Sumatra; 3-Java and Bali; 4-Lesser Sundas; 5-Kalimantan; 6-Sulawesi; 7-Maluku and Irian; 8-General topics. 1980, 1981. (P)

5059 **Government of Indonesia (1985).** A review of policies affecting the sustainable development of forest lands in Indonesia. Executive summary. Jakarta, Department of Forestry, State Ministry of Population, Environment and Development, Department of the Interior and IIED.

5060 **Government of Indonesia (1985).** Forest policies in Indonesia: the sustainable development of forest lands. Vol. 1: The discussion document. Jakarta, Department of Forestry, State Ministry of Population, Environment and Development, Department of the Interior and IIED. 24p., annexes.

5061 **Government of Indonesia (1985).** Forest policies in Indonesia: the sustainable development of forest lands. Vol. 2: A review of issues. Jakarta, Department of Forestry, State Ministry of Population, Environment and Development, Department of the Interior and IIED. 150p., annexes.

5062 **Government of Indonesia (1985).** Forest policies in Indonesia: the sustainable development of forest lands. Vol. 3: Background papers. Jakarta, Department of Forestry, State Ministry of Population, Environment and Development, Department of the Interior and IIED. 142p., annexes.

5063 **Government of Indonesia (1985).** Forest policies in Indonesia: the sustainable development of forest lands. Vol. 4: Recommended strategies. Jakarta, Department of Forestry, State Ministry of Population, Environment and Development, Department of the Interior and IIED. 11p.

5064 **Hanson, J. (1982).** Peranan biji dalam pelestarian tumbuhan. (The role of seeds in the conservation of plant genetic resources.) *Bul. Kebun Raya*, 5(1): 1-4. In (En). (E/G)

5065 **Jacobs, M. (1978).** Botanie en natuurbehoud in Indonesie. *Panda*, 14: 137-142. Du.

5066 **Jacobs, M. (1978).** Forests for people.....once? *Tigerpaper*, V(4): ? Illus. (T)

5067 **Jacobs, M. (1983).** The spirit of Bali. *Flora Males. Bull.*, 36: 3920-3925.

5068 **Jacobs, M., Boo, T.J.J. de (1982).** Conservation literature on Indonesia. Leiden, Rijksherbarium. 274p. Selected annotated bibliography. (B/P/T)

5069 **Kartawinata, K., Adisoemarto, S., Riswan, S., Vayda, A.P. (1981).** The impact of man on a tropical forest in Indonesia. *Ambio*, 10(2-3): 115-119. Illus., map. Deforestation.

5070 **Katoh, K. (1985).** The vanishing tropical rain forests of Indonesia. *Tukar-Menukar*, 4: 39-45. Illus.

5071 **Nooteboom, H.P., ed. (1984).** Will all primary forests in Indonesia be destroyed? *Flora Males. Bull.*, 9(1): 42. Tropical forest.

5072 **Reid, R. (1988).** Mango report. *Threatened Pl. Newsl.*, 19: 12. Illus. Report on the joint IBPGR/IUCN/WWF project on wild *Mangifera*. (E/T)

5073 **Rifai, M.A. (1983).** Plasma nutfah, erosi genetika dan usaha pelestarian tumbuhan obat Indonesia. (Germplasm, genetic erosion and the conservation of Indonesian medicinal plants.) *BioIndonesia*, 9: 15-28. In (En). (E)

INDONESIA

5074 **Sandy, I.M., Tantra, I.G.M., Kartawinata, K. (1984).** National parks and land use policy. *In* McNeely, J.A., Miller, K.R., *eds*. National parks, conservation and development. The role of protected areas in sustaining society. Washington, D.C., Smithsonian Institution Press. 233-236. (P)

5075 **Sastrapradja, D.S. & S., Adisoemarto, S. (1983).** Biosphere reserves in Indonesia. *BioIndonesia*, 9: 57-70. (In). Illus. (P)

5076 **Sastrapradja, D.S., Prana, M.S. (1980).** Pandangan tentang konservasi dan peranan kebun raya dalam bidang konservasi. (The viewpoint on conservation and the role of botanic gardens in conservation.) *Bul. Kebun Raya*, 4(6): 175-182. In (En). (G/T)

5077 **Sastrapradja, S. & D.S., Soetisna, U. (1987).** Botanic gardens in the process of national development: the case of Indonesia. *In* Bramwell, D., Hamann, O., Heywood, V., Synge, H., *eds*. Botanic gardens and the World Conservation Strategy. London, Academic Press. 217-226. (Sp).

5078 **Sastrapradja, S. (1983).** Activities on plant genetic resources in Indonesia. *In* Jain, S.K., Mehra, K.L., *eds*. Conservation of tropical plant resources. Howrah, Botanical Survey of India. 205-210.

5079 **Sastrapradja, S., Adisoemarto, S., Kartawinata, K., Tarumingkeng, R.C. (1980).** The conservation of forest animal and plant genetic resources. *BioIndonesia*, 7: 1-42. (In). (E)

5080 **Schelpe, E.A. (1986).** *Coelogyne multiflora*, a rare species from Indonesia. *AOS Bull.*, 55(6): 576-577. (T)

5081 **Secrett, C. (1986).** The environmental impact of transmigration. *The Ecologist*, 16(2/3): 77-88. Illus.

5082 **Setyodiwiryo, K. (1959).** Nature protection in Indonesia. *Proc. 9th Pac. Sci. Congr.*, 7: 18-20.

5083 **Sharp, T. (1984).** ASEAN nations tackle common environmental problems. *Ambio*, 13(1): 45-46. Illus.

5084 **Sormin, B. (1985).** The School of Environmental Conservation Management, Bogor, Indonesia. *In* Thorsell, J.W., *ed*. Conserving Asia's natural heritage. The planning and management of protected areas in the Indomalayan Realm. Proc. of the 25th working session of IUCN's CNPPA. Gland, IUCN. 130-131.

5085 **Sudarsan, A., Sastrapradja, S., Rifai, M.A. (1987).** The National Committee for Germplasm Conservation and the safeguarding of Indonesian endangered plants and animals. *In* Santiapillai, C., Ashby, K.R., *eds*. Proceedings of the Symposium on the Conservation and Management of Endangered Plants and Animals. Bogor, Indonesia, 18-20 June 1986. Bogor, SEAMEO-BIOTROP. 15-20. (E/T)

5086 **Sukendar (1983).** Berlomba dengan waktu. (A race against time.) *Bul. Kebun Raya*, 6(1): 15-17. In (En). (T)

5087 **Sumardja, A. (1981).** First five national parks in Indonesia. *Parks*, 6(2): 1-4. Illus., map. (P)

5088 **Sumardja, E. (1985).** The development of a protected area system for Indonesia in terms of representative coverage of ecotypes. *In* Thorsell, J.W., *ed*. Conserving Asia's natural heritage. The planning and management of protected areas in the Indomalayan Realm. Proc. of the 25th working session of IUCN's CNPPA. Gland, IUCN. 69-74. (P)

5089 **Sumardja, E.A. (1984).** *In situ* conservation of tropical forest protection and nature conservation. *In* Unesco-UNEP. Conservation, science and society. Paris, Unesco. 271-275. Contributions to the First International Biosphere Reserve Congress, Minsk, Byelorussia/USSR, 26 September - 2 October 1983. (P)

INDONESIA

5090 **Sumardja, E.A., Harsono, MacKinnon, J. (1984).** Indonesia's network of protected areas. *In* McNeely, J.A., Miller, K.R., *eds*. National parks, conservation and development. The role of protected areas in sustaining society. Washington, D.C., Smithsonian Institution Press. 214-223. Tables. (P)

5091 **Syam, A., Tarumingkeng, R.C., Wiriosoepartho, A.S., Natawiria, D. (1978).** Nature conservation and wildlife protection in Indonesia: a review. *In* Proceedings of the 8th World Forestry Congress, Jakarta, 16-28 October 1978: Forestry for quality of life. II, 4.

5092 **Tantra, I.G.M. (1978).** The tree flora and its possible utilization of some nature reserves in Indonesia. *In* Proceedings of the 8th World Forestry Congress, Jakarta, 16-28 October 1978: Forestry for quality of life. II, 5. (E/P)

5093 **Tassi, F. (1983).** Dall'isola di Bali un messaggio di speranza a favore della natura. *Nat. Montagna*, 30(3): 32. It. (L)

5094 **WWF-Indonesia (1978).** Endangered species of trees. *Conservation Indonesia*, 2(4): 4. Newsletter of WWF Indonesia Programme; lists 9 Indonesian trees. (T)

5095 **Wachtel, P.S. (1983).** Rainforest programme bears fruit. *IUCN Bull.*, 14(10-12): 127-128. Illus.

5096 **Waluyo, E.B. (1987).** The agricultural system of the Dawan people in relation to conservation of sandalwood. *In* Santiapillai, C., Ashby, K.R., *eds*. Proceedings of the Symposium on the Conservation and Management of Endangered Plants and Animals. Bogor, Indonesia, 18-20 June 1986. Bogor, SEAMEO-BIOTROP. 199-203. Map. (E/T)

5097 **Weevers-Carter, W. (1978).** Nature conservation in Indonesia. Jakarta, Pt Intermasa. 86p. Col. illus., maps.

5098 **Whitten, A.J. (1987).** Indonesia's transmigration program and its role in the loss of tropical rain forests. *J. Conservation Biology*, 1(3): 239-246. Includes tables of statistics on forest cover and conversion.

5099 **Whitten, A.J., Haeruman, H., Alikodra, H.S., Thohari, M. (1987).** Transmigration and the environment in Indonesia: the past, present and future. Gland, IUCN. 44p. Illus.

INDONESIA - Irian Jaya

5100 **Craven, I., Fretes, Y. de (1987).** The Arfak Mountains Nature Conservation Area, Irian Jaya. Management plan, 1988-1992. Bogor, WWF Project No. 3770. 175p. Maps. (P)

5101 **Diamond, J.M. (1984).** Biological principles relevant to protected area design in the New Guinea region. *In* McNeely, J.A., Miller, K.R., *eds*. National parks, conservation and development. The role of protected areas in sustaining society. Washington, D.C., Smithsonian Institution Press. 330-332. (P)

5102 **Diamond, J.M. (1986).** The design of a nature reserve system for Indonesian New Guinea. *In* Soule, M.E., *ed*. Conservation biology: the science of scarcity and diversity. Sunderland, Massachusetts, Sinauer Associates. 485-503. (P)

5103 **FAO (1980).** National conservation plan for Indonesia. Bogor, UNDP/FAO National Parks Development Project Ins/78/061. 8 vols. Field report. 1-Introduction; 2-Sumatra; 3-Java and Bali; 4-Lesser Sundas; 5-Kalimantan; 6-Sulawesi; 7-Maluku and Irian; 8-General topics. 1980, 1981. (P)

5104 **Johns, R.J. (1986).** The instability of the tropical ecosystem in New Guinea. *Blumea*, 31(2): 341-371. Illus. Tropical forest.

5105 **Petocz, R. (1984).** Irian Jaya : nature reserve design in a pristine environment. *Parks*, 9(3/4): 8-12. (P)

INDONESIA - Irian Jaya

5106 **Petocz, R. (1986).** Irian Jaya, the other side of New Guinea: biological resources and rationale for a comprehensive protected area design. *In* South Pacific Commission. Report of the Third South Pacific National Parks and Reserves Conference, Apia, 1985. Vol. II: Collected key issues and case study papers. Noumea, SPC. 237-252. (P)

5107 **Petocz, R.G. (1984).** Conservation and development in Irian Jaya: a strategy for rational resource utilization. Bogor, WWF/IUCN Conservation for Development Programme in Indonesia. 279p. Illus., map. Deforestation (P)

5108 **Schodde, R. (1973).** General problems of fauna conservation in relation to the conservation of vegetation in New Guinea. *In* Costin, A.B., Groves, K.H., eds. Nature conservation in the Pacific. Proc. Symposium A-10, XII Pacific Science Congress, Canberra, August-September 1971. IUCN and Australian Nat. Univ. Press. 123-144. Map.

INDONESIA - Java

5109 **FAO (1980).** National conservation plan for Indonesia. Bogor, UNDP/FAO National Parks Development Project Ins/78/061. 8 vols. Field report. 1-Introduction; 2-Sumatra; 3-Java and Bali; 4-Lesser Sundas; 5-Kalimantan; 6-Sulawesi; 7-Maluku and Irian; 8-General topics. 1980, 1981. (P)

5110 **Fernandes, E., Nair, P. (1986).** An evaluation of the structure and function of tropical homegardens. *Agric. Systems*, 21: 1-14. Sustainable system of agriculture. (E)

5111 **Macdonald, I.A.W., Frame, G.W. (1988).** The invasion of introduced species into nature reserves in tropical savannas and dry woodlands. *Biol. Conserv.*, 44(1-2): 67-93. Invasive species. Includes 5 case studies. (P)

5112 **Matra, I.B. (1984).** Keynote address: the Balinese view of nature. *In* McNeely, J.A., Miller, K.R., eds. National parks, conservation and development. The role of protected areas in sustaining society. Washington, D.C., Smithsonian Institution Press. 212-213.

5113 **Michon, G. (1983).** Village forest gardens in West Java. *In* Huxley, P., ed. Plant research and agroforestry. Nairobi, ICRAF. (E)

5114 **Sangat-Roemantyo, H. (1987).** Some ethnobotanical aspects and conservation strategy of several medicinal plants. *In* Santiapillai, C., Ashby, K.R., eds. Proceedings of the Symposium on the Conservation and Management of Endangered Plants and Animals. Bogor, Indonesia, 18-20 June 1986. Bogor, SEAMEO-BIOTROP. 93-100. BIOTROP Spec. Publ. No. 30. (E/T)

5115 **Soejono, Ir. (1989).** The Purwodadi Botanic Garden, East Java, Indonesia. *Bot. Gard. Conserv. News*, 1(5): 33-36. Illus., map. Includes list of 15 threatened endemics of East Java. (G/T)

5116 **Soemaatmadja, S.A.S. (1982).** Kelestarian hutan jati di wilayah kerja Perum Perhutani. (Teak forest conservation in Perum Perhutani's working territories.) *Jutta Rimba*, 8(53): 9-17. In (En). Illus. (E)

5117 **Sukardjo, S. (1978).** Mengenal hutan Gunung Tilu. (Gunung Tilu forest, West Java.) *Bul. Kebun Raya*, 3(5): 153-156. In (En). Recommends area should become a nature reserve. (P)

5118 **Sukardjo, S. (1987).** Conservation of the marine life of mangrove forests, estuaries and wetland vegetation in the Cimanuk Nature Reserve. *In* Santiapillai, C., Ashby, K.R., eds. Proceedings of the Symposium on the Conservation and Management of Endangered Plants and Animals. Bogor, Indonesia, 18-20 June 1986. Bogor, SEAMEO-BIOTROP. 35-52. Map. (P)

5119 **Sutisna, U. (1982).** The genebank at the National Biological Institute Bogor. *IBPGR Region. Committ. Southeast Asia Newsl.*, 6(3): 11-12. (E/G)

INDONESIA - Java

5120 **UNDP/FAO (1978).** Proposed Gunung Gede-Pangrango National Park. Management plan. 1979/80-1983/84. Nature Conservation and Wildlife Management Project INS/73/013. Bogor, UNDP/FAO. 96p. Maps, appendices. (P)

INDONESIA - Kalimantan

5121 **Achmad, A., Ismail, Wirawan, N. (1986).** Checklist of plants of Kutai National Park. Bogor, WWF/IUCN. 13p. (P)

5122 **Bailes, C.P. (1984).** Orchids of Borneo and their conservation. *Malay. Orchid Rev.*, 18: 20-27. Illus., col. illus. (T)

5123 **Bompard, J.M. (1988).** Wild *Mangifera* species in Kalimantan (Indonesia) and in Malaysia. Final report on the 1986-1988 collecting missions. WWF Project No. 3305. Rome, IBPGR, IUCN-WWF. 50p. Maps. (E/G/T)

5124 **Bompard, J.M., Kostermans, A.J.G.H. (1985).** Preliminary results of an IUCN/WWF sponsored project for conservation of wild *Mangifera* species in situ in Kalimantan (Indonesia). Montpellier, Laboratoire de Botanique Tropicale and Bogor, BIOTROP. Col. illus., maps. Mimeo. (E/P/T)

5125 **Davidson, J. (1987).** Conservation planning in Indonesia's transmigration programme. Case studies from Kalimantan. Gland, IUCN/UNEP. 128p. Illus., maps.

5126 **FAO (1980).** National conservation plan for Indonesia. Bogor, UNDP/FAO National Parks Development Project Ins/78/061. 8 vols. Field report. 1-Introduction; 2-Sumatra; 3-Java and Bali; 4-Lesser Sundas; 5-Kalimantan; 6-Sulawesi; 7-Maluku and Irian; 8-General topics. 1980, 1981. (P)

5127 **Giesen, W. (1987).** Danau Sentarum Wildlife Reserve, inventory, ecology and management guidelines. Zeist, The Netherlands, World Wildlife Fund. 284p. Includes checklist of plants. (P)

5128 **Mackie, C. (1984).** The lessons behind East Kalimantan's forest fires. *Borneo Res. Bull.*, 16(2): 63-74.

5129 **Nooteboom, H.P., ed. (1987).** Report of the 1982-1983 Bukit Raya expedition. Leiden, Rijksherbarium. 93p. Includes checklist of plants. (P)

5130 **Suselo, T.B., Riswan, S. (1987).** Compositional and structural pattern of lowland mixed dipterocarp forest in the Kutai National Park, East Kalimantan. *In* Kostermans, A.J.G.H., *ed*. Proceedings of the Third Round Table Conference on Dipterocarps. Jakarta, Unesco, UNEP. 459-470. Map. Paper presented at conference held at Mulawarman Univ., Samarinda, Kalimantan, 16-20 April 1985. (P)

5131 **WWF Indonesia Programme (1979).** Proposed management plan Kutai Nature Reserve, East Kalimantan, Indonesia. Bogor, WWF Indonesia Programme. 51p. Illus., map. (P)

5132 **Wirawan, N. (1984).** Can we afford to lose more of the rain forest in Kutai? A survey to the south-west corner of Kutai National Park. Bogor, WWF/IUCN. 20p. Mimeo. Map. (P)

5133 **Wirawan, N. (1984).** Good forests within the burned forest area in East Kalimantan. Bogor, WWF. 12p.

5134 **Wirawan, N. (1987).** Good forest within the burned forest area in East Kalimantan. *In* Kostermans, A.J.G.H., *ed*. Proceedings of the Third Round Table Conference on Dipterocarps. Jakarta, Unesco/UNEP. 413-425. Maps. Paper presented at conference held at Mulawarman Univ., Samarinda, Kalimantan, 16-20 April 1985. (P)

INDONESIA - Kalimantan

5135 **Wirawan, N. (1987).** The significance of the Kutai National Park for conservation and study of dipterocarps. *In* Kostermans, A.J.G.H., *ed.* Proceedings of the Third Round Table Conference on Dipterocarps. Jakarta, Unesco/UNEP. 389-411. Maps. Paper presented at conference held at Mulawarman Univ., Samarinda, Kalimantan, 16-20 April 1985. (P)

INDONESIA - Lesser Sunda Islands

5136 **FAO (1980).** National conservation plan for Indonesia. Bogor, UNDP/FAO National Parks Development Project Ins/78/061. 8 vols. Field report. 1-Introduction; 2-Sumatra; 3-Java and Bali; 4-Lesser Sundas; 5-Kalimantan; 6-Sulawesi; 7-Maluku and Irian; 8-General topics. 1980, 1981. (P)

INDONESIA - Sulawesi

5137 **Satjapradja, O., Mas'ud, F. (1978).** Kegiatan resetlmen penduduk dataran Palolo sebagai salah satu usaha untuk pengamanan hutan di daerah Sulawesi tengah. (Resettlement in the Plain of Palolo as an effort to protect the natural forests in central Sulawesi.) *Lapor. Lembaga Penelitian Hutan,* 283: 20p. In (En). Map.

5138 **Sidiyasa, K., Tantra, G.M. (1984).** Analisis vegetasi Cagar Alam Poboya, Sulawesi Tengah. (The vegetation analysis of Poboya Nature Reserve, Central Sulawesi.) *Lapor. Pusat Penelitian Hutan* , 451: 26p. In (En). Map. (P)

5139 **Simonson, D. (1987).** Morowali rainforest expedition 1985. Final report. Unknown. 23p. Includes section on rattan conservation. (E/P)

5140 **Sumardja, E.A., Tarmudji, Wind, J. (1984).** Nature conservation and rice production in the Dumoga area, North Sulawesi, Indonesia. *In* McNeely, J.A., Miller, K.R., *eds.* National parks, conservation and development. The role of protected areas in sustaining society. Washington, D.C., Smithsonian Institution Press. 224-227. (P)

5141 **WWF (1980).** Morowali Nature Reserve: a plan for conservation. WWF. Illus., maps. (P)

5142 **Whitten, A.J., Mustafa, M., Henderson, G.S. (1987).** The ecology of Sulawesi. Yogyakarta, Gadjah Mada Univ. Press. 777p. Illus., col. illus., maps. Rain forest; deforestation. (E/P)

INDONESIA - Sumatra

5143 **Budiawan (1989).** Gunung Leuser National Park and the problem of encroachment. *Tigerpaper,* 16(2): 27-32. (P/T)

5144 **FAO (1980).** National conservation plan for Indonesia. Bogor, UNDP/FAO National Parks Development Project Ins/78/061. 8 vols. Field report. 1-Introduction; 2-Sumatra; 3-Java and Bali; 4-Lesser Sundas; 5-Kalimantan; 6-Sulawesi; 7-Maluku and Irian; 8-General topics. 1980, 1981. (P)

5145 **Fowlie, J.A. (1985).** Malaya revisited, 28: *Paphiopedilum liemianum* and *Paphiopedilum tonsum* on a limestone ridge in northern Sumatra. *Orchid Digest,* 49(3): 85-90. Illus., maps. (T)

5146 **Fowlie, J.A. (1985).** Malaya revisited, 30: *Paphiopedilum chamberlainianum* on vestigial limestone outcrops of Bukit Tinggi in Sumatra. *Orchid Digest,* 49(5): 164-169. Illus. (T)

5147 **Fowlie, J.A. (1985).** Malaya revisited, 31: *Paphiopedilum victoria-mariae* (Sanders ex Masters) Rolfe refound in Sumatra at high elevation on andesite lava cliffs. *Orchid Digest,* 49(6): 204-207. Illus. (T)

5148 **Jafarsidik, Y.S., Tantra, I.G.M. (1979).** Inventarisasi flora suaka alam Isaul-Isau Pasemah, Palembang. (Flora inventory of Isau-Isau Pasemah Nature Reserve, Palembang.) *Lembaga Penelitian Hutan,* 8: 13p. In (En). Map. (P)

INDONESIA - Sumatra

5149 **Jones, D.T. (1987).** Rare plant profile no. 1: *Merrillia caloxylon* (Rutaceae). *Bot. Gard. Conserv. News*, 1(1): 38-42. Illus., map. (E/G/T)

5150 **McNeely, J.A. (1978).** Siberut: conservation of Indonesia's island paradise. *Tigerpaper*, 5(2): 16-20. Illus. (P)

5151 **Meijer, W. (1981).** Sumatra as seen by a botanist. *Indonesia Circle*, 25: 17-27. Tropical forests. (P)

5152 **Meijer, W. (1985).** Saving the world's largest flower. *Nat. Geographic*, 168(1): 136-140. (T)

5153 **Satjapradja, O. (1979).** Usaha resettlement penduduk di Malancan, kecamatan Siberut utara untuk kelestarian hutan di kepulauan Mentawai Sumatera barat. (Resettlement in the area of Malancan, within the district of N. Siberut for protecting the natural forest in Mantawi Arch., W. Sumatra.) *Lembaga Penelitian Hutan*, 22: 2p. In (En). Map. (P)

5154 **Seibert, S. (1989).** The dilemma of a dwindling resource: rattan in Kerinci, Sumatra. *Principes*, 33(2): 79-87. (E/T)

5155 **UNDP/FAO (1981).** Kerinci-Seblat proposed national park. Preliminary management plan 1982-1987. UNDP/FAO National Park Development Project INS/78/061. Bogor, FAO. 54p., appendices. Field report. (P)

5156 **WWF Indonesia Programme (1980).** Saving Siberut: a conservation master plan. Bogor, WWF. 134p. Illus., map. (P)

5157 **Whitten, A.J., Damanik, S.J., Anwar, J., Hisyam, N. (1984).** The ecology of Sumatra. Yogyakarta, Gadjah Mada Univ. Press. 583p. Illus., col. illus., maps. Covers vegetation types, flora and effects of disturbance. (P)

IRAN (ISLAMIC REPUBLIC OF)

5158 **Edmondson, J.R., Miller, A.G., Parris, B.S. (1980).** Plants of the Khabr va Ruchoun protected area, S. Iran. *Not. Roy. Bot. Gard. Edinburgh*, 38(1): 111-124. Maps. (P)

5159 **Firouz, E., Hassinger, J.D., Ferguson, D.A. (1970).** The wildlife parks and protected regions of Iran. *Biol. Conserv.*, 3(1): 37-45. Illus., map. (P)

5160 **Mossadegh, A. (1978).** Conservation de la vegetation naturelle dans la zone Caspienne en Iran. *In* Proceedings of the 8th World Forestry Congress, Jakarta, 16-28 October 1978: Forestry for quality of life. ii, 5p. Fr.

5161 **Mossadeghi, S. (1978).** Planning for the conservation of Iran's natural resources. *In* Proceedings of the 7th World Forestry Congress, Buenos Aires, 4-18 October 1972, vol. 3: Conservation and recreation. Buenos Aires, Instituto Forestal Nacional. 3684-3688.

5162 **Wendelbo, P. (1978).** Endangered flora and vegetation, with notes on some results of protection. *In* IUCN. Ecological guidelines for the use of natural resources in the Middle East and South West Asia. IUCN Publications New Series No. 34. Morges, IUCN. 189-195. (T)

IRAQ

5163 **Nasser, M.H. (1984).** Forests and forestry in Iraq: prospects and limitations. *Commonw. Forest. Rev.*, 63(4): 299-304.

5164 **Sefik, Y. (1981).** Forests of Iraq. *Istanbul Univ. Orman Fak. Dergisi, A.*, 31(1): 43-47. (Tu).

IRELAND

5165 **An Foras Forbartha (1981).** Areas of scientific interest in Ireland. National heritage inventory. Dublin, An Foras Forbartha. 166p. Illus. (P)

5166 **Anon. (1983).** Symposium on conservation in British and Irish botanic gardens. *Threatened Pl. Newsl.*, 12: 8-9. (G/T)

5167 **Bassett, J.A., Curtis, T.G.F. (1985).** The nature and occurrence of sand-dune machair in Ireland. *Proc. Roy. Irish Acad.*, 85B(1): 1-20.

5168 **Bellamy, D. (1986).** The wild boglands - Bellamy's Ireland. Dublin, Country House.

5169 **Carter, R.W.G., Hamilton, A.C., Lowry, P. (1981).** The ecology and present status of *Otanthus maritimus* on the gravel barrier at Lady's Island, Co. Wexford. *Irish Nat. J.*, 20(8): 329-331. Map. (T)

5170 **Cooper, A. (1984).** Application of multivariable methods to the conservation management of hazel scrub in northeast Ireland. *Biol. Conserv.*, 30(4): 341-357.

5171 **Cross, J.R. (1981).** The establishment of *Rhododendron ponticum* in the Killarney oakwoods, S.W. Ireland. *J. Ecol.*, 69(3): 807-824. Illus. (P)

5172 **Cross, J.R. (?)** The distribution and status of *Pyrola rotundifolia* L. in Ireland. *Ir. Nat. J.*, 22(1): 16-20.

5173 **Curtis, T.G.F., McGough, H.N. (1988).** The Irish Red Data Book 1: Vascular plants. Dublin, Wildlife Service Ireland. 168p. Col. illus., maps. (T)

5174 **Gilder, P., Jackson, M., Warren, A., eds (1980).** Mullach Mor N.P., the Burren, Co. Clare. A discussion of the size, location, management and character of national parks in the Burren. Discussion Papers in Conservation No. 27. London, University College. 46p. Maps. (P)

5175 **Ireland (Eire) Forest & Wildlife Service (1974).** Report of the Minister for Lands. Dublin, Dept of Lands. Includes forests, forestry & wildlife conservation. (E)

5176 **Jeffrey, D.W. (1977).** North Bull Island Dublin Bay - a modern coastal natural history. Dublin, Royal Dublin Society. 158p. (P)

5177 **Jeffrey, D.W., ed. (1984).** Nature conservation in Ireland; progress and problems. Dublin, Royal Irish Academy. 175p. Proceedings of a Seminar, 24-25 February 1983.

5178 **Kelly, D., Kirby, E.N. (1982).** Irish native woodlands over limestone. *J. Life Sci. Roy. Dublin Soc.*, 3: 181-198.

5179 **Kelly, D.L. (1981).** The native forest vegetation of Killarney, south-west Ireland: an ecological account. *J. Ecol.*, 69(2): 437-472. Illus., map. (P)

5180 **Lamb, J.G.D. (1984).** Woodfield Bog - an urgent case for preservation. *Irish Nat. J.*, 21(8): 363-364.

5181 **Mills, S. (1984).** Unique Irish bog may be dug up. *New Sci.*, 101(1401): 7. (P/T)

5182 **O'Connell, C., ed. (1987).** The IPCC guide to peatlands. Dublin, Irish Peatland Conservation Council. 102p.

5183 **O'Connell, C.A. (1986).** The future of Irish raised bogs. Dublin, Environment Awareness Bureau. Resource Environment Guide No. 7.

5184 **O'Connell, M. (1981).** The phytosociology and ecology of Scragh Bog, Co. Westmeath. *New Phytol.*, 87: 139-187. (P)

5185 **Pullin, A. (1986).** The status, habitat, and species association of the fen violet *Viola persicifolia* in western Ireland. *BES Bull.*, 27(1): 15-19. (T)

5186 **Reynolds, J. (1984).** Vanishing Irish boglands. *WWF News*, Spring 1984.

IRELAND

5187 **Richards, A.J. (1972).** The code of conduct: a list of rare plants. *Watsonia*, 9(1): 67-72. (L/T)

5188 **Roper, L. (1979).** Where rarities flourish: plants in Irish gardens. *Country Life*, 166(4279): 92-94. Illus. (G/T)

5189 **Ryan, J.B., Cross, J.R. (1984).** Conservation of peatlands in Ireland. *In* Proceedings of the 7th. International Peat Congress, Dublin.

5190 **Taylor, J., Smith, R. (1980).** Power in the peatlands. *New Sci.*, 88(1230): 644-646. Illus.

5191 **Van Eck, H., Govers, A., Lemaire, A., Schaminee, J. (1984).** Irish bogs. A case for planning. Nijmegen, Catholic University.

5192 **Watts, W.A. (1988).** The role of the Academy in the field of natural sciences and in conservation. *Proc. Roy. Ir. Acad.*, 88B(5): 61-67.

5193 **Webb, D.A. (1983).** The flora of Ireland in its European context. *J. Life Sc. R. Dub. Soc.*, 4: 143-160. Maps. (P/T)

5194 **Webb, D.A., Scannell, M.J.P. (1983).** Flora of the Connemara and the Burren. Dublin, Royal Dublin Society and Cambridge, Cambridge University Press. (P)

5195 **Webb, R. (1984).** The conservation of trees and woods in Ireland. *Arboricult. J.*, 8(3): 235-244. Illus. (L)

5196 **White, J. (1985).** *Limosella aquatica* L. and the vegetation of exposed mud at the Geargh, Co. Cork (H3). *Irish Nat. J.*, 21(12): 509-515. (T)

5197 **White, J. (1985).** The Geargh woodland, Co. Cork. *Irish Nat. J.*, 21(9): 391-396.

5198 **Wyse Jackson, P. (1984).** Irish rare plant conservation in the Trinity College Botanic Gardens, Dublin. *In* Jeffrey, D.W., *ed.* Nature conservation in Ireland; progress and problems. Proceedings of a Seminar, 24-25 February 1983. Dublin, Royal Irish Academy. 95-97. (G/T)

ISRAEL

5199 **Alon, A. (1973).** Saving the wild flowers of Israel. *Biol. Conserv.*, 5: 150-151. (T)

5200 **Avishai, M. (1989).** The Jerusalem and University Botanical Garden and plant conservation. *Bot. Gard. Conserv. News*, 1(5): 17-21. Illus. (G)

5201 **Clark, B. (1985).** Protecting our flowers - by computer. *Israel - land and nature*, 10(4): 156-160. Illus., map. (T)

5202 **Dafni, A., Agami, M. (1976).** Extinct plants of Israel. *Biol. Conserv.*, 10: 49-52. List of 26 species. (T)

5203 **Galil, J. (1981).** The *Gonocytisus pterocladus* - a plant in urgent need of protection. *Israel Land Nature*, 6(4): 145-147. Illus. (T)

5204 **IUCN Threatened Plants Committee Secretariat (1980).** First preliminary draft of the list of rare, threatened and endemic plants for the countries of North Africa and the Middle East. Kew, IUCN Threatened Plants Committee Secretariat. 170p. (T)

5205 **Naveh, Z. (1971).** The conservation of ecological diversity of Mediterranean ecosystems through ecological management. *In* Duffey, E., Watt, A.S., *eds.* The scientific management of animal and plant communities for conservation. The 11th Symposium of the BES, University of East Anglia, Norwich, 7-9 July 1970. Oxford, Blackwell. 605-622.

5206 **Noy-Meir, Anikster, Y., Waldman, M., Ashri, A. (1988).** Population dynamics research for *in situ* conservation: wild wheat in Israel. *Pl. Genet. Resource. Newsl.*, 75/76: 9-11. Illus. *Triticum dicoccoides*. (E)

ISRAEL

5207 **Pollak, G. (1984).** [The problems of threatened plants on Hamra and Kurkar.] *Rotem*, 13: 56-68. He (En). Illus. (T)

5208 **Rabinovitz, D. (1981).** Nature conservation and environmental protection in the Negev Desert: a challenge for Israel in the 1980's. Pamphlet No. 62. London, Anglo-Israel Association. 16p. Based on talk to Anglo-Israel Assoc., 15 Oct. 1980. (P)

5209 **Salomon, H., Shmida, A. (1985).** Computerized geographical mapping of biological species. *In* Glaeser, P.S., *ed.* The role of data in scientific progress. Holland, Elsevier. 77-79.

5210 **Zohary, D. (1983).** Wild genetic resources of crops in Israel. *Israel J. Bot.*, 32(2): 97-127. (E)

5211 **Zohary, M. (1959).** Wild life protection in Israel (flora and vegetation). *In* Animaux et vegetaux rares de la Region Mediterraneenne. Proceedings of the IUCN 7th Technical Meeting, 11-19 September 1958, Athens, vol.5. Brussels, IUCN. 199-202.

5212 **Zohary, M., Wood, H. (1975).** [Bouquet of protected wild flowers.] Tel Aviv, Nature Conservation Authority. 79p. He. Coloured plates of 37 species. (T)

ITALY

5213 **Abrami, A. (1977).** Le nuove competenze amministrative regionali in materia di interventi per la protezione della natura. (New administrative powers of the regions as regards measures for nature conservation.) *Ital. Forest. Mont.*, 32(6): 245-253. It (Fr).

5214 **Allavena, S. (1978).** Circeo National Park: reclaiming a rich heritage. *Parks*, 3(3): 3-5. Illus., map. (P)

5215 **Anchisi, E., Bernini, A., Cartasegna, N., Polani, F. (1985).** *Flora Protetta dell'Italia Settentrionale* Gruppo Naturalistico Oltrepo' Pavese. It., Col. Illus., maps. (L/T)

5216 **Anon. (1972).** Specie della Flora italiana meritevoli di protezione (Gruppo di Lavoro per la Floristica, Societa Botanica Italiana). *Inform. Bot. Ital.*, 4(1): 12-13. It (Fr, En, Ge). Lists 41 species in need of protection. (T)

5217 **Anon. (1977).** La riserva naturale orientata di Campolino, aspetti naturalistici e selvicolturali, piano di gestione. (Natural reserve of Campolino, nature and forestry aspects, plants management.) Rome, Ministerio dell' Agricoltura e delle Foreste. 102p. It. Illus., maps. (P)

5218 **Anon. (1982).** Il Parco dei Monti della Tolfa - un parco da vivere. Provincia di Roma, Assessorato Sport e Turismo. 31p. It. Illus. (P)

5219 **Arillo, A., Balletto, E., Cagnolaro, L., Orsino, F. (1975).** Proposte di riserve naturali in Liguria. Individuazione delle aree di maggior interesse faunistico, floristico e vegetazionale. Publ. Ist. Bot. Hanbury Univ. Genova No. 201. 7-58. It. Extract from vol.1 of Atti del V Simposio Nazionale sulla Conservazione della Natura, Istituto di Zoologia dell' Universita di Bari, 22-27 April 1975. (P)

5220 **Arrigoni, P.V., nardi, E., Raffaelli, M. (1985).** La vegetazione del parco naturale Maremma (Toscana). Firenze, Italy, Universita degli Studi di Firenze, Dipartimento di Biologia Vegetale. 39p. It (En). Maps. (P)

5221 **Bagnaresi, U. (1979).** Foreste tra conservazione e produzione. *Nat. Montagna*, 26(2): 5-10. It. Illus. (E)

5222 **Barberis, G., Mariotti, M. (1983).** Ricerche floristiche sulle spiagge liguri. *Arch. Bot. Biogeogr. Ital.*, 57(3-4): 154-170 (1981 publ. 1983). It (En).

5223 **Bardicci, G. (1979).** The national parks of Italy. *Natur- und Nationalparke*, 16(62): 10-12, 2-4, 23-25. (Ge, Fr). Map. (P)

ITALY

5224 **Barker, M.L. (1982).** Comparison of parks, reserves and landscape protection in three countries of the eastern Alps. *Envir. Conserv.*, 9(4): 275-285. Maps. (P)

5225 **Benvenuti, V. (1974).** La foresta nel quadro della conservazione dell'ambiente. (The forest within the framework of conservation of the environment.) Rome, Ministerio dell'Agricoltura e delle Foreste. 15p. It. Illus.

5226 **Bernetti, G. (1978).** L'assestamento forestale nelle riserve e nei parchi. (Forest management in nature reserves and parks.) *Montanaro d'Italia - Monti e Boschi*, 29(2): 61-66. It. (P)

5227 **Bortolotti, L. (1975).** Sulle leggi per la protezione della flora emanate dalle Regioni a statuto speciale e ordinario dalle Province autonome. *Boll. Soc. Bot. Ital.*, 7(2): 132-139. It. (L/T)

5228 **Bortolotti, L. (1978).** Le leggi regionali di tutela della flora spontanea. *Nat. Montagna*, 25(2): 5-8. It. Illus.

5229 **Bruynseels, G. (1983).** Flore et vegetation du parc national du Grand Paradis. *Naturalistes Belges*, 64(5): 183-197. Fr. Illus., maps. (P)

5230 **Calabri, G. (1978).** L'organizzazione per la difesa die boschi italiani dagli incendi: problemi e prospettive. (Organization of the protection of Italian forests against fires: problem and prospects.) *Ann. Accad. Ital. Sci. Forest.*, 27: 125-144. It.

5231 **Capodarca, V. (1983).** Toscana, cento alberi da salvare. (Tuscany, one hundred trees to preserve.) Firenze, Vallecchi Editore. 271p. It. Illus., col. illus., maps. (T)

5232 **Contarini, E. (1980).** Salviamo i "Gessi" di Brisighella: una proposta di protezione. *Nat. Montagna*, 26(1): 49-55. It. Illus., maps.

5233 **Contoli, L. (1980).** La promozione di un Parco naturale in rapporto alla painificazione nella Comunita montana "Monti della Tolfa". *Inform. Bot. Ital.*, 12(1): 10-14. It. (P)

5234 **Corbetta, F., Zanotti Censoni, A.L., Zarrelli, R. (1983).** Antropizzazione e depauperamento floristico-vegetazionale nella "Bassa" Bolognese. *Arch. Bot. Biogeogr. Ital.*, 57(3-4): 113-132 (1981 publ. 1983). It (En). Maps.

5235 **Corti, R. (1959).** Specie rare o minacciate della flora Mediterranea in Italia. *In* IUCN. Animaux et vegetaux rares de la Region Mediterraneenne. Proceedings of the IUCN 7th Technical Meeting, 11-19 September 1958, Athens, vol. 5. Brussels, IUCN. 112-129. It. (T)

5236 **Council of Europe (1981).** National activities: Italy. *Council of Europe Newsl.*, 81(6/7): 3. New reserves on Latium Coast and R. Sele. (P)

5237 **Council of Europe (1984).** Italy: more protected areas. *Council of Europe Newsl.*, 84(3): 3. (P)

5238 **Cseko, G. (1976).** Olaszorszag elso nemzeti parkja - a Gran Paradiso. (Gran Paradiso - the first national park in Italy.) *Buvar*, 31(1): 30-32. Hu. Illus. (P)

5239 **Filipello, S. (1979).** Projets, problemes et aboutissements de la conservation de la flore et de la vegetation en Italie. Proceedings of the 2nd OPTIMA meeting, 23-29 May 1977. *Webbia*, 34(1): 63-69. Fr. (T)

5240 **Filipello, S., Gardini Peccenini, S., Bergamo, S., eds (1979).** Repertorio delle specie della flora italiana sottoposte a vincolo di protezione nella legislazione nazionale e regionale. Collana del Programma Finalizzato "Promozione della qualita dell' ambiente," AQ/1/10. Pavia, Consiglio Nazionale delle Ricerche. It. (T)

5241 **Filipello, S., Gardini-Peccenini, S. (1985).** The Italian peninsular and alpine regions. *In* Gomez-Campo, C., *ed.* Plant conservation in the Mediterranean Area. Dordrecht, Dr W. Junk. 71-88. Illus. Maps. (T)

299

ITALY

5242 **Filipello, S., ed. (1981).** Problemi scientifici e tecnici della conservazione del patrimonio vegetale. OPTIMA Leaflet No.114. Pavia, Consiglio Nazionale delle Ricerche. 146p. It (En). Proceedings of a conference, 18-19 December 1979, Firenze. (T)

5243 **Fozzer, F., Gradi, A. (1978).** Conservation des resources genetiques de la foret Italienne. *In* Proceedings of the 8th World Forestry Congress, Jakarta, 16-28 October 1978: Forestry for quality of life. Verona. II, 9. Fr (En).

5244 **Framarin, F. (1982).** Guarding paradise. *Parks,* 7(1): 5-7. Illus. (P)

5245 **Francalancia, C., Orsomando, E. (1979).** Piante dei Monti Coscerno e Civitella (Appennino centrale) rare o interessanti per la Regione Umbria. *Arch. Bot. Biogeogr. Ital.,* 55(4): 130-142. It (En). (T)

5246 **Gardini Peccenini, S. (1984).** Flora da proteggere. Indagine su alcune specie vegetali minacciate o rare in Italia. Pavia, Istituto di Botanico e Orto Botanico, Universita di Pavia. 248p. It. Illus., maps. (G/L/P/T)

5247 **Gellini, R., Pantani, F., Grossoni, P., Bussotti, F. Barolani, E., Rinallo, C. (1985).** Further investigation on the causes of disorder of the coastal vegetation in the park of San Rossore (central Italy). *Europ. J. For. Path.,* 15(3): 145-157. (Fr, Ge). Illus. (P)

5248 **Giacomini, V. (1976).** Uomini e foreste. (Man and the forest.) *Ital. For. Mont.,* 31(5): 161-172. It (Fr).

5249 **Giannini, R., Screm, E. (1977).** Soprassuoli e rinnovazione naturale nella Riserva di Campolino (Abetone). (Forest stand and natural regeneration in Campolino Reserve [Abetone].) *Ann. Accad. Ital. Sci. Forest.,* 26: 305-324. It (En). Illus. (P)

5250 **Grey-Wilson, C. (1985).** Plants in peril. *Kew Mag.,* 2(1): 232-234. *Saxifraga florulenta.* (T)

5251 **Kammer, F. (1982).** Entwurf einer Vegetationskarte des Monte-Argentario-Gebietes (Toscana, Italien). Freiburg im Breisgau. Ge.

5252 **Moggi, G. (1972).** Specie della flora italiana meritevoli di protezione. *Imform. Bot. Ital.,* 4(1): 10-13. It. (T)

5253 **Monti, G. (1979).** Macromiceti rari o nuovi del Monte Pisano (Toscana nord-occidentale). (Rare and new macromycetes from Pisano Mountain [northwestern Tuscany].) *Micologia Italiana,* 8(3): 19-22. It. Illus. (T)

5254 **Monti, G. (1984).** Un convegno dei Lincei su "Parchi e aree protette in Italia". *Nat. Montagna,* 31(1): 75-76. It. (P)

5255 **Morandini, R. (1978).** Survey and analysis for forestry and soil protection in Italy. *Landscape Plann.,* 5(2-3): 281-292. Map.

5256 **Nano, G.M., Bicchi, C., Frattini, C., Gallino, M. (1979).** Wild Piedmontese plants, Part 2. A rare chemotype of *Tanacetum vulgare* abundant in Piedmont Italy. *Planta Med.,* 35(3): 270-274.

5257 **Neshatav, V.Yu., Yurkovskaya, T.K. (1980).** Krupnomasshtabnoe Kartirovanie rastitel'nosti okhranyaemykh territorii v Italii. (Large-scale mapping of the vegetation of protected territories in Italy.) *In* Geobotan. Kartograf. Leningrad. 65-75. Rus. (P)

5258 **Pacioni, G. (1978).** Entita micologiche del Parco Nacionale de Circeo. (Notes on mycological flora of Circeo National Park.) *Micol. Italiana,* 7(2): 39-45. Illus., map. *Scleroderma meridionale, Gyrophragmium dunalii.* (P)

5259 **Padula, M. (1978).** La difesa del patrimonio vegetale. *Nat. Montagna,* 25(3): 33-38. It. Illus.

ITALY

5260 **Pantani, F., Barbolani, E., Del Panta, S., Gellini, R. (1986).** Sulla deposizione acida nella foresta di Vallombrosa. (Acid deposition in the forest of Vallombrosa.) *Inform. Bot. Ital.*, 17(1-3): 75-86. It (En). Acid rain.

5261 **Pedrotti, F. (1971).** Censimento dei biotopi di relevante interesse vegetazionale meritevoli di conservazione in Italia, 2 vols. Camerino, Societa Botanic Italiana. It. Site descriptions, threats, proposed protection, maps. Vol. 2 publ. 1979. (P)

5262 **Pedrotti, F., Brilli-Cattarini, A.J.B. (1976).** Vegetazione e ambiente delle Marche e relativi problemi di Salvaguardia. (Vegetation and environment in the Marche and related problems of protection.) *Giorn. Bot. Ital.*, 110(6): 383-399. It (En).

5263 **Peyronel, B. (1973).** Consederazione su una legge regionale per la conservazione della flora: Italia. *Inf. Bot. Ital.*, 5(2): 151-154. It. (L/T)

5264 **Pirone, G. (1980).** Il "Faggeto" di Moliterno. *Nat. Montagna*, 27(1): 37-47. It. Map.

5265 **Pratesi, F. (1978).** Forestry and afforestation: positive and negative effects on natural environment with aim at environmental conservation. *In* Proceedings of the 8th World Forestry Congress, Jakarta, 16-28 October 1978: Forestry for quality of life. II, 4.

5266 **Raimondo, F.M. (1981).** Le specie della flora Italiana accantonate in biotopi in pericolo. (The species of Italian flora situated in threatened biotopes.) *In* Problemi scientifici e tecnici della conservazione del patrimonio vegetale, Firenze, 18-19 December 1979. Pavia, Consiglio Nazionale delle Ricerche. 103-125. It (En). Maps. (T)

5267 **Raimondo, T. (1978).** Il patrimonio vegetale e la sua degradazione. *Nat. Montagna*, 25(3): 19-31. It.

5268 **Region Marche, ed. (1979).** Flora protetta dell Marche. Region Marche. 96p. It. Illus., maps. (L/T)

5269 **Region Veneto, ed. (1975).** Fauna inferiore flora e funghi natura da Salvare. Region Veneto. 71p. It. Illus. Describes 48 protected species.

5270 **Sacchi, C.F. (1979).** The coastal lagoons of Italy. *In* Jefferies, R.L., Davy, A.J., *eds.* Ecological processes in coastal environments. British Ecological Society Symposia 19. Oxford, Blackwell Scientific Publications. 593-601. (Fr). Map.

5271 **Santucci, D. (1978).** L'oasi delle valli di Argenta e di Marmorta. *Nat. Montagna*, 25(4): 13-25. It. Illus., col. illus.

5272 **Sauro, L.R. & G. (1981).** Monte Baldo e Cansiglio-Cavallo. Riflessioni su proposte di Parchi Naturali Regionali. *Nat. Montagna*, 28(4): 89-94. It. Map. (P)

5273 **Societa Botanica Italiana (1975).** Aufruf zum Schutze der Italienischen Flora. (Appeal for protection of the Italian flora.) *Willdenowia*, 7(3): 537-538. Ge. Lists 43 protected species. (T)

5274 **Sonnino, P.F. (1975).** Protezione delle flora alpina e legislazione. *Natura e Montagna* (Italy), 22(2): 41-47. It. (L/T)

5275 **Spada, F. (1978).** Nuove segnalazioni per la flore del Parco Nazionale d'Abruzzo. (New findings of flora in the Abruzzo National Park.) *Arch. Bot. Biogeog. Ital.*, 54(3/4): 154-162. It. Illus., map. (P)

5276 **Tomei, P.E., Amadei, L., Garbari, F. (1986).** Donnees distributives de quelques angiospermes rares de la region Mediterraneenne d'Italie. *Atti Soc. Tosc. Sci. Nat., Mem., Ser. B*, 92: 207-240. Fr. Illus., maps. (T)

5277 **Tomei, P.E., Garbari, F. (1978).** Il padule di Bientina, le Cerbaie e il lago di Sibolla. *Nat. Montagna*, 25(4): 27-33. It. Illus. maps.

ITALY

5278 **Toschi, A. (1959).** Etablissement des reserves pour la protection de la faune et de la flore en Italie. *In* IUCN. Animaux et vegetaux rares de la Region Mediterraneenne. Proceedings of the IUCN 7th Technical Meeting, 11-19 September 1958, Athens, vol. 5. Brussels, IUCN. 58-63. Fr. (P)

5279 **Tucci, G.F., Pizzolongo, P. (1979).** Una pianta rara *Primula palinuri* Petagna. *Nat. Montagna*, 26(1): 15-20. It. Illus. (T)

5280 **Varotto, S. (1977).** Carta dell'uso attuale del suolo del Parco Naturale della Maremma. (Present land use of the Maremma Nature Park). *Rivista Agric. Subtrop. Trop.*, 71(7-9/10-12): 237-245. It. (P)

5281 **Zatta, T. (1979).** Il parco del Medio Mincio. *Nat. Montagna*, 26(1): 22-26. It. Illus., map. (P)

ITALY - Sardinia

5282 **Arrigoni, P.V. (1971).** Nuovi reperti di alcune species rare o notevoli della flora sarda. (New records for some rare or interesting species in Sardinia.) *Giorn. Bot. Ital.*, 105(4): 177-178. It. (T)

5283 **Gardini Peccenini, S. (1984).** Flora da proteggere. Indagine su alcune specie vegetali minacciate o rare in Italia. Pavia, Istituto di Botanico e Orto Botanico, Universita di Pavia. 248p. It. Illus., maps. (G/L/P/T)

5284 **Valsecchi, F., ed. (1988).** Biotopi di Saregna. Guide a dodici aree di rilevante interesse botanico. Sassari, Societa, Botanico Italiene. 308p. It. Col. illus., maps.

ITALY - Sicily

5285 **Anon. (1984).** Endemic fir gets protection. *Oryx*, 18(2): 107. (T)

5286 **Gardini Peccenini, S. (1984).** Flora da proteggere. Indagine su alcune specie vegetali minacciate o rare in Italia. Pavia, Istituto di Botanico e Orto Botanico, Universita di Pavia. 248p. It. Illus., maps. (G/L/P/T)

5287 **Leon, C. (1983).** Saving Sicily's relict *Abies*. *Threatened Pl. Newsl.*, 12: 3-4. (T)

5288 **Raimondo, P.M. (1983).** On the natural history of the Madonie mountains. *In* 4th. Optima Meeting, 6-14 June 1983: 20-31. Ref. to threatened *Abies nebrodensis*. (T)

5289 **Riggio, S., Massa, B. (1974).** Problemi di conservazione della natura in Sicilia. 1. Contributo. *Atti IV Simp. Naz. Conservazione Nat. Bar.*, 2: 299-425. It.

5290 **Sajeva, M., Orlando, A.M. (1987).** Conservazione *ex situ* di piante succulente minacciate da Estinzione. *Boll. Soc. Ital. Ecol.*, 8(5): 270. It. (G/T)

JAMAICA

5291 **Braatz, S.M. (1981).** Draft environmental profile on Jamaica. Washington, D.C., Department of State, U.S. Man and the Biosphere Secretariat. 97p. Maps, tables, appendices. (L)

5292 **Douglas, J. (1989).** Jamaica to develop national parks. *Biol. Conserv. Newsl.*, 80: 1-2. (P)

5293 **Kelly, D.L. (1988).** The threatened flowering plants of Jamaica. *Biol. Conserv.*, 46(3): 201-216. Maps. Includes list of threatened species. (T)

5294 **McCaffrey, D., Field, R.M., et al. (1988).** Jamaica, country environmental profile. USAID. 358p. Many authors. Maps, tables. (L)

JAPAN

5295 **Anon. (1981).** Pollen bank for fruit trees. *Chron. Hort.*, 21(1): 5. At the Fruit Tree Research Station, Yalabe, Ibaraki, Japan. (E/G)

JAPAN

5296 **Anon. (1989).** Research on the plant species and communities in urgent need of protection in Japan. Summary report of the Plant Species Research Subcommittee. 899 species of wild plants in danger. *J. Nat. Conserv. Soc. Japan*, 327. Analysis of threatened plant statistics; text summaries on 5 species. (T)

5297 **Baltus, J. (1983).** Narodni park Aso. *Ziva*, 31(3): 92-93. Cz. Illus. (P)

5298 **Briggs, S.A. (1964).** U.S. bombing range may endanger Japanese bird and plant species. *Atlantic Nat.*, 19(2): 118-119. (T)

5299 **Chen, F.S. (1977).** [Comparative study of the forest conservation systems between Taiwan and Japan: 2]. *T'ai-wan Lin Yeh Taiwan For. J.*, 3(8): 26-28. Ch.

5300 **Chen, F.S. (1977).** [Comparative study of the forest conservation systems between Taiwan and Japan: 3]. *T'ai-wan Lin Yeh Taiwan For. J.*, 3(9): 22-26, 28. Ch.

5301 **Fushimi, T. (1982).** [Studies on the relation between the forest composition and the conservation of forest land on the basin of the River Shigenobu. 2. The road net as fundamental installation]. *Bull. Ehime Univ. Forest*, 19: 1-11. Ja (En). Maps.

5302 **Hatakeyama, S. (1981).** Provenances of Sakhalin fir in conservation of gene resource in Japan. *In* XVII IUFRO World Congress, Japan 1981, International Union of Forest Research Organizations, Japanese IUFRO Congress Council. 125-128. (E/T)

5303 **Ino, H. (1978).** Forest conservation and management in Japan. *In* Proceedings of the 7th World Forestry Congress, Buenos Aires, 4-18 October 1972, vol. 3: Conservation and recreation. Buenos Aires, Instituto Forestal Nacional. 3567-3582. Sp, Fr (En). (E)

5304 **Ito, S. (1984).** [Transition of forest vegetation caused by human activity on the montane zone in central Japan - the comparison of the higher plant species distribution between the primary forest and the secondary forest]. *Bulletin of the Nagoya University Forests*, 1984 (7): 99-148. Ja (En). Illus., maps.

5305 **Kashiwadani, H. (1979).** 3 rare species of lichens from the Kii Peninsula Japan. *Mem. Nat. Sci. Mus.* (Tokyo), 12: 213-218. (T)

5306 **Kumazaki, M. (1977).** The utilization of forests and environmental conservation. Tokyo, Nihon Ringyo Gijutsu Kyokai. 202p. Ja. Illus.

5307 **Lin, C.H. (1978).** Conservation of natural resources in foreign countries. *Taiwan For. J.*, 4(6): 24-25, 46. Illus.

5308 **Miyawaki, A. (1977).** The national park system in Japan. *In* Filion, L., Villeneuve, P., *eds.* Ecological reserves and nature conservation. Rapport, Conseil Consultatif des Reserves Ecologiques, Quebec, No. 1: 20-21. Papers presented at a conference at Iles-de-la-Madeleine, Quebec, 17-18 June 1976. (P)

5309 **Miyawaki, A., Tuxen, R., eds (1977).** Vegetation science and environmental protection. Tokyo. 576p. (Ge).

5310 **Moyer, J., Higuchi, H., Matsuda, K., Hasegawa, M. (1985).** Threat to unique terrestrial and marine environments and biota in a Japanese National Park. *Envir. Conserv.*, 12(4): 293-301. (P/T)

5311 **Nature Conservation Society of Japan, WWF-Japan (1987).** [The list of plants important for conservation (the primary edition). Angiospermae: Monocotyledonae]. Tokyo, NACS-Japan and WWF-Japan. 25p. Ja. (T)

5312 **Nature Conservation Society of Japan, WWF-Japan (1987).** [The list of plants important for conservation (the primary edition). Angiospermae: Corypetalae]. Tokyo, NACS-Japan and WWF-Japan. 27p. Ja. (T)

JAPAN

5313 **Nature Conservation Society of Japan, WWF-Japan (1987).** [The list of plants important for conservation (the primary edition). Angiospermae: Sympetalae]. Tokyo, NACS-Japan and WWF-Japan. 24p. Ja. (T)

5314 **Nature Conservation Society of Japan, WWF-Japan (1987).** [The list of plants important for conservation (the primary edition). Pteridophyta]. Tokyo, NACS-Japan and WWF-Japan. 18p. Ja. (T)

5315 **Nature Conservation Society of Japan, WWF-Japan (1989).** [The conditions of important wild plant species in Japan]. Tokyo, NACS-Japan and WWF-Japan. 320p. Ja. Illus., maps. Plant Red Data Book. (T)

5316 **Nishiguchi, C. (1976).** Shinrin to ningen. (Forest and people.) Tokyo, Sanyusha. 391p. Illus.

5317 **Numata, M., Yoshioka, K., Kato, M. (1975).** Studies in conservation of natural terrestrial ecosystems in Japan. Part 1: Vegetation and communities. JIBP Synthesis vol. 8. Tokyo, Japanese Committee for the International Biological Program. 157p. Maps. Includes list of rare plants. (P/T)

5318 **Ogawa, M. (1978).** Forest land development and environmental problem. *In* Proceedings of the 8th World Forestry Congress, Jakarta, 16-28 October 1978: Forestry for quality of life. I, 7.

5319 **Ohyama, H., Hashizume, N. (1979).** [Analysis of tree density on greenery space (for conservation and landscaping planning of forests, parks, in Japan).] *Kenkyu Hokoku,* 47: 11-25. Ja (En). Illus. (G)

5320 **Ota, Y. (1976).** Hozoku ringyo no kenkyu. (Research on forest conservation.) Tokyo, Nihon Ringyo. 567p. Illus.

5321 **Rinyacho (1971).** Forest management in natural park areas upon the multiple use basis. S.L., Forestry Agency, Ministry of Agriculture and Forestry. 31p. Illus., maps. (P)

5322 **Rinyacho, K. (1976).** Hoanrin no jitsumu. (Management of forest reserves.) Tokyo, Chikyusha. 8, 428p. (P)

5323 **Sakurai, M. (1984).** Adjustment between nature and human activity in national parks in Japan. *In* McNeely, J.A., Miller, K.R., *eds.* National parks, conservation and development. The role of protected areas in sustaining society. Washington, D.C., Smithsonian Institution Press. 479-485. Map. (P)

5324 **Shidei, T., ed. (1976).** Forest protection. Tokyo, Asakura Shoten. vi, 236p. Illus.

5325 **Shimizu, T., Satomi, N. (1976).** A preliminary list of the rare and critical vascular plants of Japan, 2 parts. *J. Fac. Liberal Arts, Shinshu Univ. Nat. Sci.,* 10: 3-16; 11: 43-54. Annotated list of ferns, gymnosperms, monocotyledons and a number of dicotyledons; distribution details for Hokkaido, Honshu, Kyushu and Shikoku. (T)

5326 **Suganuma, T. (1981).** [Some problems on the nature conservation of the Hakusan National Park]. *J. Phytogeogr. Taxon.,* 29(2): 119-120. Ja. (P)

5327 **Takatsuki, S. (1980).** Ecological studies on effect of Sika deer (*Cervus nippon*) on vegetation: 2. The vegetation of Akune Island, Kagoshima Prefecture, with special reference to grazing and browsing effect of Sika deer. *Ecol. Rev.,* 19(3): 123-144. Illus.

5328 **Takehara, H. (1978).** Problems in forestry research related to environmental conservation. *In* Proceedings of the 7th World Forestry Congress, Buenos Aires, 4-18 October 1972, vol. 3: Conservation and recreation. Buenos Aires, Instituto Forestal Nacional. 4200-4217. (Sp, Fr).

5329 **Uchimura, E. (1981).** Survey of natural forests for conservation of genetic resources in Japan. *In* XVII IUFRO World Congress, Japan 1981. Ibaraki, International Union of Forest Research Organisations. 120-124.

JAPAN - Hokkaido

5330 **Simmons, I.G. (1981).** The balance of environmental protection and development in Hokkaido, Japan. *Envir. Conserv.*, 8(3): 191-198. Illus., maps. (P)

JAPAN - Kazan Retto

5331 **Okutomi, K., ed. (1982).** Conservation reports of the Minami-Iwojima Wilderness Area. Tokyo, Nature Conservation Bureau, Environment Agency of Japan. 403p. Ja (En). See in particular H. Ohba on vascular plants, with floristic analyses and distribution maps of selected species, 61-143; and H. Okutomi, H. Ohba, *et al.* on the endemic flora and fauna, 393-403. (P)

JAPAN - Ogasawara-Shoto

5332 **Degener, O. & I. (1968).** Natural history of the Bonin Islands. *Phytologia*, 21(2): 97-99.

5333 **Kashiwadani, H., Nakanishi, M. (1980).** A note on rare species of the Graphidaceae (lichens) from the Bonin Islands. *Mem. Nat. Sci. Mus. Kokuritsu Kagaku Hakubutsu-Kan*, 11: 21-26. (Ja). Illus. (T)

5334 **Kobayashi, M. (1988).** The wild orchids from Ogasawara Islands. *J. Phytogeog. Taxon.*, 36: 17-26. Distribution and conservation status. (T)

5335 **Nicholson, E.M., Douglas, G.L. (1970).** Conservation of oceanic islands. *In* IUCN Publications New Series, No. 17. IUCN Eleventh Technical Meeting, New Delhi 1969, Vol. I. 200-211.

5336 **Numata, M. (1969).** Ecological background and conservation of Japanese islands. *Micronesica*, 5(2): 295-302.

5337 **Ono, M., Kobayashi, S., Kawakubo, N. (1986).** Present situation of endangered plant species in the Bonin (Ogasawara) Islands. *Ogasawara Res.*, 12: 32p. (Ja). Illus., col. illus., maps. (T)

5338 **Shimozono, F., Iwatsuki, K. (1986).** Botanical gardens and the conservation of an endangered species in the Bonin Islands. *Ambio*, 15(1): 19-21. Illus., col. illus., map. (G/T)

5339 **Tannowa, T., Yoshida, A., Woolliams, K.R. (1976).** Tentative list of rare and endangered plants of the Ogasawara Islands. *Notes Waimea Arbor.*, 3(2): 10-12. (T)

5340 **Tokyo Metropolitan Government (1969).** Survey report on nature conservation of Bonin Islands. Tokyo, Tokyo Metropolitan Governmnent. 2 vols. Vol 1 (1969) includes flora conservation; Vol. 2 (1970). (T)

5341 **Woolliams, K.R. (1974).** Plant collecting trip to the Ogasawara Islands. *Bull. Pacific Trop. Bot. Gard.*, 4(20: 23-28. Includes observations on populations of rare plants. (T)

5342 **Woolliams, K.R. (1976).** Tentative list of rare and endangered plants of the Ogasawara Islands. *Notes Waimea Arbor.*, 3(2): 10-12. (T)

5343 **Woolliams, K.R. (1978).** Observations on the flora of the Ogasawara Islands. *Notes Waimea Arbor.*, 5(2): 2-10. Data on 18 mostly threatened plant species. (T)

5344 **Woolliams, K.R. (1983).** Ogasawara Islands: news from Hahajima. *Notes Waimea Arbor.*, 10(1): 4-5. Notes on 4 rare or threatened plants. (T)

5345 **Woolliams, K.R. (1987).** Ogasawara Islands flora update. *Notes Waimea Arbor.*, 14(2): 2-7. Illus. Includes discussion on cultivation of threatened plants at Waimea Arboretum, and update of conservation status of some species in the wild. (G/T)

5346 **Woolliams, K.R., et al. (1977).** Tentative list of rare and endangered plants of Ogasawara Islands. *Notes Waimea Arbor.*, 3(2): 10-12. (T)

JAPAN - Ogasawara-Shoto

5347 **Yoshida, A., Tannawa, T. (1977).** Endangered plant species of the Ogasawara Islands. *Notes Waimea Arbor.*, 3(2): 8-12. Tentative list of 31 'endangered'; 17 'rare' and 6 'depleted' taxa. (T)

JAPAN - Ryukyu Islands

5348 **Fosberg, F.R. (1960).** Vegetation. *In* Doan, D.B., *et al., eds.* Military geology of the Miyako Archipelago, Ryukyu-Retto. HQ U.S. Army Pacific. 167-187. Discusses man's impact on vegetation.

5349 **Fosberg, F.R. (1960).** Vegetation. *In* Foster, H.L., *et al., eds.* Military geology of Ishigaki-Shima, Ryukyu-Retto. HQ U.S. Army Pacific. 51-84. Discusses man's impact on vegetation.

5350 **Nicholson, E.M., Douglas, G.L. (1970).** Conservation of oceanic islands. *In* IUCN Publications New Series, No. 17. IUCN Eleventh Technical Meeting, New Delhi 1969, Vol. I. 200-211.

5351 **Numata, M. (1969).** Ecological background and conservation of Japanese islands. *Micronesica*, 5(2): 295-302.

5352 **U.S. Civil Administration of the Ryukyu Islands (1953).** Ryukyu Islands forest situation. U.S. Civil Administration of the Ryukyu Islands Special Bull. No. 2. Details of threats to native forests.

JORDAN

5353 **Hatough, A.M., Al-Eisawi, D.M., Disi, A.M. (?)** The effect of conservation on wild-life in Jordan. Mimeo. Dept of Biological Sciences, Faculty of Science, University of Jordan. 19p. Tables, illus., map. (P)

5354 **IUCN Threatened Plants Committee Secretariat (1980).** First preliminary draft of the list of rare, threatened and endemic plants for the countries of North Africa and the Middle East. Kew, IUCN Threatened Plants Committee Secretariat. 170p. (T)

KAMPUCHEA

5355 **Suran, C. (1985).** Intervention de la delegation du Kampuchea. *In* Thorsell, J.W., *ed.* Conserving Asia's natural heritage. The planning and management of protected areas in the Indomalayan Realm. Proc. of the 25th working session of IUCN's CNPPA. Gland, IUCN. 23-25. Fr. (P)

5356 **Talbot, L.M., Challinor, D. (1973).** Conservation problems associated with development projects on the Mekong River system. *In* Costin, A.B., Groves, K.H., *eds.* Nature conservation in the Pacific. Proc. Symposium A-10, XII Pacific Science Congress, Canberra, August-September 1971. IUCN and Australian Nat. Univ. Press. 271-283. Map.

KENYA

5357 **Ament, J.G. (1975).** The vascular plants of Meru National Park, Kenya: 1. A preliminary survey of the vegetation. *J. East Africa Nat. Hist. Soc. Nat. Mus.*, 154: 1-10. (P)

5358 **Ament, J.G., Gillett, J.B. (1975).** The vascular plants of Meru National Park, Kenya: 2. Checklist of the vascular plants recorded. *J. East Africa Nat. Hist. Soc. Nat. Mus.*, 154: 11-30. (P)

5359 **Anon. (1983).** New national park for Kenya. *Oryx*, 17:141. Kakamega Forest to be a national park. (E/P)

KENYA

5360 **Anon. (1984).** Endangered resources for development: strategy conference for the management and protection of Kenya's plant communities: forests, woodlands, bushlands, savannahs and aquatic communities. National Environment and Human Settlements Secretariat. iii, 21p. Illus. (T)

5361 **Bally, P.R.O. (1968).** The Mutomo Hill plant sanctuary in Kenya. *In* Hedberg, I. & O., *eds.* Conservation of vegetation in Africa south of the Sahara. Symposium Proceedings, at 6th plenary meeting of AETFAT, Uppsala. *Acta Phytogeogr. Suec.*, 54: 164-166. Illus. First plant sanctuary in Africa financed by the WWF. Lists plants recorded on Mutomo Hill. (P)

5362 **Beentje, H. (1989).** Mount Kenya, Kenya. *Trop. Afr. Bot. Gard. Bull.*, 1: 21. "Centres of plant diversity" data sheet. (P)

5363 **Casebeer, R.L. (1975).** Summaries of statistics and regulations pertaining to wildlife, parks and reserves in Kenya. FAO Report. no. KEN:71/526 Project working document No. 8. II, 46p. (P)

5364 **Gillett, J.B. (1979).** Possibilities and needs for conservation of plant species and vegetation in Africa. Appendix: Preliminary lists of rare and threatened species in African countries. Kenya. *In* Hedberg, I., *ed.* Systematic botany, plant utilization and biosphere conservation. Stockholm, Almqvist & Wiksell International. 93-94. Examples of taxa threatened in major vegetation types, and includes E:11, V:20, R:4, I:1. (T)

5365 **Grey-Wilson, C. (1984).** Plants in peril: 1. *Kew Mag.*, 1(2): 92-93. *Saintpaulia ionantha.* (T)

5366 **Hamilton, A.C. (1982).** Environmental history of East Africa. A study of the Quaternary. London, Academic Press. 328p.

5367 **Kokwaro, J.O. (1988).** Conservation status of Kakamenga Forest in Kenya: the easternmost relic of the equatorial rain forests of Africa. *In* Goldblatt, P., Lowry, P.P., *eds.* Modern systematic studies in African botany. Proc. Eleventh Plenary Meeting, AETFAT, Missouri Bot. Gard., 10-14 June 1985. Missouri, Missouri Bot. Gard. 471-489. Maps. Plant lists. Tropical forest.

5368 **Lamprey, H. (1978).** The Integrated Project on Arid Lands (IPAL). *Nat. Resources*, 14(4): 2-11.

5369 **Lamprey, H.F. (1981).** Kenya: seeking remedies for desert encroachment: Unesco's Integrated Project in Arid Lands (IPAL) reviewed. *Span*, 24(2): 53-56. Illus., map.

5370 **Lamprey, H.F., Yussuf, H. (1981).** Pastoralism and desert encroachment in northern Kenya. *Ambio*, 10(2-3): 131-134. Illus., map.

5371 **Lawton, R.M. (1975).** The conservation and management of tropical moist forest remnants, with particular reference to Kenya. Surbiton, Land Resources Development Centre. 10p.

5372 **Ledec, G. (1987).** Effects of Kenya's Bura irrigation settlement project on biological diversity and other conservation concerns. *J. Conservation Biology*, 1(3): 247-258. Tana River riverine forest threatened due to fuelwood shortage.

5373 **Lucas, G.Ll. (1968).** Kenya. *In* Hedberg, I. & O., *eds.* Conservation of vegetation in Africa south of the Sahara. Symposium Proceedings, at 6th plenary meeting of AETFAT, Uppsala. *Acta Phytogeogr. Suec.*, 54: 152-163. Maps. Geography, climate, vegetation description. Includes plant associations already protected and proposed areas for protection by national park status plus lists of plants recorded in areas. (P/T)

5374 **Lusigi, W.J. (1981).** New approaches to wildlife conservation in Kenya. *Ambio*, 10(2-3): 87-92. Illus., maps.

KENYA

5375 **Lusigi, W.J. (1984).** Mt. Kulal Biosphere Reserve: reconciling conservation with local human population needs. *In* Unesco-UNEP. Conservation, science and society. Paris, Unesco. 459-469. Contributions to the First International Biosphere Reserve Congress, Minsk, Byelorussia/USSR, 26 September - 2 October 1983. (P)

5376 **Mungai, G.M., Gillett, J.B., Eagle, C.F. (1980).** Plant species in Kenya: survival or extinction. Nairobi, Bulletin of Wildlife Clubs of Kenya. 6p. Illus. Lists over 20 species as threatened. (T)

5377 **O'Keefe, P. (1983).** The causes, consequences and remedies of soil erosion in Kenya. *Ambio*, 12(6): 302-305. Diagrams.

5378 **Pertet, F. (1984).** Kenya's experience in establishing coastal and marine protected areas. *In* McNeely, J.A., Miller, K.R., *eds*. National parks, conservation and development. The role of protected areas in sustaining society. Smithsonian Institution Press, Washington, D.C. 101-108. Map. (P)

5379 **Raadts, E. (1979).** Eine neue und eine seltene *Kalanchoe* aus Kenia (Ost-Africa). (Notes on a new and a rare *Kalanchoe* from Kenya (East Africa).) *Willdenowia*, 9(2): 285-287. Ge. Illus. (T)

5380 **Robertson, S.A. (1988).** Kenya strategy. *Threatened Pl. Newsl.*, 19: 5-6. Recommendations of the plenary session on the State of the Environment Report for UNEP.

5381 **Rodgers, W.A. (1983).** A note on the distribution and conservation of *Oxystigma msoo* Harms (Caesalpiniaceae). *Bull. Jard. Bot. Nat. Belg.*, 53(1-2): 161-164. (Fr). (T)

5382 **Simmons, N.M. (1984).** Problems and progress in establishing biosphere reserves in northern regions. *In* Unesco-UNEP. Conservation, science and society. Paris, Unesco. 59-64. Contributions to the First International Biosphere Reserve Congress, Minsk, Byelorussia/USSR, 26 September - 2 October 1983. (P)

5383 **Synott, T.J. (1981).** A report on the status, importance and protection of the montane forests. IPAL Technical Report No. D-2a. ii, 57p. Integrated Project in Arid Lands. In the Marsabit District of Kenya.

5384 **Taiti, S. (1977).** Ecological criteria of land evaluation for wildlife. *Misc. Soil Paper, Kenya Soil Survey*, 11: 61-72. A guide to the types of survey required for wildlife conservation planning.

5385 **Unesco/UNEP (1982).** Integrated Project In Arid Lands (IPAL): Technical Report No. A-3. Proceedings of a Scientific Seminar, Nairobi, 24-27 November, 1980. Man and the Biosphere Programme Project 3. Rome, FAO. 146p. Impact of human activities and land use practices on grazing lands.

5386 **Watterson, G.G., comp. (1963).** Conservation of nature and natural resources in modern African states. Report of a Symposium organized by CCTA and IUCN, held under the auspices of FAO and Unesco, Arusha, Tanganyika, September 1961. IUCN Publications New Series No. 1. Morges, IUCN. 367p. Illus.

5387 **Wenner, C.-G. (1983).** Soil conservation in Kenya. *Ambio*, 12(6): 305-307. Col. illus.

5388 **Western, D. (1984).** Amboseli National Park: human values and the conservation of a savanna ecosystem. *In* McNeely, J.A., Miller, K.R., *eds*. National parks, conservation and development. The role of protected areas in sustaining society. Washington, D.C., Smithsonian Institution Press. 93-100. Map. (P)

5389 **Williams, J.G., Arlott, N. (1981).** A field guide to the national parks of East Africa. London, Collins. 336p. Illus., col. illus. Includes maps of some of the parks. (P)

KIRIBATI

5390 **Dawson, E.Y. (1959).** Changes in Palmyra Atoll and its vegetation through the activities of man, 1913-1958. *Pacific Nat.*, 1(2): 1-51.

KIRIBATI

5391 **Degener, O., Gillaspy, E. (1955).** Canton Island, South Pacific. *Atoll Res. Bull.*, 41: 1-51. Lists ornamental and useful species; revegetation attempt.

5392 **Fosberg, F.R. (1977).** An irresponsible scientific expedition. *Atoll Res. Bull.*, 219: 4-5. Fire damage to vegetation. (T)

5393 **Garnett, M.C. (1983).** A management plan for nature conservation in the Line and Phoenix Islands. Part 1: Description. Government of Kiribati. viii, 318, 131p. Illus. (P)

5394 **McIntire, E.G. (1961).** Canton Island (Phoenix Islands). Riverside, California, University of California. 42p.

5395 **SPREP (1980).** Kiribati. Country Report No. 7. Noumea, New Caledonia, South Pacific Commission. 7p.

5396 **Smith, S.V., Henderson, R.S., eds (1978).** Phoenix Islands Report 1. *Atoll Res. Bull.*, 221: 183p.

5397 **Stoddart, D.R. (1971).** Conservation of the Phoenix Islands, central Pacific Ocean. London, Dept of Education and Science, and Foreign and Commonwealth Office. 20p. Report to the Southern Zone Research Committee, Royal Society.

5398 **Wodzicki, K. (1973).** Problems of vanishing plants and animals. *In* South Pacific Commission. Papers, Regional Symposium on Conservation of Nature Reefs and Lagoons, Noumea, 1971. Noumea, SPC. 217-223. (T)

KOREA, DEMOCRATIC PEOPLE'S REPUBLIC OF

5399 **Anon. (1978).** Conservation of nature in the Democratic People's Republic of Korea. *Envir. Policy Law*, 4: 181-183. Illus.

5400 **Lee, T.B. (1980).** [Conservation of threatened plants in Korea]. *Bull. Kwanak Arbor.*, 3: 190-196. Ko (En). Includes notes on plant re-introductions. (T)

KOREA, REPUBLIC OF

5401 **Choi, K.-C., Kim, C.-H. Lee, Y.-N., Wn, P.-O., Yoon, I.B. (1981).** Rare and endangered species of animals and plants of Republic of Korea. Korean Assoc. for Conservation of Nature. 293p. Lists 118 plant taxa, including widespread non-endemic species. (T)

5402 **Hyun, S.K. (1978).** The conservation of forest plant and animal genetic resources in Korea. *In* Proceedings of the 8th World Forestry Congress, Jakarta, 16-28 October 1978: Forestry for quality of life. II, 11. (E)

5403 **Kim, Y.N., Hong, S.G., Cho, T.H. (1977).** [The effect of soil hardness on tree growth for the management of trees in Seoul Children's Park.] *J. Korean Forest. Soc.*, 36: 47-55. Ko. Protected forest site. (P)

5404 **Lee, T.B. (1979).** [Distribution of *Berchemia berchemiaefolia* and an investigation for its conservation in Korea]. *Korean J. Pl. Taxon.*, 9(1-2): 1-6. Ko (En). Illus. (T)

5405 **Lee, T.B. (1980).** Rare and endangered species in the area of Mt Sorak. *Bull. Kwanak Arbor.*, 3: 197-201. Mentions 12 taxa with notes on distribution. (Ko). (T)

5406 **Lee, T.B. (1980).** [Conservation of threatened plants in Korea]. *Bull. Kwanak Arbor.*, 3: 190-196. Ko (En). Includes notes on plant re-introductions. (T)

5407 **Lee, T.B. (1984).** Endemic and rare plants of Mt. Sorak. *Bull. Kwanak Arbor.*, 5: 1-6. Enumeration of 114 vascular plant taxa of which 65 are endemic; 5 taxa are 'endangered', 12 taxa are 'rare'. (T)

5408 **Lee, Y.-N. (1981).** Plants. *In* Choi, K.-C., *et al., eds.* Rare and endangered species of animals and plants of Republic of Korea. Seoul, Korean Association for Conservation of Nature. 154-271, 277-280. (T)

LAO PEOPLE'S DEMOCRATIC REPUBLIC

5409 **Sayer, J. (1983).** Nature conservation priorities in Laos. *Tigerpaper*, 10(3): 10-14. Illus., map. (P)

5410 **Talbot, L.M., Challinor, D. (1973).** Conservation problems associated with development projects on the Mekong River system. *In* Costin, A.B., Groves, K.H., *eds.* Nature conservation in the Pacific. Proc. Symposium A-10, XII Pacific Science Congress, Canberra, August-September 1971. IUCN and Australian Nat. Univ. Press. 271-283. Map.

LEBANON

5411 **IUCN Threatened Plants Committee Secretariat (1980).** First preliminary draft of the list of rare, threatened and endemic plants for the countries of North Africa and the Middle East. Kew, IUCN Threatened Plants Committee Secretariat. 170p. (T)

LESOTHO

5412 **Beverly, A. (1978).** A survey of *Aloe polyphylla*. *Veld & Flora*, March 1978: 24-27. (T)

5413 **Beverly, A.C. (1979).** My quest for *Aloe polyphylla*. *Cact. Succ. J.* (U.S.), 51: 3-7. (T)

5414 **Codd, L.E. (1968).** Regional synthesis - Discussion. *In* Hedberg, I. & O., *eds.* Conservation of vegetation in Africa south of the Sahara. Symposium Proceedings, at 6th plenary meeting of AETFAT, Uppsala. *Acta Phytogeogr. Suec.*, 54: 257-260. Overview of conservation in southern Africa. (P)

5415 **Guillarmod, A.J. (1968).** Lesotho. *In* Hedberg, I. & O., *eds.* Conservation of vegetation in Africa south of the Sahara. Symposium Proceedings, at 6th plenary meeting of AETFAT, Uppsala. *Acta Phytogeogr. Suec.*, 54: 253-256. Illus. Vegetation, present protection (vegetation & individual species) and recommended protection. (P/T)

5416 **Guillarmod, A.J. (1979).** Lesotho. *In* Hedberg, I., *ed.* Systematic botany, plant utilization and biosphere conservation: Proc. of a sym. held in Uppsala in commemoration of the 500th ann. of the Univ. Stockholm, Almqvist & Wiksell Int'l. 101.

5417 **Guillarmod, A.J. (1979).** Possibilities and needs for conservation of plant species and vegetation in Africa. Appendix: Preliminary lists of rare and threatened species in African countries. Lesotho. *In* Hedberg, I., *ed.* Systematic botany, plant utilization and biosphere conservation. Stockholm, Almqvist & Wiksell International. 101. Contains five species and three genera: E:6, R:1, I:1. (T)

5418 **Talukdar, S. (1983).** The conservation of *Aloe polyphylla* endemic to Lesotho. *Bothalia*, 14(3-4): 985-989. (Fr). Illus., maps. (T)

5419 **Talukdar, S., Schmitz, M.O. (1982).** The conservation of *Aloe polyphylla* indigenous in Lesotho. *In* AETFAT Synopses 10th Congress: the origin, evolution and migration of African floras, Pretoria, 18-23 January 1982. 52. (T)

LIBERIA

5420 **Anon. (1979).** Liberia - dramatic about-turn. *IUCN Bull.*, 10(5): 33.

5421 **Curry-Lindahl, K. (1969).** Report to the Government of Liberia on conservation, management and utilization of wildlife resources. Morges, IUCN. 31p. Includes suggested areas for national parks and nature reserves. (P)

5422 **Jeffrey, S. (1978).** How Liberia uses wildlife. *Oryx*, 14(2): 168-173. Illus. Mainly fauna, but national forests mapped. (P)

5423 **Lamotte, M. (1983).** The undermining of Mount Nimba. *Ambio*, 12(3-4): 174-179. Illus., col.illus., maps. (P)

LIBERIA

5424 **Thorne, J.M. (1979).** Possibilities and needs for conservation of plant species and vegetation in Africa. Appendix: Preliminary lists of rare and threatened species in African countries. Liberia. *In* Hedberg, I., *ed.* Systematic botany, plant utilization and biosphere conservation. Stockholm, Almqvist & Wiksell International. 88. Includes short list of example species and genera. (T)

5425 **Verschunen, J. (1982).** Hope for Liberia. *Oryx*, 16(5): 421-427. Illus. Proposals are made for wildlife conservation measures, with reference to the setting up of a Forest Development Authority in 1976.

5426 **Voorhoeve, A.G. (1968).** Liberia. *In* Hedberg, I. & O., *eds.* Conservation of vegetation in Africa south of the Sahara. Symposium Proceedings, at 6th plenary meeting of AETFAT, Uppsala. *Acta Phytogeogr. Suec.*, 54: 74-76. Illus. Describes vegetation and recommends sites for protection. (P)

5427 **Wachtel, P.S. (1983).** Rainforest programme bears fruit. *IUCN Bull.*, 14(10-12): 127-128. Illus.

LIBYAN ARAB JAMAHARIYA

5428 **Anon. (1981).** [Kouf National Park for the conservation of natural resources and wildlife, Baida, Libya]. Libya, Arab Center for the Study of Arid Zones and Dry Land (A.C.S.A.D.). 28p. Ara (En). Col. illus., map. Popular guide. (P)

5429 **Ben Saad, A.A., Darrat, A.A. (1979).** A note on the flora and its exploitation and conservation in Libya. *In* Hedberg, I., *ed.* Systematic botany, plant utilization and biosphere conservation. Stockholm, Almqvist & Wiksell International. 139-140. (T)

5430 **IUCN Threatened Plants Committee Secretariat (1980).** First preliminary draft of the list of rare, threatened and endemic plants for the countries of North Africa and the Middle East. Kew, IUCN Threatened Plants Committee Secretariat. 170p. (T)

LIECHTENSTEIN

5431 **Baltisberger, M. (1981).** *Myosotis rehsteineri* Wartm. im Ruggeller Riet (FL). (*Myosotis rehsteineri* Wartm. in wetlands near Ruggell (FL).) *Ber. Geobot. Inst. ETH, Stiftung Rubel*, 48: 161-163. Ge. (T)

5432 **Broggi, M.F. (1977).** Nature conservation and landscape management in Liechtenstein. *Parks*, 2(3): 14-16. A short descriptive account of the history of nature conservation in Liechtenstein and habitat degradation.

5433 **Broggi, M.F., Waldburger, E. (1984).** Rote Liste der gefahrdeten und seltenen Gefasspflanzen des Furstentums Liechtenstein. *Ber. Bot.-Zool. Ges. Liechtenstein-Sargans-Werdenberg*, 13: 7-40. Ge. (T)

LUXEMBOURG

5434 **Council of Europe (1984).** Benelux: nature conservation and protection. *Council of Europe Newsl.*, 84(2): 2.

5435 **Faber, P. (1974).** L'arboretum de la Foret D'Anven. *Bull. Soc. Nat. Luxembourg*, 79: 1-29. Fr. Lists species. (G)

5436 **Fabri, R. (1983).** *Bromus grossus* s.l. et *B. secalinus* en Belgique et au Grand-Duche de Luxembourg. *Bull. Soc. Roy. Bot. Belg.*, 116(2): 207-223. Fr (En). Illus., maps. (T)

5437 **Reichling, L. (1981).** *In* Luxembourg geschutzte Pflanzen. Ubersicht sowie Anleitung zum Kennenlernen der in Luxemburg geschutzten wildwachsenden Pflanzenarten, 2nd Ed. Luxembourg, Natura Luxumburger Liga fur Natur- und Umweltschutz. 47p. Ge. Col. illus., distribution maps. Outlines the law; describes ecology and threats of plants protected. (L/T)

MADAGASCAR

5438 **Andriamampianina, J. (1979).** Madagascar. *In* Hedberg, I., *ed.* Systematic botany, plant utilization and biosphere conservation: Proc. of a sym. held in Uppsala in commemoration of the 500th ann. of the Univ. Stockholm, Almqvist & Wiksell Int'l. 103.

5439 **Andriamampianina, J. (1984).** Nature reserves and nature conservation in Madagascar. *In* Jolly, A., Oberbe, P., Albignac, R., *eds.* Madagascar. Oxford, Pergamon Press. 219-227. Map. (G/P/T)

5440 **Anon. (1983).** Conservation progress in Madagascar. *Oryx*, 17: 142.

5441 **Aubreville, A. (1971).** La destruction des forets et des sols en pays tropical. Le cas de Madagascar. *Adansonia, ser. 2,* 11(1): 5-39. Fr. Tropical rain forest. (T)

5442 **Barre, V., Sayer, J. (1986).** Hope for Madagascar? *IUCN Bull.,* 17(1-3): 29.

5443 **Baumer, M. (1978).** La conservation et la valorisation des resources ecologiques dans les iles des Comores, de Maurice, de la Reunion, des Seychelles. Paris, Agence de Cooperation Culturelle et Technique. 92p. Fr.

5444 **Bernardi, L. (1974).** Problemes de conservation de la nature les iles de l'Ocean Indien. 1. Meditations a propos de Madagascar. *Saussurea,* 5: 37-47. Fr.

5445 **Cabanis, Y., Chabouis, L. & F. (1969).** Vegetaux et groupements vegetaux de Madagascar et des Mascareignes, 4 vols. Tananarive, Bureau Pour le Developpement de la Production Agricola. Fr. Illus., maps. Published 1969-1970.

5446 **Cribb, P.J. (1983).** Orchids and conservation in tropical Africa and Madagascar. *Bothalia,* 14(3 & 4): 1013-1014. (Fr). (T)

5447 **Hardy, D.S. (1987).** Madagascar: the rescued ones. *Aloe,* 24(2): 40-41. Illus., col. illus. (T)

5448 **Helfert, M.R., Wood, C.A. (1986).** Shuttle photos show Madagascar erosion. *Geotimes,* 31(3):4-5. (P)

5449 **Hillerman, F. (1981).** Madagascar revisited: part 2. The disappearing forest. *Amer. Orchid Soc. Bull.,* 50(6): 669-673. Col. illus. Deforestation.

5450 **Humbert, H. (1946).** La protection de la nature a Madagascar. *Rev. Int. Bot. Appl. Agric. Trop.,* 26: 358. Fr.

5451 **IUCN (1972).** Comptes rendus de la conference internationale sur la conservation de la nature et de ses ressources a Madagascar. IUCN Publications New Series No. 36. Morges, IUCN. 239p. Fr. See esp. papers on the Didiereaceae thickets of southern Madagascar, 145-151, and the floristic significance of Madagascar, 139-142.

5452 **IUCN Commission on National Parks and Protected Areas (1986).** Review of the protected areas system in the Afrotropical realm. Gland, Switzerland and Cambridge, IUCN and UNEP. 259p. Illus., 5 maps. (P)

5453 **IUCN Threatened Plants Committee Secretariat (1980).** The botanic gardens list of Madagascan succulents. Kew, IUCN Botanic Gardens Conservation Co-ordination Body Report No. 2. 21p. Lists 235 succulents, most endemic to Madagascar. (G/T)

5454 **IUCN Threatened Plants Unit (1985).** The botanic gardens list of succulent euphorbias and aloes of Africa (including Madagascan succulents update). Kew, IUCN Botanic Gardens Co-ordinating Body Report No. 15. Lists plants in cultivation in botanic gardens. (G/T)

5455 **James, G. (1988).** Some observations of a population of *Nepenthes madagascariensis* in Madagascar. *Carnivorous Pl. Newsl.,* 17: 102-103. Illus. (T)

MADAGASCAR

5456 **Jenkins, M.D., ed. (1987).** Madagascar. An environmental profile. Gland, Switzerland and Cambridge, IUCN/UNEP/WWF. 374p. Compiled by IUCN Conservation Monitoring Centre, Cambridge. Maps. French edition in prep. (E/L/P/T)

5457 **Jolly, A. (1980).** A world like our own: man and nature in Madagascar. New Haven, Yale University Press. xvi, 272p. Illus., col. illus.

5458 **Jolly, A. (1982).** Island of the dodo is down to its last few native species. *Smithsonian*, 13(3): 94-103. (T)

5459 **Jolly, A. (1987).** Madagascar: a world apart. *National Geographic*, 171(2): 148-183.

5460 **Jolly, A., Oberle, P., Albignac, E.R., eds (1984).** Madagascar. Oxford, Pergamon Press. (E/P/T)

5461 **Keraudren, M. (1968).** Madagascar. *In* Hedberg, I. & O., *eds.* Conservation of vegetation in Africa south of the Sahara. Symposium Proceedings, at 6th plenary meeting of AETFAT, Uppsala. *Acta Phytogeogr. Suec.*, 54: 261-265. Fr. Illus. Lists protected reserves. (P)

5462 **Rakotovao, L., Barre, V., Sayer, J. (1988).** L'equilibre des ecosystemes forestiers a Madagascar. IUCN, Tropical Forest Programme Series. No.6. Actes d'un seminaire international. 344p.

5463 **Randrianarijaona, P. (1983).** The erosion of Madagascar. *Ambio*, 12(6): 308-311. Col. illus.

5464 **Rauh, W. (1979).** Problems of biological conservation in Madagascar. *In* Bramwell, D., *ed.* Plants and islands. London, Academic Press. 405-421. (Sp). Illus. maps. (P/T)

5465 **Rauh, W. (1985).** The succulent vegetation of Madagascar. VI. *Cact. Succ. J.* (U.S.), 57(5): 217-219. Illus. (T)

5466 **Tucker, C.J., Townshend, J.R.G., Goff, T.E. (1985).** African land-cover classification using satellite data. *Science*, 227(4685): 369-375. Maps.

5467 **Wachtel, P.S. (1983).** Rainforest programme bears fruit. *IUCN Bull.*, 14(10-12): 127-128. Illus.

5468 **Wilson, J.M., Stewart, P.D., Fowler, S.V. (1988).** Ankarana - a rediscovered nature reserve in northern Madagascar. *Oryx*, 22(3): 163-171. Illus., maps. Mainly fauna, but includes section on forest exploitation and the future. (P)

MALAWI

5469 **Anon. (1983).** Lake Malawi: a national park. *Oryx*, 17: 141. (P)

5470 **Chapman, J.D. (1968).** Malawi. *In* Hedberg, I. & O., *eds.* Conservation of vegetation in Africa south of the Sahara. Symposium Proceedings, at 6th plenary meeting of AETFAT, Uppsala. *Acta Phytogeogr. Suec.*, 54: 215-224. Describes geography, climate and vegetation, and surveys communities already protected and those requiring further protection. Table of forest patches in need of protection from fire. (P/T)

5471 **Chapman, J.D. (1981).** Conservation of vegetation and its constituent species in Malawi. *Nyala*, 6(2): 125-132 (1980 publ. 1981). Map.

5472 **Dowsett-Lemaire, F., Dowsett, R.J. (1988).** Threats to the evergreen forests of southern Malawi. *Oryx*, 22(3): 158-162. Illus., maps. Review of vegetation types, conservation status and action recommended.

5473 **Edwards, I. (1985).** Conservation of plants on Mulanje Mountain, Malawi. *Oryx*, 19: 86-90. Illus.

MALAWI

5474 **Hall-Martin, A.J., Fuller, N.G. (1975).** Observations on the phenology of some trees and shrubs of the Lengwe National Park, Malawi. *J. S. Afr. Wildl. Manage. Assoc.*, 5(1): 83-86. (P)

5475 **Hough, J. (1984).** An approach to an integrated land use system on Michiru Mountain, Malawi. *Parks*, 9(3/4): 1-3.

5476 **Kombe, A.D.C. (1984).** The role of protected areas in catchment conservation in Malawi. *In* McNeely, J.A., Miller, K.R., *eds.* National parks, conservation and development. The role of protected areas in sustaining society. Washington, D.C., Smithsonian Institution Press. 115-117. (P)

5477 **National Fauna Preservation Society of Malawi (1978).** Zomba Plateau, Mulunguzi Nature Trail. Limbe, National Fauna Preservation Society of Malawi. 30p. Illus.

5478 **National Fauna Preservation Society of Malawi (1979).** Chingwe's Hole Nature Trail, Zomba Plateau. Limbe, National Fauna Preservation Society of Malawi. 51p. Illus.

5479 **Nicholson, H.B. (1977).** The cedars of south and central Africa. A conservation tragedy. *Trees S. Afr.*, 29(2/4): 61-63. (T)

5480 **Phillips, E. (1980).** The price of development. *Soc. Malawi J.*, 33(1): 19-25.

MALAYSIA

5481 **Anon. (1981).** Conservation of the Malaysian rain forests. *J. Dendrol.*, 1(3&4): 126-127.

5482 **Bahrin, T.S. (1981).** The utilization and management of land resources in Malaysia. *Geojournal*, 5(6): 557-561. Deforestation. Summary of the development of agriculture, forest exploitation and mining, describing the accompanying environmental degradation.

5483 **Bompard, J.M. (1988).** Wild *Mangifera* species in Kalimantan (Indonesia) and in Malaysia. Final report on the 1986-1988 collecting missions. WWF Project No. 3305. Rome, IBPGR, IUCN-WWF. 50p. Maps. (E/G/T)

5484 **Bompard, J.M., Kostermans, A. (1986).** Local knowledge too late to save wild mangoes? *IUCN Bull.*, 17(1-3): 26. (E/T)

5485 **Chai, P.P.K., Choo, N.C. (1983).** Conservation of forest genetic resources in Malaysia with special reference to Sarawak. *In* Jain, S.K., Mehra, K.L., *eds.* Conservation of tropical plant resources. Howrah, Botanical Survey of India. 39-47. (E)

5486 **Christensen, B. (1983).** Mangroves - what are they worth? *Unasylva*, 35(139): 2-15. Illus. Includes report on Malaysia's mangrove management.

5487 **Cranbrook, Earl of, ed. (1988).** Key environments: Malaysia. Oxford, Pergamon Press, and Gland, Switzerland, IUCN. 317p. Illus., maps.

5488 **Cross, M. (1989).** Dispute grows over Malaysia's forests. *New Sci.*, 121(1650): 27. Deforestation.

5489 **Davison, G.W.H. (1982).** How much forest is there? *Malay. Naturalist*, 35(1/2): 11-12. Tropical forest. (T)

5490 **Dhanarajan, G. (1984).** National parks and world conservation strategy. *Malay. Naturalist*, 37(3): 7-10. (P)

5491 **Hamzah, M.B., Awang, K., Rusli, M. (1983).** Issues in Malaysian forestry. *Malays. Forest.*, 46(4): 409-424. Includes section on conservation; tropical rain forest.

MALAYSIA

5492 **Hing, L.W., Salleh Mohammed Nor (1989).** Malaysia - a responsible supplier of tropical timber products. *In* McDermott, M.J., *ed.* The future of the tropical rain forest. Proceedings of an international conference at St. Catherine's College, Oxford, 27-28 June 1988. Oxford, Oxford Forestry Institute. 44-51. (E)

5493 **Jabil, D.M. (1983).** Problems and prospects in tropical rainforest management for sustained yield. *Malays. Forest.*, 46(4): 398-408. (E)

5494 **Jin-Eong, O. (1984).** Aquaculture, forestry and conservation in Malaysian mangroves. *In* McNeely, J.A., Miller, K.R., *eds.* National parks, conservation and development. The role of protected areas in sustaining society. Washington, D.C., Smithsonian Institution Press. 165-169. (P/T)

5495 **Khoo, G. (1981).** Malaysia's forest policy. *Nat. Malaysiana*, 6(3): 4-9. Col. illus.

5496 **Khoo, G. (1982).** Logging of our forest. *Nat. Malaysiana*, 7(2): 36-39. Col. illus. Deforestation. (E)

5497 **Khoo, G. (1983).** Destroy not our wild genetic resource. *Nat. Malaysiana*, 8(2): 4-7. Col. illus. (E/G)

5498 **Kiew, R. (1983).** Conservation of Malaysian plant species. *Malay. Naturalist*, 37(1): 2-5. (T)

5499 **Lee, D. (1980).** The sinking ark: environmental problems in Malaysia and Southeast Asia. Kuala Lumpur, Heinemann. 85p.

5500 **Lovejoy, T.E. (1985).** Rehabilitation of degraded tropical forest lands. *Environmentalist*, 5(1): 13-20. Illus. Tropical forest.

5501 **Lowry, J.B. (1971).** Conserving the forest - a phytochemical view. *Malay. Nat. J.*, 24(3-4): 225-230.

5502 **Ong, J.E. (1982).** Mangroves and aquaculture in Malaysia. *Ambio*, 11(5): 252-257. Illus., map. (P)

5503 **Rajanaidu, N. (1981).** Oil palm genetic resources: current methods of conservation. *Pl. Genet. Resource. Newsl.*, no. 48: 25-30. En (Fr, Sp). (E/T)

5504 **Reid, R. (1988).** Mango report. *Threatened Pl. Newsl.*, 19: 12. Illus. Report on the joint IBPGR/IUCN/WWF project on wild *Mangifera*. (E/T)

5505 **Rubeli, K. (1989).** Pride and protest in Malaysia. *New Sci.*, 124(1687): 49-52. Col. illus. Rain forest; deforestation.

5506 **Sastry, C.B., Srivastava, P.B.L., Manap Ahmad, A., eds (1977).** A new era in Malaysian forestry. Serdang, Universiti Pertanian Malaysia Press. vii, 323p. Illus. Conservation and management.

5507 **Sharp, T. (1983).** Malaysia's environment in danger. *Ambio*, 12(5): 275-276. Illus.

5508 **Sharp, T. (1984).** ASEAN nations tackle common environmental problems. *Ambio*, 13(1): 45-46. Illus.

5509 **WWF-Malaysia (1983).** State conservation strategies: the way to sustainable development. *IUCN Bull.*, suppl. 4: 7-8. Map.

5510 **Wycherley, P.R. (1969).** Conservation in Malaysia: a manual on the conservation of Malaysia's renewable natural resources. IUCN New Series, Suppl. Paper No. 22. Morges, IUCN. 207p. Illus., map.

5511 **Zakaria, M.B., Mohamad, B.B. (1988).** Botanical gardens and protected forests in Malaysia: country report. *In* Santiapillai, C., Ashby, K.R., *eds.* Proceedings of the Symposium on the Conservation and Management of Endangered Plants and Animals. Bogor, Indonesia, 18-20 June 1986. Bogor, SEAMEO-BIOTROP. 231-232. (G/P)

MALAYSIA - Peninsular Malaysia

5512 **Aiken, S.R., Leigh, C.H. (1984).** A second national park for Peninsular Malaysia? The Endau-Rompin controversy. *Biol. Conserv.*, 29(3): 253-276. Maps. (P)

5513 **Anon. (1988).** Endau-Rompin may become a park at last. *Oryx*, 22(3): 137-138. (P)

5514 **Anon. (1988).** Malaysia ...and bad news. *Threatened Pl. Newsl.*, 19: 8-9. Deforestation of Mata Ayer Forest Reserve. (P)

5515 **Anon. (1988).** Malaysia ...good news. *Threatened Pl. Newsl.*, 19: 8. Map. Declaration of 92,000 ha of Endau-Rompin as a state park. (P/T)

5516 **Anon. (1988).** Malaysian logging threatens Thai reserve. *Oryx*, 22(1): 54. Full report in *Bangkok Bird Club Bull.*, 4(8). (P)

5517 **Anon. (1989).** Mangrove forests in danger of extinction. *Malay. Naturalist*, 42(2-3): 40-42. (P)

5518 **Bahari, J., Tahir, M. (1984).** Some notes on the coconut germplasm prospection and collection in Peninsular Malaysia. *IBPGR Region. Committ. Southeast Asia Newsl.*, 8(2): 7-8. (E/G)

5519 **Bullock, J.A. (1978).** A contribution to the estimation of litter production and tree loss in Pasoh Forest Reserve. *Malay. Nature J.*, 30(2): 363-365. Illus. (P)

5520 **Burges, A. (1978).** The conservation of Pasoh Forest. *Malay. Nature J.*, 30(2): 445-447. (P)

5521 **Dransfield, J. (1982).** *Pinanga cleistantha*, a new species with hidden flowers. *Principes*, 26(3): 126-129. (T)

5522 **Dransfield, J., Kiew, B.H. (1983).** Portraits of threatened plants 1: *Maxburretia rupicola* (Ridley) Furtado, Palmae. *Malay. Naturalist*, 37(1): 6-7. Illus. (L/T)

5523 **Dransfield, J., Kiew, R. (1987).** An annotated checklist of palms at Ulu Endau, Johore, Malaysia. *Malay. Nature J.*, 41(2-3): 257-265. (P)

5524 **Eong, O.J., Khoon, G.W. (1979).** The Pantai Acheh Forest Reserve. *Malay. Naturalist*, (Mar-Jun 1979): 14-16. Illus. (P)

5525 **Guan, S.L. (1987).** Conservation of the mango and its relatives in Peninsular Malaysia. Forest Research Institute of Malaysia. World Wildlife Fund Malaysia Project MAL 80/85. 29p. (E/T)

5526 **Heang, K.B. (1979).** The projected environmental impact assessment for the Ambang Forest Reserve and water catchment area. *Malay. Naturalist*, Mar-Jun: 3-6. Maps. (P)

5527 **Johns, A.D. (1988).** Effects of 'selective' timber extraction on rain forest structure and composition and some consequences for frugivores and folivores. *Biotropica*, 20: 31-37. Extracting 3.3% of trees in dipterocarp forest destroyed 50.9% of all tree taxa.

5528 **Jones, D.T. (1984).** Portraits of threatened plants, 4. *Maclurodendron magnificum* Hartley, 5. *Melicope suberosa* Stone. *Malay. Naturalist*, 37(4): 4-6. Illus. (T)

5529 **Jones, D.T. (1987).** Rare plant profile no. 1: *Merrillia caloxylon* (Rutaceae). *Bot. Gard. Conserv. News*, 1(1): 38-42. Illus., map. (E/G/T)

5530 **Jones, D.T. (1987).** Rimba Ilmu Universiti Malaya - preserving the natural plant heritage of the world's oldest rain forest. *In* Bramwell, D., Hamann, O., Heywood, V., Synge, H., *eds*. Botanic gardens and the World Conservation Strategy. London, Academic Press. 345-346. (G)

MALAYSIA - Peninsular Malaysia

5531 **Khan, M.K. bin M. (1985).** Peninsular Malaysia's efforts to create a system of nature reserves. *In* Thorsell, J.W., *ed.* Conserving Asia's natural heritage. The planning and management of protected areas in the Indomalayan Realm. Proc. of the 25th working session of IUCN's CNPPA. Gland, IUCN. 19-22. Map. (P)

5532 **Kiew, R. (1983).** Portraits of threatened plants. (*Ilex praetermissa* Kiew and *Didymocarpus primulina* Ridley). *Malay. Naturalist*, 37(2): 6-7. (T)

5533 **Kiew, R. (1985).** Illegal plant collection in Malaya. *Malay. Naturalist*, 38(3): 44-45. Illus. An international syndicate preys on rare plants especially *Nepenthes rajah*, now rare in Kinabalu Park. (P/T)

5534 **Kiew, R. (1985).** The limestone flora of the Batu Luas area, Taman Negara. *Malay. Naturalist*, 38(3): 30-38. Illus., map. (P/T)

5535 **Kiew, R. (1987).** Notes on the natural history of the Johore banana, *Musa gracilis* Holttum. *Malay. Nature J.*, 41(2-3): 239-248. Illus., map. (P/T)

5536 **Kiew, R. (1987).** The herbaceous flora of Ulu Endau, Johore-Pahang, Malaysia, including taxonomic notes and descriptions of new species. *Malay. Nature J.*, 41(2-3): 201-234. Includes section on the conservation status of Ulu Endau area; checklist and desciptions of new species. (P)

5537 **Kiew, R. (1988).** Portraits of threatened plants 14. *Johannesteijsmannia lanceolata* J.Dransfield. *Malay. Naturalist*, 42(1): 14-15. Illus. (T)

5538 **Kiew, R. (1988).** Portraits of threatened plants 15. *Johannesteijsmannia magnifica* J.Dransfield. *Malay. Naturalist*, 42(1): 15-16. Illus. (T)

5539 **Kiew, R. (1989).** Collecting endangered palms in Peninsular Malaysia. *Principes*, 33(2): 94-95. (T)

5540 **Kiew, R. (1989).** Lost and found - *Begonia eiromischa* and *Begonia rajah*. *Nature Malaysiana*, 14(2): 64-67. Col. illus., map. (T)

5541 **Kiew, R., Dransfield, J. (1987).** The conservation of palms in Malaysia. *Malay. Naturalist*, 41(1): 24-31. List of palms with conservation categories for Pen. Malaysia; result of WWF 3325. (T)

5542 **Kiew, R., Parris, B.S., Madhavan, S., Edwards, P.J., Wong, K.M. (1987).** The ferns and fern-allies of Ulu Endau, Johore, Malaysia. *Malay. Nature J.*, 41(2-3): 191-200. Illus. Includes checklist. (P)

5543 **Kim, T.W. (1977).** Plant inventory analysis of Pasoh Forest Reserve. *Bull. Seoul Nat. Univ. Forest*, 13: 55-65. Illus. (P)

5544 **Latiff, A. (1987).** Portraits of threatened plants: 13. *Pterisanthes pulchra* Ridl. *Malay. Nat.*, 41(2): 25-26. Illus. (T)

5545 **Latiff, A. (1989).** Portraits of threatened plants 19. *Ampelocissus floccosa* (Ridl.) Galet. *Malayan Naturalist*, 42(2-3): 3-5. Illus. (T)

5546 **Malayan Nature Society (1986).** Endau-Rompin expedition. *Malay. Naturalist*, 40(2): 44p. Illus., col. illus., maps. Includes review of the Malaysian Heritage and Scientific Expedition 1985 and various articles on botany of Endau-Rompin. (P)

5547 **Malayan Nature Society (1987).** The Malaysian Heritage and Scientific Expedition: Endau-Rompin 1985-1986. *Malay. Nat. J.*, 41(2-3): 446p. Illus., maps. Results of expedition, including papers on flora, fauna, geology. (P)

5548 **Malayan Nature Society (1988).** Proposed Endau-Rompin National Park, Malaysia. *Tigerpaper*, 15(3): 17. (P)

5549 **Marshall, A.G. (1973).** Conservation in west Malaysia: the potential for international cooperation. *Biol. Conserv.*, 5: 133, 138.

MALAYSIA - Peninsular Malaysia

5550 **Meijer, W. (1983).** *Rafflesia* rediscovered. *Malay. Naturalist*, 36(4): 21-27. Col. illus., map. (T)

5551 **Mohamad, B.B. (1987).** Management of conservation areas in Peninsular Malaysia: problems and prospects. *In* Santiapillai, C., Ashby, K.R., *eds.* Proceedings of the Symposium on the Conservation and Management of Endangered Plants and Animals. Bogor, Indonesia, 18-20 June 1986. Bogor, SEAMEO-BIOTROP. 121-130. (P/T)

5552 **Ng, F. (1984).** Portraits of threatened plants, 6. *Musa gracilis* Holttum. *Malay. Naturalist*, 38(1): 9-10. Illus. (T)

5553 **Ng, F.S.P. (1979).** Gunung Jerai - mountain resort of Kedah. *Nat. Malay.*, 3(4): 4-9. Col. illus.

5554 **Ng, F.S.P. (1984).** Portraits of threatened plants, 7. *Maingaya malayana* Oliv. *Malay. Naturalist*, 38(2): 6. Illus. (T)

5555 **Ng, F.S.P., Low, C.M. (1982).** Check list of endemic trees of the Malay Peninsula. Kepong, Forest Research Institute. 94p. Lists 654 trees endemic to the Malay Peninsula, including Peninsular Thailand; rarity is based on numbers of herbarium specimens. (T)

5556 **Parris, B.S., Edwards, P.J. (1987).** A provisional checklist of ferns and fern allies in Taman Negara, Peninsular Malaysia. *Malay. Naturalist*, 41(2): 5-10. (P)

5557 **Pong, T.Y. (1979).** The Sungei Menyala Forest Reserve, Negeri Sembilan - a conservation project for the MNS? *Malay. Naturalist*, Mar-Jun.: 2-3. Map. (P)

5558 **Putz, F.E. (1978).** A survey of virgin jungle reserves in Peninsular Malaysia. Research Pamphlet No. 73. Malaysia, Forest Research Institute. iv, 87p. Maps. (P)

5559 **Richards, P.W. (1978).** Pasoh (Forest Reserve) in perspective (Malaysia). *Malay. Nature J.*, 30(2): 145-148. (P)

5560 **Rubeli, K. (1979).** Taman Negara. *Tigerpaper*, 6(4): 6. (P)

5561 **Soepadmo, E. (1983).** Forest and man: an ecological appraisal. Kuala Lumpur, University of Malaya. 43p. Maps. Tropical forest. Deforestation. Includes map of existing forest areas. (E)

5562 **Srivastava, P.B.L., Sani Bin Shaffie, A. (1979).** Effect of final felling on natural regeneration in *Rhizophora* dominated forests of Matang Mangrove Reserve. *Pertanika*, 2(1): 34-42. Illus., map. (P)

5563 **Steenis, C.G.G.J. van (1969).** The urgent need of permanent forest reserves in Malaya. *Flora Males. Bull.*, 23: 1694-1697. Tropical forest. (P)

5564 **WWF-Malaysia (1987).** Park saved from road and gets a plan. *Oryx*, 21(2): 72-73. (P)

5565 **Weber, A. (1988).** Portraits of threatened plants 14. *Phyllagathis stonei* A.Weber. *Malay. Naturalist*, 41(3-4): 4-5. Illus. (T)

5566 **Weber, A. (1988).** Portraits of threatened plants 15. *Phyllagathis magnifica* A.Weber. *Malay. Naturalist*, 41(3-4): 5. Illus. (T)

5567 **Wong, K.M. (1987).** The bamboos of the Ulu Endau area, Johore, Malaysia. *Malay. Nature J.*, 41(2-3): 249-256. Illus., maps. Includes section on conservation status of *Racemobambos*. (P/T)

5568 **Wong, K.M., Saw, L.G., Kochummen, K.M. (1987).** A survey of the forests of the Endau-Rompin area, Peninsular Malaysia: principal forest types and floristic notes. *Malay. Nature J.*, 41(2-3): 125-144. Illus., map. (P)

MALAYSIA - Peninsular Malaysia

5569 **World Conservation Monitoring Centre (1988).** Peninsular Malaysia: conservation of biological diversity. Cambridge, World Conservation Monitoring Centre and Gland, IUCN. 23p. Briefing document prepared by the IUCN Tropical Forest Programme. Includes summary of vegetation; deforestation. Maps. (P)

5570 **Yussof, S., Kiew, R., Lim, W.H., Ibrahim, N. (1987).** A preliminary checklist of orchids from Ulu Endau, Johore, Malaysia. *Malay. Nature J.*, 41(2-3): 235-237. (P)

MALAYSIA - Sabah

5571 **Bacon, A.J. (1986).** The paphiopedilums of Sabah and the problems of conservation. *In* Rao, A.N., *ed.* Proceedings of the 5th ASEAN Orchid Congress Seminar, Singapore, 1-3 August 1984. Singapore Parks and Recreation Department, Ministry of National Development. 132-133. (T)

5572 **Bailes, C. (1985).** Kinabalu - mountain of contrasts. *Kew Mag.*, 2(2): 273-284. Illus., maps. (P)

5573 **Bailes, C.P. (1984).** Orchids of Borneo and their conservation. *Malay. Orchid Rev.*, 18: 20-27. Illus., col. illus. (T)

5574 **Beaman, J.H., Adam, J. (1983).** Observations on *Rafflesia* in Sabah. *Sabah Society J.*, 7(3): 208-212. Col. illus. (T)

5575 **Beaman, J.H., Regalado, J.C., jr (1988).** Checklist of vascular plants of the Bukit Hampuan area at the southeast base of Mount Kinabalu. East Lansing, Dept of Botany and Plant Pathology, Michigan State Univ. 21p. Illus. (P)

5576 **Beaman, R.S. & J.H., Marsh, C.W., Woods, P.V. (1985).** Drought and forest fires in Sabah in 1983. *Sabah Society J.*, 8(1): 10-30. Illus., col. illus., maps. 1 million hectares of tropical forest were burned of which 85% was previously logged-over.

5577 **Burrough, J.B. (1978).** Cabbages, conservation and copper. *In* Luping, D.M., Wen, C., Dingley, E.R., *eds.* Kinabalu summit of Borneo. Kota Kinabalu, Sabah, The Sabah Society. 75-87. Illus. (P)

5578 **Chim, L.T. (1979).** Problems relating to studies on natural regeneration and afforestation in rain forests of Sabah. *Trop. Agric. Res. Ser.*, 12: 25-30.

5579 **Cockburn, P.F. (1973).** Further case studies in selecting and allocating land for nature conservation: Sabah, virgin jungle reserves and other conservation areas. *In* Costin, A.B., Groves, K.H., *eds.* Nature conservation in the Pacific. Proc. Symposium A-10, XII Pacific Science Congress, Canberra, August-September 1971. IUCN and Australian Nat. Univ. Press. 72-81. Map. (P)

5580 **Fah, L.Y. (1987).** A preliminary survey of *Mangifera* species in Sabah. Sandakan, Forest Research Centre. Project No. 3305/MAL 75. (E/P/T)

5581 **Fowlie, J.A. (1985).** Malaya revisited, 29: Rediscovering the habitat of *Paphiopedilum dayanum* on serpentine cliffs of Mount Kinabalu in eastern Malaysia (formerly north Borneo). *The Orchid Digest*, 49(4): 124-129. Illus., maps. (T)

5582 **Grell, E., Schmude, N.F. Haas-von, Lamb, A., Bacon, A. (1988).** Re-introducing *Paphiopedilum rothschildianum* to Sabah, north Borneo. *Amer. Orchid Soc. Bull.*, 57(11): 1238-1246. Illus. (T)

5583 **Hepburn, A.J., Lamb, A. (1983).** *Glenniea philippinensis*, Sapindaceae: a rare fruit tree from Tenom, Sabah. *Malay. Naturalist*, 36(3): 9-10. Illus. (T)

5584 **Ismail, G. (1988).** Conservation of the giant *Rafflesia* in Sabah, Malaysia. *Trends Ecol. Evol.*, 3(12): 316-317. (T)

5585 **Jacobson, S.K. (1989).** Mass media and park management: reaching the public beyond the park boundary. *The Environmentalist*, 9(2): 131-137. Illus., map. Refers to media coverage of conservation problems of Kinabalu Park. (P)

MALAYSIA - Sabah

5586 **Jones, D.T. (1987).** Rare plant profile no. 1: *Merrillia caloxylon* (Rutaceae). *Bot. Gard. Conserv. News*, 1(1): 38-42. Illus., map. (E/G/T)

5587 **Kiew, B.H. (1976).** A survey of the proposed Sungai Danum National Park, Sabah. 37p. Maps. Includes checklist of plants. (P)

5588 **Kiew, R. (1985).** Illegal plant collection in Malaya. *Malay. Naturalist*, 38(3): 44-45. Illus. An international syndicate preys on rare plants especially *Nepenthes rajah*, now rare in Kinabalu Park. (P/T)

5589 **Luping, D.M., Wen, C., Dingley, E.R. (1978).** Kinabalu: summit of Borneo. Sabah Society Monograph 1978. Kota Kinabalu, Sabah, The Sabah Society. 486p. Maps. Illus., col. illus. (P/T)

5590 **Phillipps, A. (1988).** A guide to the parks of Sabah. Sabah Parks Publication No. 9. Kota Kinabalu, Sabah, Sabah Parks Trustees. 58p. Col. illus., maps. Popular guide. (P)

5591 **Phillipps, C. (1984).** Current status of mangrove exploitation, management and conservation in Sabah. *In* Proc. As. Symp. Mangr. Env. Res. & Manag., 1984. 809-820. Maps. Mangrove. (P)

MALAYSIA - Sarawak

5592 **Anderson, J.A.R., Jermy, A.C., Cranbrook, Earl of (1983).** Gunung Mulu National Park: a management and development plan. London, Royal Geographical Society. viii, 345p. Maps. (P)

5593 **Bailes, C.P. (1984).** Orchids of Borneo and their conservation. *Malay. Orchid Rev.*, 18: 20-27. Illus., col. illus. (T)

5594 **Chai, P.P.K. (1978).** Conservation in Sarawak. *Tigerpaper*, 5(4): 102.

5595 **Chai, P.P.K. (1979).** Bako National Park. *Nat. Malaysiana*, 4(4): 4-11. Col. illus., map. (P)

5596 **Chai, P.P.K., Choo, N.C. (1983).** Conservation of forest genetic resources in Malaysia with special reference to Sarawak. *In* Jain, S.K., Mehra, K.L., *eds*. Conservation of tropical plant resources. Howrah, Botanical Survey of India. 39-47. (E)

5597 **Collins, N.M., Holloway, J.D., Proctor, J. (1984).** Notes on the ascent and natural history of Gunung Api, a limestone mountain in Sarawak. *Sarawak Mus. J.*, 33(54): 219-234. Illus., map. Includes checklist of plants. (P)

5598 **Hanbury-Tenison, A.R., Jermy, A.C. (1979).** The RGS expedition to Gunong Mulu, Sarawak, 1977-78. *Geogr. J.*, 145(2): 175-191. Maps. (P)

5599 **Hanbury-Tenison, R. (1980).** Mulu: the rain forest. London, Weidenfeld and Nicolson. xi, 179p. Illus., map. (P)

5600 **Heang, K.B., Kiew, R. (1982).** The Samunsam Wildlife Sanctuary: home of the Proboscis monkey. *Nat. Malaysiana*, 7(1): 4-11. Col. illus. (P)

5601 **Kavanagh, M. (1985).** Parks and sanctuaries planning in Sarawak. *In* Thorsell, J.W., *ed*. Conserving Asia's natural heritage. The planning and management of protected areas in the Indomalayan Realm. Proc. of the 25th working session of IUCN's CNPPA. Gland, IUCN. 86-97. Map. (P)

5602 **Kiew, R. (1978).** Floristic components of the ground flora of a tropical lowland rain forest at Gunung Mulu National Park, Sarawak. *Pertanika*, 1(2): 112-119. (P)

5603 **Morshidi, A.H.K. (1977).** The development of national parks in Sarawak. *Malay. Forest.*, 40(3): 138-143. Map. (P)

5604 **Wood, J.J. (1984).** New orchids from Gunung Mulu National Park, Sarawak. *Kew Bull.*, 39(1): 73-98. Illus. (P)

MALAYSIA - Sarawak

5605 **World Conservation Monitoring Centre (1988).** Sabah and Sarawak: conservation of biological diversity. Cambridge, World Conservation Monitoring Centre and Gland, IUCN. 25p. Briefing document prepared by the IUCN Tropical Forest Programme. Includes summary of vegetation; deforestation. Maps. (P)

MALI

5606 **Anon. (1983).** La reserve de la biosphere "Boucle de Baoule" (Mali). *In*: L'Amenagement et la gestion des reserves de la biosphere en Afrique Soudano-Sahelienne; 14-84. (P)

5607 **Boudet, G. (1979).** Quelques observations sur les fluctuations du couvert vegetal sahelien au Gourma malien et leurs consequences pour une strategie de gestion sylvopastorale. *Bois Forets Trop.*, 184: 31-44. Illus.

5608 **Geerling, C. (1984).** The Boucle du Baoule, Mali. *In* Unesco-UNEP. Conservation, science and society. Paris, Unesco. 178-184. Contributions to the First International Biosphere Reserve Congress, Minsk, Byelorussia/USSR, 26 September - 2 October, 1983. (P)

5609 **Jaeger, P. (1968).** Mali. *In* Hedberg, I. & O., *eds.* Conservation of vegetation in Africa south of the Sahara. Symposium Proceedings, at 6th plenary meeting of AETFAT, Uppsala. *Acta Phytogeogr. Suec.*, 54: 51-53. Fr. Describes vegetation zones, plant sites and species needing protection. (T)

5610 **Kabala, M.M. (1983).** Le concept des reserves de la biosphere. *In* L'Amenagement et la Gestion des Reserves de la Biosphere en Afrique Soudano-Sahelienne: 7-13. Fr. (P)

5611 **Republique de Mali. Comite National Malien du Programme sur l'homme et la biosphere. (1983).** L'amenagement et la gestion des reserves de la biosphere en Afrique. Soudano-Sahelienne. Rapport de la session de formation sur: Programme sur l'homme et la biosphere, Unesco. 172p. Fr. Maps. (L/P)

5612 **Warshall, P. (1989).** Mali biological diversity assessment. Washington, D.C., Agency for International Development. 95p. (T)

MALTA

5613 **Anon. (1982).** Maltese succulents and conservation. *Kakti u Sukkulenti Ohra*, 24: 13-15. (T)

5614 **Haslam, S.M., Sell, P.D., Wolseley, P.A. (1977).** A flora of the Maltese Islands. Msida, Malta University Press. lxxi, 560p. Includes conservation, lviii.

5615 **Lanfranco, E. (1981).** Suggestions on the conservation of the unique flora associated with the Gozo Citadel. *Soc. Stud. Cons. Nat.* 3p.

5616 **Lanfranco, E. (1982).** Maltese succulents and conservation. *Kakti u Sukkulenti Ohra*, 24: 13-15.

5617 **Lanfranco, E. (1983).** Safeguarding our environment. 3. Malta's plant life. *The Teacher*, (3): 23. Illus. (T)

5618 **Schembri, P.J., Lanfranco, E., Farrugia, P., Schembri, S., Sultana, J. (1987).** Localities with conservation value in the Maltese Islands. Environment Division, Ministry of Education. 27p. (P/T)

5619 **Schembri, P.J., Sultana, J., eds (1989).** Red Data Book for the Maltese Islands. Malta, Department of Information. 142p. Col. illus. (T)

MARSHALL ISLANDS

5620 **Anderson, J.A. (1972).** Return to Eniwetok. *Micronesian Reporter*, 20(3): 28-32. Describes impact on vegetation of 1954 explosion of hydrogen bomb.

MARSHALL ISLANDS

5621 **Chamberlain, P. (1972).** Micro planning. *Micronesian Reporter*, 20(2): 33-43.

5622 **Fosberg, F.R. (1956).** Vegetation. *In* H.Q. US Army Forces Far East. Military Geography of the Northern Marshalls. 185-220.

5623 **Fosberg, F.R. (1961).** Typhoon effects on individual species of plants. *In* Blumenstock, D.I., *ed.* A report on typhoon effects upon Jaluit Atoll. *Atoll Res. Bull.*, 75: 1-105.

5624 **U.S. Congress, Office of Technology Assessment (1987).** Integrated renewable resource management for U.S. insular areas. Washington, D.C., U.S. Govt Printing Office. 443p. Illus., maps. OTA-F-325. Includes description of vegetation, remaining coverage, threatened species. (P/T)

MARTINIQUE

5625 **Anon. (1979).** Martinique and its Parc Naturel Regional. *Caribbean Conserv. News*, 1(16): 20-22. (P)

5626 **Fiard, J.P., Association des Amis du Parc Naturel Regional (1979).** La foret martiniquaise: presentation et propositions de mesures de protection. Fort de France, Parc Naturel Regional Ex-Caserne Bouille. 70p. Fr. Illus. maps. (P)

5627 **Portecop, J. (1979).** Phytogeographie, cartographie ecologique et amenagement dans une ile tropicale; le cas de la Martinique. *Doc. Cart. Ecol. Univ. Grenoble*, 21: 1-78. Fr. Map, 1:75,000.

5628 **Sastre, C. (1978).** Plantes menacees de Guadeloupe et de Martinique. 1. Especes altitudinales. *Bull. Mus. Natn. Hist. Nat.*, Paris, 3e ser., 519: 65-93. Fr (En). Description of vegetation, sheets on 13 rare and threatened species with illustrations and habitat photographs. (T)

5629 **Sastre, C., Mestoret, L. (1978).** Plantes rares ou menacees de Martinique. *Le courrier du parc naturel regional de la Martinique*, 2: 20-22. Fr. (T)

MAURITANIA

5630 **Adam, J.G. (1968).** La Mauritanie. *In* Hedberg, I. & O., *eds.* Conservation of vegetation in Africa south of the Sahara. Symposium Proceedings, at 6th plenary meeting of AETFAT, Uppsala. *Acta Phytogeogr. Suec.*, 54: 49-51. Fr. Map. Describes vegetation types, protected plants, communities andplant species needing protection. (L/T)

5631 **Hedberg, I. & O., eds. (1968).** Conservation of vegetation in Africa south of the Sahara. *Acta Phytogeogr. Suec.* (Uppsala), 54: 320p. Col. illus., maps. Proceedings of a symposium held at the 6th Plenary meeting of the "Association pour l'Etude Taxonomique de la Flore d'Afrique Tropicale" (AETFAT) in Uppsala, Sept. 12th-16th, 1966. (P/T)

MAURITIUS

5632 **Anon. (1988).** Recovery on Round Island. *Species*, 10: 20. Recovery of vegetation after rabbit eradication; introduced weeds present threat. (T)

5633 **Anon. (1988).** Round Island rebound. *Oryx*, 22(1): 52. Full report in *World Birdwatch*, 9(3). (T)

5634 **Anon. (1989).** Rare plants return to Mauritius. *Bot. Gard. Conserv. News*, 1(4): 5. (G/T)

5635 **Baumer, M. (1978).** La conservation et la valorisation des resources ecologiques dans les iles des Comores, de Maurice, de la Reunion, des Seychelles. Paris, Agence de Cooperation Culturelle et Technique. 92p. Fr.

MAURITIUS

5636 **Bullock, D., North, S. (1984).** Round Island in 1982. *Oryx*, 18: 36-41. Illus. 20km off the north coast of Mauritius. Vegetation changes in the previous seven years, since goats and rabbits were eradicated. (T)

5637 **Bullock, D., North, S., Greig, S. (1983).** Round island expedition 1982. Final report. St Andrews, Department of Botany, University of St Andrews. ii, 98p. Illus., maps. 20 km off north coast of Mauritius.

5638 **Cabanis, Y., Chabouis, L. & F. (1969).** Vegetaux et groupements vegetaux de Madagascar et des Mascareignes, 4 vols. Tananarive, Bureau Pour le Developpement de la Production Agricola. Fr. Illus., maps. Published 1969-1970.

5639 **Cadet, L.J.T. (1984).** Plantes rares ou remarquables des Mascareignes. Paris, Agence de Cooperation Culturelle et Technique. 132p. Fr. Illus., col. illus. (E/T)

5640 **Cadet, Th. (1984).** Liste et commentaires sur les plantes en danger des Mascareignes. Paris, Conservatoire et Jardins Botaniques de Nancy. Mimeo. 20p. Fr. Includes notes on local uses. (E/T)

5641 **Douglas, G. (1987).** Embryo culture of *Hyophorbe amaricaulis* from Mauritius. *Bot. Gard. Conserv. News*, 1(1): 46. Illus. (G/T)

5642 **Friedman, F. (1988).** Fleurs rares des iles Mascareignes. L'ile aux images, Ile Maurice. 31p. Illus. (T)

5643 **Lesouef, J.-Y. (1983).** Compte-rendu de la ire mission de sauvetage des elements les plus menaces de la flore des Mascareignes (La Reunion, Maurice, Rodrigues). Brest, WWF-France and Conservatoire Botanique du Stangalarc'h. 46, 15p. Illus. (T)

5644 **Lorence, D.H., Sussman, R.W. (1988).** Diversity, density, and invasion in a Mauritian wet forest. *Monogr. Syst. Bot. Missouri Bot. Gard.*, 25: 187-204. Illus. Discusses threat of invasive plants to native wet forests.

5645 **Owadally, A.W. (1979).** Possibilities and needs for conservation of plant species and vegetation in Africa. Appendix: preliminary lists of rare and threatened species in African countries. Mauritius and Rodrigues. *In* Hedberg, I., *ed.* Systematic botany, plant utilization and biosphere conservation. Stockholm, Almqvist & Wiksell International. 103. Contains 34 species: E:12, V:2, R:18, I:2. (T)

5646 **Parnell, J., Jackson, P.W., Cronk, Q. (1986).** A paradise about to be lost. *New Sci.*, 112(1528): 44-47. Col. illus. (T)

5647 **Parnell, J.A.N., Cronk, Q., Jackson, P. Wyse, Strahm, W. (1989).** A study of the ecological history, vegetation and conservation management of Ile aux Aigrettes, Mauritius. *J. Trop. Ecol.*, 5: 355-374. Maps. (P/T)

5648 **Strahm, W. (1984).** Ile aux Aigrettes. *Threatened Pl. Newsl.*, 13: 8-9. (P/T)

5649 **Strahm, W. (1985).** Mauritius success with rare plant propagation. *WWF Monthly Report*, Nov.: 261-263.

5650 **Strahm, W. (1985).** Preserving rare plants in Mauritius. *WWF Monthly Rep.*, June: 127-130. (T)

5651 **Strahm, W., Nicoll, M. (1986).** Mauritian palm - saved? Postcript. *Threatened Pl. Newsl.*, 16: 13. (T)

5652 **Synge, H. (1989).** Komt de cafeinevrije koffie straks van Mauritius? *Panda*, 11: 11-13. Du. Col. illus., map. (T)

5653 **Tirvengadum, D.D. (1980).** On the possible extinction of *"Randia" heterophylla* a Rubiaceae of great taxonomic interest from Rodrigues Island. *Mauritius Inst. Bull.*, 9(1): 1-21. (Fr). Illus., map. (T)

MAURITIUS

5654 **Vaughan, R.E. (1968).** Mauritius and Rodriguez. *In* Hedberg, I. & O., *eds.* Conservation of vegetation in Africa south of the Sahara. Symposium Proceedings, at 6th plenary meeting of AETFAT, Uppsala. *Acta Phytogeogr. Suec.,* 54: 265-272. Illus., map, tables. Discusses national reserves and conservation of species in Mauritius, and rare endemics in Rodriguez, plus ownership and management. (P/T)

5655 **Vinson, J. (1964).** Sur la disparition progressive de la flore et de la faune de l'Ile Ronde. *Proc. Roy. Soc. Arts Sci. Mauritius,* 2: 247-261. Fr.

5656 **Wyse Jackson, P. (1987).** Rare Mauritian plant and animal link-up. *Bot. Gard. Conserv. News,* 1(1): 7. Report on educational display at Jersey Zoo. (G/T)

5657 **Wyse Jackson, P., Cronk, Q., Parnell, J. (1986).** Mauritian palm - saved? *Threatened Pl. Newsl.,* 16: 12-13. Illus. (T)

5658 **Wyse Jackson, P., Cronk, Q., Parnell, J., Strahm, W. (1987).** Rare plant propagation in Mauritius. *In* Bramwell, D., Hamann, O., Heywood, V., Synge, H., *eds.* Botanic gardens and the World Conservation Strategy. London, Academic Press. 353-355. (G/T)

5659 **Wyse Jackson, P., Parnell, J.A.N., Cronk, Q.C.B. (1986).** Report of the Dublin/Cambridge expedition to Mauritius for plant conservation, 1985. Dublin. 12p. (T)

5660 **Wyse Jackson, P.S., Cronk, Q., Parnell, J.A.N. (1988).** Notes on the regeneration of two rare Mauritian endemic trees. *Tropical Ecology,* 29: 98-106. Illus. (T)

5661 **Wyse Jackson, P.S., Parnell, J.A.N., Cronk, Q.C.B., Strahm, W. (1986).** Propagation of endangered Mauritian plants for conservation. Report for IUCN. 16p. (G/T)

5662 **Wyse Jackson, P.S., Strahm, W., Cronk, Q.C.B., Parnell, J.A.N. (1988).** The propagation of endangered plants in Mauritius. *Moorea,* 7: 35-45. Map. (G/T)

MAURITIUS - Rodrigues

5663 **Anon. (1989).** Rare plants return to Mauritius. *Bot. Gard. Conserv. News,* 1(4): 5. (G/T)

5664 **Beyer, R.I. (1988).** Propagating *Ramosmania heterophylla* (Rubiaceae) at Kew, one of the world's rarest plants. *Bot. Gard. Conserv. News,* 1(3): 40-41. Illus. (G/T)

5665 **Cabanis, Y., Chabouis, L. & F. (1969).** Vegetaux et groupements vegetaux de Madagascar et des Mascareignes, 4 vols. Tananarive, Bureau Pour le Developpement de la Production Agricola. Fr. Illus., maps. Published 1969-1970.

5666 **Cadet, Th. (1984).** Liste et commentaires sur les plantes en danger des Mascareignes. Paris, Conservatoire et Jardins Botaniques de Nancy. Mimeo. 20p. Fr. Includes notes on local uses. (E/T)

5667 **Friedman, F. (1988).** Fleurs rares des iles Mascareignes. L'ile aux images, Ile Maurice. 31p. Illus. (T)

5668 **Gade, D., (1985).** Man and nature on Rodrigues: tragedy of an island common. *Envir. Conserv.,* 12(3): 207-216.

5669 **IUCN (1986).** IUCN's list of top 24 threatened animals and plants. *Biol. Conserv. Newsl.,* 44: 1-6. (T)

5670 **Lesouef, J.-Y. (1983).** Compte-rendu de la ire mission de sauvetage des elements les plus menaces de la flore des Mascareignes (La Reunion, Maurice, Rodrigues). Brest, WWF-France and Conservatoire Botanique du Stangalarc'h. 46, 15p. Illus. (T)

5671 **McHugh, D. (1986).** World's rarest tree threatened. *Endangered Species Tech. Bull.,* 3(8): 3. (T)

MAURITIUS - Rodrigues

5672 **Owadally, A.W. (1979).** Possibilities and needs for conservation of plant species and vegetation in Africa. Appendix: preliminary lists of rare and threatened species in African countries. Mauritius and Rodrigues. *In* Hedberg, I., *ed.* Systematic botany, plant utilization and biosphere conservation. Stockholm, Almqvist & Wiksell International. 103. Contains 34 species: E:12, V:2, R:18, I:2. (T)

5673 **Strahm, W. (1989).** Plant Red Data Book for Rodrigues. Konigstein, Koeltz Scientific Books. Includes detailed accounts of all endemics and threatened non-endemic species, incl. conservation categories; introduction on geography, botany and conservation problems. (T)

5674 **Tirvengadum, D.D. (1983).** L'exemple de l'Ile Rodrigues (Mascareignes) destruction d'une flore insulaire. *Compt. Rend. Seanc. Soc. Biogeogr.*, 59(2): 213-222. Fr (En). (T)

5675 **Vaughan, R.E. (1968).** Mauritius and Rodriguez. *In* Hedberg, I. & O., *eds.* Conservation of vegetation in Africa south of the Sahara. Symposium Proceedings, at 6th plenary meeting of AETFAT, Uppsala. *Acta Phytogeogr. Suec.*, 54: 265-272. Illus., map, tables. Discusses national reserves and conservation of species in Mauritius, and rare endemics in Rodriguez, plus ownership and management. (P/T)

MEXICO

5676 **Anderson, E.F. (1982).** A meeting on the cactus trade. *Cact. Succ. J.* (U.K.), 54(2): 82-85. Illus. (E/T)

5677 **Anderson, E.F. (1987).** Black magic in cactus country. *Garden* (U.S.), 11(5): 2-5, 32. Col. illus. (T)

5678 **Anon. (1979).** Especies en peligro de extincion. *Macpalxochitl*, 79: 3-4. Sp. (T)

5679 **Anon. (1986).** Mexican cacti exports decline. *Traffic (USA)*, 6(4): 12-13. (E/T)

5680 **Bolderl, R. (1984).** [A visit to the site of *Strombocactus disciformis* (De Candolle) Britton & Rose and *Lophophora diffusa* (Croizat) H. Bravo]. *Kakt. And. Sukk.*, 35(8): 177. Ge. Illus. (T)

5681 **Branigin, W. (1989).** North America's largest rain forest faces destruction. *Washington Post*, July 17: A17-A19.

5682 **Brumback, W. (1983).** Endangered species. *Amer. Assoc. Bot. Gard. Arbor. Inc. Newsl.*, 106: 5. (L/T)

5683 **Bussey, W. (1989).** Salvaging orchids in Mexico. *Amer. Orchid Soc. Bull.*, 58(1): 36-41. Col. illus. (T)

5684 **Calderon de Rzedowski, G. (1987).** *Tigridia martinezii* una especie nueva de iridaceas del estado de Hidalgo (Mexico). *Bol. Soc. Bot. Mexico*, 47: 3-6. (T)

5685 **California Native Plant Society (1982).** Endangered Species Act. *Calif. Native Pl. Soc. Bull.*, 12(3): 4. *Amsinka grandiflora, Camissonia benitensis*. (L/T)

5686 **Campbell, F. (1986).** Mexican cacti exports decline. *TRAFFIC* (U.S.A.), 6(4): 12-13. Trade. (E/T)

5687 **Campbell, F.T. (1983).** Controlling the trade in plants: a progress report. *Garden* (New York), 7(4): 2-5, 32. Illus. Effectiveness of CITES. (E/T)

5688 **Campbell, F.T. (1983).** Mexican cacti: a case study in plant trade problems. *TRAFFIC* (U.S.A.), 5(1): 12. (L)

5689 **Cornett, J.W. (1985).** Reading the fan palms - we know where these trees grow, but we're not sure how they got there. *Natural History*, 94(10): 64-73. Illus. (T)

MEXICO

5690 **Corona Nava Esparza, V., Yanez L., L. (1983).** Estudia de dos problaciones de *Cephalocereus senilis* en la Barranca de Metztitlan, Hgo. *Cact. Suc. Mex.*, 28(4): 75-80. Sp (En). (T)

5691 **Dummler, H. (1977).** Ein Reisebericht: vom 12. Internationalen Seminar uber Nationalparke und gleichwertige Schutzgebiete com 2. August bis 1. September 1977 in Kanada, USA und Mexico. *Naturschutz- und Naturparke*, 87: 1-8. Ge. Illus., map. (P)

5692 **Dunn, A.T. (1987).** Population dynamics of the Tecate cypress. *In* Elias, T.S., *ed.* Conservation and management of rare and endangered plants. Proceedings from a conference, Sacramento, California, 5-8 November 1986. Sacramento, California Native Plant Society. 367-376. Maps. *Cupressus quadalupensis* var. *forbesii* occurs with a number of widespread and rare chaparral species. Frequent fires could threaten its (T)

5693 **Estrada, A., Coates-Estrada, R. (1983).** Rain forest in Mexico: research and conservation at Los Tuxtlas. *Oryx*, 17: 201-204. Illus., map.

5694 **Ewell, P.T., Poleman, T.T. (1980).** Uxpanapa: reacomodo y desarrollo agricola en al Mexicana. 282p. Veracruz, Xalapa, Instituto nacional de investigaciones sobre recursos bioticos. Translated from English edition under the title: Uxpanapa: agricultural development in the Mexican tropics. Maps.

5695 **Ezcurra, E. (1984).** Planning a system of biosphere reserves. *In* Unesco-UNEP. Conservation, science and society. Paris, Unesco. 85-92. Contributions to the First International Biosphere Reserve Congress, Minsk, Byelorussia/USSR, 26 September - 2 October 1983. (P)

5696 **Ffolliott, P.F., Halffter, G. (1980).** Social and environmental consequences of natural resources policies, with special emphasis on biosphere reserves. *In* USDA Forest Service. Proceedings of the International Seminar, 8-13 April, 1980, Durango, Mexico. General technical report. Fort Collins, Rocky Mountain Forest and Range Experiment Station. 56p. (P)

5697 **Fundacao Zoobotanica do Rio Grande do Sul (1977).** Relatorio 1976. Porto Alegre, Fundacao Zoobotanica do Rio Grande do Sul. 64p. Por. Illus.

5698 **Hagsater, E. (1976).** Orchids and conservation in Mexico. *Orch. Rev.*, 84: 39-42. (T)

5699 **Hagsater, E. (1982).** *Encyclia kienastii*, una especie en peligro. (*Encyclia kienastii*, an endangered species.) *Orquidea* (Mexico), 8(2): 355-362. Sp (En). Illus.; col. illus. (T)

5700 **Halffter, G. (1981).** The Mapimi Biosphere Reserve: local participation in conservation and development. *Ambio*, 10(2-3): 93-96. Illus., maps. (P)

5701 **Halffter, G. (1984).** Biosphere reserves: the conservation of nature for man. *In* Unesco-UNEP. Conservation, science and society. Paris, Unesco. 450-457. Contributions to the First International Biosphere Reserve Congress, Minsk, Byelorussia/USSR, 26 September - 2 October 1983. (P)

5702 **Hendrickson, D.A., Minckley, W.L. (1985).** Cienegas: vanishing climax communities of the American Southwest. *Desert Plants*, 6(3): 131-176. Illus., maps. Wetlands.

5703 **Howard, T.M. (1981).** Current status of some endangered Mexican *Hymenocallis* species. *Pl. Life*, 37(1-4): 157-158. (T)

5704 **Hunt, D.R. (1982).** The conservation status of Mexican mammillarias: a preliminary assessment. *Cact. Succ. J.* (U.K.), 44(4): 87-88. (T)

5705 **Hurtado, J.A.Z. (1982).** Estudios ecologicas en el valle semiarido de Zapotitlan, Puebla. il. Clasificacion numerica de la vegetacion basada en atributos binarios de presencia o ausencia de las especies. *Biotica*, 7(1): 99-117. Sp (En). Illus., maps. (T)

MEXICO

5706 **IUCN Threatened Plants Unit (1983).** Rare, threatened and insufficiently known endemic cacti of Mexico, with commonly used synonyms of Mexican endemic cacti. Kew, IUCN Botanic Gardens Conservation Co-ordinating Body Report No. 13. 9p. (T)

5707 **IUFRO (1979).** Reunion del grupo tematico S5.03 "Proteccion de la Madera" de la Union Internacional del Instituto de Investigacion Silvicola (IUFRO) organizada por INIREB-LACITEMA, en Xalapa, Ver., Mexico, Julio de 1978. *Biotica*, 4(4) Suppl. 11p. Sp.

5708 **Janka, H. (1981).** La economia forestal comunal: una alternativa para el tropico humdeo? (Communal forest economy: an alternative for the humid tropics?). Publicacion especial - Instituto Nacional de Investigaciones Forestales, Sept. 1981, 26. 55-64. Sp. Conservation of natural resources in forested areas.

5709 **Kurlansky, M. (1985).** Woman in love with a jungle. *Int. Wildlife*, 15(5): 34-39. Lacandon rainforest.

5710 **Lau, A.B. (1979).** *Mammillaria neopalmeri* Craig. *Cact. Succ. J.* (U.S.), 51(6): 282-283. Illus. (T)

5711 **Lau, A.B. (1985).** New varieties and forms of *Mammillaria glassii* Foster. *Cact. Succ. J.* (U.S.), 57(5): 196-198. Illus., maps. (T)

5712 **Luca, P. de, Sabato, S., Vazquez, T.M. (1982).** Distribution and variation of *Dioon edule* (Zamiaceae). *Brittonia*, 34(3): 355-362. Illus., map. (T)

5713 **McLaughlin, S. (1988).** Salvage, propagation and re-establishment of *Mammillaria thornberi*. Progress report. Tucson Water, Tucson, Arizona. 18p. (G/T)

5714 **Muller, H. (1982).** Kakteenfrevel in Mexiko. *Kakt. And. Sukk.*, 33(2): 42-43. Ge. Illus. (T)

5715 **Munoz Pizarro, C. (1975).** Especies vegetales que se extinguen en nuestro pais. *In* Capurro, L., Vergara, R., eds. Present y futuro del medio humano. Mexico, Edit. Cont. CECSA. Capitulo XI. 161-179. Sp. (T)

5716 **Murrieta, X. (1978).** [Exploitation of the floristic resources of the arid zones]. *Cienc. Interam.*, 19(1): 10-17. Sp. Illus. Sonoran Desert, Mexico and N. America.

5717 **Nabhan, G. (1980).** *Ammobroma sonorae*, an endangered parasitic plant in extremely arid North America. *Desert Pl.*, 2(3): 188-196. Illus., col. illus. (T)

5718 **Nabhan, G.P., Greenhouse, R., Hodgson, W. (1986).** At the edge of extinction: useful plants of the border states of the United States and Mexico. *Arnoldia*, 46(3): 36-46. Illus. (E)

5719 **Oldfield, S. (1982).** Conservation news. *Cact. Succ. J.* (U.K.), 44(3): 69. (L/T)

5720 **Orellana, R. (1988).** *Ex situ* studies on five threatened species on the Yucatan Peninsula, Mexico. *Bot. Gard. Conserv. News*, 1(2): 20-22. Illus., map. (G/T)

5721 **Pina, I. (1980).** Rare and threatened Agavaceae and Cactaceae of Mexico. Sociedad Mexicana Cactologia. (T)

5722 **Puig, H., Bracho, R., Sosa, V. (1987).** Affinites phytogeographiques de la foret tropicale humide de montagne de la reserve MAB "El Cielo" de Domez-Farais, Tamaulipas, Mexique. *Compt. Rend. Soc. Biogeogr.*, 63(4): 115-140. Fr (Sp). Includes extensive list of species. (P)

5723 **Rinetti, L. (1982).** Il Parco Nazionale "Iztaccihuatl-Popocatepetl" (Messico). *Nat. Montagna*, 29(3): 13-17. Sp. Illus, map. (P)

5724 **Rushforth, K. (1986).** Mexico's spruces - rare members of an important genus. *Kew Mag.*, 3(3): 119-124. Illus. (T)

MEXICO

5725 **Rzedowski, J. (1979).** Deterioro de la flora. *In* Memorias sobre Problemas Ambientales en Mexico. Instituto Politecnico Nacional, Escuela de Ciencias Biologicas. 51-57. Sp.

5726 **Sanchez-Mejorada, H. (1982).** Informe sobre la reunion de Tucson para analizar el comercio de cactaceas. *Cact. Suc. Mex.,* 27(4): 90-96. Sp (En). (L/T)

5727 **Sanchez-Mejorada, H. (1982).** Mexico's problems and programmes monitoring trade in common and endangered cacti. *Cact. Succ. J.* (U.K.), 44(2): 36-38. (L/T)

5728 **Sanchez-Mejorada, H. (1982).** Problemas en el control del comercio de las Cactaceas. *Cact. Suc. Mex.,* 27(2): 27-33. Sp (En). (T)

5729 **Sanchez-Mejorada, H. (1983).** Mexican developments in conservation since the 16th IOS Congress. *IOS Bull.,* 4(2): 63-64. International Organization for Succulent Plant Study. (L/P)

5730 **Sanchez-Mejorada, H. (1987).** Cultivation of rare cacti in Mexico. *In* Bramwell, D., Hamann, O., Heywood, V., Synge, H., *eds.* Botanic gardens and the World Conservation Strategy. London, Academic Press. 271-275. (Sp). (G/T)

5731 **Sanchez-Mejorada, H., Anderson, E.F., Taylor, N.P. & R. (1986).** Succulent plant conservation studies and training in Mexico: Stage 1, Part 1: May-June 1986. Assessment of individual species in northeastern Mexico and initial training of conservation specialists. WWF-U.S. 158p. (T)

5732 **Sanchez-Mejorada, R.H. (1987).** Observaciones sobre el estado de conservacion de doce especies de cac taceas amenazadas del noreste de Mexico. (Notes on conservation status of 12 threatened cacti species from NE Mexico.) *Cact. Suc. Mex.,* 32(3): 61-71. Sp (En). Illus. (T)

5733 **Schreier, K. (1983).** Sind die Tage der Kakteen Mexikos gezahlt? *Kakt. And. Sukk.,* 34(3): 66-68. Ge. (T)

5734 **Schutzman, B., Vovides, A., Dehgan, B. (1988).** Two new species of *Zamia* (Zamiaceae, Cycadales) from southern Mexico. *Bot. Gaz.,* 149(3): 347-360. Gives conservation details. (T)

5735 **Tangley, L. (1988).** A new era for biosphere reserves. *BioScience,* 38(3): 148-155. (P)

5736 **Tjaden, W. (1981).** Conservation? *Cact. Succ. J.* (U.K.), 36(2): 63. Cacti collecting in Mexico.

5737 **Toledo, V.M. (1985).** Criterios fitogeograficos para la conservacion de la flora de Mexico. *In* Gomez, L.D., *ed.* Memorias del Simposio de Biogeografia de Mesoamerica. Sp.

5738 **U.S. Department of the Interior. Fish and Wildlife Service (1983).** Endangered and threatened wildlife and plants: proposed rule to determine *Frankenia johnstonii* (Johnston's frankenia) to be an Endangered species. *Fed. Register,* 48(132): 31414-31417. (T)

5739 **U.S. Department of the Interior. Fish and Wildlife Service (1984).** Endangered and threatened wildlife and plants: final rule to determine *Frankenia johnstonii* (Johnston's frankenia) to be an Endangered species. *Fed. Register,* 49(153): 31418-31421. (L/T)

5740 **U.S. Department of the Interior. Fish and Wildlife Service (1984).** Endangered and threatened wildlife and plants: proposal to determine *Mammillaria thornberi* (Thornber's fishhook cactus) to be a Threatened species. *Fed. Register,* 49(80): 17551-17555. (T)

5741 **U.S. Department of the Interior. Fish and Wildlife Service (1984).** Endangered status for three western plants. *Endangered Species Tech. Bull.,* 9(9): 6-7. (L/T)

MEXICO

5742 **U.S. Department of the Interior. Fish and Wildlife Service (1985).** CITES news: Service announces proposals for changes in CITES Appendices. *Endangered Species Tech. Bull.*, 10(1): 11. Proposals by U.S., prepared by TRAFFIC (U.S.), to move *Ceratozamia* from Appendix II to Appendix I. (L/T)

5743 **U.S. Department of the Interior. Fish and Wildlife Service (1985).** Endangered and threatened wildlife and plants: proposal to determine *Coryphantha robbinsorum* (Cochise pincushion cactus) to be a Threatened species. *Fed. Register*, 50(44): 9083-9086. (L/T)

5744 **U.S. Department of the Interior. Fish and Wildlife Service (1985).** Endangered and threatened wildlife and plants: proposal to determine *Tumamoca macdougalii* to be an Endangered species. *Fed. Register*, 50(97): 20806-20810. (L/T)

5745 **U.S. Department of the Interior. Fish and Wildlife Service (1985).** Four western species proposed for protection. *Endangered Species Tech. Bull.*, 10(6): 3-6. (T)

5746 **U.S. Department of the Interior. Fish and Wildlife Service (1986).** Endangered and threatened wildlife and plants: determination of *Tumamoca macdougalii* to be Endangered. *Fed. Register*, 51(82): 15906-15911. (L/T)

5747 **U.S. Department of the Interior. Fish and Wildlife Service (1986).** Endangered and threatened wildlife and plants: determination of Threatened status for *Coryphantha robbinsorum. Fed. Register*, 51(6): 952-956. (L/T)

5748 **Vovides, A.P. (1986).** Trade and habitat destruction threaten Mexican cycads. *TRAFFIC* (U.S.A.), 6(4): 13. (P/T)

5749 **Vovides, A.P., Gomez-Pompa, A. (1977).** The problems of threatened and endangered plant species of Mexico. *In* Prance, G.T., Elias, T.S., *eds.* Extinction is forever. Proceedings of a symposium at the New York Botanical Garden, 11-13 May 1976. New York, New York Botanical Garden. 77-88. Illus., map, graphs. Vegetational history, endangered ecosystems, biosphere reserves and present work on endangered species. (L/P/T)

5750 **WWF (1986).** Succulent plant conservation studies and training in Mexico: stage 1, part 1: May-June 1986. Assessment of individual species in northeastern Mexico and initial training of conservation specialists. WWF-U.S. 106p. Map. (T)

5751 **Wilkerson, S.J. (1985).** The Usumacinta river: troubles on a wild frontier. *Nat. Geogr.*, 168(4): 514-543. Proposed hydroelectric dams may destroy rain forest between Mexico & Guatemala.

5752 **Withner, C.L. (1977).** Threatened and endangered species of orchids. *In* Prance, G.T., Elias, T.S., *eds.* Extinction is forever. Proceedings of a symposium at the New York Botanical Garden, 11-13 May 1976. New York, New York Botanical Garden. 314-322. Discusses "Plan Chiapas", a nature protection plan in southern Mexico, specifically concerned with orchids. (T)

MEXICO - Baja California Peninsula

5753 **Barry, W.J. (1981).** San Luis Island. *Fremontia*, 9(1): 11. Illus. (T)

5754 **Medeiros, J.L. (1982).** San Luis Island. *Fremontia*, 10(1): 27. Illus.

5755 **Nabhan, G.P. (1989).** Plants at risk in the Sonoran Desert: an international concern. *Agave*, 3(3): 14-15. Map. (T)

5756 **Nabhan, G.P., Monarque, E.S., Olwell, P., Warren, P., Hodgson, W., Gallindo-Duarte, C., Bittman, R., Anderson, S. (1989).** A preliminary list of plants at risk in the Sonoran Desert of the U.S. and Mexico. *Agave*, 3(3): 15. (T)

5757 **Oberbauer, T. (1986).** Baja California's Pacific Island jewels. *Fremontia*, 14(1): 3-5. Illus.

MEXICO - Baja California Peninsula

5758 **Perez O., J.F. (1982).** Especies amenazadas y en peligro de extincion de la peninsula de Baja California. *Publ. Espec. Inst. Nacion. Invest. Forest. Mexico*, 37: 62-67. Sp. (T)

5759 **Rowley, G.D. (1983).** Cactus hunting in Baja California. *Aloe*, 20(1): 10-13. Illus. (T)

MEXICO - Chiapas

5760 **Anon. (1980).** Montes Azules Biosphere Reserve, Chiapas. *Nat. Resources*, 16(2): 21. (P)

5761 **Hartmann, W.L. (1979).** Nochmals: der "Chiapas-Plan". *Orchidee*, 30(6): 238-240. Ge. Illus.

5762 **Zamora Serrano, C. (1977).** *Pinus strobus* var. *chiapensis*, una especie en peligro de extincion en el estado de Chiapas. (*Pinus strobus* var. *chiapensis*, a species in danger of extinction in the State of Chiapas.) *Cienc. Forest.*, 2(8): 3-23. Sp. Illus. (T)

MEXICO - Chihuahua

5763 **Johnston, M.C. (1983).** *Anoda henricksonii* (Malvaceae), a new species from the southern Chihuahuan Desert region. *Phytologia*, 53(7): 451-453. (T)

5764 **Waller, R.H., Riskind, D.H., eds (1977).** Transactions of the Symposium on the Biological Resources of the Chihuahuan Desert Region, United States and Mexico. Sul Ross University, Alpine, Texas, 17-18 October 1974. National Park Service Trans. and Proc. Series No. 3. Washington, D.C., U.S. Dept of the Interior. xxii, 658p. Maps.

MEXICO - Durango

5765 **Montana, C. (1984).** Ecological and socio-economic research in the Mapimi Biosphere Reserve. *In* Unesco-UNEP. Conservation, science and society. Paris, Unesco. 520-533. Contributions to the First International Biosphere Reserve Congress, Minsk, Byelorussia/USSR, 26 September - 2 October 1983. Illus., map. (P)

5766 **Nabhan, G.P., Monarque, E.S., Olwell, P., Warren, P., Hodgson, W., Gallindo-Duarte, C., Bittman, R., Anderson, S. (1989).** A preliminary list of plants at risk in the Sonoran Desert of the U.S. and Mexico. *Agave*, 3(3): 15. (T)

MEXICO - Jalisco

5767 **Iltis, H., Doebley, J., Guzman, M., Pazy, B. (1979).** *Zea diploperennis* (Gramineae): a new teosinte from Mexico. *Science*, 203: 186-187. (E/T)

5768 **Iltis, H.H. (1980).** The 3rd University of Wisconsin/University of Guadalajara teosinte expedition to the Sierra de Manantlan, Jalisco, Mexico: 28 December 1979-21 January 1980. Madison, University of Wisconsin-Madison. 62p. Includes nature preservation.

5769 **Nault, L.R., Findley, W.R. (1982).** Update on perrenial corn discovery. *Crops and Cereals Mag.*, March: 10-13. (T)

MEXICO - Sinaloa

5770 **Nabhan, G.P., Monarque, E.S., Olwell, P., Warren, P., Hodgson, W., Gallindo-Duarte, C., Bittman, R., Anderson, S. (1989).** A preliminary list of plants at risk in the Sonoran Desert of the U.S. and Mexico. *Agave*, 3(3): 15. (T)

MEXICO - Sonora

5771 **McLaughlin, S. (1988).** Salvage, propagation and re-establishment of *Mammillaria thornberi*. Progress report. Tucson Water, Tucson, Arizona. 18p. (G/T)

MEXICO - Sonora

5772 **Nabhan, G., Valenciano, D. (1989).** A modest proposal: restoring the Sonoran Desert at Barnes Butte Bajada. *Agave,* 3(3): 3-5. Illus., col. illus. Desert degradation problems and restoration ecology.

5773 **Nabhan, G.P. (1989).** Plants at risk in the Sonoran Desert: an international concern. *Agave,* 3(3): 14-15. Map. (T)

5774 **Nabhan, G.P., Monarque, E.S., Olwell, P., Warren, P., Hodgson, W., Gallindo-Duarte, C., Bittman, R., Anderson, S. (1989).** A preliminary list of plants at risk in the Sonoran Desert of the U.S. and Mexico. *Agave,* 3(3): 15. (T)

MEXICO - Tamaulipas

5775 **U.S. Department of the Interior. Fish and Wildlife Service (1989).** Endangered and threatened wildlife and plants; withdrawal of the proposed rule to list *Boerhavia mathsiana* as Endangered. *Fed. Register,* 54(124): 27413-27414. (L/T)

MEXICO - Veracruz

5776 **Anon. (1986).** Proyecto heroes y Martires de Veracruz. Instituto Nicaraguense de Recursos Naturales y del Ambiente. 108p. Sp.

5777 **Rappole, J., Ramos, M. (1985).** Forested crater could be refuge for rare species. *WWF Monthly Rep.,* Oct.: 231-234. Rainforests of Tuxtla Mountains.

5778 **Vovides, A.P. (1978).** Practical conservation problems of a new botanic garden in Mexico. *Gartn.-Bot. Brief,* 58: 39. (Ge). Xalapa. Abstract only. (G)

5779 **Vovides, A.P. (1979).** Mexico: practical conservation problems of a new botanic garden. *In* Synge, H., Townsend, H., *eds.* Survival or extinction. Proceedings of a conference held at the Royal Botanic Gardens, Kew, 11-17 September 1978. Kew, Bentham-Moxon Trust. 117-123. New botanic garden at Xalapa. (G/T)

MEXICO - Yucatan

5780 **Garcia, R.D., Olmsted, I. (1987).** Listado floristico de la Reserva Sian Ka'an. Quintano Roo, Mexico, Puerto Morelos. 71p. Sp. Illus. (P)

MICRONESIA

5781 **Buck, M. (1984).** The precious forests of Ponape and Kosrae. *Glimpses of Micronesia,* 24(3): 33-37.

5782 **Cheatham, N.H. (1975).** Land development: its environmental impact in Micronesia. *Micronesian Reporter,* 23(3): 7-11.

5783 **Coolidge, H.J., comp. (1948).** Conservation in Micronesia. Washington, D.C., National Research Council. 70p.

5784 **Fosberg, F.R. (1953).** The naturalized flora of Micronesia and World War II. *In* Proc. Eighth Pacific Science Congress, 4: 229-234. Quezon, National Research Council of the Philippines. 1954. Introduced species.

5785 **Hosokawa, T. (1973).** On the tropical rainforest conservation to be proposed in Micronesia. *In* Planned utilization of the lowland tropical forests. Pacific Science Association Symposium. Cipayung, Bogor, Java. 150-164. (T)

5786 **Knott, N.P. (1973).** Further case studies in selecting and allocating land for nature conservation: Micronesia, a multiple land-capability inventory method. *In* Costin, A.B., Groves, K.H., *eds.* Nature conservation in the Pacific. Proc. Symposium A-10, XII Pacific Science Congress, Canberra, August-September 1971. IUCN and Australian Nat. Univ. Press. 61-66.

MICRONESIA

5787 **Otobed, D.O. (1975).** Conservation priorities in Micronesia. *In* Force, R.W., Bishop, B., *eds*. The impact of urban centers in the Pacific. Honolulu, Hawaii, Pacific Science Association. 73-79.

5788 **U.S. Congress, Office of Technology Assessment (1987).** Integrated renewable resource management for U.S. insular areas. Washington, D.C., U.S. Govt Printing Office. 443p. Illus., maps. OTA-F-325. Includes description of vegetation, remaining coverage, threatened species. (P/T)

5789 **Wenkham, R. (1971).** Micronesian parks: a proposal. *Micronesian Reporter*, 19(3): 9-22. Illus. (P)

5790 **Wenkham, R., Brower, K. (1975).** Introduction: towards oceanic parks for Micronesia - a proposal. *In* Brower, K., *ed*. Micronesia: island wilderness. San Francisco, Friends of the Earth. 161p. (P)

5791 **Wilson, P.T. (1976).** Conservation problems in Micronesia. *Oceans*, 9(3): 34-41.

MONGOLIA

5792 **Alexandrowicz, Z. (1984).** Nowe dane o ochronie przyrody w Mongolii. (New data on nature conservation in Mongolia.) *Chronmy Przyr. Ojczysta*, 40(2): 55-58. Pol. Illus.

5793 **Gal, Zh. (1981).** [Nekotorye soobshchestva Dolin Ozer kak ob'ekty okhrany tsenofondov]. *BNMAU Shinzhlekh Ukhaany Akad.*, 5: 23-28. Mong (Rus). (T)

5794 **Gubanov, I.A. (1982).** Zametki o redkikh rasteniyakh Mongolii. (Notices on rare plants of Mongolia.) *Byull. Mosk. Obshch. Ispyt. Prir., Biol.*, 87(1): 122-129. Rus. (T)

5795 **Gunin, P.D., Fedorova, I.T. (1983).** The Gobi Desert in Mongolia. *Arid Lands Newsl.*, 19: 2-9. Illus., maps.

5796 **Guricheva, N.P., Buevich, Z.G., Sukhoverko, R.V. (1984).** [Effect of the reservation conditions on the *Stipa baicalensis* steppes of Eastern Khangai.] *Bot. Zhurn.*, 69(5): 636-647. Rus (En). Illus. Literature review.

MOROCCO

5797 **Comte, M.C. (1980).** Making social forestry work. *Ceres*, 13(2): 41-44. An UNDP/FAO project on the management and improvement of forest grazing ground in Morocco. (E)

5798 **IUCN Threatened Plants Committee Secretariat (1980).** First preliminary draft of the list of rare, threatened and endemic plants for the countries of North Africa and the Middle East. Kew, IUCN Threatened Plants Committee Secretariat. 170p. (T)

5799 **Jackson, P. (1986).** National park for wetlands in Morocco. *WWF Monthly Report*, March: 69-71. (P)

5800 **Mathez, J., Quezel, P., Raynaud, C. (1985).** The Maghreb countries. *In* Gomez-Campo, C., *ed*. Plant conservation in the Mediterranean Area. Dordrecht, Dr W. Junk. 141-157. Illus., maps. (T)

5801 **Mekouar, M.A. (1984).** Foret et environment en droit marocain. (Forest and environment Moroccan law.) *In* Prieur, M., *ed*. Forets et environnement: en droit compare et international. Paris, Presses Universitaires de France. 183-198. Fr. (L)

5802 **Mellado, J. (1989).** S.O.S. Souss: argan forest destruction in Morocco. *Oryx*, 23(2): 87-93. Illus., map. (T)

5803 **Sauvage, C. (1959).** Au sujet de quelques plantes rares et menacees de la flore du Maroc. *In* IUCN. Animaux et vegetaux rares de la Region Mediterraneenne. Proceedings of the IUCN 7th Technical Meeting, 11-19 September 1958, Athens, vol. 5. Brussels, IUCN. 156-158. Fr. (T)

MOZAMBIQUE

5804 **Bruton, M.N. (1981).** Major threat to the coastal dune forest in Maputoland. *Naturalist* (S.Afr.), 25(1): 26-27. Illus. (T)

5805 **Grandvaux Barbosa, L.A., Mogg, A.O.D. (1968).** Mocambique. *In* Hedberg, I. & O., *eds.* Conservation of vegetation in Africa south of the Sahara. Symposium Proceedings, at 6th plenary meeting of AETFAT, Uppsala. *Acta Phytogeogr. Suec.*, 54: 224-232. Illus., map. Discusses present protection of animals and vegetation and suggests areas for further protection. (P/T)

5806 **Osborne, R. (1987).** *Encephalartos ferox. Encephalartos*, 9: 14-21. Illus. Data sheet; includes conservation. (T)

NAMIBIA

5807 **Berry, C. (1979).** Trees and shrubs of the Etosha National Park. Directorate of Nature Conservation. 161p. En (Af, Ge, Her). Illus., col. illus., map. (P)

5808 **Giess, W., Tinley, K.L. (1968).** South West Africa. *In* Hedberg, I. & O., *eds.* Conservation of vegetation in Africa south of the Sahara. Symposium Proceedings, at 6th plenary meeting of AETFAT, Uppsala. *Acta Phytogeogr. Suec.*, 54: 250-253. Map. Present and suggested protection. List of protected trees. (P/T)

5809 **Loope, L.L., Sanchez, P.G., Tarr, P.W., Loope, W.L., Anderson, R.L. (1988).** Biological invasions of arid land nature reserves. *Biol. Conserv.*, 44(1-2): 95-118. Invasive species. Includes 5 case studies. (P)

5810 **Nussey, W. (1979).** The Namib-Naukluft Park. *Afr. Wildl.*, 33(1): 25-28. Illus., col. illus. (P)

5811 **Smirnova, E.S., Dementeva, V.S. (1980).** *Welwitschia mirabilis.* Moskva, Nauka. 163-168. Rus. Illus. (G/T)

NAURU

5812 **Manner, H.I., Thaman, R.R., Hassall, D.C. (1984).** Phosphate mining induced vegetation changes on Nauru Island. *Ecology*, 65(5): 1454-1465. Illus., maps. Forest.

NEPAL

5813 **Acker, F. (1981).** Saving Nepal's dwindling forests. *New Sci.*, 90(1248): 92-94. Illus.

5814 **Anon. (1988).** So much yet so little: endangered plants of Nepal. *Bull. King Mahendra Trust Nat. Conserv.*, 2(1). Illus. (T)

5815 **Aswal, B.S., Mehrotra, B.N. (1983).** *Delphinium uncinatum* Hook.f. et Thoms. (Ranunculaceae) and *Lilium wallichianum* Schultes f. (Liliaceae): two rare finds from north-west Himalaya. *J. Econ. Taxon. Bot.*, 3(3): 773-775 (1982 publ. 1983). (T)

5816 **Bajracharya, D. (1983).** Deforestation in the food/fuel context and political perspectives of Nepal. *Mountain Research and Development*, 3: 227-240.

5817 **Baltus, J. (1978).** [Nature conservation in Nepal]. *Pamatky Prir.*, 4:244-247. Cz. Illus.

5818 **Bhatt, D.D. (1981).** Nepal Himalaya and change. *In* Lall, J.S., *ed.* The Himalaya: aspects of change. New Delhi, India International Centre, Oxford University Press. 253-277. (P/T)

5819 **Coburn, B. (1982).** Alternative energy sources for Sagarmatha National Park. *Parks*, 7(1): 16-18. Illus. (E/P)

5820 **Coburn, B.A. (1983).** Managing a Himalayan World Heritage Site. *Nat. Resources*, 19(3): 20-25. Illus. (P)

NEPAL

5821 **Dhungel, S.K. (1982).** A glimpse of Sagarmatha: world's highest national park. *Tigerpaper*, 9(2): 11-14, Illus. (P)

5822 **Dinerstein, E. (1979).** An ecological survey of the Royal Karnali-Bardia Wildlife Reserve, Nepal: 1. Vegetation, modifying factors and successional relationships. *Biol. Conserv.*, 15(2): 127-150. Maps. (P)

5823 **Fleming, R.L. (1969).** Nepal fauna and flora: comments on present status. *IUCN Bull.*, n.s., 2(13): 108. (T)

5824 **Gilmour, D.A. (1988).** Not seeing the trees for the forest: a re-appraisal of the deforestation crisis in two hill districts of Nepal. *Mtn Res. Dev.*, 8(4): 343-350.

5825 **Israel, S., Sinclair, T., eds (1987).** Indian wildlife: Sri Lanka, Nepal. Hong Kong, APA Productions. 362p. Col. illus. Provides overview of vegetation and conservation problems.

5826 **Jefferies, B. (1982).** Sagarmatha National Park: the impact of tourism in the Himalayas. *Ambio*, 11(5): 274-281. Illus., col. illus., maps. (P)

5827 **Jefferies, B.E. (1984).** The Sherpas of Sagarmatha: the effects of a national park on the local people. *In* McNeely, J.A., Miller, K.R., *eds*. National parks, conservation and development. The role of protected areas in sustaining society. Washington, D.C., Smithsonian Institution Press. 473-478. Map. (P)

5828 **Khadka, R.B. (1983).** Mountain flora and their conservation in Nepal. *In* Jain, S.K., Mehra, K.L., *eds*. Conservation of tropical plant resources. Howrah, Botanical Survey of India. 132-141.

5829 **King Mahendra Trust for Nature and Conservation and International Centre for Integrated Mountain Development (1986).** People and protected areas in the Hindu Kush-Himalaya. Kathmandu. Proceedings of the International Workshop on the Management of National Parks and Protected Areas in the Hindu Kush-Himalaya, 6-11 May 1985, Kathmandu, Nepal. Nepal, KMTNC. 189p. (P)

5830 **Lall, J.D.S., ed. (1981).** The Himalaya: aspects of change. New Delhi, India International Centre, Oxford Univ. Press. 253p. Col. illus. Maps. (P/T)

5831 **Lehmkuhl, J.F., Upreti, R.K., Sharma, U.R. (1988).** National parks and local development: grasses and people in Royal Chitwan National Park, Nepal. *Envir. Conserv.*, 15(2): 143-148. (P)

5832 **Martens, J. (1982).** Forests and their destruction in the Himalayas of Nepal. *Plant Research and Development*, 15: 66-96. Deforestation.

5833 **McNeely, J.A. (1985).** Man and nature in the Himalaya: what can be done to ensure that both can prosper. Switzerland, IUCN. 13p. Paper presented to the International Workshop on the Management of National Parks and Protected Areas in the Hindukush-Himalaya, Kathmandu, Nepal, 6-11 May 1985.

5834 **Mishra, H.R. (1982).** Balancing human needs and conservation in Nepal's Royal Chitwan Park. *Ambio*, 11(5): 246-251. Illus., col. illus. (P)

5835 **Myers, N. (1986).** Environmental repercussions of deforestation in the Himalayas. *J. World Forest Res. Manage.*, 2: 63-72.

5836 **National Committee for Man and the Biosphere, Research Centre for Applied Science and Technology, Nepal (1988).** A study of some rare and endemic plant species. Progress report. Unpublished. 41p. Submitted to Unesco. (T)

5837 **Nelson, D.O., Laban, P., Shrestha, B.D., Kandel, G.P. (1980).** A reconnaissance inventory of the major ecological land units and their watershed condition in Nepal. FAO Report No. FO:NEP/74/020; Project Field Document WP/17. Rome, FAO. ix, 292p. Illus., maps.

NEPAL

5838 **Nepal National Committee for Man and the Biosphere (1986).** Status of environmental knowledge in Nepal. Annotated bibliography. MAB/Nepal Publication Series No. 2-3. Kathmandu, Nepal National Committee for Man and the Biosphere. 200p. (B)

5839 **Rana, P.S.J.B., Pandey, N.R., Mishra, H.R. (1982).** Conservation for development: an introduction to the King Mahendra Trust for Nature Conservation. Kathmandu, Nepal, KMTNC. 34p. Illus.

5840 **Rieger, H.C. (1981).** Man versus mountain: the destruction of the Himalayan ecosystem. *In* Lall, J.S., *ed.* The Himalaya: aspects of change. India International Centre, New Delhi, Oxford University Press. 351-376. Many figures and tables demonstrating mechanisms of erosion. (P/T)

5841 **Roberts, J. (1982).** The Machapuchare Wildlife Conservation Project in Nepal. *Tigerpaper*, 9(4): 18-20. Illus. (P)

5842 **Sattaur, O. (1987).** Trees for the people. *New Sci.*, 115(1577): 58-62. Illus., col. illus. Afforestation projects.

5843 **Sharma, L.R. (1981).** Necessity and method of forest conservation in Nepal. *Indian Sci. Congr. Assoc. Proc.*, 68(3): 163. Abstract.

5844 **Simons, R. (1984).** Ten years later: the Smithsonian international experience since the Second World Parks Conference. *In* McNeely, J.A., Miller, K.R., *eds.* National parks, conservation and development. The role of protected areas in sustaining society. Washington, D.C., Smithsonian Institution Press. 712-718. Map. Outlines projects in Nepal and Barro Colorado Island, Panama.

5845 **Singh, J.S., Pandey, U., Tiwari, A.K. (1984).** Man and forests: a central Himalayan case study. *Ambio*, 13(2): 80-86. Illus., maps. Deforestation. (T)

5846 **Upreti, B. (1985).** The development of a protected area system in Nepal. *In* Thorsell, J.W., *ed.* Conserving Asia's natural heritage. The planning and management of protected areas in the Indomalayan Realm. Proc. of the 25th working session of IUCN's CNPPA. Gland, IUCN. 78-81. Map. (P)

5847 **Wallace, M. (1985).** Community forestry in Nepal: too little, too late? Kathmandu, Agricultural Development Council.

NETHERLANDS

5848 **Anon. (1977).** [Forestry and nature conservation]. *Netherl. Bosbouw Tijdschr.*, 49(7/8): 219-252. Du. Illus.

5849 **Bakker, P.A. (1979).** Vegetation science and nature conservation. *In* Werger, M.J.A., *ed.* The study of vegetation. The Hague, Junk. 247-288. Maps.

5850 **Barneveld, J.C. van (1981).** [The weed garden.] *In* Netherlands Plantenziektenkundige Dienst, Jaarboek, 1979. Wageningen. 97-102. Du. Weed garden at Wageningen contains species that would otherwise be extinct in Netherlands. (T)

5851 **Burton, J.F. (1981).** Among Limburg's quaking bogs: De Groote Peel Nature Reserve. *Country Life*, 170(4383): 636-637. Illus. (P)

5852 **Council of Europe (1984).** Benelux: nature conservation and protection. *Council of Europe Newsl.*, 84(2): 2.

5853 **Donselaar, J. van (1970).** De Nederlanse natuurbescherming gezien in internationaal verband-botanie. (Dutch nature conservation in the context of international botany.) *In* Kramer, J.C. van de, *et al.*, *eds.* Het veerstoorde evenwicht. Utrecht, Oosthoek. 231-244. Du. Describes important botanical sites.

5854 **Goor, C.P. van, Jagt, J.L. van der, Poel, A.J. van der (1980).** [Forest reserves]. *Nederl. Bosbouw Tijdschr.*, 52(3): 49-76. Du. (P)

NETHERLANDS

5855 **Goosen, M.G. (1978).** Het middellange termijn plan voor de boswachterijen van het staatsbosbeheer in Nederland. (Medium-term planning for conservancies of the State Forest Administration in the Netherlands.) *Nederl. Bosbouw Tijdschr.*, 7/8: 225-245. Du. Illus.

5856 **Hendriks, J.L.T. (1978).** De populier en het natuurbehoud. (Poplars and nature conservation.) *Populier*, 15(1): 16-20. Du. Illus.

5857 **Jagt, J.L.V.D. (1980).** Bosreservaten: doelstellingen, inrichting en beheer. (Forest reserves: objectives, organization, and management.) *Nederl. Bosbouw Tijdschr.*, 52(3): 64-69. Du. Illus. (P)

5858 **Joenje, W. (1979).** Plant succession and nature conservation of newly embanked tidal flat in the Lauwerszeepolder. *In* Jefferies, R.L., Davy, A.J., *eds*. Ecological processes in coastal environments. British Ecological Society Symposia 19. Oxford, Blackwell Scientific Publications. 617-634. (Fr). Illus., map.

5859 **Koningen, H.C., Zanen, G.C.N. van (1981).** Het oeverlandenreservaat aan de Amstelveense Poel. *Natura* (Netherlands), 78(11): 363-372. Du. Col. illus.

5860 **Laan, D. van der (1985).** Changes in the flora and vegetation of the coastal dunes of Voorne (The Netherlands) in relation to environmental changes. *Vegetatio*, 61(1/3): 87-95. Maps.

5861 **Maarel, E. van der (1971).** Plant species diversity in relation to management. *In* Duffey, E., Watt, A.S., *eds*. The scientific management of animal and plant communities for conservation. The 11th Symposium of the BES, University of East Anglia, Norwich, 7-9 July 1970. 45-63. Map.

5862 **Maarel, E. van der (1979).** Environmental management of coastal dunes in the Netherlands. *In* Jefferies, R.L., Davy, A.J., *eds*. Ecological processes in coastal environments. British Ecological Society Symposia 19. Oxford, Blackwell Scientific. 543-570. (Fr). Map.

5863 **Meijden, R. van der, Holverda, W.J. (1987).** Nieuwe vondsten van zeldzame planten in 1985 en 1986. (New records of rare plants in 1985 and 1986.) *Gorteria*, 13: 221-242. Du (En). (T)

5864 **Mennema, J. (1975).** Threatened and protected plants in the Netherlands. *Naturopa*, 22: 10-13. (T)

5865 **Mennema, J. (1975).** Zeldzame planten tellen. (Census of rare plants.) *Levende Nat.* 78(2): 29-31. Du. (T)

5866 **Mennema, J. (1979).** Nieuwe vondsten van zeldzame planten in Nederland, hoofdzakelijk in 1977. (Newly found and rare plants in the Netherlands mainly in 1977.) *Gorteria*, 9(6): 208-227. Du (En). (T)

5867 **Mennema, J., Holverda, W.J. (1980).** Nieuwe vondsten van zeldame planten in Nederland, hoofdzakelijk in 1979. (New records of rare plants in the Netherlands, mainly in 1979.) *Gorteria*, 10(5/6): 81-100. Du (En). (T)

5868 **Mennema, J., Holverda, W.J. (1982).** Nieuwe vondsten van zeldzame planten in Nederland, hoofdzakelijk in 1980. (New records of rare plants in the Netherlands, mainly in 1980.) *Gorteria*, 10(11/12): 189-213. Du (En). Native species, adventitious species and plants that escaped from cultivation. (T)

5869 **Mennema, J., Holverda, W.J. (1982).** Nieuwe vondsten van zeldzame planten in Nederland, hoofdzakelijk in 1981. (New records of rare plants in the Netherlands, mainly in 1981.) *Gorteria*, 11(6): 123-141. Du (En). List of native and adventitious species and plants that escaped from cultivation. (T)

5870 **Mennema, J., Ooststroom, S.J. van (1977).** Nieuwe vondsten van zeldzame planten in Nederland, hoofdzakelijk in 1975. (New and rare plants in the Netherlands, mainly in 1975.) *Gorteria*, 8(8): 135-156. Du (En). (T)

NETHERLANDS

5871 **Mennema, J., Ooststroom, S.J. van (1977).** Nieuwe vondsten van zeldzame planten in Nederland, hoofdzakelijk in 1976. (New and rare plants in the Netherlands, mainly in 1976.) *Gorteria*, 8(12): 219-240. Du (En). (T)

5872 **Mennema, J., Ooststroom, S.J. van (1979).** Nieuwe vondsten van zeldzame planten in Nederland, hoofdzakelijk in 1978. (New localities of rare plant species in the Netherlands found during 1978.) *Gorteria*, 9(11/12): 347-364. Du. Map. (T)

5873 **Mennema, J., Quene-Boterenbrood, A.J., Plate, C.L., eds (1980).** Atlas van de Nederlandse Flora 1. Uitgestorven en zeer zeldzame planten. (Extinct and very rare species.) Kosmos, Amsterdam. English edition by Junk, The Hague. 226p. Du (En). Maps. Contains conservation data for over 300 vascular plant species (native and introduced). (T)

5874 **Mennema, J., Quene-Boterenbrood, A.J., Plate, C.L., eds (1985).** Atlas van de Nederlandse Flora 2. Zeldzame en vrij zeldzame planten. (Rare and very rare plants.) Utrecht, Bohn, Scheltema and Holkema. 349p. Du (En). Maps. Contains conservation data for over 250 vascular plant species. (T)

5875 **Munckhof, P.J.J. van den (1980).** Klimopwaterranonkels in Noord-Limburg: hoe lang nog? (deel 3). *Natuur. Maandblad*, 69(1): 15-21. Du (En). Illus., map. Two planned nature reserves to protect *Ranunculus hederaceus*. (T)

5876 **Nijland, G., Veen, H.E. van de (1978).** Upgrading of natural values in the central Netherlands. Graveland, Stichting Kritisch Faunabeheer. 16p. Du. Illus. Recommendation for national park. (P)

5877 **Oosterveld, P. (1977).** Welk bosbeheer heeft wat met natuurbeheer to maken. (Relations between forestry and nature conservation.) *Nederl. Bosbouw Tijdschr.*, 49(4): 163-170. Du.

5878 **Ploeg, S.W.F. van der, Vlijm, L. (1978).** Ecological evaluation, nature conservation and land use planning with particular reference to methods used in the Netherlands. *Biol. Conserv.*, 14(3): 197-221. Map.

5879 **Poel, A.J. van der (1980).** Aanwijzing van bosreservaten en het wetenschappelijk onderzoek. (Assignment of forest reserves and scientific investigation.) *Netherl. Bosbouw Tijdschr.*, 52(3): 70-73. Du. Illus. (P)

5880 **Polen, B. van (1981).** Brabant achter de coulissen: een blik op het natuurbehoud. (Brabant behind the scenes, a look at its nature conservation.) Eindhoven, Netherlands, Bura Boeken. 221p. Du. Illus., map.

5881 **Quene-Boterebrood, A.J., Mennema, J. (1973).** Zeldzame Nederlandse plantesoorten. S'Gravenhage. 110p. Du.

5882 **Quene-Boterenbrood, A.J. (1974).** Een 'tussenrapport' over zeldzame Nederlandse plantesoorten. (An interim report of rare Dutch plant species.) *Natuur en Landschap*, 28: 297-308. Du. (T)

5883 **Read, M. (1989).** Grown in Holland? Brighton, Fauna and Flora Preservation Society. 12p. Illus., maps. European bulb trade. (E/G/L/T)

5884 **Reinink, K. (1979).** Ongewone planten van "De Konigsheide" bij Arnhem. (Unusual and rare plants at "De Konigsheide" in the municipality of Arnhem.) *Gorteria*, 9(11/12): 369-371. Du (En). (T)

5885 **Sikkel, D. (1980).** Bosreservaten in Nederland. (Forest reserves in the Netherlands.) *Nederl. Bosbouw Tijdschr.*, 52(5): 121-124. (P)

5886 **Sipkes, C. (1978).** Verrassende groeiplaatsen van orchideeen in ons land. *Levende Nat.*, 81(4): 153-160. Du. Illus.

5887 **Weijden, W.J. van der, Keurs, W.J. ter, Zande, A.N. van der (1978).** Nature conservation and agricultural policy in the Netherlands. *Ecol. Quart.*, 4: 317-335.

NETHERLANDS

5888 **Westhoff, V. (1956).** De veraming van flora en vegetatie. (The impoverishment of the flora and vegetation.) *In* Gedenkboek 50 jaar Natuurmonumenten. 151-184. Du.

5889 **Westhoff, V. (1976).** Die Verarmung der Niederlandischen Gefasspflanzenflora in den letzten 50 Jahren und ihre teilweise Erhaltung in Naturreservaten. (The decline of the Dutch vascular plant flora during the past 50 years and the contribution of nature reserves to its conservation.) *Schr.-R. Vegetationskunde*, 10: 63-73. Ge. (P/T)

5890 **Westhoff, V. (1979).** Bedrohung und Erhaltung seltener Pflanzengesellschaften in den Niederlanden. *In* Wilmans, O., Tuxen, R., *eds.* Werden und Vergehen von Pflanzengesellschaften. Vaduz, Liechstenstein. 285-313. Ge.

5891 **Westhoff, V., Weeda, E.J. (1984).** De achteruitgang van de Nederlandse flora sinds het begin van deze eeuw. (The decline of the Dutch flora since the beginning of the first century.) *Natuur en Milieu*, 8(8): 8-17. Du.

5892 **Wijnands, D.O. (1980).** *Silene viscariopsis* Bornm. *Bull. Bot. Tuinen Wageningen*, 4: 10. Du. (T)

5893 **Wijnands, D.O. (1980).** Bedreigde Nederlandse waterplanten. (Threatened Dutch water plants.) *Bull. Bot. Tuinen Wageningen*, 5: 5-9. Du. Illus., map. (T)

5894 **Wijnands, D.O. (1981).** Bedreigde Nederlandse waterplanten. (Threatened Dutch water plants.) *Bull. Arbor. Waasland*, 4(1): 38-42, 48-50. Du (En). (T)

5895 **Zeven, A.C. (1980).** Botanical gardens, public parks, road sides, historical gardens, historical arable fields, nature reserves and private gardens as genebanks of cultivated crops. *Misc. Pap. Landbouwhogesch.* (Wageningen), 19: 433-438. (E/G/P)

5896 **Zonneveld, I.S. (1978).** De plaats van de groveden in het natuurbehoud. (The place of *Pinus sylvestris* in nature conservation.) *Nederlands Bosbouw Tijdschrift*, 4: 112-114. Du.

NEW CALEDONIA

5897 **Aubert de la Rue, E. (1958).** Man's influence on tropical vegetation. *Proc. Ninth Pacific Science Congr.*, 20: 81-94. Includes case studies on metal ore mining in New Caledonia and phosphate mining which threatens limestone forest on Walpole and Makatea.

5898 **Barrau, J. (1958).** Beware this invasive noxious weed. *South Pacific Bull.*, 8(3): 7. *Cryptostegia grandiflora* (Asclepiadaceae).

5899 **Barrau, J., Devambez, L. (1957).** Quelques resultats inattendus de l'acclimatation en Nouvelle-Caledonie. *Terre et Vie*, 104(4): 324-334. Effects of introduced deer and plants on vegetation.

5900 **Catala, R.L.A. (1953).** Protection de la nature en Nouvelle-Caledonie. (Protection of the nature of New Caledonia.) *Proc. Seventh Pacific Science Congress*, 4: 674-679.

5901 **Dawson, J.W. (1981).** The species-rich, highly endemic serpentine flora of New Caledonia. *Tuatara*, 25(1): 1-6.

5902 **Given, D.R. (1981).** Flora of offshore and onlying islands. *In* XIII International Botanical Congress, Sydney, Australia, 21-28 August 1981: abstracts. Sydney, Australian Academy of Science. 215.

5903 **Guillaumin, A. (1953).** Mesures de conservation a prendre pour la sauvegarde de la flore de la Nouvelle-Caledonie. *In* Proc. Seventh Pacific Science Congress, 4: 674. Fr. Abstract.

5904 **Guillaumin, A. (1970).** L'evolution de la flore Neo-Caledonienne. *J. Soc. Oceanistes*, 9(9): 79-85. Fr.

NEW CALEDONIA

5905 **Guillaumin, A. (1970).** Le santal en Nouvelle-Caledonie. *J. Agric. Trop. Bot. Appl.*, 17(7-9): 340-341. Fr.

5906 **Hurlimann, H. (1953).** Etude sur la structure des forets de la Nouvelle-Caledonie: experiences et propositions. *Et. Melan., n.s.*, 5(7): 55-68. Fr. Includes plant conservation.

5907 **Hurlimann, H. (1959).** Need for a conservation park in New Caledonia. *In* Proc. Ninth Pacific Science Congress, 7. 50. (P/T)

5908 **Hurlimann, H. (1960).** Un parc de conservation botanique en Nouvelle-Caledonie. *J. Soc. Oceanistes*, 16: 110-112. Fr. Threatened plants have been transplanted. (G/P/T)

5909 **Lucas, G. (1980).** Deux cas remarquables de taxa menaces: Cyprinodontidae nord-africains, *Araucaria* neo-caledoniens. *Compte R. Seances Soc. Biogeogr.*, 56(489): 51-52. Fr. (T)

5910 **MacDaniels, L.H. (1952).** New Caledonia: a warning. *Cornell Plantations*, 8: 40-44. Discusses threat of uncontrolled burning of forests and grassland.

5911 **Parrat, J. (1971).** Destruction et defense de la couverture vegetale en Nouvelle-Caledonie. *In* Plessis, J., Cahiers du Pacifique. Colloque Regional sur la Protection de la Nature - Recifs et Lagons, Commission du Pacific Sud. Noumea. SPC/RSCN/WP 16 (597/71). 1-6. Fr.

5912 **Sachet, M.-H. (1957).** The vegetation of Melanesia: a summary of the literature. *Proc. Eighth Pacific Science Congress*, 4: 35-47. (B)

5913 **Schmid, M. (1981).** Fleurs et plantes de Nouvelle-Caledonie. Les Editions du Pacifique. 164p. Col. illus., map. Includes section on human impact on vegetation. (T)

5914 **Schmid, M. (1982).** Endemisme et speciation en Nouvelle-Caledonie. *Compte Rendu Seances Soc. de Biogeographie*, 58(2): 52-60. Includes statistics for this centre of plant endemism.

5915 **Shineberg, D. (1967).** They came for sandalwood: a study of the sandalwood trade in the south-west Pacific, 1830-1865. Melbourne, Melbourne Univ. Press. Impact of logging on Isle of Pines. (E)

5916 **Veillon, J.M. (1971).** La flore Neo-Caledonienne, son originalite, sa vulnerabilite face aux problemes de degradation et de pollution. Noumea, Commission du Pacifique Sud, Colloque Regional sur la Protection de la Nature - Recifs et Lagons, 4-14 Aout 1971. SPC/RSCN WP 23 (633/71): 1-5. Fr. (T)

5917 **Virot, R. (1954).** Le probleme de la protection de la nature en Nouvelle-Caledonie. *Eighth Congr. Int. Bot. Rapp. Comm.*, 21-27, 14-144. Discusses human threats to flora and conservation measures needed. (T)

5918 **Virot, R. (1956).** La vegetation Canaque. *Mem. Mus. Nation. Hist. Nat., serie B. Botanique*, 7: 1-398. Invasive species and other potential threats discussed; extensive bibliography. (B)

HUNTER, MATTHEW AND WALPOLE ISLANDS

5919 **Aubert de la Rue, E. (1958).** Man's influence on tropical vegetation. *Proc. Ninth Pacific Science Congr.*, 20: 81-94. Includes case studies on metal ore mining in New Caledonia and phosphate mining which threatens limestone forest on Walpole and Makatea.

NEW ZEALAND

5920 **Aitken, R. (1979).** National parks down under. *Town and Country Planning*, 48(9): 303-304. Illus. (P)

NEW ZEALAND

5921 **Allen, R.B. (1978).** Scenic reserves of Otago Land District. Wellington, Department of Lands and Survey. xxviii, 322p. Maps. Includes plant lists. (P)

5922 **Anon. (1977).** Environment and conservation organisations of New Zealand. *Hort. New Zealand*, 4: 7.

5923 **Anon. (1977).** Whakarewarewa State Forest Park. *New Zealand Wildlife*, 7(53): 1-10. (P)

5924 **Anon. (1978).** Conference on conservation of high mountain resources. Lincoln College, 1977. *Rev. Tussock Grassl. Mount. Lands Inst.*, 37: 87-93.

5925 **Anon. (1978).** Environment and conservation organisations of New Zealand. *Hort. New Zealand*, 6: 13-14.

5926 **Anon. (1978).** Guides and policies in the exercise of the Reserves Act 1977. No. 2. Leasing of reserves and other rights of occupation. Reserves Series No. 3. Wellington, N.Z., Department of Lands and Survey. 34p. (L/P)

5927 **Anon. (1979).** Ruahine State Forest Park. *New Zealand Wildlife*, 8(58): 22-27. Illus., map. (P)

5928 **Anon. (1980).** Castle Hill Nature Reserve management plan. Christchurch, The Commissioner of Crown Lands. 33p. Map. (P/T)

5929 **Anon. (1983).** Preserving New Zealand's ecosystems. *Search*, 14(7-8): 176. (L)

5930 **Anon. (1984).** Mooring at Snares Islands, New Zealand: a case of reluctant compromise. *Oryx*, 8: 6-7.

5931 **Anon. (1987).** Geothermal allocations concern council. *Nature Conservation Council Newsl.*, 66: 3. (T)

5932 **Anon. (1987).** New Zealand forest protected at last. *Oryx*, 21(3): 193. Report based on *Forest and Bird*, 17(4). Establishment of Paparoa National Park and legal protection for North Westland Wildlife Corridor. (P)

5933 **Atkinson, I.A.E. (1973).** Protection and use of the islands in Hauraki Gulf Maritime Park. *Proc. New Zealand Ecol. Soc.*, 20: 103-114. (P)

5934 **Australian and New Zealand Association for the Advancement of Science (1979).** A vanishing heritage: the problem of endangered species and their habitats. Wellington, Nature Conservation Council. 273p. (T)

5935 **Beever, R.E. (1982).** Submission on proposed forest park at Cornwallis. *Newsl. Auckland Bot. Soc.*, 37(2): 1-10. (P)

5936 **Beever, R.E. (1983).** Management of Kirk's Bush, Papakura. *Newsl. Auckland Bot. Soc.*, 38(1): 6-8. Letter signed for J. Mackinder, President. (P)

5937 **Benham, S. (1983).** Poor knights brush lily. *Garden* (London), 108(9):377. Col. illus. *Xeronema callistemon*. (T)

5938 **Bickerstaff, R. (1981).** A rare species can be yours! *Orchids New Zealand*, 6(5): 114-115. (T)

5939 **Burns, C.W. (1984).** Protected areas and introduced species in New Zealand. *In* McNeely, J.A., Miller, K.R., *eds*. National parks, conservation and development. The role of protected areas in sustaining society. Washington, D.C., Smithsonian Institution Press. 403-411. Map. (P)

5940 **Burrows, C.J., ed. (1974).** Handbook to the Arthurs Pass National Park. 3rd revised ed. Christchurch. Arthurs Pass National Park Board. 104p. Illus., maps. (P)

5941 **Clarkson, B.D. (1985).** The vegetation of the Kaitake Range, Egmont National Park, New Zealand. *New Zealand J. Bot.*, 23(1): 15-31. Illus., maps. (P)

NEW ZEALAND

5942 **Connor, H.E. (1979).** Genetics in conservation and biosphere reserve selection. *In* N.Z. Man and Biosphere Report No. 2: 31-34. (P)

5943 **Costin, A.B., Groves, R.H., eds (1973).** Nature conservation in the Pacific. Proceedings of Symposium A-10, XII Pacific Science Congress, August-September 1971, Canberra, Australia. IUCN New Series No. 25. Morges, IUCN and Canberra, Australian National University Press. 337p. Illus., maps.

5944 **Cumberland, K.B. (1963).** Man's role in modifying island environments in the southwest Pacific, with special reference to New Zealand. *In* Fosberg, F.R., *ed.* Man's place in the island ecosystem: a symposium. Honolulu, Bishop Museum Press. 186-206.

5945 **Dingwall, P. (1982).** New Zealand: saving some of everything. *Ambio*, 11(5): 296-301. Illus., col. illus. (P/T)

5946 **Dingwall, P.R. (1983).** Recent progress in the management of rare plants and their habitats in protected areas. *In* Given, D.R., *ed.* Conservation of plant species and habitats. Wellington, Nature Conservation Council. 67-82. Illus. Gives lists of threatened plants and protected areas in New Zealand. (P/T)

5947 **Dingwall, P.R. (1984).** Moving toward a representative system of protected areas in New Zealand. *In* McNeely, J.A., Miller, K.R., *eds.* National parks, conservation and development. The role of protected areas in sustaining society. Washington, D.C., Smithsonian Institution Press. 386-393. Maps, illus. (P)

5948 **Dingwall, P.R., Miers, K.H. (1979).** Biosphere reserves: the selection and protection of equivalent reserves in New Zealand. *In* N.Z. Man and Biosphere Report No. 2: 19-30. (P)

5949 **Druce, A.P., Simpson, M.J.A. (1979).** Plants of the Gouland Downs and Perry Pass along the Heaphy Track north west Nelson Forest Park. *Inform. Ser. New Zealand*, 80: 15. Illus., maps. (P)

5950 **Edmonds, A.S. (1979).** The clearance of the lowlands: the passing of New Zealand's lowland indigenous forests. *Ann. Roy. N.Z. Instit. Hort. J.*, 7: 65-74.

5951 **Fisher, A.B. (1972).** In search of a rare plant in Stewart Island. *Forest and Bird*, 185: 21-13. Illus. (T)

5952 **Forde, M.B. (1986).** Plant genetic resources in New Zealand. *J. Roy. Soc. New Zealand*, 16(1): 101-110. Includes tables of the forms in which various economic crop plant gene resources are kept in botanic gardens. (E/G)

5953 **Forde, M.B., Burdon, R.D., Dawes, S.N., Dunbier, M.W., Given, D.R., Gray, M. (1985).** Report of the *ad hoc* committee on conservation of plant genetic resources in New Zealand. *Proc. Roy. Soc. New Zealand.*, 113: 117-133. (E)

5954 **Frankel, E. (1978).** Bibliography of the Great Barrier Reef Province. Canberra, Great Barrier Reef Marine Park Authority. Lists 4444 publications. (B)

5955 **Given, D.R. (1974).** Documentation and protection of rare taxa - the Cinderella of botanical conservation. *In* Pap. Ann. Conf. N.Z. Ecol. Soc. (T)

5956 **Given, D.R. (1975).** *Celmisia spedeni* G.Simpson and *Celmisia thomsoni* Cheeseman -two rediscovered species. *New Zealand J. Bot.*, 13: 547-556. (T)

5957 **Given, D.R. (1975).** Conservation of rare and threatened plant taxa in New Zealand - some principles. *Proc. New Zealand Ecol. Soc.*, 22: 1-6. Map. Includes remarks on Philip Island. (T)

5958 **Given, D.R. (1976).** A register of rare and endangered indigenous plants in New Zealand. *New Zealand J. Bot.*, 14(2): 135-149. Lists 314 taxa under consideration for threatened status. (T)

5959 **Given, D.R. (1976).** Endangered plants and photography. *New Zealand Camera*, 23(6): 2-4. (T)

NEW ZEALAND

5960　**Given, D.R. (1976).** Endangered plants. *New Zealand Nat. Herit.*, 7(102): 2838-2841. Illus. (T)

5961　**Given, D.R. (1977).** Rare and endangered forest plants. *New Zealand Foresters Forestry Handbook*, 1977: 43-46. (T)

5962　**Given, D.R. (1977).** Tropical ferns and fern allies of thermal areas. *Forest and Bird*, August: 5-8. (T)

5963　**Given, D.R. (1978).** "If you want to keep them, give them away!" *N.Z. Dairy Exporter*, 53: 76-7. (T)

5964　**Given, D.R. (1978).** The preservation of rare flora. *In* Proceedings of the Conference of High Mountain Resources, Lincoln College, 9-15 November 1977. Wellington, Dept of Lands and Survey. 294-295. (T)

5965　**Given, D.R. (1978).** Tropical pteridophytes of geothermal sites in the Taupo-Rotorua region. *In* Broadlands geothermal power development. Environmental impact report. Submissions. Wellington, Commission for the Environment. 17p. (T)

5966　**Given, D.R. (1979).** Endangered New Zealand plants. *Canterbury Envir. J.*, 4(7): 7-9. (T)

5967　**Given, D.R. (1979).** Rare and endangered plants with particular reference to high mountains. *In* N.Z. Man and Biosphere Report No. 2: 59-63. (T)

5968　**Given, D.R. (1979).** Register of threatened plant taxa of New Zealand. *In* Commission for the Environment. Biological resources workshop, Wellington, 12-13 September 1979. 23-30. (T)

5969　**Given, D.R. (1979).** Species at risk in Canterbury. *Canterbury Bot. Soc. J.*, 13: 54-55. (T)

5970　**Given, D.R. (1979).** Threatened New Zealand plants. *New Zealand Envir.*, 24: 4-9. (T)

5971　**Given, D.R. (1979).** Threatened plants in New Zealand. *In* Nature Conservation Council. A vanishing heritage. The problem of endangered species and their habitats. 22-45. Proceedings of the 49th Congress of the Australian and New Zealand Assoc. for the Advancement of Science. Part 1: Botany session. (T)

5972　**Given, D.R. (1981).** Flora of offshore and onlying islands. *In* XIII International Botanical Congress, Sydney, Australia, 21-28 August 1981: abstracts. Sydney, Australian Academy of Science. 215.

5973　**Given, D.R. (1981).** Rare and endangered plants of New Zealand. Wellington, Reed. vi, 154p. Illus., col. illus., maps. (T)

5974　**Given, D.R. (1981).** Threatened plants of New Zealand: documentation in a series of islands. *In* Synge, H., *ed.* The biological aspects of rare plant conservation. Chichester, Wiley. 67-79. Proceedings of International Conference, King's College, Cambridge, 14-19 July 1980. Maps. (T)

5975　**Given, D.R. (1983).** Monitoring and science - the next stage in New Zealand. Pacific Science Association 15th Congress, Dunedin, New Zealand. Abstracts 1. p.83. Abstract only.

5976　**Given, D.R. (1983).** Monitoring and science - the next stage in threatened plant conservation in New Zealand. *In* Given, D.R. *ed.* Conservation of plant species and habitats. Wellington, Nature Conservation Council. 83-102 Illus. Gives lists of threatened species. (P/T)

5977　**Given, D.R. (1984).** Documentation and assessment of geothermal habitats for management. *In* Dingwall, P.R., *comp.* Protection and parks, essays in the preservation of natural values in protected areas. Dept of Lands and Survey. 15-24. Includes rare species. (T)

NEW ZEALAND

5978 **Given, D.R. (1986).** Priceless plants. *N.Z. Gardener*, 7-8.

5979 **Given, D.R. (1988).** Rare and endangered plants - protection and conservation. *In* Proceedings of the New Zealand Parks and Recreation Admin. Conference, 1987. 151-165. (T)

5980 **Given, D.R., Kelly, G.C. (1976).** Rare and local plants of the West Coast Beech Project Area. *Beech Res. News*, 4: 16-24. (T)

5981 **Given, D.R., Sykes, W.R., Williams, P.A., Wilson, C.M., comps (1987).** Threatened and local plants of New Zealand - a revised checklist. Christchurch, Botany Division, DSIR. 17p. (T)

5982 **Given, D.R., comp. (1976).** Threatened plants of New Zealand: a register of rare and endangered plants of the New Zealand Botanical Region. Christchurch, DSIR. Loose-leaf series of detailed double-paged datasheets with maps on 50 selected threatened species. Supplements in 1977, 1978. (T)

5983 **Greenwood, R.M., Skipworth, J.P. (1979).** Keebles bush, Palmeston north - a lowland forest remnant. How should it be managed? *New Zealand J. Ecol.*, 2: 91-92.

5984 **Hosking, G.P., Hutcheson, J.A. (1988).** Mountain beech (*Nothofagus solandri* var. *cliffortioides*) decline in the Kaweka Range, North Island, New Zealand. *New Zealand J. Bot.*, 26(3): 393-400. Illus. (T)

5985 **Howard, W.E. (1964).** Modification of New Zealand's flora by introduced mammals. *Proc. New Zealand Ecol. Soc.*, 11: 59-62. (T)

5986 **Howden, C. (1978).** Native Plant Protection Act, 1934. *Hort. New Zealand*, 6: 10-11. (L)

5987 **IUCN Threatened Plants Unit (1983).** The botanic gardens list of rare and threatened plants of New Zealand. Kew, IUCN Botanic Gardens Conservation Co-ordination Body Report No. 8. Lists plants in cultivation in botanic gardens. (G/T)

5988 **Kerby, R. (1983).** *Elingamita johnsonii. The Plantsman*, 5(1): 32-34. Illus. (T)

5989 **Knox, G.A. (1973).** Conservation and research on the offshore islands of New Zealand. *In* Costin, A.B., Groves, K.H., *eds*. Nature conservation in the Pacific. Proc. Symposium A-10, XII Pacific Science Congress, Canberra, August-September 1971. IUCN and Australian Nat. Univ. Press. 229-239. Maps.

5990 **Mackinder, J. (1982).** Missing dicots from the Auckland flora. *Newsl. Auckland Bot. Soc.*, 37(1): 4-5. (T)

5991 **Majstrik, V. (1978).** Narodni parky a ochrana prirody na Novem Zelande. (National parks and conservation of nature in New Zealand.) *Ziva*, 16(1): 7-11. Cz. Illus. Map. (P)

5992 **Manning, J. (1977).** Development of scenic reserves as urban amenities. *Hort. New Zealand*, 5: 10-12. (P)

5993 **Mark, A.F. (1977).** Vegetation of Mount Aspiring National Park, New Zealand. *Nat. Parks Sci. Ser.*, 2: 79p. Col. illus., maps. Plant lists. (P)

5994 **Mark, A.F. (1978).** Permanent photographic points for following vegetation changes. *Tussock Grasslands Mountain Lands Inst. Rev.*, 37: 38-45. Illus. Mount Aspiring National Park vegetation survey. (P)

5995 **Mark, A.F. (1985).** The botanical component of conservation in New Zealand. *New Zealand J. Bot.*, 23(4): 789-810. Illus., map.

5996 **Mark, A.F., Baylis, G.T.S. (1982).** Further studies on the impact of deer on Secretary Island, Fiordland, New Zealand. *New Zealand J. Ecol.*, 5: 67-75. The continuing effects of red deer on the rain forest of Secretary Island. (P)

NEW ZEALAND

5997 **Martin, W. (1941).** Preservation of native plants. *J. Roy. New Zealand Inst. Hort.*, 11: 43-46. (T)

5998 **Mascarenhas, A.F., comp.** (?) Biological diversity and genetic resources techniques and methods: tissue culture directory for Asia-Pacific. CSC Technical Publication Series No. 249. London, Commonwealth Science Council. 12p. (E/G)

5999 **McCaskill, L.W. (1974).** Scenic reserves of Canterbury. Wellington, New Zealand Department of Lands and Survey. Illus., map. (P)

6000 **McCaskill, L.W. (1975).** Notes on *Ranunculus paucifolius* (Castle Hill Buttercup) Reserve for Department of Lands and Survey, Christchurch. Mimeo. 5p. Includes list of other plants in Castle Hill Reserve. (P/T)

6001 **McCaskill, L.W. (1979).** Castle Hill Reserve for the preservation of flora and fauna - *Ranunculus paucifolius*. *In* N.Z. Man and Biosphere Report No. 2. 64-66. (P/T)

6002 **McCaskill, L.W. (1982).** The Castle Hill buttercup (*Ranunculus paucifolius*): a story of preservation. Review - Tussock Grasslands and Mountain Lands Institute. 1982(41). 38-48. Illus., maps. (P/T)

6003 **McKerchar, N.D., Dingwall, P.R. (1984).** Identifying the essential scientific needs of protected area managers. *In* Unesco-UNEP. Conservation, science and society. Paris, Unesco. 320-330. Contributions to the First International Biosphere Reserve Congress, Minsk, Byelorussia/USSR, 26 September - 2 October 1983. (P)

6004 **Molloy, B.P.J. (1971).** Possibilities and problems for nature conservation in a closely settled area. *Proc. New Zealand Ecol. Soc.* 1971, 18: 25-27. Illus., maps, list of protected areas in the Canterbury region. (P)

6005 **Molloy, L.F. (1984).** The reservation of commercially important lowland forests in New Zealand. *In* McNeely, J.A., Miller, K.R., *eds.* National parks, conservation and development. The role of protected areas in sustaining society. Washington, D.C., Smithsonian Institution Press. 394-402. Maps. (E/P)

6006 **Moore, L.B. (1983).** *Pseudowintera axillaris*. *Newsl. Auckland Bot. Soc.*, 38(1): 10. (T)

6007 **Morrah, P. (1981).** Mount Lees Reserve, a gift to New Zealand. *Hort. New Zealand,* 10: 12-15. Illus. (P)

6008 **Nature Conservation Council (1984).** Collection of native orchids could endanger wild populations. *Newsl. New Zealand Native Orchid Group,* 9: 2-4. Collectors. (L/P/T)

6009 **Nature Conservation Council (1984).** Collection of native orchids could endanger wild populations. *Orchids New Zealand,* 9(6): 138-140. (T)

6010 **Nature Conservation Council (1984).** Threatened plants programme. *Nat. Conserv. Council Newsl.,* 56, Oct-Nov 1984: 6. (T)

6011 **Nelson, R. (1979).** Deer and resulting devastation in New Zealand. Wellington, Royal Forest & Bird Protection Society of N.Z. iv, 71p. Illus.

6012 **New Zealand Ecological Society (1978).** The future of west coast forests and forest industries. *New Zealand J. Ecology,* 1: 166-172. Ref. in *Forest. Abs.,* 41: 100 (1980).

6013 **New Zealand Forest Service (1976).** Tararua State Forest Park. Information Series No. 73. Wellington, New Zealand Forest Service. 27p. Illus. (P)

6014 **New Zealand Forest Service (1978).** New Zealand Forest Service Seminar. Management proposals for state forests of the Rangitoto and Hauhungaroa Ranges, central North Island, Taupo, 23-30 March 1978. Wellington, New Zealand Forest Service. (P)

NEW ZEALAND

6015 **New Zealand Forest Service (1978).** Submissions on the west Taupo State forests. Conference: Seminar on Management Proposals for State Forests of the Rangitoto and Hauhungaroa Ranges, Central North Island, N.Z., 1978. Wellington, New Zealand Forest Service. 67p.

6016 **New Zealand Wildlife Service (1983).** Species at risk. New Zealand Wildlife Service. 17p. Includes half-page accounts on 6 plants. (T)

6017 **Newsome, A.E. (1973).** The adequacy and limitations of flora conservation for fauna conservation in Australia and New Zealand. *In* Costin, A.B., Groves, R.H., *eds.* Nature conservation in the Pacific. Proc. Symposium A-10, XII Pacific Science Congress, Canberra, August-September 1971. IUCN and Australian Nat. Univ. Press. 93-110. Illus., maps. (P)

6018 **Nicholls, J.L. (1977).** Rare indigenous forest plants. *New Zealand J. Forest.*, 22(1): 155-161. (T)

6019 **Nordmeyer, A.H. (1978).** Protection forestry. *New Zealand J. Forest.*, 23(2): 169-172.

6020 **O'Connor, K.F. (1982).** The implications of past exploitation and current developments to the conservation of South Island tussock grasslands. *New Zealand J. Ecol.*, 5: 97-107. (E/P)

6021 **O'Connor, K.F., Ackley, K.A. (1981).** New Zealand's waitaki: a planning prospective for regional resources. *Ambio*, 10(2-3): 142-147. Illus., maps.

6022 **OECD (1981).** Environmental policies in New Zealand. Paris, OECD. 78p.

6023 **Paterson, G. (1983).** The role of horticulture and botanic gardens. *In* Given, D.R., *ed.* Conservation of plant species and habitats. Wellington, N.Z., Nature Conservation Council. 59-66. Illus. (G/T)

6024 **Paterson, G. (1984).** An effort from "Down Under". *Threatened Pl. Newsl.*, 13: 12. (G/T)

6025 **Pekelharing, C.J., Reynolds, R.N. (1983).** Distribution and abundance of browsing mammals in Westland National Park in 1978, and some observations on their impact on the vegetation. *New Zealand J. Forest. Sci.*, 13(3): 247-265. Maps. (P)

6026 **Purdie, A.W. (1985).** *Chordospartium muritai* (Papilionaceae) - a rare new species of New Zealand tree broom. *New Zealand J. Bot.*, 23: 157-161. Illus., map. (T)

6027 **Sage, B. (1979).** New Zealand's forests in danger. *New Sci.*, 82(1149): 31-33. Illus.

6028 **Seddon, M. (1982).** *Todea barbara. Newsl. Auckland Bot. Soc.*, 37(2): 13. (T)

6029 **Simpson, P. (1986).** The rehabilitation of endangered plant species on the Three Kings Islands with particular reference to *Tecomanthe*. *In* Wright, A.E., Beever, R.E., *eds.* The offshore islands of northern New Zealand. Proceedings of a symposium convened by Offshore Island Research Group in Australia, 10-13 May 1983. Wellington, Department of Lands and Survey. 187-195. (T)

6030 **Thomas, G.M., Ogden, J. (1983).** The scientific reserves of Auckland University: 1. General introduction to their history, vegetation, climate and soils. *Tane*, 29: 143-161. Maps. Reserves at Swanson, Huapai, Oratia, Anawhata and Leigh. (P)

6031 **Thompson, K. (1979).** A case for the conservation and reservation of New Zealand's peatlands. *In* Hamilton, L.S., Hodder, A.P.W., *eds.* Proc. of a Symposium on New Zealand's peatlands, Hamilton, 23-24 Nov. 1978. 111-119. Wetlands.

6032 **Thompson, K. (1979).** Towards a better understanding of New Zealand's biological resources: some recent approaches, some recommendations and some problems. *In* Proceedings of the Biological Resources Workshop, Wellington, 12-13 September 1979. Wellington, Commission for the Environment. 119-144.

NEW ZEALAND

6033 **Thompson, K. (1983).** The status of New Zealand's wetlands in 1983. *New Zealand Envir.*, 37: 3-7.

6034 **Thompson, K. (1983).** The status of New Zealand's wetlands; still a question of conflicting interests. *In* Given, D.R., *ed.* Conservation of plant species and habitats. Wellington, Nature Conservation Council. 103-116. Illus. (P)

6035 **Timmins, S.M., Clarkson, B.D., Shaw, W.B., Atkinson, I.A.E. (1984).** Mapping native vegetation using Landsat data. *New Zealand J. Sci.*, 27(4): 389-397. Illus.

6036 **Trussell, D. (1982).** History in an antipodean garden. *The Ecologist*, 12(1): 32-42. Illus. Deforestation.

6037 **Warburton, T.H. (1978).** Control of our shrinking heritage. *Roy. New Zealand Inst. Hort. Ann. J.*, 6: 112-114.

6038 **Wardle, P. (1978).** Regeneration status of some New Zealand conifers with particular reference to *Libocedrus bidwillii* in Westland National Park. *New Zealand J. Bot.*, 16(4): 471-477. (P)

6039 **Wardle, P. (1979).** Plants and landscape in Westland National Park. *Nat. Parks Sci.*, ser. no. 3: 168p. Illus., map. (P)

6040 **Williams, G.R., Given, D.R., comps (1981).** The Red Data Book of New Zealand: rare and endangered species of endemic terrestrial vertebrates and vascular plants. Wellington, N.Z., Nature Conservation Council. 175p. Maps. (T)

6041 **Williams, P.A. (1983).** Woody weeds and native vegetation in New Zealand: a conservation problem. Pacific Science Assoc. Congress, Dunedin, 1-11 February 1983. 15(1/2): 256.

6042 **Wilson, C.M., Given, D.R. (1988).** Priorities for threatened plant conservation in New Zealand. Christchurch, Botany Division. 55p. Data sheets on 21 taxa. (T)

6043 **Wilson, C.M., Given, D.R. (1989).** Threatened plants of New Zealand. Wellington, DSIR. viii, 151p. Col. illus. Data on over 95 threatened plants; distribution maps. (T)

6044 **Wilson, H.D. (1976).** Vegetation of Mount Cook National Park, New Zealand. *Nat. Parks Sci.*, 1: 138p. Illus., maps. (P)

6045 **Wilson, H.D. (1978).** Field guide: wild plants of Mount Cook National Park. Christchurch, N.Z., Field Guide Publication. 294p. Illus. (P)

6046 **Wisheart, P. (?)** Forgotten habitats matter. The Scout Association of New Zealand. 21p. Illus.

6047 **Wright, A.E. (1983).** Conservation status of the Three Kings Islands endemic flora in 1982. *Rec. Auckland Insk. Mus.*, 20: 175-184. Map. (T)

6048 **Wynn, G. (1979).** Pioneers, politicians and the conservation of forests in early New Zealand (legislation). *J. Hist. Geogr.*, 5(2): 171-188. Illus., map. (L)

NEW ZEALAND - Antipodes Islands

6049 **Knox, G.A. (1973).** Conservation and research on the offshore islands of New Zealand. *In* Costin, A.B., Groves, K.H., *eds.* Nature conservation in the Pacific. Proc. Symposium A-10, XII Pacific Science Congress, Canberra, August-September 1971. IUCN and Australian Nat. Univ. Press. 229-239. Maps.

NEW ZEALAND - Auckland Islands

6050 **Campbell, D.J., Rudge, M.R. (1976).** A case for controlling the distribution of the tree daisy (*Olearia lyallii*) Hook. f. in its type locality, Auckland Islands. *Proc. New Zealand Ecol. Soc.*, 23: 109-115. Illus., map. (T)

NEW ZEALAND - Auckland Islands

6051 **Campbell, D.J., Rudge, M.R. (1978).** Reply to: Goats on Auckland Islands. *New Zealand J. Bot.*, 16(2): 293-296. Deals with feral goat problem.

6052 **Knox, G.A. (1973).** Conservation and research on the offshore islands of New Zealand. *In* Costin, A.B., Groves, K.H., *eds.* Nature conservation in the Pacific. Proc. Symposium A-10, XII Pacific Science Congress, Canberra, August-September 1971. IUCN and Australian Nat. Univ. Press. 229-239. Maps.

6053 **Wardle, P., Moar, N.T., Given, D.R. (1978).** Goats on Auckland Islands. *New Zealand J. Bot.*, 16(2): 291-292. Feral goats are a threat to the flora.

NEW ZEALAND - Bounty Islands

6054 **Knox, G.A. (1973).** Conservation and research on the offshore islands of New Zealand. *In* Costin, A.B., Groves, K.H., *eds.* Nature conservation in the Pacific. Proc. Symposium A-10, XII Pacific Science Congress, Canberra, August-September 1971. IUCN and Australian Nat. Univ. Press. 229-239. Maps.

NEW ZEALAND - Campbell Island

6055 **Brockie, R.E., Loope, L.L., Usher, M.B., Hamann, O. (1988).** Biological invasions of island nature reserves. *Biol. Conserv.*, 44(1-2): 9-36. Includes 6 case studies. (P)

6056 **Knox, G.A. (1973).** Conservation and research on the offshore islands of New Zealand. *In* Costin, A.B., Groves, K.H., *eds.* Nature conservation in the Pacific. Proc. Symposium A-10, XII Pacific Science Congress, Canberra, August-September 1971. IUCN and Australian Nat. Univ. Press. 229-239. Maps.

6057 **McKerchar, N.D.R., Devine, W.T. (1984).** The Campbell Island story: the management challenge of sub-Antarctic islands. *In* McNeely, J.A., Miller, K.R., *eds.* National parks, conservation and development. The role of protected areas in sustaining society. Washington, D.C., Smithsonian Institution Press. 376-385. Maps. (P)

6058 **Meurk, C.D. (1982).** Regeneration of subantarctic plants on Campbell Island following exclusion of sheep. *New Zealand J. Ecol.*, 5(5): 51-58. Illus., map.

NEW ZEALAND - Chatham Islands

6059 **Devine, W.T. (1982).** Nature conservation and land-use-history of the Chatham Islands, New Zealand. *Biol. Conserv.*, 23(2): 127-140. Illus., map.

6060 **Given, D., Williams, P. (1985).** Conservation of Chatham Island: flora and vegetation. Botany Division, DSIR, New Zealand. Reprinted with amendments March 1985. Contains 44 plant Red Data Sheets. (T)

6061 **Kelly, G.C. (?)** Distribution and ranking of remaining areas of indigenous vegetation in the Chatham Islands. Department of Land and Surveys Resource Inventory. In prep. Map with extended legend. (T)

6062 **Knox, G.A. (1973).** Conservation and research on the offshore islands of New Zealand. *In* Costin, A.B., Groves, K.H., *eds.* Nature conservation in the Pacific. Proc. Symposium A-10, XII Pacific Science Congress, Canberra, August-September 1971. IUCN and Australian Nat. Univ. Press. 229-239. Maps.

NEW ZEALAND - Kermadec Islands

6063 **McCombs, P. (1987).** Cousteau and the capture of paradise. *Washington Post*, G1-G6. Describes programme to eradicate invasive species on Raoul Island.

6064 **Sykes, W.R. (1969).** The effect of goats on vegetation of the Kermadec Islands. *Proc. New Zealand Ecol. Soc.*, 16: 13-16. (T)

NEW ZEALAND - South Island

6065 **Anon. (1988).** Swamp forests cause concern. *Oryx*, 22(1): 60. Logging of *Dacryocarpus dacrydioides* forests, South Westland. (P)

NICARAGUA

6066 **D'Arcy, W.G. (1977).** Endangered landscapes in Panama and Central America: the threat to plant species. *In* Prance, G.T., Elias, T.S., *eds.* Extinction is forever. Proceedings of a symposium at the New York Botanical Garden, 11-13 May 1976. New York, New York Botanical Garden. 89-104. Maps. The flora and deforestation today. (P/T)

6067 **Eremeev, A.G., Zhivotiagin, I.F., Shishkov, E.V. (1981).** [Forest resources of the Nicaraguan Republic]. *Lesnoe Khoz.*, 6: 69-72. Rus. Illus.

6068 **Ryan, D.A. (1978).** Recent development of national parks in Nicaragua. *Biol. Conserv.*, 13(3): 179-182. (Sp). (P)

NIGER

6069 **Gillet, H., Fabreques, B.P. de (1982).** Quelques arbres utiles, en voie de disparition, dans le centre-est du Niger. *Terre Vie*, 36(3): 465-470. Fr (En). Includes *Khaya senegalensis, Terminalia avicennioides.* (T)

6070 **Kinako, P.D.S. (1977).** Conserving the mangrove forest of the Niger Delta. *Biol. Conserv.*, 11(1): 35-39. Map.

6071 **Newby, J. (1984).** The role of protected areas in saving the Sahel. *In* McNeely, J.A., Miller, K.R., *eds.* National parks, conservation and development. The role of protected areas in sustaining society. Washington, D.C., Smithsonian Institution Press. 130-136. Map. (P)

6072 **Newby, J. (1986).** Niger plans wildlife protection. *WWF Monthly Report*, July 1986: 167-168. Illus. (T)

6073 **Newby, J. (1986).** The fabulous lake of wild sorghum. *WWF News*, 40: 7. Air Tenere region, Sahara. Gene source. (E/T)

NIGERIA

6074 **Afolayan, T.A. (1978).** Effects of fire on the vegetation and soils in Kainji Lake National Park, Nigeria. *In* Hyder, D.N., *ed.* Proceedings of the First International Rangeland Congress, Denver, 14-18 August 1978. Denver, Society for Range Management. 55-59. (P)

6075 **Afolayan, T.A. (1978).** The effect of fires (burning treatments) on the vegetation in Kainji Lake National Park, Nigeria. *Oikos*, 31: 376-382. (Rus). (P)

6076 **Ajayi, S.S., Milligan, K.R.N., Ayeni, J.S.O., Afolayan, T.A. (1981).** A management programme for Kwiambana Game Reserve, Sokoto State, Nigeria. *Biol. Conserv.*, 20(1): 45-57. Maps. (P)

6077 **Anadu, P.A. (1987).** Progress in the conservation of Nigeria's wildlife. *Biol. Conserv.*, 41(4): 237-251. Map. (L)

6078 **Ayeni, J.S.O. (1985).** Strategies for quick and effective conservation and recovery of the diminishing wildlife resources of Nigeria. *Nigerian Field*, 50(1-2): 13-20.

6079 **Chapman, J.D. (1982).** Conservation of Afromontane forest: Ngel Nyaki Forest Reserve. *Nigerian Field*, 47(1-3): 133. (P)

6080 **Charter, J.R. (1968).** Nigeria. *In* Hedberg, I. & O., *eds.* Conservation of vegetation in Africa south of the Sahara. Symposium Proceedings, at 6th plenary meeting of AETFAT, Uppsala. *Acta Phytogeogr. Suec.*, 54: 91-94. Describes vegetation, outlines protected areas. (P)

NIGERIA

6081 **Egunjobi, L. (1989).** Nigeria and the advancing desert. *The Environmentalist*, 9(1): 31-37. Illus., map. Desertification.

6082 **Gbile, Z.O., Ola-Adams, B.A., Soladoye, M.O. (1978).** Endangered species of the Nigerian flora. *Nigerian J. For.*, 8(1,2): 14-20. (T)

6083 **Gbile, Z.O., Ola-Adams, B.A., Soladoye, M.O. (1981).** List of rare species of the Nigerian flora. *Research Paper (Forest Series)* 47. Ibadan, Forest Research Institute of Nigeria. 35p. (T)

6084 **Jibirin Jia, M.O.N. (1971).** The Yankari Game Reserve: 1955-1970. *In* Happold, D.C.D., *ed.* Wildlife conservation in West Africa. Proceedings of a Symposium held at the University of Ibadan, Nigeria, 2 April 1970. Morges, IUCN. 27-30. (P)

6085 **Kaliduu, O., ed. (1988).** A national conservation strategy for Nigeria. Nigeria, Wildlife Conservation Division and Reprographics Unit, v, 7 5p. Illus., maps. (L/T)

6086 **Kio, P.R.O. (1983).** Management potentials of the tropical high forest with special reference to Nigeria. *In* Sutton, S.L., Whitmore, T.C., Chadwick, A.C., *eds.* Tropical rain forest: ecology and management. Oxford, Blackwell Scientific. 445-455. Special Publications Series of the British Ecological Society No. 2.

6087 **Lowe, R.G. (1984).** Forestry and forest conservation in Nigeria. *Commonw. Forest. Rev.*, 63(2): 129-136. An outline is given of the development of the timber industry in Nigeria over the past 80 years.

6088 **Nowakowski, M. (1978).** Rezerwat Yankari jako przyklad zabiegow o ochrone przyrody na terenie Nigerii. (The Yankari Nature Reserve as an example of the protective measures in nature conservation applied in Nigeria.) *Chronmy Przyr. Ojczysta*, 34(6): 68-71. Po. Illus., map. (P)

6089 **Oguntala, A.B. (1978).** The changing structure of Nigerian tropical rain forests. *Newsl. Int. Union Forest Res. Org.*, 1: 2-3.

6090 **Oguntala, A.B. (1980).** A proposal for recreation use of Nigeria's Sapoba Forest Reserve. *Parks*, 5(1): 15-17. Illus., map. (P)

6091 **Okali, D.U.U. (1979).** The Nigerian rainforest project - an overview. *Nat. Resources*, 15(4): 31-33. MAB Project 1.

6092 **Ola-Adams, B.A. (1977).** Conservation of genetic resources of indigenous forest tree species in Nigeria: possibilities and limitations. *Forest Genetic Resources Inf.*, 7: 1-9. (E/G)

6093 **Ola-Adams, B.A., Iyamabo, D.E. (1977).** Conservation of natural vegetation in Nigeria. *Envir. Conserv.*, 4(3): 217-226. Illus.

6094 **Osborne, R. (1988).** *Encephalartos barteri*. *Encephalartos*, 14: 8-16. Illus., map. Data sheet; includes conservation. (G/T)

6095 **Oyewole, S.O., Harris, B. (1979).** Nigeria. *In* Hedberg, I., *ed.* Systematic botany, plant utilization and biosphere conservation. Proc. of a sym. held in Uppsala in commemoration of the 500th ann. of the Univ. Stockholm, Almqvist & Wiksell Int'l. 92.

6096 **Rajanaidu, N., Ooi, S.C., Lawrence, M.J. (1981).** Conservation of oil palm genetic resources and genotype x environment interaction. *In* Fourth International Sabrao Congress, 4-8 May 1981, at Universiti Kebangsaan Malaysia and Federal Hotel, Kuala Lumpur. Gene banks. (E)

6097 **Sanford, W.W. (1969).** Conservation of West African orchids I: Nigeria. *Biol. Conserv.*, 1(2): 148-150. (T)

6098 **Sanford, W.W. (1980).** Humid savannah of north-western Nigeria. *Nat. Resources*, 16(2): 15-17. Map.

NORFOLK ISLAND

6099 **Anon. (1967).** Norfolk Island. *IUCN Bull.*, 16(4): 25-26. Rain forest on Mt Pitt and Mt Bates threatened by encroaching road projects. (P)

6100 **Anon. (1986).** Found again. *Species*, 6: 20. Rediscovery of *Abutilon julianae*. (T)

6101 **Anon. (1988).** Recovery of a *Hibiscus*. *Species*, 10: 21. (T)

6102 **Anon. (1989).** Restoring a ravished island. *The Environmentalist*, 9(2): 142-144. Illus. Philip Island. (T)

6103 **Australian National Parks and Wildlife Service (1984).** Plan of management Norfolk Island National Park and plan of management Norfolk Island Botanic Garden. Canberra, ANPWS. 112p. Maps. (G/P)

6104 **Australian National Parks and Wildlife Service (1989).** Philip Island revised draft plan of management. Australian National Parks and Wildlife Service. 128p. Includes annotated list of species; introduced species eradication programmes. (T)

6105 **Coyne, P. (1983).** Revegetation attempt on Philip Island, South Pacific. *Threatened Pl. Newsl.*, 12: 14. Map. (T)

6106 **Given, D.R. (1975).** Conservation of rare and threatened plant taxa in New Zealand - some principles. *Proc. New Zealand Ecol. Soc.*, 22: 1-6. Map. Includes remarks on Philip Island. (T)

6107 **Green, P.S. (1979).** Observations on the phytogeography of the New Hebrides, Lord Howe Island and Norfolk Island. *In* Bramwell, D., *ed.* Plants and islands. London, Academic Press. 41-53. (Sp). Map.

6108 **Green, P.S. (1985).** Refound: on South Sea isle. *Threatened Pl. Newsl.*, 15: 21. (T)

6109 **Green, P.S., Hicks, J. (1989).** Norfolk Island Botanic Garden: a new conservation garden. *Bot. Gard. Conserv. News*, 1(5): 31-32. Illus., map. (G)

6110 **McMichael, D.F., Talbot, F.H. (1970).** Conservation of islands and coral reefs of the Great Barrier Reef system, the islands of the Coral Sea, and Norfolk and Lord Howe Islands. *Micronesia*, 5(2): 493-496.

6111 **Melville, P.S. (1970).** Notes relating to the floras of Norfolk and Lord Howe islands, I. *J. Arnold Arb.*, 51(2): 204-220. Illus. (P/T)

6112 **Melville, R. (1969).** The endemics of Philip Island. *Biol. Conserv.*, 1: 170-172. (T)

6113 **Sykes, W.R., Atkinson, I.A.E. (1988).** Rare and endangered plants of Norfolk Island. Wellington, DSIR, Botany Division. (T)

6114 **Turner, J.S., Smithers, C.N., Hoogland, R.D. (1968).** Conservation of Norfolk Island. Australian Conservation Foundation Special Publ. No. 1. Norfolk Island, Australian Conservation Foundation. 41p. Includes checklist of plants on the islands with notes on local distribution, frequency, habitats; chapters on conservation problems and recommendations. Illus., map.

NORTHERN MARIANA ISLANDS

6115 **Anon. (1958).** The vegetation of Micronesia. Engineer Intelligence Study No. 257. Washington, D.C., U.S. Geological Survey, Military Geology Branch. 160p. Includes causes of denudation of vegetation in all island groups.

6116 **Anon. (1985).** CNMI Northern islands win preservation. *Pacific Science Assoc. Inf. Bull.*, 7(6): 57-58. Commonwealth of Northern Mariana Islands. Maug, Uracas, Asuncion and Guguan set aside for conservation; Sariguan Island affected by introduced rats and goats. (P)

NORTHERN MARIANA ISLANDS

6117 **Anon. (1985).** Northern Marianas slated for preservation. *Coastal Views* (Saipan, Marianas), 7(2): 1, 3, 10. Illus. Describes Maug, Uracas, Asuncion and Guguan, to be given protection. (P)

6118 **DiSalvatore, B. (1981).** The goat men of Aguijan. *Islands*, 1(1): 86-92. Introduced species, especially goats, destroy vegetation.

6119 **Fosberg, F.R. (1960).** The vegetation of Micronesia: 1. General descriptions, the vegetation of the Marianas Islands, and a detailed consideration of the vegetation of Guam. *Bull. Amer. Mus. Nat. Hist.*, 119: 75p. Illus.

6120 **Hoffman, C.W. (1950).** Saipan: the beginning of the end. Washington, D.C., U.S. Marine Corps. 286p. Impact of military operations.

6121 **Hoffman, C.W. (1951).** The seizure of Tinian. Washington, D.C., U.S. Marine Corps. 169p. Impact of military operations.

6122 **Spoehr, A. (1954).** Saipan: the ethnology of a war-devastated island. *Fieldiana: Anthropol.*, 41: 1-379. Discusses the impact of military activities on the vegetation.

6123 **U.S. Congress, Office of Technology Assessment (1987).** Integrated renewable resource management for U.S. insular areas. Washington, D.C., U.S. Govt Printing Office. 443p. Illus., maps. OTA-F-325. Includes description of vegetation, remaining coverage, threatened species. (P/T)

NORWAY

6124 **Anon. (1978).** Landskapsvern og naturvern i skogen: veiledande retningslinjer i skogbruket. (Care of the landscape and nature conservation in the forest: guidelines for forestry.) *Norsk Skogbruk*, 24(2): 3-14, 31-43. No.

6125 **Bakkevig, S. (1981).** Vegetasjon og naturgrunnlag i Tveitaneset Naturreservat, Rogaland. (Vegetation and ecology at Tveitaneset nature conservation area, Rogaland county, western Norway). *Blyttia*, 39(3): 107-113. No (En.) Illus., maps. (P)

6126 **Council of Europe (1981).** National activities: Norway. *Council of Europe Newsl.*, 81(6/7): 3. Jotunheim Nat. Park. (P)

6127 **Elven, R. (1979).** [Areas deserving botanical protection in Roros, Sor-Trondelag County, central Norway.] *Rapp. Nongl. Norske Vidensk. Selsk., Bot.*, 6: 1-158. No. (P)

6128 **Elven, R., Aarhus, A. (1984).** A study of *Draba cacuminum* (Brassicaceae). *Nordic J. Bot.*, 4(4): 425-441. Illus., maps. (T)

6129 **Frisen, R. (1988).** [Cooperation for protection of species and biotopes in the Nordic countries]. *Sven. Bot. Tidskr.*, 82(6): 499-501. Sw (En). (P)

6130 **Frislid, R. (1977).** Skog og villmark. Friluftsomrader og skogreservater pa statens grunn. (Forest and wilderness areas. Recreation areas and forest reserves on state land.) Oslo, Luther Forlag. 128p. No. (P)

6131 **Gjerlaug, H.C. (1975).** Liste over truede oglleler sjeldne planter i Norge, karsporeplanter og froplanter. Oslo. No. Includes a list of rare and threatened plants. (T)

6132 **Gjerlaug, H.C. (1977).** Liste over antatt utdodde, truete, sarbare og sjeldne plantearter i Norge. Oslo. 7p. No. (T)

6133 **Halvorsen, R. (1982).** Sjeldne og sarbare plantearter i Sor-Norge: 5. Strandtistel (*Eryngium maritimum*). (Rare and threatened plant species in south Norway: 5. *Eryngium maritimum.*) *Blyttia*, 40(3): 163-173. No (En). Illus., map. (T)

6134 **Halvorsen, R. (1984).** Sikring av sor-norske forekomster for nasjonalt truete plantearter: tilbakeblikk og presentasjon av en arbeidsplan. (Conservation of southern Norwegian stations for regionally threatened vascular plants: retrospect and plans for the future.) *Blyttia*, 42(4): 130-137. No (En). (T)

NORWAY

6135 **Halvorsen, R., Evje, G.A., Iversen, I. (1984).** Sjeldne og sarbare plantearter i Sor-Norge: 6. Dvergtistel *Cirsium acaule*. (Rare and threatened plants in south Norway: 6. *Cirsium acaule*.) *Blyttia*, 42(4): 143-148. No. (T)

6136 **Halvorsen, R., Fagernaes, K.E. (1980).** Rare and threatened plant species in south Norway: 2. *Glyceria plicata*. *Blyttia*, 38(3): 127-132. (T)

6137 **Halvorsen, R., Fagernaes, K.E. (1980).** Rare and threatened plant species in south Norway: 3. *Hornungia petraea*. *Blyttia*, 38(4): 171-180. (T)

6138 **Halvorsen, R., Fagernaes, K.E. (1980).** Sjeldne og sarbare plantearter i Sor-Norge: 1. Kubjelle (*Pulsatilla pratensis*). (Rare and threatened plant species in south Norway: 1. *Pulsatilla pratensis*.) *Blyttia*, 38(1): 3-8. No (En). Illus., maps. (T)

6139 **Hoiland, K. (1984).** Russearve, *Moehringia laterifolia*, en truet plante in Norge? (*Moehringia lateriflora*, an endangered species in the Norwegian flora?). *Blyttia*, 42:149-156. No. Illus. (T)

6140 **Hoiland, K. (1988).** Vern og forvaltning av sjeldne planter i Norge. (Protection and conservation of threatened plants in Norway.) *Svensk Bot. Tidskr.*, 82(6): 385-388. Sw (En). (T)

6141 **Karlsson, L. (1983).** [Notes on the vascular plant flora in the southwestern part of the Padjelanta National Park]. *Sven. Bot. Tidskr.*, 77(4): 217-220. Sw (En). Illus. (P)

6142 **Krohn, O. (1982).** Skogbruk og naturvern. (Forestry and nature conservation). Oslo, Norges Naturvernforbund, Gyldendal Norsk Forlag. 122p. No. Illus., maps.

6143 **Mejstrik, V. (1980).** Ochrana prirody a narodni parky v Norsku. *Ziva*, 28(5): 164-166. Cz. Illus., map.

6144 **Miljovernodepartementet (1976).** Oversikt over omrader og forekomster i Norge som er fredet eller vernet etter naturvernloven, samt omrader og forekomster som er administrativt fredet. Norwegian Ministry of Environmental Affairs. No. (List of areas and species in Norway protected by the Nature Conservation Act, and areas and objects protected by administrative regulations). (L/P)

6145 **Nilsson, O. (1979).** Threatened plants in the Nordic countries. *In* Hedberg, I., *ed.* Systematic botany, plant utilization and biosphere conservation: Proceedings of a symposium held in Uppsala in commemoration of the 500th ann. of the Univ. Stockholm, Almqvist & Wiksell International. 72-75. (T)

6146 **Norderhaug, M. (1984).** Trua planter og dyr i Noreg. Statens Naturvernrad. No (En). (T)

6147 **Okland, T. (1984).** Utsatte plantearter i Vestfold fylke - en oversikt. (Exposed vascular plants in Vestfold County, SE Norway - a survey). *Blyttia*, 42: 167-172. No (En). Illus. Species in decline due to human activities. (T)

6148 **Osthagen, H. (1984).** Fredete plantearter i Norge. (Plant species protected in Norway.) *Blyttia*, 42(4): 157-162. No (En). Col. illus., map. (L/T)

6149 **Schumacher, T., Ostmode, K. (1978).** Floristiske bidrag fra Rondane Nasjonal Park. (The flora of Rondane National Park, Central Norway.) *Blyttia*, 36(4): 193-195. No. Illus. (P)

6150 **Stormer, P., Torkelsen, A.-E. (1979).** Fra plantelivet i Skultrevassasen skogreservat i Drangedal. (Botanical observations from the Skultrevassasen Forest Reserve in Drangedal, southern Norway.) *Blyttia*, 37(1): 25-37. No. Illus., map. (P)

6151 **Vader, W. (1983).** Regionalmuseenes rolle i naturvernforskning og utredningsarbeid i Norge. (The importance of regional museums for research and development in nature conservation in Norway.) *Mem. Soc. Fauna Flora Fenn.*, 59(3): 101-105. No (En).

NORWAY

6152 **Vuokko, S. (1983).** Uhatut Kasvimme. (Our threatened plants.) Suomen Luonnonsuojelum Tuki Oy, Helsinki. 96p. Fi. Col. illus. Covers Nordic region. (T)

6153 **Wielgolaski, F.E. (1971).** IBP Ecosystems studies in Norway. *Biol. Conserv.*, 4(1): 71-72.

6154 **Yndgaard, F. (1982).** A documentation system for the Nordic Gene Bank. *Pl. Genet. Resource. Newsl.*, 49: 34-36. (Sp, Fr). (G)

OMAN

6155 **Astric, C. (1979).** Territoire Francais des Afars et des Issacs. *In* Hedberg, I., *ed.* Systematic botany, plant utilization and biosphere conservation. Proc. of a sym. held in Uppsala in commemoration of the 500th ann. of the Univ. Stockholm, Almqvist & Wiksell Int'l. 93.

6156 **Lewis, G. (1985).** Plants in peril. *Kew Mag.*, 2(4): 380-382. *Ceratonia oreothauma* identified as an endangered species. (T)

6157 **Miller, A.G., Morris, M. (1988).** Plants of Dhofar, the southern region of Oman: traditional, economic and medicinal uses. Muscat, Oman, The Office of The Advisor for Conservation of the Environment, Diwan of Royal Court, Sultanate of Oman. 361p. Col. illus. (E/T)

PAKISTAN

6158 **Afzal, Z., et al. (1987).** Genetic conservation activities in Pakistan. *FAO/IBPGR Plant Genetic Resources News*, 70: 33-34. Collectors. Seed bank. Collection of cereals, lentils & rice. (E)

6159 **Aswal, B.S., Mehrotra, B.N. (1983).** *Delphinium uncinatum* Hook.f. et Thoms. (Ranunculaceae) and *Lilium wallichianum* Schultes f. (Liliaceae): two rare finds from north-west Himalaya. *J. Econ. Taxon. Bot.*, 3(3): 773-775 (1982 publ. 1983). (T)

6160 **Cawardine, M., ed. (1986).** The nature of Pakistan. A guide to conservation and development issues: no. 1. Gland, Switzerland, IUCN Conservation for Development Centre and Lahore, WWF-Pakistan. 72p. Col. illus., maps. (E/L/P)

6161 **Chaudhry, I. (1978).** Problems of wildlife conservation in the developing countries, with special reference to Pakistan. *In* Proceedings of the 7th World Forestry Congress, Buenos Aires, 4-18 October 1972, vol. 3: Conservation and recreation. Buenos Aires, Instituto Forestal Nacional. 4013-4016. (Sp, Fr)

6162 **Khan, A.A., Hussain, M. (1985).** The development of a protected area system in Pakistan in terms of representative coverage of ecotypes. *In* Thorsell, J.W., *ed.* Conserving Asia's natural heritage. The planning and management of protected areas in the Indomalayan Realm. Proc. of the 25th working session of IUCN's CNPPA. Gland, IUCN. 60-68. (P)

6163 **Khan, S.M. (1977).** Ecological changes in the alpine pasture vegetation at Paya (Kaghan) due to complete protection from grazing. *Pakistan J. Forest.*, 27(3): 139-142.

6164 **Nawaz, M. (1985).** National parks reserves for Pakistan's North West Frontier Province. *Parks*, 10(1): 6-7. Map. Illus. (P)

6165 **Rasul, G. (1984).** Wild almond: an endangered species. *WWF-Pakistan Newsl.*, 3(3): 8-10. Illus. (E/T)

6166 **Siddiqui, I.A. (1984).** Plants of medicinal value found in Baluchistan. *WWF-Pakistan Newsl.*, 3(3): 5-7. (E)

6167 **Singh, P. (1983).** Extended distribution of an endemic species: *Tribulus rajasthanensis* Bandari & Sharma. *Bull. Bot. Surv. India*, 24(1-4): 237-238 (1982 publ. 1983). (T)

PALAU

6168 **Canfield, J.E. (1981).** Palau: diversity and status of the native vegetation of a unique Pacific island ecosystem. *Newsl. Hawaii. Bot. Soc.*, 20: 14-20.

6169 **Faulkner, D. (1981).** Palau: a pattern of islands. *Oceans*, 14(4): 36-43. Management plans threaten island.

6170 **Johnson, S.P. (1972).** Palau: conservation frontier of the Pacific. *Nat. Parks Conserv. Mag.*, 46(4): 12-17.

6171 **Johnson, S.P., Pierrepont, S. (1972).** Palau and a Seventy Islands Tropical Park. *Nat. Parks Conserv. Mag.*, 46(4): 12-17. Illus., map. Refers to proposed Seventy Islands Tropical Park. (P)

6172 **Johnson, S.P., Pierrepont, S. (1972).** Palau and a Seventy Islands Tropical Park. *Nat. Parks Conserv. Mag.*, 46(7): 4-8. Illus., map. Refers to proposed Seventy Islands Tropical Park. (P)

6173 **Johnson, S.P., Pierrepont, S. (1972).** Palau and a Seventy Islands Tropical Park. *Nat. Parks Conserv. Mag.*, 46(8): 9-13. Illus., map. Refers to proposed Seventy Islands Tropical Park. (P)

6174 **Kluge, P.F. (1986).** Palau: problems in the Pacific. *Smithsonian*, 17(6): 44-55.

6175 **Kochi, J.S. (1971).** Objectives and importance of conservation. *Atoll Res. Bull.*, 148: 21-22.

6176 **Nicholson, E.M., Douglas, G.L. (1970).** Conservation of oceanic islands. *In* IUCN Publications New Series, No. 17. IUCN Eleventh Technical Meeting, New Delhi 1969, Vol. I. 200-211.

6177 **Owen, R.P. (1978).** Conservation is for everyone. *Micronesian Reporter*, 26(3): 16-20. (T)

6178 **U.S. Congress, Office of Technology Assessment (1987).** Integrated renewable resource management for U.S. insular areas. Washington, D.C., U.S. Govt Printing Office. 443p. Illus., maps. OTA-F-325. Includes description of vegetation, remaining coverage, threatened species. (P/T)

PANAMA

6179 **Anon. (1979).** Panama declares national park in canal zone. *Envir. Conserv.*, 6(3): 186. (P)

6180 **Anon. (1987).** Darien. The road less-travelled. IUCN Special Report: Central America. *IUCN Bull.*, 18(10-12): 8-9. Illus. (P)

6181 **D'Arcy, W.G. (1977).** Endangered landscapes in Panama and Central America: the threat to plant species. *In* Prance, G.T., Elias, T.S., *eds.* Extinction is forever. Proceedings of a symposium at the New York Botanical Garden, 11-13 May 1976. New York, New York Botanical Garden. 89-104. Maps. The flora and deforestation today. (P/T)

6182 **Foster, R.B., Brokaw, N.V.L. (1982).** Structure and history of the vegetation of Barro Colorado Island (semideciduous forest, Panama). *In* Leigh, E.G., jr, Rand, A.S., Windsor, D.M., *eds.* The ecology of a tropical forest: seasonal rhythms and long-term changes. Washington, D.C., Smithsonian Institution Press. 67-81. Illus., maps. (P)

6183 **La Bastille, A. (1973).** An ecological survey of the proposed Volcano Baru National Park, Republic of Panama. IUCN Occas. Papers No. 6. Switzerland, IUCN. 77p. Illus. (P)

6184 **Lamlein, J. (1984).** Indian Forest Reserve in Panama. *Threatened Pl. Newsl.*, 13: 6. (P)

PANAMA

6185 **Leigh, E.G., jr, Rand, A.S., Windsor, D.M., eds. (1982).** The ecology of a tropical forest: seasonal rhythms and long term changes. Washington, D.C., Smithsonian Institution Press, 1982. Illus., maps. (P)

6186 **Mittermeier, R.A., Milton, K. (1978).** Proposal for an island national park in Panama. *Oryx*, 14(4): 343-344. Map. Isla Coiba. (P)

6187 **Morales, R., Barborak, J.R., MacFarland, C. (1984).** Planning and managing a multi-component biosphere reserve, the case of the La Amistad/Talamanca Range/Bocas de Toro Wildlands Complex of Cosa Rica and Panama. *In* Unesco-UNEP. Conservation, science and society. Paris, Unesco. 168-177. Contributions to the First International Biosphere Reserve Congress, Minsk, Byelorussia/USSR, 26 September - 2 October, 1983. Map. (P)

6188 **Navarro, J.C., Fletcher, R. (1988).** Preserving Panama's parks. *The Nature Conservancy Mag.*, 38: 20-24. (P)

6189 **Read, R.M. (1985).** *Aechmea strobilina* rediscovered? *J. Bromeliad Soc.*, 35(3): 104-105. (T)

6190 **Rubinoff, I., Smythe, N. (1982).** A jungle kept for study. *New Sci.*, 95(1319): 495-499. Illus., col. illus. (P)

6191 **Simons, R. (1984).** Ten years later: the Smithsonian international experience since the Second World Parks Conference. *In* McNeely, J.A., Miller, K.R., *eds.* National parks, conservation and development. The role of protected areas in sustaining society. Washington, D.C., Smithsonian Institution Press. 712-718. Map. Outlines projects in Nepal and Barro Colorado Island, Panama.

6192 **Smythe, N. (1985).** The park that protects the Panama Canal. *Oryx*, 18(2): 42-46. Illus., map. (P)

6193 **Steven, D. de, Putz, F.E. (1985).** Mortality rates of some rain forest palms in Panama. *Principes*, 29(4): 162-165.

6194 **Tarr, B. (1984).** Endangered species: *Peristeria elata* taken from *The Orchid World*, Volume 5 (1914). *Orchid Rev.*, 92(1083): 33. (T)

6195 **Torres, H. (1987).** La Amistad Biosphere Reserve. Conservation in the name of friendship. IUCN Special Report: Central America. *IUCN Bull.*, 18(10-12): 6. Illus. (P)

6196 **Torres, H., Hurtado, L., eds (1987).** Parque international de La Amistad - plan general de manejo y desarrollo. Madrid, Instituto de Cooperacion Iberoamericano. (P)

6197 **Torres, H., Hurtado, L., eds (1987).** Reserva la Biosfera de La Amistad: una estrategia para su conservacion y desarrollo. Turrialba, Costa Rica, Tropical Agricultural Research and Training Center. (P)

6198 **Woolliams, K.R. (1985).** Endangered *Heliconia*: how serious a problem? *Notes Waimea Arbor.*, 12(1): 5-8. (T)

6199 **Wright, R.M., Houseal, B., Leon, C. de (1985).** Kuna Yala: indigenous biosphere reserve in the making? *Parks*, 10(3): 25-26. Illus. (P)

PAPUA NEW GUINEA

6200 **Cruttwell, Rev. N. (1983).** Mount Gahavisuka Provincial Park, Papua New Guinea. *Austral. Parks. Recr.*, Feb: 20-23. Illus. (P)

6201 **Cruttwell, Rev. N.E.G. (1984).** Lipizauga Botanical Sanctuary, Mount Gahavisuka Provincial Park. *Rhododendron Not. Rec.*, 1: 147-155. Illus. (G/P)

6202 **Cruttwell, Rev. N.E.G., Siga, B. (1987).** Lipizauga Botanical Sanctuary, Papua New Guinea. *Bot. Gard. Conserv. News*, 1(1): 10-11. Map. (G/P)

PAPUA NEW GUINEA

6203 **Diamond, J.M. (1984).** Biological principles relevant to protected area design in the New Guinea region. *In* McNeely, J.A., Miller, K.R., *eds.* National parks, conservation and development. The role of protected areas in sustaining society. Washington, D.C., Smithsonian Institution Press. 330-332. (P)

6204 **Eaton, P. (1984).** Environmental policies to combat deforestation. UN/ESCAP. 17p. Mimeo. Tropical forest, deforestation.

6205 **Eaton, P. (1985).** Customary land tenure and conservation in Papua New Guinea. *In* McNeely, J.A., Pitt, D., *eds.* Culture and conservation: the human dimension in environmental planning. London, Croom Helm. 181-192.

6206 **Eaton, P. (1986).** Grassroots conservation: wildlife management areas in Papua New Guinea. Land Studies Centre Report 86/1. University of Papua New Guinea. v, 101p. Illus. (P)

6207 **Gagne, W.C., Gressitt, J.L. (1982).** Conservation in New Guinea. *In* Gressitt, J.L., *ed.* Biogeography and ecology of New Guinea. The Hague, Junk. 945-966. Illus., maps. (P)

6208 **Gare, N. (1986).** The making of a national park. *Parks*, 11(4): 13-18. Illus. (P)

6209 **Given, D.R. (1981).** Flora of offshore and onlying islands. *In* XIII International Botanical Congress, Sydney, Australia, 21-28 August 1981: abstracts. Sydney, Australian Academy of Science. 215.

6210 **Gorio, S. (1980).** Papua New Guinea involves its people in national park development. *Tigerpaper*, 7(1): 9-11. Illus., map. (P)

6211 **Heylingers, P.C. (1967).** Vegetation and ecology of Bougainville and Buka Islands. *In* CSIRO (Australia). Lands of Bougainville and Buka Islands, Territory of Papua and New Guinea. Melbourne, CSIRO. 121-145. Maps.

6212 **Hilton, R.G.B., Johns, R.J. (1984).** The future of forestry in Papua New Guinea. *Commonw. Forest. Rev.*, 63(2): 103-106. A discussion of exploitation of the natural rainforest, its productivity, and the role of plantations as a resource for the forest industry. (E)

6213 **Howcroft, N.H.S. (1988).** From the back of the bushhouse. *Orchadian*, 9: 126-128. Conservation of orchids and cultivation centres. (G/T)

6214 **Hoyle, M.A. (1977).** Forestry and conservation in Papua New Guinea. *Tigerpaper*, 4(4): 10-12. Illus.

6215 **Johns, R.J. (1986).** The instability of the tropical ecosystem in New Guinea. *Blumea*, 31(2): 341-371. Illus. Tropical forest.

6216 **Johns, R.J., Stevens, P.F. (1971).** Mount Wilhelm flora: a checklist of the species. Botany Bulletin No. 6. Papua New Guinea, Dept of Forests. 61p.

6217 **Kores, P. (1977).** Papua New Guinea's orchids: an exploited resource. *Sci. New Guinea*, 5: 51-66. Paper presented to the PNG Botanical Society, Wau Ecology Institute, October 1977; describes the threats to orchids. (E/T)

6218 **Kwapena, N. (1984).** Wildlife management by the people. *In* McNeely, J.A., Miller, K.R., *eds.* National parks, conservation and development. The role of protected areas in sustaining society. Washington, D.C., Smithsonian Institution Press. 315-321. Maps. (P)

6219 **Lamb, K.P., Gressitt, J.L., eds (1976).** Ecology and conservation in Papua New Guinea: a symposium held at the Wau Ecology Institute on 4th November 1975. Wau Ecology Institute Pamphlet No. 2.

6220 **McIntosh, D.H. (1962).** The effect of man on the forests of the highlands of eastern New Guinea. *In* Symposium on the Impact of Man on Humid Tropics Vegetation, Goroka, Papua-new Guinea, September, 1960. Canberra, A.J. Arthur, Commonwealth Govt. Printer. 123-126. Tropical rain forest.

PAPUA NEW GUINEA

6221 **Morauta, L., Pernetta, J., Heaney, W., eds (1982).** Traditional conservation in Papua New Guinea. Port Moresby, Papua New Guinea Office of Environment and Conservation. xii, 392p. Illus., map. (E)

6222 **Robbins, R.G. (1972).** Vegetation and man in the south-west Pacific and Papua New Guinea. *In* Ward, R.G., *ed*. Man in the Pacific islands: essays on geographical change in the Pacific islands. London, Clarendon Press. x, 339p.

6223 **Routley, R. & V. (1980).** Destructive forestry in Melanesia and Australia. *Ecologist*, 10(1-2): 56-67. Illus.

6224 **Saulei, S.M. (1984).** Natural regeneration following clear-fell logging operations in the Gogol Valley, Papua New Guinea. *Ambio*, 13(5-6): 351-354. Illus., charts.

6225 **Schodde, R. (1973).** General problems of fauna conservation in relation to the conservation of vegetation in New Guinea. *In* Costin, A.B., Groves, K.H., *eds*. Nature conservation in the Pacific. Proc. Symposium A-10, XII Pacific Science Congress, Canberra, August-September 1971. IUCN and Australian Nat. Univ. Press. 123-144. Map.

6226 **Singh, B. (1984).** Keynote address: the Oceanian Realm. *In* McNeely, J.A., Miller, K.R., *eds*. National parks, conservation and development. The role of protected areas in sustaining society. Washington, D.C., Smithsonian Institution Press. 310-314. (P)

6227 **Specht, R.L., Roe, E.M., Boughton, V.H., eds. (1974).** Conservation of major plant communities in Australia and Papua New Guinea. Australian J. Bot. Supp. Series 7. 667p. Detailed assessment of conservation status of major plant communities. (T)

6228 **Viner, A.B. (1984).** Environmental protection in Papua New Guinea. *Ambio*, 13(5-6): 342-344. Illus.

6229 **Womersley, J.S., comp. (1974).** Conservation of primitive, rare, and endangered species. *In* Specht, R.L., *et al*., *eds*. Conservation of major plant communities in Australia and Papua New Guinea. Australian J. Bot. Suppl. No. 7: 594. (T)

PAPUA NEW GUINEA - Bougainville

6230 **Heylingers, P.C. (1967).** Vegetation and ecology of Bougainville and Buka Islands. *In* CSIRO (Australia). Lands of Bougainville and Buka Islands, Territory of Papua and New Guinea. Melbourne, CSIRO. 121-145. Maps.

PARAGUAY

6231 **Gauto, R. (1989).** Private conservation programs in Paraguay. *Conservation Biology*, 3(2): 120-122.

6232 **Rios, E., Zardini, E. (1989).** Conservation of biological diversity in Paraguay. *Conservation Biology*, 3(2): 118-120. Map. (P)

6233 **Stutz, L.C. (1983).** Etudes floristiques de divers stades secondaires des formations forestieres du haut Parana (Paraguay) oriental. *Candollea*, 38(2): 541-573. Fr (En). Maps. (P)

PERU

6234 **Aguilar, D.P.R. (1986).** Yanachaga-Chemillen: futuro parque nacional en la Selva Central del Peru. *Bol. Lima*, 45: 7-21. (P)

6235 **Anon. (1977).** Peru. Supreme Decree No. 158-77-AG approving the wild flora and fauna conservation regulations, 31 March 1977. *Food Agric. Legis.*, 26(2): 111-114. (L/T)

6236 **Anon. (1981).** New protected areas of Peru. *IUCN Bull.*, 12(3-4): 17. (P)

PERU

6237 **Anon. (1987).** *In situ* conservation in Peru: a case study 1. *Forest Genetic Resources Information,* 15: 5-22. Maps. (E/P)

6238 **Bennett, D.E. (1988).** Conserving Peruvian orchids. *Amer. Orchid Soc. Bull.,* 57(11): 1247-1249. Col. illus. (T)

6239 **Denevan, W.M. (1973).** Development and the imminent demise of the Amazon rain forest. *Profess. Geogr.,* 25: 130-135.

6240 **Denevan, W.M., Treacy, J.M., Alcorn, C., Padoch, J., Denslow, J., Paitan, S.F. (1984).** Indigenous agroforestry in the Peruvian Amazon: Bora Indian management of swidden fallows. *Interciencia,* 9: 346-357. Sustainable agriculture system in tropics. (E)

6241 **Dourojeanni, M.J., Ponce, C.F. (1978).** Los parques nacionales del Peru. Coleccion "La naturaleza en Iberoamerica". Madrid, INCAFO & CIC. 224p. Col. illus., map. (P)

6242 **Ferreyra, R. (1977).** Endangered species and plant communities in Andean and coastal Peru. *In* Prance, G.T., Elias, T.S., *eds.* Extinction is forever. Proceedings of a symposium at the New York Botanical Garden, 11-13 May 1976. New York, New York Botanical Garden. 150-157. Graphs. Vegetation regions and associated endangered plants. (T)

6243 **Gentry, A.H. (1977).** Endangered plant species and habitats of Ecuador and Amazonian Peru. *In* Prance, G.T., Elias, T.S., *eds.* Extinction is forever. Proceedings of a symposium at the New York Botanical Garden, 11-13 May 1976. New York, New York Botanical Garden. 136-149. Maps. An assessment of threatened and endangered habitats. (P/T)

6244 **Goodland, R.J., Irwin, H.S. (1977).** Amazonian forest and cerrado: development and environmental conservation. *In* Prance, G.T., Elias, T.S., *eds.* Extinction is forever. Proceedings of a symposium at the New York Botanical Garden, 11-13 May 1976. New York, New York Botanical Garden. 214-233. Maps, graphs. (P/T)

6245 **Gutierrez, H.F., Ramirez, Q.O. (1980).** La regeneracion naturel en el Bosque von Humboldt. (Natural regeneration in the Von Humboldt National Forest.) *Rev. Forest. Peru,* 9(1): 11-18. (P)

6246 **Jordan, C.F., ed. (1987).** Amazonian rain forests. Ecosystem disturbance and recovery. Ecological Studies No. 60. New York, Springer-Verlag. 133p.

6247 **Nano, C.O. (1980).** Las cactaceas y la conservacion de la flora. *Bol. Lima,* 7: 40-44. Sp (Ge, En). Illus., map. (T)

6248 **Peters, C.M., Balick, M.J., Kahn, F., Anderson, A.B. (1989).** Oligarchic forests of economic plants in Amazonia: utilization and conservation of an important tropical resource. *Conservation Biology,* 3(4): 341-349. Illus. (E)

6249 **Ponce del Prado, C.F. (1984).** Objectives for managing biosphere reserves; the Peruvian example. *In* Unesco-UNEP. Conservation, science and society. Paris, Unesco. 125-132. Contributions to the First International Biosphere Reserve Congress, Minsk, Byelorussia/USSR, 26 September - 2 October 1983. Tables. (P)

6250 **Prance, G.T. (1977).** The phytogeographic subdivisions of Amazonia and their influence on the selection of biological reserves. *In* Prance, G.T., Elias, T.S., *eds.* Extinction is forever. Proceedings of a symposium at the New York Botanical Garden, 11-13 May 1976. New York, New York Botanical Garden. 195-213. Maps. (P)

6251 **Rios, M.A., Ugaz, J., Vasquez, P.A., Ghersi, F., Parro-C., R.M. (1985).** Centro de Datos para la Conservacion del Peru. *Diversidata,* 2(4): 1-8. Sp. (T)

6252 **Saavedra, C, Freitas, S. de (1989).** Manu - two decades later. *WWF Reports,* June/July 1989: 6-9. (P)

PERU

6253 **Salati, E., Vose, P.B.F. (1984).** Amazon basin: a system in equilibrium. *Science*, 225(4658): 129-138. Amazonia.

6254 **Vasquez, P.G., Ugaz, J. (1988).** Conservando los manglares Peruanos. *Diversidata*, 5(2): 1-2. Sp.

6255 **Vasquez, R., Gentry, A.H. (1989).** Use and misuse of forest-harvested fruits in the Iquitos area. *Conservation Biology*, 3(4): 350-361. Illus. (E/T)

6256 **Veening, W. (1989).** Is er hoop voor Amazonie? *Panda*, 11: 8-10. Du. Col. illus., maps. Deforestation.

PHILIPPINES

6257 **Andrews, S., Lewis, G. (1984).** Plants in peril. 3. *Strongylodon macrobotrys*. *Kew Mag.*, 1(4): 188-190. Illus. (T)

6258 **Baconguis, S.R., Abrigo, D.P. (1980).** Erosion control and economic potential of tiger grass in the northwestern part of Camarines Sur. *Canopy*, 6(9): 4-5. Illus. (E)

6259 **Cox, R. (1988).** The conservation status of biological resources in the Philippines. Cambridge, IUCN Conservation Monitoring Centre. 68p. Maps. Includes discussion on threats to plant life; vegetation. (P/T)

6260 **Eakle, T.W. (1978).** Ecology, conservation and preservation defined (forests). *Canopy*, 4(6): 8-9.

6261 **Gutierrez, H.G. (1974).** The endemic flowering plant species of the Philippines. Philippine National Herbarium. 242p. Bound manuscript. List of 5221 endemic taxa, assigned to earlier IUCN numerical system, 0-4, to indicate degree of threat. Taxonomy rather dated. (T)

6262 **Guzman, E.D. de (1975).** Conservation of vanishing timber species in the Philippines. *In* Williams, J.T., *et al.*, *eds*. S.E.Asian plant genetic resources. Proc. symposium on S.E.Asian plant genetic resources, Kopo, Cisarua, Bogor, Indonesia, 20-22 March 1975. Bogor, IBPGR. 198-204. (E/T)

6263 **IBPGR (1984).** Establishment of a coconut genetic resources center in the Philippines. *IBPGR Region. Committ. Southeast Asia Newsl.*, 8(2): 12-14. Map. (E/G)

6264 **Lechoncito, J. (1985).** The development of a protected area system in the Philippines in terms of representative coverage of ecotypes. *In* Thorsell, J.W., *ed*. Conserving Asia's natural heritage. The planning and management of protected areas in the Indomalayan Realm. Proc. of the 25th working session of IUCN's CNPPA. Gland, IUCN. 82-85. (P)

6265 **Lerio, E.V. (1983).** There is a need to teach forest conservation in elementary and high schools. *Canopy*, 8(10): 3-5. Illus.

6266 **MacKenzie, D. (1988).** Uphill battle to save Filipino trees. *New Sci.*, 118(1619): 42-43. Illus., map. Deforestation.

6267 **Madulid, D.A. (1975).** Soil and sand binding grasses in the Philippines and its conservation value. *Acta Manilana, ser. A*, 14(22): 76-161. Illus., map.

6268 **Madulid, D.A. (1982).** Plants in peril. *Filipinas J. Sci. Cult.*, 3: 8-16. Refers to trade in plants, as recommended by CITES. (E/L/T)

6269 **Madulid, D.A. (1985).** The botanist's role in Philippine conservation. *Enviroscope*, 5(1): 2.

6270 **Madulid, D.A. (1987).** Plant conservation in the Philippines. *In* Santiapillai, C., Ashby, K.R., *eds*. Proceedings of the Symposium on the Conservation and Management of Endangered Plants and Animals. Bogor, Indonesia, 18-20 June 1986. Bogor, SEAMEO-BIOTROP. 21-26. (T)

PHILIPPINES

6271 **Mamicpic, N.G. (1982).** National and regional genebank in the Philippines. *IBPGR Region. Committ. Southeast Asia Newsl.*, 6(3): 8-9. Illus. (E/G)

6272 **Mamicpic, N.G., Movillon, M.M. (1980).** Collection, evaluation and conservation of winged bean germplasm in the Philippines. *In* Philippine Council for Agriculture and Resources Research. The winged bean. Los Banos, Laguna. 58-62. Papers presented in the 1st International Symposium on developing the potentials of the winged bean. January 1978. (E)

6273 **Mendoza, V.B. (1979).** Deteriorating forest environment: a serious ecological problem. *Canopy*, 5(9): 3, 14. Illus.

6274 **Myers, N. (1988).** Environmental degredation and some economic consequences in the Philippines. *Envir. Conserv.*, 15: 205-214.

6275 **Myers, N. (1988).** Tropical forests: shifted cultivators in the Philippines. *Plants Today*, 3: 6.

6276 **Pancho, J.V. (1975).** Vanishing plants of the Philippines with special reference to potential ornamental plants. *In* Williams, J.T., *et al.*, eds. S.E.Asian plant genetic resources. Proc. symposium on S.E.Asian plant genetic resources, Kopo, Cisarus, Bogor, Indonesia, 20-22 March 1975. Bogor, IBPGR. 231-234. (E/T)

6277 **Panot, I.A. (1983).** Floristic composition of Mt. Pulog. *Canopy*, 9(9): 6. Illus. Centre of plant diversity.

6278 **Pasig, S.D., Garcia, P.R. (1983).** Save a national park from dying: the Bicol National Park. *Canopy*, 9(9): 8-10. Illus. (P)

6279 **Penafiel, S.R. (1987).** Status of research and development work in eastern Visayas, Philippines on two endangered plant species. *In* Santiapillai, C., Ashby, K.R., *eds.* Proceedings of the Symposium on the Conservation and Management of Endangered Plants and Animals. Bogor, Indonesia, 18-20 June 1986. Bogor, SEAMEO-BIOTROP. 115-119. Map. (T)

6280 **Porter, G., Ganapin, D.J. (1989).** Resources, population, and the Philippines' future. Washington, D.C., World Resources Institute, WRI Publications Brief. 4p. Summary of more detailed case study; includes map of logging operations and useful statistics on deforestation.

6281 **Quisumbing, E. (1962).** The vanishing species of plants in the Philippines. *In* Symposium on the Impact of Man on Humid Tropics Vegetation, Goroka, Papua-New Guinea, September 1960. Canberra, A.J. Arthur, Commonwealth Govt Printer. 344-349. (T)

6282 **Quisumbing, E. (1967).** Philippine species of plants facing extinction. *Araneta J. Agric.*, 14(3): 135-162. Lists about 100 taxa at risk, including non-endemics. (T)

6283 **Ramoran, E.B. (1980).** Conservation and management of Philippine mossy forests. *Canopy*, 6(8): 8-9. Laguna Forest Research Institute.

6284 **Reyes, M.R. (1978).** Natural forests need attention. *Canopy*, 4(12): 3. Illus. Tropical forest.

6285 **Reyes, M.R. (1982).** Intensifying conservation of our dipterocarp forest. *Canopy*, 8(1): 15. Illus. (T)

6286 **Serrano, R.C. (1978).** Preserving the Philippine mangrove swamps. *Canopy*, 4(7): 6-7.

6287 **Sharp, T. (1984).** ASEAN nations tackle common environmental problems. *Ambio*, 13(1): 45-46. Illus.

6288 **Siebert, S.F. (1984).** Conserving tropical rainforests: the case of Leyte Mountains National Park, Philippines. *Mountain Res. Devel.*, 4(3): 272-276. Illus. Maps. (P)

PHILIPPINES

6289 **Soriano, M.A. (1981).** Bicol National Parks will disappear in 5 years. *The Peninsula News*, 41(18): 1. (P)

6290 **Tan, B.C., Fernando, E., Rojo, J.P. (1986).** An updated list of endangered Philippine plants. *Yushania*, 3(2): 1-5. (T)

6291 **Terborgh, J., Winter, B., Hauge, P., Parkinson, J. (1984).** Conservation priorities in the Philippine archipelago. Princeton, New Jersey, Princeton University.

6292 **Tolentino, A.S., jr (1984).** How to protect coastal and marine ecosystems: lessons from the Philippines. *In* McNeely, J.A., Miller, K.R., *eds.* National parks, conservation and development. The role of protected areas in sustaining society. Washington, D.C., Smithsonian Institution Press. 160-164. (P)

6293 **Tolentino, E.P. (1979).** National parks, equivalent reserves, and forest recreation: the Philippines. *In* Nelson, J.G. & S.H., Needham, R.D., Scace, R.C., *eds.* The Canadian national parks: today and tomorrow. Conference II: ten years later. Vol I; II. 809-816. (P)

6294 **Valmayor, R.V. (1980).** The regional gene bank. *Newsl. Regional Comm. SE Asia*, 5(1): 8-9. Gene bank. (G)

6295 **Yao, C.E., Ulep, E.V. (1981).** Almaciga plantation: antidote to Almaciga extinction. *Canopy*, 7(11): 11-12. Illus. *Agathis philippinensis*. (E/L/T)

6296 **Zamora, P.M. (1980).** Philippine mangroves: assessment of status, environmental problems, conservation and management strategies. *Bakawan: Newsl. Nat. Mangrove Committee (NRMC)*, 1: 2.

PHILIPPINES - Luzon

6297 **Gruezo, W.S., Fernando, E.S. (1985).** Notes on *Phoenix hanceana* var. *philippinensis* in the Batanes Islands, Philippines. *Principes*, 29(4): 170-176. Illus. (T)

PHILIPPINES - Mindanao

6298 **Raber, D.S. (1977).** Notes on the ecology of the Sulu Archipelago. *Pterocarpus*, 3(1): 33-41. Map.

PHILIPPINES - Palawan

6299 **Anon. (1980).** Palawan almacigas might yet be saved. *Canopy*, 6(9): 2. *Agathis dammara*.

6300 **Clad, J., Vitug, M.D. (1988).** Palawan's forests appear doomed in a power struggle. The politics of plunder. *Far East. Econ. Rev.*, 24 November: 48-52. Illus. Deforestation.

6301 **Pido, M.D. (1988).** A proposed parks/protected area system for Palawan island groups, Philippines. *Tigerpaper*, 15(3): 21-30. Maps. (P)

6302 **Pido, M.P. (1986).** Palawan: on natural resources conservation. *Canopy*, 12(2): 1,4-5. Illus.

6303 **Quinnell, R., Balmford, A. (1986).** Palawan forest research area 1984. Final report. An expedition from Cambridge University. Cambridge, University of Cambridge. 57p.

6304 **Quinnell, R., Balmford, A. (1988).** A future for Palawan's forests? *Oryx*, 22(1): 30-35. Illus., maps. Deforestation. (P)

PITCAIRN - Henderson Island

6305 **Anon. (1983).** An island at risk. *Oryx*, 17(3): 109. Incipient danger in proposed development.

6306 **Fosberg, F.R. (1984).** Henderson Island saved. *Envir. Conserv.*, 11(2): 183-184. (T)

6307 **Fosberg, F.R., Sachet, M.-H. (1983).** Henderson Island threatened. *Envir. Conserv.*, 10(2): 171-173. Threatened desecration of unique biota, including 10 endemic plant taxa; later averted by British government. (T)

6308 **Fosberg, F.R., Sachet, M.-H., Stoddart, D.R. (1983).** Henderson Island (southeastern Polynesia): summary of current knowledge. *Atoll Res. Bull.*, 272: 53p. Illus., maps. Lists native and endemic species.

6309 **Serpell, J. (1983).** Desert island risk. *New Sci.*, 98(1356): 320. Describes the importance of Henderson Island and threats to its flora and fauna; reprinted in *Threatened Pl. Newsl.*, 11: 14, 1983. (T)

6310 **Serpell, J., Collar, N., Davis, S., Wells, S. (1983).** Submission to the Foreign and Commonwealth Office on the future conservation of Henderson Island in the Pitcairn Group. WWF-UK, IUCN, ICBP. 27p. Mimeo.

PITCAIRN - Pitcairn Island

6311 **Randall, J.E. (1973).** Expedition to Pitcairn. *Oceans*, 6(2): 12-21.

POLAND

6312 **Adamczyk, B. (1979).** Apel o utworzenie rezerwatu stepowego w Wolce Leszczanskiej kolo Chelma. (An appeal for the establishment of a nature reserve to safeguard the steppe vegetation at Wolka Leszczanska near Chelm.) *Chronmy Przyr. Ojczysta*, 35(3): 59-60. Pol. (P)

6313 **Anon. (1976).** [Special issue on the Swietokrzyski National Park]. *Sylwan*, 120(4): 1-72. Pol. (P)

6314 **Anon. (1981).** Polskie towarzystwo botaniczne w obronie przyrody. (Polish Botanical Association in defence of nature.) *Wiadomosci Bot.*, 25(4): 295-302. Pol. (P)

6315 **Antczak, A. (1979).** Rezerwatowa ochrony jodly pospolitej (*Abies alba* Mill.) w Polsce. (Protection of silver fir (*Abies alba* Mill.) in reserves in Poland.) *Sylwan*, 123(5): 37-47. Pol (En). Illus., map. (P/T)

6316 **Aulak, W.A., et al. (1977).** Ochrona lasu: podr ecznik dla student ow wydzia ow le snych adademii rolniczych. (Protection of forest.) Warszawa, Panstwowe Wydawn. Rolnicze i le Sne. 530p. Pol. Illus. (P)

6317 **Baron, H. (1980).** Vascular flora of the Rybnic coal region, Silesia, Poland. *Pol. Ecol. Stud.*, 6(4): 585-592. (Pol). Species that should be protected are mentioned. (T)

6318 **Bartyzel, R. (1980).** Przeglad badan naukowych dotyczacych Pieninskiego Parku Narodowego. (A review of scientific studies on the Pieniny National Park.) *Chronmy Przyr. Ojczysta*, 36(4): 17-24. Pol (En). Illus. (P)

6319 **Bartyzel, R. (1985).** Czy Male Pieniny beda parkiem narodowym? (Will the Lesser Pieniny Mts be proclaimed a national park?) *Chronmy Przyr. Ojczysta*, 41(2): 57-59. Pol. (P)

6320 **Bialobok, S. (1977).** [Protection of trees]. *Nasze Drzewa Lesne*, 5: 550-553. Pol (En). Illus., map. *Picea abies*.

POLAND

6321 **Bialobok, S. (1979).** Ochrona drzew. (Protection of trees.) *In* Bialobok, S., *ed.* Brzozy: *Betula* L. Warsaw, Polska Akademia Nauk. 381-388. Pol (En). Map. Lists protected trees. (L/T)

6322 **Bogucka, A. (1979).** Problemy ksztaltowania obrzeza Pieninskiego Parku Narodowego na tle warunkow srodowiska przyrodniczego. (The problems of the formation of the borders of the Pieniny National Park on the background of the conditions of its natural environment.) *Ochr. Przyr.*, 42: 217-243. Pol (En). Maps. (P)

6323 **Boratynska, K. (1978).** Stanowisko zimosiolu polnocnego *Linnaea borealis* w wojewodztwie leszczynskim. (A locality of the twinflower *Linnaea borealis* in the province of Leszno.) *Chronmy Przyr. Ojczysta*, 34(6): 43-45. Pol. Maps. (T)

6324 **Boratynski, A. (1978).** Sosna blotna (*Pinus uliginosa* Neumann) w rezerwacie Bledne Skaly w gorach stolowych. (*Pinus uliginosa* Neumann in the Bledne Skaly Reserve in Stolowe Mountains.) *Arboretum Kornickie*, 23: 261-267. Pol (En, Rus). (P)

6325 **Bosiak, A. (1979).** Lasy ochronne nad Kanalem Augustowskim. (Protective forests on Kanal Augustowski.) *Chronmy Przyr. Ojczysta*, 35(5): 28-36. Pol (En). Map. (P)

6326 **Braun, J. (1978).** Nowy rezerwat "Milechowy" w poludniowo-zachodniej czesci Gor Swietokrzyskich. ("Milechowy", a new nature reserve in the southwestern part of the Swietokrzyskie [Holy Cross] Mts.) *Chronmy Przyr. Ojczysta*, 34(5): 69-72. Pol. (P)

6327 **Broda, J. (1979).** Rola lasu w ochronie srodowiska naturalnego w Polsce w ujeciu historycznym (od konca *XVIII* w.). (The role of forest in the conservation of the natural environment in Poland in a historical approach [from the end of the 18th century].) *Sylwan*, 123(11): 1-17. Pol (Rus, En).

6328 **Broz, E. (1983).** Notatki florystyczne z Gor Swietokrzyskich: Czesc 3. (Floristic notes from the Swietokrzyskie Mountains [central Poland]: part 3.) *Fragm. Flor. Geobot.*, 27(4): 607-617 (1981 publ. 1983). Pol (En). Maps. (T)

6329 **Broz, E. (1985).** Roslinnosc rezerwatu stepowego "Polana Polichno" koto Pinczowa oraz uwagi dotyczace jej ochrony. (The "Polana Polichno" nature reserve near Pinczow safeguarding steppe vegetation and some remarks on its protection). *Chronmy Przyr. Ojczysta*, 41(6): 22-35. Pol. Illus., maps. (P)

6330 **Broz, E., Cieslinski, S. (1976).** Rezerwat modrzewia polskiego Ciechostowice w Gorach Swietokrzyskich. (The Ciechostowice Nature Reserve of the Polish larch *Larix polonica* RAC. in the Holy Cross Mountains.) *Ochr. Przyr.*, 41: 155-178. (En). (P)

6331 **Broz, E., Przemyski, A. (1985).** Nowe stanowiska rzadkich gatunkow roslin naczyniowych z lasow Wyzyny Srodkowomalopolskiej. (New localities of rare species of vascular plants from the forest of the Central Little Poland Upland). *Fragm. Flor. Geobot.*, 29(1): 19-30 (1983 publ. 1985). Pol (En). Map. (T)

6332 **Budzynski, W., Chielnik, H. (1978).** [Conservation and renovation work on parks in the Bydgoszcz Province]. *Sluzba Rolna*, 6: 10-12. Pol. Illus. (P)

6333 **Capecki, Z. (1983).** Charakterystyka zdrowotnosci i zagrozenia lasow Karpackich w Polsce. (Characteristics of the health and threats to the Carpathian forests in Poland.) *Prac. Inst. Badaw. Lesn.*, 616-620: 27-54. Pol (Ge, Rus). Illus., maps.

6334 **Capecki, Z., Zwolinski, A. (1984).** Charakterystyka zagrozenia lasow Karkonoskiego Parku Narodowego. (Character of the threat to forests in the Karkonosze National Park.) *Sylwan*, 128(8): 1-21. Pol (En). Illus., maps. (P)

6335 **Celinski, F., Wika, S. (1978).** Proba nowego spojrzenia na stosunki fitosocjologiczne rezerwatu "Parkowe" w Zlotym Potoku kolo Czestochowy. (A new look at phytosociological conditions in the "Parkowe" reservation in Zloty Potok near Czestochowa.) *Fragm. Florist. Geobot.*, 24(2): 277-301. Pol (En). Illus., maps. (P)

POLAND

6336 **Celinski, F., Wojterski, T. (1978).** Zespoly lesne Masywu Babiej Gory. (Forest associations of the Babia Gora-Massif.) *Prace Kom. Biol. Poznan. Tow. Przy. Nauk. Mat.-Przyrod.*, 48: 60p. Pol (En). Illus. Includes protection of forests. (P)

6337 **Chojnacki, J.M., Mroz, W.J. (1984).** Wplyw antropogennych zmian stosunkow wodnych na roslinnosc rezerwatu Las Bielanski w Warszawie. (Effect of anthropogenous changes of water conditions on the vegetation in the forest reserve Las Bielanski in Warsaw.) *Wiadomosci Ekologiczne*, 30(2): 167-192. Pol (En). Illus., maps. (P)

6338 **Cichy, D. (1978).** Problemy ksztalcenia ogolnego w zakresie ochrony i ksztaltowania srodowiska w Polsce. (The problems of general education in the field of environmental conservation and management in Poland.) *Chronmy Przyr. Ojczysta*, 34(3): 17-24. Pol (En).

6339 **Cieslak, M., Gacka-Grzesikiewicz, E., Lubelska, T. (1979).** System obszarow Chronionych. (The system of protected regions in Chelmo voivodship.) *Przyroda Polska*, 4: 9-11. Pol. Col. illus. (P/T)

6340 **Cwiklinski, E., Oziminski, K. (1984).** Walory i funkcje Minskiego Obszaru Chronioniego Krajobrazu w wojewodztwie Siedleckim. (The qualities and functions of the Minsk area of protected landscape in the province of Siedke.) *Chronmy Przyr. Ojczysta*, 40(1): 12-34. Pol (En). Illus., maps. (P)

6341 **Czarnecka, B. (1979).** Charakterystyka geobotaniczna Rezerwatu Lesnego Jarugi na Rostoczu s'rodkowym. (Geobotanical characteristics of the Jarugi Forest Reserve in Central Roztocze.) *Ann. Univ. Mariae Curie-Sklodowska, C.*, 33: 309-331 (1978 publ. 1979). Pol. Illus., maps. (P)

6342 **Czeczuga, B. (1979).** Rezerwat Gorbacz zagrozony! (Gorbacz preserve in danger!) *Przyr. Polska*, 4: 6-8. Pol. Col. illus. Peatlands. (P)

6343 **Czekalski, M. (1985).** Godny ochrony park wiejski w Dobrej w wojewodztwie Opolskim. (The rural park at Dobra in the province of Opole deserves protection). *Chronmy Przyr. Ojczysta*, 41(4): 57-64. Pol. Illus. (P)

6344 **Czekalski, M. (1987).** Kalina koralowa (*Viburnum opulus* L.): krzew crzsciowo chroniony. (European cranberry-bush [*Viburnum opulus* L.]: a shrub protected in part by conservation laws). *Roczn. Sekc. Dendrol. Pol. Tow. Bot.*, 36: 73-81 (1984-1985 publ. 1987). Cz (En). Illus. (L)

6345 **Czubinski, Z., Gawlowska, J., Zabierowski, K. (1977).** Rezerwaty przyrody w Polsce. (Nature reserves in Poland.) Warszawa, Krakow, Panstwowe Wydawnictwo Naukowe. Pol. (P)

6346 **Czyzewska, K. (1979).** Zaleczanski Park Krapbrazowy. (Zalecze Landscape Park.) *Przyr. Polska*, 11: 20-21. Pol. Col. illus. (P)

6347 **Denisiuk, Z. (1978).** Wielkopolski Park Narodowy- stan aktualny i perspektywy rozwoju. (The Wielkopolski (major Poland) National Park - its present state and possible future development.) *Chronmy Przyr. Ojczysta*, 4: 5-20. Pol (En). Illus., map. (P)

6348 **Denisiuk, Z., Dziewalski, J., Madziara-Borusiewicz, K. (1983).** Niektore problemy rezerwatu Romanka w Beskidzie Zywieckim. (Some problems of the Romanka Nature Reserve in the Beskid Zywiecki Mountain Range. *Chronmy Przyr. Ojczysta*. 39(3): 15-29. Pol (En). Illus. Map. (P)

6349 **Drewniak, S. (1984).** Pomniki, rezerwaty przyrody, parki narodowe. *Przyr. Polska*, 3: 34-35. Pol. The list of National Parks, Reserves and native monuments in Poland. (P)

6350 **Drzal, M. (1978).** Wspolczesny stan ochrony przyody dorzecza Pilicy. (The present state of protection of the Pilicy River Basin.) *Stud. Osrod. Dok. Fizjogr.*, 6: 313-317. Pol (En).

POLAND

6351 **Drzal, M., Kleczkowski, A.S. (1978).** Koncepcja ochrony srodowiska przyrodniczego dorzecza Pilicy. (The protection of the natural environment of the River Pilicy Basin.) *Stud. Osrod. Dok. Fizjogr.*, 6: 319-340. Pol (En)

6352 **Dswonko, Z., Zemanek, B. (1976).** Rezerwat modrzewia polskiego Ciechostowice w Gorach Swietokrzyskich. (The vegetation of the reserve Gora Sobien near Monastarzec [Polish Eastern Carpathians].) *Ochr. Przyr.*, 41: 179-204. Pol (En). Illus. (P)

6353 **Dzieciolowski, R., Sokolowski, A. (1978).** The conservation of forest animal and plant genetic resources in Poland. *In* Proceedings of the 8th World Forestry Congress, Jakarta, 16-28 October 1978: Forestry for quality of life. I, 12. (E)

6354 **Dziewolski, J. (1972).** Naturalne zmiany w strukturze drzewostanow Pieninskiego Parku Narodowego w okresie 32 lat (1936-1968). (Natural changes in the structure of the woodstands of the Pieniny National Park over a 32-year period. [1936-1968].) *Ochr. Przyr.*, 37: 263-283. Pol (En). Illus. (P)

6355 **Dziewolski, J. (1978).** Nieznane stanowiska jodly *Abies alba* w lasach Tatrzanskiego Parku Narodowego. (The localities of the fir *Abies alba* hitherto unknown in the forests of the Tatry National Park.) *Chronmy Przyr. Ojczysta*, 3: 62-66. Pol. Illus., map. (P)

6356 **Dzwonko, Z., Zemanek, B. (1976).** Rezerwat modrzewia polskiego Ciehchostowice w Gorach Swietokrzyskich. (The vegetation of the reserve Gora Sobien near Monastarzec [Polish Eastern Carpathians].) *Ochr. Przyr.*, 41: 179-204. Pcl (En). Illus. (P)

6357 **Falinska, K. (1979).** Modifications of plant populations in forest ecosystems and their ecotones. *Pol. Ecol. Stud.*, 5(1): 89-150. (Pol). Illus. Bialowieza National Park. (P)

6358 **Falinski, J.B. (1972).** Podstawy i formy eksploracji naukowej Bialowieskiego Parku Narodowego. (Basis and forms of scientific exploration of the Bialowieza National Park.) *Ochr. Przyr.*, 37: 7-55. Pol (En). Illus., maps. (P)

6359 **Falinski, J.B., ed. (1976).** Synantropizacja szaty roslinnej VI. Wymieranie skladnikow flory polskiej i jego przyczyny. (Synanthropization of plant cover VI. The decline and extinction of native plant species in Poland.) *Phytocoenosis*, 5(3/4): 157-409. Pol (En). 22 research papers. (T)

6360 **Faltynowicz, W. (1978).** Rzadsze gatunki roslin naczyniowych nadlesnictwa Przymuszewo w Borach Tucholskich. (Rare vascular plants in the Przymuszewo-Forest [Bory Tucholskie-Region] in West Pomerania.) *Badan. Fizjogr. Pol. Zachod.*, B, 30: 193-197. Pol (En). Map. (T)

6361 **Faltynowicz, W., Szmeja, J. (1978).** O potrzebie utworzenia stref ochronnych wokol jezior lobeliowych. (On the need for providing protective zones round *Lobelia*-lakes.) *Chronmy Orzyr. Ojczysta*, 34(5): 5-17. Pol (En). Illus., map. (P)

6362 **Fijalkowski, D., Kseniak, M. (1977).** Park Jakubowice Murowane. (The Park Jakubowice Murowane.) *Ann. Univ. Mariae Curie-Sklodowska, C*, 31: 79-85 (1976 publ. 1977). Pol (Rus, En). Map. Lists rare species. (P/T)

6363 **Fijalkowski, D., Kseniak, M. (1977).** Stan parkow Wiejskich regionu Lubelskiego. (Sostoyanie sel'skikh parkov Lyublinskogo raiona [PNR].) (State of the national parks in the Lubelski Region.) *Czlow. i Srodow*, 1(4). Pol (Rus, En). (P)

6364 **Fijalkowski, D., Kseniak, M. (1978).** Park Jablonna kolo Lublina. (Park at Jablonna near Lublin.) *Ann. Univ. Mariae Curie-Sklodowska, C*, 32: 179-184 (1977 publ. 1978). Pol (En). Lists rare species. (P/T)

6365 **Fijalkowski, D., Kseniak, M. (1978).** Park dworski w Snopkowie kolo Lublina. (A mansion park in Snopkow near Lublin.) *Ann. Univ. Mariae Curie-Sklodowska, C*, 33: 361-367. Pol (Rus, En). Illus., map. (P)

POLAND

6366 **Fijalkowski, D., Kseniak, M. (1978).** Park palacowy w Tomaszowicach. (The Palace Park in Tomaszowice.) *Ann. Univ. Mariae Curie-Sklodowska, C*, 33: 369-374. Pol (Rus, En). Map. (P)

6367 **Fijalkowski, D., Sokolowski, A., Puszkar, L. (1978).** Rzadsze rosliny synantropijne w otoczeniu Zakladow Azotowych w Pulawach. (Rarer synantropic plants in the environs of nitrogen works in Pulawy.) *Ann. Univ. Mariae Curie-Sklodowska, C*, 32: 97-105 (1977 publ. 1978). Pol (En). Map. (T)

6368 **Filkowa, B. (1978).** Osobliwosci dendrologiczne Ojcowskiego Parku Narodowego. (Dendrological rarities in the Ojcow National Park.) *Chronmy Przyr. Ojczysta*, 4: 59-64. Pol. Illus., map. (P/T)

6369 **Garvel, T. (1984).** W sprawie parku wiejskiego w Porebie Wielkiej. (The rural park at Poreba Wielka.) *Chronmy Przyr. Ojczysta*, 40(1): 57-59. Pol. (P)

6370 **Gaulowska, J. (1982).** Ochrona zasobow roslin leczniczych i przemyslowych. Badania prowadzone w Zakladzie Ochrony Przyrody i Zasobow Naturalnych PAN w Krakowie. (Conservation of the resources of medicinal and industrial plants. Polish Academy of Sciences in Cracow.) *Chronmy Przyr. Ojczysta*, 38: 20-31. Pol (En). Illus. (E/T)

6371 **Gawel, T. (1978).** Park w Tarnowie-Gumniskach jako obiekt dydaktyczny. (The park in Tarnow-Gumniska as an educational object.) *Chronmy Przyr. Ojczysta*, 34(5): 52-55. Pol. Illus. (P)

6372 **Glazek, T. (1975).** Roslinnosc rezerwatu acheologicznego Krzemionki Opatowskie kolo Ostrowca Swietokrzyskiego. (The vegetation of the "Krzemionki Opatowskie" archaeological reserve near Ostrowiec Swietokrzyski.) *Ochr. Przyr.*, 40: 139-162. Pol (En). Illus., map. (P)

6373 **Glazek, T. (1984).** Rezerwat Stepoury Gory Pinczowskie w wojewodztwie kieleckim. (The "Gory Pincoczowskie" nature reserve in the province of Kielce safeguarding steppe vegetation.) *Chronmy Przyr. Ojczysta*, 40(5-6): 5-13. Pol (En). Illus., map. (P)

6374 **Goldyn, H. (1978).** Nowe stanowiska rzadszych roslin naczyniowych w polnocnej czesci Wysoczyzny Leszozynskiej. (New localities of rare vascular plants in the northern part of the Leszno-Plateau.) *Badan. Fizjogr. Pol. Zachod., B*, 30: 199-202. Pol (En). Map. (T)

6375 **Gorska-Zajaczkowska, M., Wojtowicz, W. (1985).** Stanowisko roslin chronionych w Borach Tucholskich na Pomorzu Zachodnim. (The locality of protected plants in Bory Tucholskie [Tuchola coniferous forests] in Western Pomerania.) *Chronmy Przyr. Ojczysta*, 41(5): 54-56. Pol. (L)

6376 **Grodzinska, K. (1979).** Mapa zbiorowisk roslinnych rezerwatu Przelom Bialki pod Krempachami (Pieninski Pas Skalkowy). (Map of plant communities in the nature reserve of "Przelom Bialki pod Krempachami" [Bialka river gorge at Krempachy, Pieniny Klippen Belt].) *Ochr. Przyr.*, 42: 29-73. Pol (En). Illus., maps. (P)

6377 **Grodzinska, K., Olaczek, R., eds (1985).** Zagrozenie parkow Narodowych w Polsce. (National parks in danger). Warszawa, Polska Akademia Nauk Komitet Ochrony Przyrody. (Polish Academy of Sciences Committee for Nature Conservation.) 156p. Pol (En). Illus., maps. (P)

6378 **Herbich, J. (1975).** Problem zachowania rezerwatow lesnych w okolicy Opalenia nad dolna Wisla. (Problems of preserving forest reserves near Opalenie on the Lower Vistula.) *Ochr. Przyr.*, 40: 113-138. Pol (En). Illus., maps. (P)

6379 **Holeksa, J (1986).** Zaslugujacy na ochrone Kompleks lesny na zachodnim Krancu Karpat Polskich. (The sylvan complex at the western border of the Polish Carpathian Mts deserves protection.) *Chronmy Przyr. Ojczysta*, 42(3): 5-16. Pol (En). Illus., map.

POLAND

6380 **Holeska, J. (1979).** Godne ochrony zrodliska na peryferiach Puszczy Bialowieskiej. (Safeguarding the headwaters in the periphery of the Bialowieza Primeval Forest Reserve.) *Chronmy Przyr. Ojczysta*, 35(3): 33-40. Pol (En). Illus. (P)

6381 **Jakubowska-Gabara, J., Kucharski, L. (1983).** Rzeka Rawka, projektowany rezerwat przyrody. (The Rawka stream: a projected nature reserve.) *Chronmy Przyr. Ojczysta*, 39(1-2): 17-28. Pol (En). Illus., maps. (P)

6382 **Jakuczun, H. (1978).** Osobliwosci florystyczne Siwianskich Turni w Tatrzanskim Parku Narodowym. (Floristic rarities of the Siwianskie Peaks in the Tatra National Park.) *Chronmy Przyr. Ojczysta*, 34(6): 36-39. Pol. Illus. (P/T)

6383 **Jakuczun, H. (1978).** Stanowiska obuwicka pospolitego *Cypripedium calceolus* w Tatrzanskim Parku Narodowym. (The locality of the lady's slipper, *Cypripedium calceolus* in the Tatry National Park.) *Chronmy Przyr. Ojczysta*, 34(3): 59-62. Pol. Illus., map. (P/T)

6384 **Janecki, J. (1978).** Rezerwat Siedliszcze w poludniowej czesci Kotliny Dubienki. (Siedliszcze Nature Reserve in the southern part of the Dubienka valley.) *Chronmy Przyr. Ojczysta*, 34(3): 68-73. Pol. Maps. (P)

6385 **Jasiewicz, A. (1981).** Wykaz gatunkow rzadkich i zagrozonych flory polskiej. (List of rare and endangered plants from the Polish flora.) *Fragm. Flor. Geobot.*, 27(3): 401-414. Pol (En). Over 400 threatened taxa listed. (T)

6386 **Jasnowska, J., Jasnowski, M. (1977).** Zagrozone gatunki flory torfowisk. (Endangered plant species in the flora of peatbogs.) *Ojczysta "Zagrozone Gatunki Flory Torjowisk*, 33(4): 5-14. Pol (En). Illus. (P/T)

6387 **Jasnowska, J., Jasnowski, M. (1977).** [Orchidaceae of the peat bog reserve "Bagno Chlopiny" in the Mysliborz Lakeland]. *Zesz. Nauk. Akad. Roln. Szczecin*, 61: 163-184. Pol (En). (P)

6388 **Jasnowski, M., Friedrich, S. (1979).** Znaczenie i zadania Puszczy Bukowej kolo Szczecina a potrzeby jej ochrony. (The importance and tasks of Puszcza Bukowa (primeval beechwoods) near Szczecin and the need for their protection.) *Chronmy Przyr. Ojczysta*, 35(1): 15-27. Pol (En). Illus., map. (P)

6389 **Jasnowski, M., Jasnowska, J., Kowalski, W., Markowski, S., Radomski, J. (1972).** Warunki siedliskowe i szata roslinna torfowiska nakredowego w rezerwacie Tchorzyno na Pojezierzu Mysliborskim. (Habitat conditions and vegetation of the peat bog on chalk substratum in the reserve Tchorzyno in the Mysliborz lake region.) *Ochr. Przyr.*, 37: 157-232. Pol (En). Illus., map. (P)

6390 **Jasnowsksa, J. (1977).** [A threatened stand of the protected orchid *Orchis militaris* L. near Lake Bedzin, the Mysliborz Lakeland]. *Zesz. Nauk Akad. Roln. Szczecin*, 53: 85-93 (1976 publ. 1977). Pol (En). Maps. (T)

6391 **Jedrzejko, K. (1983).** W sprawie utworzenia rezerwatu torfowiskowego Bagna im. Prof. Bronislawa Szafrana W Zaglebiu Dabrowskim. (Notes on the "Bagna" peat bog reserve to be established in the Dabrowa Coal Basin in honour of Professor Branislaw Szafran.) *Chronmy Przyr. Ojczysta*, 39(5): 63-68. Pol. Maps. (P)

6392 **Joseph-Tomaszewska, E. (1982).** W sprawie ochrony i zagospodarowania Zespolu Jurajskich Parkow Krajobrazowych w wojewodztwie katowickim. (On the protection and management of the complex of the Jurassic Landscape Parks in the province of Katowice.) *Chronmy Przyr. Ojczysta*, 38(1-2): 76-82. Pol. Maps. (P)

6393 **Jost-Jakubowska, B. (1979).** Flora i roslinnosc projektowanego rezerwatu lesnego "Rokinciny" kolo Lodzi. (Flora of the projected forest reservation "Rokinciny" near Lodz.) *Acta Univ. Lodz., Bot.*, 27: 17-38. Pol (En, Rus). (P)

6394 **Kaminski, R. (1983).** Aldrowanda pecherzykowata *Aldrovanda vesiculosa* ginaca roslina w Polsce. (*Aldrovanda vesiculosa*, a vanishing plant in Poland.) *Chronmy Przyr. Ojczysta*, 39(4): 20-24. Pol (En). Illus., maps. (T)

POLAND

6395 **Kepczynski, K., Zaluski, T. (1982).** Rosliny rzadziej spotykane w okolicach Wloclawka. 2. (Rare plants in the vicinity of Wlockawka. 2.) *Biologia*, 24(53): 39-53. Pol (Ge). (T)

6396 **Kinasz, W. (1976).** Ekologiczne podstawy urzadzenia lak w Peininskim Parku Narodowym. (Ecological basis of the management of the meadows of the Pieniny National Park.) *Ochr. Przyr.*, 41: 77-118. Pol (En). Maps. (P)

6397 **Klarowski, R. (1978).** Utworzono Mazurski Park Krajobrazowy. (The establishment of the Mazurian Landscape Park.) *Chronmy Przyr. Ojczysta*, 3: 49-51. Pol. Map. (P)

6398 **Klarowski, R. (1978).** Zagrozenie rezerwatu na rzece Paslece. (The nature reserve on the river Pasleka is in danger.) *Chronmy Przyr. Ojczysta*, 3: 67-68. Pol. (P)

6399 **Klorowski, R. (1984).** O ochrone przelomu rzeki Lyny kolo Olsztyna. (An appeal for the protection of the Lyna river gorge near Olsztyn.) *Chronmy Przyr. Ojczysta*, 40(2): 41-45. Pol. Illus., maps. (P)

6400 **Klosowski, S., Checinska-Rybak, A. (1984).** Gozdzikowate. (Carnation family.) *Przyroda Polska*, 6: 23-25. Pol. Col. illus. Protected plants. (L/T)

6401 **Klosowski, S., Checinska-Rybak, A. (1984).** Rosliny nowo objete ochrona gatunkowa. *Przyroda Polska*, 3: 22-25. Pol. Col. illus. Plant species protected in Poland. (L/T)

6402 **Kloss, M., Wilpiszewska, I. (1983).** O roslinnosci niewielkich zaglebien bezodplywowych okolic Mikolajek i potrzebie ich ochrony. (Vegetation in small hollows without outflow in the environs of Mikolajki and on the need for their protection.) *Chronmy Przyr. Ojczysta*, 39(4): 25-29. Pol (En).

6403 **Knapik, M. (1978).** O ochrone Garbu Tenczynskiego i jego okolic. (An appeal for the protection of the Tenczynek Hummock and its environs.) *Chronmy Przyr. Ojczysta*, 34(5): 87-90. Pol. Illus., map. (P)

6404 **Kolago, C. (1979).** Parki narodowe wobec naporu urbanizacyjnego. (National parks faced with the pressure of urbanization.) *Chronmy Przyr. Ojczysta*, 35(1): 5-14. Pol (En). Illus. (P)

6405 **Kornas, J. (1961).** The extinction of the association *Spergoletum-Lolietum remoti* in flax cultures in the Gorce (Polish Western Carpathian Mountains). *Bull. Acad. Polonaise Sci.*, Cl. II-9(1):37-40. (P/T)

6406 **Kornas, J. (1971).** Changements recents de la flore polonaise. *Biol. Conserv.*, 4(1): 43-47. Fr.

6407 **Kornas, J. (1971).** Uwagi o wspolczesnym wymieraniu niektorych gatunkow roslin synantropijnych w Polsce. (Recent decline of some synanthropic plant species in Poland.) *Mater. Zakt. Ritosoc. Stos. U.W.*, 27: 51-64. Pol. (T)

6408 **Kornas, J. (1986).** Zmiany roslinnosci segetalnej w Gorcach w ostatnich 35 latach. (Changes of segetal vegetation in the Gorce Mts (Polish Western Carpathians) during the last 35 years.) *Institute of Botany of the Jagiellonian University*, 15: 8-26. Pol (En). Illus. Sum.

6409 **Kornas, J., et al. (1976).** Wymieranie skladnikow flory polskiej i jego przyczyny. (The causes of plant decline in the Polish flora.) *Phytocoenosis (Warsaw)*, 5: 161-410. Pol.

6410 **Kostyniuk, M. (1961).** Our plant protection. Breslau, Poland. 202p. (T)

6411 **Kostyniuk, M.A., Marczek, E. (1961).** Nasze ro sliny chronione. (Our protected plants.) Warsaw, Panstwowe Wydawnictwo Naukowe. 201p. Pol (En). Illus. (L/T)

POLAND

6412 **Kozlowski, J.M. (1984).** Threshold approach to the definition of environmental capacity in Poland's Tatry National Park. *In* McNeely, J.A., Miller, K.R., *eds.* National parks, conservation and development. The role of protected areas in sustaining society. Washington, D.C., Smithsonian Institution Press. 450-462. Maps. (P)

6413 **Kozlowski, S. (1973).** Program ochrony krajobrazu Polski i jego pierwsze realizacje. (The plan for safeguarding Poland's landscape and its initial accomplishment.) *Ochr. Przyr.*, 38: 61-83. Pol (En). (P)

6414 **Kozlowski, S. (1979).** Paki Krajobrazowe. (Landscape parks.) *Przyroda Polska*, 7-8: 14-15. Pol. Col. illus. (P)

6415 **Krysztofik, E. (1978).** Lata kleski w Puszczy Jodlowej. (The years of calamity in the Primeval Fir Forest.) *Chronmy Przyr. Ojczysta*, 34(5): 39-46. Pol.

6416 **Laszek, C. (1985).** Rezerwat przyrody Wolica w dolinie Utraty. (The "Wolica" nature reserve in the Utrata river valley). *Chronmy Przyr. Ojczysta*, 41(5): 51-53. Pol. Map. (P)

6417 **Lesniak, A. (1983).** Aktualny stan ochrony lasu w Polsce. (Current state of forest protection in Poland.) *Postepy Nauk Rolniczych*, 30(6): 71-80. Pol.

6418 **Lisiewska, M. (1979).** Flora macromycetes Swietokrzyskiego Parku Narodowego. (Flora of macromycetes of the Swietokrzyski National Park.) *Acta Mycol.*, 15(1): 21-43. Rus, (En.) Illus. (P)

6419 **Lomnicki, A. (1971).** The management of plant and animal communities in the Tatra Mountains National Park. *In* Duffey, E., Watt, A.S., *eds.* The scientific management of animal and plant communities for conservation. The 11th Symposium of the BES, University of East Anglia, Norwich, 7-9 July 1970. Blackwell. 599-604. (P)

6420 **Marasimink, M., Janecki, J., Jezierski, W., Krol, T., Szwajgier, W. (1982).** Rezerwat geologiczno-stepowy w Zmudzi w wojewodztwie chetmskim. (The nature reserve at Zmudz in the province of Chetmsk safeguarding geological phenomena and steppe vegetation.) *Chronmy Przyr. Ojczysta*, 38(6): 46-58. Pol (En). Illus., map. (P/T)

6421 **Marcinkiewicz, J. (1987).** SOS Poland. *Environment Now*, 1: 48-51. Illus., map. Acid rain.

6422 **Matsenko, A.E., Pen'kos-Mirkova, G. (1980).** Ob okhrane prirode v. Pol'ske. (On protected nature in Poland.) *Byull. Glavn. Bot. Sada* (Moscow), 116: 72-78. Rus. Map. (P)

6423 **Matuszkiewicz, A. & W. (1975).** Mapa zbiorowisk roslinnych Karkonoskiego Parku Narodowego. (Carte de la vegetation du Park National de Karkonosze [sudetes occidentales].) *Ochr. Przyr.*, 40: 45-112. Pol (Fr). (P)

6424 **Michal, I. (1979).** Dynamika prirozenych lesu Bialovezskeho hvozdu 1. *Ziva*, 27(2): 44-47. Cz.

6425 **Michal, I. (1979).** Dynamika prirozenych lesu Bialovezskeho hvozdu 2. *Ziva*, 27(3): 88-92. Cz. Illus.

6426 **Michalik, S. (1974).** Antropogeniczne przemiany szaty roslinnej Ojcowskiego Parku Narodowego od poczatkow XIX wieku do 1960 roku. (The changes induced by man in the vegetation of the Ojcow National Park from the beginning of the 19th century to 1960.) *Ochr. Przyr.*, 39: 65-154. Pol (En). Illus., maps. (P)

6427 **Michalik, S. (1978).** Ochrona Bieszczadow zachodnich w swietle waloryzacji przyrodniczej. (Protection of Western Bieszczady Mountains in the light of the evaluation of natural resources [including the flora].) *Kosmos, Ser. A., Biol.*, 27(4): 383-391. Pol. (P)

POLAND

6428 **Michalik, S. (1978).** Rosliny naczyniowe Ojcowskiego Parku Narodowego. (Vascular plants of the Ojcow National Park.) *Ochr. Przyr., Stud. Nat., ser. A,* 16: 171p. Pol (En). Illus., maps. (P)

6429 **Michalik, S. (1979).** Przestrzenna i ekologiczna koncepcja ochrony szaty roslinnej centralnej czesci wyzyny krakowskiej. (Spatial and ecological conception of the conservation of vegetation in the central part of the Cracow upland.) *Ochr. Przyr.,* 42: 75-91. Pol (En). Maps.

6430 **Michalik, S. (1979).** Zagadnienia ochrony zagrozonych gatunkow roslin w Polsce. (Some problems of the conservation of threatened plant species in Poland.) *Ochr. Przyr.,* 42: 11-28. Pol (En). (T)

6431 **Mielcarski, C. (1978).** Stanowiska roslin rzadszych i chronionych, wystepujacych w zbiorowiskach lesnych okolic Czeszewa nad Warta i Lutynia. (Localities of rare and protected plants occurring in the forest communities in the neighbourhood of Czeszewo on the rivers Warta and Lutynia.) *Badan. Fizjogr. Pol. Zachod., B,* 30: 207-208. Pol (En). (T)

6432 **Mielnicka, B., Warkowska, H. (1979).** Proba okreslenia pojemnosci turystycznej parkow narodowych na przykladzie Babiogorskiego Parku Narodowego. (An attempt at defining the tourist capacity of national parks on the example of the Babia Gora National Park.) *Ochr. Przyr.,* 42: 279-296. Pol (En). Map. (P)

6433 **Milewski, W. (1979).** Biebrzanski skarb natury. (Nature treasure at Biebrza.) *Przyroda Polska,* no. 11: 15-17. Pol. Col. illus.

6434 **Mirek, Z.L. (1976).** The extinction of flax weed *Camelina alyssum* in Poland. *Phytocoenosis,* 5(3/4) : 227-236. Map. (T)

6435 **Molski, B.A. (1979).** The relationship between the national reserves and the activities of botanic gardens in plant genetic resource conservation. *In* Synge, H., Townsend, H., *eds.* Survival or extinction. Proceedings of a conference held at the Royal Botanic Gardens, Kew, 11-17 September 1978. Kew, Bentham-Moxon Trust. 53-62. (E/G/P/T)

6436 **Moser, M. (1979).** Uber einige neue oder seltene Agaricales-Arten aus dem Pieniny und aus Biescziade, Polen. (Some new or rare Agaricales species from the Pieniny and the Biescziade, Poland.) *Sydowia,* 8: 268-275. Ge. Illus. (T)

6437 **Okolow, C. (1986).** The Bialowieza primeval forest - the pearl of European forests. *Parks,* 11(2-3): 6-10. (P)

6438 **Olaczek, R. (1972).** Parki wiejskie ostoja rodzimej flory lsnej. (Rural parks as the refuges of the indigenous sylvan flora). *Chronmy Przyr. Ojczysta,* 28(2): 5-22. Pol (En). Illus. Map. (P/T)

6439 **Olaczek, R. (1978).** Chronione i rzadkie skladniki flory dorzecza Pilicy. (Protected and rare components of flora in the Pilica catchment basin.) *Stud. Osrod. Dok. Fizjogr.,* 6: 165-180. Pol (En). (T)

6440 **Olaczek, R. (1979).** [Human pressure effect on reserves, national parks and areas of the protected landscape in the perspective of the year 2000.] *Zeszyty Probl. Post. Nauk Rolnyczych, K.* 217: 302-318. Pol (En, Rus). (P/T)

6441 **Olaczek, R. (1982).** Synanthropization of phytocoenoses. *Memorabilia Zool.,* 37: 93-112. (Pol, Rus). (P/T)

6442 **Olaczek, R. (1983).** O wspoltczesnym rozumieniu ochrony gatunkowej roslin. (On modern c Somprehension of plant species protection.) *Przyroda Polska,* 11: 3-6. Pol (En). Col. illus. (T)

6443 **Olaczek, R., Piotrowska, H. (1979).** Mapa roslinnosci Wolinsliego Parku Narodowego. (Map of the vegetation of the Wolin National Park.) *Przyroda Polska,* 7-8: 9-11. Pol. Col. illus. (P)

POLAND

6444 **Olaczek, R., Piotrowska, H. (1980).** [The role of the Wolin National Park in the protection of nature on the west part of the Baltic sea coast]. *Zakl. Ochr. Przyr. Polsk. Akad. Nauk*, R43: 29-53. Pol (En). Illus. Maps. (P/T)

6445 **Olaczek, R., Sowa, R. (1971).** Roslinnosc lasu jodlowo-bukowego rezerwatu Galkow pod Lodzig. (Vegetation of the fir-beech forest reserve Galkow, near Lodz.) *Ochr. Przyr.*, 36: 131-169. Pol (En). Illus., maps. (P)

6446 **Olaczek, R., Sowa, R. (1976).** Wymieranie flory rodzimej w obszarze zurbanizowanym na przykladzie rezerwatu lesnego "polesie konstantinowskie" w Lodzi. (Decline of the native flora in urban areas using the example of the "Polesie Konstantinowskie" Forest Reserve in Lodz.) *Phytocoenosis*, 5(3/4). Warszawa-Bialowieza. Materialy Sympozjum w Krakowie 8-10.VI. 1976. Pol (En). (P/T)

6447 **Olaczek, R., Sowa, R. (1980).** Flora rezerwatu lesnego "polesie konstantinowskie" w lodzi. (Flora of the "Polesie Konstantinowskie" Forest Reserve in Lodz.) *Spraw. Czynn. Posied. Nauk.*, R. 34(11): 1-5. Pol. (T)

6448 **Olaczek, R., Zarzycki, K., eds (1988).** Problemy ochrony Polskiej przyrody. (Problems of conservation in Poland.) Warsaw, Polish Scientific Publishers. 113p. Pol (En).

6449 **Palczynski, A. (1984).** Peat bogs of Biebrza Valley - a proposed biosphere reserve. *In* Unesco/UNEP. Conservation, science and society. Paris, Unesco. 297-299. Contributions to the First International Biosphere Reserve Congress, Minsk, Byelorussia/USSR, 26 September - 2 October 1983. (P)

6450 **Parusel, J.B. (1985).** Przyroda rezerwatu florystycznego Ochojec w Katowicach. (The nature in the "Ochojec" floristic reserve in Katowice.) *Chronmy Przyr. Ojczysta*, 41(3): 52-55. Pol. Map. (P)

6451 **Pawlowska, S. (1953).** Rosliny endemiczne w Polsce i ich ochrona. (Les especes endemiques en Pologne et leur protection.) *Ochr. Przyr.*, 21: 1-33. Pol (Fr). Illus., maps. (T)

6452 **Pedrotti, F. (1980).** Il Parco nazionale di Bialowieza. *Nat. Montagna*, 27(3): 177-187. It. Illus. (P)

6453 **Piasecki, K. (1983).** Projektowany rezerwat przyrody Zabuze w wodewodztwie bialskopodlaskim. ("Zabuze": the proposed nature reserve in the province of Biata Podlaska.) *Chronmy Przyr. Ojczysta*, 39(4): 42-50. Pol. Illus., maps. (P)

6454 **Piekos-Mirkowa, H. (1978).** Dramatyozne losy azalii pontyjskiej, zwanej rozanecznikium zoltym, jedynego gatunku, ktory dziko rosnie w Polsce. (The dramatic story of the pontic azalea, the only species of azalea growing wild in Poland.) *Kwiaty*, 4: 6-7. Pol. Illus. *Rhododendron luteum* protected in Rzeszow Province. (T)

6455 **Piekos-Mirkowa, H. (1980).** [Distribution of rare taxa of vascular plants in the area of the Tatra Mts.] *Chronmy Przyr. Ojczysta*, 36(3): 34-44. Pol (En). (T)

6456 **Piekos-Mirkowa, H., Mirek, Z. (1978).** O rzadkich lub dotychczas z obszaru Tatr nie znanych gatunkach roslin naczyniowych. (On rare or hitherto unlisted vascular plant species in the Tatra Mts.) *Fragm. Flor. Geobot.*, 24(3): 363-368. Pol (En). (T)

6457 **Plackowski, R. & B. (1985).** Rzadkie gatunki roslin w projektowanym rezerwacie Mlyn Wojcik kolo Radomska. (Rare plant species growing in the projected "Mlyn Wojcik" nature reserve near Radomsko.) *Chronmy Przyr. Ojczysta*, 41(3): 56-59. Pol. Map. (P/T)

6458 **Plackowski, R. (1984).** Ginace torfowisko w okolicy Belchatowa. (A vanishing peat-bog in the vicinity of Belchatow.) *Chronmy Przyr. Ojczysta*, 40(2): 32-34. Pol.

POLAND

6459 **Polakowski, B., Dziedzio, J., Polakowska, E. (1973).** Szata roslinna rezerwatu przyrody Jezioro Lukniany na Pojezierzu Mazurskim. (Vegetation of the nature reserve Jezioro Lukniany in Mazurian Lake region [N.E. Poland].) *Ochr. Przyr.*, 38: 85-114. Pol (En). Illus., map. (P)

6460 **Poznanska, Z. (1978).** Dziewiecsil poplocholistny *Carlina onopordifolia* i problem jego ochrony w Polsce. (The carline thistle, *Carlina onopordifolia* Bess., and its protection in Poland.) *Chronmy Przyr. Ojczysta*, 34(5): 18-27. Pol (En). Illus., map. (T)

6461 **Razmierczakowa, R., Rams, B. (1976).** Proba okreslenia optymalnej wielkosci zbioru kopytnika pospolitego *Asarum europaeum* L. na wybranej powierzchni lesnej w Ojcowskim Parku Narodowym. *Ochr. Przyr.*, 41: 231-248. Pol (En). (An attempt at the determination of the resources of *Asarum europaeum* L. and of the optimum amount of harvest from a chosen experimental forest area in the Ojcow National Park). (P)

6462 **Rejewski, M., Olesinska, H. (1974).** Zaslugajace na ochrone olesy i legi nad Jeziorem Rakutowskim na Kujawach. (Wet alderwoods and riverside carrs on Rakutowskie Lake in the region of Kujawy which deserve protection.) *Ochr. Przyr.*, 39: 173-199. Pol (En). Maps. (P)

6463 **Rostanski, K., Sendek, A., Jedrzejko, K. (1980).** [The yew reservation Zadni Gaj near Geszyn, Poland]. *Pr. Nauk Univ. Slask. Katowicach*, 375: 81-96. Pol. *Taxus baccata.* (P)

6464 **Rutkowski, L. (1982).** Nowe stanowiska rzadszych roslin w okolicach Szczecinka. (New locations of rare vascular plants in the vicinity of Szczecinek.) *Biologia*, 24(53): 107-116. Pol (Ge). (T)

6465 **Salata, B. (1979).** Grzyby wyzsze Rezerwatu Lesnego Jta k Lukowa. (Higher fungi of the Forest Reserve Jata near Lukow.) *Ann. Univ. Mariae Curie-Sklodowska*, C33: 127-148 (1978 publ. 1979). Pol (En, Fr). Map. (P)

6466 **Sembrat, K., et al. (1972).** Przyroda rezerwatu Muszkowicki Las Bukowy w wojewodztwie wroclawskim. (Nature of the Muszkowicki Las Bukowy reserve in the Wroclaw district.) *Ochr. Przyr.*, 37: 57-156. Pol (En). Illus., maps. (P)

6467 **Sherpinski, Z. (1979).** Forest protection problems in Poland. *Gorsko Stopanstvo*, 35(10): 42-46.

6468 **Sicinski, J.T. (1987).** Agrorezerwaty - forma czynnej ochrony przyrody. (Agroreserves - a form of active nature conservation.) *Chronmy Przyr. Ojczysta*, 43(5-6): 31-36. Pol (En).

6469 **Skibinski, S. (1979).** Problemy ochrony srodowiska przyrodniczego w okolicach Chelma. (Problems of the natural environment in the surroundings of Chelm.) *Chronmy Przyr. Ojczysta*, 35(6): 24-29. Pol.

6470 **Sokol, S., Szczepka, M.Z. (1985).** Zabytkowy park w Swierklancu zagrozony. (The monumental park at Swierklaniec is in danger.) *Chronmy Przyr. Ojczysta*, 41(6): 60-64. Pol. Illus. (P)

6471 **Sokolowski, A.W. (1974).** Projekt racjonalnej sieci rezerwatow przyrody w wojewodztwie Bialostokim. (Project of a rational network of nature reserves in the province of Bialystok.) *Ochr. Przyr.*, 39: 155-172. Pol (En). Maps. (P)

6472 **Sokolowski, A.W. (1976).** Projekt uzupelniajacej sieci rezerwatow przyrody w Puszczy Bialowieskiej. (The project of the supplementary nature reserves network in the Bialowieza forest.) *Ochr. Przyr.*, 41: 119-154. Pol (En). Illus., maps. (P)

6473 **Sokolowski, A.W. (1979).** Waloryzacja przyrodnicza projektowanych rezerwatow Puszczy Boreckiej. (Evaluation of natural qualities of the proposed nature reserves in the Borecka Primeval Forest.) *Chronmy Przyr. Ojczysta*, 35(4): 15-25. Pol (En). Illus., map. (P)

POLAND

6474 **Sokolowski, A.W. (1981).** Flora roslin naczyniowych Bialowieskiego Parku Narodowego. (Vascular plants of the Bialowieski National Park.) *Fragm. Florist. Geobot.*, 27(1-2): 51-131. Pol (En). Maps. (P)

6475 **Sokolowski, A.W., Piotrowska, M. (1985).** Projekt parku krajobrazowego w dolinie Narwi. (The project for a landscape park in the Narew River valley.) *Chronmy Przyr. Ojczysta*, 41(5): 5-18. Pol (En). Illus., col. illus., map. (P)

6476 **Spiewakowski, E.R. (1977).** K voprosu ob okhrane rastitel'nogo pokrova Srednego Pomor'ya. (Z zagadnien ochrony szaty roslinnej Pomorza Srodkowego.) (The problem of protecting the plant cover of the Middle Pomeronian Region.) *Koszalin Stud. Mater.*, 1: 117-130. Rus (Pol).

6477 **Stachurski, M., Stachurska, E. (1979).** Aktualny stan rezerwatow stepowych i florystycznych w okolicach Miechowa. (The present state of the steppe vegetation and floristic reservations in the environs of Miechow.) *Chronmy Przyr. Ojczysta*, 35(1): 28-40. Pol (En). Illus., map. (P)

6478 **Staszkiewicz, J. (1972).** Dolnoreglowe rezerwaty lesne Beskidu Sadeckiego. (The Lower Subalpine Forest Reserve of the Beskid Sadecki Region.) *Ochr. Przyr.*, 37: 233-262. Pol (Fr). Illus., map. (P)

6479 **Stojny, Wladyslaw (1972).** Plant protection in Poland. Warszawa, Panstwowe Wydawn. Rolnicze i Lesne. 166p. (T)

6480 **Sudnik-Wojcikowska, B. (1983).** Rzadkie i interesujace gatunki roslin naczyniowych z obszaru Wielkiej Warszawy. (Rare and interesting vascular plants in the Greater Warsaw area.) *Fragm. Flor. Geobot.*, 27(4): 565-576 (1981 publ. 1983). Pol (En). (T)

6481 **Szadkowska-Izydorek, M. (1985).** O ochrone projektowanego rezerwatu buczyny nizowej w okolicach Ustki. (An appeal in favour of the protected nature reserve safeguarding the beech stands typical of lowlands in the environs of Ustka.) *Chronmy Przyr. Ojczysta*, 41(4): 51-54. Pol. Illus., map. (P)

6482 **Szadkowska-Izydorek, M., Kaminska, A. (1978).** Projectowany rezerwat sosny Pomnikowe kolo Czluchowa. (The proposed nature reserve of the "Monumental Pines" near Czluchow.) *Chronmy Przyr. Ojczysta*, 34(6): 40-43. Pol. Illus., map. (P)

6483 **Szafer, W. (1958).** Chronione w Polsce gatunki roslin. (Plant species protected in Poland.) *Zakl. Ochrony Przyrody PAN, Wyd.*, 14: 1-108. Pol. Illus. (T)

6484 **Szczesny, T. (1983).** Obszary chronione w Polsce w swietle systemu i zasad klasyfikacji miedzynarodowej. (Natural areas protected in Poland in the light of the system and principles of international classification.) *Chronmy Przyr. Ojczysta*, 39(1-2): 5-16. Pol (En). Col. illus. (P)

6485 **Szotkowski, P. (1978).** Interesujace obiekty i obszary przyrodnicze w okolicy Glogowka na Slasku Opolskim. (Some interesting protected objects and areas in the vicinity of Glogowek near Opole in Silesia.) *Chronmy Przyr. Ojczysta*, 34(6): 63-67. Pol. Illus. Protection of landscape. (P)

6486 **Szukiel, E. (1982).** Wplyw przegeszczenia jeleni na odnowienia w lasach bieszczadzkich. (Impact of excessive red deer density upon forest regeneration in the Bieszczady Mountains.) *Sylwan*, 26(1-3): 41-47. Pol. (P)

6487 **Szymanski, B., Tym, Z. (1975).** Bieszczadzki Park Narodowy. (The Bieszczady National Park.) *Sylwan*, 119(9): 25-36. Pol (En, Rus). Map. (P)

6488 **Trzcinska-Tacik, H., Wasylikowa, K. (1982).** [History of the synanthropic changes of flora and vegetation of Poland]. *Memorabilia Zoologica*, 37: 47-69. Pol, Rus (En). Effects of human habitation.

POLAND

6489 **Wiecko, E. (1978).** The Bialowieza Forest. A nature protection centre of world significance. *In* Proceedings of the 7th World Forestry Congress, Buenos Aires, 4-18 October 1972, vol. 3: Conservation and recreation. Buenos Aires, Instituto Forestal Nacional. 4152-4158. (Sp, Fr). (P)

6490 **Wiecko, E. (1978).** [Protected areas in Poland]. *Lesnic. Warsz. Szk. Gl. Gospod. Wiejsk.*, 24: 97-121. Pol (En). Map. (P)

6491 **Wierchowska, E. (1985).** [A locality of the false orchid, *Chamaeorchis alpina*, in the Tatra National Park]. *Chronmy Przyr. Ojczysta*, 41(6): 50-51. Pol (En). (P/T)

6492 **Wilczkiewicz, M. (1984).** Las w Wojborzu zasluguje na ochrone. (The Wolborz forest deserves protection.) *Chronmy Przyr. Ojczysta*, 40(2): 30-31. Pol.

6493 **Wojterski, T. (1977).** La protection de la nature en Pologne. *In* Filion, L., Villeneuve, P., *eds.* Ecological reserves and nature conservation. Rapport, Conseil Consultatif des Reserves Ecologiques, Quebec, 1: 18-34. Fr. Papers presented at a conference at Iles-de-la-Madeleine, Quebec, 17-18 June 1976. (P)

6494 **Wojterskiego, T., ed. (1976).** Roslinnosc rezerwatu 'Debina' pod Wagrowem w Wielkopolsce. (Vegetation of the 'Debina' Reserve in Wielkopolska Region.) *Badan. Fizjogr. Pol. Zachod., B*, 29: 243p. Pol. Illus. Contains papers on the flora, ecology and scientific importance of the reserve. (P)

6495 **Wrobel, J. (1979).** W Slowinskim Parku Narodowym. (In the Slowinski National Park.) *Przyroda Polska*, 6: 11-13. Pol. Col. illus., map. (P)

6496 **Wyrzykiewicz, M. (1982).** Rezerwat brzozy niskiej *Betula humilis* bezpowrotnie stracony. (A nature reserve safeguarding the birch *Betula humilis* has been lost for ever.) *Chronmy Przyr. Ojezysta*, 38(6): 95-96. Pol. (P/T)

6497 **Zabawski, J. (1978).** Godne ochrony torfowisko w okolicy Swiecia nad Wisla. (The peat bog in the environs of Swiecie on the Vistula deserves protection.) *Chronmy Przyr. Ojczysta*, 34(5): 83-87. Pol. Map. (P)

6498 **Zarzycki, K. (1976).** Male populacje pieninskich roslin reliktowych i endemicznych, ich zagrozenie i problemy ochrony. (Small populations of relict and endemic plant species of the Pieniny range [West Carpathian Mts], and their conservation.) *Ochr. Pryzyr.*, 41: 7-75. Pol (En). Illus., maps. (P/T)

6499 **Zarzycki, K., Wojewoda, W. (1986).** Lista roslin wymierajacych i zagrazonych w Polsce (List of threatened plants in Poland.) Warszawa, Panstwowe Wydawnictwo Naukowe (Polish Scientific Publishers). 128p. Pol(En). (T)

6500 **Zyber, G.M. (1979).** Obszary chronionego krajobrazu w Koszalinskiem. *Przyroda Polska*, no. 11: 12-13. Pol. Col. illus.

PORTUGAL

6501 **Beliz, J.M. (1983).** A Serra de Monchique. Flora e vegetacao. Lisbon, National Parks and Country Heritage Authority. 86-92. Por (En). (T)

6502 **Dray, A.M. (1985).** Plantas a Proteger em Portugal Continental. Servico Nacional de Parques, Reservas e Conservacao da Natureza, Lisboa. 56p. (T)

6503 **Franco, J.D.A., Malato-Beliz, J., Mota, M. (1984).** Portugal as a genetic reserve of aromatic and medicinal plants. *In* EUCARPIA International Symposium on Aromatic Plants, Oeiras, Portugal. 39-46. Maps. Includes species list. (E)

6504 **Gomez-Campo, C. (1985).** The Iberian Peninsula. *In* Gomez-Campo, C., *ed.* Plant conservation in the Mediterranean area. Dordrecht, Dr W. Junk. 47-70. Illus. Maps. (T)

6505 **Malato-Beliz, J. (1986).** O Barrocal Algarvio. Flora e Vegetacao da Amendoeira (Loule). Colecao Parques Naturais No. 17. Lisboa, Servico Nacional de Parques, Reservas e Conservacao da Natureza. 51p. Por. Col. illus. Maps. (P)

PORTUGAL

6506 **Mendes, L. (1978).** Le parc national de Peneda Geres. (National park of Peneda Geres.) *In* Proceedings of the 7th World Forestry Congress, Buenos Aires, 4-18 October 1972, vol. 3: Conservation and recreation. Buenos Aires, Instituto Forestal Nacional. 4139-4142. (P)

6507 **Paiva, J.A.R. (1981).** Mata da Margaraca e sua conversao em reserva. *Anuar. Soc. Brot.*, 47: 49-66. Por. (P)

6508 **Pinto da Silva, A.R. (?)** Plantas em perigo: as armerias. *Bol. Comis. Nac. Ambiente*, 6 (7-9, anexo): 2p. (?1980). Por. Illus. (T)

6509 **Pinto da Silva, A.R. (?)** Plantas em perigo: azereiro *Prunus lusitanica* L. *Bol. Comis. Nac. Ambiente*, 5 (3, anexo 1): 2p. (?1980). Por. Illus. (T)

6510 **Pinto da Silva, A.R. (?)** Plantas em perigo: feto-de-botao *Woodwardia radicans* (L.) Smith. *Bol. Comis. Nac. Ambiente*, 5 (11-12, anexo): 2p. (?1980). Por. Illus. (T)

6511 **Pinto da Silva, A.R. (?)** Plantas em perigo: jascones. *Bol. Comis. Nac. Ambiente*, 5 (4, anexo): 2p. (?1980). Por. Illus. (T)

6512 **Pinto da Silva, A.R. (?)** Plantas em perigo: loendreira ou adelfeira *Rhododendron ponticum* L. subsp. *baeticum* (Boissier & Reuter) Handel-Mazzetti. *Bol. Comis. Nac. Ambiente*, 5 (5-6, anexo 2): 2p. (?1980). Por. Illus. (T)

6513 **Pinto da Silva, A.R. (?)** Plantas em perigo: pinheiro-silvestre *Pinus sylvestris* L. *Bol. Comis. Nac. Ambiente*, 5 (1, anexo): 2p. (?1980). Por. Illus. (T)

6514 **Pinto da Silva, A.R. (?)** Plantas em perigo: samouco ou faia *Myrica faya* Aiton. *Bol. Comis. Nac. Ambiente*, 6 (4-6, anexo 1): 2p. (?1980). Por. Illus. (T)

6515 **Sainz-Ollero, H., Hernandez-Bermejo, J.E. (1981).** Sintesis corologica de las dicotiledoneas endemicas de la Peninsula Iberica e Islas Baleares. Madrid, Coleccion Monografica INIA Num. 31. 111p. Sp (En). (T)

6516 **Salvo, E. (1985).** Libro Rojo de los helechos de la Peninsula Iberica y Baleares. *Quercus*, 18: 29. Sp. Illus. (T)

6517 **Tavares, C.N. (1959).** Protection of the flora and plant communities in Portugal. *In* IUCN. Animaux et vegetaux rares de la Region Mediterraneenne. Brussels, IUCN. 86-94. Proceedings of the IUCN 7th Technical Meeting, 11-19 September, 1958, Athens, vol.5. (T)

6518 **Vasconcellos, J. de C. (1950).** Proteccao a flora do Geres. *Agron. Lusit.* 12(4): 611-617. Por (Fr). (T)

PORTUGAL - Azores

6519 **Dias, M.H.P., Gama, M.I.D., da (1982).** Arvores e arbustos do Parque das Furnas (S. Miquel-Acores). (Trees and shrubs of the Furnas Park [Sao Miguel, Azores].) *Anuar. Soc. Broter., Coimbra A Soc.*, 48: 15-23. Por. (P)

6520 **Gomes, M.B.A. (1989).** The Faial Botanic Garden, Azores. *Bot. Gard. Conserv. News*, 1(4): 28-31. Illus., map. Includes list of endemic taxa in cultivation and their conservation status. (G/T)

6521 **IUCN Threatened Plants Committee Secretariat (1980).** First preliminary draft of the list of rare, threatened and endemic plants for the countries of North Africa and the Middle East. Kew, IUCN Threatened Plants Committee Secretariat. 170p. (T)

6522 **IUCN Threatened Plants Committee Secretariat (1981).** The botanic gardens list of rare and threatened plants of Macaronesia (excluding the Cape Verde Islands). Kew, IUCN Botanic Gardens Conservation Co-ordination Body Report No. 4. Lists plants in cultivation in botanic gardens. (G/T)

PORTUGAL - Azores

6523 **Kammer, F. (1979).** The influence of man on the vegetation of Macaronesia. *In* Berichte der Internationalen Symposien der Internationalen Vereinigung fur Vegetatsionskunde Herausgegeben von Reinhold Tuxen, 20-23 March 1978. 601-616. Illus. (T)

6524 **Kammer, F. (1982).** Beitrage zu einer kritischen Interpretation der rezenten und fossilen Gefasspflanzenflora und Wirbeltierfauna der Azoren, des Madeira-Archipels, der Ilhas Selvagens, der Kanarischen Inseln und der Kapverdischen Inseln, mit einem Ausblick auf Artenschwund in Makaronesien. Freiburg im Breisgau. 179p. Ge.

6525 **Malato-Beliz, J. (1983).** Proposition pour l'elaboration d'un catalogue des plantes endemiques, menacees et disparues en Macaronesie. *In* Comunicacoes apresentadas ao 2 Congresso Internacional pro flora Macaronesica, Funchal, 19-25 de Junho de 1977. Funchal. 437-440. Fr. (T)

6526 **Pinto da Silva, A.R. (?)** Plantas em perigo: feto-de-folha-de-hera *Asplenium hemionitis* L. *Bol. Comis. Nac. Ambiente,* 5 (2, anexo 1): 1p. (?1980). Por. Illus. (T)

6527 **Sjogren, E. (1973).** Conservation of natural plant communities in Madeira and of the Azores. *In* Proc. 1 Intern. Congress pro Flora Macaronesica. 148-153.

6528 **Sjogren, E. (1984).** Acores flores. (Azores flowers). Horta Faial, Azores, Direcao Regional de Turismo. Por (En, Fr, Ge). Col. illus. Includes list of threatened plants. (T)

PORTUGAL - Madeira

6529 **Bramwell, D., Montelongo, V., Navarro, B., Ortega, J. (1982).** Informe sobre la conservacion de los bosques y la flora de la Isla de Madeira. Jardin Botanico "Viera y Clavijo". Sp (Por). Report to International Dendrology Society and IUCN, outlining proposals for a protected areas system on Madeira. (P)

6530 **Bramwell, D., Synge, H. (1983).** A conservation project in Madeira. *Int. Dendrol. Soc. Year Book,* 1982: 73-74. (E/P/T)

6531 **Costa Neves, H. (1988).** Saving Madeira's laurel forest. *Threatened Pl. Newsl.,* 20: 4-5. Illus. (T)

6532 **IUCN Threatened Plants Committee Secretariat (1980).** First preliminary draft of the list of rare, threatened and endemic plants for the countries of North Africa and the Middle East. Kew, IUCN Threatened Plants Committee Secretariat. 170p. (T)

6533 **IUCN Threatened Plants Committee Secretariat (1981).** The botanic gardens list of rare and threatened plants of Macaronesia (excluding the Cape Verde Islands). Kew, IUCN Botanic Gardens Conservation Co-ordination Body Report No. 4. Lists plants in cultivation in botanic gardens. (G/T)

6534 **Kammer, F. (1979).** The influence of man on the vegetation of Macaronesia. *In* Berichte der Internationalen Symposien der Internationalen Vereinigung fur Vegetatsionskunde Herausgegeben von Reinhold Tuxen, 20-23 March 1978. 601-616. Illus. (T)

6535 **Kammer, F. (1982).** Beitrage zu einer kritischen Interpretation der rezenten und fossilen Gefasspflanzenflora und Wirbeltierfauna der Azoren, des Madeira-Archipels, der Ilhas Selvagens, der Kanarischen Inseln und der Kapverdischen Inseln, mit einem Ausblick auf Artenschwund in Makaronesien. Freiburg im Breisgau. 179p. Ge.

6536 **Malato-Beliz, J. (1977).** Consideracoes sobre a protecao da flora e da vegetacao na Madeira. *Natur. Pais.,* 3: 1-11. Por. (T)

PORTUGAL - Madeira

6537 **Malato-Beliz, J. (1983).** Considerations sur la protection de la flore et de la vegetation a Madere. *In* Comunicacoes apresentadas ao 2 Congresso Internacional pro flora Macaronesica, Funchal, 19-25 de Junho de 1977. Funchal. 353-364. Fr. (T)

6538 **Malato-Beliz, J. (1983).** Proposition pour l'elaboration d'un catalogue des plantes endemiques, menacees et disparues en Macaronesie. *In* Comunicacoes apresentadas ao 2 Congresso Internacional pro flora Macaronesica, Funchal, 19-25 de Junho de 1977. Funchal. 437-440. Fr. (T)

6539 **Sjogren, E. (1973).** Conservation of natural plant communities in Madeira and of the Azores. *In* Proc. 1 Intern. Congress pro Flora Macaronesica. 148-153.

PORTUGAL - Salvage Islands

6540 **Brockie, R.E., Loope, L.L., Usher, M.B., Hamann, O. (1988).** Biological invasions of island nature reserves. *Biol. Conserv.*, 44(1-2): 9-36. Includes 6 case studies. (P)

6541 **IUCN Threatened Plants Committee Secretariat (1980).** First preliminary draft of the list of rare, threatened and endemic plants for the countries of North Africa and the Middle East. Kew, IUCN Threatened Plants Committee Secretariat. 170p. (T)

6542 **IUCN Threatened Plants Committee Secretariat (1981).** The botanic gardens list of rare and threatened plants of Macaronesia (excluding the Cape Verde Islands). Kew, IUCN Botanic Gardens Conservation Co-ordination Body Report No. 4. Lists plants in cultivation in botanic gardens. (G/T)

6543 **Kammer, F. (1979).** The influence of man on the vegetation of Macaronesia. *In* Berichte der Internationalen Symposien der Internationalen Vereinigung fur Vegetatsionskunde Herausgegeben von Reinhold Tuxen, 20-23 March 1978. 601-616. Illus. (T)

6544 **Kammer, F. (1982).** Beitrage zu einer kritischen Interpretation der rezenten und fossilen Gefasspflanzenflora und Wirbeltierfauna der Azoren, des Madeira-Archipels, der Ilhas Selvagens, der Kanarischen Inseln und der Kapverdischen Inseln, mit einem Ausblick auf Artenschwund in Makaronesien. Freiburg im Breisgau. 179p. Ge.

PUERTO RICO

6545 **Anon. (1987).** *Goetzea elegans* on endangered species lists. *Solanaceae Newsl.*, 2(4): 68-69. (T)

6546 **Ayensu, E.S., DeFilipps, R.A. (1978).** Endangered and threatened plants of the United States. Washington, D.C., Smithsonian Institution and World Wildlife Fund Inc. 403p. Lists 90 'Extinct', 839 'Endangered' and 1211 'Threatened' taxa for the continental U.S. Also covers Hawaii, Puerto Rico and Virgin Is. (T)

6547 **Birdsey, R.A., Miller, W.F., Clark, J., Smith, R.H. (1984).** Forest area estimates from Landsat MSS (Multispectral scanner) and forest inventory plot data. U.S. Department of Agriculture, Forest Service Research Paper SO No.211. Washington, D.C., U.S. Department of Agriculture, Forest Service, Southern Forest Experiment Station. 7p. Illus., maps.

6548 **Elugo, A., Brown, S. (1981).** Ecological monitoring in the Luquillo Forest Reserve. *Ambio*, 10(2-3): 102-107. Illus., maps. (P)

6549 **Figueroa-Colon, J.C., et al. (1984).** Directices para la evaluacion de areas naturales en Puerto Rico. Commonwealth of Puerto Rico, Department of Natural Resources. 69p. Sp. (T)

PUERTO RICO

6550 **Little, E.L., jr, Woodbury, R.O. (1976).** Trees of the Caribbean National Forest, Puerto Rico. U.S. Department of Agriculture, Forest Service Res. Pap., ITF-20. Washington, D.C., U.S. Govt Print. Off. 27p. (Sp). (P)

6551 **Little, E.L., jr, Woodbury, R.O. (1980).** Rare and endemic trees of Puerto Rico and the Virgin Islands. U.S. Department of Agriculture, Forest Service Conserv. Res. Rep., 27. Washington, D.C., U.S. Govt Print. Off. 26p. (T)

6552 **Martorell, L.F., Loigier, A.H., Woodbury, R.O. (1981).** Catalogo de los nombres vulgares y cientifico de las plantas de Puerto Rico. Boletin 263, Universidad de Puerto Rico. Rio Piedras, Puerto Rico, Recinto de Mayaguez, Estacion Experimental Agricola. 231p. Sp.

6553 **Quevedo, V. (1987).** Data center targets Puerto Rico's rare species for conservation. *TNC International News*, (Fall 1987): 4. (T)

6554 **Robinson, A.F., jr, ed. (1982).** Endangered and threatened species of the southeastern United States including Puerto Rico and the Virgin Islands. Washington D.C., U.S. Department of Agriculture, Forest Service. Illus., maps, col. illus. (T)

6555 **Stafford, R. (1989).** Wheeler's peperomia and 'ohai: two tropical rarities in the Center's charge. *Plant Conservation*, 4(3): 2-3. Illus. Rare plants covered by the Center of Plant Conservation's programme. (T)

6556 **U.S. Congress, Office of Technology Assessment (1987).** Integrated renewable resource management for U.S. insular areas. Washington, D.C., U.S. Govt Printing Office. 443p. Illus., maps. OTA-F-325. Includes description of vegetation, remaining coverage, threatened species. (P/T)

6557 **U.S. Department of the Interior. Fish and Wildlife Service (1984).** Endangered and threatened wildlife and plants: proposed Endangered status for *Buxus vahlii* (Vahl's boxwood). *Fed. Register*, 49(136): 28580-28583. (L/T)

6558 **U.S. Department of the Interior. Fish and Wildlife Service (1984).** Endangered and threatened wildlife and plants: proposed Endangered status for *Goetzea elegans* (Beautiful goetzea). *Fed. Register*, 49(118): 24903-24906. (L/T)

6559 **U.S. Department of the Interior. Fish and Wildlife Service (1984).** Four plants in danger of extinction. *Endangered Species Tech. Bull.*, 9(8): 1, 5 Illus. (L/T)

6560 **U.S. Department of the Interior. Fish and Wildlife Service (1984).** Four plants proposed for listing. *Endangered Species Tech. Bull.*, 9(7): 1, 4-5. Illus. (L/T)

6561 **U.S. Department of the Interior. Fish and Wildlife Service (1985).** Endangered and threatened wildlife and plants: final rule to determine *Buxus vahlii* as an Endangered species. *Fed. Register*, 50(156): 32572-32575. (L/T)

6562 **U.S. Department of the Interior. Fish and Wildlife Service (1985).** Endangered and threatened wildlife and plants: final rule to determine *Goetzea elegans* (Beautiful goetzea) as an Endangered species. *Fed. Register*, 50(76): 15564-15567. (T)

6563 **U.S. Department of the Interior. Fish and Wildlife Service (1985).** Endangered and threatened wildlife and plants: final rule to determine *Zanthoxyllum thomasianum* to be an Endangered species. *Fed. Register*, 50(245): 51867-51870. (L/T)

6564 **U.S. Department of the Interior. Fish and Wildlife Service (1985).** Four plants given Endangered Species Act protection. *Endangered Species Tech. Bull.*, 10(5): 1,10. (L/T)

6565 **U.S. Department of the Interior. Fish and Wildlife Service (1986).** Endangered and threatened wildlife and plants: proposed Endangered status for *Banara vanderbiltii. Fed. Register*, 51(69): 12455-12457. (L/T)

PUERTO RICO

6566 **U.S. Department of the Interior. Fish and Wildlife Service (1986).** Endangered and threatened wildlife and plants: proposed Endangered status for *Peperomia wheeleri. Fed. Register*, 51(69): 12457-12460. (L/T)

6567 **U.S. Department of the Interior. Fish and Wildlife Service (1989).** Proposed listings - July 1989. Palo de Rosa (*Ottoschulzia rhodoxylon*). *Endangered Species Tech. Bull.*, 14(8): 5. (L/T)

6568 **Wadsworth, F.H. (1974).** Management of mountain habitat on a densely populated tropical island. Session 7. Paper 19. *In* IUCN Second World Conference on National Parks. Yellowstone and Grand Teton National Parks, U.S.A., 18-27 September 1972. Switzerland, IUCN. 212-219. Map.

6569 **Weaver, P.L. (1981).** The forest resources of Puerto Rico. U.S. Department of Agriculture Forest Service, Southern Forest Experiment Station. 19-20. 1980 Annual Letter, Institute of Tropical Forestry, Rio Piedras, Puerto Rico (June 1981).

6570 **Weaver, P.L. (1989).** Rare trees in the Colorado Forest of Puerto Rico's Luquillo Mountains. *Natural Areas J.*, 9(3): 169-173. (T)

6571 **Woodbury, R.O., et al. (1975).** Rare and endangered plants of Puerto Rico: a committee report. Commonwealth of Puerto Rico, U.S.D.A. Soil Conservation Service and Dept. of Natural Resources. 85p. Lists 515 rare and endangerd species of endemic and non-endemic plants, with their habitat, distribution and threat. (T)

REUNION

6572 **Baumer, M. (1978).** La conservation et la valorisation des resources ecologiques dans les iles des Comores, de Maurice, de la Reunion, des Seychelles. Paris, Agence de Cooperation Culturelle et Technique. 92p. Fr.

6573 **Cabanis, Y., Chabouis, L. & F. (1969).** Vegetaux et groupements vegetaux de Madagascar et des Mascareignes, 4 vols. Tananarive, Bureau Pour le Developpement de la Production Agricola. Fr. Illus., maps. Published 1969-1970.

6574 **Cadet, Th. (1984).** Liste et commentaires sur les plantes en danger des Mascareignes. Paris, Conservatoire et Jardins Botaniques de Nancy. Mimeo. 20p. Fr. Includes notes on local uses. (E/T)

6575 **Castillon, J.-B. (1984).** Orchidees rares ou nouvelles de l'ile de la Reunion. *Orchidophile*, 15(61): 589-591. Fr. (T)

6576 **Doumenge, C., Renard, Y. (1989).** La conservation des ecosystemes forestiers de l'ile de la Reunion. Cambridge, IUCN. 95p. Illus., maps. (L/P)

6577 **Friedman, F. (1988).** Fleurs rares des iles Mascareignes. L'ile aux images, Ile Maurice. 31p. Illus. (T)

6578 **Lavergne, R. (1982).** Orchidees menacees de la Reunion. *Orchidophile*, 13(50): 17-21. Fr. Illus. (T)

6579 **Lesouef, J.-Y. (1983).** Compte-rendu de la ire mission de sauvetage des elements les plus menaces de la flore des Mascareignes (La Reunion, Maurice, Rodrigues). Brest, WWF-France and Conservatoire Botanique du Stangalarc'h. 46, 15p. Illus. (T)

6580 **Lesouef, J.Y. (1988).** Rare plant profile No.2. The rescue of *Ruizia cordata* and the possible extinction of *Astiria rosea. Bot. Gard. Conserv. News*, 1(2): 36-39. Illus., map. (T)

6581 **Rivals, P. (1986).** La Reunion. *In* Hedberg, I. & O., *eds*. Conservation of vegetation in Africa south of the Sahara. Symposium Proceedings, at 6th plenary meeting of AETFAT, Uppsala. *Acta Phytogeogr. Suec.*. 54: 272-275. Fr. Lists species needing protection and discusses vegetation types requiring protective measures. (P/T)

REUNION

6582 **Valck, D. (1988).** Garden profile no. 3. Conservatoire et Jardin Botanique de Mascarin, Reunion Island, Indian Ocean. *Bot. Gard. Conserv. News*, 1(3): 50-53. Includes conservation programme. (G/T)

ROMANIA

6583 **Baicu, T. (1976).** Citeva probleme ecologice ale protoctiei plantelor. (Some ecological problems of plant protection.) *Prod. Veg. Cereale Plante Teh*, 28(12): 33-38. Rum.

6584 **Banarescu, P., Nalbant, T., Oarcea, Z. (1979).** Viitonue Parc National Cheile Nerei - Beusnita. (The future national park Cheile Nerei - Beusnita.) *Ocrot. Nat.*, 23(2): 99-103. Rum. Col. illus., map. (P)

6585 **Bleahu, M. (1979).** Conservarea padurii si turismul. Citeva observatii pe marginea unui proiect de parc national in Muntii Bihor. *Rev. Padur. Ind. Lemn. Celul. Hirtie Silvic. Expl. Padur.*, 94(4): 231-234. Rum. A national park project in the Bihor Mts in relation to forest conservation and tourism. (P)

6586 **Borza, A. (1941).** Die Pflanzenwelt Rumaniens und ihr Schutz. (The flora of Romania and its protection.) *Ber. Deutsch. Bot. Ges.*, 59(5): 153-168. Ge. (T)

6587 **Boscaiu, N. (1975).** La protection de la flore dans les Carpates Roumains. *In* Jordanov, D., *et al., eds.* Problems of Balkan flora and vegetation. Sofia, Bulgarian Academy of Sciences. 428-430. Fr. Proceedings of the 1st International Symposium on Balkan Flora and Vegetation, 7-14 June 1973. (T)

6588 **Boscaiu, N. (1975).** Probleme le conservanii vegetatiei alpine si subalpine. *Ocrot. Nat.*, 19(1): 17-23. Rum.

6589 **Boscaiu, N. (1979).** Integrarea fitocenotica si constituirea rezervatiilor botanice. (L'integration phytocenotique et la constitution de reserves botaniques.) *Ocrot. Nat.*, 23(2): 105-110. Rum (Fr). Col. illus. (P)

6590 **Boscaiu, N. (1985).** Criterii pentru constituirea si gestiunea ecologica a rezervatiilor botanice. (Criteres pour la constitution et la gestion ecologique des reserves botaniques). *Ocrot. Nat.*, 29(2): 126-135. Rum (Fr). (P)

6591 **Boscaiu, N. (1986).** Importanta sistematicii pentru protectia florei. (L'importance de la systematique pour la protection de la flore). *Ocrot. Nat.*, 30(2): 117-120. Rum (Fr).

6592 **Botnariuc, N., Toniuc, N. (1980).** Rezervatia biosferei Repetek. (La reserve de la biosphere Repetek.) *Ocrot. Nat.*, 24(1): 35-37. Rum (Fr). (P)

6593 **Botnariuc, N., Toniuc, N., Boscaiu, N. (1985).** Parcul national Retezat la cinci decenii de la infiintare. (Le parc national de Retezat a cinq decennies d'existence). *Ocrot. Nat. Mediul. Inconjurat.*, 29(1): 5-11. Rum (Fr). (P)

6594 **Burduja, C., Vidrascu, P. (1984).** L'utilite de conservation et d'extention des plantes decoratives rares. *Lucr. Grad. Bot. Bucuresti*, 1883-84: 295-300. Fr (Rum). Illus. (T)

6595 **Cioaca, A. (1986).** Gradina Zmeilor (Iudetul Salaj): monument al naturii. (Le monument naturel "Gradina Zmeilor" (Le Jardin des Dragons - district de Salaj): monument de la nature). *Ocrot. Nat.*, 30(2): 124-129. Rum (Fr). Illus., map. (G)

6596 **Cosma, I. (1978).** Ocrotirea naturii si turismul. (La protection de la nature et le tourisme.) *Ocrot. Nat.*, 1: 13-20. Rum (Fr).

6597 **Council of Europe (1983).** Rumania. *Council of Europe Newsl.*, 83(8-9): 4. Need for protection of Danube Delta.

6598 **Enescu, V. (1979).** Probleme actuale ale conservarii resurselor genetice forestiere in Romania. (Current problems in the conservation of forest-tree genetic resources in Romania.) *Probl. Genet. Teor. Aplic.*, 11(2): 129-136. Rum (En). (E)

ROMANIA

6599 **Fuhn, I.E., Cristurean, I. (1977).** Situatia actuala a rezervatiei naturale Padurea Hagieni. (La situation actuelle de la reserve naturelle de la foret de Hagieni.) *Ocrot. Nat.*, 21(2): 103-110. Rum (Fr). Illus., map. (P)

6600 **Giurgiu, V. (1978).** Conservarea padurilor. (Conserving forest.) Bucuresti, Editura Ceres. 307. Rum. Illus.

6601 **Grigore, S. (1975).** Flora si vegetatia Rezervatiei de Saraturi de la Dinias Judetul Timis. (Flora und Vegetation des Salzboden - Naturschutzgebiet Dinias, Kries Timis.) *Tibiscus*, 1975: 49-68. Rum (Ge). Illus. (P)

6602 **Grigore, S., Coste, I. (1975).** Flora Rezervatiei naturale Valea Mare - Moldova Noua (Banat). (Die Flora des Naturschutzgebietes Valea Mare - Moldova Noua [Banat].) *Tibiscus*, 1975: 69-81. Rum (Ge). Illus., map. (P)

6603 **Horeanu, C. (1978).** [La sauvegarde de *Schivereckia podolica* (Bess.) Andrz. par transplantation dans la station Ripiceni]. *Ocrot. Nat.*, 1: 43-46. Rum (Fr). Illus. (T)

6604 **Horeanu, C., Borcea, M. (1982).** Ceahlaul: viitor parc national. (Le Ceahlau: futur parc national.) *Ocrot. Nat.*, 26(1-2): 20-23. Rum (Fr). Illus., col. illus. (P)

6605 **Ianovici, V. (1979).** Protectia mediului inconjurator si prezervarea cadrului biologic in Tarile. Balcanice. (La protection de l'environnement et la conservation du cadre biologique dans les pays Balkaniques.) *Ocrot. Nat.*, 23(1): 5-7. Rum (Fr). Conference at Bucharest on 18-22 September 1978.

6606 **Ioanid, V. (1979).** Aspecte ale relatiei sistematizare: protectia mediului inconjurator. (Aspects de la relation amenagement territorial: protection de l'environnement.) *Ocrot. Nat.*, 23(1): 15-22. Rum (Fr).

6607 **Jordanov, D., et al., eds (1975).** Problems of the Balkan flora and vegetation. Sofia, Bulgarian Academy of Sciences. Proceedings of the First International Symposium on Balkan Flora and Vegetation, Vorna, 7-14 June 1973. 441p. (T)

6608 **Karacsonyi, C. (1985).** Rezervatiile naturale de pe Nisipurile din nord-vestul Romaniei. (Naturschutzgebiete auf den Sandgebieten im nordwesten Rumaniens.) (That we preserve all the sandy areas of north-west Rumania.) *Ocrot. Nat.*, 29(2): 119-125. Rum (Ge). Illus., map. (P)

6609 **Kirby, K.J., Heap, J.R. (1984).** Forestry and nature conservation in Romania. *Quart. J. Forest.*, 78(3): 145-155. Maps.

6610 **Kuzmanov, B. (1981).** Balkan endemism and the problem of species conservation, with particular reference to the Bulgarian flora. *Bot. Jahrb. Syst.*, 102(1-4): 255-270. Illus., maps. (T)

6611 **Malos, C. A. (1982).** Argumente botanice pentru constituirea rezervatiei Muntele Piatra Closani. (Arguments botaniques pour la constitution de la reserve du Mont Piatra Closanilor.) *Ocrot. Nat. Mediul. Inconjurat.*, 26(1-2): 75-81. Rum (Fr). Illus., map. (P)

6612 **Manoliu, A. (1981).** Ocrotirea naturii in R.S.S. Lituaniana. (La protection de la nature dans la R.S.S.) *Ocrot. Nat.*, 25(1): 110-112. Rum. Map.

6613 **Marossy, A. (1977).** Padurea ou bujori (*Paeonia officinalis* L. ssp. *banatica* (Roch.) Soo) de pe Dealul Pacau (Jud. Bihor.) (La foret a pivoine [*Paeonia officinalis* L. ssp. *banatica* (Roch.) Soo] de Dealul Pacau [dep. de Bihor].) *Ocrot. Nat.*, 21(2): 127-129. Rum (Fr). (T)

6614 **Miclaus, V., Szabo, A.T. (1979).** Bodenokologische Bedingungen seltener Pflanzen aus S.R. Rumanien. *Tofieldia calyculata* (L.) Wahlbg. *Not. Bot. Hort Agrobot. Cluj-Napoca*, 10: 105-113. Rum. Maps.

ROMANIA

6615 **Mihailescu-Firea, S. (1979).** Plante rare din flora R.S.R., cultivate in Gradina Botanica din Bucuresti. (Rare plants of the Romanian flora cultivated at Bucharest Botanic Garden.) *Culegere de studii si articole de Biologie.* Gradina Botanica Iasi, 1: 189-194. Rum (Fr). Describes 70 rare taxa in cultivation. (G/T)

6616 **Milescu, I. (1978).** Ideea de conservare a padurilor in legislatia romaneasca. (The idea of forest conservation in Romanian laws.) *Rev. Padur. Ind. Lemn. Celul. Hirtie Silvic. Expl. Padur.*, 93(4): 175-179. Rum. (L)

6617 **Mititelu, D., Burduja, L. (1984).** Dealul Ciorsaci-Falticeni, un monument al naturii. (La colline de Corsaci-Fasticeni, monument naturel.) *Ocrot. Nat.*, 28(2): 116-117. Rum (Fr). (P)

6618 **Mititelu, D., Navrotescu, T. (1983).** Pentru o rezervatie botanica ba Saca-Todireni (Judetul Vaslui). (Pour une reserve botanique a Saca-Todireni (Dep. Vaslui).) *Ocrat. Nat. Mediul. Inconjurat.*, 27(2): 139-141. Rum (Fr). (P/T)

6619 **Moldovan, I., Pazmany, D., Szabo, A., Chirca, E., Leon, C. (1984).** List of rare, endemic and threatened plants in Romania (I.). *Not. Bot. Hort. Agrobot. Cluj-Napoca*, 14: 5-16. Rom (En). (T)

6620 **Moldovan, I., Szabo, A.T., Pazmany, D. Chirca, E. (1986).** Computer-aided herbarium records (COHER). Endemics included in the list of endangered species (IUCN - 1983) in the herbarium of the Agronomy Institute Cluj-Napoca (CLA). *Not. Bot. Hort. Agrobot. Cluj-Napoca*, 16: 99-105. (T)

6621 **Morariu, I. (1981).** *Spiraea crenata* L., its ecology and protection in Brasov. *Rev. Roum. Biol., Biol. Veg.*, 26(1): 65-67. (T)

6622 **Muica, C., Paraschivu, G. (1980).** Rezervatia forestiera "Cotul cu Aluni" (Jud. Gorj). (La reserve forestiere "Cotul cu Aluni" [Dep. de Gorj].) *Ocrot. Nat.*, 24(1): 43-47. Rum (Fr). (P)

6623 **Oprea, I.V. & V., Purdela, L. (1982).** Rezervatii forestiere din sud-vestul Romaniei si vegetatia acestora. (Wald Rezervate aus dem sudwesten Rumaniens und ihre Vegetation.) *Ocrot. Nat.*, 26(1-2): 97-99. Rum (Ge). (P)

6624 **Oprea, I.V. & V., Purdela, L. (1983).** Analiza floristica a Parcului National "Semenic-Cheile Carasului", proiectat. (Floristic analysis of the National Park of Semenic-Mountains furrows of Caras.) *Contrib. Bot.*, 1983: 33-37. Rum (Fr). Maps. (P)

6625 **Oprea, I.V., & V. (1984).** Obiective ocrotite in sud-vestul Romaniei. (Protected plants in south-western Romania.) *Natura* (Romania), 35(2): 23-31. Rum. Illus., distribution maps.

6626 **Peterfi, S., Toniuc, N., Boscaiu, N. (1978).** Cinci decenii de activitate pentru ocrotirea naturii. (Cinquante annees d'activite de protection de la nature.) *Ocrot. Nat.*, 1: 7-12. Rum (Fr).

6627 **Peterfi, S., et al. (1977).** Noi initiative ale Comisiei Monumentelor Naturii pentru conservarea genofondului Romaniei. *In* Opris, T., *ed.* Ocrot. Nat. Maramuresene. Cluj-Napoca, Subcomisia Monumentelor. 7-48. Rum.

6628 **Plamada, E. (1979).** Specii arctic-alpine din Masivul Fagaras noi sau rare in Brioflora Romaniei. (Arctic-alpine species new or rare for the Bryoflora of Romania, in the Fagaras Mountains.) *Contrib. Bot.*: 45-50. Rum (Fr). Illus. (T)

6629 **Ploaie, G., Stefureac, T.I. (1983).** O viitoare rezervatie botanica si peisagistica: Piatra Tirnovului (Judetul Vilcea). (Une reserve botanique et paysagiste en perspective: le mont Piatra Tirnovului (Dep. Vilcea).) *Ocrat. Nat.*, 27(2): 135-138. Rum (Fr). Illus., map. (P/T)

ROMANIA

6630 **Popovici, D., Ciubotariu, C. (1979).** Contributii la stabilerea modului de ocrotire a rezervatiei naturale Ponoare - Bosanci, judetul Suceava. (Contributions concernant la modalite de protection de la reserve scientifique naturelle de Ponoare-Bosanci dep. de Suceava.) *Ocrot. Nat.*, 23(2): 173-177. Rum (Fr). (P)

6631 **Puscariu, V. (1978).** Insemnatatea primului congres al naturalistor din Romania (aprilie 1928), pentru ocrotirea naturii. (L'importance du premier Congres des naturalistes de Roumanie - (Avril 1928) pour la protection de la nature.) *Ocrot. Nat.*, 1: 21-24. Rum (Fr).

6632 **Puscariu, V., Boscaiu, N. (1981).** Viitorul parc national al Muntilor Apuseni: 1. (Le futur parc national des monts Apuseni: 1.) *Ocrot. Nat.*, 25(2): 165-178. Rum (Fr). Illus., map. (P)

6633 **Puscariu, V., Boscaiu, N. (1982).** Viitorul parc national al Muntilor Apuseni: 2. Propuneri pentru organizarea si function area parcului. (Le futur parc national des monts Apuseni: 2. Propositions concernant l'organisation et le fonctionnement du parc.) *Ocrot. Nat.*, 26(1-2): 5-13. Rum (Fr). Illus, col. illus. (P)

6634 **Racz, G., Dogaru, M.T. (1982).** Conservarea genofondului natural la specii de plante recoltate in scopuri medicinale. (The conservation of the natural genepool of plants used for therapeutic purposes.) *Ocrot. Nat.*, 26(1-2): 14-19. Rum (En). Illus., col. illus. (E)

6635 **Ratiu, F. (1979).** Importanta fitoistorica a rezervatiei de la Valea Morii: dealul Feleacului. (Die phytohistorische Bedeutung des Naturschutzgebiets Valea Morii: Feleac Hugel.) *Ocrot. Nat.*, 23(1): 31-34. Rum (Ge). Illus., map. (P)

6636 **Ratiu, F., Gergely. I. (1974).** Associatii vegetale noi si rare pentru tara noastra. (New and rare plant associations in Romania.) *Biologia*, 19(2): 7-15. Rum (Rus.) (T)

6637 **Resmerata, I. (1983).** Conservarea dinamica a naturii. Bucharest, Editura stiintific si Enciclopedica. Rum.

6638 **Resmerita, I. (1971).** Statium noi cu plante rare din Romania. (Status of some rare plants in Romania.) *Stud. Cerc. Biol., Ser. Bot.*, 23(6): 491-493. Rum. 12 rare taxa described. (T)

6639 **Schrott, L. (1979).** [The need to protect some rare taxa, due to their disappearance in the flora of the province of Timis]. *Tisbiscus Stiinte Nat.*, 11-16. Rum (Fr). (T)

6640 **Stancu, R., Deaconu, Gh., Richiteanu, A., Stancu, S. (1977).** Aspecte ale dezvoltarii legislatiei de ocrotire a naturii in Romania. (Aspects du developpement de la legislation concernant la protection de la nature en Roumanie.) *Ocrot. Nat.*, 21(2): 95-101. Rum (Fr). Illus. (L)

6641 **Stanescu, V., Parascan, D. (1979).** Padurea si protectia mediului inconjurator. Problematica si obiective generale. (Problems and general objectives for forest and environmental protection.) *Rev. Padur. Ind. Lemn. Celul. Hirtie Silvic. Expl. Padur.*, 94(4): 202-205. Rum. (P)

6642 **Stefureac, T.I., Panzaru, G. (1978).** Statiunea du *Cochlearia pyrenaica* DC. de la Silhoi (Maramures) si ocrotirea sa. (La station avec *Cochlearia pyrenaica* DC: de Silhoi (Maramures) et sa sauvegarde.) *Ocrot. Nat.*, 1: 39-42. Rum (Fr). Illus., map. (T)

6643 **Stoiculescu, C. (1978).** Arbori seculari si de mari dimensiuni din Oltenia, propusi pentru ocrotire. (Arbres seculiers de grandes dimensions de la region d'Oltenie recommandes pour protection.) *Ocrot. Nat.*, 1: 55-58. Rum (Fr). Illus. (T)

6644 **Szabo, A.T. (1981).** Problems of genetic erosion in Transylvania, Romania. *Kulturpflanze*, 29: 47-62. Maps. (E/G)

ROMANIA

6645 **Toniuc, N., Boscaiu, N. (1978).** Parcurile nationale in contextul culturii contemporane. (Les parcs nationaux dans le contexte de la culture contemporaine.) *Sargetia*, 11-12: 61-66 (1975-1976 publ. 1978). Rum (Fr). (P)

6646 **Toth, K. (1981).** Parcul National Kiskunsag. (Parc National Kiskunsag.) *Ocrot. Nat.*, 25(1): 87-89. Rum (Fr). Illus. (P)

6647 **Turnock, D. (1988).** Woodland conservation: the emergence of rational land use policies in Romania. *GeoJournal*, 17(3): 413-433.

6648 **Vicol, E.C., Georgescu, A. (1979).** Caracterizarea ecologica a reservatiei naturale Cheile Turzii (Jud. Cluj). (Okologische Charakterisierung des Naturreservats "Cheile Turzii" [Kreis Cluj].) *Contrib. But. Univ. Cluj*, 73-75. Rum (Ge). (P)

RWANDA

6649 **Anon. (1989).** Rwanda Arboretum. *Trop. Afr. Bot. Gard. Bull.*, 1: 12-13. Map. Lists native species. (E/G)

6650 **Deuse, P. (1968).** Rwanda. *In* Hedberg, I. & O., eds. Conservation of vegetation in Africa south of the Sahara. Symposium Proceedings, at 6th plenary meeting of AETFAT, Uppsala. *Acta Phytogeogr. Suec.*, 54: 125-127. Fr. Climate, vegetation, protected flora, and associations for protection. (T)

6651 **INEAC (1954).** Carte des Sols et de la Vegetation du Congo, du Rwanda et du Burundi. Brussels, INEAC. Fr. A series of vegetation and soil maps covering Zaire Rwanda and Burundi in c. 25 parts, published between 1954 and c. 1970. Each map is accompanied by a descriptive memoir, & several maps are to dif. scales.

6652 **Lebrun, J. (1955).** Esquisse de la vegetation du Parc National de la Kagera. Fasc. 2 of Exploration du Parc National de la Kagera Mission J. Lebrun (1937-1938). Bruxelles, Publ. Inst. Parcs Nat. Congo Belge. 89p. Fr. Illus. (P)

6653 **Weber, B. (1985).** Le Parc National des Volcans biosphere reserve: cooperation between conservation and development. *Parks*, 10(3): 19-21. Illus. (P)

SAMOA

6654 **Holloway, C.W., Floyd, C.H. (1975).** A national parks system for Western Samoa. Suva, Fiji, U.N. Development Advisory Team. vi, 71p. Illus. (P)

6655 **Trotman, I.G. (1979).** Western Samoa launches a national park program. *Parks*, 3(4): 5-8. Illus., map. A paper with same title in *Tigerpaper*, 6(4): 11-14. (P)

6656 **Watters, R.F. (1960).** The nature of shifting cultivation: a review of recent research. *Pacific Viewpoint*, 1: 59-99. Impact of shifting cultivation.

6657 **Whistler, W.A. (1983).** Vegetation and flora of the Aleipata Islands, Western Samoa. *Pacific Sci.*, 37(3): 227-249. Extent and condition of vegetation.

SAO TOME AND PRINCIPE

6658 **IUCN (?)** Conservacao e desenvolvimento sustentado dos ecossistemas florestais na Republica Democratica de Sao Tome e Principe. IUCN, Tropical Forest Programme Series. c.200p. In prep.

6659 **Exell, A.W. (1968).** Principe, S. Tome and Annobon. *In* Hedberg, I. & O., eds. Conservation of vegetation in Africa south of the Sahara. Symposium Proceedings, at 6th plenary meeting of AETFAT, Uppsala. *Acta Phytogeogr. Suec.*, 54: 132-134. Includes lists of main plants of interest, and areas proposed for conservation. (P/T)

SAO TOME AND PRINCIPE

6660 **Monad, T. (1960).** Notes botaniques sur iles de Sao Tome et de Principe. *Bull. IFAN*, 22A: 19-94.

SAUDI ARABIA

6661 **Abulfatih, H.A., Nasher, A.K. (1988).** Rare and endangered succulent plants in southwestern Saudi Arabia. *Arab Gulf J. Sci. Res. B. Agric. Biol. Sci.*, 6(3): 399-408. (Ara). Map. (T)

6662 **Anon. (1981).** New desert park in Saudi Arabia. *Threatened Pl. Commit. Newsl.*, 8: 15. Near Jeddah, close to the King Abdul Aziz University. (P)

6663 **Hajrah, H.H., Amer, H.I., Zahran, M.A., Naguib, I. (1980).** Preliminary ecological studies on Jeddah Desert Park. *In* Saudi Biological Society. Fourth Symposium on the Biological Aspects of Saudi Arabia. 10-13 March 1980. Riyad, Riyad University Press. 118. (Ara). Abstract. (P)

6664 **Nader, I. (1984).** Ein Nationalpark im Hochland des sudwestlichen Saudi-Arabiens. *Nat. Mus.* (Frankfurt), 114(2): 34-45. Ge. Illus., map. (P)

6665 **Vincent-Barwood, A. (1980).** A park for 'Asir. *Aramco World Mag.*, 31(5): 22-23. Saudi Arabia's first national park. (P)

SENEGAL

6666 **Adam, J.G. (1968).** Senegal. *In* Hedberg, I. & O., *eds.* Conservation of vegetation in Africa south of the Sahara. Symposium Proceedings, at 6th plenary meeting of AETFAT, Uppsala. *Acta Phytogeogr. Suec.*, 54: 65-69. Fr. Map. Describes vegetation types, protected plant communities, and areas and plant species proposed for protection. (P)

6667 **Dupuy, A.R. (1971).** Le Parc National du Niokolo-Koba. Premier des grand parcs nationaux de la Republique du Senegal. *In* Happold, D.C.D., *ed.* Wildlife conservation in West Africa. Proceedings of a Symposium held at the University of Ibadan, Nigeria, 2 April 1971. Morges, IUCN. 21-25. (P)

6668 **Mekouar, M.A., Wenger, E. (1984).** Foret et environnement en droit senegalais. (Forest and environment in Senegalese law.) *In* Prieur, M., Forets et environnement: en droit compare et international. Paris, Presses Universitaires de France: 165-182. Fr. (L)

SEYCHELLES

6669 **Baumer, M. (1978).** La conservation et la valorisation des resources ecologiques dans les iles des Comores, de Maurice, de la Reunion, des Seychelles. Paris, Agence de Cooperation Culturelle et Technique. 92p. Fr.

6670 **Edward, N. (1978).** Medusa with a death wish. *Country Life*, 164(4240): 1098. Illus. (T)

6671 **Jeffrey, C. (1968).** Seychelles. *In* Hedberg, I. & O., *eds.* Conservation of vegetation in Africa south of the Sahara. Symposium Proceedings, at 6th plenary meeting of AETFAT, Uppsala. *Acta Phytogeogr. Suec.*, 54: 275-279. Illus. Includes Felicite Is., Praslin Is., Curieuse Is., Mahe Is., and Aldabra islands. Present and future conservation measures. (P/T)

6672 **Lionnet, G. (1986).** The romance of a palm: Coco de Mer. Mauritius, l'ile aux images editions. 95p. Illus. *Lodoicea maldivica.* (T)

6673 **Polunin, N., Proctor, J. (1973).** The Vallee de Mai, Praslin, Seychelles. *Biol. Conserv.*, 5(4): 314-316. Lists some of the endemic plants. (P)

SEYCHELLES

6674 **Procter, J. (1974).** The endemic flowering plants of the Seychelles: an annotated list. *Candollea*, 29(2): 345-387. Lists 72 endemic species with the previous IUCN numerical system (0-4) to indicate degree of threat. (T)

6675 **Procter, J. (?)** Conservation in the Seychelles. Report of the conservation adviser, 1970. Mahe, Government Printer. 35p.

6676 **Seychelles Government (1971).** Conservation policy in the Seychelles. Seychelles Government White Paper. 10p. (P/T)

6677 **Swabey, C. (1970).** The endemic flora of the Seychelles Islands and its conservation. *Biol. Conserv.*, 2(3): 171-177. (T)

6678 **Warman, S., Todd, D. (1984).** A biological survey of Aride Island Nature Reserve, Seychelles. *Biol. Conserv.*, 28(1): 51-71. Illus., maps. (P/T)

SEYCHELLES - Coralline Islands

6679 **Beamish, H.H. (1967).** Saving Aldabra. *The Listener*, 78: 605-606. (T)

6680 **Beamish, H.H. (1970).** Aldabra alone. *Int. J. Envir. Stud.*, 1(2): 168. (T)

6681 **Brockie, R.E., Loope, L.L., Usher, M.B., Hamann, O. (1988).** Biological invasions of island nature reserves. *Biol. Conserv.*, 44(1-2): 9-36. Includes 6 case studies. (P)

6682 **Coblentz, B.E., Vuren, D.V. (1987).** Effects of feral goats (*Capra hircus*) on Aldabra Atoll. *Atoll Res. Bull.*, 306: 1-6. (P)

6683 **Gaymer, E.D.T. (1966).** Aldabra: the case for conserving this coral atoll. *Oryx*, 8: 348-352. (T)

6684 **Gould, M.S., Swingland, I.R. (1980).** The tortoise and the goat: interactions on Aldabra Island. *Biol. Conserv.*, 17(4): 267-279. Illus., maps.

6685 **Jeffrey, C. (1968).** Seychelles. *In* Hedberg, I. & O., *eds.* Conservation of vegetation in Africa south of the Sahara. Symposium Proceedings, at 6th plenary meeting of AETFAT, Uppsala. *Acta Phytogeogr. Suec.*, 54: 275-279. Illus. Includes Felicite Is., Praslin Is., Curieuse Is., Mahe Is., and Aldabra islands. Present and future conservation measures. (P/T)

6686 **Stoddart, D.R. (1968).** The conservation of Aldabra. *Geog. Journ.*, 134(4): 471-486. Illus. (T)

6687 **Stoddart, D.R. (1971).** The settlement, development and conservation of Aldabra. *Phil. Trans. R. Soc. Lond, B*, 60(836): 611-628. (T)

6688 **Stoddart, D.R., Ferrari, J.D.M. (1983).** Aldabra atoll: a stunning success story for conservation. *Nat. Resources*, 19(1): 20-28. Illus., map. (P)

6689 **Stoddart, D.R., Savy, S. (1983).** Aldabra: island of giant tortoises. *Ambio*, 12(3/4): 180-185. Illus., maps.

6690 **Stoddart, D.R., ed. (1967).** Ecology of Aldabra Atoll, Indian Ocean. *Atoll Res. Bull.*, 118: 144p. Illus. List of endemic plant species, bibliography of Aldabra. (B/T)

SIERRA LEONE

6691 **Cole, N.H. (1980).** The Gola Forest in Sierra Leone: a remnant primary tropical rain-forest in need of conservation. *Envir. Conserv.*, 7(1): 33-40. Map. (P)

6692 **Cole, N.H. Ayodele (1979).** Sierra Leone. *In* Hedberg, I., *ed.* Systematic botany, plant utilization and biosphere conservation: Proc. of a sym. held in Uppsala in commemoration of the 500th ann. of the Univ. Stockholm, Almqvist & Wiksell Int'l. 88.

SIERRA LEONE

6693 **Davies, A.G. (1987).** The Gola Forest Reserves, Sierra Leone: wildlife conservation and forest management. Cambridge, IUCN. 126p. IUCN Tropical Forest Programme. (P)

6694 **Morton, J.K. (1968).** Sierra Leone. *In* Hedberg, I. & O., *eds.* Conservation of vegetation in Africa south of the Sahara. Symposium Proceedings, at 6th plenary meeting of AETFAT, Uppsala. *Acta Phytogeogr. Suec.*, 54: 72-74. Describes vegetation and outlines sites of importance to plant conservation. (P)

6695 **White, J.A. (1972).** Report to the Government of Sierra Leone on forest inventory of the Gola Forest Reserves. Rome, FAO. vi, 50p. Illus. map. (P)

SINGAPORE

6696 **Holttum, R.E. (1982).** *Diplazium prescottianum* (Wall. ex Hook.) Bedd: a Singapore fern now possibly extinct. *Gard. Bull. Singapore*, 35(1): 65-68. Illus. (T)

6697 **Sharp, T. (1984).** ASEAN nations tackle common environmental problems. *Ambio*, 13(1): 45-46. Illus.

SOLOMON ISLANDS

6698 **Boutilier, J.A. (1981).** The nature, scope, and impact of the tourist industry in the Solomon Islands. *In* Force, R.W., Bishop, B., *eds.* Persistence and change. Honolulu, Pacific Science Association. 37-50.

6699 **Cribb, P.J., Campbell, J., Dennis, G. (1985).** *Paphiopedilum* in the Solomon Islands: the rediscovery of "*P. dennisii*". *Orchid Review*, 93(1098): 130-131. (T)

6700 **Glaser, T. (1987).** Solomon Islands: paradise lost and found. *The Courier*, 102: 44-51.

6701 **Hansell, J.F.R., Wall, J.R.D. (1976).** Land resources of the Solomon Islands. Land Resources Study No. 18. Surbiton, ODA. 8 vols. Maps.

6702 **Hoyle, M.A. (1978).** Forestry and conservation in the Solomon Islands and the New Hebrides. *Tigerpaper*, 5(2): 21-24.

6703 **Kalkman, C., ed. (1983).** People unite against Unilever. *Flora Males. Bull.*, 36: 3916-3917. Deforestation.

6704 **Schenk, J.R. (1984).** A conservation strategy for forest utilization in Solomon Islands. Solomon Islands, Physical Planning Division. 62p. Illus., maps, forest. (P)

SOMALIA

6705 **Bally, P.R.O. (1968).** Somali Republic South. *In* Hedberg, I. & O., *eds.* Conservation of vegetation in Africa south of the Sahara. Symposium Proceedings, at 6th plenary meeting of AETFAT, Uppsala. *Acta Phytogeogr. Suec.*, 54: 145-148. Maps. Proposals for future conservation discussed. (P/T)

6706 **Barbier, C. (1985).** Further notes on *Livistona carinensis* in Somalia. *Principes*, 29(4): 151-155. Illus. (T)

6707 **Douthwaite, R.J. (1987).** Lowland forest resources and their conservation in southern Somalia. *Envir. Conserv.*, 14: 29-35.

6708 **Hemming, C.F. (1968).** Somali Republic North. *In* Hedberg, I. & O., *eds.* Conservation of vegetation in Africa south of the Sahara. Symposium Proceedings, at 6th plenary meeting of AETFAT, Uppsala. *Acta Phytogeogr. Suec.*, 54: 141-145. Map. Protection problems and laws are discussed. (L/P/T)

6709 **Lewis, G. (1985).** Plants in peril. *Kew Mag.*, 2(4): 380-382. *Ceratonia oreothauma* identified as an endangered species. (T)

SOMALIA

6710 **Madgwick, J. (1989).** Somalia's threatened forests. *Oryx*, 23(2): 94-101. Illus., maps. Tropical forest.

6711 **Trager, J.N. (1983).** The salvation of an endangered *Euphorbia*. *Cact. Succ. J.* (U.K.), 55(6): 246-247. Illus. (T)

6712 **Verdcourt, B. (1968).** French Somaliland. *In* Hedberg, I. & O., *eds.* Conservation of vegetation in Africa south of the Sahara. Symposium Proceedings, at 6th plenary meeting of AETFAT, Uppsala. *Acta Phytogeogr. Suec.*, 54: 140-141. Discusses geography, climate and vegetation. Lists desert fringe species characteristic of area.

SOUTH AFRICA

6713 **Ackermann, D.P. (1979).** The reservation of wilderness areas in South Africa. *S. Afr. Forest J.*, 108: 2-4. (P)

6714 **Anon. (1967).** Some protected wild flowers of the Cape Province. Cape Town, Dept of Nature Conservation of the Cape Provincial Administration. Unpaginated. With colour paintings of 244 protected species. (T)

6715 **Anon. (1979).** Looking for the last forests. *African Wildlife*, 33(2): 13. Col. illus.

6716 **Anon. (1981).** Conservation of germplasm. *Forum Bot.*, 19(6): 73. Annual Report of the Botanical Research Institute, South Africa. Includes collections of *Lagenaria*, *Sorghum*, *Pennisetum*, *Citrullus*. (E)

6717 **Anon. (1981).** Conservation of threatened plants. *Forum Bot.*, 19(6): 73-74. Annual report of the Botanical Research Institute, South Africa. (T)

6718 **Anon. (1982).** 10th International AETFAT Congress. The Palmiet River dam scheme. *Veld & Flora*, 68(1): 7-8. Map.

6719 **Anon. (1983).** Forestry and resource conservation. What is the role of the forester in South Africa? *S. Afr. Forest. J.*, 126: 12-14.

6720 **Anon. (1983).** Rediscovered! The black orchid. *S. Afr. Gard. Home*, (April): 53, 55. Col. illus. *Disa bodkinii*. (T)

6721 **Anon. (1983).** The conservation framework. *J. Dendrol.*, 3(3 & 4): 171-172. (L)

6722 **Anon. (1984).** Endangered species. *AABGA Newsl.*, 113: 3. (T)

6723 **Anon. (1984).** Our wonderful ericas. *Veld & Flora*, 70(suppl., 5(2)): i-iii. Protected erica species in South Africa. (T)

6724 **Anon. (1984).** Report to SAAB on the activities of the Flora Conservation Committee of the Botanical Society of South Africa for 1983. *Forum Bot.*, 22(6): 55-60. (T)

6725 **Bands, D.P. (1977).** Planning a wilderness area. *S. Afr. Forest. J.*, 103: 22-27. (P)

6726 **Bigalke, R.C. (1983).** Forestry and resource conservation. What is the role of the forester in South Africa. *S. Afr. Forest. J.*, 1983(126): 12-14.

6727 **Bliss-Guest, P. (1983).** Environmental stress in the East African Region. *Ambio*, 12(6): 290-295. Col. illus, maps.

6728 **Bossi, L. (1984).** Mapping Cape Fynbos vegetation with the aid of Landsat imagery. *Veld & Flora*, 70(1): 31-33. Illus.

6729 **Broomberg, B. (1980).** Giant's Castle Nature Reserve. *Trees S. Afr.*, 32(2): 49-50. (P)

6730 **Broomberg, B. (1981).** Preliminary survey of the trees of Empisini Nature Reserve. *Trees S. Afr.*, 33(2): 55-56. (P)

SOUTH AFRICA

6731 **Bunton, P.H. (1986).** Nature conservation laws in the Cape. *Naturalist* (S. Afr.), 26(2): 35-36. (L)

6732 **Codd, L.E. (1968).** Regional synthesis - Discussion. *In* Hedberg, I. & O., eds. Conservation of vegetation in Africa south of the Sahara. Symposium Proceedings, at 6th plenary meeting of AETFAT, Uppsala. *Acta Phytogeogr. Suec.*, 54: 257-260. Overview of conservation in southern Africa. (P)

6733 **Coetzee, B.J. (1983).** Phytosociology, vegetation structure and landscapes of the Central District, Kruger National Park, South Africa. *Dissert. Bot.*, 69: xxiii, 456p. Illus., map. (P)

6734 **Cottrell, M.J. (1983).** Palmiet Nature Reserve: an urban venture. *Park Admin.*, 36(2): 26-28. Illus. (P)

6735 **Cowley, J. (1987).** Plants in peril: 2. *Kew Mag.*, 4(4): 199-201. *Gladiolus aureus.* (T)

6736 **Cowling, R.M., Pierce, S.M. (1985).** Southern Cape coastal dunes - an ecosystem lost? *Veld & Flora*, 71(4): 99-103.

6737 **Cowling, R.M., Pierce, S.M., Moll, E.J. (1986).** Conservation and utilization of South Coast Renosterveld, an endangered South African vegetation type. *Biol. Conserv.*, 37(4): 363-377. Map.

6738 **Cunningham, A.B., Nichols, G. (1984).** *Guettarda speciosa* - tenuous foothold on the South African mainland. *Veld & Flora*, 70(3): 90-92. Endangered species in South Africa. (T)

6739 **Dellatola, L. (1980).** Saving rare plants. *S. Afr. Panorama.*, August: 42-43. (T)

6740 **Dellatola, L. (1984).** Save this heritage! *S. Afr. Panorama*, 29(2): 34-39.

6741 **Donald, D.G.M. (1982).** The control of *Pinus pinaster* in the Fynbos biome. *S. Afr. Forest. J.*, 123: 3-7. Illus.

6742 **Duncan, G. (1981).** *Moraea loubseri* Goldbl.: saved through cultivation. *Veld & Flora*, 67(1): 18-19. (G/T)

6743 **Duncan, G. (1983).** *Moraea aristata. Veld & Flora*, 69(4): 143-144. Illus. Description and notes on cultivation. (G/T)

6744 **Everard, D.A. (1988).** Threatened plants of the eastern Cape: a synthesis of collection records. *Bothalia*, 18(2): 271-277. En (Af). Map. (T)

6745 **Gill, K. (1982).** Legal protection for trees: Forest Amendment Act 1982. *Trees S. Afr.*, 33(4): 100-101. (L/T)

6746 **Goldblatt, P. (1981).** Moraeas: one lost, one saved. *Veld & Flora*, 67(1): 19-21. Illus. (T)

6747 **Hall, A.V. (1979).** Republic of South Africa. *In* Hedberg, I., ed. Systematic botany, plant utilization and biosphere conservation: Proc. of a sym. held in Uppsala in commemoration of 500th ann. of the Univ. Stockholm, Almqvist & Wiksell Int'l. 100-101.

6748 **Hall, A.V. (1981).** Information handling for Southern Africa's rare and endangered species survey. *In* Morse, L.E., Henifin, M.S., eds. Rare plant conservation: geographical data organization. New York, The New York Botanical Garden. 167-184. (T)

6749 **Hall, A.V. (1987).** Threatened plants in the Fynbos and Karoo biomes, South Africa. *Biol. Conserv.*, 40(1): 29-52. (T)

6750 **Hall, A.V., Ashton, E.R. (1983).** Threatened plants of the Cape Peninsula. Bolus Herbarium, Threatened-Plants Research Group. 26p. Lists 174 Cape Peninsula plants, endemic and non-endemic (world categories): Ex:5, E:25, V:28, R:50, I:40, K:26. (T)

SOUTH AFRICA

6751 **Hall, A.V., Rycroft, H.B. (1979).** South Africa: the conservation policy of the National Botanic Gardens and its regional gardens. *In* Synge, H., Townsend, H., *eds.* Survival or extinction. Proceedings of a conference held at the Royal Botanic Gardens, Kew, 11-17 September 1978. Kew, Bentham-Moxon Trust. 125-134. Map. (G/T)

6752 **Hall, A.V., Veldhuis, H.A. (1985).** South African Red Data Book: plants - Fynbos and Karoo biomes. South African National Scientific Programmes Report No. 117. Pretoria, CSIR. 160p. (P/T)

6753 **Hall, A.V., ed. (1984).** Conservation of threatened natural habitats. South African National Scientific Programmes Report No. 92. Pretoria, CSIR Foundation for Research Development. ix, 185p. Map.

6754 **Hall, H. (1983).** Notes on some rare stapeliads from Namaqualand. *Asklepios*, 27: 38-44. Illus., col. illus. (T)

6755 **Hartmann, H.E.K. (1981).** Ecology, distribution and taxonomy in Mesembryanthemaceae as a basis for conservation decisions. *In* Synge, H., *ed.* The biological aspects of rare plant conservation. Chichester, Wiley. 297-303. Illus. Proceedings of International Conference, King's College, Cambridge, 14-19 July 1980. (T)

6756 **Hilliard, O.M., Burtt, B.L. (1987).** The botany of the southern Natal Drakensberg. *Ann. Kirstenbosch Bot. Gard.*, 15: 253p.

6757 **IUCN (1986).** IUCN's list of top 24 threatened animals and plants. *Biol. Conserv. Newsl.*, 44: 1-6. (T)

6758 **IUCN/SSC TRAFFIC Group (?)** Black Stinkwood threatened in South Africa. *IUCN/SSC TRAFFIC Group Bull.*, 1(6): 3. (T)

6759 **Jaarsveld, E. van, Duncan, G. (1983).** *Freylinia visseri.* A plant which has become extinct in the wild, but which has been saved from total obliteration through cultivation. *Veld & Flora* (Kirstenbosch), 69(1): 2-3. Illus., col. illus. (G/T)

6760 **Jarman, M.L. (1982).** The Fynbos Biome Project. *In* AETFAT Synopses 10th Congress: the origin, evolution and migration of African floras, Pretoria, 18-23 January 1982. 69.

6761 **Kruger, F.J. (1981).** Conservation: South African heathlands. *In* Specht, R.L., *ed.* Heathlands and related shrublands: analytical studies. Ecosystems of the World 9B. Amsterdam, Elsevier. 231-234.

6762 **Kruger, P.R. (1978).** *Tinnea barbata*, a rare and unusual labiate. *Veld & Flora* (Kirstenbosch), 64(4): 112-113. Illus. (T)

6763 **Kundaeli, J.N. (1983).** Making conservation and development compatible. *Ambio*, 12(6): 326-331. Col. illus.

6764 **Lawder, M. (1981).** Conservation by cultivation. *Veld & Flora*, 67(1): 30-31. Illus. (G)

6765 **Low, A.B. (1988).** Conservation priority survey of the Cape Flats: the first four months. *Veld & Flora* (Kirstenbosch), 74(3): 83.

6766 **Lubke, R.A., Everard, D.A., Jackson, S. (1986).** The biomes of the eastern Cape with emphasis on their conservation. *Bothalia*, 16(2): 251-261. (Af). Maps.

6767 **Macdonald, I.A.W., Graber, D.M., DeBenedetti, S., Groves, R.H., Fuentes, E.R. (1988).** Introduced species in nature reserves in Mediterranean-type climatic regions of the world. *Biol. Conserv.*, 44(1-2): 37-66. Invasive species. Includes 7 case studies. (P)

6768 **McDonald, D. (1985).** Helicopter reconnaissance of some south-western Cape vegetation. *Veld & Flora*, 71(2): 37-38. Illus.

SOUTH AFRICA

6769 **McDowell, C. (1987).** Bid to save *Protea odorata*. *Veld & Flora* (Kirstenbosch), 72(4): 98-101. Illus., map. (T)

6770 **McMahan, L. (1984).** Cynthia Giddy's nursery for cycads. *Garden* (New York), 8(4): 6-7 Illus. S.Afican C.Giddy grows and sells rare S.African cycads to relieve the collecting pressure on wild populations. (G/T)

6771 **Mentis, M.T. (1984).** Monitoring in South African grasslands. South African National Scientific Programmes Report No. 91. Pretoria, CSIR Foundation for Research Development, Council for Scientific and Industrial Research. viii, 53p.

6772 **Moll, E.J., Bossi, L. (1984).** Assessment of the extent of the natural vegetation of the fynbos biome of South Africa. *S. Afr. J. Sci.*, 80(8): 355-358. Maps. Endangered hatitat.

6773 **Mooney, H.A. (1988).** Lessons from Mediterranean-climate regions. *In* Wilson, E.O., *ed.* Biodiversity. Washington, D.C., National Academy of Science Press. 157-165. Discussion of biodiversity of tropical and temperate zones.

6774 **Netshiungani, E.N., Wyk, A.E. van, Linger, M.T. (1981).** The holy forest of the Vhavenda. *Veld & Flora* (Kirstenbosch), 67(2): 51-52. Illus., col. illus., map.

6775 **Nichols, G.R. (1983).** Durban Parks Department establishes a natural environmental conservation section. *Park Admin.*, 36(2): 29-30. Illus. (P)

6776 **Noble, R.G. (1974).** An evaluation of the conservation status of aquatic biotypes (in South Africa). *Koedoe*, 17: 71-83.

6777 **Oberholster, C. (1983).** The orchids of the Vryheid Hill Game Reserve. *Veld & Flora* (Kirstenbosch), 69(2): 39-41. Col. illus. (P)

6778 **Oldfield, S. (1983).** South Africa launches plants campaign. *Threatened Pl. Newsl.*, 12: 2.

6779 **Oldfield, S. (1984).** Rare, wild succulents enter trade? *Oryx*, 18: 48-49. *Haworthia* spp. (T)

6780 **Olivier, M.C. (1983).** An annotated systematic checklist of the Angiospermae of the Cape Receife Nature Reserve, Port Elizabeth. *J. S. Afr. Bot.*, 49(2): 161-174. Maps. (P)

6781 **Olivier, M.C. (1986).** Valley Bushveld: an endangered veld type. *Veld & Flora* (Kirstenbosch), 72(2): 49-50. Map.

6782 **Olivier, W. (1982).** Jim Holmes and a Cape flora santuary. *Veld & Flora* (Kirstenbosch), 68(3): 88-90. Col. illus. (G/T)

6783 **Osborne, R. (1987).** *Encephalartos ferox*. *Encephalartos*, 9: 14-21. Illus. Data sheet; includes conservation. (T)

6784 **Osborne, R. (1987).** South African cycads - a bibliography. *Encephalartos*, 11: 8-13. (B)

6785 **Poynton, J.C. (1983).** Values, purpose and strategy in park services: some biological considerations. *Park Admin.*, 36(1): 33-38. (P)

6786 **Puzo, B. (1978).** Patterns of man-land relationships. *In* Werger, M.J.A., *ed.* Biogeography and ecology of Southern Africa. The Hague, W. Junk. 1049-1112. (P)

6787 **Rebelo, A.G., Holmes, P.M. (1988).** Commercial exploitation of *Brunia albiflora* (Bruniaceae) in South Africa. *Biol. Conserv.*, 45(3): 195-207. (T)

6788 **Richardson, D.M., Manders, P.T. (1988).** Reflections on the fynbos. *S. Afr. J. Sci.*, 84(11): 875-876.

SOUTH AFRICA

6789 **Roberts, C.P.R. (1982).** Environmental implications of the proposed Palmiet River water and power development projects. *Veld & Flora*, 68(1): 4-6.

6790 **Rooyen, N. van (1983).** Die plantegroei van die Roodeplaatdam-Natuurreservaat. 1. 'n voorlopige plantspesielys. (The vegetation of the Roodeplaat Dam Nature Reserve. 1. A preliminary plant species list.) *S. Afr. J. Bot.*, 2(2): 105-114. (P)

6791 **Rooyen, N. van (1983).** Die plantegroei van die Roodeplaatdam-Natuurreservaat. 2. Die plantgemeenskappe. (The vegetation of the Roodeplaat Dam Nature Reserve. 2. The plant communities). *S. Afr. J. Bot.*, 2(2): 115-125. Af (En). Illus., maps. (P)

6792 **Rupert, A. (1983).** Flora '83: conservation through education. *Veld & Flora* (Kirstenbosch), 69(3): 66-67. En, Af. (T)

6793 **Sargeant, P. (1985).** *Brunsvigia orientalis* - a disappearing flower. *Veld & Flora*, 71(suppl. 6(2)): i-ii. Illus. (T)

6794 **Scheepers, J.C. (1982).** The status of conservation in South Africa. *J. S. Afr. Biol. Soc.*, 23: 64-71. Map.

6795 **Scheepers, J.C. (1983).** Progress with vegetation studies in South Africa. *In* Killick, D.J.B., *ed.* The Proceedings of the 10th AETFAT congress held at the CSIR Conference Centre, Pretoria, Republic of South Africa, from 19-23 January 1982. *Bothalia*, 14(3/4). 683-690. Includes a good bibliography. (B)

6796 **Scheepers, J.C. (1983).** The present status of vegetation conservation in South Africa. *In* Killick, D.J.B., *ed.* The Proceedings of the Xth AETFAT congress held at the CSIR Conference Centre, Pretoria, Republic of South Africa, from 19-23 January 1982. *Bothalia*, 14(3/4). 991-995. (T)

6797 **Simmonds, M. (1985).** The fynbos and the frogs. *Oryx*, 19: 104-108. Illus. Map.

6798 **Skinner, G. (1981).** Some interesting trees in Van Staden's Wild Flower Reserve. *J. Dendrol*, 1(1&2): 36-37. Illus. Near Port Elizabeth, South Africa. (P)

6799 **South Africa. Department of Environment Affairs (1983).** Nature conservation. *Ann. Rep. Dep. Environ. Affairs* (S.Afr.), 1981-82: 217-218. (Af).

6800 **Taylor, H.C. (1980).** Weed research and veld conservation: some impressions on the Proceedings of the Third National Weeds Conference of South Africa, 1979. *Veld & Flora* (Kirstenbosch), 66(3): 85-87. Col. illus.

6801 **Van Wyk, B.-E. & C.M., Novellie, P.A. (1988).** Flora of the Zuurberg National Park. 2. An annotated checklist of ferns and seed plants. *Bothalia*, 18(2): 211-232.

6802 **Van Wyk, G.N. (1983).** Flora legislation. *Veld & Flora* (Kirstenbosch), 69(3): 71-72. (L/T)

6803 **Watson, H.K., MacDonald, I.A.W. (1983).** Vegetation changes in the Hluhluwe-Umfolozi Game Reserve Complex from 1937 to 1975. *Bothalia*, 14(2): 265-269. Maps. (P)

6804 **Werger, M.J.A., ed. (1978).** Biogeography and ecology of Southern Africa. Monographiae Biologicae 31. The Hague, W. Junk. 2 vols. xv, 1439p. Illus. maps. (P)

6805 **Whateley, A., Porter, R.N. (1983).** The woody vegetation communities of the Hluhluwe-Corridor-Umfolozi Game Reserve Complex. *Bothalia*, 14(3/4): 745-758. (Fr). Maps. Presented at the AETFAT Symposium on the Origin, Evolution and Migration of African Floras, Pretoria, Jan. 19-23, 1982. (P)

6806 **Wightman, D.R.D. (1981).** Koeberg and conservation. *Veld & Flora* (Kirstenbosch), 67(2): 48.

SOUTH AFRICA - Cape Province

6807 **Anon. (1979).** *Sesbania* tot onkruid verklaar. *Newsl. Soc. Protect. Envir.*, 9(2): 2. Af. Illus. *Sesbania punicea.*

6808 **Anon. (1983).** Flora '83 exhibitors - uitstallers. *Veld & Flora* (Kirstenbosch), 69(3): 83-84, 86-89, 91-95, 97-99, 101-103, 105. (Af). Illus., col. illus. (T)

6809 **Anon. (1983).** Fynbos: feite en vrae. *Veld & Flora* (Kirstenbosch), 69(3): 72, 74. Af. Col. illus.

6810 **Anon. (1983).** New parks for South Africa. *Oryx*, 17: 141. West Coast at Langebaan Lagoon, Cape Province and Vaalbos near Berkeley West in Kalahari Sand Belt. (P)

6811 **Anon. (1983).** Only one per cent of Cape conserved. *Forum Bot.*, 21(8): 60-61. 38th Annual Report of Cape Department of Nature and Environmental Conservation. (P)

6812 **Anon. (1983).** Some notes on the early history of flora conservation. *Veld & Flora* (Kirstenbosch), 69(3): 71. (T)

6813 **Anon. (1983).** The on-going battle against aliens. *Veld & Flora*, 69(3): 75, 77. (T)

6814 **Anon. (1984).** South African reserve to be a military target? *Oryx*, 18: 3. De Hoop Provincial Nature Reserve, South Cape coast. (P)

6815 **Boucher, C. (1981).** Autecological and population studies of *Orothamnus zeyheri* in the Cape of South Africa. *In* Synge H., *ed.* The biological aspects of rare plant conservation. Chichester, Wiley. 343-353. Illus. Proceedings of International Conference, King's College, Cambridge, 14-19 July 1980. (T)

6816 **Boucher, C. (1982).** The Kogelberg State Forest and environs - a paradise for Cape flora. *Veld & Flora* (Kirstenbosch), 68(1): 9-11. (P)

6817 **Brand, J.G., Hoven, H.J. van der, Ferguson, G., Petersen, C.H. (1984).** Taffelberg (Sudafrika). (Table Mountain (South Africa).) *Gard. Landsch.*, 94(11): 39-42. Af (En, Ge). Illus, maps.

6818 **Campbell, B., Gubb, A., Moll, E. (1980).** The vegetation of the Edith Stephens Cape Flats Flora Reserve. *J. S. Afr. Bot.*, 46(4): 435-444. (Af). Map. (P)

6819 **Codd, L.E. (1968).** Regional synthesis - Discussion. *In* Hedberg, I. & O., *eds.* Conservation of vegetation in Africa south of the Sahara. Symposium Proceedings, at 6th plenary meeting of AETFAT, Uppsala. *Acta Phytogeogr. Suec.*, 54: 257-260. Overview of conservation in southern Africa. (P)

6820 **Cole, M.M. (1984).** Economic opportunities and environmental conflicts in the south-eastern Cape. *Naturalist* (S.Afr.), 28(1): 27-36. Illus.

6821 **Cowling, R., McKenzie, B. (1979).** The conservation status of the Keiskamma River Mouth Region. *East. Cape Nat.*, 66: 29-30. Illus., map.

6822 **Cowling, R.M. (1980).** The coastal dune ecosystems of the Humansdorp District -a plea for their conservation. *East. Cape Nat.*, 70: 25-28. Illus., map.

6823 **Duncan, G. (1980).** Skaars inheemse plante van die Saldanha-Langebaan veld. *Veld & Flora* (Kirstenbosch), 66(2): 49-51. Af (En). Col. illus. (T)

6824 **Gawith, E.L., Schmeidler III, N.J. (1987).** Rare plant conservation in the Southwestern Cape. *In* Elias, T.S., *ed.* Conservation and management of rare and endangered plants. Proceedings from a conference, Sacramento, California, 5-8 November 1986. Sacramento, California Native Plant Society. 283-287. Illus., map. Includes overview of Fynbos flora and case study of *Witsenia maura* Thunb. (T)

SOUTH AFRICA - Cape Province

6825 **Geldenhuys, R. (1977).** Threatened species: habitat conservation and their survival. Department of Nature and Environmental Conservation, Provincial Administration of the Cape of Good Hope. 48p. Illus. (T)

6826 **Giliomee, J. (1979).** Rare flowers of the New Year. *African Wildlife*, 33(6): 9-13. Illus., col. illus. (T)

6827 **Hall, A.V. (1983).** Threatened plants at the south-western corner of Africa. *Bothalia*, 14(3 & 4): 981-984. (Fr). Whole volume covers the proceedings of the 10th AETFAT congress: the origin, evolution and migration of African floras, Pretoria, 18-23 January 1982. (T)

6828 **Holmes, J.L. (1983).** *Gladiolus stokoei. Veld & Flora* (Kirstenbosch), 69(2): 41-42. Col. illus. (T)

6829 **Kemp, M. (1986).** *Encephalartos latifrons. Encephalartos*, 8: 8-15. Illus. Includes note on conservation status. (T)

6830 **Kemp, M. (1987).** *Encephalartos arenarius. Encephalartos*, 11: 4-7. Illus. Includes section on conservation. (T)

6831 **Kemp, M. (1988).** *Encephalartos altensteinii. Encephalartos*, 13: 8-17. Illus. Data sheet; includes conservation. (T)

6832 **Kruger, F.J. (1979).** The Hottentots Holland: newest and largest of the Forestry Department's nature reserves. *Afr. Wildl.*, 33(4): 14-16. (P)

6833 **Low, A.B. (1979).** Whither Cape flats? *Veld & Flora* (Kirstenbosch), 65(3): 83-87. Af. Illus., col. illus., map.

6834 **Luckhoff, H.A. (1981).** Enlarged Marloth Nature Reserve and Swellendam hiking trail opened. *Veld & Flora*, 67(3): 67-69. (P)

6835 **Massyn, W. (1983).** A floral wonderland. *S. African Panorama*, 28(8): 48-50. Col. illus. Van Stadens Wild Flower Reserve. (P/T)

6836 **McKenzie, B. (1976).** Suggested management plan for the indigenous vegetation of Orange Kloof, Table Mountain, based on a phytosociological survey. *S. Afr. Forest. J.*, 99: 1-6. Illus. (P/T)

6837 **Milewski, A.V. (1978).** Habitat of threatened Proteaceae endemic to western Cape coastal flats. *J. S. Afr. Bot.*, 44(1): 55-65. (Af). (T)

6838 **Milewski, A.V. (1978).** Habitat of threatened species of *Serruria* and *Protea* endemic to Western Cape Coastal Flats. *J. S. Afr. Bot.*, 44(4): 363-371. (Af). Illus. (T)

6839 **Moll, E.J. (1981).** Is Fynbos being burnt to death? *Staavia dodii* as an example. *Forum Bot.*, 19(2): 40. Abstract only. Paper presented at SAAB Annual Congress, January 1981. (T)

6840 **Moll, E.J. (1981).** Table Mountain conservation: fiction or reality? *Forum Bot.*, 19(2): 41. Abstract only. Paper presented at SAAB Annual Congress, January 1981. (T)

6841 **Moll, E.J., Gubb, A.A. (1981).** Aspects of the ecology of *Staavia dodii* in the south western Cape of South Africa. *In* Synge, H., *ed*. The biological aspects of rare plant conservation. Chichester, Wiley. 331-342. Illus., map. Proceedings of International Conference, King's College, Cambridge, 14-19 July 1980. (T)

6842 **Moll, E.J., McKenzie, B., McLachlan, D., Campbell, B.M. (1978).** A mountain in a city - the need to plan the human usage of the Table Mountain National Monument, South Africa. *Biol. Conserv.*, 13(2): 117-131. Illus. (P)

6843 **Myles, C.F. (1979).** Baakens Valley: fauna and flora or freeway and fumes? *East Cape Nat.*, no. 66: 22-23.

SOUTH AFRICA - Cape Province

6844 **Oliver, I.B. (1989).** The cultivation of *Lithops salicola* L. Bolus a xerophyte with a vulnerable status. *Bot. Gard. Conserv. News*, 1(5): 25-27. Illus. (G/T)

6845 **Olivier, M.C. (1979).** An annotated systematic check list of the Angiospermae of the Worcester Veld Reserve. *J. S. Afr. Bot.*, 45(1): 49-62. (Af). (P)

6846 **Palmer, A.R. (1981).** Comparative total floristic survey of three east Cape nature reserves. *Forum Bot.*, 19(2): 32. Abstract only. Paper presented at SAAB Annual Congress, January 1981. (P/T)

6847 **Parker, D. (1979).** Whither the conservation of our fynbos? *Veld & Flora*, 65(2): 59. (T)

6848 **Parker, D. (1982).** The western Cape lowland fynbos. What is there left to conserve?! *Veld & Flora* (Kirstenbosch), 68(4): 98-99, 101. Maps. (T)

6849 **Parker, D. (1984).** The conservation of De Hoop. *Veld & Flora* (Kirstenbosch), 70(1): 26-27. (P)

6850 **Parker, R.D.M. (1983).** De Hoop: a heritage under threat. *Veld & Flora* (Kirstenbosch), 69(3): 69. (Af). Col. illus. (P)

6851 **Penzhorn, B.L. (1977).** Toevoegings tot die blomplantlys van die Bergkwagga Nasionale Park. (Additions to the check list of flowering plants of the Mountain Zebra National Park.) *Koedoe*, 20: 203-204. Af (En). (P)

6852 **Peterson, C. (1988).** Re-establishment of indigenous flora at the Camps Bay Reservoir, Table Mountain Nature Reserve. *Veld & Flora* (Kirstenbosch), 74(3): 96-100. (P)

6853 **Peterson, C.H. (1983).** The vegetation of the Western Table of Table Mountain. *Veld & Flora* (Kirstenbosch), 69(2): 50-53. Col. illus. (T)

6854 **Plessis, J.L. du (1978).** Blomme van die Bontebok Nasionale Park. *Veld & Flora* (Kirstenbosch), 64(4): 104-105. Af. Col. illus. (P)

6855 **Robertson, C. (1980).** New botanical centre for Fernkloof Nature Reserve. *Veld & Flora* (Kirstenbosch), 66(1): 22-23. Illus. (P)

6856 **Robertson, C. (1980).** Vogelgat Nature Reserve. *Veld & Flora* (Kirstenbosch), 66(3): 72-75. Illus., col. illus. (P)

6857 **Rooyen, M.W. van, Grobbelaar, N., Theron, G.K. (1979).** Phenology of the vegetation in the Hester Malan Nature Reserve in the Namaqualand broken veld: 2. The therophyte population. *J. S. Afr. Bot.*, 45(4): 433-452. (Af). (P)

6858 **Rooyen, M.W. van, Theron, G.K., Grobbelaar, N. (1979).** Phenology of the vegetation in the Hester Malan Nature Reserve in the Namaqualand broken veld. *J. S. Afr. Bot.*, 45(3): 279-293. (Af). (P)

6859 **Roux, A. le, Schelpe, E.A.C.L.E. (1981).** Namaqualand and Clanwilliam; South African wild flower guide: 1. Botanical Society of South Africa. 173p. Illus., col illus., map. Project of Cape Dept of Nature & Environmental Conservation. Includes list of protected and endangered flora. (T)

6860 **Rycroft, H.B. (1968).** Cape Province. *In* Hedberg, I. & O., eds. Conservation of vegetation in Africa south of the Sahara. Symposium Proceedings, at 6th plenary meeting of AETFAT, Uppsala. *Acta Phytogeogr. Suec.*, 54: 235-239. Illus. Lists numerous protected areas and recommends further protective measures. (L/P/T)

6861 **Small, J.G.C., Garner, C.J. (1980).** Gibberellin and stratification required for the germination of *Erica junionia*, an endangered species. *Zeitschr. Pflanzenphysiol.*, 99(2): 172-182. (G/T)

6862 **Strauss, H. (1978).** The Karoo National Park: a gain amidst losses. *Afr. Wildlife*, 32(6): 12-13. Col. illus. (P)

SOUTH AFRICA - Cape Province

6863 **Tansley, S.A. (1988).** The status of threatened Proteaceae in the Cape Flora, South Africa, and the implications for their conservation. *Biol. Conserv.,* 43(3): 227-239. (T)

6864 **Taylor, H.C. (1977).** Aspects of the ecology of the Cape of Good Hope Nature Reserve in relation to fire and conservation. Stellenbosch, S.A., Botanical Research Unit. (P)

6865 **Taylor, H.C. (1983).** The vegetation of the Cape of Good Hope Nature Reserve. *Bothalia,* 14(3 & 4): 779-784. Illus., maps. About 40 spp. endemic, rare or endangered to varying degrees. (P/T)

6866 **Taylor, H.C. (1984).** A vegetation survey of the Cape of Good Hope Nature Reserve. I. The use of association-analysis and Braun-Blanquet methods. *Bothalia,* 15(1/2): 245-258. (Af). Maps. (P)

6867 **Taylor, H.C. (1984).** A vegetation survey of the Cape of Good Hope Nature Reserve. II. Descriptive account. *Bothalia,* 15(1/2): 259-291. Illus., maps. (P)

6868 **Taylor, H.C. (1985).** An analysis of the flowering plants and ferns of the Cape of Good Hope Nature Reserve. *S. Afr. J. Bot.,* 51(1): 1-13. (Af). Maps. (P)

6869 **Taylor, H.C., Macdonald, S.A. & I.A.W. (1985).** Invasive alien woody plants in the Cape of Good Hope Nature Reserve. II. Results of a second survey from 1976 to 1980. *S. Afr. J. Bot.,* 51(1): 21-29. (Af). Maps. (P)

6870 **Taylor, H.C., Macdonald, S.A. (1985).** Invasive alien woody plants in the Cape of Good Hope Nature Reserve. I. Results of a first survey in 1966. *S. Afr. J. Bot.,* 51(1): 14-20. (Af). Maps. (P)

6871 **Van Rooyen, M.W., Theron, G.K., Grobbelaar, N. (1979).** Phenology of the vegetation in the Hester Malan Native Reserve in the Namaqualand Broken Veld: 1. General observations. *J. S. Afr. Bot.,* 45(3): 279-293. (Af). Illus. (P)

6872 **Van Rooyen, N., Van Rensburg, D.J., Theron, G.K., Bothma, J.D.P. (1988).** A check list of flowering plants of the Kalahari Gemsbok National Park. *Koedoe,* 31: 115-135. (P)

6873 **Van Wyk, B.-E., Novellie, P.A., Van Wyk, C.M. (1988).** Flora of the Zuurberg National Park. 1. Characterization of major vegetation units. *Bothalia,* 18(2): 211-220. En (Af). Illus., map. (P)

6874 **Werger, M.J.A., Coetzee, B.J. (1977).** A phytosociological and phytogeographical study of Augrabies Falls National Park, Republic of South Africa. *Koedoe,* 20: 11-51. Illus., map. (P)

6875 **Wyk, P. van (1974).** Trees of the Kruger National Park. Cape Town, Purnell. 2 vols. xxv, 597p. Illus. (P)

SOUTH AFRICA - Natal

6876 **Anon. (1978).** Do not mine Mapelane. *African Wildlife,* 32(4): 38-40. Col. illus. (P)

6877 **Bayer, A.W., Bigalke, R.C., Crass, R.S. (1968).** Natal. *In* Hedberg, I. & O., eds. Conservation of vegetation in Africa south of the Sahara. Symposium Proceedings, at 6th plenary meeting of AETFAT, Uppsala. *Acta Phytogeogr. Suec.,* 54: 243-247. Illus. Present and suggested protection of vegetation types. (L/P/T)

6878 **Codd, L.E. (1968).** Regional synthesis - Discussion. *In* Hedberg, I. & O., eds. Conservation of vegetation in Africa south of the Sahara. Symposium Proceedings, at 6th plenary meeting of AETFAT, Uppsala. *Acta Phytogeogr. Suec.,* 54: 257-260. Overview of conservation in southern Africa. (P)

6879 **Cunningham, T. (1988).** Over exploitation of medicinal plants in Natal/Kwazulu: root causes. *Veld & Flora (Kirstenbosch),* 74(3): 85-87. Col. illus. Medicinal plants. (E/T)

SOUTH AFRICA - Natal

6880 **Freire, M.S.B. (1983).** Experiencia de revegetacao nas dunao costeiras do Natal. (Revegetation experiments in the coastal dunes of Natal.) *Brasil Florest.*, 13(53): 35-42. Por (En). Illus.

6881 **Garland, I. (1981).** The dilemma of survival: the heritage of Natal forests and their value to the environment. *Trees S. Afr.*, 33(2): 30-37. Illus.

6882 **Harvey, P.H. (1983).** The new Formosa Nature Reserve: Estcourt. *Trees S. Afr.*, 35(1-2): 24-25. (P)

6883 **Lehmiller, D.J. (1987).** Plant conservation in Natal. *Herbertia*, 43(2): 60-62. Illus. (T)

6884 **Macdonald, I.A.W., Frame, G.W. (1988).** The invasion of introduced species into nature reserves in tropical savannas and dry woodlands. *Biol. Conserv.*, 44(1-2): 67-93. Invasive species. Includes 5 case studies. (P)

6885 **Moll, E.J. (1977).** A plea for the Gwalaweni Forest, Zululand. *Trees. S. Afr.*, 29(1): 17-23. Illus.

6886 **Moll, E.J. (1977).** The vegetation of Maputoland - a preliminary report of the plant communities and their present and future conservation status. *Trees S. Afr.*, 29(2/4): 31-58. Illus. (T)

6887 **Moll, E.J. (1978).** A plea for the Ngoye Forest. *Trees S. Afr.*, 30(3): 63-71. Illus. (P)

6888 **Nicholson, H.B. (1978).** A new nature reserve and developing arboretum in southern Natal. *Trees S. Afr.*, 30(1): 9-17. Illus. (G/P)

6889 **Nicholson, H.B. (1982).** The forests of the Umtamvuna River Reserve. *Trees S. Afr.*, 34(1&2): 2-10. Illus., maps. (P)

6890 **Osborne, R. (1987).** *Encephalartos ghellinckii. Encephalartos*, 12: 16-23. Illus., map. Data sheet; includes conservation. (G/T)

6891 **Saleh, F. (1982).** Rare plant find. *Gartn.-Bot. Brief*, 72: 6. *Asclepias concinna.* (T)

6892 **Weisser, P.J., Garland, I.F., Drews, B.K. (1982).** Dune advancement 1937-1977 at the Mlalazi Nature Reserve, Mtunzini, Natal, South Africa, and a preliminary vegetation-succession chronology. *Bothalia*, 14(1): 127-130. Illus. (P)

6893 **Weisser, P.J., Ward, C.J. (1982).** Destruction of the *Phoenix/Hibiscus* and *Barringtonia racemosa* communities at Richards Bay, Natal, South Africa. *Bothalia*, 14(1): 123-125. Illus. (T)

6894 **Wyk, A.E. van (1981).** Vir 'n groter Umtamvuna - natuurreservaat. *J. Dendrol.*, 1(3&4): 106-108. Af. Map. (P)

SOUTH AFRICA - Orange Free State

6895 **Codd, L.E. (1968).** Regional synthesis - Discussion. *In* Hedberg, I. & O., eds. Conservation of vegetation in Africa south of the Sahara. Symposium Proceedings, at 6th plenary meeting of AETFAT, Uppsala. *Acta Phytogeogr. Suec.*, 54: 257-260. Overview of conservation in southern Africa. (P)

6896 **Leistner, O.E. (1973).** Trees and shrubs of the Willem Pretorias Game Reserve. Nat. Conserv. Misc. Publ. No. 1. Orange Free State Provincial Administration. 27p. Illus. (P)

6897 **Oliver, I.B. (1989).** The cultivation of *Lithops salicola* L. Bolus a xerophyte with a vulnerable status. *Bot. Gard. Conserv. News*, 1(5): 25-27. Illus. (G/T)

6898 **Roberts, B.R. (1968).** The Orange Free State. *In* Hedberg, I. & O., eds. Conservation of vegetation in Africa south of the Sahara. Symposium Proceedings, at 6th plenary meeting of AETFAT, Uppsala. *Acta Phytogeogr. Suec.*, 54: 247-250. Describes vegetation types and recommends areas for protection. (P)

SOUTH AFRICA - Prince Edward Islands

6899 **Heymann, G., Erasmus, T., Huntley, B.J., Liebenberg, A.C., Retief, G. de F., Condy, P.R., Westhuysen, O.A. van der (1987).** An environmental impact assessment of a proposed emergency landing facility on Marion Island - 1987. South African National Scientific Programmes Report No. 140-1987. Pretoria, CSIR. 209p. Illus., col. illus., maps. Includes checklist of plants, overview of vegetation and threats.

SOUTH AFRICA - Transkei

6900 **Osborne, R. (1987).** *Encephalartos ghellinckii. Encephalartos*, 12: 16-23. Illus., map. Data sheet; includes conservation. (G/T)

6901 **Wyk, A.E. van (1981).** Vir 'n groter Umtamvuna - natuurreservaat. *J. Dendrol.*, 1(3&4): 106-108. Af. Map. (P)

SOUTH AFRICA - Transvaal

6902 **Anon. (1978).** Een Lavin Nature Reserve. *African Wildlife*, 32(5): 16-17. Col. illus. (P)

6903 **Anon. (1978).** Nature reserves. *In* Transvaal Provincial Administration, Nature Conservation Division. Twelfth Annual Report 1976-1977. 8-15. Col. illus. (P)

6904 **Anon. (1979).** Leave the park: coal mining in the Kruger National Park. *African Wildlife*, 33(5): 18-20. Illus. (P)

6905 **Berry, A. (1980).** The Melville Koppies Nature Reserve. *Veld & Flora* (Kirstenbosch), 66(2): 43-45, 47-49. Col. illus. (P)

6906 **Bredenkamp, G.J., Lambrechts, A. van W. (1979).** A check list of ferns and flowering plants of the Suikerbosrand Nature Reserve. *J. S. Afr. Bot.*, 45(1): 25-47. (Af). (P)

6907 **Burger, G. (1980).** Visit to the Transvaal Provincial Nature Reserve "Doorndraaidam". *Trees S. Afr.*, 32(2): 51-56. (P)

6908 **Codd, L.E. (1968).** Regional synthesis - Discussion. *In* Hedberg, I. & O., *eds.* Conservation of vegetation in Africa south of the Sahara. Symposium Proceedings, at 6th plenary meeting of AETFAT, Uppsala. *Acta Phytogeogr. Suec.*, 54: 257-260. Overview of conservation in southern Africa. (P)

6909 **Coetzee, B.J. (1982).** Vegetation of the Rustenburg Nature Reserve. *Park Admin.*, 35(4): 11, 13, 15, 17. (P)

6910 **Fourie, S.P. (1982).** Threatened euphorbias in the Transvaal. *Aloe*, 19(4): 111-125. Illus., col. illus., maps, key. (L/T)

6911 **Fourie, S.P. (1985).** Threatened euphorbias in the Transvaal. *Euphorbia J.*, 2: 75-90. (T)

6912 **Fourie, S.P. (1986).** The Transvaal, South Africa, threatened plants programme. *Biol. Conserv.*, 37(1): 23-42. Maps. (T)

6913 **Killick, D.J.B. (1968).** Transvaal. *In* Hedberg, I. & O., *eds.* Conservation of vegetation in Africa south of the Sahara. Symposium Proceedings, at 6th plenary meeting of AETFAT, Uppsala. *Acta Phytogeogr. Suec.*, 54: 239-243. Discusses vegetation, present protection within various reserves, protection of individual species, and suggested protection. (P/T)

6914 **Macdonald, I.A.W., Frame, G.W. (1988).** The invasion of introduced species into nature reserves in tropical savannas and dry woodlands. *Biol. Conserv.*, 44(1-2): 67-93. Invasive species. Includes 5 case studies. (P)

SOUTH AFRICA - Transvaal

6915 **Schijff, H., van der (1969).** A check list of the vascular plants of the Kruger National Park. Pretoria, Univ. of Pretoria, N.R. 53. 100p. (P)

6916 **Van Rooyen, N. (1978).** [A supplementary list of plant species for the Kruger National Park of the Pafuri area, South Africa]. *Koedoe*, 21: 37-46. Af. (P)

6917 **Zunckel, K. (1987).** Humilis on the move. *Encephalartos*, no. 12: 24. Illus. Short account of re-introduction of *Encephalartos humilis*. (G/T)

SPAIN

6918 **ADENA (1984).** ADENA, presenta la compana: salvemos a las que nos sal van a nosotros. *Panda* (WWF), 6: 20-23. Sp. Col. illus. (T)

6919 **Anon. (1979).** Nature in Spain. Barcelona, Servicio de Publicidad e Informacion de Turismo, Secretariat de Estado de Turismo. Col. illus. National parks and nature reserves. (P)

6920 **Anon. (1982).** A new botanic garden for Spain. *Threatened Pl. Commit. Newsl.*, 10: 14. In Cordoba. Conservation goals outlined. (E/G/T)

6921 **Anon. (1984).** ADENA presenta una campaa internacional de proteccion de las plantas. *Quercus*, 14: 40. Sp. Outlines 6 project proposals. (T)

6922 **Aparcio Martinez, A., Silvestre Domingo, S. (1987).** Flora del Parque Natural de la Sierra de Grazalema. Seville, Junta de Andalucia, Agencia de Media Ambiente. 303p. Sp. Illus. (P)

6923 **Aritio, L.B. (1979).** Parques nacionales espanoles. Madrid, Incafo. 192p. Sp. Col. illus. (P)

6924 **Barreno, E., et al. (1984).** Listado de Plantas Endemicas, Raras o Amenazadas de Espana-Espana peninsular. *Informacion Ambiental*. Conservacionismo de Espana. No. 3: I-XIII. Sp. (T)

6925 **Blanca Lopez, G,, Cueto Romero, M. (1984).** *Crepis pygmaea* L. (Compositae) en el sur de la Peninsula Iberica. (*Crepis pygmaea* L. (Compositae) in the southern Iberian Peninsula.) *An. Jard. Bot. Madrid*, 41(2): 341-350. Sp (En). Illus, maps. *Crepis granatensis*. (T)

6926 **Blanca, G., Diaz de la Guardia, C., Ortiz, M., Valle, F. (1986).** Flora medicinal de la provincia de Jaen. Nota 1. *Blancoana*, 4: 41-47. Sp (En). Chorology, ecology and medicinal use of 22 taxa: 8 threatened. (E/T)

6927 **Blas Aritio, L. (1975).** El parque nacional de Covadonga. *Vida Silvestre*, 16: 226-237. Sp. Col. illus. (P)

6928 **Blas Aritio, L. (1975).** El probleme de la proteccion a la naturaleza. *Vida Silvestre*, 13: 3-9. Sp. Col. illus.

6929 **Bolos, A. de (1959).** A propos de quelques plantes menacees dans le nord-est d'Espagne. *In* IUCN 7th Tech. Meeting, vol. 5:102. Fr. (T)

6930 **Cabezudo, B. (1979).** Plantas de la Reserva Biologica de Donana (Huelva): 2. *Lagascalia*, 8(2): 167-181 (1978 publ. 1979). Sp (En). (P)

6931 **Camavasa Castillo, J.M., Folch Ciullen, R., Masalles Saumeu, R.M., eds (1979).** El patrimonio naturai de la Comarca de Barcelona. Medidas necesarias para su proteccion y conservacion. Los recursos renovables terrestres. Barcelona, Corporacion Metropolitana. 269p. Sp. Illus., maps.

6932 **Carrasco-Munoz de Vera, C. (1977).** Les parcs nationaux d'Espagne. *Biol. Conserv.*, 11(1): 5-11. Fr (En). (P)

SPAIN

6933 **Costa, M., Garcia Carrascosa, M., Monzo, F., Peris, J.B., Stubing, G., Valero, E. (1984).** Estado actual de la flora y fauna marinas en al litoral de la comunid ad Valencia. Castellon de la Plana, Publicaciones del Excelentisimo Ayuntamiento. 209p. Sp. Col. illus. (The present state of marine flora and fauna on the Valencia coast.) (T)

6934 **Council of Europe (1981).** National activities: Spain. *Council of Europe Newsl.*, 81(6/7): 3. Creation of Alto Tajo Nat. Park. (L/P)

6935 **Cuellar Carrasco, L., Catalan Bachiller, G. (1978).** Conservacion de los recursos geneticos forestales, animales y vegetales en Espana. (Conservation of forest genetic resources, both animal and vegetable, in Spain.) *In* Proceedings of the 8th World Forestry Congress, Jakarta, 16-28 October 1978: Forestry for quality of life. II, 9. Sp (En). (E)

6936 **Dalda Gonzalez, J. (1976).** Conservacion de las comunidades climax. *Vida Silvestre*, 19: 152-167. Sp. Col. illus.

6937 **De Viedma, M.G. (1976).** Nature conservation in Spain: a brief account. *Biol. Conserv.*, 9(3): 181-190.

6938 **Fernandez-Galiano, E. (1971).** Problemes de la conservation de la vegetation et de la flore en Espagne. *Boissiera*, 19: 81-86. Fr. (T)

6939 **Fisher, R.C., Hooper, M.D., Warren, A., eds (1979).** Proposals for the biological management of the Parque Nacional de la Montana de Covadonga, Asturias, Spain. Discussion Papers in Conservation No.25. London, University College. 44p. (P)

6940 **Garcia Dory, M.A. (1977).** Covadonga National Park, Asturias, Spain. Its history, conservation interest and management problems. *Biol. Conserv.*, 11(2): 79-85. (Sp). (P)

6941 **Garcia Novo, F. (1979).** The ecology of vegetation of the dunes Donana National Park (south-west Spain). *In* Jefferies, R.L., Davy, A.J., eds. Ecological processes in coastal environments. British Ecological Society Symposia 19. Oxford, Blackwell Scientific Publications. 571-592. En (Fr). Illus. (P)

6942 **Garcia-Gonzalez, A., Goldsmith, F.B. (1988).** Cantabrican Mountain beechwoods: a survey and the case for their conservation. *Biol. Conserv.*, 45(2): 121-134. Illus. (P)

6943 **Gardner, M.F. (1986).** Plants in peril, 8. *Kew Mag.*, 3(3): 140-142. Illus. *Vella pseudocytisus.* (T)

6944 **Goday, R. (1959).** Algunas Especies raras o relictas que deben protejerse en la Espana mediterranea. *In* IUCN 7th Tech. Meeting. 5: 95-101. Sp. (T)

6945 **Gomez-Campo, C. (1978).** Studies on Cruciferae: 6. Geographical distribution and conservation status of *Boleum* Desv., *Guiraoa* Coss. and *Euzomodendron* Coss. *An. Inst. Bot. A.J. Cavanilles*, 35: 165-176. (Sp). Maps. (T)

6946 **Gomez-Campo, C. (1979).** Proteccion de especies vegetales amenazadas en Espana. *In* II Semana de Biologia, Conferencias - coloquio sobre investigaciones biologicas, 1979. Serie Universitaria, Botanica, 87. Madrid, Fundacion Juan March. 25-33. Sp. (T)

6947 **Gomez-Campo, C. (1984).** La flora Espanola. Una Estrategia para su conservacion. *Panda* (WWF), (5): 3-8. Sp. Col. illus. (T)

6948 **Gomez-Campo, C. (1985).** The Iberian Peninsula. *In* Gomez-Campo, C., ed. Plant conservation in the Mediterranean area. Dordrecht, Dr W. Junk. 47-70. Illus. Maps. (T)

6949 **Gomez-Campo, C., ed. (1987).** Libro rojo de especies vegetales amenazadas de espana peninsular e islas baleares. ICONA, Ministerio de Agricultura, Pesca y Alimentacion. 676p. Sp. Col illus., maps. Red Data Book. (T)

6950 **Herrera, C.M. (1987).** Distribucion, ecologia y conservacion de *Atropa baetica* Willk. (Solanaceae) en la Sierra de Cazorla. *An. Jard. Bot. Madrid*, 43(2): 387-398. Sp (En). (T)

6951 **Heywood, V. (1984).** Spain. *Threatened Pl. Newsl.*, 13: 13. Report on a threatened plants list for Spain. (T)

6952 **ICONA (1984).** Conservation in Spain. XVI General Assembly of the IUCN (summary), Madrid. 10. (P/T)

6953 **Kalutskii, K.K. (1978).** [Forests and national parks in Spain]. *Lesnoe Khoz.*, 10: 91-94. Rus. (P)

6954 **Larramendi, A. (1984).** Lluvia acida, la muerte viene del cielo. (Acid rain, death comes from the sky.) *Panda* (WWF), 6: 11-14. Sp. Col. illus. (T)

6955 **Lazare, J.-J., Miradalles, J., Villar, L. (1987).** *Cypripedium calceolus* L. (Orchidaceae) en el Pirineo. *An. Jardin Bot. Madrid*, 43(2): 375-382. Sp (En). Illus., map. Rediscovery in Pyrenees; distribution data. (T)

6956 **Marin del Valle, A. (1983).** La region extremena como ejemplo de los factores que intervienen en la distribucion de las plantas. *Quercus*, 26: 24-26. Sp. Illus. Map. (T)

6957 **Marraco Solana, S. (1978).** La proteccion de la natureleza y el caracter social del bosque. (Protection of nature and the social character of forests.) *In* Proceedings of the 7th World Forestry Congress, Buenos Aires, 4-18 October 1978, vol. 3: Conservation and recreation. 3677-3683. Sp (En, Fr).

6958 **Merino, J., Martin, V.A. (1981).** Biomass, productivity and succession in the scrub of the Donana Biological Reserve in southwest Spain. *Tasks for Vegetation Science*, 4: 197-203. (P)

6959 **Molero-Mesa, J. (1984).** El patrimonio vegetal de Sierra Nevada y su proteccion. *Panda* (WWF), 2(6): 3-10. Sp. Col. illus. (T)

6960 **Molero-Mesa, J., Perez-Raya, F. (1987).** La Flora de Sierra Nevada. Avance sobre el catalogo floristico nevadense. Granada, Universidad de Granada. 397p. Sp. Includes a checklist of the flora, with annotations of taxa threatened. (T)

6961 **Montoya, O.J.M. (1983).** [Alternative uses and conservation of coppices of *Quercus pyrenaica* Willd.] *Bol. Estac. Centr. Ecol.* (Madrid), 12(23): 35-42. Sp (En). Illus.

6962 **Moreno, M. (1983).** [*Iberis grossii* Pau: a little known species of the Andalusian flora.] *An. Jard. Bot. Madrid*, 40(1): 53-61. Sp (En). Illus. (T)

6963 **Nelson, C., McClintock, D., Small, D. (1985).** The natural habitat of *Erica andevalensis* in south-western Spain. *Kew Mag.*, 2(3): 324-330. Illus., maps. (T)

6964 **Novo, F.G. (1977).** The effects of fire on the vegetation of Donana National Park. Gen. Techn. Report No. WO-3. Washington, D.C., U.S. Department of Agriculture. 318-325. (P)

6965 **Ortiz Valbuena, A. (1984).** A proposito de la presencia de *Atropa baetica* Willk. en la provincia de Cuenca (Espana). (Occurrence of *Atropa baetica* Willk. in the province of Cuenca, Spain.) *An. Jard. Bot. Madrid*, 41(1): 161-165. Sp (En). (T)

6966 **Ortuna Medina, F. (1974).** Los parques nacionales in Espana. *Vida Silvestre*, 12: 222-228. Sp. Col. illus. (P)

6967 **Ortuna Medina, F. (1983).** La politica forestal en Espana. (Forest policy in Spain.) *In* Publicacion especial - Instituto Nacional de Investigaciones Forestales. Mexico The Instituto (41): 27-37. Sp. (L)

6968 **Ortuno, F., Pena, J. de la (1976).** Reservas y cotos nacionales de caza: 1. Region Pirenaica. Coleccion Naturaleza Espanola: 2. Madrid, INCAFO. 255p. Sp. Col. illus., maps. (P)

SPAIN

6969 **Ortuno, F., Pena, J. de la (1977).** Reservas y cotos nacionales de caza: 2. Region Cantabrica. Coleccion Naturaleza Espanola: 3. Madrid, INCAFO. 253p. Sp. Col. illus., maps. (P)

6970 **Ortuno, F., Pena, J. de la (1978).** Reservas y cotos nacionales de caza: 3. Region Central. Coleccion Naturaleza Espanola: 4. Madrid, INCAFO. 253p. Sp. Col. illus., maps. (P)

6971 **Peris, J.B., Stubing, G. (1984).** *Lavatera mauritanica* subsp. *davaei* en las Islas Columbretes (Castellon). *Ann. Jard. Bot. Madrid*, 41(2): 455-456. Sp. (T)

6972 **Prieto, A., Tejada, H. de (1979).** Aspectos forestales de la Provincia de Madrid. Madrid, Servicios de Extension Cultural y Divulgacion de la Diputacion Provincial de Madrid. 247p. Sp. Illus.

6973 **Rivas-Goday, S. (1959).** Algunas especies raras o relicticas que deben protegerse en la Espaa Mediterranea. *In* IUCN. Animaux et vegetaux rares de la Region Mediterraneenne. Proceedings of the IUCN 7th Technical Meeting, 11-19 September, 1958, Athens, vol. 5. Brussels, IUCN. 95-101. Sp. Briefly describes 38 threatened plant species. (T)

6974 **Rivas-Martinez, S. (1971).** Bases ecologicas para la conservacion de la vegetacion. *Rev. Cienc.*, 36(2). 6p. Sp. (T)

6975 **Sainz-Ollero, H., Hernandez-Bermejo, J.E. (1979).** Experimental reintroductions of endangered plant species in their natural habitats in Spain. *Biol. Conserv.*, 16(3): 195-206. Map. (T)

6976 **Sainz-Ollero, H., Hernandez-Bermejo, J.E. (1981).** Sintesis corologica de las dicotiledoneas endemicas de la Peninsula Iberica e Islas Baleares. Madrid, Coleccion Monografica INIA Num. 31. 111p. Sp (En). (T)

6977 **Salvo, E. (1985).** Libro Rojo de los helechos de la Peninsula Iberica y Baleares. *Quercus*, 18: 29. Sp. Illus. (T)

6978 **Tamames, R., et al., eds (1984).** El libro de la naturaleza. Madrid, El Pais. 304p. Sp. A handbook of environmental issues in Spain, including many articles about flora and vegetation, its destruction and protection.

SPAIN - Balearic Islands

6979 **Barreno, E., et al., eds (1984).** Listado de plantas endemics, raras o amenezadas de Espana: Islas Baleares. (List of endemic, rare or threatened plants in Spain: Balearic Islands.) *Informacion Ambiental*, No. 3 Conservacionismo en Espana: XIV-XVII. Sp (En). (T)

6980 **Cardona, M., Contandriopoulos, J. (1977).** L'edemisme dans les fleurs insulaires mediterraneennes. *Mediterranea*, 2: 49-77. Fr. (T)

6981 **Cardona, M.A. (1978).** Consideracions sobre l'endemisme i l'origen de la flora de les illes Balears. *Bull. Inst. Cat. Hist. Nat.*, 44(3): 7-15. Ca (Sp). (T)

6982 **Goldsmith, F.B. (1974).** An assessment of the nature conservation value of Majorca. *Biol. Conserv.* 6(2): 79-83.

6983 **Gomez-Campo, C., ed. (1987).** Libro rojo de especies vegetales amenazadas de espana peninsular e islas baleares. ICONA, Ministerio de Agricultura, Pesca y Alimentacion. 676p. Sp. Col illus., maps. Red Data Book. (T)

6984 **Sainz-Ollero, H., Hernandez-Bermejo, J.E. (1981).** Sintesis corologica de las dicotiledoneas endemicas de la Peninsula Iberica e Islas Baleares. Madrid, Coleccion Monografica INIA Num. 31. 111p. Sp (En). (T)

6985 **Salvo, E. (1985).** Libro Rojo de los helechos de la Peninsula Iberica y Baleares. *Quercus*, 18: 29. Sp. Illus. (T)

SPAIN - Canary Islands

6986 **Anon. (1978).** Botanisher Garten der Universitat Erlangen-Nurnberg: Informationsblatt Canarengewachshaus. Zur Vegetation der Canarischen Inseln. *Gartn.-Bot. Brief*, 58: 42-43. Ge. Illus., map. (G)

6987 **Anon. (1978).** Proyecto de parque natural "Dunas de Maspalomas". (Dunes of Maspalomas). *Aguayro*, 105: 8-13. Sp. (P)

6988 **Anon. (1981).** Protection for remaining laurel forest in Gran Canaria, Canary Islands. *Threatened Pl. Commit. Newsl.*, 8: 20. (P)

6989 **Arozarena Villar, A., Hernan Valero, M., Vega Hidalgo, J.A. (1976).** Parque Nacional de la Caldera de Taburiente Isla de la Palma. Monografias 12. Madrid, Ministerio de Agricultura, Instituto Nacional para la Conservacion de la Naturaleza. 162p. Sp. Col. illus., maps. (P)

6990 **Barreno, E., et al., eds (1984).** Listado de plantas endemics, raras o amenezadas de Espana: Islas Baleares. (List of endemic, rare or threatened plants in Spain: Balearic Islands.) *Informacion Ambiental*, No. 3 Conservacionismo en Espana: XIV-XVII. Sp (En). (T)

6991 **Blas Aritio, L. (1977).** El parque nacional de Timanfaya. *Vida Silvestre*, 21: 17-32. Sp. Col. illus. (P)

6992 **Blas Aritio, L. (1977).** El parque nacional de la Caldera de Taburiente. *Vida Silvestre*, 23: 153-168. Sp. Col. illus., map. (P)

6993 **Blas Aritio, L. (1977).** El parque nacional del Teide. *Vida Silvestre*, 22: 85-100. Sp. Col. illus. (P)

6994 **Bramwell, D. & Z.I. (1974).** Wild flowers of the Canary Islands. London, Stanley Thornes. 261p. Illus., keys, descriptions, mostly of the endemics; also describes areas of botanical interest. Spanish edition as Flores silvestres de las Islas Canarias, 2nd Ed., Madrid, 1983. Also German edition.

6995 **Bramwell, D. (1979).** A local botanic garden: its role in plant conservation. *In* Synge, H., Townsend, H., *eds*. Survival or extinction. Proceedings of a conference held at the Royal Botanic Gardens, Kew, 11-17 September 1978. Kew, Bentham-Moxon Trust. 47-52. (G)

6996 **Bramwell, D. (1983).** The Jardin Botanico "Viera y Clavijo" and its role in the conservation of the Macaronesian flora. *In* Comunicacoes apresentadas ao 2 Congresso Internacional pro flora Macaronesica, Funchal, 19-25 de Junho de 1977. Funchal. 365-371. (G/T)

6997 **Bramwell, D. (1987).** The role of the Jardin Botanico Canario "Viera y Clavijo" in the conservation of endangered Canarian endemics. *In* Bramwell, D., Hamann, O., Heywood, V., Synge, H., *eds*. Botanic gardens and the World Conservation Strategy. London, Academic Press. 175-181. (Sp). (G/T)

6998 **Bramwell, D., Beltram, W., Montelongo, V., Rios, C. (1986).** Plan especial de proteccion de los espacios naturales de Gran Canaria. *Bot. Macaronesica*, 15: 72p. Sp (En). Col. illus., maps. (1985 publ. 1986).

6999 **Bramwell, D., Rodrigo Perez, J. (1984).** Prioridades para la conservacion de la diversidad genetica en la flora de las Islas Canarias. *Bot. Macaronesica*, 10: 3-17 (1982 publ. 1984). Sp. (E/G/T)

7000 **Brochmann, C. (1984).** Hybridization and distribution of *Argyranthemum coronopifolium* (Asteraceae, Anthemideae) in the Canary Islands. *Nordic J. Bot.*, 4(6): 729-736. Illus., maps. (T)

7001 **Brochmann, C. (1987).** Kanarioyene - mer enn sol, badeliv og diskoteker? (Canary Islands - anything else than sun, bathing, and discotheques?) *Blyttia*, 45: 2-11. No (En). Discusses threats to species and habitats. Illus. (T)

SPAIN - Canary Islands

7002 **Cabildo Insular de Gran Canaria (1987).** Politica de restauracion del medio natural. Gran Canaria, La Seccion de Medio Ambient del Excmo, Cabildo Insular de Gran Canaria. Unpaginated. Sp. Col. illus., maps.

7003 **Dickson, J.H., Rodriguez, J.C., Machado, A. (1987).** Invading plants at high altitudes on Tenerife especially in the Teide National Park. *Bot. J. Linn. Soc.*, 95(3): 155-179. Illus., maps. (P)

7004 **Diez, E.B. (1972).** Impresiones botanicas, con algunas citas zoologicas, de un viaje a la Isla del Hierro. *Vieraea Fol. Sc. Biol. Can.: 10-24.* Sp (En). Reports an expedition to Hierro island (Canaries) and botanical and zoological observations made.

7005 **IUCN (1973).** Focus on conservation in the Canary Islands. *IUCN Bull.*, n.s., 4(6): 23.

7006 **IUCN (1986).** IUCN's list of top 24 threatened animals and plants. *Biol. Conserv. Newsl.*, 44: 1-6. (T)

7007 **IUCN Threatened Plants Committee Secretariat (1980).** First preliminary draft of the list of rare, threatened and endemic plants for the countries of North Africa and the Middle East. Kew, IUCN Threatened Plants Committee Secretariat. 170p. (T)

7008 **IUCN Threatened Plants Committee Secretariat (1981).** The botanic gardens list of rare and threatened plants of Macaronesia (excluding the Cape Verde Islands). Kew, IUCN Botanic Gardens Conservation Co-ordination Body Report No. 4. Lists plants in cultivation in botanic gardens. (G/T)

7009 **Kammer, F. (1972).** Erganzungen zu O. Eriksson: check-list of vascular plants of the Canary Islands (1971). *Cuad. Bot. Canar.*, 16: 47-49. Ge.

7010 **Kammer, F. (1976).** Biogeography of man on the vegetation of the island of Hierro. *In* Kunkel, G., *ed.* Biogeography and ecology in the Canary Islands. The Hague, Junk. 327-346. Illus., maps.

7011 **Kammer, F. (1979).** The influence of man on the vegetation of Macaronesia. *In* Berichte der Internationalen Symposien der Internationalen Vereinigung fur Vegetatsionskunde Herausgegeben von Reinhold Tuxen, 20-23 March 1978. 601-616. Illus. (T)

7012 **Kammer, F. (1982).** Beitrage zu einer kritischen Interpretation der rezenten und fossilen Gefasspflanzenflora und Wirbeltierfauna der Azoren, des Madeira-Archipels, der Ilhas Selvagens, der Kanarischen Inseln und der Kapverdischen Inseln, mit einem Ausblick auf Artenschwund in Makaronesien. Freiburg im Breisgau. 179p. Ge.

7013 **Kunkel, G. (1971).** Our heritage - past and future. *The Canary Islands Sun*, 6 March: 3-4. Illus. (T)

7014 **Kunkel, G. (1978).** La vida vegetal del Parque Nacional de Timanfaya, Lanzarote, Islas Canarias. *Nat. Hispanica*, 15: 94p. Sp. Illus., col. illus., map. (P)

7015 **Kunkel, G., ed. (1975).** Inventario de los recursos naturales renovables de la Provincia de las Palmas. Las Palmos de Gran Canaria, Excmo. Cabildo Insular. 156p. Sp. Maps. Results of IUCN/WWF Project 817, undertaken by Asociacion Canaria para la Defensa de la Naturaleza.

7016 **Malato-Beliz, J. (1983).** Proposition pour l'elaboration d'un catalogue des plantes endemiques, menacees et disparues en Macaronesie. *In* Comunicacoes apresentadas ao 2 Congresso Internacional pro flora Macaronesica, Funchal, 19-25 de Junho de 1977. Funchal. 437-440. Fr. (T)

7017 **Milburn, M. (1984).** Dragon's blood in east and west Africa and the Canary Islands. *Africa* (Rome, Instituto Italo-Africano), 39(3): 486-493. (T)

SPAIN - Canary Islands

7018 **Ortega Gonzales, C.I. (1984).** Micropropagacion de *Lotus berthelotii* Masf. (Leguminosae), un endemismo canario en peligro de extincion. *Bot. Macaronesica,* 10: 18-25 (1982). Sp (En). (G/T)

7019 **Ortuna Medina, F. (1975).** El parque nacional de Timanfaya. *Vida Silvestre,* 13: 18-37. Sp. Col. illus. (P)

7020 **Ortuna Medina, F. (1975).** El parque nacional del Teide. *Vida Silvestre,* 15: 150-165. Sp. Col. illus. (P)

7021 **Reiss, J. (1980).** Der Sukkulentenbusch auf Teneriffa. Ein naturlicher Standort seltener Sukkulenten. (The succulent [euphorbia] bush at Teneriffe, a natural site of rare succulents.) *Kakt. And. Sukk.,* 31(2): 45-50. Ge. Illus., maps. (T)

7022 **Rodriguez, C.S. (1985).** Jardines botanicos estrategia mondial para conservacion. Las Palmas de Gran Canaria, Instituto Nacional para la Conservacion de la Naturaleza (ICONA). 32p. Sp. Illus. Maps. Prepared as an excursion guide for the international botanical symposium. (G/P/T)

7023 **Santos Guerra, A. (1974).** Plantas en vias de extincion en las Islas Canarias. Canary Islands, Instituto Nacional de Investigaciones Agrarias Centro Regional de Canarias. 6p. Sp. Presented at Convencion de Asociaciones de Amigos de la Naturaleza, Pamplona, 1974. (T)

7024 **Taylor, N. (1979).** Identifying a rare *Ceropegia. Cact. Succ. J.* (U.K.), 41(2): 31-33. Illus. (T)

SRI LANKA

7025 **Abeywickrama, B.A. (1983).** Threatened or endangered plants of Sri Lanka and the status of their conservation measures. *In* Jain, S.K., Mehra, K.L., *eds.* Conservation of tropical plant resources. Howrah, Botanical Survey of India. 11-18. (L/T)

7026 **Alwis, L. (1984).** River basin development and protected areas in Sri Lanka. *In* McNeely, J.A., Miller, K.R., *eds.* National parks, conservation and development. The role of protected areas in sustaining society. Washington, D.C., Smithsonian Institution Press. 178-182. (P)

7027 **Amerasinghe, S.R. (1979).** Coast conservation. *Tigerpaper,* 6(4): 22-24. Illus.

7028 **Ariyaratna, N.T. (1980).** Hopes afresh for a national park. *Tigerpaper,* 7(1): 8. Illus. Catchment area of Walawe River was designated a national park in 1972, but has undergone disturbances. (P)

7029 **Bharathie, K.P. (1979).** Man and Biosphere reserves in Sri Lanka. *Sri-Lanka Forest.,* 14(1-2): 37-40. Map. 36 reserves listed. (P)

7030 **Cramer, L.H. (1977).** The significance of the indigenous flora in the areas of the Mahaweli complex. *Sri Lanka Forest.,* 13(1-2): 9-17. (E/T)

7031 **Crusz, H. (1973).** Nature conservation in Sri Lanka (Ceylon). *Biol. Conserv.,* 5: 199-208.

7032 **Erdelen, W. (1988).** Forest ecosystems and nature conservation in Sri Lanka. *Biol. Conserv.,* 43(2): 115-135. Maps. (P)

7033 **FAO Regional Office for Asia and the Pacific (1985).** Dipterocarps of South Asia. Bangkok, FAO. 321p. (Rapa Monograph 1985/4). (E/T)

7034 **Fernando, R., Samarsinghe, S.W.R. de A. (1988).** Forest conservation and the Forestry Master Plan for Sri Lanka. A review. Colombo, The Wildlife and Nature Protection Society. 76p.

SRI LANKA

7035 **Gunatilleke, C.V.S. & I.A.U.N. (1980).** The floristic composition of Sinharaja: a rain forest in Sri Lanka with special reference to endemics. *Sri Lanka Forester*, 14(3/4): 170-179. Illus., map. Includes rare endemic species. (P/T)

7036 **Gunatilleke, C.V.S. & I.A.U.N. (1985).** Phytosociology of Sinharaja: a contribution to rain forest conservation in Sri Lanka. *Biol. Conserv.*, 31(1): 21-40. Maps. (P)

7037 **Gunatilleke, C.V.S. & I.A.U.N. (1987).** Rare woody species of Sinharaja rain forest in Sri Lanka. *In* Kostermans, A.J.G.H., *ed.* Proceedings of the Third Round Table Conference on Dipterocarps. Jakarta, Unesco-MAB and UNEP. 519-530. Paper presented at conference held at Mulawarman Univ., Samarinda, Kalimantan, April 1985. (P/T)

7038 **Gunatilleke, C.V.S. & I.A.U.N., Sumithraarachchi, B. (1987).** Woody endemic species of the wet lowlands of Sri Lanka and their conservation in botanic gardens. *In* Bramwell, D., Hamann, O., Heywood, V., Synge, H., *eds.* Botanic gardens and the World Conservation Strategy. London, Academic Press. 183-194. (Sp). Map. (G/P/T)

7039 **Gunatilleke, C.V.S. (1978).** Sinharaja today. *Sri Lanka Forest.*, 13(3/4): 57-64. Maps, tropical forest. (P)

7040 **Gunatilleke, C.V.S., Wijesundara, D.S.A. (1982).** *Ex-situ* conservation of woody plant species in Sri Lanka. *Loris*, 16(2): 73-79. Principally a list of endemic woody plants at Peradeniya Botanic Garden. (G/T)

7041 **Gunatilleke, I.A.U.N. & C.V.S. (1983).** Conservation of natural forests in Sri Lanka. *Sri Lanka Forest.*, 16(1-2): 39-56. (E/P)

7042 **Gunatilleke, I.A.U.N. & C.V.S. (1984).** Distribution of endemics in the tree flora of a lowland hill forest in Sri Lanka. *Biol. Conserv.*, 28(3): 275-285. Map. Hill forest Hinidumkanda.

7043 **Gunatilleke, I.A.U.N. & C.V.S. (?)** Threatened woody endemics of the wet lowlands of Sri Lanka and their conservation. *Biol. Conserv.*, in press. (T)

7044 **Hoffmann, Th.W. (1985).** The Sinharaja forest. *Sri Lanka Wildlife*, 1: 25-27. Illus. (P)

7045 **Hoffmann, Th.W. (1985).** The Sinharaja forest. *Sri Lanka Wildlife*, 2: 29-34. Illus. (P)

7046 **Israel, S., Sinclair, T., eds (1987).** Indian wildlife: Sri Lanka, Nepal. Hong Kong, APA Productions. 362p. Col. illus. Provides overview of vegetation and conservation problems.

7047 **Jayasuriya, A.H.M. (1984).** Flora of Ritigala Strict Nature Reserve. *Sri Lanka Forest.*, 16(3/4): 61-156. Illus., maps. Annotated list; includes discussion on rare endemics. (P/T)

7048 **Kostermans, A.J.G.H. (1973).** A forgotten Ceylonese cinnamon-tree (*Cinnamomum capparu-coronde* Bl.). *Ceylon J. Sci. (Bio. Sci.)*, 10(2): 119-121. (T)

7049 **Kostermans, A.J.G.H., ed. (1987).** Proceedings of the Third Round Table Conference on Dipterocarps. Jakarta, Unesco. 657p. Papers presented at an international conference held at Mulawarman Univ., Samarinda, E. Kalimantan, 16-20 April 1985. Includes conservation. (E/T)

7050 **McNeely, J.A. (1987).** How dams and wildlife can co-exist: natural habitats, agriculture, and major water resource development projects in tropical Asia. *J. Conservation Biology*, 1(3): 228-238. Maps. (P)

7051 **Moyle, P.B., Senanayake, F.R.R. (1982).** Wildlife conservation in Sri Lanka: a Buddhist dilemma. *Tigerpaper*, 9(4): 1-4.

7052 **Perera, W.R.H. (1978).** A position paper on conservation of the Mahaveli catchment area. *Sri Lanka Forest.*, 13(3/4): 75-85.

SRI LANKA

7053 **Poore, D. (1979).** The values of tropical moist forest ecosystems and the environmental consequences of their removal. *Sri-Lanka Forest.*, 14(1-2): 15-36.

7054 **Senanayake, F.R., et al. (1977).** Habitat values and endemicity in the vanishing rain forests of Sri Lanka. *Nature*, 265: 351-354.

7055 **Silva, A. de (1980).** The impact of Buddhism on the conservation of the fauna and flora in ancient Sri Lanka. *Tigerpaper*, 7(4): 21-25. Illus.

7056 **Sumithraarachchi, D.B. (1986).** Conservation of orchids in Sri Lanka. *In* Rao, A.N., *ed.* Proceedings of the 5th ASEAN Orchid Congress Seminar, Singapore, 1-3 August, 1984. Singapore Parks and Recreation Department, Ministry of National Development. 140-144. (T)

7057 **Werner, W.L. (1981).** From far and near. A plea for the conservation of three unique forests in Sri Lanka. *Loris*, 15(6): 330-331. Swamp forest south east of Bulatsinhala, Rilagala (west of Dimbula), and plateau between Madugoda and Mabiyangana.

7058 **Wijayadasa, K.H.J. (1983).** Formulating a Sri Lankan strategy. *IUCN Bull.*, suppl. 4: 3. Illus.

7059 **Zoysa, N. de, Raheem, R. (1987).** Sinharaja - a rainforest in Sri Lanka. Colombo, March for Conservation. 92p. Illus. (P)

ST HELENA

7060 **Anon. (1981).** St. Helena. *Threatened Pl. Commit. Newsl.*, 8: 12-13. Account of efforts to rescue endangered flora. (T)

7061 **Cronk, Q.C.B. (1981).** *Senecio redivivus* and its successful conservation in St. Helena. *Envir. Conserv.*, 8(2): 125-126. Illus. Now known as *Lachanodes arborea.* (T)

7062 **Cronk, Q.C.B. (1983).** The decline of the redwood *Trochetiopsis erythroxylon* on St. Helena. *Biol. Conserv.*, 26(2): 163-174. Illus. (T)

7063 **Cronk, Q.C.B. (1987).** The plight of the St Helena olive. *Bot. Gard. Conserv. News*, 1(1): 30-32. Illus. (T)

7064 **Fay, M. (1989).** *Nesiota elliptica* - the St Helena olive; new moves to safeguard its future. *Bot. Gard. Conserv. News*, 1(4): 7. Illus. (G/T)

7065 **Goodenough, S. (1983).** A botanic gardener on St. Helena. *Threatened Pl. Newsl.*, 12: 10-11. (G/T)

7066 **Goodenough, S. (1983).** Conservation of the endemic flora of Saint Helena: a plant propagation project and recommendations for the conservation of the endemic flora of the island. Kew, Royal Botanic Gardens. 13p. (T)

7067 **Goodenough, S. (1984).** St. Helena: so far so good. *Threatened Pl. Newsl.*, 13: 9-10. (G/T)

7068 **Goodenough, S. (1985).** St. Helena and its endemic plants - a conservation success. The best known unfamiliar land in the world. *Kew Mag.*, 2(4): 369-379. Illus., maps. (T)

7069 **Goodenough, S. (1986).** St Helena: a conservation success. *Gartn. Bot. Brief*, 87: 10-11. (T)

7070 **Williamson, M. (1984).** St. Helena ebony tree saved. *Nature*, 309(5969): 581. (T)

ST HELENA - Ascension Island

7071 **Cronk, Q.C.B. (1980).** Extinction and survival in the endemic vascular flora of Ascension Island. *Biol. Conserv.*, 17(3): 207-219. Illus., map. (T)

TRISTAN DA CUNHA ISLANDS

7072 **Wace, N.M., Holdgate, M.W. (1976).** Man and nature in the Tristan da Cunha Islands. IUCN Monograph No. 6. Morges, IUCN. 114p. Illus., maps.

SUDAN

7073 **Ahmed, A.A. (1983).** Forest reserves and woodland savanna regeneration on the sub-Saharan massif of Jebel Marra, Democratic Republic of the Sudan. *Vegetatio*, 54(2): 65-78. Illus., maps.

7074 **Bari, E.A. (1968).** Sudan. *In* Hedberg, I. & O., *eds.* Conservation of vegetation in Africa south of the Sahara. Symposium Proceedings, at 6th plenary meeting of AETFAT, Uppsala. *Acta Phytogeogr. Suec.*, 54: 59-64. Figs. Describes vegetation, enclosed areas and areas proposed for conservation.

7075 **Geerling, C. (1985).** The status of the woody species of the Sudan and Sahel zones of West Africa. *Forest Ecol. Manage.*, 13: 247-255. (T)

7076 **Gorse, J. (1987).** Desertification in the Sahelian and Sudanian zones of West Africa. *Unasylva*, 37: 2-18.

7077 **Jenkin, R.N., Howard, W.J., Thomas, P., Abell, T.M.B., Deane, G.C. (1977).** Forestry development prospects in the Imatong Central Forest Reserve, southern Sudan. Land Resource Study No. 28. Land Resources Division, Ministry of Overseas Development. 27, 217p. Illus., map. (P)

7078 **Osborne, R. (1988).** *Encephalartos barteri*. *Encephalartos*, 14: 8-16. Illus., map. Data sheet; includes conservation. (G/T)

7079 **Wickens, G. (1979).** Possibilities and needs for conservation of plant species and vegetation in Africa. Appendix: Preliminary lists of rare and threatened species in African countries. Sudan. *In* Hedberg, I., *ed.* Systematic botany, plant utilization and biosphere conservation. Stockholm, Almqvist & Wiksell International. 85-88. Contains 258 species and infraspecific taxa. (T)

SURINAME

7080 **Boxman, O., Graaf, N.R. de, Hendrison, J., Jonkers, W.B.J., Poels, R.L.H., Schmidt, P., Sang, R.T.L. (1985).** Towards sustained timber production from tropical rain forests in Suriname. *Netherlands J. Agric. Sci.*, 33: 125-132. (E)

7081 **Schulz, J.P. (1968).** Nature preservation in Suriname - a review of the present status. Paramaribo, Suriname Forest Service. Principally covers the protected areas. (P)

7082 **Schulz, J.P., Mittermeier, R.A., Reichart, H.A. (1977).** Wildlife in Surinam. *Oryx*, 14(2): 133-144. Illus. Mainly fauna, but notes on 8 nature reserves. (P)

SVALBARD AND JAN MAYEN ISLANDS

7083 **Ministry of Environment (1981).** Environmental regulations for Svalbard. Ministry of Environment. (Revised edition). (L)

SWAZILAND

7084 **Anon. (1983).** Conservation in Swaziland. *Oryx*, 17: 141. (P)

7085 **Codd, L.E. (1968).** Regional synthesis - Discussion. *In* Hedberg, I. & O., *eds.* Conservation of vegetation in Africa south of the Sahara. Symposium Proceedings, at 6th plenary meeting of AETFAT, Uppsala. *Acta Phytogeogr. Suec.*, 54: 257-260. Overview of conservation in southern Africa. (P)

7086 **Compton, R.H. (1968).** Swaziland. *In* Hedberg, I. & O., *eds.* Conservation of vegetation in Africa south of the Sahara. Symposium Proceedings, at 6th plenary meeting of AETFAT, Uppsala. *Acta Phytogeogr. Suec.*, 54: 256-257. Vegetation, present and recommended protection. (P)

SWAZILAND

7087 **Kemp, E.S. (1979).** Possibilities and needs for conservation of plant species and vegetation in Africa. Appendix: Preliminary lists of rare and threatened species in African countries. Swaziland. *In* Hedberg, I., *ed.* Systematic botany, plant utilization and biosphere conservation. Stockholm, Almqvist & Wiksell International. 101-103. Contains 155 species and infraspecific taxa: E:2, V:16, R:137. (T)

7088 **Reilly, E. & T. (1979).** The political alternative. *African Wildlife*, 33(4): 18-21. Col. illus.

7089 **Reilly, T.E. (1985).** The Mlilwane story: a history of nature conservation in the Kingdom of Swaziland and fund raising appeal. Mbabane, Swaziland, Mlilwane Wildlife Sanctuary Trust. 84p. Col. illus. (P)

SWEDEN

7090 **Agren, C. (1983).** Skogsdoden har natt Sverige! (Tree death has reached Sweden.) *Sveriges Nat.*, 7: 7-9. Sw (En).

7091 **Akesson, S. (1979).** Avvagningar mellan skogsbruk och naturvard. (Balancing the requirements of forestry and conservation.) *Sveriges Skogsvardsf. Tidskr.*, 77(1): 83-87. Sw.

7092 **Andersson, B., Landin, B. (1978).** Vanerskargardarna - skyddsvarda for bade batliv och naturvard. (The Vaner Islands are worth protecting for boating and nature.) *Sveriges Nat.*, 69(3): 143-146. Sw (En). Illus., col. illus. (P)

7093 **Andersson, F. (1981).** The Swedish coniferous forest project. *Ambio*, 10(2-3): 126-129. Illus., map.

7094 **Andersson, L., Appelquist, T. (1982).** Brunbraken, *Asplenium adulterinum*, funnen i norra Vastgoterland. *Sven. Bot. Tidskr.*, 76: 306-310. Sw (En). Illus. (T)

7095 **Anon. (1951).** Naturvard i statens skogar. (Protection of nature in the State forests.) Stockholm, Domanverket. 30p. Sw (En). Illus. Report on forest conservation measures taken by Domanverket.

7096 **Anon. (1985).** Preliminar lista over hotade karlvaxter i Sverige. (A preliminary list of threatened vascular plants in Sweden.) *Svensk Bot. Tidskr.*, 79(5): 362-366. Sw (En). (T)

7097 **Aronsson, M. (1979).** Det relikta odlingslandskapet i mellersta Kalmar lan. (Relicts of the old pastoral landscape in E. Smaland, Sweden.) *Svensk Bot. Tidskr.*, 73(2): 97-114. Sw. Illus., col. illus.

7098 **Arvidsson, L., Skoog, L. (1984).** Svaveldioxidens inverkan pa lavfloran i Goteborgsomradet. (Effect of sulphur dioxide air pollution on the distribution of lichens in the Goteborg area, SW Sweden.) *Svensk Bot. Tidskr.*, 78(3): 137-144. Sw (En). Maps.

7099 **Bergenstrahle, A., Milberg, P. (1985).** Gucusko i Lule lappmark. *Svensk Bot. Tidskr.*, 79(2): 84. Sw (En). Summary tells that *Cypripedium calceolus* was found for the first time in 1984 in Lule lappmark, N. Sweden. (T)

7100 **Bjorkback, F., Imby, L., Lidberg, R., Sjostrom, I., Osterdahl, L. (1976).** Nagot um brunkullans (*Nigritella nigra*) utbredning och ekologi i Sverige. Exempel pa ADB-anpassad katalogisering och bearbetning. *Fauna och Flora*, 2/1976: 49-60. Sw. Illus., col. illus. Maps. (T)

7101 **Bjorkback, F., Lundqvist, J. (1982).** Aktion Brunkulla: ett botaniskt WWF-projekt. (The *Nigritella* orchid project.) *Svensk Bot. Tidskr.*, 76(4): 215-228. Sw (En). Illus., col. illus., maps. *Nigritella nigra.* (T)

7102 **Brakenhielm, S. (1980).** Naturvardens behov av landskaps floror. (Nature conservation and the need for province floras.) *Svensk Bot. Tidskr.*, 73(6): 517-520 (1979 publ. 1980). Sw (En).

SWEDEN

7103 **Council of Europe (1984).** New nature reserve at Tandovala. *Council of Europe Newsl.*, 84(4): 4. (P)

7104 **Council of Europe (1984).** Sweden: water and nature protection act. *Council of Europe Newsl.*, 84(3): 4. (L)

7105 **Council of Europe (1984).** Sweden: wetlands in danger in the south. *Council of Europe Newsl.*, 84(2): 3.

7106 **Curry-Lindahl, K. (1984).** Conservation in Sweden: problems and progress. *Oryx*, 18(4): 202-209. Illus.

7107 **Deltin, M. (1974).** Naturvard i jamtlands lan. *Sveriges Nat.*, 3/1974: 111-114. Sw. Illus. Maps. (P/T)

7108 **Eckerberg, K. (1988).** Clearfelling and environmental protection - results of an investigation in Swedish forests. *J. Envir. Manage.*, 27(3): 237-256.

7109 **Egerstedt, O. (1983).** Riksdagen: Nyheter our natur- och miljodvard. (The Riksdag: new developments concerning nature conservancy and environment protection.) *Sveriges Nat.*, 4: 7. Sw (En). 2,100 hectares of forest and wetlands, in the Tinas district, lower reaches of Dalalv river.

7110 **Elven, R., Aarhus, A. (1984).** A study of *Draba cacuminum* (Brassicaceae). *Nordic J. Bot.*, 4(4): 425-441. Illus., maps. (T)

7111 **Emanuelsson, U. (1980).** Den ekologiska betydelsen av mekanisk paverkan pa vegetation i fjallterrang. (Mechanical impact on vegetation in the Tometrask area.) *Fauna Flora* (Stockholm), 1: 37-42. Sw (En). Illus.

7112 **Engelmark, O. (1981).** Forest history of Muddus National Park, northern Sweden. *Wahlenb. Scripta Bot. Umensia*, 7: 33-38. Illus., map. (P)

7113 **Engelmark, O. (1984).** Forest fires in the Muddus National Park (northern Sweden) during the past 600 years. *Canad. J. Bot.*, 62(5): 893-898. (Fr). Illus., maps. (P)

7114 **Faxen, L. (1974).** Fillsta-tufferna. *Sveriges Nat.*, 3: 115-118. Sw. Illus. (P/T)

7115 **Floravardskommitten for Karlvaxter 1985. (1985).** Preliminar lista over hotade karlvaxter i Sverige. (A preliminary list of threatened vascular plants in Sweden.) *Svensk Bot. Tidskr.* 79: 362-366. Covers approx. 400 taxa. (T)

7116 **Floravardskommitten for Lavar (1987).** Preliminar lista over hotade lavar i Sverige. *Svensk Bot. Tidskr.* 81:237-256. Sw (En). Illus. (T)

7117 **Frisen, R. (1988).** [Cooperation for protection of species and biotopes in the Nordic countries]. *Sven. Bot. Tidskr.*, 82(6): 499-501. Sw (En). (P)

7118 **Fritz, O. (1989).** Flytsvaltung, *Luronium natans*, funnen i Halland 1988. (A find of *Luronium natans* in the Province of Halland, south-west Sweden). *Sven. Bot. Tidsk.*, 83(2): 65-136. Sw (En). (T)

7119 **Gustafsson, L.-A. (1974).** Var radd om Ostgottaslattens orkideangar! *Sveriges Nat.*, 3/74: 125-130. Sw. Illus. Maps. (T)

7120 **Gustafsson, L.-A. (1976).** Projekt Linne samlar llandets botanister till kraftinsats. (Project Linnaeus unites the botanists of Sweden in a powerful effort). *Fauna & Flora*, 71(5): 189-201. Sw (En). Illus. (T)

7121 **Gustafsson, L.-A. (1978).** Hundlokelasbraken och bondskog. (*Botrychium virginianum* and farm forests.) *Sveriges Nat.*, 69(7): 425-429. (En). Illus. Plants in Jamtland seen on farm while researching for Project Linnaeus inventory. (T)

7122 **Gustafsson, L.-A., Nilsson, O. (1976).** Projekt Linne: Klatt och Kung Karls spira. *Sartyrick ur Sveriges Natur*, 2/76: 65-68. Sw. Illus. Maps. (T)

SWEDEN

7123 **Hagner, M. (1977).** Hur skall genresurser i inforda tradarter bevaras. (How are the gene resources of imported tree species to be conserved.) *Rapp. Upps. Inst. Skogsgen.*, 24: 113-119. Sw (En). (E)

7124 **Hamdahl, B. (1983).** Den nya vattenlagen och naturvarden. (The new Water Act and nature conservation.) *Sveriges Nat.*, 5: 3-4. Sw (En). (L)

7125 **Hamdahl, B. (1984).** Fjallskogarna maste raddas! (Save the mountain forests!) *Sveriges Nat.*, 2: 7-8. Sw (En).

7126 **Hedberg, I. (1985).** Sweden: international work in plant conservation. *Threatened Pl. Newsl.*, 14: 21-22.

7127 **Herrmann, J. (1979).** Den ekologiska grundsynen - synad i grunden. *Sveriges Nat.*., 6: 251-253. Sw. Illus. (T)

7128 **Hjelm, K. (1983).** Tandovala - reservatet antligen a hamn! (Tandovala reserve secured at last!) *Sveriges Nat.*, 5: 4. Sw (En). (P)

7129 **Holmstedt, S. (1978).** Nedre Dalalven - nu kravs det en ordentlig satsning pa bevarande! (The Lower Dalalven Valley: preservation is urgent.) *Sveriges Nat.*, 69(6): 387-390. Sw (En). Illus. Threatened by modern forestry. (P)

7130 **Ingelog, T. (1979).** Vedkraft - ett miljovanligt alternativ? *Sveriges Nat.*, 6: 245-250. Sw. Illus. (T)

7131 **Ingelog, T. (1981).** Floravard i Skogsbruket. Del. 1. (Plant conservation and forestry.) Jonkoping, Skogsstyrelsen. 153p. Sw. Illus. Lists 278 native and threatened forest species of flowering plants, ferns, mosses, fungi and lichens. (T)

7132 **Ingelog, T. (1988).** Floralaget i Sverige. (Conservation and status of the Swedish flora.) *Svensk Bot. Tidskr.*, 82(6): 376-378. Sw (En). (T)

7133 **Ingelog, T., Gustafsson, L., Thor, G. (1984).** Skyddsvarda Skogsvaxter i Sverige, Floravard i skogsbruket. Del. 3 - Fotoflora i farg. Jonkoping, Skogsstyrelsen. 64p. Sw. Col. illus., maps. Briefly describes 73 vascular plants threatened by forestry. (T)

7134 **Ingelog, T., Thor, G., Gustafsson, L. (1984).** Floravard i Skogsbruket. Del 2 - Artdel. Jonkoping, Skogsstyrelsen. 407p. Sw. (T)

7135 **Jaakkola, S. (1983).** Skolig fjarranalys: Forskningsprogram for skoglig fjarranalys 1982-1990. (Remote sensing analysis applied to forestry: research program for 1982-1990.) *Tidskr. Kungl.*, 122(3): 101-132. Sw (En). Illus.

7136 **Jarai, M. (1975).** Termeszetvedelem Svedorszagban. (Nature protection in Sweden.) *Buvar*, 30(5): 203-208. Sw. Illus.

7137 **Johnson, G. (1978).** Naturvarden i planeringen - forslag till agerande. *Skanes Nat.*, 66(1): 1-10. Sw. Illus., map.

7138 **Kardell, L. (1979).** Sydbillingens plata: historien om hur en skovlad skog pa hundra ar blir ett naturreservat. *Fauna Flora* (Stockholm), 74(2): 79-88. Sw. Illus., maps. (P)

7139 **Kardell, L. (1980).** Skogsmarkens bar och svampar - en hotad resurs. (Are the berries and mushrooms of the forest area a threatened resource.) *Tidskr. Sverig. Skogsvards.*, 78(5): 5-19. Sw. Illus. (E)

7140 **Karlsson, L. (1983).** [Notes on the vascular plant flora in the southwestern part of the Padjelanta National Park]. *Sven. Bot. Tidskr.*, 77(4): 217-220. Sw (En). Illus. (P)

7141 **Kers, L.E. (1978).** Rommehed raddat - tack vare en svamp! (Rommehed saved -thanks to a fungus!) *Sveriges Nat.*, 69(2): 55-57. Sw (En). Land set aside for nature reserve. (P)

SWEDEN

7142 **Kopp, H. (1980).** Environment protection in Sweden. *Nature and National Parks,* 17(65): 12-14. Illus. (P)

7143 **Kraft, J. (1979).** Upprop om sallsynta vaxter pa Kullaberg. (List of rare plants at Kullaberg.) *Sven. Bot. Tidskrift,* 73(6): 435-436. Sw. (T)

7144 **Landell, N.E. (1979).** Hur gick det med Sjaunja-debatten? *Sveriges Nat.,* 7: 274-280. Sw. Illus., maps. National park. (P)

7145 **Larsson, T. (1977).** Nature conservation in Sweden: recent developments and legislative changes. *Biol. Conserv.,* 11(2): 129-143. (L)

7146 **Lindberg, P.S. (1980).** Toppjungfrulin, *Polygala comosa,* en vaxt pa Tillbakagang. (*Polygala comosa* threatened on the Swedish mainland.) *Sven. Bot. Tidskr.,* 74(3): 213-219. Sw. Illus., map. (T)

7147 **Lofgren, L. (1988).** Projekt hotade urbogavaxter. (Threatened plants in the commune of Arboga, C. Sweden.) *Sven. Bot. Tiddskr.,* 82(6): 446-448. Sw (En). Illus. (T)

7148 **Lojtnant, B., ed. (1984).** Spredningsokogi. (Dispersal ecology.) Denmark, Naturfredningsradet og Fredningsstyrelsen, 84: 111p. Da, En. Illus., maps. Covers island biogeography and design of nature reserves for threatened species, using examples from Denmark and Sweden. (P/T)

7149 **Merker, A. (1985).** Den vilda floran som gerensurs i vaxtforadlingen. (The wild flora as a genetic resource in plant breeding.) *Sven. Bot. Tidskr.,* 79: 65-72. Sw (En). (E/G)

7150 **Nilsson, C., Grelsson, G. (1979).** Nagra sallsynta inslag i norra Norrlands storalvflora. (Some rare members of the flora of the larger rivers of Norrland Province [Sweden].) *Sven. Bot. Tidskr.,* 73(2): 89-95. Sw (En). Illus. (T)

7151 **Nilsson, I.N. (1984).** Nagra synpunkter pa obiogeografisk teori och naturvardsplanering. *In* Lojtnant, B., *ed.* Spredningsokologi. (Dispersal ecology). Denmark, Naturfredningsradet og Fredningsstyrelsen, 84: 11-13. Covers the theory of island biogeography and planning for nature conservation. Da (En). (P)

7152 **Nilsson, O. (1972).** Bevarandet av utrotningshotade svenska vaxter - en vadjan om deltagande i ett nystartat projekt. *Sven. Bot. Tidskr.,* 66: 291-293. Sw. (T)

7153 **Nilsson, O. (1976).** The records system of Uppsala Botanic Garden. *In* Simmons, J.B., *et al., eds.* Conservation of threatened plants. NATO Conference Series 1: Ecology, vol. 1. New York and London, Plenum Press. 89-93. Proc. Conference on the Functions of Living Plant Collections in Conservation and Conservation-Orientated Research and Public Education, Kew, 2-6 September 1975. (G)

7154 **Nilsson, O. (1978).** Project Linne - the preservation of species of Swedish plants threatened with extinction. *Current Sweden,* 201: 2-9. Map. (T)

7155 **Nilsson, O. (1979).** Threatened plants in the Nordic countries. *In* Hedberg, I., *ed.* Systematic botany, plant utilization and biosphere conservation: Proceedings of a symposium held in Uppsala in commemoration of the 500th ann. of the Univ. Stockholm, Almqvist & Wiksell International. 72-75. (T)

7156 **Nilsson, O. (1981).** Project Linnaeus: assessing Swedish plants threatened with extinction. *In* Synge, H., *ed.* The biological aspects of rare plant conservation. Chichester, Wiley. 105-112. Proceedings of International Conference, King's College, Cambridge, 14-19 July 1980. (T)

7157 **Nilsson, O. (1982).** *Astragalus arenarius. Sven. Bot. Tidskr.,* 76: 374. Sw. (T)

7158 **Nilsson, O., Gustafsson, L. (1976).** Projekt Linne rapporterar 1-13. *Sven. Bot. Tidskr.,* 70: 166-175. Sw (En). Illus., maps (T)

7159 **Nilsson, O., Gustafsson, L. (1977).** Projekt Linne rapporterar 14-28. *Sven. Bot. Tidskr.,* 70(3): 211-224. Sw (En). (T)

SWEDEN

7160 **Nilsson, O., Gustafsson, L. (1977).** Projekt Linne rapporterar 29-48. *Sven. Bot. Tidskr.*, 71(1): 3-22. Sw (En). (T)

7161 **Nilsson, O., Gustafsson, L. (1977).** Projekt Linne rapporterar 49-63. *Sven. Bot. Tidskr.*, 71(3): 205-224. Sw (En). (T)

7162 **Nilsson, O., Gustafsson, L. (1977).** Projekt Linne rapporterar 64-79. *Sven. Bot. Tidskr.* 72(1): 1024. Sw (En). (T)

7163 **Nilsson, O., Gustafsson, L. (1978).** Projekt Linne rapporterar 80-92. *Sven. Bot. Tidskr.*, 72(3): 189-204. Sw (En). (T)

7164 **Nilsson, O., Gustafsson, L. (1979).** Projekt Linne rapporterar 106-120. *Sven. Bot. Tidskr.*, 73(5): 353-372. Sw (En). Illus., col. illus., map. (T)

7165 **Nilsson, O., Gustafsson, L. (1979).** Projekt Linne rapporterar 93-105. *Sven. Bot. Tidskr.*, 73(1): 71-85. Sw (En). Illus. (T)

7166 **Nilsson, O., Gustafsson, L. (1982).** Projekt Linne rapporterar 121-132. *Sven. Bot. Tidskr.*, 76(2): 135-145. Sw (En). (T)

7167 **Nilsson, O., Gustafsson, L. (1985).** Projekt Linne: slutrapport. (Final report from Project Linnaeus.) *Sven. Bot. Tidskr.*, 79(5): 319-328. Sw (En). Includes an index of reports in *Sven. Bot. Tidskr.* 1976-1982 on the status of rare, endangered and/or extinct vascular taxa; total 143. (B/T)

7168 **Nilsson, O., Gustafsson, L., Karlsson, T. (1982).** Projekt Linne rapporterar 133-143. *Sven. Bot. Tidskr.*, 76(5): 273-284. Sw (En). Illus., maps. (T)

7169 **Nordstrom, H. (1983).** The Linnaeus Project for plant species at risk. *Sveriges Nat. Arsbook*, (1983): 106-115. (T)

7170 **Olsson, R. (1982).** Ett strot steg mot utarmning! (Forests: a big step towards impoverishment.) *Sveriges Nat.*, 4: 12-15. Sw (En). Illus.

7171 **Regnander, J., Nilsson, L., Jennersten, K.L., Fransson, Y., Staal, E., Bostrom, C. (1978).** Skogsskydd. (Forest protection.) *Skogen*, 7: 10-21. Illus.

7172 **Rogstadius, M. (1978).** Far vi lamma nagra blommor. (Saying it with flowers.) *Sveriges Nat.*, 69(4): 297-299. (En). Illus. Regulations concerning protected plants in Sweden. (L/T)

7173 **Selander, S. (1957).** Det Levande Landskapet i Sverige. (The living landscape of Sweden.) 2nd Ed. Stockholm. 492p. Sw. Describes main vegetation types and effects of human interference.

7174 **Simonsson, P. (1978).** Skuleskogen - var nasta nationalpark? (Skuleskogen - our next national park?) *Sveriges Nat.*, 69(5): 333-338. Sw (En). Illus., map. Mountain flora and virgin forest. (P)

7175 **Soderstrom, L. (1983).** Threatened and rare bryophytes in spruce forests of central Sweden. *Sven. Bot. Tidskr.*, 77(1): 4-12. (T)

7176 **Stahl, P. (1988).** Hotade vaxter i ostra Gastrikland. (Threatened plants in S.E. Gastrikland, a province in C. Sweden.) *Svensk Bot. Tidskr.*, 82(6): 393-400. Sw (En). Illus., col illus., map. (T)

7177 **Svensson, R. (1988).** [Flora conservation in the Swedish landscape]. *Sven. Bot. Tidskr.*, 82(6): 485-486. (T)

7178 **Thor, G., Ingelog, T. (1985).** Projekt Linne i ny skepnad. (A new emphasis for Project Linnaeus). *Sven. Bot. Tidskr.*, 79(5): 357-361. Sw (En). Describes a new organisation for the conservation of flora and fauna in Sweden. (T)

7179 **Tyler, C. (1984).** Calcareous fens in South Sweden. Previous use, effects of management and management recommendations. *Biol. Conserv.*, 30(1): 69-89. Map.

SWEDEN

7180 **Vuokko, S. (1983).** Uhatut Kasvimme. (Our threatened plants.) Suomen Luonnonsuojelum Tuki Oy, Helsinki. 96p. Fi. Col. illus. Covers Nordic region. (T)

7181 **Wahlberg, S. (1979).** God created, Linnaeus arranged: Project Linnaeus, an effort to save that good work for the future. *In* Synge, H., Townsend, H., *eds.* Survival or extinction. Proceedings of a conference held at the Royal Botanic Gardens, Kew, 11-17 September 1978. Kew, Bentham-Moxon Trust. 25-30. (T)

7182 **Westman, G. (1985).** Klibbal som relikt i mellersta Norrland. (Relict occurrences of *Alnus glutinosa* in the northern Swedish inland.) *Svensk Bot. Tidskr.*, 79(1): 51-64. Sw (En). (T)

7183 **Widen, B. (1987).** Population biology of *Senecio integrifolius* (Compositae), a rare plant in Sweden. *Nordic. J. Bot.*, 1(6): 687-704. Maps. (T)

7184 **van Dijk, G. (1981).** Het beheer in een aantal Zuidzweedse natuurterreinen, grotendeels botanische reservaten. *Levende Nat.*, 83(4): 159-168. Du; illus. (P)

SWITZERLAND

7185 **Aeberhard, T. (1978).** Berucksichtigung des Naturschutzes bei Gewasserverbauungen. *In* Naturschutzinspektorat des Kantons Bern Bericht 1977. *Mitt. Naturforsch. Ges. Bern*, 35: 129-145. Ge. Illus.

7186 **Aeschimann, D., Roguet, D. (1986).** L'Ete suisse au Jardin botanique: rocailles et plantes protegees. *Mus. Geneve*, 267: 10-13. Fr. Illus. (G)

7187 **Anon. (1977).** Schutz- und Wohlfahrtsfunktionen des Waldes. (Protective and social welfare functions of forests.) *Bericht. Eidgenoss. Anst. Forstl. Versuchsw.*, 177: 50p. Arbeitsgemeinschaft fur den Wald. Ge.

7188 **Anon. (1979).** La haie, Bale, Protection de la Nature. *Bull. Ligue Suisse Prot. Nat.*, Special edition: 48p. Fr.

7189 **Bach, R., et al. (1976).** Durch den Schweizerischen Nationalpark. (Through the Swiss National Park.) Basel, Schweizerischer Bund fur Naturschutz. 258p. Ge. Illus., maps. (P)

7190 **Becherer, A. (1972).** Erloschene Arten der Schweizer Flora. *Ber. Schweiz. Bot. Ges.*, 82: 300-301. Ge. Extinct species in the Swiss flora. (T)

7191 **Bossert, A. (1978).** Teichbau im Naturschutzgebiet Wengimoos. *In* Naturschutzinspektorat des Kantons Bern Bericht 1977. *Mitt. Naturforsch. Ges. Bern*, 35: 145-152. Ge. Illus.

7192 **Bressond, B., Oggier, P.A., Catzeflis, F. (1977).** Etude botanique de la reserve de Pouta-Fontana Grone. *Bull. Murithienne*, 94: 85-117. Fr. Illus., map. (P)

7193 **Bucher, J.B. (1984).** Bemerkungen zum Waldsterben und Umweltschutz in der Schweiz. (Observations on forest dieback and environmental protection in Switzerland.) *Forstwiss. Zentralbl.*, 103(1): 16-27. Ge (En). Maps.

7194 **Bucher, J.B., Kaufmann, E., Landolt, W. (1984).** [Forest damages in Switzerland -1983. (1) Interpretation of the Sanasilva enquiry and spruce needle analyses from the viewpoint of emission protection.] *Schweiz. Zeitschr. Forstw.*, 135(4): 271-287. Ge (Fr). Illus., maps. Pollution, forests.

7195 **Burkhard, H.P. (1984).** Das Waldsterben als Herausforderung an den Politiker. (Forest dying as challenge to politicians.) *Schweiz. Zeitschr. Forstw.*, 135(4): 321-328. Ge (Fr). Pollution.

7196 **Candolle, R. de (1980).** The protection of trees in Switzerland. *Int. Dendrol. Soc. Yearbook*, (1979): 12-15. (T)

7197 **Chappuis, J.B. (1987).** Proteger la Nature. Ligne Suisse pour la Protection de la Nature. 156p. Fr. Guide pratique pour la protection de la nature et du paysage au niveau communal.

SWITZERLAND

7198 **Council of Europe (1984).** Switzerland: peat bogs menaced. *Council of Europe Newsl.*, 84(1): 4.

7199 **Ewald, K. (1982).** Naturschutz und Forstwirtschaft. (Nature protection and forest management.) *J. Forest. Suisse*, 133(1): 29-36. Ge (Fr). Illus.

7200 **Fehlmann, S.R. (1979).** Geschutzte Pflanzen. Zurich, Silva. 124p. Ge. Col. illus. (L/T)

7201 **Hofle, H.H. (1976).** Holzernte und Umweltschutz - ein unlosbarer Konflikt. (Wood harvesting and protection of the environment.) *Schweizerische Zeitschr. ft fur Forstwesen*, 127(6): 373-403. Ge (Fr). A review of the effects and types of damage caused by wood harvesting to social welfare and the forest ecosystem, including effects on wildlife.

7202 **Kaufmann, E., Bucher, J.B., Landolt, W., Jud, B., Hoffmann, C. (1984).** Waldschaden in der Schweiz - 1983. (III). (Forest damages in Switzerland - 1983. (III).) *Schweiz. Zeitschr. Forstw.*, 135(10): 817-831. Ge (Fr). Illus., maps.

7203 **Kessler, E. (1976).** Grundlagen fur die Ausscheidung von Schutzgebieten in der Schweiz. (Criteria for designating areas as protected areas in Switzerland.) *Natur und Landschaft*, 51(5): 143-149. Ge. (P)

7204 **Klein, A. (1980).** Die Vegetation an Nationalstrassenboschungen der Nordschweiz und ihre Eignung fur den Naturschutz. (The vegetation on motorway verges in northern Switzerland and its suitability for nature protection purposes.) *Veroff. Geobot. Inst. Rubel*, 72: 75p. (Fr, En).

7205 **Klein, A. (1981).** Die Vegetation an Nationastrassenboschungen der Nordschweiz und ihre Eignung fur den Naturschutz. (The vegetation on motorway verges in northern Switzerland and its suitability for nature protection purposes.) *Ber. Geobot. Inst. Eidgenoss. Techn. Hochsch. Stift. Rubel*, no. 48: 28 (1980 publ. 1981). Ge.

7206 **Klotzli, F. (1969).** Zur okologie schweizerischer bruckwalder unter besonderer berucksichtigung des waldreservates moos bei birmensdorf und Katzensees. *Ber. Geobot. Inst. Eidgenoss. Techn. Hochsch.* (Zurich), 39: 56-123. Ge. Illus. Map. (T)

7207 **Klotzli, F. (1978).** Zur Bewaldungsfahigkeit von Mooren der Schweiz. (Afforestation capacity of bogs in Switzerland.) *Telma*, 8: 183-192. Ge.

7208 **Koeppel, H.D. (1977).** Das Naturschutzzentrum Aletschwald. *Naturschutz- und Naturparke*, 86: 45-52. Ge. Illus., map.

7209 **Koeppel, H.D. (1979).** The Aletschwald native conservation and study center. *Parks*, 4(1): 8-10. Illus., maps.

7210 **Krebs, E. (1981).** Die Walderhaltungspolitik. (Forest conservation policy.) *J. Forest. Suisse*, 132(12): 1053-1058. Ge.

7211 **Landolt, E. (1982).** Geschutzte Pflanzen in der Schweiz. (Protected plants in Switzerland.) 3rd Ed. Basel, Schweizerischer Bund fur Naturschutz. 215p. Ge. Describes over 150 protected taxa with colour illus.; detailed introduction with vegetation descriptions and legislation details in each Canton; French edition also available, published in 1970. (L/T)

7212 **Landolt, E., Fuchs, H.-P., Heitz, C., Sutter, R. (1982).** Bericht uber die gefahrdeten und seltenen Gefasspflanzenarten der Schweiz ("Rote Liste"). (Report on threatened and rare vascular plants of Switzerland ["Red List"]. *Ber. Geobot. Inst. Eidgenoss. Techn.* 49: 195-218 (1981). Ge (En). (T)

7213 **Ligue Suisse pour la Protection da la Nature (1975).** Inventaire des plantes protegees par la loi federale et les lois cantonales, dresse en novembre 1974, revise octobre 1975. Mimeo. Fr. (L/T)

7214 **Moor, M. von (1942).** Die Pflanzengesellschaften der Freiberge (Berner Jura). *Bericht. Schweiz. Bot. Gesellsch.*, 1942(52): 363-422. Ge. (P/T)

SWITZERLAND

7215 **Moor, M. von (1968).** Die Pflanzenwelt schweizerischer Flussauen. Sonderabdruck aus *Bauhinia. Zeitschr. Basl. Bot. Gesellsch.*, 4(1): 31-46. Ge. (P/T)

7216 **Muller, W. (1979).** Bedeutung, Schutz und Plfege von Hecken. (Significance, protection and care of hedges.) Birmensdorf, Schweizerisches Landeskomitee fur Vogelschutz. 12p. Ge. Illus.

7217 **Naf, E. (1981).** Zur Entstehung und Erhaltung von Mooren und Streuwiesen im Reusstal. (Development and conservation of wetlands in the Reuss Valley.) *Ber. Geobot. Inst. Eidgenoss. Techn. Hochsch. Stift. Rubel*, 48: 28 (1980 publ. 1981). Ge.

7218 **Parish, M. (1980).** The Swiss National Park. *Hort. Northwest*, 7(4): 66-68. (P)

7219 **Peter, R., Rueger, J., Keller, H. (1983).** Orchideenbiotope im Kanton Aargau (CH) - Ihr Schutz und Pflegemassnahmen durch die Arbeitsgruppe Einheimische Orchideen Aarau AGEO-Aarau. (Conservation and improvement of orchid biotopes in the Kanton Aargau (CH) by the Arbeitsgruppe Einheimische Orchideen Aarau AGEO-Aarau.) *Mitteilungsbl. Arbeitskr. Heim. Orch. Baden-Wuerttemberg*, 15(3): 331-350. Ge (En). Illus. (T)

7220 **Plackowki, R. (1983).** O ochronie gatunkowej roslin i ginieciu rzadkich taksonow w florze szwajcarii. (On the conservation of plant species and the vanishing of their rare taxa in Switzerland.) *Chronmy Przyr. Ojczysta*, 39(5): 87-92. Pol. Illus. (T)

7221 **Ritter, M., Waldis, R. (1983).** Vue d'ensemble des perils menacent la flore vegetale et ruderale: avec la liste rouge de la flore vegetale et ruderale. Fr. *Contributions a la protection de la nature en suisse*, No. 5. Ligue suisse pour la protection de la nature (LSPN). (T)

7222 **Rohrer, N., Chappuis, J.-B. (1982).** Coquelicots et bluets. Numero Special 1/1982. Basel, Ligue Suisse pour la Protection de la Nature. 25p. Fr. Col. illus. (E/T)

7223 **Schmalz, K.L. (1978).** Das Naturschutzgebiet Aegelsee-Moor auf dem Bergli, Gemeinde Diemtigen. *In* Naturschutzinspektorat des Kantons Bern Bericht 1977. *Mitt. Naturforsch. Ges. Bern*, 35: 176-185. Ge. Illus.

7224 **Schnyder, H. (1979).** Plan der geschutzten Naturobjekte im Kanton Luzern. (Plan for nature protection sites in the Canton of Lucerne.) *J. Forest. Suisse*, 130(7): 515-520. Ge. Illus. (P)

7225 **Steinlin, H. (1984).** Forstwirtschaft und Naturschutz - Spannung oder Ausgleich. (Forestry and nature protection - tension or balance.) *Schweiz. Zeitschr. Forstw.*, 135(2): 81-98. Ge (Fr).

7226 **Stockli, P.P. (1981).** Denkmalpflege im Freiraum. (Conservation of "Open Space" monuments.) *Anthos*, 20(2): 1-2. Ge (Fr, En). Illus. (P)

7227 **Waldis-Meyer, R. (1978).** Die Verarmung der Unkrautflora und einige Gedanken zu ihrer Erhaltung (mit besonderer Berucksichtigung des Wallis). (Impoverishment of the weedflora and some thoughts on its conservation [with special regard to Le Valais].) *Mitt. Ver. Forstl. Standortsk. Forstpfl.*, 26: 70-71. Ge. (T)

7228 **Weck, B. de (1981).** Rapport de la Ligue fribourgeoise pour la protection de la nature pour l'annee 1980/81. *Bull. Soc. Fribourg, Sci. Nat.*, 70(1-2): 30-31. Fr. (P)

7229 **Wendelberger, G. (1979).** Das Waldreservat "Les Follateres" ob Fully (Wallis) -eine pflanzensoziologische Studie. (The "Les Follateres" Forest Reserve near Fully (Wallis) - a study of plant ecology.) *Ber. Geobot. Inst. Eidgenoss. Techn. Hochsch. Stift. Rubel*, 46: 117-144. Ge. Illus., maps. (P)

7230 **Werner, P., Bressoud, B., Delarze, R. (1983).** Situation des plantes rares et de leurs milieux en Valais. *Bull. Murithienne*, 100: 195-211. Fr. Maps. (T)

7231 **Yerly, M. (1978).** Flore et vegetation du Parc National et de l'Engadine; problemes d'equilibre entre faune et vegetation, egalement dans la Reserve du Vanil-Noir. *Bull. Soc. Fribourg, Sci. Nat.*, 67(1): 3-5. Fr. Illus. (P)

SYRIA

7232 **IUCN Threatened Plants Committee Secretariat (1980).** First preliminary draft of the list of rare, threatened and endemic plants for the countries of North Africa and the Middle East. Kew, IUCN Threatened Plants Committee Secretariat. 170p. (T)

TAIWAN, PROVINCE OF

7233 **Anon. (1980).** [The rare and threatened plants of Taiwan.] Keelung, Taiwan Provincial Keelung Junior High School. 100p. Ch. Col. illus. (T)

7234 **Chen, F.S. (1977).** [Comparative study of the forest conservation systems between Taiwan and Japan: 2]. *T'ai-wan Lin Yeh Taiwan For. J.*, 3(8): 26-28. Ch.

7235 **Chen, F.S. (1977).** [Comparative study of the forest conservation systems between Taiwan and Japan: 3]. *T'ai-wan Lin Yeh Taiwan For. J.*, 3(9): 22-26, 28. Ch.

7236 **Chou, C. (1983).** Ecological conservation of mangrove forests in Taiwan. *Pac. Sci. Congr. Proc.*, 15(1-2): 42.

7237 **Hsu, K.-S., ed. (1982).** [The rare and threatened plants of Taiwan.] Keelung, Taiwan Provincial Junior High School. 100p. Ch. Lists 37 ferns, 8 gymnosperms, 293 angiosperms, with codes for habitat loss, overcollecting and restricted geographical ranges; includes non-endemic threatened plants; many colour plates. (T)

7238 **Hsu, W.S. (1978).** [How to draw up plans for national park recreation and its management (in Taiwan).] *Taiwan Forest. J.*, 4(3): 16-20. Ch. (P)

7239 **Huang, T.-C., Hsu, S.-J. (1982).** [The problems of Taiwan's mangrove forest]. *Quart. J. Chinese Forest.*, 15(3): 77-83. Ch (En). (T)

7240 **IUCN (1986).** IUCN's list of top 24 threatened animals and plants. *Biol. Conserv. Newsl.*, 44: 1-6. (T)

7241 **Kiang, Y.T., Antonovics, J., Wu, L. (1979).** The extinction of wild rice (*Oryza perennis formosana*) in Taiwan. *J. Asian Ecology*, 1: 1-9.

7242 **Lin, P.N. (1978).** [A review on the speedy afforestation of mountain conservation areas in Taiwan in the last years.] *Taiwan For. J.*, 4(3): 26-34. Ch. Illus. (P)

7243 **Lin, P.N. (1978).** [Examining recent achievements in speedy afforestation on mountainous reserves in Taiwan.] *Taiwan For. J.*, 4(4): 14-15. Ch. Illus. (P)

7244 **Liu, T., Hsu, K.-S. (1971).** [The rare and threatened plants and animals of Taiwan.] *Quart. J. Chinese Forest.*, 4(4): 89-96. Ch. (T)

7245 **McHenry, T.J.P., Yaw-Yuan Lin (1984).** Taiwan's first national parks. *Parks*, 9(3/4): 13-16. (P)

7246 **Ministry of Interior (Taiwan) (1985).** The Taiwan Nature Conservation Strategy. Taipei, Ministry of Interior. 39p.

7247 **Su, H.-J. (1980).** [Studies on the rare and threatened forest plants of Taiwan.] *Bull. Exp. Forest Nation. Taiwan Univ.*, 125: 165-205. Ch (En). (T)

TANZANIA, UNITED REPUBLIC OF

7248 **Anon. (1985).** [Problems in conservation of *Acacia tortilis* in Serengeti Reserve.] *Rotem*, 14: 88-89. He. (P/T)

7249 **Barnes, R.F.W. (1985).** Woodland changes in Ruaha National Park (Tanzania) between 1976 and 1982. *Afr. J. Ecol.*, 23(4): 215-221. (Fr). Maps. (P)

7250 **Clutton-Brock, T.H., Gillett, J.B. (1979).** A survey of forest composition in the Gombe National Park, Tanzania. *Afr. J. Ecol.*, 17: 131-158. Illus., map. (P)

TANZANIA, UNITED REPUBLIC OF

7251 **Greenway, P., Vesey-Fitzgerald, D. (1969).** The vegetation of Lake Manyara National Park. *J. Ecol.*, 57: 127-149. (P)

7252 **Greenway, P., Vesey-Fitzgerald, D. (1972).** Annotated check-list of plants occurring in Lake Manyara National Park. *J. East Afr. Nat. Hist. Soc.*, 28: 1-29. (P)

7253 **Grey-Wilson, C. (1984).** Plants in peril: 1. *Kew Mag.*, 1(2): 92-93. *Saintpaulia ionantha.* (T)

7254 **Hamilton, A.C. (1982).** Environmental history of East Africa. A study of the Quaternary. London, Academic Press. 328p.

7255 **Hamilton, A.H. (?)** Forest conservation in the East Usambara Mountains, Tanzania. IUCN, Tropical Forest Programme Series. c.200p. In prep.

7256 **Lovett, J. (1985).** Endemic plants of the Tanzanian forest. *WWF Monthly Report*, April: 77-81.

7257 **Lovett, J. (1985).** The moist forests of eastern Tanzania. *Swara*, 8(5): 8-9.

7258 **Lovett, J. (1986).** A future for the Tanzanian forest. *WWF News*, 39: 2. (P)

7259 **Lovett, J. (1986).** Upland Tanzania. *Threatened Pl. Newsl.*, 16: 9-10. Tropical forest, montane. (T)

7260 **Lovett, J.C. (1986).** The Eastern Arc forests of Tanzania. *Kew Mag.*, 3(2): 83-87. Illus. (T)

7261 **Lovett, J.C. (1988).** Practical aspects of moist forest conservation in Tanzania. *In* Goldblatt, P., Lowry, P.P., eds. Modern systematic studies in African botany. Proc. Eleventh Plenary Meeting, AETFAT, Missouri Bot. Gard., 10-14 June 1985. Missouri, Missouri Bot. Gard. 491-496. Tropical forest.

7262 **Lovett, J.C., Bridson, D.M., Thomas, D.W. (1988).** A preliminary list of the moist forest angiosperm flora of Mwanihana Forest Reserve, Tanzania. *Ann. Missouri Bot. Gard.*, 75(3): 874-885. (P)

7263 **Lovett, J.C., Norton, G.W. (1989).** Afromontane rainforest on Malundwe Hill in Mikumi National Park, Tanzania. *Biol. Conserv.*, 48(1): 13-19. (P)

7264 **Macdonald, I.A.W., Frame, G.W. (1988).** The invasion of introduced species into nature reserves in tropical savannas and dry woodlands. *Biol. Conserv.*, 44(1-2): 67-93. Invasive species. Includes 5 case studies. (P)

7265 **Mascarenhas, A. (1983).** Ngorongoro: a challenge to conservation and development. *Ambio*, 12(3-4): 146-152. Col. illus., maps. (P)

7266 **Mashalla, S.K. (1988).** The human impact on the natural environment of the Mbeya Highlands, Tanzania. *Mtn Res. Dev.*, 8(4): 283-288.

7267 **McNaughton, S.J. (1979).** Grazing as an optimization process: grass-ungulate relationships in the Serengeti. *Amer. Nat.*, 113(5): 691-703. (P)

7268 **Mshigeni, K.E. (1979).** Exploitation and conservation of biological resources in Tanzania. *In* Hedberg, I., ed. Systematic botany, plant utilization and biosphere conservation. Stockholm, Almqvist & Wiksell International. 140-143. Illus. (E)

7269 **Owen, J.S. (1969).** Development and consolidation of Tanzania national parks. *Biol. Conserv.*, 1(2): 156-158. (P)

7270 **Polhill, R.M. (1968).** Tanzania. *In* Hedberg, I. & O., eds. Conservation of vegetation in Africa south of the Sahara. Symposium Proceedings, at 6th plenary meeting of AETFAT, Uppsala. *Acta Phytogeogr. Suec.*, 54: 166-178. Illus., maps. Describes physiography, climate, vegetation plus plant conservation at present with proposals for the provinces. Species list included. (P/T)

TANZANIA, UNITED REPUBLIC OF

7271 **Rodgers, W.A. (1983).** A note on the distribution and conservation of *Oxystigma msoo* Harms (Caesalpiniaceae). *Bull. Jard. Bot. Nat. Belg.*, 53(1-2): 161-164. (Fr). (T)

7272 **Rodgers, W.A., Hall, J.B., Mwasumbi, L., Vollesen, K. (1984).** The conservation values and status of the Kimbosa Forest Reserve, Tanzania. Dar es Salaam, University of Dar es Salaam. 84p. Mimeo. Includes checklist. (P)

7273 **Rodgers, W.A., Homewood, K.M. (1979).** The conservation of the east Usambara Mountains, Tanzania: a review of biological values and land use pressures. Dar es Salaam, University of Dar es Salaam. 45p. Maps.

7274 **Rodgers, W.A., Homewood, K.M. (1982).** Biological values and conservation prospects for the forests and primate populations of the Uzungwa Mountains, Tanzania. *Biol. Conserv.*, 24: 285-304.

7275 **Rogers, G.K. (1987).** Thriving on the windowsill but endangered in the wild. *Missouri Bot. Gard. Bull.*, 75(2): 7. (T)

7276 **Schmidt, W. (1985).** Landschafts- und Vegetationsgliederung im Nordosten des Serengeti Nationalparks (Tanzania). (Landscape and vegetation classification in the northeastern Serengeti National Park, Tanzania.) *Phytocoenologia*, 13(1): 139-156. Ge (En). Illus., maps. (P)

7277 **Vollesen, K. (1980).** Annotated check-list of the vascular plants of the Selous Game Reserve, Tanzania. *Op. Bot.*, 59: 1-117. (P)

7278 **Watterson, G.G., comp. (1963).** Conservation of nature and natural resources in modern African states. Report of a Symposium organized by CCTA and IUCN, held under the auspices of FAO and Unesco, Arusha, Tanganyika, September 1961. IUCN Publications New Series No. 1. Morges, IUCN. 367p. Illus.

7279 **White, F. (1979).** Some interesting and endangered plant species in the West Usambara Mountains, Tanzania. *Not. Forest Herb. Univ. Oxford*, 1. 5p. Mimeo. (T)

7280 **Williams, J.G., Arlott, N. (1981).** A field guide to the national parks of East Africa. London, Collins. 336p. Illus., col. illus. Includes maps of some of the parks. (P)

7281 **Wingfield, R.C. (1979).** Possibilities and needs for conservation of plant species and vegetation in Africa. Appendix: Preliminary lists of rare and threatened species in African countries. Tanzania. *In* Hedberg, I., *ed.* Systematic botany, plant utilization and biosphere conservation. Stockholm, Almqvist & Wiksell International. 95-99. Contains about 390 endemic species and infraspecific taxa. (T)

THAILAND

7282 **Anon. (1988).** Malaysian logging threatens Thai reserve. *Oryx*, 22(1): 54. Full report in *Bangkok Bird Club Bull.*, 4(8). (P)

7283 **Anon. (1988).** Nam Choan inquiry: the environmental dilemma of the decade. *The Nation*, 18 April 1988: 36p. Special issue. (P)

7284 **Anon. (1988).** Thailand's controversial dam plan resurrected. *Oryx*, 22(1): 54. Nam Choan Dam threat. (P)

7285 **Arbhabhirama, A., Phantumvanit, D., Elkington, J., Ingkasuwan, P. (1987).** Thailand: natural resources profile. Bangkok, Thailand Development Research Institute. 310p. Illus., maps. Status, trends and key issues documented for natural resources, including wildlife and forests. (L/P)

7286 **Bain, J.R., Humphrey, S.R. (1982).** A profile of the endangered species of Thailand, 2 vols. Gainesville, USA, Office of Ecological Services. Lists 53 plant taxa rare or threatened in Thailand, with 3 data sheets and 14 dot maps. (T)

THAILAND

7287 **Banziger, H. (1988).** How wildlife is helping to save Doi Suthep: Buddhist sanctuary and national park of Thailand. *In* Hedberg, I., *ed.* Proceedings of the Symposium on Systematic Botany - A Key Science for Tropical Research and Documentation, Sweden, 14-17 Sept. 1987. Stockholm, Almqvist & Wiksell. 255-267. Illus., col. illus. (P/T)

7288 **Boonkird, S.A., Fernandes, E.C.M., Nair, P.K.R. (1985).** Forest villages: an agroforestry approach to rehabilitating forest land degraded by shifting agriculture. *Agroforestry Systems*, 2: 87-102. (E)

7289 **Brockelman, W.Y. (1987).** Nature conservation. *In* Arbhabhirama, A., Phantumvanit, D., Elkington, J., Ingkasuwan, P., *eds.* Thailand natural resources profile. Bangkok, Thailand Development Research Institute.

7290 **Chapman, E.C., Sabhasri, S., eds (1983).** Natural resource development and environmental stability in the highlands of northern Thailand. Proceedings of a workshop organised by Chiang Mai University and the United Nations University - Thailand, November 1979. *Mountain Research and Development*, 3(4): 309-431.

7291 **Chettamart, S. (1985).** Preparing a management plan for Khao Yai Park: the process involved and the lessons learned. *In* Thorsell, J.W., *ed.* Conserving Asia's natural heritage. The planning and management of protected areas in the Indomalayan Realm. Proc. of the 25th working session of IUCN's CNPPA. Gland, IUCN. 162-165. (P)

7292 **Chomchalow, S. (1986).** The national genebank of Thailand. *IBPGR Region. Committ. Southeast Asia Newsl.*, 10(1): 4-6. (E/G)

7293 **Davis, S. (1988).** Nam Choan dam threat. *Threatened Pl. Newsl.*, 19: 9. Proposals for dam will flood 223 sq. km of riverine forest. (P)

7294 **Davis, S. (1988).** Nam Choan update. *Threatened Pl. Newsl.*, 20: 10. Map. (P)

7295 **Dobias, R.J. (1982).** The Shell guide to the national parks of Thailand. Bangkok, The Shell Co. of Thailand Ltd. 137p. Illus., col. illus., maps. (P)

7296 **Dobias, R.J. (1987).** A demonstration project for integrating park conservation with rural development at Khao Yai National Park, Thailand. *Parks*, 12(1): 17-18. (P)

7297 **DuPuy, D. (1984).** Flowers of the Phu Luang wildlife sanctuary. *Kew Mag.*, 1(2): 75-84. Illus. Map. (P/T)

7298 **DuPuy, D.J. (1983).** The wildlife sanctuary of Phu Luang, Thailand and its rich orchid flora. *Orchid Rev.*, 91(1082): 366-371. Col. illus. (P)

7299 **Elliott, S. (1988).** Thai forest wins reprieve from dam. *Oryx*, 22(4): 191-192. (P)

7300 **Ewins, P.J., Bazely, D.R. (1989).** Jungle law in Thailand's forests. *New Sci.*, 124(1691): 42-46. Col. illus. Deforestation. (L)

7301 **IBPGR (1980).** The Thai National Rice Seed Storage Laboratory for Genetic Resources. *IBPGR Region. Committ. Southeast Asia Newsl.*, 5(1): 14. Seed bank. (E)

7302 **Jones, D.T. (1987).** Rare plant profile no. 1: *Merrillia caloxylon* (Rutaceae). *Bot. Gard. Conserv. News*, 1(1): 38-42. Illus., map. (E/G/T)

7303 **Kempf, E. (1987).** Dam controversy growing in Thailand. *WWF News*, 46: 6, 8.

7304 **Khemnark, C, (1984).** Using research to guide improvements to deteriorating conditions in Thailand: the experience of Sakaerat environmental research station. *In* Unesco-UNEP. Conservation, science and society. Paris, Unesco. 354-359. Contributions to the First International Biosphere Reserve Congress, Minsk, Byelorussia/USSR, 26 September - 2 October 1983. (P)

7305 **Lohmann, L. (1989).** Forestry in Thailand: the logging ban and its consequences. *Ecologist*, 19(2): 76-77. Tropical forest; deforestation.

THAILAND

7306 **Maxwell, J.F. (1977).** Vascular flora of Sam Lan Forest Park Muang District, Saraburi Province, Thailand. *Thai For. Bull.*, 10: 47-110. Illus. (P)

7307 **Maxwell, J.F. (1980).** Vegetation of Khao Khieo Game Sanctuary Chonburi Province, Thailand. *Nat. Hist. Bull. Siam Soc.*, 28: 9-24. Illus. (P)

7308 **Maxwell, J.F. (1989).** The vegetation of Doi Sutep-Pui National Park, Chiang Mai Province, Thailand. *Tigerpaper*, 15(4): 6-14. Illus., map. (P)

7309 **Ng, F.S.P., Low, C.M. (1982).** Check list of endemic trees of the Malay Peninsula. Kepong, Forest Research Institute. 94p. Lists 654 trees endemic to the Malay Peninsula, including Peninsular Thailand; rarity is based on numbers of herbarium specimens. (T)

7310 **Niyomdham, C. (1989).** A preliminary study on flora of Khao Noi Chuchi (Khao Pra Baang Khraam Non-hunting Area): Krabi Province. Report submitted to WCI, Thailand. Checklist of flora with descriptions. (P)

7311 **Phantumvanit, D., Sathirathai, K.S. (1987).** Pressure on forest - the Thai experience. *Forest News*, 1(3): 1-5. Tropical forest; deforestation; sustainable use. Supplement to *Tigerpaper*, 19(4), 1987. (P)

7312 **Polprasid, P. (1982).** Genetic resources of *Salacca* palms in Thailand. *IBPGR Region. Committ. Southeast Asia Newsl.*, 6(4): 2-11. Illus. (E)

7313 **Pradhan, U.C. (1975).** Conservation of eastern Himalayan orchids. Problems and prospects. Part I. *Orch. Rev.*, 83: 314-317. (T)

7314 **Pradhan, U.C. (1975).** Conservation of eastern Himalayan orchids. Problems and prospects. Part II. *Orch. Rev.*, 83: 345-347. (T)

7315 **Pradhan, U.C. (1975).** Conservation of eastern Himalayan orchids. Problems and prospects. Part III. *Orch. Rev.*, 83: 374. (T)

7316 **Round, P.D. (1989).** Monitoring the conservation status of species and habitats in Thailand. *Tigerpaper*, 16(3): 20-27. Illus.

7317 **Sagarik, R. (1983).** Some ideas on conservation and development of rarer orchid species of Thailand. *In* Robinson, R., *ed.* Proceedings of the 8th Australian Orchid Conference, Townsville, 27 August 1983-4 September 1983. Townsville Orchid Society. 34-36. (G/T)

7318 **Santiapillai, C., Ashby K.R., eds (1987).** Proceedings of the Symposium on the Conservation and Management of Endangered Plants and Animals. Bogor, Indonesia, 18-20 June 1986. Bogor, SEAMEO-BIOTROP. 246p. Maps. (E/G/L/T)

7319 **Sharp, T. (1984).** ASEAN nations tackle common environmental problems. *Ambio*, 13(1): 45-46. Illus.

7320 **Singhapant, S. (1978).** National parks of Thailand. *In* Proceedings of the 8th World Forestry Congress. Jakarta, 16-28 October 1978: Forestry for quality of life. II, 10. (P)

7321 **Smitinand, T. (1968).** Some rare and vanishing plants of Thailand. IUCN Publications New Series No. 10. Morges, IUCN. 344-346. (T)

7322 **Smitinand, T., ed. (1977).** Plants of Khao Yai National Park. Thailand, Friends of Khao Yai National Park. 73p. Col. illus. Map. (P)

7323 **Sribhibhadh, P., Juntarogool, R., Mead, E., eds (1987).** Consider the costs : a position paper on the Nam Choan Dam. Bangkok, Wildlife Fund Thailand. 47p. Col. illus., maps. (P)

THAILAND

7324 **Suvanakorn, Ph., Dobias, R. (1985).** Using economic incentives to improve park protection: a case study from Thailand. *In* Thorsell, J.W., *ed.* Conserving Asia's natural heritage. The planning and management of protected areas in the Indomalayan Realm. Proc. of the 25th working session of IUCN's CNPPA. Gland, IUCN. 177-182. (P)

7325 **Talbot, L.M., Challinor, D. (1973).** Conservation problems associated with development projects on the Mekong River system. *In* Costin, A.B., Groves, K.H., *eds.* Nature conservation in the Pacific. Proc. Symposium A-10, XII Pacific Science Congress, Canberra, August-September 1971. IUCN and Australian Nat. Univ. Press. 271-283. Map.

7326 **Topark-Ngarm, A., Manidool, C., Chankam, S., Tiyawalee, D. (1981).** Species evaluation and conservation of forage legumes in Thailand. *IBPGR Region. Commit. Southeast Asia Newsl.,* 5(3): 11-12. (E)

7327 **UNDP/FAO (1981).** National parks and wildlife management. Thailand. A review of the nature conservation programmes and policies of the Royal Forest Department. Project Working Document THA 77/003. Bangkok, UNDP/FAO. Maps. (L/P/T)

7328 **Vejaboosakorn, S. (1985).** The development of a protected area system for Thailand in terms of representative coverage of ecotypes. *In* Thorsell, J.W., *ed.* Conserving Asia's natural heritage. The planning and management of protected areas in the Indomalayan Realm. Proc. of the 25th. Working Session of IUCN's CNPPA. Gland, IUCN. 75-77. (P)

7329 **Yingvanasiri, T. (1982).** *In situ* conservation of *Pinus merkusii* in Thailand. *IBPGR Region. Committ. Southeast Asia Newsl.,* 6(3): 11. (E)

TOGO

7330 **Osborne, R. (1988).** *Encephalartos barteri. Encephalartos,* 14: 8-16. Illus., map. Data sheet; includes conservation. (G/T)

TONGA

7331 **Dahl, A.L. (1978).** Environmental and ecological report on Tonga. Part 1: Tongatapu. Noumea, New Caledonia, South Pacific Commission.

7332 **Rinke, D. (1986).** The status of wildlife in Tonga. *Oryx,* 20: 146-151. Human impact on vegetation.

7333 **Straatmans, W. (1964).** Dynamics of some Pacific island forest communities in relation to the survival of the endemic flora. *Micronesica,* 1(1-2): 113-122.

7334 **Tongilava, S.L. (1979).** Development and management of marine parks and reserves in the Kingdom of Tonga. *In* Second South Pacific Conference on National Parks and Reserves, Proceedings, vol. 1. 148-152. Mostly about reefs. (P)

TRINIDAD AND TOBAGO

7335 **Adams, C.D., Baksh, Y.S. (1981).** What is an endangered plant? *Living World,* 1981-1982: 9-14. Journal of the Trinidad and Tobago Field Naturalists Club. Includes distribution tables of 648 threatened non-endemic species and of 215 endemic species, with criteria for ranking the selected species. (T)

7336 **Anon. (1979).** From around the Caribbean: Trinidad & Tobago. *Caribbean Conserv. News,* 1(17): 26. Discovery of rare orchid in Trinidad: *Notylia augustifolia.* (T)

7337 **Anon. (1983).** Caribbean mangroves to be protected. *Oryx,* 17: 143. (L/P)

7338 **Bacon, P.R., French, R.P., eds (1972).** The wildlife sanctuaries of Trinidad and Tobago. Trinidad and Tobago, Wildlife Conservation Committee, Ministry of Agriculture, Lands and Fisheries. 80p. Illus., map.

TRINIDAD AND TOBAGO

7339 **Thelen, K.D., Faizool, S., eds (1980).** Policy for the establishment and management of a national park system in Trinidad and Tobago. Port of Spain, Forestry Division, Ministry of Agriculture. 26p. (P)

TUNISIA

7340 **Blanc, C.P., Gaultier, T., Figier, J., Memmi, L., Muller, H.P. (1978).** Plaidoyer pour la conservation du Djebel Zaghouan. *Bull. Soc. Sci. Nat. Tunisie*, 13: 3-4. Fr.

7341 **Hollis, G.E. (1981).** Les programmes de conservation de la nature et de gestion des eaux du Parc National de l'Ichkeul. Discussion Papers in Conservation No. 31. London, University College. 31p. Fr. Maps. (P)

7342 **IUCN Threatened Plants Committee Secretariat (1980).** First preliminary draft of the list of rare, threatened and endemic plants for the countries of North Africa and the Middle East. Kew, IUCN Threatened Plants Committee Secretariat. 170p. (T)

7343 **Mathez, J., Quezel, P., Raynaud, C. (1985).** The Maghreb countries. *In* Gomez-Campo, C., *ed.* Plant conservation in the Mediterranean Area. Dordrecht, Dr W. Junk. 141-157. Illus., maps. (T)

7344 **Novikoff, G., Skouri, M. (1981).** Balancing development and conservation in pre-Saharan Tunisia. *Ambio*, 10(2-3): 135-141. Illus., map.

7345 **Pottier-Alapetite, G. (1959).** Especes vegetales rares ou menacees de Tunisie. *In* IUCN. Animaux et vegetaux rares de la Region Mediterraneenne. Proceedings of the IUCN 7th Technical Meeting, 11-19 September 1958, Athens, vol. 5. Brussels, IUCN. 135-139. (T)

TURKEY

7346 **Akesen, A. (1977).** Turkiye'deki ulusal parklarin acik hava rekreasyonu yonunden nitelikleri ve sorunlari (ornek Uludag uludsal parki). (Characteristics and problems of the national parks in Turkey with respect to outdoor recreation.) *Istanbul Univ. Orman Fak. Dergisi, A.*, 27(2): 294-339. Tu (En). Illus., maps. (P)

7347 **Altan, T., Fahrenhorst, B. (1988).** Der Export von Geophyten aus der Turkei und seine Folgen. (The export of geophytes from Turkey and its consequences.). *Natur und Landschaft*, 63(1): 21-25. Ge. Illus. (E)

7348 **Ayasligil, Y. (1987).** Der Koprulu Kanyon Nationalpark. Seine Vegetation und ihre Beeinflussung durch den Menschen. Technische Universitat Munchen-Weihenstephan. Ge (En). Illus., maps. (P)

7349 **Ball, M (1980).** *Iris susiana*: an endangered species? *New Zealand Iris Soc. Bull.*, 92: 13-15. Illus. (T)

7350 **Baytop, A., Demiriz, H. (1980).** Rare plants and endemics in Turkey-in-Europe. *Istanbul Univ. Fen Fak. Mec. Seri B*, 45: 109-111. Brief floristic account; lists 29 rare and threatened taxa. (T)

7351 **Baytop, T. (1959).** Les plantes rares de l'Anatolie et les precautions prises en vue et leur protection. *In* IUCN. Animaux et vegetaux rares de la Region Mediterraneenne. Proceedings of the IUCN 7th Technical Meeting, 11-19 September, 1958, Athens, vol. 5. Brussels, IUCN. 197-198. Fr. (T)

7352 **Birand, H. (1959).** La vegetation Anatolienne et la necessite de sa protection. *In* IUCN. Animaux et vegetaux rare de la Region Mediterraneenne. Proceedings of the IUCN 7th Technical Meeting, 11-19 September, 1958, Athens, vol. 5. Brussels, IUCN. 192-196. Fr. (T)

7353 **Boydak, M. (1985).** The distribution of *Phoenix theophrasti* in the Datca Peninsula, Turkey. *Biol. Conserv.* 32: 129-135. Illus., map. (G/P/T)

TURKEY

7354 **Boydak, M. (1987).** A new occurrence of *Phoenix theophrasti* in Kumluca-Karaoz, Turkey. *Principes*, 31(2): 89-95. Illus., maps. Includes population data and conservation measures required for a species originally presumed endemic to Crete. (T)

7355 **Chown, M. (1985).** Rare plants piled high and sold cheap. *New Sci.*, 105(1446): 10. Illus. (E/G/L/T)

7356 **Council of Europe (1984).** Turkey: national parks law. *Council of Europe Newsl.*, 84(2): 4. (L/P)

7357 **Davis, P.H. (1985).** Why is the flora of Turkey interesting and important? *Kew Mag.*, 2(4): 357-367. Illus. Map (T)

7358 **Demiriz, H. (1973).** Endemiten der Umgebung von Istanbul. *In* Istanbul: Universitesi, Orman Fakultesi. Kazdagi goknari ve Turkiye florasi uluslararasi Sympozyomu bilolirileri Istanbul 1973. 147-150. Ge.

7359 **Demiriz, H. (1981).** Nature conservation in Turkey. *Threatened Pl. Committ. Newsl.*, 7: 21-22.

7360 **Demiriz, H., Baytop, T. (1985).** The Anatolian Peninsula. *In* Gomez-Campo, C., *ed.* Plant conservation in the Mediterranean area. Dordrecht, Dr W. Junk. 113-121. Illus., map. (T)

7361 **Efe, A. (1988).** *Liquidambar orientalis. Threatened Pl. Newsl.*, 20: 9-10. Illus. Exploitation and conservation efforts. (E/T)

7362 **Ekim, T., Koyuncu, M., Erik, S., Guner, A., Yildiz, B., Vural, M. (1984).** [Taxonomic and ecological investigations on the economic geophytes of Turkey.] *In* Arastirma Gurubu Proje. No.: T.B.A.G. -490A. Ankara. Tu (En). (E/G/T)

7363 **Ekim, T., Koyuncu, M., Erik, S., Ilarslan, R., eds (1989).** Turkiye'nin Tehlike Altindaki Nadir ve Endemik Bitki Turleri. (List of rare, threatened and endemic plants in Turkey.) Ankara, Turkiye Tabiatini Koruma Dernegi (Turkish Association for the Conservation of Nature and Natural Resources). Serie No: 18. 227p. Tu (En). (T)

7364 **Gurpinar, T. (1981).** The seven lakes. *Naturopa*, 38: 29-30. Illus. (P)

7365 **IUCN Commission on Ecology (1968).** Proceedings of a Technical Meeting on Wetland Conservation, Ankara, Bursa, Istanbul, 9-16 October 1967. IUCN Publications New Series No. 12. Morges, IUCN. 274p. Maps.

7366 **Ianovici, V. (1979).** Protectia mediului inconjurator si prezervarea cadrului biologic in Tarile. Balcanice. (La protection de l'environnement et la conservation du cadre biologique dans les pays Balkaniques.) *Ocrot. Nat.*, 23(1): 5-7. Rum (Fr). Conference at Bucharest on 18-22 September 1978.

7367 **Jordanov, D., et al., eds (1975).** Problems of the Balkan flora and vegetation. Sofia, Bulgarian Academy of Sciences. Proceedings of the First International Symposium on Balkan Flora and Vegetation, Vorna, 7-14 June 1973. 441p. (T)

7368 **Michael, J.B., Almond, L.A. (1976).** Sites and flowers of west Turkey. *The Rock Garden*, 19(3): 243-254. (T)

7369 **Read, M. (1988).** Bulb trade in Turkey. *Species*, 10: 40. (E/T)

7370 **Read, M. (1989).** Grown in Holland? Brighton, Fauna and Flora Preservation Society. 12p. Illus., maps. European bulb trade. (E/G/L/T)

7371 **Tezcan, S., et al. (1979).** Halic Master Plani Butunu Icinde Eyup Bolge Parke Onerisi. (The Eyup Regional Park proposal within the general framework of the Golden Horn Master Plan.) *Istanbul Univ. Orman Fak. Dergisi, A.*, 29(2): 1-30. Tu (En). Maps. (P)

TURKEY

7372　**Turkoz, N. (1968).** Conservation of natural resources: the example of forest management in Turkey. *In* Proc. Tech. Meeting on Wetland Conservation, 9-16 October, 1967. Ankara-Bursa-Istanbul. IUCN Publications New Series No. 12. Morges, IUCN Commission on Ecology. 23-24.

7373　**Uner, N. (1968).** Soil and water problems in Turkey from the standpoint of nature conservation. *In* IUCN Commission on Ecology. Proceedings of a Technical Meeting, Ankara-Bursa-Istanbul, 9-16 October 1967. 17-22.

7374　**Vardar, Y., Oflas, S., Oguz, G. (1970).** Tabiati degerlendirmenin onemi ve kum yiginlarinda kendiliginden beliren vegetasyona ait bir gozlem. (The importance of nature conservation and observations on the vegetation of sand dunes.) *Ege Univ. Fen Fak. Ilmi Raporlar Ser.*, 109: 3-10. Tu. Illus.

TURKS AND CAICOS ISLANDS

7375　**Sauleda, R.P., Adams, R.M. (1979).** *Encyclia inaguensis* Nash ex Britton and Millspaugh - a rare orchid from the Bahama Islands and the Caicos group. *Bull. Amer. Orchid Soc.*, 48(3): 257-260. Illus. (T)

UGANDA

7376　**Edroma, E.L. (1977).** Wirkungen von Grosswild-Bestanden und Branden auf die Vegetation im Ruwenzori-Nationalpark. (Effects of big game and fires on the vegetation of the Ruwenzori National Park.) *Oberhess. Naturwiss. Zeitschr.*, 43: 39-54. Ge (En). (P)

7377　**Hamilton, A.C. (1982).** Environmental history of East Africa. A study of the Quaternary. London, Academic Press. 328p.

7378　**Howard, P. (1986).** Conserving tropical forest wildlife in Uganda. *WWF Monthly Report,* July: 185-191. Illus., map. (T)

7379　**Katende, A.B. (1976).** The problem of plant conservation and the endangered plant species in Uganda. *In* Miege, J., Stork, A.L., eds. Comptes Rendus de la VIIIe Reunion de l'AETFAT, 2 vols. Proceedings of the 8th plenary meeting of AETFAT in Geneva 16-21 September, 1974. *Boissiera*, 24a, b: 451-456. (T)

7380　**Katende, A.B. (1979).** Uganda. *In* Hedberg, I., ed. Systematic botany, plant utilization and biosphere conservation: Proc. of a sym. held in Uppsala in commemoration of the 500th ann. of the Univ. Stockholm, Almqvist & Wiksell Int'l. 93.

7381　**Katsigouzi, C., Bass, S. (1983).** Uganda: conservation for rebuilding a country. *IUCN Bull.*, suppl. 4: 6. Illus., map.

7382　**Kayanja, F., Douglas-Hamilton, I. (1984).** The impact of the unexpected on the Uganda National Parks. *In* McNeely, J.A., Miller, K.R., eds. National parks, conservation and development. The role of protected areas in sustaining society. Washington, D.C., Smithsonian Institution Press. 87-92. (P)

7383　**Lock, J.M. (1977).** The vegetation of Ruwenzori National Park, Uganda. *Bot. Jahrb. Syst.*, 98(3): 372-448. En (Ge). (P)

7384　**Musoke, M.B. (1980).** Overbrowsing of *Capparis tomentosa* bushes by goats in Ruwenzori National Park, Uganda. *Afr. J. Ecol.*, 18(1): 7-10. (Fr). (P/T)

7385　**Osborne, R. (1988).** *Encephalartos barteri. Encephalartos*, 14: 8-16. Illus., map. Data sheet; includes conservation. (G/T)

7386　**Osmaston, H.A. (1968).** Uganda. *In* Hedberg, I. & O., eds. Conservation of vegetation in Africa south of the Sahara. Symposium Proceedings, at 6th plenary meeting of AETFAT, Uppsala. *Acta Phytogeogr. Suec.*, 54: 148-151. Associations and plants already protected, and those requiring protection. (P/T)

UGANDA

7387 **Smart, N.O.E., Hatton, J.C., Spence, D.H.N. (1985).** The effect of long-term exclusion of large herbivores on vegetation in Murchison Falls National Park, Uganda. *Biol. Conserv.*, 33(3): 229-245. Maps. (P)

7388 **Spence, D.H.N., Angus, A. (1971).** African grassland management - burning and grazing in Murchison Falls National Park, Uganda. *In* Duffey, E., Watt, A.S., *eds.* The scientific management of animal and plant communities for conservation. The 11th Symposium of the BES, University of East Anglia, Norwich, 7-9 July 1970. Oxford, Blackwell. 319-331. Illus., map. (P)

7389 **Struhsaker, T. (1987).** Forestry issues and conservation in Uganda. *Biol. Conserv.*, 39(3): 209-234.

7390 **Williams, J.G., Arlott, N. (1981).** A field guide to the national parks of East Africa. London, Collins. 336p. Illus., col. illus. Includes maps of some of the parks. (P)

UNION OF SOVIET SOCIALIST REPUBLICS

7391 **Abele, G.P., Miezite, I. Ya (1982).** Zapovednik Krustkalny. (Krustkaln Nature Reserve.) Latv, SSR, Zinatne. (Seriya "Flora Okhranyaemykh Territorii Latvii"). Rus. (P)

7392 **Abele, G.T., Miezite, I. Ya (1982).** Zapovednik Krustkalny. (Krustkaln Nature Reserve.) Flora Okhranyaemykh Territorii Latvii. Riga, Zinatne. 108p. Rus. Illus. (P)

7393 **Abramchuk, A.V., Gorchakowskii, P.L. (1980).** [Formation and anthropogenic degradation of meadow plant communities in the Trans-Ural forest-steppe region]. *Sov. J. Ecol.*, 11(1): 11-20. Translated from *Ekologiya*, (1980). 11(1): 22-34.

7394 **Agal'tsova, V.A. (1979).** Sokhranenie memorial'nykh lesoparkov. USSR, Lesnaya Prom.-St', 1980. Rus. Memorial parks at Mikhailovskoe, Boldino, Tarkhany, Ostaf'evo and elsewhere. (P)

7395 **Akademiya Nauk SSSR (1978).** Rastitel'nyi mir okhranyaemykh territorii. (The plant world of protected territories.) Riga, Zinatne. 168p. Rus. (P)

7396 **Akhundov, G.F., Gozina, E.E., Prilipko, L.I. (1978).** [Rare species from the natural flora of the Nakhichevan ASSR]. *Byull. Glavn. Bot. Sada* (Moscow), 108: 54-62. Rus. (T)

7397 **Aksenova, N.A., Frolova, L.A. (1984).** O kulture redkikh vidov drevesnykh rastenii SSSR v Botanicheskom Sadu MGU im. M.V. Lomonosova. (Cultivation of rare species of woody plants of the USSR in the botanical garden of the M.V. Lomonosov Moscow State University. *Byull. Glavn. Bot. Sada* (Moscow), 133: 85-91. Rus. (G/T)

7398 **Albertina, S. (1978).** Dabas un apkartejas vides aizsardzibaviens no svarigakajiem sisdienas uzdevumiem. (Nature and environment conservation - an urgent task [of forestry specialists].) *Mezsaimnieciba un Mezrupnieciba*, 1: 22-24. (Rus).

7399 **Alekperov, U.K., Troitskaya, E.A. (1983).** Sessiya Nauchnogo Soveta AN SSSR po probleme "Biologicheskie osnovy ratsional'nogo ispol'zovaniya, preobrazovaniya i okhrany rastitel'nogo mira" ("Biological foundations of the rational utilization, transformation and protection of the plant world.") *Bot. Zhurn.*, 68(9): 1289-1292. Rus.

7400 **Alekseev, V.D., ed. (1982).** Rastitelnyi pokrov Dagestana i ego okhrana: Mezhvuz. Nauch. temat. sb. Makhachkala, Dag. Un-t. 136p. Rus. Illus.

7401 **Alekseeva, L.M. (1976).** Rare species of vascular plants of Kunashir Island. *In* "Preservation of nature in the Far East". Vladivostok, Far East Scientific Center of the Soviet Academy of Sciences. 93-99. (T)

UNION OF SOVIET SOCIALIST REPUBLICS

7402 **Amirkhanov, A.M. (1978).** (Patterns of altitudinal zonation of vegetation of the North-Osetian National Park.) *Byull. Mosk. Obshch. Ispyt. Prir., Biol.,* 83(3): 136-142. Rus (En). (P)

7403 **Andreev, G.N., Golovkin, B.N. (1978).** [Introduction as a method of conservation of rare and disappearing plant species from the far north and high mountain areas (of the USSR)]. *Byull. Glavn. Bot. Sada,* 109: 3-6. Rus. (T)

7404 **Andreev, L.N., ed. (1986).** Introduktsiya i okhrana rastenii v SSSR i SShA. (Introduction and protection of plants in the USSR and USA.) Moskva, Nauka. 129p. Rus. Maps. (P/T)

7405 **Andreev, V.N., ed. (1987).** Krasnaya kniga Yakutskaya ASSR: redkie i nakhodyashchiesya pod ugrozoi ischeznoveniya vidy rastenii. (Red Book of the Yakut ASSR: rare and threatened species of plants). Novosibirsk, Nauka. 247p. Rus. Maps. (P/T)

7406 **Andronova, N.N., Gaisenok, S.N., Kucheneva, A.E., Grebennikova, V. (1981).** K spisku redkikh i ischezayushchikh rastenii Kabinin gradskoi oblasti (Prostrel lugovoi). *In* Puti adapt. rast. pri introduktsii na severe. Petrozavodsk. 37-40. Rus. Note in *Ref. Zhurn., Biol.,* 8(2): V450 (1982).

7407 **Anon. (1980).** Khorologiya flory Latviiskoi SSR: redkie vidy rastenii II gruppy okhrany. (Chorology of the flora of the Latvian SSR: rare species of plants of the group of protection [i.e. needing protection].) Latv. SSR, Zinatne. Rus. (T)

7408 **Anon. (1980).** Nauchnye osnovy razmeshcheniya prirodnykh rezervatov v Sverdlovskoi oblasti. (Scientific conditions for the location of reserves in the Sverdlovsk Region.) U.S.S.R., Izd. Ural'skogo Nauch. Tsentra AN SSSR. 10p. Rus. (P)

7409 **Anon. (1981).** Okhrana prirody Prichernomor'ya. (Protection of nature in the Black Sea Region.) U.S.S.R., Lesnaya Prom-st'. Rus.

7410 **Anon. (1982).** Inventarizatsiya, metody issledovaniya i okhrana redkikh rastitel'nykh soobshchestv: materialy 1-7. Vses. kong., Moskva, 29 okt.-2 noyab., 1981. VNII okhrany prirody i zapoved. dela SSSR. Rukopsis'dep. v Viniti No. 1059-83 Dep. Moskva. 317p. Making an inventory, methods of investigation and the protection of rare plant associations: materials of the 17th All Union Congress, Moscow, 29 Oct-2 Nov. 1981. Rus. (T)

7411 **Anon. (1982).** Lakhemaaskii Natsional'nyi Park: nauchnye trudy po okhrane prirody. (Lakhemaaski National Park: scientific works on the protection of nature.) *Uch. Zap. Tart. Un-ta,* 575: 3-78. Rus. (P)

7412 **Anon. (1982).** Okhrana redkikh rastitel'nykh soobshchestva: sb. nauch. tr. VNII okhrany prirody i zapoved. dela. (The protection of rare plant associations: collection of scientific works of VNII on the protection of nature and nature reserves.) Moskva, VNII. 80p. Rus (En). (P/T)

7413 **Anon. (1982).** Okhrana, izuchenie i obogashchenie rastitel'nogo mira sb 9. (Protection, study and enrichment of the plant world: part 9.) Kiev, Izd-vo Kiev. Un-ta. Rus.

7414 **Anon. (1983).** Okhrana flory rechnykh dolin v Pribaltiiskikh Respublikakh. (Protection of the flora of the river valleys of the Baltic Republics.) Latv, SSR, Zinatne. Rus. (T)

7415 **Anon. (1983).** Sezonnaya i raznogodichnaya dinamika rastitel'nogo pokrova v zapovednikakh RSFSR: sb. nauch. tr. TsNIL okhat. Kh-va i zapoved. (Seasonal and annual dynamic of the plant cover in the nature reserves of the RSFSR.) Moskva. 134p. Rus. Illus. (P)

7416 **Anon. (1983).** Zapovedniki SSSR. (Nature reserves of the U.S.S.R.) U.S.S.R., Lesnaya Prom-st'. Rus. Includes climate, physical and biological information. (P)

UNION OF SOVIET SOCIALIST REPUBLICS

7417 **Anon. (1984).** Green medicine for the biosphere. *Trees*, 44(1): 10-12. (E/L)

7418 **Anon. (1984).** Obshchie problemy okhrany rastitel'nosti: materialy Vses. soveshch. Okhrana rastit. mira sev. regionov, Syktyvkar, 7-9 Sent. 1982. T.I. (General problems of the protection of plants: papers of the All Union Conference for the protection of the plant world of the North Region.) Syktyvkar, Komi Fil. AN SSSR. 175p. Rus.

7419 **Aralova, N.S., Cherkasova, M.V. (1980).** Antropogennaya transformatsiya flory sosudistykh rastenii. (Anthropogenic transformation of the flora of vascular plants.) *In* Okhrana redk. rast. i fitotsenozov. Moskva. 40-48. Rus (En). From the Red Book of the USSR. (T)

7420 **Aralova, N.S., Zykov, K.D. (1984).** USSR nature reserves: their role in environmental education. *In* Unesco-UNEP. Conservation, science and society. Paris, Unesco. 573-576. Contributions to the First International Biosphere Reserve Congress, Minsk, Byelorussia/USSR, 26 September - 2 October 1983. (P)

7421 **Aristov, Yu.V., Kotova, T.V. (1979).** O kartograficheskom materiale v "Krasnoi knige SSSR: redkie i nakhodya-shchesya pod ugrozoi ischeznoveniya vidy zhivotnykh i rastenii." (A review of the cartographical material in the "Red Book of the U.S.S.R.: rare and disappearing animal and plant species".) Rus. (T)

7422 **Astanin, L., Blagosklonov, K. (1983).** Okhrana prirody. (The protection of nature.) U.S.S.R., Progress. Rus.

7423 **Astanin, L.P., Blagosklonov, K.N. (1983).** Conservation of nature. Moscow, Progress Publishers, 149p. English translation of the revised Russian text. (L/P/T)

7424 **Azovskii, M.G. (1981).** Nakhodki redkikh dlya vostochnoi Sibiri Pribrezhno-vodnykh i vodnykh rastenii po trasse Baikalo-Ámurskoi magistrali. (The finds of rare east Siberian littoral and aquatic plants on the Baikalo-Amur route.) *Bot Zhurn.*, 66(8): 1218-1220. Rus.

7425 **Balabas, G.M., Satsyperova, I.F., Sinitskii, V.S. (1981).** Ratsional'noe ispol'zovanie, vosproizvodstvo i okhrana ofitsial'nykh travyanistykh lekarstvennykh rastenii lesnoi zony SSSR. (Rational usage, reproduction and protection of herbaceous medicinal plants.) *Rast. Resursy*, 17(3): 325-337. Rus. (E)

7426 **Bannikov, A.G., Bogdanov, B.N. (1973).** Conservation as a long-term development tool. *In* IUCN. Twelfth technical meeting, Banff, Alberta, Canada, 12-15 September 1972. 121-130. Morges, IUCN.

7427 **Bassiev, T.U., et al. (1981).** [Red data book of the Northern Osetia.] Ordjonikidze. 88p. Rus. (T)

7428 **Bazilevich, N.I., Gilamanov, T.G. (1984).** Conceptual-balance models of natural and semi-natural ecosystems of the central-chernozyom biosphere reserve and their analysis. *In* Unesco-UNEP. Conservation, science and society. Paris, Unesco. 347-350. Contributions to the First International Biosphere Reserve Congress, Minsk, Byelorussia/USSR, 26 September - 2 October 1983. Table. (P)

7429 **Beloussova, L. (1977).** Endangered plants of the U.S.S.R. *Biol. Conserv.*, 12(1): 1-11. (T)

7430 **Beloussova, L., Denisova, L. (1981).** The USSR Red Data Book and its compilation. *In* Synge, H., *ed.* The biological aspects of rare plant conservation. Chichester, Wiley. 93-99. Refers to Borodin, A.M. *et al.*, *The Red Data Book of USSR*. 1st ed., 1978. (T)

7431 **Beloussova, L.S., Denisova, L.V. (1974).** Okhrana redkikh vidov rastenii v SSSR. (The protection of rare varieties of plants in the Soviet Union.) Moskva, Vniitetsel'khoz. 68p. Rus. (T)

UNION OF SOVIET SOCIALIST REPUBLICS

7432 **Beloussova, L.S., Denisova, L.V. (1980).** Botanicheskie zakazniki SSSR. (Botanical orders of the USSR.) Rukopis dep v VINITI no. 5010-80 Dep. Moskva, MSKh SSSR. 222p. Rus. (L)

7433 **Beloussova, L.S., Denisova, L.V., Nikitina, S.V. (1979).** [Rare plants of the U.S.S.R.] Moscow, Lesnaya Promyshlennost Pub. c.150p. Rus. (T)

7434 **Beloussova, L.S., Denisova, L.V., Nikitina, S.V. (1986).** [Rare plants of the world and their protection.] Moscow, International System of Scientific and Technical Information on Agriculture and Forestry. Rus. (T)

7435 **Bespalova, A.E. (1978).** [Regeneration of woody and shrubby plant species in protected plantings of the semidesert area]. *Lesnoe Khoz.*, 9: 51-54. Rus. Illus. (G)

7436 **Bezgodov, A.G. (1984).** Redkie rasteniya zapovednika "Basegi". (Rare plants of the "Basegi" Nature Reserve.) *In* Fiz. - geogr. osnovy razvitiya i razmeshch. proizvodit. sil. Nechernozem. Urala. Perm. 150-156. Rus. (P)

7437 **Bobrovskii, R.V. (1984).** Okhrana rastitel'nykh soobshchestv i redkikh vidov v Vologodskoi oblasti. (Protection of the plant associations and rare species in the Volgogod Region.) *In* Obshch. probl. okhrany rastit. Materialy Vses. soveshch. Okhrana rastit. mira sev. regionov, Syktyvkar, 7-9 sent., 1982. T.I. Syktyvkar. 60-62. Rus. (T)

7438 **Borodin, A.M., Isakov, Y., Krinitsky, V.V. (1984).** The system of natural protected areas in the USSR: biosphere reserves as part of this system. *In* Unesco-UNEP. Conservation, science and society. Paris, Unesco. 221-228. Contributions to the First International Biosphere Reserve Congress Minsk, Byelorussia/USSR, 26 September - 2 October 1983. Tables. (P)

7439 **Borodin, A.M., et al., eds (1985).** Krasnaya kniga SSSR: redkie i nakhodyashchiesya pod ugrozoi ischeznoveniya vidy zhivotnykh i rastenii. - izdanie vtoroe [2]: tom pervyi - vtoroi [1-2]. 2nd. Ed. Moscow, Lesnaya Promyshlennost. Rus. Illus., col. illus., maps. Vol. 1 - Animals: Vol. 2 - Plants. (T)

7440 **Chernaia, G.A. (1978).** [Findings of rare aquatic plants in the northern Donets River]. *Ukr. Bot. Zhurn.*, 35(5): 476-478. Uk (En, Rus). Map. (T)

7441 **Chertovskoi, V.G., ed. (1984).** Izuchenie i ockhrana rastitel'nosti severa. (Study and protection of plants of the north). Syktyvkar, Komi Fil. AN SSSR. 144p. Rus. Illus.

7442 **Chervonnyi, M.G. (1974).** Beregite les: ob organizatsii okhrany lesov i otvetstvennosti za lesonarusheniya. (Protect the forest: on the organization of protection of the forests and responsibilities for forest infringements.) Moscow, Moskovskii Rabochii. 58p. Rus.

7443 **Chupakhin, V.M., Zykov, K.D., Krinitskii, V.V. (1983).** Sarny-Chelekskii gornyi biosfernyi zapovednik. *Priroda Ezhem. Pop. Estest. Zhurn. Akad. Nauk*, 12: 48-55. Rus. Col. illus. (P)

7444 **Cinovskis, R. (1977).** [A collection of rare and valuable species of forest trees and shrubs in Cesis]. *Daildarznieciba*, 11: 16-18. Latv. (E/G)

7445 **Danilov, V.I., Nukhimovskaya, Yu.D., Shtilmark, F.R. (1983).** Problemy sozdraniya stepnykh zaponednikov v RSFSR. (Problems of reservation establishment in the steppe zone of the Russian Federation.) *Byull. Mosk. Obshch. Ispyt. Prir., Biol.*, 88(6): 92-104. Rus. (P)

7446 **Dengubenko, A.V. (1979).** [Some rare plants of West Pamir.] *Isw. Akad. Nauk Tadzh. SSR, Biol. Nauk*, 4: 88-89. Rus. (T)

7447 **Denisova, L.V. (1974).** [Rare and disappearing plants of the USSR.] Moskva, Lesnakila promyshlennost. 149p. Rus. (T)

UNION OF SOVIET SOCIALIST REPUBLICS

7448 **Denisova, L.V., Belousova, L.S. (1974).** Redkie i ischezaiushchie rasteniia SSSR. Moscow, Lesnaia Promyshlennost. 149p. Rus. Col. illus.

7449 **Denisova, L.V., Levichev, I.G., Nikitina, S.V., Tikhomirov, V.N. (1981).** Ob izuchennosti flory i rastitel'nosti zapovednikov Sovetskogo soynza. (Study of the flora and vegetation of nature reserves in the Soviet Union.) *Byull. Mosk. Obsch. Ispyt. Prir., Biol.,* 86(3): 135-145. Rus. Tables of numbers of publications from different divisions of botany in the nature reserves of the USSR. (P)

7450 **Didukh, Ya.P. (1978).** [Biomorphological structure of flora in the Yalta Mountain Forest State Reservation]. *Ukr. Bot. Zhurn.,* 35(5): 470-475. Uk (En, Rus). Illus. (P)

7451 **Didukh, Ya.P. (1981).** [Vegetation of the "Agarmysh" reservation.] *Ukr. Bot. Zhurn.,* 38(2): 96-101. Uk (En, Rus). Illus. (P)

7452 **Diuriagina, G.P. (1983).** [Problems of studying rare plants in botanical gardens.] *Byull. Glavn. Bot. Sada,* 1983 (129): 49-55. Rus. (G)

7453 **Dmitrieva, A.A. (1979).** [Conservation of natural flora of the Batumi coast and the ravine of Adzharistkali River]. *Byull. Glavn. Bot. Sada* (Moscow), 112: 46-48. Rus. (T)

7454 **Dolukhanov, A.G. (1980).** O zapovednikakh v svyazi s problemoi okhrany rastitel'nogo mira gornykh stran. (On the reserves and the problem of protection of the plant world in the mountain countries.) *Bot. Zhurn.,* 65(7): 1037-1040. Rus. (P/T)

7455 **Dulepova, B.I., Umanskaia, N.V. (1979).** [New habitats of plant species rare for Dauria]. *Bot. Zhurn.,* 64(8): 1199-1200. Rus. (T)

7456 **Dyrenkov, S.A., Krasnitskii, A.M. (1982).** Osnovnye funktsii zapovednykh territorii i ikh otrazhenie v rezhime okhrany lesnykh ekosistem. (Main functions of reservations and reflection on these functions in the policy of forest ecosystem protection.) *Byull. Mosk. Obshch. Ispyt. Prir., Biol.,* 87(6): 105-115. Rus. (P)

7457 **Dyuryagina, G.P. (1982).** Method of introduction of rare and vanishing plants. *Bot. Zhurn.* (Leningr.), 67(5): 679-687. (T)

7458 **Egorova, E.M, Chernianeva, A.M. (1980).** [Rare and disappearing species of plants from Sakhalin and Kurile Islands plant conservation.] *Byull. Glavn. Bot. Sada* (Moskva), 116: 64-72. Rus. (T)

7459 **Elias, T.S. (1983).** Rare and endangered species of plants: the Soviet side. *Science,* 219(4580): 19-23. Illus. (T)

7460 **Emel'ianova, V.G. (1975).** Okhrana zapovednikov, zakaznikov, pamiatnikov prirody. (The protection of nature preserves, game reservations, and natural monuments.) Moscow, Iuridicheskaia Literatura. 69p. (P)

7461 **Eremeeva, G.E. (1979).** Biologicheskie osobennosti i okhrana nekotorykh reliktovykh i redkikh rastenii vodoemov Priamur'ya. *Nauch. Trud. Kuibyshev. Gos. Ped. In-t.,* 229: 92-97.

7462 **Esipov, V.M. (1983).** Chatkal'skii gorno-lesnoi zapovednik. U.S.S.R., Uzbekistan. Rus.

7463 **Faiziev, A. (1985).** Sredneaziatskie endemichnye rasteniya v Yuzhnom Kyzylkume i voprosy ikh okhrany. (Middle Asian endemic plants in South Kyzylkum and problems of their protection). *In* Ekol. geogr. osnovy kompleks. osvoeniya Yuzh. Kyzylkuma. Tashkent. 31-42. Rus. (T)

7464 **Fedyaeva, V.V. (1977).** Material for the Red Book from the Rostov Oblast. *Izv. Sev-Kavk. Nauchn. Tsentra Vyssh. Shk. Estestv. Nauki,* 5(1): 105-108. (T)

7465 **Franklin, J.F., Krugman, S.L. (1979).** Selection, management and utilization of biosphere reserves. USDA, Forest Serv., Gen. Tech. Rep., PNW-82: 308p. Proc. of US/USSR Symp., Moscow May 1976. Illus. (P)

UNION OF SOVIET SOCIALIST REPUBLICS

7466 **Gagnidze, R.I., Zurebiani, B.G., Mukbaniani, M.V., Chelidze, D.T. (1978).** [New and rare species of the flora of Svanetia]. *Zam. Sist. Geogr. Rast.*, 35: 38-54. Geo (Rus). (P/T)

7467 **Gal, Zh. (1981).** [Nekotorye soobshchestva Dolin Ozer kak ob'ekty okhrany tsenofondov]. *BNMAU Shinzhlekh Ukhaany Akad.*, 5: 23-28. Mong (Rus). (T)

7468 **Gavva, I.A., Krinitsky, V.V., Yazan, Y.P. (1984).** Development of nature reserves and national parks in the U.S.S.R. *In* McNeely, J.A., Miller, K.R., *eds.* National parks, conservation and development. The role of protected areas in sustaining society. Washington, D.C., Smithsonian Institution Press. 463-465. (P)

7469 **Gavva, I.A., Yazan, Y.P. (1984).** Approaches to creation of an effective network of reserves in the USSR. *In* Unesco-UNEP. Conservation, science and society. Paris, Unesco. 33-36. Contributions to the First International Biosphere Reserve Congress, Minsk, Byelorussia/USSR, 26 September - 2 October 1983. (P)

7470 **Geideman, T.S., Nikolaeva, L.P., Simonov, G.P. (1980).** Konspekt flory zapovednika "Kodry". Mold, SSR, Shtiintsa. Rus.

7471 **Gensiruk, S.A. (1979).** Ratsional'noe prirodopol' zovanie. (Rational use of natural resources.) Moscow, Lesnaia Promyshlennost. 308p. Rus. Illus., maps.

7472 **Gerasimov, I.P. (1984).** Geosystem monitoring and its realization in biosphere reserves. *In* Unesco-UNEP. Conservation, science and society. Paris, Unesco. 435-438. Contributions to the First International Biosphere Reserve Congress, Minsk, Byelorussia/USSR, 26 September - 2 October 1983. Table. (P)

7473 **Giriaev, D.M. (1977).** [Protection of forests (from damage caused by man and fires) in a nationwide cause]. *Lesnoe Khoz.*, 11: 72-77. Rus.

7474 **Gogina, E.E. (1979).** USSR: the policies of botanic gardens and their activities in the conservation of threatened plants. *In* Synge, H., Townsend, H., *eds.* Survival or extinction. Proceedings of a conference held at the Royal Botanic Gardens, Kew, 11-17 September 1978. Kew, Bentham-Moxon Trust. 141-147. (G/T)

7475 **Golubev, V.N. (1977).** K metodike kolicheskvennogo izucheniya redkikh i ischezayushchikh rastenii flory Kryma. (On the methods of quantitative study of rare and vanishing plants in the Crimean flora.) *Byull. Gos. Nikit. Bot. Sada*, 1(32): 11-15. (En). (T)

7476 **Golubev, V.N. (1982).** [Procedure of ecological and biological studies of rare and vanishing plants in natural vegetation communities.] *Byull. Gos. Nikit. Bot. Sada*, 1982 (47): 11-16. Rus (En). (T)

7477 **Golubev, V.N., Golubeva, I.V. (1986).** The forest of *Juniperus excelsa* with a rare fern species *Cheilanthes persica* in south coast of the Crimea. *Bull. State Nikita Bot. Gard.*, 59: 12-13. (T)

7478 **Golubev, V.N., Korzhenevsky, V.V. (1986).** On the description methods of rare plant communities to be entered into the Green Book. *Bull. State Nikita Bot. Gard.*, 59: 5-8. (T)

7479 **Golubev, V.N., Kossykh, V.M. (1980).** The population-quantitative analysis of endemic, rare and threatened plants of the Crimea. *In* Dietmar Behnke, H., *ed.* Second International Congress of Systematic and Evolutionary Biology, the University of British Columbia, Vancouver, Canada, 17-24 July 1980. 214. Abstract. (T)

7480 **Golubeva, I.V. (1982).** Ob adventivnykh rasteniyakh zapovednika "Mys Martyan". (On adventive plants of the nature reserve "Cape Martyan".) *Byull. Gos. Nikit. Bot. Sada*, 49: 13-16. Rus (En). (P)

7481 **Gonchar, M.T. (1980).** Zembya: nasha kormilitsa. Lvov, Izd. L'vovskogo Universiteta. Rus.

UNION OF SOVIET SOCIALIST REPUBLICS

7482 **Gorchakovskii, P.L., Lalayan, N.T. (1982).** Relict black-alder forests of the Kazakh undulating plain and changes caused by human activity. *Sov. J. Ecol.*, 12(4): 201-212 (1981 publ. 1982). Description is given of unique relict *Alnus glutinosa* communities in the Bayan Aul mountain forest massif.

7483 **Gorchakovskii, P.L., Shurova, E.A., Kirshin, I.K. (1982).** Redkie i ischezaiuschchie rasteniia Urala i Priural'ia. (Rare and vanishing plants of the Urals and the Ural Mountain Region.) Moskva Nauka. 208p. Rus. Illus., maps. (T)

7484 **Gorchakovskii, P.L., ed. (1982).** Izuchenie i osvoenie flory i rastitel'nosti vysokogorii. 8. Vses. soveshch. Tez. dokl. 4: Rastitel'nye resursy, okhrana i ratsional'noe ispol'zovanie rastitelnogo mira vysokogorii. Sverdlovsk. 64p. Rus. Protection of the high altitude flora of U.S.S.R.

7485 **Gunin, P.D., Neronov, V.M., Veyisov, S.V. (1982).** Razvitie seti biosfernykh zapovednikov v Azii. (Development of biosphere reserves network in Asia.) *Probl. Osvoen. Pustyn'* (Turkm. SSR), 4: 7-24. Rus (En). Maps. (P)

7486 **Hryhora, I.M. (1981).** [Forest bog mak and the necessity of its conservation]. *Ukr. Bot. Zhurn.*, 38(5): 24-27. Uk (En, Rus). Illus.

7487 **Isakov, Yu.A. (1984).** Chto predstavlyayut soboi biosfernye zapovedniki? (What are "biosphere reserves?"). *Byull. Mosk. Obshch. Ispyt. Prir., Biol.*, 89(4): 20-26. Rus (En). (P)

7488 **Isakov, Yu.A., Krinitsky, V.V. (1980).** [A system of nature conservation territories in the USSR: structure and development trends.] *Izvest. Akad. Nauk SSSR, Ser. Geogr.*, 3: 46-52. Rus. (P)

7489 **Ishankuliev, M. (1983).** Flora Repetekskogo biosfernogo zapovednika. (The flora of the Repetet Biosphere Reserve.) Ashkhabad, Ylym. 37p. Rus. (P)

7490 **Ivanova, M.M. (1978).** [New and rare species in the flora of the Upper Angara Valley]. *Bot. Zhurn.*, 63(12): 1721-1730. Rus (En). (T)

7491 **Izrael, Y. (1984).** The concept of ecological monitoring and biosphere reserves. *In* Unesco-UNEP. Conservation, science and society. Paris, Unesco. Vol. 2, Annex. II, 1-8. (P)

7492 **Izrael, Y., Rovinsky, F.Y., Filippova, L.M. (1984).** Integrated monitoring of background pollution and its ecological effects in biosphere reserves. *In* Unesco-UNEP. Conservation, science and society. Paris, Unesco. 404-410. Contributions to the First International Biosphere Reserve Congress, Minsk, Byelorussia/USSR, 26 September - 2 October 1983. (P)

7493 **Izrael, Y., Rovinsky, F.Z., Gorokhov, W.A. (1984).** Ecological monitoring and biosphere reserves in the USSR. *In* McNeely, J.A., Miller, K.R., *eds.* National parks, conservation and development. The role of protected areas in sustaining society. Washington, D.C., Smithsonian Institution Press. 466-469. (P)

7494 **Kalinin, S.D., Lapina, L.G. (1979).** [Reservations, herbaria and botanical gardens of the central part of the middle and lower Volga area]. *Byull. Glavn. Bot. Sada* (Moscow), 112: 81-83. Rus. (G/P)

7495 **Kamelin, R.V. (1978).** Printsipy otbora redkikh vidov rastenii dlya krasnoi knigi. (Principles for the selection of rare species of plants for the Red Book.) *In* Latvia. Akademiya Nauk Latviiskoi SSSR. Institut Biologii Rastitel'nyi mir okhranyaemykh territorii Riga, Zinatne. 60-67. Rus. (T)

7496 **Karpenko, A.S., Stavrova, N.I. (1980).** [Protecting the plant world in the non-Chernozem Zone]. Leningrad, Nauka. 112p. Rus. Illus., map. (T)

7497 **Karpenko, A.S., Vetsel', N.K. (1980).** Soveshchanie "Problemy okhrany pamyatnikov prirody y sostave pamyatnikov istorii i kultury" (28-31 viii 1979, Pskov). (On the conference "Problems in protection of natural relics as parts of the historical and cultural monuments" [28-31 August 1979, Pskov].) *Bot. Zhurn.*, 65(11): 1665-1668. Rus. (P)

UNION OF SOVIET SOCIALIST REPUBLICS

7498 **Karpisonova, R.A. (1979).** Rare species of herbaceous plants from broadleaved forests of the USSR in the Central Botanical Garden. *Byull. Glavn. Bot. Sada* (Moscow), 112: 54-59. Illus. (G/T)

7499 **Kasach, A.E. (1979).** [Rare herbaceous plants from western areas of the Gorno-Badakhshan region and possibilities of their introduction]. *Probl. Bot.*, 41(2): 170-175. Rus. (T)

7500 **Kharkevich, S.S. (1979).** [Conservation of the (Soviet) Far East flora]. *Priroda*, 8: 63. Rus. (T)

7501 **Kharkevich, S.S., Kachura, N.N. (1981).** Redkie vidy rastenii Sovetskogo dal'nego vostoka i ikh okhrana. (Rare plant varieties of the Soviet Far East and their protection.) Moscow, Nauka. 230p. Rus. Illus., maps. (T)

7502 **Khokhlov, A.M., Solod'ko, A.S. (1979).** Caucasian biosphere reservation and problems of nature conservation in north-western Caucasus. *Priroda*, 2: 58-69. Illus., map. Includes plants. (P)

7503 **Khrapko, O.V. (1979).** [Analysis and ways of conservation of rare and vanishing species of the Far East flora]. *Byull. Glavn. Bot. Sada* (Moscow), 112: 50-53. Rus. Illus. (T)

7504 **Kiseleva, A.A. (1979).** [New and rare plants from the foothills of the eastern Sayan Mountains: 1]. *Izvest. Sibir. Otdel. Akad. Nauk SSR, Biol. Nauk*, 3: 59-62. Rus (En). (T)

7505 **Knyazev, M.S. (1984).** Sokhranenie nekotorykh redkikh vidov rastenii Urala v antropogennykh landshaftakh. (Preservation of some rare species of plants of the Urals, in anthropogenic landscapes.) *In* Ekol. aspekty optimiz. tekhnogen. landshaftov. Sverdlovsk. 36-39. Rus. (T)

7506 **Knystautas, A. (1987).** The natural history of the USSR. London, Century. 224p. Col. illus., maps.

7507 **Kolesnikov, B.P. (1979).** The problems of the conservation of the plant world in the USSR. *In* Lebedev, D.V., ed. Proceedings of the International Botanical Congress, Leningrad, 3-10 July 1975. Leningrad, Komarov Botanical Institute. 92-104. (T)

7508 **Kolesnikov, B.P., Semenova-Tyan-Shanskaya, A.M., Parfenov, V.I., Boch, M.S. (1979).** Okhrana rastitel'nogo mira v SSSR (obzor issledovanii). (Protection of plant world in the U.S.S.R. [survey of studies].) *Bot. Zhurn.*, 64(7): 1051-1064. Rus. (T)

7509 **Komarov, B. (1978).** The destruction of nature in the Soviet Union. London, Pluto Press. 150p.

7510 **Komendar, V.I., Solod'ko, A.S. (1982).** K okhrane redkikh vidov flory Severo-Zapadnogo Kavkaza i Karpat. (On the protection of rare species of the flora of north-west Caucasus and the Carpathians.) *In* Izuch i osvoenie flory i rastit. vgsokogorii. 8 Vses. soveshch. Tez. dokl. 4: Rastit. resursy okhrana i rats. ispol'z rastit. mira vysokogorii. Sverdlovsk. 26. Rus. (T)

7511 **Kondratiuk, I., Ivashyn, D.S., Burda, R.I. (1978).** [Study of the flora and vegetation of the Stanitsa-Lugansk branch of the Lugansk State Reservation]. *Introd. Aklim. Rosl. Ukr. Akad. Nauk. URSR*, 13: 18-22. Uk (Rus). (P)

7512 **Kondratiuk, I., Ivashyn, D.S., Kharkhota, H.I., Chupryna, T.T. (1978).** [Useful plants of the Provalskaya Steppe and prospects of their utilization]. *Introd. Aklim. Rosl. Ukr. Akad. Nauk. URSR*, 13: 3-6. Uk (Rus). Conservation aspects. (E)

7513 **Konstantinova, N.A. (1978).** [The rare liverwort species *Sphenolobopsis pearsonii* (Spruce) Schust. in Khibiny Mountains]. *Bot. Zhurn.*, 63(7): 1032-1035. Rus. (T)

UNION OF SOVIET SOCIALIST REPUBLICS

7514 **Korovina, O.N. (1978).** Materialy dlya *"Krasnoi Knigi"*. (Material for "the Red Book".) *Trudy Prikl. Bot. Genet. Selek.*, 62(1): 50-54. Rus. (T)

7515 **Korovina, O.N. (1980).** Organizatsiya zapovednikov i zakaznikov v SSSR: osnova sokhraneniya populyatsii dikikh sorodichei kul'turnykh rastenii. (Establishment of reserves and protected natural lands of the USSR: the basis for conservation of populations of cultivated plant relatives.) *Trudy Prikl. Bot. Genet. Selek.*, 68(3): 145-150. Rus (En). (E/P)

7516 **Kosykh, V.M., Korzhenevskii, V.V. (1978).** [New and rare species of the flora of the Crimea]. *Byull. Glavn. Bot. Sada* (Moscow), 108: 28-30. Rus. (T)

7517 **Kosykh, V.M., Korzhenevskii, V.V. (1979).** [Some rare and neglected ferns of Crimea]. *Bot. Zhurn.*, 64(8): 1197-1199. Rus. *Cheilanthes persica, Notholaena marantae, Phyllitis scolopendrium, Anogramma leptophylla, Asplenium germanicum* ssp. *heuflerii*. (T)

7518 **Kotliakov, V.M., Suprunenko, I. (1979).** [Creation of high mountain glacial national parks]. *Izv. Akad. Nauk SSSR*, 5: 25-32. Rus. Illus., map. (P)

7519 **Kotlyarov, I.I. (1977).** Zashchitnaya rol' gornykh Okhotskogo poberezh'ya. (The protective role of the montane forests of the Okhotsk coast.) *Lesnoe Khoz.*, 1977(1): 44-46. Rus. (T)

7520 **Kotlyarov, I.I. (1980).** [Improving the conservation and regeneration of scrub pine (*Pinus pumila*) thickets (in the USSR).] *Lesnoe Khoz.*, 1980(8): 29-31. Rus. (T)

7521 **Kotsuenko, V.D. (1979).** Rare and endangered plants of the flora of the Adygei Autonomous Oblast Russian-SFSR USSR and problems of their protection. *In* Reports of the 2nd regional student scientific conference of North Caucasus Universities, Russian-SFSR, USSR. *Izv Sev-Kavk Nauchn Tsentra Vyssh Shk Estestv Nauki*, 7(1): 103. (T)

7522 **Kovalev, P.V., Pugach, E.A., Shevernozhuk, R.G. (1980).** [Tasks in the conservation and rational utilization of the forest gene pool]. *Lesnoe Khoz.*, 1980(4): 64-67. Rus. (E)

7523 **Kovda, V., Kerzhentsev, A., Zablotskaya, L. (1984).** The biosphere reserve on the Oka river in central Russia. *In* U.S.S.R. Academy of Sciences. Man and biosphere. Moscow, Nauka. 188-196. Map. (P)

7524 **Kovshar, A.F., Ivashchenko, A.A. (1980).** Zapovednik Aksu-Dzabagly. (The Aksu-Dzabagly Nature Reserve.) KazSSR, Kainar. Rus. (P)

7525 **Kozhevnikov, Yu P. (1983).** O predgor'yakh i gorakh Byrranga. (On the foothills and mountains of Byrrang.) *Priroda Ezhem. Pop. Estest. Zhurn. Akad. Nauk SSSR.*, 7: 36-41. Rus. Col. illus.

7526 **Krasnitskii, A.M. (1978).** Istoricheskie i funktsional'nye predposylki organizatsii gosudarstvennoi zapovednoi sluzhby. (Historical and functional prerequisites for the organization of a state native reserve service.) *Byull. Mosk. Obshch. Ispyt. Prir., Biol.*, 83(4): 152-161. Rus. (P)

7527 **Krasnitskii, A.M. (1978).** Osnovye zadachi spetsializatsii zapovednogo dela. (Foundation problems of specialization in the nature reserve business.) *In* Latvia. Akademiya Nauk Latviiskoi SSR. Institut Biologii. Rastitel'nyi mir okhranyaemykh territorii. Riga, Zinatne. 29-33. Rus. Includes nature reserve organization. (P)

7528 **Krasnitskii, A.M. (1983).** Problemy zapovednogo dela. (The problems of nature reserves: their organization in the U.S.S.R.) Moscow, Lesnaya Prom-st'. Rus. (P)

7529 **Krasnov, N.A. (1979).** Materialy k analizu flory Volzhsko-Kamskogo Zapovednika. (Materials for the analysis of Volga-Kama Reservation flora.) *Bot. Zhurn.*, 64(10): 1481-1485. Rus. (P)

UNION OF SOVIET SOCIALIST REPUBLICS

7530 **Krauklis, A.A., Mikhailev, Yu.P. (1979).** Rastitel'nost' Taigi i okhrana sredy. (Taiga vegetation and nature conservation.) *Bot. Zhurn.*, 64(11): 1674-1681. Boreal forest region south of the tundra.

7531 **Krivolutskk, D.A., ed. (1984).** Vliyanie promyshlennykh predpriyatii na okruzhayushchuyu sredy. Vses. shk. Zvenigorod, 4-8 dek., 1984: tez. dokl. (Influence of industrial undertakings on the surrounding environment.) Pushchino. 240p. Rus. Illus.

7532 **Kryuchkov, V.V. (1984).** Strategiya okhrony prirody severa. (Strategy for the protection of nature in the north.) *Prir. Ezhem. Pop. Estest. Zhurn. Akad. Nauk*, 1: 38-50. Rus. Col. illus., map. North of the USSR. Map of nature reserves in Kolskoi Peninsula. (P)

7533 **Kucheneva, G.G. & A.E. (1983).** Kurshskaya kosa. (The Kurshskaya Spit.) *Prir. Ezhem. Pop. Estest. Zhurn. Akad. Nauk*, 8: 44-52. Rus. Illus., col. illus., map.

7534 **Kucherov, E.V. (1985).** Ob okhrane i obogashchenii genofonda rastenii na Yuzhnom Urale. (On protection and enrichment of the gene fund in South Urals.) Ufa, In-t Biol. 15p. Rus.

7535 **Kucherov, E.V., Muldashev, A.A., Galeeva, A.K. (1987).** Okhrana redkikh vidov rastenii na Yuzhnom Urale. (Protection of rare species of plants in the South Urals.) Moskva, Nauka. 202p. Rus. Illus., maps. (T)

7536 **Kudinov, K.A., Kostyleva, N.I., Saksonov, S.V. (1984).** Rasteniya Zhigulevskogo zapovednika, redkie dlya flory SSSR. (Flora of the Zhigulevsk Nature Reserve, rare for the flora of the USSR.) *In* Probl. okhrany genofonda i upr. ekosistemami v zapoved. step. i pustyn. zon. Tez. dokl. Vses. soveshch., Askaniya-Nova, 21-25 maya 1984.Moskva. 138-143. Rus. (P)

7537 **Kuliev, A.N., Starkov, V.G. (1982).** O nekotorykh redkikh soobshchestvakh Polyarnogo Urala. (On some rare associations of the Polar Urals.) *In* Okhrana redk. rastit. soobshchestv. Moskva. 17-23. Rus (En). (T)

7538 **Kurentsova, G.E. (1981).** Sosudistye rasteniya ostrovov Dal'nevostochnogo gosudarstvennogo morskogo Zapovednika. (Vascular plants of the islands of the Far East State Marine Sea nature reserves). *In* Tsvetkov. rast. o-vov. Dal'nevost. mor.Zapoved. Vladivostok. 34-61. Rus. (P)

7539 **Kurnaev, S.F. (1980).** Tul'skie zaseki, ikh priroda nauchnoe i khozyaistvennoe znachenie. (Tula tree felling, its nature, scientific and agricultural significance.) *Prir. Ezhem. Pop. Estest. Zhurn. Akad. Nauk*, 3: 86-99. Illus., col. illus., maps.

7540 **Kuvaev, V.B., Denisova, L.V., Poshkurlat, A.P. (1981).** O printsipakh okhrany dikorastushchikh poleznykh rastenii (na primere lekarstvennykh rastenii). (On the principles of preserving wild growing useful plants [the example of medicinal plants].) *Rast. Resursy*, 17(2): 272-281. Rus. Map. (E)

7541 **L'vov, L. (1980).** Materialy dlya *"Zelenoi knigi Dagestana"*. (Materials for the Green Book of Dagestan.) *In* Rastitel'nyi pokrov. Dagestana i ego okhrana. Makhachkala. 11-19. Rus. (T)

7542 **L'vov, P.L. (1979).** Conservation of rare plants and phytocoenoses in Dagestan. *Priroda*, 3: 80-87. Illus. (T)

7543 **Laivins, M., Kreile, V. (1980).** Augu sugu kartesana aizsargajamas teritorijas. (Mapping plants on conservation territories.) *Mezsaimnieciba un Mezrupnieciba*, 1980(3): 29-30. Latv (Rus). Illus.

7544 **Lanyi, G. (1979).** A sivatag varazsa: Turkmen tudosok eredmenyei a homoksivatag hasznositasaban. *Buvar*, 34(1): 4-14. Hu. Illus., map. Rezervatumok a Karakumban.

7545 **Lapin, P.I. (1975).** Our endangered environment - the Russian view: rare and endangered plant species in the USSR. *Gard. J.*, 25(6): 171-175. (T)

UNION OF SOVIET SOCIALIST REPUBLICS

7546 **Lapin, P.I. (1977).** Contributions of botanical gardens of the USSR to enrichment of plant resources of the country. *Biol. Bull. Aca. Sci. USSR*, 4(5): 557-568. (G)

7547 **Lapin, P.I., ed. (1984).** Rol' introduktsii v sokranenii genofonda redkikh i ischezayushchikh vidov rastenii. (The role of introductions into the gene fund collections of rare and disappearing species of plants.) Moskva, Nauka, 197p. Rus. Illus. (G/P/T)

7548 **Latvia: Akademiya Nauk Latviiskoi SSR. Sovet Botanicheskikh Sadov Pribaltiiskogo Regiona (1977).** Botanicheskie sady pribaltiki: okhrana rastenii. (Horti botanici Baltici: conservatio plantarum.) (Botanic gardens of the Baltic: the protection of plants.) Riga, Zinatne. 226p. Rus. Illus., maps. (G/T)

7549 **Lenisova, L.V., Levichev, I.G., Nikitina, S.V. (1978).** O botanicheskoi izuchennosti okhranyaemykh territorii Sovetskogo Soyuza. (On botanical studies of protected territories of the Soviet Union.) *In* Latvia. Akademiya Nauk Latviiskoi SSR. Institut Biologii. Rastitel'nyi mir okhranyaemykh territorii. Riga, Zinatne. 83-87. Rus. (P)

7550 **Leon, C. (1981).** Red Data Books for the USSR. *Threatened Pl. Commit. Newsl.*, 8: 18-19. (T)

7551 **Luks, I. (1978).** [Classification of Orchidaceae species in the Crimean flora according to their rarity]. *Byull. Gos. Nikit. Bot Sada*, 3: 15-18. Rus (En). Plant conservation. (T)

7552 **Luks, Yu.A., Kryukova, I.V. (1979).** [The USSR Red Data Book; the rare and disappearing animal and plant species, 1978.] *Bot. Zhurn.*, 64(12): 1825-1832. Rus. A review. (T)

7553 **Luks, Yu.A., Privalova, L.A., Kryukova, I.V. (1980).** Zapovednye rasteniya Kryma. (Protection of plants in the Crimea.) USSR, Tavriya. Rus. (T)

7554 **Makarov, F.N. (1978).** [Kinds of felling and conservation of mountain forests]. *Lesnaia Prom.*, 10: 20. Rus.

7555 **Makhmedov, A.M. (1980).** Rod *Salvia* (Lamiaceae) material dlya Krasnoi Knigi SSSR. (The genus *Salvia* (Lamiaceae): the material for the Red Book of the USSR.) *Bot. Zhurn.*, 65(8): 1208-1211. Rus. Maps. (T)

7556 **Malyshev, L.I. (1980).** Soveshchanie po aktual'nym voprosam okhrany rastitel'nogo mira Sibiri (Novosibirsk, 4-5 Maya 1979). (Conference on the urgent problems in protection of the Siberian plant world [Novosibirsk, 4-5 May 1979].) *Bot. Zhurn.*, 65(5): 758-760. Rus. (T)

7557 **Malyshev, L.I., Sobolevskaya, K.A. (1983).** O programme "Biologiya nuzhdayushchikhsya v Gosudarstvennoi okhrane rastenii Sibiri". (On the programme "Biology of Siberian plants for state conservation".) *Bot. Zhurn.*, 68(4): 541-545. Rus. (T)

7558 **Mamaev, S.A. (1980).** [Preservation and rational use of the natural resources of the Ural Mountains.] Sverdlovsk, Academy of Sciences of the USSR, Ural Scientific Center. Vol. 4: Flora and Fauna.

7559 **Mamaev, S.A. (1986).** Problemy i dostizheniya okhrany genofonda rastenii na Urale. (Problems and achievements in the protection of the gene pool of plants in the Urals.) *Byull. Glavn. Bot. Sada* (Moscow), 140: 37-41. Rus.

7560 **Mamaev, S.A., ed. (1980).** Nauchnye osnovy razmeshcheniya prirodnykh, rezervatov Sverdlovskoi oblasti. (Scientific foundations of the siting of nature reserves of the Sverdlovsk Region.) Sverdlovsk, Uralsk. Nauch. Tsentr. 152p. Rus. Illus. (P)

UNION OF SOVIET SOCIALIST REPUBLICS

7561 **Marek, S., Koziol, E. (1980).** Materialy do rozmieszczenia rzadkick roslin torfowiskowych w wojewodztuie koszalinskim. (New data on the distribution of rare bog and water plants in the Koszalin District.) *Badan Fizjogr. Pol. Zachod., B,* 31: 181-185 (1979 publ. 1980). Pol (En). Maps. (T)

7562 **Martusova, E.G. (1981).** [Vegetation of the Baikal State Reservation and its study]. *Okhr. Ras. Mira Sib.,* 1981: 110-116. Rus. (P)

7563 **Mel'nikova, A.B. (1979).** [New and rare species of the flora from Khekhtsir.] *Byull. Glavn. Bot. Sada,* 113: 63-66. Rus. (T)

7564 **Mel'nikova, A.B. (1983).** [New data on the flora of Bolshekherhtsirsk Reserve.] *Bot. Zhurn.,* 68(7): 932-938. Rus. (P)

7565 **Migunova, E.S. (1983).** [Problems of steppe afforestation and nature conservation in the "Lesnoi zhurnal" ("Forestry journal").] *Lesnoe Khoz.,* (8): 72-73. Rus.

7566 **Mikheev, A.D. (1979).** Nekotorye voprosy okhrany botanicheskikh ob'ektov raiona Kavkazskikh Mineral'nykh Vod. (Some problems of the conservation of plants in the region of the Caucasian mineral waters.) *Byull. Mosk. Obshch. Ispyt. Prir., Biol.,* 84(2): 101-110. Rus (En). (T)

7567 **Mikheeva, N.N. (1979).** [Areas of endemic and rare steppe species of Altai]. *Bot. Mater. Gerb. Inst. Bot. Akad. Nauk Kazakhshoi SSR,* 11: 22-38. Rus. Maps. (T)

7568 **Mirimanian, K.P. (1976).** Measures for protection and rational land utilization in arid mountain regions. *In* Proceedings of an International Meeting on Ecological Guidelines for the use of Natural Resources in the Middle East and South West Asia, Persepolis, Iran, 24-30 May 1975. IUCN Publ. New Series No. 34. Morges, IUCN. 224-226. (P)

7569 **Molchanov, E.F., Shcherbatyuk, L.K., Golubev, V.N., Kosykh, V.M. (1983).** Aktual'nye voprosy sovershenstvovaniya seti zapovednykh territorii v Krymu. (Urgent problems of improving reserved territory in the Crimea.) *Byull. Gos. Nikit. Bot. Sada,* 52: 5-9. Rus (En). (P)

7570 **Molchanov, E.F., ed. (1980).** Izuchenie prirodnykh kompleksov yuzhnogo berega Kryma v svyazi s ikh okhranoi. (Study of natural complexes of the Crimean south coast as related to their conservation.) *Trudy Gos. Nikit. Bot. Sada,* 81: 96p. Rus (En).

7571 **Muizharaya, E.Ya., Plaudis, A.A., Kazaka, R.M., Limberva, R.E. (1983).** Semennoe razmnozhenie redkikh vidov rastenii v natsional'nom parke "Gauya" s tselo sokhraneniya genofonda flory. (Seed reproduction of rare species of plants in the national park "Gauya" with the whole preservation of the flora's gene fund.) *In* Okhrana flory rech. dolin v Pribalt. resp. Riga. 86-88. Rus. (P/T)

7572 **Nagalevskii, V.Ya. (1980).** O solonchakovoi rastitel'nosti del'ty Kubani, ee okhrane i ratsional'nom ispol'zovanii. (On the salt marsh vegetation of the Kuban Delta, its protection and its rational utilization.) *Izv. Sev. Kavkaz. Nauch. Tsentra Vyssh. Shkoly Estestv. Nauk.,* 2: 76-77. Rus.

7573 **Nasimovich, A.A. (1979).** Okhrana prirody v Berezinskom Zapovednike. (Conservation of nature in the Berezinsk Nature Reserve.) *Byull. Mosk. Obshch. Ispyt. Prir., Biol.,* 84(3): 111-113. (P)

7574 **Nasimovich, A.A., Isakov, Y. (1985).** Preservation of model ecosystems in reserves, problems and their possible solutions. *In* Unesco-UNEP. Conservation, science and society. Paris, Unesco. 265-270. Contributions to the First International Biosphere Reserve Congress, Minsk, Byelorussia/USSR, 26 September - 2 October 1983. Illus. (P)

UNION OF SOVIET SOCIALIST REPUBLICS

7575 **Natkevichaite-Ivanauskene, M. (1979).** Otlichitel'nye rasteniya i fitotsenozy otdel'nykh landshaftov Litvy i vopros otbora trebuyushchikh okhrany botanicheskikh ob'ektov. (Distinctive plants and the phytocoenoses of ... Lithuania and the question of selection of the required protection of plants.) *In* Formir. rastitel'n. pokrova pri optimiz. landschafta Materialy 2-i Vses. Shkoly Kaunas, 1979. Vilnyus. 63-65. Rus.

7576 **Nepomilueva, N.I., Alekseeva, R.N. (1984).** Zonal'no-provintsial'nyi analiz botanicheskikh okhranyaemykh v Komi SSSR. (Zonal provincial analysis of botanical protected regions in the Komi SSSR.) *In* Obshch. probl. okhrany rastit. Materialy Vses. soveshch. Okhrana rastit. mira sev. regionov, Syktyvkar, 7-9 sent., 1982. T.I. Syktyvkar. 52-59. Rus.

7577 **Nikitina, S.V. (1980).** [Rare and endangered plants of the Soviet Union and their preservation]. Moscow, Vses. Nauchno-Issledovatel'skii Inst. Inform. Tekhniko-Ekon. Issledovanii Sel'skomu Khoz. 53p. Rus. (T)

7578 **Nikolajenko, W. (1978).** Walderneuerung in der UDSSR zum Schutze der Umwelt. (Forest renovation in the USSR for environmental protection.) *Int. Zeitschr. Landwirtsch.*, 6: 589-591. Illus.

7579 **Odnoralov, V.S. (1978).** [Introduction of felling methods for nature conservation.] *Lesnaia Promyshl.*, 10: 7-8. Rus. Illus.

7580 **Odnoralov, V.S. (1978).** [Nature - conservation methods of felling for mountain forests]. *Lesnoe Khoz.*, 10: 21-24. Rus.

7581 **Olsauer, J. (1978).** [Conservation of nature in the U.S.S.R.]. *Pamatky Prir.*, 5: 312-313. Cz. Illus.

7582 **Orlov, A.Ya., Abaturov, Yu.D., Pis'merov, A.V. (1980).** [The last area of virgin spruce forests in the southern taiga on the Russian plain.] *Lesovedenie*, 4: 38-45. Rus (En).

7583 **Parfenov, V.I., Liakavitsius, A.A., Kozlovskaia, N.V., Vinaev, G.V., Iankiavitsenee, R.L., Baliavitsenee, U.U., Lazdayckaite, Z., Lapele, M. (1987).** [Rare and threatened plant species of Byelorussia and Lithuania.] Akademiya Nauk Byelorusskoi SSR. Minsk, Nauka i Tekhnica. 352p. Rus. (T)

7584 **Platnikova, L.S. (1984).** Okhrana dendroflory SSSR i ee zadachi. (The protection of the dendroflora of the USSR and its aims.) *In* Lapin, P.I., *ed.* Rol' introduktsii v sokhranenii genofonda redkikh i izchezayushchikh vidov rastenii. Moskva, Nauka. 21-68. Rus. (E/T)

7585 **Plotnikov, V.V. (1978).** Genezis malykh ozer Il Istmenskogo Zapovednika (Predlesostepnoe Zaural Iste). (Development of small lakes in the Il Istmen Nature Reserve [Transural forest-steppe].) *Trudy Inst. Ekol. Rast. Zhivotn*, 108: 15-34. (P)

7586 **Plotnikova, L.S. (1980).** Introduktsiya redkikh vidov drevesnykh rastenii SSSR v Glavnom Botanicheskom Sadu AN SSSR. (Introduction of rare species of woody plants of the USSR into Main Botanical Garden, Moscow.) *In* U.S.S.R. AN SSSR Glavnyi Botanicheskii Sad. Introduktsiya drevesnykh rastenii. Moskva, Nauka. 35-48. Rus. (G/T)

7587 **Plotnikova, L.S. (1983).** Dendroflora zapovednikov S.S.S.R. (Dendroflora of nature reserves in the USSR.) *In* Tez. dokl. 7 Delegat. s'ezda Vses. botan. o-va, Donetsk, 11-14 maya, 1983. Leningrad. 284-285. Rus. (P)

7588 **Polyakova, G.A. (1979).** Rekreatsiya i degradatsiya lesnykh biogeotsenozov. (Recreation and the degradation of forest biogeocoenoses.) *Lesovedinie*, 3: 70-80. (En).

7589 **Ponomarev, A.N., Kamelin, R.V., Dem'yanova, E.I. (1983).** Konspekt flory Troitskogo lesostepnogo zapovnednika. (Conspectus of the flora of the Troitsk Forest Steppe Nature Reserve [Ural Region].) Perm', Perm. Un.t. 76p. (Rukopis' dep. v VINITI, no. 5987-83 Dep.) Rus. (P)

UNION OF SOVIET SOCIALIST REPUBLICS

7590 **Priimak, D.P. (1980).** Bayanaul Zapovednik. (Bayanaul Nature Reserve.) KazSSR, Kazakhstan. Rus. (P)

7591 **Prilipko, L.I., Gozina, E.E. (1978).** Rare species of the natural flora of Talysh deserving protection. *Byull. Glavn. Bot. Sada* (Moscow), 108: 62-68. (T)

7592 **Pronin, M.I. (1977).** [Effect of recreation on tree stands and fauna in forest parks]. *Lesnoe Khoz.*, 10: 68-70. Rus. (P)

7593 **Punonin, A.I., Gruzdev, G.S. (1983).** Programmnye lesa i voprosy okhrany prirody: sbornik nauchnykh trudov. (The programmed forest and questions concerning nature conservation.) Moskva, TSKhA. 103p. Rus. Illus.

7594 **Radzhi, A.D. (1982).** K okhrane endemikov i rekikh vidov vysokogornogo Dagestana. (On the protection of endemic and rare species of high mountains of Dagestan.) *In* Izuch i osvoenie flory i rastit. vgsokogorii. 8 Vses. soveshch. Tez. dokl. 4: Rastit. resursy okhrana i rats. ispol'z rastit. mira vysokogorii. Sverdlovsk. 42. Rus. (T)

7595 **Reimers, N.F., Shtil'mark, F.R. (1978).** Osobo okhranyaemye prirodnye territorii. (Specially protected natural territories.) Moskva, Nauka. 295p. Rus. (P)

7596 **Roost, V. (1986).** Orchids in the U.S.S.R.: conservation, research and cultivation. *Amer. Orchid. Soc. Bull.*, 55(6): 591-594. Illus., col. illus., map. (T)

7597 **Rusanov, G.M. (1983).** [Present state of naturally vegetated lands in the outfall area of the Volga River offshore and prospects of their further changes.] *Byull. Mosk. Obshch. Ispyt. Prir., Otdel Biol.*, 88(5): 10-21. Rus (En).

7598 **Rysin, L.P., Savel'eva, L.I. (1978).** [Forest reservations (conservation)]. *Lesnoe Khoz.*, 12: 18-21. Rus.

7599 **Semenova-Tyan-Shanskaya, A.M. (1981).** Rezhim okhrany rastitel'nogo pokrova zapovednykh territorii. (The regime of protection of the vegetational cover in the reserve territories.) *Bot. Zhurn.*, 66(7): 1060-1067. Rus. (P)

7600 **Seredin, R.M. (1980).** [Protection-requiring rare and vanishing higher cryptogamous & gymnospermous species and a class of monocotyledonous flowering plants as material for the Red Book of northern Caucasus, Ciscaucasia & the Dagestan ASSR, Russian SFSR, USSR.] *Izv. Sev-Kauk. Nauchn. Tcentra Vyssh Shk Estestv. Nauk.*, 2: 90-94. Rus. (T)

7601 **Sevel'eva, L.I. (1980).** [Forest reserves as a form of environmental protection, using the Moscow, U.S.S.R., district as an example]. *Lesovedenie*, 4: 29-37. Rus (En). (P)

7602 **Shcherbova, M.A., Shemetova, N.S. (1981).** New sites of discovery of rare species of plants in the Khabarovsk territory. *Isv. Sibirsk. Otdel. Akad. Nauk SSSR, Ser. Biol. Nauk*, 10: 7-9. Rus (En). Illus. (T)

7603 **Shelyag-Sosonko, Yu.R., Didukh, I. (1978).** [Essay of flora and vegetation of the Yalta Mountain Forest State Reservation: 2. Flora]. *Bot. Zhurn.*, 63(9): 1285-1301. Rus (En). Illus., map. (P)

7604 **Shelyag-Sosonko, Yu.R., Didukh, Ya.P. (1980).** Yaltinshii Gorno-Lesnoi Gosudarstvennyi Zapovednik: botaniko-geograficheskii ocherk. (The Yalta Mountain Forest State Nature Reservation.) Kiev, Naukova Dumka. 183p. Rus. Col. illus., map. (P)

7605 **Shelyag-Sosonko, Yu.R., Didukh, Ya.P., Zhizhin, N.P. (1982).** Elementarnaya flora i problema okhrany vidov. (Elementary flora and the problem of species protection.) *Bot. Zhurn.*, 67(6): 842-852. Rus. Map. (T)

7606 **Shelyag-Sosonko, Yu.R., Sytnik, K.M., et al. (1980).** [Ukraine, White Russia, Moldavia. The protection of important botanical regions.] Kiev, Naukova Dumka. 332-370. Rus. Includes lists of threatened plants by region. (P/T)

UNION OF SOVIET SOCIALIST REPUBLICS

7607 **Sheveloukha, V.S. (1984).** Nature conservation in the USSR. *In* Unesco-UNEP. Conservation, science and society. Paris, Unesco. Vol. 2, Annex I, 1-8. Contributions to the First International Biosphere Reserve Congress, Minsk, Byelorussia/USSR, 26 September - 2 October 1983. (P)

7608 **Shilov, M.P. (1980).** [Mapping of the distribution of vanishing plant species for conservation purposes]. *Byull. Mosk. Obshch. Ispyt. Prir. Otdel. Biol.*, 85(4): 89-91. Maps. Klyazma river valley. (T)

7609 **Shtarker, V.V. (1988).** Flora yuzhnogo i yugo-vostochnogo makrosklonov glavnogo mezhdurech'ya zapovednika "Stolby". (Flora of S. and S.E. slopes of the State "Interriverine" Reserve "Stolby". Krasnoyarsk Region.) *Trudy Gos. Zapoved. Stolby*, no. 15: 3-7. Rus. (P/T)

7610 **Shul'kina, T.V. (1982).** [Distribution of endemic and rare *Campanula* species of the USSR flora and prospects of their introduction.] *Byull. Glavn. Bot. Sada*, 1982(124): 39-43. Rus. Maps. (T)

7611 **Shur-Bagdasarian, E.F., Avanesov, A.A. (1979).** [Change of vegetation on eroded pastures of steppes under reservation conditions]. *Biol. Zhurn. Arm.*, 32(6): 506-512. Rus (Arm, En). Illus. (P)

7612 **Shvedchikova, N.K. (1983).** Onovykh i redkikh vidakh flory Kryma. (On new and rare species in the Crimean flora.) *Byull. Mosk. Obshch. Ispyt. Prir. Biol.*, 88(2): 122-128. Rus (En). 22 rare species noted. (T)

7613 **Sirko, A.V. (1978).** K flore i ekologii sumchatykh gribov Il Istmenskogo Zapovednika. (Flora and ecology of ascomycetes in the Il Istmen Nature Reserve.) *Trudy Inst. Ekol. Rast. Zhivotn.*, 108: 53-67. Rus. (P)

7614 **Skripchinskii, V.V. (1980).** Okhrana prirodnoi flory i rastitel'nosti Stavropolskoi vozvyshenosti. (The protection of natural flora and vegetation of the Stavropol Hills.) *In* Stepi i luga Stavropolsk Kraya. Stavropol'. 49-64. Rus.

7615 **Skvortsov, A.K. (1988).** Krasnaya kniga SSSR i okhrana redkikh vidov rastenii. ("Red Book" of the USSR and protection of rare plant species.) *Bot. Zhurn.*, 73(2): 282-288. Rus. (L/T)

7616 **Skvortsova, L.S. (1980).** Study and conservation of plants from the Crimean flora. *Polezn. Rast. Prir. Flory Ispol'z. Nar. Khoz.*, 145-147. Rus. (T)

7617 **Smolonogov, E.P. (1978).** [Basic principles of the management of protection forests in Sverdlovsk province]. *Trudy Inst. Ekol. Rast. Zhiv.*, 118: 46-61. Rus. (P)

7618 **Sobolevskaya, K.A. (1983).** Problemy introduktsii ischezayushchikh vidov prirodnoi flory Sibiri. (The problems of the introduction of the disappearing species in the natural Siberian flora.) *Izv. Sibir. Otdel. Akad. Nauk, Biol. Nauk*, 10(2): 3-9. Rus (En). (T)

7619 **Sobolevskaya, K.A., ed. (1988).** Bioekologicheskie osobennosti rastenir sibiri, nuzhdayushchikhsya v okhrane. (Biological peculiarities of plants of Siberia, in need of protection.) Novosibirsk, Naika. 220p. Rus. Illus., maps. (T)

7620 **Sokolov, V. (1981).** The biosphere reserve concept in the U.S.S.R. *Ambio*, 10(2-3): 97-101. Illus., map. (P)

7621 **Sokolov, V. (1984).** Man and Biosphere. *Science in USSR*, 4 (1984): 22-35. Col. illus. (P/T)

7622 **Sokolov, V. (1985).** The system of biosphere reserves in the USSR. *Parks*, 10(3): 6-8. Illus. (P)

7623 **Sokolov, V.E., Borodin, A.M., Zykov, K.D. (1983).** Novaya "*Krasnaya kniga SSSR*". *Priroda Ezhen. Pop. Estest. Zhurn. Akad. Nauk*, 2: 36-39. Rus. Col. illus. Review of 2nd edition of the *Red Book of the USSR* (1st edition published 1978). (T)

440

UNION OF SOVIET SOCIALIST REPUBLICS

7624 **Somermaa, A.L. (1979).** Species list and ecological groups in corticolous lichens. *Estonian Contrib. Internat. Biol. Programme*, 12: 25-29. Vooremaa Forest Ecology Station. (P)

7625 **Stavrovskaya, L.A. (1982).** Redkie i izchezayushchie vidy rastenii Berezinskogo zapovednika i mery ikh okhrany. (Rare and disappearing species of plants of the Berezinsky Nature Reserve and measures for their protection.) *Zapoved. Belorussii.* (Minsk), 6: 24-31. Rus. (P/T)

7626 **Stepanova, N.T. (1977).** Griby poryadka Aphyllophorales v lesakh Il Istmenskogo Gosudarstvennogo Zapovednika im. V.I. Lenina. (Fungi of the order Aphyllophorales in the forests of the Il Istmenskii State Nature Reserve.) *Trudy Inst. Ekol. Rast. Zhivotn.*, 107: 3-22. Rus. (P)

7627 **Stoiko, S.M. (1983).** [Ecological bases of conservation of rare, unique and typical phytocoenoses.] *Bot. Zhurn.*, 68(11): 1574-1583. Rus.

7628 **Strazdaite, Yu. (1979).** Puti sokhraneniya mestoobitanii redkikh vidov rastenii Litvy v protsesse formirovaniya landschafta. (Ways of preserving the habitat of rare species of plants of Lithuania with the process of forming the landscape.) *In* Formir. rastitel'n. pokrova pri optimiz. landshafta. Materialy 2-i Vses Shkoly Kaunas, 1979. Vilnyus. 137-140. Rus. (T)

7629 **Syroechkovskii, E.E., Rogacheva, E.E., Klokov, K.B. (1982).** Taezhnoe prirodopol'zovanie. (Taiga natural resources/possibilities.) U.S.S.R., Lesnaya Prom-st'. Rus. Illus.

7630 **Syroechkovskii, E.E., ed. (1983).** Okhrana i ratsional'noe ispol'zovanie biologicheskikh resursov Krainego Severa. (The protection and rational utilization of the biological resources of the Far North.) U.S.S.R., Kolos. Nauch. Trudy Vaskhnil. Rus. (P)

7631 **Takhtajan, A.L. (1981).** Rare and vanishing plants of the USSR to be protected. Leningrad, Nauka. 202p. (T)

7632 **Takhtajan, A.L., ed. (1975).** [Red Book: native plant species to be protected in the USSR.] Leningrad, Nauka. 201p. Rus. (T)

7633 **Tikhomirov, V.N. (1981).** Regional rare plant conservation schemes in the USSR. *In* Synge, H., *ed.* The biological aspects of rare plant conservation. Chichester, Wiley. 101-104. (T)

7634 **Tikhomirov, V.N. (1984).** Osebennosti okhrany rastitel'nogo pokrova na malykh zapovednykh territoriyakh (na primere zapovednika "Galich'ya gora"). (Peculiarities of plant cover protection on small reserved territories (the example of the "Galichya Gora" reservation).) *Byull. Mosk. Obshch. Ispyt. Prir., Biol.*, 89(4): 27-35. Rus (En). (P)

7635 **Tikhomirov, V.N., Shcherbakov, A.V., Denisova, L.V. (1985).** Osnovnye istochniki po flore zapovednikov Sovetskogo Soyuza. (Main sources of information on the flora of the USSR reserves.) *Byull. Mosk. Obshch. Ispyt. Prir., Biol.*, 90(5): 119-136. Rus. Bibliography arranged by regions. (B/P/T)

7636 **Tille, I. (1983).** Zur Erhaltung der Wildformen von Kulturpflanzen in Naturschutzgebieten der Sowjetunion. (On the conservation of cultivated plants' wild forms in nature reserves of the U.S.S.R.) *Arch. Naturschutz Landschaftsforsch.*, 23(2): 113-116. Ge. (E/P/T)

7637 **Tkachenko, V.S., Chuprina, T.T., Baklanov, O.V. (1979).** The Steppe Reserve "Proval Istskaya Step" (present state and the problems of research). *Ukr. Bot. Zhurn.*, 36(4): 352-356. En, Rus. (P)

7638 **Tobias, A.V. (1981).** New and rare species of Discomycetes for Leningrad Oblast Russian-SFSR USSR. *Vestn. Leningr. Univ. Biol.*, 1: 112-115. (T)

7639 **Trass, H. (1978).** (New and rare taxa of Cladoniaceae in the lichen-flora of the U.S.S.R.) *Folia Cryptogamica Estonica*, 11: 1-6. Rus (En). Illus. (T)

UNION OF SOVIET SOCIALIST REPUBLICS

7640 **Troitskaya, E.A., Stepanova, E.G. (1988).** [On the session of the Scientific Council of the USSR Academy of Sciences on the problem "Biological Basis of Rational Utilization, Transformation and Protection of the Plant World"]. *Bot. Zhurn.*, 73(5): 776-780. Rus. (T)

7641 **U.S.S.R. Akademiya Nauk SSSR (1980).** Zapovedniki Belorussii: issledovaniya: vyp. 4. (Nature reserves of Byelorussia: investigations: part 4.) BSSR, Uradzhai. Rus. (P)

7642 **U.S.S.R. Akademiya Nauk SSSR. Nauchnyi Sovet po probleme "Introduktsiya i Akklimatizatsiya Rastenii" (1983).** Redkie i ischezayushchie vidy prirodnoi flory SSR, kul'tiviruemye v botanicheskikh sadakh i drugikh introduktsionnykh tsentrakh strany. (Rare and disappearing species of the natural flora of the USSR cultivated in botanic gardens and others introduced into the central region.) Moskva, Nauka. 301p. Rus (En). (G/T)

7643 **U.S.S.R. Ministerstvo Sel'skogo Khozyaistva SSSR. Glavnoe Upravlenie po Okrane Prirody, Zapovednikam, Lesnomu i Okhotnich'emu Khozyaistvu (1978).** Krasnaya kniga SSSR: redkie i nakhodyashchiesya pod ugrozoi ischeznoveniya vidy zhivotnykh i rastenii. (Red Data Book of U.S.S.R.: rare and endangered species of animals and plants.) Moskva, Izdatel'stvo Lesnaya Promyshlennost'. Rus. Illus., maps. (T)

7644 **Ul'yanova, T.N. (1980).** Novye i redkie dlya flory Ostrova Iturup rasteniya (kratkoe soobshchenie). (New rare plants occurring in the Iturup Island flora; a short report.) *Trudy Prikl. Bot. Genet. Selek.*, 68(3): 42-44. Rus (En). (T)

7645 **Vardanyan, Zh.A. (1979).** [Arid vaika light forest and ways of their restoration (conservation measures)]. *Biol. Zhurn. Arm.*, 32(1): 51-56. Rus (Arm, En).

7646 **Varlygina, T.I., Kulikova, G.G. (1981).** O rabochem soveshchanii Kommissii po okrhane rastenii. (On the working conference of the Commission of Plant Conservation.) *Byull. Mosk. Obsch. Ispyt. Prir., Biol.*, 86(2): 130-131. Rus. (T)

7647 **Varsanof'eva, V.A. (1980).** Vremena goda v Pechorskom krae. U.S.S.R. Nauka. (Seriya "Chelovek i Okruzhayushchaya Sreda".) Pechoro-Ylychskii Nature Reserve, North Urals. (The season in the Pechora Territory.) Rus. (P)

7648 **Vasil'chenko, I.T., Konnov, A.A. (1986).** Archevye Srednei Azii: ocherednye zadachi ikh izucheniya, okhrany i ratsional'nogo ispol'zovaniya. (Juniper forests of Middle Asia: the immediate tasks of their study, protection and rational use.) *Bot. Zhurn*, 71(4): 554-556. Rus.

7649 **Verechaka, T.V., Kulikova, G.C. (1986).** [Valuable areas of living nature in the Moscow region]. Gugk. Rus. (P/T)

7650 **Viktorov, S.V. (1978).** [Conservation et regeneration de la couverture vegetale dans la region d'Ustjurt]. *Probl. Osvoen. Pustyn, Turkm. SSR*, 4: 77-82. Rus (Fr).

7651 **Vinogradov, B.V. (1984).** Aerospace studies of protected natural areas in the U.S.S.R. *In* Unesco-UNEP. Conservation, science and society. Paris, Unesco. 439-448. Contributions to the First International Biosphere Reserve Congress, Minsk, Byelorussia/USSR, 26 September - 2 October 1983. Table. (P)

7652 **Vinogradov, V., Chernyavskaya, S. (1984).** The oldest reserve in the Soviet Union. *In* U.S.S.R. Academy of Sciences. Man and biosphere. Moscow, Nauka Publishers. 173-178. Illus., map. Lower Volga delta. (P)

7653 **Vinogradov, V.N. (1977).** [Forest science and the conservation of environment]. *Vestn. Sel'skokhos. Nauki*, 12: 123-131. Rus.

7654 **Volkov, A.D. (1985).** The Red Data Book of the Karelia ASSR. Rare and endangered plant and animal species. Kareliya, Petrozavodsk. 181p. Illus., col. illus. (T)

7655 **Vorontsov, A.I., Isaev, A.S. (1979).** Novye zadachi lesozashchity. (New tasks in forest protection.) *Lesovedenie*, 6: 3-11. Rus (En).

UNION OF SOVIET SOCIALIST REPUBLICS

7656 **Vorontsova, L.I., Vasilyeva, V.D., Kuliyev, A.N., Lomikina, G.A. (1988).** Zadachi klassifikatsii redkikh rastitel'nykh soobshchestv v svyazi s ikh okhranoi. (The aims of rare plant community classification in connection with their protection.) *Bot. Zhurn.*, 73(5): 733-740. Rus. (T)

7657 **Vrishch, D.L. (1976).** [Beautiful flowering herbaceous species of the Primorski Krai and their protection.] *Bot. Zhurn. Leningr.*, 61(1): 121-130. Rus. (T)

7658 **Weiner, D.R. (1988).** Models of nature: ecology, conservation and cultural revolution in Soviet Russia. Bloomington, Indiana, Indiana University Press. 312p.

7659 **Yadrov, A.A., Golubev, V.N. (1980).** [Conservation of the genetic diversity of subtropical fruit crops]. *Byull. Glavn. Bot. Sada*, 115: 67-70. Rus. Recommends botanic gardens and other horticultural institutions to pay particular attention to establishing collections of useful species. (E)

7660 **Yazan, Yu.P., ed. (1982).** Kompleksnye biogeotsenoticheskie issledovaniya v Tsentral'no - lesnom Zapovednike za 50 let: Tez. dokl. nauch. konf., 27-29 iyulya, 1982. (Complex bio-geocoenotic investigations in the Central Forest Reserve after 50 years: reports of a scientific conference, 27-29 July, 1982.) Moskva, 70p. Rus. Illus. (P)

7661 **Yurtsev, B.A. (1983).** Rastitelnyi pokrov polyarnoi bezlesnoi oblasti: problemy izucheniya i okhrany. (Plant cover of glade, treeless regions: problems of study and protection.) *In* Probl. ekol. polyar. obl. Tr. Shk.: Seminara, Murmansk, apr., 1980. M. 34-45. Rus.

7662 **Yurtsev, B.A. (1984).** Voprosy okhrany redkikh vidov i rastitel'nykh soobshchesti Chukotskoi tundry. (Problems of protection of rare species and the vegetation associations of the Chukotsk tundra.) *In* Obshch. probl. okhrany rastit. Materialy Vses. soveshch. Okhrana rastit. mira sev. regionov, syktyvkar, 7-9 sent., 1982. T.I. Syktyvkar. 131-135. Rus.

7663 **Zhdanov, A.A., Levshin, L.V. (1958).** Okhrana lesnykh i vodnykh bogatstv v SSSR. (Protection of forest and wetland sites in the USSR.) Moskva, Gosiurizdat. 49p.

7664 **Zhirmunsky, A. (1984).** The reserve in Peter the Great Bay. *In* U.S.S.R. Academy of Sciences. Man and biosphere. Moscow, NAUKA Publishers. 197-201. Map. (P)

7665 **Zhirmunsky, A.V. (1979).** [Reservation in Peter the Great Bay]. *Priroda*, 8: 52-61. Rus. Illus., map. Includes plant conservation. (P)

7666 **Zimina, R.P. (1978).** The main features of the Caucasian natural landscapes and their conservation, U.S.S.R. *Arctic Alp. Res.*, 10(2): 479-488.

7667 **Zubrebiani, B.G., Bolkvadze, I.I. (1979).** (Rare species of plants for Lower Svanetia flora.) *Soobshch. Akad. Nauk Gruz. SSR*, 95(1): 165-167. Geo (En, Rus). Ref. in *Agricola*, 79(10-11). (T)

7668 **Zvereva, G.A. (1979).** [New habitats of discovery of rare plants over the area of the southern part of the Krasnoyarsk territory]. *Izvest. Sibir. Otdel. Akad. Nauk. SSR, Biol. Nauk.*, 3: 56-59. Abstr. in *Agricola*, 80(10). (T)

ARCTIC U.S.S.R.

7669 **Budnikova, G.P. (1980).** Flora vodoemov Kemerovskoi oblasti i voprosy ee okhrany. (Flora of the reservoirs of the Kemerovsk Region and the question of its protection.) *In* Priroda i ekon. Kuzbassa Tez. dokl. k predstoyashch. 18i nauch konf. po itogam n-i raboty in-ta Ser. estest.-geogr. nauk. Novokuznetsk. 88-90. Rus.

7670 **Fuller, W.A., Kevan, P.G., eds (1970).** Productivity and conservation in northern circumpolar lands. Proceedings of a conference, Edmonton, Alberta 15-17 October 1969. IUCN New Series No. 16. Morges, IUCN. 344p. Illus., map.

7671 **Sokolov, V.E., Chernov, Y.I. (1983).** Arctic ecosystems: conservation and development in an extreme environment. *Nat. Resources*, 19(3): 2-9. Illus., maps.

ASIATIC U.S.S.R.

7672 **Amelchenko, V.P., Agafonov, G.I., Ignatenko, N.A. (1986).** Redkie i ischezayushchie rasteniya Tomskoi oblasti v Sibirskom Botanicheskom Sadu. (Rare and disappearing plants of the Tomsk Oblast in the Siberian Botanic Garden in Tomsk). *Byull. Glavn. Bot. Sada* (Moscow), 141: 58-61. Rus. (G/T)

7673 **Anon. (1978).** Aktual'nye voprosu okhrany na Dal'nem Vostoke. (Urgent issues on the protection of the Soviet Far East.) U.S.S.R., Izd. Dal'nevost. Nauch., Tsentra AN SSSR. 160p. Rus.

7674 **Anon. (1981).** Rasteniya Dal'nevostochnogo Morskogo Zapovednika i ikh okhrana. (Plants of the Far Eastern Sea Nature Reserve and their protection.) U.S.S.R., Izd. Dal'nevost. Nauch. Tsentra AN SSSR. Rus. (P)

7675 **Bol'shakov, N.M. (1984).** [New and rare vascular plants for the Angara-Tunguska floristic area (Krasnoyarsk Territory).] *Bot. Zhurn.*, 69(7): 963-965. Rus. (T)

7676 **Borovskii, V.M. (1979).** O vliyanii khozyaistvennoi deyatel'nosti cheloveka na izmenenie prirodnoi obstanovki. (The influence of the agricultural activities of men on changes in the natural conditions.) *Izv. Akad. Nauk Kaz. SSR. Biol.*, 6: 40-44. Rus.

7677 **Charkevitch, S.S., Katchura, N.N. (1981).** [Rare plant species of the Soviet Far East.] Moscow, Nauka. 230p. Rus. Illus., maps. Species accounts. (T)

7678 **Far East Scientific Center of the Soviet Academy of Sciences (1976).** Preservation of nature in the Far East. Vladivostok, Far East Scientific Center of the Soviet Academy of Sciences.

7679 **Gorelova, T.G., Kamelin, R.V. (1978).** [Present state of rare and endemic species of Badkhys vegetation]. *Izv. Akad. Nauk Turkm. SSR, Biol. Nauk*, 4: 26-33. Rus (En). Plant conservation aspects. (T)

7680 **Grey-Wilson, C. (1985).** Plants in peril: 6. *Kew Mag.*, 2(3): 330-331. *Iris winogradowii.* (T)

7681 **Kamakhina, G.L. (1983).** O flore ushchelii uchastka Babazo Kopetdagskogo Zapovednika. (On the canyon's flora of Babazo area of the Kopetdag Reserve.) *Izv. Akad. Nauk Turkm. SSR, Biol. Nauk.*, 3: 31-34. Rus (En). (P)

7682 **Kasiev, K.S. (1982).** K voprosy okhrany rastitel'nosti pribrezhnoi zony ozera Issyk-Kul. (On the problem of vegetation of the lake-side zone of the Issyk-Kul Lake.) *In* Izuch i osvoenie flory i rastit. vysokogorii 8 Vses. soveshch. Tez. dokl. 4: Rastit. resursy okhrana i rats. ispol'z. rastit. mira vysokogorii. Sverdlovsk. 23. Rus.

7683 **Kharkevich, S.S., Man'ko, Yu.I., Vasil'ev, N.G., Zhivotchenko, V.I. (1983).** Sozdat' Dzhugdzhurskii Zapovednik. (The creation of the Dzhugdzhur Nature Reserve.) *Priroda Ezhem. Pop. Estest. Zhurn. Akad. Nauk*, 4: 34-43. Rus. Illus., col. illus., map. Situated on the Sea of Okhotsk. (P)

7684 **Khrapko, O.V. (1976).** Rare and vanishing species of herbaceous perennials of the southern Maritime region. *In* Preservation of nature of the Far East. Vladivostok, Far East Scientific Center of the Soviet Academy of Sciences. 100-104. (T)

7685 **Kohlein, F. (1979).** Lilien und andere Seltenheiten aus dem fernen Osten der USSR. (Lilies and other rare plants from the Soviet Far East.) *Staudengarten*, 2: 27-29. Ge. (T)

7686 **Krasovskaya, L.S., Levichev, I.G. (1986).** Flora Chatkal'skogo zapovednika. (Flora of the Chatkal Nature Reserve). Tashkent, Fan. 176p. Rus. (P)

7687 **Kurentsova, G.E., Kharkevich, S.S. (1975).** Objectives of preservation and utilization of rare plant species of the Soviet Far East. *Bull. Main Bot. Gard. Soviet Acad. Sci.*, 95: 77-84. (T)

ASIATIC U.S.S.R.

7688 **Malyshev, L.I., Peshkova, G.A. (1979).** Nuzhdayutsya v okhrane: redkie i ischezayushchie rasteniya Tsentral'noi Sibiri. (To be protected: rare and disappearing plants of central Siberia.) Novosibirsk, Nauka. 173p. Rus. Illus., maps. (T)

7689 **Malyshev, L.I., Sobolevskaya, K.A., eds (1980).** [Rare and endangered plant species of Siberia]. Novosibirsk, Nauka. 223p. Rus. Illus., maps. Describes over 300 species. (T)

7690 **Malyshev, L.I., Sobolevskaya, K.A. (1984).** Okhrana genofonda rastenii v stepnykh ekosistemakh severnoi Azii. (Protection of the gene fund of plants in the steppe ecosystem of north Asia.) In Probl. okhrany genofonda i upr. ekosistemami v zapoved. step. i pustyn. zon. Tez. dokl. Vses. soveshch., Askaniya-Nova, 21-25 maya 1984. Moskva. 147-151. Rus.

7691 **Mironova, L.N. (1977).** Wild growing species of iris in the Maritime region and their preservation and geographical distribution. In Flora of the Far East (biology, utilization, preservation). Vladivostok, Far East Scientific Center of the Soviet Academy of Sciences. 59-61.

7692 **Naumenko, A.T. (1978).** Scientific objectives of preservation and reproduction of Kamchatkan *Abies*. In Botanical investigations of the Far East. Vladivostok, Far East Scientific Center of the Soviet Academy of Sciences.

7693 **Pautova, V.N., Galimulin, M.G. (1980).** Discoveries of rare eastern Siberia species of higher aquatic plants. *Bot. Zhurn. Leningr.*, 65(7): 1020-1022. (T)

7694 **Rodin, L.E., Miroshnichenko, I. (1978).** [Ecological basis of conservation of desert plant resources]. *Probl. Osvoen, Pustyn, Turkm. SSR*, 6: 10-14. Rus (En).

7695 **Rustamov, I.G., Klyushkin, E.A. (1980).** Botanicheskie issledovaniya v zapovednikakh i okhrana rastitel'nosti pustyn' Turkmenistana. (Botanical studies in nature reserves and the protection of the Turkmenistan desert vegetation.) In Vopr. izuch flory i fauny Turkenistana. Ashkhabad. 3-12. Rus. (P)

7696 **Safarov, I.S. (1986).** Redkie i ischezayushchie vidy dendroflory vostochnogo Zakavkaz'ya i ikh okhrana. (Rare and vanishing species of the eastern Transcaucasian dendroflora and their protection.) *Bot. Zhurn.*, 71(1): 102-106. Rus. Map. (T)

7697 **Samoilov, T.V. (1980).** Nekotorie redkie i izchezayushchie vidy prirodnoi flory dal'nevostochnogo regiona v usloviyakh kultury dendroparka. (Some rare and disappearing species of the native flora of the Far Eastern Region and in cultivation in an arboretum.) In Izuch i ispolz rastitel'n resursov Sakhalina i Yuga Primor'ya. Yuzhno-Sakkhalinsk. 127-135. Rus. (G/T)

7698 **Seledets, V.P. (1983).** Okhrana fitogenofonda v prirodookhrannykh kompleksakh Dal'nego Vostoka. (The protection of the plant gene pool of protected natural complexes of the [Soviet] Far East.) In Tez. dokl. 7 Delegat. S'ezda Vses. botan. o-va, Donetsk, 11-14 maya. Leningrad. (?1984). Rus. (P)

7699 **Semkin, B.I., Borzova, L.M. (1986).** Sravnitel 'nyi analiz spiskov vodov sosudistykh rastenir ostrovov Dal 'nevostochnogo gosudarstvennogo morskogo zapovednika. (A comparative analysis of the lists of vascular plant species from islands Far Eastern State Marine Reserve.) *Bot. Zhurn.*, 71(5): 652-657. Rus. Map. (P)

7700 **Sirodova, T.V. (1984).** Osobennosti flory Kurgal'dzhinskogo zapovednika, ee okhrana. (Peculiarities of the flora of the Kurgal'dzhin Nature Reserve, its protection.) In Probl. okhrany genofonda i upr. ekosistemami v zapoved. step. i pustyn. zon. Tez. dokl. Vses. soveshch., Askaniya-Nova, 21-25 maya 1984. Moskva. 177-178. Rus. (P)

7701 **Smirnov, E.N., Podushko, M.B., Vasil'ev, N.G. (1981).** Sikhote-Alinskii biosfernyi Zapovednik. (Sikhote-Alin Biosphere Reserve.) *Priroda Ezhem. Pop. Estest. Zhurn. Akad. Nauk*, 3: 32-45. Rus. Col. illus., map. (P)

445

ASIATIC U.S.S.R.

7702 **Sobolevskaya, K.A. (1980).** Rare and disappearing plants of Siberia. Novosibirsk, Nauka. 223p. (T)

7703 **Sokolov, V.E., Syroechkovskii, E.E., eds (1985).** Zapovedniki SSSR: zapovedniki Dal'nego Vostoka. (Nature reserves of the USSR: reserves of the Far East.) Moskva, Mysel. 319p. Rus. Illus., col. illus. Map. Describes 13 reserves. (P/T)

7704 **Sukhova, G.V. (1980).** Okhrana i obogashchenie rastitel'nogo mira Turkmenskoi SSR. (The protection and enrichment of the plant world of the Turkmen SSR.) *Izv. Akad. Nauk Turkm. SSR, Biol. Nauk,* 3: 94-95. Rus.

7705 **Tugushi, K.L. (1980).** [Reliable conservation of valuable gene pool of beech *Fagus orientalis* and fir *Abies,* virgin forests, Caucasus]. *Lesnoe Khoz.,* 1980(11): 34-37. Rus. (E)

7706 **Urusov, V.M. (1976).** Coniferales of the maritime region - coenosis, condition and ways of their preservation. *In* Preservation of nature of the Far East. Vladivostok, Far East Scientific Center of the Soviet Academy of Sciences. 37-45.

7707 **Urusov, V.M. (1984).** V primor'e neobkhodim natsional'nyi prirodnyi park. (The need for a national nature park in coastal regions.) *Priroda Ezhem. Pop. Estest. Zhurn. Akad. Nauk,* 7: 57-65. Rus. Col. illus. (P)

7708 **Vlasova, N.V. (1984).** [New and rare species of the flora of Southern Yakutia.] *Bot. Zhurn.,* 69(8): 1102-1104. Rus. (T)

7709 **Vorob'ev, D.F. (1969).** Rare plant species of the Maritime region and the Amur River region. *In* Botany of the Far East; to commemorate the 100th birthday of V.L. Komarov (1869-1969). Vladivostok, Far East filial V.L. Komarov of the Soviet Academy of Sciences. 119-123. (T)

7710 **Voskanyan, V.E. (1982).** Flora i rastitelnost' vysokogorii Armyanskoi SSR, ikh okhrana i ratsional'noe ispol'zovanie. (Flora and vegetation of the high mountains of the Armenian SSR, their protection and rational utilization.) *In* Izuch i osvoenie flory i rastit. vysokogorii. 8 Vses. soveshch. Tez. dokl. 4: Rastit. resursy, okhrana i rats. ispol'z rastit. mira vysokogorii. Sverdlovsk. 11. Rus. (E)

7711 **Vyshin, I.B. (1983).** [Rare species of vascular plants from central Sikhote-Alin Range and their conservation needs]. *Komarovskie Chteniia,* (30): 29-39. Rus. Maps. (P/T)

7712 **Yakubov, V.V. (1983).** Nakhodki redkikh i novykh dlya Kamchatskoi oblasti vidov sosudistykh rastenii v Kronotskom Gosudarstvennom Zapovednike. (Finds of rare and new species of vascular plants for the Kamchatskaya District in the Kronotsky State Reserve.) *Bot. Zhurn.,* 68(5): 678-679. Rus. (P)

7713 **Yakubov, V.V. (1983).** Taksonomicheskii sostav sosudistykh rastenii Kronotskogo gosudarstvennogo zapovednika. (Taxonomic structure of the vascular plants of the Kronotskaya National Park.) *In* Okhrana zhivoi prirody. Tez. Vses. konf. mol. uchenykh, noyab. Moskva. 226-228. Rus. (P/T)

EUROPEAN U.S.S.R.

7714 **Abele, G. (1979).** Krustkalnskii Zapovednik. (Krustkaln Nature Reserve.) *Nauki i Tekhn.* (Riga), 7: 23-25. Rus. (P)

7715 **Abele, G. et al. (1986).** Khorologiya flory Latviiskoi SSR: perspektivnye dlya okhrany vidy rastenii. (Chorology of the flora of the Latvian SSR: perspective for the protection of plant species). Riga, Zinatne. 110p. Rus. Illus. (T)

7716 **Abele, G.T. (1978).** Inventarizatsiya okhranyaemykh i redkikh vidov rastenii v Latviiskoi SSR. (Inventory of protected and rare species of plants in the Latvian SSR.) *In* Latvia. Akademiya Nauk Latviiskoi SSSR. Institut Biologii. Rastitel'nyi mir okhranyaemykh territorii. Riga, Zinatne. 72-76. Rus. (T)

EUROPEAN U.S.S.R.

7717 **Agadzhanov, S.D., Rabotina, E.N. (1984).** Redkie i endemichnye vidy flory Apsheronskogo poluo strova i voprosy ikh okhrany. (Rare and endemic species of plants of the flora of the Apsheran Peninsula and problems of their protection.) *In* Biol. produktiv. polez. rast. flory Kobystana i Apsheron. p-va. Baku. 29-37. Rus. (T)

7718 **Aleksieva, T.S. (1980).** A disappearing nut species (Turkish Filbert, *Corylus colurna*), a grove of wild ancient trees, Kyustendil region. Conservatior aspects. Ovoshtarstvo. *Sofia, Tsentralen Suvet na Natsionalniia Agrarno-Promishlen Suinz*, 59(11): 37-38. (T)

7719 **Andreev, G.N., et al. (1979).** [The animals and plants of the Murmansk Region which are rare and require protection.] Murmansk. 160p. Rus. (T)

7720 **Andronova, N., et al. (1981).** K voprosy vozobnovlenii redkikh i ischezayushchikh rastenii Kaliningradskoi oblasti. (On the problem of conserving rare and disappearing plants of the Kaliningrad Region.) *In* Ekol. i okhrana rast. Nechernozem. Zony RSFSR. Ivanovo. 50-53. Rus. (T)

7721 **Baranova, E.V., Baranov, M.P., Tikhonova, O.A. (1984).** Materialy k flore Nizhne-Svirskogo gosudarstvennogo zapovednika. (On the flora of the Lower-Svir Reserve.) *Vestn. Leningrad. Univ., Biol.*, 3: 105-108. Rus (En). (P)

7722 **Blagoveshchenskii, V., Pchelkin, Yu., Rakov, N., Shustov, V. (1981).** Itogi ucheta redkikh i ischezayushchikh rastenii v Ul'yanovskoi oblasti. (Taking stock of the rare and disappearing plants in the Ulyanov Region.) *In* Ekol. i okhrana rast. Nechernozem. zony RSFSR. Ivanovo. 53-55. Rus. (T)

7723 **Borodina, N.V., Talmatova, L.V., Medvedeva, L.V., Tereshkin, I.S. (1982).** Dopolnenie k flore Mordovskogo gosudarstvennogo zapovednika. (Addition to the flora of the Mordovskaya State Nature Reserve.) *In* Rast. i sreda. Saransk. 5-12. Rus. (P)

7724 **Chiguryaeva, A.A., Ivanova, R.D., Michurin, V.G., Milovidova, I.B. (1984).** Redkie i ischezayushchie vidy rastenii prirodnoi flory Saratovskoi oblasti. (Rare and disappearing species of plants of the native flora of Saratov Region). *In* Vopr. botan. Yugo-vost.: Flora Rastit. Fiziol. Saratov. 49-78. Rus. (T)

7725 **Chiguryaeva, A.A., Milovidova, I.B., Michurin, V.G. (1983).** Botanicheskie ob'ekty Saratovskoi oblasti, podlezhashchie okhrane. (Plants of the Saratov Region, destined for protection.) *In* Tez. dokl. Y Delegat. S'ezda Vses. botan. o-va, Donetsk, 11-14 Maya, 1983. Leningrad. 328-329. Rus. (T)

7726 **Chiguryaeva, A.A., ed. (1979).** Okhranyaemye rasteniya Saratovskoi oblasti. (Protected plants of the Saratov Region.) Saratov, Privolsh. Kn. Isd-vo. 120p. Rus. Illus. (T)

7727 **Danilov, V.I., Kulikova, G.G., Nitina, S.V., Novikov, V.S. (1980).** O nekotorykh botunickeskikh ob'ektakh v priokskoi polose Moskovskoi oblasti nuzhdayushchikhsya i okhrane. *In* Okhrana redk. rast. i fitotsenozov. (Moskva). 22-29. Rus (En).

7728 **Danilov, V.I., Kulikova, G.G., Novikov, V.S., Tikhomirov, V.N. (1983).** Botanicheskie ob'ekty vdoline reki Osetr, nuzhdayushchiesya v okhrane. (Botanic objects in the Osertr river valley requiring protection.) *Byull. Mosk. Obshch. Ispyt. Prir., Biol.*, 88(4): 150-157. Rus. (T)

7729 **Denisova, L.V., ed. (1980).** Okhrana redkikh rastenii i fitotsenozov: sb. nauch. tr. VNII okhrana prirody i zapovedn dela. Moskva. 93p. Rus. Illus.

7730 **Dyrenkov, S.A., Fedorchuk, V.N., Mel'nitskaya, G.B. (1980).** Redkie lesnye rastitel'nye soobshchestva Leningradskoi oblasti, nuzhdayushchiesya v okhrane. (Rare forest communities to be protected in the Leningrad region.) *Ref. Zhurn. Biol.*, 65(8): 1202-1208. Rus. (P)

EUROPEAN U.S.S.R.

7731 **Eustratov, U.I., Tarshis, L.G. & G.I., Vasfilova, E.S. (1983).** Vnutrividovaya izmenchivost' i okhrana travyanistykh rastenii Srednego Urala. *In* Okhrana zhivoi prirody. Tez. Vses. konf. mol. uchenykh noyab. Moskva. 59-60. Rus. (P)

7732 **Fedotov, V.V. (1984).** Redkie i ischezayushchie vidy obshchesoyuznogo spiska vo flore Pechoro-Ilychskogo gosudarstvennogo zapovednika. *In* Obshch. probl. okhrany rastit. Materialy Vses. soveshch. Okhrana rastit. mira sev. regionov, syktyvkar, 7-9 sent., 1982. T.I. Syktyvkar. 118-122. Rus.

7733 **Gladkov, V.P. (1978).** Printsipy vydeleniya, otsenki zonirovaniya territorii i opredeleniya ustoichivosti geokompleksov Prirodnogo Parka y Komi ASSR. *In* Latvia. Akademiya Nauk Latviiskoi SSSR. Institut Biologii. Rastitel'nyi mir okhranyaemykh territorii. Riga, Zinatne. 38-43. Rus. Komi ASSR.

7734 **Gogina, E.E., Novikov, V.S., Skvortsov, A.K., Tikhomirov, V.N. (1981).** O kadastre botanicheskikh ob'ektov nuzhdayushchikhsya v okhrane na territorii Moskovskoi oblasti. (Plants to be protected in the Moscow district.) *Bot. Zhurn.*, 66(4): 595-600. Rus. (T)

7735 **Golubev, V.N., Molchanov, E.F. (1978).** [Systematic instructions on the population quantity and eco-biological studies of rare, disappearing and endemic plants of the Crimea.] Yalta, Nikita Botanical Garden. Rus. (T)

7736 **Grin, A.M., Utekhin, V.D. (1981).** Tsentral'no Chernozemnyi biosfernyi zapovednik: geosistemyi monitoring v biosfernom zapovednike. (The central Black Earth Centre Biosphere Nature Reserve: geosystems for monitoring in the Biosphere Nature Reserve.) *Priroda Ezhem. Pop. Estest. Zhurn. Akad. Nauk*, 9: 30-34. Rus. Col. illus., map. (P)

7737 **Gubina, E.M. (1983).** Redkie i ischezayushchie vidy drevesnoi flory Kavkaza v botanicheskikh sadakh SSSR. (Rare and disappearing species of the flora of the Caucasus in botanical gardens of the USSR.) *In* Okhrana zhivoi prirody. Tez. Vses. Konf. mol. uchenykh, noyab. Moskva. 41-42. Rus. (G/T)

7738 **Gushchina, E.G. (1978).** Redkie i ischezayushchie rasteniya Ryazanskoi lesostepi i perspektivy ikh okhrany. (Rare and disappearing plants of the Ryazan Wooded Steppe and perspectives of their protection.) *In* Latvia. Akademiya Nauk Latviiskoi SSR. Institut Biologii. Rastitel'nyi mir okhranaemykh territorii. Riga, Zinatne. 76-79. Rus. (T)

7739 **Hang, V. (1980).** Punane raamat ei saa kunagi valmis. (The Red Data Book will never be finished.) *Eesti Loodus*, 23(12): 773-776. Es (Rus, En). Illus., col. illus. (T)

7740 **Ignatenko, O.S. (1978).** Okhrana rastitel'nosti v Tsentral'no Chernozemmom zapovednike im. Prof. V.V. Alexina. (Protection of vegetation of the Prof. V.V. Alekhin Black Earth Region Nature Reserve.) *In* Probl. vzaimodeistviya cheloveka s okhruzh. Srednoi. Materialy Vses. soveshch. Kursk. 151-153. Rus. (P)

7741 **Ignatenko, O.S., Krasnityskii, A.M. (1981).** Tsentral'no - Chernozemnyi biosferyi Zapovednik: Zapovednik im. V.V. Alekhina prirodnoe yadro biosfernogo zapovednika. (The Central Black Earth Region Biosphere Nature Reserve: the V.V. Alekhin Nature Reserve, the natural kernel of the Biosphere Reserve.) *Priroda Ezhem. Pop. Estest. Zhurn Akad. Nauk*, 9: 35-39. Rus. Col. illus. (P/T)

7742 **Ignatenko, O.S., Sobakinskikh, V.D. (1983).** Nekotorye itogi okhrany rastitel'nosti Streletskoi stepi. (Some results of the protection of vegetation of the Streletsk Steppe.) *In* U.S.S.R. AN SSSR. Lab. Lesovedeniya, MOIP. Ekologo-tsenoticheskie i geograficheskie osobennosti rastitel'nosti. Moskva, Nauka. 99-106. Rus. (T)

7743 **Ignatenko, V.I., Parfenov, P.V. (1983).** Parawnal'naya kharaktarystyka vodnai flory i raslinnastsi azer Byarezinskaha biyasfernaha zapavednika. (Comparative characteristics of the aquatic flora and vegetation of the lakes of the Berezinsk Biosphere Reserve.) *Vestsi Akad. Navuk BSSR, Biyal. Navuk*, 3: 16-23. Be (En, Rus). (P)

EUROPEAN U.S.S.R.

7744 **Il'minskikh, N.G., Dimitriev, A.V. (1981).** Neudachnaya kniga o prirode Chuvashii: *Priroda Chuvashii i ee okhrana.* (On the nature of Chuvackia and its conservation.) Cheboksary, Chuvashskoe Knizhnoe Izdatel'stvo. 168p. Rus.

7745 **Isachenkov, V.A., ed. (1983).** Rastitel'nyi pokrov Pskovskoi oblasti i voprosy ego okhrany mezhvuz. sb. nauch. tr. (Plant cover of the Pskov Region and questions of its protection: inter-university collected scientific papers.) Leningrad, Gos. Ped. In-t. 98p. Rus. Illus. (T)

7746 **Kalda, A.A. (1979).** Nekotorye aspekty okhrany rastitel'nosti okul'turennykh territorii. (Some aspects of the protection of plants of cultivated areas.) *Uchen Zap. Tartus. Un-ta,* 475: 62-66. Rus (En).

7747 **Karpenko, A.S., Stavrova, N.I. (1980).** Printsipy i metody kartografirovaniya okhranyaemykh botanicheskikh ob'ektov (na primere Nechernozem'ya). (Principles and methods of mapping protected plant species [exemplified by the non-chernozem zone].) *Bot. Zhurn.,* 65(8): 1192-1202. Rus. Maps. (L/T)

7748 **Korobtsova, Z.V. (1984).** Sokhrani rastenie. (Protected plants.) Saratov, Privolzh. Kn. Izd-vo. 152p. Rus. Illus. (T)

7749 **Kosykh, V.M. (1978).** Chislennost' i struktura populyatsii nekotorykh redkikh i ischezayushchikh vidov flory Kryma. (Size and structure of populations of some rare and vanishing species in the Crimean flora.) *Trudy Gos. Nikit. Bot. Sada,* 74: 85-90. Rus (En). (T)

7750 **Kosykh, V.M. (1982).** Struktura populyatsii nekotorykh redkikh vidov Yaltinskogo gorno-lesnogo Zapovednika. (Structure of some rare species populations in Yalta Mountain Forest Reservation.) *Trudy Nikit. Bot. Sada,* 87: 72-79. Rus (En). (P/T)

7751 **Kosykh, V.M., Golubev, V.N. (1983).** Sovremennoe sostoyanie populyatsii redkikh ischezayushchikh i endemichnykh rastenii gornogo Kryma. (Present condition of the population of rare and disappearing plants of the mountains of the Crimea.) Rukopis' dep. v VINITI, no. 3360-3383 Dep. Yalta, Gos. Nikit. Botan. Sad. 118p. Rus. Illus. (T)

7752 **Kucherov, E.V., ed. (1982).** Redkie i ischezayushchie vidy poleznykh rastenii Bashkirii i puti ikh okhrany. (Rare and disappearing species of useful plants of Bashkir and the means of their protection.) Ufa, In-t Biol. 103p. Rus. Illus. (E/T)

7753 **Kukk, U. (1983).** Harju rajooni taimhàraldused ja nende Kaitse. (Redkie rasteniya i ikh okhrana v Khar'yuskom raione.) (Rare plants of Harju district and their conservation.) *Lesovod. Issled.* (Tallinn), 18: 77-90. Es (En, Rus). (T)

7754 **Kulikova, G.G., Novikov, V.S., Tikhomirov, V.N., Variygina, T.I. (1983).** Opyt razrabotki sistemy okhranyaemykh prirodnykh territorii Moskovskoi oblasti. (Experiment on the cultivation system of protected natural territories of the Moscow Region.) *In* Tez. dokl. Delegat. S'ezda Vses. botan. o-va, Donetsk, 11-14 Maya, 1983. Leningrad. 299-300. Rus. (P)

7755 **Kulikova, G.G., Tiknomirov, V.N. (1984).** Rabota botanicheskikh sadov tsentra Evropeiskoi chasti SSSR po okhrane rastenii i rastitel'nykh soobshchestv. (The work of botanic gardens of the centre of the European Section of the USSR in the protection of plants and plant associations.) *Byull. Glavn. Bot. Sada* (Moscow), 133: 91-92. Rus. (G)

7756 **Kuzenkova, L.Ya (1979).** Botanicheskie ob'ekty basseina r. Pakhry i ikh okhrana. (Plants of the basin of the River Pakhra and their protection.) *In* Priroda i prirod protsessy na territorii podmoskov'ya. Moskva. 71-83. Rus.

7757 **Kuznetsov, L.A. (1979).** Informatsiya ob okhranyaemykh rastenii Leningradskoi oblasti. (Information on the protected plants of the Leningrad Region.) *In* Sistematika, anatomiya i ekol. rast. Evrop. chasti SSSR. Leningrad. 10-116. Rus. (T)

7758 **Litvinskaya, S.A., Tilka, A.P., Filimonova, R.G. (1983).** [Rare and threatened plants of Kuban.] Krasnodar, Krasnodarskoye knizhnoe izdatelstvo. 159p. Rus. (T)

EUROPEAN U.S.S.R.

7759 **Luks, Yu.A., Krukyova, I.V. (1973).** [Valuable, rare and vanishing plants of Crimean flora urgently requiring protection.] *Bot. Zhurn.*, 58(1): 97-106. Rus. (T)

7760 **Luks, Yu.A., Privalova, L.A., Kryukova, I.V., eds (1975).** [Catalogue of rare, vanishing and extinct plants of the Crimean flora recommended for preservation.] *Bull. Gos. Nikit. Bot. Sada*, 3(28): 13-20 and appendix of 174p. Rus (En). (T)

7761 **Lyubchenko, V.M., Yatsenko, N.P. (1980).** Okhrana genofonda flory i rastitel'nosti v zone Kanevskogo Zapovednika. (Protection of the gene pool of flora and vegetation in the Kanev Nature Reserve Zone.) *Ochrana Izuch. Obogashch. Rast. Mira* (Kiev), 7: 9-18. Rus. (P)

7762 **Malyshev, L.I. (1980).** Strateliya i taktika okhrany flory. (Strategy and tactics for flora protection.) *Bot. Zhurn.*, 65(6): 875-887. Rus. (T)

7763 **Matsenko, A.E., Kulikova, G.G., Kamysheva, N.P. (1988).** Osobennosti organizatsii okhrany prirodnykh territorii Podmoskov'ya. (Peculiarities of the organization of the protection of natural regions in the Moscow region.) *Byull. Glavn. Bot. Sada*, (Moskva), 148: 61-66. Rus. (P)

7764 **Matveev, V.I., Plaksina, T.I. (1983).** Flora vodoemov Zhigulevskogo zapovednika im. L.L. Sprygina. (Flora of the reservoires of the L.L.Sprygin State Nature Reserve at Zhigulev.) *In* Probl. ratsional. ispol'z. i okhrany prirod. kompleksa Samar. Luki. Kuibyshev. 56-58. Rus. (P)

7765 **Mavrishchev, V.V., et al. (1988).** Lesotipologicheskie kompleksy Berezinskogo biosfernogo zapovednika kak ob'ekty litotsenstichesogo monitoringa. (Forest type complexes of the Berezinsky Biosphere Nature Reserve as objects of phytocoenotic monitoring.) *Botanika* (Minsk), 29: 117-133. Rus. Map. (P)

7766 **Mikheev, A.V., Gladkov, N.A., Galushin, V.M. (1981).** Okhrana prirody: uchebnik dlya studentov ped.-in-tov. (The protection of nature: a textbook for students of pedagogical institutes.) Izd. 2.e., USSR, Kolos. Rus.

7767 **Minibaev, R.G., Nazirova, Z.M. (1986).** [Some results and perspectives on the study of rare and disappearing species of plants of Bashkir]. *In* Region. florist. issled. i metod. prepokavaniya botan. distsiplin. Krasnodar. 83-87. Rus. (T)

7768 **Minyaev, N.A., Konechnaya, G.Yu. (1976).** Flora tsentral'no-lesnogo gosudarstvennogo. Leningrad, Izdatel'stvo Otdelenie. 102p. Rus. Illus., maps.

7769 **Mironenko, O.N. (1984).** Nuzhdayushchiesya v okhrane rasteniya i sitotsenozy Arkhangel'skoi oblasti. (Plants in need of protection and the cytocoenosis of the Arkhangel Region.) *In* Probl. okhrany prirody v basseine Belogo mor'ya. Murmansk. 64-70. Rus. (T)

7770 **Mironova, T.I., Slepyan, E.I. (1982).** Priroda Leningradskoi oblasti i ee okhrana. (Nature of the Leningrad region and its protection.) Leningrad, Lenizdat. Rus.

7771 **Moskovskoe Obshchestro Ispytatelei Prirody (1984).** Sosoyanie i perspektivy issledovaniya flory srednei polosy Evropeiskoi chasti SSSR: materialy soveshchaniya. dekabr' 1983 g. (Condition and perspective of the study of the flora of the central region of European part of U.S.S.R. material from the conference Dec. 1983.) Moskva, MOIP. 89p. Rus. Map.

7772 **Nepomilueva, N.I. (1981).** O sokhranenii taezhnykh landshaftov na Evropeiskoi severo-vostoke. (Conservation of boreal coniferous forest landscapes in the north-east of Europe.) *Bot. Zhurn.*, 66(11): 1616-1622. Rus. Map.

7773 **Nepomilueva, N.I., Lashchenkova, A.N. (1978).** Okhrana flory i rastitel'nosti Prirodnogo Parka Komi ASSR. (The protection of the flora and vegetation of the National Park of the Komi ASSR.) *In* Latvia. Akademiya Nauk Latviiskoi SSSR. Institut Biologii: Rastitel'nyi mir okhranyaemykh territorii. Riga, Zinatne. 43-47. Rus. (P)

EUROPEAN U.S.S.R.

7774 **Newcombe, L.F. (1985).** Protected natural territories in the Crimean USSR. *Envir. Conserv.*, 12(2): 147-155. Illus. (P/T)

7775 **Nikitin, V.V., Krasikova, N.S. (1978).** *Homalodiscus ochradeni*, a rare disappearing plant, requiring protection. *Izv. Akad. Nauk Turkm. SSR, Biol. Nauk*, 6: 71-73. Illus. (T)

7776 **Nikolaenko, V. (1978).** [Regeneration of forest resources and environmental conservation]. *Mezhdun. Sel'Skokhoz. Zhurn.*, 6: 80-82. Rus. Illus.

7777 **Olovyannikova, I.N. (1977).** [Effect of 1972 drought on the vegetation of Solonetz complex in the north Caspian area]. *Byull. Mosk. Obsch. Ispyt. Prir. Biol.*, 82(6): 63-73. Rus.

7778 **Plaksina, T.I. (1986).** Novye dannye redkikh rasteniyakh yugo-vostoka evropeiskoi chasti SSSR i ikh okhrana. (New data on some rare plants of the south-east of the European part of the USSR and their protection.) *Bot. Zhurn.*, 71(5): 695-702. Rus. Illus., maps. (T)

7779 **Poluyakhtov, K.K., ed. (1981).** Biologicheskie osnovy povysheniya produktivnosti i okhrany rastitel'nykh soobshchestv Povolzh'yoi Mezhvuz. sb. (Biological foundations of increased productivity and protection of the plant associations of the Volga Region.) Gorkii, Gos. Ped. In-t. 113p. Rus. Illus.

7780 **Ronkonen, N.I. (1979).** Nekotorye voprosy okhrany lekarstvennykh rastenii. (Some questions on the protection of medicinal plants.) *In* Ekologiya, produktivn i biokhim. sostav lekarstv. i yagod. rast. lesov i bolot Karelii. Petrozavodsk. 36. Karelia. Rus. (E)

7781 **Ronkonen, N.I. (1984).** Okhrana redkie rastenii Yuzhnoi Karelii. (The protection of rare plants of South Karelia.) *In* Obshch. probl. okhrany rastit. Materialy Vses. soveshch. Okhrana rastit. mira sev. regionov, Syktyvkar, 7-9 sent., 1982. T.I. Syktyvkar. 104-107. Rus. (T)

7782 **Ronkonen, N.I., Andreev, K.A. (1978).** Rastitel'nost' Karelii i mery po organizatsii ee okhrany. (The vegetation of Karelia and measures for the organization of its protection.) *In* Latvia. Akademiya Nauk Latviiskoi SSSR. Institut Biologii. Rastitel'nyi mir okhranyaemykh territorii. Riga, Zinatne. 48-51. Rus.

7783 **Roshchevsky, M.P., et al., eds (1982).** [Rare animals and plants of the Komi ASSR in need of protection.] Syktyvkar. 152p. Rus. (T)

7784 **Rusyaeva, G.G. (1986).** [The influence of anthropogenic factors on the composition of the herbaceous shrub cover]. Sverdlovsk, Il'm. Gos. Zapoved., 67p. Rus. Illus. (P)

7785 **Rysina, G.P. (1984).** Opyt vosstanovleniya populyatsii okhranyaemykh rastenii v Podmoskov'e. (Experiment on the renewal of the population of protected plants of the Moscow Region.) *Byull. Glavn. Bot. Sada* (Moscow), 133: 81-85. Rus. (T)

7786 **Salo, K. (1986).** Kivatsu, luonnosuojelalue Karjalan ASNT: ssa. (Kivatsu, nature reserve in the Karelian Autonomic Socialist Republic.) *Luonnon Tutkija*, 90(2): 100-106. Fi (En). Illus., map. (P)

7787 **Schmidt, V.M., Sergienko, V.G. (1986).** Vidy vysshikh rastenii rekomenduemye k okhrane na territorii zapada Nenetskogo avtonomnogo okruga. (Recommendations to the protection of vascular plants species in the western part of Nenetski Autonomous District.) *Vestn. Leningrad. Univ., Biol.*, 1: 100-102. Rus.

7788 **Semagina, R.N. (1983).** Condition of natural regeneration of *Taxus baccata* in the Black Sea area forests of the Caucasian reserve. *Byull. Mosk. Obshch. Ispyt. Prir. Otd. Biol.*, 88(4): 146-149. Rus (En). (P/T)

EUROPEAN U.S.S.R.

7789 **Sergienko, V.G. (1983).** Perspekivy okhrany genofonda Arkhangelskoi oblasti na primere flory Kanina. (Perspectives of protection of the gene pool of the Arkhangel Region, especially of the flora of Kanin.) *In* Okhrana zhivoi prirody. Tez. Vses. konf. mol. uchenykh noyal. Moskva. 188-190. Rus. (T)

7790 **Sergienko, V.G. (1984).** Voprosy okhrany rastitel'nykh kompleksov v raione kontakta lesa i tundry na Kanine. (Problems of protection of the vegetation complexes in the region of contact of forest and tundra at Kanin.) *In* Obshch. probl. okhrany rastit. Materialy Vses. soveshch. Okhrana rastit. mira sev. regionov, Syktyvkar, 7-9 sent., 1982. T.I. Syktyvkar. 43-46. Rus.

7791 **Shatko, V.G. (1984).** Okhranyaemye vidy prirodnoi flory Kryma v Moskve. (Protected species of the Crimean natural flora in Moscow). *Byull. Glavn. Bot. Sada* (Moscow), 130: 67-74. Rus. Illus. (G/T)

7792 **Shelyag-Sosonko, Yu.R., Didukh, Ya.P., Molchanov, E.F. (1985).** Gosudarstvennyi zapovednik "Mys Mart'yan". (The State Nature Reserve "Mys Mart'yan".) Kiev, Naukova Dumka. 256p. Rus. Illus., col. illus., maps. In the central region of the south coast of Crimea. (P)

7793 **Shelyag-Sosonko, Yu.R., Sytnik, K.M., et al. (1980).** [Ukraine, White Russia, Moldavia. The protection of important botanical regions.] Kiev, Naukova Dumka. 332-370. Rus. Includes lists of threatened plants by region. (P/T)

7794 **Shevchenko, G.T. (1983).** Nekotorye biologicheskie osobennosti redkikh i endemichnykh rastenii severnogo Kavkaza. (Some biological peculiarities of rare and endemic plants of North Caucasus.) *In* Vopr-o, okhrana i ratsional. ispol'z. prirod. rastit. resursov. Stavropol'. 107-119. Rus. (T)

7795 **Shilov, M.P. (1983).** Sostoyanie populyatsii i puti okhrany vodyanogo orekha v Ivanovskoi i Vladimirskoi oblastiyakh. (The condition of the population and the route to the protection of [an aquatic nut] of the Ivanov and Vladimir Regions.) *In.* Izuch. redk. i okhranyaem. nidov travyanist. rast. Moskva. 61-64. Rus.

7796 **Shkhagapsoev, S.Kh., Abramova, T.I. (1987).** K okhrane skal'no-osypnoi rastitel'nosti v Kabardino-Balkarskom Vysokogornom Zapovednike. (On the protection of rock scree vegetation of the Kabardino-Balkar Nature Reserve.) *In* Priroda mal. okhranyaem territorir. Voronezh. 84-90. Rus. (P/T)

7797 **Shmidt, V.M., Simacheva, E.V. (1984).** Materialy k okhrane flory Arkhangel'skoi oblasti. (Materials concerning the protection of the flora of the Arkhangel region.) *Vestn. Leningrad. Univ., Biol.,* 9: 50-54. Rus (En). (T)

7798 **Shmidt, V.M., Simacheva, E.V. (1984).** Problemy okhrany genofonda flory Arkhangel'skoi oblasti. (Problems of protection of the gene pool of the flora of the Arkhangel Region.) *In* Obshch. probl. okhrany rastit. Materialy Vses. soveshch. okhrana rastit. mira sev. regionov, Syktyvkar, 7-9 sent., 1982. T.I. Syktyvkar. 108-112. Rus.

7799 **Shvedchikova, N.K. (1983).** Unikal'nye rastitel'nye soobshchestva vostochnogo Kryma. (Unique plant communities of the East Crimea.) *Biol. Nauki* (Moscow), 7(235): 83-86. Rus (En).

7800 **Shvedchikova, N.K. (1983).** [Pine-juniper forests in the eastern Crimea.] *Byull. Mosk. Obshch. Ispyt. Prir. Biol.,* 88(4): 125-134. Rus (En). Protected species. (T)

7801 **Simachev (1980).** Biologicheskie osnovy okhrany redkikh reliktovykh vidov vysshikh rastenii Leningradskoi oblasti na primere *Pulsatilla vernalis, Viscaria alpina, Oxytropis sordida. Bot. Zhurn.,* 65(5): 725-737. Illus., map. Biological foundations of protection of rare relict species of higher plants of Leningrad District with special reference to ... *Pulsatilla vernalis, Viscaria alpina, Oxytropis sordida.*

EUROPEAN U.S.S.R.

7802 **Simachev, V.I., Simacheva, E.V. (1985).** Botanicheskoe obosnovanie predlozhenir po organizatsii okhrany nekotorykh urochishch Leningradskoi oblasti. (Botanical grounds for suggestions on the organization of reservation of some landscape places of Leningrad region.) *Vestn. Leningrad. Univ. Biol.*, 24: 21-35. Rus (En). (P)

7803 **Skvortsov, A.K., Tikhomirov, V.N. (1986).** Redkie ischezayushchie i nuzhdayushchiesya v okhrane vidy sosudistykh rastenii moskovskoi oblasti. (Species of vascular plants rare, disappearing and in need of protection in Moscow Region). *Byull. Mosk. Obshch. Ispyt. Prir., Biol.*, 91(6): 111-118. Rus (En). Covers 261 species. (T)

7804 **Smirnova, A.D. (1982).** Okhranyaemye rasteniya Gor'kovskoi oblasti. (Protected plants of the Gorki Region.) Gor'kii, Volgo-Vyat. Kn. Izd-vo. 96p. Rus. Illus. (T)

7805 **Sukhikh, V.I. (1979).** Distantsionnye metody zondirovaniya v lesnom khozyaistve okhrane prirody. (Remote scanning methods in forestry and nature protection.) *Lesnoe Khoz.*, 3: 41-45. Rus.

7806 **Tabaka, L.V., Baroninya, V.K. (1979).** Floristicheskaya struktura intensivno ispol'zuemoi (rekreatsionnoi) zony. (The floristic structure of an intensively exploited [recreational] zone.) *In* Latvia. AN LatvSSR. Inst. Biologii. Flora i rastitel'nost' Latviiskoi SSR. Severo - Vidzemskii Geobotanicheskii raion. Riga, Zinatne. 103-107. Rus.

7807 **Tabaka, L.V., ed. (1983).** Okhrana flory rechnykh dolin v Pribaltiiskikh respublikakh. (Protection of the flora of river valleys in the Baltic Republics.) Riga, Zinatne. 102p. Rus. Illus.

7808 **Tamilova, L.I. (1977).** (Experience in the introduction of some endemic and relict plants from the Urals.) *Trudy Inst. Ekol. Rast. Zhivotn. Ural Nauchn. Tsentr.*, 123-131. Rus. (T)

7809 **Tikhomirov, V.N., ed. (1983).** Izuchenie redkikh i okhranyaemykh vidov travyanistykh rastenii. (The study of rare and protected species of herbaceous plants.) U.S.S.R., Geogr. O-va S.S.S.R. 96p. Rus. (T)

7810 **Tumanov, V.K., ed. (1983).** Zllenyi shum. (Green noise: problems and protection of nature in the Volga region.) Kuibyshev, Kuibyshev. Kn. Izd.-vo. Rus.

7811 **U.S.S.R. Akademiya Nauk S.S.S.R. Laboratoriya Lesovedeniya (1980).** Biogeotsenologicheskie osnovy sozdaniya prirodnykh zakaznikov na primera zakaznika "Verkhnyaya Moskva-Reka". (Biogeocoenotic foundations for the creation of nature reserves, for example the Upper Moscow River Reserve.) Moskva, Nauka. 173p. Rus. Illus. (P)

7812 **Vardanyan, Zh.A. (1987).** Redkie i ischezayushchie vidy dendroflory Armenii v Erevanskom botanicheskom sadu. (Rare and disappearing species of the dendroflora of Armenia in the Erevan Botanical Garden.) *Byull. Glavn. Bot. Sada* (Moscow), 146: 72-77. Rus. (G/T)

7813 **Voronezh. Universitet (1977).** Materialy k poznaniyu prirody Galich'ei Gory. (Materials for the recognition of the natural life of Galich'ya Gora.) Voronezh, Izdatel'stvo Voronezhskogo Universiteta. 143p. Rus. Maps.

7814 **Zozulin, G.M., Pashkov, G.D., Abramova, T.I., Stepnin, G.I., Fedyaeva, V.V. (1977).** [Material for the Red Book from the Rostov Oblast]. *Izv. Severo-Kavk. Nauch. Tsentra Vyssh. Shkoly Estest. Nauk.*, 5(1): 105-108. Rus. Abstr. in *Biol. Abstr.*, 67(4): 20998 (1979). (T)

U.S.S.R. - Armenia S.S.R.

7815 **Abovyan, Yu.I., Bunatyan, E.G., Davtyan, L.V. (1983).** Okhrana prirodnykh resursov Armyanskoi SSR. (The protection of natural resources of the Armenian SSR.) Arm, SSR, Aiastan. Rus. (P)

U.S.S.R. - Armenia S.S.R.

7816 **Akademiya Nauk Arm. SSR. Botanicheskii Institut. (1979).** [List of rare and disappearing species of the flora of Armenia.] Erevan. 27p. Arm. (T)

7817 **Arevshatyan, I.G. (1980).** Nekotorye novye i redkie vidy flory Armenii sem. Fabaceae. (Some new and rare species of Fabaceae family in Armenian flora.) *Biol. Zhurn. Arm.,* 33(5): 505-508. Rus (Arm, En). (T)

7818 **Barsegyan, A.M. (1980).** Redkie i ischezayushchie rastitelnye formatsii Armenii i ikh okhrana. (Rare and disappearing plant formations in Armenia and their conservation). *Biol. Zhurn. Arm.,* 12(5): 515-521. Arm (En). (T)

7819 **Eramian, E.N., Galstian, M.G. (1982).** [New and rare species of Armenian flora]. *Biol. Zhurn. Armen.,* 35(5): 412-414. Rus. Includes list. (T)

7820 **Gabrielyan, E.T. (1981).** The conservation of rare and threatened species and types of vegetation in Armenia. (Conservacion de especies y tipos de vegetacion raros y amenazados en Armenia.) *An. Jard. Bot.* (Madrid), 37(2): 773-778 (1980 publ. 1981). (Sp). (T)

7821 **Gabrielyan, E.T., Gusyan, K.E. (1980).** Novye i redkie rody i vidy iz severnoi Armenii. (New and rare genera and species of northern Armenia.) *Biol. Zhurn. Arm.,* 33(5): 535-537. Arm. (T)

7822 **Gabrielyan, E.T., Tamanian, K.C. (1982).** [New and rare species of the flora of Armenia]. *Biol. Zhurn. Armen.,* 35(3): 227-229. Rus. (T)

7823 **Khandzhian, N.S. (1982).** [Rare species of the genus *Tanacetum* L. from Armenia]. *Biol. Zhurn. Armen.,* 35(1): 72-74. Rus. (T)

7824 **Khandzhian, N.S. (1984).** [New and rare plants for the Armenian flora]. *Biol. Zhurn. Armen.,* 37(5): 430-432. Rus. (T)

7825 **Khurshudyan, P.A., Barsegyan, A.M., Afrikyan, K.G. (1980).** Nuzhdayushchiesya v okhrane botanicheskie ob'ekty Sevanskogo Natsional'nogo Parka. (Plants of the Sevan National Park in need of protection.) *Biol. Zhurn. Arm.,* 33(1): 12-19. Rus (Arm, En). Map. (P/T)

7826 **Oganesyan, M.E. (1980).** Nekotorye kriticheskie redkie vidy kolskol'chikovykh (Campanulaceae) iz yuzhno Zakavkaz'ya. (Some critical and rare Campanulaceae species of southern Transcaucasia.) *Biol. Zhurn. Arm.,* 33(12): 496-504. (Arm, Rus, En). Maps. (T)

7827 **U.S.S.R. Akademiya Nauk Arm. SSR Botanicheskii Institut (1979).** Spisok redkikh i ischezayushchikh vidov flory Armenii. (List of rare and disappearing species of the flora of Armenia.) Yerevan. 27p. La, Rus, Arm. (T)

7828 **Voskanyan, V.E., Arutiunian, M.G., Gukasian, A.G. (1984).** [State and conservation of carpet phytocoenoses with the domination of *Campanula tridentata* Schreb. in the Armenian SSR]. *Biol. Zhurn. Armen.,* 37(4): 281-287. Rus (Arm, En).

U.S.S.R. - Azerbaydzhan S.S.R.

7829 **Asakov, K.S. (1984).** [Some rare trees and bushes of the Nakhichevan ASSR.] Doklady - Akademiya nauk Azerbaidzhanskoi SSR. *Baku "Elm",* 40(11): 83-85. Az (En, Rus). (T)

7830 **Askerov, A.M. (1981).** [Rare and vanishing species of Pteropsida plants from Azerbaydzhan and their conservation]. *Byull. Glavn. Bot. Sada,* 1981(122): 85-90. Rus. (T)

7831 **Ismikhanova, A.A. (1983).** [Valuable relict tree species in the Talysh Mountains.] *Lesnoe Khoz.* (Moskva), (11): 35-36. Rus. (T)

U.S.S.R. - Azerbaydzhan S.S.R.

7832 **Kapinos, G.E., Ibadov, O.V., Abdullaeva, I.K. (1982).** [Rare and disappearing *Tulipa* species of the Azerbaydzhan flora.] *Byull. Glavn. Bot. Sada,* 1982 (125): 44-49. Rus. (T)

7833 **Liatifova, A.Kh., Evstratova, O.I. (1984).** [Analysis of the flora of the S.M. Kirov Kyzylgach reservation.] Izvestiya Akademiya Nauk Azerbaidzhanskoi SSR. Seria biologicheskikh nauk. Baku "Elm", 1: 34-40. Rus (Az). (P)

7834 **Mamedov, R.M. (1984).** [Rare and disappearing species of the genus *Muscari* growing in the Kuba-Khachmas zone and their biological peculiarities]. *Izvest. Akad. Nauk Azerbaydzhan. SSR,* (1): 41-43. Az (Rus). (T)

7835 **Safarov, I.S. (1982).** Rastitel'nye soobshchestva s redkimi vidami rastenii v Azerbaidzhanskoi SSR. (Plant associations of rare species of plants in the Azerbaydzhan SSR.) *In* Okhrana redk. rastit. soobshchestv. Moskva. 67-75. Rus (En). (T)

7836 **Shatko, V.G., Mironova, L.I. (1983).** Kvoprosy izuchenii redkikh, ischezayushchikh i endemichnykh vidov rastenii Karadagskogo Zapovednika. (On the problems of the study of rare, disappearing and endemic species of the vegetation of the Karadag Nature Reserve.) *In* Okhrana zhivoi prirody. Tez. Vses. Konf. mol. uchenykh, noyab. Moskva. 213-214. Rus. (P/T)

7837 **Shatko, V.G., Mironova, L.P. (1986).** Sostoyanie populyatsii nekotorykh redkikh rastenii v Karadagskom Gosudarstvennom Zapovednike. (Condition of the population of some rare plants in the Karadag State Nature Reserve). *Byull. Glavn. Bot. Sada* (Moscow), 141: 61-67. Rus. Map. (P/T)

U.S.S.R. - Byelorussian S.S.R.

7838 **Akademiya Nauk Ukrainskoi SSR. Institut Botaniki im. N.G. Kholodnogo (1980).** Ukrainy, Belorussii Moldavii: okhrana vazhneishikh botanisheskikh ob'ektov. (The Ukraine, Byelorussia and Moldavia: the protection of important plants.) Kiev, Naukova Dumka. 389p. Rus. Illus., col. illus., map. (T)

7839 **Bibikov, Yu.A. (1986).** Okhranyaemye rasteniya Krupskogo raiona Minskoi oblasti. (Protected plants of the Krup region of Minsk Oblast.) *Vestn. Belorus. Gos. Univ.,* 1: 34-38. Rus. (L/T)

7840 **Bibikov, Yu.A., Zubkevich, G.I., Sautkina, T.A. (1980).** Redkiya rasliny pawdeneva-zakhodnyai chastki Beloruskaha paazer'ya. (Rare plants in the south-west part of the Byelorussian Poozerie.) *Vestsi Akad. Navuk BSSR, Biyal. Navuk.,* 6: 20-24. Rus (En). (T)

7841 **Boiko, A.V., Loznukho, I.V. (1981).** Bioekologicheskikh osobennosti rastitel'nykh kompleksov Pripyatskogo zapovednika. (Bioecological peculiarities of the plant complex of the Pripyat Nature Reserve.) BSSR, Nauka i Tekhnika. Rus. (P)

7842 **Geltman, V.S. (1985).** Rastitel'nost' Pripyatskogo zapovednika. *Zapoved. Belorussii,* 10: 9-20. Rus. Vegetation of Pripyat Nature Reserve. (P)

7843 **Kazlouskaia, N.V., Bulat, V.S. (1980).** [Representation of the Byelorussian flora on reservation territories.] *Vest. Akad. Navuk BSSR, Ser. Biol. Navuk,* 1980 (5): 5-9. Be (En, Rus). (P)

7844 **Kim, G.A. (1984).** Printsipy okhrany lugovoi rastitel'nosti v B.S.S.R. (The principles of protection of meadow plants in the Byelorussian SSR.) *In* Tez. dokl. 7 Delegat. S'ezda Vses. botan. o-va, Donetsk, 11-14 Maya, 1983. Leningrad. 296-297. Rus.

7845 **Klakotskaya, T.N. (1983).** Dopolnenie i spisku flory Pripyatskogo landshaftno-gidrologicheskogo Zapovednika i ego okrestnostei. (Additional lists of the flora of the Pripyat Landscape Hydrological Nature Reserve and its environs.) *Zapoved. Belorussii* (Minsk), 7: 41-47. Rus. (P)

U.S.S.R. - Byelorussian S.S.R.

7846 **Kovalenko, G.G., Kozlov, V., Parfenov, V.I. (1984).** Nature conservation in the Byelorussian SSR. *In* Unesco-UNEP. Conservation, science and society. Paris, Unesco. 190-194. Contributions to the First International Biosphere Reserve Congress, Minsk, Byelorussia/USSR, 26 September - 2 October 1983. (P)

7847 **Kozlov, V., et al., eds (1981).** [Red Data Book of the Byelorussian SSR. Rare and endangered species of animals and plants.] Minsk, Byelorussian Soviet Encyclopedia. 288p. Be, Rus. Editors of the plant section: V.I. Parfenov and N.V. Koslovsckaya. (T)

7848 **Kozlovskaya, N.V., Bulat, V.S. (1980).** Representatywnasts' flory Belarusi na zapavednykh terytoryyakh. (Representation of the Byelorussian flora within protected areas.) *Vestsi Akad. Navuk BSSR, Biyal Navuk*, 5: 5-9. Rus (En). (P)

7849 **Kozlovskaya, N.V., Simonowich, L.G., Blazhevich, R.I., Vynaev, G.V., Tret'iakov, D.I. (1979).** [Species of Carpathian origin rare for the Soviet Union flora and new for the Byelorussian SSR flora]. *Dokl. Akad. Nauk Bel. SSR*, 23(10): 933-936. Rus (En). (T)

7850 **Krinitskii, V.V. (1978).** Printsipy organizatsii okhrany i izucheniya prirodnykh kompleksov gosudarstvennykh zapovednikov. (Principles of organization and study of natural complexes of state nature reserves.) *In* Sovrem. zadachi gos. zapovednikov les zony evrop. chasti SSSR. Materialy nauch. konf. Domzheritsy, Minsk, 1975. 3-10. Rus. (P)

7851 **Martsinkevich, G.I. (1977).** Ispol'zovanie prirodnykh resursov i okhrana prirody. (The utilization of natural resources and the protection of nature.) Minsk, Izd-vo im. V.I. Lenina. 198p. Rus.

7852 **Misiewicz, J. (1985).** Berezynski rezerwat biosfery. (The biosphere reserve in the Berezina river valley.) *Chronmy Przyr. Ojczysta*, 41(6): 65-67. Pol. Map. (P)

7853 **Parfenov, V.I. (1978).** Problemy ispol'zovaniya i okhrany rastitel'nogo mira Belorussii. (Problems of utilization and protection of the plant world of Byelorussia.) Minsk, Nauka i Tekhn. 104p. Rus.

7854 **Parfenov, V.I. (1983).** Berezinsky Biosphere Reserve: an example in a temperate mixed forest. *Nat. Resources*, 19(2): 26-35. Illus., maps. (P)

7855 **Parfenov, V.I., Kudin, M.V. (1983).** Berezinskiy Biosfernyi Zapovednik. (Berezinsky Biosphere Reserve.) *Priroda, Ezhem. Pop. Estest. Zhurn. Akad. Nauk SSSR*, 6: 2-11. Rus. Illus., col. illus., map. (P)

7856 **Parfenov, V.I., Rykovskii, G.F., Vynaev, G.V. (1982).** Tearetychnyya pryntsypy arhanizatsyi setki akhowvaemykh pryrodnykh terytoryi Belarusi. (The theoretical fundamentals for the arrangement of a system of protected areas in the Byelorussian S.S.R.) *Vestsi Akad. Navuk BSSR, Biyal Navuk*, 6: 6-13. Be (En). Map. (P)

7857 **Rinkus, E. (1979).** [Berezina State Reserve]. *Mezsaimnieciba un Mezrupnieciba*, 1: 26-28. Latv (Rus). (P)

7858 **Rykovskii, G.F. (1977).** [Discovery of a rare species - *Sphagnum molle* Sull. - in the northern part of Byelorussia]. *Bot. Issled Beloruss. Otd. Vses Bot. O-va*, 19: 152-154. Rus. (T)

7859 **Rykovskii, G.F. (1980).** Mokhoobraznye Berezinskogo Biosfernogo Zapovednika. (Bryophyta of the Berezina Biosphere Reserve.) Minsk, Nauka I Tekhnika. 133p. Rus. (P)

7860 **Stavrovskaya, L.A. (1986).** Flora shirokolistvennykh lesov Berezinskogo biosfernogo zapovednika. (Flora of broad-leaved forests of the Berezinsky Biosphere Reserve). *Zapoved. Belorussii* (Minsk), 10: 60-64. Rus. (P)

U.S.S.R. - Byelorussian S.S.R.

7861 **Suschenyz, L.M., Parfenov, V.I., Vinayev, G.V., Rykovsky, G.F. (1984).** Scientific principles of designing a system of nature reserves in Byelorussian USSR. *In* Unesco-UNEP. Conservation, science and society. Paris, Unesco. 121-124. Contributions to the First International Biosphere Reserve Congress, Minsk, Byelorussia/USSR, 26 September - 2 October 1983. (P)

7862 **Yurkevich, I.D., ed. (1981).** Zapovedniki Belorussii: vyp.5. (09). (Nature reserves of white Russia: part 5.) BSSR, Uradshai. Rus. (P)

7863 **Yurkomich, I.D., ed. (1982).** Zapovedniki Belorussii: sb. vyp. 7. (Nature reserves of Byelorussia: part 7.) Rus. (P)

U.S.S.R. - Estonia S.S.R.

7864 **Anon. (1979).** Metroloogiaalase too ulesannetest ning korraldamisest eesti nsv metsamajanduse ja looduskaitse ministeeriumi susteemis. *Mets, Puit, Paber. Les, Drevesina, Bumago. Informatsiooniseeria,* 6(9): 18-22. Es. On the work of the meteorology service in forestry and nature protection in Estonia.

7865 **Anon. (1986).** Siiraid soove murdeeas rahvuspargile. (Fifteen years of the Lahemaa National Park.) *Eesti Loodus,* 6: 338-358. Es (En, Rus). Illus., col. illus. (P)

7866 **Arun, M. (1982).** Okhrana prirody v Estonskoi SSR. (Protection of nature in Estonia.) Est SSR, Periodika. Rus.

7867 **Flint, V. (1979).** Punane raamat. (The Red Data Book.) *Eesti Loodus,* 22(6): 365-369. Rus (En). Red Data Book of Estonia - method of compilation. (T)

7868 **Herman, M. (1982).** Otepaa maastikukaitseala. (The Otepaa Landscape Reserve.) *Eesti Loodus,* 25(1): 9-17. Es (Rus, En). Illus., col. illus., map. (P)

7869 **Herman, M., Kaljumae, H., Paakspuu, V. (1983).** Elu matsalus soltub tagamaadest. (Life in the Matsalu State Nature Reserve depends on the hinterland.) *Eesti Loodus,* 26(10): 634-640. Es (En). Illus., col. illus. (P)

7870 **Herman, M., Loopmann, A., Ranniku, V. (1982).** Soakaitsealad: milleks ja kuhu? (Wetland reserves: their purpose and future?) *Eesti Loodus,* 25(10): 626-632. Es (Rus, En). Illus., col. illus., map. (P)

7871 **Jeeser, M. (1981).** Kumme aastat Hiiumaa laidude riiklikku maastikukaitseala. (Ten years of the Hiiumaa Islets State Landscape Reserve.) *Eesti Loodus,* 24(10): 626-632. Es (Rus, En). Illus., col. illus. (P)

7872 **Kalda, A. (1981).** Human impact on the plant cover of Lahemaa National Park. *In* Laasimer, L., *ed.* Anthropogenous changes in the plant cover of Estonia. Tartu, Academy of Sciences of the ESSR 32-45. (P)

7873 **Kalda, A. (1982).** Lahemaa rahvuspargi taimkate Rakvere rajooni taimkatte Etalonina. (Vegetation cover of the Lahemaa National Park as a standard of the vegetation cover in the Rakvere district [Estonian SSR].) *In* Loodusvarade kasutamine ja keskkonnakaitse: Teaduslik-praktiline konverents 11. Ja 12. Tallinn, ENSV Teaduste Akadeemia Tallinna Botaanikaaed. 118-121. Es (Rus). (P)

7874 **Kukk, U. (1982).** Iseloomulikke jooni Rakvere rajooni taimestikust ja selle Kaitsest. (Typical characters of vegetation and their conservation in the Rakvere district [Estonian SSR].) *In* Loodusvarade kasutamine ja keskkonnakaitse: Teaduslik-praktiline konverents 11. Ja 12. Tallinn, ENSV Teaduste Akadeemia Tallinna Botaanikaaed. 121-124. Es (Rus). Maps.

7875 **Kumari, E. (1982).** [Red Data Book of the Estonian SSR.] Tallinn, Valgus. 248p. Es (Rus, En). (T)

7876 **Kuulpak, H. (1980).** Laiendati looduskaitsealuseid maid. (Nature conservation areas are extended.) *Eesti Loodus,* 23(2): 73-74. Rus (En). (P)

U.S.S.R. - Estonia S.S.R.

7877 **Kuulpak, H. (1983).** Kaitsealuseid taimeliike on nuud poole rohkem. (The number of protected plant species has doubled.) *Eesti Loodus*, 26(7): 431-435. Es (Rus, En). Illus., col. illus. (L/T)

7878 **Laasimer, L. (1981).** Anthropogenous changes of plant communities and problems of conservation. *In* Laasimer, L., *ed.* Anthropogenous changes in the plant cover of Estonia. Tartu, Academy of Sciences of the Estonian S.S.R. 18-31.

7879 **Laasimer, L., ed. (1981).** Anthropogenous changes in the plant cover of Estonia. Tartu, Academy of Sciences of the Estonian S.S.R. 163p.

7880 **Leht, M. (1979).** Taimeliigid punastes raamatutes. (Plant species in the Red Data Books.) *Eesti Loodus*, 22(7): 420-426. Es (title), Rus (text). Col. illus., map. Red Data Book of the Estonian SSR. (T)

7881 **Luik, H. (1984).** Meie vabariik sai looduskaitsefondi. (The Estonian SSR now has a nature conservation fund.) *Eesti Loodus*, 27(5): 282-289. Es (En, Rus). Illus., col. illus.

7882 **Mazing, V.V. (1978).** Problemy sokhraneniya bolot (na primere Estonskoi SSR). (Problems of the preservation of wetlands, for example of the Estonian SSR.) *Tartu Ulik. Toimet.*, 458(1): 90-92. Rus.

7883 **Miilmets, A. (1983).** Matsalu, looduskaitseala ja margala. (Matsalu: a nature reserve and wetland.) *Eesti Loodus*, 27(11): 690-695. Es (En). Illus., col. illus., maps. (P)

7884 **Paal, J., Herman, M. (1986).** Uus looduskaitseala. (A new nature reserve.) *Eesti Loodus*, 1: 10-15. Es (En, Rus). Illus., col. illus., map. Oostriku Wetland Reserve on the Pandivere Upland. (P)

7885 **Pork, K. (1981).** [Anthropogenous dynamics of meadows in recent decades. Protection of meadow communities]. *In* Laasimer, L. *ed.* Anthropogenous changes in the plant cover of Estonia. Tartu, Academy of Sciences of the ESSR, Publications Advisory Committee. 46-63. Es (Rus, En).

7886 **Pungas, K. (1984).** Uljaste Kaitse ja puhkepaigana. (Uljaste as a conservation and recreation area.) *Eesti Loodus*, 27(6): 362-366. Es (En, Rus). Illus., map. (P)

7887 **Randlane, T. (1978).** New lichen species to the Estonian lichen-flora from the Vudumae Nature Reserve (Island Saaremaa). *Folia Cryptogamica Estonica*, 11: 7-8. (Rus). (P)

7888 **Ranniku, V. (1983).** Karula maastikukaitseala. (The Karula Landscape Nature Reserve.) *Eesti Loodus*, 26(6): 338-345. Es. Col. illus., map. (P)

7889 **Rebassoo, H. (1980).** Botaaniline uunikum. (A botanically unique reserve.) *Eesti Loodus*, 23(5): 293-300. Rus (En). Col. illus., map. Vilsandi State Nature Reserve, Estonia. (P)

7890 **Rebassoo, H., Viires, H. (1979).** Eesti taimharuldusi. (Rare plants of Estonia). Tallinn, Valgus. 1 portfolio. Es (Rus, En). Pictorial work. (T)

7891 **Reitalu, M. (1981).** Hanuldaste taimede kodu. (Home of rare plants.) *Eeesti Leodus*, 24(5): 279-287. Rus, En. Illus., col. illus. Viidumae State Nature Reserve, Estonia. (P/T)

7892 **Teder, Kh.O. (1983).** [Single system of forestry and nature conservation in Estonia.] *Lesnoe Khoz.*, (10): 9-12. Rus.

7893 **Trass, H. (1985).** Problems of the plant communities protection (for example from Estonia). *In* International Symposium "Protection of Natural Areas and the Genetic Fund they Contain. Project no. 8 on the Programme "Man and the Biosphere" (MAB) of Unesco, Sofia, 23-28.09. 1985. (En). (P)

U.S.S.R. - Estonia S.S.R.

7894 **Valk, U., Eilart, J., comps (1981).** Eesti metsad. (Forests of Estonia.) Tallinn, Estonian SSR, Valgus. 307, 29p. Es (Rus, Ge, En). Illus. Notes on tree species, wildlife and minor forest produce. (E)

7895 **Vilbaste, H. (1982).** Raba looduskaitseala. (The Nigula Bog State Nature Reserve.) *Eesti Loodus*, 25(11): 696-707. Es (Rus, En). Col. illus., map. (P)

7896 **Zobel, M. (1980).** Kaitsta okosusteeme. (Ecosystems require protection.) *Eesti Loodus*, 23(5): 301-308. (Rus, En). Illus., col. illus. Vilsandi Island, Estonia.

U.S.S.R. - Georgia S.S.R.

7897 **Asieshvili, L.V. (1979).** USSR: rare and protected decorative plants of the wild flora of Georgia and their cultivation in the Tbilisi Botanic Garden. *In* Synge, H., Townsend, H., *eds*. Survival or extinction. Proceedings of a conference held at the Royal Botanic Gardens, Kew, 11-17 September 1978. Kew, Bentham-Moxon Trust. 153-156. Illus. (G/T)

7898 **Dolukhanov, A.G. (1987).** K voprosam okhrany i izucheniya rastitel'nogo mira v Lagodekhskom i drugikh zapovednikakh Gruzii. (On the problems of protection and study of the plant world of the Lagodekhi Reservation and other reservations of Georgia.) *Bot. Zhurn.*, 72(10): 1405-1412. Rus. (P)

7899 **Katcharava, V.Ja., ed. (1982).** [Red Data Book of the Georgian SSR. Rare and endangered species of animals and plants.] Tbilisi, Sabcota Sakartvelo. 256p. Rus. Editor of the plant section: N.N. Ketzkhoveli. (T)

7900 **Kolakovskii, A.A., Yabrova, V.S. (1980).** Rasteniya Pitsunda-Myuserskogo Zapovednika. (Plants of the Pitsunda-Myus Nature Reserve.) *Novye Knigi SSSR*, 46: 77. Rus. (P)

7901 **Zaikonnikova, T.I. (1979).** *Sorbus velutina* (Albov) Schneid. (Rosaceae) - a vanishing species from the Caucasus. *Bot. Zhurn.*, 64(9): 1345-1348. (T)

U.S.S.R. - Kazakhstan S.S.R.

7902 **Andreeva, E.I. (1979).** [A rare species of a soil-surface lichen from the desert areas of Kazakhstan.] *Bot. Mater. Gerb. Inst. Bot. Akad. Nauk Kazakhskoi SSR*, ll: 78-80. Rus. (T)

7903 **Anon. (1980).** Okhrana rastitel'nogo mira Kazakhstana. (The protection of the plant world of Kazakhstan.) Kaz, SSR, Nauka. Rus.

7904 **Anon. (1981).** Krasnaya kniga Kazakhskoi SSR: redkie i nakhodyashchiesya pod ugrozoi ischezayushchiya vidy zhivotnykh i rastenii: ch. 2. Rasteniya. (A review: Red Data Book of the Kazakh SSR: rare and endangered animal and plant species: part 2. Plants.) Alma-Ata, Nauka Kaz, SSR. 263p. Rus. Illus. (T)

7905 **Anon. (1982).** Okhrana okruzhayushchei sredy v Kazakhstan: t.1. (The protection of the environment in Kazakhstan, vol. 1.) Kaz SSR, Nauka. Rus. (P)

7906 **Anon. (1983).** Okhrana okruzhayushchei sredy v Kazakhstan: t.2. (Protection of the environment of Kazakhstan, vol. 2.) Kaz SSR, Nauka. Rus.

7907 **Baimukhambetova, Zh.U. (1984).** Okhrana redkikh i ischezayushchikh vidov flory gor ulytau. (The protection of rare and disappearing species of plants of the mountains of Ulytau.) *Izv. Akad. Nauk Kaz. SSR, Biol.*, 6: 78-79. Rus.

7908 **Bel'gibaev, M.E. (1981).** Problemy okhrany prirody Severnogo i Tsentral'nogo Kazakhstana. (Problems of the protection of nature in north and central Kazakhstan.) *Izv. Akad. Nauk. Kaz. SSR, Biol.*, 6: 73-76. Rus.

7909 **Bijaschev, G.S., Bajtenov, M.S. (1981).** [Red Data Book of Kazakh SSR. Rare and endangered species of animals and plants. Part 2. Plants.] Alma-Ata, Nauka. 260p. Rus. Illus., maps. (T)

U.S.S.R. - Kazakhstan S.S.R.

7910 **Bizhanova, G. (1983).** K metodike sostavleniya kart okhrany pastbishchnykh ugodii. (A case study of maps compiling methods of the range lands conservation.) *Probl. Osvoen. Pust. Turkm. SSR,* 5: 54-57 Rus (En). Maps. The myunkum lands.

7911 **Byashev, G.S., Bykov, B.A. (1982).** O chem govorit "Krasnaya Kniga Kazakhskoi SSR"? (Who speaks of "The Red Data Book of the Kazakh SSR"?) *Izv. Akad. Nauk Kaz. SSR, Biol.,* 2: 4-10. Rus. Maps. (T)

7912 **Bykov, B.A., ed. (1987).** Okhrana redkikh vidov rastenii i rastitel'nosti Kazakhstana. (The protection of rare species of plants and vegetation of Kazakhstan.) Alma-Ata, Nauka. 83p. Rus. (T)

7913 **Karmysheva, N.K. (1973).** Flora i rastitel'nost' zapovednika Aksu-Dzhabagly: (Talasskii Alatau). (Flora and vegetation of the Aksu-Dzhabagly Nature Reservation: Talasskii Alatau.) Alma-Ata, Nauka. 176p. Rus. Illus., maps. (P)

7914 **Khrokov, V.V. (1980).** Zapovednik Kurgal'dzhino. (The Kurgal'dzhino Nature Reserve.) KazSSR, Kainar. Rus. (P)

7915 **Lyashenko, N.V. (1979).** Results of introduction of some rare species of flora of the Kazakh-SSR USSR. *Izv. Akad. Nauk Kaz. SSR Ser. Biol. Nauk,* 17(3): 13-20. (T)

7916 **Nechaeva, N.T., ed. (1984).** Resursy biosfery pustyn' Srednei Azii i Kazakhstana. Sovremennye Problemy Biosfery. (Resources of the biosphere of the deserts of Middle Asia and Kazakhstan. Contemporary problems of the biosphere.) Moskva, Nauka. 191p. Rus. (E)

7917 **Pryadko, G.F., Pyakina, A.K., Ban'kovskii, L.V. (1983).** Redkie rasteniya gor Ermentau, rekomenduemye dlya zapovednoi okhrany. (Rare plants of the Ermentau Mountains recommended for official protection.) *Izv. Akad. Nauk Kaz. SSR, Biol.,* 3: 64-66. Rus. (T)

7918 **Siderova, T.V. (1985).** Redkie rasteniya Kurgal'dzhinskogo zapovednika. (Rare plants of the Kurgal'dzhinsk Nature Reserve.) *In* Izuch. i okhrana zapoved. ob'ektov. Alma-Ata. 84. Rus. (P/T)

7919 **Sinitsyn, G.S. (1980).** K organizatsii i rezhimu Kapchagaiskogo Botanicheskogo Zakaznika. (On the organization and regime of the Kapchagai Botanical Reserve.) *Izv. Akad. Nauk. Kaz. SSR, Biol.,* 4: 77-79. Rus (Ka). *Ephedra equisetina, Ferula iliensis, Celtis caucasica.* (P/T)

7920 **Smetana, N.G., et al. (1982).** Nauzumskii zapovednik. (Nauzumskii Nature Reserve.) Kaz, SSR, Kainar. Rus. (P)

7921 **Vintergoller, B.A. (1976).** [Rare plants of Kazakhstan.] Alma-Ata, Nauka. 199p. Rus. (T)

7922 **Winterholler, B.A. (1979).** USSR: rare and threatened plants and their conservation in the botanic gardens of Kazakhstan. *In* Synge, H., Townsend, H., *eds.* Survival or extinction. Proceedings of a conference held at the Royal Botanic Gardens, Kew, 11-17 September 1978. Kew, Bentham-Moxon Trust. 149-151. (G/T)

7923 **Zhaparova, N.K. (1986).** Redkie i ischezayushchie rasteniya proektiruemykh Karatauskogo i Betpak-Dalinskogo gosudarstvennykh zapovednikow. (Rare and disappearing plants of the projected Kara Tau and Bet-Pak-Dala Reserves.) *Izv. Akad. Nauk Kaz. SSR, Biol.,* 1(133): 13-17. Rus (Ka). (P/T)

U.S.S.R. - Kirghizia S.S.R.

7924 **Aidarova, R.A. (1982).** Sostoyanie okhrany rastitel'nogo pokrova v Kirgizii i zadachi dal'neishikh nauchnykh issledovanii. (The condition of the plant cover of Kirghizia and problems of distant scientific studies.) *In* Rastit. resursy gor Kirgizii. Frunze. 133-137. Rus.

U.S.S.R. - Kirghizia S.S.R.

7925 **Arbaeva, Z.S., Bazhetskaya, A.A. (1982).** O sostayanii resursov poleznykh rastenii Respubliki i merakh po ratsional'nomu ikh ispol'zovaniyu, okhrane i vosproizvodstvu. *In* Rastit. resursy gor Kirgizii. Frunze. 90-104. On the condition of resources of useful plants of the Republic and of the measures for the rationalization of their utilization, protection and reproduction. Rus. (E)

7926 **Assorina, I.A. (1981).** [Rare and vanishing species of herbaceous plants from Kirghizia and experience of their introduction]. *Byull. Glavn. Bot. Sada,* 1981(122): 90-94. Rus. (T)

7927 **Tkachenko, V.I., Assorina, I.A. (1978).** [Rare and endangered plant species of the Kirghizia wild flora.] Frunze. 128p. Rus. (T)

7928 **Vorob'eva, M.G. & G.G. (1983).** Zapovednye territorii i ikh rol' v okhrane rastitel'nosti Kirgizii. (Protected territories and their role in the protection of the vegetation of Kirgizia.) *In* Rastit. resursy gor Kirgizii. Frunze. 137-149. Rus. (P)

U.S.S.R. - Latvia S.S.R.

7929 **Abele, G. (1981).** Krustalnu Rezervata flora un vegetacija. (Vegetation and the flora of the Krustalnu Reserve.) *Mezsaimnieciba un Mezrupnieciba,* 1981(3): 11-13. Latv (Rus). (P)

7930 **Abele, G.T., Limbena, R.E. (1979).** Flora zapovednoi zony "Nurmizhi". (Flora of the "Nurmizh" Protected Zone.) *In* Latvia. AN LatvSSR. Inst. Biologii. Flora i rastitel'nost' Latviiskoi SSR. Severo-Vidzemskii Geobotanicheskii raion. Riga, Zinatne. 100-103. Rus. (P)

7931 **Andrusaitis, G., ed. (1985).** Latvijas PSR Sarkana gramata: retas un iznikstosas dzivnieku un augu sugas. (Red Data Book of the Latvian SSR: rare and endangered species of animals and plants.) Riga, Zinatne. 526p. Latv, Rus (En). Col. illus., maps. (T)

7932 **Baroniya, V.K. (1985).** Novye mestonakhozhdeniya okhranyaemykh vidov rastenii. (New locations of protected species of plants: Eastern Latvia. *In* Tabaka, L.V., *ed.* Flora i rastitel'nost' Latviiskoi SSR: Vostochno-Latvii geobotanicheskii raion. Riga, Zinatne. 135-142. Rus. (T)

7933 **Bergkhol'tsas, I.I, Skriba, G.V. (1982).** Natsional'nyi Park "Gauya". (The "Gauya" National Park.) Moskva, Lesnaya Promyshlennost'. 168p. Rus (En, Fr, Ge). Col. illus., maps. (P)

7934 **Birkmane, K. (1974).** [Protected plants in Latvia.] Riga, Zinatne. 58p. Rus. (T)

7935 **Damberga, R., Kviese, D., Ledeboka, G. (1982).** [Flora of "Rochi" reservation of the "Gauja" National Park in the Latvian SSR]. *Trudy Latv. Sel'Skokhoz. Akad.,* 194: 41-52. Rus. (P)

7936 **Devichev, I.G., Krasovskaya, L.S. (1978).** O ponyatii "redkii vid" primenitel'no k malym territoriyam. (On the meaning of "rare plant" used in the small territories.) *In* Latvia Akademiya Nauk Latviiskoi SSSR. Institute Biologii Rastitel'nyi mir okhranyaemykh territorii. Riga, Zinatne. 67-72. Rus. (T)

7937 **Fatare, I. Ya., Gavrilova, G.B. (1985).** Redkie vidy rastenii. (Rare species of plants: Eastern Latvia.) *In* Tabaka, L.V., *ed.* Flora i rastitel'nost' Latviiskoi SSR: Vostochno-Latviiskii geobotanicheskii raion. Riga, Zinatne. 142-154. Rus. (T)

7938 **Kivi, V. (1977).** Redkie i okhranyaemye dekorativnye rasteniya dikorastushchei flory Estonskoi SSR v kulture Botanicheskogo Sada Tartuskogo Gosudarstvennogo Universiteta. (Rare decorative plants of Estonian SSR in cultivation in Tartu State University.) *In* Latvia AN LatvSSR. Sovet Bot. Sadov Pribaltiki Regiona. Botanischskie sady Pribaltiki: okhrana rastenii. Riga, Zinatne. 36-41. Rus. (G/T)

U.S.S.R. - Latvia S.S.R.

7939 **Kondratowicz, R., Czekalski, M. (1986).** Lotewski Park Narodowy "Gauja". (The Gauja National Park in Latvia.) *Chronmy Przyr. Ojczysta*, 41(4): 70-77. Pol. Illus., maps. (P)

7940 **Laivina, S., & M. (1981).** Grinu rezervata augu sabiedribu struktura un vides faktori. (Structure of plant associations and environment factors in the 'Sliteres' Reserve.) *Mezsaimnieciba un Mezrupnieciba*, 1981(3): 16-21. Latv (Rus). Illus. (P)

7941 **Luchinskene, A. (1977).** Redkie dikorastushchie tsvetochnye rasteniya dlye sadov i parkov Litovskoi SSR. (Rare native flowering plants for gardens and parks of the Lithuanian SSR.) *In* Latvia. AN LatvSSR. Sovet Bot. Sadov Pribaltiiskogo Regiona. Botanicheskie sady Pribaltiki: okhrana rastenii. Riga, Zinatne. 208-214. Rus. (G/T)

7942 **Maksin'sh, A.A. (1978).** [Gauja National Park (Latvian SSR).] *Lesnoe Khoz.*, 6: 54-57. Rus. (P)

7943 **Melluma, A.Kh. (1978).** Funktsionalnoe zonirovanie natsional'nogo Parka "Gauya" kak osnova dlya razrabotki programmy differentsirovannoi okhrany prirody. (Functional zoning of the Gauya Nat. Pk as a foundation for the exploitation of the programme of differentiation of the protection of nature.) *In* Latvia. Akademiya Nauk Latviiskoi SSR. Institut Biologii. Rastitel'nyi mir okhranyaemykh territorii. Riga, Zinatne. 12-19. Rus. (P)

7944 **Pirags, D. (1985).** Preservation of forest species gene pool in the Latvian SSR. *In* International Symposium: Protection of Natural Areas and the Genetic Fund They Contain. Sofia. 170. (E/G/T)

7945 **Rasin'sh, A., Kucheneva, G., Andronova, N., Kireeva, E. (1977).** Redkie i ischezayushchie vidy rastenii Kaliningradskoi oblasti. (Rare and disappearing species of plants of the Kaliningrad Region.) *In* Latvia. AN LatvSSR. Sovet Bot. Sadov Pribaltiiskogo Regiona. Botanicheskie sady Pribaltiki: okhrana rastenii. Riga, Zinatne. 113-116. Rus. (T)

7946 **Ripa, A.K. (1979).** Voprosy ratsional'nogo ispol'zovaniya Vosproizvodstva i okhrany klyukvy v latviiskoi SSR. (Problems of rational use, reproduction and protection of *Vaccinium oxycoccus* in the Latvian SSSR.) *Rast. Resursy*, 15(1): 70-75. Rus.

7947 **Tabaka, L., Fatare, I. (1981).** Sistema okhranyaemykh ob'ektov prirody i ee rol' v sokhranenii genofonda flory v Latviiskoi SSR. (Nature conservation and its role in the protection of the gene pool of the flora of the Latvian SSR.) *In* Ekol. i okhrana rast. Nechernozem. zony RSFSR. Ivanovo. 43-50. Rus. (T)

7948 **Tabaka, L., Klavina, G. (1981).** Floristisko petijumu nozime aisargajamo teritoriju augu valsts genetiska fonda saglabasana. (Importance of floristic studies in the preservation of the gene pool of reservation territories [Latvian SSR].) *Mezsaimnieciba un Mezrupnieciba*, 1981(3): 9-10. Latv (Rus). (P)

7949 **Tabaka, L.V., Kļyavinya, G.B. (1980).** O novykh i redkikh vidakh rastenii Latviiskoi SSR. (On the new and rare species of the plants from the Latvian SSR.) *Bot. Zhurn.*, 65(12): 1799-1801. Rus. (T)

7950 **Tsinovskis, R., Zvirgzd, A., Knape, D. (1977).** Okhranyaemye dendrologicheskie redkosti sel'skikh raionov Latviiskoi SSR. (Protected tree and shrub rarities of rural regions of the Latvian SSR.) *In* Latvia. AN LatvSSR. Sovet Bot. Sadov Pribaltiiskogo Regiona. Botanicheskie sady Pribaltiki: okhrana rastenii. Riga, Zinatne. 137-157. Rus. (T)

7951 **Vimba, E. (1979).** Papildinajumi Sliteres Rezervata augstako augu florai. (New habitats of rare plant species in the Slitere Reserve.) *Mezsaimnieciba un Mezrupnieciba*, 1: 23-25. Latv (Rus). Illus. (P/T)

U.S.S.R. - Latvia S.S.R.

7952 **Zvirgzd, A., Tsinovskis, R., Knape, D. (1977).** Itogi inventarizatsii starykh sel'skikh parkov Latvii. *In* Latvia. AN LatvSSR. Sovet Bot. Sadov Pribaltiiskogo Regiona. Botanicheskie sady Pribaltiki: okhrana rastenii. Riga, Zinatne. 117-136. Maps. Old rural parks of Latvia. (P)

U.S.S.R. - Lithuania S.S.R.

7953 **Apalia, D. (1977).** [New locations of 18 rare species of flora of the Lithuanian SSR and their ecological characteristics]. *Darb. Tr. Ser., C Ser. V Liet. Tsr. Mokslu Adad.*, 2: 17-26. Rus (En). Maps. (T)

7954 **Balyavichene, Yu.Yu., Lazdauskaite, Zh.P. (1978).** Okhrana rastitel'nosti Litovskogo Natsional'nogo Parka "Aukshtaitiya". (The protection of plants of the Lithuanian National Park "Aukshtaitiya".) *In* Latvia. Akademiya Nauk Latviiskoi SSSR. Institut Biologii Rastitel'nyi mir okhranyaemykh territorii. Riga, Zinatne. 19-23. Rus. (P)

7955 **Gudanavichyus, S. (1977).** Rasteniya landschaftnogo Zapovednika Esya, podlezhashchie okhrane. (Vegetation of the landscape of the River Esya Nature Reserve, as a subject for protection.) *In* Latvia. AN Latv SSR. Sovet Bot. Sadov Pribaltiiskogo Regiona. Botanicheskie sady Pribaltiki: okhrana rastenii. Riga, Zinatne. 104-105. Rus. (P)

7956 **Jankevicius, et al. (1981).** [Red Data Book of the Lithuanian SSR. Rare and endangered species of animals and plants.] Vilnius, Mosklas. 84p. Rus (Li). Plant section prepared by G. Baleviciene and K. Balevicius. (T)

7957 **Kairiukstis, L.A., Valius, M.I., Bumblauskis, T.T. (1983).** Standardprogramm komplexer Untersuchungen von Naturschutzgebieten am Beispiel des Naturschutzgebietes Zuvintas, Litauische S.S.R. (A standard programme of complex researches in nature reserves using the examples of the nature reserve Zuvintas, Lithuanian S.S.R.) *Arch. Naturschutz Landschaftsforsch.*, 23(1): 41-48. Ge. Map. (P)

7958 **Lithuania. Akademiya Nauk Litovskoi SSR. Institut Botaniki (1979).** Formirovanie rastitel'nogo pokrova pri optimizatsii landshafta: materialy 2-i Vses. Shkoly, Kaunas, 10-14 Sentyabr' 1979. (Formation of the plant cover for the optimization of the landscape: material of the second All-Union School, Kaunas, 10-14 September 1979.) Vil'nyus, In-t Botan. AN Lit. SSR. 206p. Rus. Illus.

7959 **Lithuania. Akademiya Nauk Litovskoi SSR. Otdel' Geografii Geograficheskoe Obshchestvo Litovskoi SSR. (1979).** Geograficheskii ezhegodnik 17. Kraeustroistvo i okhrana prirody. (The Geographical Yearbook 17. Land arrangement and nature protection.) Vilnyus, AN Litovskoi SSR. 253p. Li (Rus). Illus., maps. (P)

7960 **Marchyulensis, V.I. (1983).** Itogi i zadachi izucheniya rastitel'nykh resursov Litovskoi SSR. (Results and challenges of investigation of plant resources of the Lithuanian S.S.R.) *Rast. Resursy*, 19(1): 3-7. Rus. (E)

7961 **Tuciene, A. (1984).** [Vegetation of the "Cepkeliai" reserve. 4. Woodless and open woodland-associations of a fen.] *Trud. Akad. Nauk Litovsk. SSR* (Vilnius), ser. V, 3: 20-29. Rus (Li). Wetlands. (P)

U.S.S.R. - Moldavia S.S.R.

7962 **Akademiya Nauk Ukrainskoi SSR. Institut Botaniki im. N.G. Kholodnogo (1980).** Ukrainy, Belorussii Moldavii: okhrana vazhneishikh botanisheskikh ob'ektov. (The Ukraine, Byelorussia and Moldavia: the protection of important plants.) Kiev, Naukova Dumka. 389p. Rus. Illus., col. illus., map. (T)

7963 **Geideman, T.S. (1980).** O sokhranenii genofonda dikorastushchikh rastenii Moldavii. (On the preservation of the genofund of wild plants of Moldavia.) *In* Genetich. prohl. zagryazneniya okruzh. sredy na territorii Mold. SSR Zasedanie Sekts. genetich. aspektov probl. Chelovek i biosfera. Tez, Dokl. Kishinev. 43-45. Rus.

U.S.S.R. - Moldavia S.S.R.

7964 **Geideman, T.S., Chebotar', A.A. (1982).** Redkie vidy flory Moldavii: biologiia, ekologiia, geografiia. (Rare varieties of Moldavian flora.) Kishinev Shtiintsa. 102 p. Rus (Fr). Maps. (T)

7965 **Geideman, T.S., Nikolaeva, L.P. (1986).** Sovremennoe sostoyanie i okhrana flory moldavii. *Isv. Akad. Nauk Moldav. SSR, Biol. Khim.*, 3: 17-21. Rus. (T)

7966 **Geideman, T.S., Simonov, G.P. (1978).** Kharakterististika rastitel'nosti budeschchego prirodnogo parka Moldavii. (Characteristics of vegetation of a future natural park in Moldavia.) *Izr. Akad. Nauk Moldav. SSR, Biol. Khim*, 4: 5-12. Rus. (P)

7967 **Geideman, T.S., et al. (1982).** [Rare species of the flora of Moldavia.] Kishinev. 80p. Rus. Includes biological, ecological and geographical data. (T)

7968 **Kononov, V.N., Shabanova, G.A. (1978).** [New and rare species of the flora of Moldavia and their protection]. *Bot. Zhurn.*, 63(6): 908-912. Rus. (T)

7969 **Kononov, V.N., Shabanova, G.A., Mol'kova, I.F., Danilova, L.V. (1979).** Travyanistaya rastitel'nost' lesnykh polyan Feteshtskogo zapovednika. (Herbaceous vegetation of forest clearings of the Fetesht Nature Reserve.) *In* Sovrem. zadachi okhrani i ratsionae'n ispol'z. flory Moldavii. Kishvev. 33-36. Rus. (P)

7970 **Konovov, V. (1978).** [Strict conservation of flora in the Moldavian SSR]. *Sel'skoe Khoz. Moldavii*, 7: 60-61. Rus. (T)

7971 **Kravchuk, I. (1978).** [Acceleration of the creation of national and natural parks]. *Sel'skoe Khoz. Moldavii*, 8: 58-59. Rus. (P)

7972 **Lysikov, V.N., Buyukli, P.I., Kotel'nikova, L.K. (1980).** [Conservation of the genetic resources of cultivated plants in Moldavia]. *In* [Genetic problems of pollution of the surrounding environment in the territory of the Moldavian SSR]. Kishinev, Akademiya Nauk Moldavskoi SSR. 47-49. Rus. Conference of the Section of the Genetic Aspects of the Problems of Man and the Biosphere: collected papers. (E/G)

7973 **Moldavia. Moldavskoe Obschestvo Okrany Prirody (1978).** Redkie i ischezayushchie rasteniya Moldavii. (Rare and disappearing plants of Moldavia.) Kishinev, Timpul. 27p. Rus. Col. illus. (T)

7974 **Nikolaeva, L. (1980).** Alder *Alnus glutinosa* and *Alnus incana*, rare species in the Moldavian SSR. *Sel'sk. Khoz. Mold.*, 12: 55-56. (T)

7975 **Postolake, G.G., Golban, V.A. (1978).** [Rare plants of the forest area with a predominance of oak and an admixture of birch]. *Izv. Akad. Nauk Moldav. SSR, Biol. Khim.*, 5: 82. Rus. (T)

7976 **Rud', G., Zhadan, V., Bondarenko, V. (1978).** Some problems of conservation and regeneration of forests. *Sel'skoe Khoz. Moldavii*, 5: 32-34.

7977 **Shelyag-Sosonko, Yu.R., Sytnik, K.M., et al. (1980).** [Ukraine, White Russia, Moldavia. The protection of important botanical regions.] Kiev, Naukova Dumka. 332-370. Rus. Includes lists of threatened plants by region. (P/T)

7978 **Vitko, K.R., Istratii, A.I., Railyan, A.F. (1984).** Kompleksnaya otsenka rastitel'nosti okhranyaemykh territorii na primere Selishtskogo zakaznika lekarstvennykh rastenii (Moldauskaya S.S.R.).(Complex evaluation of vegetation on protected territories illustrated by the Selishte reservation of drug plants (Moldavian S.S.R.)). *Izv. Akad. Nauk Moldau. SSR, Biol. Khim.*, 1: 11-15. Rus. (G/P)

7979 **Abramchuk, A.V., Gorchakovskii, P.L. (1982).** Ob okhrane geneticheskikh resursov flory gornykh lugov Urala. (On the protection of the genetic resources of the flora of mountain meadows of the Urals.) *In* Izuch i osvoenie flory i rastit. vgsokogorii. 8 Vses. soveshch. Tez. dokl. 4: Rastit. resursy okhrana i rats. ispol'z rastit. mira vysokogorii. Sverdlovsk. 3. Rus.

7980 **Alekseeva, L.M. (1983).** [New and rare species for Kunashir Island.] *Byull. Glavn. Bot. Sada*, 127: 25-28. Rus. Illus. (T)

7981 **Anon. (1984).** Botanicheskie issledovaniya v zapovednikakh RSFSR: sb. nauch. tr. Tsentr. n-i. lab. okhotn. kh-va i zapovednikov. (Botanical investigations in nature reserves of the R.S.F.S.R.) (Moskva). 145p. Rus. Illus.

7982 **Bazilevich, N.I., Shmakova, E.I. (1984).** [Productivity of the completely reserved meadow steppe in the Central-Chernozem Biosphere Reserve.] *Byull. Mosk. Obshch. Ispyt. Prir.*, 89(4): 94-107. Rus (En). Illus. Meadow steppes. (P)

7983 **Belaya, G.A. (1983).** Novye vidy dlya flory Ussuriiskogo zapovednika. (New species for the flora of the Ussuriysk Reserve.) *Bot. Zhurn.*, 68(10): 1426-1427. Rus. (P)

7984 **Boikov, T.G. (1982).** Redkie i ischezayushchie vidy rastenii Zapadnogo Zabaikal'ya. (Rare and disappearing species of plants of western Baikal Region.) *In* Biol. resursy Zabaikal'ya i ikh okhrana. Uban-Ude. 19-46. Rus. (T)

7985 **Chugunov, Yu.D., ed. (1981).** Tsvetkovye rasteniya ostrovov Dal'nevostochnogo morskogo zapovednika. (Flowering plants of the islands of the Far Eastern Sea Nature Reserve.) Vladivostok, In-t-Biol. Morya. 155p. Rus. Illus. (P)

7986 **Eustratov, U.I., Tarshis, L.G. & G.I., Vasfilova, E.S. (1983).** Vnutrividovaya izmenchivost' i okhrana travyanistykh rastenii Srednego Urala. *In* Okhrana zhivoi prirody. Tez. Vses. konf. mol. uchenykh noyab. Moskva. 59-60. Rus. (P)

7987 **Fliagina, I.A. (1981).** Flora of the Sikhote-Alin State Reservation. *Byull. Glavn. Bot. Sada*, 1981(122): 59-63. (P)

7988 **Golovanov, V.D., et al., eds (1988).** Red Data Book of RSFSR, plants. Academy of Sciences of USSR, Botanical Institut of V.L. Komarov, Moskow, Rosagropromizdat. Rus. Col. illus., maps. (T)

7989 **Gorovai, P.G., Boiko, E.V. (1981).** Konspekt flory ostrova Furugel'ma. *In* Tsvetkov. rast. o-vov. Dal'nevost. mor.Zapoved. Vladivostok. 62-80. Rus. Abstr. in *Ref. Zhurn., Biol.*, 2(2): V694 (1983).

7990 **Grigor'evskaya, A.Ya, Kartashova, N.D. (1987).** [Plants of the Sokol'skoye mountain and their protection (on the question of the organization of new sections of the nature reserve "Galich'ya Gora" and of the study of vegetation of the North Donets Relict Region)]. *In* Priroda mal. okhranyaem territorii. Voronezh. 54-76. Rus. (P)

7991 **Hoey-Smith, J.R.P. van (1973).** Two very rare conifers *Microbiota decussata* and *Cathaya argyrophylla*. *IDS Year Book 1972*: 50-54. Illus. (T)

7992 **Ignatov, M.S. (1984).** [Discoveries of rare plants in the Moscow Region.] *Byull. Glavn. Bot. Sada*, (131): 86-89. Rus. (T)

7993 **Kamyshev, N.S. (1979).** Kratkaya istoriya stanovleniya i razvitiya zapovednika Galich'ya Gora i zadachi ego botanicheskogo issledovaniya. (A short history of the formation and development of Galich'ya Gora and the problems of its botanical investigation.) *In* Izuch zapovedn. landshaftov Galich'ei gory. Voronezh. 6-11. Rus.

7994 **Kononov, V.N., Tanfil'ev, V.G. (1982).** [New and rare plants in the flora of the Stavropol territory]. *Nov. Sistem. Vyssh. Rast. Akad. Nauk SSSR, Bot. Inst.*, 19: 196-199. Rus. (T)

U.S.S.R. - R.S.F.S.R.

7995 **Kozhevnikov, A.E., Gorshkov, A.Yu. (1984).** New and rare species of vascular plants of the flora of Kamchatka Peninsula. *Bot. Zhurn.*, 69(11): 1555-1562. Rus. Maps. (T)

7996 **Kubaev, V.B., Shelgunova, M.L. (1982).** Redkie i kharakternye rastitelnye soobshchestva yuzhnoi chasti Yakutii. (Rare and characteristic associations of the southern region of Yakutia.) *In* Okhrana redk. rastit. soobshchestv. Moskva. 10-17, 76. Rus (En). (T)

7997 **Kuvaev, V.B., Denisova, L.V., Beloussova, L.S., Nikitina, S.V. (1979).** Krasnaya kniga SSSR i okhrana redkikh i ischezayushchikh vidov rastenii sovetskogo Dal'nego Vostoka. (The Red Book of the USSR and the protection of rare and disappearing species of plants of the Soviet Far East.) *In* 14- i Tikhookean, nauch. kongr., Khabarovsk, 1979. Kom. A. Sekts. A.2. Tez. dokl. M.48-49. Rus. (T)

7998 **Kuzmin, G.F., Petrovskii, E.E. (1980).** Vydelenie ob'ektov okhrany dlya bolot Sakhalina. (The isolation of objectives of the protection of the wetlands of Sakhalin.) *Trudy Darwinsk. Gos. Zapovednika*, 15: 74-82. Rus.

7999 **Makarov, V.V., Nedoluzhko, V.A., Urusov, V.M. (1982).** [Additions to the flora of the "Kedrovaia Pad" reservation]. *Byull. Glavn. Bot. Sada*, 1982(123): 47-51. Rus. (P)

8000 **Malyshev, L.I. (1979).** Sokhranenie flory na izolirovannykh territoriyakh. (Preservation of the flora in isolated territories.) *In* 14-i Tikhookean. Nauch. Kongr. Khabarovsk, 1979. Kom A. Sekts. A.2. Tez. dokl. M. 58-59. Rus.

8001 **Masing, V. (1980).** Kellele seda soist tihnikut Sosval vaja on? 1. (Who needs the swampy thickets on the Sosva River?: part 1.) *Eesti Loodus*, 23(1): 43-48. Rus (En). Illus., col. illus. The Konda-Sosva State Nature Reserve in West Siberia (1934-1952). (P)

8002 **Mazing, V.V. (1979).** Ob okhrane redkikh vidov (na primere bolotnykh rastenii). (On the protection of rare species, especially of wetland plants.) *Trudy Darvinsk. Gos. Zapov.*, 15: 64-68. Rus. (T)

8003 **Milov, V.M. (1979).** O nekotorykh okhranyaemykh vidakh rastenii Leningradskoi oblasti. (On some protected species of plants of the Leningrad Region.) *Nauch. Trud. Leningr. S.-Kh. In-ta*, 370: 94-98. Rus. (T)

8004 **Novikov, V.S., Oktiabreva, N.B., Tikhomirov, V.N. (1979).** [*Dentaria bulbifera* L., a new species for the flora of the Moscow region]. *Biol. Nauki* (Moscow), 8: 80-82. Rus. Includes conservation.

8005 **Pechenyuk, E.V. (1982).** Rare aquatic plants in the Khoper State Reserve Russian-SFSR USSR. *Bot. Zh. (Leningr.)*, 67(5): 647-651. (P/T)

8006 **Polozhii, A.V., Revushkin, A.S., Vydrina, S.N. (1982).** Ob okhrane rastitel'nogo mira vysokogornoi Altae Sayanskoi provintsii. (On the protection of the plant world of the high mountains in the Altai Sayan Province.) *In* Izuch. i osvoenie flory i rastit. vysokogorii. 8 Vses. soveshch. Tez. dokl 4: Rastit. resursy, okhrana i rats. ispol'z rastit. mira vysokogorii. Sverdlovsk. p.38. Rus.

8007 **Polyakova, G.A., Flerov, A.A. (1982).** O redkikh; ne ukazannykh dlya Moskovskoi oblasti rasteniyakh. (On some rare plants and the plants that are not indicated for the Moscow District.) *Bot. Zhurn.*, 67(11): 1543-1544. Rus. (T)

8008 **Pridnya, M.V. (1981).** Network of protected natural areas in the western Caucasus and the role of the Caucasus Reserve. *Sov. J. Ecol.*, 12(6): 322-327 (1981 publ. 1982). Map. (P)

8009 **Rakova, M.V. (1980).** O redkom dal'nevostochnom vide Fialki *Viola rossii* (Violaceae). (On the rare species *Viola rossii* (Violaceae) from the Far East.) *Bot. Zhurn.*, 65(7): 994-1000. Illus. (T)

U.S.S.R. - R.S.F.S.R.

8010 **Samarin, V.P. (1983).** Osobo redkie endemy i relikty flory tsvetkovykh Chelyabinskoi oblasti i osnovnye puti ikh okhrany. (Especially rare endemics and relicts of the flora of flowering plants of the Chelyabinsk Region and basic methods for its protection.) *In* Flora i rastit. Urala i puti ikh okhrany. Chelyabinsk. 3-15. Rus. (T)

8011 **Savel'eva, L.I. (1980).** [Forest reserves as a form of nature conservation, as exemplified by the Moscow region]. *Lesovedenie*, 4: 29-37. Rus (En). (L/P)

8012 **Semenova-Tyan-Shanskaya, A.M., Ivashintsov, A.D. (1980).** Priroda Leningradskoi oblasti i ee okhrana. (Nature of the Leningrad Region and its protection.) *Prirode Ezhem. Pop. Estest. Zhurn. Akad. Nauk*, 6: 80-89. Rus. Illus., map.

8013 **Seredin, R.M. (1981).** [Data for the Red Book of northern Caucasia, Ciscaucasia and Dagestan ASSR, Russian SFSR, USSR]. *Izv. Severo-Kavk. Nauch. Tsentra Vyssh. Shkoly. Estest. Nauk.*, 1: 78-85. Rus. (T)

8014 **Sergienko, V.G. (1983).** [New and rare species for the north-western part of the Kirov region]. *Nov. Sistem. Vys. Rast. Akad. Nauk SSSR, Bot. Inst.*, 20: 201-202. Rus.

8015 **Shilov, M.P. (1982).** [Distribution of certain rare and vanishing plant species in Ivanovo Oblast and Vladimir Oblast Russian-SFSR USSR and the state of their populations.] *Biol. Nauki.* (Moscow), (4): 58-62. Rus. (T)

8016 **Slizik, L.N. (1978).** Redkie i tsennye vidy derevyanistykh lian Primorskoi kraya vozmozhnosti ikh okhrany i vosproizvodstva. (Rare and valuable species of woody lianas of the Primorsk Region and the means of their protection and reproduction.) *In* Aktualn. vopr. okhrany prirody na Dal'n. Vost. Vladivostok. 47-55. Rus. (T)

8017 **Smirnova, A.D. (1979).** Opyt organizatsii okhrany bolotnykh ekosistem (na primere Gor'kovskoi oblasti). (An experiment in the organization of the protection of wetland ecosystems, especially of the Gorki Region.) *Trudy Darvinsk. Gos. Zapovednika*, 15: 69-73. Rus.

8018 **Trufanova, E.R. (1979).** Botanicheskie ob'ekty v nizov'e reki Leny, podlezhashchie okhrane. (Plants in the lower reaches of the River Lena, entitled to protection.) *In* Okhrana prirody yakutii. Yakutsk. 20-23. Rus. (T)

8019 **Tsaregradskaya, A.P. (1979).** Lekarstvennye rasteniya Yakutii i ikh okhrana. (The medicinal plants of Yakutia and their protection.) *In* Okhrana prirody Yakutii. Yakutsk. 20-26. Rus. (E)

8020 **Turkov, V.G. (1979).** Dinamika rastitel'nogo pokrova Visimskogo zapovednika v protsesse khozyaistvennogo osvoeniya ego territorii (XVII-XXVV). (The dynamic of the plant cover of Visim Nature Reserve in the process of agricultural production in its territory.) *Trudy in-ta Ekol. Rast. i Zhivotnyk Uralsk. Nauch. Tsentr. AN SSSR*, 128: 34-50. Rus. (P)

8021 **Ukhacheva, V.N. (1984).** Floristicheskii spisok Uzyanskogo uchastka Bashkirskogo zapovednika. (The list of species of Uzyanski part of Bashkir Reserve.) *Vestn. Leningrad Univ. Biol.*, 9: 42-50. Rus (En). Central part of southern Ural. (P)

8022 **Zabrodin, V.A., ed. (1981).** Floristickeskie issledovaniya v zapovednikakh RSFSR. (Floristic studies in nature reserves in the RSFSR.) Moskva. 153p. Rus. Illus. (P)

8023 **Zaikonnikova, T.I. (1979).** *Sorbus velutina* (Albov) Schneid. (Rosaceae) - a vanishing species from the Caucasus. *Bot. Zhurn.*, 64(9): 1345-1348. (T)

8024 **Zarubin, S.I., Neshta, I.D., Malova, A.N., Donskova, A.A., Karavaeva, Z.I. (1983).** Redkie i ischezayushchie vidy flory Tyumenskoi oblasti. (Rare and vanishing species of the flora of the Tymen region.) *Bot. Zhurn.*, 68(9): 1265-1269. Rus. In West Siberia. (T)

U.S.S.R. - R.S.F.S.R.

8025 **Zhirnova, T.V., Alekseev, IU.E., Chechetkin, E.V. (1984).** [Second addition to the list of vascular plants from the Bashkir reservation]. *Biologicheskie Nauki: Nauchnye Doklady Vysshei Shkoly,* 1984 (4): 68-72. Rus (En). (P)

8026 **Zolotukhin, N.I., Zolotukhina, I.B., Marina, L.V. (1986).** [Flowering plants of the Altai Nature Reserve on the upper high altitude boundary]. *In* Ekosistemy ekstremal. uslovii sredy v zapoved. RSFSR. [Moskva]. 74-80. Rus. (P/T)

U.S.S.R. - Tadzhikistan S.S.R.

8027 **Rasulova, M.R., Yunusov, S.Yu., Ashurov, A.A. (1981).** Voprosy okhrany flory i rastitel'nosti Tadzhikistana. (Questions of protection of the flora and vegetation of Tadzhikistan.) *Okhr. Prirody Tadzhikistana* (Dushanbe), 2: 60-71. Rus.

8028 **Yunushov, S., Kamelin, R.V. (1980).** [Materials for the Red Data Book of the Tadzhik SSR. Rare and endangered animals and plants.] Dushanbe. 32p. Rus. (T)

U.S.S.R. - Turkmenistan S.S.R.

8029 **Abramova, T.I., Shkhagapsoev, S.Kh. (1984).** Endemichnye reliktovye i redkie vidy rastenii Kabardino-Balkarskogo vysokogornogo zapovednika. (Endemic relict and rare plant species in the Kabardino-Balkarian High Mountain Reservation.) *Byull. Mosk. Obshch. Ispyt. Prir., Biol.,* 89(2): 114-118. Rus (En). (P/T)

8030 **Babaev, A.G., Durdyev, K. (1982).** [Brief physical-geographic characteristics of the western Kopet Dagh]. *Priroda Zapadn. Kopetd.,* 1982: 7-19. Rus. Includes vegetation and conservation problems. (P)

8031 **Babaev, A.G., ed. (1985).** Ocherki prirody i khozyaistva Turkmenistana. (Essays on the nature and economy of Turkmenistan). Ashkhabad, Ylym. 251p. Rus. (E/G/T)

8032 **Berkutenko, A.N., Kamelin, R.V. (1975).** Opredelitel' rastenii Repetekskogo Zapovednika. (Determining the plants of the Repetek Reserve.) Ashkabad, Ylym. 109p. Rus. Illus. (P)

8033 **Frantskevich, N.A. (1978).** [Wild relatives of cultivated plants and their conservation in the basin of Ai-Dere River (Kara-Kala District of the Turkmen SSR)]. *Biul. Vses. Inst. Rastenievodstva,* 81: 86-91. Rus. (E)

8034 **Gudkova, E.P., Seifullin, E.M., Chopanov, P.M. (1982).** [Conspects of the flora of the Western Kopet Dagh]. *Prirod. Zapadn. Kopetdaga,* 1982: 38-119. Rus. Conservation problems. (P)

8035 **Gunin, P.D., Veyisov, S.V. (1987).** Repetekskii biosfernyi zapovednik kak regional'nyi ekologicheskii tsentr Vostochnyk Karakumov. (Repetek Biosphere Reserve as regional ecological centre of the East Karakum.) *Probl. Osvoen. Pustyn'* (Turkm. SSR), 5: 54-61. Rus (En). (P)

8036 **Ishankuliev, M.I., Kuz'menko, V.D., Atakhanov, O.A. (1983).** [Summary of flora of the Eradzhi reservation (East Kara-Kum).] *Izvest. Akad. Nauk Turkmen. SSR,* (6): 66-73. Rus (En). (P)

8037 **Kamakhina, G.L. (1985).** Ob endemichnykh i redkikh vidakh uchaska Babazo Kopetdagskogo zapovedrika. (On state of endemic and rare plants on Babazo Plot of Kopetdag Reserve.) *Izv. Akad. Nauk Turkm. SSR, Biol. Nauk,* 3: 6-13. Rus (En). (P/T)

8038 **Kamelin, R.V., Kurbanov, D. (1987).** O nekotorykh ischezayushchikh i redkikh rasteniyakh zapadnykh nizkogorii Turkmenii. (On some endangered and rare plants of the west low mountains of Turkmenistan). *Bot. Zhurn.,* 17(3): 397-402. Rus. (T)

U.S.S.R. - Turkmenistan S.S.R.

8039 **Kerbabaev, B.B., Ishchenko, L.E. (1985).** Puti sokhraneniya genofonda mestnoi flory. (Means of preserving the gene pool of the local flora.) *Izv. Akad. Nauk Turkm. SSR, Biol.* Nauk, 4: 3-9. Rus. Gives list of Turkmenistan plants cultivated in the Botanic Garden of the Academy of Sciences of the Turkmen SSR, Ashkhabad. (G)

8040 **Kogan, S. (1978).** [Water bodies conservation in the Turkmenistan.] *Izv. Akad. Nauk Turkm. SSR, Biol. Nauk*, 4: 84-91. Rus (En). Includes aquatic plants.

8041 **Kulibaba, V.V., Neshataeva, G.I., Skalon, A.V. & N.V., Kazakova, T. (1982).** [Analysis of the present state of the Tugai ecosystem of Sumbar River Area.] *In* Priroda Zapadnogo Kopetdaga. 133-145. Rus. Illus. Vegetation associations and nature conservation problems.

8042 **Levin, G.M., Mishchenko, A.S., Petrova, E.F., Chopanov, P., Esenova, K. (1978).** Some rare wild plants of Turkmenistan. *Izv. Akad. Nauk Turkm. SSR, Biol. Nauk*, 4: 48-52. Rus (En). Includes aspects of conservation. (T)

8043 **Mesheheryakov, A.A. (1980).** Novye i redkie rasteniya dlya flory Turkmenii. (De plantis pro flora Turcomania novis et raris.) (New and rare plants for the flora of Turkmenia.) *Novosti Sist. Vyssh. Rast.*, 17: 239-240. Rus. *Arnebia baldschuanica, Crucianella chlorostachys, Astragalus spinescens.* (T)

8044 **Mizgireva, O.F. (1978).** Mandragora Turcomanica Mizgir. *Izv. Akad. Nauk Turkm. SSR, Biol. Nauk*, 4: 54-55. Rus (En). Plant conservation aspects. (T)

8045 **Nikitin, V.V. (1978).** Disappearing rare plants of Kopet Dagh and the problem of restoring their area. *Izv. Akad. Nauk Turkm. SSR, Biol. Nauk*, 2: 3-9. Rus (En). (T)

8046 **Nikitin, V.V. (1979).** Rasteniya flory Turkmenii, predlagaemye v *Krasnuyu knigi Turkmenskoi SSR.* (List of the Turkmenistan plants included in the *Turkmen SSR Red Data Book.*) *Izv. Akad. Nauk Turkm. SSR, Biol. Nauk*, 6: 3-11. (T)

8047 **Nikitin, V.V. (1979).** Redkie i ischezayushchie rasteniya flory Turkmenii v "Krasnoi knige SSR". (Rare and disappearing plants of the Turkmenian flora in the "Red data of the U.S.S.R.") *Bot. Zhurn.*, 64(12): 1799-1807. (T)

8048 **Nikitin, V.V., Klyushkin, E.A. (1975).** [Plant species of the Turkmen SSR to be entered in the Red Book.] *Izv. Akad. Nauk Turkm. SSR, Ser. Biol. Nauk.*, 2: 73-76. Rus. (T)

8049 **Nikitin, V.V., Krasikova, N.S. (1981).** [Rare, little known plant *Dendrostellera turkmenorum.*] *Izv. Akad. Nauk Turkm. SSR, Biol. Nauk*, 3: 58-60. Rus. (T)

8050 **Nikitin, V.V., Kurbandurdyev, M. (1978).** Endemic and rare plants in the Central Kopet Dagh Reserve. *Izv. Akad. Nauk Turkm. SSR, Biol. Nauk*, 1: 10-22. Rus (En). (P/T)

8051 **Nikitin, V.V., Muradov, K.M., Kliushkin, E.A. (1978).** Endemic and rare species of the Turkmenistan flora. *Izv. Akad. Nauk Turkm. SSR, Biol. Nauk*, 4: 11-25. Illus. (T)

8052 **Proskuriakova, G.M. (1978).** Juniper stands in Turkmenia and problems of their conservation. *Izv. Akad. Nauk Turkm. SSR, Biol. Nauk*, 4: 34-41. Rus. (T)

8053 **Razumovskii, S.M., Neshataeva, G.I. (1982).** [Ecological fundamentals of the restoration of natural vegetation of Western Kopet Dagh]. *Prir. Zapadn. Kopetd.*, 1982: 120-132. Rus. Illus. Conservation problems.

8054 **Rustamov, A.K. (1978).** [Red Book of the Turkmen SSR]. *Izv. Akad. Nauk Turkm. SSR, Biol. Nauk*, 4: 8-10. Rus (En). (T)

8055 **Rustamov, A.K. (1983).** Kratkie itogi, sostoyanie i zadachi okhrany prirody v Turkmenistane: okhrana prirody: kompleksnaya zadacha. (Brief results: of the condition and aims of the protection of nature in Turkmenistan; the protection of nature; combined aims.) *Izv. Akad. Nauk Turkm. SSR, Biol. Nauk*, 5: 16-23. Rus. (P)

U.S.S.R. - Turkmenistan S.S.R.

8056 **Rustamov, A.K., Babaev, A.G., Klyushkin, E.A. (1978).** [La conservation de la nature au Turkmenistan et ses perspectives]. *Probl. Osvoen. Pustyn, Turkm. SSR,* 4: 3-7. Rus (Fr).

8057 **Sokolov, V., Gunin, P. (1984).** The Repetek reserve: the first desert biosphere reserve in the USSR. *In* USSR Academy of Sciences. Man and biosphere. Moscow, Nauka Publishers. 179-187. Maps. (P)

U.S.S.R. - Ukrainian S.S.R.

8058 **Akademiya Nauk Ukrainskoi SSR. Institut Botaniki im. N.G. Kholodnogo (1980).** Ukrainy, Belorussii Moldavii: okhrana vazhneishikh botanisheskikh ob'ektov. (The Ukraine, Byelorussia and Moldavia: the protection of important plants.) Kiev, Naukova Dumka. 389p. Rus. Illus., col. illus., map. (T)

8059 **Andrienko T.L., Pryadko E.L. (1989).** Fitots enotychna reprezentatyunist' bolotnykh pryodno-zapovidnykh ob'ektiv Ukrayiny. (Phytocenotic representation of the bog nature reserve objects of the Ukraine.) *Ukr. Bot. Zhurn.,* 46(1): 77-80. Uk (En, Rus). (P)

8060 **Andrienko, T.L. (1980).** Distribution, ecology and phytocenology of rare boggy species of the family Orchidaceae in the Ukraine. Tallin, Akademiia Nauk Estonskoi SSR. 57-60. (T)

8061 **Andrienko, T.L. (1983).** Roslynnist' zakaznyka "Horodnyts'kyi" (Zhytomyrs'ke Polissya). (Vegetation of the reservation "Gordodnitsky" [Zhitomir Polessie].) *Ukr. Bot. Zhurn.,* 40(2): 107-111. Uk (En, Rus). Illus. (P)

8062 **Andrienko, T.L., Partyka, L. Ya (1984).** Rosslynnist' ta florystychni osoblyvosti zakaznyka "Nechymne" (Volyns'ka Oblast'). (Vegetation and floristic peculiarities of the "Nechimnoe" Reservation [Volyn Region].) *Ukr. Bot. Zhurn.,* 41(1): 90-94. Uk (En, Rus). Illus., map. (P)

8063 **Andrienko, T.L., Popovich, S.Yu., Shelyagsosonko, Yu.R. (1986).** Polesskii gosudarstvennyi zapovednik: rastitel'nyi mir. (Polesye State Nature Reserve in Ukraine in Zhitomir Obl. MAB). Kiev, Naukova Dumka. 203p. Rus. Illus., col. illus., maps. (P)

8064 **Andrienko, T.L., Priadko, O.I., Udra, I.K., Sheliag-Sosonko, R. (1981).** [Vegetation of the "Babka" forest reservation in the Kiev Polesye]. *Ukr. Bot. Zhurn.,* 38(2): 91-95. Uk (En, Rus). Illus. Conservation of rare species. (P)

8065 **Andrienko, T.L., Shelyag-Sosonko, Yu.R., Ustimenko, P.M. (1982).** Lisova roslynnist' terytoriyi zaproektovanoho Mezyns'koho Pryrodnoho Parku. (Forest plants in the territory of the planned natural Mezin Park.) *Ukr. Bot. Zhurn.,* 39(2): 74-81. Uk (En, Rus). Map. (P)

8066 **Anon. (1980).** Chervona kniha Ukrayinskoy RSR. (Krasnaya kniga Ukrainskoi SSR.) (The Red Data Book of the Ukrainian SSR.) Kyyiv, Naukova Dumka. 499p. Uk. Illus. (T)

8067 **Anon. (1985).** Fitosfera ta yiyi okhorana. (Phytosphere and its protection.) *Ukr. Bot. J.,* 42(2): 49-82. Uk.

8068 **Antonyuk, N.E. (1980).** [Conservation of rare forest species of the flora of the Ukrainian SSR in the Central Republican Botanical Garden of the Ukrainian SSR Academy of Sciences]. *Polezn. Rast. Prir. Flory Ispol'z. Nar. Khoz.,* 150-154. Rus. Illus. (G/T)

8069 **Antonyuk, N.E. (1982).** [Rare plants of the Ukraine in cultivation.] Kiev, Naukova Dumka. 212p. Rus. (G/T)

8070 **Antonyuk, N.E., Borodina, R.N., Sobko, V.G., Skvortsova, L.S. (1982).** [Rare plants of the flora of the Ukraine in cultivation.] Kiev, Central Republic Botanic Gardens. 212p. Rus. Based on information received from 94 botanic gardens and contains data on 117 native threatened plant species cultivated in the U.S.S.R. (G/T)

U.S.S.R. - Ukrainian S.S.R.

8071 **Badalov, P.P. (1978).** [Rare species of *Fraxinus* in the steppe zone of the Ukraine and their importance for forestry]. *Lesovodstvo i Agrolesomelioratsiia*, 50: 69-77. Rus. Illus. (T)

8072 **Burda, R.I. (1984).** Flory osobo okhranyaemykh territorii na yugo-vostoke U.S.S.R. (Floras, especially of the protected territories of the South Eastern Ukrainian SSR.) *In* Probl. okhrany genofonda i upr. ekosistemami v zapoved. step. i pustyn'. zon. Tez. dokl. Vses. soveshch., Askaniya-Nova, 21-25 maya, 1984. Moskva. 14-18. Rus.

8073 **Cherevchenko, T.M. (1982).** Orchids in protected cultivation of the Central Republican Botanical Garden of the Ukrainian SSR Academy of Sciences. *Introd. Aklim. Rosl. Ukr.*, 20: 71-75. (G)

8074 **Chopik, V.I. (1970).** [Scientific grounds for protection of the Ukrainian flora rare species.] *Ukr. Bot. Zhurn.*, 27: 693-704. Rus (En). Includes a list of 185 spp. in need of protection. (T)

8075 **Chopik, V.I. (1971).** [Flora and technical progress.] *Bot. J. Acad. Sci. USSR*, 57: 281-290. Rus (En.) Presents a survey of the literature on changes of flora and disappearance of species in the Ukraine. (T)

8076 **Chopik, V.I. (1978).** Redkie i ischezayushchie rasteniya Ukrainy: spravochnik. (Rare and threatened plants of the Ukraine.) Kiev, Naukova Dumka. 211p. Illus., col. illus., maps. (T)

8077 **Chopik, V.I., Pogrebennik, V.P., Nechitailo, V.A., Yatsenko, N.P., Bortnyak, N.N. (1984).** K okhrane genofonda Stepnykh nidov Flory Srednego Pridneprov'ya Ukrainskoi S.S.R. *In* Probl. okhrany genofonda i upr. ekosistemami v zapoved. step. i pustyn'. zon. Tez. dokl. Vses. soveshch., Askaniya-nova, 21-25 maya, 1984. Moskva. 62-65. Rus. Genetic resources. (E)

8078 **Chupryna, T.T. (1978).** [Some biological characteristics of rare species of *Stipa* introduced into the Donets Basin]. *Introd. Aklim. Rosl. Ukr. Akad. Nauk. USSR*, 13: 23-25. Uk (Rus). (T)

8079 **Deliamure, S.L., Ena, V.G., Mishnev, V.G., Molchanov, E.F. (1978).** [Problems of the protection of rare organisms and unique natural complexes in the southern part of the Ukrainian SSR]. *Byull. Gos. Nikit. Bot. Sada*, 3: 5-11. Rus. (T)

8080 **Didukh, Ya.P., Sukhoi, I.B. (1984).** Suchasnyi stan roslynnoho pokryvu lisovoho masyvu Bannyi Yar (Sums'ka oblast') ta ioho okhorona. (The present state of the plant cover in the Banny Yar Forest Massif (The Sumy Region) and problems of its protection.) *Ukr. Bot. Zhurn.*, 41(5): 70-73. Uk (En, Rus). Map. (P/T)

8081 **Dubina, D.V. (1987).** Roslynnist' Prydunais'kykh ozer ta yiyi okhorona. (Vegetation of Danube estuaries and its protection.) *Ukr. Bot. Zhurn.*, 44(6): 77-81. Uk (En, Rus). Proposal to create Danube Lakes Nature Reserve to join Danube Biosphere Reserve. (P)

8082 **Dubyna, D.V. (1980).** [Conservation and enrichment of Nymphaeaceae species from the natural flora of the Ukraine]. *Pol. Rast. Prirod. Flory Ispol. Narod. Khoz.*, 1980: 147-150. Rus.

8083 **Dubyna, D.V. (1986).** Funktsional'ne zonuvannya terytoriyi zaproektovanoho Nizhn'odniprovs'koho pryrodnoho natsional'noho parku "Khersons'ka oblast". (Functional zonation of the territory of the planned Lower-Dnieper Natural National Park - Kherson Region). *Ukr. Bot. Zhurn.*, 43(3): 94-98. Uk (En, Rus). Map. (P)

8084 **Dubyna, D.V. (1986).** Roslynnist' vodoim dolyny r. Pivlennyi Buh, yiyi florystychni osoblyvosti ta okhorona. (Vegetation of water bodies of the Yuzhny Bug River valley, its floristic peculiarities and protection). *Ukr. Bot. Zhurn.*, 43(6): 64-69. Uk (En, Rus). Map.

U.S.S.R. - Ukrainian S.S.R.

8085 **Ena, A.V., Dubonos, V.N. (1982).** Karadagskii Zapovednik. (Karadag Nature Reserve.) *Prir. Ezhem. Pop. Estest. Zhurn. Akad. Nauk*, 9: 80-87. Rus. Col. illus., map. (P)

8086 **Genov, A.P. (1984).** [Conservation of small forests in steppe ravines in the Left-Bank Ukraine.] *Introd. Akklim. Rast.*, 1: 36-38. Rus.

8087 **Genova, L.F. (1984).** [Some rare and unique phytocenoses in the branches of the Ukrainian State Reserve "Khomutovskaia Steppe" and "Kamennye Nogily".] *Introd. Akklim. Rast.*, 1: 38-39. Rus. (P)

8088 **Gensiruk, S.A., Nizhnik, M.S., Mitsenko, V.O. (1982).** Ekoloho-ekonomichni aspekty pryrodo-korystuvannya. (Ecological and economic aspects of the use of nature.) Kyyiv, Naukova Dunka. 174p. Uk. Col. illus.

8089 **Gladun, Ya.D., Gladun, M.I. (1983).** Poshyrennya i zapasi naivazhlyvishykh likars'kykh roslyn L'vivs'koyi oblasti. (Distribution and reserves of imported drug plants of the Lvov Region.) *Ukr. Bot. Zhurn.*, 40(5): 15-18. Uk (En, Rus). Maps. Medicinal. (E/T)

8090 **Golubets, M.A., Zaverukha, B.V. (1987).** Suchasnyi stan vyvchenosti, vykorystannya, popovnennya ta zberezhennya henofondu flory URSR. (Modern state of the study, use, supplement and preservation of the gene pool of the Ukrainian SSR flora). *Ukr. Bot. Zhurn.*, 44(2): 1-8. Uk (En, Rus).

8091 **Gorelova, L.N. (1986).** [The condition and perspective of the protection of species of plants of the middle stream of the North Donets River included in the "Redbook of the USSR" and the "Redbook of the Ukrainian SSR".] *In* Flora i rastit. Ukrainy. Kiev. 19-22. Rus. (T)

8092 **Gusev, A.A. (1986).** Tsentral'nochernozemnomu gosudarstvennomu biosfernomu zapovedniku imeni professora V.V. Alekhina 50 let. (The Prof. V.V. Alekhin Central Chernozem State Biosphere Reserve is 50 years old.) *Byull. Mosk. Obshch. Ispyt. Prir., Biol.*, 91(2): 159-161. Rus. (P)

8093 **Hanzha, R.V., Liuryn, I.B., Shevchneko, M.T. (1983).** [Principles of production, nature conservation and evaluation of ecological systems]. *Ukr. Bot. Zhurn.*, 40(5): 9-11, 14. Uk (En, Rus).

8094 **Hizatullin, S.K. (1978).** [Problems of preserving forests on flooded river banks (in the Ukrainian SSR)]. *Lisove Hospodarstvo*, 3: 6-7. Uk.

8095 **Holubiev, V.M. (1984).** [Urgent problems of ecological studies and nature conservation]. *Ukr. Bot. Zhurn.*, 41(2): 66-75. Uk (En, Rus). Illus.

8096 **Horb, V.K. (1984).** Natural populations of *Syringa josikaea* Jacq.f. in the Ukrainian Carpathians. *Ukr. Bot. Zhurn.*, 41(4): 62-64. Uk (En, Rus). (T)

8097 **Iashchenko, P.T. (1984).** Biomorphological spectrum of the flora of the area of Shatsk lakes. *Ukr. Bot. Zhurn.*, 41(5): 73-77. Uk (En).

8098 **Iashchenko, P.T., Andrienko, T.L., Shelyag-Sosonko, Yu.R., Stoiko, S.M. (1983).** [Vegetation cover of the planned Shatskiy National Park]. *Ukr. Bot. Zhurn.*, 40(4): 71-76. Uk (En, Rus). Illus., maps. (P)

8099 **Ignatenko, O.S., Semenova-Tyan-Shanskaya, A.M. (1979).** Okhrana redkikh vidov flory Tsentral'no - Chernozemnogo Zapovednika. (Protection of rare species of Central Chernozem Reservation flora.) *Bot. Zhurn.*, 64(12): 1816-1824. Rus. (P/T)

8100 **Ivashin, D.S., Ganzha, R.V., Stasilyunas, O.A., Golova, T.P., Litvinova, M.D. (1985).** Ridkisni roslyny pivdenno - skhidnoyi chastyny livoberezhnoho lisostepu Ukrayiny. (Rare plants of the South Eastern part of the left bank forest steppe of the Ukraine.) *Ukr. Bot. Zhurn.*, 42(1): 71-75. Uk (En, Rus). Map. (T)

U.S.S.R. - Ukrainian S.S.R.

8101 **Ivashin, D.S., Isaeva, R.Ya, Kuznetsova, P.I., Maslova, V.R., Nikolaeva, E.S. (1981).** Reliktovi ta endemichni roslyny dolyny r. Sivers'kyi Donets' u yiyi nyzhnii techiyi. (Relict and endemic plants of the lower S. Donets Valley and problems of their preservation.) *Ukr. Bot. Zhurn.*, 38(5): 60-64. Uk (Rus, En). Illus. (T)

8102 **Ivchenko, I.S. (1983).** Suchasnyi stan okhorony ridkisnykh i znykayuchykh vydiv dendroflory Ukrayiny. (Present day studies in rare and disappearing species of the Ukraine dendroflora.) *Ukr. Bot. Zhurn.*, 40(3): 81-87. Uk (En, Rus). Map. (T)

8103 **Kokhno, N.A. (1986).** Redkie vidy dendroflory SSSR v Tsentral'nom Respulilikanskom Botanicheskom Sadu AN USSR. (Rare species of the dendroflora of the USSR in the Central Republican Botanic Garden of the Ukrainian SSR, Kiev). *Byull. Glavn. Bot. Sada* (Moscow), 141: 55-57. Rus. (G/T)

8104 **Komendar, V.I. (1988).** Problemy okhorony fitohenofondu Karpat. (Problems of protection of the Carpathian phytogene pool) *Ukr. Bot. Zhurn.*, 45(1): 1-6. Uk (En, Rus). (L/T)

8105 **Kondratiuk, I.M., Burda, R.I. (1979).** [Donets Basin flora as an object for protection]. *Introd. Aklim. Rosl. Ukr. Akad. Nauk. URSR*, 15: 36-42. Uk.

8106 **Kondratyuk, E.N., Burda, R.I. (1981).** Stan i perspektyvy okhorony vydiv flory Donbasu, Zanesenykh do *Chervonoyi knyhy Ukrayins'koyi RSR*. (State and perspectives of protection of plant species in the Donets Basin included in the *Red Data Book of the Ukrainian SSR*.) *Ukr. Bot. Zhurn.*, 38(5): 1-7, 23. Uk (Rus, En). Maps. (T)

8107 **Kondratyuk, E.N., Burda, R.I. (1987).** Okhorona roslyn na pivdennomu skhodi Ukrayiny. (Protection of plants in the South-East of the Ukraine.) *Ukr. Bot. Zhurn.*, 44(5): 85-89. Uk (En, Rus). Map. (T)

8108 **Kostyl'ov, O.V., Movchan, Ya.I., Osychniuk, V.V., Solomakha, V.A. (1984).** Main associations of steppe vegetation in the "Khomutovskaia steppe" reservation. *Ukr. Bot. Zhurn.*, 41(6): 12-17. Uk (En, Rus). (P)

8109 **Kostylev, A.V. (1987).** Roslynnist' zaproektovanono zapovidnyka "Elanets'kyi". (Vegetation of the planned nature reservation "Elanetsky"). *Ukr. Bot. Zhurn.*, 44(2): 77-81. Uk (En, Rus). (P)

8110 **Kosykh, V.M., Lisovskii, B.V., Shapovalova, E.E. (1982).** [Genetic populations of some rare plants from the Ai-Petri and Yalta Yailas]. *Trudy Gos. Nikit. Bot. Sada*, 86: 80-88. Rus (En). (T)

8111 **Koval'chuk, S.I., Kl'ots, O.M. (1984).** [Discoveries of *Euonymus nana* Bieb. in Podolia (Khmlnik Region).] *Ukr. Bot. Zhurn.*, 41(4): 69-71. Uk (En, Rus). (T)

8112 **Koval'chuk, S.I., Zadorozhnyi, M.A. (1986).** Novi zapovidni ob'yekty Khmel'nyts'koho Prydnistrov'ya. (New reserve objects of the Dniester Area in the Khmelnitsky Region.) *Ukr. Bot. Zhurn.*, 43(4): 92-93. Uk (En, Rus). (P)

8113 **Krasnova, A.N., Kuz'michev, A.I. (1987).** Stan okhorony ridkisnykh ta endemichnykh vydiv roslyn zapovidnyka "Askaniya-Nova". (State of protection of rare and endemic plant species in the "Askania Nova" reserve). *Ukr. Bot. Zhurn.*, 44(3): 77-80. Uk (En, Rus). (P/T)

8114 **Kucheryavaya, L.F. (1983).** Flora vysshikh vodnykh i pribrezhno - vodnykh makrofitov Kanevskogo zapovednika i ee okhrana. (Flora of higher water and coastal macrophytes of the Kanev Nature Reserve and their protection.) *Okhrana Izuch. Obogashch. Rast. Mira* (Kiev), 10: 19-32. Rus. Kanev State Nature Reserve. (P)

8115 **Kuznetsova, P.L., Nikolaeva, O.S., Dika, M.P. (1979).** [Flora and vegetation of the Kremenian Forest]. *Ukr. Bot. Zhurn.*, 36(1): 58-61. Uk (En).

U.S.S.R. - Ukrainian S.S.R.

8116 **Latyshenko, M.D. (1980).** Nauchnye i prakticheskie zadachi okhrany botanicheskikh ob'ektov v zakaznikakh USSR. (Scientific and practical tasks for the protection of plants named for the Ukrainian SSR.) *Okhrana Izuch Obogashch. Rast. Mira* (Kiev), 7: 109-111. Rus (En).

8117 **Lyna, A.L., Reshetniak, T.A. (1978).** [Rare and unique coniferous species introduced to the Ukraine]. *Byull. Glavn. Bot. Sada* (Moscow), 108: 69-71. Rus. (T)

8118 **Lypa, A.L. (1978).** [Nature reservations of the Ukraine, their modern state, tasks and prospects for botanical studies]. *Ukr. Bot. Zhurn.*, 35(5): 513-515. Uk (En, Rus). (P)

8119 **Malinovskii, K.A. (1981).** Okhorona ridkisnykh vydiv vysokohirnoyi flory Ukrayinskykh Karpat. (Protection of rare species of alpine flora of the Ukrainian Carpathians.) *Ukr. Bot. Zhurn.*, 38(4): 63-67, 73. Uk (En, Rus). (T)

8120 **Marchenko, P.D. (1979).** [Powdery mildew fungi (Erysiphaceae) new and rare for the Ukraine.] *Ukr. Bot. Zhurn.*, 36(4): 360-366. Uk (En, Rus).

8121 **Matveyeva, E.P. (1979).** Askaniya-Nova segodnya. (Askania-Nova today.) *Bot. Zhurn.*, 64(8): 1201-1205. Rus.

8122 **Milkina, L.I. (1979).** O sistemnom podkhode v organizatsii prirodno-zapovednogo fonda na primere Ukrainskikh Karpat. (On the systematic approach to the organization of nature reserve resources at, for example, the Carpathian Mountains.) *Bot. Zhurn.*, 64(2): 199-210. Rus (En). Map. (P)

8123 **Milkina, L.I. (1984).** [Rare broadleaved forest communities of the Carpathian State Nature Reservation.] *Ukr. Bot. Zhurn.*, 41(4): 10-14. Uk (En, Rus). Illus. (P)

8124 **Osychniuk, V.V. (1979).** [Some characteristics of the protection regime in parts of the Ukrainian Steppe reserve]. *Ukr. Bot. Zhurn.*, 36(4): 347-352. (En, Rus). (P)

8125 **Osychniuk, V.V., Kostyl'ov, O.V., Movchan, I., Solomakha, V.A. (1984).** [Floristic classification of vegetation from the "Khomutovskaia Step" Reservation]. *Ukr. Bot. Zhurn.*, 41(2): 11-16. Uk (En, Rus). (P)

8126 **Ovsyannikova, E.S., Skrypko, G.S., Cherevko, S.P., Yatsenko, A.V. (1980).** [Problems of the genetic conservation of plant material in Zaporozh'e Province]. *In* Genetic problems of pollution of the surrounding environment in the territory of the Moldavian SSR. Kishinev, Akademiya Nauk Moldavskoi SSR. 49-52. Rus. Conference of the Section of the Genetic aspects of the Problems of Man and the Biosphere: collected papers. (E)

8127 **Panova, L.N. (1978).** Introduktsiya vidov roda *Betula* u Botanicheskom Parke "Askaniya-Nova". *Byull. Glavn. Bot. Sada* (Moscow), 108: 18-22. Rus. (P)

8128 **Parakhons'ka, N.O., Tkachenko, V.S. (1984).** [Changes in the floristic composition of the Mikhailovskaia Tselina under reservation conditions. *Ukr. Bot. Zhurn.*, 41(5): 13-16. Uk (En, Rus). (P/T)

8129 **Pogrebennik, V.P., et al. (1987).** Redkie i ischezayushchie rasteniya v urochishchie "Shandrovskii Les". (Rare and disappearing plants in "Shandrovskii Les" in Kiev Region.) *Okhrana Izuch. Obogashch. Rast. Mira* (Kiev), 14: 13-18. Rus. (T)

8130 **Popovich, S.Yu. (1983).** [Exogenic changes of forest vegetation in the Polesskii State Reservation]. *Ukr. Bot. Zhurn.*, 40(4): 77-81. Uk (En, Rus). (P)

8131 **Popovich, S.Yu. (1983).** [Floristic discoveries on the territory of the Polesskii State Reservation]. *Ukr. Bot. Zhurn.*, 40(6): 94-98. Uk (En, Rus). Maps. (P)

8132 **Popovich, S.Yu., Andrienko, T.L. (1982).** Roslynnist' ozera Hropa ta ioho naukova tsinnist'. (Plants of Lake Gropa and its scientific significance.) *Ukr. Bot. Zhurn.*, 39(4): 92-95. Uk (Rus, En). Illus. Lake protection in the Ukrainian Carpathians.

U.S.S.R. - Ukrainian S.S.R.

8133 **Popovich, S.Yu., Balashev, L.S. (1983).** Pryrodni ta antropohenni zminy roslynnoho pokryvu bolit Polis'koho derzhavnoho zapovidnyka. (Natural and anthropogenic changes in plant cover of the Polessie State Reservation bogs.) *Ukr. Bot. Zhurn.*, 40(3): 86-92. Uk (En, Rus). (P)

8134 **Pryadko, E.I. (1983).** Funktsional'ne zonuvannya terytoriyi zaproektovanoho Dniprovs'koho pryrodnoho natsional'noho parku. (Functional zonation of the territory of the planned Dnieper Natural National Park.) *Ukr. Bot. Zhurn.*, 40(5): 85-89. Uk (En, Rus). Maps. (P)

8135 **Pryadko, E.I. (1986).** Analiz flory zaproektovanoho Dniprovs'koho pryrodnoho natsional'noho parku. (Analysis of flora of the planned Dnieper Natural National Park.) *Ukr. Bot. Zhurn.*, 43(3): 102-106. Uk (En, Rus). (P)

8136 **Semenikhina, K.A. (1979).** New habitats of rare aquatic species in water bodies of the Desna River floodplains. *Ukr. Bot. Zhurn.*, 36(3): 214-218. Uk (En, Rus). Map. (T)

8137 **Shcherbakova, O.N. (1984).** Redkie rasteniya lesnogo zakaznika "Stradchanskii les". (Rare plants of the forest reserve "Stradchansk Forest".) *In* Izuch. i okhrana zapoved. ob'ektov. Alma-Ata. 101-102. Rus. (P/T)

8138 **Shelyag-Sosonko, Yu.R., Andrienko, T.L., Udra, I.F. (1979).** Roslynnist "Nadsluchanskoi Shveitsariyi" tsinnoyi pam'yatky pryrody polissya. (Vegetation of "Nasluchanskaya Shveitsaria" a valuable monument of the polessie [a region of swamps and forests] nature.) *Ukr. Bot. Zhurn.*, 36(6): 578-583. Uk (Rus, En). Illus., map. (P)

8139 **Shelyag-Sosonko, Yu.R., Didukh, Ya.P. (1978).** Outline of the flora and vegetation of the Yalta Mountain Forest State Reservation 2. *Flora. Bot. Zhurn. Leningr.*, 63(10): 1430-1439. (P)

8140 **Shelyag-Sosonko, Yu.R., Dubyna, D.V. (1984).** Flora Gosudarstvennogo zapovednika "Dunaiskie plavni". (Flora of The Danube Marshes State Reserve.) *Bot. Zhurn.*, 69(5): 654-661. Rus. Illus.; maps. (P)

8141 **Shelyag-Sosonko, Yu.R., Dubyna, D.V. (1984).** [Present state and prospects of studying the higher aquatic flora and vegetation of the Ukraine]. *Ukrains'kyi Botanichnyi Zhurnal*, 41(2): 1-11. Uk (Rus, En). Includes plant conservation.

8142 **Shelyag-Sosonko, Yu.R., Sytnik, K.M., et al. (1980).** [Ukraine, White Russia, Moldavia. The protection of important botanical regions.] Kiev, Naukova Dumka. 332-370. Rus. Includes lists of threatened plants by region. (P/T)

8143 **Shelyag-Sosonko, Yu.R., ed. (1987).** Zelenaya kniga Ukrainskoi SSR: redkie, ischezayushchie i tipichnye, nuzhdayushchiesya v okhrane rastitel'nye soobshchestva. Kiev, Naukova Dumka. 213p. Rus. Illus., col. illus., maps. Green Book of the Ukrainian SSR: rare, disappearing and characteristic plant associations, in need of protection. (T)

8144 **Shevchenka, T.H. (1978).** Okhorona, vivchennia ta zbagachennia roslinnogo svitu. (Conservation, study and enrichment of the vegetable kingdom.) Kiev, Vyshcha Shkola. 1974-1978. 5 vols. Uk. Illus.

8145 **Sokko, V.G., Nefedova, O.N., Dubenets, T.G. (1985).** [Taxonomy and geographical analysis of Orchidaceae species of Ukraine in relation with their introduction and protection.] *Introd. Akklim. Rast.* (Kiev), 4: 33-36. Rus. (T)

8146 **Stoiko, S.M. (1980).** [Preservation of nature in the Ukrainian Carpathians and adjacent territories.] Kiev, Naukova dumka. 261p. Rus. (P)

8147 **Stoiko, S.M., Saik, D.S., Sukharyuk, D.D., Tasenkevich, L.A. (1985).** Karpatskii gosudarstvennyi zapovednik i neobkhodimost' uluchsheniya ego territorial'noi struktury. (The Carpathian State Reserve and the necessity of improvement of its territorial structure.) *Bot. Zhurn.*, 70(10): 1418-1425. Rus. Illus. Map. (P)

U.S.S.R. - Ukrainian S.S.R.

8148 **Stoiko, S.M., Zhizhin, N.P., Tasenkevich, L.A. (1986).** Fitosozolonichni osnovy stvorennya merezhi natsional'nykh rehional'nykh parkiv v Ukrayinskykh Karpatakh. (Phytosociological grounds for creation of a network of national and regional parks in the Ukrainian Carpathians.) *Ukr. Bot. Zhurn.*, 43(6): 55-58. Uk (En, Rus). Map. (P)

8149 **Sytnik, K.M., et al. (1980).** [Preservation of the most important botanical species of the Ukraine.] Kiev, Naukova Dumka. 390p. Rus. (P)

8150 **Sytnik, K.M., et al., eds (1979).** [Red Data Book of the Ukrainian SSR.] Kiev, Academy of Sciences of the Ukrainian SSR. Naukova Dumka. 497p. Rus (En). Illus., maps. Covers 151 vascular plant species. (T)

8151 **Terlets'kyi, V.K. (1984).** [New habitats of rare species in Western Polesye.] *Ukr. Bot. Zhurn.*, 41(5): 92-94. Uk (En, Rus). (T)

8152 **Tkachenko, V.S. (1980).** Struktura roslynnoho pokryvu Zapovidnyka "Proval'skyi Step" za danymy krupnomasshtabnoho geobotanichnoho kartuvannya. (Structure of the "Provalskaya Steppe" Reservation plants according to the data of large scale geological mapping.) *Ukr. Bot. Zhurn.*, 37(6): 20-26. Uk (En, Rus). Map. (P)

8153 **Tkachenko, V.S. (1984).** [Nature of the meadow steppe reservation "Mikhailovskaia Virgin Land" and the forecast of its development under reservation conditions]. *Bot. Zhurn.*, 69(4): 448-457. Rus (En). Illus. Literature review. Sumy Region. (P)

8154 **Tkachenko, V.S., Genov, A.P. (1986).** Florotsenotichna kharakterystyka zaproponovanoho Kal'mius'koho derzhavnoho zakaznika. (Florocenotic characteristic of the Kalmius State Reserve.) *Ukr. Bot. Zhurn.*, 43(5): 92-96. Uk (En, Rus). Maps. (P)

8155 **Tkachenko, V.S., Genov, A.P., Movchan, Ya.I. (1987).** Florotsenotychna kharakterystyka Kryvoluts'koho Kretofitnoho stepu na Donbasi ta neobkhidnist' ioho zapovidannya. (Florocenotic characteristic of the Krivolutsk Cretophytic Steppe in the Donbas and the necessity of its reservation). *Ukr. Bot. Zhurn.*, 44(4): 70-75. Uk (En, Rus). Map. (P/T)

8156 **Tkachenko, V.S., Genov, A.P., Parakhonskaya, N.A. (1987).** Heobotanichna otsinka pryrodnykh uhid' okolynts' deyakykh stepovykh zapovidnykiv AN URSR i neobknidnist' yikh okhorony. *Ukr. Bot. Zhurn.*, 44(3): 66-72. (Geobotanical estimation of natural lands in outskirts of certain steppe reserves of the Academy of Sciences of Ukrainian SSR and the necessity of their protection). Uk (En, Rus). Map. (P)

8157 **Tykhonenko, Yu.Ya. (1981).** [New and rare mycoflora of the Ukrainian-SSR USSR Uredinales species.] *Ukr. Bot. Zh.*, 38(4): 77-79.

8158 **Ulychna, K.O., Partyka, L.Y. (1972).** Rare species of bryoflora in the Ukraine and necessity of their protection. *Ukr. Bot. Zhurn.*, 29: 581-585. (T)

8159 **Ustimenko, P.M. (1983).** Roslynnist' lisovoho masyvu "Vetykyi Lis" (Chernihivs'ke polissya). (Plants of the "Veliky Les" Forest Massif [Chernigov Polessie].) *Ukr. Bot. Zhurn.*, 40(3): 92-9. Uk (En, Rus). Map.

8160 **Ustimenko, P.M. (1984).** [Floristic discoveries on the territory of the planned Mezin Natural National Park.] *Ukr. Bot. Zhurn.*, 41(4): 64-67. Uk (En, Rus). (P/T)

8161 **Ustimenko, P.M. (1986).** Funktsional'ne zonuvannya terytoriyi zaproektovanoho Mezyns'koho pryrodnoho natsional'noho parku. (Functional zonation of the territory of the planned Mezin Natural National Park.) *Ukr. Bot. Zhurn.*, 43(3): 99-102. Uk (En, Rus). Map. (P)

8162 **Ustimenko, P.M. (1986).** Merezha pryrodnykh natsional'nykh parkiv Ukrayins'koho Polissya. (A net of natural national parks of Ukrainian Polessie.) *Ukr. Bot. Zhurn.*, 43(4): 42-43. Uk (En, Rus). (P)

U.S.S.R. - Ukrainian S.S.R.

8163 **Vavrysh, P.O., Sobko, V.H. (1984).** [Rare population of *Cypripedium calceolus* L. on the Volyn upland]. *Ukr. Bot. Zhurn.*, 41(2): 86-88. Uk (En, Rus). Illus., maps. (T)

8164 **Vetrova, Z.I. (1979).** (New and rare for the Ukrainian SSR taxa of Euglenophyta from water bodies of the Transcarpathian area.) *Ukr. Bot. Zhurn.*, 36(3): 252-255. Uk (En, Rus). Illus. (T)

8165 **Zaeto, Z.S., Stefanik, V.I., Solodkova, T.I. (1981).** Stepovi dilyanky radyans'koyi Bukovyny shcho potrebuyut' okhorony. (Steppe sites of the Soviet Bukovina which should be protected.) *Ukr. Bot. Zhurn.*, 38(5): 64-67. Uk (Rus, En). (P)

8166 **Zaverukha, B.V., Protopopova, V.V., Andrienko, T.L. (1982).** Po stranitsam *Krasnoi knigi Ukrainskoi SSR*: vysshie rasteniya. (From the pages of the Red Data Book of the Ukrainian SSR: higher plants.) [UkrSSR], Naukova Dumka. (T)

8167 **Zelinka, S.V. (1979).** Tsennye botanicheskie ob'ekty Zapadnoi Podolii i neobkhodimost' ikh okhrany. (Important plants of Western Podolia and the need for protection.) *In* Aktualn. vopr. sovrem. botaniki. Kiev. 65-69. Rus.

8168 **Zelinka, S.V., Balashev, L.S., Shimanskaya, V.E. (1984).** Bolotni zakaznyyky zakhidnoho Podillya. (Bog reservations of Western Podolia). *Ukr. Bot. Zhurn.*, 41(6): 77-81. Uk. (En, Rus). Illus., map. (P)

8169 **Zhizhin, N.P., Zagul'skii, M.N., Kagalo, A.A. (1987).** Poshyrennya ta okhorona ridkisnykh vydiv n Voronyakakh (Volyno-Podillya URSR). (Distribution and protection of rare species in Voronyaki [The Volyn-Podolias Territory of the Ukrainian SSR].) *Ukr. Bot. Zhurn.*, 44(6): 73-77. Uk (En, Rus). Map. (L/T)

U.S.S.R. - Uzbekistan S.S.R.

8170 **Anon. (1983).** Krasnaya kniga Uzbekskoi SSSR: redkie i ischezayushchie vidy zhivotnykh i rastenii: t.2 Rasteniya. (Red Data Book of the Uzbek.SSR: rare and threatened species of animals and plants: vol. 2, plants.) UzSSR, Fan. Rus. (T)

8171 **Belolipov, I.V. (1975).** [Goals of botanical gardens of the Uzbekistan Soviet Socialist Republic for the protection of plants of the native flora of central Asia.] *Byull. Glavn. Bot. Sada*, 95: 88-89. Rus. (G)

8172 **Kenig, G.F. (1985).** Ob okhrane endemichnykh drevesnykh rastenii v doline Syrdari. (On protection of endemic species on the Syrdarya Valley.) *Probl. Osvoen. Pustyn* (Turkm.SSR) 5: 79-81. Rus. (G/T)

8173 **Ketskhoveli, N.I., Gagnidze, R.I. (1979).** Sostoyanie i perspektivy izucheniya rastitel'nykh resursov Gruzii. (The conditions and perspectives of the study of plant resources of Georgia.) *In* Sostoyanie i perspektivy izuch. i ispol'z prirod. rastitel'n resurvov SSSR. Tashkent. 95-97. Rus.

8174 **Nabiev, M.M., ed. (1984).** [Red data book of the Uzbek SSR. Rare and endangered species of animals and plants. Vol. 2 - plants.] Tashkent, Fan. 151p. Rus (with Uzbek and English preface). (T)

8175 **Nikalaev, F.I., et al. (1983).** Ekologiya rastenii i zhivotnykh zapovednikov Uzbekistana. (Ecology of plants and animals in nature reserves of Uzbekistan.) U.S.S.R., Fan. Rus. Book review in *Novge Kingi SSSR*, 24: 116 (1982). (P)

8176 **Ratseka, V.I., ed. (1980).** Zapovednye territorii Uzbekistana. (Protected territories of Uzbekistan.) Tashkent, Uzbekistan. 70p. Rus. Col. illus., map. (P)

8177 **Vernik, R.S., Mailun, Z.A., Khamidov, G.Kh., Mukhamedzhanova, F.I., Talipov, K.I. (1984).** Orekhovye lesa Uzbekistana i puti ikh okhrany. (Walnut woods of Uzbekistan and the means of their protection.) *Uzb. Biol. Zhurn*, 3: 30-32. Rus.

UNITED ARAB EMIRATES

8178 **Khan, M.I.R. (1982).** Mangrove forest of the United Arab Emirates. *Pakistan J. Forest.*, 32(2): 36-39. Illus. (E/P)

UNITED ARAB EMIRATES

8179 **Khan, M.I.R. (1982).** Need for conservation and development of natural mangrove forest of U.A.E. *Pakistan J. Forest.*, 32(4): 127-129.

8180 **Khan, M.I.R. (1982).** Status of mangrove forests in the United Arab Emirates. *Bull. Emirates Nat. Hist. Group* (Abu Dhabi), 17: 15-17. (P)

UNITED KINGDOM

8181 **Adams, W.M. (1981).** A plea for partial conservation. *Ecos*, 2(4): 16-21. (L)

8182 **Allen, D.E. (1980).** The early history of plant conservation in Britain. *Leicester Lit. Phil. Soc.*, 72: 35-50.

8183 **Anon. (1975).** Wildlife, the law and you: an explanation of the Wild Creatures and Wild Plants Act, 1975. London, British Museum (Natural History). Unpaged. (L/T)

8184 **Anon. (1982).** Plant thieves threaten rare species. *New Sci.*, 95(1312): 5. Illus. Lizard Orchids (*Himantoglossum hiricinum*) and Sundew (*Drosera anglica*). Collectors. (T)

8185 **Anon. (1982).** Seeds to save: guide to the suppliers of tree and shrub seed to the UK. *GC & HTJ*, 193(23): 31, 33, 35-38. (G/T)

8186 **Anon. (1983).** A conservation/development programme for UK forestry. *In* Anon. The conservation and development programme for the UK: a response to the World Conservation Strategy. London, Kogan Page. 215-224.

8187 **Anon. (1983).** Porridge-eating fungus saves endangered orchids. *New Sci.*, 98(1359): 540. Illus. (T)

8188 **Anon. (1983).** Symposium on conservation in British and Irish botanic gardens. *Threatened Pl. Newsl.*, 12: 8-9. (G/T)

8189 **Anon. (1983).** Threat to rare buttercup. *Times*, 5 May. Illus. *Ranunculus ophioglossifolius* has the smallest nature reserve in the world: Badgeworth in Gloucestershire. (P/T)

8190 **Anon. (1984).** Further discoveries of the Fen Violet (*Viola persicifolia* Schreber) at Wicken Fen, Cambridgeshire. *Watsonia*, 15(2):122-123. (T)

8191 **Anon. (1984).** The problem of primrose picking. *Oryx*, 18(2): 70-71. Collectors.

8192 **Anon. (1986).** Protecting the countryside: the Government's consultative proposals for landscape conservation orders. Department of the Environment, Scottish Office, Welsh Office. 13p. (L)

8193 **Anon. (1988).** Exotic aquatic threatens UK reserves. *Oryx*, 22(1): 49-50. *Crassula helmsii* from Australia. (P)

8194 **Anon. (1988).** Scottish primrose takes the high road to extinction. *New Sci.*, 117(1602): 41. Illus. (T)

8195 **Bailey, R.H., Dunning, R.J. (1984).** The Society and the conservation of *Cyclamen*. *Cyclamen Soc. J.*, 8(1): 9-10. (T)

8196 **Bellamy, D. (1986).** A heritage for all reasons; Upper Teesdale, Cumbria. *Country Life*, 8 May: 1240-1243. Illus. A plea that Upper Teesdale should be given protection. (P)

8197 **Black, D. (1975).** Conservation begins at home. How the World Wildlife Fund is saving wild Britain. *Wildlife*, 17(12): 546-548. Col. illus. Illus. Plant surveys in Scotland. (T)

8198 **Blacksell, M. (1982).** The spirit and purpose of national parks in Britain. *Parks*, 6(4): 14-17. Illus. (P)

UNITED KINGDOM

8199 **Botanical Society of the British Isles (1982).** Code of conduct for the conservation of wild plants. London, British Museum (Nat. Hist.). Pamphlet. (T)

8200 **Bradshaw, M.E. (1981).** Monitoring grassland plants in Upper Teesdale, England. *In* Synge, H., *ed.* The biological aspects of rare plant conservation. Chichester, Wiley. 241-251. Proceedings of International Conference, King's College, Cambridge, 14-19 July 1980. Discussion on plant demography as aid to rare plant conservation. (T)

8201 **Brickell, C.D. (1979).** The RHS Conservation Conference. *Garden* (London), 104(4): 161-171. The practical role of gardens in the conservation of rare and threatened plants. (G/T)

8202 **Briggs, J.D., Tandy, C. (1988).** The Montgomery canal: its aquatic plants and their conservation. *BSBI News*, 49: 28-30. Illus. (T)

8203 **British Museum (Natural History) (1981).** Threatened plants. British Museum Publ. 836. London, The Yale Press. 12p. (T)

8204 **Brockie, R.E., Loope, L.L., Usher, M.B., Hamann, O. (1988).** Biological invasions of island nature reserves. *Biol. Conserv.*, 44(1-2): 9-36. Includes 6 case studies. (P)

8205 **Brookes, B.S. (1981).** The discovery, extermination, translocation and eventual survival of *Schoenus ferrugineus* in Britain. *In* Synge, H., *ed.* The biological aspects of rare plant conservation. Chichester, Wiley. 421-428. Illus. Proceedings of International Conference, King's College, Cambridge, 14-19 July 1980. (T)

8206 **Brooks, A. (1981).** Waterways and wetlands: a practical conservation handbook, rev. ed. Reading, British Trust for Conservation Volunteers. 186p. Illus.

8207 **Brooks, A., Follis, A., comps (1980).** Woodlands: a practical conservation handbook. Reading, British Trust for Conservation Volunteers. 187p.

8208 **Bullard, E.R., Shearer, H.D.H., Day, J.D., Crawford, R.M.M. (1987).** Survival and flowering of *Primula scotica* Hook. *J. Ecol.*, 75(3): 589-602. Map. (T)

8209 **Christie, S.J. (1986).** Peatlands of Northern Ireland - a review with recommendations. Belfast, Ulster Trust for Nature Conservation.

8210 **Council for Nature (1978).** A directory of natural history and nature conservation films. 3rd ed. London, Council for Nature. 65p.

8211 **Council for Nature (1979).** Directory of lecturers in natural history and nature conservation. 2nd ed. London, Council for Nature. 22p.

8212 **Crompton, G. (1981).** Surveying rare plants in eastern England. *In* Synge, H., *ed.* The biological aspects of rare plant conservation. Chichester, Wiley. 117-124. (T)

8213 **Cruickshank, J.G., Wilcock, D.N. (1982).** Northern Ireland environment and nature resources. Belfast, The Queen's University of Belfast and The New University of Ulster.

8214 **Crumley, J., Fairley, R., Grant, J., Smith, R. (1987).** The Cairngorms. *Environment Now*, January 1987: 54-60. Col. illus., map. (P)

8215 **Dalyell, T. (1981).** Wildlife and midnight oil. *New Sci.*, 90(1258): 786. U.K. Wildlife and Countryside Bill. (L)

8216 **Davies, J.A. (1983).** *Cotoneaster integerrimus*: a step away from extinction? *BSBI Welsh Bull.*, 37: 10-11. (T)

8217 **Dring, M.J., Frost, L.C. (1971).** Studies of *Ranunculus ophioglossifolius* in relation to its conservation at Badgeworth Nature Reserve, Gloucestershire, England. *Biol. Conserv.*, 4(1): 48-56. (P/T)

UNITED KINGDOM

8218 **Duffey, E. (1971).** The management of Woodwalton Fen: a multidisciplinary approach. *In* Duffey, E., Watt, A.S., *eds.* The scientific management of animal and plant communities for conservation. Oxford, Blackwell Scientific Publications. 581-597. From the 11th symposium of the British Ecological Society, Univ. of East Anglia, Norwich, July 7-9, 1970. Wetlands. (P)

8219 **Duffey, E., Watt, A.S., eds (1971).** The scientific management of animal and plant communities for conservation. The 11th symposium of the British Ecological Society, Univ. of East Anglia, Norwich, July 7-9, 1970. Oxford, Blackwell Scientific Publications. 652p. (P)

8220 **Dunlop, D.J., Christie, S.J. (1987).** Peatlands in Northern Ireland. *Natural World.*

8221 **Ellis, E.A., Perring, F., Randall, R.E. (1977).** Britain's rarest plants. Norwich, Jarrold. 41p. Illus. (T)

8222 **Esslemont, H. (1985).** Plant portraits: *Iris winogradowii. The Rock Garden. J. Scottish Rock Garden Club,* XIX(3), 76: 275-283. Illus. (G/T)

8223 **Farrell, L. (1984).** Eyes on orchids. *Natural World,* summer 1984. Illus. 12 orchids qualify for inclusion in the Red Data Book. (T)

8224 **Farrell, L. (1987).** Notices (others). Rare plants: their conservation and legislation. *BSBI News,* 45: 27. (L/T)

8225 **Fitter, R. (1970).** Danger! Rare British flora. *World Wildlife News,* Summer: 3. (T)

8226 **Fitter, R. (1982).** Newly protected plants in Britain. *Oryx,* 16: 317. (T)

8227 **Flower, C. (1976).** Those other wetlands. *Wildlife Mag.,* 18(9): Illus. Focus on European Wetlands Year. (T)

8228 **Foster, J., Phillips, A., Steele, R. (1982).** Limited choices: protected areas in the United Kingdom. IUCN: World National Parks Congress; 16 October 1982: 1-27. Gives lists of protected areas in England, Wales, Scotland and Northern Ireland. (P)

8229 **Foster, J., Phillips, A., Steele, R. (1984).** Protected areas in the United Kingdom: an approach to the selection, establishment, and management of natural and scenic protected areas in a densely populated country. *In* McNeely, J.A., Miller, K.R., *eds.* National parks, conservation and development. The role of protected areas in sustaining society. Smithsonian Institution Press, Washington, D.C. 426-437. (P)

8230 **Frost, L.C. (1981).** The study of *Ranunculus ophioglossifolius* and its successful conservation at the Badgeworth Nature Reserve, Gloucestershire. *In* Synge, H., *ed.* The biological aspects of rare plant conservation. Chichester, Wiley. 481-489. Illus. (P/T)

8231 **Fry, G. (1987).** Acid rain and the management of protected areas. *Ecos,* 8(1): 40-44. Illus. Acid rain. (P)

8232 **Fry, G.L.A., Cooke, A.S. (1987).** Focus on nature conservation: no. 7. Acid deposition and its implications for nature conservation in Britain. Peterborough, Nature Conservancy Council, 59p. Maps. Acid rain.

8233 **Fuller, R.M. (1987).** The changing extent and conservation interest of lowland grasslands in England and Wales: a review of grassland surveys 1930-84. *Biol. Conserv.,* 40: 281-300.

8234 **Gibbons, R.B. (1980).** Britain's disappearing plants. *Garden,* 105(3): 99-102. (T)

8235 **Giddens, C., Bristow, H., Allen, N., eds (1988).** The flora and fauna of Exmoor National Park. A provisional check-list. Minehead, Exmoor Natural History Society. 272p. Annotated list; includes lower plants and fungi. (P)

8236 **Gillie, O. (1982).** Found: a lily from the Ice Age. *Sunday Times,* Jan. 1982. Illus. (T)

UNITED KINGDOM

8237 **Godfrey, P.J., Alpert, P. (1985).** Racing to save the coastal heaths. *Nat. Conserv. News*, 35(4): 10-14.

8238 **Goode, D. (1981).** The threat to wildlife habitats. *New Sci.*, 22 Jan. Illus. Wildlife and Countryside Bill not strong enough. (T)

8239 **Grainger, A. (1981).** The shape of woods to come. *New Sci.*, 90(1247): 26-28. Illus. (G)

8240 **Great Britain. Department of the Environment (1976).** Report on the implementation of the Convention on International Trade in Endangered Species of Wild Fauna and Flora in the United Kingdom. London, Department of the Environment. 98p. (L)

8241 **Green, B. (1981).** Countryside conservation: the protection and management of amenity ecosystems. London, George Allen & Unwin. xiii, 249p. Illus., maps. (P)

8242 **Green, B.H. (1981).** A policy on introductions to Britain. *In* Synge, H., *ed*. The biological aspects of rare plant conservation. Chichester, Wiley. 403-412. Proceedings of International Conference, King's College, Cambridge, 14-19 July 1980. Paper covers implications of introductions for conservation.

8243 **Greenwood, E.F., Gemmell, R.P. (1978).** Derelict industrial land as a habitat for rare plants in south Lancashire and west Lancashire England UK. *Watsonia*, 12(1): 33-40. (P/T)

8244 **Greig-Smith, J., Sagar, G.R. (1981).** Biological causes of local rarity in *Carlina vulgaris*. *In* Synge, H., *ed*. The biological aspects of rare plant conservation. Chichester, Wiley. 389-400. Proceedings of International Conference, King's College, Cambridge, 14-19 July 1980. (T)

8245 **Gritten, R. (1988).** Invasive plants in the Snowdonia National Park. *Ecos*, 9(1): 17-22. Illus. Invasive species. (P)

8246 **Grove, R.H. (1981).** The use of disguise in nature conservation: the evidence from 3 case studies. Discussion Papers in Conservation No. 32. London, University College. 38p.

8247 **Harkness, C.E. (1983).** Mapping changes in the extent and nature of woodlands in national parks in England and Wales. *Arboricult. J.*, 7(4): 309-319. Maps. (P)

8248 **Harley, J.L. (1984).** The flora and vegetation of Britain: origins and changes - the facts and their interpretation. London, Academic Press. 209p. Illus., maps.

8249 **Harvey, H.J., Meredith, T.C. (1981).** Ecological studies of *Peucedanum palustre* and their implications for conservation management at Wicken Fen, Cambridgeshire. *In* Synge, H., *ed*. The biological aspects of rare plant conservation. Chichester, Wiley. 365-378. Proceedings of International Conference, King's College, Cambridge, 14-19 July 1980. Paper covers studies at Wicken Fen Nature Reserve. (P)

8250 **Harvey, H.J., Meredith, T.C. (1981).** The biology and conservation of milk-parsley *Peucedanum palustre* at Wicken Fen. *Nat. Cambridge*, 24: 38-42. Illus. (T)

8251 **Hepper, F.N., ed. (1982).** The Royal Botanic Gardens, Kew: gardens for science and pleasure. London, HMSO. vii, 195p. Illus., col. illus. Includes chapter "Kew & Plant Conservation". (G/T)

8252 **Hodgson, J.G. (1986).** Commonness and rarity in plants with special reference to the Sheffield flora. *Biol. Conserv.*, 36(3): 197-314. Parts I, II, III, IV. (T)

8253 **Hodgson, J.G. (1989).** What is happening to the British flora: an investigation of commonness and rarity. *Pl. Today*, 2(1): 26-32. Illus. (T)

8254 **Hope-Simpson, J.F. (1975).** Mendip rare plants in the context of their wider distribution. *Somerset Trust Nat. Conserv. Ann. Rep.*, 10: 17-22 (1974 publ. 1975). Map. (T)

UNITED KINGDOM

8255 **Horley, J.L., Lewis, D.H. (1985).** The flora and vegetation of Britain. Origin and changes. London, Academic Press. 228p.

8256 **Huntley, B. (1981).** The past and present vegetation of the Caenlochan National Nature Reserve, Scotland. II. Palaeoecological investigations. *New Phytol.*, 87(1): 189-222. Illus., maps. (P)

8257 **Hywel-Davies, J., Thom, V. (1986).** The Macmillan guide to Britain's nature reserves. London, Macmillan. 780p. Illus., col. illus., maps. Hardback edition published 1984. (P)

8258 **Ingham, R. (1984).** Orchid site saved. *Shropshire Wildlife*, 60. Illus. West Telford. (P)

8259 **Johnson, B. (1983).** The conservation and development programme for the UK: a response to the World Conservation Strategy. An overview - resourceful Britain. London, Kogan Page. 104p.

8260 **Kemp, E.E. (1979).** A phytosociological layout for locally endangered species. *In* Synge, H., Townsend, H., *eds*. Survival or extinction. Proceedings of a conference held at the Royal Botanic Gardens, Kew, 11-17 September 1978. Kew, Bentham-Moxon Trust. 135-139. University Botanic Garden, Dundee, Scotland. (G/T)

8261 **Kirby, K.J. (1984).** Forestry operations and broadleaf woodland conservation. *Focus Nat. Conserv.*, 8: 59p.

8262 **Kirby, K.J. (1984).** Scottish birchwoods and their conservation: a review. *Trans. Bot. Soc. Edinb.*, 44(3): 205-218. Maps.

8263 **Kirby, K.J., Peterken, G.F., Spencer, J.W., Walker, G.J. (1984).** Inventories of ancient semi-natural woodland. *Focus Nat. Conserv.*, 6: 67p.

8264 **Lear, M. (1983).** Conserving our rare trees. *Arboricult. J.*, 7(1): 17-37. Illus. Includes documentation. (G/T)

8265 **Lewington, J. (1982).** The Wildlife and Countryside Act, 1981. *Progress, ADAS*, 20(1): 12-13. (L/T)

8266 **Ling, K.A., Ashmore, M.R. (1987).** Focus on nature conservation: No. 19. Acid rain and trees: an appraisal of the evidence for damage to native tree species by air pollution and acid precipitation in the United Kingdom. Peterborough, Nature Conservancy Council. 78p. Maps.

8267 **Lovejoy, D. (1983).** Trees, landscape and conservation. *Trees*, 43(1): 7-11. Illus. (E/T)

8268 **Mills, S. (1986).** Roads run over conservation. *New Sci.*, 109(1496): 44-48. Col. illus. Sites of Special Scientific Interest are threatened by road schemes. (P)

8269 **Milne-Redhead, E., ed. (1963).** The conservation of the British flora. BSBI Conference Report No. 8. London, BSBI. 90p. (T)

8270 **Moore, J.A. (1988).** *Lamprothamnium*: a pioneer in the conservation of the aquatic environment. *BSBI News*, 49: 48.

8271 **Moore, N. (1982).** What parts of Britain's countryside must be conserved. *New Sci.*, 93(1289): 147-149. Illus. (L/P)

8272 **Moore, P. (1985).** The death of the elm. *New Sci.*, 107(1466): 32-34. Illus.

8273 **Morton, J.K. (1982).** Preservation of endangered species by transplantation. *Canad. Bot. Ass. Bull.*, 15(3): 32. Refers to *Schoenus ferrugineus*. (T)

8274 **Mundey, G.R. (1976).** Lost ghost of the woods. *Wildlife*, 18(8): 352-353. Col. illus. Ghost Orchid (*Epipogium aphyllum*) may be less rare than supposed. (T)

8275 **Nature Conservancy Council (1982).** Wildlife, the law and you. London, NCC. 15p. (L)

UNITED KINGDOM

8276 **Nature Conservancy Council (1984).** Nature conservation in Great Britain. Shrewsbury, NCC. 15p. Illus., maps. (L/P)

8277 **Nectoux, F. (1895).** Timber! An investigation of the UK tropical timber industry. London, Friends of the Earth. (E)

8278 **North, R. (1983).** Wild Britain: the Century book of marshes, fens and broads. London, Century. 192p. Illus., col. illus., maps.

8279 **O'Connor, F.B. (1983).** The conservation and development programme for the UK. A response to the world conservation strategy. London, Kogan Page. 496p. Illus.

8280 **O'Riordan, T. (1983).** Preserving the heritage landscapes of Broadland. *Landscape Design*, 8/83: 14-16. Illus. Maps. Should the Broads (Broadland) be a national park? (P)

8281 **Oldfield, S. (1983).** Threatened plants for sale in UK. *TRAFFIC Bull.*, 5(2): 21. Trade. (L/T)

8282 **Oldfield, S. (1986).** Orchid conservation. *BSBI News*, 42: 25-26. (L)

8283 **Parker, D.M. (1981).** The re-introduction of *Saxifraga cespitosa* to north Wales. *In* Synge, H., *ed.* The biological aspects of rare plant conservation. Chichester, Wiley. 506-508. Proceedings of International Conference, King's College, Cambridge, 14-19 July 1980.

8284 **Parry, M., Bruce, A., Harkness, C. (1981).** The plight of British moorlands. *New Sci.*, 28 May: 350-351. Illus. (T)

8285 **Perring, F. (1976).** Wetland flora in danger. *Conserv. Rev.* 13: Illus. (T)

8286 **Perring, F., Randall, R.E. (1981).** Britain's endangered plants. Norwich, Jarrold. 42p. Illus. Describes 42 taxa. (T)

8287 **Perring, F., ed. (1945).** The flora of a changing Britain, BSBI Conference Report No. 11. Hampton, Classey. 157p. Examines factors responsible for change in the British flora since 1945 and predicts futher changes. (T)

8288 **Perring, F.H. (1971).** Rare plant recording and conservation in Great Britain. *Boissiera*, 19: 73-79. (T)

8289 **Perring, F.H. (1975).** Problems of conserving the flora of Britain. *BSBI News*, 9: 18-24.

8290 **Perring, F.H. (1977).** Wild flowers and the law. *The Garden*, 102(5): 202-203. (L)

8291 **Perring, F.H. (1978).** Rare vascular plant species in cultivation. *Nat. Wales*, 16(2): 139. (G/T)

8292 **Perring, F.H., Farrell, L. (1983).** British Red Data Books: 1. Vascular plants, 2nd Ed. Lincoln, RSNC. 99p. Identifies more than 300 rare and threatened taxa, 17.6% of the native flora, and summarizes status of and threats to each species. (T)

8293 **Perring, F.H., Randall, R.E. (1982).** Britain's endangered plants. Norwich, Jarrold Colour Publication. 42p. (T)

8294 **Perring, F.H., Walters, S.M. (1971).** Conserving rare plants in Britain. *Nature*, 229(5284): 375-377. (T)

8295 **Peterken, G.F. (1981).** Woodland conservation and management. London, New York, Chapman and Hall. xv, 328p. Illus., maps.

8296 **Phillips, A. (1985).** Socio-economic development in the "National Parks" of England and Wales. *Parks*, 10(1): 1-5. Illus. Tables. (P)

UNITED KINGDOM

8297 **Pigott, C.D. (1984).** The flora and vegetation of Britain: ecology and conservation. *In* Harley, J.L., Lewis, D.H., *eds*. Proceedings of a Symposium: The flora and vegetation of Britain: origins and changes - the facts and their interpretation, 19 May 1984. *New Phytologist*, 98(1): 119-128. (T)

8298 **Preston, C.D., Whitehouse, H.L.K. (1986).** The habitat of *Lythrum hyssopifolia* L. in Cambridgeshire, its only surviving English locality. *Biol. Conserv.*, 35(1): 41-62. (T)

8299 **Prince, S.D., Hare, D.R. (1981).** *Lactuca saligna* and *Pulicaria vulgaris* in Britain. *In* Synge, H., *ed*. The biological aspects of rare plant conservation. Chichester, Wiley. 379-388. Maps. Proceedings of International Conference, King's College, Cambridge, 14-19 July 1980. Both species are on the edge of their range in Britain. (T)

8300 **Pritchard, N.M. (1972).** Where have all the gentians gone? *Trans. Bot. Soc. Edin.*, 41: 279-291. (T)

8301 **Putwain, P.D., Gillham, D.A. (1988).** Restoration of heather moorland. *Landscape Design*, 172: 51-56. Illus., col. illus., maps. (L)

8302 **Rabinowitz, D., Cairns, S., Dillon, T. (1986).** Seven forms of rarity and their frequency in the flora of the British Isles. *In* Soule, M.E., *ed*. Conservation biology: the science of scarcity and diversity. Sunderland, Massachussetts, Sinauer Assoc. 183-204. (T)

8303 **Rackham, O. (1976).** Trees and woods in the British landscape. London, Dent. 204p. Illus., maps. Describes woodland evolution, management and conservation.

8304 **Ranwell, D.S. (1981).** Introduced coastal plants and rare species in Britain. *In* Synge, H., *ed*. The biological aspects of rare plant conservation. Chichester, Wiley. 413-419. Proceedings of International Conference, King's College, Cambridge, 14-19 July 1980. (T)

8305 **Ratcliffe, D.A. (1973).** Safeguarding wild plants. *In* Green, P.S., *ed*. Plants: wild and cultivated. A conference on horticulture and field botany, RHS and BSBI, 2-3 Sept. 1972. BSBI, 18-24.

8306 **Ratcliffe, D.A. (1979).** The role of the Nature Conservancy Council in the conservation of rare and threatened plants in Britain. *In* Synge, H., Townsend, H., *eds*. Survival or extinction. Proceedings of a conference held at the Royal Botanic Gardens, Kew, 11-17 September 1978. Kew, Bentham-Moxon Trust. 31-35. (P/T)

8307 **Ratcliffe, D.A., ed. (1977).** A nature conservation review. The selection of biological sites of national importance to nature conservation in Britain. Cambridge, Cambridge Univ. Press. 2 vols. 401, 320p. 1-vegetation communities; 2-site accounts, with evaluations of national conservation importance. (P)

8308 **Rawes, M., Welch, D. (1972).** Trials to recreate floristically rich vegetation by plant introduction in the northern Pennines, England. *Biol. Conserv.*, 4(2): 135-140. (P)

8309 **Riley, H., Ferry, B. (1986).** Fighting for the beaches of Dungeness. *New Sci.*, 110(1511): 46-47, 50-52. Illus., col. illus., maps. Dune system with numerous rare species is threatened by gravel extraction and nuclear power plant construction. (T)

8310 **Rothschild, M. (1980).** Britain's disappearing plants ?-further reflections. *Garden*, 105(7): 296. (T)

8311 **Rowell, T.A., Walters, S.M., Harvey, H.J. (1982).** The rediscovery of the Fen Violet, *Viola persicifolia* Schreber, at Wicken Fen, Cambidgeshire. *Watsonia*, 14(2): 183-184. (T)

8312 **Sands, T. (1981).** Wildlife and Countryside Bill. *BSE News*, 33: 2-6. (L)

UNITED KINGDOM

8313 Sayers, C.D., Gaman, J.H. (1972). Gene banks: a case study with Teesdale species. *J. Roy. Hort. Soc.*, 97: 488-491. (G)

8314 Shoard, M. (1980). The theft of the countryside. London, Temple Smith. 272p.

8315 Silverside, A.J. (1981). The conservation of weed communities of sandy soils in Britain (Phytosociological survey). *In* Schwabe-Braun, A., ed. Vegetation als anthropo-okologischer Gegenstand (Rinteln, 5-8 April 1971). Gefahrdete Vegetation und ihr Erhaltung (Rinteln, 27-30 March 1972). Vaduz, J. Cramer. 511-513.

8316 Simmons, J.B.E. (1986). Conservation and the living collections, Royal Botanic Gardens, Kew. *Kew Mag.*, 3(1): 39-48. Illus. (G)

8317 Simons, P. (1988). The day of the rhododendron. *New Sci.*, 119(1620): 50-55. Col. illus. *Rhododendron ponticum* as an invasive species.

8318 Smith, M. (1979). Creating specialized habitats in a garden. *In* Synge, H., Townsend, H., eds. Survival or extinction. Proceedings of a conference held at the Royal Botanic Gardens, Kew, 11-17 September 1978. Kew, Bentham-Moxon Trust. 219-221. Describes creation of sand dune and *Sphagnum* bog habitats at University of Bristol Botanic Garden. (G)

8319 Smith, M. (1982). Growing back from the brink of extinction. *Guardian*, 11/82. 62 British plants covered by 1981 Wildlife and Countryside Act. (T)

8320 Smith, M. (1982). How to save the forests of Snowdonia. *New Sci.*, 95(1312): 14-17. Illus. Map. (P/T)

8321 Smith, P.H. (1984). The distribution, status and conservation of *Juncus balticus* Willd. in England. *Watsonia*, 15: 15-26. Illus. (T)

8322 Somerville, A.H., Small, G. (1981). *Trichomanes speciosum. Trans. Bot. Soc. Edinb.*, 43(4): (T)

8323 Thompson, D. (1987). Battle of the bog. *New Sci.*, 113(1542): 41-45. Peatlands in Scotland.

8324 Torrance, W.G. (1968). Saving our floral heritage. Derby, U.K., Harper & Sons. 36p. (T)

8325 Tubbs, C. (1982). The New Forest: conflict and symbiosis. *New Sci.*, 95(1312): 10-13. Illus. Map. Grazing by ponies and cattle helps to maintain unique habitats. But there are too many domestic animals. Can economy and wildlife be reconciled? (T)

8326 Tully, C. (1987). Conserving Broadland. *Environment Now*, January 1987: 36-39. Col. illus., map.

8327 U.K. Committee for International Nature Conservation (1979). Wildlife introductions to Great Britain: the introduction, reintroduction and restocking of species in Great Britain: some policy implications for nature conservation. London, Nature Conservancy Council. 32p.

8328 Vines, G. (1983). Science can make a meadow. *New Sci.*, 99(1371): 486-487. Illus. Recreating plant-rich meadows. (T)

8329 Walters, S.M. (1979). The eastern England rare plant project in the University Botanic Garden, Cambridge. *In* Synge, H., Townsend, H., eds. Survival or extinction. Proceedings of a conference held at the Royal Botanic Gardens, Kew, 11-17 September 1978. Kew, Bentham-Moxon Trust. 37-46. Illus., maps. (G/T)

8330 Ward, L.K. (1981). The demography, fauna and conservation of *Juniperus communis* in Britain. *In* Synge, H., ed. The biological aspects of rare plant conservation. Chichester, Wiley. 319-329. Maps.

UNITED KINGDOM

8331 **Warner, N. (1981).** *Selinum carvifolia* in Cambridgeshire. *Nat. Cambridge*, 24: 42-43.

8332 **Wells, D.A. (1981).** The protection of British rare plants in nature reserves. *In* Synge, H., *ed.* The biological aspects of rare plant conservation. Chichester, Wiley. 475-480. Proceedings of International Conference, King's College, Cambridge, 14-19 July 1980. (P/T)

8333 **Wells, T.C.E. (1981).** Population ecology of terrestrial orchids. *In* Synge, H., *ed.* The biological aspects of rare plant conservation. Chichester, Wiley. 281-295. Illus. Proceedings of International Conference, King's College, Cambridge, 14-19 July 1980. (T)

8334 **Witt, J., Witt, J. (1979).** Endangered plants and botanic gardens. *Univ. Wash. Arbor. Bull.*, 42(2): 15-22. Illus., col. illus. (G/T)

8335 **Witt, J., Witt, J. (1979).** Endangered plants and botanic gardens: 2. *Univ. Wash. Arbor. Bull.*, 42(3): 2-6. Illus. (G/T)

8336 **Wood, J., Clements, M., Muir, H. (1984).** Plants in peril. 2. *Kew Mag.*, 1(3): 139-142. Illus. *Cypripedium calceolus.* (T)

8337 **Woodings, T.L., Ratcliffe, D. (1981).** Declining populations of annual *Veronica* species in Britain: studies on seed production, germination and survival. *In* Synge, H., *ed.* The biological aspects of rare plant conservation. Chichester, Wiley. 508-511. Proc. International Conference, King's College, Cambridge, 14-19 July 1980. *Veronica verna, V. praecox, V. triphyllos* included in *British Red Data Book* (T)

8338 **Yarrow, C. (1980).** Preserving lowland woods. *Town and Country Planning*, 49(6): 190. Illus. Woodlands and tourism could support each other.

UNITED STATES

8339 **Academy of Natural Sciences of Philadelphia (1977).** Our vanishing species: should they be saved? *Frontiers*, 41(4): 10-39. (T)

8340 **Adams, W.T. (1981).** Population genetics and gene conservation in Pacific Northwest conifers. *In* Scudder, G.G.E., Reveal, J.L., *eds.* Evolution today: proceedings of the second International Congress of Systematic and Evolutionary Biology. Pittsburgh. Hunt Inst. Bot. Documentation. 401-415.

8341 **American Horticultural Society (1983).** New plants listed as Endangered. *Amer. Hort.*, 62(1): 8-9. Illus. (T)

8342 **Anderson, E.F. (1982).** A meeting on the cactus trade. *Cact. Succ. J.* (U.K.), 54(2): 82-85. Illus. (E/T)

8343 **Anderson, M.P. (1904).** The protection of our native plants. *J. New York Bot. Gard.*, 5: 71-79. (T)

8344 **Andreev, L.N., ed. (1986).** Introduktsiya i okhrana rastenii v SSSR i SShA. (Introduction and protection of plants in the USSR and USA.) Moskva, Nauka. 129p. Rus. Maps. (P/T)

8345 **Andrus, C.D. (1982).** The Endangered Species Act in transit. *Garden* (New York), 6(4): 2-6. Col. illus. (L)

8346 **Anon. (1966).** The Multiple Use-Sustained Yield Act of 1960 - a highlight in the history of forest conservation. *Program Aid*, 771: 6. Illus. (L)

8347 **Anon. (1974).** Ten percent of U.S. plants endangered. *Sci. News*, 106: 204. (T)

8348 **Anon. (1975).** How to save a wildflower: NPCA progress report on endangered plants. *Nat. Parks Conserv. Mag.*, 49(4): 10-14. (T)

8349 **Anon. (1975).** The Conservation of Wild Creatures and Wild Plants Act 1975. *Habitat*, 11(8): 1-3. (L/T)

UNITED STATES

8350 **Anon. (1976).** Endangered plants: *Bureaucratus delayus. Nat. Parks Conserv. Mag.,* 50(10): 29, 30. (L/T)

8351 **Anon. (1977).** Cactus rustling in southwest. *Field Mus. Nat. Hist. Bull.* (Chicago), 48(8): 21.

8352 **Anon. (1977).** National parks declining? *Field Mus. Nat. Hist. Bull.* (Chicago), 48(1): 6-7. (P)

8353 **Anon. (1978).** The national park system: thirteen possible additions. *Nat. Parks Conserv. Mag.,* 52(7): 20-25. Illus. (P)

8354 **Anon. (1978).** Thirteen plants added to endangered and threatened lists. *Field Mus. Nat. Hist. Bull.* (Chicago), 49(7): 3, 9. Illus. (T)

8355 **Anon. (1978).** Thirty "endangered species" candidates not in trouble after all. *Field Mus. Nat. Hist. Bull.* (Chicago), 49(9): 4. (T)

8356 **Anon. (1979).** Chapman rhododendron endangered. *Amer. Hort. Soc. News Views,* 21(5): 4. Illus. (T)

8357 **Anon. (1979).** NPCA adjacent lands update: no park is an island. *Nat. Parks Conserv. Mag.,* 53(11): 21-22. Illus. (P)

8358 **Anon. (1979).** Parks versus power plants? *Nat. Parks Conserv. Mag.,* 53(10): 25-26. (P)

8359 **Anon. (1979).** Plants and plant material. *AABGA Newsl.,* 57: 3-4. Illus. (T)

8360 **Anon. (1980).** Endangered and threatened plant species. *Field Mus. Nat. Hist. Bull.* (Chicago), 51(10): 11. (T)

8361 **Anon. (1980).** Mountain golden-heather proposed for Endangered status. *Amer. Hort. Soc. News Views,* 22(5): 7. (T)

8362 **Anon. (1980).** Where have all the prairies gone? *Missouri Bot. Gard. Bull.,* 68(2): 1, 3. (T)

8363 **Anon. (1981).** The GCA: a lifeline for plants. *Garden* (New York), 5(5): 2-3, 32. Illus. Garden Club of America - efforts to save species in jeopardy. (T)

8364 **Anon. (1982).** Cycad Society seed bank. *Cycad Newsl.,* 5(1): 15. (G/T)

8365 **Anon. (1982).** Endangered species. *AABGA Newsl.,* 97: 4. (L/T)

8366 **Anon. (1982).** Research: Wingra Fen and its orchids. *Arbor. News Univ. Wisc.,* 31(4): 11. (T)

8367 **Anon. (1982).** Scientists for the Endangered Species Act. *Ass. Syst. Collect. Newsl.,* 10(2): 17. (L/T)

8368 **Anon. (1982).** Silky camellia is a wild eye-catcher. *Amer. Camellia Yearbook,* 1982: 100-102. Illus. *Stewartia malacodendron.* Threatened and endangered plants of Florida from Division of Plant Industry, Gainesville. (L/T)

8369 **Anon. (1983).** Endangered species in USA. *Gartn.-Bot. Brief,* 76: 39-40. (T)

8370 **Anon. (1983).** Nature Conservancy launches Endangered Wetlands Program. *BioScience,* 33(6): 404. Illus. (P)

8371 **Anon. (1983).** New plants listed as Endangered. *Amer. Horticulturist,* 62(1): 8-9. Illus. (T)

8372 **Anon. (1983).** Two new plants proposed as Endangered. *Amer. Horticulturist,* 62(9): 5. (T)

8373 **Anon. (1984).** Endangered plant update. *Amer. Horticulturist,* 63(11): 14-15. Description of six endangered plants and eleven proposed endangered species. (T)

UNITED STATES

8374 **Anon. (1984).** Endangered species. *AABGA Newsl.*, 114: 6-7. Adapted from *Endangered Species Tech. Bull.*, 9(3) (1984). (T)

8375 **Anon. (1984).** More U.S. plants protected. *Threatened Pl. Newsl.*, 13: 14-15. (L/T)

8376 **Anon. (1985).** Endangered plant update. *Amer. Horticulturist*, 64(11): 1-4, 6-7. Illus. (T)

8377 **Anon. (1986).** Saving *Isotria medeoloides*. *Amer. Orchid Soc. Bull.*, 55(1): 16-18. Col. illus. (T)

8378 **Anon. (1986).** USA seizes rare cacti. *TRAFFIC Bull.*, 8(2): 32. (T)

8379 **Anon. (1987).** Decline of sugar maple. *Oryx*, 21(2): 121. *Acer saccharum* threatened by acid rain. Extract from *The New York Times*, 7 December 1987.

8380 **Anon. (1987).** Tree die-off in eastern U.S. *Oryx*, 21(2): 121. Pollution damage. Extract from *Science*, 22 August 1986.

8381 **Anon. (1988).** Extinction imminent for native plants. *Science*, 242(4885): 1508. (T)

8382 **Apel, J. (1979).** Naturschutz-Eindrucke in den U.S.A. *Gartn.-Bot. Brief*, 62: 4. Ge. (P)

8383 **Arnett, G.R. (1984).** Cooperation between government and the private sector: a north American example. *In* McNeely, J.A., Miller, K.R., *eds.* National parks, conservation and development. The role of protected areas in sustaining society. Washington, D.C., Smithsonian Institution Press. 534-537. (P)

8384 **Atwood, D. (1979).** Management program for plants on federal lands. *In* Wood, S.L. The endangered species: a symposium. 81-86.

8385 **Auchincloss, A. (1982).** Where the wild thyme grows. New York, Natural Resources Defense Council. 8p.

8386 **Ayensu, E.S. (1975).** Endangered and threatened orchids of the United States. *Amer. Orch. Soc. Bull.*, 44: 384-394. (T)

8387 **Ayensu, E.S. (1978).** The U.S. Red Data Book. *Garden* (New York), 2(5): 2-3. (T)

8388 **Ayensu, E.S. (1981).** Assessment of threatened plant species in the United States. *In* Synge, H., *ed.* The biological aspects of rare plant conservation. Chichester, Wiley. 19-58. Proceedings of International Conference, King's College, Cambridge, 14-19 July 1980. Maps. (T)

8389 **Ayensu, E.S. (1983).** Smithsonian Institution endangered flora information-processing: experiences and goals. *In* Jain, S.K., Mehra, K.L., *eds.* Conservation of tropical plant resources. Howrah, Botanical Survey of India. 29-38. (T)

8390 **Ayensu, E.S., DeFilipps, R.A. (1978).** Endangered and threatened plants of the United States. Washington, D.C., Smithsonian Institution and World Wildlife Fund Inc. 403p. Lists 90 'Extinct', 839 'Endangered' and 1211 'Threatened' taxa for the continental U.S. Also covers Hawaii, Puerto Rico and Virgin Is. (T)

8391 **Backiel, A., Hunt, F. (1986).** "Acid rain" and forests. An attempt to clear the air. *Amer. Forests*, 92(2): 42-47. (T)

8392 **Baker, D. (1986).** Virgin forests under fire. *Nat. Wildlife*, 24(2): 4-11. North-west forests being cut.

8393 **Barrick, F.B. (1977).** Federal and state programs on endangered plants. *In* U.S. Department of Agriculture Forest Service. Conference on endangered plants in the Southeast: proceedings. Asheville, NC, Southeast Forest Exp. Station. (T)

8394 **Barry, W.J. (1981).** Jepson Prairie: will it be preserved? *Fremontia*, 9(1): 7-11. Illus., map.

8395 **Bartgis, R.L. (1985).** Rediscovery of *Trifolium stoloniferum* Muhl. ex A.Eaton. *Rhodora*, 87(851): 425-429. Illus. (T)

UNITED STATES

8396 **Barton, F. (1978).** Rare plant sanctuary. *Herbarist*, 44: 6-14. Illus. (P/T)

8397 **Bean, M.J. (1977).** The Endangered Species Act under fire. *Nat. Parks Conserv. Mag.*, 51(6): 16-20. (L)

8398 **Bean, M.J. (1983).** The evolution of national wildlife law. Rev. and exp. ed. New York, Praeger. 449p. (L)

8399 **Beede, H.L. (1966).** America's endangered orchids. *Nat. Wildl.*, 4(1): 50-51. (T)

8400 **Beeking, R.W. (1977).** The national park system in the United States. *In* Filion, L., Villeneuve, P., *eds*. Ecological reserves and nature conservation. Rapport, Conseil Consultatif des Reserves Ecologiques, Quebec, 1. 35-40. Papers presented at a conference at Iles-de-la-Madeleine, Quebec, 17-18 June 1976. (P)

8401 **Behlen, D. (1979).** Protecting special trees. *Amer. Forests*, 85(11): 28-30. Illus.

8402 **Beimborn, W.A. (1979).** The Endangered Species Act of 1973. *In* Proceedings of the Annual Technical Meeting of the Institute of Environmental Science. 392-393. Illus. Ref. in *Agricola*, 79(8-9). (L)

8403 **Bender, M., ed. (1984).** Listing proposal for eastern plant withdrawn. [*Paronychia argyrochoma* var. *albimontana*]. *Endangered Species Tech. Bull.*, 8(11): 8. (T)

8404 **Bender, M., ed. (1984).** San Benito evening-primrose proposed as endangered. *Endangered Species Tech. Bull.*, 8(11):?. (T)

8405 **Benninghoff, W.S. (1977).** The concept of natural area and species conservation in the United States. *In* Filion, L., Villeneuve, P., *eds*. Ecological reserves and nature conservation. Rapport, Conseil Consultatif des Reserves Ecologiques, Quebec, 1. 9-13. Papers presented at a conference at Iles-de-la-Madeleine, Quebec, 17-19 June 1986. (P)

8406 **Benson, B. (1976).** Endangered Species - heads in the clouds or the sand? *Cact. Succ. J.* (U.S.), 48(5): 207-212. (L/T)

8407 **Benson, L. (1982).** The cacti of the United States and Canada. Stanford, CA, Stanford University Press. ix, 1044p. Illus., col. illus., maps, keys. Includes conservation, 242-247. (T)

8408 **Berger, T.J., Neuner, A.M. (1979).** Directory of state protected species: a reference to species controlled by non-game regulations. Lawrence, Kansas, Association of Systematics Collections. (L)

8409 **Berger, T.J., Phillips, J.D. (1977).** Index to U.S. federal wildlife regulations. Lawrence, KS, Association of Systematics Collections, Museum of Natural History, Univ. of Kansas. Loose-leaf. (L)

8410 **Berger, T.J., et al. (1979).** Directory of federally controlled species. Lawrence, KA, Association of Systematics Collections. 1v. Loose-leaf. (L)

8411 **Bernhardt, P. (1983).** Skipping over the lady's slipper. *Garden* (New York), 7(2): 14-15, 31. Illus., col. illus. (T)

8412 **Betz, R.F. (1979).** Resurrecting the prairie. *Garden* (New York), 3(4): 28-34. Illus., col. illus., map. Prairie cemeteries now being made into state nature reserves. (P)

8413 **Bleahu, M. (1984).** Protectia naturii in Statele Unite ale Americii. (The protection of nature in the U.S.A.) *Ocrot. Nat. Mediul. Inconjurat.*, 28(2): 120-125. Rum. Illus.

8414 **Boyd, W.S. (1970).** Federal protection of endangered wildlife species. *Stanford Law Review*, 22: 1289-1309. (T)

8415 **Boyles, J.L. (1979).** A long time coming - Ted Shanks Wildlife Management Area. *Missouri Conservationist*, 40(11): 4-7. Col. illus. (P)

UNITED STATES

8416 **Bratton, S.P., Owen, I., White, P.S. (1982).** The status of botanical information on national parks in the southeastern United States. *Castanea*, 47(2): 137-147. Including checklists, collections, rare and exotic species data, vegetation inventories, maps and management. (P/T)

8417 **Bratton, S.P., White, P.S. (1980).** Rare plant management - after preservation what? *Rhodora*, 82(829): 49-75. (T)

8418 **Bratton, S.P., White, P.S. (1981).** Rare and endangered plant species management: potential threats and practical problems in US national parks and preserves. *In* Synge, H., *ed.* The biological aspects of rare plant conservation. Chichester, Wiley. 459-474. Proceedings of International Conference, King's College, Cambridge, 14-19 July 1980. (P/T)

8419 **Brooks, H.T. (1979).** Reserved water rights and our national forests. *Nat. Resources J.*, 19(2): 433-443.

8420 **Brooks, R.T., Birch, T.W. (1988).** Changes in New England forests and forest owners: implications for wildlife habitat resources and management. *Trans. N. Amer. Wildl. Nat. Resour. Conf.*, 53: 78-87.

8421 **Brouillet, L. (1985).** La conservation des plantes rares: le fondement biologique. *Nat. Canad.*, 112(2): 263-273. Fr (En). (T)

8422 **Brown, D. (1975).** Wildlife habitat management needs. U.S. Department of Agriculture Forest Service Project Record No. 7626 2201. Missoula, MT, U.S. Department of Agriculture Forest Service, Equipment Development Center. iii, 10p.

8423 **Brumback, B. (1983).** Endangered species. *Amer. Assoc. Bot. Gard. Arbor. Inc. Newsl.*, 102: 9. (L/T)

8424 **Brumback, B., et al. (1981).** Rare and endangered native plant exchange. *Amer. Assoc. Bot. Gard. Arbor. Newsl.*, 81: 7-8. (G/T)

8425 **Brumback, W. (1983).** Endangered species. *Amer. Assoc. Bot. Gard. Arbor. Inc. Newsl.*, 100: 4. (L/T)

8426 **Brumback, W. (1983).** Endangered species. *Amer. Assoc. Bot. Gard. Arbor. Inc. Newsl.*, 105: 6. Proposed for protection under U.S. Endangered Species Act. (L/T)

8427 **Brumback, W. (1983).** Endangered species. *Amer. Assoc. Bot. Gard. Arbor. Inc. Newsl.*, 103: 5. 4th meeting of Conference of Parties to Convention on International trade in Endangered Species of Wild Fauna and Flora (CITES). (L/T)

8428 **Brumback, W.E. (1981).** Endangered plant species programs for botanic gardens with examples from North American institutions. Newark, Univ. of Delaware. 232p. M.Sc. thesis. (G/T)

8429 **Burdick, G. (1977).** Plant species to be protected under 1973 Endangered Species Act. *Environment Reporter*, 7(50): 1906-1907. (L/T)

8430 **Bureau of Sport Fisheries and Wildlife (1972).** Proposed Endangered Species Conservation Act of 1972. Draft environmental impact statement. Washington, D.C., The Bureau. 27p. (L)

8431 **Bureau of Sport Fisheries and Wildlife (1973).** Proposed Endangered Species Conservation Act of 1972. Final environmental impact statement. Washington, D.C., The Bureau. 309p. (L)

8432 **Burgess, R.L. (1978).** Deciduous: the changing face of eastern North America. *Frontiers*, 42(3): 9-11. Illus.

8433 **Burk, J., Sackett, M. (1978).** Ecology seminar series: the problem of endangered species. *Appalachia*, 44(7): 14-15, 18. Illus. (T)

UNITED STATES

8434 **Burk, J., Sackett, M. (1979).** Ecology seminar series: the coastal zone. *Appalachia Bull.*, 45(2): 22-24. Illus. Massachusetts Bay.

8435 **Burt, DeVere E. (1973).** The geography of extinct and endangered species in the United States. *The Explorer*, 15(3): 4-10. (T)

8436 **Byers, A.M. (1978).** The island appeal: inland islands preserved through the efforts of the conservancy. *Nat. Conserv. News*, 28(4): 16-19. Illus., col. illus.

8437 **Cain, K.R. (1978).** Beauty among the tupelos. *Soil Conservation*, 43(11): 8. Illus. Conservation of remnant of the Pocotaligo.

8438 **California Native Plant Society (1978).** Endangered species act survives? *Calif. Native Pl. Soc. Bull.*, Nov-Dec.: [1]p. (L/T)

8439 **California Native Plant Society (1980).** Conservation news: Red Rock Canyon State Park. *Calif. Native Pl. Soc. Bull.*, 10(4): 4. (P)

8440 **California Native Plant Society (1981).** Assessment of native plant vulnerability. *Calif. Native Pl. Soc. Bull.*, 11(5): 9. (T)

8441 **California Native Plant Society (1982).** Endangered Species Act reauthorization. *Calif. Native Pl. Soc. Bull.*, 12(2): 6. (L/T)

8442 **California Native Plant Society (1982).** Ernest C. Twisselmann Botanical Area (proposed). *Calif. Native Pl. Soc. Bull.*, 12(3): 4. (P)

8443 **California Native Plant Society (1982).** Trade in wild plants. *Calif. Native Pl. Soc. Bull.*, 12(2): 6. (E/L/T)

8444 **California Native Plant Society (1983).** Rare Plant Scientific Advisory Committee report. *Calif. Native Pl. Soc. Bull.*, 13(1): 2. (T)

8445 **Campbell, F.T. (1980).** Conserving our wild plant heritage. *Environment*, 22(9): 14-24.

8446 **Campbell, F.T. (1982).** The Lacey Act and the cactus collector. *Cact. Succ. J.* (U.K.), 54(5): 213-214. (L/T)

8447 **Campbell, F.T. (1983).** Carnivorous plants deserve protection. *Carniv. Pl. Newsl.*, 12(4): 96-98. Illus. (L/T)

8448 **Campbell, F.T. (1983).** Controlling the trade in plants: a progress report. *Garden* (New York), 7(4): 2-5, 32. Illus. Effectiveness of CITES. (E/T)

8449 **Campbell, F.T. (1983).** Plant trade regulations. *TRAFFIC* (U.S.A.), 5(1): 6-8. (E/L)

8450 **Caplenor, D. (1978).** Additions to the list of plants of the gorges at Fall Creek Falls. *J. Tenn. Acad. Sci.*, 53(4): 135. Fall Creek State Park. (P)

8451 **Cardozo, Y. (1982).** Sanctuary for some small cypress. *Garden* (New York), 6(2): 22-25. Illus., col. illus. (P/T)

8452 **Carl, A. (1976).** The legacy of Gardiners Island. *Conservationist*, 31(1): 18-23, 45. Col. illus.

8453 **Case, F.W., jr (1964).** Orchids of the western Great Lakes region. *Cranbrook Inst. Sci. Bull.*, 48. (T)

8454 **Chalk, D.E., Miller, S.A., Hoekstra, T.W. (1984).** Multiresource inventories: integrating information on wildlife resources. *Wildl. Soc. Bull.*, 12(4): 357-364. Forests. (L)

8455 **Cheatham, N.H.D. (1979).** Living laboratories in the desert. *Fremontia*, 6(4): 26-30. Illus.

8456 **Christian, P.H., Shelly, F.L. (1981).** News and notes: Thompson Wetlands Preserve established. *Bartonia*, 48: 43. (P)

UNITED STATES

8457 **Clark, T.D. (1984).** The greening of the south: the recovery of land and forest. Lexington, Univ. Press of Kentucky. xvi, 168p. Illus.

8458 **Coggins, G.C., Patti, S.T. (1980).** The emerging law of wildlife II: a narrative bibliography of federal wildlife law. *Harvard Environ. Law Rev.,* 4(1): 164-190. (B/L)

8459 **Coggins, G.C., Smith, D.L. (1976).** The emerging law of wildlife: a narrative bibliography. *Harvard Environ. Law Rev.,* 6: 583-618. (B/L)

8460 **Committee on Merchant Marine and Fisheries (1976).** Hearings before the Subcommittee on Fisheries and Wildlife Conservation and the Environment of the Committee on Merchant Marine and Fisheries on implementation and administration of the Endangered Species Act and its amendments; and to review the problems and issues encountered. Series No. 94. 17: 1-367. (L)

8461 **Cooper, T. (1978).** Nature - of the people, by the people, for the people. *Frontiers,* 42(4): 21-23.

8462 **Council on Environmental Quality (1981).** A summary of the legal authorities for conserving wild plants. Washington, D.C., Council on Environmental Quality. 156p. (L/T)

8463 **Countryman, W.D., Dowhan, J.J., Morse, L.E. (1981).** Regional coordination of rare plant information synthesis by the New England Botanical Club. *In* Morse, L.E., Henifin, M.S., *eds.* Rare plant conservation: geographical data organization. New York, New York Botanical Garden. 123-131. (T)

8464 **Crow, G.E., et al. (1981).** Rare and endangered vascular plant species in New England. *Rhodora,* 83(834): 259-299. (T)

8465 **Crowe, D.S. (1979).** The value of small woodlands to landscape and society. *Parks,* 4(1): 5-7. Illus.

8466 **Cutler, M.R. (1980).** Wilderness decisions: values and challenges to science. *J. Forest.,* 78(2): 74-77. (P/T)

8467 **Cutler, R.M., Jenkins, R.E. (1978).** Ecology forum, No. 28. Islands as preserves. *Nat. Conserv. News,* 28(4): 24-25.

8468 **Daiber, F. (1986).** Conservation of tidal marshes. New York, Van Nostrand Reinhold. 341p. Wetlands.

8469 **Dale, N. (1980).** A mountain park in a great city. *Fremontia,* 8(1): 7-12. Illus.

8470 **Dasmann, R.F. (1973).** Reconciling conservation and development in the coastal zone. *In* Costin, A.B., Groves, A.K., *eds.* Nature conservation in the Pacific. Proc. Symposium A-10, XII Pacific Science Congress, Canberra, August-September 1971. IUCN and Australian Nat. Univ. Press. 285-297.

8471 **Davis, R.C., ed. (1983).** Encyclopedia of American forest and conservation history. New York, Collier Macmillan Publishers. 2 vols, 780p. Illus.

8472 **DeDecker, M. (1981).** The state of the endangered mallow. *Fremontia,* 8(4): 16. (T)

8473 **DeDecker, M. (1982).** Overdraft victim. *Fremontia,* 10(1): 13. Illus. *Sidalcea covillei.* (T)

8474 **Dickenson, R.E. (1980).** The national parks today and tomorrow. *Nat. Parks Conserv. Mag.,* 54(8): 8. (P)

8475 **Dilcher, D. (1982).** Endangered Species Act. *Pl. Sci. Bull.,* 28(1): 2. (L/T)

8476 **Dolan, R., Hayden, B., Fisher, J. (1973).** A strategy management of marine and lake systems within the national park system. National Science Report No. 6. Washington, D.C., U.S. Dept of the Interior, National Park Service. 40p. Illus. Dune stabilization study. (P)

UNITED STATES

8477 **Dolan, R., Hayden, B.P., Vincent, C.L. (1974).** Shore zone land use and land cover: central Atlantic regional ecological test site. National Resources Report No. 8. Washington D.C., U.S. Geological Survey, National Park Service. v, 50p. Maps.

8478 **Dummler, H. (1977).** Ein Reisebericht: vom 12. Internationalen Seminar uber Nationalparke und gleichwertige Schutzgebiete com 2. August bis 1. September 1977 in Kanada, USA und Mexico. *Naturschutz- und Naturparke*, 87: 1-8. Ge. Illus., map. (P)

8479 **Duncan, W.H. (1973).** Endangered, rare and uncommon wildflowers found on the southern national forests. Atlanta, U.S. Department of Agriculture, Forest Service, South Region. 20p. (T)

8480 **Durrant, M. (1985).** The hazardous life of our rarest plants. *Audubon*, 87(4): 50-61. (T)

8481 **Dusek, K. (1985).** Update on our rarest pine. *American Forests*, 91(10): 26-29, 61, 63. (T)

8482 **Dyrness, C.T., et al. (1975).** Research natural area needs in the Pacific Northwest: a contribution to land-use planning. USDA For. Serv. Gen. Tech. Rep. No. PNW-38. Portland, OR, U.S. For. Serv. Pacific Northwest Forest and Range Experiment Station. 231p. (P)

8483 **Eastman, P., Bodde, T. (1982).** Endangered Species Act: more is at stake than the Bald Eagle. *BioScience*, 32(4): 246-248. (L)

8484 **Eckhardt, J.P. (1983).** The Conservancy's endangered species program. *Nat. Conserv. News*, 33(6): 14-17. (T)

8485 **Ecological Society of America (1921).** Preserves of natural conditions. Springfield, IL, Ecological Society of America, Committee on the Preservation of Natural Conditions. 31p. (P)

8486 **Elias, T.S. (1975).** Vascular plants. *In* Proceedings of the Symposium on endangered and threatened species of North America, Washington, D.C., 1974. St. Louis, MO, Wild Canid Survival and Research Center. 339p. (T)

8487 **Elias, T.S. (1977).** An overview. *In* Prance, G.T., Elias, T.S., *eds.* Extinction is forever. Proceedings of a symposium at the New York Botanical Garden, 11-13 May 1976. New York, New York Botanical Garden. 13-16. Illus. Discusses state of threatened plant knowledge in the U.S. (G/T)

8488 **Elias, T.S. (1984).** Okhrana redkikh i ischezayushchikh vidov rastenii SShA. (The protection of rare and disappearing plants of the U.S.A.) *Priroda Ezhem. Pop. Estest. Zhurn. Akad. Nauk*, 8: 88-95. Rus. Col. illus. Translated from English. (T)

8489 **Elias, T.S. (1987).** Can threatened and endangered species be maintained in botanic gardens? *In* Elias, T.S., *ed.* Conservation and management of rare and endangered plants. Proceedings from a conference, Sacramento, California, 5-8 November 1986. Sacramento, California Native Plant Society. 563-566. Examples and data from Rancho Santa Ana Botanic Garden, Claremont, CA. (G/P/T)

8490 **Ellman, P. (1975).** Let's save Ring Mountain. *Fremontia*, 3(2): 10-14. (P)

8491 **Ellmore, G.S., Phair, W.E. (1987).** Status of elm preservation in New England. *Rhodora*, 89(857): 27-33. Illus.

8492 **Endangered Species Scientific Authority (1977).** Interim charter. Requests for comments on interim charter and on criteria for permit application evaluation. *Fed. Register*, 42(32): 35799-35802. (L)

8493 **Evans, A.M. (1979).** Survival strategies of southern Appalachian pteridophytes. *ASB Bull.*, 26(2): 101. Abstract. (T)

UNITED STATES

8494 **Ewacha, J. (1985).** Vanishing cypripediums. *Amer. Orchid Soc. Bull.*, 54(10): 1194-1196. Col. illus. (T)

8495 **Fahl, R.J. (1977).** North American forest and conservation history: a bibliography. Santa Barbara, Forest History Society. 408p. (B)

8496 **Falk, D.A,, Thibodeau, F.R. (1986).** Saving the rarest. *Arnoldia*, 46(3): 3-17. Illus. (G/T)

8497 **Falk, D.A. (1988).** The Center for Plant Conservation: conserving the native plant genetic diversity of the United States. *Diversity*, 16: 20-21. Map. (G/T)

8498 **Falk, D.A., Walter, K.S. (1986).** Networking to save imperilled plants. *Garden* (U.S.), 10(1): 2-6, 32. The new Center for Plant Conservation spearheads a nationwide preservation program that draws on the efforts of U.S. botanical gardens and arboreta. (G/T)

8499 **Farnsworth, N., Soejarto, D.D. (1985).** Potential consequence of plant extinction in the United States on the current and future availability of prescription drugs. *Econ. Bot.*, 39(3): 231-240. Medicinal (E/T)

8500 **Faulf, D.A. (1987).** Endangered species conservation *ex situ*: the national view. *In* Elias, T.S., *ed.* Conservation and management of rare and endangered plants. Proceedings from a conference, Sacramento, California, 5-8 November 1986. Sacramento, California Native Plant Society. 553-561. (G/T)

8501 **Fay, J.J. (1981).** The Endangered Species Program and plant reserves in the United States. *In* Synge, H., *ed.* The biological aspects of rare plant conservation. Chichester, Wiley. 447-452. Proceedings of International Conference, King's College, Cambridge, 14-19 July 1980. (P/T)

8502 **Federal Committee on Ecological Reserves (1977).** A directory of research natural areas on Federal lands of the United States of America. USDA, Forest Service. v, 280p. (P)

8503 **Ffolliott, P.F., Halffter, G. (1980).** Social and environmental consequences of natural resources policies, with special emphasis on biosphere reserves. *In* USDA Forest Service. Proceedings of the International Seminar, 8-13 April, 1980, Durango, Mexico. General technical report. Fort Collins, Rocky Mountain Forest and Range Experiment Station. 56p. (P)

8504 **Fitzgerald, S.G. (1986).** The state of the States in plant protection. *Garden* (New York), 10(5): 2-5, 31-32. (T)

8505 **Fosberg, F.R. (1972).** Our native plants - a look to the future. *Nat. Parks Conserv. Mag.*, 46(11): 17-21.

8506 **Fox, J. (1985).** Stalking and preserving US wild plants. *BioScience*, 35(5): 276-277.

8507 **Franklin, J.F. (1977).** The biosphere reserve program in the United States. *Science*, 95: 262-267. (P)

8508 **Franklin, J.F., Krugman, S.L. (1979).** Selection, management and utilization of biosphere reserves. USDA, Forest Serv., Gen. Tech. Rep., PNW-82: 308p. Proc. of US/USSR Symp., Moscow May 1976. Illus. (P)

8509 **Fritz, E.C. (1983).** Saving species is not enough. *BioScience*, 33(5): 301. (P/T)

8510 **Frome, M. (1981).** Politics and the parks. *Nation. Parks*, 55(9-10): 28-30. (P)

8511 **Frome, M. (1981).** The national parks: a plan for the future. *Nation. Parks*, 55(11-12): 10-13. Illus. (P)

8512 **Frome, M. (1984).** Promised land: twenty years after the Wilderness Act, the challenge passes to a new generation. *Nation. Parks*, 58(1-2): 24-29. Col. illus. (L/P)

UNITED STATES

8513 **Frome, M. (1989).** Wilderness. *Nation. Parks*, 63(7-8): 34-38. Wilderness Act. (L/P)

8514 **Fryxell, P.A. (1984).** Taxonomy and germplasm resources. *Agronomy*, (24): 27-57. *Gossypium.* Germplasm genetic resources.

8515 **Fuller, D. (1985).** U.S. cactus and succulent business moves toward propagation. *TRAFFIC (U.S.A.) Bull.*, 6(2): 1, 3-5, 11. (E/G)

8516 **Fuller, D. (1986).** American ginseng: harvest and export, 1982-1984. *TRAFFIC (U.S.A.) Bull.*, 7(1): 6. Trade. (E/T)

8517 **Gade, S. (1987).** State-by-state summary of protection afforded to native orchid species. *Amer. Orchid Soc. Bull.*, 56(2): 147-163. Col. illus. (L/T)

8518 **Gaskin, J.W., Douglass, J.E., Swank, W.T. (1983).** Annotated bibliography of publications on watershed management and ecological studies at Coweeta Hydrologic Laboratory. 1934-1984. USDA Forest Service General Technical Report SE No. 30. Southeastern Forest Experiment Station. 140p. (B)

8519 **Geary, I. (1978).** The Presidio Clarkia. *Fremontia*, 5(4): 11.

8520 **Gillham, M.E., Smith, J.K. (1983).** Industry and wildlife: compromise and coexistence. *Endeavour, n.s.*, 7(4): 162-172. Illus., col. illus.

8521 **Glenn-Lewin, D.C., Landers, R.Q., eds (1978).** Fifth Midwest Prairie conference proceedings, Iowa State Univ. Ames, Aug. 22-24, 1976. Ames, Iowa State Univ. 230p.

8522 **Godfrey, M.A. (1979).** Of polygons and plants. *Nat. Conserv. News*, 29(2): 21-24. Col. illus.

8523 **Godin, V.B., Leonard, R.E. (1979).** Management problems in designated wilderness areas. *J. Soil Water Conserv.*, 34(3): 141-143. (P)

8524 **Goldberg, L.J. (1979).** A hardy orchid conservancy. *Amer. Orchid Soc. Bull.*, 48(10): 1003-1006. Col. illus. Jackson's Garden at Union College, Schenectady. (G)

8525 **Goldwater, B.M. (1984).** Conservation through consensus. *Nation. Parks*, 58(7-8): 14-15.

8526 **Goodwin, H.A., Dawson, E.P. (1971).** Status of endangered species program. *In* Trefethen, J.B., *ed.* Transactions of the 36th North American Wildlife and Natural Resource Conference. Symposium VIII. Washington, D.C., Wildlife Management Institute. (L/T)

8527 **Gordon, S.L. (1980).** Development versus environment conflicts: an analysis of the Endangered Species Act and its implications for federal public works projects. Los Angeles, CA, UCLA Architecture and Urban Planning. 54p. Masters Thesis. (L)

8528 **Grave, N.P. (1979).** National parks of the USA. *Les i Chelovek*, (1979): 165p. Illus. (P)

8529 **Graves, W.L., Kay, B.L., Williams, W.A. (1978).** Revegetation of disturbed sites in the Mojave Desert with native shrubs. *Calif. Agric.*, 32(3): 4-5.

8530 **Greenwalt, L.A., Schoning, R.W. (1976).** Guidelines to assist federal agencies in complying with section 7 of the Endangered Species Act of 1973. Memorandum FWS/AFA. Washington, D.C., U.S. Fish and Wildlife Service. (L)

8531 **Griggs, T. (1982).** Diversity and adaptation. *Fremontia*, 9(4): 7. Illus. *Orcuttia inequalis, Neostapfia colusana.* (T)

8532 **Guthrie, R.L. (1974).** A living museum. *Arbor. Newsl. W. Virg. Univ. Dep. Biol.*, 22(1): 1-7. Map. Cranesville Swamp Nature Sanctuary. (P)

8533 **Hagen, M. (1980).** The Wagar influence. *American Forests*, 86(7): 10-13, 54-55. Illus.

UNITED STATES

8534 **Halvorson, C., Linduska, J., Stebler, A. (1978).** Economic values of wildlife production in forestry areas in North America. *In* Proceedings of the 7th World Forestry Congress, Buenos Aires, 4-18 October 1972, vol. 3: Conservation and recreation. Buenos Aires, Instituto Forestal Nacional. 3921-3946. (Sp, Fr). (E)

8535 **Harmon, D., Freed, M. (1978).** The foresters role in wilderness land preservation and management. *In* Proceedings of the 7th World Forestry Congress, Buenos Aires, 4-18 October 1972, vol. 3: Conservation and recreation. Buenos Aires, Instituto Forestal Nacional. 4133-4138. (Sp, Fr).

8536 **Harrington, H.A. (1980).** The need for protection of our native cacti. *Cact. Succ. J.* (U.S.), 52(5): 224-226, 232. Illus. (T)

8537 **Harris, T.H. (1979).** A weekend at Rose Ranch. *Fremontia*, 6(4): 24-25. Illus. Conservation, wildflowers, Sierra Nevada. (T)

8538 **Hart, D. (1974).** Proceedings of a symposium on endangered and threatened species of North America, June 11-14, 1974, Washington, D.C. St. Louis, MO, Wild Canid Survival and Research Center. 339p. (T)

8539 **Hart, W.J. (1980).** Corps lands: is anybody minding the store? *Nat. Parks Conserv. Mag.*, 54(1): 16-19. Illus.

8540 **Heckscher, S. (1981).** News and notes. Philadelphia conservationists establish Fortescue Glades Wildlife Reserve. *Bartonia*, 48: 45-46. (P)

8541 **Henry, V.G. (1977).** Role of the Fish and Wildlife Service concerning endangered flora. *In* Proceedings of the USDA Forest Service conference on endangered plants in the southeast. Asheville, NC, Southeast Forest Experiment Station. (T)

8542 **Hilts, L. (1980).** National forest guide. Chicago, Rand McNally. 218p. Illus., maps. National parks and reserves, forest reserves. (P)

8543 **Holden, C. (1977).** Endangered species: review of law triggered by Tellico impasse. *Science*, 196(4297): 1426-1428. (L/T)

8544 **Holden, C. (1982).** Endangered Species Act in jeopardy. *Science*, 215(4537): 1212-1214. Illus. (L/T)

8545 **Holden, C. (1982).** Endangered Species Act reauthorized. *Science*, 216(4553): 1390. (L/T)

8546 **Holden, C. (1986).** Forest death showing up in the United States. *Science*, 233(4766): 837. Illus.

8547 **Holden, D.J., et al. (1978).** Cloning native prairie plants by tissue culture. *In* Glenn-Lewin, D.C., Landers, R.Q., *eds.* Fifth Midwest Prairie conference proceedings, Iowa State Univ., Ames, Aug. 22-24, 1976. Ames, Iowa State Univ. 92-95. (G)

8548 **Holgren, P.K. (1977).** Uses of the New York Botanical Garden's systematic collections for solution of problems of human health, food resources, environmental quality and location and utilization of natural resources. *Bull. Amer. Assoc. Bot. Gard. Arbor.*, 11(1): 2-13. (E/G)

8549 **Holmgren, A.H. (1977).** Some intermountain endemics. 55th Faculty Haven Lecture. Logan, UT, Utah State University. 12p.

8550 **Hopple, M.K. (1978).** Special places for everyone. *Conservationist*, 33(2): 25-27. Col. illus.

8551 **Howard, A. (1978).** Federal-State rare plant news. *Calif. Native Pl. Soc. Bull.*, Nov.-Dec.: [1]p. (T)

8552 **Howard, A. (1979).** Conservation briefs. *Fremontia*, 7(3): 16-19. Illus.

UNITED STATES

8553 **Howard, A. (1979).** Once more the Endangered Species Act. *Calif. Native Pl. Soc. Bull.*, 10(1): 4. (L/T)

8554 **Howard, A. (1979).** Rare plants and water conflicts east of the Sierra. *Fremontia*, 7(3): 12-14. Illus. (T)

8555 **Howard, A. (1980).** Conservation briefs. *Fremontia*, 8(2): 14-16. Illus.

8556 **Howard, A. (1981).** Native grasslands endangered at Hungry Valley. *Fremontia*, 9(1): 12. Illus. (T)

8557 **Ittner, R., et al., eds (1979).** Conference proceedings: Recreational Impact on Wildlands, October 27-29, 1978, Seattle, Washington. U.S. Forest Service No. R-6-001. U.S. Department of Agriculture Forest Service. 341p. Illus. (P)

8558 **Jackson, J.P. (1979).** Criminal case of the Marthasville Oak. *Amer. Forests.*, 85(11): 14-17, 44-45. Illus.

8559 **Jarvis, T.D. (1980).** NPS study confirms NPCA findings. *Nat. Parks Conserv. Mag.*, 54(8): 9-11. Illus. (P)

8560 **Jeffery, W. (1979).** Ginseng - Ozark gold. *Missouri Conservationist*, 40(8): 12-13. Col. illus. *Panax quinquefolium.* (E/T)

8561 **Jenkins, D.W. (1973).** List of rare and endangered plants of the U.S. Washington, D.C., Smithsonian Institution. 3p. (T)

8562 **Johnson, J., Dunning, G. (1977).** Land planning in national parks and forests: a selective bibliography. Exchange Bibliography, Council of Planning Librarians, No. 1291/1292. 68p. Annotated bibliography, 1970-1975. (B/P)

8563 **Jones, Q. (1981).** The National Plant Germplasm System. *Hort. Science*, 16(6) Sect. 1: 737-739. (G)

8564 **Jordan, W.R. (1988).** Ecological restoration: reflections on a half-century of experience at the University of Wisconsin-Madison Arboretum. *In* Wilson, E.O., ed. Biodiversity. Washington, D.C., National Academy Press. 311-316. (G/T)

8565 **Joyce, C. (1987).** Trees and lakes 'need fear no acid'. *New Sci.*, 115(1579): 21. Acid rain.

8566 **Kartesz, J.T. & R. (1977).** The biota of North America: part 1. Vascular plants. Volume 1: rare plants. Pittsburgh, Bonac. iii, 361p. Includes Endangered Species Act, 1973. (L/T)

8567 **Kartesz, J.T. (1981).** Maintaining awareness of state rare-plant lists and projects. *In* Morse, L.E., Henifin, M.S., *eds.* Rare plant conservation: geographical data organization. New York, The New York Botanical Garden. 103-107. (T)

8568 **Kassler, K. (1980).** Preserving our natural heritage. *Nat. Parks Conserv. Mag.*, 54(7): 23-24.

8569 **Keener, C.S. (1983).** Distribution and biochemistry of the endemic flora of the mid-Appalachian shale-barrens. *Bot. Rev.*, 49(1): 65-115. Maps. 18 species listed. (T)

8570 **Kimball, T.L. (1979).** What citizen conservationists need to know. Conference: North America's forests: gateway to opportunity, 1978. *Proc. Soc. Amer. Foresters*, (1979): 25-28.

8571 **Kinish, D.R. (1980).** About the Endangered Species Act. *J. Bromeliad Soc.*, 30(5): 195-196. (L/T)

8572 **Klopatek, J.M., Olson, R.J., Emerson, C.J., Joness, J.L. (1979).** Land-use conflicts with natural vegetation in the United States. *Envir. Conserv.*, 6(3): 191-199. Maps.

UNITED STATES

8573 **Kologiski, R. (1982).** Endangered species. *Amer. Assoc. Bot. Gard. Arbor. Inc. Newsl.*, 91: 3. (L/T)

8574 **Kral, R. (1983).** A report on some rare, threatened or endangered forest-related vascular plants of the south. Technical Publication R8-TP2. Atlanta, GA, U.S. Department of Agriculture. Forest Service. 2 vols. x, 1305p. Maps. (T)

8575 **Kruckeberg, A.R. (1976).** Scientific values of preserves for rare biota. *In* Andrews, R.D., Carr, R.L., Gibson, F., Lang, B.Z., Soltero, R.A., Swedberg, K.C., *eds.* Proc. Symp. on Terrestrial & Aquatic Ecol. Studies of the NW, 26-27 March. Cheney, E. Washington State Coll. 48-57. (P/T)

8576 **Krugman, S.L., Cowling, E.B. (1982).** Our national resources: basic research needs in forestry and renewable natural resources. Washington, D.C., U.S. Department of Agriculture Forest Service. viii, 35p. Illus. Forest conservation.

8577 **Kurtz, W.B., Alig, R.J., Mills, T.J. (1980).** Retention and condition of agricultural conservation program conifer planning. *J. Forest.*, 78(5): 273-276. Illus.

8578 **Lachenmeir, R.R. (1974).** The Endangered Species Act of 1973: preservation or pandemonium? *Environ. Law*, 5(1): 29-82. (L)

8579 **Lambert, E. (1979).** A national park in the Great Basin? *Nat. Parks Conserv. Mag.*, 53(9): 4-9. Illus. (P)

8580 **Lammers, T.G. (1980).** The vascular flora of Starr's Cave State Preserve. *Proc. Iowa Acad. Sci.*, 87(1): 7. Abstract of papers at 92nd Session of Iowa Acad. Sci., 18-19 April 1980, Simpson College, Indianola. Rare species. (P/T)

8581 **Lathrop, E.W. (1981).** Sensitive plants in the Cleveland National Forest. *Crossosoma*, 7(4): 1-3. (T)

8582 **Lawyer, J.I., et al. (1979).** A guide to selected current literature on vascular plant floristics for the contiguous U.S., Alaska, Canada, Greenland and the U.S. Caribbean and Pacific Islands. New York, New York Botanical Garden. 138p. Suppl. 1: Research on floristic information synthesis: a report to the Division of Natural History, Nat. Park Service, U.S. Dept of Interior. Washington, D.C. (B)

8583 **Layne, E.N. (1973).** Who will save the cacti? *Audubon*, 75(4): 4. (T)

8584 **Leedy, D.L. (1979).** An annotated bibliography on planning and management for urban-suburban wildlife. U.S. Dept of the Interior, Fish and Wildlife Service. 256p. (B)

8585 **Leisz, D., Horton, L.E. (1978).** Protecting rare plants in national forests. *Fremontia*, 5(4): 22-24. (P/T)

8586 **Lemons, T., Stout, D. (1982).** National parks legislative mandate in the United States of America. *Envir. Manage.*, 6(3): 199-207. (L/P)

8587 **Leopold, A. (1936).** Threatened species. *Am. Forests*, 42: 116-119. (T)

8588 **Lincer, J.L. (1982).** Protecting endangered species at the local government level. *Florida Sci.*, 45(Suppl. 1). 46th. Annual Meeting of the Florida Academy of Sciences, Deland, Florida, April 22-24. (L/T)

8589 **Lindsey, A.A., Escobar, L.K. (1976).** Eastern deciduous forest vol. 2. Beech-maple region. Inventory of natural areas and sites recommended as potential natural landmarks. Washington, D.C., U.S. National Park Service. xiv, 238p. (P)

8590 **Lipske, M. (1981).** Endangered species law, targeted by industry, is again in jeopardy. *Defenders*, 56(6): 37. (L/T)

8591 **Little, E.L., jr (1975).** Our rare and endangered trees. *Amer. Forest*, 81(2): 16-17, 44-45. (T)

UNITED STATES

8592 **Little, E.L., jr (1975).** Our rare and endangered trees. *Amer. Forest*, 81(7): 16-21, 55-57. (T)

8593 **Little, E.L., jr (1975).** Rare and local conifers in the United States. U.S. Department of Agriculture Forest Service Conserv. Res. Rep., 19. Washington, D.C., U.S. Govt Print. Office. 25p. (T)

8594 **Little, E.L., jr (1977).** Rare and local trees in the national forests. U.S. Department of Agriculture, Forest Service Conserv. Res. Rep., 21. Washington, D.C., U.S. Govt Print. Off. 14p. (G/T)

8595 **Litzow, M. (1979).** Threatened and endangered plant species: literature review. *Misc. Report* (Minnesota), 3: 45-47. (B/T)

8596 **Locklear, J. (1987).** Nebraska Statewide Arboretum conserves plants of the Great Plains. *Center Pl. Conserv.*, 2(1): 2, 8. Illus. (G)

8597 **Look, C.A. (1979).** Sempervirens fund: an evergreen conservation effort. *Fremontia*, 7(30): 28-29. Illus. (T)

8598 **Lucas, R.C., Kovalicky, T.J. (1981).** Opportunities for meeting future demands for wilderness. U.S. Department of Agriculture, Forest Service Forest Res. Rep., 22. Washington, D.C., U.S. Govt Print. Off. 105-107. Illus.

8599 **Lucas, R.C., Kovalicky, T.J. (1981).** The national wilderness preservation system. U.S. Department of Agriculture, Forest Service Forest Res. Rep., 22. Washington, D.C., U.S. Govt Print. Off. 100-105. Illus., map. (P)

8600 **Lyons, G. (1979).** The C.S.S.A. and conservation: a boom or a bust? *Cact. Succ. J.* (U.S.), 51(1): 9-15. Illus. (T)

8601 **Lyons, G. (1979).** The C.S.S.A. code of conduct for the conservation of succulent plants. *Cact. Succ. J.* (U.S.), 51(6): 284-285. (T)

8602 **Lyons, G. (1980).** At long last: protection for endangered cacti. *Cact. Succ. J.* (U.S.), 52(5): 229, 232. (L/T)

8603 **MacBryde, B. (1977).** Endangered plant reponsibilities of the U.S. Fish and Wildlife Service. *In* Abstracts of papers, 30th Annual Meeting, Society for Range Management, 14-17 Feb. 1977. Portland, OR. 16. (L/T)

8604 **MacBryde, B. (1977).** Plant conservation in the United States Fish and Wildlife Service. *In* Prance, G.T., Elias, T.S., *eds.* Extinction is forever. Proceedings of a symposium at the New York Botanical Garden, 11-13 May 1976. New York, New York Botanical Garden. 62-74. Discusses the role of the U.S. Government in matters of plant conservation and its commitments. (L/P/T)

8605 **MacBryde, B. (1979).** Plant conservation in North America: developing structure. *In* Hedberg, I., *ed.* Systematic botany, plant utilization and biosphere conservation. Stockholm, Almqvist & Wiksell International. 105-109. Map. (T)

8606 **MacBryde, B. (1979).** Plants for all seasons: conservation as if nature mattered. *Nat. Conserv. News*, 29(2): 9-11. Map.

8607 **MacBryde, B. (1979).** U.S. possibilities for plant conservation internationally. *In* Hedberg, I., *ed.* Systematic botany, plant utilization and biosphere conservation. Stockholm, Almqvist & Wiksell International. 151-154.

8608 **MacBryde, B. (1980).** Guest column: notice of review. *Garden* (New York), 4(6): 2-3, 32. Illus. Fish and Wildlife Service reviewing 3,000 plants for possible protection. (L/T)

8609 **MacBryde, B. (1981).** Information needed to use the Endangered Species Act for plant conservation. *In* Morse, L.E., Henifin, M.S., *eds.* Rare plant conservation: geographical data organization. New York, New York Botanical Garden. 49-55. (L/T)

UNITED STATES

8610 **MacBryde, B. (1983)**. Assessment of vulnerable native plants updated. *Endangered Species Tech. Bull.*, 8(12): 1, 6. Illus. (L/T)

8611 **MacBryde, B. (1984)**. Assessing U.S. threatened plants. *Threatened Pl. Newsl.*, 13: 13-14. (L/T)

8612 **MacBryde, B., McMahan, L. (1980)**. Legal protection for rare plants. *Amer. Univ. Law Rev.*, 29(3): 515-569. (L/T)

8613 **MacKnight, J.C. (1986)**. Botanical gardens and conservation education. *AABGA Public Gardens*, 1(1): 18-19. Illus. (G)

8614 **Malloch, B.S. (1978)**. Mapping rare plants at the Geysers. *Fremontia*, 5(4): 30-32. (T)

8615 **Malyshev, L.I., Plotnikova, L.S. (1982)**. Voprosy okhrany rastitel'nogo mira v SShA (materialy III Sovetsko-Amerikanskoi Botanicheskoi Ekspeditsii po Okhrane Rastitel'nogo Mira). (Problems in protection of plants in the USA [the 3rd Soviet-American botanical expedition on plant conservation]). *Bot. Zhurn.*, 67(3): 373-381. Rus.

8616 **Manheim, B.S., Bean, M.J. (1984)**. Undermining the plant protection effort. *Garden* (New York), 8(4): 2-5. Col. illus. Since 1981, only 13 of nearly 1800 eligible plant species have been listed for protection. (G/L/T)

8617 **Manning, R. (1979)**. RARE II: wilderness elimination. *Appalachia*, 42(3): 60-73. Illus. Roadless Areas Review and Evaluation - U.S. Forest Service.

8618 **Manning, R.E. (1979)**. Strategies for managing recreational use of national parks. *Parks*, 4(1): 13-15. Illus. (P)

8619 **Martin, T. (1980)**. How can we protect southwestern national parks? *Nat. Parks & Conserv. Mag.*, 54(3): 4-9. Illus., map. (P)

8620 **Marvinney, S. (1976)**. Swamps, bogs and marshes: safeguarding a fragile resource. *Conservationist*, 30(6): 11-111. Illus. Wetlands.

8621 **Mathews, J.F. (1977)**. Definition and classification of Endangered and Threatened plant species (1). *In* Conference on endangered plants in the Southeast: Proceedings, May 11-13, 1976, Asheville, NC. Asheville, NC, U.S. Department of Agriculture Forest Service, Southeastern Forest Experiment Station. 2-7. (T)

8622 **Matyshev, L.I., Plotnikova, L.S. (1982)**. Voprosy okhrany rastitel'nogo mira v SShA: materialy III Sovetsko-Amerikanskoi Botanicheskoi Ekspeditsii po Okhrane Rastitel'nogo Mira. (Problems in protection of plants in the U.S.A.; the 3rd Soviet-American botanical expedition on plant conservation.) *Bot. Zhurn.*, 67(3): 373-381. Rus.

8623 **May, R. (1979)**. *Sclerocactus polyancistrus*: its growth, distribution and cultivation. *Cact. Succ. J.* (U.K.), 51(5): 228-232. Illus. (G/T)

8624 **Mayolo, K.A. de (1987)**. Opportunities for involvement in endangered plant education. *In* Elias, T.S., *ed.* Conservation and management of rare and endangered plants. Proceedings from a conference, Sacramento, California, 5-8 November 1986. Sacramento, California Native Plant Society. 619-622.

8625 **McClintock, E., Danielson, M. (1975)**. Three thistles - two of them rare. *Fremontia*, 2(4): 27. (T)

8626 **McClure, J.A. (1977)**. Congress takes a look at the Endangered Species Act of 1973. *Frontiers*, 41(4): 35-36. (L/T)

8627 **McCrone, H. G. (1976)**. America's cacti are endangered and threatened. *Cact. Succ. J.* (U.S.), 48: 119. (T)

UNITED STATES

8628 **McCurdy, D.R. (1977).** Overview of the National Wilderness Preservation System. *J. Forest.*, 75(5): 260-261. (P)

8629 **McMahan, L. (1980).** Legal protection for rare plants. *Amer. Univ. Law Review*, 29(3): 515-569. (L/T)

8630 **McMahan, L. (1981).** The trade, biology and management of American ginseng *Panax quinquefolius.* Washington, D.C., International Convention Advisory Commission. 26p. Appendices. (E/L)

8631 **McMahan, L. (1981).** U.S. exports and imports of cacti, 1977-1979. Washington, D.C., TRAFFIC (U.S.A.). 65p. (Trade). (E/L/T)

8632 **McMahan, L. (1982).** Lacey Act strengthened. *TRAFFIC (U.S.A.) Newsl.*, 3(4): 2. Trade. (L)

8633 **McMahan, L. (1982).** U.S. is major importer of orchids. *TRAFFIC (U.S.A.) Newsl.*, 4(2): 7. (L)

8634 **McMahan, L. (1983).** U.S. exports of American ginseng soar. *TRAFFIC (U.S.A.) Newsl.*, 4(2): 2. (E/L)

8635 **McMahan, L. (1984).** What is protection? An overview of state conservation laws for plants. *Tennessee Conservationist*, 50(2): 5-7. (L)

8636 **McMartin, B. (1977).** A new life for the High Peaks Wilderness. *Conservationist*, 32(2): ii-iv. Illus. (P)

8637 **McRae, J.F. (1979).** A lily sanctuary at the Rae Selling Berry Botanic Garden. *Lily Yearbook*, 32: 96-100. (G)

8638 **Meagher, T.R., Antonovics, J., Primack, R. (1978).** Experimental ecological genetics in *Plantago*: 3. Genetic variation and demography in relation to survival of *Plantago cordata*, a rare species. *Biol. Conserv.*, 14(4): 243-257. Map. (T)

8639 **Medeiros, J.L. (1982).** Drainage pond or refuge? *Fremontia*, 10(1): 36. (P)

8640 **Miller, J.R. (1978).** A simple economic model of endangered species preservation in the United States. *J. Envir. Econ. Manage.*, 5: 292-300. (T)

8641 **Miller, R. (1986).** Predicting rare plant distribution in the southern Appalachians of the south-eastern U.S.A. *J. Biogeogr.*, 13: 293-311. Maps. (T)

8642 **Miller, S.F. (1977).** Wild flowers, The Endangered Species Act, and you. *Flower and Garden*, 21(12): 12-16. (L/T)

8643 **Minckler, L.S. (1979).** The challenge of forest management. *Nat. Parks Conserv. Mag.*, 53(9): 16-19. Illus.

8644 **Mohlenbrock, R.H. (1983).** Where have all the wildflowers gone? New York, Macmillan, and London, Collier Macmillan. 239p. (T)

8645 **Mohlenbrock, R.H. (1985).** Plant protection is back on track. *Garden* (U.S.), 9(4): 22-27. U.S. protects 15 more plants in 1984. (L/T)

8646 **Mohlenbrock, R.H. (1986).** A banner year for plant protection. *Garden* (U.S.), 10(4): 20-23. Illus.

8647 **Mohlenbrock, R.H. (1987).** Plant protection '86: running hard to stay in place. *Garden* (U.S.), 11(4): 16-21. Illus., col. illus. (L)

8648 **Mohlenbrock, R.H. (1988).** Preparing an RDB plant list. *Species*, 10: 18. Background to SSC project to prepare a Red Data Book. (T)

8649 **Mooney, E.C. (1979).** Enchanted rock: land of legend and life. *Nat. Conserv. News*, 29(2): 18-20. Illus., col. illus.

UNITED STATES

8650 **Moore, G. (1977).** The deflowering of the Endangered Species Act. *Horticulture*, 55(5): 37-39. (L/T)

8651 **Morse, L. (1988).** Rare plants of Appalachian bedrock. *Nat. Conserv. Mag.*, 38(2): 28-30. (T)

8652 **Morse, L.E. (1981).** The Nature Conservancy and rare plant conservation in the United States. *In* Synge, H., *ed.* The biological aspects of rare plant conservation. Chichester, Wiley. 453-457. Proceedings of International Conference, King's College, Cambridge, 14-19 July 1980. (P/T)

8653 **Morse, L.E., Lamson, M., Tryon, A.F., Walton, R.R. (1981).** Specimen locality indexing by the New England Botanical Club. *In* Morse, L.E., Henifin, M.S., *eds*. Rare plant conservation: geographical data organization. New York, New York Botanical Garden. 185-192.

8654 **Moser, R. (1979).** Anza-Borrego Desert State Park. *Envir. Southwest*, 484: 14-16. Illus. (P)

8655 **Mosher, L. (1983).** The continuing "emasculation" of the EPA. *Ambio*, 12(3-4): 216-217. Illus. United States Environmental Protection Agency. (L)

8656 **Mountjoy, J.H. (1979).** Broom - a threat to native plants. *Fremontia*, 6(4): 11-15. Illus., maps. Invasive species. (T)

8657 **Moyseenko, H.P. (1981).** Limiting factors and pitfalls of environmental data management: some considerations in developing the information system for the State Natural Heritage Programs. *In* Morse, L.E., Henifin, M.S., *eds*. Rare plant conservation: geographical data organization. New York, New York Botanical Garden. 237-253. The work of The Nature Conservancy.

8658 **Muir, J.J. (1978).** Split Rock Gulf and the hart's-tongue fern. *Conservationist*, 32(6): 27-29. Illus., col. illus.

8659 **Murrieta, X. (1978).** [Exploitation of the floristic resources of the arid zones]. *Cienc. Interam.*, 19(1): 10-17. Sp. Illus. Sonoran Desert, Mexico and N. America.

8660 **Nabhan, G. (1979).** Endangered plants of southwestern Indian agriculture. *In* Proceedings of the 23rd Annual Meetings of the Arizona-Nevada Academy of Science, 13-14 April 1979. Arizona State University Temple. *J. Ariz. Acad. Sci.*, 14 suppl.: 48. (E/T)

8661 **Nagy, J., Calef, C.E. (1979).** ESUSA: U.S. endangered species distribution file. Upton, NY, Brookhaven National Laboratory. vi, 64p. (T)

8662 **Nash, R. (1978).** Wilderness management: a contradiction in terms? Forest, Wildlife and Range Experiment Station Contrib. No. 122. Idaho, University of Idaho Wilderness Research Center. 18p. Illus. (P)

8663 **Nees, M. (1983).** Preservation of Pickerington Marsh. *Ohio J. Sci.*, 83(2): 67. Abstract.

8664 **Nelson, B.B., Arndt, R.E. (1980).** Eastern states endangered plants. Alexandria, VA, U.S. Dept of Interior, Bureau of Land Management. 109p. (T)

8665 **Nelson, B.B., Taylor, S.E. (1980).** Endangered and threatened species and related habitats in five southeastern states. Alexandria, VA, U.S. Dept. Interior, Bureau of Land Management. 104p. (T)

8666 **Nickerson, N.H., Thibodeau, F.R. (1983).** Destruction of *Ammophila breviligulata* by pedestrian traffic: quantification and control. *Biol. Conserv.*, 27(3): 277-287. (T)

8667 **Nilsson, G. (1983).** The endangered species handbook. Washington, D.C., Animal Welfare Institute. See in particular M.Bean on questions and answers about the Endangered Species Act (114-116) & section on state endangered species programs with useful table on no. of protected plants in each state (119-120). (L/T)

UNITED STATES

8668 **Norse, E., Rosenbaum, K., et al. (1986).** Conserving biological diversity in our national forests. Washington, D.C., The Wilderness Society. 116p. (P)

8669 **Odell, R., ed. (1978).** The Endangered Species Law is under scrutiny (Part I). Conservation Foundation Letter. 8p. (L)

8670 **Odum, W.E. (1982).** Environmental degradation and the tyranny of small decisions. *BioScience*, 32(9): 728-729. (L)

8671 **Olson, S. (1978).** University receives Federal grant: information sought on propagation of endangered plant species. *Minn. Hort.*, 106(3): 70-71. (G/T)

8672 **Olson, T.E., Knopf, F.L. (1987).** An exotic spreads. *Wildlife Soc. Bull.*, 14: 492-493. *Elaeagnus angustifolia* threatens native plants.

8673 **Olwell, P. (1984).** Endangered species. *Newsl. Amer. Assoc. Bot. Gard. Arbor.*, 119: 4. (T)

8674 **Owens, J.M., Nelson, J. (1981).** Easements to protect rare plants. *Fremontia*, 8(4): 22-23. (L/T)

8675 **Palmer, M.E. (1987).** A critical look at rare plant monitoring in the United States. *Biol. Conserv.*, 39(2): 113-127. (T)

8676 **Palmer, R. (1978).** Why not move the tarweed? *Fremontia*, 5(4): 15-16.

8677 **Palmer, W.D. (1975).** Endangered species protection: a history of congressional action. *Envir. Affairs*, 4: 255-293. (L)

8678 **Parcells, S.J. (1980).** How long the tallgrass...? *Nat. Parks Conserv. Mag.*, 54(4): 4-8. Illus.

8679 **Pardue, J.W., Olson, R.J., Burgess, R.L. (1981).** Locating critical natural features information for environmental planning. *In* Morse, L.E., Henifin, M.S., *eds.* Rare plant conservation: geographical data organization. New York, New York Botanical Garden. 69-75. Refers to work at the Oak Ridge National Laboratory, Tennessee.

8680 **Parr, P.D., Taylor, F.G. (1979).** Plant species on the Department of Energy - Oak Ridge Reservation that are rare, threatened or of special concern. *J. Tenn. Acad. Sci.*, 54(3): 100-102. (P/T)

8681 **Peterson, C. (1987).** Scenic sites under siege. *National Wildlife*, 25(4): 44-46. Air pollution in national parks. (P)

8682 **Peterson, J. (1981).** Preserving the world's wonders. *Ambio*, 10(1): 36-38. Illus. (P)

8683 **Peterson, J.S. (1982).** 1981 Rocky Mountain Regional Rare Plant Conference. *Green Thumb*, 39(1): 24-26. Illus. (T)

8684 **Peterson, M., et al. (Dynamic Corp., MD.) (1982).** The effects of air pollution and acid rain on fish, wildlife and their habitats: critical habitats of threatened and endangered species. FWS Report OBS-80/40.11, June 83, (57).

8685 **Phelps, O.A. (1976).** Slippers of the queen. *Conservationist*, 30(6): 37. Col. illus. *Cypripedium reginae* - protected plant. (L/T)

8686 **Pickering, J. (1989).** Conservation efforts boost hopes for rare clover. *Center for Plant Conserv.*, 4(2): 3. Running buffalo clover. (T)

8687 **Pittillo, J.D. (1979).** Southern Appalachian endangered and threatened flowering plants. *ASB Bull.*, 26(2): 101. Abstract. (T)

8688 **Plotnikova, L.S. (1982).** Okhrana rastenii v SSHA. (The protection of plants in the U.S.A.) *Bot. Zhurn.*, 67(3): 381-385. Rus. (T)

UNITED STATES

8689 **Poole, J.M. (1987).** Ashy dogwood (*Thymophylla tephroleuca*) recovery plan. Albuquerque, New Mexico, U.S. Fish and Wildlife Service. 46p. Map. (T)

8690 **Postel, S. (1984).** Air pollution, acid rain and the future of forests. 5. *Amer. Forests*, 90(11): 46-47. Pollution.

8691 **Powell, A. (1979).** Recreation or preservation? AMC examines its role in marketing the back country. *Appalachia Bull.*, 45(1): 18-21. Illus.

8692 **Primack, R.B. (1980).** Phenotypic variation of rare and widespread species of *Plantago*. Conference: Symposium on Rare and Endangered Species in New England 1979. *Rhodora*, 82(829): 87-95. Illus. (T)

8693 **Pyle, R.M. (1980).** Management of nature reserves. *In* Soule, M.E., Wilcox, B.A., *eds.* Conservation biology: an evolutionary and ecological perspective. Massachusetts, Sinauer Associates. 319-327. (P)

8694 **Quigg, P.W. (1978).** Protecting natural areas - an introduction to the creation of national parks and reserves. New York, National Audubon Society. 44p. (P)

8695 **Radford, A.E. (1977).** Natural area classification system: a standardization scheme. *In* U.S. Department of Agriculture, Forest Service. Conference on endangered plants in the southeast: proceedings. Asheville, NC, Southeast Forest Experiment Station. 95-101. (P)

8696 **Radford, A.E. (1978).** Natural area classification system: a standardization scheme for basic inventory of species, community, and habitat diversity. *In* Conference, National Symposium on Classification, Inventory and Analysis of Fish and Wildlife Habitat, Phoenix, 1977. FWS/OBS, 78/76. 243-280. Illus.

8697 **Range, J.D. (1977).** Legislative history and overview of congressional action to date. *Frontiers*, 41(4): 33-34. (L)

8698 **Raven, P.H. (1982).** Oversight hearings on Endangered Species Act. *Bull. Amer. Assoc. Bot. Gard. Arbor.*, 16(3): 109-114. (L)

8699 **Raven, P.H. (1983).** The importance of preserving species. *Fremontia*, 11(1): 9-12. Illus. (L/T)

8700 **Rawinski, T.J. (1986).** Vandalism of small whorled pogonia. *Endangered Species Tech. Bull.*, 11(12): 6. (T)

8701 **Reichenbacher, F.W. (1985).** Conservation of southwestern agaves. *Desert Pl.*, 7(2): 103-106, 88. Maps. (T)

8702 **Reveal, J.L. (1981).** Notes on endangered buckwheats (*Eriogonum*: Polygonaceae) with three newly described from the western United States. *Brittonia*, 33(3): 441-448. Illus. (T)

8703 **Rickard, W.H., Rogers, L.E. (1983).** Industrial land-use and the conservation of native biota in the shrub-steppe region of western North America. *Envir. Conserv.*, 10(3): 205-211. Illus., maps.

8704 **Ripley, S.D. (1975).** Report on endangered and threatened plant species of the United States. House Document No. 94-51. 94th. Congress, 1st. Session. (L/T)

8705 **Robinson, A.F. (1980).** The impact of the Endangered Species Act of 1973 on private landowners includes plants. *In* Annual For. Symposium, Division of Continuing Education. Baton Rouge, LA, Louisiana State University. 5-9. (L/T)

8706 **Robinson, A.F., Horton, L.E., Schlatterer, E. (1981).** Forest Service programs and activities for the conservation of rare and sensitive plant species. *In* Morse, L.E., Henifin, M.S., *eds.* Rare plant conservation: geographical data organization. New York, New York Botanical Garden. 207-216. (T)

UNITED STATES

8707 **Robinson, A.F., jr, ed. (1982).** Endangered and threatened species of the southeastern United States including Puerto Rico and the Virgin Islands. Washington D.C., U.S. Department of Agriculture, Forest Service. Illus., maps, col. illus. (T)

8708 **Rogers, G. (1987).** Missouri grows rarest plants of its region. *Center for Plant Conserv.*, *2(1): 2, 8.* Illus. (G/T)

8709 **Romspert, A. (1982).** Comparison of two sensitive species of the locoweed, *Astragalus* in the Algadones Dunes. *Crossosom,* 8(2): 7-9. (T)

8710 **Rosenberg, R.H. (1980).** Federal protection of unique environment interests: endangered and threatened species. *North Carolina Law Rev.*, 58: 491-559. (L/T)

8711 **Rosencranz, A. (1986).** The acid rain controversy in Europe and North America: a political analysis. *Ambio*, 15(1): 19-21.

8712 **Rowley, G. (1973).** Save the succulents! A practical step to aid conservation. *Cact. Succ. J.* (U.S.), 45(1): 8-11.

8713 **Runte, A. (1977).** The national park idea: origins and paradox of the American experience. *J. Forest Hist.*, 21(2): 64-75. Illus. (P)

8714 **Runte, A. (1979).** National parks: the American experience. Lincoln, University of Nebraska Press. (P)

8715 **Schaefer, P. (1977).** Watershed forests. *Conservationist*, 31(5): i-v. Illus.

8716 **Schnell, D. (1980).** *Drosera linearis. Carniv. Pl. Newsl.*, 9(1): 16-18. Illus., col. illus. Declining species. (T)

8717 **Scott, J. (1986).** Native plants and the nursery trade. *Amer. Hort.*, 65(6): 27-32. Illus., col. illus. Overcollecting of native plants. (G)

8718 **Seneres, A. (1980).** A new conservation program. *Four Seasons*, 6(2): 23. Regional Parks Botanic Garden. (G)

8719 **Seneres, A. (1980).** Notes on rare and endangered plants. *Four Seasons*, 6(1): 22. Regional Parks Botanic Garden, Berkeley. (G/T)

8720 **Shands, W.E., Hagenstein, P.R., Roche, M.T. (1979).** National forest policy. Washington, D.C., Conservation Foundation. ix, 37p.

8721 **Shaw, J. (1980).** Conservation and endangered native plants; reaching and teaching the public. *Bull. Amer. Assoc. Bot. Gard. Arbor.*, 44(3): 65-68. (G/T)

8722 **Shearer, T.D., Coggeshall, A.D. (1978).** Today's trend in forest preserve management. *Conservationist*, 33(2): 14-17. Col. illus. (P)

8723 **Shreve, D., et al. (1978).** The Endangered Species Act and energy facility planning: compliance and conflict. Upton, NY, Brookhaven National Laboratory, Biomedical and Environmental Assessment Division, National Center for Analysis of Energy Systems, BNL 50841. 51p. (L)

8724 **Smith, E.L. (1980).** Plant conservation and the Endangered Species Act. *In* Feret, P.P., Sharik, T.L., *eds.* Dendrology in the eastern deciduous forest biome. Proceedings, September 11-13, 1979. Blacksburg, VA, Virginia Polytechnic Institute. (L/T)

8725 **Smith, M.N. (1979).** Report from the Escaped Exotics Committee. *Fremontia*, 6(4): 18-19.

8726 **Smithsonian Institution (1975).** Report on endangered and threatened plant species of the United States. Committee on Merchant Marine and Fisheries. Committee on Merchant Marine and Fisheries, Serial No. 94-A. Washington, D.C., U.S. Govt Printing Office. 200p. (L/T)

UNITED STATES

8727 **Society of American Foresters (1984).** New forests for a changing world. Proceedings of the 1983 convention of the Society of American foresters, Portland, Oregon, October 16-20. Bethesda, Md. 578p.

8728 **Solheim, W.G. (1976).** Endangered Species Act of 1973 - Public Law 93-205. *Pl. Sci. Bull.*, 22(3): 31-32. Some notes on the Act. (L)

8729 **Soucy, D.S. (1979).** Saving our native orchids. *Amer. Horticulturist*, 58(5): 21, 43. Col. illus. (T)

8730 **Speth, G. (1980).** The Sisyphus syndrome: acid rain and public responsibility. *Nat. Parks Conserv. Mag.*, 54(2): 12-17. Illus., map. Acid rain.

8731 **Stanley, J. (1980).** Nurse logs and horsetails. *GC & HTJ*, 187(6): 18-19. Illus., map.

8732 **Stanley, J. (1982).** Lonesome pine. *GC & HTJ*, 191(24): 29. Illus. (P/T)

8733 **Starkey, E.E., Franklin, J.F., Matthews, J.W. (1979).** Ecological research in national parks of the Pacific Northwest. San Francisco, Conference on Scientific Research in the National Parks 1979. 142p. Illus., maps. (P)

8734 **Stebbins, J. (1981).** Sedge no longer endangered. *Fremontia*, 9(3): 28-29. *Carex whitneyi.* (T)

8735 **Stebbins, J.C., Smith, J.R., Holeman, J.R. (1981).** *Carex whitneyi* Olney (Cyperaceae): not endangered. *Madrono*, 28(3): 190-191. (T)

8736 **Stout, M., Painter, W. (1985).** Saving endangered species. Implementation of the U.S. Endangered Species Act in 1984. Washington, D.C., Defenders of Wildlife. 16p. Reviews one year's record on funding, listing, and recovery and makes recommendations for future progress. (L)

8737 **Stutz, H.C. (1979).** The meaning of 'rare' and 'endangered' in the evolution of western shrubs. *In* Wood, S.L., *ed.* The endangered species: a symposium. 119-128. (T)

8738 **Styron, C.E., ed. (1976).** Symposium on rare, endangered, and threatened biota of the southeast. *ASB Bull.*, 23: 137-167. (T)

8739 **TIE (The Institute of Ecology) (1977).** Experimental ecological reserves: a proposed national network. Washington, D.C., U.S. Govt Printing Office. 40p. Illus., map. (P)

8740 **Taft, K.A., jr, Davis, S.E. (1978).** Integrated forest wildlife management through graphical analysis. *In* Proceedings of the 8th World Forestry Congress, Jakarta, 16-28 October 1978: Forestry for food. II, 12.

8741 **Tans, W. (1974).** Criteria for priority ranking of biotic natural areas. *Michigan Bot.*, 13: 31-39. (P)

8742 **Teague, L.L. (1982).** Endangered now, extinct soon. *Cact. Succ. J.* (U.K.), 54(3): 122. (L/T)

8743 **Teer, J.G. (1978).** Manipulation of forestry or wildlife habitat to regulate wildlife resources. *In* Proceedings of the 7th World Forestry Congress, Buenos Aires, 4-18 October 1972, vol. 3: Conservation and recreation. Buenos Aires, Instituto Forestal Nacional. 3988-4001. (Sp, Fr).

8744 **The Nature Conservancy (1975).** Preserving our natural heritage, vol. 1: Federal activities; vol. 2: State activities. Washington, D.C., U.S. Govt Printing Office.

8745 **The Nature Conservancy (1975).** The preservation of natural diversity: a survey and recommendations. U.S. Dept. of the Interior, contract no. CX0001-5-0110. Final report. Washington, DC, The Nature Conservancy. 212p.

8746 **Thibodeau, F., Falk, D. (1986).** The Center for Plant Conservation: a new response to endangerment. *Center Pl. Conserv.*, 1(1): 14-19. Illus. (G/T)

UNITED STATES

8747 **Thomas, L.K., jr (1979).** Distribution and ecology of *Sida hermaphrodita*: a rare plant species. *Bartonia*, 46: 51-59. Maps. (T)

8748 **Thomas, L.K., jr (1980).** The impact of three exotic plant species on a Potomac Island. Scientific Monogr. Series No. 13. Washington, D.C., U.S. Department of the Interior, National Park Service. xvi, 179p. Illus., map. *Lonicera japonica, Hedera helix, Iris pseudacorus.* (T)

8749 **Thompson Campbell, F., Tarr, J. (1980).** Plant contraband. *Garden* (New York), 4(6): 8-12. Illus. Trade. (T)

8750 **Thorhaug, A. (1977).** Symposium on restoration of major plant communities in the United States. *Envir. Conserv.*, 4(1): 49-50. (P)

8751 **Towell, W.E., Poole, D.A., Kimball, T.L. (1978).** Achieving balanced use in forest conservation. *In* Proceedings of the 7th World Forestry Congress, Buenos Aires, 4-18 October 1972, vol. 3: Conservation and recreation. Buenos Aires, Instituto Forestal Nacional. 3627-3658. En, Sp, Fr.

8752 **Travis, M.S. (1977).** The Endangered Species Act of 1973. *Harvard Envir. Law Rev.*, 1: 129-142. (L)

8753 **Tryon, A.F. (1980).** Foreword to the symposium Rare and Endangered Plant Species in New England. *Rhodora*, 82(829): 1-2. (T)

8754 **Tucker, W. (1982).** Progress and privilege: America in the age of environmentalism. New York, Doubleday. 336p.

8755 **Turner, M.G., Gregg, W.P., jr (1983).** The status of scientific activities in United States biosphere reserves. *Envir. Conserv.*, 10(3): 231-237. (P)

8756 **U.S. Congress House Committee on Interior & Insular Affairs. Subcommittee on National Parks and Insular Affairs (1980).** National Parks Service's new area study program. Washington, D.C., U.S. Govt Printing Office. Maps. (P)

8757 **U.S. Congress House. Committee on Science and Technology (1985).** Tropical Forest development projects: status of environmental and agricultural research: hearing before the Subcommittee on Natural Resources, Agriculture Research and Environment of the Commit. on Science and Technology, U.S. House of Representatives, 19th Congress, 2nd session. Washington, D.C., U.S. Govt Print. Office. 242p. Illus., maps.

8758 **U.S. Council on Envrionmental Quality (1979).** Our nation's wetlands: an interagency task force report 1978. Washington, D.C., Govt Printing Office. v, 70p. Illus., col. illus.

8759 **U.S. Department of Agriculture (1978).** Activities for learning about conservation of forest resources: a guide for leaders of youth groups. PA, U.S. Department of Agriculture. 1214. 29p.

8760 **U.S. Department of Agriculture (1979).** Future challenges in renewable natural resources: proceedings of a national workshop, January 22-25, 1979, Rosslyn, Virginia. Miscellaneous Publication No. 1376. Washington, D.C., U.S. Department of Agriculture. 116p. Illus.

8761 **U.S. Department of Agriculture. Forest Service (1970).** Endangered rare and uncommon wildflowers found on the southern national forests. U.S. Department of Agriculture Forest Service, Southern Region 8. 20p. (T)

8762 **U.S. Department of Agriculture. Forest Service (1977).** Conference on endangered plants in the southeast. Forest Service Tech. Rept. SE-11. Asheville, NC, U.S. Department of Agriculture, Southeastern Forest Experimental Station. 104p. (T)

UNITED STATES

8763 **U.S. Department of Agriculture. Forest Service (1978).** Guadelupe Escarpment wilderness proposal, Lincoln National Forest. Alamagordo, U.S. Department of Agriculture Forest Service. 42p. Endangered species. (P/T)

8764 **U.S. Department of Agriculture. Forest Service (1978).** Trabuco planning unit, Cleveland National Forest, region 5. San Diego, U.S. Department of Agriculture Forest Service. 305p. (P)

8765 **U.S. Department of Agriculture. Forest Service (1979).** Alsea planning unit land-management plan, Siuslaw National Forest. Portland, OR, U.S. Department of Agriculture Forest Service. 389p. (P)

8766 **U.S. Department of Agriculture. Forest Service (1979).** Proposed regulation for national forest system planning. Washington, D.C., U.S. Department of Agriculture Forest Service. 126p. (L/P)

8767 **U.S. Department of Agriculture. Forest Service (1979).** Roadless area review and evaluation, RARE 2. (Final environmental statement). Washington, D.C., U.S. Department of Agriculture Forest Service. 755p. Illus., maps. (P)

8768 **U.S. Department of Agriculture. Forest Service (1979).** Timber resource management plan, Lakeview Federal Sustained Yield Unit, Fremont National Forest. Portland, OR, U.S. Department of Agriculture Forest Service. 335p. Endangered species. (P/T)

8769 **U.S. Department of Agriculture. Forest Service (1981).** Loss of wetlands. Forest Res. Report No. 22. Washington, D.C., U.S.D.A. Forest Service. 136-137. Illus., map.

8770 **U.S. Department of Agriculture. Forest Service (1985).** Symposium and workshop on wilderness fire. USDA Forest Service technical report INT - Intermountain forest and range experiment station. Ogden, UT, April 1985. 411p. Contains 35 individual papers on fire and management in all the major National Parks and Forest Reserves. (P)

8771 **U.S. Department of Agriculture. Soil Conservation Service (1972).** Rare and endangered plant and animal species. Environment Memorandum 5. Suppl. 1. 1-6. (T)

8772 **U.S. Department of State (1980).** The world's tropical forests: a policy, strategy, and program for the United States. Washington, D.C., U.S. Govt Print. Office. 53p.

8773 **U.S. Department of the Interior. Bureau of Land Management (1979).** Aravaipa Canyon wilderness. Phoenix, Bureau of Land Management. 40p.

8774 **U.S. Department of the Interior. Fish and Wildife Service (1983).** Proposed rule on ginseng exports. *Endangered Species Tech. Bull.*, 8(10): 5, 8. Trade. *Panax guinquefolius.* (E/L)

8775 **U.S. Department of the Interior. Fish and Wildife Service (1983).** Republication of the lists of endangered and threatened species. Final rule. U.S. Department of the Interior, Fish and Wildlife Service. 25p. Includes endangered plants, 22-25. (L/T)

8776 **U.S. Department of the Interior. Fish and Wildlife Service (1975).** Notice of review of four eastern U.S. plants. *Fed. Register*, 40(77): 17612. (L/T)

8777 **U.S. Department of the Interior. Fish and Wildlife Service (1975).** Notice of review of over 3000 vascular plants and determination of critical habitat. *Fed. Register*, 40: 27823-27924. (L/T)

8778 **U.S. Department of the Interior. Fish and Wildlife Service (1976).** Convention on International Trade in Endangered Species of Wild Fauna and Flora: proposed implementation. *Fed. Register*, 41(117): 24367-24378. (L/T)

8779 **U.S. Department of the Interior. Fish and Wildlife Service (1976).** First U.S. plants proposed as Endangered. *Endangered Species Tech. Bull.*, 1(1): 1. (L/T)

UNITED STATES

8780 **U.S. Department of the Interior. Fish and Wildlife Service (1976).** Plant listings produce conflicting views at hearings. *Endangered Species Tech. Bull.*, 1(3): 1-2. (L/T)

8781 **U.S. Department of the Interior. Fish and Wildlife Service (1976).** Proposed Endangered status for some 1700 U.S. vascular plants. *Fed. Register*, 41(117): 24524-22922. (L/T)

8782 **U.S. Department of the Interior. Fish and Wildlife Service (1976).** Proposed prohibitions on certain uses of endangered or threatened plants, permits for exceptions to such prohibitions, and related items. *Fed. Register*, 41(110): 22915-22922. (L/T)

8783 **U.S. Department of the Interior. Fish and Wildlife Service (1976).** Pros. and cons.: excerpts of testimony at endangered plant hearings. *Endangered Species Tech. Bull.*, 1(3): 3. (L/T)

8784 **U.S. Department of the Interior. Fish and Wildlife Service (1978).** Endangered and threatened species of the southeastern United States. Atlanta, GA, U.S. Fish and Wildlife Service, Region 4. 1v. (L/T)

8785 **U.S. Department of the Interior. Fish and Wildlife Service (1978).** Endangered species: Great Lakes Region. Twin Cities, MN, Fish and Wildlife Service. (L/T)

8786 **U.S. Department of the Interior. Fish and Wildlife Service (1978).** Furbish lousewort among 13 plant taxa newly listed by service for protection. *Endangered Species Tech. Bull.*, 3(7): 1, 7, 8. (L/T)

8787 **U.S. Department of the Interior. Fish and Wildlife Service (1978).** Plan advanced for resolving furbish lousewort conflict. *Endangered Species Tech. Bull.*, 3(7): 1, 5. (L)

8788 **U.S. Department of the Interior. Fish and Wildlife Service (1979).** Service lists 32 plants. *Endangered Species Tech. Bull.*, 4(11): 1, 5-8. Additions to the U.S. Endangered Species Act. (L/T)

8789 **U.S. Department of the Interior. Fish and Wildlife Service (1981).** Endangered means: there is still time. Washington, D.C., U.S. Govt. Print Off. 32p. (T)

8790 **U.S. Department of the Interior. Fish and Wildlife Service (1981).** Endangered species: the road to recovery. Washington, D.C., U.S. Govt Print Off. 11p. (T)

8791 **U.S. Department of the Interior. Fish and Wildlife Service (1981).** The 1978 and 1979 amendments to the Endangered Species Act: a discussion. *In* Morse, L.E., Henifin, M.S., *eds.* Rare plant conservation: geographical data organization. New York, New York Botanical Garden. 313-365. Includes the full text of the Act. Reprinted from *Endangered Species Technical Bulletin*, October 1978 and January 1980. (L/T)

8792 **U.S. Department of the Interior. Fish and Wildlife Service (1983).** Endangered and threatened wildlife and plants: notice of withdrawal of proposed rule for *Paronychia argyrocoma* var. *albimontana* (Silver ling) to be a Threatened species. *Fed. Register*, 48(207): 49316-49318. (L/T)

8793 **U.S. Department of the Interior. Fish and Wildlife Service (1983).** Endangered and threatened wildlife and plants: proposed Endangered status for the pedate checker-mallow (*Sidalcea pedata*) and slender-petaled mustard (*Thelypodium stenopetalum*). *Fed. Register*, 48(137): 32522-32525. (L/T)

8794 **U.S. Department of the Interior. Fish and Wildlife Service (1983).** Endangered and threatened wildlife and plants: supplement to review of plant taxa for listing as Endangered or Threatened species. *Fed. Register*, 48(229): 53640-53670. (L/T)

8795 **U.S. Department of the Interior. Fish and Wildlife Service (1983).** Export of American ginseng harvested in 1983 season. *Fed. Register*, 48(176): 40750-40752. Trade. (L/T)

UNITED STATES

8796 **U.S. Department of the Interior. Fish and Wildlife Service (1983).** Export of American ginseng harvested in 1983 season. *Fed. Register*, 48(183): 42840. Trade. (E/L/T)

8797 **U.S. Department of the Interior. Fish and Wildlife Service (1983).** Export of American ginseng harvested in 1983 season. *Fed. Register*, 48(196): 45775-45778. Trade. (E/L/T)

8798 **U.S. Department of the Interior. Fish and Wildlife Service (1983).** Final listing and recovery priority guidelines approved. *Endangered Species Tech. Bull.*, 8(10): 6-7. (L/T)

8799 **U.S. Department of the Interior. Fish and Wildlife Service (1983).** Liaison conservation directory for endangered and threatened species. 5th. ed. Washington, D.C., U.S. Fish and Wildlife Service, Office of Endangered Species. v, 122p. Map.

8800 **U.S. Department of the Interior. Fish and Wildlife Service (1983).** Listing and recovery priorities in draft guidelines. *Endangered Species Tech. Bull.*, 8(6):4, 7. (L)

8801 **U.S. Department of the Interior. Fish and Wildlife Service (1983).** Revised interagency consultation rules proposed. *Endangered Species Tech. Bull.*, 8(7): 1, 10. (L/T)

8802 **U.S. Department of the Interior. Fish and Wildlife Service (1983).** Service seeks data on ginseng status. *Endangered Species Tech. Bull.*, 8(7): 5. *Panax quinquefolius.* (E/L/T)

8803 **U.S. Department of the Interior. Fish and Wildlife Service (1983).** Three plants proposed as Endangered. *Endangered Species Tech. Bull.*, 8(10): 1, 4-5. Illus. (E/L/T)

8804 **U.S. Department of the Interior. Fish and Wildlife Service (1984).** Export of American ginseng harvested in 1983 season. *Fed. Register*, 49(54): 10123-10124. Trade. (E/T)

8805 **U.S. Department of the Interior. Fish and Wildlife Service (1984).** Final findings for export of American Ginseng harvested in 1984 season. *Fed. Register*, 49(168): 34020-34023. Trade. (E/L/T)

8806 **U.S. Department of the Interior. Fish and Wildlife Service (1984).** Protection becomes final for four plants. *Endangered Species Tech. Bull.*, 9(8): 3-5. Illus. (L/T)

8807 **U.S. Department of the Interior. Fish and Wildlife Service (1984).** Seven plants in southern U.S. proposed for listing. *Endangered Species Tech. Bull.*, 9(12): 1, 6-7. (L/T)

8808 **U.S. Department of the Interior. Fish and Wildlife Service (1985).** Controlling the take of rare cacti. *Endangered Species Tech. Bull.*, 10(1): 9. Collectors are a threat to cacti. (L/T)

8809 **U.S. Department of the Interior. Fish and Wildlife Service (1985).** List of approved recovery plans. *Endangered Species Tech. Bull.*, 10(1): 3-6. 33 plant species listed. (L/T)

8810 **U.S. Department of the Interior. Fish and Wildlife Service (1985).** Review of plant taxa for listing as Endangered or Threatened species; notice of review. *Fed. Register*, 50(188): 39526-39584. (L/T)

8811 **U.S. Department of the Interior. Fish and Wildlife Service (1986).** Endangered Species Act protection proposed for four plants. *Endangered Species Tech. Bull.*, 9(10): 3-4. (L/T)

8812 **U.S. Department of the Interior. Fish and Wildlife Service (1986).** Endangered plants update. *Amer. Horticulturist*, 65(1): 2-3. Illus. Reprinted from *Endangered Species Tech. Bull.*, Sept. 1985. (L/T)

UNITED STATES

8813 **U.S. Department of the Interior. Fish and Wildlife Service (1986).** Two plants given final Endangered Species Act protection. *Endangered Species Tech. Bull.* 11(2): 3. (L/T)

8814 **U.S. Department of the Interior. Fish and Wildlife Service (1988).** Endangered Species Act protection is proposed for nine species. *Endangered Species Tech. Bull.,* 13(5): 3-4. Illus. (L/T)

8815 **U.S. Department of the Interior. Fish and Wildlife Service (1988).** Endangered and threatened wildlife and plants; proposal to determine *Platanthera leucophaea* (Eastern prairie fringed orchid) and *Platanthera praeclara* (Western prairie fringed orchid) to be a Threatened species. *Fed. Register,* 53(196): 39621-39625. (L/T)

8816 **U.S. Department of the Interior. Fish and Wildlife Service (1988).** Listings approved for three plants. *Endangered Species Tech. Bull.,* 13(5): 5. (L/T)

8817 **U.S. Department of the Interior. Fish and Wildlife Service (1989).** Endangered & threatened wildlife and plants. Washington, D.C., U.S. Department of the Interior, Fish and Wildlife Service. Doc. No. 50 CFR 17.11 & 17.12: 34p. Current list of taxa determined as Endangered or Threatened, with distribution, status and date of listing. (L/T)

8818 **U.S. Department of the Interior. Fish and Wildlife Service (1989).** Endangered and threatened wildlife and plants; determination of Threatened status for eastern and western prairie fringed orchids. *Fed. Register,* 54(187): 39857-39862. *Platanthera leucophaea, P. praeclara.* (L/T)

8819 **U.S. Department of the Interior. Fish and Wildlife Service (1989).** Endangered and threatened wildlife and plants; proposed Threatened status for *Spiraea virginiana* (Virginia spiraea). *Fed. Register,* 54(139): 30577-30581. (L/T)

8820 **U.S. Department of the Interior. Fish and Wildlife Service (1989).** Final listing rules approved for 10 species. *Endangered Species Tech. Bull.,* 14(8): 7-8. (L/T)

8821 **U.S. Department of the Interior. Fish and Wildlife Service (1989).** Jesup's milk-vetch (*Astragalus robbinsii* var. *jesupi*) recovery plan. Washington, D.C., U.S. Department of the Interior, Fish and Wildlife Service. 26p. Illus., map. (L/T)

8822 **U.S. Department of the Interior. Fish and Wildlife Service (1989).** Proposed listings - July 1989. Virginia spiraea (*Spiraea virginiana*). *Endangered Species Tech. Bull.,* 14(8): 6-7. Illus. (L/T)

8823 **U.S. Department of the Interior. Fish and Wildlife Service (1989).** Recovery 2000: an intensive approach for restoring endangered species. *Endangered Species Tech. Bull.,* 14(8): 1, 3. Illus. (L/T)

8824 **U.S. Department of the Interior. Fish and Wildlife Service (1989).** Recovery plan for Running buffalo clover (*Trifolium stoloniferum*). Twin Citie, MN, U.S. Department of the Interior, Fish and Wildlife Service. 26p. Map. (L/T)

8825 **U.S. Department of the Interior. Forest Service (1978).** Please help protect America's past (cultural heritage, national forests). South Reg. U.S. For. Serv. 14. 6p.

8826 **U.S. Department of the Interior. National Park Service (1975).** Eastern deciduous forest, vol. 1. Southeastern evergreen and oak-pin region. Inventory of natural areas and sites recommended as potential natural landmarks. Washington, D.C., National Park Service. xiv, 206p. (P)

8827 **Van Name, W.G. (1979).** Vanishing forest reserves. New York, Arno Press. x, 190p. Illus. (P)

8828 **Vernimb, B. (1976).** Endangered and threatened rhododendron species. *Quart. Bull. Amer. Rhod. Soc.,* 30(2): 124. (T)

UNITED STATES

8829 **WWF (1974).** Symposium on Endangered and Threatened Species of North America, Washington, D.C., 1974. Proceedings of the Symposium on Endangered and Threatened Species of North America. Washington, D.C., World Wildlife Fund. 339p. (T)

8830 **Waburn, P. (1983).** Legal hassles over the West's natural heritage. *Ambio*, 12(5): 274. (L)

8831 **Waggoner, G.S. (1981).** The conservation of rare flora through the National Natural Landmarks Program. *In* Morse, L.E., Henifin, M.S., *eds.* Rare plant conservation: geographical data organization. New York, The New York Botanical Garden. 57-61. (P/T)

8832 **Wagner, F.H. (1978).** U.S. deserts. *Frontiers*, 42(3): 17-21. Illus., map.

8833 **Walker, K.G. (1979).** Summary of the endangered plant program in the Bureau of Land Management. *In* Wood, S.L., *ed.* The endangered species: a symposium. 165-170. (T)

8834 **Wallinger, R.S. (1987).** AFA: mobilizing conservation action. *Amer. Forests*, 93(1-2): 20-21, 36. Forest.

8835 **Walters, D.R. (1977).** Rare plants: a consideration. *In* Walters, D.R., *et al., eds.* Symposium proceedings: Native plants: a viable option. Special Publication No. 3. California Native Plant Society. 68-79. Illus. Symposium held in 1975. (T)

8836 **Waltrip, R. (1979).** The redwoods still need you. *Nat. Parks Conserv. Mag.*, 53(11): 4-8. Illus., map. (T)

8837 **Walz, N. (1976).** What's happening in the Pacific Northwest. *Univ. Wash. Arbor. Bull.*, 39(4): 17-21. Illus. (P/T)

8838 **Wang, J.-X. (1982).** [A brief account of the research work on "air pollution and plants" in the United States.] *Bull. Nanjing Bot. Gard. Mem. Sun Yat Sen*, 1982: 120-126. Ch (En).

8839 **Wauer, R.H. (1980).** The role of the National Park Service natural resources manager. UW/CPSU Report B-80-2. Seattle, Co-operative Park Studies Unit National Park Service, College of Forest Resources Univ. of Washington. 20p. (P)

8840 **Weedin, J. (1980).** Cactus conservation: a sticky proposition. *Chihuahuan Desert Discovery*, 7: 13-14. Illus. (L/T)

8841 **Whetstone, R.D. (1984).** Notes on *Croomia pauciflora* (Stemonaceae). *Rhodora*, 86(846): 131-137. Illus., maps. Occurrence and distribution in the southeastern United States. (T)

8842 **White, P.S., Bratton, S.P. (1981).** Monitoring vegetation and rare plant populations in US national parks and preserves. *In* Synge, H., *ed.* The biological aspects of rare plant conservation. Chichester, Wiley. 265-278. Proceedings of International Conference, King's College, Cambridge, 14-19 July 1980. (P/T)

8843 **Whitney, G.G., Somerlot, W.J. (1985).** A case study of woodland continuity and change in the American Midwest. *Biol. Conserv.*, 31(3): 265-285. Map.

8844 **Whitson, P.D. (1974).** The impact of human use upon the Chisos Basin and adjacent lands. National Park Service Sci. Monogr. Series No. 4. Washington, D.C., U.S. Govt Printing Office. ix, 92p. Maps.

8845 **Wilcove, D.S. (1988).** National forests: policies for the future. Vol. 2: Protecting biological diversity. Washington, D.C., The Wilderness Soc. 50p.

8846 **Wiley, L. (1969).** Rare wild flowers of North America. 2nd ed. rev. Portland, OR, n.p. 501p. (T)

8847 **Williams, J.D., Baker, G.S. (1976).** A review of the Endangered Species Act of 1973. *Assoc. Southeastern Biol. Bull.*, 23(3): 138-141. (L)

UNITED STATES

8848 **Williams, M. (1984).** Predicting from inventories: a timely issue. *J. Forest Hist.*, 28(2): 92-98. Illus. Timber scarcity, history of forest conservation in U.S.

8849 **Wilson, D., Brig. Gen. (1977).** Endangered species and the U.S. Corps of Engineers. *Frontiers*, 41(4): 37-38. (T)

8850 **Winkler, M.G., DeWitt, C.B. (1985).** Environmental impacts of peat mining in the United States: documentation for wetland conservation. *Envir. Conserv.*, 12(4); 317-330. (T)

8851 **Witt, J.A. (1976).** The arboretum and endangered plants. *Univ. Wash. Arbor. Bull.*, 39(4): 9-11. Illus. (G/T)

8852 **Wood, L.D. (1975).** Section 7 of the Endangered Species Act of 1973: a significant restriction for all federal activities. *Envir. Law Reporter*, 5(10): 50189-50201. (L)

8853 **Wright, H.E. (1977).** The Boundary Waters Canoe Area: a wilderness ecosystem in need of protection. *Field Mus. Nat. Hist. Bull.*, 48(7): 18-19.

8854 **Yaffee, S.L. (1982).** Prohibitive policy: implementing the Federal Endangered Species Act. Cambridge, MA, MIT Press. 239p. (L)

8855 **Yates, T. (1979).** Preserving endangered species. *Arbor. Leaves*, 21(2): 26, 28. Holden Arboretum. (G/T)

8856 **York, R., Smith, J.P., Cochrane, S. (1982).** New developments in the rare-plant program. *Fremontia*, 9(4): 11-13. Illus. (T)

8857 **Zeedyk, W.D., Farmer, R.E., jr, MacBryde, B., Baker, G.S. (1978).** Endangered plant species and wildland management. *J. Forest.*, 76(1): 31-36. Col. illus. (T)

U.S. - Alabama

8858 **Anon. (1979).** *Sarracenia oreophila. AABGA Newsl.*, 59: 8. Illus. (T)

8859 **Atwood, J.T. (1985).** The range of *Cypripedium kentuckiense. Amer. Orch. Soc. Bull.*, 54(10): 1197-1199. Illus., maps. (T)

8860 **Boschung, H., ed. (1976).** Endangered and threatened plants and animals of Alabama. *Alabama Mus. Nat. Hist. Bull.*, 2: 1-92. (T)

8861 **Freeman, J.D., Causey, A.S., Short, J.W., Haynes, R.R. (1979).** Endangered, threatened and special concern plants of Alabama. Departmental Series No. 3. Auburn, Auburn University. 25p. Col. illus. Over 300 species listed, annotated descriptions. (T)

8862 **Hardin, J.W. (1977).** The southeastern United States. *In* Prance, G.T., Elias, T.S., *eds.* Extinction is forever. Proceedings of a symposium at the New York Botanical Garden, 11-13 May 1976. New York, New York Botanical Garden. 36-40. Discusses status of state programmes within the 14 southeastern states of endangered plant species. (L/P/T)

8863 **Schnell, D. (1988).** Proposed listing of *Sarracenia rubra* ssp. *alabamensis* as an Endangered species under the U.S. Endangered Species Act. *Carniv. Pl. Newsl.*, 7: 104. (T)

8864 **Stamps, E.R., Linder, C.Y. (1973).** Wildflower conservation list for Alabama. Alabama Dept. of Conservation and Natural Resources. 15p. (T)

8865 **Thomas, J.L. (1976).** Plants. *In* Boschung, H., *ed.* Endangered and threatened plants and animals of Alabama. *Bull. Alabama Mus. Nat. Hist.*, 2: 5-12. A paper published as part of the results of a symposium sponsored by the Game and Fish Div. of the Alabama Dept of Conservation and Natural Resources and Alabama Museum of Natural History. (T)

U.S. - Alabama

8866 **U.S. Department of the Interior. Fish and Wildlife Service (1983).** Habitat description notice on green pitcher plant. *Endangered Species Tech. Bull.*, 8(9): 4, 8. (L/T)

8867 **U.S. Department of the Interior. Fish and Wildlife Service (1985).** Endangered and threatened wildlife and plants: Endangered status for *Clematis socialis. Fed. Register*, 50(235): 49970-49972. (L/T)

8868 **U.S. Department of the Interior. Fish and Wildlife Service (1985).** Endangered and threatened wildlife and plants: proposed Endangered status for *Lindera melissifolia* (Pondberry). *Fed. Register*, 50(156): 32582-32585. (L/T)

8869 **U.S. Department of the Interior. Fish and Wildlife Service (1986).** Endangered and threatened wildlife and plants: Endangered status for *Clematis socialis. Fed. Register*, 51(186): 34420. (T)

8870 **U.S. Department of the Interior. Fish and Wildlife Service (1988).** Endangered and threatened wildlife and plants; determination of Endangered status for *Trillium reliquum* (Relict trillium). *Fed. Register*, 53(64): 10879-10884. (L/T)

8871 **U.S. Department of the Interior. Fish and Wildlife Service (1988).** Endangered and threatened wildlife and plants; determination of Threatened status for *Marshallia mohrii* (Mohr's Barbara's-buttons). *Fed. Register*, 53(173): 34698-34701. (T)

8872 **U.S. Department of the Interior. Fish and Wildlife Service (1988).** Endangered and threatened wildlife and plants; proposed Endangered status for *Sarracenia rubra* ssp. *alabamensis* (Alabama canebrake pitcher-plant). *Fed. Register*, 53(77): 13230-13234. (T)

8873 **U.S. Department of the Interior. Fish and Wildlife Service (1989).** Endangered and threatened wildlife and plants; Threatened status for *Phyllitis scolopendrium* var. *americana* (American hart's-tongue). *Fed. Register*, 54(134): 29726-29730. (L/T)

8874 **U.S. Department of the Interior. Fish and Wildlife Service (1989).** Endangered and threatened wildlife and plants; proposed Threatened status for *Sagittaria secundifolia* (Kral's water-plantain). *Fed. Register*, 54(200): 42816-42820. (L/T)

8875 **U.S. Department of the Interior. Fish and Wildlife Service (1989).** Protection approved for the Alabama canebrake pitcher plant. *Endangered Species Tech. Bull.*, 14(4): 6. (T)

8876 **U.S. Department of the Interior. Fish and Wildlife Service. (1989).** Alabama leather flower (*Clematis socialis*) recovery plan. U.S. Fish and Wildlife Service. Southeast Region. (T)

U.S. - Alaska

8877 **Batten, A.R., et al. (1979).** Threatened and endangered plants in selected areas of the BLM Fortymile Planning Unit, Alaska. BLM-Alaska Tech. Report No. 3. Anchorage, AK, U.S. Dept of the Interior, Bureau of Land Management, Alaska State Office. 127p. (T)

8878 **Coster, B.A., Hewett, C.E., Smith, T.G. (1978).** Economics and policy. Alaska's forests: what does the future hold? Working group technical session. *In* Society of American Foresters. Forests for people: a challenge in world affairs. Proceedings, National Convention, Alberquerque, New Mexico, 2-6 October 1977. Washington. 283-294.

8879 **Council of Europe (1981).** National activities. Alaska: a new protected area. *Council of Europe Newsl.*, 81: 4. (P)

8880 **Eidsvik, H. (1983).** Under joint responsibility: the Kluane-Wrangell-St. Elias World Heritage Site. *Ambio*, 12(3-4): 191-196. Col. illus., map. (P)

U.S. - Alaska

8881 **Fuller, W.A., Kevan, P.G., eds (1970).** Productivity and conservation in northern circumpolar lands. Proceedings of a conference, Edmonton, Alberta 15-17 October 1969. IUCN New Series No. 16. Morges, IUCN. 344p. Illus., map.

8882 **Gardner, J.E., Nelson, J.G. (1981).** National parks and native peoples in northern Canada, Alaska and northern Australia. *Envir. Conserv.*, 8(3): 207-215. Maps. (P)

8883 **Hakala, D.R. (?)** Interpreting the Tongass National Forest. Juneau, U.S. Department of Agriculture Forest Service, Alaska Region. 81p. Illus., maps. (P)

8884 **Jordan, J. (1979).** Impacts of the expanding urban influence on forest resource management on the White Mountain National Forest. *Proc. Soc. Amer. Foresters*: 35-37. (P)

8885 **Kartesz, J.T. & R. (1977).** The biota of North America: part 1. Vascular plants. Volume 1: rare plants. Pittsburgh, Bonac. iii, 361p. Includes Endangered Species Act, 1973. (L/T)

8886 **Lawyer, J.I., et al. (1979).** A guide to selected current literature on vascular plant floristics for the contiguous U.S., Alaska, Canada, Greenland and the U.S. Caribbean and Pacific Islands. New York, New York Botanical Garden. 138p. Suppl. 1: Research on floristic information synthesis: a report to the Division of Natural History, Nat. Park Service, U.S. Dept of Interior. Washington, D.C. (B)

8887 **Lyon, J.G., George, T.L. (1979).** Vegetation mapping in the gates of Arctic National Park. *Photogramm. Engin. Rem. Sens.*, 45(6): 790. (P)

8888 **McKendrick, J.D., Mitchell, W.W. (1978).** Fertilizing and seeding oil-damaged arctic tundra to effect vegetation recovery Prudhoe Bay, Alaska. *Arctic*, 31(3): 296-304. Illus.

8889 **Muller, M. (1980).** Progress report - Phase II. Threatened and endangered plants in the Tongass National Forest, Alaska. Sitka, AK, U.S. Forest Service Rep. 30p. (P/T)

8890 **Murray, D.F. (1980).** Threatened and endangered plants of Alaska. U.S. Department of Agriculture, Forest Service and U.S. Department of the Interior, Bureau of Land Management. 59p. Illus., dot maps, 42 species. (T)

8891 **Murray, D.F., Lipkin, R. (1987).** Candidate threatened and endangered plants of Alaska with comments on other rare plants. Fairbanks, University of Alaska Museum. 76p. Illus., maps. (T)

8892 **Nagazina, J. (1979).** President Carter protects Alaska land. *Appalachia Bull.*, 45(2): 10-11. Illus. (P)

8893 **Neiland, B.J. (1978).** Rehabilitation of bare sites in interior Alaska. *Agroborealis*, 10(1): 21-25. (G)

8894 **Pain, S. (1987).** After the goldrush. *New Sci.*, 115(1574): 36-40. Illus., col. illus. Deforestation, soil erosion caused by mining.

8895 **Pain, S. (1987).** Alaska lays its wildlife on the line. *New Sci.*, 114(1558): 51-55. Illus., col. illus., map. Threats posed by Trans-Alaska Pipeline.

8896 **Rice, K.W., Vorobik, L.A. (1979).** Endangered, threatened and rare plant program inventories on Alaskan national forests (USA). *In* Proceedings of 30th Alaska Sci. Conference, Fairbanks, Alaska, USA, 19-21 September 1979. 10. Abstract. (T)

8897 **Sandgren, G.D., Noble, M.G. (1978).** A floristic survey of a subalpine meadow on Mt. Wright, Glacier Bay National Monument, Alaska. *Northwest Sci.*, 52(4): 329-336. Maps. (P)

8898 **Senner, S. (1978).** RARE (Roadless Area Review and Evaluation). II, Challenge in the Chugach (Alaska, National Forests, Wilderness Conservation). *Alaska Conserv. Rev.*, 19(2): 3-4.

U.S. - Alaska

8899 **Smith, D.K. (1985).** *Polystichum aleuticum* from Adak Island, Alaska: a second locality for the species. *Amer. Fern J.*, 75(2): 72. (T)

8900 **Swem, T., Cahn, R. (1983).** The politics of parks in Alaska. *Ambio*, 12(1): 14-19. Illus., col. illus. Map. (P)

8901 **Swem, T.R., Cahn, R. (1984).** The politics of parks in Alaska: innovative planning and management approaches for new protected areas. *In* McNeely, J.A., Miller, K.R., *eds.* National parks, conservation and development. The role of protected areas in sustaining society. Washington, D.C., Smithsonian Institution Press. 518-524. Maps. (P)

8902 **Therrien, N. (1979).** Managing the landscape. *Appalachia Bull.*, 45(1): 16-17. Illus. (P)

8903 **Therrien, N. (1979).** U.S. Forest Service recommends wilderness. *Appalachia Bull.*, 45(3): 8-9. Map. (P)

8904 **Tomanova, E. (1978).** Po vlnach aljasske reky Noatak. *Ziva*, 26(5): 170-171. Cz. Illus., map. (P)

8905 **U.S. Department of Agriculture. Forest Service (1973).** New national forests for Alaska. Anchorage, U.S. Department of Agriculture Forest Service. 244p. Illus., maps. (P)

8906 **Watkins, T.H. (1988).** Vanishing Arctic: Alaska's National Wildlife Refuge. Wilderness Society. (P)

8907 **Wayburn, E. (1984).** Annex: Alaskan conservation units of great international significance. *In* McNeely, J.A., Miller, K.R., *eds.* National parks, conservation and development. The role of protected areas in sustaining society. Washington, D.C., Smithsonian Institution Press. 525-526.

U.S. - Arizona

8908 **Anon. (1979).** Protecting Arizona's native plants by law and regulation: an interview with Richard A. Countryman. *Desert Plants*, 1(2): 61-70. Illus. (L/T)

8909 **Anon. (1984).** Endangered species. *AABGA Newsl.*, 111: 4. (T)

8910 **Anon. (1984).** Endangered species. *AABGA Newsl.*, 115: 5. (T)

8911 **Anon. (1984).** Endangered species. *AABGA Newsl.*, 116: 5. (T)

8912 **Anon. (1987).** Famous cactus disappears - and is found. *Oryx*, 21(4): 252. Report on article in *Audubon*, July 1987. *Carnegiea gigantea* illegally uprooted. (L)

8913 **Arizona Commission of Agriculture and Horticulture (1976).** Our protected native plants. AH-N. 509-Rev. 8-76-R-D. Tucson, AZ. 1p.

8914 **Arizona State Office (1980).** The BLM program in Arizona for threatened and endangered plants. *Desert Pl.*, 2(2): 115-118. Illus. (T)

8915 **Banks, V. (1980).** Arizona's cacti have met the enemy - and they are us. *Smithsonian*, 11(8): 94-103. (T)

8916 **Cornett, J.W. (1985).** Reading the fan palms - we know where these trees grow, but we're not sure how they got there. *Natural History*, 94(10): 64-73. Illus. (T)

8917 **Countryman, R.A. (1980).** Protecting Arizona's native plants by law and regulation. *Desert Pl.*, 1(1): 61-70. Illus. (L/T)

8918 **DeLamater, R., Hodgson, W. (1987).** *Agave arizonica*: an endangered species, a hybrid, or does it matter? *In* Elias, T.S., *ed.* Conservation and management of rare and endangered plants. Proceedings from a conference, Sacramento, California, 5-8 November 1986. Sacramento, California Native Plant Society. 305-309. (T)

U.S. - Arizona

8919 **Dodero, M.W. (1983).** Organ Pipe Cactus National Monument. *Environment Southwest*, 503: 4. Col. illus. (P)

8920 **Erickson, J. (1989).** The regeneration gap. *Parks*, 63(7-8): 30-33. Saguaros (*Carnegiea gigantea*) declining.

8921 **Hendrickson, D.A., Minckley, W.L. (1985).** Cienegas: vanishing climax communities of the American Southwest. *Desert Plants*, 6(3): 131-176. Illus., maps. Wetlands.

8922 **Hodgson, W., Nabhan, G., Ecker, L. (1989).** Prehistoric fields in central Arizona: conserving rediscovered agave cultivars. *Agave*, 3(3): 9-11. Illus., col. illus. (G/T)

8923 **Hodgson, W.C. (1989).** A tale of two saiyas: conserving plant lore and gene pools. *Agave*, 3(3): 12-14. Illus. *Amoreuxia palmatifida, A. gonzalezii.* (T)

8924 **House, D. (1987).** Navajo sedge (*Carex specuicola*) recovery plan. Albuquerque, New Mexico, U.S. Fish and Wildlife Service. (T)

8925 **Johnston, M.C. (1977).** The southwestern United States. *In* Prance, G.T., Elias, T.S., eds. Extinction is forever. Proceedings of a symposium at the New York Botanical Garden, 11-13 May 1976. New York, New York Botanical Garden. 60-61. Legislation regarding threatened wild plants in the southwestern U.S. (L/T)

8926 **Lehto, E. (1980).** "Extinct" wire-lettuce, *Stephanomeria schottii* (Compositae), rediscovered in Arizona after more than one hundred years. *Desert Pl.*, 1(1): 22. (T)

8927 **Loope, L.L., Sanchez, P.G., Tarr, P.W., Loope, W.L., Anderson, R.L. (1988).** Biological invasions of arid land nature reserves. *Biol. Conserv.*, 44(1-2): 95-118. Invasive species. Includes 5 case studies. (P)

8928 **McLaughlin, S. (1988).** Salvage, propagation and re-establishment of *Mammillaria thornberi*. Progress report. Tucson Water, Tucson, Arizona. 18p. (G/T)

8929 **McLaughlin, S., Mason, C.T., jr (1977).** Notes on new and rare Arizona plants. *J. Ariz. Acad. Sci.*, 12(3): 125-126. (T)

8930 **McMahan, L.R. (1987).** Rare plant conservation by state government: case studies from the western United States. *In* Elias, T.S., ed. Conservation and management of rare and endangered plants. Proceedings from a conference, Sacramento, California, 5-8 November 1986. Sacramento, California Native Plant Society. 23-31. Maps. (L/T)

8931 **Milne, J. (1987).** Conserving some of the world's tiniest cacti. *Center for Plant Conservation*, 2(1): 1, 8. Illus. (T)

8932 **Nabhan, G. (1980).** *Ammobroma sonorae*, an endangered parasitic plant in extremely arid North America. *Desert Pl.*, 2(3): 188-196. Illus., col. illus. (T)

8933 **Nabhan, G., Valenciano, D. (1989).** A modest proposal: restoring the Sonoran Desert at Barnes Butte Bajada. *Agave*, 3(3): 3-5. Illus., col. illus. Desert degradation problems and restoration ecology.

8934 **Nabhan, G.P. (1987).** Nurse plant ecology of threatened desert plants. *In* Elias, T.S., ed. Conservation and management of rare and endangered plants. Proceedings from a conference, Sacramento, California, 5-8 November 1986. Sacramento, California Native Plant Society. 377-383. Illus. Numerous desert plants depend on protective overstorey for survival; case studies of 2 cacti and *Capsicum annuum* var. *glabriusculum* (T)

8935 **Nabhan, G.P. (1989).** Plants at risk in the Sonoran Desert: an international concern. *Agave*, 3(3): 14-15. Map. (T)

8936 **Nabhan, G.P., Greenhouse, R., Hodgson, W. (1986).** At the edge of extinction: useful plants of the border states of the United States a. ' Mexico. *Arnoldia*, 46(3): 36-46. Illus. (E)

517

U.S. - Arizona

8937 **Nabhan, G.P., Monarque, E.S., Olwell, P., Warren, P., Hodgson, W., Gallindo-Duarte, C., Bittman, R., Anderson, S. (1989).** A preliminary list of plants at risk in the Sonoran Desert of the U.S. and Mexico. *Agave*, 3(3): 15. (T)

8938 **Nabhan, G.P., Reichhardt, K.L. (1983).** Hopi protections of *Helianthus anomalus*, a rare sunflower. *Southwestern Naturalist*, 28: 231-235. (T)

8939 **Pew, T. (1978).** In Arizona there's a cactus law, a cactus cop to enforce it, and a million dollar rustle. *Horticulture*, 56(1): 20-24. Collectors. (L)

8940 **Sluis, C.J., Wochok, Z.S. (1980).** In-vitro propagation of an endangered *Pediocactus paradinei* species. Annual meeting of the American society of Plant Physiologists and the Phytochemical Society of North America, Pullman, Wash., USA, 3-7 Aug. 1980. *Plant Physiol.*, 65(6 Suppl.): 36. (G/T)

8941 **Smith, E., et al. (1973).** Fossil Creek Springs. Natural area report no. 11. Tempe, AZ, Arizona Acad. of Sci. 22p. (P)

8942 **U.S. Department of Agriculture. Forest Service (1980).** Coconino. Washington, D.C., U.S. Department of Agriculture Forest Service, Southwestern Region. 28p. Illus. A recreation guide to the Coconino National Forest. (P)

8943 **U.S. Department of the Interior. Bureau of Land Management. Arizona State Office (1980).** The BLM program in Arizona for threatened and endangered plants. *Desert Plants*, 2(2): 115-118. (T)

8944 **U.S. Department of the Interior. Fish and Wildlife Service (1980).** Endangered species of Arizona and New Mexico. Albuquerque, NM, U.S. Fish and Wildlife Service. 63p. (L/T)

8945 **U.S. Department of the Interior. Fish and Wildlife Service (1983).** Comment period reopened for proposed plant. *Endangered Species Tech. Bull.*, 8(4): 4, 6. (L/P/T)

8946 **U.S. Department of the Interior. Fish and Wildlife Service (1983).** Endangered and threatened wildlife and plants: final rule to determine *Senecio franciscanus* (San Francisco Peaks groundsel) to be Threatened species and determination of its critical habitat. *Fed. Register*, 48(226): 52743-52747. Map. (L/T)

8947 **U.S. Department of the Interior. Fish and Wildlife Service (1983).** Endangered and threatened wildlife and plants: proposal to determine *Agave arizonica* to be an Endangered species. *Fed. Register*, 48(99): 4p. (T)

8948 **U.S. Department of the Interior. Fish and Wildlife Service (1983).** Endangered and threatened wildlife and plants: proposal to determine *Hedeoma diffusum* (Flagstaff pennyroyal) to be a Threatened species. *Fed. Register*, 48(126): 29929-29931. (L/T)

8949 **U.S. Department of the Interior. Fish and Wildlife Service (1983).** Endangered and threatened wildlife and plants: proposed Endangered status for *Cowania subintegra* (Arizona cliffrose). *Fed. Register*, 48(137): 32520-32522. (T)

8950 **U.S. Department of the Interior. Fish and Wildlife Service (1983).** Rulemaking actions: June 1983. Flagstaff pennyroyal proposed as Threatened. *Endangered Species Tech. Bull.*, 8(7): 3. *Hedeoma diffusum*. (L/T)

8951 **U.S. Department of the Interior. Fish and Wildlife Service (1984).** Endangered and threatened wildlife and plants: final rule to determine *Agave arizonica* (Arizona agave) to be an Endangered species. *Fed. Register*, 49(98): 21055-21058. (L/T)

8952 **U.S. Department of the Interior. Fish and Wildlife Service (1984).** Endangered and threatened wildlife and plants: final rule to determine *Cowania subintegra* (Arizona cliffrose) to be an Endangered species. *Fed. Register*, 49(104): 22326-22329. (L/T)

U.S. - Arizona

8953 **U.S. Department of the Interior. Fish and Wildlife Service (1984).** Endangered and threatened wildlife and plants: notice of 6-month extension the proposed rule for *Hedeoma diffusum*. *Fed. Register*, 49(115): 24416. (L/T)

8954 **U.S. Department of the Interior. Fish and Wildlife Service (1984).** Endangered and threatened wildlife and plants: proposal to determine *Carex specuicola* to be a Threatened species and to determine its critical habitat. *Fed. Register*, 49(71): 14406-14410. Map. (L/T)

8955 **U.S. Department of the Interior. Fish and Wildlife Service (1984).** Four western plants threatened by habitat degradation. *Endangered Species Tech. Bull.*, 9(5): 1, 8-9. Illus. (L/T)

8956 **U.S. Department of the Interior. Fish and Wildlife Service (1984).** Two Arizona plants listed as Endangered. *Endangered Species Tech. Bull.* 9(6): 8. (L/T)

8957 **U.S. Department of the Interior. Fish and Wildlife Service (1985).** Endangered and threatened wildlife and plants: determination of *Carex specuicola* to be a Threatened species with critical habitat. *Fed. Register*, 50(89): 19370-19374. (L/T)

8958 **U.S. Department of the Interior. Fish and Wildlife Service (1985).** Endangered and threatened wildlife and plants: notice of six-month extension on the proposed rule for *Mammillaria thornberi* (Thornber's fishhook cactus). *Fed. Register*, 50(111): 24241-24242. (L/T)

8959 **U.S. Department of the Interior. Fish and Wildlife Service (1985).** Endangered and threatened wildlife and plants: notice of withdrawal of proposed rules to list *Hedeoma diffusum* and *Phlox longipilosa* var. *longipilosa*. *Fed. Register*, 50(60): 12348-12350. (L/T)

8960 **U.S. Department of the Interior. Fish and Wildlife Service (1985).** Endangered and threatened wildlife and plants: proposal to determine *Coryphantha robbinsorum* (Cochise pincushion cactus) to be a Threatened species. *Fed. Register*, 50(44): 9083-9086. (L/T)

8961 **U.S. Department of the Interior. Fish and Wildlife Service (1985).** Endangered and threatened wildlife and plants: proposal to determine *Cycladenia humilis* var. *jonesii* to be an Endangered species. *Fed. Register*, 50(7): 1247-1251. (L/T)

8962 **U.S. Department of the Interior. Fish and Wildlife Service (1985).** Endangered and threatened wildlife and plants: proposal to determine *Tumamoca macdougalii* to be an Endangered species. *Fed. Register*, 50(97): 20806-20810. (L/T)

8963 **U.S. Department of the Interior. Fish and Wildlife Service (1985).** Four plants given Endangered Species Act protection. *Endangered Species Tech. Bull.*, 10(5): 1,10. (L/T)

8964 **U.S. Department of the Interior. Fish and Wildlife Service (1985).** Four western species proposed for protection. *Endangered Species Tech. Bull.*, 10(6): 3-6. (T)

8965 **U.S. Department of the Interior. Fish and Wildlife Service (1986).** Endangered and threatened wildlife and plants: determination of *Tumamoca macdougalii* to be Endangered. *Fed. Register*, 51(82): 15906-15911. (L/T)

8966 **U.S. Department of the Interior. Fish and Wildlife Service (1986).** Endangered and threatened wildlife and plants: determination of Threatened status for *Coryphantha robbinsorum*. *Fed. Register*, 51(6): 952-956. (L/T)

8967 **U.S. Department of the Interior. Fish and Wildlife Service (1989).** Endangered and threatened wildlife and plants; proposed Endangered status for *Astragalus cremnophylax* var. *cremnophylax* (Sentry milk-vetch). *Fed. Register*, 54(200): 42820-42822. Threatened by trampling at Grand Canyon. (L/P/T)

8968 **U.S. Department of the Interior. National Park Service (1979).** Colorado River management plan, Grand Canyon National Park, Arizona. Arizona, National Park Service. 359p. (P)

U.S. - Arizona

8969 **U.S. Department of the Interior. National Park Service (1979).** General management plan, wilderness recommendation and road study alternatives for the Glen Canyon National Recreation Area, Arizona and Utah. Missoula, National Park Service. 357p. (P)

8970 **U.S. Department of the Interior. National Park Service (1980).** Feral Burro management and ecosystem restoration plan, Grand Canyon National Park, Arizona. Denver, CO, National Park Service. 283p. (P)

U.S. - Arkansas

8971 **Arkansas Department of Planning (1974).** Threatened native plants and animals of Arkansas. *In* Arkansas Dept. of Planning, Arkansas Nat. Area Plan. 35-37. (T)

8972 **Arkansas Natural Heritage Commission (1977).** Arkansas plants nominated for listing as endangered: a fact sheet. Little Rock, AR, Arkansas Natural Heritage Commission. 21p. (T)

8973 **Atwood, J.T. (1985).** The range of *Cypripedium kentuckiense. Amer. Orch. Soc. Bull.*, 54(10): 1197-1199. Illus., maps. (T)

8974 **Hardin, J.W. (1977).** The southeastern United States. *In* Prance, G.T., Elias, T.S., *eds.* Extinction is forever. Proceedings of a symposium at the New York Botanical Garden, 11-13 May 1976. New York, New York Botanical Garden. 36-40. Discusses status of state programmes within the 14 southeastern states of endangered plant species. (L/P/T)

8975 **Mohlenbrock, R.H. (1985).** Mount Magazine, Arkansas. *Natural History*, 94(10): 82-85. Illus. (Ozark National Forest.)

8976 **Pell, W.F. (1981).** Classification and protection status of remnant natural plant communities in Arkansas, U.S.A. *Ark. Acad. Sci. Proc.*, 35: 55-59. (P)

8977 **Roedmer, B.J., Hamilton, D.A., Evans, K.E. (1978).** Rare plants of the Ozark Plateau: a field identification guide. St Paul, North Central Forest Experiment Station, U.S. Department of Agriculture Forest Service. 238p. Illus., map. (T)

8978 **Tucker, G.E. (1974).** Threatened native plants of Arkansas. *In* Arkansas Natural Areas Plan. Little Rock, Arkansas Department of Planning. 39-65. Descriptions, characteristics, dot maps and photos. (T)

8979 **U.S. Department of Agriculture. Forest Service (1979).** Upper Arkansas planning unit, Pike and San Isabel National Forests. Pueblo, U.S. Department of Agriculture Forest Service. 3 vols. Endangered species. (P/T)

8980 **U.S. Department of the Interior. Fish and Wildlife Service (1985).** Endangered and threatened wildlife and plants: determination of Endangered status for *Lindera melissifolia* (Pondberry). *Fed. Register*, 50(147): 27495-27500. (L/T)

8981 **U.S. Department of the Interior. Fish and Wildlife Service (1985).** Endangered and threatened wildlife and plants: proposed Endangered status for *Lindera melissifolia* (Pondberry). *Fed. Register*, 50(156): 32582-32585. (L/T)

8982 **U.S. Department of the Interior. Fish and Wildlife Service (1985).** Protection sought for four vulnerable plants. *Endangered Species Tech. Bull.*, 19(9): 1, 8. (L/T)

8983 **U.S. Department of the Interior. Fish and Wildlife Service (1986).** Endangered and threatened wildlife and plants: proposed Threatened status for *Geocarpon minimum. Fed. Register*, 51(69): 12460-12463. (L/T)

U.S. - California

8984 **Ackerman, T.L. (1981).** A survey of possible threatened and endangered plant species on the Desert National Wildlife Range. Mission Viejo, CA. 323p. (P/T)

U.S. - California

8985 **Amme, D., Havlik, N. (1987).** Assessment and management of *Arctostaphylos pallida* Eastwood. *In* Elias, T.S., *ed.* Conservation and management of rare and endangered plants. Proceedings from a conference, Sacramento, California, 5-8 November 1986. Sacramento, California Native Plant Society. 447-453. Illus. (T)

8986 **Anon. (1981).** Endangered and threatened wildlife and plants: review of plant taxa for listing as endangered or threatened species. *Lotus Newsl.*, 12: 18. *Lotus dendroides* ssp. *traskiae*, *L. argophyllus* ssp. *adsurgens*, *L. argophyllus* ssp. *niveus*. (T)

8987 **Anon. (1982).** *Cordylanthus palmatus*. *Fremontia*, 10(1): 12. (T)

8988 **Anon. (1982).** Notices. New research preserve in California. *Pl. Sci. Bull.*, 28(1): 3. Cub Creek Research Natural Area. (P)

8989 **Anon. (1983).** Ring Mountain, a preserve at last. *Fremontia*, 11(3): 29-30. (P)

8990 **Anon. (1984).** Endangered species. *AABGA Newsl.*, 116: 5. (T)

8991 **Anthrop, D.F. (1977).** Redwood National and State Parks. Happy Camp, Naturegraph Publishers. 70p. (P)

8992 **Armstrong, W.P. (1978).** Four wildflowers vanishing from northern San Diego County. *Envir. Southwest*, 480: 3-6. (T)

8993 **Armstrong, W.P. (1979).** Unicorn plants in California. *Fremontia*, 7(1): 16-22. Illus.

8994 **Armstrong, W.P. (1983).** Searching for rare plants: 2. Rediscovering the Rock Creek broomrape. *Fremontia*, 11(1): 16-18. Illus. (T)

8995 **Armstrong, W.P., Norris, L.L. (1984).** Death Valley's rare *Gilmania*. *Fremontia*, 11(4): 25-26. Illus. (T)

8996 **Barry, J. (1987).** Rare and endangered species management in the California State park system. *In* Elias, T.S., *ed.* Conservation and management of rare and endangered plants. Proceedings from conference, Sacramento, California, 5-8 November 1986. Sacramento, California Native Plant Society. 73-77. Describes habitat management techniques. (P/T)

8997 **Barry, W.J. (1981).** Native grasslands then and now. *Fremontia*, 9(1): 18. Map.

8998 **Bartel, J.A. (1987).** The Federal listing of rare and endangered plants: what is involved and what does it mean? *In* Elias, T.S., *ed.* Conservation and management of rare and endangered plants. Proceedings from a conference, Sacramento, California, 5-8 November 1986. Sacramento, California Native Plant Society. 15-22. (L/T)

8999 **Bauder, E.T. (1987).** Threats to San Diego vernal pools and a case study in altered pool hydrology. *In* Elias, T.S., *ed.* Conservation and management of rare and endangered plants. Proceedings from a conference, Sacramento, California, 5-8 November 1986. Sacramento, California Native Plant Society. 209-213. Vernal pools have a unique flora. (T)

9000 **Beauchamp, R.M. (1987).** San Clemente Island: remodelling the museum. *In* Elias, T.S., *ed.* Conservation and management of rare and endangered plants. Proceedings from a conference, Sacramento, California, 5-8 November 1986. Sacramento, California Native Plant Society. 575-578. Military and feral animals have devastated flora. Several island endemics threatened; 2 rediscoveries of "extinct" plants. (T)

9001 **Benedict, N.B. (1983).** Plant associations of subalpine meadows, Sequoia National Park. *Arctic and Alpine Research*, 15(3): 383-396. Illus., maps. (P)

9002 **Bennett, P.S., Johnson, R.R., Kunzman, M.R. (1987).** Cactus collection factors of interest to resource managers. *In* Elias, T.S., *ed.* Conservation and management of rare and endangered plants. Proceedings from a conference, Sacramento, California, 5-8 November 1986. Sacramento, California Native Plant Society. 215-223. (T)

U.S. - California

9003 **Berg, K. (1983).** The Tahquitz *Ivesia*, presumed extinct. *Fremontia*, 11(1): 13-15. Illus. (T)

9004 **Berg, K.S. (1987).** Population dynamics of *Erysimum menziesii*, a facultative biennial mustard. *In* Elias, T.S., *ed.* Conservation and management of rare and endangered plants. Proceedings from a conference, Sacramento, California, 5-8 November 1986. Sacramento, California Native Plant Society. 351-357. (T)

9005 **Blake, T.A. (1985).** Wild California. Vanishing lands, vanishing wildlife. Berkeley, CA, University of California Press. 144p.

9006 **Bowland, J.L. (1987).** Guadelupe Dunes revegetation programme. *In* Elias, T.S., *ed.* Conservation and management of rare and endangered plants. Proceedings from a conference, Sacramento, California, 5-8 November 1986. Sacramento, California Native Plant Society. 487-491. Restoration of vegetation following oil pipeline construction.

9007 **Boyd, R. (1987).** The effects of controlled burning on three rare plants. *In* Elias, T.S., *ed.* Conservation and management of rare and endangered plants. Proceedings from a conference, Sacramento, California, 5-8 November 1986. Sacramento, California Native Plant Society. 513-517. (T)

9008 **Bratt, C.C. (1987).** Point Loma lichens - now and then. *In* Elias, T.S., *ed.* Conservation and management of rare and endangered plants. Proceedings from a conference, Sacramento, California, 5-8 November 1986. Sacramento, California Native Plant Society. 289-293. (P/T)

9009 **Butcher, R.D. (1981).** The California Desert Plan. *Nation. Parks*, 55(7-8): 29-31, 34-35. Illus., col. illus., maps. Protection plan for east Mojave desert. (P)

9010 **Butterwick, M. (1987).** Bureau of Land Management's efforts to conserve *Pediocactus peeblesianus* var. *peeblesianus* (Cactaceae). *In* Elias, T.S., *ed.* Conservation and management of rare and endangered plants. Proceedings from a conference, Sacramento, California, 5-8 November 1986. Sacramento, California Native Plant Society. 257-262. (T)

9011 **California Department of Fish and Game (1980).** List of designated endangered or rare plants. Sacramento, CA, California Dept of Fish and Game. 5p. (T)

9012 **California Department of Fish and Game (1984).** Potential candidates for listing. California Dept of Fish and Game Endangered Plant Program. 4p. Listing of 103 species. (T)

9013 **California Desert Advisory Committee (1980).** California Desert conservation area. Sacramento, Department of the Interior, Bureau of Land Management. 436p.

9014 **California Native Plant Society (1973).** Inventory of rare, endangered and possibly extinct vascular plants of California. Berkeley, CA, California Native Plant Society. 27p. (T)

9015 **California Native Plant Society (1980).** Inventory of rare and endangered vascular plants of California. 2nd ed. Berkeley, CA, California Native Plant Society. 115p. (T)

9016 **California Native Plant Society (1980).** U.S. Fish and Wildlife Service acquires Antioch Dunes to protect three endangered species. *Calif. Native Pl. Soc. Bull.*, 10(3): 7. (P/T)

9017 **California Native Plant Society (1981).** California BLM begins study phase in wilderness program. *Calif. Native Pl. Soc. Bull.*, 11(6): 3. Bureau of Land Management.

9018 **California Native Plant Society (1981).** Hall Canyon Research Natural Area proposal. *Calif. Native Pl. Soc. Bull.*, 11(6): 4. San Bernardino National Forest. (P/T)

U.S. - California

9019 **California Native Plant Society (1981).** Inventory of rare and endangered vascular plants of California. Supplement. Berkeley, CA, California Native Plant Society. 56p. (T)

9020 **California Native Plant Society (1981).** Santa Monica Mountains. *Calif. Native Pl. Soc. Bull.*, 11(5): 8.

9021 **California Native Plant Society (1981).** State legislation of interest. *Calif. Native Pl. Soc. Bull.*, 11(5): 6. (L)

9022 **California Native Plant Society (1982).** Endangered Species Act. *Calif. Native Pl. Soc. Bull.*, 12(3): 4. *Amsinka grandiflora, Camissonia benitensis.* (L/T)

9023 **California Native Plant Society (1982).** Research preserve established on Lassen National Forest in California. *Calif. Native Pl. Soc. Bull.*, 12(2): 5. (P)

9024 **California Native Plant Society (1982).** The California Critical Areas Program. *Calif. Native Pl. Soc. Bull.*, 12(2): 7.

9025 **California Native Plant Society (1984).** Top 'candidates' for state listing. Berkeley, CA, California Native Plant Society, Rare Plant Program. 5p. 183 species listed and entered into computer data base system. (T)

9026 **Campbell, F.T. (1987).** The potential for permanent plant protection. *In* Elias, T.S., *ed.* Conservation and management of rare and endangered plants. Proceedings from a conference, Sacramento, California, 5-8 November 1986. Sacramento, California Native Plant Society. 7-13. (L/T)

9027 **Castagnoli, S.P., et al. (1983).** Vegetation and flora. *In* Stone, R.D., Sumidor, V.A., *eds.* The Kingston Range of California: a resource survey. Publ. No. 10. Santa Cruz, Environmental Field Program. 44-104.

9028 **Cheatham, N.H., Barry, W.J., Hood, L. (1988).** Research natural areas and related programs in California. *In* Barbour, M.G., Major, J., *eds.* Terrestrial vegetation of California. CNPS Spec. Publ. No. 9. California Native Plant Society. 75-108.

9029 **Clark, R.A., Fellers, G.M. (1987).** Rare plants at Point Reyes National Seashore. *Fremontia*, 15(1): 13-16. Illus. (P)

9030 **Clifton, G., Callizo, J. (1987).** Monitoring rare plant populations in the Knoxville area of California. *In* Elias, T.S., *ed.* Conservation and management of rare and endangered plants. Proceedings from a conference, Sacramento, California, 5-8 November 1986. Sacramento, California Native Plant Society. 397-400. (T)

9031 **Cochrane, S.A. (1985).** *Dudleya brevifolia*: California's own "living stone". *Fremontia*, 13(1): 21. Illus. (T)

9032 **Cochrane, S.A. (1987).** Endangered plants and California State laws. *In* Elias, T.S., *ed.* Conservation and management of rare and endangered plants. Proceedings from a conference, Sacramento, California, 5-8 November 1986. Sacramento, California Native Plant Society. 33-37. (L/T)

9033 **Coleman, R. (1989).** The cypripediums of California. *Fremontia*, 17(2): 17-19. (T)

9034 **Committee to Save San Bruno Mountain (1981).** Rare fauna and flora on San Bruno Mountain. *Calif. Native Pl. Soc. Bull.*, 11(6): 4. (T)

9035 **Conkle, M.T. (1987).** Electrophoretic analysis of variation in native Monterey cypress (*Cupressus macrocarpa* Hartw.). *In* Elias, T.S., *ed.* Conservation and management of rare and endangered plants. Proceedings from a conference, Sacramento, California, 5-8 November 1986. Sacramento, California Native Plant Society. 249-256. Map. (T)

9036 **Cornett, J.W. (1985).** Reading the fan palms - we know where these trees grow, but we're not sure how they got there. *Natural History*, 94(10): 64-73. Illus. (T)

U.S. - California

9037 **Cummings, E.W. (1987).** Using the California Endangered Species Act consultation provisions for plant conservation. *In* Elias, T.S., *ed.* Conservation and management of rare and endangered plants. Proceedings from a conference, Sacramento, California, 5-8 November 1986. Sacramento, California Native Plant Society. 43-50. (L/T)

9038 **DaVilla, W.B., Taylor, D.M., Stone, R.D., Willoughby, J.W. (1987).** Determining population sizes of narrowly endemic but locally common plants in the Red Hills, California. *In* Elias, T.S., *ed.* Conservation and management of rare and endangered plants. Proceedings from a conference, Sacramento, California, 5-8 November 1986. Sacramento, California Native Plant Society. 167-172. Maps. (T)

9039 **Dawson, B.E. (1987).** Development of management plans for sensitive plant species. *In* Elias, T.S., *ed.* Conservation and management of rare and endangered plants. Proceedings from a conference, Sacramento, California, 5-8 November 1986. Sacramento, California Native Plant Society. 455-459. Includes case study of the Indian Valley Brodiaea. (T)

9040 **DeBenedetti, S.H. (1987).** Management of feral pigs at Pinnacles National Monument: why and how. *In* Elias, T.S., *ed.* Conservation and management of rare and endangered plants. Proceedings from a conference, Sacramento, California, 5-8 November 1986. Sacramento, California Native Plant Society. 193-197. Feral pigs have considerable impact on native flora. (P)

9041 **DeDecker, M. (1978).** The loss of *Sidalcea covillei. Fremontia*, 5(4): 34-35. (T)

9042 **Dunn, A.T. (1985).** The Tecate cypress. *Fremontia*, 13(3): 3-7. Illus., maps. (T)

9043 **Dunn, A.T. (1987).** Population dynamics of the Tecate cypress. *In* Elias, T.S., *ed.* Conservation and management of rare and endangered plants. Proceedings from a conference, Sacramento, California, 5-8 November 1986. Sacramento, California Native Plant Society. 367-376. Maps. *Cupressus quadalupensis* var. *forbesii* occurs with a number of widespread and rare chaparral species. Frequent fires could threaten its (T)

9044 **Dunn, P. (1983).** Two federally prominent plants: San Diego's endangered species. *Environment Southwest*, 501: 8. Illus. *Podogyne ambramsii, Cordylanthus maritimus.* (T)

9045 **Dunn, P.V. (1987).** Endangered species management in southern California coastal salt marshes: a conflict or an opportunity. *In* Elias, T.S., *ed.* Conservation and management of rare and endangered plants. Proceedings from a conference, Sacramento, California, 5-8 November 1986. Sacramento, California Native Plant Society. 441-446. Illus. Management for endangered birds has threatened rare plant. (T)

9046 **Edwards, S.W. (1980).** California manzanitas in the Regional Parks Botanic Garden. *Four Seasons*, 6(2): 6-10. Rare plants. (G/T)

9047 **Elias, T.S. (1987).** Can threatened and endangered species be maintained in botanic gardens? *In* Elias, T.S., *ed.* Conservation and management of rare and endangered plants. Proceedings from a conference, Sacramento, California, 5-8 November 1986. Sacramento, California Native Plant Society. 563-566. Examples and data from Rancho Santa Ana Botanic Garden, Claremont, CA. (G/P/T)

9048 **Elias, T.S., ed. (1987).** Conservation and management of rare and endangered plants. Proceedings of a conference of the California Native Plant Society, Sacramento, California, 5-8 November 1986. Sacramento, California Native Plant Society. x, 630p. Illus., maps. (L/T)

9049 **Ellman, P. (1982).** Report from Ring Mountain. *Fremontia*, 10(2): 9-10. Illus. Includes 4 plants in CNPS inventory of rare and endangered vascular plants of California, including *Tiburon mariposa, Eriogonum caninum* and *Calamagrostis orphiditis.* (T)

U.S. - California

9050 **Evans, J.M., Bohn, J.W. (1987).** Revegetation with rare and endangered species: the role of the propagator and grower. *In* Elias, T.S., *ed.* Conservation and management of rare and endangered plants. Proceedings from a conference, Sacramento, California, 5-8 November 1986. Sacramento, California Native Plant Society. 537-545. Case studies on cultivation and transplanting of threatened plants. (G/T)

9051 **Evans, T. (1978).** BLM, custodian and planner for the desert. *Fremontia*, 6(3): 7-12. Illus., map. California Desert.

9052 **Farve, R.M. (1987).** A management plan for rare plants in the Red Hills of Tuolumne County, California. *In* Elias, T.S., *ed.* Conservation and management of rare and endangered plants. Proceedings from a conference, Sacramento, California, 5-8 November 1986. Sacramento, California Native Plant Society. 425-427. Maps. (T)

9053 **Faull, M.R. (1987).** Management of *Hemizonia arida* (Asteraceae) by the California Department of Parks and Recreation. *In* Elias, T.S., *ed.* Conservation and management of rare and endangered plants. Proceedings from a conference, Sacramento, California, 5-8 November 1986. Sacramento, California Native Plant Society. 429-439. Illus., map. Control of invasive species. (P/T)

9054 **Ferguson, H.L. (1979).** The goats of San Clemente Island. *Fremontia*, 7(3): 3-8. Illus.

9055 **Ferlatte, W.J. (1978).** Notes on two rare endemic species from the Klamath Region of northern California. *Phacelia dalesiana* (Hydrophyllaceae) and *Raillardella pringlei* (Compositae). *Madrono*, 25(3): 138. (T)

9056 **Ferreira, J., Smith, S. (1987).** Methods of increasing native populations of *Erysimum menziesii*. *In* Elias, T.S., *ed.* Conservation and management of rare and endangered plants. Proceedings from a conference, Sacramento, California, 5-8 November 1986. Sacramento, California Native Plant Society. 507-511. (G/T)

9057 **Ferren, W.R., jr, Forbes, H.C., Roberts, D.A., Smith, D.M. (1984).** The botanical resources of La Purisima Mission State Historic Park, California Publ. No. 3. S. Barbara, Department of Biological Sciences, University of California. (P)

9058 **Ford, L., Norris, K. (1989).** The University of California natural reserve system. Progress and prospects. *Fremontia*, 17(2): 11-16. (P)

9059 **Freas, K.E., Murphy, D.D. (1988).** Taxonomy and the conservation of the critically endangered Bakersfield saltbush, *Atriplex tularensis*. *Biol. Conserv.*, 46(4): 317-324. (T)

9060 **Frey, W. (1984).** The Sargent Cypress Botanical Reserve: a hammock forest. *Pacif. Hort.*, 45(3): 51-52. Illus. (P)

9061 **Greenacres Consulting Corporation (1977).** Redwood National Park proposed 48,000-acre expansion, Washington. Washington, Kenore. 109p. (P)

9062 **Griffin, J.R. (1976).** Native plant reserves at Ford Ord. *Fremontia*, 4(2): 25-28. (P)

9063 **Griggs, F.T., Jain, S.K. (1983).** Conservation of vernal pool plants in California: 2. Population biology of a rare and unique grass genus *Orcuttia*. *Biol. Conserv.*, 27(2): 171-193. (T)

9064 **Griggs, T. (1983).** Kaweah Oaks Preserve. *Fremontia*, 11(3): 25. Illus. (P)

9065 **Gunn, P. (1981).** Vernal pool habitat threatened by dam enlargement. *Calif. Native Pl. Soc. Bull.*, 11(5): 8.

U.S. - California

9066 **Hall, L.A. (1987).** Transplantation of sensitive plants as mitigation for environmental impacts. *In* Elias, T.S., *ed.* Conservation and management of rare and endangered plants. Proceedings from a conference, Sacramento, California, 5-8 November 1986. Sacramento, California Native Plant Society. 413-420. Includes case studies on rare species. (T)

9067 **Hartesvelt, R.J., Harvey, H.T., Shellhammer, H.S., Stecher, R.E. (1975).** The giant *Sequoia* of the Sierra Nevada. Washington, D.C., U.S. Dept of the Interior, National Park Service. xiv, 180p. Illus. (P/T)

9068 **Harvey, H.T., Shellhammer, H.S., Stecker, R.E. (1980).** Giant *Sequoia* ecology: fire and reproduction. Scientific Monogr. Series No. 2. Washington, D.C., U.S. Dept of the Interior, National Park Service. xxii, 182p. Illus., maps. Scientific Monogr. Series No. 2. (T)

9069 **Hastey, E.L. (1987).** Conservation of rare plants on public lands in California. *In* Elias, T.S., *ed.* Conservation and management of rare and endangered plants. Proceedings from a conference, Sacramento, California, 5-8 November 1986. Sacramento, California Native Plant Society. 51-59. (L/T)

9070 **Hatch, D.A. (1980).** Spring on the Point Reyes headlands. *Fremontia*, 8(1): 3-6. Illus. Rare plants. (P/T)

9071 **Havlik, N.A. (1987).** The 1986 Santa Cruz tarweed relocation project. *In* Elias, T.S., *ed.* Conservation and management of rare and endangered plants. Proceedings from a conference, Sacramento, California, 5-8 November 1986. Sacramento, California Native Plant Society. 421-423. (G/T)

9072 **Hayes, M., Schlising, R., Wurlitzer, H. (1979).** *Calycadenia fremontii* rediscovered. *Fremontia*, 7(1): 14-15. Illus. (T)

9073 **Heady, H.F., Zinke, P.J. (1978).** Vegetational changes in Yosemite Valley. National Park Service Occas. Papers No. 5. Washington, D.C., U.S. Dept of the Interior. Illus., maps. (P)

9074 **Heil, K.D., Brack, S. (1985).** The cacti of Carlsbad Caverns National Park. *Cact. Succ. J.* (U.S.), 57(3): 127-134. Illus. (P)

9075 **Heintz, G. (1980).** Exploring the Santa Monicas. *Fremontia*, 8(1): 12-14. Illus. (P)

9076 **Henry, M.A. (1979).** A rare grass on the Eureka dunes. *Fremontia*, 7(2): 3-6. Illus. (T)

9077 **Henry, M.A. (1983).** Plant hunting. I. A Mojave fish-hook cactus weekend. *Fremontia*, 10(4): 23-24. Illus. California, includes endangerment status. (T)

9078 **Holland, R.F. (1987).** Is *Quercus lobata* a rare plant? Approaches to conservation of rare plant communities that lack rare plant species. *In* Elias, T.S., *ed.* Conservation and management of rare and endangered plants. Proceedings from a conference, Sacramento, California, 5-8 November 1986. Sacramento, California Native Plant Society. 129-132. Riparian forests are used to illustrate the Natural Diversity Database.

9079 **Holland, R.F. (1987).** What constitutes a good year for an annual plant? Two examples from the Orcuttieae. *In* Elias, T.S., *ed.* Conservation and management of rare and endangered plants. Proceedings from a conference, Sacramento, California, 5-8 November 1986. Sacramento, California Native Plant Society. 329-333. (T)

9080 **Howald, A.M. (1987).** Strategies for protecting rare plants from oil developments: a Santa Barbara County perspective. *In* Elias, T.S., *ed.* Conservation and management of rare and endangered plants. Proceedings from a conference, Sacramento, California, 5-8 November 1986. Sacramento, California Native Plant Society. 409-411. (T)

U.S. - California

9081 **Howard, A. (1978).** Endangered species in California: federal procedures and status report. *Madrono*, 25(4): 232-233. (L/T)

9082 **Howard, A. (1980).** In search of the Yosemite onion. *Fremontia*, 8(1): 15-18. Illus.

9083 **Howard, A. (1981).** New rare plant list appears. *Calif. Native Pl. Soc. Bull.*, 11(3): 5. (T)

9084 **Howard, A., Arnold, R.A. (1980).** The Antioch dunes - safe at last? *Fremontia*, 8(3): 3-12. Illus. (P)

9085 **Hudson, D.A. (1979).** The fight to preserve Redwood National Park. *Ecol. Law Quarterly*, 7: 781-859. (P)

9086 **Jezik, L. (1978).** California fritillarias: vanishing wildlflower. *Plants Alive*, 6(4): 64-65. (T)

9087 **Jones & Stokes Associates (1987).** Sliding towards extinction: the state of California's natural heritage. San Francisco, The Nature Conservancy. 123p.

9088 **Jones, K.G. (1984).** The Nipomo dunes. *Fremontia*, 11(4): 3-10. Illus., map. Rare and endangered plants listed. (T)

9089 **Katibah, E.F., Dummer, K.J., Nedeff, N.E. (1984).** Current condition of riparian resources in the Central Valley of California. *In* Warner, R.E., Hendrix, K.M., *eds.* California riparian systems: ecology, conservation and productive management. Berkeley, CA, Univ. of California Press. 314-321. Maps.

9090 **Keil, D.J., McLeod, M.G. (1987).** Rare plants in the Arroyo de la Cruz endemic area, San Luis Obispo County, California. *In* Elias, T.S., *ed.* Conservation and management of rare and endangered plants. Proceedings from a conference, Sacramento, California, 5-8 November 1986. Sacramento, California Native Plant Society. 141-154. Maps showing distributions of 12 rare species. (T)

9091 **Knudsen, M.D. (1987).** Recovery of endangered and threatened plants in California: the Federal role. *In* Elias, T.S., *ed.* Conservation and management of rare and endangered plants. Proceedings from a conference, Sacramento, California, 5-8 November 1986. Sacramento, California Native Plant Society. 461-469. (T)

9092 **Krantz, T. (1985).** New additions to the federal endangered species list. *Fremontia*, 13(1): 22-23. Illus. (L/T)

9093 **Krantz, T. (1987).** Island biogeography and preserve design of an insular rare plant community. *In* Elias, T.S., *ed.* Conservation and management of rare and endangered plants. Proceedings from a conference, Sacramento, California, 5-8 November 1986. Sacramento, California Native Plant Society. 605-614. Relict alpine community of Big Bear Valley, San Bernardino Mountains. (P/T)

9094 **Krantz, T.P. (1981).** Rare plant species in the Big Bear Lake Basin, San Bernadino Mountain. *Crossosoma*, 7(1): 1-5. (T)

9095 **Kruckeberg, A.R. (1987).** Serpentine endemism and rarity. *In* Elias, T.S., *ed.* Conservation and management of rare and endangered plants. Proceedings from a conference, Sacramento, California, 5-8 November 1986. Sacramento, California Native Plant Society. 121-128. (T)

9096 **Larson, D.W. (1978).** The living desert reserve. *Fremontia*, 6(3): 28-29. Illus. Palm Desert, California. (P)

9097 **Lathrop, E.W., Archbold, E.F. (1980).** Plant response to utility right of way construction in the Mojave Desert. *Envir. Manag.*, 4(3): 215-226. Illus.

9098 **Lathrop, E.W., Thorne, R.F. (1985).** A new preserve on the Santa Rosa Plateau. *Fremontia*, 13(1): 15-19. Illus., maps. (P)

U.S. - California

9099 **Latting, J. (1978).** The future of the desert is being decided now. *Fremontia*, 6(3): 13-20. Illus. California Desert.

9100 **Latting, J. (1980).** The Desert Plan - protection or destruction? *Fremontia*, 7(4): 18-26. Illus. California Desert.

9101 **Ledig, F.T. (1987).** Genetic structure and the conservation of California's endemic and near-endemic conifers. *In* Elias, T.S., *ed.* Conservation and management of rare and endangered plants. Proceedings from a conference, Sacramento, California, 5-8 November 1986. Sacramento, California Native Plant Society. 587-594. Map. (T)

9102 **Leitner, B.M., deBecker, S. (1987).** Monitoring the Geyser's Panicum (*Dichanthelium lanuginosum* var. *thermale* at the Little Geysers, Sonoma County, California. *In* Elias, T.S., *ed.* Conservation and management of rare and endangered plants. Proceedings from a conference, Sacramento, California, 5-8 November 1986. Sacramento, California Native Plant Society. 391-396. (T)

9103 **Leopold, A.S. (1986).** Wild California: vanishing lands, vanishing wildlife. University of California Press. 144p.

9104 **Libby, J. (1979).** *Cupressus abramsiana* goes to court. *Fremontia*, 7(3): 15. Illus. (T)

9105 **Loope, L.L., Sanchez, P.G., Tarr, P.W., Loope, W.L., Anderson, R.L. (1988).** Biological invasions of arid land nature reserves. *Biol. Conserv.*, 44(1-2): 95-118. Invasive species. Includes 5 case studies. (P)

9106 **Lozier, L. (1987).** The California Nature Conservancy's Landowner Contract Registry Program: voluntary protection for rare plant sites. *In* Elias, T.S., *ed.* Conservation and management of rare and endangered plants. Proceedings from a conference, Sacramento, California, 5-8 November 1986. Sacramento, California Native Plant Society. 567-571. (P/T)

9107 **Macdonald, I.A.W., Graber, D.M., DeBenedetti, S., Groves, R.H., Fuentes, E.R. (1988).** Introduced species in nature reserves in Mediterranean-type climatic regions of the world. *Biol. Conserv.*, 44(1-2): 37-66. Invasive species. Includes 7 case studies. (P)

9108 **Marangio, M.S., Morgan, R. (1987).** The endangered sandhills plant communities of Santa Cruz County. *In* Elias, T.S., *ed.* Conservation and management of rare and endangered plants. Proceedings from a conference, Sacramento, California, 5-8 November 1986. Sacramento, California Native Plant Society. 267-273. Map. Inland sandhills dominated by *Pinus ponderosa* and *Arctostaphylos silvicola* support a unique flora. (T)

9109 **Martz, C. (1987).** Endangered plants along California highways: considerations for right-of-way management. *In* Elias, T.S., *ed.* Conservation and management of rare and endangered plants. Proceedings from a conference, Sacramento, California, 5-8 November 1986. Sacramento, California Native Plant Society. 79-84. Discusses management techniques for rare plants along some of California's 15,000 miles of roadway. (T)

9110 **Matteson, S.H. (1979).** Boiling point at Lassen. *Nat. Parks Conserv. Mag.*, 53(10): 4-8. Illus.

9111 **McCarten, N.F. (1987).** Ecology of the serpentine vegetation in the San Francisco Bay region. *In* Elias, T.S., *ed.* Conservation and management of rare and endangered plants. Proceedings from a conference, Sacramento, California, 5-8 November 1986. Sacramento, California Native Plant Society. 335-339. Map. Includes design of serpentine plant preserves. (P/T)

9112 **McClintock, E. (1973).** Rare and endangered - the story of San Francisco's dune tansy. *Fremontia*, 1(3): 8-10. (T)

U.S. - California

9113 **McClintock, E. (1987).** The displacement of native plants by exotics. *In* Elias, T.S., *ed.* Conservation and management of rare and endangered plants. Proceedings from a conference, Sacramento, California, 5-8 November 1986. Sacramento, California Native Plant Society. 185-188. Invasive species.

9114 **McLeod, M. (1981).** Endangered plant report. *Calif. Native Pl. Soc. Bull.*, 11(1): 3. (T)

9115 **McLeod, M.G. (1987).** Rare and endangered plant successes in San Luis Obispo County. *In* Elias, T.S., *ed.* Conservation and management of rare and endangered plants. Proceedings from a conference, Sacramento, California, 5-8 November 1986. Sacramento, California Native Plant Society. 275-278. (T)

9116 **McMahan, L.R. (1987).** Rare plant conservation by state government: case studies from the western United States. *In* Elias, T.S., *ed.* Conservation and management of rare and endangered plants. Proceedings from a conference, Sacramento, California, 5-8 November 1986. Sacramento, California Native Plant Society. 23-31. Maps. (L/T)

9117 **Medeiros, J.L. (1979).** San Luis Island: the last of the Great Valley. *Fremontia*, 7(1): 3-9. Illus., map.

9118 **Messick, T. (1984).** Conservation. *Calif. Native Pl. Soc. Bull.*, 14(3): [3]. California Native Plant Society (CNPS) contacts with IUCN. (T)

9119 **Messick, T. (1985).** Threatened plant conservation in California. *Threatened Pl. Newsl.*, 14: 24-25. (T)

9120 **Messick, T.C. (1987).** Research needs for rare plant conservation in California. *In* Elias, T.S., *ed.* Conservation and management of rare and endangered plants. Proceedings from a conference, Sacramento, California, 5-8 November 1986. Sacramento, California Native Plant Society. 99-108. (T)

9121 **Millar, C. (1986).** Gene conservation in California's forests. *Fremontia*, 14(1): 6-7. Illus. (G/T)

9122 **Mooney, H.A. (1988).** Lessons from Mediterranean-climate regions. *In* Wilson, E.O., *ed.* Biodiversity. Washington, D.C., National Academy of Science Press. 157-165. Discussion of biodiversity of tropical and temperate zones.

9123 **Muick, P.C. (1980).** Restoring habitats in Sonoma County. *Fremontia*, 8(2): 17-21. Illus.

9124 **Myatt, M.M. (1987).** Predicting the habitat geography of sensitive plants and community types. *In* Elias, T.S., *ed.* Conservation and management of rare and endangered plants. Proceedings from a conference, Sacramento, California, 5-8 November 1986. Sacramento, California Native Plant Society. 173-179. Maps. Geographic Information System to be used in Natural Diversity Database.

9125 **Myatt, R.G. (1987).** Germination and seedling establishment of the Ione Buckwheat. *In* Elias, T.S., *ed.* Conservation and management of rare and endangered plants. Proceedings from a conference, Sacramento, California, 5-8 November 1986. Sacramento, California Native Plant Society. 547-551. Includes field studies and controlled environments. (G/T)

9126 **Nabhan, G. (1980).** *Ammobroma sonorae*, an endangered parasitic plant in extremely arid North America. *Desert Pl.*, 2(3): 188-196. Illus., col. illus. (T)

9127 **Nabhan, G.P. (1987).** Nurse plant ecology of threatened desert plants. *In* Elias, T.S., *ed.* Conservation and management of rare and endangered plants. Proceedings from a conference, Sacramento, California, 5-8 November 1986. Sacramento, California Native Plant Society. 377-383. Illus. Numerous desert plants depend on protective overstorey for survival; case studies of 2 cacti and *Capsicum annuum* var. *glabriusculum*. (T)

U.S. - California

9128 **Nabhan, G.P. (1989).** Plants at risk in the Sonoran Desert: an international concern. *Agave*, 3(3): 14-15. Map. (T)

9129 **Nabhan, G.P., Monarque, E.S., Olwell, P., Warren, P., Hodgson, W., Gallindo-Duarte, C., Bittman, R., Anderson, S. (1989).** A preliminary list of plants at risk in the Sonoran Desert of the U.S. and Mexico. *Agave*, 3(3): 15. (T)

9130 **Natural Diversity Data Base (1984).** Elements with occurrence. California Department of Fish and Game. 13-31. Listing of 391 special plants, categories of endangerment and entered into computer data base system. (T)

9131 **Newton, G.A. (1987).** The ecology and management of three rare salt marsh species of Humboldt Bay. *In* Elias, T.S., *ed.* Conservation and management of rare and endangered plants. Proceedings from a conference, Sacramento, California, 5-8 November 1986. Sacramento, California Native Plant Society. 263-266. (T)

9132 **Nicola, S.J. (1987).** Government funding of research, protection and management of rare plants in California. *In* Elias, T.S., *ed.* Conservation and management of rare and endangered plants. Proceedings from a conference, Sacramento, California, 5-8 November 1986. Sacramento, California Native Plant Society. 67-72. (T)

9133 **Norris, L.L. (1987).** Status of five rare plant species in Sequoia and Kings Canyon National Parks. *In* Elias, T.S., *ed.* Conservation and management of rare and endangered plants. Proceedings from a conference, Sacramento, California, 5-8 November 1986. Sacramento, California Native Plant Society. 279-282. (P/T)

9134 **Nuorteva, M. (1979).** Preservation problems of redwoods in California. *Silva Fennica*, 13(1): 51. (T)

9135 **Oberbauer, T. A. (1978).** San Diego County and its rare plants. *Fremontia*, 5(4): 12-15. (T)

9136 **Orsak, L. (1984).** Brinkmanship on San Bruno mountain. *Garden* (U.S.), 8(5): 16-21, 32. Col. illus.

9137 **Palmer, R. (1987).** Evolutionary relationships of *Holocarpha macradenia*. *In* Elias, T.S., *ed.* Conservation and management of rare and endangered plants. Proceedings from a conference, Sacramento, California, 5-8 November 1986. Sacramento, California Native Plant Society. 295-304. Illus., maps. (T)

9138 **Parker, V.T. (1987).** Effects on wet-season management burns on chapparal vegetation: implications for rare species. *In* Elias, T.S., *ed.* Conservation and management of rare and endangered plants. Proceedings from a conference, Sacramento, California, 5-8 November 1986. Sacramento, California Native Plant Society. 233-237. (T)

9139 **Parsons, D.J. (1983).** Wilderness protection: an example from the southern Sierra Nevada, U.S.A. *Envir. Conserv.*, 10(1): 23-30. Illus. Maps. Kings Canyon National Park, Rae Lakes basin. (P)

9140 **Pavlik, B.M. (1987).** Attributes of plant populations and their management implications. *In* Elias, T.S., *ed.* Conservation and management of rare and endangered plants. Proceedings from a conference, Sacramento, California, 5-8 November 1986. Sacramento, California Native Plant Society. 311-319. (T)

9141 **Pavlik, B.M., Barbour, M.G. (1988).** Demographic monitoring of endemic sand dune plants, Eureika Valley, California. *Biol. Conserv.*, 46(3): 217-242. Illus. (T)

9142 **Pearlman, N. (1980).** Help save the Santa Rosa Mountains. *Calif. Native Pl. Soc. Bull.*, 10(5): 6.

9143 **Pitschel, B.M. (1984).** A role for botanical gardens and arboreta in preserving, restoring and reestablishing native California grasslands. *Bull. Amer. Assoc. Bot. Gard. Arbor.*, 18(2): 53-61. Illus. (G)

U.S. - California

9144 **Powel, W.R. (1975).** The CNPS rare plant project. *Fremontia*, 2(4): 14-19. (T)

9145 **Powell, W.R. (1978).** The CNPS inventory - a progress report. *Fremontia*, 5(4): 28-29.

9146 **Powell, W.R., Duncan, T., Howard, A.Q. (1981).** The California Native Plant Society Rare Plant Project. *In* Morse, L.E., Henifin, M.S., *eds.* Rare plant conservation: geographical data organization. New York, The New York Botanical Garden. 193-197. (T)

9147 **Powell, W.R., ed. (1974).** Inventory of rare and endangered vascular plants of California. Special Publication No. 1. Berkeley, California Native Plant Society. 56p. (T)

9148 **Reid, T.S., Walsh, R.C. (1987).** Habitat reclamation for endangered species on San Bruno Mountain. *In* Elias, T.S., *ed.* Conservation and management of rare and endangered plants. Proceedings from a conference, Sacramento, California, 5-8 November 1986. Sacramento, California Native Plant Society. 493-499. Maps. (T)

9149 **Reiseberg, L.H. (1988).** Saving California's rarest tree. *Center Pl. Conserv.*, 3(1): 1, 8. Catalina mahogany. (T)

9150 **Reveal, J.L. (1977).** The western United States. *In* Prance, G.T., Elias, T.S., *eds.* Extinction is forever. Proceedings of a symposium at the New York Botanical Garden, 11-13 May 1976. New York, New York Botanical Garden. 50-59. Discusses existing laws and present policies regarding endangered plants for five western states of the U.S. (L/T)

9151 **Scheid, G.A. (1987).** Habitat characteristics of willowy monardella in San Diego County: site selection for transplants. *In* Elias, T.S., *ed.* Conservation and management of rare and endangered plants. Proceedings from a conference, Sacramento, California, 5-8 November 1986. Sacramento, California Native Plant Society. 501-506. Map. (T)

9152 **Schoolcraft, G. (1987).** The designation process for a research natural area. *In* Elias, T.S., *ed.* Conservation and management of rare and endangered plants. Proceedings from a conference, Sacramento, California, 5-8 November 1986. Sacramento, California Native Plant Society. 85-90. (P)

9153 **Shelgren, M. (1983).** A poppy paradise. *Garden* (New York), 7(3): 15-17, 32. Col. illus. Antelope Valley, California Poppy Reserve. (P)

9154 **Shevock, J., Taylor, D.W. (1987).** Plant exploration in California, the frontier is still here. *In* Elias, T.S., *ed.* Conservation and management of rare and endangered plants. Proceedings from a conference, Sacramento, California, 5-8 November 1986. Sacramento, California Native Plant Society. 91-98. Map. Discusses plant exploration of California and geographic distribution of new plant taxa.

9155 **Shevock, J.R., Hennessy, L.L. (1987).** The California Natural Diversity Data Base - a common denominator. *In* Elias, T.S., *ed.* Conservation and management of rare and endangered plants. Proceedings from a conference, Sacramento, California, 5-8 November 1986. Sacramento, California Native Plant Society. 181-184. Database includes locational data on plants, animals and communities. Shows field survey form.

9156 **Shevock, J.R., Norris, L.L. (1981).** The mountain parsleys of California. *Fremontia*, 9(3): 22-25. Illus., maps. (T)

9157 **Showers, M.A. & D.W. (1981).** A field guide to the flowers of Lassen Volcanic National Park. Loomis Museum Association. 112p. Illus. (P)

9158 **Sigg, J. (1980).** The uncertain future of Golden Gate Park. *Pac. Hort.*, 41(4): 3-8. Illus. (P)

U.S. - California

9159 **Simmons, I.G., Vale, T.R. (1975).** Conservation of the California coast redwood and its environment. *Envir. Conserv.*, 2: 29-38. (P/T)

9160 **Smith, J.P. (1981).** Inventory of rare and endangered vascular plants of California. First supplement. Berkeley, California Native Plant Society. 28p. (T)

9161 **Smith, J.P. (1982).** Inventory of rare and endangered vascular plants of California. Second supplement. Berkeley, California Native Plant Society. 28p. (T)

9162 **Smith, J.P. (1987).** California: leader in endangered plant protection. *Fremontia*, 15(1): 3-7. (T)

9163 **Smith, J.P., Berg, K., eds (1988).** Inventory of rare and endangered vascular plants of California. Sacramento, California, California Native Plant Society. 168p. (T)

9164 **Smith, J.P., York, R. (1984).** Inventory of rare and endangered vascular plants of California. Berkeley, CA, California Native Plant Society. 174p. Special Publication No. 1 (3rd Ed.). (T)

9165 **Smith, J.P., et al. (1980).** Inventory of rare and endangered vascular plants of California. 2nd ed. Berkeley, California Native Plant Society. 115p. (T)

9166 **Smith, J.P., jr (1987).** California's endangered plants and the CNPS Rare Plant Program. *In* Elias, T.S., *ed.* Conservation and management of rare and endangered plants. Proceedings from a conference, Sacramento, California, 5-8 November 1986. Sacramento, California Native Plant Society. 1-6. (T)

9167 **Smith, J.P., jr, York, R. (1984).** Inventory of rare and endangered vascular plants of California, 3rd Ed. Special Publication No. 1. Berkeley, California Native Plant Society. 174p. List of 34 taxa 'presumed extinct in CA', 604 'rare or endangered in CA and elsewhere', 198 'R/E in CA, more common elsewhere', 144 'needing more info.', and 499 'limited distribution'. (T)

9168 **Smith, W.R. (1987).** Studies of the population biology of prairie bush-clover (*Lespedeza leptostachya*). *In* Elias, T.S., *ed.* Conservation and management of rare and endangered plants. Proceedings from a conference, Sacramento, California, 5-8 November 1986. Sacramento, California Native Plant Society. 359-366. Illus. (T)

9169 **Smith, Z.G., jr (1987).** Sensitive plant management in the national forests in California. *In* Elias, T.S., *ed.* Conservation and management of rare and endangered plants. Proceedings from a conference, Sacramento, California, 5-8 November 1986. Sacramento, California Native Plant Society. 61-65. Map. Discusses the National Forest Service Sensitive Plant List. (P/T)

9170 **Sorrie, B.A. (1978).** *Corallorhiza trifida* in California and Nevada. *Wasmann J. Biol.*, 36(1-2): 199-200. (T)

9171 **Stebbins, G.L. (1978).** Why are there so many rare plants in California? I. Environmental factors. II. Youth and age of species. *Fremontia*, 5(4): 6-10; 6(1): 17-20. (T)

9172 **Stebbins, G.L. (1986).** Rare plants in California's national forests: their scientific value and conservation. *Fremontia*, 13(4): 9-12. (T)

9173 **Stone, D.R., Clifton, G.L., DaVilla, W.B., Stebbins, J.C., Taylor, D.M. (1987).** Endangerment status of the grass tribe Orcuttieae and *Chamaesyce hooveri* (Euphorbiaceae) in the Central Valley of California. *In* Elias, T.S., *ed.* Conservation and management of rare and endangered plants. Proceedings from a conference, Sacramento, California, 5-8 November 1986. Sacramento, California Native Plant Society. 239-247. (T)

9174 **Stone, R.D., Sumidor, V.A., eds (1983).** The Kingston Range of California: a resource survey. Publication No. 10. Santa Cruz, Environmental Field Program. 393p.

U.S. - California

9175 **Strahan, J., Wolley, G.J. (1987).** The Ring Mountain restoration plan. *In* Elias, T.S., *ed.* Conservation and management of rare and endangered plants. Proceedings from a conference, Sacramento, California, 5-8 November 1986. Sacramento, California Native Plant Society. 477-486. Ring Mountain Preserve. (P/T)

9176 **Taylor, D.M., Palmer, R.E. (1987).** Ecology and endangerment status of *Silene invisa* populations in the central Sierra Nevada, California. *In* Elias, T.S., *ed.* Conservation and management of rare and endangered plants. Proceedings from a conference, Sacramento, California, 5-8 November 1986. Sacramento, California Native Plant Society. 321-327. Illus., map. (T)

9177 **Taylor, D.M., Stebbins, J.C., DaVilla, W.B. (1987).** Endangerment status of *Collomia rawsoniana* (Polemoniaceae), western Sierra Nevada, California. *In* Elias, T.S., *ed.* Conservation and management of rare and endangered plants. Proceedings from a conference, Sacramento, California, 5-8 November 1986. Sacramento, California Native Plant Society. 225-231. Illus., map. (T)

9178 **Taylor, D.W. (1988).** The Californian jewelflower: one of California's most endangered plants. *Fremontia*, 16(1): 18-19. *Caulanthus californicus.* (T)

9179 **Taylor, M.S. (1983).** Rare and endangered plants of Butte County, California. *Flora Buttensis*, 4(1): 59. (T)

9180 **Taylor, T.P., Erman, D.C. (1979).** The response of benthic plants to past levels of human use in high mountain lakes in Kings Canyon National Park, California, U.S.A. *J. Envir. Manage.*, 9(3): 271-278. Illus., map. (P)

9181 **Thomas, T. (1987).** Approach to rare plant management at Golden Gate National Recreation Area. *In* Elias, T.S., *ed.* Conservation and management of rare and endangered plants. Proceedings from a conference, Sacramento, California, 5-8 November 1986. Sacramento, California Native Plant Society. 471-475. (P/T)

9182 **Turner, C.E. (1980).** Help find *Sagittaria sandfordii. Calif. Native Pl. Soc. Bull.*, 10(5): 5. Rare species. (T)

9183 **Twisselmann, E.C. (1969).** Status of the rare plants of Kern County. *Calif. Native Plant Soc. Newsl.*, 5(3): 1-7. (T)

9184 **U.S. Department of Agriculture. Forest Service (1979).** Medicine Lake planning unit, Modoc, Shasta-Trinity and Klamath National Forests. Alturas, U.S. Department of Agriculture Forest Service. 235p. (P)

9185 **U.S. Department of the Interior. Bureau of Land Management (1980).** California desert conservation area. Sacramento, CA, Dept of the Interior, Bureau of Land Management. 2 vols. (P)

9186 **U.S. Department of the Interior. Fish and Wildlife Service (1983).** Endangered and threatened wildlife and plants: proposed Endangered status for *Camissonia benitensis* (San-Benito evening-primrose). *Fed. Register*, 48(211): 50126-50128. (L/T)

9187 **U.S. Department of the Interior. Fish and Wildlife Service (1983).** Endangered and threatened wildlife and plants: proposed Endangered status for the pedate checker-mallow (*Sidalcea pedata*) and slender-petaled mustard (*Thelypodium stenopetalum*). *Fed. Register*, 48(137): 32522-32525. (L/T)

9188 **U.S. Department of the Interior. Fish and Wildlife Service (1983).** Eureka Valley dunes recovery plan approved. *Endangered Species Tech. Bull.*, 8(3): 10-11. Illus. (T)

9189 **U.S. Department of the Interior. Fish and Wildlife Service (1983).** San Francisco Peaks groundsel listed as Threatened. *Endangered Species Tech. Bull.*, 8(12): 3. (L/T)

U.S. - California

9190 **U.S. Department of the Interior. Fish and Wildlife Service (1984).** Endangered and threatened wildlife and plants: determination of Endangered status for *Thelypodium stenopetalum* (Slender-petaled mustard) and *Sidalcea pedata* (Pedate checker-mallow). *Fed. Register*, 49(171): 34497-34500. (L/T)

9191 **U.S. Department of the Interior. Fish and Wildlife Service (1984).** Endangered and threatened wildlife and plants: proposal of Endangered status and critical habitat for the large-flowered fiddleneck (*Amsinckia grandiflora*). *Fed. Register*, 49(90): 19534-19538. Map. (L/T)

9192 **U.S. Department of the Interior. Fish and Wildlife Service (1984).** Endangered and threatened wildlife and plants: proposed Endangered status for the San Mateo thornmint (*Acanthomintha obovata* ssp. *duttonii*). *Fed. Register*, 49(118): 24906-24909. (L/T)

9193 **U.S. Department of the Interior. Fish and Wildlife Service (1984).** Endangered status for three western plants. *Endangered Species Tech. Bull.*, 9(9): 6-7. (L/T)

9194 **U.S. Department of the Interior. Fish and Wildlife Service (1984).** Four plants proposed for listing. *Endangered Species Tech. Bull.*, 9(7): 1, 4-5. Illus. (L/T)

9195 **U.S. Department of the Interior. Fish and Wildlife Service (1984).** Three plants proposed for listing. *Endangered Species Tech. Bull.*, 9(6): 3, 10. Illus. (L/T)

9196 **U.S. Department of the Interior. Fish and Wildlife Service (1985).** Emergency protection for Loch Lomond Coyota-thistle. *Endangered Species Tech. Bull.*, 10(9): 3. (L/T)

9197 **U.S. Department of the Interior. Fish and Wildlife Service (1985).** Endangered and threatened wildlife and plants: determination of Endangered species status for *Acanthomintha obovata* ssp. *duttonii*. *Fed. Register*, 50(181): 37858-37863. (L/T)

9198 **U.S. Department of the Interior. Fish and Wildlife Service (1985).** Endangered and threatened wildlife and plants: determination of Threatened status for *Camissonia benitensis* (San-Benito evening-primrose). *Fed. Register*, 50(29): 5755-5759. (L/T)

9199 **U.S. Department of the Interior. Fish and Wildlife Service (1985).** Endangered and threatened wildlife and plants: determination of Threatened status with critical habitat for six plants and one insect in Ash Meadows, Nevada, California; and Endangered status with critical habitat for one plant in Ash Meadows, Nevada, California. *Fed. Register*, 50(97): 20777-20815. (L/T)

9200 **U.S. Department of the Interior. Fish and Wildlife Service (1985).** Endangered and threatened wildlife and plants: determination that *Amsinckia grandiflora* is an Endangered species and designation of critical habitat. *Fed. Register*, 50(89): 19374-19378. (L/T)

9201 **U.S. Department of the Interior. Fish and Wildlife Service (1985).** Endangered and threatened wildlife and plants: emergency determination of Endangered status for Loch Lomond coyote thistle (*Eryngium constancei*). *Fed. Register*, 50(147): 31187-31190. (L/T)

9202 **U.S. Department of the Interior. Fish and Wildlife Service (1985).** Endangered and threatened wildlife and plants: proposal to determine *Cupressus abramsiana* to be an Endangered species. *Fed. Register*, 50(177): 37249-37252. (L/T)

9203 **U.S. Department of the Interior. Fish and Wildlife Service (1985).** Endangered and threatened wildlife and plants: proposed Endangered status for *Cordylanthus palmatus* (Palmate-bracted bird's beak). *Fed. Register*, 50(136): 28870-28873. (L/T)

9204 **U.S. Department of the Interior. Fish and Wildlife Service (1985).** Endangered and threatened wildlife and plants: public hearing and reopening of comment period on proposed Endangered status for *Cordylanthus palmatus* (Palmate bracted bird's beak). *Fed. Register*, 50(206): 43260-43261. (L/T)

U.S. - California

9205 **U.S. Department of the Interior. Fish and Wildlife Service (1985).** Endangered and threatened wildlife and plants: reopening of comment period on proposed Endangered status for *Cupressus abramsiana* (Santa Cruz cypress). *Fed. Register,* 50(228): 48616-48617. (L/T)

9206 **U.S. Department of the Interior. Fish and Wildlife Service (1985).** Endangered classification proposed for four plants. *Endangered Species Tech. Bull.,* 10(8): 1, 4-5. (L/T)

9207 **U.S. Department of the Interior. Fish and Wildlife Service (1985).** Four plants given Endangered Species Act protection. *Endangered Species Tech. Bull.,* 10(5): 1,10. (L/T)

9208 **U.S. Department of the Interior. Fish and Wildlife Service (1985).** Protection recommended for three plants. *Endangered Species Tech. Bull.,* 10(1): 1, 5-6. (L/T)

9209 **U.S. Department of the Interior. Fish and Wildlife Service (1985).** Regional briefs: regions 1 & 2. *Endangered Species Tech. Bull.,* 10(10): 2, 9-11. (T)

9210 **U.S. Department of the Interior. Fish and Wildlife Service (1986).** Endangered and threatened wildlife and plants: proposal to determine *Eryngium constancei* (Loch Lomond coyote thistle) to be an Endangered species. *Fed. Register,* 51(58): 10412-10415. (L/T)

9211 **U.S. Department of the Interior. Fish and Wildlife Service (1986).** Endangered and threatened wildlife and plants: proposed Endangered status for *Eriastrum densifolium* ssp. *sanctorum* (Santa Ana River woolly star) and *Centrostegia leptoceras* (Slenderhorned spineflower). *Fed. Register,* 51(68): 12180-12184. (L/T)

9212 **U.S. Department of the Interior. Fish and Wildlife Service (1989).** Proposed listings - July 1989. Five San Joaquin Valley plants. *Endangered Species Tech. Bull.,* 14(8): 4-5. Illus. (L/T)

9213 **U.S. Department of the Interior. National Park Service (1979).** Giant forest/lodgepole area, Sequoia and Kings Canyon National Parks, California. Three Rivers, National Park Service. 291p. (P)

9214 **U.S. Department of the Interior. National Park Service (1979).** Redwood National Park, California. Denver, CO, National Park Service. 2 vols. (P)

9215 **U.S. Department of the Interior. National Park Service (1980).** General management plan, Yosemite National Park, California (supplement). Denver, CO, National Park Service. 2 vols. (P)

9216 **U.S. Department of the Interior. National Park Service (1980).** Redwood National Park, Del Norte and Humboldt Counties, California. Denver, CO, U.S. Department of the Interior, National Park Service. 2 vols. (P)

9217 **Vankat, J.L. (1982).** A gradient perspective on the vegetation of Sequoia National Park, California. *Madrono,* 29(3): 200-214. Illus. (P)

9218 **Vankat, J.L., Major, J. (1978).** Vegetation changes in Sequoia National Park, California. *J. Biogeogr.,* 5(4): 377-402. Illus., map. (P)

9219 **Wagtendonk, J.W. van (?)** Refined burning prescriptions for Yosemite National Park. Washington, D.C., U.S. Govt Printers. iv, 21p. Illus., map. (P)

9220 **Warner, R.E. (1984).** Structural, floristic, and condition inventory of Central Valley riparian systems. *In* Warner, R.E., Hendrix, K.M., *eds.* California riparian systems: ecology, conservation and productive management. Berkeley, CA, Univ. of California Press. 356-374. Illus. Covers site quality.

9221 **Warner, R.E., Hendrix, K.M., eds (1984).** California riparian systems: ecology, conservation and productive management. Berkeley, Univ. of California Press. Illus., maps.

U.S. - California

9222 **Wayburn, E., McCloskey, M., Howard, B. (1984).** Redwood National Park: a case study in preserving a vanishing resource. *In* McNeely, J.A., Miller, K.R., *eds.* National parks, conservation and development. The role of protected areas in sustaining society. Washington, D.C., Smithsonian Institution Press. 503-507. (P)

9223 **Westman, W.E. (1987).** Implications of ecological theory for rare plant conservation in coastal sage scrub. *In* Elias, T.S., *ed.* Conservation and management of rare and endangered plants. Proceedings from a conference, Sacramento, California, 5-8 November 1986. Sacramento, California Native Plant Society. 133-140. Map. About 50% of plants are rare in this endangered habitat which is reduced to 10-15% of its original area (T)

9224 **Willoughby, J.W. (1987).** Effects of livestock grazing on two rare plant species in the Red Hills, Tuolumne County, California. *In* Elias, T.S., *ed.* Conservation and management of rare and endangered plants. Proceedings from a conference, Sacramento, California, 5-8 November 1986. Sacramento, California Native Plant Society. 199-208. (T)

9225 **Willy, A.G. (1987).** Feral hog management at Golden Gate National Recreation Area. *In* Elias, T.S., *ed.* Conservation and management of rare and endangered plants. Proceedings from a conference, Sacramento, California, 5-8 November 1986. Sacramento, California Native Plant Society. 189-191. Feral hogs have considerable impact on native flora. (P)

9226 **York, R.P. (1987).** A list is a list ... or is it? *In* Elias, T.S., *ed.* Conservation and management of rare and endangered plants. Proceedings from a conference, Sacramento, California, 5-8 November 1986. Sacramento, California Native Plant Society. 39-41. Discusses Federal List and California Native Plant Society lists and their legal implications. (L/T)

9227 **York, R.P. (1987).** California's most endangered plants. *In* Elias, T.S., *ed.* Conservation and management of rare and endangered plants. Proceedings from a conference, Sacramento, California, 5-8 November 1986. Sacramento, California Native Plant Society. 109-120. Illus., maps. Data sheets on 12 taxa. (T)

9228 **Zedler, J.B. (1988).** Restoring diversity in salt marshes: can we do it? *In* Wilson, E.O., *ed.* Biodiversity. Washington, D.C., National Academy Press. 317-325. Wetlands.

9229 **Zedler, P.H., Guehlotorff, K., Scheidlinger, C., Gautier, C.R. (1983).** The population ecology of a dune thistle, *Cirsium rhothophilum* (Asteraceae). *Amer. J. Bot.*, 70(10): 1516-1527. (T)

9230 **Zembal, R., Kramer, K.J. (1985).** The status of the Santa Ana River woolly-star. *Fremontia*, 13(3): 19-20. Illus. (T)

U.S. - Colorado

9231 **Colorado Native Plant Society (1978).** Working list of threatened and endangered plant species of Colorado. Colorado Native Plant Society Newsletter Suppl. (T)

9232 **Colorado Native Plant Society (1989).** Rare plants of Colorado. Estes Park, Colorado, Rocky Mountain National Park. 75p. Data on 92 threatened plants; introductory chapters on plant communities and habitats. Col. illus. (T)

9233 **Colorado Natural Heritage Inventory (1983).** Plant associations of special concern in Colorado. Denver, Colorado Natural Heritage Inventory. 14p. (T)

9234 **Colorado Natural Heritage Inventory (1983).** Plant species of special concern. Denver, Colorado Natural Heritage Inventory. 12p.

9235 **Dickenson, R.E. (1981).** Mesa Verde: a world heritage. *Nat. Parks Conserv. Mag.*, 55(9-10): 14-15. (P)

U.S. - Colorado

9236 **Ecology Consultants, Inc. (1978).** An illustrated guide to the proposed threatened and endangered plant species in Colorado. Denver, CO, U.S. Dept of the Interior, Fish and Wildlife Service. 114p. (T)

9237 **Johnston, B.C., et al. (1980).** Proposed and recommended threatened and endangered plant species of the Forest Service Rocky Mountain Region: an illustrated guide to certain species in Colorado, Wyoming, South Dakota, Nebraska and Kansas. Denver, CO, U.S. Fish and Wildlife Service. 164p. (T)

9238 **McMahan, L.R. (1987).** Rare plant conservation by state government: case studies from the western United States. *In* Elias, T.S., *ed.* Conservation and management of rare and endangered plants. Proceedings from a conference, Sacramento, California, 5-8 November 1986. Sacramento, California Native Plant Society. 23-31. Maps. (L/T)

9239 **Olmsted, C.E. (1979).** The effect of large herbivores on aspen in Rocky Mountain National Park. Ann Arbor, University Microfilms International. xiii, 141p. (P)

9240 **Reveal, J.L. (1977).** The western United States. *In* Prance, G.T., Elias, T.S., *eds.* Extinction is forever. Proceedings of a symposium at the New York Botanical Garden, 11-13 May 1976. New York, New York Botanical Garden. 50-59. Discusses existing laws and present policies regarding endangered plants for five western states of the U.S. (L/T)

9241 **Trammell, J.R. & V.M. (1986).** Knowing: growing: conserving native plants Colorado's great natural diversity. *Green Thumb*, 43(1): 17-24. Illus., map. (G/T)

9242 **U.S. Department of the Interior. Fish and Wildlife Service (1983).** A Colorado wild-buckwheat proposed with critical habitat. *Endangered Species Tech. Bull.*, 8(7): 3, 5. Illus. (L/T)

9243 **U.S. Department of the Interior. Fish and Wildlife Service (1983).** Endangered and threatened wildlife and plants: proposal to determine *Eriogonum pelinophilum* (Clay loving wild buckwheat) to be an Endangered species and to determine its critical habitat. *Fed. Register*, 48(121): 28504-28507. Map. (L/T)

9244 **U.S. Department of the Interior. Fish and Wildlife Service (1984).** Endangered and threatened wildlife and plants: final rule to determine *Eriogonum pelinophilum* to be an Endangered species and to designate its critical habitat. *Fed. Register*, 49(136): 28562-28565. Map. (L/T)

9245 **U.S. Department of the Interior. Fish and Wildlife Service (1984).** Endangered and threatened wildlife and plants: proposal to determine *Astragalus humillimus* (Mancos milkvetch) to be an Endangered species. *Fed. Register*, 49(126): 26610-26614. (L/T)

9246 **U.S. Department of the Interior. Fish and Wildlife Service (1984).** Protection becomes final for four plants. *Endangered Species Tech. Bull.*, 9(8): 3-5. Illus. (L/T)

9247 **U.S. Department of the Interior. Fish and Wildlife Service (1988).** Clay-loving wild-buckwheat (*Erigonum pelinophilum*) recovery plan. U.S. Fish and Wildlife Service. Region 6. (T)

9248 **U.S. Department of the Interior. Fish and Wildlife Service (1988).** Endangered and threatened wildlife and plants; proposal to determine *Astragalus osterhoutii* and *Penstemon penlandii* to be Endangered species. *Fed. Register*, 53(128): 25181-25185. (L/T)

9249 **U.S. Department of the Interior. Fish and Wildlife Service (1988).** Endangered and threatened wildlife and plants; public hearing and reopening of comment period on proposed Endangered status for *Astragalus osterhoutii* (Osterhout milkvetch) and *Penstemon penlandii* (Penland beardtongue). *Fed. Register*, 53(185): 37009-37010. (L/T)

U.S. - Colorado

9250 **U.S. Department of the Interior. Fish and Wildlife Service (1989).** Endangered and threatened wildlife and plants; final rule to determine *Astragalus osterhoutii* and *Penstemon penlandii* to be Endangered species. *Fed. Register*, 54(133): 29658-29663. (L/T)

9251 **U.S. Department of the Interior. National Park Service, Forest Service (1979).** Rocky Mountain National Park boundary study. Washington, D.C., Department of the Interior, National Park Service. 152p. Illus., maps. (P)

9252 **Wenger, S.R. (1978).** Flowers of Mesa Verde National Park. Colorado, Mesa Verde Museum Association. 50p. (P)

U.S. - Connecticut

9253 **Anderson, G.J. (1980).** The status of the very rare *Prunus gravesii* Small. Conference: Symposium on Rare and Endangered Species in New England, 1979. *Rhodora*, 82(829): 113-129. Illus. (T)

9254 **Countryman, W.D. (1977).** The northeastern United States. *In* Prance, G.T., Elias, T.S., *eds*. Extinction is forever. Proceedings of a symposium at the New York Botanical Garden, 11-13 May 1976. New York, New York Botanical Garden. 30-35. Legislation regarding conservation (primarily of plants and natural areas) in the northeastern U.S. (L/P/T)

9255 **Dowhan, J.J., Craig, R.J. (1976).** Rare and threatened species of Connecticut and their habitats. Report of investigations No. 6. Hartford, State Geological and Natural History Survey of Connecticut, Natural Resources Center, Dept. of Environment Protection. v, 137p. (T)

9256 **Field, K.G., Primack, R.B. (1980).** Comparison of the characteristics of rare, common and endemic plant species in New England USA. *In* The University of British Columbia. Second International Congress of Systematic and Evolutionary Biology, Vancouver, B.C. Canada, July 17-24, 1980. Vancouver, B.C., Univ. of British Columbia. (T)

9257 **Foy, L. (1977).** Protection of endangered plants in New England: lists, laws and activities. Framingham, MA, New England Wild Flower Society, Inc. (Various paging). (L/T)

9258 **Mehrhoff, L.J. (1978).** Rare and endangered vascular plant species in Connecticut. Newton Corner, MA, U.S. Fish and Wildlife Service and The New England Botanical Club. vii, 41p. Map. (T)

9259 **Mehrhoff, L.J. (1980).** Connecticut's endangered species program. *Rhodora*, 82(829): 141-144. (T)

9260 **New England Botanical Club. Natural Areas Criteria Committee (1972).** Guidelines and criteria for the evaluation of natural areas. A report prepared for the New England Natural Resources Center by the Natural Areas Criteria Committee of the New England Botanical Club Inc. (P)

9261 **Tryon, R.M., ed. (1980).** Symposium on rare and endangered plant species in New England, Cambridge, Mass., 1979. The proceedings of the symposium "Rare and Endangered Plant Species in New England," 4-5 May 1979. *Rhodora* (Cambridge, MA, New England Botanical Garden), 82(829): 237p. (T)

9262 **U.S. Department of the Interior. Fish and Wildlife Service (1988).** Endangered and threatened wildlife and plants; determination of *Agalinis acuta* (Sandplain gerardia) to be an Endangered species. *Fed. Register*, 53(173): 34701-34705. (L/T)

9263 **U.S. Department of the Interior. Fish and Wildlife Service (1989).** Sandplain gerardia (*Agalinis acuta*) recovery plan. Newton Corner, MA, U.S. Department of the Interior, Fish and Wildlife Service. 47p. Illus., map. (L/T)

U.S. - Connecticut

9264 **U.S. Environmental Protection Agency, Region 1 (1982).** New England Wetland Flora and Wetland Protection Laws. Washington, D.C., U.S. Govt Print. Off. 172p. Illus. (L/P)

U.S. - Delaware

9265 **Boone, D.D., Fenwick, G.H., Hirst, F. (1984).** The rediscovery of *Oxypolis canbyi* on the Delmara Peninsula. *Bartonia*, 50: 21-22. (T)

9266 **Dill, N.H., Tucker, A.O., Davis, J.E. (1982).** Computer access to locations of rare plants in Delaware, U.S.A. *Amer. Assoc. Adv. Sci. Abstr. Pap. Nation. Meeting*, 148:140. (T)

9267 **Hardin, J.W. (1977).** The southeastern United States. *In* Prance, G.T., Elias, T.S., *eds*. Extinction is forever. Proceedings of a symposium at the New York Botanical Garden, 11-13 May 1976. New York, New York Botanical Garden. 36-40. Discusses status of state programmes within the 14 southeastern states of endangered plant species. (L/P/T)

9268 **Naczi, R.F.C. (1984).** Rare sedges discovered and rediscovered in Delaware. *Bartonia*, 50: 31-35. (T)

9269 **Tucker, A.O., Dill, N.H., Broome, C.R., Phillips, C.E., Maciarello, M.J. (1979).** Rare and endangered vascular plant species in Delaware. Newton Corner, MA, U.S. Fish and Wildlife Service and The Society of Natural History of Delaware. x, 89p. Map. Lists 449 rare plants with annotations. (T)

9270 **U.S. Department of the Interior. Fish and Wildlife Service (1985).** Endangered and threatened wildlife and plants: proposal to determine *Oxypolis canbyi* (Canby's dropwort) to be an Endangered species. *Fed. Register*, 50(60): 12345-12348. (L/T)

9271 **U.S. Department of the Interior. Fish and Wildlife Service (1986).** Endangered and threatened wildlife and plants: determination of *Oxypolis canbyi* (Canby's dropwort) to be an Endangered species. *Fed. Register*, 51(37): 6690-6693. (L/T)

9272 **U.S. Department of the Interior. Fish and Wildlife Service (1988).** Endangered and threatened wildlife and plants; determination of *Helonias bullata* (Swamp pink) to be an Endangered species. *Fed. Register*, 53(175): 35076-35080. (L/T)

9273 **White, C.P. (1982).** Endangered and threatened wildlife of the Chesapeake Bay region. Centreville, Tidewater Publishers, 147p. Illus., maps, col. illus. (T)

U.S. - Florida

9274 **Anon. (1979).** Climatron shelters endangered species *Torreya taxifolia*. *Missouri Bot. Gard. Bull.*, 67(5): 9. (G/T)

9275 **Anon. (1981).** *Ilex buswellii* seed. *Holly Lett.*, 72: 6. Threatened with extinction in Florida. (T)

9276 **Anon. (1982).** *Torreya taxifolia*: an endangered tree species. *J. Dendrol.*, 2(3-4): 176. (T)

9277 **Anon. (1982).** Silky camellia is a wild eye-catcher. *Amer. Camellia Yearbook*, 1982: 100-102. Illus. *Stewartia malacodendron*. Threatened and endangered plants of Florida from Division of Plant Industry, Gainesville. (L/T)

9278 **Anon. (1983).** A commendable orchid conservation project. *Florida Orchidist*, 26(4): 156. Illus. *Cyrtopodium punctatum*. (T)

9279 **Anon. (1983).** Wetlands programme. *Oryx*, 17:143. 17,000 acres along the Escambia River. (P)

9280 **Anon. (1984).** Revitalising the Everglades. *Oryx*, 18(2): 111. (P)

U.S. - Florida

9281 **Austin, D.F. (1977).** Endangered species: Florida plants. *Florida Naturalist,* 50(3): 15-21. (T)

9282 **Austin, D.F. (1983).** Endangered plants: lessons in survival. *Bull. Fairchild Trop. Gard.,* 38(3): 32-35. Illus. (T)

9283 **Beck, C.H. (1967).** Rare plants of Florida. *Florida Naturalist,* 40(2): 57-60. (T)

9284 **Benzing, D. (1982).** Everglades Park bromeliads face possible danger. *J. Bromeliad Soc.,* 32(2): 75-78. Illus. (P/T)

9285 **Croat, T.B. (1984).** Rediscovery of a rare monstera (*Monstera gracilis*). *Aroideana,* 7(1): 12-13. Illus. (T)

9286 **Florida Committee on Rare and Endangered Plants and Animals (1978).** Plants. Rare and endangered biota of Florida. Gainesville, FL, Univ. Presses of Florida. 175p. (T)

9287 **Gatewood, S., Hardin, D. (1985).** La Florida: the land of flowers. *Nature Conserv. Newsl.,* 35(5): 6-12. Describes the natural communities & their endangered species. (T)

9288 **George, J.C. (1976).** The battle to save Florida's palms. *Nat. Wildl.,* 14(4): 17-19. (T)

9289 **Hardin, J.W. (1977).** The southeastern United States. *In* Prance, G.T., Elias, T.S., *eds.* Extinction is forever. Proceedings of a symposium at the New York Botanical Garden, 11-13 May 1976. New York, New York Botanical Garden. 36-40. Discusses status of state programmes within the 14 southeastern states of endangered plant species. (L/P/T)

9290 **Harmon, S.A. (1979).** An update on endangered Chapman Rhododendron. *Amer. Hort. Soc. News Views,* 21(6): 7. (T)

9291 **Hemming, E.S. (1979).** Conservation in Florida. *Amer. Nurseryman,* 149(10): 46-49.

9292 **Hendrix, G., Morehead, J. (1983).** Everglades National Park: an imperilled wetland. *Ambio,* 12(3-4): 153-157. Illus., col. illus., map. (P)

9293 **Holmes, J. (1988).** The invasion of the killer plants. *Insight,* 5 December: 58. Invasive species. (P)

9294 **Kral, R. (1979).** Endangered or threatened vascular plant species of the Florida sandscrub. *ASB Bull.,* 26(2): 102. Abstract. (T)

9295 **Langdon, K.R. (1981).** The seriously depleted cowhorn orchid *Cyrtopodium punctatum,* endangered in Florida. *Florida Dept Agric. Nematol. Circ.,* 74 Bot. 13: 2p. (T)

9296 **Lewis, R.R., Cole, D.P., eds (1977).** Proceedings of the Fourth Annual Conference on Restoration of Coastal Vegetation in Florida, Hillsborough Community College, 14 May 1977. Tampa, Hillsborough Community College. ii, 167p. Illus. Maps.

9297 **Little, E.L., jr (1976).** Rare tropical trees of south Florida. Washington, D.C., U.S. Department of Agriculture, Forest Service. Conserv. Res. Rep. 20: 20p. (T)

9298 **Loope, L.L., Avery, G.N. (1979).** A preliminary report on rare plant species in the flora of National Park Services areas of South Florida. South Florida Research Center Rept. M-548. 42p. (P/T)

9299 **McCartney, C. (1985).** Orchids of Florida: the orchids of Everglades National Park. 1. *Amer. Orch. Soc. Bull.,* 54(3): 265-276. Illus., maps. (P)

9300 **McCartney, C. (1985).** Orchids of Florida: the orchids of Everglades National Park. 2. *Amer. Orch. Soc. Bull.,* 54(4): 440-448. Illus. (P)

U.S. - Florida

9301 **McCoy, E.D. (1981).** Rare, threatened, and endangered plant species of southwest Florida and potential OCS activity impacts. Performed for National Coastal Ecosystems Team, Biological Services Program, Fish and Wildlife Service, U.S. Dept of the Interior. Washington, D.C., The Team. 83p. (T)

9302 **Morehead, J.M. (1984).** Attempts to modify significant deterioration of a park's natural resources: Everglades National Park. *In* McNeely, J.A., Miller, K.R., eds. National parks, conservation and development. The role of protected areas in sustaining society. Washington, D.C., Smithsonian Institution Press. 496-502. Maps. (P)

9303 **Norquist, C. (1984).** Savannahs and bogs of the southern U.S. U.S. threatened ecosystems. *Endangered Species Tech. Bull.,* 10(9): 4-5 (T)

9304 **Reed, N.P. (1983).** A saga that may yet lead to a legacy: Everglades National Park. *Bull. Fairchild Trop. Gard.,* 38(3): 8-15. Illus., map. (P)

9305 **Remus, J.Y. (1979).** Status of two endangered Florida species of *Chamaesyce* (Euphorbiaceae). *Florida Scient.,* 42(3): 130-136. (T)

9306 **Savage, T. (1983).** A Georgia station for *Torreya taxifolia* Arn. survives. *Florida Scientist,* 46(1): 62-64. Maps. (T)

9307 **Shalter, R., Dial, S.C. (1984).** Hammock vegetation of Little Talbot Island State Park, Florida. *Bull. Torrey Bot. Club,* 111(4): 494-497.

9308 **Stalter, R., Dial, S. (1984).** Environmental status of the stinking cedar, *Torreya taxifolia. Bartonia,* 50: 40-42. (T)

9309 **Stern, W.L. (1977).** Can we save the native plants of the Florida Keys? *Kampong Notes,* 11(1): 1-4. (T)

9310 **Tolle, P. (1984).** The Everglades and the National Park Service. *Florida Orchidiskt,* 27(1): 36-37. Illus. (P/T)

9311 **U.S. Department of the Interior. Fish and Wildlife Service (1979).** Determination that *Harperocallis flava* is an Endangered species. *Fed. Register,* 44(192): 56862-56863. (L/T)

9312 **U.S. Department of the Interior. Fish and Wildlife Service (1983).** Endangered and threatened wildlife and plants: proposal to determine *Torreya taxifolia* (Florida torreya) as an endangered species. *Fed. Register,* 48(68): 15168-15171. (L/T)

9313 **U.S. Department of the Interior. Fish and Wildlife Service (1983).** Endangered and threatened wildlife and plants: proposed Endangered status for *Cereus robinii* (Key tree cactus). *Fed. Register,* 48(147): 34483-34486. (L/T)

9314 **U.S. Department of the Interior. Fish and Wildlife Service (1983).** Rulemaking actions: May 1983. Disease threatens tree: Endangered status proposed. *Endangered Species Tech. Bull.,* 8(5). Illus. (L/T)

9315 **U.S. Department of the Interior. Fish and Wildlife Service (1984).** Endangered and threatened wildlife and plants: *Ribes echinellum* (Miccosukee gooseberry) proposed to be a Threatened species. *Fed. Register,* 49(171): 34535-34538. (L/T)

9316 **U.S. Department of the Interior. Fish and Wildlife Service (1984).** Endangered and threatened wildlife and plants: final rule to determine *Cereus robinii* (Key tree-cactus) to be an Endangered species. *Fed. Register,* 49(140): 29234-29237. (L/T)

9317 **U.S. Department of the Interior. Fish and Wildlife Service (1984).** Endangered and threatened wildlife and plants: final rule to determine *Torreya taxifolia* (Florida torreya) to be an Endangered species. *Fed. Register,* 49(15): 2783-2786. (L/T)

U.S. - Florida

9318 **U.S. Department of the Interior. Fish and Wildlife Service (1984).** Endangered and threatened wildlife and plants: proposed Endangered status for *Dicerandra immaculata* (Lakela's mint). *Fed. Register*, 49(142): 29632-29634. (L/T)

9319 **U.S. Department of the Interior. Fish and Wildlife Service (1984).** Endangered and threatened wildlife and plants: proposed Endangered status for five Florida pine rockland plants. *Fed. Register*, 49(217): 44507-44512. (L/T)

9320 **U.S. Department of the Interior. Fish and Wildlife Service (1984).** Endangered and threatened wildlife and plants: public hearing and extension of comment period on proposed Threatened status for *Ribes echinellum* (Miccosukee gooseberry). *Fed. Register*, 49(205): 41266-41267. (L/T)

9321 **U.S. Department of the Interior. Fish and Wildlife Service (1984).** Evergreen tree listed as Endangered. *Endangered Species Tech. Bull.*, 9(2): 4-5. Illus. (L/P/T)

9322 **U.S. Department of the Interior. Fish and Wildlife Service (1984).** Four plants in danger of extinction. *Endangered Species Tech. Bull.*, 9(8): 1, 5 Illus. (L/T)

9323 **U.S. Department of the Interior. Fish and Wildlife Service (1984).** Miccosukee Gooseberry proposed for listing as Threatened. *Endangered Species Tech. Bull.*, 9(9): 30. (L/T)

9324 **U.S. Department of the Interior. Fish and Wildlife Service (1984).** Protection becomes final for four plants. *Endangered Species Tech. Bull.*, 9(8): 3-5. Illus. (L/T)

9325 **U.S. Department of the Interior. Fish and Wildlife Service (1985).** Endangered and threatened wildlife and plants: Endangered and Threatened status for five Florida pine rockland plants. *Fed. Register*, 50(138): 29345-29349. *Amorpha crenulata, Euphorbia deltoidea* ssp. *deltoidea, Euphorbia garberi, Galactia smallii, Polygala smallii.* (L/T)

9326 **U.S. Department of the Interior. Fish and Wildlife Service (1985).** Endangered and threatened wildlife and plants: determination of Endangered status for two Florida mints. *Fed. Register*, 50(212): 45621-45624. (T)

9327 **U.S. Department of the Interior. Fish and Wildlife Service (1985).** Endangered and threatened wildlife and plants: determination of Endangered status for: *Cereus eriophorus* var. *fragrans* (Fragrant prickly apple). *Fed. Register*, 50(212): 45618-45620. (L/T)

9328 **U.S. Department of the Interior. Fish and Wildlife Service (1985).** Endangered and threatened wildlife and plants: determination of Threatened status for *Ribes echinellum* (Miccosukee gooseberry). *Fed. Register*, 50(138): 29338-29341. (L/T)

9329 **U.S. Department of the Interior. Fish and Wildlife Service (1985).** Endangered and threatened wildlife and plants: determination of endangered status for *Dicerandra immaculata* (Lakela's mint). *Fed. Register*, 50(94): 20212-20215. (L/T)

9330 **U.S. Department of the Interior. Fish and Wildlife Service (1985).** Endangered and threatened wildlife and plants: proposed Endangered and Threatened status for three Florida shrubs. *Fed. Register*, 50(212): 45634-45637. (L/T)

9331 **U.S. Department of the Interior. Fish and Wildlife Service (1985).** Endangered and threatened wildlife and plants: proposed Endangered status for *Chrysopsis floridana* (Florida golden aster). *Fed. Register*, 50(150): 31629-31632. (L/T)

9332 **U.S. Department of the Interior. Fish and Wildlife Service (1985).** Endangered and threatened wildlife and plants: proposed Endangered status for *Dicerandra frutescens* (Scrub balm) and *Dicerandra cornutissima* (Longspurred balm). *Fed. Register*, 50(61): 12587-12591. (L/T)

9333 **U.S. Department of the Interior. Fish and Wildlife Service (1985).** Endangered and threatened wildlife and plants: proposed Endangered status for *Lindera melissifolia* (Pondberry). *Fed. Register*, 50(156): 32582-32585. (L/T)

U.S. - Florida

9334 **U.S. Department of the Interior. Fish and Wildlife Service (1985).** Endangered and threatened wildlife and plants: proposed rule to list the Fragrant prickly-apple as an Endangered species. *Fed. Register*, 50(44): 9089-9092. *Cereus eriophorus* var. *fragrans*. (L/T)

9335 **U.S. Department of the Interior. Fish and Wildlife Service (1985).** Final protection given to four plants. *Endangered Species Tech. Bull.*, 10(11): 5. *Hoffmannseggia tenella* (Slender rush-pea). *Cereus eriophorus* var. *fragrans* (Fragrant prickly apple). *Dicerandra frutescens* (Scrub mint),. *D. cornutissima* (Longspurred mint). (L/T)

9336 **U.S. Department of the Interior. Fish and Wildlife Service (1985).** Florida plant. *Endangered Species Tech. Bull.*, 10(6): 7. (T)

9337 **U.S. Department of the Interior. Fish and Wildlife Service (1985).** Protection proposed for four plants. *Endangered Species Tech. Bull.*, 10(11): 3-4. *Deeringothamnus pulchellus*, *D. rugelii*, *Asimina tatramera* and *Serianthes nelsonii*. Illus. (L/T)

9338 **U.S. Department of the Interior. Fish and Wildlife Service (1985).** Protection sought for four vulnerable plants. *Endangered Species Tech. Bull.*, 19(9): 1, 8. (L/T)

9339 **U.S. Department of the Interior. Fish and Wildlife Service (1986).** Current status of the Chapman rhododendron. *Endangered Species Tech. Bull.*, 11(3): 8. (T)

9340 **U.S. Department of the Interior. Fish and Wildlife Service (1986).** Endangered and threatened wildlife and plants: Endangered status for *Chrysopsis floridana* (Florida golden aster). *Fed. Register*, 51(95): 17974-17977. (L/T)

9341 **U.S. Department of the Interior. Fish and Wildlife Service (1986).** Endangered and threatened wildlife and plants: Endangered status for *Lupinus aridorum* (Scrub lupin). *Fed. Register*, 51(79): 15514-15517. (L/T)

9342 **U.S. Department of the Interior. Fish and Wildlife Service (1986).** Endangered and threatened wildlife and plants: proposed Endangered or Threatened status for seven Florida scrub plants. *Fed. Register*, 51(69): 12444-12451. (L/T)

9343 **U.S. Department of the Interior. Fish and Wildlife Service (1986).** Endangered and threatened wildlife and plants: proposed Endangered status for *Warea amplexifolia* (Wide-leaf warea). *Fed. Register*, 51(95): 18010-18013. (L/T)

9344 **U.S. Department of the Interior. Fish and Wildlife Service (1988).** Endangered and threatened wildlife and plants; proposal to determine *Campanula robinsiae* (Brooksville bellflower) and *Justicia cooleyi* (Cooley's water-willow) to be an Endangered species. *Fed. Register*, 53(176): 35215-35219. (L/T)

9345 **U.S. Department of the Interior. Fish and Wildlife Service (1988).** Endangered and threatened wildlife and plants; proposal to determine Endangered status for *Liatris ohlingerae* and *Ziziphus celata*. *Fed. Register*, 53(188): 37818-37822. (L/T)

9346 **U.S. Department of the Interior. Fish and Wildlife Service (1989).** Endangered and threatened wildlife and plants; *Dicerandra christmanii* (Garrett's mint) determined to be Endangered. *Fed. Register*, 54(182): 38946. (L/T)

9347 **U.S. Department of the Interior. Fish and Wildlife Service (1989).** Endangered and threatened wildlife and plants; Endangered status for four Florida plants. *Fed. Register*, 54(143): 31190-31196. *Campanula robinsiae*, *Justicia cooleyi*, *Liatris ohlingerae*, *Ziziphus celata*. (L/T)

9348 **U.S. Department of the Interior. Fish and Wildlife Service (1989).** Endangered and threatened wildlife and plants; proposed Endangered status for *Salpingostylis coelestina* (Bartram's ixia). *Fed. Register*, 54(96): 21632-21635. (L/T)

9349 **U.S. Department of the Interior. National Park Service (1979).** Master plan, Everglades National Park, Florida. Atlanta, GA, National Park Service. 2 vols. (P)

U.S. - Florida

9350 **Wallace, S.R. (1989).** Endangered gourd thrives at Bok Tower Gardens. *Plant Conservation*, 4(3): 3, 8. Illus. (G/T)

9351 **Ward, D.B. (1974).** Rare and endangered Florida plants. *Florida Flora Newsl.*, 19. (T)

9352 **Ward, D.B., ed. (?)** Rare and endangered biota of Florida: 5. Plants. Gainesville, University Presses of Florida. 175p. Illus., maps. Series edited by P.C.H. Pritchard. Reviewed in *Threatened Pl. Commit. Newsl.*, 7: 23-24 (1981). (T)

U.S. - Georgia

9353 **Anon. (1979).** *Sarracenia oreophila*. *AABGA Newsl.*, 59: 8. Illus. (T)

9354 **Anon. (1979).** Climatron shelters endangered species *Torreya taxifolia*. *Missouri Bot. Gard. Bull.*, 67(5): 9. (G/T)

9355 **Anon. (1982).** *Torreya taxifolia*: an endangered tree species. *J. Dendrol.*, 2(3-4): 176. (T)

9356 **Anon. (1984).** Georgia's mystery tree. *Georgia Forestry*, 37(4): 6, 14. Illus. (T)

9357 **Bozeman, J.R., Rogers, G.A. (1979).** Notes on the endangered Georgia endemic, *Elliottia racemosa* Muhl. ex Elliott (Ericaceae). *ASB Bull.*, 26(2): 94. Abstract. (T)

9358 **Burbanck, M.P. (1979).** Endangered and threatened species of plants on granite outcrops of Georgia. *ASB Bull.*, 26(2): 101-102. Abstract. (T)

9359 **Georgia Botanical Society (1973).** Rare and endangered plants of Georgia. Tiger, GA, Georgia Botanical Society. (T)

9360 **Haehnle, G.G., Jones, S.M. (1985).** Geographical distribution of *Quercus oglethorpensis*. *Castanea*, 50(1): 26-31. Illus., maps. (T)

9361 **Hardin, J.W. (1977).** The southeastern United States. *In* Prance, G.T., Elias, T.S., *eds.* Extinction is forever. Proceedings of a symposium at the New York Botanical Garden, 11-13 May 1976. New York, New York Botanical Garden. 36-40. Discusses status of state programmes within the 14 southeastern states of endangered plant species. (L/P/T)

9362 **Johnson, A.S., Hillestad, H.O., Shanholtzer, S.F. & G.F. (1974).** An ecological survey of the coastal region of Georgia. National Park Service Scientific Monographs Series No. 3. Washington, D.C., U.S. Govt Printing Office. xv, 233p. Maps, illus.

9363 **Jones, S.M., Dunn, B.A. (1979).** Aspects of the biology of *Shortia galacifolia* (Diapensiaceae). *ASB Bull.*, 26(2): 92. Abstract. (T)

9364 **Leslie, K.A., Burbanck, M.P. (1979).** Vegetation of granitic outcroppings at Kennesaw Mountain, Cobb County, Georgia. *Castanea*, 44(2): 80-87. Map. Kennesaw Mountain National Battlefield Park. (P)

9365 **McCollum, J.L. (1976).** Endangered, threatened and unusual plants - protected plants list. *Outdoors in Georgia*, 5(9): 27-31. (T)

9366 **McCollum, J.L., Ettman, D.R. (1977).** Georgia's protected plants. Atlanta, Resource Planning Section, OPR, Endangered Plant Program. xxv, 64p. Illus., col. illus., maps. 58 species described, dot maps and line drawings. (L/T)

9367 **McCollum, J.L., ed. (1974).** Endangered species of Georgia. Atlanta, GA, Natural Areas Unit, Office of Planning & Research, Georgia Dept of Natural Resources. Proceedings of the 1974 Conference, sponsored by the Georgia Dept of Natural Resources. (T)

9368 **Schilling, T. (1984).** *Elliottia racemosa*. *Garden* (London), 109(9): 349-350. Col. illus. (T)

U.S. - Georgia

9369 **Stalter, R., Dial, S. (1984).** Environmental status of the stinking cedar, *Torreya taxifolia*. *Bartonia*, 50: 40-42. (T)

9370 **U.S. Department of Agriculture. Forest Service (1981).** Sensitive plants of the Francis Marion National Forest. Atlanta, GA, U.S. Department of Agriculture Forest Service, Southern Region. 12p. (P)

9371 **U.S. Department of the Interior. Fish and Wildlife Service (1983).** Endangered and threatened wildlife and plants: proposal to determine *Torreya taxifolia* (Florida torreya) as an endangered species. *Fed. Register*, 48(68): 15168-15171. (L/T)

9372 **U.S. Department of the Interior. Fish and Wildlife Service (1983).** Habitat description notice on green pitcher plant. *Endangered Species Tech. Bull.*, 8(9): 4, 8. (L/T)

9373 **U.S. Department of the Interior. Fish and Wildlife Service (1984).** Endangered and threatened wildlife and plants: final rule to determine *Torreya taxifolia* (Florida torreya) to be an Endangered species. *Fed. Register*, 49(15): 2783-2786. (L/T)

9374 **U.S. Department of the Interior. Fish and Wildlife Service (1984).** Evergreen tree listed as Endangered. *Endangered Species Tech. Bull.*, 9(2): 4-5. Illus. (L/P/T)

9375 **U.S. Department of the Interior. Fish and Wildlife Service (1985).** Endangered and threatened wildlife and plants: determination of Endangered status for *Lindera melissifolia* (Pondberry). *Fed. Register*, 50(147): 27495-27500. (L/T)

9376 **U.S. Department of the Interior. Fish and Wildlife Service (1985).** Endangered and threatened wildlife and plants: proposal to determine *Oxypolis canbyi* (Canby's dropwort) to be an Endangered species. *Fed. Register*, 50(60): 12345-12348. (L/T)

9377 **U.S. Department of the Interior. Fish and Wildlife Service (1985).** Endangered and threatened wildlife and plants: proposed Endangered status for *Lindera melissifolia* (Pondberry). *Fed. Register*, 50(156): 32582-32585. (L/T)

9378 **U.S. Department of the Interior. Fish and Wildlife Service (1985).** Endangered and threatened wildlife and plants: proposed Endangered status for *Scutellaria montana*. *Fed. Register*, 50(219): 46797-46799. (L/T)

9379 **U.S. Department of the Interior. Fish and Wildlife Service (1985).** Protection sought for four vulnerable plants. *Endangered Species Tech. Bull.*, 19(9): 1, 8. (L/T)

9380 **U.S. Department of the Interior. Fish and Wildlife Service (1986).** Endangered and threatened wildlife and plants: determination of *Oxypolis canbyi* (Canby's dropwort) to be an Endangered species. *Fed. Register*, 51(37): 6690-6693. (L/T)

9381 **U.S. Department of the Interior. Fish and Wildlife Service (1986).** Endangered and threatened wildlife and plants: determination of Endangered status for *Scutellaria montana* (Large-flowered skullcap). *Fed. Register*, 51(119): 22521-22524. (L/T)

9382 **U.S. Department of the Interior. Fish and Wildlife Service (1988).** Endangered and threatened wildlife and plants; determination of *Helonias bullata* (Swamp pink) to be an Endangered species. *Fed. Register*, 53(175): 35076-35080. (L/T)

9383 **U.S. Department of the Interior. Fish and Wildlife Service (1988).** Endangered and threatened wildlife and plants; determination of Endangered status for *Ptilimnium nodosum*. *Fed. Register*, 53(188): 37978-37982. (L/T)

9384 **U.S. Department of the Interior. Fish and Wildlife Service (1988).** Endangered and threatened wildlife and plants; determination of Endangered status for *Trillium reliquum* (Relict trillium). *Fed. Register*, 53(64): 10879-10884. (L/T)

U.S. - Georgia

9385 **U.S. Department of the Interior. Fish and Wildlife Service (1988).** Endangered and threatened wildlife and plants; determination of Threatened status for *Marshallia mohrii* (Mohr's Barbara's-buttons). *Fed. Register*, 53(173): 34698-34701. (T)

9386 **U.S. Department of the Interior. Fish and Wildlife Service (1989).** Endangered and threatened wildlife and plants; determination of Endangered status for *Rhus michauxii* (Michaux's sumac). *Fed. Register*, 54(187): 39854-39857. (L/T)

9387 **U.S. Department of the Interior. Fish and Wildlife Service (1989).** Endangered and threatened wildlife and plants; proposed Endangered status for *Rhus michauxii* (Michaux's sumac). *Fed. Register*, 54(4): 441-445. (L/T)

9388 **U.S. Department of the Interior. Fish and Wildlife Service (1989).** Endangered and threatened wildlife and plants; proposed Threatened status for *Sagittaria secundifolia* (Kral's water-plantain). *Fed. Register*, 54(200): 42816-42820. (L/T)

9389 **U.S. Department of the Interior. National Park Service (1979).** Preliminary status of rare plants in Great Smoky Mountains National Park. Management Report No. 25. Atlanta, GA, Southeast Regional Office. 46p. (P/T)

U.S. - Hawaii

9390 **Albert, H. (1986).** Structure of a disturbed forest community replanted with *Eucalyptus robusta* on Wai'alae Ridge, Oahu, Hawaii. *Newsl. Hawaii. Bot. Soc.*, 25(2): 60-69. *Eucalyptus* is invading beyond regenerated area. (P)

9391 **American Horticultural Society (1983).** New plants listed as Endangered. *Amer. Hort.*, 62(1): 8-9. Illus. (T)

9392 **Amerson, A.B. (1971).** The natural history of French Frigate Shoals, northwestern Hawaiian Islands. *Atoll Res. Bull.*, 150: 383p. Threats to Tern Island vegetation by naval air facilities and invading weeds. (P)

9393 **Amerson, A.B., Shelton, P.C. (1976).** The natural history of Johnston Atoll, central Pacific Ocean. *Atoll Res. Bull.*, 192: 479p. Terrestrial vegetation has been heavily disturbed by man. (P)

9394 **Anon. (1954).** Biological control in the Hawaiian Islands. *Pacific Science Assoc. Inf. Bull.*, 6(2): 9-10. Biological control of invasive species. (P)

9395 **Anon. (1973).** Museum miscellany. *Ka 'Elele*, 116-118: 4. Hawaiian threatened plants have been planted at the Bishop Museum, Honolulu in hopes of preserving the species. (G/T)

9396 **Anon. (1977).** The greening of Kauai. *Marathon World*, 14(2): 10-13. Concerns the Pacific Tropical Botanical Gardens role in conserving threatened species. (G)

9397 **Anon. (1978).** Hide-and-seek orchid found. *Bishop Mus. News and Ka 'Elele*, 5(2): 3. (P/T)

9398 **Anon. (1979).** *Kokia cookei* - extinction or survival? *Notes Waimea Arbor.*, 6(1): 2-5. Discusses tissue culture experiments. (G/T)

9399 **Anon. (1979).** Rare and endangered species planted at Waimea Arboretum. *Notes Waimea Arbor.*, 6(2): 7-10. (G/T)

9400 **Anon. (1979).** Tree fern logging on Hawaii. *Oryx*, 15(2): 127.

9401 **Anon. (1981).** *Kokia cookei*: progress report. *Notes Waimea Arbor.*, 8(1): 8. Several graftings of *Kokia cookei* onto *K. drynarioides* have been planted at Waimea. (G/T)

9402 **Anon. (1981).** Conservation of Hawaii's coastal plants. *Notes Waimea Arbor.*, 8(1): 14. Conference on Conserving Hawaii's Coastal Ecosystems, 27 March 1981. (P)

U.S. - Hawaii

9403 **Anon. (1985).** Kauai: the garden island. *Hawaii*, 2(1) (Issue No. 3): 10-15. Threat posed by increased tourism to a relatively unspoiled environment. (P)

9404 **Anon. (1985).** Koke'e logging: "maintenence"? *Elepaio*, 45(12): 132-134. Logging in Koke'e State Park threatens the most diverse mesic forests of Hawaii. (P/T)

9405 **Anon. (1985).** Ohi'a woodchipping double talk. *Elepaio*, 45(12): 132-134. Continuation of logging threatens the last lowland tropical forest. (P)

9406 **Ayensu, E.S., DeFilipps, R.A. (1978).** Endangered and threatened plants of the United States. Washington, D.C., Smithsonian Institution and World Wildlife Fund Inc. 403p. Lists 90 'Extinct', 839 'Endangered' and 1211 'Threatened' taxa for the continental U.S. Also covers Hawaii, Puerto Rico and Virgin Is. (T)

9407 **Baker, J. (1983).** Research on endangered Hawaiian species at the Hawaii Field Research Center. *Newsl. Hawaii. Bot. Soc.*, 20: 5-8. (T)

9408 **Baldwin, P.H., Fagerlund, G.O. (1943).** The effect of cattle grazing on koa reproduction in Hawaii National Park. *Ecology*, 24: 118-122. (P/T)

9409 **Barrau, J. (1960).** The sandalwood tree. *South Pacific Bull.*, 10(4): 39, 63. Recounts the history of decimation; lists species of *Santalum* in Oceania.

9410 **Bender, M., ed. (1984).** Carter's panicgrass listed as endangered. *Endangered Species Tech. Bull.*, 8(11): 7. (L/T)

9411 **Berger, A.J. (1975).** Hawaii's dubious distinction. *Defenders*, 50(6): 491-496. Deforestation; introduced species. (P)

9412 **Berger, A.J. (1977).** Aloha means goodbye. *National Wildlife*, 15(1): 28-35. Deforestation. (P)

9413 **Brockie, R.E., Loope, L.L., Usher, M.B., Hamann, O. (1988).** Biological invasions of island nature reserves. *Biol. Conserv.*, 44(1-2): 9-36. Includes 6 case studies. (P)

9414 **Bryan, L.W. (1973).** Ahinahina. *Newsl. Hawaii. Bot. Soc.*, 12(1): 1-2. (T)

9415 **Carlquist, S. (1965).** Island life: a natural history of the islands of the world. New York, Natural History Press. 451p. Origin, evolution and adaptations of island flora and fauna; Galapagos and Hawaiian Islands well covered.

9416 **Carlquist, S. (1970).** Hawaii: a natural history. New York, Natural History Press. 463p.

9417 **Carlquist, S. (1982).** Hawaii: a museum of evolution. *The Nature Conservancy News*, 32(3): 4-11. Includes threats to flora.

9418 **Carlquist, S. (1983).** Hawaii: a museum of evolution. *Bulletin Pacific Tropical Botanical Garden*, 13(2): 33-39. Includes threats to flora.

9419 **Carlson, N.K. (1973).** The Kamehameha Schools - Bernice Pauahi Bishop Estate and the forests of the Big Island. *Newsl. Hawaii. Bot. Soc.*, 12(3): 16-19.

9420 **Carson, H.L. (1982).** A cloudy future. *Natural History*, 91(12):72. Threats to rain forests.

9421 **Carson, H.L. (1982).** Hawaii: showcase of evolution, an introduction. *Natural History*, 91(12): 16-18.

9422 **Caum, E.L. (1936).** Notes on the flora and fauna of Lehua and Kaula islands. *Occas. Pap. B.P. Bishop Mus.*, 11(21): 3-17. Introduced and invasive species.

9423 **Char, W. (1976).** Field studies of the *Sesbania* complex on the island of Hawaii. *Bull. Pacif. Trop. Bot. Gard.*, 6(2): 41. (T)

U.S. - Hawaii

9424 **Char, W.P., Balakrishnan, N. (1979).** Ewa Plains botanical survey. U.S. Department of the Interior Contract Report. Honolulu, Hawaii, University of Hawaii at Manoa. U.S. Department of the Interior Contract Report. (T)

9425 **Chock, A.K. (1963).** Kokee. *Newsl. Hawaii. Bot. Soc.*, 2(3): 37-39. Threats from introduced and invasive species on Kokee.

9426 **Christensen, C. (1979).** Propagating Kauai's *Brighamia*. *Bull. Pac. Trop. Bot. Gard.*, 9(1): 2-4. (G/T)

9427 **Clapp, R.B., Kridler, E., Fleet, R.B. (1977).** The natural history of Nihoa Island, northwestern Hawaiian Islands. *Atoll Res. Bull.*, 207: 147p. (T)

9428 **Clapp, R.B., Wirtz, W.O. (1975).** The natural history of Lisianski Island, northwestern Hawaiian Islands. *Atoll Res. Bull.*, 186: 196p.

9429 **Conry, P. (1988).** MAS conservation project: recovery of the endangered tree *Serianthes nelsonii. Koko's Call*, 5(8): 1-2. (T)

9430 **Cooray, R.G. (1974).** Stand structure of a montane rain forest on Mauna Loa, Hawaii. Honolulu, Hawaii, University of Hawaii, Island Ecosystems IRP/IBP Hawaii, Technical Report No. 44. 98p. Includes notes on threat caused by introduced species. (T)

9431 **Cory, C. (1986).** Saving endangered plants. *Notes Waimea Arbor.*, 13(1): 5-6. (G/T)

9432 **Craine, C. (1985).** Dangerous and endangered species: a political update on native ecosystems. *Newsl. Hawaii. Bot. Soc.*, 14(1): 13-18. (T)

9433 **Cranwell, L.M. (1984).** Lehua Maka Noe, an endangered bog. *Newsletter Hawaiian Botanical Society*, 23: 3-6.

9434 **Croft, L., Hemmes, D.E., Macneil, J.D. (1976).** Puukohola Heiau National Historic Site plant survey. *Newsl. Hawaii. Bot. Soc.*, 15(4-5): 81-94. (P/T)

9435 **Cuddihy, L.W., Anderson, S.J., Stone, C.P., Smith, C.W. (1986).** A botanical baseline study of forests along the East Rift and Hawaii Volcanoes National Park adjacent to Kahaualea. University of Hawaii Coop. Nat. Park Resources Stud. Unit Tech. Report 61: 180p. (P)

9436 **Degener, O. & I. (1963).** Kaena Point, Oahu. *Newsl. Hawaii. Bot. Soc.*, 2(6): 77-79. (T)

9437 **Degener, O. & I. (1966).** Yes, thank you; we love ferns. *Phytologia*, 13(7): 449-452. Mentions areas where native flora is threatened. (T)

9438 **Degener, O. & I. (1971).** *Pritchardia* and *Cocos* in the Hawaiian Islands. *Phytologia*, 21(5): 320-326. (T)

9439 **Degener, O. & I. (1971).** *Sophora* in Hawaii. *Phytologia*, 21(6): 411-415. (T)

9440 **Degener, O. & I. (1972).** *Wikstroemia pulcherrima* var. *petersonii* Deg. & Deg., from Hawaii. *Phytologia*, 24(2): 151-154. (T)

9441 **Degener, O. & I. (1973).** *Santalum paniculatum* var. *chartaceum* Deg. & Deg. *Phytologia*, 27(3): 145-147. (T)

9442 **Degener, O. & I. (1974).** To save a rare naupaka. *Newsl. Hawaii. Bot. Soc.*, 13(4): 16. (T)

9443 **Degener, O. & I. (1975).** Silverswords and the Blue Data Book. *Not. Waimea Arbor.*, 2(1): 3-6. (T)

9444 **Degener, O. & I. (1976).** *Wikstroemia perdita* Deg. & Deg., an extinct(?) endemic of a paradise lost by exotic primates. *Phytologia*, 34(1): 28-32. (T)

U.S. - Hawaii

9445 **Degener, O. & I. (1977).** *Hibiscadelphus* number KK-HX-1: an international treasure in Hawaii. *Phytologia*, 35(5): 385-396. (P/T)

9446 **Degener, O. & I., Hormann, H. (1969).** *Cyanea carlsonii* Rock and the unnatural distribution of *Sphagnum palustre* L. *Phytologia*, 19(1): 1-3. (T)

9447 **Degener, O. (1977).** Help save the dwindling endemic flora of the Hawaiian Islands at least as herbarium specimens for museums of the world. *Phytologia*, 37(4): 281-284. (T)

9448 **Degener, O., Sunada, K. (1976).** *Argyroxiphium kauense*, the Kau silversword. *Phytologia*, 33(3): 173-177. (T)

9449 **Devaney, D.M., Kelly, M., Lee, P.J., Motteler, L.S. (1976).** Kaneohe: a history of change (1778-1950). Honolulu, Hawaii, Bernice P. Bishop Museum. 271p. (T)

9450 **Donaghho, W.R. (1970).** Destruction of virgin ohia and koa forest on Hawaii by the Division of Forestry. *Elepaio*, 30(7): 67. Deforestation.

9451 **Doria, J.J. (1979).** Haleakala's silversword has a chance. *Nat. Parks Conserv. Mag.*, 53(12): 14-16. Illus. (P/T)

9452 **Duefrene, P. (1984).** The top of Mauna Kea. *Aloha*, 7(4): 62-67. Effect of introduced species; overgrazing.

9453 **Dworsky, S. (1986).** Two in the tropics. *Horticulture*, 64(3): 56-62. Pacific Tropical Botanical Garden and Allerton Gardens conserve rare species. (G/T)

9454 **Elliott, M.E., Hall, E.M. (1977).** Wetlands and wetland vegetation of Hawaii. Fort Shafter, U.S. Army Corps of Engineers, Pacific Ocean Division. 344p. Reports on threats to wetlands.

9455 **Ellshoff, Z.E. (1986).** Symposium on control of introduced plants in native ecosystems of Hawaii: summary of presentations. *Newsl. Hawaii. Bot. Soc.*, 25(3): 79-88.

9456 **Elton, C.S. (1958).** The ecology of invasions by animals and plants. London, Methuen. 181p. Introduced species. (P)

9457 **Fay, J.J. (1978).** Hawaii: extinction unmerciful. *Garden*, 2(4): 22-27. Illus. (T)

9458 **Fay, J.J. (1980).** Endangered and threatened wildlife and plants: proposed endangered status for the 'Ewa Plains 'akoko (*Euphorbia skottsbergii* var. *kalaeloana*). *Fed. Reg.*, 45(171): 58166-58168. (L/T)

9459 **Fay, J.J. (1982).** Endangered and threatened wildlife and plants: determination that *Euphorbia skottsbergii* var. *kalaeloana* ('Ewa Plains 'akoko) is an endangered species. *Fed. Reg.*, 47(164): 36846-36849. (L/T)

9460 **Fay, J.J. (1988).** One man's preserve in Hawaii. *Species*, 10: 26-27. Establishment of 100 acre nature reserve on Kauai containing several threatened plants. (P/T)

9461 **Flanders, G. (1985).** Preserving Hawaii's heritage. *Hawaii*, 2(3): 22-25.

9462 **Flynn, D. (1987).** The Pacific Tropical Botanical Garden: conserving Hawaii's unique flora. *Center Plant Conserv. Newsl.*, 2(3): 3. (G)

9463 **Fosberg, F.R. (1954).** Vanishing island floras and vegetation. IUCN Technical Meeting, Caracas, 1952 (Reports). 5p.

9464 **Fosberg, F.R. (1963).** Disturbance in island ecosystems. *In* Gressitt, J.L., ed. Pacific basin biogeography: a symposium. Honolulu, Hawaii, Museum Press. 557-561.

9465 **Fosberg, F.R. (1967).** Some ecological effects of wild and semi-wild exotic species of vascular plants. *In* IUCN. Towards a new relationship of man and nature in temperate lands, Part III. Changes due to introduced species. Morges, Switzerland, IUCN. 98-109.

U.S. - Hawaii

9466 **Fosberg, F.R. (1971).** Endangered island plants. *Bull. Pac. Trop. Bot. Gard.,* 1(3): 1-7. (P/T)

9467 **Fosberg, F.R. (1972).** The axis deer problem. *Elepaio,* 32(9): 86-88.

9468 **Fosberg, F.R. (1975).** The deflowering of Hawaii. *Nat. Parks Conserv. Mag.,* 49(10): 4-10. (P)

9469 **Fosberg, F.R. (1976).** Endangered species in Hawaii, effect on other resources management. A response. *Newsl. Hawaii. Bot. Soc.,* 15(1): 14-21. (T)

9470 **Fosberg, F.R., Degener, I. (1969).** Hawaii's vanishing native plants. *Honolulu Star Bull.,* 30 December. (T)

9471 **Fosberg, F.R., Herbst, D. (1975).** Rare and endangered species of Hawaiian vascular plants. *Allertonia,* 1(1): 72p. Estimates 70% of flora is threatened; lists 1186 taxa, of which 273 'extinct', 800 'endangered'. (P/T)

9472 **Frome, M. (1986).** Hawaii's heritage remains at risk. *Defenders,* 61(5): 18-19, 44.

9473 **Gagne, W. (1975).** Hawaii's tragic dismemberment. *Defenders Wildl. News,* 50: 461-470.

9474 **Gagne, W.C. (1983).** Nihoa: biological gem of the northwest Hawaiian Islands. *Ka 'Elele,* 10(7): 3-5. (T)

9475 **Gagne, W.C. (1986).** Hawaii's botanic gardens: panacea or Pandora's box in the conservation of Hawaii's native flora. *Newsl. Hawaii. Bot. Soc.,* 25(1): 7-10. (G)

9476 **Gardiner, J.M. (1979).** Silverswords and greenswords from Hawaii. *The Garden* (U.K.), 104(2): 50-54. (T)

9477 **Garnett, W.G. (1989).** Rare plant profile no. 4. *Hibiscus brackenridgei* subsp. *mokuleianus. Bot. Gard. Conserv. News,* 1(4): 47-48. Illus. (G/T)

9478 **Gerrish, G., Mueller-Dombois, D. (1980).** Behavior of native and non-native plants in two tropical rain forests on Oahu, Hawaiian Islands. *Phytocoenologia,* 8(2): 237-295. Introduced species.

9479 **Gerum, S. (1983).** In search of the "wild" bananas. *Not. Waimea Arbor.,* 10(1): 9-11. Illus., map.

9480 **Giffin, J. (1977).** Ecology of the feral pig on Hawai'i Island. *Elepaio,* 37(12): 140-142. Introduced species.

9481 **Gilmartin, A.J. (1970).** First colloquium on rare and endangered species of Hawaii. *Assoc. Trop. Biol. Newsl.,* 22: 1-4. (T)

9482 **Gon, S.M. (1987).** The dunes of Mo'omomi. *Nat. Conserv. News,* 37(1): 14-17.

9483 **Gormley, R. (1984).** Molokai-on the edge. *Aloha,* 7(1): 20-27.

9484 **Gosnell, M. (1976).** The island dilemma. *Internat. Wildl.,* 6(5): 24-35. (T)

9485 **Gourou, P. (1963).** Pressure on island environment. *In* Fosberg, F.R., *ed.* Man's place in the island ecosystem: a symposium. Honolulu, Bishop Museum Press. 207-225.

9486 **Gustafson, R.J. (1979).** Hawaii's unique and vanishing flora - the genesis of an exhibit. *Terra,* 18(2): 3-9. (T)

9487 **Harrison, B.C. (1972).** The vegetation of Waihoi Valley, East Maui. *In* Kjargaard, J.I., *ed.* Scientific report of the Waihoi Valley Project. University of Hawaii. 94-136.

9488 **Hartt, C.E., Neal, M.C. (1940).** The plant ecology of Mauna Kea, Hawaii. *Ecology,* 21(2): 237-266.

U.S. - Hawaii

9489 **Hatheway, W.H. (1952).** Composition of certain native dry forests: Mokuleia, Oahu, Territory of Hawaii. *Ecol. Monogr.*, 22: 153-168. Introduced species.

9490 **Hawaii Volcanoes National Park (1974).** National Park Service silversword restoration project proposal. Hawaii Volcanoes National Park, National Park Service, U.S. Department of the Interior. 15p. (P/T)

9491 **Herbst, D. (1972).** Ohai, a rare and endangered Hawaiian plant. *Bull. Pacif. Trop. Bot. Gard.*, 2(3): 58. (T)

9492 **Herbst, D. (1977).** Endangered Hawaiian plants. *Newsl. Hawaii. Bot. Soc.*, 16(1-2): 22-29. (T)

9493 **Herbst, D., Fay, J.J. (1981).** Proposal to list *Panicum carteri* (Carter's panicgrass) as an endangered species and determine its critical habitat. *Fed. Register*, 46(20): 9976-9979. Map. (L/T)

9494 **Higashino, P.K., Cuddihy, L.W., Anderson, S.J., Stone, C.P. (1988).** Bryophytes and vascular plants of Kipahulu Valley, Haleakala National Park. Cooperative National Park Resources Studies Unit Technical Report No. 65. Honolulu, University of Hawaii at Manoa. 63p. Map. (P/T)

9495 **Higashino, P.K., Cuddihy, L.W., Anderson, S.J., Stone, C.P. (1988).** Checklist of vascular plants of Hawaii Volcanoes National Park. Cooperative National Park Resources Studies Unit Technical Report No. 64. Honolulu, University of Hawaii at Manoa. 81p. Map. (P/T)

9496 **Hirano, R.T. (1973).** Preservation of the Hawaiian flora. *Arbor. Bot. Gard. Bull.*, 7(1): 10-11. (T)

9497 **Hodel, D. (1980).** Notes on *Pritchardia* in Hawaii. *Principes*, 24(2): 65-81. (T)

9498 **Hodel, D.R. (1985).** A new *Pritchardia* from South Kona, Hawaii. *Principes*, 29(1): 31-34. Illus., notes on habitat and population size. (T)

9499 **Holden, C. (1985).** Hawaiian rainforest being felled. *Science*, 228: 1073-1074. Deforestation. (T)

9500 **Holing, D. (1987).** Hawaii: Eden of endemism. *Nat. Conserv. News*, 37(1): 6-13. Illus. (P)

9501 **Hosmer, R.S. (1910).** Kahoolawe Forest Reserve. *Hawaiian For. Agr.*, 7: 264-267. (P)

9502 **Howard, R.A. (1962).** Hawaii - a botanical and horticultural opportunity. *Garden* (U.S.), 12(6): 223-226.

9503 **IUCN Threatened Plants Unit (1985).** The botanic gardens list of rare and threatened species of the Hawaiian Islands. Kew, IUCN Botanic Gardens Co-ordinating Body Report No. 14. 21p. Lists plants in cultivation in botanic gardens. (G/T)

9504 **Jacobi, J.D. (1978).** Vegetation map of the Kau Forest Reserve and adjacent lands, island of Hawaii. Resource Bulletin PSW-16. Berkeley, CA, Pacific Southwest Forest and Range Experiment Station. 1p. Map. Outlines areas of introduced shrub-dominated community. (P)

9505 **Jacobi, J.D. (1981).** Vegetation changes in a subalpine grassland in Hawaii following disturbance by feral pigs. Cooperative National Park Resources Studies Unit, Technical Report 41. University of Hawaii at Manoa. 23p. (P)

9506 **Jenkins, D.W., Ayensu, E.S. (1975).** One-tenth of our plant species may not survive. *Smithsonian*, 5(10): 92-96. Includes discussion on threatened *Rollandia* and *Argyroxiphium*. (T)

9507 **Johnson, M. (1986).** *Brighamia citrina* var. *napaliensis*. *Kew Mag.*, 3(2): 68-72. Col. illus., map. (T)

U.S. - Hawaii

9508 **Juvik, J.O. & S.P. (1984).** Mauna Kea and the myth of multiple use endangered species and mountain management in Hawaii. *Mountain Res. Devel.*, 4(3): 191-202. Illus. Map. (P/T)

9509 **Kikukawa, H.H., LeBarron, R.K. (1971).** Ohia-lehua. *Aloha Aina*, 2(2): 12-13. Describes threat of grazing to *Metrosideros collina* and *Acacia koa* forests. (T)

9510 **Kimura, B.Y., Nagata, K.M. (1980).** Hawaii's vanishing flora. Honolulu, Oriental Publ. Co. 88p. Illus., col. illus. Details on 56 threatened and candidate species. (P/T)

9511 **King, J. (1978).** Hawaii's wildlife - legacy and stewardship. *Elepaio*, 38(11): 122-125. Mentions effects of grazing and introduced species. (T)

9512 **King, W. (1971).** Hawaii: haven for endangered species? *Nat. Parks Conserv. Mag.*, 45(10): 9-13. (T)

9513 **Kirch, P.V. (1982).** The impact of the prehistoric Polynesians on the Hawaiian ecosystem. *Pacific Science*, 36(1): 1-14.

9514 **Knapp, R. (1975).** Vegetation of the Hawaiian Islands. *Newsl. Hawaii. Bot. Soc.*, 14(5): 95-121. Includes changes to vegetation since European discovery.

9515 **Kobayashi, H.K. (1973).** Present status of the ahinahina or silversword, *Argyroxiphium sandwicense* DC. on Haleakala, Maui. *Newsl. Hawaii. Bot. Soc.*, 12(4): 23-26. (P/T)

9516 **Kobayashi, H.K. (1974).** Preliminary investigations on insects affecting the reproductive stage of the silversword (*Argyroxiphium sandwicense* DC., Compositae), Haleakala Crater, Maui, Hawaii. *Proc. Hawaii Ent. Soc.*, 21(3): 397-402. (P/T)

9517 **Konishi, T., Kondo, N., Yoshida, A. (1979).** *Kokia cookei*: extinction or survival? *Not. Waimea Arbor.*, 6(1): 7-14. Illus. (T)

9518 **Kores, P. (1979).** A review of the literature on Hawaiian orchids. *Newsl. Hawaii. Bot. Soc.*, 18(3-5): 34-55. (G/T)

9519 **Kramer, R. (1969).** We're botching conservation! Do you care? *Elepaio*, 29(11): 98-101. Includes effects of feral grazing animals.

9520 **Lamoureux, C.H. (1968).** Should the axis deer be introduced to the island of Hawaii? *Elepaio*, 29(2): 10-15. Introduced species.

9521 **Lamoureux, C.H. (1973).** Conservation problems in Hawaii. *In* Costin, A.B., Groves, K.H., *eds*. Nature conservation in the Pacific. Proc. Symposium A-10, XII Pacific Science Congress, Canberra, August-September 1971. IUCN and Australian Nat. Univ. Press. 315-319.

9522 **Lamoureux, C.H. (1976).** Endangered species in Hawaii, effect on other resource management: a response. *Newsl. Hawaii. Bot. Soc.*, 15(1): 14-21.

9523 **Landgraf, L.K. (1973).** Mauna Kea and Mauna Loa silversword: alive and perpetuating. *Bull. Pac. Trop. Bot. Gard.*, 3(4): 64-66. (T)

9524 **Lasseter, J.S., Gunn, C.R. (1979).** *Vicia menziesii* Sprengel (Fabaceae) rediscovered: its taxonomic relationships. *Pacific Science*, 33(1): 85-101. (T)

9525 **LeBarron, R.K. (1970).** Saving Hawaii's forests. *Aloha Aina*, 1(1): 7-8.

9526 **LeBarron, R.K. (1971).** A forester's point of view. *Aloha Aina*, 2(2): 6-11. Effect of grazing on *Gardenia remyi*. (T)

9527 **LeBarron, R.K. (1971).** Kahoolawe. *Aloha Aina*, 2(2): 16-20. Describes threats to vegetation.

9528 **Lindsey, R. (1986).** Hawaii issue: how much tourism is too much? *The New York Times*, 13 March: A10.

U.S. - Hawaii

9529 **Little, H.P. (1984).** The Nature Conservancy of Hawaii's endangered forest and bird project. *In* McNeely, J.A., Miller, K.R., *eds.* National parks, conservation and development. The role of protected areas in sustaining society. Washington, D.C., Smithsonian Institution Press. 355-358. (P/T)

9530 **Long, Senator O.E. (1960).** Sheep destruction of woodland. *Bull. Conserv. Council Hawaii*, 1(2): 4.

9531 **Loope, L.L., Hamann, O., Stone, C.P. (1988).** Comparative conservation biology of oceanic archipelagoes: Hawaii and Galapagos. *BioScience*, 38(4): 272-282. Illus., maps. Includes discussion on introduced and invasive species. (P)

9532 **Loope, L.L., Stone, C.P. (1984).** Introduced vs. native species in Hawaii: a search for solutions to problems of island biosphere reserves. *In* Unesco/UNEP. Conservation, science and society. Paris, Unesco. 283-288. Contributions to the First International Biosphere Reserve Congress, Minsk, Byelorussia/USSR, 26 September-2 October 1983. (P)

9533 **Mack, J. (1975).** Hawaii's first natural area reserve. *Defenders*, 50(6): 500-503. (P)

9534 **Mangenot, G. (1963).** The effects of man on the plant world. *In* Fosberg, F.R., *ed.* Man's place in the island ecosystem: a symposium. Honolulu, Bishop Museum Press. 117-132. Reprinted 1965.

9535 **Matthiessen, P., Wenkham, R. (1970).** Kipahulu - from cinders to the sea. *Audubon*, 14p. Kipahulu Valley rain forests, Maui.

9536 **McIntire, E.G. (1961).** Hawaiian Islands, with special reference to Kaneohe Bay, Oahu; South Point, Hawaii; Waimea District, Kauai. Riverside, California, University of California. 73p.

9537 **McKinney, J. (1983).** Haleakala and Hawaii volcanoes: from the sublime to the ridiculous. *Islands*, 3(2): 20-32. Discusses problem of introduced species. (P)

9538 **McKinney, J. (1985).** Bombs away, Kahoolawe. *Islands*, 5(1): 10. Destruction caused by 1985 military operations.

9539 **McKinney, J. (1986).** Kauai: a journey through the Garden Isle. *Islands*, 6(2): 38-59.

9540 **Mitchell, F. (1981).** Mouflon sheep and Kau silverswords. *Not. Waimea Arbor.*, 8(1): 6. (T)

9541 **Mitchell, F. (1982).** *Acacia koa* FAD and *Vicia menziesii. Not. Waimea Arbor.*, 9(1): 6-7. (T)

9542 **Mohlenbrock, R.H. (1983).** Where have all the wildflowers gone? New York, Macmillan, and London, Collier Macmillan. 239p. (T)

9543 **Montgomery, S. (1972).** Feral animals. *Botanical Society*, 11(2): 13-16. Impact of introduced species on native vegetation.

9544 **Mooney, H.A., Drake, J.A., eds (1986).** Ecology of biological invasions of North America and Hawaii. New York, Springer-Verlag.

9545 **Morales, P. (1981).** The rain forests of Hawaii. *Pacific Discovery*, 35(5): 1-11. Illus.

9546 **Mueller-Dombois, D. (1967).** Ecological relations in the alpine and subalpine vegetation of Mauna Loa, Hawaii. *J. Indian Bot. Soc.*, 46(6): 403-411. Discusses effects of grazing animals and introduced plants.

9547 **Mueller-Dombois, D. (1973).** Some aspects of island ecosystems analysis. Honolulu, Hawaii, University of Hawaii. 26p. Island Ecosystems IRP/IBP Hawaii, Technical Report No. 19. Includes threat caused by introduced species.

U.S. - Hawaii

9548 **Mueller-Dombois, D. (1980).** The ohia dieback phenomenon in the Hawaiian rain forest. *In* Cairns, J., *ed.* The recovery process in damaged ecosystems. Ann Arbor, Michigan, Ann Arbor Science Publishers, Inc. 153-161.

9549 **Mueller-Dombois, D. (1981).** Understanding Hawaiian forest ecosystems: the key to biological conservation. *In* Mueller-Dombois, D., Bridges, K.W., Carson, H.L., *eds.* Island ecosystems: biological organization in selected Hawaiian communities. US/IBP Synthesis Series No. 15. Stroudsburg, PA, Hutch. Ross. 502-520. Illus. (P)

9550 **Mueller-Dombois, D. (1983).** Population death in Hawaiian plant communities: a causal theory and its successional significance. *Tuexenia*, 3: 117-130.

9551 **Mueller-Dombois, D. (1984).** Classification and mapping of plant communities: a review with emphasis on tropical vegetation. *In* Woodwell, G.M., *ed.* The role of terrestrial vegetation in the global carbon cycle: measurement by remote sensing. Chichester, Wiley. 21-88. Deforestation; vegetation mapping.

9552 **Mueller-Dombois, D. (1984).** Ohi'a dieback in Hawaii: 1984 synthesis and evaluation. Hawaii Botanical Science Paper No. 45. Honolulu, Hawaii, University of Hawaii. 44p.

9553 **Mueller-Dombois, D., Krajina, V.J. (1968).** Comparison of east-flank vegetations on Mauna Loa and Mauna Kea, Hawaii. *Proc. Symposium Recent Advances in Tropical Ecology*, 2: 508-520. Discusses impact of grazing cattle. (T)

9554 **Mueller-Dombois, D., Spatz, G. (1972).** The influence of feral goats on the lowland vegetation in Hawaii Volcanoes National Park. *Phytocoenologia*, 3(1): 1-29. (P)

9555 **Mueller-Dombois, D., ed. (1973).** Island ecosystems stability and evolution subprogram. IBP/IRP Technical Report No.2. Honolulu, Hawaii, University of Hawaii. 262p. Covers threats to native vegetation.

9556 **Mull, M.E. (1975).** Comments on silversword planting project, draft environmental assessment, June 1974 to superintendent G.Bryan, Hawaii Volcanoes National Park, from Mae E. Mull, 31 July 1974. *Elepaio*, 36(4): 45-47. (P/T)

9557 **Mull, M.E. (1978).** Question: should wild sheep be allowed to roam free on Mauna Kea? *Elepaio*, 38(10): 117.

9558 **Munro, G.C. (1933).** Preserving the rare plants of Hawaii. *Bernice P. Bishop Mus. Spec. Publ.*, 21: 26-27. (T)

9559 **Munro, G.C. (1952).** Na Laau Hawaii. *Elepaio*, 13(6): 39-43. Discusses attempts to conserve native vegetation. (T)

9560 **Munro, G.C. (1952).** Revisiting the island of Lanai in 1952. *Elepaio*, 12(10): 62-64. (T)

9561 **Munro, G.C. (1955).** Preserving the rare plants of Hawaii. *Elepaio*, 15(10): 57-58. (T)

9562 **Munro, G.C. (1970).** Axis deer on Molokai and Lanai, circa 1952. *Elepaio*, 31(2): 15-17. Introduction of deer and effects on vegetation.

9563 **Myrhe, S.B. (1970).** Kahoolawe. *Newsl. Hawaii. Bot. Soc.*, 9(4): 21-27. Effects of introduced species.

9564 **Nelson, R.E. (1967).** Records and maps of forest types in Hawaii. U.S. Forest Service Resource Bulletin PSW-8. 22p.

9565 **Nelson, R.E. (1971).** Hawaii's forest resource needs, production potentials, and constraints. Proc. Twelfth Pacific Science Congress. 1. 118.

9566 **Nelson, R.E., Hornibrook, E.M. (1962).** Commercial uses and volume of Hawaiian tree fern. Pacific Southwest Forest and Range Experiment Station, Technical Paper 73. 10p. (T)

U.S. - Hawaii

9567 **Nisbet, I.C.T. (1976).** Pacific follies, or the ravishing of Hawaii. *Tech. Rev.*, 78: 8-9. (T)

9568 **Norris, R. (1986).** The last interstate battle. *Audubon*, 88(6): 46, 48-51.

9569 **Obata, J.K. (1985).** Another noxious melastome? *Oxyspora paniculata. Newsl. Hawaii. Bot. Soc.*, 24: 25-26.

9570 **Obata, J.K. (1985).** The declining forest cover of the Ko'olau summit. *Newsl. Hawaii. Bot. Soc.*, 24: 41-42. (T)

9571 **Obata, J.K. (1986).** The demise of a species: *Urera kaalae. Newsl. Hawaii. Bot. Soc.*, 25(2): 74-75. (T)

9572 **Obata, J.K. (1988).** Rare, threatened and endangered native flora of O'ahu. *Newsl. Hawaii. Bot. Soc.*, 27(2): 39-82. (T)

9573 **Oberhansley, F.R. (1953).** Some conservation problems in Hawaii National Park. *Proc. Seventh Pacific Science Congress*, 4: 652-657. (P)

9574 **Ord, W.M. (1962).** Preservation of plants and wildlife in Hawaii. *Elepaio*, 22(10): 75-77.

9575 **Parman, T. (1975).** An autecological review of *Sophora chrysophylla* in Hawaii. *Newsl. Hawaii. Bot. Soc.*, 14(3): 40-49. (T)

9576 **Perlman, S. (1978).** A rare Hawaiian orchid. *Bull. Pacif. Trop. Bot. Gard.*, 8(1): 19. (T)

9577 **Perlman, S.P. (1979).** *Brighamia* in Hawaii. *Bull. Pacif. Trop. Bot. Gard.*, 9(1): 1-2. (T)

9578 **Powell, E. (1985).** The Mauna Kea silversword, a species on the brink of extinction. *Newsl. Hawaii. Bot. Soc.*, 24: 44-57. (T)

9579 **Powell, G.A. (1982).** Palmae collection. *Notes Waimea Arbor.*, 9(1): 4-5. Illus. (G/T)

9580 **Pung, E. (1971).** Forestry saves koai'a. *Aloha Aina*, 2(2): 25-26. (T)

9581 **Ralph, C.J. (1978).** Hawaiian plant on endangered species list. *Elepaio*, 38(12): 142-143. (T)

9582 **Ralph., C.J., Pearson, A.P., Phillips, D.C. (1981).** Observations on the life history of the endangered Hawaiian vetch (*Vicia menziesii*) (Fabaceae) and its use by birds. *Pac. Sci.*, 34(2): 83-92 (1980 publ. 1981). Illus. (T)

9583 **Rauh, W. (1981).** *Brighamia insignis*, a curious succulent of the lobelia family, from the Hawaiian Islands. *Cact. Succ. J.* (U.S.), 53(5): 219-220. (T)

9584 **Reeser, D.W. (1976).** Successful goat control at Hawaii Volcanoes. *Parks*, 1(2): 14-15. (P/T)

9585 **Richmond, G.B. (1965).** Naturalization of Java *Podocarpus* in Hawaii rain forest. Research Note PSW-76. Berkeley, CA, U.S. Forest Service, Pacific Southwest Forest and Range Experiment Station. Introduced species.

9586 **Scowcroft, P.G. (1971).** Koa: monarch of Hawaiian forests. *Newsl. Hawaiian Bot. Soc.*, 10(3): 23-26. Discusses threat of cattle, logging, weeds and fungal diseases to *Acacia koa.* (T)

9587 **Scowcroft, P.G. (1983).** Tree cover changes in mamane (*Sophora chrysophylla*) forests grazed by sheep and cattle. *Pacific Sci.*, 37(2): 109-119. (T)

9588 **Scowcroft, P.G., Sakai, H.F. (1984).** Stripping of *Acacia koa* bark by rats on Hawaii and Maui. *Pacific Sci.*, 38(1): 80-86. Threat to Hawaii's most valuable timber. (T)

U.S. - Hawaii

9589 **Smathers, G.A. (1969).** Plant succession and recovery in the 1959 Kilauea Iki devastation area, Hawaii Volcanoes National Park. Hawaii, National Park Service, Office of Natural Science Studies. Annual report 1968: 59-72. (P)

9590 **Smathers, G.A., Mueller-Dombois, D. (1974).** Invasion and recovery of vegetation after a volcanic eruption in Hawaii. National Park Service Sci. Monogr. Series No. 5. Washington, D.C., U.S. Govt Printing Office. xiv, 129p. Illus., maps.

9591 **Smith, R. (1989).** Hawaiian paradise - for plants. *Garden*, 13(6): 12-15. Waimea Garden cultivates imperiled Hawaiian plants. (G/T)

9592 **Smithsonian Institution (1975).** Report on endangered and threatened plant species of the United States. Committee on Merchant Marine and Fisheries. Committee on Merchant Marine and Fisheries, Serial No. 94-A. Washington, D.C., U.S. Govt Printing Office. 200p. (L/T)

9593 **Sorensen, J. (1977).** *Andropogon virginicus* (Broomsedge). *Newsl. Hawaiian Bot. Soc.*, 16(1-2): 7-22. Invasive species threat.

9594 **Spatz, G., Mueller-Dombois, D. (1972).** The influence of feral goats on koa (*Acacia koa* Gray) reproduction in Hawaii Volcanoes National Park. Honolulu, University of Hawaii. Island Ecosystems IRP/IBP Hawaii, Tech. Report No. 3: 16p. (P/T)

9595 **Spatz, G., Mueller-Dombois, D. (1973).** The influence of feral goats on koa tree reproduction in Hawaii Volcanoes National Park. *Ecology*, 54(4): 870-876. (P/T)

9596 **Spence, G.E., Montgomery, S.L. (1976).** Ecology of the dryland forest at Kanepu'u Island of Lanai. *Newsl. Hawaiian Bot. Soc.*, 15(4-5): 62-80. Discusses conservation.

9597 **St. John, H. (1947).** The history, present distribution, and abundance of sandalwood on Oahu, Hawaiian Islands: Hawaiian Plant Studies 14. *Pacif. Sci.*, 1(1): 5-20. Maps, illus. (T)

9598 **St. John, H., Corn, C.A. (1981).** Rare endemic plants of the Hawaiian Islands, Book 1. Honolulu, Department of Land and Natural Resources, Division of Forestry and Wildlife. 70p. 68 threatened taxa giving status and threats. (T)

9599 **Stafford, R. (1989).** Wheeler's peperomia and 'ohai: two tropical rarities in the Center's charge. *Plant Conservation*, 4(3): 2-3. Illus. Rare plants covered by the Center of Plant Conservation's programme. (T)

9600 **Stemmermann, L. (1980).** Observations on the genus *Santalum* (Santalaceae) in Hawaii. *Pacific Sci.*, 34(1): 41-54. Relict populations remain after habitat destruction and exploitation. (T)

9601 **Stine, P.A. (1986).** Refuge established for endangered Hawaiian forest birds. *Endangered Species Tech. Bull.*, 11(1): 5. Hakalau Forest National Wildlife Refuge will protect several threatened plants, including *Clermontia, Cyanea, Gouldia* and *Platydesma*. (P/T)

9602 **Stone, C.P. (1983).** Research on exotic and endangered species in Hawaiian National Parks USA. *Pac. Sci. Congr. Proc.*, 15(1-2). 228p. South Pacific Science Assoc. 15th Congress, Dunedin, New Zealand, Feb. 1-11, 1983. (T)

9603 **Stone, C.P., Scott, J.M. (1985).** Hawaii's native ecosystems: importance, conflicts, and suggestions for the future. *In* Stone, C.P., Scott, J.M., *eds*. Hawaii's terrestrial ecosystems: preservation and management. Honolulu, University of Hawaii Press. 495-534.

9604 **Stone, C.P., Scott, J.M., eds (1985).** Hawaii's terrestrial ecosystems: preservation and management. Honolulu, University of Hawaii Press.

U.S. - Hawaii

9605 **Tabata, R.S. (1980).** The native coastal plants of Oahu, Hawaii. *Newsl. Hawaiian Bot. Soc.,* 19: 2-44. Threats to native species discussed in detail; conservation measures needed. (T)

9606 **Tagawa, T.K. (1976).** Endangered species in Hawaii: effect on other resource management. *Newsl. Hawaiian Bot. Soc.,* 15(1): 7-14. (T)

9607 **Taketa, K.H. (1987).** Hawaii's islands of life: a campaign to stem the tide of extinction. *The Nature Conservancy News,* 37(1): 4-5.

9608 **The Nature Conservancy (1983).** The Nature Conservancy in Hawaii. *Nature Conserv. News,* 32(3): 18-23. Discussion of unique areas preserved in Hawaii. (P)

9609 **Titcomb, M. (1969).** Axis deer: welcome or not? *Elepaio,* 30(6): 52-54. Threat to East Molokai forests.

9610 **Titcomb, M. (1969).** The axis deer: impending threat to the Big Island. *Elepaio,* 30(3): 21-25.

9611 **Togawa, T.K. (1976).** Endangered species in Hawaii, effect on other resource management. *Newsl. Hawaii. Bot. Soc.,* 15(1): 7-14. (T)

9612 **Tomich, P.Q. (1965).** A question of values. *Elepaio,* 25(7): 54-55. Discusses impact of rabbits on Manana Island.

9613 **Tomich, P.Q. (1969).** Mammals in Hawaii. Honolulu, Bishop Museum Special Publ. No. 57. Introduced species which now pose a threat to native vegetation.

9614 **Tomich, P.Q. (1972).** The feral goat in Hawaii, with particular reference to problems in national parks. *In* Mueller-Dombois, D., *ed.* Island ecosystems stability and evolution subprogramme. Honolulu, University of Hawaii. IBP/IRP Tech. Report No. 2: 203-204. (P)

9615 **Tomich, P.Q., Wilson, N., Lamoureux, C.H. (1968).** Ecological factors on Manana Island, Hawaii. *Pacific Sci.,* 22: 352-368. Discusses impact of rabbits.

9616 **U.S. Department of the Interior. Fish and Wildlife Service (1979).** Service lists 32 plants. *Endangered Species Tech. Bull.,* 4(11): 1, 5-8. Additions to the U.S. Endangered Species Act. (L/T)

9617 **U.S. Department of the Interior. Fish and Wildlife Service (1980).** 'Ewa Plains 'akoko proposed as endangered. *Endangered Species Tech. Bull.,* 5(10): 5-6. (L/T)

9618 **U.S. Department of the Interior. Fish and Wildlife Service (1982).** Endangered and threatened wildlife and plants: determination that *Euphorbia skottsbergii* var. *kalealoana* (Ewa Plains Akoko) is an Endangered species. *Fed. Register,* 47(164): 36846-36849. (L/T)

9619 **U.S. Department of the Interior. Fish and Wildlife Service (1983).** Endangered and threatened wildlife and plants: proposed Endangered status and critical habitat for *Gouania hillebrandii. Fed. Register,* 48(174): 40407-40411. Maps. (L/T)

9620 **U.S. Department of the Interior. Fish and Wildlife Service (1983).** Endangered and threatened wildlife and plants: proposed Endangered status and critical habitat for *Kokia drynarioides* (hau-hele'ula). *Fed. Register,* 48(177): 40920-40923. Maps. (L/T)

9621 **U.S. Department of the Interior. Fish and Wildlife Service (1983).** Endangered and threatened wildlife and plants: rule to list *Panicum carteri* (Carter's panicgrass) as an Endangered species and determine its critical habitat. *Fed. Register,* 48(198): 46328-46332. (L/T)

9622 **U.S. Department of the Interior. Fish and Wildlife Service (1983).** Three plants proposed as Endangered. *Endangered Species Tech. Bull.,* 8(10): 1, 4-5. Illus. (E/L/T)

U.S. - Hawaii

9623 **U.S. Department of the Interior. Fish and Wildlife Service (1984).** Endangered and threatened wildlife and plants: determination of Endangered status and critical habitat for *Kokia drynarioides* (Koki'o). *Fed. Register*, 49(234): 47397-47401. (L/T)

9624 **U.S. Department of the Interior. Fish and Wildlife Service (1984).** Endangered and threatened wildlife and plants: final rule to list *Bidens cuneata* and *Schiedea adamantis* as Endangered species. *Fed. Register*, 49(34): 6099-6102. (L/T)

9625 **U.S. Department of the Interior. Fish and Wildlife Service (1984).** Endangered and threatened wildlife and plants: final rule to list *Gouldia hillebrandii* as an Endangered species and to designate its critical habitat. *Fed. Register*, 49(219): 44753-44757. (L/T)

9626 **U.S. Department of the Interior. Fish and Wildlife Service (1984).** Endangered and threatened wildlife and plants: proposed Endangered status and critical habitat for *Gardenia brighamii* Mann (Na'u or Hawaiian gardenia). *Fed. Register*, 49(199): 40058-40062. (L/T)

9627 **U.S. Department of the Interior. Fish and Wildlife Service (1984).** Hawaiian Gardenia proposed for Endangered listing. *Endangered Species Tech. Bull.*, 9(11): 5-7. (L/T)

9628 **U.S. Department of the Interior. Fish and Wildlife Service (1984).** Protection given to two rare Hawaiian plants. *Endangered Species Tech. Bull.*, 9(3): 1-7. (L/T)

9629 **U.S. Department of the Interior. Fish and Wildlife Service (1984).** Two animals and one plant added to list of threatened and endangered species. *Endangered Species Tech. Bull.*, 9(12): 5-6. (L/T)

9630 **U.S. Department of the Interior. Fish and Wildlife Service (1985).** Endangered and threatened wildlife and plants: determination of Endangered status for *Gardenia brighamii* (Na'v or Hawaiian gardenia) and withdrawal of proposed designation of critical habitat. *Fed. Register*, 50(162): 33728-33731. (L/T)

9631 **U.S. Department of the Interior. Fish and Wildlife Service (1985).** Endangered and threatened wildlife and plants: proposed Endangered status for *Abutilon menziesii* (Ko'oloa 'ula). *Fed. Register*, 50(136): 28876-28878. (L/T)

9632 **U.S. Department of the Interior. Fish and Wildlife Service (1985).** Endangered and threatened wildlife and plants: proposed Endangered status for *Achyranthes rotundata*. *Fed. Register*, 50(77): 15764-15767. (L/T)

9633 **U.S. Department of the Interior. Fish and Wildlife Service (1985).** Endangered and threatened wildlife and plants: proposed Endangered status for *Argyroxiphium sandwicense* var. *sandwicense* ('ahinahina or Maura Kea silversword). *Fed. Register*, 50(44): 9092-9095. (L/T)

9634 **U.S. Department of the Interior. Fish and Wildlife Service (1985).** Endangered and threatened wildlife and plants: proposed Endangered status for *Santalum freycinetianum* Guad. var. *lanaiense* Rock (Lanai sandalwood or 'iliahi). *Fed. Register*, 50(44): 9086-9089. (L/T)

9635 **U.S. Department of the Interior. Fish and Wildlife Service (1985).** Endangered and threatened wildlife and plants: proposed rule to determine Endangered status for *Hibiscadelphus distans* (Kauai hau kuahiwi). *Fed. Register*, 50(136): 28873-28876. (L/T)

9636 **U.S. Department of the Interior. Fish and Wildlife Service (1985).** Endangered and threatened wildlife and plants: proposed rule to determine Endangered status for *Scaevola coriacea* (Dwarf naupaka). *Fed. Register*, 50(136): 28878-28881. (L/T)

U.S. - Hawaii

9637 **U.S. Department of the Interior. Fish and Wildlife Service (1985).** Endangered and threatened wildlife and plants: public hearing and reopening of comment period on proposed Endangered status for *Achyranthes rotundata*. *Fed. Register*, 50(137): 28959-28960. (L/T)

9638 **U.S. Department of the Interior. Fish and Wildlife Service (1985).** Endangered and threatened wildlife and plants: public hearing and reopening of comment period on proposed Endangered status for *Argyroxiphium sandwicense* var. *sandwicense* (Mauna Kea silversword). *Fed. Register*, 50(110): 24001-24002. (L/T)

9639 **U.S. Department of the Interior. Fish and Wildlife Service (1985).** Endangered classification proposed for four plants. *Endangered Species Tech. Bull.*, 10(8): 1, 4-5. (L/T)

9640 **U.S. Department of the Interior. Fish and Wildlife Service (1985).** Protection sought for four vulnerable plants. *Endangered Species Tech. Bull.*, 19(9): 1, 8. (L/T)

9641 **U.S. Department of the Interior. Fish and Wildlife Service (1985).** Rare Hawaiian tree listed as Endangered. *Endangered Species Tech. Bull.*, 10(1): 1. (L/T)

9642 **U.S. Department of the Interior. Fish and Wildlife Service (1986).** Two plants given final Endangered Species Act protection: Lanai sandalwood or 'iliahi. *Endangered Species Tech. Bull.*, 11(2): 3. (L/T)

9643 **U.S. Department of the Interior. National Park Service (1978).** General management plan, Haleakala National Park, Hawaii. San Francisco, National Park Service. 180p. Endangered species. (P/T)

9644 **Vitousek, P.M., Loope, L.L., Stone, C.P. (1987).** Introduced species in Hawaii: biological effects and opportunities for ecological research. *Trends Ecol. Evol.*, 2(7): 224-227. Illus. Effects of exotic plants and animals. (P)

9645 **Vogl, R.J. (1971).** General ecology of northeast outer slopes of Haleakala Crater, East Maui, Hawaii. *Contrib. Nat. Conserv.*, 6: 1-8. Recommends protection.

9646 **Vogl, R.J., Henrickson, J. (1971).** Vegetation of the alpine bog on East Maui, Hawaii. *Pacific Sci.*, 25(4): 475-483. Grazing may have eliminated some endemic *Lobelia* and *Argyroxiphium*. (T)

9647 **Wagner, J.P. (1985).** The "scandalwood". *Hawaii*, 2(2) (Issue No. 4): 51-52. Exploitation of sandalwood has had devastating effects. (T)

9648 **Wagner, W.H. (1950).** Ferns naturalized in Hawaii. *B.P.Bishop Mus. Occ. Papers*, 20(8): 95-121. Historical data on establishment of invasive ferns.

9649 **Wagner, W.H. (1981).** Ferns in the Hawaiian Islands. *Fiddlehead Forum*, 8(6): 43-44. Includes threatened ferns, and ferns introduced for commercial reasons. (E/T)

9650 **Warner, R.E. (1960).** A forest dies on Mauna Kea. *Pacific Discovery*, 13(2): 6-14. Feral sheep have destroyed "mamane" forest dominated by *Sophora chrysophylla*. Illus. (T)

9651 **Warner, R.E. (1961).** The problem of native forest destruction in Hawaii. *Tenth Pacific Sci. Congr. Abstr.*, 251-252.

9652 **Warner, R.E., ed. (1968).** Scientific report of the Kipahulu Valley expedition. The Nature Conservancy. 184p. Includes status and threats to vegetation.

9653 **Warshauer, F.R. (1977).** The Kalapana extension of Hawaii Volcanoes National Park: its variety, vegetation, and value. *Newsl. Hawaiian Bot. Soc.*, 16(3-4): 57-60. (P)

9654 **Warshauer, F.R., Jacobi, J.D. (1982).** Distribution and status of *Vicia menziesii* Spreng. (Leguminosae): Hawaii's first officially listed endangered plant species. *Biol. Conserv.*, 23(2): 111-126. Maps. (T)

U.S. - Hawaii

9655 **Wenkham, R. (1967).** A Kauai national park. *National Parks Mag.*, 41(234): 4-8. Proposal to establish protected area. (P)

9656 **Whiteaker, L.D. (1983).** The vegetation and environment of the Crater District of Haleakala National Park. *Pacific Sci.*, 37(1): 1-24. (P)

9657 **Wichman, C. (1978).** Limahuli Valley botanical survey. *Bull. Pacific Trop. Bot. Gard.*, 8(1): 1-6. Rare endemics found in PTBG satellite garden. (G/T)

9658 **Winner, W.E., Mooney, H.A. (1985).** Ecology of SO2 resistance. V. Effects of volcanic SO2 on native Hawaiian plants. *Oecologia*, 66(3): 387-393.

9659 **Woolliams, K. (1972).** A report on the endangered species. *Bull. Pacific Trop. Bot. Gard.*, 2(3): 46-49. Data on propagation of 12 species. (G/T)

9660 **Woolliams, K. (1972).** Propagation of endangered tropical plants. *Bull. Pacific Trop. Bot. Gard.*, 2(1): 17-20. (G/T)

9661 **Woolliams, K. (1987).** Garden profile no. 1: Waimea Arboretum and Botanical Garden. *Bot. Gard. Conserv. News*, 1(1): 33-37. Illus. Includes conservation of threatened species. (G/T)

9662 **Woolliams, K., Degener, O. & I. (1980).** *Kokia cookei* Deg. ...then there were two!. *Notes Waimea Arbor.*, 7(1): 2-3. (T)

9663 **Woolliams, K., Degener, O. & I. (1980).** Cooke's kokia again. *Notes Waimea Arbor.*, 7(2): 8-9. (T)

9664 **Woolliams, K.R. (1974).** Endangered species now established in the grounds of Pacific garden. *Bull. Pacific Trop. Bot. Gard.*, 4(2): 33. Lists 26 species. (G/T)

9665 **Woolliams, K.R. (1975).** Propagation (*Sesbania tomentosa*). *Notes Waimea Arbor.*, 2(2): 7-8. (G/T)

9666 **Woolliams, K.R. (1975).** The propagation of Hawaiian endangered species. *Newsl. Hawaii Bot. Soc.*, 14(4): 59-68. (T)

9667 **Woolliams, K.R. (1976).** Propagation (*Chenopodium pekeloi*). *Notes Waimea Arbor.*, 3(1): 5-6. (G/T)

9668 **Woolliams, K.R. (1976).** The propagation of Hawaiian endangered species. *In* Simmons, J.B., *et al.*, *eds.* Conservation of threatened plants. NATO Conference Series 1: Ecology, vol. 1. New York and London, Plenum Press. 73-83. Proc. Conference on the Functions of Living Plant Collections in Conservation and Conservation-Orientated Research and Public Education. (G/T)

9669 **Woolliams, K.R. (1978).** Propagation of some endangered Hawaiian plants at Waimea Arboretum. *Notes Waimea Arbor.*, 5(1): 3-4. *Sophora, Mezoneuron, Lepechinia.* (G/T)

9670 **Woolliams, K.R. (1979).** *Kokia cookei*: extinction or survival? *Notes Waimea Arbor.*, 6(1): 2-5. (T)

9671 **Woolliams, K.R. (1980).** Oahu yellow hibiscus found. *Notes Waimea Arbor.*, 7(1): 9, 12. (T)

9672 **Woolliams, K.R. (1981).** *Kokia cookei* - progress report. *Notes Waimea Arbor.*, 8(1): 8. (G/T)

9673 **Woolliams, K.R. (1981).** *Serianthes nelsonii* - an up-date. *Notes Waimea Arbor.*, 8(1): 8-9. First discovery in 1980 on Rota I. (T)

9674 **Woolliams, K.R. (1982).** *Kokia cookei*: more good news. *Notes Waimea Arbor.*, 9(1): 3-4. (G/T)

9675 **Young, R.A., Popenoe, P. (1916).** Saving the kokia tree. *J. Heredity*, 7(1): 24-28. (T)

U.S. - Idaho

9676 **Chambers, K.L. (1977).** The northwestern United States. *In* Prance, G.T., Elias, T.S., *eds*. Extinction is forever. Proceedings of a symposium at the New York Botanical Garden, 11-13 May 1976. New York, New York Botanical Garden. 45-49. General status of programmes in the Pacific northwest. (L/T)

9677 **Crawford, R.C. (1980).** Ecological investigations and management implications of six northern Idaho endemic plants on the proposed endangered and threatened lists. Moscow, ID, University of Idaho. Masters Thesis. (T)

9678 **Eidemiller, B.J. (1976).** Threatened and endangered plant inventory report for the Shoshone District Bureau of Land Management, Idaho. Irvine, CA, Univ. of Calif., Dept Ecol. & Evol. Biol. 56p. (T)

9679 **Eidemiller, B.J. (1977).** Endangered and threatened plant inventory, Sun Valley ES Area. Report for the Shoshone District Bureau of Land Management, Idaho. Irvine, CA, Univ. of Calif., Dept Ecol. & Evol. Biol. 33p. (T)

9680 **Henderson, D.M., Johnson, F.D., Packard, P., Steele, R. (1977).** Endangered and threatened plants of Idaho - A summary of current knowledge. College of Forestry, Wildlife and Range Sciences Bulletin 21. Moscow, ID, University of Idaho Forest, Wildlife and Range Experiment Station. 161p. Updated in 1983, dot maps and annotations. (T)

9681 **Idaho Natural Areas Council (1983).** Vascular plant species of concern in Idaho. 1983 status change and additions. University of Idaho FWR Bull. No. 34. 1981. Addition. Idaho Natural Areas Council. 20p. (T)

9682 **Idaho Natural Areas Council. Rare and Endangered Plants Technical Committee (1979).** Species of concern in Idaho. Idaho Natural Areas Council. n.p. (T)

9683 **Lesica, P., Allendorf, F., Leary, R., Bilderback, D. (1988).** Lack of genetic diversity within and among populations of an endangered plant, *Howellia aquatilis. Conserv. Biol.,* 2(3): 275-282. En (Sp). Map. (T)

9684 **McMahan, L.R. (1987).** Rare plant conservation by state government: case studies from the western United States. *In* Elias, T.S., *ed.* Conservation and management of rare and endangered plants. Proceedings from a conference, Sacramento, California, 5-8 November 1986. Sacramento, California Native Plant Society. 23-31. Maps. (L/T)

9685 **Phillips, T.A. (1976).** Threatened and endangered plant species location and site description. Idaho, USFS Field Notes on file, Sawtooth Nat. Forest, Twin Falls. 19p. (T)

9686 **Rosentreter, R. (1979).** Endangered, threatened, and uncommon plants inventory report for the Boise District, Bureau of Land Management. Boise, ID, The District? 69 leaves. (T)

9687 **Rosentreter, R. (1983).** Sensitive and uncommon plants inventory report for the Boise District Bureau of Land Management. Boise, ID, Bureau of Land Management. 92p. (T)

9688 **Steele, R., Brunsfeld, S., Henderson, D., Holte, K., Johnson, F., Packard, P. (1981).** Rare and endangered vascular plant species of concern by the Plant Technical Committee of the Idaho Natural Areas Council, Supplement: 1983 Status changes and addition to Bulletin 34. Moscow, Idaho, College of Forestry, Wildlife and Range Sciences Bulletin 34, Univ. of Idaho Forest, Wildlife and Range Experiment Station. (T)

9689 **Steele, R.W. (1975).** A directory of disjunct and endemic plants of central and southern Idaho. Univ. of Idaho, College of Forestry, Infor. Ser. No. 9. 1-26.

9690 **U.S. Congress Senate Committee on Energy & Natural Resources (1979).** River of no return wilderness proposals. Washington, D.C., U.S. Govt Printing Office. Illus., maps. (L/P)

U.S. - Idaho

9691 **U.S. Department of Agriculture. Forest Service (1979).** Hells Canyon National Recreation Area land and resource management plan. Baker, OR, U.S. Department of Agriculture Forest Service. 418p. (P)

9692 **Willey, R. (1978).** Endangered and threatened plant inventory report. Coeur d'Alene, ID, Bureau of Land Management, Coeur d'Alene District. 34p. (T)

U.S. - Illinois

9693 **Bowles, M.L., ed. (1981).** Endangered and threatened vertebrate animals and vascular plants of Illinois. Springfield, Illinois Dept. Conservation. (T)

9694 **Ebinger, J.E., ed. (1981).** Illinois endangered and threatened plant species. *In* Bowles, M.L., *ed.* Endangered and threatened vertebrate animals and vascular plants of Illinois. Springfield, Illinois Dept of Conservation. 72-186. (T)

9695 **Illinois Nature Preserves Commission (1976).** Endangered, vulnerable, rare and extirpated vascular plants in Illinois. Interim list of species. Rockford, IL, Illinois Dept. of Conserv. 13p. (T)

9696 **Illinois Nature Preserves Commission (1976).** Preliminary list of extinct, rare and endangered plants in Illinois. Rockford, IL, Illinois Nature Preserves Commission. (T)

9697 **Madany, M. (1977).** Looking for 'unimproved' land: the Illinois natural areas inventory. *Field Mus. Nat. Hist. Bull.* (Chicago), 48(6): 18-23. Illus.

9698 **McFall, D.W. (1984).** Vascular plants of the Manito Gravel Prairie, Tazewell County, Illinois. *Trans. Illinois State Acad. Sci.*, 77(1/2): 9-14. Maps. (P/T)

9699 **Mohlenbrock, R.H. (1977).** The midwestern United States. *In* Prance, G.T., Elias, T.S., *eds.* Extinction is forever. Proceedings of a symposium at the New York Botanical Garden, 11-13 May 1976. New York, New York Botanical Garden. 41-44. Discusses the status of legal protection for threatened and endangered species of plants which occur in the midwestern states. (L/T)

9700 **Natural Land Institute (1978).** Semi-final list of endangered and threatened plants. Rockford, IL, Natural Land Institute. n.p. (T)

9701 **Paulson, G.A. (1976).** Preliminary list of extinct, rare, and endangered plants in Illinois. Illinois Natural Preserves Commission, Natural History Survey, Ill. State Mus. 7p. (T)

9702 **Paulson, G.A., Schwegman, J. (1976).** Endangered, vulnerable, rare and extirpated vascular plants in Illinois - Interim list of species. Rockford and Springfield, Illinois Nature Preserves Commission and Department of Conservation. 189p. and appendices. Descriptions and dot maps. (T)

9703 **Peck, J.H. (1979).** Compilatory list of the rare, threatened, and endangered plants of the five states of the Upper Mississippi River: Illinois, Iowa, Minnesota, Missouri, and Wisconsin. Contributions from University of Wisconsin-La Crosse Herbarium No. 26. La Crosse, WI, Univ. of Wisconsin-La Crosse. 48p. (T)

9704 **Price, S., Rogers, G. (1988).** Rare plant profile no. 3. *Boltonia decurrens* (Compositae): a midwestern U.S. riverine species in decline. *Bot. Gard. Conserv. News*, 1(3): 46-49. Illus., map. (T)

9705 **Sheviak, C.J. (1978).** Semi-final list of endangered and threatened plants. Rockford, IL, Natural Land Institute. n.p. Photocopy. (T)

9706 **U.S. Department of the Interior. Fish and Wildlife Service (1985).** Endangered and threatened wildlife and plants: proposal to determine *Lespedeza leptostachya* to be a Threatened species. *Fed. Register*, 50(235): 49967-49969. (L/T)

U.S. - Illinois

9707 **U.S. Department of the Interior. Fish and Wildlife Service (1988).** Endangered and threatened wildlife and plants; determination of Threatened status for *Asclepias meadii* (Mead's milkweed). *Fed. Register*, 53(170): 33992-33995. (L/T)

9708 **U.S. Department of the Interior. Fish and Wildlife Service (1988).** Endangered and threatened wildlife and plants; determination of Threatened status for *Boltonia decurrens* (Decurrent false aster). *Fed. Register*, 53(219): 45858-45861. (L/T)

9709 **U.S. Department of the Interior. Fish and Wildlife Service (1988).** Endangered and threatened wildlife and plants; determination of Threatened status for *Cirsium pitcheri. Fed. Register*, 53(137): 27137-27141. (L/T)

9710 **U.S. Department of the Interior. Fish and Wildlife Service (1988).** Endangered and threatened wildlife and plants; determination of Threatened status for *Hymenoxys acaulis* var. *glabra* (Lakeside daisy). *Fed. Register*, 53(121): 23742-23745. (L/T)

9711 **U.S. Department of the Interior. Fish and Wildlife Service (1988).** Endangered and threatened wildlife and plants; proposal to determine *Boltonia decurrens* (Decurrent false aster) to be a Threatened species. *Fed. Register*, 53(37): 5598-5601. (L/T)

U.S. - Indiana

9712 **Anon. (1980).** Bill to expand Indiana Dunes clears house without fight. *Nat. Parks Conserv. Mag.*, 54(1): 16-19. (L/P)

9713 **Bacone, J.A., Hedge, C.L. (1980).** A preliminary list of endangered and threatened vascular plants in Indiana. *Proc. Indiana Acad Sci.*, 89: 359-371. (T)

9714 **Barnes, W.B. (1975).** Rare and endangered plants in Indiana. Indianapolis, IN, Division of Nature Preserves, Indiana Dept. of Natural Resources. 12p. Mimeographed. (T)

9715 **Crovello, T.J., Keller, C. (1981).** The Indiana Biological Survey and rare plant data: an unending synthesis. *In* Morse, L.E., Henifin, M.S., *eds*. Rare plant conservation: geographical data organization. New York, New York Botanical Garden. 133-147. (T)

9716 **Gray, L.M., Bacone, J.A. (1980).** A floristic inventory of Hemlock Bluff Nature Preserve, Jackson County, Indiana. *Proc. Indiana Acad. Sci.*, 89: 372-379. (P)

9717 **Lindsey, A.A., ed. (1966).** Natural features of Indiana. Indianapois, IN, Indiana Academy of Sciences. 597p. (P)

9718 **Lindsey, A.A., et al. (1969).** Natural areas in Indiana and their preservation. Lafayette, IN, Indiana Natural Areas Survey. 594p. (P)

9719 **Mohlenbrock, R.H. (1977).** The midwestern United States. *In* Prance, G.T., Elias, T.S., *eds*. Extinction is forever. Proceedings of a symposium at the New York Botanical Garden, 11-13 May 1976. New York, New York Botanical Garden. 41-44. Discusses the status of legal protection for threatened and endangered species of plants which occur in the midwestern states. (L/T)

9720 **U.S. Department of the Interior. Fish and Wildlife Service (1988).** Endangered and threatened wildlife and plants; determination of Threatened status for *Asclepias meadii* (Mead's milkweed). *Fed. Register*, 53(170): 33992-33995. (L/T)

9721 **U.S. Department of the Interior. Fish and Wildlife Service (1988).** Endangered and threatened wildlife and plants; determination of Threatened status for *Cirsium pitcheri. Fed. Register*, 53(137): 27137-27141. (L/T)

U.S. - Iowa

9722 **Carroll, S., Miller, R.L., Whitson, P.D. (1984).** Status of four orchid species at Silver Lake Fen complex. *Proc. Iowa Acad. Sci.*, 91(4): 132-139. Illus., maps. Wetlands. (T)

9723 **Gilly, C., McDonald, M. (1936).** Rare and unusual plants from southeastern Iowa. *Proc. Iowa Acad. Sci.*, 43: 143-149. (T)

9724 **Mohlenbrock, R.H. (1977).** The midwestern United States. *In* Prance, G.T., Elias, T.S., *eds.* Extinction is forever. Proceedings of a symposium at the New York Botanical Garden, 11-13 May 1976. New York, New York Botanical Garden. 41-44. Discusses the status of legal protection for threatened and endangered species of plants which occur in the midwestern states. (L/T)

9725 **Peck, J.H. (1979).** Compilatory list of the rare, threatened, and endangered plants of the five states of the Upper Mississippi River: Illinois, Iowa, Minnesota, Missouri, and Wisconsin. Contributions from University of Wisconsin-La Crosse Herbarium No. 26. La Crosse, WI, Univ. of Wisconsin-La Crosse. 48p. (T)

9726 **Roosa, D.M., Eilers, L.J. (1978).** Endangered and threatened Iowa vascular plants. State Preserves Advisory Board Special Report 5. Des Moines, State Conservation. 93p. Descriptions and Iowa Endangered Species Act. (T)

9727 **Thorne, R.F. (1953).** Notes on rare Iowa plants. *Iowa Acad. Sci.*, 60: 260-274. (T)

9728 **Thorne, R.F. (1956).** Notes on rare Iowa plants II. *Iowa Acad. Sci.*, 63: 214-227. (T)

9729 **U.S. Department of the Interior. Fish and Wildlife Service (1985).** Endangered and threatened wildlife and plants: proposal to determine *Lespedeza leptostachya* to be a Threatened species. *Fed. Register*, 50(235): 49967-49969. (L/T)

9730 **U.S. Department of the Interior. Fish and Wildlife Service (1988).** Endangered and threatened wildlife and plants; determination of Threatened status for *Asclepias meadii* (Mead's milkweed). *Fed. Register*, 53(170): 33992-33995. (L/T)

9731 **Vander Zee, D. (1979).** The vascular flora of Gitchie Manitou State Preserve, Lyon County, Iowa. *Proc. Iowa Acad. Sci.*, 86(2): 66-75. Illus., maps. (P)

9732 **Watson, W.C. (1983).** Survey of seven endangered plant species in Iowa. *Proc. Iowa Acad. Sci.*, 90(1): 56. Abstract. Includes *Asclepias lanuginosa*. (T)

9733 **Zee, D. van der (1979).** The vascular flora of Gitchie Manitou State Preserve, Lyon County, Iowa. *Proc. Iowa Acad. Sci.*, 86(2): 66-75. Map. (P)

U.S. - Kansas

9734 **Baird, J. (1985).** A return to prairie. *Geogr. Mag.*, 57(11): 582-583. Semi-arid zones, farming systems and soil conservation. (T)

9735 **Johnston, B.C., et al. (1980).** Proposed and recommended threatened and endangered plant species of the Forest Service Rocky Mountain Region: an illustrated guide to certain species in Colorado, Wyoming, South Dakota, Nebraska and Kansas. Denver, CO, U.S. Fish and Wildlife Service. 164p. (T)

9736 **McGregor, R.L. (1977).** Rare native vascular plants of Kansas. Lawrence, Kansas, Technical Publications of the State Biological survey of Kansas 5. 44p. 114 species described. (T)

9737 **Mohlenbrock, R.H. (1977).** The midwestern United States. *In* Prance, G.T., Elias, T.S., *eds.* Extinction is forever. Proceedings of a symposium at the New York Botanical Garden, 11-13 May 1976. New York, New York Botanical Garden. 41-44. Discusses the status of legal protection for threatened and endangered species of plants which occur in the midwestern states. (L/T)

9738 **U.S. Department of the Interior. Fish and Wildlife Service (1988).** Endangered and threatened wildlife and plants; determination of Threatened status for *Asclepias meadii* (Mead's milkweed). *Fed. Register*, 53(170): 33992-33995. (L/T)

U.S. - Kentucky

9739 **Atwood, J.T. (1985).** The range of *Cypripedium kentuckiense. Amer. Orch. Soc. Bull.,* 54(10): 1197-1199. Illus., maps. (T)

9740 **Babcock, J.V. (1977).** Endangered plants and animals of Kentucky. Institute for Mining and Minerals Research Publ. no. IMMR25-GR4-77. Lexington, KY, Univ. of Kentucky. 128p. (T)

9741 **Baskin, J.M. & C.C. (1977).** *Leavenworthia torulosa* Gray: an endangered plant species in Kentucky. *Castanea,* 42(1): 15-17. (T)

9742 **Baskin, J.M. & C.C. (1984).** Rediscovery of the rare Kentucky endemic *Solidago shortii* T. & G. in Fleming and Nicholas Counties. *Trans. Kentucky Acad. Sci.,* 45(3/4): 159. (T)

9743 **Baskin, J.M. & C.C. (1985).** A floristic study of a cedar glade in Blue Licks Battlefield State Park, Kentucky. *Castanea,* 50(1): 19-25. Illus. *Juniperus virginiana.*

9744 **Branson, B.A., et al. (1981).** Endangered, threatened and rare animals and plants of Kentucky. *Trans. Kentucky Acad. Sci.,* 42: 77-89. (T)

9745 **Chester, E.S. (1982).** Notes on some rare Kentucky USA vascular plants including 3 additions to the state flora. *Trans. Kentucky Acad. Sci.,* 43(1-2): 91. (T)

9746 **Cranfill, R., Medley, M.E. (1985).** Taxonomy, distribution and rarity of *Leavenworthia* and *Lesquerella* (Brassicaceae) in Kentucky. *Sida: contributions to botany* (Dallas, TX, Mahler), 11(2): 189-199. *Lesquerella* taxonomy, keys, geographical distribution, endangered species. (T)

9747 **Hardin, J.W. (1977).** The southeastern United States. *In* Prance, G.T., Elias, T.S., *eds.* Extinction is forever. Proceedings of a symposium at the New York Botanical Garden, 11-13 May 1976. New York, New York Botanical Garden. 36-40. Discusses status of state programmes within the 14 southeastern states of endangered plant species. (L/P/T)

9748 **Kentucky Department of Fish and Wildlife Resources (?)** A list of rare, threatened, and endangered flora in Kentucky. Frankfort, KY, Kentucky Dept of Fish and Wildlife Resources. (T)

9749 **Parker, W., Dixon, L. (1980).** Endangered and threatened wildlife of Kentucky, North Carolina, South Carolina and Tennessee. Raleigh, NC, North Carolina Agricultural Extension Service. 116p. (T)

9750 **U.S. Department of the Interior. Fish and Wildlife Service (1984).** Endangered and threatened wildlife and plants: proposed Endangered status for *Solidago shortii* (Short's goldenrod). *Fed. Register,* 49(198): 39873-39876. (L/T)

9751 **U.S. Department of the Interior. Fish and Wildlife Service (1984).** Kentucky plant proposed as Endangered. *Endangered Species Tech. Bull.,* 9(11): 3-4. (L/T)

9752 **U.S. Department of the Interior. Fish and Wildlife Service (1985).** Endangered and threatened wildlife and plants: Endangered status for *Solidago shortii* (Short's goldenrod). *Fed. Register,* 50(172): 36085-36089. (L/T)

9753 **U.S. Department of the Interior. Fish and Wildlife Service (1988).** Endangered and threatened wildlife and plants; determination of Threatened status for *Solidago albopilosa* (White-haired goldenrod). *Fed. Register,* 53(67): 11612-11615. (L/T)

U.S. - Louisiana

9754 **Atwood, J.T. (1985).** The range of *Cypripedium kentuckiense. Amer. Orch. Soc. Bull.,* 54(10): 1197-1199. Illus., maps. (T)

9755 **Balogh, P. (1976).** Notes on rare or endangered species of Louisiana. *In* A vegetational survey Barksdale Air Force Base-an environmental assessment. 343. (T)

U.S. - Louisiana

9756 **Craig, N.J., Turner, R.E., Day, J.W., jr (1980).** Wetland losses and their consequences in coastal Louisiana. *Zeitschr. Geomorph., Suppl.*, 34: 224-241. Illus. (P)

9757 **Curry, M.G. (1976).** Rare vascular plants of Louisiana. Metairie, Louisiana, VTN Louisiana Inc. (T)

9758 **Haehnle, G.G., Jones, S.M. (1985).** Geographical distribution of *Quercus oglethorpensis*. *Castanea*, 50(1): 26-31. Illus., maps. (T)

9759 **Hardin, J.W. (1977).** The southeastern United States. *In* Prance, G.T., Elias, T.S., *eds.* Extinction is forever. Proceedings of a symposium at the New York Botanical Garden, 11-13 May 1976. New York, New York Botanical Garden. 36-40. Discusses status of state programmes within the 14 southeastern states of endangered plant species. (L/P/T)

9760 **Meeks, G., jr (1980).** A Louisiana swamp story. *Planning (ASPO)*, 46(2): 12-15. (P)

9761 **U.S. Department of the Interior. Fish and Wildlife Service (1985).** Endangered and threatened wildlife and plants: proposed Endangered status for *Lindera melissifolia* (Pondberry). *Fed. Register*, 50(156): 32582-32585. (L/T)

U.S. - Maine

9762 **Conkling, P.W., Drury, W.H., Leonard, R.E., eds (1984).** People and islands: resource management issues for islands in the Gulf of Maine. Maine Island Institute. Illus.

9763 **Countryman, W.D. (1977).** The northeastern United States. *In* Prance, G.T., Elias, T.S., *eds.* Extinction is forever. Proceedings of a symposium at the New York Botanical Garden, 11-13 May 1976. New York, New York Botanical Garden. 30-35. Legislation regarding conservation (primarily of plants and natural areas) in the northeastern U.S. (L/P/T)

9764 **Deis, R. (1979).** Maine's endangered plants. *Down East*, May: 48-51. (T)

9765 **Eastman, L.M. (1978).** Rare and endangered vascular plant species in Maine. Cambridge, MA, New England Botanical Club. 33p. (T)

9766 **Eastman, L.M. (1980).** Abstract: the rare and endangered species in Maine. *Rhodora*, 82(829): 191-192. (T)

9767 **Field, K.G., Primack, R.B. (1980).** Comparison of the characteristics of rare, common and endemic plant species in New England USA. *In* The University of British Columbia. Second International Congress of Systematic and Evolutionary Biology, Vancouver, B.C. Canada, July 17-24, 1980. Vancouver, B.C., Univ. of British Columbia. (T)

9768 **Foy, L. (1977).** Protection of endangered plants in New England: lists, laws and activities. Framingham, MA, New England Wild Flower Society, Inc. (Various paging). (L/T)

9769 **Gawler, S.C. (1982).** An annotated list of Maine's rare vascular plants. Augusta, ME, State Planning Office. 68p. (T)

9770 **Gawler, S.C. (1982).** Arethus (*Arethusa bulbosa* L.): a rare orchid in Maine and its relevance to the Critical Areas Program. Maine Critical Areas Program Planning Report No. 76. Augusta, Maine State Planning Office. 75p. (P/T)

9771 **Gawler, S.C., Waller, D.M., Menges, E.S. (1987).** Environmental factors affecting establishment and growth of *Pedicularis furbishiae*, a rare endemic of the St John River Valley, Maine. *Bull. Torrey Bot. Club*, 114(3): 280-292. (T)

9772 **Lannon, M.M.C.S. (1978).** Three Thousand Islands, their future? *Nat. Conserv. News*, 28(4): 12-15. Illus., col. illus., map. Maine coast.

U.S. - Maine

9773 **Maine State Planning Office Critical Areas Program (1981).** Rare vascular plants of Maine. Augusta, ME, State Planning Office. 656p. (T)

9774 **New England Botanical Club. Natural Areas Criteria Committee (1972).** Guidelines and criteria for the evaluation of natural areas. A report prepared for the New England Natural Resources Center by the Natural Areas Criteria Committee of the New England Botanical Club Inc. (P)

9775 **Tryon, R.M., ed. (1980).** Symposium on rare and endangered plant species in New England, Cambridge, Mass., 1979. The proceedings of the symposium "Rare and Endangered Plant Species in New England," 4-5 May 1979. *Rhodora* (Cambridge, MA, New England Botanical Garden), 82(829): 237p. (T)

9776 **Tyler, H.R., jr, Gawler, S.C. (1980).** The botanical aspect of Maine's critical areas program. *Rhodora*, 82(829): 207-225. Map. (P)

9777 **U.S. Environmental Protection Agency, Region 1 (1982).** New England Wetland Flora and Wetland Protection Laws. Washington, D.C., U.S. Govt Print. Off. 172p. Illus. (L/P)

9778 **Vickery, B.St.John & P.D. (1983).** New England notes. Note on the status of *Agalinis maritima* (Raf.) Raf. in Maine. *Rhodora*, 85(842): 267-269. (T)

U.S. - Maryland

9779 **Broome, C.R., Tucker, A.O., Reveal, J.L., Dill, N.H. (1979).** Rare and endangered vascular plant species in Maryland. Newton Corner, MA, U.S. Fish and Wildlife Service. 64p. List of 237 species and annotations. (T)

9780 **Hardin, J.W. (1977).** The southeastern United States. *In* Prance, G.T., Elias, T.S., *eds.* Extinction is forever. Proceedings of a symposium at the New York Botanical Garden, 11-13 May 1976. New York, New York Botanical Garden. 36-40. Discusses status of state programmes within the 14 southeastern states of endangered plant species. (L/P/T)

9781 **Klockner, W. (1986).** Rediscovered plant is almost lost again. *TNC (Maryland Chapter) Newsl.*, 10(2): 5. (T)

9782 **Little, S., Mohr, J.J. (1979).** Reestablishing understory plants in overused wooded areas of Maryland State Parks. U.S. Department of Agriculture, Forest Service Res. Pap., 431. Washington, D.C., U.S. Govt Print. Off. 9. Illus. (P)

9783 **Maryland Natural Heritage Program (1984).** Threatened and endangered plants and animals of Maryland. Annapolis, Maryland, Maryland Dept of Natural Resources. 476p. Proceedings of symposium held 3-4 September 1981, including 7 papers on endangered plants and habitats. (T)

9784 **Morse, L.E. (1980).** Working list of special plants. Arlington, VA, Maryland Natural Heritage Program. 15p.

9785 **Reveal, J.L., Broome, C.R. (1981).** Minor nomenclatural and distributional notes on Maryland vascular plants with comments on the State's proposed endangered and threatened species. *Castanea*, 46(1): 50-82. (T)

9786 **Reveal, J.L., Broome, C.R. (1982).** Comments on Maryland's proposed endangered and threatened vascular plants. *Castanea*, 47(2): 191-200. (T)

9787 **Riefner, R.E. (1981).** Notes on some proposed rare and endangered vascular plant species in Maryland. *Phytologia*, 47(5): 397-403. (T)

9788 **Riefner, R.E., jr, Hill, S.R. (1983).** Notes on infrequent and threatened plants of Maryland including new state records. *Castanea*, 48(2): 117-137. 47 species suggested for monitoring on the Maryland list of endangered and threatened plants, and 27 changes in status from the previously published list. (T)

U.S. - Maryland

9789 **Sipple, W.S., Klockner, W.A. (1980).** A unique wetland in Maryland. *Castanea*, 45(1): 60-69. Map.

9790 **U.S. Department of the Interior. Fish and Wildlife Service (1985).** Endangered and threatened wildlife and plants: proposal to determine *Oxypolis canbyi* (Canby's dropwort) to be an Endangered species. *Fed. Register*, 50(60): 12345-12348. (L/T)

9791 **U.S. Department of the Interior. Fish and Wildlife Service (1986).** Endangered and threatened wildlife and plants: determination of *Oxypolis canbyi* (Canby's dropwort) to be an Endangered species. *Fed. Register*, 51(37): 6690-6693. (L/T)

9792 **U.S. Department of the Interior. Fish and Wildlife Service (1988).** Endangered and threatened wildlife and plants; determination of *Helonias bullata* (Swamp pink) to be an Endangered species. *Fed. Register*, 53(175): 35076-35080. (L/T)

9793 **U.S. Department of the Interior. Fish and Wildlife Service (1989).** Sandplain gerardia (*Agalinis acuta*) recovery plan. Newton Corner, MA, U.S. Department of the Interior, Fish and Wildlife Service. 47p. Illus., map. (L/T)

9794 **Walton, S. (1982).** Chesapeake Bay: threats to ecological stability. *BioScience*, 32(11): 843-844. Map.

9795 **White, C.P. (1982).** Endangered and threatened wildlife of the Chesapeake Bay region. Centreville, Tidewater Publishers, 147p. Illus., maps, col. illus. (T)

U.S. - Massachusetts

9796 **Brumback, W.E. (1986).** Endangered plants at the Garden in the Woods: problems and possibilities. *Arnoldia*, 46(3): 33-35. Illus. The garden of the New England Wild Flower Society. (G/T)

9797 **Coddington, J., Field, K.G. (1978).** Candidate species for a list of rare or declining vascular plants of Massachusetts. Cambridge, MA, New England Botanical Club. 52p. (T)

9798 **Coddington, J., Field, K.G. (1978).** Rare and endangered vascular plant species in Massachusetts. Newton Corner, MA, U.S. Fish and Wildlife Service, New England Botanical Club. ii, 62p. Map. Annotated list of 243 species with country distributions, threats and habitats. (T)

9799 **Countryman, W.D. (1977).** The northeastern United States. *In* Prance, G.T., Elias, T.S., *eds*. Extinction is forever. Proceedings of a symposium at the New York Botanical Garden, 11-13 May 1976. New York, New York Botanical Garden. 30-35. Legislation regarding conservation (primarily of plants and natural areas) in the northeastern U.S. (L/P/T)

9800 **Field, K.G., Coddington, J. (1980).** Rare plant species in Massachusetts. *Rhodora*, 82(829): 151-162. (T)

9801 **Field, K.G., Primack, R.B. (1980).** Comparison of the characteristics of rare, common and endemic plant species in New England USA. *In* The University of British Columbia. Second International Congress of Systematic and Evolutionary Biology, Vancouver, B.C. Canada, July 17-24, 1980. Vancouver, B.C., Univ. of British Columbia. (T)

9802 **Fisher, M.N., Buttrick, S.C. (1980).** Massachusetts Natural Heritage Program. *Rhodora*, 82(829): 227-237.

9803 **Foy, L. (1977).** Protection of endangered plants in New England: lists, laws and activities. Framingham, MA, New England Wild Flower Society, Inc. (Various paging). (L/T)

9804 **Levering, D.F. (1988).** The changing flora of the Boston Harbor Islands. *Arnoldia*, 48(3): 18-23.

U.S. - Massachusetts

9805 **McDonnell, M.J. (1981).** Trampling effects on coastal dune vegetation in the Parker River National Wildlife Refuge, Massachusetts, USA. *Biol. Conserv.*, 21(4): 289-301. Map. (P)

9806 **New England Botanical Club. Natural Areas Criteria Committee (1972).** Guidelines and criteria for the evaluation of natural areas. A report prepared for the New England Natural Resources Center by the Natural Areas Criteria Committee of the New England Botanical Club Inc. (P)

9807 **Nicholson, R.G. (1986).** To the Arks with rabbitbane: plant conservation at the Arnold Arboretum. *Arnoldia*, 46(3): 23-25. (G/T)

9808 **Sorrie, B.A. (1983).** Native plants for special considerations in Massachusetts. Boston, Massachusetts Natural Heritage Program, Div. of Fisheries and Wildlife. 14p. List of 250 rare plant species. (T)

9809 **Tryon, R.M., ed. (1980).** Symposium on rare and endangered plant species in New England, Cambridge, Mass., 1979. The proceedings of the symposium "Rare and Endangered Plant Species in New England," 4-5 May 1979. *Rhodora* (Cambridge, MA, New England Botanical Garden), 82(829): 237p. (T)

9810 **U.S. Department of Agriculture. Soil Conservation Service (1975).** Threatened species of Massachusetts. Amherst, MA, U.S. Department of Agriculture Soil Conservation Service. (T)

9811 **U.S. Department of the Interior. Fish and Wildlife Service (1988).** Endangered and threatened wildlife and plants; determination of *Agalinis acuta* (Sandplain gerardia) to be an Endangered species. *Fed. Register*, 53(173): 34701-34705. (L/T)

9812 **U.S. Department of the Interior. Fish and Wildlife Service (1989).** Sandplain gerardia (*Agalinis acuta*) recovery plan. Newton Corner, MA, U.S. Department of the Interior, Fish and Wildlife Service. 47p. Illus., map. (L/T)

9813 **U.S. Environmental Protection Agency, Region 1 (1982).** New England Wetland Flora and Wetland Protection Laws. Washington, D.C., U.S. Govt Print. Off. 172p. Illus. (L/P)

9814 **Wheeler, R.P. (1975).** Ashumet Holly Reservation and Wildlife Sanctuary, East Falmouth, Massachusetts. *In* Holly Society of America. Proceedings of the 52nd meeting, 11-15 November 1975. 16-17. (P)

U.S. - Michigan

9815 **Barnes, B.V. (1989).** Newly discovered birch ranks among rarest of rare. *Center for Plant Conservation*, 4(2): 1, 8. Murray birch, *Betula alleghaniensis*. (T)

9816 **Beaman, J.H. (1977).** Commentary on endangered and threatened plants in Michigan. *Michigan Bot.*, 16(3): 110-122. (T)

9817 **Beaman, J.H., et al. (1985).** Endangered and threatened vascular plants in Michigan: 2. Third biennial review proposed list. *Michigan Bot.*, 24(3): 99-116. (T)

9818 **Crispin, S.R. (1980).** Nature preserves in Michigan 1920-1979. *Mich. Bot.*, 19(3): 99-242. (P)

9819 **Fernald, M.L. (1935).** Critical plants of the upper Great Lakes region of Ontario and Michigan. *Rhodora*, 37: 197-222, 238-262, 272-301, 324-341. (T)

9820 **Janke, R.A., McKaig, D., Raymond, R. (1978).** Comparison of presettlement and modern upland boreal forests on Isle Royale National Park. *For. Sci.*, 24(1): 115-121. (P)

9821 **Kohring, M.A. (1981).** Saving Michigans (U.S.A.) railroad strip prairies. *Ohio Biol. Surv. Biol. Not.* 15: 150-151. Illus. Maps.

U.S. - Michigan

9822 **Michigan. Department of Natural Resources (?)** Michigan's endangered and threatened species program, 1976-1978. Department of Natural Resources. 30p. (T)

9823 **Mohlenbrock, R.H. (1977).** The midwestern United States. *In* Prance, G.T., Elias, T.S., *eds.* Extinction is forever. Proceedings of a symposium at the New York Botanical Garden, 11-13 May 1976. New York, New York Botanical Garden. 41-44. Discusses the status of legal protection for threatened and endangered species of plants which occur in the midwestern states. (L/T)

9824 **Schmaltz, N.J. (1979).** Academia gets involved in Michigan forest conservation. *Michigan Academician*, 12(1): 25-46.

9825 **Toner, M. (1985).** Is this the chestnut's last stand? *Nat. Wildlife*, 23(6): 24-27. Michigan's efforts to save this species. (T)

9826 **U.S. Department of the Interior. Fish and Wildlife Service (1988).** Endangered and threatened wildlife and plants; determination of Threatened status for *Cirsium pitcheri. Fed. Register*, 53(137): 27137-27141. (L/T)

9827 **U.S. Department of the Interior. Fish and Wildlife Service (1988).** Endangered and threatened wildlife and plants; determination of Threatened status for *Solidago houghtonii* (Houghton's goldenrod). *Fed. Register*, 53(137): 27134-27137. (L/T)

9828 **U.S. Department of the Interior. Fish and Wildlife Service (1988).** Endangered and threatened wildlife and plants; proposal to determine *Iris lacustris* (Dwarf lake iris) to be a Threatened species. *Fed. Register*, 52(233): 46334-46336. (L/T)

9829 **U.S. Department of the Interior. Fish and Wildlife Service (1989).** Endangered and threatened wildlife and plants; Endangered status for *Mimulus glabratus* var. *michiganensis* (Michigan monkey flower). *Fed. Register*, 54(189): 40454-40458. (L/T)

9830 **U.S. Department of the Interior. Fish and Wildlife Service (1989).** Endangered and threatened wildlife and plants; Threatened status for *Phyllitis scolopendrium* var. *americana* (American hart's-tongue). *Fed. Register*, 54(134): 29726-29730. (L/T)

9831 **Voss, E.G. (1965).** Some rare and interesting aquatic vascular plans of northern Michigan, with special reference to Cusino Lake (Schoolcraft Co.). *Michigan Bot.*, 4: 11-25. (T)

9832 **Voss, E.G. (1972).** The state of things. *Michigan Acad.*, 5(1): 1-7. (T)

9833 **Wagner, W.H., Voss, E.G., Beaman, J.H., Bourdo, E.A., Case, F.W., Churchill, J.A., Thompson, P.W. (1977).** Endangered, threatened, and rare vascular plants in Michigan. *Michigan Bot.*, 16: 99-110. (T)

9834 **Yakes, N. (1977).** Plants. *In* A guide to the recorded distribution of endangered, threatened and rare species in Michigan. Denver, CO, Fish and Wildlife Reference Service, Denver Public Library. 55-61. Annotated reference list and directory of local nature organizations and field stations. (B/T)

U.S. - Minnesota

9835 **Herman, K.D. (1981).** Special plant species in the Chippewa National Forest, Minnesota: Final Report. Kalamazoo, MI, Gove Associates. 159p.

9836 **Johnson, A.G., Pauley, S.S. (1967).** Rare native trees of Minnesota. *Minnesota Sci.*, 23(4): 27-31. (T)

9837 **Minnesota Deptartment of Natural Resources (1975).** The uncommon ones: animals and plants which merit special consideration and management. St. Paul, MN, Minnesota Dept of Natural Resources. 32p. (T)

U.S. - Minnesota

9838 **Mohlenbrock, R.H. (1977).** The midwestern United States. *In* Prance, G.T., Elias, T.S., *eds.* Extinction is forever. Proceedings of a symposium at the New York Botanical Garden, 11-13 May 1976. New York, New York Botanical Garden. 41-44. Discusses the status of legal protection for threatened and endangered species of plants which occur in the midwestern states. (L/T)

9839 **Morley, T. (1972).** Rare or endangered plants of Minnesota with the counties in which they have been found. Minneapolis, MN, Univ. of Minn., Bot. Dept. 13p. Mimeographed. (T)

9840 **Morley, T. (1978).** Distribution and rarity of *Erythronium propullans* of Minnesota, with comments on certain distinguishing features. *Phytologia*, 40(5): 381-389. Map. (T)

9841 **Moyle, J.B. (1974).** Minnesota animals and plants in need of special consideration with suggestions for management. Spec. Publ. 104. MN Div. Fish & Wildlife Service. 26p. (T)

9842 **Peck, J.H. (1979).** Compilatory list of the rare, threatened, and endangered plants of the five states of the Upper Mississippi River: Illinois, Iowa, Minnesota, Missouri, and Wisconsin. Contributions from University of Wisconsin-La Crosse Herbarium No. 26. La Crosse, WI, Univ. of Wisconsin-La Crosse. 48p. (T)

9843 **Ray, J. (1978).** Saving Minnesota's natural diversity (preservation of areas which are rare or which represent the pre-European settlement landscape). *Minn. Hortic.*, 106(8): 228-232. Map. (P)

9844 **U.S. Department of the Interior. Fish and Wildlife Service (1985).** Endangered and threatened wildlife and plants: proposal to determine *Erythronium propullans* (Minnesota trout lily) to be an Endangered species. *Fed. Register*, 50(86): 18893-18895. (L/T)

9845 **U.S. Department of the Interior. Fish and Wildlife Service (1985).** Endangered and threatened wildlife and plants: proposal to determine *Lespedeza leptostachya* to be a Threatened species. *Fed. Register*, 50(235): 49967-49969. (L/T)

9846 **U.S. Department of the Interior. Fish and Wildlife Service (1986).** Endangered and threatened wildlife and plants: determination of Endangered status for *Erythronium propullans* (Minnesota trout lily). *Fed. Register*, 51(58): 10521-10523. (L/T)

9847 **U.S. Department of the Interior. National Park Service (1980).** Master plan, Voyageurs National Park, Minnesota. Denver, CO, National Park Service. 328p. (P)

9848 **U.S. Department of the Interior. National Park Service (1980).** Wilderness recommendation, Voyageurs National Park, St. Louis & Koochiching Counties, Minnesota. Denver, CO, Dept of the Interior, National Park Service. (P)

U.S. - Mississippi

9849 **Atwood, J.T. (1985).** The range of *Cypripedium kentuckiense. Amer. Orch. Soc. Bull.*, 54(10): 1197-1199. Illus., maps. (T)

9850 **Hardin, J.W. (1977).** The southeastern United States. *In* Prance, G.T., Elias, T.S., *eds.* Extinction is forever. Proceedings of a symposium at the New York Botanical Garden, 11-13 May 1976. New York, New York Botanical Garden. 36-40. Discusses status of state programmes within the 14 southeastern states of endangered plant species. (L/P/T)

9851 **McDearman, W. (1981).** Threatened, rare and plants of special concern in the Desoto National Forest, Mississippi, U.S.A. *J. Missouri Acad. Sci.*, 26(suppl.): 13. Abstract. (P/T)

U.S. - Mississippi

9852 **Miller, W.F., Carter, B.D. (1979).** Rational land use decision-making: the Natchez State Park. *Rem. Sens. Envir.*, 8(1): 25-38. Illus. (P)

9853 **Pullen, T.M. (1975).** Rare and endangered plant species in Mississippi. Department of Biology, University of Mississippi. (T)

9854 **U.S. Department of the Interior. Fish and Wildlife Service (1985).** Endangered and threatened wildlife and plants: determination of Endangered status for *Lindera melissifolia* (Pondberry). *Fed. Register*, 50(147): 27495-27500. (L/T)

9855 **U.S. Department of the Interior. Fish and Wildlife Service (1985).** Endangered and threatened wildlife and plants: proposed Endangered status for *Lindera melissifolia* (Pondberry). *Fed. Register*, 50(156): 32582-32585. (L/T)

9856 **U.S. Department of the Interior. Fish and Wildlife Service (1985).** Protection sought for four vulnerable plants. *Endangered Species Tech. Bull.*, 19(9): 1, 8. (L/T)

U.S. - Missouri

9857 **Anon. (1987).** Rare *Boltonia* rediscovered by Missouri botanical staff. *Wildflower*, 3(2): 43. (T)

9858 **Dunn, D.B., Knauer, D.F. (1975).** Plant introductions by waterfowl to Mingo National Wildlife Refuge, Missouri. *Trans. Missouri Acad. Sci.*, 9: 27-28. (P)

9859 **Holt, F.T., et al. (1974).** Rare and endangered species of Missouri. Jefferson City, MO, Missouri Dept of Conservation. 77p. (T)

9860 **Hornberger, K.L. (1981).** Rare and endangered species of Roaring River State Park, Barry County, Missouri. *Trans. Missouri Acad. Sci.*, 15: 61-63. (P/T)

9861 **Lewis, W.H. (1981).** Wild American ginseng: *Panax quinquefolium* L. - conserve this forest resource in Missouri. Jefferson City, Missouri Dept of Conservation. 4p. Illus. (E)

9862 **Missouri Department of Conservation (1974).** Rare and endangered species of Missouri. Jefferson City, MO, Missouri Dept of Conservation. 77p. (T)

9863 **Mohlenbrock, R.H. (1977).** The midwestern United States. *In* Prance, G.T., Elias, T.S., *eds.* Extinction is forever. Proceedings of a symposium at the New York Botanical Garden, 11-13 May 1976. New York, New York Botanical Garden. 41-44. Discusses the status of legal protection for threatened and endangered species of plants which occur in the midwestern states. (L/T)

9864 **Morgan, S. (1984).** Select rare and endangered plants of Missouri. Jefferson City, Missouri, Missouri Dept of Conservation. 29p. (T)

9865 **Nordstrom, G.R., Pflieger, W.L., Sadler, K.C., Lewis, W.H. (1977).** Rare and endangered species of Missouri. St. Louis, U.S. Department of Agriculture, Missouri Department of Conservation. 130p. (T)

9866 **Orzell, S.L. (1983).** Notes on rare and endangered Missouri fen plants. *Trans. Missouri Acad. Sci.*, 17: 67-71. Wetlands. (T)

9867 **Orzell, S.L. (1984).** Additional notes on rare, endangered and unusual Missouri fen plants. *Trans. Missouri Acad. Sci.*, 18: 13-16. (T)

9868 **Peck, J.H. (1979).** Compilatory list of the rare, threatened, and endangered plants of the five states of the Upper Mississippi River: Illinois, Iowa, Minnesota, Missouri, and Wisconsin. Contributions from University of Wisconsin-La Crosse Herbarium No. 26. La Crosse, WI, Univ. of Wisconsin-La Crosse. 48p. (T)

9869 **Pickering, J. (1989).** A collection of rare species from Missouri and surrounding states, displayed at the Missouri Botanical Garden. Missouri Botanical Garden. 4p. For "The Genetics of Rare Plant Conservation: A Conference on Integrated Strategies for Conservation and Management 9-11 March 1989". (G/T)

U.S. - Missouri

9870 **Price, S., Rogers, G. (1988).** Rare plant profile no. 3. *Boltonia decurrens* (Compositae): a midwestern U.S. riverine species in decline. *Bot. Gard. Conserv. News*, 1(3): 46-49. Illus., map. (T)

9871 **Roedmer, B.J., Hamilton, D.A., Evans, K.E. (1978).** Rare plants of the Ozark Plateau: a field identification guide. St Paul, North Central Forest Experiment Station, U.S. Department of Agriculture Forest Service. 238p. Illus., map. (T)

9872 **Rogers, G. (1987).** Missouri grows rarest plants of its region. *Center for Plant Conserv., 2(1): 2, 8.* Illus. (G/T)

9873 **Skinner, B.R., Probasco, G.E., Samson, F.B. (1983).** Environmental requirements of three threatened plants on limestone glades in southern Missouri. *Biol. Conserv.*, 25(1): 63-73. Map. *Penstemon cobaea, Stenosiphon linifolius, Centaurium texense.* (T)

9874 **U.S. Department of the Interior. Fish and Wildlife Service (1985).** Endangered and threatened wildlife and plants: determination of Endangered status for *Lindera melissifolia* (Pondberry). *Fed. Register*, 50(147): 27495-27500. (L/T)

9875 **U.S. Department of the Interior. Fish and Wildlife Service (1985).** Endangered and threatened wildlife and plants: proposed Endangered status for *Lindera melissifolia* (Pondberry). *Fed. Register*, 50(156): 32582-32585. (L/T)

9876 **U.S. Department of the Interior. Fish and Wildlife Service (1985).** Protection sought for four vulnerable plants. *Endangered Species Tech. Bull.*, 19(9): 1, 8. (L/T)

9877 **U.S. Department of the Interior. Fish and Wildlife Service (1986).** Endangered and threatened wildlife and plants: proposal to determine *Lesquerella filiformis* (Missouri bladderpod) to be an Endangered species. *Fed. Register*, 51(66): 11874-11877. (L/T)

9878 **U.S. Department of the Interior. Fish and Wildlife Service (1988).** Endangered and threatened wildlife and plants; determination of Threatened status for *Asclepias meadii* (Mead's milkweed). *Fed. Register*, 53(170): 33992-33995. (L/T)

9879 **U.S. Department of the Interior. Fish and Wildlife Service (1988).** Endangered and threatened wildlife and plants; determination of Threatened status for *Boltonia decurrens* (Decurrent false aster). *Fed. Register*, 53(219): 45858-45861. (L/T)

9880 **U.S. Department of the Interior. Fish and Wildlife Service (1988).** Endangered and threatened wildlife and plants; proposal to determine *Boltonia decurrens* (Decurrent false aster) to be a Threatened species. *Fed. Register*, 53(37): 5598-5601. (L/T)

9881 **Vassar, J.N., Henke, G.A., Blakeley, C. (1981).** Prairie restoration in north-central Missouri, U.S.A. *Ohio Biol. Surv. Biol. Not.* 15: 197-199. Illus. Maps.

9882 **Wilson, J.H., Lewis, W.H. (1979).** The Missouri ginseng conservation program. *In* Proceedings of the First National Ginseng Conference. Kentucky, Governor's Council on Agriculture. 35-42. Maps. *Panax quinquefolium.* (E)

U.S. - Montana

9883 **Carlson, C.E., Dewey, J.E. (1971).** Environmental pollution by fluorides in Flathead National Forest and Glacier National Park. Missoula, MT, U.S. Department of Agriculture, Forest Service, Division of State and Private Forestry, Forest Insect and Disease Branch. 57p. Illus. (P)

9884 **Chadwick, D.H. (1980).** Cutting Glacier to size. *Nat. Parks Conserv. Mag.*, 54(8): 12-17. Illus. map. Glacier National Park. (P)

U.S. - Montana

9885 **Chambers, K.L. (1977).** The northwestern United States. *In* Prance, G.T., Elias, T.S., *eds.* Extinction is forever. Proceedings of a symposium at the New York Botanical Garden, 11-13 May 1976. New York, New York Botanical Garden. 45-49. General status of programmes in the Pacific northwest. (L/T)

9886 **Johnson, F.D., Crawford, R.C. (1978).** Ecology and distribution of six species of sensitive plants. Report to U.S. Department of Agriculture Forest Service. Missoula, MT. n.p.

9887 **Lesica, P., Allendorf, F., Leary, R., Bilderback, D. (1988).** Lack of genetic diversity within and among populations of an endangered plant, *Howellia aquatilis. Conserv. Biol.*, 2(3): 275-282. En (Sp). Map. (T)

9888 **Lesica, P., Moore, G., Peterson, K.M., Rumely, J.H. (1984).** Vascular plants of limited distributions. Monograph No. 2. *Montana Acad. Sci.*, 43: 1-61. (T)

9889 **McMahan, L.R. (1987).** Rare plant conservation by state government: case studies from the western United States. *In* Elias, T.S., *ed.* Conservation and management of rare and endangered plants. Proceedings from a conference, Sacramento, California, 5-8 November 1986. Sacramento, California Native Plant Society. 23-31. Maps. (L/T)

9890 **Rousch, K. (1987).** Pine Butte Swamp: a plum of a preserve. *Nat. Conserv. Mag.*, 37(4): 18-21. (P)

9891 **Schuyler, A.E. (1980):** *Carex chordorrhiza* in Glacier National Park, Montana. *Rhodora*, 82(831): 519. (P)

9892 **U.S. Department of Agriculture. Forest Service (1979).** Land-management plan, Bull River - Clark Fork planning unit, Kootenai National Forest. Libby, U.S. Department of Agriculture Forest Service. 216p. (P)

9893 **U.S. Department of Agriculture. Forest Service (1979).** Land-management plan, Keeler planning unit, Kootenai National Forest. Missoula, U.S. Department of Agriculture Forest Service. 177p. (P)

9894 **U.S. Department of the Interior. Fish and Wildlife Service (1980).** Management of Charles M. Russell National Wildlife Refuge. Denver, U.S. Dept of the Interior. Fish and Wildlife Service. 219p. (P)

9895 **Watson, T.J. (1976).** An evaluation of putatively threatened or endangered species from the Montana flora. Univ. of Montana. 31p. (T)

9896 **Wiersma, G.B., Davidson, C.I., Breckenridge, R.P., Binda, R.E., Hull, L.C., Herrmann, R. (1984).** Integrated global monitoring in mixed forest biosphere reserves. *In* Unesco-UNEP. Conservation, science and society. Paris, Unesco. 395-4-3. Contributions to the First International Biosphere Reserve Congress, Minsk, Byelorussia/USSR, 26 September - 2 October 1983. Map. (P)

9897 **Wiersma, G.B., Davidson, C.I., Mizell, S.A., Breckenridge, R.P., Binda, R.E., Hull, L.C., Herrmann, R. (1984).** Integrated monitoring in mixed forest biosphere reserves. Washington, U.S. Department of Agriculture, Forest Service General Te chnical Report INT No. 173. Intermountain Forest and Range Experiment Station. 29-39. Illus., maps. (P)

U.S. - Nebraska

9898 **Johnston, B.C., et al. (1980).** Proposed and recommended threatened and endangered plant species of the Forest Service Rocky Mountain Region: an illustrated guide to certain species in Colorado, Wyoming, South Dakota, Nebraska and Kansas. Denver, CO, U.S. Fish and Wildlife Service. 164p. (T)

U.S. - Nebraska

9899 **Mohlenbrock, R.H. (1977).** The midwestern United States. *In* Prance, G.T., Elias, T.S., *eds.* Extinction is forever. Proceedings of a symposium at the New York Botanical Garden, 11-13 May 1976. New York, New York Botanical Garden. 41-44. Discusses the status of legal protection for threatened and endangered species of plants which occur in the midwestern states. (L/T)

9900 **U.S. Department of Agriculture. Soil Conservation Service (1975).** Threatened and endangered species of vascular plants in Nebraska. Lincoln, U.S. Department of Agriculture Soil Conservation Service. (T)

9901 **U.S. Department of the Interior. Fish and Wildlife Service (1986).** Endangered and threatened wildlife and plants: proposal to determine *Penstemon haydenii* to be an Endangered species. *Fed. Register*, 51(82): 15929-15932. (L/T)

U.S. - Nevada

9902 **Beatley, J.C. (1977).** Endangered plant species of the Nevada Test Site, Ash Meadows, and South Central Nevada; threatened species; addendum. Department of Biological Sciences, University of Cincinnati. 150p. Dot map, photos and descriptions. (T)

9903 **Bender, M., ed. (1984).** Eight more Ash Meadows species proposed as endangered. *Endangered Species Tech. Bull.*, 8(11): 8-10. (L/T)

9904 **Cochrane, S. (1979).** Status of endangered and threatened plant species on Nevada Test Site - a survey. Parts 1 and 2. Goleta, CA, EG&G Inc. EGG 1183-2356. S-646-R. Appendix C: Collection records for the taxa considered. (T)

9905 **Gunn, P. (1982).** Vernal pools and rare plants of Haystack Mountain. *Fremontia*, 9(4): 3-10. Illus. (T)

9906 **Loope, L.L., Sanchez, P.G., Tarr, P.W., Loope, W.L., Anderson, R.L. (1988).** Biological invasions of arid land nature reserves. *Biol. Conserv.*, 44(1-2): 95-118. Invasive species. Includes 5 case studies. (P)

9907 **Mooney, M., Pinzl, A., comps (1984).** Sensitive plant list for Nevada. Reno, Nevada, March 1984, Threatened and Endangered Plant Workshop. 8p. (T)

9908 **Mozingo, H., William, M. (1980).** Threatened and endangered plants of Nevada; an illustrated manual. Portland, Oregon, U.S. Fish and Wildlife Service. (T)

9909 **Nevada Threatened and Endangered Plant Workshop (1983).** Summary of February 11, 1983 Meeting. Nevada Threatened and Endangered Plant Workshop. 8p. (T)

9910 **Northern Nevada Native Plant Society, Rare Plant Committee (1978).** Nevada's T/E plant map book. Carson City, NV, Nevada State Museum. 1 v. Chiefly maps. 1978? (T)

9911 **Pinzl, A. (1979).** Nevada's endangered plant program. *Fremontia*, 7(3): 21-22. Illus. (T)

9912 **Pinzl, A. coord. (1983).** Sensitive plant list for Nevada. Carson City, Nevada Threatened and Endangered Plant Workshop. 7p. Summary of February 11, 1983 meeting. (T)

9913 **Reveal, J.L. (1977).** The western United States. *In* Prance, G.T., Elias, T.S., *eds.* Extinction is forever. Proceedings of a symposium at the New York Botanical Garden, 11-13 May 1976. New York, New York Botanical Garden. 50-59. Discusses existing laws and present policies regarding endangered plants for five western states of the U.S. (L/T)

9914 **Rhoads, W.A., Cochrane, S.A., Williams, M.P. (1978).** Status of endangered and threatened plant species on Nevada Test Site - a survey, Part 2: Threatened species. Goleta, California, Santa Barbara Operations, EG & G Inc. 148p. Descriptions and dot maps. (T)

U.S. - Nevada

9915 **Rhoads, W.A., Williams, M.P. (1977).** Status of endangered and threatened plant species on Nevada Test Site - a survey, Part 1: Endangered species. Goleta, California, Santa Barbara Operations, EG & G. Inc. 102p. Descriptions, dot maps and photos. (T)

9916 **Rogers, B.S., Tiehm, A. (1979).** Vascular plants of the Sheldon National Wildlife Refuge. Portland, U.S. Fish and Wildlife Service. iv, 87p. Illus., maps. (P)

9917 **Schwartz, A. (1984).** Bright future for a desert refugium. *Garden* (New York), 8(3): 26-29. Col. illus. Ash meadows, Mojave Desert. 13,000 acres protected. (P/T)

9918 **U.S. Department of the Interior. Fish and Wildlife Service (1984).** Ash Meadows becomes nature preserve. *Endangered Species Tech. Bull.*, 9(3): 5. Illus. Desert wetland ecosystem. (P)

9919 **U.S. Department of the Interior. Fish and Wildlife Service (1985).** Endangered and threatened wildlife and plants: comment period for public hearing for *Eriogonum ovalifolium* var. *williamsiae* and *Cupressus abramsiana* (Santa Cruz cypress). *Fed. Register*, 50(211): 45443-45444. (L/T)

9920 **U.S. Department of the Interior. Fish and Wildlife Service (1985).** Endangered and threatened wildlife and plants: proposed Endangered status for *Eriogonum ovalifolium* var. *williamsiae* (Steamboat buckwheat). *Fed. Register*, 50(177): 37252-37255. (L/T)

9921 **U.S. Department of the Interior. Fish and Wildlife Service (1985).** Endangered and threatened wildlife and plants: reopening of comment period on proposed Endangered status for *Eriogonum ovaliofolium* var. *williamsiae* (Steamboat buckwheat). *Fed. Register*, 50(228): 48617-?. (L/T)

9922 **U.S. Department of the Interior. Fish and Wildlife Service (1985).** Final protection for seven plants and an insect in Ash Meadows. *Endangered Species Tech. Bull.*, 10(6): 1-7. (L/T)

9923 **U.S. Department of the Interior. Fish and Wildlife Service (1985).** Protection recommended for three plants. *Endangered Species Tech. Bull.*, 10(1): 1, 5-6. (L/T)

9924 **Villa-Lobos, J.L. (1985).** Desert oasis saved. *Threatened Pl. Newsl.*, 15: 17-18. Ash Meadows National Wildlife Refuge. (P/T)

9925 **Williams, M. (1979).** Eighteen Nevada rare plants listed. *Madrono*, 26(1): 50. (L/T)

U.S. - New Hampshire

9926 **Anon. (1980).** Endangered and threatened species determinations. *AABGA Newsl.*, 71: 6. (L/T)

9927 **Countryman, W.D. (1977).** The northeastern United States. *In* Prance, G.T., Elias, T.S., *eds.* Extinction is forever. Proceedings of a symposium at the New York Botanical Garden, 11-13 May 1976. New York, New York Botanical Garden. 30-35. Legislation regarding conservation (primarily of plants and natural areas) in the northeastern U.S. (L/P/T)

9928 **Crow, G.E., Feltes, T. (1982).** New England's rare, threatened and endangered plants. Washington, D.C. U.S. Department of the Interior. Fish and Wildlife Service. x, 130p. Illus., col. illus., maps.

9929 **Crow, G.E., Storks, I.M. (1980).** Rare and endangered plants of New Hampshire: a phytogeographic viewpoint. *Rhodora*, 82(829): 173-189. (T)

9930 **Dunlop, D.A., Crow, G.E. (1985).** Rare plants of coastal New Hampshire. *Rhodora*, 87(852): 487-501. (T)

9931 **Dunlop, D.A., Crow, G.E. (1985).** The vegetation and flora of the Searbrook Dunes with special reference to rare plants. *Rhodora*, 87(852): 471-486. Illus., maps. (T)

U.S. - New Hampshire

9932 **Field, K.G., Primack, R.B. (1980).** Comparison of the characteristics of rare, common and endemic plant species in New England USA. *In* The University of British Columbia. Second International Congress of Systematic and Evolutionary Biology, Vancouver, B.C. Canada, July 17-24, 1980. Vancouver, B.C., Univ. of British Columbia. (T)

9933 **Foy, L. (1977).** Protection of endangered plants in New England: lists, laws and activities. Framingham, MA, New England Wild Flower Society, Inc. (Various paging). (L/T)

9934 **Graber, R.E., Brewer, L.G. (1985).** Changes in the population of the rare and endangered plant *Potentilla robbinsiana* Oakes during the period 1973-1983. *Rhodora*, 87(852): 449-458. (T)

9935 **Graber, R.E., Crow, G.E. (1982).** Hiker traffic on and near the habitat of Robbins cinquefoil, an endangered plant species. *Bull. New Hampshire Agric. Exper. Stat.*, 1982(522): 10. Illus. *Potentilla robbinsiana*, in the White Mountains of New Hampshire. (T)

9936 **Hodgdon, A.R. (1973).** Endangered plants of New Hampshire: a selected list of endangered species. *Forest Notes*, 114: 2-6. (T)

9937 **Kimball, K.D. (1985).** Progress in the Robbins' cinquefoil (*Potentilla robbinsiana*) recovery program. *Endangered Species Tech. Bull.*, 10(5): 6-11. Illus. Forest, montain, alpine. (T)

9938 **New England Botanical Club. Natural Areas Criteria Committee (1972).** Guidelines and criteria for the evaluation of natural areas. A report prepared for the New England Natural Resources Center by the Natural Areas Criteria Committee of the New England Botanical Club Inc. (P)

9939 **Storks, I.M., Crow, G.E. (1978).** Rare and endangered vascular plant species in New Hampshire. Newton Corner, MA, U.S. Fish and Wildlife Service and The New England Botanical Club. iv, 66, 10p. Map. List of species with annotations. (T)

9940 **Tryon, R.M., ed. (1980).** Symposium on rare and endangered plant species in New England, Cambridge, Mass., 1979. The proceedings of the symposium "Rare and Endangered Plant Species in New England," 4-5 May 1979. *Rhodora* (Cambridge, MA, New England Botanical Garden), 82(829): 237p. (T)

9941 **U.S. Department of the Interior. Fish and Wildlife Service (1985).** Endangered and threatened wildlife and plants: proposal to determine *Astragalus robbinsii* var. *jesupii* to be an Endangered species. *Fed. Register*, 50(244): 51718-51722. (L/T)

9942 **U.S. Environmental Protection Agency, Region 1 (1982).** New England Wetland Flora and Wetland Protection Laws. Washington, D.C., U.S. Govt Print. Off. 172p. Illus. (L/P)

U.S. - New Jersey

9943 **Buckley, P.A. (1978).** The sum of the parts: preservation of the New Jersey Pine Barrens. *Frontiers*, 42(2): 13-15. (P)

9944 **Conservation and Environmental Studies Center, Inc. (1979).** Inventory of endangered, threatened, rare and undetermined status flora in the Passaic River Basin. New York, U.S. Army Corps of Engineers. 32p. (T)

9945 **Cook, B., Baxter, P. (1984).** Two new nature preserves in Pine Barrens. *Bartonia*, 50-61. (P/T)

9946 **Cook, R.T. (1981).** News and notes. Natural areas protected by The Nature Conservancy in Pennsylvania and New Jersey. *Bartonia*, 48: 44-45. (P)

U.S. - New Jersey

9947 **Countryman, W.D. (1977).** The northeastern United States. *In* Prance, G.T., Elias, T.S., *eds.* Extinction is forever. Proceedings of a symposium at the New York Botanical Garden, 11-13 May 1976. New York, New York Botanical Garden. 30-35. Legislation regarding conservation (primarily of plants and natural areas) in the northeastern U.S. (L/P/T)

9948 **Cromartie, W.J., ed. (1980).** Endangered and threatened plants and animals of New Jersey: proceedings of the second annual Pine Barrens research conference, Stockton State College, Pomona, NJ, 31 March-April 1979. Pomona, NJ, Stockton Center for Environmental Research. 82p. (T)

9949 **Ehrenfeld, J.G. (1983).** The effects of changes in land-use on swamps of the New Jersey Pine Barrens. *Biol. Conserv.*, 25(4): 353-375. Map. (T)

9950 **Fairbrothers, D.E. (1980).** Plants. *In* Cromartie, W.J., *ed.* Endangered and threatened plants and animals of N.J. Proc. of the second annual Pine Barrens Research Conference. Pomona, NJ, Stockton Center for Environmental Research. 12-19. (T)

9951 **Fairbrothers, D.E., Hough, M.Y. (1973).** Rare or endangered vascular plants of New Jersey. *New Jersey State Mus. Sci. Not.*, 14: 1-53. (T)

9952 **Good, R.E., Good, N.F. (1984).** The Pinelands National Reserve: an ecosystem approach to management. *BioScience*, 34(3): 169-173. Map. (L/P)

9953 **Hales, D.F. (1984).** The Pinelands National Reserve: an approach to cooperative conservation. *In* Unesco-UNEP. Conservation, science and society. Paris, Unesco. 553-564. Contributions to the First International Biosphere Reserve Congress, Minsk, Byelorussia/USSR, 26 September - 2 October 1983. (P)

9954 **Kantor, R.A., Pillsbury, M.K. (1976).** Upland living resources: Endangered and rare vegetation. A staff working paper. Trenton, NJ, NJ Dept of Environmental Protection. (T)

9955 **Letcher, G.R. (1979).** New Jersey's pine barrens: strategies for protecting a critical area. *J. Soil Water Conserv.*, 34(5): 211-214. Illus.

9956 **Little, S. (1974).** Wildflowers of the Pine Barrens and their niche requirements. *New Jersey Outdoors*, 1: 16-18.

9957 **Roman, C.T., Zampella, R.A., Jaworski, A.Z. (1985).** Wetland boundaries in the New Jersey pinelands: ecological relationships and delineation. *Water Resources Bull.*, 21(6): 1005-1012. Conservation areas. (P)

9958 **Smith, A.P. (1972).** Survival and seed production of transplants of *Dionaea muscipula* in the new Jersey Pine Barrens. *Torreya*, 99: 145-146. (G/T)

9959 **Snyder, D.B., Vivian, V.E. (1975).** Rare and endangered vascular plant species of New Jersey. *Proc. Roch. Acad. Sci.*, 12(4): 400. (T)

9960 **Snyder, D.B., Vivian, V.E. (1981).** Rare and endangered vascular flora of New Jersey. *Bull. New Jersey Acad. Sci.*, 26(2): 71. (T)

9961 **Synder, D. (1986).** Rare New Jersey plant species rediscovered. *Bartonia*, 52: 44-48. (T)

9962 **U.S. Department of the Interior. Fish and Wildlife Service (1988).** Endangered and threatened wildlife and plants; determination of *Helonias bullata* (Swamp pink) to be an Endangered species. *Fed. Register*, 53(175): 35076-35080. (L/T)

9963 **U.S. Department of the Interior. Fish and Wildlife Service; Heritage Conservation and Recreation Service; National Park Service. (1980).** Comprehensive management plan for the Pinelands National Reserve, New Jersey. Philadelphia, Pennsylvania, Department of the Interior, Heritage Conservation and Recreation Service. 508p. (P)

U.S. - New Jersey

9964 **Volk, J.M. (1987).** Is Pine Barren stream ecology threatened? *Bartonia*, 53: 63-64.

U.S. - New Mexico

9965 **Anon. (1984).** Endangered species. *AABGA Newsl.*, 116: 5. (T)

9966 **Fletcher, R., Isaacs, B., Knight, P., Martin, W., Sabo, D., Spellenberg, R., Todsen, T. (1984).** A handbook of rare and endemic plants of New Mexico. Albuquerque, Univ. of New Mexico Press. 291p. List of over 130 species, ink drawings, descriptions and dot maps. (T)

9967 **Guthrie, P. (1987).** Lavaland. *National Parks*, 61(5-6): 22-27. Proposed El Malpais National Monument. (P)

9968 **Harrington, W. (1980).** Endangered species protection and water resource development. Los Alamos, NM, Los Alamos Scientific Laboratory. 60p. (T)

9969 **Hennessy, J.T., Gibbens, R.P., Tromble, J.M., Cardenas, M. (1983).** Vegetation changes from 1935 to 1980 in Mesquite dunelands and former grasslands on southern New Mexico. *J. Range Manage.*, 36(3): 370-374. Illus.

9970 **Irving, R.S. (1979).** *Hedeoma todsenii* (Labiatae), a new and rare species from New Mexico. *Madroño*, 26(4): 184-187. Illus. (T)

9971 **Johnston, M.C. (1977).** The southwestern United States. *In* Prance, G.T., Elias, T.S., eds. Extinction is forever. Proceedings of a symposium at the New York Botanical Garden, 11-13 May 1976. New York, New York Botanical Garden. 60-61. Legislation regarding threatened wild plants in the southwestern U.S. (L/T)

9972 **McMahan, L.R. (1987).** Rare plant conservation by state government: case studies from the western United States. *In* Elias, T.S., *ed.* Conservation and management of rare and endangered plants. Proceedings from a conference, Sacramento, California, 5-8 November 1986. Sacramento, California Native Plant Society. 23-31. Maps. (L/T)

9973 **Milne, J. (1987).** Conserving some of the world's tiniest cacti. *Center for Plant Conservation*, 2(1): 1, 8. Illus. (T)

9974 **Nabhan, G.P. (1987).** Nurse plant ecology of threatened desert plants. *In* Elias, T.S., *ed.* Conservation and management of rare and endangered plants. Proceedings from a conference, Sacramento, California, 5-8 November 1986. Sacramento, California Native Plant Society. 377-383. Illus. Numerous desert plants depend on protective overstorey for survival; case studies of 2 cacti and *Capsicum annuum* var. *glabriusculum.* (T)

9975 **Nabhan, G.P., Greenhouse, R., Hodgson, W. (1986).** At the edge of extinction: useful plants of the border states of the United States and Mexico. *Arnoldia*, 46(3): 36-46. Illus. (E)

9976 **New Mexico Plant Protection Advisory Committee (1983).** A handbook of rare and endemic plants of New Mexico. Albuquerque, Univ. of New Mexico Press. 291p. (T)

9977 **Olwell, P., Cully, A., Knight, P., Brack, S. (1987).** *Pediocactus knowltonii* recovery efforts. *In* Elias, T.S., *ed.* Conservation and management of rare and endangered plants. Proceedings from a conference, Sacramento, California, 5-8 November 1986. Sacramento, California Native Plant Society. 519-522. Illus., map. Results of efforts to re-introduce cultivated plants to the wild. (G/T)

9978 **Peirce, P. (1975).** Endangered and threatened cacti of New Mexico. Albuquerque, NM. (T)

9979 **U.S. Department of the Interior. Fish and Wildlife Service (1980).** Endangered species of Arizona and New Mexico. Albuquerque, NM, U.S. Fish and Wildlife Service. 63p. (L/T)

U.S. - New Mexico

9980 **U.S. Department of the Interior. Fish and Wildlife Service (1984).** Endangered and threatened wildlife and plants: correction. *Fed. Register*, 49(127): 49639. *Eriogonum gypsophilum.* (L/T)

9981 **U.S. Department of the Interior. Fish and Wildlife Service (1984).** Endangered and threatened wildlife and plants: proposal to determine *Astragalus humillimus* (Mancos milkvetch) to be an Endangered species. *Fed. Register*, 49(126): 26610-26614. (L/T)

9982 **U.S. Department of the Interior. Fish and Wildlife Service (1984).** Endangered and threatened wildlife and plants: proposal to determine *Cirsium vinaceum* to be a Threatened species and to determine its critical habitat. *Fed. Register*, 49(90): 20735-20739. (L/T)

9983 **U.S. Department of the Interior. Fish and Wildlife Service (1984).** Endangered and threatened wildlife and plants: proposal to determine *Erigeron rhizomatus* (Rhizome fleabane) to be a Threatened species. *Fed. Register* 49(80): 17548-17551. (L/T)

9984 **U.S. Department of the Interior. Fish and Wildlife Service (1984).** Four western plants threatened by habitat degradation. *Endangered Species Tech. Bull.*, 9(5): 1, 8-9. Illus. (L/T)

9985 **U.S. Department of the Interior. Fish and Wildlife Service (1985).** Endangered and threatened wildlife and plants: final rule to determine *Astragalus humillimus* to be Endangered. *Fed. Register*, 50(124): 26568-26572. (L/T)

9986 **U.S. Department of the Interior. Fish and Wildlife Service (1985).** Endangered and threatened wildlife and plants: final rule to determine *Erigeron rhizomatus* to be a Threatened species. *Fed. Register,* 50(81): 16680-16682. (L/T)

9987 **U.S. Department of the Interior. Fish and Wildlife Service (1985).** Endangered and threatened wildlife and plants: final rule to determine *Erigeron rhizomatus* to be a Threatened species. *Fed. Register*, 50(81): 16680-16682. (L/T)

U.S. - New York

9988 **Art, H.W. (1976).** Ecological studies of the sunken forest, Fire Island National Seashore, New York. National Park Service Sci. Monogr. Series No. 7. Washington, D.C., Govt Print. Office. xv, 237p. Illus., maps. (P)

9989 **Clemants, S.E. (1986).** New York State rare plants: spring 1986 status report. New York State Department of Environmental Conservation and The Nature Conservancy. v, 26p. (T)

9990 **Countryman, W.D. (1977).** The northeastern United States. *In* Prance, G.T., Elias, T.S., *eds.* Extinction is forever. Proceedings of a symposium at the New York Botanical Garden, 11-13 May 1976. New York, New York Botanical Garden. 30-35. Legislation regarding conservation (primarily of plants and natural areas) in the northeastern U.S. (L/P/T)

9991 **Fairbrothers, D.E. (1979).** Endangered, threatened and rare vascular plants of the Pine Barrens and their biogeography. *In* Forman, R.T.T., *ed.* Pine Barrens. New York, Academic Press. 395-405. (T)

9992 **Forest, H.S. (1978).** The American chestnut in New York. *Conservationist*, 33(1): 25-26. Col. illus.

9993 **Frankel, E. (1984).** Rare, endangered, threatened and protected native ferns and fern allies of New York State. *Fiddlehead Forum*, 11(5): 25. (T)

9994 **Hill, D.B. (1985).** Forest fragmentation and its implications in central New York. *Forest Ecol. Manage.*, 12(2): 113-128. Land use and fragmentation of forests. (T)

9995 **Keller, J.E. (1980).** Adirondack wilderness. New York, Syracuse University Press. 747p. (P)

U.S. - New York

9996 **Leopold, D.J., Reschke, C., Smith, D.S. (1988).** Old-growth forests of Adirondack Park, New York. *Nat. Areas J.*, 8(3): 166-189. (P)

9997 **McMullen, J.M. (1986).** New York's endangered, threatened, and rare plant species. *Clintonia*, 1(6): 1-3. (T)

9998 **Mitchell, R.S. (1979).** Preliminary lists of rare, endangered and threatened plant species in New York State. New York State Museum Leaflet No. 21. New York, State Education Department. 15p. (T)

9999 **Mitchell, R.S., Sheviak, C.J. (1981).** Rare plants of New York State. New York State Museum Bulletin No. 445. Albany, The University of the State of New York, The State Education Department. viii, 96p. Illus. Reprinted 1984. (T)

10000 **Mitchell, R.S., Sheviak, C.J., Dean, J.K. (1980).** Rare and endangered vascular plant species in New York State. Newton Corner (Ma.), U.S. Fish and Wildlife Service and Albany, The State Botanist's Office, New York State Museum. 38p. (T)

10001 **New York State (1974).** List of protected native plants in New York State. New York State Dept. Environ. Conserv. 2p. (T)

10002 **Rechlin, M.A. (1973).** Recreational impact in the Adirondack High Peaks Wilderness. Ann Arbor, University of Michigan. 64p. (P)

10003 **Stauffer, M.R. (1975).** Inventory of rare and endangered plants of New York. *Proc. Roch. Acad. Sci.*, 12(4): 400. (T)

10004 **Terrie, P. (1985).** Forever wild. Philadelphia, Temple University Press. 209p. Environmental aesthetics and Adirondack Forest Preserve. (P)

10005 **Thompson, L. (1971).** The future of the Adirondacks. *Appalachia*, 36(14): 196-201. Maps.

10006 **U.S. Department of the Interior. Fish and Wildlife Service (1988).** Endangered and threatened wildlife and plants; determination of *Agalinis acuta* (Sandplain gerardia) to be an Endangered species. *Fed. Register*, 53(173): 34701-34705. (L/T)

10007 **U.S. Department of the Interior. Fish and Wildlife Service (1988).** Endangered and threatened wildlife and plants; determination of *Helonias bullata* (Swamp pink) to be an Endangered species. *Fed. Register*, 53(175): 35076-35080. (L/T)

10008 **U.S. Department of the Interior. Fish and Wildlife Service (1989).** Endangered and threatened wildlife and plants; Threatened status for *Phyllitis scolopendrium* var. *americana* (American hart's-tongue). *Fed. Register*, 54(134): 29726-29730. (L/T)

10009 **U.S. Department of the Interior. Fish and Wildlife Service (1989).** Sandplain gerardia (*Agalinis acuta*) recovery plan. Newton Corner, MA, U.S. Department of the Interior, Fish and Wildlife Service. 47p. Illus., map. (L/T)

10010 **Zander, R.H. (1975).** Protected plants in New York State. *Not. Clinton Herb.*, 2.

10011 **Zander, R.H. (1976).** Floristics and environmental planning in western New York and adjacent Ontario: distribution of legally protected plants and plant sanctuaries. Buffalo, NY, Buffalo Society of Natural Sciences. 47p. (L/P/T)

U.S. - North Carolina

10012 **Anon. (1980).** Endangered and threatened species determinations. *AABGA Newsl.*, 71: 6. (L/T)

10013 **Au, S.-F. (1974).** Vegetation and ecological processes on Shackleford Bank, North Carolina. National Park Service Sci. Monogr. Series No. 6. Washington, D.C., U.S. Govt Printing Office. xii, 86p. Illus., map.

U.S. - North Carolina

10014 **Cooper, J.E., et al., eds (1977).** Endangered and threatened plants and animals of North Carolina. Proceedinga of the Symposium on Endangered and Threatened Biota of North Carolina. 1 biological concerns; Meredith College, Raleigh, Nov. 7-8, 1975. Raleigh, NC, North Carolina State Museum of Natural History. 444p. (T)

10015 **Godfrey, P.J. & M.M. (1976).** Barrier Island ecology of Cape Lookout National Seashore and vicinity, North Carolina. National Park Service Sci. Monogr. Series No. 9. Washington, D.C., U.S. Govt Print. Off. ix, 160p. Illus., maps. (P)

10016 **Hardin, J.W. (1977).** The southeastern United States. *In* Prance, G.T., Elias, T.S., *eds.* Extinction is forever. Proceedings of a symposium at the New York Botanical Garden, 11-13 May 1976. New York, New York Botanical Garden. 36-40. Discusses status of state programmes within the 14 southeastern states of endangered plant species. (L/P/T)

10017 **Hardin, J.W., et al. (1977).** North Carolina endangered and threatened vascular plants. *In* Cooper, J.E., *et al., eds.* Endangered and threatened plants and animals of NC. Proceedings of the Symposium on Endangered and Threatened Biota of NC. Raleigh, NC State Mus. of Nat. Hist. 56-142. Symposium held at Meredith College, Raleigh, Nov. 7-8, 1975. (T)

10018 **Hicks, M., Davison, P. (1989).** Some rare, endemic and disjunct liverworts in North Carolina. *Castanea*, 54(4): 255-261. (T)

10019 **Jones, S.M., Dunn, B.A. (1979).** Aspects of the biology of *Shortia galacifolia* (Diapensiaceae). *ASB Bull.*, 26(2): 92. Abstract. (T)

10020 **Lamm, L. (1984).** Two state endangered plants rediscovered. *Wild Flower* (North Carolina Wild Flower Preservation Society), Spring: 12. (T)

10021 **Massey, J., Otte, D., Atkinson, T., Whetstone, R. (1983).** An atlas and illustrated guide to the threatened and endangered vascular plants of the mountains of North Carolina and Virginia. Asheville, North Carolina, U.S. Department of Agriculture, Forest Service, Southeastern Forest Experiment Station. 218p. 45 species listed with descriptions, habitat details, ink drawings, dot maps. (T)

10022 **Moore, J.K., Bell, C.R. (1974).** The North Carolina Botanical Garden: a natural garden of native plants. *Amer. Hort.*, 53(5): 23-29. (G)

10023 **Moore, K. (1980).** Conservation through propagation: a status report and request for assistance from the North Carolina Botanical Garden. *Newsl. Alabama Wildfl. Soc.* 19: 4-6. (G)

10024 **Norquist, C. (1984).** Savannahs and bogs of the southern U.S. U.S. threatened ecosystems. *Endangered Species Tech. Bull.*, 10(9): 4-5 (T)

10025 **North Carolina Endangered Species Committee (1973).** Preliminary list of endangered plant and animals species in North Carolina. Raleigh, NC, North Carolina Dept of Natural and Economic Resources. 24p. (T)

10026 **Parker, W., Dixon, L. (1980).** Endangered and threatened wildlife of Kentucky, North Carolina, South Carolina and Tennessee. Raleigh, NC, North Carolina Agricultural Extension Service. 116p. (T)

10027 **Phillips, H. (1984).** Spreading the native-plant idea. *Garden* (New York), 8(3): 2-8. Col. illus. N. Carolina Bot. Gard. encouraging gardeners to buy propagated native plants. (G/T)

10028 **Pittillo, J.D. (1974).** Rare vascular plants of western North Carolina. Cullowhee, NC, Western Carolina Univ. 5p. (T)

10029 **Richardson, C.J. (1983).** Pocosins: vanishing wastelands or valuable wetlands. *BioScience*, 33(10): 626-633. Illus., maps. Nutrient-poor, freshwater, evergreen shrub bogs.

U.S. - North Carolina

10030 **Sutter, R. (1982).** Protection strategy for mountain golden heather. *Wild Flower* (Newsletter of the North Carolina Wild Flower Preservation Society), Fall 1982: 29-31. (T)

10031 **Sutter, R.D. (1982).** Is Venus fly trap (*Dionaea muscipula*) an endangered species? *ASB Bull.*, 29(2): 86. (T)

10032 **Sutter, R.D., Mansberg, L., Moore, J. (1983).** Endangered, threatened and rare plant species of North Carolina: a revised list. *A.S.B. Bull.*, 30(4): 153-163. (T)

10033 **U.S. Department of the Interior. Fish and Wildlife Service (1980).** National wildlife refuge on the Currituck Outer Banks, Currituck County, N. Carolina. Newton Corner, MA, Dept of the Interior, Fish and Wildlife Service. 367p. (P)

10034 **U.S. Department of the Interior. Fish and Wildlife Service (1984).** Endangered and threatened wildlife and plants: proposed Threatened status for *Solidago spithamaea* (Blue ridge goldenrod). *Fed. Register*, 49(142): 29629-29632. (L/T)

10035 **U.S. Department of the Interior. Fish and Wildlife Service (1984).** Four plants in danger of extinction. *Endangered Species Tech. Bull.*, 9(8): 1, 5 Illus. (L/T)

10036 **U.S. Department of the Interior. Fish and Wildlife Service (1985).** Endangered and threatened wildlife and plants: determination of Endangered status for *Lindera melissifolia* (Pondberry). *Fed. Register*, 50(147): 27495-27500. (L/T)

10037 **U.S. Department of the Interior. Fish and Wildlife Service (1985).** Endangered and threatened wildlife and plants: determination of Threatened status for *Solidago spithamaea* (Blue ridge goldenrod). *Fed. Register*, 50(60): 12306-12309. (L/T)

10038 **U.S. Department of the Interior. Fish and Wildlife Service (1985).** Endangered and threatened wildlife and plants: proposal to determine *Oxypolis canbyi* (Canby's dropwort) to be an Endangered species. *Fed. Register*, 50(60): 12345-12348. (L/T)

10039 **U.S. Department of the Interior. Fish and Wildlife Service (1985).** Endangered and threatened wildlife and plants: proposed Endangered status for *Lindera melissifolia* (Pondberry). *Fed. Register*, 50(156): 32582-32585. (L/T)

10040 **U.S. Department of the Interior. Fish and Wildlife Service (1985).** Protection sought for four vulnerable plants. *Endangered Species Tech. Bull.*, 19(9): 1, 8. (L/T)

10041 **U.S. Department of the Interior. Fish and Wildlife Service (1986).** Endangered and threatened wildlife and plants: determination of *Oxypolis canbyi* (Canby's dropwort) to be an Endangered species. *Fed. Register*, 51(37): 6690-6693. (L/T)

10042 **U.S. Department of the Interior. Fish and Wildlife Service (1986).** Endangered and threatened wildlife and plants: proposed Endangered status for *Lysimachia asperulifolia. Fed. Register*, 51(69): 12451-12455. (L/T)

10043 **U.S. Department of the Interior. Fish and Wildlife Service (1988).** Endangered and threatened wildlife and plants; determination of *Helonias bullata* (Swamp pink) to be an Endangered species. *Fed. Register*, 53(175): 35076-35080. (L/T)

10044 **U.S. Department of the Interior. Fish and Wildlife Service (1988).** Endangered and threatened wildlife and plants; determination of Endangered status for *Sarracenia rubra* ssp. *jonesii* (Mountain sweet pitcher plant). *Fed. Register*, 53(190): 38470-38474. (L/T)

10045 **U.S. Department of the Interior. Fish and Wildlife Service (1988).** Endangered and threatened wildlife and plants; determination of Threatened status for *Liatris helleri. Fed. Register*, 52(223): 44397-44401. (L/T)

10046 **U.S. Department of the Interior. Fish and Wildlife Service (1988).** Endangered and threatened wildlife and plants; proposed Endangered status for *Sarracenia rubra* ssp. *jonesii* (Mountain sweet pitcher plant). *Fed. Register*, 53(27): 3901-3905. (L/T)

U.S. - North Carolina

10047 **U.S. Department of the Interior. Fish and Wildlife Service (1989).** Endangered and threatened wildlife and plants; Small anthered bittercress determined to be Endangered. *Fed. Register*, 54(182): 38947-38950. (L/T)

10048 **U.S. Department of the Interior. Fish and Wildlife Service (1989).** Endangered and threatened wildlife and plants; Threatened status of *Hexastylis naniflora* (Dwarf-flowered heartleaf). *Fed. Register*, 54(71): 14964-14967. (L/T)

10049 **U.S. Department of the Interior. Fish and Wildlife Service (1989).** Endangered and threatened wildlife and plants; determination of Endangered status for *Rhus michauxii* (Michaux's sumac). *Fed. Register*, 54(187): 39854-39857. (L/T)

10050 **U.S. Department of the Interior. Fish and Wildlife Service (1989).** Endangered and threatened wildlife and plants; proposed Endangered status for *Geum radiatum* and *Hedyotis purpurea* var. *montana*. *Fed. Register*, 54(139): 30572-30576. (L/T)

10051 **U.S. Department of the Interior. Fish and Wildlife Service (1989).** Endangered and threatened wildlife and plants; proposed Endangered status for *Rhus michauxii* (Michaux's sumac). *Fed. Register*, 54(4): 441-445. (L/T)

10052 **U.S. Department of the Interior. Fish and Wildlife Service (1989).** Proposed listings - July 1989. Two southern Appalachian plants. *Endangered Species Tech. Bull.*, 14(8): 4. *Geum radiatum, Hedyotis purpurea* var. *montana*. (L/T)

10053 **U.S. Department of the Interior. National Park Service (1982).** General management plan, Great Smoky Mountains National Park, North Carolina, Tennessee. Denver, CO, U.S. Department of the Interior. National Park Service. 2 vols. (P)

U.S. - North Dakota

10054 **Hansen, P.L., Hoffman, G.R., Bjugstad, A.J. (1984).** The vegetation of Theodore Roosevelt National Park, North Dakota: a habitat type classification. Fort Collins, CO, U.S. Dept of Agriculture, Forest Service, Rocky Mountain Forest and Range Experiment Station. 35p. Illus., map. (P)

10055 **Luoma, J.R. (1985).** Twilight in pothole country. *Audubon*, 87(5): 66-85. Devastation of North Dakota prairie wetland.

10056 **Mohlenbrock, R.H. (1977).** The midwestern United States. *In* Prance, G.T., Elias, T.S., *eds*. Extinction is forever. Proceedings of a symposium at the New York Botanical Garden, 11-13 May 1976. New York, New York Botanical Garden. 41-44. Discusses the status of legal protection for threatened and endangered species of plants which occur in the midwestern states. (L/T)

10057 **U.S. Department of Agriculture. Soil Conservation Service (1972).** Rare and endangered plant species in North Dakota. Bismarck, ND, U.S. Department of Agriculture Soil Conservation Service. (T)

U.S. - Ohio

10058 **Andreas, B., Burns, J., Cusick, A., Emmitt, D., Marshall, J., Spooner, D. (1984).** Ohio endangered and threatened vascular plants: abstracts of state-listed taxa. Columbus, Ohio, Ohio Dept of Natural Resources. 635p. 367 species described, including habitat, threats, dot maps. (T)

10059 **Carr, W.R. (1983).** Vascular plants of Bigelow (Ichuckery) Cemetery State Nature Reserve in northern Madison County, Ohio, U.S.A. *Ohio Biol. Surv. Biol. Not.*, 15. Illus., maps. (P)

10060 **Cooperrider, T.S. ed. (1982).** Endangered and threatened plants of Ohio. *Ohio Biol. Surv. Biol. Not.*, 16: 92p. 821 species listed with descriptions. (T)

U.S. - Ohio

10061 **Cusick, A.W., Spooner, D.M., Andreas, B.K., Anderson, D.M. (1983).** Status of Ohio plants considered for federal listing. *Ohio J. Sci.*, 83(2): 8. Abstract. (L/T)

10062 **Easterly, N.W. (1979).** Rare and infrequent plant species in the oak openings of northwestern Ohio. *Ohio J. Sci.*, 79(2): 51-58. (T)

10063 **Easterly, N.W. (1981).** Floristic notes on the Irwin Prairie State Nature Reserve and Schwamberger Prairie Reserve in northwestern Ohio, U.S.A. *Ohio Biol. Surv. Biol. Not.*, 15:146-147. (P)

10064 **Easterly, N.W. (1983).** Endangered and threatened plant species of Schwamberger Preserve, Lucas County, Ohio. *Ohio J. Sci.*, 83(3): 97-102. Map. (P/T)

10065 **Gambill, W.G., jr (1969).** Our vanishing native flora dilemma in southern Ohio. *In* Proc. Int. Bot. Congr. XI: 67. Seattle. Abstract. (T)

10066 **Jones, C.H. (1943).** Studies in Ohio floristics. II. Rare plants of Ohio. *Castanea*, 8(5-6): 81-108. (T)

10067 **Mohlenbrock, R.H. (1977).** The midwestern United States. *In* Prance, G.T., Elias, T.S., *eds.* Extinction is forever. Proceedings of a symposium at the New York Botanical Garden, 11-13 May 1976. New York, New York Botanical Garden. 41-44. Discusses the status of legal protection for threatened and endangered species of plants which occur in the midwestern states. (L/T)

10068 **Ohio Department of Natural Resources (1981).** Rare species of native Ohio wild plants. Columbus, OH, Ohio Dept. of Natural Resources, Div. of Natural Areas and Preserves. 20p. (T)

10069 **Parsons, B. (1982).** Rare plant conservation. *Arbor. Leaves*, 24(2): 10-11. Illus. (G/L/T)

10070 **Spooner, D.M. (1981).** Distributional records of some previously rare Ohio plants. *Ohio J. Sci.*, 81 (Apr. Prog. Abstr.): 20. Ohio Acad. Sci., 90th Ann. Meeting, 24-26 April 1981, College of Wooster. (T)

10071 **Spooner, D.M., Cusick, A.W., Andreas, B., Anderson, D. (1983).** Notes on Ohio vascular plants previously considered for listing as federally endangered or threatened species. *Castanea*, 48(4): 250-258. (L/T)

10072 **Stuckey, R.L., Roberts, M.L. (1977).** Rare and endangered aquatic vascular plants of Ohio: an annotated list of the imperiled species. *Sida*, 7(1): 24-41. (T)

10073 **Transeau, E.N. (1981).** The vanishing prairies of Ohio, U.S.A. *Ohio Biol. Surv. Biol. Not.*, 15: 61-62. Illus. Maps.

10074 **U.S. Department of the Interior. Fish and Wildlife Service (1984).** Endangered and threatened wildlife and plants: proposed Endangered status for *Solidago shortii* (Short's goldenrod). *Fed. Register*, 49(198): 39873-39876. (L/T)

10075 **U.S. Department of the Interior. Fish and Wildlife Service (1985).** New publications: Ohio endangered and threatened vascular plants. *Endangered Species Tech. Bull.*, 10(8): 8. (T)

10076 **U.S. Department of the Interior. Fish and Wildlife Service (1988).** Endangered and threatened wildlife and plants; determination of Threatened status for *Hymenoxys acaulis* var. *glabra* (Lakeside daisy). *Fed. Register*, 53(121): 23742-23745. (L/T)

U.S. - Oklahoma

10077 **Anon. (1983).** Endangered species. *AABGA Newsl.*, 107: 3. (T)

U.S. - Oklahoma

10078 **Crockett, J.J., et al. (1975).** Herbaceous plants. *In* Rare and endangered vertebrates and plants of Oklahoma. Stillwater, Rare and Endangered Species of Oklahoma Committee and U.S. Department of Agriculture, Soil Conservation Service. (T)

10079 **Johnston, M.C. (1977).** The southwestern United States. *In* Prance, G.T., Elias, T.S., *eds*. Extinction is forever. Proceedings of a symposium at the New York Botanical Garden, 11-13 May 1976. New York, New York Botanical Garden. 60-61. Legislation regarding threatened wild plants in the southwestern U.S. (L/T)

10080 **Lewis, J.C., ed. (1975).** Rare and endangered vertebrates and plants of Oklahoma. Stillwater, OK, Rare and Endangered Species of Oklahoma Committee. 44p. (T)

10081 **Magrath, L.K., et al. (1978).** New, rare and infrequently collected plants in Oklahoma. Herbarium Publication No. 2. Durant, OK, Herbarium, Southeastern Oklahoma State University. 120p. (T)

10082 **Mahler, W.F. (1981).** Notes on rare Texas and Oklahoma USA plants. *Sida Contrib. Bot.*, 9(1): 76-86. (T)

10083 **Oklahoma Committee on Rare and Endangered Species, eds (1975).** Rare and endangered vertebrates and plants of Oklahoma. U.S. Department of Agriculture/SCS. 44p. (T)

10084 **Roedmer, B.J., Hamilton, D.A., Evans, K.E. (1978).** Rare plants of the Ozark Plateau: a field identification guide. St Paul, North Central Forest Experiment Station, U.S. Department of Agriculture Forest Service. 238p. Illus., map. (T)

10085 **Smola, N.E., Teate, J.L. (1975).** Trees, shrubs, vines and other woody plants. *In* Rare and endangered vertebrates and plants of Oklahoma. Stillwater, U.S. Department of Agriculture, Soil Conservation Service. (T)

10086 **Taylor, R.J. & C.E. (1977).** The rare and endangered *Hypoxis longii* Fernald. (Amaryllidaceae) in Oklahoma. *Bull. Torrey Bot. Club*, 104(3): 276. (T)

10087 **Taylor, R.J. & C.E. (1977).** Those endangered species. *Oklahoma Observer*, 9(12): 12. (T)

10088 **Taylor, R.J. & C.E. (1978).** An annotated list of rare or infrequently collected vascular plants that grow in Oklahoma. Publ. Herb. S.E. Okla. State Univ. No. 2. 15-121. (T)

10089 **U.S. Department of the Interior. Fish and Wildlife Service (1985).** Endangered and threatened wildlife and plants: notice of withdrawal of proposed rules to list *Hedeoma diffusum* and *Phlox longipilosa* var. *longipilosa*. *Fed. Register*, 50(60): 12348-12350. (L/T)

10090 **U.S. Department of the Interior. National Park Service (1979).** Chickasaw National Recreation Area, Oklahoma. Santa Fe, National Park Service. 2 vols. (P)

10091 **Zanoni, T.A., Gentry, J.L., Tyrl, R.J., Risser, P.G. (1979).** Endangered and threatened plants of Oklahoma. Norman, University of Oklahoma, Department of Botany and Microbiology. 64p. 26 species listed. Illus., maps. (T)

U.S. - Oregon

10092 **Anon. (1988).** Rare buckwheat rediscovered after 87 years. *Bull. Native Plant Soc. Oregon*, 21(9): 90.

10093 **Anon. (1988).** Wire-lettuce progress. *Oryx*, 22(3): 148. Re-introduction to wild. (G/T)

10094 **Bornholdt, M., (1985).** Rare flower focuses fight to save an Oregon wetland. *Audubon Action*, 3(5): 12. (T)

U.S. - Oregon

10095 **Chambers, K.L. (1977).** The northwestern United States. *In* Prance, G.T., Elias, T.S., *eds.* Extinction is forever. Proceedings of a symposium at the New York Botanical Garden, 11-13 May 1976. New York, New York Botanical Garden. 45-49. General status of programmes in the Pacific northwest. (L/T)

10096 **Chambers, K.L., Siddall, J.L. (1976).** Provisional list of the rare, threatened and endangered plants of Oregon. Oregon Rare and Endangered Plant Species Task Force. (T)

10097 **Franklyn, A.L., et al. (1980).** Rare, threatened and endangered plant survey. Supplement. Burns, OR, Bureau of Land Management, Burns District. (T)

10098 **Gruber, E.H., et al. (1979).** Rare, threatened and endangered plant survey, 1979. Burns, OR, Bureau of Land Management, Burns District. 303p. (T)

10099 **Jessup, D. (1976).** Oregon forest protection survey report. Portland, Oregon Forest Protection Association. 74p. Illus., maps.

10100 **Johnson, J.M. (1980).** Handbook of uncommon plants in the Salem BLM District. Salem, OR, Salem Bureau of Land Management. 291p. (T)

10101 **Kruckeberg, A.R. (1980).** Why worry about rare plants? Pullman, WA, University of Washington. 24p. Abstract of papers presented at the symposium of Threatened and Endangered Plants, a West Coast Perspective, 25-27 July, 1980. Ashland, Oregon, Oregon State College. (T)

10102 **Lesica, P., Allendorf, F., Leary, R., Bilderback, D. (1988).** Lack of genetic diversity within and among populations of an endangered plant, *Howellia aquatilis. Conserv. Biol.*, 2(3): 275-282. En (Sp). Map. (T)

10103 **McMahan, L.R. (1987).** Rare plant conservation by state government: case studies from the western United States. *In* Elias, T.S., *ed.* Conservation and management of rare and endangered plants. Proceedings from a conference, Sacramento, California, 5-8 November 1986. Sacramento, California Native Plant Society. 23-31. Maps. (L/T)

10104 **McNeil, R.C., Zobel, D.B. (1980).** Vegetation and fire history of a ponderosa pine-white fir forest in Crater Lake National Park. *Northwest Sci.*, 54(1): 30-46. Illus., map. (P)

10105 **Meinke, R.J. (1982).** Threatened and endangered vascular plants of Oregon: an illustrated guide. Portland, Oregon, U.S. Fish and Wildlife Service. (T)

10106 **Mitchell, R. (1979).** A checklist of the vascular plants in Abbott Creek Research Natural Area, Oregon. Research Note No. 341. U.S. Department of Agriculture Forest Service. Illus. map. (P)

10107 **Siddall, J.L. (1977).** Field checking progress report to botanists on provisional list of rare, threatened and endangered plants in Oregon. Taken from the 1976 conference and worksheets submitted to the Oregon Rare and Endangered Plant Project. 23p. (T)

10108 **Siddall, J.L. (1977).** Provisional list of rare, threatened and endangered plants in Oregon. Portland, OR, Oregon Rare and Endangered Plant Project. 22p. (T)

10109 **Siddall, J.L. (1981).** The role of the volunteer in gathering rare plant information. *In* Morse, L.E., Henifin, M.S., *eds.* Rare plant conservation: geographical data organization. New York, New York Botanical Garden. 95-102. Refers to work in compiling the Oregon threatened plant list.

10110 **Siddall, J.L., Chambers, K.L., Wagner, D.H. (1979).** Rare, threatened and endangered vascular plants in Oregon - an interim report. Salem, Natural Area Preserves Advisory Committee. 109p. Over 600 species listed with descriptions. (T)

U.S. - Oregon

10111 **Soper, C., Kagan, J., Yamamoto, S. (1983).** Rare and endangered plants and animals of Oregon. Portland, OR, The Nature Conservancy. (T)

10112 **Thomas, J.W., ed. (1979).** Wildlife habitats in managed forests: the Blue Mountains of Oregon and Washington. Agriculture Handbook No. 553. U.S. Department of Agriculture, Forest Service. 512p. Illus., col. illus., maps.

10113 **U.S. Department of Agriculture. Forest Service (1978).** Desolation Planning Unit, Umatilla, Wallowa-Whitman and Malheur National Forests, Umatille. Union and Grant Countries, Oregon. Portland, OR, U.S. Department of Agriculture Forest Service, Pacific Northwest Region. viii, 550p. Illus., maps. (P)

10114 **U.S. Department of Agriculture. Forest Service (1978).** Land-management plan for Deschutes National Forest and portions of Willamette and Umpqua National Forests. Portland, OR, U.S. Department of Agriculture Forest Service. 635p. (P)

10115 **U.S. Department of Agriculture. Forest Service (1978).** Oregon Dunes National Recreation Area: management plan (supplement). Corvallis, OR, U.S. Department of Agriculture Forest Service. 6p. (P)

10116 **U.S. Department of Agriculture. Forest Service (1979).** Desolation Planning Unit, Umatilla, Wallowa-Whitman and Malhew National Forests. Portland, OR, U.S. Department of Agriculture Forest Service. 559p. (P)

10117 **U.S. Department of Agriculture. Forest Service (1979).** Elgin Planning Unit land-management plan, Umatilla, Union, and Wallowa Counties, Oregon. Portland, OR, U.S. Department of Agriculture Forest Service. 584p. (P)

10118 **U.S. Department of Agriculture. Forest Service (1979).** Hells Canyon National Recreation Area land and resource management plan. Baker, OR, U.S. Department of Agriculture Forest Service. 418p. (P)

10119 **U.S. Department of Agriculture. Forest Service (1979).** Heppner Planning Unit land-management plan, Umatilla National Forest. Portland, OR, U.S. Department of Agriculture Forest Service. 494p. (P)

10120 **U.S. Department of Agriculture. Forest Service (1979).** Rogue - Illinois land management plan, Josephine, Curry and Coos Counties, Oregon. Grants Pass, U.S. Department of Agriculture Forest Service. xiv, 370p. Illus., maps.

10121 **Walsh, S.J. (1980).** Coniferous tree species mapping using Landsat data. *Remote Sens. Envir.*, 9(1): 11-16. Illus., maps. Crater Lake National Park. (P)

U.S. - Pennsylvania

10122 **Buker, W.E. (1975).** Rare western Pennsylvania plants. Pittsburgh, PA, Carnegie Museum, Section of Plants. 4p. (T)

10123 **Cook, R.T. (1981).** News and notes. Natural areas protected by The Nature Conservancy in Pennsylvania and New Jersey. *Bartonia*, 48: 44-45. (P)

10124 **Countryman, W.D. (1977).** The northeastern United States. *In* Prance, G.T., Elias, T.S., *eds.* Extinction is forever. Proceedings of a symposium at the New York Botanical Garden, 11-13 May 1976. New York, New York Botanical Garden. 30-35. Legislation regarding conservation (primarily of plants and natural areas) in the northeastern U.S. (L/P/T)

10125 **Genoways, H.H., Brenner, F.J., eds (1985).** New publications: species of special concern in Pennsylvania. *Endangered Species Tech. Bull.*, 10(5): 12. (T)

10126 **Gress, E.M. (1925).** Preservation of wild flowers in Pennsylvania. *Bull. Pennsylvania Dept. Agric.*, 8(3): 1-17. (T)

10127 **Rhoads, A.F. (1984).** Rare Pennsylvania plants. *Bartonia*, 50: 61-63. (T)

U.S. - Pennsylvania

10128 **Schweiger, L.J. (1979).** New threat to environment: acid rain on Pennsylvania forests, waters. *Pennsylvania Forests*, 69(2): 6-8. Illus., map.

10129 **Wherry, E.T. (1975).** Rare plants of southeastern Pennsylvania. *Bartonia*, 44: 22-26. 1975-1976. (T)

10130 **Wiegman, P.G. (1979).** Rare and endangered vascular plant species in Pennsylvania. Newton Corner, MA, U.S. Fish and Wildlife Service and Western Pennsylvania Conservancy. 93p. List of species and bibliography. (B/T)

U.S. - Rhode Island

10131 **Church, G.L. (1980).** Plant conservation concerns in Rhode Island. *Rhodora*, 82(829): 145-149. (T)

10132 **Church, G.L., Champlin, R.L. (1978).** Rare and endangered vascular plants species in Rhode Island. Newton Corner, MA, U.S. Fish and Wildlife Service, New England Botanical Club. ii, 17p. 120 species listed and annotated. (T)

10133 **Countryman, W.D. (1977).** The northeastern United States. *In* Prance, G.T., Elias, T.S., *eds*. Extinction is forever. Proceedings of a symposium at the New York Botanical Garden, 11-13 May 1976. New York, New York Botanical Garden. 30-35. Legislation regarding conservation (primarily of plants and natural areas) in the northeastern U.S. (L/P/T)

10134 **U.S. Department of the Interior. Fish and Wildlife Service (1988).** Endangered and threatened wildlife and plants; determination of *Agalinis acuta* (Sandplain gerardia) to be an Endangered species. *Fed. Register*, 53(173): 34701-34705. (L/T)

10135 **U.S. Department of the Interior. Fish and Wildlife Service (1989).** Sandplain gerardia (*Agalinis acuta*) recovery plan. Newton Corner, MA, U.S. Department of the Interior, Fish and Wildlife Service. 47p. Illus., map. (L/T)

U.S. - South Carolina

10136 **Forsythe, D.M., Ezell, W.B., jr, eds (1979).** Proceedings of the first South Carolina Endangered Species Symposium. Charleston, Nongame-Endangered Species Section, South Carolina Wildlife and Marine Resources Department and The Citadel. 201p. 147 species listed.

10137 **Gaddy, L.L. (1979).** The hollies of the proposed Congaree Swamp National Monument. *Holly Letter*, 64: 4-6. Illus. (P)

10138 **Gaddy, L.L., Rayner, D.A. (1980).** Rare or overlooked? Recent plant collections from the coastal plain of South Carolina. *Castanea*, 45(3): 181-184. (T)

10139 **Haehnle, G.G., Jones, S.M. (1985).** Geographical distribution of *Quercus oglethorpensis*. *Castanea*, 50(1): 26-31. Illus., maps. (T)

10140 **Hardin, J.W. (1977).** The southeastern United States. *In* Prance, G.T., Elias, T.S., *eds*. Extinction is forever. Proceedings of a symposium at the New York Botanical Garden, 11-13 May 1976. New York, New York Botanical Garden. 36-40. Discusses status of state programmes within the 14 southeastern states of endangered plant species. (L/P/T)

10141 **Jones, S.M., Dunn, B.A. (1979).** Aspects of the biology of *Shortia galacifolia* (Diapensiaceae). *ASB Bull.*, 26(2): 92. Abstract. (T)

10142 **Kohlsaat, T., South Carolina Wildlife and Marine Resource Department (1971).** Working list of endangered and threatened plant species of South Carolina. Columbia, SC, South Carolina Wildlife and Marine Resource Dept. (T)

10143 **Martin, C.E., Christy, E.J., McLeod, K.W. (1977).** Changes in the vegetation of a South Carolina swamp following cessation of thermal pollution. *J. Elisha Mitchell Scientific Soc.*, 93(4): 173-176. Wetlands.

U.S. - South Carolina

10144 **Parker, W., Dixon, L. (1980).** Endangered and threatened wildlife of Kentucky, North Carolina, South Carolina and Tennessee. Raleigh, NC, North Carolina Agricultural Extension Service. 116p. (T)

10145 **Porcher, R.D. (1981).** Endangered and threatened vascular plants of the Francis Marion National Forest, South Carolina: part 2. *ASB Bull.*, 28(2): 83. Abstr. of paper, 42nd Ann. Meeting of the Assoc. of Southeastern Biologists, Knoxville, Texas. (P/T)

10146 **Porcher, R.D. (1981).** Endangered and threatened vascular plants of the Francis Marion National Forest, South Carolina: part 1. *ASB Bull.*, 28(2): 82-83. Abstr. of paper, 42nd Ann. Meeting of the Assoc. of Southeastern Biologists, Knoxville, Texas. (P/T)

10147 **Rayner, D.A. (1979).** Native vascular plants: endangered, threatened or otherwise in jeopardy in South Carolina. *Bull. South Carolina Mus.*, 4: 1-22. (T)

10148 **Rodgers, C.L., et al. (1979).** Status report: native vascular plants endangered, threatened, or otherwise in jeopardy. *In* Forsythe, D.M., Ezell, W.B., *jr, eds.* Proceedings of the first South Carolina endangered species symposium, Nov. 11-12, 1979, Charleston, SC. Charleston, SC, The Citadel. 26-32. (T)

10149 **Schilling, T. (1984).** *Elliottia racemosa. Garden* (London), 109(9): 349-350. Col. illus. (T)

10150 **Sutter, R.D. (1982).** Is Venus fly trap (*Dionaea muscipula*) an endangered species? *ASB Bull.*, 29(2): 86. (T)

10151 **U.S. Department of the Interior. Fish and Wildlife Service (1984).** Endangered and threatened wildlife and plants: *Ribes echinellum* (Miccosukee gooseberry) proposed to be a Threatened species. *Fed. Register*, 49(171): 34535-34538. (L/T)

10152 **U.S. Department of the Interior. Fish and Wildlife Service (1984).** Endangered and threatened wildlife and plants: public hearing and extension of comment period on proposed Threatened status for *Ribes echinellum* (Miccosukee gooseberry). *Fed. Register*, 49(205): 41266-41267. (L/T)

10153 **U.S. Department of the Interior. Fish and Wildlife Service (1984).** Miccosukee Gooseberry proposed for listing as Threatened. *Endangered Species Tech. Bull.*, 9(9): 30. (L/T)

10154 **U.S. Department of the Interior. Fish and Wildlife Service (1985).** Endangered and threatened wildlife and plants: determination of Endangered status for *Lindera melissifolia* (Pondberry). *Fed. Register*, 50(147): 27495-27500. (L/T)

10155 **U.S. Department of the Interior. Fish and Wildlife Service (1985).** Endangered and threatened wildlife and plants: determination of Threatened status for *Ribes echinellum* (Miccosukee gooseberry). *Fed. Register*, 50(138): 29338-29341. (L/T)

10156 **U.S. Department of the Interior. Fish and Wildlife Service (1985).** Endangered and threatened wildlife and plants: proposal to determine *Oxypolis canbyi* (Canby's dropwort) to be an Endangered species. *Fed. Register*, 50(60): 12345-12348. (L/T)

10157 **U.S. Department of the Interior. Fish and Wildlife Service (1985).** Endangered and threatened wildlife and plants: proposed Endangered status for *Lindera melissifolia* (Pondberry). *Fed. Register*, 50(156): 32582-32585. (L/T)

10158 **U.S. Department of the Interior. Fish and Wildlife Service (1985).** Protection sought for four vulnerable plants. *Endangered Species Tech. Bull.*, 19(9): 1, 8. (L/T)

10159 **U.S. Department of the Interior. Fish and Wildlife Service (1986).** Endangered and threatened wildlife and plants: determination of *Oxypolis canbyi* (Canby's dropwort) to be an Endangered species. *Fed. Register*, 51(37): 6690-6693. (L/T)

U.S. - South Carolina

10160 **U.S. Department of the Interior. Fish and Wildlife Service (1988).** Endangered and threatened wildlife and plants; determination of *Helonias bullata* (Swamp pink) to be an Endangered species. *Fed. Register*, 53(175): 35076-35080. (L/T)

10161 **U.S. Department of the Interior. Fish and Wildlife Service (1988).** Endangered and threatened wildlife and plants; determination of Endangered status for *Ptilimnium nodosum*. *Fed. Register*, 53(188): 37978-37982. (L/T)

10162 **U.S. Department of the Interior. Fish and Wildlife Service (1988).** Endangered and threatened wildlife and plants; determination of Endangered status for *Sarracenia rubra* ssp. *jonesii* (Mountain sweet pitcher plant). *Fed. Register*, 53(190): 38470-38474. (L/T)

10163 **U.S. Department of the Interior. Fish and Wildlife Service (1988).** Endangered and threatened wildlife and plants; determination of Endangered status for *Trillium reliquum* (Relict trillium). *Fed. Register*, 53(64): 10879-10884. (L/T)

10164 **U.S. Department of the Interior. Fish and Wildlife Service (1988).** Endangered and threatened wildlife and plants; proposed Endangered status for *Sarracenia rubra* ssp. *jonesii* (Mountain sweet pitcher plant). *Fed. Register*, 53(27): 3901-3905. (L/T)

10165 **U.S. Department of the Interior. Fish and Wildlife Service (1989).** Endangered and threatened wildlife and plants; Threatened status of *Hexastylis naniflora* (Dwarf-flowered heartleaf). *Fed. Register*, 54(71): 14964-14967. (L/T)

10166 **U.S. Department of the Interior. Fish and Wildlife Service (1989).** Endangered and threatened wildlife and plants; determination of Endangered status for *Rhus michauxii* (Michaux's sumac). *Fed. Register*, 54(187): 39854-39857. (L/T)

10167 **U.S. Department of the Interior. Fish and Wildlife Service (1989).** Endangered and threatened wildlife and plants; proposed Endangered status for *Rhus michauxii* (Michaux's sumac). *Fed. Register*, 54(4): 441-445. (L/T)

10168 **Zingmark, R.G., et al. (1979).** Status report: Lower plants. *In* Forsythe, D.M., Exell, W.B., jr., *eds.* Proceedings of the First SC Endangered Species Symposium, Nov. 11-12, 1976, Charleston, SC. Charleston, SC, The Citadel. 22-25. (T)

U.S. - South Dakota

10169 **Bock, J.H. & C.E. (1984).** Effect of fires on woody vegetation in the pine-grassland ecotone of the southern Black Hills. *Amer. Midl. Nat.*, 112(1): 35-42. Prescribed burns, Wind Cave National Park, South Dakota. (P)

10170 **Johnston, B.C., et al. (1980).** Proposed and recommended threatened and endangered plant species of the Forest Service Rocky Mountain Region: an illustrated guide to certain species in Colorado, Wyoming, South Dakota, Nebraska and Kansas. Denver, CO, U.S. Fish and Wildlife Service. 164p. (T)

10171 **Mohlenbrock, R.H. (1977).** The midwestern United States. *In* Prance, G.T., Elias, T.S., *eds.* Extinction is forever. Proceedings of a symposium at the New York Botanical Garden, 11-13 May 1976. New York, New York Botanical Garden. 41-44. Discusses the status of legal protection for threatened and endangered species of plants which occur in the midwestern states. (L/T)

10172 **Schumacher, C.M. (1975).** Endangered plants of South Dakota, preliminary draft. Huron, SD, South Dakota Endangered Species Committee, U.S. Dept. of Agriculture. (T)

10173 **Schumacher, C.M. (1977).** Endangered and threatened species - plants of South Dakota. Huron, SD, South Dakota Endangered Species - Plant Committee. 4p. (T)

10174 **Schumacher, C.M. (1979).** Status of endangered and threatened plants in South Dakota. Technical Notes, Environment, 9. Huron, U.S. Department of Agriculture Soil Conservation Service. (T)

U.S. - South Dakota

10175 **U.S. Department of Agriculture. Forest Service (1979).** Norbeck Wildlife Preserve land-management plan, Black Hills National Forest. Custer, U.S. Department of Agriculture Forest Service. 119p. (P)

10176 **U.S. Department of the Interior. National Park Service (1982).** Badlands National Park, South Dakota. Denver, CO, U.S. Department of the Interior, National Park Service. 138p. (P)

U.S. - Tennessee

10177 **Anon. (1979).** *Sarracenia oreophila. AABGA Newsl.,* 59: 8. Illus. (T)

10178 **Anon. (1979).** Tennessee purple coneflower endangered. *Amer. Hort. Soc. News Views,* 21(6): 6. (T)

10179 **Atwood, J.T. (1985).** The range of *Cypripedium kentuckiense. Amer. Orch. Soc. Bull.,* 54(10): 1197-1199. Illus., maps. (T)

10180 **Baskin, J.M. & C.C. (1982).** Effects of vernalization and photoperiod on flowering in *Echinacea tennesseensis,* an endangered species. *J. Tennessee Acad. Sci.,* 57(2): 53-56. (T)

10181 **Bratton, S.P. (1979).** Preliminary status of rare plants in Great Smoky Mountains National Park. Research/Resources Management Report No. 25. Gatlinburg, Tenn., U.S. Department of the Interior, National Park Service, Southeast Region. 46p. Map. (P/T)

10182 **Bratton, S.P. (1981).** Information storage and population monitoring within Great Smoky Mountain National Park. *In* Morse, L.E., Henefin, M.S., *eds.* Rare plant conservation: geographical data organization. New York, New York Botanical Garden. 63-68. (P/T)

10183 **Bratton, S.P., Hickler, M.G., Graves, J.H. (1978).** Visitor impact on backcountry campsites in the Great Smoky Mountains National Park. *Envir. Manage.,* 2(5): 431-441. Illus. (P)

10184 **Committee for Tennessee Rare Plants (1978).** The rare vascular plants of Tennessee. *J. Tenn. Acad. Sci.,* 53(4): 128-133. (T)

10185 **Devine, H.A., jr, Borden, F.Y., Turner, B.J. (1976).** A simulation study of the Cades Cove visitor vehicle flow (Gt Smokey Mts National Park). National Park Service Occas. Papers No. 4. Washington, D.C., U.S. Govt Printers. vi, 25p. Map. (P)

10186 **Farmer, R.E. (1975).** Rare, endangered and threatened flora propagation newsletter. Norris, TN, Tennessee Valley Authority. n.p. (G/T)

10187 **Goff, F.G., et al. (1975).** Rare, endangered and endemic taxa and their habitats of the East Tennessee Development District. Publ. No. 684. Oak Ridge National Laboratory Environmental Sci. Div. 32p. (T)

10188 **Hardin, J.W. (1977).** The southeastern United States. *In* Prance, G.T., Elias, T.S., *eds.* Extinction is forever. Proceedings of a symposium at the New York Botanical Garden, 11-13 May 1976. New York, New York Botanical Garden. 36-40. Discusses status of state programmes within the 14 southeastern states of endangered plant species. (L/P/T)

10189 **Herrmann, R., Bratton, S. (1977).** Great Smoky Mountains National Park as a biosphere reserve: a research/monitoring perspective. Management Report No. 23. Atlanta, GA, U.S. Department of the Interior. National Park Service, Southeast Region. 38p. (P)

10190 **Horn, G.S. van, Williams, L.G. (1981).** Notes and news. New county records for endangered and threatened species in Tennessee. *Castanea,* 46(4): 343-345. *Talinum tenerifolium, Lilium philadelphicum, Lymnophyllus fraseri, Lycopodiun alopecuroides, Hydrastis canadensis.* (T)

U.S. - Tennessee

10191 **Johnson, W.C., Bratton, S.P. (1978).** Biological monitoring in Unesco biosphere reserves with special reference to the Great Smoky Mountains National Park. *Biol. Conserv.*, 13(2): 105-115. (P)

10192 **Lindsay, M.M., Bratton, S.P. (1979).** Grassy balds of the Great Smoky Mountains: their history and flora in relation to potential management. *Envir. Manag.*, 3(5): 417-430. Illus. (P)

10193 **Lindsay, M.M., Bratton, S.P. (1979).** The vegetation of grassy balds and other high elevation disturbed areas in the Great Smoky Mountains National Park. *Bull. Torrey Bot. Club*, 106(4): 264-275. Illus., maps. (P)

10194 **Lindsay, M.M., Bratton, S.P. (1980).** The rate of woody plant invasion on two grassy balds. *Castanea*, 45(2): 75-87. Illus., maps. Great Smoky Mountains National Park. (P)

10195 **Man, L.K., Patrick, T.S., DeSelm, H.R. (1985).** A checklist of the vascular plants of the Department of Energy Oak Ridge Reservation. *J. Tenn. Acad. Sci.*, 60(1): 8-13. Maps. Montane forest. (P)

10196 **Mellichamp, T.L. (1982).** *Astilbe crenatiloba*: extinct or non-existent? *ASB Bull.*, 29(2): 72. (T)

10197 **Parker, W., Dixon, L. (1980).** Endangered and threatened wildlife of Kentucky, North Carolina, South Carolina and Tennessee. Raleigh, NC, North Carolina Agricultural Extension Service. 116p. (T)

10198 **Parr, P.D. (1984).** Endangered and threatened plant species on the Department of Energy Oak Ridge Reservation: an update. *J. Tenn. Acad. Sci.*, 59(4): 65-68. (T)

10199 **Quarterman, E. (1973).** Endangered species of plants in middle Tennessee. Nashville, TN, Vanderbilt Univ., Biol. Dept. (T)

10200 **Sharp, A.J. (1974).** Rare plants of Tennessee. *Tennessee Conserv.*, 40: 20-21. (T)

10201 **U.S. Department of the Interior. Fish and Wildlife Service (1983).** Habitat description notice on green pitcher plant. *Endangered Species Tech. Bull.*, 8(9): 4, 8. (L/T)

10202 **U.S. Department of the Interior. Fish and Wildlife Service (1983).** Tennessee coneflower. *Endangered Species Tech. Bull.*, 8(7): 7-8, 12. Illus. (T)

10203 **U.S. Department of the Interior. Fish and Wildlife Service (1984).** Endangered and threatened wildlife and plants: proposed Endangered status for *Pityopsis ruthii* (Ruth's golden aster). *Fed. Register*, 49(225): 45766-45769. (L/T)

10204 **U.S. Department of the Interior. Fish and Wildlife Service (1984).** Endangered and threatened wildlife and plants: proposed Threatened status for *Solidago spithamaea* (Blue ridge goldenrod). *Fed. Register*, 49(142): 29629-29632. (L/T)

10205 **U.S. Department of the Interior. Fish and Wildlife Service (1984).** Four plants in danger of extinction. *Endangered Species Tech. Bull.*, 9(8): 1, 5 Illus. (L/T)

10206 **U.S. Department of the Interior. Fish and Wildlife Service (1985).** Endangered and threatened wildlife and plants: determination of Endangered status for *Pityopsis ruthii* (Ruth's golden aster). *Fed. Register*, 50(138): 29341-29345. (T)

10207 **U.S. Department of the Interior. Fish and Wildlife Service (1985).** Endangered and threatened wildlife and plants: determination of Threatened status for *Solidago spithamaea* (Blue ridge goldenrod). *Fed. Register*, 50(60): 12306-12309. (L/T)

10208 **U.S. Department of the Interior. Fish and Wildlife Service (1985).** Endangered and threatened wildlife and plants: proposed Endangered status for *Scutellaria montana. Fed. Register*, 50(219): 46797-46799. (L/T)

U.S. - Tennessee

10209 **U.S. Department of the Interior. Fish and Wildlife Service (1985).** Listings become final for two animals and seven plants. *Endangered Species Tech. Bull.*, 10(8): 1,5-6. (L/T)

10210 **U.S. Department of the Interior. Fish and Wildlife Service (1986).** Endangered and threatened wildlife and plants: determination of Endangered status for *Scutellaria montana* (Large-flowered skullcap). *Fed. Register*, 51(119): 22521-22524. (L/T)

10211 **U.S. Department of the Interior. Fish and Wildlife Service (1988).** Endangered and threatened wildlife and plants; determination of Endangered status for *Arenaria cumberlandensis*. *Fed. Register*, 53(121): 23745-26369. (L/T)

10212 **U.S. Department of the Interior. Fish and Wildlife Service (1989).** Endangered and threatened wildlife and plants; Threatened status for *Phyllitis scolopendrium* var. *americana* (American hart's-tongue). *Fed. Register*, 54(134): 29726-29730. (L/T)

10213 **U.S. Department of the Interior. Fish and Wildlife Service (1989).** Endangered and threatened wildlife and plants; proposed Endangered status for *Geum radiatum* and *Hedyotis purpurea* var. *montana*. *Fed. Register*, 54(139): 30572-30576. (L/T)

10214 **U.S. Department of the Interior. Fish and Wildlife Service (1989).** Proposed listings - July 1989. Two southern Appalachian plants. *Endangered Species Tech. Bull.*, 14(8): 4. *Geum radiatum, Hedyotis purpurea* var. *montana*. (L/T)

10215 **U.S. Department of the Interior. National Park Service (1979).** Stones River National Battlefield and Cemetery, Munfreesboro, Tennessee. Denver, CO, National Park Service. (P)

10216 **U.S. Department of the Interior. National Park Service (1980).** Stones River National Battlefield & Cemetery, Murfreesboro, Tennessee. Denver, CO, National Park Service. (P/T)

10217 **White, A.J. (1978).** Range extension of the proposed endangered plant, *Heterotheca ruthii* (Compositae) (Tennessee). *Castanea*, 43(4): 263. (T)

10218 **White, P.S. (1981).** Rarity: the case for vascular plants at Great Smoky Mountains National Park. *A.S.B. Bull.*, 28(2): 84. Abstract of paper, 42nd. Annual Meeting of the Assoc. of Southeastern Biologists, Knoxville, Tennessee. (P/T)

10219 **Wofford, B.E. (1976).** The Tennessee list of possibly extinct, endangered, threatened and special concern vascular plants. Knoxville, TN, Univ. of Tennessee, Dept. of Botany. 12p. (T)

U.S. - Texas

10220 **Anon. (1980).** Two new plants listed as endangered. *Amer. Horticulturist*, 8 (November 1980). *Spiranthes parksii, Callirhoe scabriuscula*. (L/T)

10221 **Beaty, H.E., et al. (1983).** Endangered, threatened and watch lists of plants of Texas. Austin, Texas Organization for Endangered Species. 7p. Lists habitats and reason for status. (T)

10222 **Bennett, B.C. (1985).** Notes on the Bromeliaceae of Big Bend National Park. *J. Bromeliad Soc.*, 35(1): 24-25. (P)

10223 **Christenson, E.A. (1988).** Conservation of *Spiranthes parksii*: a beginning. *Orchid Rev.*, 96(1135): 148-149. Illus., col. illus. (T)

10224 **Corin, C.W. (1988).** Black lace cactus recovery plan. *Endangered Species Tech. Bull.*, 13(5): 7-8. Illus. (T)

10225 **Everitt, J.H, Alaniz, M.A. (1979).** Propagation and establishment of 2 rare and endangered native plants from southern Texas. *J. Rio Grande Valley Hort. Soc.*, 33: 133-136. (G/T)

U.S. - Texas

10226 **Everitt, J.H. (1978).** Rare and endangered native plants of ornamental value in southern Texas. *J. Rio Grande Valley Hort. Soc.*, 32: 21-25. (T)

10227 **Gardener, S., O'Brien, R. (1988).** Slender rush-pea (*Hoffmannseggia tenella*) recovery plan. Albuquerque, New Mexico, Fish and Wildlife Service. 38p. (T)

10228 **Gehlbach, F.R., Polley, H.W. (1982).** Relict trout lilies *Erythronium mesochoreum* in central Texas: a multivariate analysis of habitat for conservation. *Biol. Conserv.*, 22(4): 251-258. (T)

10229 **Genoways, H.H., Baker, R.J., eds (1979).** Biological investigations in the Guadalupe Mountains National Park, Texas. Proceedings of a Symposium held at Texas Tech. University, Lubbock, Texas, April 4-5, 1975. National Parks Service Proc. & Trans Series No.4. Washington, D.C., U.S. Govt Printing Office. xvii, 442p. Illus., maps. (P)

10230 **Johnston, M.C. (1974).** Rare and endangered plants native to Texas. Austin, Rare Plant Study Center, University of Texas. 12p. (T)

10231 **Johnston, M.C. (1977).** The southwestern United States. *In* Prance, G.T., Elias, T.S., eds. Extinction is forever. Proceedings of a symposium at the New York Botanical Garden, 11-13 May 1976. New York, New York Botanical Garden. 60-61. Legislation regarding threatened wild plants in the southwestern U.S. (L/T)

10232 **Mahler, W.F. (1981).** Notes on rare Texas and Oklahoma USA plants. *Sida Contrib. Bot.*, 9(1): 76-86. (T)

10233 **McMahan, L.R. (1987).** Rare plant conservation by state government: case studies from the western United States. *In* Elias, T.S., ed. Conservation and management of rare and endangered plants. Proceedings from a conference, Sacramento, California, 5-8 November 1986. Sacramento, California Native Plant Society. 23-31. Maps. (L/T)

10234 **Nabhan, G.P., Greenhouse, R., Hodgson, W. (1986).** At the edge of extinction: useful plants of the border states of the United States and Mexico. *Arnoldia*, 46(3): 36-46. Illus. (E)

10235 **Norquist, C. (1984).** Savannahs and bogs of the southern U.S. U.S. threatened ecosystems. *Endangered Species Tech. Bull.*, 10(9): 4-5 (T)

10236 **Price, S. (1980).** Where the rainbow waits for the rain. *Nat. Parks Conserv. Mag.*, 54(11): 12-17. Illus., col. illus., map. Big Bend National Park, Texas. (P)

10237 **Rare Plant Study Center (1974).** Rare and endangered plants native to Texas. Austin, TX, Univ. of Texas. 12p. (T)

10238 **Riskind, D. (1975).** Provisional list of Texas' threatened and endangered plant species. Austin, TX, Texas Organization of Threatened and Endangered Species, Texas Parks and Wildlife Dept. (T)

10239 **Terrell, E.E., Emery, W.H.P., Beaty, H.E. (1978).** Observations on *Zizania texana* (Texas wildrice), an endangered species. *Bull. Torrey Bot. Club*, 105(1): 50-57. (T)

10240 **U.S. Department of the Interior. Fish and Wildlife Service (1982).** Endangered and threatened wildlife and plants: determination of *Spiranthes parksii* (Navasota ladies'-tresses) to be an Endangered species. *Fed. Register*, 47(88): 19539-19542. (L/T)

10241 **U.S. Department of the Interior. Fish and Wildlife Service (1983).** Endangered and threatened wildlife and plants: proposal to determine *Dyssodia tephroleuca* (Ashy dogweed) to be an Endangered species. *Fed. Register*, 48(142): 33501-33503. (L/T)

10242 **U.S. Department of the Interior. Fish and Wildlife Service (1983).** Endangered and threatened wildlife and plants: proposal to determine *Styrax texana* (Texas snowbells) to be an Endangered species. *Fed. Register*, 48(197): 46086-46088. (L/T)

U.S. - Texas

10243 **U.S. Department of the Interior. Fish and Wildlife Service (1983).** Endangered and threatened wildlife and plants: proposed rule to determine *Frankenia johnstonii* (Johnston's frankenia) to be an Endangered species. *Fed. Register*, 48(132): 31414-31417. (T)

10244 **U.S. Department of the Interior. Fish and Wildlife Service (1983).** Three plants proposed as Endangered. *Endangered Species Tech. Bull.*, 8(10): 1, 4-5. Illus. (E/L/T)

10245 **U.S. Department of the Interior. Fish and Wildlife Service (1984).** Endangered and threatened wildlife and plants: final rule to determine *Dyssodia tephroleuca* (Ashy dogweed) to be an Endangered species. *Fed. Register*, 49(140): 29232-29234. (L/T)

10246 **U.S. Department of the Interior. Fish and Wildlife Service (1984).** Endangered and threatened wildlife and plants: final rule to determine *Frankenia johnstonii* (Johnston's frankenia) to be an Endangered species. *Fed. Register*, 49(153): 31418-31421. (L/T)

10247 **U.S. Department of the Interior. Fish and Wildlife Service (1984).** Endangered and threatened wildlife and plants: final rule to determine *Styrax texana* (Texas snowbells) to be an Endangered species. *Fed. Register*, 49(199): 40036-40038. (L/T)

10248 **U.S. Department of the Interior. Fish and Wildlife Service (1984).** Endangered and threatened wildlife and plants: proposal to determine *Hoffmanseggia tenella* to be an Endangered species. *Fed. Register*, 49(226): 45884-45887. (L/T)

10249 **U.S. Department of the Interior. Fish and Wildlife Service (1984).** Endangered status for three western plants. *Endangered Species Tech. Bull.*, 9(9): 6-7. (L/T)

10250 **U.S. Department of the Interior. Fish and Wildlife Service (1984).** Endangered status given to Texas plant. *Endangered Species Tech. Bull.*, 9(11): 1, 12. (L/T)

10251 **U.S. Department of the Interior. Fish and Wildlife Service (1984).** Protection becomes final for four plants. *Endangered Species Tech. Bull.*, 9(8): 3-5. Illus. (L/T)

10252 **U.S. Department of the Interior. Fish and Wildlife Service (1985).** Endangered and threatened wildlife and plants: listing *Hoffmanseggia tenella* as an Endangered species. *Fed. Register*, 50(212): 45614-45618. (L/T)

10253 **U.S. Department of the Interior. Fish and Wildlife Service (1985).** Endangered and threatened wildlife and plants: proposal to determine *Hymenoxys texana* to be an Endangered species. *Fed. Register*, 50(44): 9095-9097. (L/T)

10254 **U.S. Department of the Interior. Fish and Wildlife Service (1985).** Final protection given to four plants. *Endangered Species Tech. Bull.*, 10(11): 5. *Hoffmannseggia tenella* (Slender rush-pea). *Cereus eriophorus* var. *fragrans* (Fragrant prickly apple). *Dicerandra frutescens* (Scrub mint),. *D. cornutissima* (Longspurred mint). (L/T)

10255 **U.S. Department of the Interior. Fish and Wildlife Service (1986).** Endangered and threatened wildlife and plants: determination of Endangered status for *Hymenoxys texana*. *Fed. Register*, 51(49): 8681-8683. (L/T)

10256 **U.S. Department of the Interior. Fish and Wildlife Service (1986).** Endangered and threatened wildlife and plants: proposed Endangered staus for *Lesquerella pallida* (White bladderpod). *Fed. Register*, 51(68): 12184-12187. (L/T)

10257 **U.S. Department of the Interior. Fish and Wildlife Service (1988).** Endangered and threatened wildlife and plants; determination of *Abronia macrocarpa* (Large-fruited sand verbena) to be an Endangered species. *Fed. Register*, 53(188): 37975-37978. (L/T)

U.S. - Texas

10258 **U.S. Department of the Interior. Fish and Wildlife Service (1988).** Endangered and threatened wildlife and plants; determination of *Echinocereus chisoensis* var. *chisoensis* (Chisos Mountain hedgehog cactus) to be a Threatened species. *Fed. Register*, 53(190): 38453-38456. (L/T)

10259 **U.S. Department of the Interior. Fish and Wildlife Service (1988).** Endangered and threatened wildlife and plants; determination of *Quercus hinckleyi* (Hinckley oak) to be a Threatened species. *Fed. Register*, 53(166): 32824-32827. (L/T)

10260 **U.S. Department of the Interior. Fish and Wildlife Service (1988).** Endangered and threatened wildlife and plants; notice of six-month extension of the proposed rule for *Boerhavia mathisiana* (Mathis spiderling). *Fed. Register*, 53(135): 26616-26617. (L/T)

10261 **U.S. Department of the Interior. Fish and Wildlife Service (1989).** Endangered and threatened wildlife and plants; withdrawal of the proposed rule to list *Boerhavia mathsiana* as Endangered. *Fed. Register*, 54(124): 27413-27414. (L/T)

10262 **U.S. Department of the Interior. National Park Service (1979).** Master plan, Big Bend National Park, Texas. Santa Fe, National Park Service. 251p. (P)

10263 **Waller, R.H., Riskind, D.H., eds (1977).** Transactions of the Symposium on the Biological Resources of the Chihuahuan Desert Region, United States and Mexico. Sul Ross University, Alpine, Texas, 17-18 October 1974. National Park Service Trans. and Proc. Series No. 3. Washington, D.C., U.S. Dept of the Interior. xxii, 658p. Maps.

U.S. - Utah

10264 **Cottam, W.P. (1929).** Man as a biotic factor illustrated by recent floristic and physiographic changes at Mountain Meadows, Washington County, Utah. *Ecology*, 10: 361-363.

10265 **Loope, L.L., Sanchez, P.G., Tarr, P.W., Loope, W.L., Anderson, R.L. (1988).** Biological invasions of arid land nature reserves. *Biol. Conserv.*, 44(1-2): 95-118. Invasive species. Includes 5 case studies. (P)

10266 **Madany, M.H., West, N.E. (1984).** Vegetation of two relict mesas in Zion National Park. *J. Range Manage.*, 37(5): 456-461. Illus. Forest. (P)

10267 **McMahan, L.R. (1987).** Rare plant conservation by state government: case studies from the western United States. *In* Elias, T.S., *ed.* Conservation and management of rare and endangered plants. Proceedings from a conference, Sacramento, California, 5-8 November 1986. Sacramento, California Native Plant Society. 23-31. Maps. (L/T)

10268 **Mutz, K.M. (1980).** Conflict and coevolution of endemics and energy resources in the Uinta Basin of Utah. *Bot. Soc. Amer. Misc. Ser. Pub.*, 158: 79. Abstr. in Botany 80: abstracts of papers to be presented at the University of British Columbia, Vancouver, 12-16 July 1980.

10269 **Nabhan, G.P., Reichhardt, K.L. (1983).** Hopi protections of *Helianthus anomalus*, a rare sunflower. *Southwestern Naturalist*, 28: 231-235. (T)

10270 **Reveal, J.L. (1977).** The western United States. *In* Prance, G.T., Elias, T.S., *eds.* Extinction is forever. Proceedings of a symposium at the New York Botanical Garden, 11-13 May 1976. New York, New York Botanical Garden. 50-59. Discusses existing laws and present policies regarding endangered plants for five western states of the U.S. (L/T)

10271 **U.S. Department of the Interior. Fish and Wildlife Service (1984).** Endangered and threatened wildlife and plants: proposal to determine *Asclepias welshii* (Welsh's milkweed) to be an Endangered species and to designate its critical habitat. *Fed. Register*, 49(110): 23399-23402. Map. (L/T)

U.S. - Utah

10272 **U.S. Department of the Interior. Fish and Wildlife Service (1984).** Endangered and threatened wildlife and plants: proposal to determine *Erigeron maguirei* var. *maguirei* (Maguire daisy) to be an Endangered species. *Fed. Register*, 49(146): 30211-30214. (L/T)

10273 **U.S. Department of the Interior. Fish and Wildlife Service (1984).** Endangered and threatened wildlife and plants: proposal to determine *Primula maguirei* (Maguire primrose) to be a Threatened species. *Fed. Register*, 49(73): 14771-14774. (L/T)

10274 **U.S. Department of the Interior. Fish and Wildlife Service (1984).** Endangered and threatened wildlife and plants: proposal to determine *Townsendia aprica* to be an Endangered species. *Fed. Register*, 49(104): 22352-22355. (L/T)

10275 **U.S. Department of the Interior. Fish and Wildlife Service (1984).** Endangered and threatened wildlife and plants: public hearing and reopening of comment period on proposed Endangered with critical habitat status for *Asclepias welshii* (Welsh's milkweed). *Fed. Register*, 49(172): 34879. (L/T)

10276 **U.S. Department of the Interior. Fish and Wildlife Service (1984).** Four plants in danger of extinction. *Endangered Species Tech. Bull.*, 9(8): 1, 5 Illus. (L/T)

10277 **U.S. Department of the Interior. Fish and Wildlife Service (1984).** Four plants proposed for listing. *Endangered Species Tech. Bull.*, 9(7): 1, 4-5. Illus. (L/T)

10278 **U.S. Department of the Interior. Fish and Wildlife Service (1984).** Four western plants threatened by habitat degradation. *Endangered Species Tech. Bull.*, 9(5): 1, 8-9. Illus. (L/T)

10279 **U.S. Department of the Interior. Fish and Wildlife Service (1985).** Endangered and threatened wildlife and plants: determination of Endangered status for *Erigeron maguirei* var. *maguirei* (Maguire daisy). *Fed. Register*, 50(172): 36089-36092. (L/T)

10280 **U.S. Department of the Interior. Fish and Wildlife Service (1985).** Endangered and threatened wildlife and plants: final rule to determine *Primula maguieri* to be a Threatened species. *Fed. Register*, 50(62): 33731-33734. (L/T)

10281 **U.S. Department of the Interior. Fish and Wildlife Service (1985).** Endangered and threatened wildlife and plants: proposal to determine *Cycladenia humilis* var. *jonesii* to be an Endangered species. *Fed. Register*, 50(7): 1247-1251. (L/T)

10282 **U.S. Department of the Interior. Fish and Wildlife Service (1985).** Endangered and threatened wildlife and plants: proposal to determine *Glaucocarpum suffrutescens* to be an Endangered species with critical habitat. *Fed. Register*, 50(172): 36118-36122. (L/T)

10283 **U.S. Department of the Interior. Fish and Wildlife Service (1985).** Endangered and threatened wildlife and plants: public hearing and extension of comment period on proposed Endangered status with critical habitat for *Glaucocarpum suffrutescens* (Toad-flax cress). *Fed. Register*, 50(213): 45846. (L/T)

10284 **U.S. Department of the Interior. Fish and Wildlife Service (1985).** Endangered and threatened wildlife and plants: rule to determine *Townsendia aprica* (Last chance townsendia) to be a Threatened species. *Fed. Register*, 50(162): 33734-33737. (L/T)

10285 **U.S. Department of the Interior. Fish and Wildlife Service (1985).** Protection recommended for three plants. *Endangered Species Tech. Bull.*, 10(1): 1, 5-6. (L/T)

10286 **U.S. Department of the Interior. Fish and Wildlife Service (1985).** Seven species receive protection. *Endangered Species Tech. Bull.*, 10(10); 3-4. (L/T)

10287 **U.S. Department of the Interior. Fish and Wildlife Service (1985).** Utah plant, the Jones cycladenia, proposed for listing. *Endangered Species Tech. Bull.*, 10(2): 3,12. *Cycladenia humilis* var. *jonesii.* (L/T)

U.S. - Utah

10288 **U.S. Department of the Interior. Fish and Wildlife Service (1986).** Endangered and threatened wildlife and plants: proposal to determine *Eriogonum humivagans* to be an Endangered species. *Fed. Register*, 51(66): 11880-11883. (L/T)

10289 **U.S. Department of the Interior. Fish and Wildlife Service (1986).** Endangered and threatened wildlife and plants: proposed determination of Endangered status for: *Pediocactus despainii* (San Raphael cactus). *Fed. Register*, 51(59): 10560-10563. (L/T)

10290 **U.S. Department of the Interior. Fish and Wildlife Service (1986).** Endangered and threatened wildlife and plants: rule to determine *Cycladenia humilis* var. *jonesi* to be a Threatened species. *Fed. Register*, 51(86): 16526-16530. (L/T)

10291 **U.S. Department of the Interior. Fish and Wildlife Service (1988).** Endangered and threatened wildlife and plants; proposed delisting of *Astragalus perianus* (Rydberg milkvetch). *Fed. Register*, 53(196): 39626-39627. (T)

10292 **U.S. Department of the Interior. National Park Service (1979).** General management plan, wilderness recommendation and road study alternatives for the Glen Canyon National Recreation Area, Arizona and Utah. Missoula, National Park Service. 357p. (P)

10293 **U.S. Department of the Interior. National Park Service (1982).** Capitol Reef National Park, Utah. Denver, CO, U.S. Department of the Interior, National Park Service. 143p. (P)

10294 **Welsh, S.H., Chatterley, L.M. (1985).** Utah's rare plants revisited. *Great Basin Nat.*, 45(2): 173-237. (T)

10295 **Welsh, S.L. (1976).** Proposed threatened, endangered, presumed extinct or extinct and disjunct relict plants in the Cedar City and Richfield districts, Utah. Final report. n.p., Bureau of Land Management. 205p. (T)

10296 **Welsh, S.L. (1977).** Endangered and threatened plant species of the Central Coal Lands, Utah. Final Report. Interagency Task Force on Coal, U.S. Geological Survey. 422p. (T)

10297 **Welsh, S.L. (1978).** Endangered and threatened plants of Utah: a reevaluation. *Great Basin Nat.*, 38(1): 1-18. (T)

10298 **Welsh, S.L. (1979).** Endangered and threatened plants of Utah: a case study. *In* Wood, S.L., *ed.* The endangered species: a symposium. 69-80. (T)

10299 **Welsh, S.L. (1979).** Illustrated manual of proposed endangered and threatened plants of Utah. Denver, Colorado, Denver Federal Center, U.S. Fish and Wildlife Service. 318p. 148 species listed, with ink drawings, descriptions and dot maps. (T)

10300 **Welsh, S.L., Atwood, N.D., Reveal, J.L. (1975).** Endangered, threatened, extinct, endemic and rare or restricted Utah vascular plants. *Great Basin Nat.*, 35(4): 327-376. (T)

10301 **Wood, S.L., ed. (1979).** The endangered species: a symposium. Great Basin Naturalist Memoirs No. 3. Provo, UT, Brigham Young Univ. 171p. Great Basin Naturalist Memoirs no. 3. (T)

U.S. - Vermont

10302 **Anon. (1980).** Endangered and threatened species determinations. *AABGA Newsl.*, 71: 6. (L/T)

10303 **Countryman, W.D. (1977).** The northeastern United States. *In* Prance, G.T., Elias, T.S., *eds.* Extinction is forever. Proceedings of a symposium at the New York Botanical Garden, 11-13 May 1976. New York, New York Botanical Garden. 30-35. Legislation regarding conservation (primarily of plants and natural areas) in the northeastern U.S. (L/P/T)

U.S. - Vermont

10304 **Countryman, W.D. (1978).** Rare and endangered vascular plant species in Vermont. Newton Corner, MA, U.S. Fish and Wildlife Service, New England Botanical Club. 68p. Map. Species listed and annotated. (T)

10305 **Countryman, W.D. (1980).** Vermont's endangered plants and the threats to their survival. *Rhodora*, 82(829): 163-171. (T)

10306 **Egler, F.E. (1985).** Observations of the changing vegetation of Camel's Hump, Vermont, in relation to acid deposition. *Phytologia*, 57(3): 182-205. Acid rain.

10307 **Field, K.G., Primack, R.B. (1980).** Comparison of the characteristics of rare, common and endemic plant species in New England USA. *In* The University of British Columbia. Second International Congress of Systematic and Evolutionary Biology, Vancouver, B.C. Canada, July 17-24, 1980. Vancouver, B.C., Univ. of British Columbia. (T)

10308 **Foy, L. (1977).** Protection of endangered plants in New England: lists, laws and activities. Framingham, MA, New England Wild Flower Society, Inc. (Various paging). (L/T)

10309 **New England Botanical Club. Natural Areas Criteria Committee (1972).** Guidelines and criteria for the evaluation of natural areas. A report prepared for the New England Natural Resources Center by the Natural Areas Criteria Committee of the New England Botanical Club Inc. (P)

10310 **Tryon, R.M., ed. (1980).** Symposium on rare and endangered plant species in New England, Cambridge, Mass., 1979. The proceedings of the symposium "Rare and Endangered Plant Species in New England," 4-5 May 1979. *Rhodora* (Cambridge, MA, New England Botanical Garden), 82(829): 237p. (T)

10311 **U.S. Department of the Interior. Fish and Wildlife Service (1985).** Endangered and threatened wildlife and plants: proposal to determine *Astragalus robbinsii* var. *jesupii* to be an Endangered species. *Fed. Register*, 50(244): 51718-51722. (L/T)

10312 **U.S. Environmental Protection Agency, Region 1 (1982).** New England Wetland Flora and Wetland Protection Laws. Washington, D.C., U.S. Govt Print. Off. 172p. Illus. (L/P)

10313 **Vermont (1974).** Protection of endangered species. Vermont Statutes Annotated. Title 13, Chapter 79. 22-23. (L)

10314 **Zika, P.F., Dann, K.T. (1985).** Rare plants on ultramafic soils in Vermont. *Rhodora*, 87(851): 293-304. Maps. (T)

U.S. - Virginia

10315 **Acken, D. van (1981).** Geplante "Wildnis" der Shenandoah National Park in Virginia, U.S.A. *Gartn.-Bot. Brief*, 68: 37-43. Ge (En). (P)

10316 **Bartgis, R.L., Lang, G.E. (1984).** Marl Wetlands in eastern West Virginia: distribution, rare plant species and recent history. *Castanea.*, 49(1): 17-25. Map. (E/T)

10317 **Core, E.L. (?)** Rare and endangered plant species in West Virginia. Charleston, West Virginia Dept of Agriculture. 7p. (T)

10318 **Fortney, R.H., et al. (1978).** Rare and endangered species of West Virginia, a preliminary report. Vol. 1: Vascular Plants. Charleston, WV, West Virginia Dept of Natural Resources. (T)

10319 **Hardin, J.W. (1977).** The southeastern United States. *In* Prance, G.T., Elias, T.S., *eds.* Extinction is forever. Proceedings of a symposium at the New York Botanical Garden, 11-13 May 1976. New York, New York Botanical Garden. 36-40. Discusses status of state programmes within the 14 southeastern states of endangered plant species. (L/P/T)

U.S. - Virginia

10320 **Harwill, A.M., jr (1975).** Some new and very local populations of rare species in Virginia. *Castanea*, 38: 305-307. (T)

10321 **Linzey, D.W., ed. (1979).** Endangered and threatened plants and animals of Virginia. Blacksburg, Virginia Polytechnic Institute and State University. (T)

10322 **Massey, J., Otte, D., Atkinson, T., Whetstone, R. (1983).** An atlas and illustrated guide to the threatened and endangered vascular plants of the mountains of North Carolina and Virginia. Asheville, North Carolina, U.S. Department of Agriculture, Forest Service, Southeastern Forest Experiment Station. 218p. 45 species listed with descriptions, habitat details, ink drawings, dot maps. (T)

10323 **Mazzeo, P.M. (1972).** An illustrated guide to the ferns and fern allies of Shenandoah National Park, Virginia. Luray, Shenandoah Natural History Association. 52p. Illus. (P)

10324 **Mazzeo, P.M. (1974).** *Betula uber:* what is it and where is it? *Castanea*, 39(3): 273-278. (T)

10325 **Mazzeo, P.M. (1974).** Trees of Shenandoah National Park in the Blue Ridge Mountains of Virginia. Luray, Shenandoah Natural History Association. 80p. (P)

10326 **Mazzeo, P.M. (1977).** *Betula uber* - the preservation of its germplasm. *Virg. J. Sci.*, 28(2): 74. (T)

10327 **Mazzeo, P.M. (1979).** Trees of Shenandoah National Park. Luray, Shenandoah Natural History Association. (P)

10328 **Ogle, D.W. (1980).** An unusual locality for rare plants in southwestern Virginia. *Castanea*, 45(4): 243-247. (T)

10329 **Ogle, D.W. (1989).** Rare vascular plants of the Clinch River Gorge area in Russell County, Virginia. *Castanea*, 54(2): 105-110. (T)

10330 **Ogle, D.W., Mazzeo, P.M. (1976).** *Betula uber*, the Virginia round-leaf birch, rediscovered in southwest Virginia. *Castanea*, 41(3): 248-256. (T)

10331 **Porter, D. (1980).** Rare and endangered vascular plant species in Virginia. Washington, DC, U.S. Fish and Wildlife Service. 52 leaves. (T)

10332 **Porter, D.M. (1979).** Vascular plants. *In* Linzey, D.W., *ed.* Endangered and threatened plants and animals of Virginia. Blacksburg, Virginia Polytechnic Institute and State University. 31-122. 333 species listed, dot maps and descriptions. (T)

10333 **Sharik, T.L., Ford, R.H., Feret, P.P. (1980).** The status of the endangered Virginia round-leaf birch, *Betula uber* (Ashe) Fernald. *Virg. J. Sci.*, 30(2): 58 (1979). Abstract. (T)

10334 **U.S. Department of the Interior. Fish and Wildlife Service (1985).** Endangered and threatened wildlife and plants: proposal to determine *Iliamna corei* (Peter's mountain mallow) to be an Endangered Species. *Fed. Register*, 50(170): 35584-35587. (L/T)

10335 **U.S. Department of the Interior. Fish and Wildlife Service (1985).** Protection sought for four vulnerable plants. *Endangered Species Tech. Bull.*, 19(9): 1, 8. (L/T)

10336 **U.S. Department of the Interior. Fish and Wildlife Service (1986).** Endangered and threatened wildlife and plants: determination of *Iliamna corei* (Peter's mountain mallow) to be an Endangered species. *Fed. Register*, 51(91): 17343-17346. (L/T)

10337 **U.S. Department of the Interior. Fish and Wildlife Service (1988).** Endangered and threatened wildlife and plants; determination of *Helonias bullata* (Swamp pink) to be an Endangered species. *Fed. Register*, 53(175): 35076-35080. (L/T)

U.S. - Virginia

10338 **U.S. Department of the Interior. Fish and Wildlife Service (1988).** Endangered and threatened wildlife and plants; proposal to determine *Arabis serotina* (Shale Barren rock cress) to be an Endangered species. *Fed. Register*, 53(222): 46479-46482. (L/T)

10339 **U.S. Department of the Interior. Fish and Wildlife Service (1989).** Endangered and threatened wildlife and plants; "*Arabis serotina*" (Shale barren rock cress) determined to be an Endangered species. *Fed. Register*, 54(133): 29655-29658. (L/T)

10340 **Uttall, L.J. (1972).** Endangered status of Virginia flowering plants. Blacksburg, VA, Virginia Polytechnic Institute and State Univ., Biol. Dept. 6p. (T)

10341 **Walton, S. (1982).** Chesapeake Bay: threats to ecological stability. *BioScience*, 32(11): 843-844. Map.

10342 **White, C.P. (1982).** Endangered and threatened wildlife of the Chesapeake Bay region. Centreville, Tidewater Publishers, 147p. Illus., maps, col. illus. (T)

U.S. - Washington

10343 **Anon. (1981).** Mount St. Helens proposed as new National Monument. *Nation. Parks*, 55(6): 24. (P)

10344 **California Native Plant Society (1981).** Washington State Natural Heritage Program. *Calif. Native Pl. Soc. Bull.*, 11(5): 9. (P)

10345 **Chambers, K.L. (1977).** The northwestern United States. *In* Prance, G.T., Elias, T.S., *eds*. Extinction is forever. Proceedings of a symposium at the New York Botanical Garden, 11-13 May 1976. New York, New York Botanical Garden. 45-49. General status of programmes in the Pacific northwest. (L/T)

10346 **Denton, M., et al. (1977).** A working list of rare, endangered, threatened and endemic vascular plant taxa for Washington. Seattle, WA, Univ. of Washington, Dept. of Botany. Revised 1978 + 1981, Washington Natural Heritage Program, Olympia, Washington. (T)

10347 **Franklin, J.F. (1982).** Ecosystem studies in the Hoh River drainage, Olympic National Park. *In* Starkey, E., Franklin, J., Matthews, J., *eds*. Ecological research in the national parks of the Pacific NW. Proc. 2nd Conf. on Scientific Research in the National Parks, San Francisco, Calif., Nov. 1979. Illus., maps. (P)

10348 **Hemstrom, M.A., Franklin, J.F. (1982).** Fire and other disturbances of the forests in Mount Rainier National Park. *Quaternary Research*, 18(1): 32-51. Illus., maps. (P)

10349 **Kruckeberg, A.R. (1974).** Partial inventory of rare, threatened, and unique plants of Washington. *In* Dyrness, C.T., *et al*. Research natural area needs in the Pacific Northwest - a contribution to land-use planning. Report on natural area needs, 29 November - 1 December 1973 at Wemme, Oregon. 311-319. (T)

10350 **Lesica, P., Allendorf, F., Leary, R., Bilderback, D. (1988).** Lack of genetic diversity within and among populations of an endangered plant, *Howellia aquatilis*. *Conserv. Biol.*, 2(3): 275-282. En (Sp). Map. (T)

10351 **McMahan, L.R. (1987).** Rare plant conservation by state government: case studies from the western United States. *In* Elias, T.S., *ed*. Conservation and management of rare and endangered plants. Proceedings from a conference, Sacramento, California, 5-8 November 1986. Sacramento, California Native Plant Society. 23-31. Maps. (L/T)

10352 **Sauer, R.H., Mastrogiuseppe, J.D., Smookler, P.H. (1979).** *Astragalus columbianus* (Leguminosae) - rediscovery of an extinct species. *Brittonia*, 31(2): 161-264. (T)

10353 **Sheehan, M. (1984).** Endangered, threatened and sensitive vascular plants of Washington. Olympia, WA, Natural Heritage Program. 29p. (T)

U.S. - Washington

10354 **Sheehan, M., Schuller, R. (1981).** An illustrated guide to the endangered, threatened and sensitive vascular plants of Washington. Olympia, WA, Natural Heritage Program. 328p. (T)

10355 **Thomas, J.W., ed. (1979).** Wildlife habitats in managed forests: the Blue Mountains of Oregon and Washington. Agriculture Handbook No. 553. U.S. Department of Agriculture, Forest Service. 512p. Illus., col. illus., maps.

10356 **U.S. Department of Agriculture. Forest Service (1979).** Canal front planning unit land-management plan, Olympic National Forest. Portland, OR, U.S. Department of Agriculture Forest Service. 391p. (P)

10357 **Washington Natural Heritage Program (1981).** Endangered, threatened and sensitive vascular plants of Washington. Olympia, WA, Washington Natural Heritage Program. 26p. (T)

U.S. - West Virginia

10358 **Clarkson, R.B., Evans, D.K., Fortney, R.H., Grafton, B., Rader, L. (1981).** Rare and endangered vascular plant species in West Virginia. Newton Corner, MA, U.S. Fish and Wildlife Service. (T)

10359 **U.S. Department of the Interior. Fish and Wildlife Service (1986).** Endangered and threatened wildlife and plants: proposal to determine *Trifolium stoloniferum* to be an Endangered species. *Fed. Register*, 51(46): 8217-8219. (L/T)

10360 **U.S. Department of the Interior. Fish and Wildlife Service (1988).** Endangered and threatened wildlife and plants; proposal to determine *Arabis serotina* (Shale Barren rock cress) to be an Endangered species. *Fed. Register*, 53(222): 46479-46482. (L/T)

U.S. - Wisconsin

10361 **Beyer, R.I. (1982).** Mitchell Park Conservatory, Milwaukee. *Threatened Pl. Commit. Newsl.*, 10: 13. Comments on conservation activities. (G)

10362 **Bosley, T.R. (1978).** Loss of wetlands on the west shore of Green Bay. *Trans. Wisc. Acad. Sci. Arts Lett.*, 66: 235-245. Wetlands.

10363 **Botanical Club of Wisconsin (1973).** Endangered species of vascular plants in Wisconsin. *Bot. Club. Wisconsin Newsl.*, 5(3): n.p. (T)

10364 **Hartley, T.G. (1959).** Notes on some rare plants of Wisconsin. *Trans. Wisconsin Acad. Sci. (Arts and Lett.)* 48: 57-64. (T)

10365 **Linterreur, L.J. (1969).** Endangered species of trees. *Wisconsin Conserv. Bull.*, 34(5): 14-15. (T)

10366 **Mohlenbrock, R.H. (1977).** The midwestern United States. *In* Prance, G.T., Elias, T.S., *eds.* Extinction is forever. Proceedings of a symposium at the New York Botanical Garden, 11-13 May 1976. New York, New York Botanical Garden. 41-44. Discusses the status of legal protection for threatened and endangered species of plants which occur in the midwestern states. (L/T)

10367 **Peck, J.H. (1979).** Compilatory list of the rare, threatened, and endangered plants of the five states of the Upper Mississippi River: Illinois, Iowa, Minnesota, Missouri, and Wisconsin. Contributions from University of Wisconsin-La Crosse Herbarium No. 26. La Crosse, WI, Univ. of Wisconsin-La Crosse. 48p. (T)

10368 **Read, R.H. (1976).** Endangered and threatened vascular plants in Wisconsin. Scientific Areas Prevention Council Technical Bulletin 92. Madison, Department of Natural Resources. 58p. 268 taxa included with photos and descriptions. (T)

U.S. - Wisconsin

10369 **U.S. Department of the Interior. Fish and Wildlife Service (1985).** Endangered and threatened wildlife and plants: proposal to determine *Lespedeza leptostachya* to be a Threatened species. *Fed. Register,* 50(235): 49967-49969. (L/T)

10370 **U.S. Department of the Interior. Fish and Wildlife Service (1988).** Endangered and threatened wildlife and plants; determination of Threatened status for *Asclepias meadii* (Mead's milkweed). *Fed. Register,* 53(170): 33992-33995. (L/T)

10371 **U.S. Department of the Interior. Fish and Wildlife Service (1988).** Endangered and threatened wildlife and plants; determination of Threatened status for *Cirsium pitcheri. Fed. Register,* 53(137): 27137-27141. (L/T)

10372 **U.S. Department of the Interior. Fish and Wildlife Service (1988).** Endangered and threatened wildlife and plants; determination of Threatened status for *Oxytropis campestris* var. *chartacea* (Fassett's locoweed). *Fed. Register,* 53(188): 37970-37972. (L/T)

10373 **U.S. Department of the Interior. Fish and Wildlife Service (1988).** Endangered and threatened wildlife and plants; proposal to determine *Iris lacustris* (Dwarf lake iris) to be a Threatened species. *Fed. Register,* 52(233): 46334-46336. (L/T)

10374 **Wisconsin. Department of Natural Resources (1981).** Endangered and threatened plants. *In* Wisconsin Administrative Code, Chapter NR27. Madison, WI, The Dept. 315-(334-3). (T)

10375 **Zimmerman, J.H., Iltis, H.H. (1961).** Conservation of rare plants and animals. *Wisconsin Acad. Rev.,* (Winter): 7-11. (T)

U.S. - Wyoming

10376 **Chase, A. (1986).** Playing God in Yellowstone: the destruction of America's first national park. New York, Atlantic Monthly Press. 446p. (P)

10377 **Clark, T.W., Dorn, R.D., eds (1979).** Rare and endangered vascular plants and vertebrates of Wyoming. Available from R.D.D., Box 1471, Cheyenne, Wyoming. 78p. (T)

10378 **Dorn, R.D. (1976).** Wyoming plants proposed for endangered or threatened status. Report given to Bureau of Land Management, Rawlins, Wyoming. (T)

10379 **Dorn, R.D. (1977).** Rare and endangered species. *In* Dorn, R.D., *ed.* Manual of the vascular plants of Wyoming, Vol.2. New York, Garland. 1394-1400. (T)

10380 **Dorn, R.D. (1980).** Illustrated guide to special interest vascular plants of Wyoming. U.S. Fish and Wildlife Service and Bureau of Land Management. 67p. Lists 13 plants considered of special interest, with dot maps and illus. (T)

10381 **Johnston, B.C., et al. (1980).** Proposed and recommended threatened and endangered plant species of the Forest Service Rocky Mountain Region: an illustrated guide to certain species in Colorado, Wyoming, South Dakota, Nebraska and Kansas. Denver, CO, U.S. Fish and Wildlife Service. 164p. (T)

10382 **McMahan, L.R. (1987).** Rare plant conservation by state government: case studies from the western United States. *In* Elias, T.S., *ed.* Conservation and management of rare and endangered plants. Proceedings from a conference, Sacramento, California, 5-8 November 1986. Sacramento, California Native Plant Society. 23-31. Maps. (L/T)

10383 **Midgley, G.W. (1980).** Yellowstone: majestic world wilderness. *Habitat (Australia),* 8(2): 24-25. Illus. (P)

10384 **Reveal, J.L. (1977).** The western United States. *In* Prance, G.T., Elias, T.S., *eds.* Extinction is forever. Proceedings of a symposium at the New York Botanical Garden, 11-13 May 1976. New York, New York Botanical Garden. 50-59. Discusses existing laws and present policies regarding endangered plants for five western states of the U.S. (L/T)

U.S. - Wyoming

10385 **Romme, W.H. (1982).** Fire and landscape diversity in subalpine forests of Yellowstone National Park. *Ecological Monographs*, 52(2): 199-221. Maps, forests. (P)

10386 **Shaw, R.J. (1981).** Plants of Yellowstone and Grand Teton National Parks. Salt Lake City, Utah, Wheelwright Press. 159p. Col. illus. (P)

UNITED STATES MINOR OUTLYING ISLANDS

U.S. - Howland Island and Baker Island

10387 **U.S. Congress, Office of Technology Assessment (1987).** Integrated renewable resource management for U.S. insular areas. Washington, D.C., U.S. Govt Printing Office. 443p. Illus., maps. OTA-F-325. Includes description of vegetation, remaining coverage, threatened species. (P/T)

U.S. - Midway Islands

10388 **Anon. (1942).** Midway plants. *Scientific American*, 167: 170. Laysan vegetation was destroyed by rabbits in 1903.

10389 **Apfelbaum, S.I., Ludwig, J.P. & C.E. (1983).** Ecological problems associated with disruption of dune vegetation dynamics by *Casuarina equisetifolia* L. at Sand Island, Midway Atoll. *Atoll Res. Bull.*, 261: 1-19. Invasive species.

10390 **U.S. Congress, Office of Technology Assessment (1987).** Integrated renewable resource management for U.S. insular areas. Washington, D.C., U.S. Govt Printing Office. 443p. Illus., maps. OTA-F-325. Includes description of vegetation, remaining coverage, threatened species. (P/T)

VANUATU

10391 **Gilpin, M.E., Diamond, J.M. (1980).** Subdivision of nature reserves and the maintenance of species diversity. *Nature*, 285(5766): 567-568. (P)

10392 **Green, P.S. (1979).** Observations on the phytogeography of the New Hebrides, Lord Howe Island and Norfolk Island. *In* Bramwell, D., *ed.* Plants and islands. London, Academic Press. 41-53. (Sp). Map.

10393 **Hoyle, M.A. (1978).** Forestry and conservation in the Solomon Islands and the New Hebrides. *Tigerpaper*, 5(2): 21-24.

10394 **Jenks, J.A. (1988).** The forests of Vanuatu. *Tigerpaper Forest News*, 2(3): 10-11. Statistics on forest cover and progress with forest inventory.

10395 **Lee, K.E. (1972).** Proposal for a forest sanctuary at Erromangi Is., New Hebrides. Royal Society, Southern Zone Research Committee. Mimeo. Concerns conservation of *Agathis obtusa*.

10396 **Marshall, A.G. (1973).** A start to nature conservation in the New Hebrides. *Biol. Conserv.*, 5(1): 67-69. Four areas proposed Reef Island Reserve; Duck Lake Reserve, Efate; Botanical and Ornithologica Garden, Efate; Kauri Forest Sanctuary, Erromango. (P)

10397 **Sachet, M.-H. (1957).** The vegetation of Melanesia: a summary of the literature. *Proc. Eighth Pacific Science Congress*, 4: 35-47. (B)

VENEZUELA

10398 **Anon. (1983).** Caribbean mangroves to be protected. *Oryx*, 17: 143. (L/P)

10399 **Bisbal, F. (1988).** [Human impact on habitat in Venezuela]. *Interciencia*, 13(5): 226-232. Sp.

VENEZUELA

10400 **Denevan, W.M. (1973).** Development and the imminent demise of the Amazon rain forest. *Profess. Geogr.*, 25: 130-135.

10401 **Ferrer Veliz, E., Smith, R.F., Chavez Perez, A. (1971).** Esquema general de los aspectos dasonomicos ecologicos y el subsector economico forestal en la region centro occidental. Barquisimeto, Fundeco Division de Programacion Economica. 220p. Por. Illus., maps.

10402 **Garcia, J.R. (1984).** Waterfalls, hydro-power, and water for industry: contributions from Canaima National Park, Venezuela. *In* McNeely, J.A., Miller, K.R., *eds.* National parks, conservation and development. The role of protected areas in sustaining society. Washington, D.C., Smithsonian Institution Press. 588-591. Map. (P)

10403 **Gondelles, A., R., Garcia A., J.R., Steyermark, J. (1977).** Los parques nacionales de Venezuela. Coleccion "La naturaleza en Iberoamerica". Madrid, INCAFO & CIC. 224p. Sp. Col. illus., map. (P)

10404 **Goodland, R.J., Irwin, H.S. (1977).** Amazonian forest and cerrado: development and environmental conservation. *In* Prance, G.T., Elias, T.S., *eds.* Extinction is forever. Proceedings of a symposium at the New York Botanical Garden, 11-13 May 1976. New York, New York Botanical Garden. 214-233. Maps, graphs. (P/T)

10405 **Hadley, M., ed. (1986).** Rain forest regeneration and management. Report of a workshop. *Int. Union Biol. Sci.*, 18: 68p.

10406 **Hamilton, L.S., Steyermark, J., Veillon, J.P., Mondolfi, E. (1976).** Conservacion de los bosques humedos de Venezuela. 2da edicion. Caracas, Sierra Club-Consejo de Bienestar Rural. 181p. Sp. (T)

10407 **Hamilton, L.S., ed. (1976).** Tropical rain forest use and preservation: a study of problems and practices in Venezuela. International Series No. 4. San Francisco, Sierra Club, Office of International Affairs. 115p. (T)

10408 **Jordan, C.F., ed. (1987).** Amazonian rain forests. Ecosystem disturbance and recovery. Ecological Studies No. 60. New York, Springer-Verlag. 133p.

10409 **Prance, G.T. (1977).** The phytogeographic subdivisions of Amazonia and their influence on the selection of biological reserves. *In* Prance, G.T., Elias, T.S., *eds.* Extinction is forever. Proceedings of a symposium at the New York Botanical Garden, 11-13 May 1976. New York, New York Botanical Garden. 195-213. Maps. (P)

10410 **Salati, E., Vose, P.B.F. (1984).** Amazon basin: a system in equilibrium. *Science*, 225(4658): 129-138. Amazonia.

10411 **Steyermark, J.'(1973).** Prservamos las cumbres de la Peninsula de Paria. Defensa de la Naturaleza. *Ao*, 2(6): 33-35. Sp.

10412 **Steyermark, J. (1979).** Plant refuge and dispersal centres in Venezuela: their relict and endemic element. *In* Larsen, K., Holm-Nielsen, L.B., *eds.* Tropical botany. London, Academic Press. 185-221. Includes an appendix of endemic and relict species in the coastal Cordillera de Serrania del Interior. (T)

10413 **Steyermark, J. (1982).** Relationships of some Venezuelan forest refuges with lowland tropical floras. *In* Prance, G.T., *ed.* Biological diversification in the tropics. Proceedings of the 5th International Symposium of the Association for Tropical Biology held at Macuto Beach, Caracas Venezuela, 1979. 182-220. List of endemics in each of the 5 principal forest refuges.

10414 **Steyermark, J.A. (1977).** Future outlook for threatened and endangered species in Venezuela. *In* Prance, G.T., Elias, T.S., *eds.* Extinction is forever. Proceedings of a symposium at the New York Botanical Garden, 11-13 May 1976. New York, New York Botanical Garden. 128-135. General review of the extent and occurrence of endemism and floristic relationships in Venezuela. (L/P/T)

VENEZUELA

10415 **Steyermark, J.A. (1984).** The flora of the Guyana highland: endemicity of the genetic flora of the summits of the Venezuela Tepuis. *Taxon*, 33: 371-372.

10416 **Veillon, J.P. (1976).** Deforestations in the western llanos of Venezuela since 1950 to 1975. *In* Hamilton, L.S., *ed.* Tropical rain forest use and preservation: a study of problems and practices in Venezuela. Int. Ser. No. 4. San Francisco, Sierra Club. 97-110.

VIET NAM

10417 **Cuc, L.T. (1989).** The current issues of natural conservation in Vietnam. *Tigerpaper*, 16(3): 7-10.

10418 **Haager, J. (1978).** Setkani v narodnim parku Cuc Phuong: expedice orchidea 3. *Ziva*, 26(6): 223-225. Cz. Illus. (P)

10419 **IUCN (1985).** Vietnam counts forest and species loss after years of war and colonial rule. *Tigerpaper*, 12(1): 17. (T)

10420 **Kempf, E. (1986).** The re-greening of Vietnam. *WWF News*, 41: 4-5.

10421 **McNeely, J.A. (1987).** Conserving Vietnam's biological diversity: how to manage the nation's living resources to support development. International conference on ecology in Vietnam, Mohonk Mountain House, New Platz, New York, 28-30 May 1987. Gland, Switzerland, IUCN. 18p. Maps. (L/P)

10422 **Morris, G.E., comp. (1987).** News of Nam Cat Tien. *Garrulax*, 2: 3-5. Map. (P)

10423 **Pfeiffer, E.W. (1984).** The conservation of nature in Viet Nam. *Envir. Conserv.*, 11(3): 217-221. Illus.

10424 **Phan Ke Loc (1983).** Preliminary results of the discovery of some rare and endangered plant species to be protected in Viet Nam. Ha Noi, Report in the symposium of the national workshop on rational use of resources and environmental protection, 19-23 November 1983. (T)

10425 **Phan Ke Loc (1989).** Map showing the distribution of rare and endangered plant species to be protected in Viet Nam. Ha Noi, Report in the Symposium on the National Atlas of Viet Nam, 26-27 November, 1983. With list of 568 species.

10426 **Quy, V. (1985).** Rare species and protection measures proposed for Vietnam. *In* Thorsell, J.W., *ed.* Conserving Asia's natural heritage. The planning and management of protected areas in the Indomalayan Realm. Proc. of the 25th working session of IUCN's CNPPA. Gland, IUCN. 98-102.

10427 **Richards, P.W. (1984).** The forests of South Viet Nam in 1971-72: a personal account. *Envir. Conserv.*, 11(2): 147-153. Illus. (P)

10428 **Talbot, L.M., Challinor, D. (1973).** Conservation problems associated with development projects on the Mekong River system. *In* Costin, A.B., Groves, K.H., *eds.* Nature conservation in the Pacific. Proc. Symposium A-10, XII Pacific Science Congress, Canberra, August-September 1971. IUCN and Australian Nat. Univ. Press. 271-283. Map.

10429 **Tam, T.Q. (1988).** A preliminary list of epiphyte orchids at Nam Cat Tien Forest Reserve. *Garrulax*, 4: 10. (P/T)

10430 **Trung, T. van (1985).** The development of a protected area system in Vietnam. *In* Thorsell, J.W., *ed.* Conserving Asia's natural heritage. The planning and management of protected areas in the Indomalayan Realm. Proc. of the 25th working session of IUCN's CNPPA. Gland, IUCN. 30-31. Condensed from an original paper presented in French. (P)

10431 **Trung, T. van (1986).** The forest reserve of Nam Cat Tien in southern Vietnam. *Garrulax*, 1: 3-6. (P/T)

VIET NAM

10432 **Vidal, J.E. (1983).** Consequences ecologiques de la defoliation chimique au Viet Nam. (Ecological consequences of chemical defoliation in Viet Nam.) *Bull. Soc. Bot. France Lett. Bot.*, 130(4-5): 363-369. Fr (En). Illus.

10433 **Westing, A. (1975).** Postwar forestry in North Vietnam. *World Wood*, (1975): 26-28. Illus. Deforestation.

10434 **Westing, A.H. & C.E. (1981).** Endangered species and habitats of Viet Nam. *Envir. Conserv.*, 8(1): 59-62. (T)

10435 **Westing, A.H. (1971).** Ecological effects of military defoliation in the forests of S. Vietnam. *BioScience*, 21(17): 893, 896.

VIRGIN ISLANDS, U.S.

10436 **Anon. (1979).** The Virgin Islands Conservation Society. *Caribbean Conserv. News*, 1(18): 14-15.

10437 **Ayensu, E.S., DeFilipps, R.A. (1978).** Endangered and threatened plants of the United States. Washington, D.C., Smithsonian Institution and World Wildlife Fund Inc. 403p. Lists 90 'Extinct', 839 'Endangered' and 1211 'Threatened' taxa for the continental U.S. Also covers Hawaii, Puerto Rico and Virgin Is. (T)

10438 **Knausenberger, W.I., Matuszak, J.M., Thomas, T.A. (1987).** Herbarium of the Virgin Islands National Park: consolidation and curation of a reference collection. St. Thomas, U.S. Virgin Islands, Virgin Islands Resource Management Cooperative. 51p. (Biosphere Reserve Research Report No. 18A.) Checklist; mentions rare and threatened species. (P/T)

10439 **Little, E.L., jr, Woodbury, R.O. (1980).** Rare and endemic trees of Puerto Rico and the Virgin Islands. U.S. Department of Agriculture, Forest Service Conserv. Res. Rep., 27. Washington, D.C., U.S. Govt Print. Off. 26p. (T)

10440 **Robinson, A.F., jr, ed. (1982).** Endangered and threatened species of the southeastern United States including Puerto Rico and the Virgin Islands. Washington D.C., U.S. Department of Agriculture, Forest Service. Illus., maps, col. illus. (T)

10441 **Rogers, C.S., Teytaud, R. (1988).** Marine and terrestrial ecosystems of the Virgin Islands National Park and Biosphere Reserve. St. Thomas, Virgin Islands, U.S. Dept of the Interior, National Park Service, and Virgin Islands Resource Managemant Cooperative, and Island Resources Foundation. 112p. Illus., maps. Includes list of threatened species; vegetation. (P/T)

10442 **U.S. Congress, Office of Technology Assessment (1987).** Integrated renewable resource management for U.S. insular areas. Washington, D.C., U.S. Govt Printing Office. 443p. Illus., maps. OTA-F-325. Includes description of vegetation, remaining coverage, threatened species. (P/T)

10443 **U.S. Department of the Interior. Fish and Wildlife Service (1984).** Endangered and threatened wildlife and plants: proposed Endangered status for *Buxus vahlii* (Vahl's boxwood). *Fed. Register*, 49(136): 28580-28583. (L/T)

10444 **U.S. Department of the Interior. Fish and Wildlife Service (1985).** Endangered and threatened wildlife and plants: final rule to determine *Zanthoxyllum thomasianum* to be an Endangered species. *Fed. Register*, 50(245): 51867-51870. (L/T)

10445 **Woodbury, R.O., Little, E.L., jr (1979).** Flora of Buck Island Reef National Monument (U.S. Virgin Islands). U.S. Dept Agric. Forest Serv. Res. Pap., no. ITF-19. 27p. (Sp). Illus. (P)

WALLIS AND FUTUNA ISLANDS

10446 **Morat, P., Veillon, J.-M. (1985).** Contribution a la connaissance de la vegetation et de la flore de Wallis et Futuna. *Adansonia*, 3: 259-329. Fr. Deforestation.

YEMEN

10447 **Hepper, F.N., Wood, J.R.I. (1979).** Were there forests in the Yemen? *Proc. Seminar for Arabian Studies*, 9: 65-69. Illus.

YEMEN, DEMOCRATIC

YEMEN, DEMOCRATIC - Socotra

10448 **Cronk, Q. (1986).** Plant protection for Socotra. *WWF Monthly Report*, Feb.: 43-46. (T)

10449 **Gwynne, M.D. (1968).** Socotra. *In* Hedberg, I. & O., *eds.* Conservation of vegetation in Africa south of the Sahara. Symposium Proceedings, at 6th plenary meeting of AETFAT, Uppsala. *Acta Phytogeogr. Suec.*, 54: 179-185. Illus., map. Describes vegetation and makes recommendations of the three regions which are worthy of conservation. (P)

10450 **WWF (1986).** Plant protection for Socotra. *WWF Monthly Report*, February 1986: 43-49. Illus. Lists endemic plants. (E/P/T)

YUGOSLAVIA

10451 **Akeroyd, J.R., Preston, C.D. (1980).** Observations on two narrowly endemic plants, *Moehringia minutiflora* Bornm. and *Silene viscariopsis* Bornm., from Prilep, Yugoslavia. *Biol. Conserv.*, 19: 223-233. Illus,. map. (T)

10452 **Akeroyd, J.R., et al. (1979).** Macedonian floristic conservation expedition 1976. Cambridge, Univ. of Cambridge, Botany School.

10453 **Androic, M. (1982).** Aktualna problematika zastite suma u nas. (Present problems of forest protection in our country.) *Sum. List*, 106(11-12): 441-452. Se (En).

10454 **Anko, B. (1978).** O novih gozdnih rezervatih v sloveniji. (New forest reserves in Slovenia.) *Varstvo Narvave*, 11: 57-63. Se (En). Illus. (P)

10455 **Anon. (1984).** Naucen sobir Florata i vegetacijata na Jugoslavija i problemot na nivnata zaxtita. (Symposium on the flora and

vegetation in Yugoslavia and their protection.) Skopje, 2-4 June 1982. Maked. akad. nauk. umetn., Oddel. biol. med. nauk. 162p. Se. (T)

10456 **Baumann, S., Pradler, H.-W. (1976).** Juwel im Hinterland der Adriakuste: das Naturschutzgebiet des Krka-Tales. *Naturschutz- und Naturparke*, 83: 33-34. Ge. Illus.

10457 **Bedalov, M., Gazi-Baskova (1981).** Einige seltene Pflanzen der jugoslawischen Flora auf den Kornaten-Inseln. *Linz. Biol. Beitr.*, 13(2): 131-141. Ge (En.). (T)

10458 **Bertovic, S., Kamenarovic, M., Kevo, R. (1961).** The protection of nature in Croatia. Zagreb.

10459 **Blecic, V. (1957).** Endemicne i retke biljke u Srbiji. (Endemic and rare plants in Serbia.) *Zast. Prir.*, 9: 1-6. Sc. (T)

10460 **Brelih, S., Gregori, J. (1980).** Redke in ogrozene zivelskev Sloveniji. (Rare and endangered living creatures of Slovenia.). Ljubliana, Prirido slovni muzej Slovenije. 253p. Se. (T)

10461 **Brkic, D., Barbalic, L. (1982).** Rezultati istrazivanja o Prilagodavanju nekih vrsta ljekovitih biljaka na promjenjene ekoloske uvjet·. (Adaptation of some medicinal plant species to changed ecological conditions.) *Sumarski List*, 106(11/12): 463-470. Se (En). (E)

10462 **Broz, V. (1963).** Rad na zastiti retke i ugrozene flore. (The protection of rare and threatened flora.) *Zast. Prir.*, 26: 125-130. Se. (T)

10463 **Broz, V. (1978).** Zastita retkih i ugrozenih biljnih vrsta. (Protection of rare and endangered plants.) *Yugoslav J. for Promoting the Quality of Life*, 3: 33-35. Se. (T)

YUGOSLAVIA

10464 **Colic, D.B. (1978).** Neophodnost daljeg odrzanja i zastite podrucja Biogradske fore kao nacionalnog parks. (Action needed to maintain and protect the Belgrad mountain area as a national park.) *Posebna Izd. Republ. Zavod za Zastitu Prirode SR Srbije*, 11: 9-13. Se, Rus, Ge. (P)

10465 **Colic, D.B. (1978).** Report on current activities in the protection of nature and national parks (1977). *Posebna Izd.-Republ. Zavod za Zastitu Prirode SR Srbije*, 11: 159-164. (Se, Ge, Fr). (P)

10466 **Em, H. (1984).** [Ash community along the lower Vardar River: Periploco-Fraxinetum angustifoliae-pallisae as. nov. (Last remnants of an important monument to nature).] *Sumarski Pregled*, 32(1/4): 3-17. Mac (Ge). Illus. *Fraxinus pallisae.*

10467 **Erker, R. (1979).** Redke drevesne vrste v nasih parkih in nasadih. (Rare tree species in Slovenian parks and plantations.) *Gozdarski Vestnik*, 37(2): 58-62. Se. (P/T)

10468 **Fukarek, P. (1959).** Arbres et arbustes rares et menaces de la flore de Yougoslavie. *In* IUCN. Animaux et vegetaux rares de la Region Mediterraneenne. Proceedings of the IUCN 7th Technical Meeting, 11-19 September 1958, Athens, vol. 5. Brussels, IUCN. 159-165. Fr. (T)

10469 **Godicl, L. (1981).** The protection of rare plants in nature reserves and national parks in Yugoslavia. *In* Synge, H., *ed.* The biological aspects of rare plant conservation. Chichester, Wiley. 491-502. Map. Proceedings of International Conference, King's College, Cambridge, 14-19 July 1980. (P/T)

10470 **Godicl, L. (1985).** Seltene und gefahrdete Pflanzen ind ihr Schutz in Yugoslawien. (Rare and endangered plants and their protection in Yugoslavia.) *Stapfia*, 14: 77-84. Ge. Map. (T)

10471 **Ianovici, V. (1979).** Protectia mediului inconjurator si prezervarea cadrului biologic in Tarile. Balcanice. (La protection de l'environnement et la conservation du cadre biologique dans les pays Balkaniques.) *Ocrot. Nat.*, 23(1): 5-7. Rum (Fr). Conference at Bucharest on 18-22 September 1978.

10472 **Jancovic, M.M., Stevanovic, V. (1982).** [Problem of endangering and protection of flora and vegetation of SR Sebia]. *Maked. akad. nauk. unmetn., Oddel. biol. med. nauk.*, contrib. 3(1): 41-58. (T)

10473 **Jaric, Z. (1978).** Nacionalni park kao faktor regionalnog razvoja. (The national park as a factor of regional development.) *Posebna Izd.-Republ. Zavod za Zastitu Prirode SR Srbije*, 11: 133-158. Se (Fr, Rus). Illus. (P)

10474 **Jovicecic, G. (1978).** Razoj zakonodavstva o zastiti prirode Crne Gore. (Legal developments on nature protection in Montenegro.) *Glasn. Republ.*, 11: 139-148. Se (En). (L)

10475 **Klotzli, F. (1980).** Technicheskaya storona okhrany prirody na bolotakh. (Technischer Naturschutz in Mooren.) *Poroc. Vzhodnoalp-Dinar. Dr. Preuc. Veget.*, 14: 199-209 (1979 publ. 1980). Se. Swamps.

10476 **Kmecl, M. (1979).** Gozdarstvo v varstvu okolja na slovenskem. (Relation of forestry to environmental protection in Slovenia.) *Gozdarski Vestnik*, 37(5): 214-218. Se (Ge). Illus.

10477 **Kopasz, M., Szalay-Marzso, L. (1974).** Nemzetkozi Termeszetvedelmi Konferencia Jugoszlaviaban. (International conference for the protection of nature in Yugoslavia.) *Buvar*, 29(1): 55-56. Se. Illus.

10478 **Krasulja, S. (1978).** Rad na zastiti prirode u SR Srbiji sa posebnim osvrtom na nacionalne i regionalne prirodne parkove. (Report on nature conservation in Serbia with special attention given to national and regional nature parks.) *Posebna Izd.-Republ. Zavod za Sastitu Prirod. SR Srbije*, 11: 169-178. Se. (P)

YUGOSLAVIA

10479 **Kuzmanov, B. (1981).** Balkan endemism and the problem of species conservation, with particular reference to the Bulgarian flora. *Bot. Jahrb. Syst.,* 102(1-4): 255-270. Illus., maps. (T)

10480 **Majer, D. (1983).** Uredajne osnove za sume Nacionalnog parka "Mljet" od 1875. do 1980, godine. (Management bases for forests of the Mljet National Park from 1875 to 1980.) *Sumarski list,* 107(1/2): 71-95. Se (En). Illus., maps. (P)

10481 **Micevski, K. et al. (1982).** [Flora and vegetation in Yugoslavia and their protection.] Symposium, Macedonian Acad. of Sciences and Art, Skopje. Se. (P/T)

10482 **Micevski, Lj., Manevski, Lj. (1984).** [Ecological and floristic changes caused by degradation of the *Quercus sessilis* forests in the Jakupica mountain area.] *Sumarski Pregled,* 32(5/6): 19-29. Mac (Fr). Illus. (T)

10483 **Mihajlovic, I. (1978).** Derdapsko podrucje i njegore karakteristike od opsteg drustvenog znacaja. *Sumarstvo,* 31(2-3): 64-71. Se. Illus., map.

10484 **Movcan, J. (1982).** Development and economics in Plitvice National Park. *Ambio,* 11(5): 282-285. Illus., col. illus. (P)

10485 **Movcan, J. (1984).** National park development and its economics: experience from Plitvice National Park, Yugoslavia. *In* McNeely, J.A., Miller, K.R., *eds.* National parks, conservation and development. The role of protected areas in sustaining society. Washington, D.C., Smithsonian Institution Press. 442-445. (P)

10486 **Peterlin, S., Skoberne, P., Wraber, T. (1985).** Na poti k botanicni "Rdneci Knjigi" za Slovenijo. (Towards the plant "Red Data Book" of Slovenia. *Biol. Vestn.,* 33(2): 61-72. Se (En). (T)

10487 **Pevalek, I. (1959).** Sur les plantes rares et menacees de la region Mediterraneenne de la Yougoslavie. *In* IUCN. Animaux et vegetaux rares de la Region Mediterraneenne. Proceedings of the IUCN 7th Technical Meeting, 11-19 September 1958, Athens, vol. 5. Brussels, IUCN. 166-167. Fr. (T)

10488 **Plavsic-Gojkovic, N. (1972).** Zasticene biljne vrste u SR Hrvatskoj. (Protected plant species in the Socialist Republic of Croatia.) *Mala Hort. Bibl.* 2. 68p. Se. Illus., maps. (L/T)

10489 **Plavsic-Gojkovic, N. (1976).** Seltene Geschutzte Pflanzen der Sozialistischen Republik Kroatien (Jugoslawien). (Rare protected plants of the Croatia, Yugoslavia.) *Poljopr. Znan. Smotra,* 36(46): 61-71. Ge (Se). Illus., maps. (L/T)

10490 **Prpic, B. (1980).** Problematika Motovunske sume s prijedlogom rjesenja. (Present problems of the Motovun Forest in Istria and proposed solution.) *Sum. List.,* 104(5-6): 189-200. Se (En). Illus., map.

10491 **Pulevic, V. (1983).** Zasticene biljne vrste u SR Crnoj Gori. (Protected plant species in SR Montenegro (Yugoslavia). *Glasn. Republ.,* 16: 33-54. Se (En). (P/T)

10492 **Sadar, V. (1979).** Varstvo okolja z vidika pravilne rabe gozdnega prostora. (Environmental protection from the viewpoint of the correct utilization of forest space.) *Gozdarski Vestnik,* 37(5): 205-213. Sl (Ge). Illus.

10493 **Savjetovanje o novijim dostignucima u zastiti bilja, Zagreb (1969).** Zbornik radova. Savjetovanje o nivijim dostignucima u zastiti bilja (povodom 60-godisnjice Instituta za zastitu bilja Poljoprivrednog fakulteta u Zagrebu). Zagreb, 13.-14. II 1969. Beograd, Savez poljoprivrednih inzenjera i tehnicara SR Hrvatske-Sekcija za zastitu bilja. 240p. Se.

10494 **Sicevski, K. (1978).** [Rare and unknown species in the flora of Macedonia]. *God. Zborn. Biol.,* 31: 151-165. Mac (Ge). (T)

YUGOSLAVIA

10495 **Skoberne, P. (1984).** Zavarovane rastline. (Protected plants.) Skofja Loka. 32p. Se. Colour photographs with brief distribution and conservation notes for 28 species protected in Slovenia. (L/T)

10496 **Stajic, S. (1978).** Kriterijumi za turisticko aktivirenje Sarplanine kao nacionalnog parka. (Criteria for the activation of tourism in the Sar mountain area as a national park.) *Posebna Izd. Republ. Zavod za Zastitu Prirode SR Srbije*, 11: 99-131. Se (En, Rus). Illus. (P)

10497 **Strgar, V. (1975).** O varstvu blagajevega volcina na Slovenskem. (On the protection of *Daphne blagayana* Freyer in Slovenia.) *Varstvo Narvave*, 8: 67-70. Se (En). Illus. (T)

10498 **Strgar, V. (1979).** Trying to conserve the rare and endangered *Degenia. In* Synge, H., Townsend, H., *eds.* Survival or extinction. Proceedings of a conference held at the Royal Botanic Gardens, Kew, 11-17 September 1978. Kew, Bentham-Moxon Trust. 211-214. Map. (T)

10499 **Sugar, I. (1982).** Rad na "crvenoj knjiziJugoslavie" i znazenje zastite Biljnih i zivotinjskih vrsta za zastitu prirode. *In* Micevski, K., *et al.*, *eds.* [Flora and vegetation in Yugoslavia and their protection]. Skopje, Macedonian Acad. Sci. & Arts. 147-150. Se. (P)

10500 **Tortic, M. (1977).** Two rare polypores from Lindtner's collection, new for Yugoslavia. *Glasn. Muz. Beogradu*, 32: 35-40. (Se). Map. *Chaetoporellus latitans, Rigidoporus undatus.* (T)

10501 **Wojterski, T. (1971).** Parki Narodowe Jugoslawii. (National parks of Yugoslavia.) *Ochr. Przyr.*, 36: 11-129. Se (En). Illus., maps. (P)

10502 **Wraber, T. (1965).** Nekaj misli o varstvu narave, posebej se rastlinstva. (Some ideas about nature protection, especially of plants.) *Varstvo Narave* (Ljubljana), 2-3: 75-88. Se (En). Discusses 56 species protected in Slovenia. (T)

10503 **Wraber, T., Skoberne, P. (1989).** Rdeci seznam ogrozenih praprotnic in semenk SR Slovenije. (The Red List of threatened vascular plants in Socialist Republic of Slovenia.) *Varstvo Narave*, 14-15. 429p. Se (En). Maps. (T)

10504 **Zukrigl, K. (1979).** (Sootnoshenie mezhdu fitotsenologei okhranoi prirody i ratsionalnym ispolzovaniem landshaftov.) Die Beziehungen der Vegetationskunde zu Naturschutz und Landschaftspflege. *Poroc. Vzhodnoalp. - Dinar. Dr. Preuc. Veget.*, m 14: 417-429. Ge (Sln, It).

ZAIRE

10505 **INEAC (1954).** Carte des Sols et de la Vegetation du Congo, du Rwanda et du Burundi. Brussels, INEAC. Fr. A series of vegetation and soil maps covering Zaire Rwanda and Burundi in c. 25 parts, published between 1954 and c. 1970. Each map is accompanied by a descriptive memoir, & several maps are to dif. scales.

10506 **IUCN (?)** La conservation des ecosystemes forestiers du Zaire. IUCN, Tropical Forest Programme Series. c.200p. In prep.

10507 **Malaisse, F. (1979).** Zaire. *In* Hedberg, I., *ed.* Systematic botany, plant utilization and biosphere conservation: Proc. of a sym. held in Uppsala in commemoration of the 500th ann. of the Univ. Stockholm, Almqvist & Wiksell Int'l. 92.

10508 **Olschowy, G. (1976).** Zaire und seine Nationalparke. *Naturschutz- und Naturparke*, 80: 33-42. Ge. Illus. (P)

10509 **Tirziu, D. & E. (1978).** Aspecte privind conservarea naturii in Republica Zair. (Aspects concerning nature preservation in the Republic of Zaire.) *Natura* (Rumania), 29(2): 13-19. Rum.

ZAMBIA

10510 **Davidson, J. (1984).** Zambia: conserving for progress. *Countryside Commis. News*, 8: 6. Illus.

10511 **Fanshawe, D.B. (1961).** Evergreen forest relics in Northern Rhodesia. *Kirkia*, 1: 20-24.

10512 **Fanshawe, D.B. (1978).** Conservation of vegetation in Zambia. *In* Proceedings of the 7th World Forestry Congress, Buenos Aires, 4-18 October 1972, vol. 3: Conservation and recreation. Buenos Aires, Instituto Forestal Nacional. 4126-4132. En (Sp, Fr).

10513 **Government of Zambia/IUCN (1985).** The National Conservation Strategy for Zambia. Gland, Switzerland, IUCN. 96p. Illus. Maps. (P/T)

10514 **Hans, A.S., Mwambetania, F. (1978).** Conservation of genetic resources of Zambian forest trees. *In* Proceedings of the 8th World Forestry Congress, Jakarta, 16-28 October 1978: Forestry for quality of life. II, 5. (E)

10515 **Lawton, R.M. (1967).** The conservation and management of the riparian evergreen forests of Zambia. *Comm. For. Rev.*, 46(3): 223-232.

10516 **White, F. (1968).** Zambia. *In* Hedberg, I. & O., *eds.* Conservation of vegetation in Africa south of the Sahara. Symposium Proceedings, at 6th plenary meeting of AETFAT, Uppsala. *Acta Phytogeogr. Suec.*, 54: 208-215. Lists protected areas and discusses need for further protection. (P/T)

ZIMBABWE

10517 **Anon. (1987).** Botanical reserves threatened. *Oryx*, 21(2): 117. Extract from *Zimbabwe Wildlife*, 46. (P)

10518 **Child, G. (1984).** Managing wildlife for people in Zimbabwe. *In* McNeely, J.A., Miller, K.R., *eds.* National parks, conservation and development. The role of protected areas in sustaining society. Washington, D.C., Smithsonian Institution Press. 118-123. Tables; map. (P)

10519 **Du Toit, R.F. (1984).** Some environmental aspects of proposed hydro-electric schemes on the Zambezi River, Zimbabwe. *Biol. Conserv.*, 28(1): 73-87. Maps.

10520 **Kappeyne, J. (1984).** An evaluation of Landsat as a method for monitoring indigenous forests in Matabeleland. *Zimbabwe Sci. News*, 18(1/2): 17-19. Illus.

10521 **Kimberley, M.J. (1975).** Plant protection legislation in Rhodesia. *Excelsa*, 5: 3-16. (L/T)

10522 **Kimberley, M.J. (1980).** Specially protected plants in Zimbabwe. *Excelsa*, 9: 53-54 (1979 publ. 1980). (T)

10523 **Kimberley, M.J., comp. (982).** Seed bank. *Aloe, Cact. Succ. Soc. Zimbabwe Quart. Newsl.*, 52: 33-34. (G)

10524 **Taylor, R.D. (1979).** Biological conservation in Matusadona National Park, Kariba. *Trans. Rhodesia Sci. Assoc.*, 59(5): 30-40. (P)

10525 **Thomson, P.J. (1975).** The role of elephants, fire and other agents in the decline of a *Brachystegia boehmii* woodland. *J. S. Afr. Wildl. Manage. Assoc.*, 5(1): 11-18. Illus.

10526 **Tomlinson, D.N.S. (1980).** Nature conservation in Rhodesia: a review. *Biol. Conserv.*, 18(3): 159-177. Maps.

10527 **Weir, J.S. (1971).** The effect of creating additional water supplies in a Central African national park. *In* Duffey, E., Watt, A.S., *eds.* The scientific management of animal and plant communities for conservation. The 11th Symposium of the BES, University of East Anglia, Norwich, 7-9 July 1970. Oxford, Blackwell. 367-385. Map. (P)

ZIMBABWE

10528 **Wild, H. (1968).** Rhodesia. *In* Hedberg, I. & O., *eds*. Conservation of vegetation in Africa south of the Sahara. Symposium Proceedings, at 6th plenary meeting of AETFAT, Uppsala. *Acta Phytogeogr. Suec.*, 54: 202-207. Map. Survey of areas already protected and needs for further protection. (P/T)

10529 **Wild, H., Muller, T. (1979).** Possibilities and needs for conservation of plant species and vegetation in Africa. Appendix: Preliminary lists of rare and threatened species in African countries. Rhodesia. *In* Hedberg, I., *ed*. Systematic botany, plant utilization and biosphere conservation. Stockholm, Almqvist & Wiksell International. 99-100. Contains 84 species and infraspecific taxa: E:18, V:26, R:40. (T)

10530 **Wilson, L.K.S. (1983).** The conservation trust: yesterday, today and tomorrow. *Zimbabwe Sci. News*, 17(2): 27-29. Illus.

INDEX OF PLANT NAMES

The Index of Plant Names refers to those names used in the WCMC–TPU plants database. Numbers refer to the sequential numbers used throughout the bibliography and not to page numbers.

Plants are indexed to those publications which cover one, or a small number of, species in detail. The index, therefore, will not take the user to every publication which refers to a given species. For example, a Red Data Book containing lists and data sheets on many species would be cross–referenced to the subject area of 'Threatened Plants' (indicated by the code 'T' after the citation) rather than cross–referenced to all the individual species it contains.

Most of the species listed are threatened, either nationally or on a global scale. Occasionally, widespread species are included if they are threatened in part of their range.

INDEX OF FAMILIES

The Index of Families refers to family names as used in the WCMC–TPU database. Numbers relate to the sequential reference numbers used throughout the bibliography and not to page numbers.

Families are indexed to those publications which deal with one, or a small number of, plant families, or to works covering individual species within a given family. The index will not, however, take the user to every publication which refers to that family.

INDEX OF PLANT NAMES

INDEX OF PLANT NAMES

INDEX OF PLANT NAMES

INDEX OF PLANT NAMES

INDEX OF PLANT NAMES

INDEX OF PLANT NAMES

INDEX OF PLANT NAMES

INDEX OF PLANT NAMES

INDEX OF PLANT NAMES

INDEX OF PLANT NAMES

INDEX OF PLANT NAMES

INDEX OF PLANT NAMES

INDEX OF PLANT NAMES

INDEX OF PLANT NAMES

INDEX OF PLANT NAMES

INDEX OF PLANT NAMES

INDEX OF PLANT NAMES

INDEX OF PLANT NAMES

INDEX OF PLANT NAMES

INDEX OF PLANT NAMES

INDEX OF FAMILIES

INDEX OF FAMILIES

INDEX OF FAMILIES

GEOGRAPHICAL INDEX

GEOGRAPHICAL INDEX

GEOGRAPHICAL INDEX

GEOGRAPHICAL INDEX

644

GEOGRAPHICAL INDEX